Graduate Texts in Physics

Series Editors

Kurt H. Becker, NYU Polytechnic School of Engineering, Brooklyn, USA

Jean-Marc Di Meglio, Matière et Systèmes Complexes, Bâtiment Condorcet, Université Paris Diderot, Paris, France

Sadri Hassani, Department of Physics, Illinois State University, Normal, USA

Morten Hjorth-Jensen, Department of Physics, Blindern, University of Oslo, Oslo, Norway

Bill Munro, Okinawa Institute of Science and Technology Graduate University, Onna-son, Japan

Richard Needs, Cavendish Laboratory, University of Cambridge, Cambridge, UK

William T. Rhodes, Department of Computer and Electrical Engineering and Computer Science, Florida Atlantic University, Boca Raton, USA

Susan Scott, Australian National University, Acton, Australia

H. Eugene Stanley, Department of Physics, Center for Polymer Studies, Boston University, Boston, MA, USA

Martin Stutzmann, Walter Schottky Institute, Technical University of Munich, Garching, Germany

Andreas Wipf, Institute of Theoretical Physics, Friedrich-Schiller-University Jena, Jena, Germany

Graduate Texts in Physics publishes core learning/teaching material for graduate- and advanced-level undergraduate courses on topics of current and emerging fields within physics, both pure and applied. These textbooks serve students at the MS- or PhD-level and their instructors as comprehensive sources of principles, definitions, derivations, experiments and applications (as relevant) for their mastery and teaching, respectively. International in scope and relevance, the textbooks correspond to course syllabi sufficiently to serve as required reading. Their didactic style, comprehensiveness and coverage of fundamental material also make them suitable as introductions or references for scientists entering, or requiring timely knowledge of, a research field.

Simon Širca · Martin Horvat

Computational Methods in Physics

Compendium for Students

Third Edition

 Springer

Simon Širca
Faculty of Mathematics and Physics
University of Ljubljana
Ljubljana, Slovenia

Martin Horvat
Faculty of Mathematics and Physics
University of Ljubljana
Ljubljana, Slovenia

ISSN 1868-4513   ISSN 1868-4521   (electronic)
Graduate Texts in Physics
ISBN 978-3-031-68568-2   ISBN 978-3-031-68566-8   (eBook)
https://doi.org/10.1007/978-3-031-68566-8

Dedicated to our parents

Preface

This book evolved from short homework instructions for courses in computational physics at the Faculty of Mathematics and Physics, University of Ljubljana. The feedback received from the students was used to gradually supplement the assignments by classroom presentations and additional material on the web. The heritage of these sessions represented a basis onto which we attempted to span a richer manifold and to elucidate the "exercises" from the mathematical, physical, as well as programming and computational viewpoints. The somewhat spartan instructions thus evolved into a much more general textbook aimed primarily at third- and fourth-year physics students, and at Ph.D. students as an aid for all courses with a mathematical physics tinge. It was one of our local goals to modestly interweave physics and mathematics studies, and this is why the book steers between mathematical rigidity and more profane perspectives of numerical methods, while it tries to preserve the colorful content of the field of mathematical physics.

We were driven by the realization that physics students are often insufficiently prepared to face various obstacles they encounter in numerical solutions or modeling of physical problems. Only a handful of them truly know how something can "actually be computed" or how their work can be efficiently controlled and its results reliably checked. Everyone can solve the matrix system $Ax = b$, but almost no one has an idea of how to estimate the corresponding uncertainty and how to relate this estimate to the possible true error. Students and established researchers alike use explicit integrators of differential equations indiscriminately until they happen to examine closely the solutions of a problem as simple as $\ddot{x} = -x$. The direness of the situation is compounded by many freely available or commercial tools giving a false impression that virtually all problems can be solved by a single keystroke. In parts of the text where basic approaches are discussed, we therefore insist on seemingly redundant numerical details while, on the other hand, we do strive to present at least some of the "serious" methods and illustrate them by manageable examples. Apart from trying to be pedagogic, our guideline has been to include topics and items which are not provided by default in popular software or are rarely discussed elsewhere. The book oscillates between all these extremes: it tries to be neither fully elementary nor encyclopedically complete, but at any rate representative.

The book is structured exactly with such gradations in mind: additional, "noncompulsory" chapters and sections are marked with stars ★ and can be read by particularly motivated students or used as reference. Similarly, the ⊙ symbols denote simpler tasks in the end-of-chapter problems, while more demanding ones are marked by the symbols ⊕. The main text is peppered with examples terminated by the symbol ◁. The purpose of the Appendices is not merely to remove the superfluous contents from the main text, but to enhance programming efficiency. The sour apples we force our reader to bite are the lack of detailed derivations and references to formulas placed in remote parts of the text, although we tried to design the chapters as self-contained units. This style requires more concentration and consultation with literature on the reader's part, but makes the text more concise. In turn, the book does call for an inspired course tutor. In a typical one-semester course, she may hand-pick and fine-tune a dozen or so end-of-chapter problems and supply the necessary background, while the students may peruse the book as a convenient point of departure for work.

The end-of-chapter problems should resonate well with the majority of physics students. We scooped up topics from the most varied disciplines and tried to embed them into the framework of the book. Chapters are concluded by relatively long lists of references, with the intent that the book will be useful also as a stepping stone for further study and as a decent vademecum.

In spite of all care, mistakes do occur. We shall be grateful to all readers turning our attention to any error they might spot, no matter how relevant. The Errata is maintained at the book's web page `http://cmp.books.fmf.uni-lj.si` which also contains the data files needed in some of the problems.

Updates in This Edition

In the process of selecting the material to be included in the new edition, we tried to persevere in our fundamental creed: include either topics or methods which are not universally known—calibrated against our target audience, of course—or topics which are "on the market" for anyone to see and choose from, yet for some unknown reason they seem to have been overlooked or are hard to come by in a succinct manner.

By far the largest new contribution to this book is Chap. 3 on numerical integration and differentiation, both of which were previously treated only in an appendix. At least the former segment, the problem of quadrature, may look like common knowledge, but the proper ways in which individual quadrature formulas should and can be used in order to attain fast and reliable integral estimates—for instance, the surprisingly effective trapezoidal rule—certainly do not seem to be ingrained in modern students' minds and have therefore warranted a more dedicated and comprehensive exposition. The same applies to differentiation: the mantra that it should not be performed numerically is widely spread, yet obviously one can do much better than simply evaluate $\Delta f / \Delta x$!

Section 4.7 on eigenvalue problems with sparse matrices has been made more "hands-on" by including the description of the Arnoldi and Lanczos algorithms. Section 5.3 on transformations with orthogonal polynomials now contains new material on Laguerre and Hermite polynomials, as well as rational Chebyshev functions, all of which are needed to support the new matter in Chap. 12 on spectral methods. Section 7.8 and Appendix D present optimal (Kalman) filtering; according to our experience, physics students indeed do get exposed to this truly well-established methodology, but rarely venture beyond the linear model and certainly fail to reach for the powerful extensions of unscented and dual filtering. Chapter 8 on initial-value problems for ODEs has been enriched by a subsection on Hermite integration of gravitational many-body problems and supplemented by Appendix E on the computation of Poincaré maps and Appendix G on the regularization of trajectories in orbital mechanics. Section 9.8 on singular Sturm–Liouville problems, previously just a "stub", now brings a discussion on the classification of singular endpoints and its relation to spectra and spectral densities. Section 10.11 on conservation laws and high-resolution methods for PDEs has been augmented by a description of high-resolution slope-limiter methods, in particular the Kurganov–Tadmor second-order and Kurganov–Levy third-order schemes, as well as by an introduction to the (weighted) essentially non-oscillatory schemes, important tools in the study of shocks and discontinuities in diffusion-advection processes. Section 12.1 on spectral representation of spatial derivatives contains new material on Laguerre and Hermite spectral derivatives, both of which are needed in Sect. 12.7 devoted to problems with ODEs and PDEs on semi-infinite and infinite domains. In Sect. 12.6, we introduce strong stability preserving methods and special time-integration techniques which can be used in spectral methods where large temporal stepsizes must always be weighted against the discretization of the spatial operator. Section 13.7 brings an introduction to phase retrieval, which is the process of recovering both the modulus and the phase of a complex-valued input signal from measurements that provide only the intensities.

We have slightly revised the rest of the text at too many places to enumerate them individually and tried to fix the typos that resisted being captured in the second edition. Every single figure has been redrawn in the hope to improve readability and better to convey its primary message.

Ljubljana, Slovenia

Simon Širca
Martin Horvat

Acknowledgments

The text of the original Slovenian edition was scrutinized by two physicists, Profs. Alojz Kodre and Tomaž Prosen, as well as three mathematicians, Associate Professors Emil Žagar, Marjetka Krajnc, and Gašper Jaklič. We thank them; as we have stated in the preface to the first edition, navigating between the Scylla and Charybdis of these reviewers allowed us to emerge as better sailors and to arrive happily, after years of roaming the stormy seas, to our Ithaca.

In processing the literature for our presentation of regularization in orbital mechanics in this new edition, we had the pleasure to communicate with Profs. James Binney and Scott Tremaine, who kindly supplied valuable details on Burdet–Heggie and Kustaanheimo–Stiefel methods. We also thank Prof. Mario Spera who provided us with his N-body codes which we used to cross-check our own computations. We are also grateful to Prof. John Pryce, who helped us to understand a few intricate technical aspects of endpoint classification in singular Sturm–Liouville problems.

We wish to express our gratitude to Prof. Claus Ascheron, former Senior Editor at Springer, for his effort in preparation and advancement of the first and second editions of this book. His work has now been taken over by Dr. Zachary Evenson, Senior Editor at Springer Nature, whom we have found just as communicative and forthcoming, and we wish to thank him for his goodwill and patience.

Contents

Chapter 1
Basics of Numerical Analysis

Abstract After reviewing the basic properties of floating-point representation, examples of typical use of expressions in finite-precision arithmetic are given, along with illustrations of common programming pitfalls. Selected classes of function approximation are shown next: optimal (minimax) and rational (Padé) approximations, as well as approximations of evolution operators for Hamiltonian systems. Emphasis is given to the efficient calculation of the Padé approximants and preservation of unitarity. Fundamental techniques of power and asymptotic expansions and asymptotic analysis are discussed: Taylor and Laurent series, treatment of divergent asymptotic series, asymptotic analysis of integrals by the Laplace method and the stationary-phase approximation. The treatment of differential equations involving large parameters is elucidated on the example by the WKB method in quantum mechanics. Tests of convergence of series are listed, followed by a presentation of efficient acceleration techniques for their summation. The technique of series reversion is introduced. Examples and Problems include the system of interacting electric dipoles, integral of the Gaussian function, computation of Airy and Bessel functions, and calculation of the Coulomb scattering amplitude.

1.1 Introduction

An ever increasing amount of computational work is being relegated to computers, and often we almost blindly assume that the obtained results are correct. At the same time, we wish to accelerate individual computation steps and improve their accuracy. Numerical computations should therefore be approached with a good measure of skepticism. Above all, we should try to understand the meaning of the results and the precision of operations between numerical data.

A prudent choice of appropriate algorithms is essential (see, for example, [1, 2]). In their implementation, we should be aware that the compiler may have its own "will" and has no "clue" about mathematical physics. In order to learn more about the essence of the computation and its natural limitations, we strive to simplify complex operations and restrict the tasks of functions to smaller, well-defined domains. It is also worthwhile to measure the execution time of programs: large fluctuations in these well measurable quantities without modifications in running conditions typically point to a poorly designed program or a lack of understanding of the underlying problem.

1.1.1 Finite-Precision Arithmetic

The key models for computation with real numbers in finite precision are the *floating-point* and *fixed-point arithmetic*. A real number x in floating-point arithmetic with base β is represented by the approximation

$$\mathrm{fl}(x) = \pm(d_0 + d_1\beta^{-1} + d_2\beta^{-2} + \cdots + d_{p-1}\beta^{-(p-1)}) \cdot \beta^e \equiv \pm d_0.d_1\ldots d_{p-1} \cdot \beta^e ,$$

where $\{d_i\}_{i=0}^{p-1}$, $d_i \in \{0, 1, \ldots, \beta - 1\}$, is a set of p integers and the exponent e is within $[e_{\min}, e_{\max}]$. The expression $m = d_0.d_1\ldots d_{p-1}$ is called the *significand* or *mantissa*, while $f = 0.d_1\ldots d_{p-1}$ is its *fractional part*. Here we are mostly interested in binary numbers ($\beta = 2$) which can be described by the fractional part f alone if we introduce two classes of numbers. The first class contains *normal* numbers with $d_0 = 1$; these numbers are represented as $\mathrm{fl}(x) = 1.f \cdot 2^e$, while the number zero is defined separately as $\mathrm{fl}(0) = 1.0 \cdot 2^{e_{\min}-1}$. The second class contains *subnormal* numbers, for which $d_0 = 0$. Subnormal numbers fall in the range between the number zero and the smallest positive normal number $2^{e_{\min}}$. They can be represented in the form $\mathrm{fl}(x) = 0.f \cdot 2^{e_{\min}}$. Data types with single ("`float`") and double ("`double`") precision, as well as algorithms for computation of basic operations between them are defined by the IEEE 754 standard; see Table 1.1, the details in Appendix B, as well as [3, 4].

Computations in fixed-point arithmetic (in which numbers are represented by a *fixed* value of e) are faster than those in floating-point arithmetic, and become relevant when working with a restricted range of values. They becomes useful on

Table 1.1 The smallest and largest exponents and approximate values of some important numbers representable in single- and double-precision floating-point arithmetic in base two, according to the IEEE 754 standard. Only positive values are listed

Precision	Single ("`float`")	Double ("`double`")
$e_{\max}$	127	1023
$e_{\min} = 1 - e_{\max}$	-126	-1022
Smallest normal number	$\approx 1.18 \times 10^{-38}$	$\approx 2.23 \times 10^{-308}$
Largest normal number	$\approx 3.40 \times 10^{38}$	$\approx 1.80 \times 10^{308}$
Smallest representable number	$\approx 1.40 \times 10^{-45}$	$\approx 4.94 \times 10^{-324}$
Machine precision, ε_M	$\approx 1.19 \times 10^{-7}$	$\approx 2.22 \times 10^{-16}$
Format size	32 bits	64 bits

very specific architectures where large speed and small memory consumption are crucial (for example, in GPS devices or CNC machining tools). In scientific and engineering work, floating-point arithmetic dominates.

The elementary binary operations between floating-point numbers are addition, subtraction, multiplication, and division. We denote these operations by

$$+ : x \oplus y , \qquad - : x \ominus y , \qquad \times : x \otimes y , \qquad / : x \oslash y .$$

Since floating-point numbers have finite precision, the results of the operations $x + y$, $x - y$, $x \times y$, and x/y, computed with exact values of x and y, are not identical to the results of the corresponding operations in finite-precision arithmetic, $x \oplus y$, $x \ominus y$, $x \otimes y$, and $x \oslash y$. One of the key properties of finite-precision arithmetic is the non-associativity of addition and multiplication,

$$x \oplus (y \oplus z) \neq (x \oplus y) \oplus z , \qquad x \otimes (y \otimes z) \neq (x \otimes y) \otimes z .$$

This has important consequences, as demonstrated by the following Example.

Example By writing a simple program in C or C++ you can convince yourself that in the case $\texttt{float x} = \texttt{1e9}$; $\texttt{float y} = -\texttt{1e9}$; $\texttt{float z} = \texttt{1}$; with the GNU compiler $\texttt{c++}$ and option $-\texttt{O2}$ you obtain $(\texttt{x} \oplus \texttt{y}) \oplus \texttt{z} = 1$, while $\texttt{x} \oplus (\texttt{y} \oplus \texttt{z}) = 0$. (Other compilers may behave differently.) ◁

Example The effects of rounding errors can also be neatly demonstrated [5] by observing the results of three algebraically identical, but numerically different ways of computing the values of the polynomial $y(x) = (1 - x)^8$ in the vicinity of $x = 1$. Let us denote the results of the three methods by $p(x)$, $q(x)$, and $r(x)$. We first compute

$$p(x) = (1 - x)^8$$

by using the standard power function $\texttt{pow(1-x, 8)}$ available in C or C++. The same polynomial can also be expanded as

$$q(x) = 1 - 8x + 28x^2 - 56x^3 + 70x^4 - 56x^5 + 28x^6 - 8x^7 + x^8 ,$$

which we compute by simple multiplication, for example, $x^3 = x \cdot x \cdot x$. Finally, we compute $r(x)$ just like $q(x)$, each term in a row, but evaluate the powers x^n by using the $\texttt{pow(x, n)}$ function. Figure 1.1 shows the results of the three methods (note the change of scale on the horizontal and vertical axes). ◁

Binary operations in floating-point arithmetic are thus performed with errors which depend on the arguments, the type of operation, the programming language, and the compiler. The precision of floating-point arithmetic, called the *machine precision* or *machine epsilon*, and denoted by ε_M, is defined as the difference between the representation of the number 1 and the representation of the nearest larger number. In single precision, we have $\varepsilon_M = 2^{-23} \approx 1.19 \cdot 10^{-7}$, while in double precision

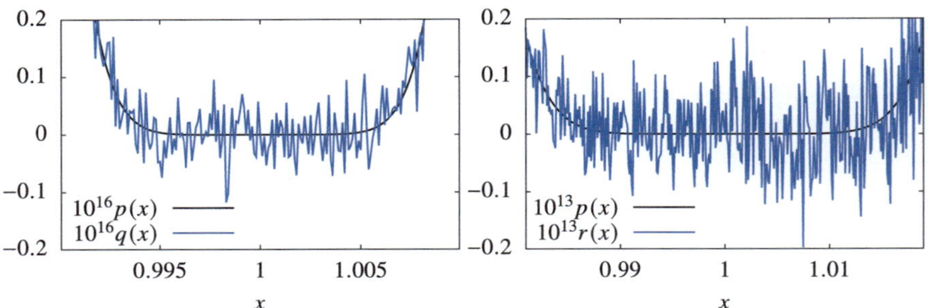

Fig. 1.1 Computation of $(1 - x)^8$ in the vicinity of $x = 1$ in double-precision floating-point arithmetic. [LEFT] Formula $p(x) = (1 - x)^8$ by using `pow(1-x,8)` and formula $q(x) = 1 - 8x + 28x^2 - \cdots$ by using simple multiplications of x. [RIGHT] Computation of $r(x)$ with formula for $q(x)$, but by using `pow` functions for the powers

$\varepsilon_M = 2^{-52} \approx 2.22 \cdot 10^{-16}$. The precision of the arithmetic can also be defined as the largest ε for which $\mathrm{fl}(1 + \varepsilon) = 1$, but the value of ε obtained in this manner depends on the rounding method. The IEEE standard prescribes rounding to the nearest representable result: in this case $\varepsilon \approx \varepsilon_M/2$, which is known as *unit round-off*. For arbitrary data or result of a basic binary operation between normal numbers we have

$$|\mathrm{fl}(x) - x| \le \frac{\varepsilon_M}{2} \, |x| \, , \qquad |\mathrm{fl}(x \circ y) - x \circ y| \le \frac{\varepsilon_M}{2} \, |x \circ y| \, ,$$

where $\circ$ denotes a basic binary operation. The floating-point numbers therefore behave as exact values perturbed by a relative error:

$$\mathrm{fl}(x) = x(1 + \delta) \, , \qquad |\delta| \le \frac{\varepsilon_M}{2} \, .$$

When operations are performed between such numbers, a seemingly small perturbation δ can greatly increase the relative error of the result, leading to a *loss of significant digits* (one or more incorrect digits). As an example, we subtract two different numbers $x = \mathrm{fl}(x)(1 + \delta_1)$ and $y = \mathrm{fl}(y)(1 + \delta_2)$:

$$x - y = (\mathrm{fl}(x) - \mathrm{fl}(y))(1 + h) \, , \qquad h = \frac{\mathrm{fl}(x)\delta_1 - \mathrm{fl}(y)\delta_2}{\mathrm{fl}(x) - \mathrm{fl}(y)} \, .$$

The relative error is bounded by $|h| \le \varepsilon_M \max\{|x|, |y|\}/|x - y|$ and can become very large if x and y are nearby.

Considering the effects of small perturbations also helps us to analyze the relative errors of other basic operations and the precision of more complex algorithms. Special libraries like GMP [6] do allow for computations with almost arbitrary precision, but let us not use this possibility absent-mindedly: do not try to out-smart an unstable algorithm solely by increasing the arithmetic precision!

For a detailed description of how floating- or fixed-point arithmetic behave in typical algorithms, see [4, 7–9]; for details on floating-point data format according to the IEEE 754 standard see Appendix B. We conclude this Section by advices for a stable and precise solution of a few seemingly trivial tasks [4, 10].

Branch statements In programs, we tend to use simple branch statements like

$$
\texttt{if } (\, |x_k - x_{k-1}| < \varepsilon_{\text{abs}} \,) \texttt{ then } \ldots , \tag{1.1}
$$

where, for example, $\varepsilon_{\text{abs}} = 10^{-6}$. But for $\text{fl}(x_k) = 10^{50}$ the nearest representable number in double precision is $\varepsilon_{\text{M}} \, \text{fl}(x_k) \approx 10^{34}$ away. By using an algorithm which returns x_k in double precision, the condition (1.1) will never be met, except in the trivial case of equal x_k and x_{k-1} (for example, due to rounding to zero). It is much better to use

$$
\texttt{if } (\, |x_k - x_{k-1}| < \varepsilon_{\text{abs}} + \varepsilon_{\text{rel}} \, \max\{x_k, x_{k-1}\} \,) \texttt{ then } \ldots ,
$$

where $\varepsilon_{\text{rel}} \approx \varepsilon_{\text{M}}$. Similarly, we repeatedly encounter sign-change tests like

$$
\texttt{if } (\, f_k f_{k-1} < 0 \,) \texttt{ then } \ldots . \tag{1.2}
$$

If $f_k = 10^{-200}$ and $f_{k-1} = -10^{-200}$, we expect $f_k f_{k-1} = -10^{-400} < 0$, but in double precision we get an underflow $\text{fl}(-10^{-400}) = 0$ and the statement (1.2) goes the wrong way. If $f_k = 10^{200}$ and $f_{k-1} = -10^{200}$, we get an overflow. Let us rather check only the sign of the arguments! In C or C++ this can be accomplished by using the function `template < typename T > int sign(constT & val) {returnint((val > 0) − (val < 0)); }`, and then comparing

$$
\texttt{if } (\, \texttt{sign}(f_k) \neq \texttt{sign}(f_{k-1}) \,) \texttt{ then } \ldots .
$$

Roots of the quadratic equation The quadratic equation $ax^2 + bx + c = 0$ has the roots

$$
x_{\pm} = \frac{-b \pm \sqrt{b^2 - 4ac}}{2a} = \frac{2c}{-b \mp \sqrt{b^2 - 4ac}} ,
$$

related by $x_+ x_- = c/a$. In most cases the absolute value of one of the roots is larger than the other and is computed with a larger relative precision in floating-point arithmetic. Once b and $\sqrt{b^2 - 4ac}$ are known, it is therefore preferable from the numerical viewpoint to first compute the larger root x_{max} and then use $x_{\text{min}} = c/(ax_{\text{max}})$ to compute the smaller one. In a similar spirit we compute $\sqrt{1 + x^2} - 1$ for $|x| \ll 1$ where subtraction and potential loss of significant digits can be avoided by rewriting the expression as

$$
\sqrt{1 + x^2} - 1 = \frac{(\sqrt{1 + x^2} - 1)(\sqrt{1 + x^2} + 1)}{\sqrt{1 + x^2} + 1} = \frac{x^2}{\sqrt{1 + x^2} + 1} .
$$

Area of triangle Heron's formula $S = \sqrt{d(d-a)(d-b)(d-c)}$ for the area of a triangle with sides $a \geq b \geq c$, where $d = (a+b+c)/2$, is very sensitive to round-off errors, in particular when one of the angles is larger than $90°$ and $a \approx b + c$. In such cases it is advisable to use the following formula which is accurate to $\approx 10\,\varepsilon_{\mathrm{M}}$:

$$S = \frac{1}{4}\sqrt{[a+(b+c)][c-(a-b)][c+(a-b)][a+(b-c)]}\;.$$

Magnitude of complex number, ratio of complex numbers The magnitude (absolute value) of a complex number $z = x + iy$ is $|z| = (x^2 + y^2)^{1/2}$. Squaring and addition of large numbers, which both may lead to overflows, can be avoided by using the formula

$$|z| = \begin{cases} |x|\sqrt{1 + (|y|/|x|)^2} \;; & 0 < |y| < |x|\,, \\ |y|\sqrt{1 + (|x|/|y|)^2} \;; & 0 < |x| < |y|\,, \\ |x|\sqrt{2} & ;\ \text{otherwise}\,. \end{cases}$$

We can exploit a similar trick to avoid overflows when computing the ratio of complex numbers $(a+ib)/(c+id) = (ac+bd)/(c^2+d^2) + i(bc-ad)/(c^2+d^2)$. If this formula is applied directly, overflow may occur even though the correct result is within the allowed range. A safer way to compute the ratio is

$$\frac{a+ib}{c+id} = \begin{cases} \dfrac{a+b(d/c)}{c+d(d/c)} + i\,\dfrac{b-a(d/c)}{c+d(d/c)} \;; & |d| < |c|\,, \\[2ex] \dfrac{b+a(c/d)}{d+c(c/d)} - i\,\dfrac{a-b(c/d)}{d+c(c/d)} \;; & |d| \geq |c|\,. \end{cases}$$

Natural logarithm To compute $\log(1+x)$ at $0 \leq x < 3/4$ we recommend

$$\log(1+x) = \begin{cases} x & ;\ 1 \oplus x = 1\,, \\ \dfrac{x\log(1+x)}{(1+x)-1} & ;\ \text{otherwise}\,, \end{cases} \tag{1.3}$$

which has an error smaller than $5\varepsilon_{\mathrm{M}}$. Such a precise calculation finds its uses in economics for computation of interest rates where wrong results literally cost money. Let us assume we have some funds A and a small interest rate x for a short period of time. After n periods we have $A' = A(1+x)^n$. If $x \ll 1$, errors can accumulate in computing A' for $n \gg 1$. It is preferable to use the formula $A' = A\exp(n\log(1+x))$ and resort to (1.3).

Average of two numbers Even a simple expression like the arithmetic mean of two floating-point numbers, $x = (a+b)/2$, may overflow, and one should use $x = a + (b-a)/2$ or $a/2 + b/2$ instead.

1.2 Approximation of Expressions

Approximation is one of the key concepts of numerical analysis [11]. In this book we only refer to approximations of scalar functions and expressions involving operators, and therefore only discuss these two examples.

1.2.1 Optimal (Minimax) and Almost Optimal Approximations

The optimal approximation of degree n of a continuous function f on the interval $[a, b]$ is defined as the polynomial $p_n^\star$ for which $E_n^\star \in \mathbb{R}$ exists such that

$$E_n^\star = \max_{a \leq x \leq b} \left| p_n^\star(x) - f(x) \right| \leq \max_{a \leq x \leq b} |p_n(x) - f(x)| , \tag{1.4}$$

where p_n is any polynomial of degree n. We are therefore seeking a polynomial $p_n^\star$ that minimizes the maximal error with respect to the function f. Such a polynomial represents an *optimal* or *minimax approximation*. The curve $p_8^\star$ in Fig. 1.2 (left) is the optimal approximation of the function $e^x \sin(3\pi x)$ by a polynomial of degree eight ($n = 8$), while the corresponding curve in the right panel is the error of the approximation.

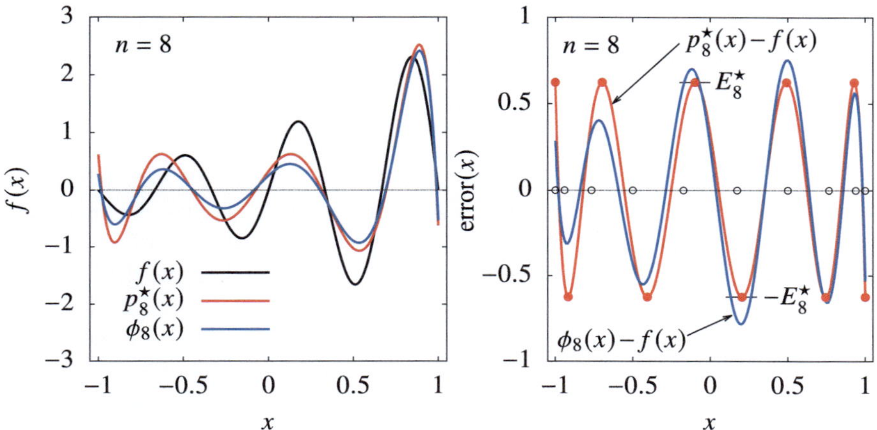

Fig. 1.2 Approximation of $f(x) = e^x \sin(3\pi x)$ by an optimal polynomial and by a Chebyshev expansion. [LEFT] The function f, the optimal polynomial $p_8^\star$ of degree eight ($n = 8$), and the expansion in terms of Chebyshev polynomials $\phi_8(x) = \sum_{k=0}^{8} \hat{\tau}_k T_k(x)$. [RIGHT] Error of the approximations. The full circles denote $(n + 2) = 10$ characteristic points at which the absolute value of the error $\pm E_n^\star = \pm |p_n^\star - f|$ is maximal, and between which the error changes its sign $(n + 1)$-times. The empty circles denote the points at which $T_{n+1}(x) = \pm 1$

The optimal polynomial approximation of a continuous function f on $[a, b]$ can be computed by the iterative algorithm due to Remez (see e.g. [12]). The basis for this procedure is the Borel equi-oscillation theorem which states that a degree-n polynomial $p_n^\star$ is the optimal approximation of f precisely when $(n + 2)$ distinct points x_i exist, arranged as $a \leq x_0 < x_1 < \cdots < x_{n+1} \leq b$, for which

$$p_n^\star(x_i) - f(x_i) = \lambda(-1)^i E_n^\star, \qquad i = 0, 1, \ldots, n + 1,$$

and where λ has a fixed value of $+1$ or -1. The extremes $\pm E_n^\star$ of the error of the optimal approximation $p_n^\star$ therefore occur at the points x_i with the signs flipping back and forth (Fig. 1.2 (right)). This implies that the coefficients of $p_n(x) = \sum_{j=0}^{n} a_j x^j$ and the parameter E_n must fulfill the system of equations

$$\sum_{j=0}^{n} a_j x_i^j - f(x_i) = (-1)^i E_n, \qquad i = 0, 1, \ldots, n + 1, \tag{1.5}$$

$$\sum_{j=0}^{n} a_j x_i^j - f(x_i) = \text{extreme}. \tag{1.6}$$

If f is differentiable, we can rewrite the last equation as

$$\sum_{j=1}^{n} j a_j x_i^{j-1} - f'(x_i) = 0. \tag{1.7}$$

The iteration is started with an initial guess for x_i. A good choice for x_i are the points at which the values of the Chebyshev polynomials T_{n+1} are ± 1,

$$x_i = \cos\left(\frac{i\pi}{n + 1}\right), \qquad i = 0, 1, \ldots, n + 1.$$

By solving the system (1.5) we obtain $n + 1$ coefficients a_j and the parameter E_n, which are the first approximations for the coefficients of the optimal polynomial and the maximum error of $p - f$. These parameters are then used to solve (1.6) or (1.7): we seek the points x_i at which $p - f$ has an extremum. This is the most difficult step in the procedure, especially when it needs to be automated. The new points x_i are then again used in (1.5), and we repeat the procedure until the difference between the consecutive values of E_n drops below a desired tolerance. At the end of the iteration the $\{a_j\}_{j=0}^{n}$ are the coefficients of the optimal polynomial $p_n^\star$, and $E_n = E_n^\star$.

Actually, the power basis $\{1, x, x^2, \ldots\}$ of $p_n^\star$ is a bad choice. The condition number of the matrix corresponding to the system (1.5) deteriorates exponentially with increasing n. Instead of polynomials, rational functions $p(x) = P_n(x)/Q_m(x)$ can be plugged into the Remez' procedure, but the system of equations for the

$(n + 1) + (m + 1)$ coefficients and the errors E becomes non-linear; it can be solved by linearization [13]. In the context of high-order optimal approximations, Chebyshev polynomials are also an attractive option [14].

Even without the Remez' algorithm, the Chebyshev polynomials lead us to an almost identical goal. Instead of searching for the optimal polynomial $p_n^\star$, we may be satisfied by finding an approximation q_n for which

$$\varepsilon = \max_{a \leq x \leq b} \left| p_n^\star(x) - q_n(x) \right|$$

with some small ε. Such an approximation is called *almost optimal* or *near-minimax*. Often it is much easier to find than the true optimal approximation. The most important among them is the approximation by Chebyshev polynomials

$$f(x) \approx \phi_n(x) = \sum_{k=0}^{n} \hat{\tau}_k T_k(x) \,, \tag{1.8}$$

where the coefficients $\hat{\tau}_k$ are given in (5.43). We can switch between the intervals $[a, b]$ and $[-1, 1]$ (domain of Chebyshev polynomials) by the transformations

$$t \mapsto x = 2\frac{t - a}{b - a} - 1 \,, \qquad x \mapsto t = \frac{1 - x}{2} a + \frac{1 + x}{2} b \,.$$

The summation of the series (1.8) can be terminated at the same n that would be used in the optimal approximation. The error due to the series truncation is then determined by the term $\hat{\tau}_{n+1} T_{n+1}(x)$, which has interchanging extreme values of $\pm\hat{\tau}_{n+1}$ at $(n + 2)$ points on $[-1, 1]$ and closely resembles the genuine optimal approximation (see Fig. 1.2 (left and right)).

Chebyshev polynomials have many pleasing and useful properties which we exploit heavily in function and signal transformations (Chap. 5) and spectral methods for partial differential equations (Chap. 12). In the context of optimal approximations, an important property of Chebyshev polynomials is their point-wise orthogonality (5.42). There exists a whole class of polynomials which are orthogonal on a discrete set of points, but only Chebyshev polynomials on $[-1, 1]$ oscillate uniformly and can be generated by so few numerical operations (see recurrence (5.41)).

The shape of the optimal approximation depends on the norm in which its error is measured. Equation (1.4) defines the *uniform* optimal approximation in the "max"-norm $\| \cdot \|_\infty$. The computation of the optimal approximation for an arbitrary function, in particular if it is given only at specific points, can be cumbersome. The main nuisance is the large sensitivity of the Remez' algorithm to individual values $f(x_i)$ that strongly deviate from the average (e.g. outliers in discrete-time signals). In physics we often resort to the optimal approximation in the Euclidean norm (A.3). In this case, the optimal approximation of the function f on $[a, b]$ is a function $p^\star$ for which $E^\star \in \mathbb{R}$ exists such that

$$E^\star = \int_a^b \left[p^\star(x) - f(x) \right]^2 \, \mathrm{d}x \leq \int_a^b \left[p(x) - f(x) \right]^2 \, \mathrm{d}x \tag{1.9}$$

for any polynomial p of degree n (compare this to Eq. (1.4)). The form (1.9) is less sensitive to outliers. We are of course referring to the method of least squares which we most often encounter in its discrete form, where we search for

$$\min_{p \in P_n} \sum_i \left[p(x_i) - f(x_i) \right]^2 .$$

1.2.2 Rational (Padé) Approximation

Suppose that we wish to approximate a function f with the power expansion

$$f(z) = \sum_{k=0}^{\infty} c_k z^k . \tag{1.10}$$

A Padé approximation of f is a rational function with a numerator of degree L and denominator of degree M, determined such that its power expansion matches the series (1.10) up to including the power $L + M$. In other words, if we can find a polynomial P_L of degree L and Q_M of degree M such that

$$f(z) = \frac{P_L(z)}{Q_M(z)} + O(z^{L+M+1}) , \qquad Q_M(0) = 1 ,$$

then

$$[L/M]_f(z) = \frac{P_L(z)}{Q_M(z)} = \frac{a_0 + a_1 z + a_2 z^2 + \cdots + a_L z^L}{b_0 + b_1 z + b_2 z^2 + \cdots + b_M z^M} , \qquad b_0 = 1 , \tag{1.11}$$

defines a Padé approximation of order (L, M) of the function f. The coefficients a_k and b_k can be determined by equating the approximation $[L/M]_f$ with the power series for f and reading off the coefficients of the same powers of x:

$$\left(b_0 + b_1 z + \cdots + b_M z^M \right) \left(c_0 + c_1 z + \cdots \right) = a_0 + a_1 z + \cdots + a_L z^L + O(z^{L+M+1}) .$$

By comparing the terms with powers $z^{L+1}, z^{L+2}, \ldots, z^{L+M}$ we obtain a system of equations for the coefficients b_k of the denominator Q_M, while by comparing terms with powers $z^0, z^1, \ldots, z^L$ we obtain explicit equations for the coefficients a_k of the numerator P_L. For example, with $L = M = 3$, we get

$$
\begin{pmatrix} c_3 & c_2 & c_1 \\ c_4 & c_3 & c_2 \\ c_5 & c_4 & c_3 \end{pmatrix} \begin{pmatrix} b_1 \\ b_2 \\ b_3 \end{pmatrix} = \begin{pmatrix} -c_4 \\ -c_5 \\ -c_6 \end{pmatrix}, \qquad \begin{pmatrix} a_0 \\ a_1 \\ a_2 \\ a_3 \end{pmatrix} = \begin{pmatrix} c_0 & 0 & 0 & 0 \\ c_1 & c_0 & 0 & 0 \\ c_2 & c_1 & c_0 & 0 \\ c_3 & c_2 & c_1 & c_0 \end{pmatrix} \begin{pmatrix} 1 \\ b_1 \\ b_2 \\ b_3 \end{pmatrix}.
$$

Since $b_0 = 1$, the first line of the equation on the right tells us that the zeroth coefficient of P_L is equal to the zeroth coefficient of the power expansion of the function, $a_0 = c_0$. The bulk of the work is hidden in the matrix system on the left. Large Padé systems of this type are solved by robust algorithms like Gauss elimination with complete pivoting because, in most cases, we are interested in relatively low degrees L and M and because accuracy takes precedence over speed. In the sense of properties mentioned in the following, *diagonal Padé approximations*, in which $L = M$, are the most efficient.

Example (Adapted from [15], p. 4.) The function

$$
f(z) = \sqrt{\frac{1 + z/2}{1 + 2z}} = \sum_{k=0}^{\infty} c_k z^k = 1 - \frac{3}{4} z + \frac{39}{32} z^2 - \frac{267}{128} z^3 + \frac{7563}{2048} z^4 - \cdots ,
$$

$$(1.12)$$

has the first two diagonal Padé approximations

$$
[1/1]_f(z) = \frac{1 + \frac{7}{8} z}{1 + \frac{13}{8} z}, \qquad [2/2]_f(z) = \frac{1 + \frac{17}{8} z + \frac{61}{64} z^2}{1 + \frac{23}{8} z + \frac{121}{64} z^2}.
$$

(Compute them!) The comparison of two power expansions and these Padé approximations is shown in Fig. 1.3, revealing a most pleasant property: if some power series converges to a function with a convergence radius ρ, that is, for all $|z| < \rho$ and $0 < \rho < \infty$, then an appropriately chosen Padé approximation converges to f for all $z \in \mathcal{R}$, where the domain $\mathcal{R}$ is larger than the domain defined by $|z| < \rho$. The function f in Example (1.12) has the limit $\lim_{z \to \infty} f(z) = 1/2$ and its power series has a convergence radius of only $\rho = 1/2$. On the other hand, both lowest-order diagonal Padé approximations are stable at infinity. Moreover, when the order of the approximation is increased, the correct limit $1/2$ is approached rapidly: $\lim_{z \to \infty} [1/1]_f(z) = 7/13 \approx 0.5385$ and $\lim_{z \to \infty} [2/2]_f(z) = 61/121 \approx 0.5041$. In other words, the Padé approximation tells us something about how the function behaves outside of the convergence radius of its power series, and ensures better asymptotics. ◁

Example Take the function $f(z) = e^z$, for which the lowest-order Padé approximations are listed in Table I.1. This table shows the Padé approximation in the context of solving partial differential equations and also summarizes the leading error terms. The Padé approach is also fruitful in approximating the time evolution of Hamiltonian operators: see Example (1.18) below. ◁

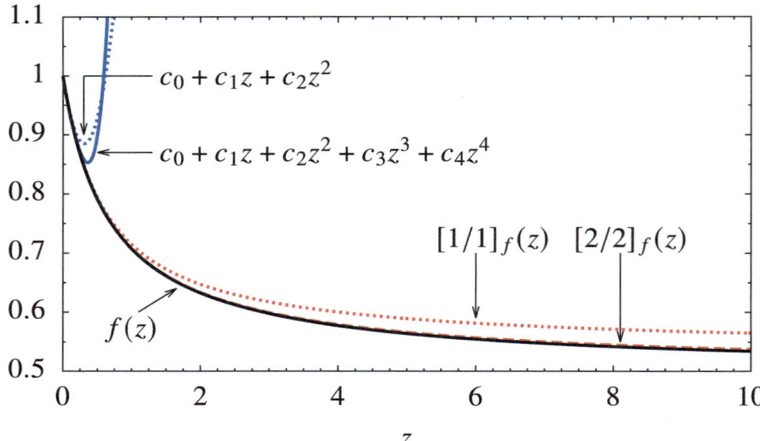

Fig. 1.3 Comparison of power expansions of degree two and four for the function (1.12) and the diagonal Padé approximations $[1/1]_f$ and $[2/2]_f$. The curves for f and $[2/2]_f$ are barely distinguishable in this plot

In certain cases the Padé approximation of f does not exist. For example, for $f(z) = 1 + z^2$ we would try to set up the $[1/1]$ expansion $(a_0 + a_1 z)/(b_0 + b_1 z) = 1 + z^2 + O(z^3)$ and compare the coefficients with equal powers of z to obtain $a_0 = b_0$, $a_1 = b_1$ and $a_0 = 0$. This leads to $[1/1]_f = (0 + a_1 z)/(0 + b_1 z) = 1 \neq 1 + z^2 + O(z^3)$. One can also encounter problems if the denominator of the Padé approximation is zero when $z = 0$, or to degeneracies when the numerator and the denominator both end up having less than the allowed degree. For details on negotiating these exceptional issues, see [15, 16].

Padé approximations have numerous important transformation and invariance properties, of which *duality* and *unitarity* are the most relevant. Duality connects the Padé approximations for reciprocal functions:

$$g(z) = \{f(z)\}^{-1} \text{ and } f(0) \neq 0 \quad \Leftrightarrow \quad [L/M]_g(z) = \left\{[M/L]_f(z)\right\}^{-1} \quad \forall L, M \,.$$

In physics applications, unitarity is even more important, in particular in the theory of scattering matrices. Assume that $f(z) = \sum_{k=0}^{\infty} c_k z^k$ is unitary, so that $f(z) f^*(z) = 1$, and that $[M/M]_f$ is its diagonal Padé approximation. Then we also have

$$[M/M]_f(z) [M/M]_f^*(z) = 1 \,,$$

where the symbol $*$ denotes complex conjugation of the coefficients of the approximation (not of the argument z). In approximations of time evolution of Hamiltonian operators [see Eq. (1.17)], preservation of unitarity is crucial.

Summation of Series by Using Padé Approximations (Wynn's ϵ-Algorithm)

Summation of series is the topic of Sect. 1.4, but here we wish to discuss an important method based on the Padé approximation. We have just witnessed how this approximation can be used to extend the convergence radius of a power series. But this very same approximation can be used to accelerate the summation of a series within its convergence radius. Assume that a series (in the region where it converges) can be approximated by a rational function f which is analytic in this region and has the form

$$f(z) = c_0 + c_1 z + c_2 z^2 + \cdots = \frac{P_L(z)}{Q_M(z)} \equiv [L/M]_f(z) \,,$$

where $P_L(z) = a_0 + a_1 z + \cdots + a_L z^L$ and $Q_M(z) = 1 + b_1 z + \cdots + b_M z^M$. An efficient procedure to compute the diagonal approximations at a given z can be obtained by transformations between the values of $[L/M]_f(z)$ for different L and M. A connection between these approximations can be established which is known as the Wynn's ϵ-algorithm [17]. It can be written as a recurrence

$$\epsilon(n, m + 1) = \epsilon(n + 1, m - 1) + \frac{1}{\epsilon(n + 1, m) - \epsilon(n, m)} \,. \tag{1.13}$$

We start the recurrence with $\epsilon(n, -1) = 0$ for $\forall n \geq 0$ and $\epsilon(n, 0)$ which contain the partial sums

$$\epsilon(n, 0) = S_n = \sum_{k=0}^{n} c_k z^k \,, \qquad n \geq 0 \,.$$

The initial state of the algorithm is given by the first two columns of the $\epsilon(n, m)$ table:

$$
\begin{array}{llllll}
\epsilon(0, -1) = 0 & \underline{\epsilon(0, 0) = S_0} & \epsilon(0, 1) & \underline{\epsilon(0, 2)} & \epsilon(0, 3) & \underline{\epsilon(0, 4)} \cdots \\
\epsilon(1, -1) = 0 & \underline{\epsilon(1, 0) = S_1} & \epsilon(1, 1) & \underline{\epsilon(1, 2)} & \epsilon(1, 3) & \epsilon(1, 4) \cdots \\
\epsilon(2, -1) = 0 & \epsilon(2, 0) = S_2 & \epsilon(2, 1) & \epsilon(2, 2) & \epsilon(2, 3) & \epsilon(2, 4) \cdots \\
\epsilon(3, -1) = 0 & \epsilon(3, 0) = S_3 & \epsilon(3, 1) & \epsilon(3, 2) & \epsilon(3, 3) & \epsilon(3, 4) \cdots \\
\vdots & \vdots & \vdots & \vdots & \vdots & \vdots
\end{array}
$$

By repeated use of (1.13) we obtain all other entries, and these are related to Padé approximations of various orders. The diagonal approximations $[p/p]_f(z)$ at a given z can be read off from the underlined entries with even indices m:

$$[p/p]_f(z) = \epsilon(0, 2p) \,, \qquad p = 0, 1, \ldots \,.$$

We stop the algorithm when the value of the denominator of the fraction in (1.13) drops below a prescribed tolerance; since the convergence is very fast, this can be set to $\approx \varepsilon_M$. Some authors recommend the Wynn procedure as the best general algorithm to accelerate the summation of any slowly converging series [18].

As an exercise, compare the extremely slow convergence of the partial sums $S_n = \sum_{k=0}^{n}(-1)^k/(k+1)$ with $\lim_{n\to\infty} S_n = \log 2$ (second column of the Wynn table) to the much faster convergence of the entries $\epsilon(0,0)$, $\epsilon(0,2)$, $\epsilon(0,4)$, ... in the first row of the table!

Example The Wynn procedure also enables us to evaluate asymptotic (divergent) series in the sense of Sect. 1.3.2. The exponential integral $f(z) = \text{Ei}(z)$ for large positive z can be represented by the asymptotic series

$$\text{Ei}(z) = \int_z^{\infty} \frac{e^{-t}}{t}\, dt \sim \frac{e^{-z}}{z}\left[1 - \frac{1!}{z} + \frac{2!}{z^2} - \frac{3!}{z^3} + \cdots \right], \qquad z \to \infty . \qquad (1.14)$$

The partial sums of (1.14),

$$S_n = \frac{e^{-z}}{z} \sum_{k=0}^{n} \frac{k!}{(-z)^k}, \qquad n \geq 0, \qquad (1.15)$$

do not converge for any z, as the upper two curves in Fig. 1.4 (left) indicate for $z = 2$ and $z = 4$. (At even higher z, the minimum of the error would just shift to the right and downwards.) The lower two curves in the same Figure show the error when the series is evaluated by the Wynn algorithm. ◁

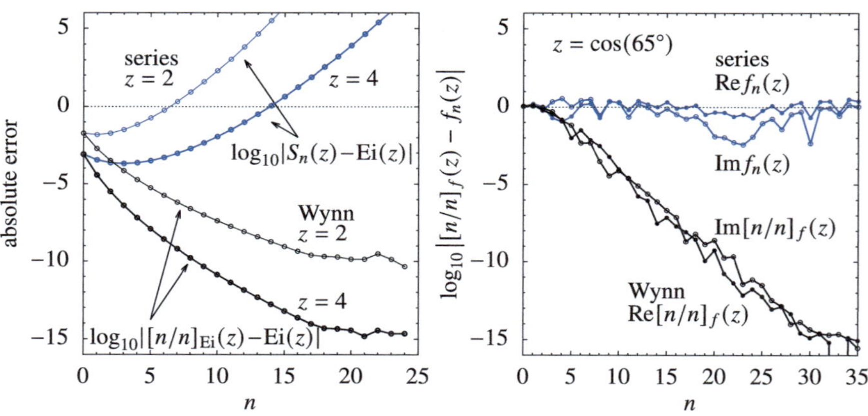

Fig. 1.4 Evaluation of asymptotic series by the Wynn algorithm. [LEFT] Errors of the partial sums (1.15) of (1.14) and of the Padé approximation $[n/n]_{\text{Ei}}$ with respect to the exact values Ei(2) and Ei(4) as a function of the index n. [RIGHT] Errors of the partial sums of (1.16) and of $[n/n]_f$ with respect to the exact amplitude (1.73) at $\theta = 65°$

Example A physically more convincing use of the Wynn's method for divergent series is described in Problem 1.6.5. There, the function f represents the Coulomb scattering amplitude with a divergent power series in $z = \cos\theta$:

$$f(\theta) = \frac{1}{2\,\mathrm{i}\,k} \sum_{l=0}^{\infty} (2l+1)\, P_l(\cos\theta) \left(\mathrm{e}^{2\,\mathrm{i}\,\sigma_l} - 1\right) . \qquad (1.16)$$

The errors of the partial sums of (1.16) and of the Wynn approximations with respect to the exact amplitude are shown in Fig. 1.4 (right).   ◁

Approximation of the Evolution Operator for a Hamiltonian System

The time evolution of a quantum Hamiltonian system from time t to time $t + \Delta t$ (for example, of a wave-function $\Psi(x, t)$ governed by the non-stationary Schrödinger equation) is given by

$$\Psi(x, t + \Delta t) = \mathrm{e}^{-\mathrm{i}\,H\Delta t/\hbar}\,\Psi(x, t) . \qquad (1.17)$$

In difference methods for partial differential equations (Chap. 10) we prefer to approximate the evolution operator $\exp[-\mathrm{i}\,H\Delta t/\hbar]$ by low-order unitary approximations. If the order of the approximation is too low, the solution can become noisy and meaningless even on short time scales. If H does not depend on time, a good way to improve the accuracy of the difference method in its temporal part is to approximate the exponential operator by a diagonal Padé approximation

$$\mathrm{e}^z = \sum_{k=0}^{\infty} c_k z^k = \frac{1 + a_1 z + \cdots + a_M z^M}{1 + b_1 z + \cdots + b_M z^M} + O\left(z^{2M+1}\right) , \qquad (1.18)$$

where $c_k = 1/k!$, and the coefficients a_m and b_m are complex in general. The coefficients at a chosen order M are computed by the procedure described on page 10. The numerator and the denominator of (1.18) can then be factorized as [15]

$$\mathrm{e}^z = \prod_{s=1}^{M} \left(\frac{1 - z/z_s^{(M)}}{1 + z/z_s^{*(M)}}\right) + O\left(z^{2M+1}\right) , \qquad (1.19)$$

where $z_s^{(M)}$ are the zeros of the numerator, and $z_s^{*(M)}$ the zeros of the denominator (which are just their complex conjugates). For example, these zeros up to $M = 3$, in double precision, are

$$z_1^{(1)} = -2 \, ,$$

$$z_{1,2}^{(2)} = -3 \pm i\sqrt{3} \, ,$$

$$z_1^{(3)} = -4.6443707092521712 \, ,$$

$$z_{2,3}^{(3)} = -3.6778146453739144 \pm i\, 3.5087619195674433 \, .$$

At the lowest order ($M = 1$) the expression (1.19) with $z = -i\,H\,\Delta t/\hbar$ simplifies to

$$e^{-iH\Delta t/\hbar} = \frac{1 - \frac{1}{2}i\,H\,\Delta t/\hbar}{1 + \frac{1}{2}i\,H\,\Delta t/\hbar} + O\left(\Delta t^3\right) \, .$$

In the context of (1.17) this can be read as

$$\left[1 + \frac{i\,H\,\Delta t}{2\hbar}\right]\Psi(x, t + \Delta t) = \left[1 - \frac{i\,H\,\Delta t}{2\hbar}\right]\Psi(x, t)$$

(see Problem 10.12.9). At higher orders, we get a product operator

$$e^{-iH\Delta t/\hbar} \approx \prod_{s=1}^{M} E_s^{(M)} \, , \qquad E_s^{(M)} = \frac{1 + (i\,H\,\Delta t/\hbar)/z_s^{(M)}}{1 - (i\,H\,\Delta t/\hbar)/z_s^{*(M)}} \equiv \frac{R_s^{(M)}}{L_s^{(M)}} \, ,$$

which allows us to compute the time evolution as

$$\Psi(x, t_{n+1}) = E_M^{(M)} \cdots E_2^{(M)} E_1^{(M)} \Psi(x, t_n) \, , \qquad t_{n+1} - t_n = \Delta t \, .$$

In a practical implementation, this implies a sequence of steps

$$L_1^{(M)}\Psi(x, t_{n+1/M}) = R_1^{(M)}\Psi(x, t_n) \, ,$$

$$L_2^{(M)}\Psi(x, t_{n+2/M}) = R_2^{(M)}\Psi(x, t_{n+1/M}) \, ,$$

$$\cdots \quad \cdots$$

$$L_M^{(M)}\Psi(x, t_{n+1}) = R_M^{(M)}\Psi(x, t_{n+(M-1)/M}) \, .$$

Note the symbolic notation: $t_{n+1} = t_n + \Delta t$ for each n, but the time advance in each line ($t_n \to t_{n+1/M}$, $t_{n+1/M} \to t_{n+2/M}$, ...) is not equidistant. If H does not depend on time, the sequence of steps is arbitrary. The band structure of matrices L and R depends on the spatial discretization of H: if it contains the kinetic energy operator $-(\hbar^2/2m)\partial^2/\partial x^2$, which is discretized in the form (10.11) on an equidistant spatial mesh, the matrices are tridiagonal; if it is discretized as in (10.12), they are pentadiagonal. Details can be found in [19]; the methods for time-dependent Hamiltonians are discussed in [20].

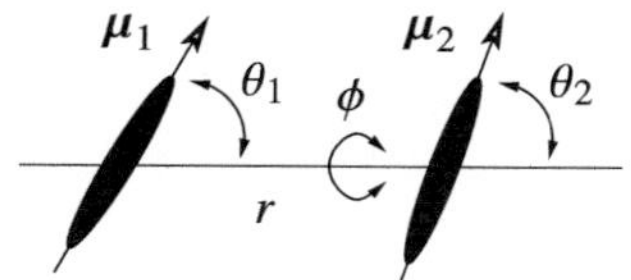

Fig. 1.5 Freely rotating electric dipoles

1.3 Power and Asymptotic Expansion, Asymptotic Analysis

Power expansion is a tool to describe the behavior of a function in the vicinity of a specific point, while asymptotic expansion and analysis are languages in which we express the dependence of integrals and solutions of differential equations on parameters governing their behavior in the limit of very small or very large values. The basic notation (symbols O, o and $\sim$) is given in Appendix A.1.

Example A nice instance of asymptotic analysis can be found in the study of the system of two freely rotating electric dipoles (Fig. 1.5) with an interaction of the form

$$V(r, \Omega) = \frac{\mu_1 \mu_2}{4\pi\varepsilon_0 r^3}\, F(\Omega)\,, \qquad \Omega = (\theta_1, \theta_2, \phi)\,,$$

where $F(\Omega) = -2\cos\theta_1 \cos\theta_2 + \sin\theta_1 \sin\theta_2 \cos\phi$. One encounters such systems in studies of orientation of nitroaniline molecules in zeoliths. The relevant quantity is the interaction energy V weighted by the Boltzmann factor $\exp(-V/kT)$ and averaged over all possible orientations of the dipoles Ω,

$$\left\langle V e^{-V/kT} \right\rangle = \frac{\mu_1 \mu_2}{4\pi\varepsilon_0 r^3} \frac{\int d\Omega\, F(\Omega)\, e^{\lambda F(\Omega)}}{\int d\Omega\, e^{\lambda F(\Omega)}}\,, \qquad \lambda = \frac{\mu_1 \mu_2}{4\pi\varepsilon_0 r^3 kT}\,,$$

where $d\Omega = \sin\theta_1 \sin\theta_2\, d\theta_1\, d\theta_2\, d\phi$. By using standard collections of integrals [21, 22] the triple integrals can be reduced to a single integration [23],

$$\left\langle V e^{-V/kT} \right\rangle = -\frac{\mu_1 \mu_2}{4\pi\varepsilon_0 r^3} \frac{d}{d\lambda} \log K(\lambda)\,, \qquad K(\lambda) = \frac{8\pi}{\sqrt{3}\,\lambda} \int_1^2 \frac{\sinh(\lambda x)}{\sqrt{x^2 - 1}}\, dx$$

We are interested in the behavior of $K(\lambda)$ for large values of the parameter λ, i.e. at fixed r and low temperatures T. By asymptotic analysis (see page 25) we get

$$K(\lambda) \sim \frac{4\pi}{3} \frac{e^{2\lambda}}{\lambda^2} \left(1 + \frac{2}{3\lambda} + \frac{1}{\lambda^2} + \frac{22}{9\lambda^3} + \cdots \right) \tag{1.20}$$

or

$$\frac{d}{d\lambda} \log K(\lambda) = \frac{1}{K}\frac{dK}{d\lambda} \sim 2 - \frac{2}{\lambda} - \frac{2}{3\lambda^2} + \cdots\,, \qquad \lambda \to \infty\,. \tag{1.21}$$

The terms in the expansion (1.21) have clear physical meanings. The leading term $+2$ does not depend on T and reflects the attraction of the dipoles in the state with $\theta_1 = \theta_2 = 0$ that has the lowest energy, $\langle V \exp(-V/kT) \rangle = -\mu_1 \mu_2 / 2\pi \varepsilon_0 r^3$. The second term (proportional to T) expresses the average potential energy of a pair of anisotropic two-dimensional oscillators [23]. The third term (proportional to T^2) corresponds to the non-harmonic part of the potential.   ◁

1.3.1   Power Expansion

Assume that a real function f is at least $(n+1)$-times differentiable in the vicinity of the point x_0. The nth order *power (Taylor) expansion* of f is defined as

$$f(x) = \sum_{k=0}^{n} \frac{f^{(k)}(x_0)}{k!} (x - x_0)^k + R_n(x) \tag{1.22}$$

with the remainder

$$R_n(x) = \int_{x_0}^{x} \frac{f^{(n+1)}(t)}{n!} (x - t)^n \, dt = \frac{f^{(n+1)}(x^\star)}{(n+1)!} (x - x_0)^{n+1} \,, \qquad x^\star \in [x_0, x] \,.$$

In the last step we have used the mean value theorem [24]: for any continuous function g on the interval $[a, b]$ there exists a point $a \leq x^\star \leq b$ such that

$$\int_{a}^{b} g(x) \, dx = g(x^\star)(b - a) \,,$$

and $g(x^\star)$ is the average value of g. If at x_0 all derivatives of f exist and are finite, f can be expanded in the vicinity of x_0 in an infinite series. This series converges on the interval $(x_0 - r, x_0 + r)$ where r is the *convergence radius of the series*. It is given by the formulas (1.56).

The extension of the Taylor series to the complex plane and inclusion of negative powers of the arguments is called the *Laurent expansion*:

$$f(z) = \sum_{k \in \mathbb{Z}} a_k (z - z_0)^k \,, \qquad a_k = \frac{1}{2\pi i} \oint_\gamma \frac{f(z)}{(z - z_0)^{k+1}} \, dz \,.$$

Here γ is an arbitrary closed contour encircling z_0 in the positive sense. According to the coefficients $\{a_{-k}\}_{k \in \mathbb{N}}$ of the negative powers, we have three distinct cases. The function f is *analytic* if all these coefficients are zero; it is *meromorphic* if there exists a minimal N such that $a_{-k} = 0$ for $\forall k \geq N$; if such N does not exist, we say that f has an *essential singularity* at z_0.

1.3.2 Asymptotic Expansion

The asymptotic expansion of a function is a series, the partial sum of which is an approximation of this function in the regime where the parameter describing the asymptotics becomes large or small. The classical example (see e.g. [25]), is the gamma function. For small x it has the Laurent expansion

$$\Gamma(x) = \frac{1}{x} - \gamma + \left(\frac{\gamma^2}{2} + \frac{\pi^2}{12}\right)x + \cdots , \qquad 0 < |x| < 1 , \tag{1.23}$$

where $\gamma \approx 0.577216$ is the Euler constant. The expansion (1.23) converges for all x satisfying $0 < |x| < 1$, and both the left and the right side are functions of x. On the other hand, for large positive x, we have the expansion

$$\Gamma(x) \sim e^{-x} x^x \sqrt{\frac{2\pi}{x}} \left(1 + \frac{1}{12x} + \frac{1}{288x^2} + \cdots\right) , \qquad x \to \infty , \tag{1.24}$$

which does not converge for *any* x! The right side of (1.24) represents an *asymptotic expansion* of the function $\Gamma(x)$ in the limit $x \to \infty$ and is not a function of x. It is merely a sequence of approximations for the function. The error due to the truncation of the series at order n and fixed x does *not* go to zero when n is increased, but it does vanish when $x \to \infty$ at fixed n.

An asymptotic series may converge or diverge. Colloquially, an asymptotic series usually means a divergent series. If the series converges, it can be summed up to an arbitrary term. But it makes no sense to sum a divergent asymptotic series; rather, we use such a series to obtain as good as possible an approximation to the function f. We sum the series only until the sequence of partial sums appears to converge, or stop the summation with the term just before the smallest one. (It turns out that in many asymptotic series the truncation error does not exceed the first omitted term.)

For a more general discussion of asymptotics at an arbitrary point x_0 we choose a sequence of functions $\{\phi_k\}_{k=0}^{\infty}$ with the property $\phi_{k+1}(x) = o(\phi_k(x))$ in the limit $x \to x_0$ an $\phi_k(x) \neq 0$ in the neighborhood of x_0, excluding x_0 itself. The sum $\sum_{k=0}^{\infty} c_k \phi_k(x)$ is called the *general asymptotic expansion* of f if

$$f(x) = \sum_{k=0}^{n} c_k \phi_k(x) + o(\phi_n(x)) , \qquad x \to x_0 . \tag{1.25}$$

The limit point x_0 can be finite or infinite. With the chosen sequence of functions $\{\phi_k\}$, the expansion coefficients in (1.25) are given by

$$c_k = \lim_{x \to x_0} \frac{f(x) - \sum_{m=0}^{k-1} c_m \phi_m(x)}{\phi_k(x)} , \qquad k = 0, 1, \ldots, n . \tag{1.26}$$

Example (see [25], p. 30). We seek an asymptotic expansion of the function

$$f(x) = \frac{1}{x} + \mathrm{e}^{-x}\left(1 - \frac{1}{x}\right)^{-1} , \qquad x \to \infty , \tag{1.27}$$

based on the sequence of functions $\{\phi_k(x)\}_{k=0}^{\infty} = \{1, x^{-1}, x^{-2}, \ldots\}$. We compute the coefficients c_k of (1.25) by using (1.26): we get $c_0 = \lim_{x\to\infty} f(x)/\phi_0(x) = 0$ and $c_1 = \lim_{x\to\infty}(f(x) - c_0\phi_0(x))/\phi_1(x) = 1$, while $c_k = 0$ for $k \geq 2$. Based on the chosen sequence $\{\phi_k(x)\}_{k=0}^{\infty}$, the function f has the expansion

$$f(x) \sim c_1\phi_1(x) = \frac{1}{x} , \qquad x \to \infty . \tag{1.28}$$

The same function f can have another asymptotic expansion if a different sequence $\{\phi_k\}_{k=0}^{\infty}$ is chosen. If we select $\{\phi_k\}_{k=0}^{\infty} = \{1, x^{-1}, \mathrm{e}^{-x}, \mathrm{e}^{-x}x^{-1}, \mathrm{e}^{-x}x^{-2}, \ldots\}$, we get for the same f as before the coefficients $c_0 = 0$ and $c_k = 1$ for $k \geq 1$, thus

$$f(x) \sim \frac{1}{x} + \sum_{k=1}^{\infty} \mathrm{e}^{-x}x^{1-k} , \qquad x \to \infty .$$

By reading this example backwards we realize that different functions may correspond to the same asymptotic expansion. Based on the sequence $\{1, x^{-1}, x^{-2}, \ldots\}$ both (1.27) and $f(x) = 1/x$ have identical expansions, namely (1.28). ◁

1.3.3 Asymptotic Analysis of Integrals by Integration by Parts

For an arbitrary function f and positive $m \in \mathbb{N}$ let f_m represent its mth derivative and F_m its mth indefinite integral,

$$f_0 = f , \qquad f_m = \frac{\mathrm{d}^m f}{\mathrm{d}x^m} , \qquad \frac{\mathrm{d}F_m}{\mathrm{d}x} = F_{m-1} .$$

Suppose we are interested in the asymptotic behavior of the integral

$$I(\lambda) = \int_a^b g(\lambda, x)h(\lambda, x)\,\mathrm{d}x ,$$

where the asymptotics is determined by the parameter λ. Let g be at least n-times differentiable and let h be integrable. By using integration by parts, $I(\lambda)$ can be written as the sum

$$I_n(\lambda) = \sum_{k=0}^{n-1} s_k(\lambda) + R_n(\lambda) \,,$$

where the terms s_k and the remainder R_n are

$$s_k(\lambda) = (-1)^k \left[g_k(\lambda, b) H_{k+1}(\lambda, b) - g_k(\lambda, a) H_{k+1}(\lambda, a) \right] \,,$$

$$R_n(\lambda) = (-1)^n \int_a^b g_n(\lambda, x) H_n(\lambda, x) \, \mathrm{d}x \,.$$

If g is $(n+1)$-times continuously differentiable, we have $R_n = s_n + R_{n+1}$, which can be used to estimate the value of the remainder in two cases [26].

1. If g and h are real and the products $g_n H_n$ and $g_{n+1} H_{n+1}$ have constant and equal signs on $[a, b]$, then R_n has the same sign as s_n and opposite than R_{n+1}, and $|R_n| \leq |s_n|$. Example:

$$g(\lambda, x) = (1 + \lambda x)^{-1} \,, \qquad h(x) = \exp(-x) \,,$$

where $0 \leq (-1)^m R_m \leq (-1)^m s_m$. In the case $a = 0$ and $b = \infty$ we are dealing with the Euler integral $I(\lambda) = \int_0^\infty e^{-x}(1 + \lambda x)^{-1} \, \mathrm{d}x$, which can be represented as a finite alternating series and the remainder

$$I(\lambda) = \sum_{k=0}^{n} (-1)^k k! \, \lambda^k + (-\lambda)^{n+1}(n+1)! \int_0^\infty \frac{e^{-x}}{(1 + \lambda x)^{n+2}} \, \mathrm{d}x \,,$$

and which is closely related to the exponential integral Ei (see Fig. 1.6).

2. If g is real, $|H_{n+1}|$ an increasing function of x and both g_n and g_{n+1} have constant and equal signs on $[a, b]$, or if g is real, $|H_{n+1}|$ is a decreasing function of x, and both g_n and g_{n+1} have constant but opposite signs on $[a, b]$, we have $|R_n| \leq 2|s_n|$.

If λ is complex, we rename $x \mapsto x/\lambda$; thus $g(x) = (1 + x)^{-1}$ and $h(\lambda, x) = \lambda^{-1} \exp(-x/\lambda)$. Then for $\operatorname{Re} \lambda > 0$, the functions g_n, g_{n+1} and H_{n+1} correspond to the second criterion of item 2 above, and we have $|R_n| \leq 2|s_n|$. At any rate, the remainder R_n is on the order of the first omitted term, $R_n = O(s_n)$.

The approach described above is particularly useful when h can be integrated easily, like in the case of $h(\lambda, x) = \exp(\lambda x)$ when $H_m(\lambda, x) = h(x)/\lambda^m$. This is the foundation of the asymptotic expansion of Laplace and Fourier integrals

$$L(\lambda) = \int_a^b e^{-\lambda x} \phi(x) \, \mathrm{d}x \,, \qquad F(\lambda) = \int_a^b e^{\mathrm{i}\lambda x} \phi(x) \, \mathrm{d}x \,,$$

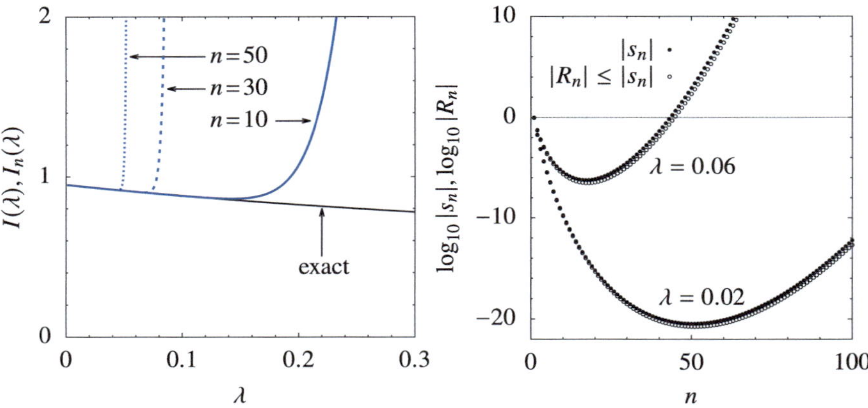

Fig. 1.6 Computation of the $I(\lambda) = \int_a^b e^{-x}(1 + \lambda x)^{-1}\,dx$ on $[a, b] = [0, 3]$ at small λ by asymptotic series. [LEFT] Divergent behavior of partial sums as a function of λ at $n = 10$, 30, and 50. [RIGHT] The size of the first omitted term s_n and the remainder R_n as a function of n at $\lambda = 0.02$ and 0.06: we see that $|R_n| \leq |s_n|$. (The error $|I(\lambda) - I_n(\lambda)|$ has the same qualitative behavior.) We stop the summation when we reach the minimum in this graph. This point also determines the smallest error one can achieve at a given λ

in the limit $\lambda \to \pm\infty$, which are hard to compute by other means. The asymptotic expansion of the Fourier integral is

$$F_n(\lambda) = \sum_{k=0}^{n-1} \left(\frac{i}{\lambda}\right)^{k+1} \left[e^{i\lambda a}\phi^{(k)}(a) - e^{i\lambda b}\phi^{(k)}(b)\right] + R_n(\lambda) \,.$$

The remainder after the truncation of the series to n terms,

$$R_n(\lambda) = \left(\frac{i}{\lambda}\right)^n \int_a^b \phi^{(n)}(x)e^{i\lambda x}\,dx \,,$$

has an upper limit when dealing with finite intervals $[a, b]$. By integrating the remainder by parts one more time, we get [26]

$$|R_n(\lambda)| \leq \lambda^{-n-1}\left[\left|\phi^{(n)}(a)\right| + \left|\phi^{(n)}(b)\right| + \int_a^b \left|\phi^{(n+1)}(x)\right|\,dx\right] = O(\lambda^{-n-1}) \,.$$

1.3.4 Asymptotic Analysis of Integrals by the Laplace Method

Here we analyze integrals of the form

$$I(\lambda) = \int_a^b \phi(x)e^{-\lambda h(x)}\,dx \tag{1.29}$$

in the limit of large positive λ, where ϕ and h are real functions of a real variable x. Assume that h has a global minimum at one of the internal points ξ of the interval $[a, b]$, thus $h'(\xi) = 0$ and $h''(\xi) > 0$. Therefore $\exp(-\lambda h(x))$ reaches its maximum at ξ and in its vicinity we may expect the largest contribution to the integral (1.29). We expand h in the Taylor series around ξ,

$$h(x) = h(\xi) + \frac{1}{2}h''(\xi)(x - \xi)^2 + O\left((x - \xi)^3\right), \tag{1.30}$$

while we take simply $\phi(x) \approx \phi(\xi)$. When these are used in the integral (1.29) and its integration limits are extended to $[-\infty, +\infty]$, we obtain

$$I(\lambda) \approx \int_a^b \phi(\xi)\,e^{-\lambda[h(\xi)+h''(\xi)(x-\xi)^2/2]}\,dx \approx \phi(\xi)\,e^{-\lambda h(\xi)} \int_{-\infty}^{\infty} e^{-\lambda h''(\xi)x^2/2}dx\;.$$

This is the Laplace approximation, which is the leading term in the asymptotic expansion

$$I(\lambda) = e^{-\lambda h(\xi)}\left[\phi(\xi)\sqrt{\frac{2\pi}{\lambda h''(\xi)}} + O\left(\frac{1}{\lambda}\right)\right], \qquad \lambda \to \infty. \tag{1.31}$$

If h reaches its maximum only at $x = a$ and $h'(a) > 0$, the leading term in the asymptotic expansion becomes

$$I(\lambda) = e^{-\lambda h(a)}\left[\frac{\phi(a)}{h'(a)}\frac{1}{\lambda} + O\left(\frac{1}{\lambda^2}\right)\right], \qquad \lambda \to \infty, \tag{1.32}$$

while if it reaches its minimum only at $x = b$ and $h'(b) < 0$, we have

$$I(\lambda) = e^{-\lambda h(b)}\left[-\frac{\phi(b)}{h'(b)}\frac{1}{\lambda} + O\left(\frac{1}{\lambda^2}\right)\right], \qquad \lambda \to \infty. \tag{1.33}$$

The asymptotics of the integrals of the form (1.29), where the dominant contributions originate in the minima of h, is given by expressions (1.31)–(1.33) [25]. Similar formulas can be derived for the case where $\exp(\lambda h(x))$ appears in the integral (1.29) instead of $\exp(-\lambda h(x))$.

Example We seek the leading term in the asymptotic expansion of the integral

$$I(\lambda) = \int_0^1 \frac{\exp(-\lambda[1 + x(1-x)])}{\sqrt{x^2 + 1}}\, dx\,, \qquad \lambda \to \infty\,.$$

In this case $h(x) = 1 + x(1-x)$ and $\phi(x) = 1/\sqrt{x^2 + 1}$. The function h reaches its minimum at both extreme points of the interval, $x = a = 0$ and $x = b = 1$, at which $h(a) = h(b) = 1$, $h'(a) = 1$, $h'(b) = -1$, $\phi(a) = 1$ and $\phi(b) = 1/\sqrt{2}$. The asymptotic expansion is therefore given by the sum of (1.32) and (1.33):

$$I(\lambda) = e^{-\lambda}\left[\left(1 + \frac{1}{\sqrt{2}}\right)\frac{1}{\lambda} + O\left(\frac{1}{\lambda^2}\right)\right]\,, \qquad \lambda \to \infty\,.$$

As an exercise, see how the integral behaves in the limit $\lambda \to -\infty$. ◁

For better Laplace approximations, we need a more general expansion [27]. Assume that h has only one minimum on the interval $[a, b]$, at $x = a$; otherwise, we split the whole interval on suitable subintervals. Assume that h in the vicinity of $x = a$ (in the limit $x \searrow a$) can be represented as

$$h(x) \sim h(a) + \sum_{s=0}^{\infty} a_s (x - a)^{s+\alpha}\,, \tag{1.34}$$

where $\alpha \in \mathbb{R}$ and $\alpha > 0$, $a_0 \neq 0$, while ϕ can be represented as

$$\phi(x) \sim \sum_{s=0}^{\infty} b_s (x - a)^{s+\beta-1}\,, \tag{1.35}$$

where $b_0 \neq 0$ and we require $\mathrm{Re}\,\beta > 0$ for the constant $\beta \in \mathbb{C}$. Let h' and ϕ be continuous around $x = a$ (except perhaps at $x = a$ itself). Under these assumptions, if the integral $I(\lambda)$ absolutely converges for all large enough λ, we have the asymptotic expansion

$$I(\lambda) \sim e^{-\lambda h(a)} \sum_{s=0}^{\infty} \Gamma\left(\frac{s+\beta}{\alpha}\right) \frac{c_s}{\lambda^{(s+\beta)/\alpha}}\,, \qquad \lambda \to \infty\,. \tag{1.36}$$

The coefficients c_s in the expansion (1.36) can be expressed by the coefficients a_k and b_k for $k \leq s$. Here we write the first three,

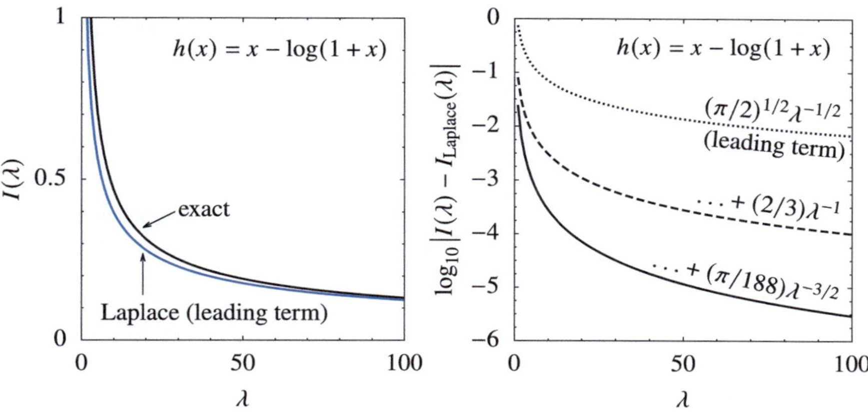

Fig. 1.7 Asymptotic behavior of $I(\lambda) = \int \phi(x) \exp[-\lambda h(x)]\,dx$, where $\phi(x) = 1$ and $h(x) = x - \log(1+x)$. [LEFT] Exact dependence on λ and the leading Laplace approximation. [RIGHT] The error of the approximation. The function h has a minimum at $a = 0$, where it can be expanded as $h(x) = x^2/2 - x^3/3 + x^4/4 - \cdots$. By comparison with (1.34) and (1.35) we get $\alpha = 2$, $a_s = (-1)^s/(s+2)$, $\beta = 1$, $b_0 = 1$, and $b_s = 0$, $s \geq 1$. From here we get the coefficients $c_0 = 1/\sqrt{2}$, $c_1 = 2/3$, and $c_2 = \sqrt{2}/12$ of (1.36)

$$
\begin{aligned}
c_0 &= \frac{b_0}{\alpha a_0^{\beta/\alpha}}, \\[2mm]
c_1 &= \left[\frac{b_1}{\alpha} - \frac{(\beta+1)a_1 b_0}{\alpha^2 a_0}\right] a_0^{-(\beta+1)/\alpha}, \\[2mm]
c_2 &= \left[\frac{b_2}{\alpha} - \frac{(\beta+2)a_1 b_1}{\alpha^2 a_0} + \left\{(\alpha+\beta+2)a_1^2 - 2\alpha a_0 a_2\right\} \frac{(\beta+2)b_0}{2\alpha^3 a_0^2}\right] a_0^{-(\beta+2)/\alpha},
\end{aligned}
\tag{1.37}
$$

while the procedure to compute any c_k can be found in [27] and [28]. The estimate of the error due to the truncation of (1.36) to a finite number of terms is discussed by [29] in Sect. 3.9. The general Laplace method is illustrated in Fig. 1.7 and by the following Example.

Example Let us revisit the calculation of the average interaction energy of two electric dipoles and its asymptotic behavior at low temperatures (1.20). We use the Laplace method to analyze the integral

$$
K(\lambda) = \frac{8\pi}{\sqrt{3}\lambda} \int_1^2 \frac{\sinh(\lambda x)}{\sqrt{x^2-1}}\,dx = \frac{4\pi}{\sqrt{3}\lambda}\left[\underbrace{\int_1^2 \frac{e^{\lambda x}}{\sqrt{x^2-1}}\,dx}_{I_1(\lambda)} - \underbrace{\int_1^2 \frac{e^{-\lambda x}}{\sqrt{x^2-1}}\,dx}_{I_2(\lambda)}\right]
$$

in the limit $\lambda = \mu_1\mu_2/(4\pi\varepsilon_0 r^3 kT) \to \infty$. By using $x \mapsto -x$ the first term can be rewritten as

$$I_1(\lambda) = \int_1^2 \frac{e^{\lambda x}}{\sqrt{x^2-1}}\,dx = \int_{-2}^{-1} \frac{e^{-\lambda x}}{\sqrt{x^2-1}}\,dx\ ,$$

so that $h(x) = x$ and $\phi(x) = 1/\sqrt{x^2-1}$. The function h has a minimum at $x = a = -2$, as required by the assumptions for the expansion (1.36). Since h is so simple, its expansion (1.34) has only two terms,

$$h(x) = x = h(a) + \sum_{s=0}^{\infty} a_s(x-a)^{s+\alpha} = -2 + a_0(x-(-2))^{0+\alpha} + 0 + 0 + \cdots\ ,$$

from which we read off $\alpha = 1$, $a_0 = 1$, and $a_s = 0$ for $s \geq 1$. We expand ϕ around a in the Taylor series and compare it to the expansion (1.35),

$$\phi(x) = \frac{1}{\sqrt{x^2-1}} = \frac{1}{\sqrt{3}} + \frac{2(x-a)}{3\sqrt{3}} + \frac{(x-a)^2}{2\sqrt{3}} + \cdots = \sum_{s=0}^{\infty} b_s(x-a)^{s+\beta-1}\ ,$$

from which we infer $\beta = 1$, $b_0 = 1/\sqrt{3}$, $b_1 = 2/(3\sqrt{3})$, and $b_2 = 1/(2\sqrt{3})$. By using the coefficients a_s and b_s we compute c_s by (1.37) and use them in the expansion (1.36). We get $c_0 = b_0$, $c_1 = b_1$, $c_2 = b_2$. Finally, we have $-\lambda h(a) = 2\lambda$, thus

$$I_1(\lambda) = e^{2\lambda}\left[\frac{1}{\sqrt{3}\lambda} + \frac{2}{3\sqrt{3}\lambda^2} + \frac{1}{\sqrt{3}\lambda^3} + \cdots\right]\ ,$$

from which (1.20) follows. The integral $I_2(\lambda)$ is already in the appropriate form, with the function $h(x) = x$ reaching its minimum at $x = a = 1$. Since $I_2(\lambda)$ behaves like $e^{-\lambda}$ it is negligible in comparison to $I_1(\lambda)$ in the limit $\lambda \to \infty$.   ◁

1.3.5 Stationary-Phase Approximation

The method of *stationary-phase approximation* is used to deduce the asymptotic series for the integrals of the form

$$I(\lambda) = \int_a^b \phi(x)\,e^{i\lambda h(x)}\,dx\ , \tag{1.38}$$

where h is a real function of a real variable x. The integrand thus contains the exponential function with an imaginary argument, and we are interested in the behavior

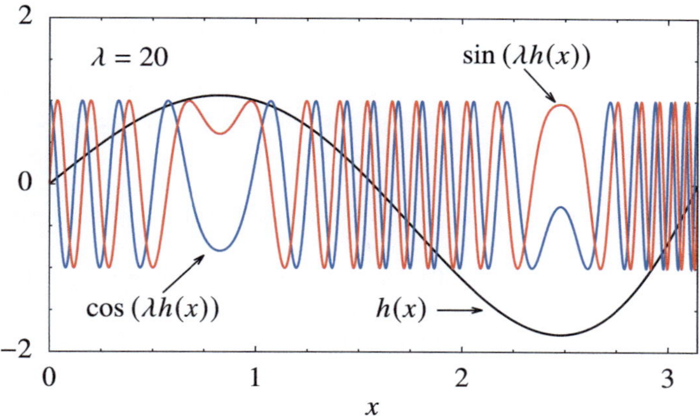

Fig. 1.8 The basic idea of stationary-phase approximation. At large λ the function $\exp(i\lambda h(x))$ in (1.38) rapidly oscillates around zero. In the regions of large variations of $h(x)$ the contributions to the integral therefore largely cancel out, while there is much less cancellation where the variation is small (near maxima and minima)

of the integral at $|\lambda| \gg 1$. In order to compute $I(\lambda)$ for negative arguments, we use the symmetry $I(\lambda)^\star = I(-\lambda)$.

The approximation can be established by realizing that the leading contribution to $I(\lambda)$ for $\lambda \to \infty$ comes from the integral over the points at which the *phase function h is stationary*, that is, $h'(x) = 0$. Assume that h has only one minimum ($h''(\xi) > 0$) or one maximum ($h''(\xi) < 0$) on the interval $[a, b]$, at ξ. We insert the expansion (1.30) into (1.38) and obtain

$$I(\lambda) \approx \int_a^b \phi(\xi)\, e^{i\lambda[h(\xi)+h''(\xi)(x-\xi)^2/2]} dx = \phi(\xi)\, e^{i\lambda h(\xi)} \int_a^b e^{i\lambda h''(\xi)(x-\xi)^2/2} dx \ .$$

We extend the integral on the right to the whole real axis and obtain the leading term of the stationary-phase approximation:

$$I(\lambda) \sim \phi(\xi) \sqrt{\frac{2\pi}{\lambda |h''(\xi)|}} \ \exp\left\{ i\left[\lambda h(\xi) + \frac{\pi}{4} \text{sign}(h''(\xi)) \right] \right\} \ , \tag{1.39}$$

where we have assumed $h''(\xi) \neq 0$ and used $\int_{-\infty}^\infty e^{ix^2} dx = \sqrt{\pi}\, e^{i\pi/4}$.

Example We are interested in the leading asymptotic term of the integral

$$I(\lambda) = \int_0^\pi \phi(x)\, e^{i\lambda h(x)} dx \ , \qquad h(x) = \sin(2x)\, e^{x^2/10} \ , \qquad \phi(x) = \frac{1}{\sqrt{x^2+1}} \ ,$$

in the limit $\lambda \to \infty$. On $[0, \pi]$, the function h has a maximum at $\xi_1 \approx 0.8266$ and a minimum at $\xi_2 \approx 2.4776$ (see Fig. 1.8). At these points, $h''(\xi_1) \approx -4.0841$ and $h''(\xi_2) \approx 7.2550$. The asymptotics of the integral is determined by two contributions of the form (1.39) which, at any $\lambda \gg 1$, are computed for $\xi = \xi_1$ and $\xi = \xi_2$, and then summed. With $\lambda = 20$, for example, we obtain

$$I(20) \approx (-0.0487 + 0.3566\,\mathrm{i}) + (-0.4814 + 0.2781\,\mathrm{i}) = -0.5301 + 0.6347\,\mathrm{i}\,,$$

while $I(20) \approx -0.5290 + 0.6280\,\mathrm{i}$ by precise numerical integration. The leading-order stationary-phase approximation to the complex value of the integral thus leads to the error in the modulus of $\approx 0.5\,\%$ and in the phase of $\approx 0.2°$. $\triangleleft$

Higher terms of the stationary-phase approximation can be obtained by the generalization of this method [27]. Assume that h has a finite number of stationary points (zeros of $h'(x)$) on the integration interval. We split it into subintervals such that there is one stationary point at the lower edge of every subinterval. Let $[a, b]$ be such a subinterval, and let h be monotonously increasing on $[a, b]$, thus $h'(x) > 0$ for $x \in [a, b]$ (in the opposite case, transform $x \mapsto -x$). Assume that h has the form

$$h(x) = h(a) + (x - a)^\alpha h_1(x)\,, \qquad h_1(a) \neq 0\,,$$

where h_1 is smooth on $[a, b]$ and $\alpha \geq 1$. Let ϕ have the form

$$\phi(x) = (x - a)^{\beta-1}\phi_1(x)\,,$$

where ϕ_1 is smooth on $[a, b]$ and $\beta \in (0, 1]$. Then the asymptotic expansion of the integral (1.38) in the limit $\lambda \to \infty$ is

$$I(\lambda) = \mathrm{e}^{\mathrm{i}\lambda h(a)}\left[\sum_{n=0}^{N-1} a_n \left(\frac{\mathrm{i}}{\lambda}\right)^{(n+\beta)/\alpha}\right] + R_N^{(1)} - \mathrm{e}^{\mathrm{i}\lambda h(b)}\left[\sum_{n=0}^{M-1} b_n \left(\frac{\mathrm{i}}{\lambda}\right)^{n+1}\right] + R_M^{(2)}\,,$$

where $R_N^{(1)} = O(\lambda^{-(N+\beta)/\alpha})$ and $R_M^{(2)} = o(\lambda^{-M})$ are the remainders due to the truncation of the series (see [27] for details). By introducing new variables $t^\alpha = h(x) - h(a)$ the coefficients a_n can be determined as

$$a_n = \frac{1}{\alpha n!}\,\Gamma\left(\frac{n+\beta}{\alpha}\right)\left(\frac{\mathrm{d}}{\mathrm{d}t}\right)^n\left[\left(\frac{x-a}{t}\right)^{\beta-1}\phi_1(x)\,\frac{\mathrm{d}x}{\mathrm{d}t}\right]\Bigg|_{t=0}\,,$$

and the coefficients b_n as

$$b_n = \left(\frac{1}{h'(x)}\frac{\mathrm{d}}{\mathrm{d}x}\right)^n\left[\frac{\phi(x)}{h'(x)}\right]\Bigg|_{x=b}\,.$$

Only few instances of functions h and ϕ allow for a simple calculation of the coefficients a_n and b_n. We typically let this work be done by programs for symbolic computation like MATHEMATICA [30]. In connection to the integration of rapidly oscillating functions see also Sect. 3.4.

General integrals along a contour C in the complex plane

$$I(\lambda) = \int_C g(z)\, e^{\lambda f(z)} dz\,, \qquad \lambda \to \infty\,,$$

where f and g are analytic, can be computed by means of the *method of steepest descent* and by the *saddle-point method*, which are both similar to the Laplace method in spirit, but technically more complicated. For further information, we refer the reader to [27] and [31].

1.3.6 Differential Equations with Large Parameters

Asymptotic approaches are also applicable to the analysis of differential equations. For a physicist, second-order homogeneous equations

$$y''(x) + p(x, \lambda)y'(x) + q(x, \lambda)y(x) = 0 \tag{1.40}$$

in the limit $\lambda \to \infty$ may be particularly relevant. By using the ansatz $y(x) = z(x)\exp\left(-\frac{1}{2}\int p(x,\lambda)\,dx\right)$ Eq. (1.40) can be put into the standard form

$$z''(x) + h(x, \lambda)z(x) = 0\,, \qquad h(x, \lambda) = q(x, \lambda) - \tfrac{1}{2}p'(x, \lambda) - \tfrac{1}{4}p^2(x, \lambda)\,. \tag{1.41}$$

Assume that $h(x, \lambda)$ has the Laurent expansion

$$h(x, \lambda) = \lambda^{2k} \sum_{n=0}^{\infty} h_n(x)\lambda^{-n}\,, \tag{1.42}$$

where k is a positive integer and $h_0 \neq 0$. By using this expansion, a large class of problems can be treated, in spite of the seemingly restrictive character of the leading term $\sim \lambda^{2k}$ in (1.42). Equation (1.40) has two types of solutions [26].

The first type of the solution has the form

$$z(x, \lambda) = A(x, \lambda)\, e^{S(x,\lambda)}\,, \tag{1.43}$$

where we have introduced the amplitude function $A(x, \lambda)$ and the action function $S(x, \lambda)$. They are defined as series in the parameter λ:

$$A(x, \lambda) = \sum_{n=0}^{\infty} a_n(x)\lambda^{-n} , \qquad S(x, \lambda) = \lambda^k \sum_{n=0}^{k-1} b_n(x)\lambda^{-n} . \tag{1.44}$$

When we insert (1.43) in (1.41) and collect the terms with powers λ^{2k-n}, we get

$$\left(b_0'\right)^2 + h_0 = 0 , \tag{1.45}$$

$$2b_0'b_m' + h_m + \sum_{n=1}^{m-1} b_n'b_{m-n}' = 0 , \qquad m = 1, 2, \ldots, k - 1 , \tag{1.46}$$

where $'$ denotes the derivative with respect to x. This is a system of differential equations for the coefficient functions b_n of the action $S(x, \lambda)$. Since $h_0 \neq 0$, squaring in (1.45) implies two possible signs for the derivative of the leading coefficient, $b_0' = \pm\sqrt{-h_0}$. These possibilities correspond to two linearly independent solutions of (1.41), as expected for a second-order equation. We then use the computed b_n in the equations for the coefficient functions a_n:

$$2a_0'b_0' + a_0 \left(b_0'' + h_k + \sum_{n=1}^{k-1} b_n'b_{k-n}' \right) = 0 , \tag{1.47}$$

$$2a_n'b_0' + \sum_{m=0}^{n} a_{n-m} A_m + 2 \sum_{m=1}^{n} a_{n-m}'b_m' + a_{n-k}'' = 0 , \qquad n = 1, 2, \ldots .$$

The functions h_n, a_n, and b_n are zero if the subscripts are outside of their ranges required by Eqs. (1.42) and (1.44), thus $h_{-n} = a_{-n} = b_{-n} = b_{k-1+n} = 0$ for $\forall n \in \mathbb{N}$. We have also introduced

$$A_m = b_m'' + h_{k+m} + \sum_{l=m+1}^{k-1} b_l'b_{k+m-l}' .$$

The second type of the solution of Eq. (1.41) has the form

$$z(x, \lambda) = e^{Z(x,\lambda)} , \qquad Z(x, \lambda) = \lambda^k \sum_{n=0}^{\infty} c_n(x)\lambda^{-n} , \tag{1.48}$$

where k is a positive integer. When the ansatz (1.48) is used in (1.41) and the terms with equal powers of λ are combined, we obtain

$$\left(c_0'\right)^2 + h_0 = 0 \,, \tag{1.49}$$

$$2c_0'c_n' + h_n + \sum_{m=1}^{n-1} c_m'c_{n-m}' = 0 \,, \qquad n = 1, 2, \ldots k-1 \,, \tag{1.50}$$

$$2c_0'c_n' + h_n + \sum_{m=1}^{n-1} c_m'c_{n-m}' + c_{n-k}'' = 0 \,, \qquad n = k, k+1, \ldots \,, \tag{1.51}$$

while the functions c_{-n} vanish for $\forall n \in \mathbb{N}$. By determining the phase of the leading term (the derivative of which has two possible dependencies, $c_0' = \pm\sqrt{-h_0}$) this procedure gives us two linearly independent solutions of (1.41).

For subscripts $0 \le n \le k-1$ the system of Eqs. (1.45) and (1.46) for the functions b_n is the same as the system (1.49) and (1.50) for the functions c_n, so $b_n = c_n$ for $0 \le n \le k-1$. By comparing (1.43) to (1.48) it becomes clear that the amplitude function of the first-type solution is just a formal expansion of the remainder of the action of the second-type solution,

$$A(x, \lambda) = \exp\left(\sum_{n=k}^{\infty} c_n(x)\lambda^{k-n} \right) \,,$$

with functions $\{c_n\}_{n=k}^{\infty}$ determined by (1.51).

We have assumed that $h(x, \lambda)$ as a function of λ has a pole of even degree at infinity, $h(x, \lambda) = O(\lambda^{2k})$. If the pole has an odd degree, $h(x, \lambda) = O(\lambda^{k})$, the solutions (1.43) and (1.48) are not valid. In this case we can introduce a new asymptotic parameter $\lambda' = \lambda^{1/2}$ and use λ' in the formulas derived above.

The procedure described here is called the WKB (Wentzel–Kramers–Brillouin) method. The special case $k = 1$ coincides with the problems of the Schrödinger equation for a particle of mass m in an one-dimensional potential V,

$$-\frac{\hbar^2}{2m} \frac{\partial^2}{\partial x^2} \psi(x) + \left[V(x) - E\right]\psi(x) = 0 \,. \tag{1.52}$$

In the limit of large energies ($E \to \infty$) or small Planck constant ($\hbar \to 0$) we speak of a *semi-classical* approach. The asymptotic parameter λ then represents high energies $E = \lambda^2$ or the smallness of $\hbar = \lambda^{-1}$. The WKB method is illustrated by the following Example adapted from [25]; for a detailed exposition see [32–34].

Example The classical example of the WKB method in quantum mechanics is the calculation of the particle's wave-function in a space with linear potential [33]. In Eq. (1.52) this means $V(x) = \lambda^2 x$, and it can be rewritten as

$$z''(x) - \lambda^2 x z(x) = 0 \,. \tag{1.53}$$

by a suitable change of variables. This equation is of the form (1.41) with $h(x, \lambda) = -\lambda^2 x$, but let us pretend for a moment that h is still general and has

the expansion (1.42). To solve (1.53), we use the ansatz (1.43). First, we determine
the leading coefficient of the action, b_0. From (1.45) it follows that

$$b_0'(x) = \pm\sqrt{-h_0(x)}\,, \qquad b_0''(x) = \mp\frac{h_0'(x)}{2\sqrt{-h_0(x)}}\,, \qquad b_0(x) = \pm\int_{x^\star}^x \sqrt{-h_0(t)}\,\mathrm{d}t\,.$$

The lower integration limit $x^\star$ will be determined in the following. We compute
the coefficient a_0 of the amplitude function (1.44) by using (1.47), $2a_0'b_0' + a_0b_0'' +
a_0h_1 = 0$. This is a differential equation for a_0:

$$\frac{\mathrm{d}a_0}{a_0} = \left(-\frac{1}{4}\frac{h_0'}{h_0} \mp \frac{h_1}{2\sqrt{-h_0}}\right)\mathrm{d}x\,.$$

We use $h_0'/h_0 = (\log h_0)'$, and obtain

$$a_0(x) = \frac{1}{[h_0(x)]^{1/4}}\exp\left\{\mp\frac{1}{2}\int_{x^\star}^x \frac{h_1(t)}{\sqrt{-h_0(t)}}\,\mathrm{d}t\right\}\,.$$

Since $k = 1$, the series (1.44) for $S(x, \lambda)$ contains only the term $\lambda^1 b_0\lambda^0 = \lambda b_0$. The
final structure of the solution to the leading order in λ is therefore simply $z(x, \lambda) =
a_0(x)\exp\{\lambda b_0(x)\}$, but its precise form still depends on the sign of the coefficient
function h_0 from the expansion (1.42).

The point $x^\star$, in which h_0 has a simple zero, $(h_0(x^\star) = 0, h_0'(x^\star) \neq 0)$, is called
the *turning* or *transition point*, since the physical character of the solution changes
at this point. In the region where $h_0 > 0$, the expressions written above yield two
linearly independent oscillatory solutions

$$z_{\mathrm{osc}}^{\pm}(x, \lambda) \approx \frac{1}{[h_0(x)]^{1/4}}\exp\left\{\pm\mathrm{i}\lambda\int_{x^\star}^x \sqrt{h_0(t)}\,\mathrm{d}t \pm \frac{\mathrm{i}}{2}\int_{x^\star}^x \frac{h_1(t)}{\sqrt{h_0(t)}}\,\mathrm{d}t\right\}\,,$$

while in the region with $h_0 < 0$ we get exponentially increasing or decreasing solutions

$$z_{\mathrm{exp}}^{\pm}(x, \lambda) \approx \frac{1}{[|h_0(x)|]^{1/4}}\exp\left\{\pm\lambda\int_{x^\star}^x \sqrt{|h_0(t)|}\,\mathrm{d}t \mp \frac{1}{2}\int_{x^\star}^x \frac{h_1(t)}{\sqrt{|h_0(t)|}}\,\mathrm{d}t\right\}\,.$$

In order for the WKB analysis to be valid, some authors require that the whole
function h, not just its leading term h_0, should have a zero at the turning point. It
turns out that it is very hard to formulate an asymptotic analysis of the WKB type if
h has a zero in the region being discussed while h_0 does not. The explicit demand

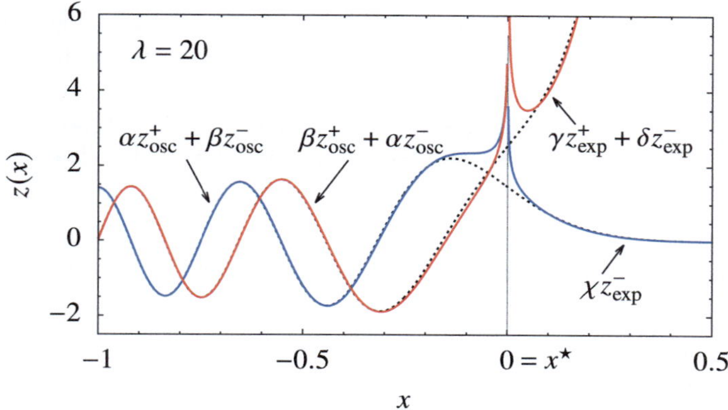

Fig. 1.9 The solution of $z''(x) - \lambda^2 x z(x) = 0$ with $\lambda = 20$ in the WKB approximation. The exact solutions (dashed curves) are given by the Airy functions Ai and Bi (Problem 1.6.2). The coefficients $\alpha, \beta, \gamma, \delta$ and χ in the linear combinations $z_{\text{osc}}^\pm$ and $z_{\text{exp}}^\pm$ are determined such that the WKB solutions match the exact solutions far from the turning point $x^\star = 0$

that h_0 has a simple zero can thus be understood as a necessary condition for the applicability of the WKB method.

Let us reconsider Eq. (1.53). From (1.42) we read off $h_0(x) = -x$ and $h_n(x) = 0$ for $n \geq 1$. In the exponents of $z_{\text{osc}}^\pm(x, \lambda)$ and $z_{\text{exp}}^\pm(x, \lambda)$ only the first term appears, and the turning point is $x^\star = 0$. In its vicinity the WKB approximation fails (see Fig. 1.9). The solution of (1.53) in the WKB approximation to the left of the turning point ($x < x^\star$ and $h_0(x) > 0$) is a linear combination of the solutions z_{osc}^+ and z_{osc}^-. The solution on the right ($x > x^\star$ and $h_0(x) < 0$) is a linear combination of z_{exp}^+ and z_{exp}^-. ◁

The method described above can be generalized to the case when h has a zero $x^\star$ of degree p. Assume that h is analytic at $x^\star$ and that in the vicinity of $x^\star$, in the limit $\lambda \to \infty$, it has the asymptotic expansion

$$h(x, \lambda) \sim C \lambda^{2k}(x - x^\star)^p , \qquad C \in \mathbb{R} . \tag{1.54}$$

By substituting $x - x^\star = |C \lambda^{2k}|^{-1/(2+p)} t$ and by using (1.54) we rewrite (1.41) as

$$\frac{d^2 z(t)}{dt^2} + s t^p z(t) = 0 , \qquad s = \text{sign}(C) .$$

This equation is valid near $t = 0$. Its solutions are known [35] and can be expressed in terms of the Bessel functions of the first and second kind [21] as

$$z(t) = \sqrt{t} \begin{cases} C_1 J_{1/2q}(t^q/q) + C_2 Y_{1/2q}(t^q/q) \; ; \; s = +1 , \\ C_1 I_{1/2q}(t^q/q) + C_2 K_{1/2q}(t^q/q) \; ; \; s = -1 , \end{cases}$$

where $q = \frac{1}{2}(p + 2)$. In seeking the solutions over a larger region, the constants C_1 and C_2 can be determined such that the solution near the turning point matches the solution far from the turning point in amplitude and phase.

For details on the formulation and use of asymptotic series see [26, 27]. Connections of asymptotic series to special functions are discussed in the classic work [29].

1.4 Summation of Finite and Infinite Series

Physical quantities are often represented as infinite or finite series

$$S = \sum_{k=0}^{\infty} a_k , \qquad S_n = \sum_{k=0}^{n} a_k , \qquad a_k \in \mathbb{R} \text{ or } \mathbb{C} .$$

The sum S_n of the first $n + 1$ terms of S is the *nth partial sum* of S. The series S *converges* if the sequence $\{S_n\}$ converges, i.e. if for any $\varepsilon > 0$ a $\kappa \in \mathbb{N}$ can be found such that for each $p \in \{0\} \cup \mathbb{N}$ we have $n > \kappa \implies |S_{n+p} - S_n| < \varepsilon$. A convergent sequence converges to a finite limit and this limit is its one and only cluster point. The series S is said to *diverge* if the sequence $\{S_n\}$ has a limit at infinity, has multiple cluster points or has no cluster points at all. Sometimes we carelessly interpret divergence as "convergence" to infinity.

General properties of series and summation methods are treated by the theory of summability [36]. Further reading on modern techniques of symbolic summation of series to closed forms can be found in [37, 38].

1.4.1 Tests of Convergence

Tests of convergence are procedures used to identify sufficient conditions for convergence of infinite series $\sum_{k=0}^{\infty} a_k$. In many tests, we disregard the signs of the terms (if $a_k \in \mathbb{R}$) or their phases (if $a_k \in \mathbb{C}$) and only use their absolute values. This simplification is based on the Cauchy inequality $|\sum_k a_k| \leq \sum_k |a_k|$, from which we infer that a series converges if the corresponding series with absolute values of terms converges (*absolute convergence*). The necessary condition for the convergence of any series is $\lim_{k\to\infty} a_k = 0$.

Comparison test For a given sequence $\{a_k\}_{k\in\mathbb{N}_0}$, where all $a_k \geq 0$, we find a sequence $\{b_k\}_{k\in\mathbb{N}_0}$. If there exists a $N \in \mathbb{N}_0$ such that $0 \leq a_k \leq b_k$ for all $k > N$ and the series $\sum_k b_k$ converges, the series $\sum_k a_k$ also converges. If at some $N \in \mathbb{N}_0$ we find $0 \leq b_k \leq a_k$ for all $k > N$ and the series $\sum_k b_k$ diverges, then the series $\sum_k a_k$ also diverges.

Quotient and Cauchy square-root test In the quotient test we observe the upper limit of the quotient of the consecutive terms of the series,

$$\rho = \limsup_{k \to \infty} \left| \frac{a_{k+1}}{a_k} \right| ,$$

while in the square-root test, we look at the upper limit of the square roots,

$$\rho = \limsup_{k \to \infty} |a_k|^{1/k} .$$

The series (absolutely) converges if $\rho < 1$, and (absolutely) diverges if $\rho > 1$. In the case $\rho = 1$ the test is inconclusive (the series may either converge or diverge).

Integral test Assume we have a sequence $\{a_k\}_{k \in \mathbb{N}}$ where all $a_k \geq 0$, and there exists a continuous monotonously decreasing function f such that $f(k) = a_k$ for all $k \geq 1$. Then the series $\sum_{k=1}^{\infty} a_k$ and the integral $\int_1^{\infty} f(x)\, dx$ either both converge or both diverge. In the case of convergence, the difference

$$R_n = S - S_n = \sum_{k=n+1}^{\infty} a_k$$

satisfies

$$\int_{n+1}^{\infty} f(x)\, dx \leq R_n \leq \int_n^{\infty} f(x)\, dx .$$

Kummer's and Raabe's test To perform the Kummer's test, we need a sequence $\{a_k\}_{k \in \mathbb{N}_0}$, $a_k > 0$, and a sequence $\{b_k\}_{k \in \mathbb{N}_0}$, $b_k > 0$, from which we form the limit

$$\rho = \lim_{k \to \infty} \left(b_k \frac{a_k}{a_{k+1}} - b_{k+1} \right) .$$

The series $\sum_k a_k$ converges if $\rho > 0$, while it diverges if $\rho < 0$ and the series $\sum_k 1/b_k$ diverges. If $\rho = 0$ the criterion is useless. In the case $b_k = k$ we obtain the Raabe's test, where we observe the limit

$$\rho = \lim_{k \to \infty} \left(k \frac{a_k}{a_{k+1}} - k - 1 \right) = \lim_{k \to \infty} \left[k \left(\frac{a_k}{a_{k+1}} - 1 \right) \right] - 1 .$$

The series $\sum_k a_k$ converges if $\rho > 0$, and diverges if $\rho < 0$. In the case $\rho = 0$ the convergence or divergence can not be ascertained.

Limit comparison test To a sequence $\{a_k\}_{k \in \mathbb{N}_0}$, $a_k > 0$, we find a sequence $\{b_k\}_{k \in \mathbb{N}_0}$, $b_k > 0$, such that the limit $\rho = \lim_{k \to \infty} a_k/b_k$ exists. If ρ is finite and $\rho \neq 0$, then both $\sum_k a_k$ and $\sum_k b_k$ either converge or diverge.

Leibniz's test for alternating series An important class of real series is represented by alternating series $\sum_{k=0}^{\infty}(-1)^k a_k$ where $a_k \geq 0$ (the consecutive terms change signs). If a_k decrease monotonically and $\lim_{k\to\infty} a_k = 0$ holds true, the alternating series converges. The remainder can be bounded as $|S - S_n| \leq a_n$.

Most of the enumerated tests are adapted for analytic work, but very often we can also use them numerically to determine with large certainty whether a given series diverges or converges.

We are also interested in the convergence of the power series

$$\sum_{k=0}^{\infty} a_k (z - z_0)^k \,, \qquad a_k, z, z_0 \in \mathbb{C}\,, \tag{1.55}$$

which is used to describe functions around a point z_0. Let us consider only absolute convergence and define the largest disk (circular region of points z in the complex plane) $\{z : |z - z_0| \leq r\}$, within which the series (1.55) absolutely converges. The disk radius r is the *convergence radius* and can be computed as

$$r = \left(\limsup_{k\to\infty} |a_k|^{1/k}\right)^{-1} \qquad \text{or} \qquad r = \limsup_{k\to\infty} \left|\frac{a_k}{a_{k+1}}\right|\,. \tag{1.56}$$

1.4.2 Summation of Series in Floating-Point Arithmetic

In floating-point arithmetic, the summation of real series $\sum_{k=0}^{n} a_k$ implies rounding errors. In particular for series with $n \to \infty$, precision is of utmost importance. Substantial work has been done in the minimization of summation errors (see [7, 39–41]). Here we list three most widely used summation methods that do not require more than $O(n)$ of operations.

Simple recursive summation Assume that we have the values $\{a_k\}_{k=0}^{n}$ and wish to compute their sum $S = \sum_{k=0}^{n} a_k$. Most obviously, this can be accomplished by computing $\hat{S} = (\ldots(((a_0 \oplus a_1) \oplus a_2) \oplus a_3) \ldots \oplus a_{n-1}) \oplus a_n$, in a loop

> **Input**: real numbers $a_0, a_1, \ldots, a_n$
> $\hat{S} = a_0$;
> **for** $k = 1$ **step** 1 **to** n **do**
> $\quad | \quad \hat{S} = \hat{S} + a_k$;
> **end**
> **Output**: $\hat{S}$ is the numerical sum of numbers a_k

The deviation of the numerical sum $\hat{S}$ from the exact sum S strongly depends on how a_k are sorted. If they are unsorted, we have

$$|S - \hat{S}| \leq \frac{\varepsilon_M}{2} n \sum_{k=0}^{n} |a_k| + O(\varepsilon_M^2) , \tag{1.57}$$

where ε_M is the arithmetic precision (see page 2) [40]. In simple summation one can therefore expect a loss of up to $\log_{10} n$ significant digits. The estimate for the upper limit of the error (not the error itself) is smallest when the terms are sorted as $|a_k| \leq |a_{k+1}|$. Sorting requires at least $O(n \log n)$ additional operations.

Kahan's algorithm A much better procedure to sum a series, by which the effect of rounding errors is greatly diminished, was proposed by Kahan [42]:

Input: real numbers $a_0, a_1, \ldots, a_n$
$\hat{S} = a_0$;
$c = 0$;
for $k = 1$ **step** 1 **to** n **do**
$\quad$ $y = a_k - c$;
$\quad$ $t = \hat{S} + y$;
$\quad$ $c = (t - \hat{S}) - y$; // do not omit brackets
$\quad$ $\hat{S} = t$;
end

Output: $\hat{S}$ is the numerical sum of numbers a_k

Algebraically, the value of c is zero, but in finite arithmetic it represents a large part of the lost precision when summing $t = \hat{S} + y$. It is added to the sum in the next step and by doing this, it compensates the rounding error from the previous step. The deviation of the numerical sum from the exact one satisfies

$$|S - \hat{S}| \leq \left(\varepsilon_M + O\left(n \varepsilon_M^2\right) \right) \sum_{k=0}^{n} |a_k| . \tag{1.58}$$

According to (1.58), Kahan's summation is more precise than simple summation for $n \varepsilon_M / 2 \leq 1$. In practice, this actually applies to even larger n (see Fig. 1.10). In the implementation of the algorithm we should make sure that the compiler does not simplify it, since the essence of its strength is hidden in the rules of floating-point arithmetic. In C and C++ variables should be declared `volatile`.

Recursive summation of pairs Summation is an associative and commutative operation between real numbers. Algebraically, the order of summation is thus irrelevant, and this fact is exploited by the Linz's procedure [43]. In the first step we sum the consecutive pairs of terms and obtain a new series. In this series we again sum the consecutive pairs and repeat this ($r = \lceil \log_2 n \rceil$)-times, until we are left with only one term, which represents the final sum:

Input: real numbers $a_0, a_1, \ldots, a_{n-1}$, where $n = 2^r, r \in \mathbb{N}$
$m = m' = n/2$;
for $k = 0$ **step** 1 **to** $m - 1$ **do**
$\quad | \quad S_{0,k} = a_{2k} + a_{2k+1}$;
end
for $j = 1$ **step** 1 **to** $r - 1$ **do**
$\quad | \quad m' = m'/2$;
$\quad | \quad$ **for** $k = 0$ **step** 1 **to** $m' - 1$ **do**
$\quad | \quad\quad | \quad S_{j,k} = S_{j-1,2k} + S_{j-1,2k+1}$;
$\quad | \quad$ **end**
end
Output: $\hat{S} = S_{r-1,0}$ is the numerical sum of numbers a_k

Because each term a_k in the sum is touched only r-times, the deviation of the sum $\hat{S}$
from the exact value S is much smaller than in simple recursive summation:

$$\left| S - \hat{S} \right| \le \frac{\varepsilon_M}{2} \log_2 n \sum_{k=0}^{n-1} |a_k|$$

(compare to (1.57)). For the intermediate sums $S_{j,k}$ we need additional computer
memory to store $n/2$ real numbers, which is a bit wasteful compared to the simple
and Kahan's summation which require only $O(1)$ of memory. Linz's algorithm can
be improved by compensating the numerical error and selecting the pairs in a more
intricate manner. For details, consult [44].

In all three methods we specified the upper bounds for $|S - \hat{S}|$; for a given set of
numbers $\{a_k\}$, all methods may be equally precise. In general, we recommend the
Linz's method unless pairs can not be formed or this does not make much sense (for
example, for relatively short series). On the other hand, the Kahan's method, which
is both simple and precise, never fails to enchant (see Fig. 1.10). For very precise
summation, we resort to more sophisticated but slower methods like *distillation
algorithms* described in [39, 45, 46].

1.4.3   Acceleration of Convergence

The convergence of the partial sums $S_n = \sum_{k=0}^{n} a_k$ to the limit $S = \lim_{n\to\infty} S_n$ may
be slow. By "slow" we mean its leading behavior to be $|S_n - S| = O(n^{-p})$ (power) or
$|S_n - S| = O((\log n)^{-p})$ (logarithmic) where $p > 0$ is the convergence order. Slow
convergence is not desired since it implies large numerical costs and a potential
accumulation of rounding errors.

We speak of "fast" convergence when it is better than "slow" according to the
definition given above. Ideally, one would like to have exponential (geometric) con-
vergence $|S_n - S| = O(a^n)$ where $a \in [0, 1)$. In many cases, convergence can be
accelerated by transforming the original series into another series that converges

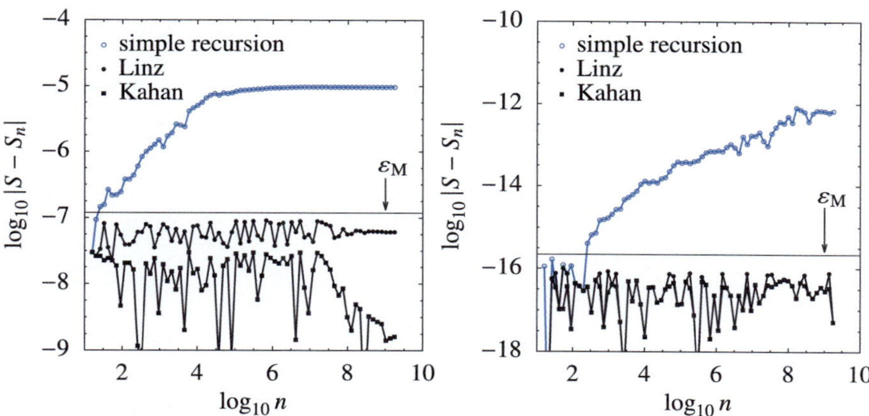

Fig. 1.10 Rounding errors in summing series with many terms. Shown is the absolute error of the numerical partial sums $S_n = \sum_{k=0}^{n}(-1)^k/(k+1)$ with respect to the limiting value $S_\infty = \log 2$. [LEFT] Summation in single-precision arithmetic. [RIGHT] Summation in double-precision arithmetic. The horizontal lines correspond to $\varepsilon_M = 1.19 \cdot 10^{-7}$ (left) and $\varepsilon_M = 2.22 \cdot 10^{-16}$ (right) — see page 2

more rapidly. In the following, we describe a few basic approaches. A modern introduction to convergence acceleration with excellent examples and many hints can be found in [47]; for a more detailed review, see [48].

Richardson extrapolation Assume that we already know the order of convergence for a series $S = \sum_{k=0}^{\infty} a_k$ so that for its partial sums $S_n = \sum_{k=0}^{n} a_k$ we have

$$S = S_n + \frac{\alpha}{n^p} + O(n^{-r}), \qquad r > p > 0.$$

We think of the "value of the series" as being the value of the partial sum plus a correction with the known leading-order behavior α/n^p. By transforming

$$T_n^{(1)} = \frac{2^p S_{2n} - S_n}{2^p - 1} = S_{2n} + \frac{S_{2n} - S_n}{2^p - 1}$$

the term α/n^p can be eliminated and the terms $T_n^{(1)}$ give us a better estimate for the sum, for which we obtain $S = T^{(1)} + O(n^{-r})$. The very same trick can be repeated — until this makes sense — by forming new sequences,

$$T_n^{(2)} = \frac{2^{p+1} T_{2n}^{(1)} - T_n^{(1)}}{2^{p+1} - 1}, \qquad T_n^{(3)} = \frac{2^{p+2} T_{2n}^{(2)} - T_n^{(2)}}{2^{p+2} - 1}, \qquad \cdots.$$

Richardson's procedure is an example of a *linear extrapolation method* X, in which for partial sums S_n and T_n of two series we have $X(\lambda S_n + \mu T_n) = \lambda X(S_n) + \mu X(T_n)$. It is efficient if the partial sums S_n behave like polynomials in some sequence

h_n, that is, $S_n = S + c_1 h_n^{p_1} + c_2 h_n^{p_2} + \cdots$ or $S_{n+1} = S + c_1 h_{n+1}^{p_1} + c_2 h_{n+1}^{p_2} + \cdots$, where the ratio h_{n+1}/h_n is constant. If this condition is not met, linear extrapolation may become inefficient or does not work at all. In such cases we resort to *semi-linear* or *non-linear extrapolation* [47].

Aitken's method Aitken's method is one of the classical and most widely used ways to accelerate the convergence by non-linear extrapolation. Assume that we have a sequence of partial sums S_n with the limit $S = \lim_{n\to\infty} S_n$. We transform the sequence S_n into a new sequence

$$T_n^{(1)} = S_n - \frac{(S_{n+1} - S_n)^2}{S_{n+2} - 2S_{n+1} + S_n}, \qquad n = 0, 1, 2, \ldots, \tag{1.59}$$

where the fraction should be evaluated exactly in the given form in order to minimize rounding errors. (Check that the transformed sequence (1.59) is identical to the column $\epsilon(n, 2)$ of the Wynn's table (1.13).) We repeat the process by using $T_n^{(1)}$ instead of S_n to form yet another, even more accelerated sequence $T_n^{(2)}$, and proceed thus until it continues to make sense as far as the rounding errors are concerned. Figure 1.11 (left) shows the comparison of convergence speeds for the unaccelerated partial sums S_n and the accelerated sequences $T_n^{(1)}$, $T_n^{(2)}$, and $T_n^{(3)}$.

Aitken's method is optimally suited for acceleration of linearly convergent sequences, for which $\lim_{n\to\infty}(S_{n+1} - S)/(S_n - S) = a$ with $-1 \le a < 1$. Such sequences originate in numerous numerical algorithms based on finite differences. In some cases, we apply Aitken's formula to triplets of partial sums S_{n+p}, S_n, and S_{n-p}, where $p > 1$, because sometimes the geometric convergence of a series only becomes apparent at larger p; see also Sect. 2.3 and [49].

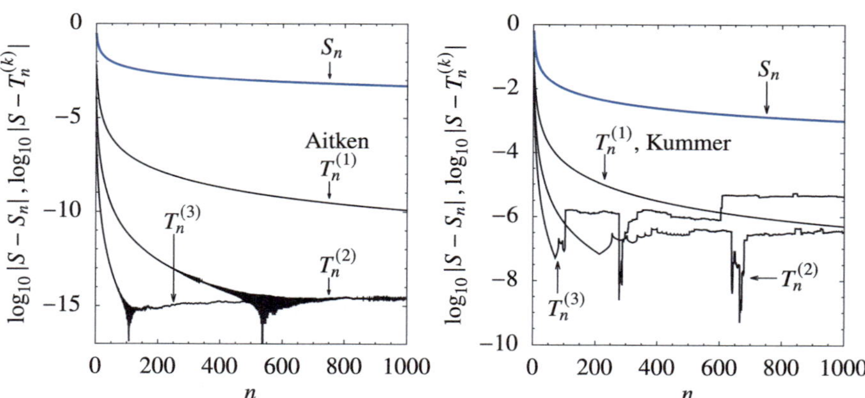

Fig. 1.11 Acceleration of convergence of partial sums. [LEFT] Aitken's method for the sum $S_n = \sum_{k=0}^{n}(-1)^k/(k+1)$ with the limit $S = \lim_{n\to\infty} S_n = \log 2$. Shown is the acceleration of this series with very slow (logarithmic) convergence by three-fold repetition of the Aitken's method. For typical series usually a single step of (1.59) suffices. [RIGHT] Kummer's acceleration of the sums $S_n = \sum_{k=1}^{n} 1/k^2$ with the limit $S = \lim_{n\to\infty} S_n = \pi^2/6$ by using the auxiliary series $\sum_{k=1}^{\infty} 1/(k(k+1))$

Kummer's acceleration The basic idea of the Kummer's method of summing a convergent series $S = \sum_k a_k$ is to subtract from it another (auxiliary) convergent series $B = \sum_k b_k$ with the known limit B, such that

$$\lim_{k \to \infty} \frac{a_k}{b_k} = \rho \neq 0 .$$

Then the original series can be transformed to

$$T = \sum_k a_k = \rho \sum_k b_k + \sum_k (a_k - \rho b_k) = \rho B + \sum_k \left(1 - \rho \frac{b_k}{a_k}\right) a_k . \qquad (1.60)$$

The convergence of the series on the right is faster than the convergence of that on the left since $(1 - \rho b_k / a_k)$ tends to zero when $k \to \infty$. An example is the sum $S = \sum_{k=1}^{\infty} 1/k^2 = \pi^2/6$ from which we subtract $B = \sum_{k=1}^{\infty} 1/(k(k+1)) = 1$, thus $\rho = \lim_{k \to \infty} k(k+1)/k^2 = 1$. We use the terms a_k and b_k, as well as ρ and B, in Eq. (1.60), and get the transformed partial sum

$$T_n^{(1)} = 1 + \sum_{k=1}^{\infty} \left(1 - \frac{k^2}{k(k+1)}\right) \frac{1}{k^2} ,$$

which has a faster convergence than the original series. Again, the procedure can be invoked repeatedly (see [50] and Fig. 1.11 (right)).

1.4.4 Alternating Series

In alternating series the sign of the terms flips periodically,

$$S = a_0 - a_1 + a_2 - a_3 + \ldots = \sum_{k=0}^{\infty} (-1)^k a_k , \qquad S_n = \sum_{k=0}^{n} (-1)^k a_k .$$

In physics such examples can be encountered e.g. in electro-magnetism in problems with oppositely charged particles or currents flowing in opposite directions. An example is the calculation of the electric potential $U(x, y, z)$ of charges of opposite signs lying next to each other at distances a along the x-axis:

$$U(x, y, z) \propto \sum_{k=-\infty}^{\infty} \frac{(-1)^k}{\sqrt{(x + ka)^2 + y^2 + z^2}} .$$

Making a monotonous series alternate For realistic physics problems, the results of series summation may be unpredictable. Simple recursive summation may suffer from large rounding errors. On the other hand, the acceleration of alternating series is typically more efficient than the acceleration of series with exclusively positive (or exclusively negative) terms. A monotonous sequence can be transformed into an alternating one by using the Van Wijngaarden's trick:

$$\sum_{k=0}^{\infty} a_k = \sum_{k=0}^{\infty} (-1)^k b_k \,, \qquad b_k = \sum_{j=0}^{\infty} 2^j a_{2^j(k+1)-1} \,.$$

Euler's transformation One of the oldest ways to accelerate the convergence of an alternating sequence by a linear combination of its terms is the *Euler transformation*. We rewrite the original sum $S = \sum_k (-1)^k a_k$ and its partial sum as

$$S = \sum_{k=0}^{\infty} (-1)^k \frac{\Delta^k a_0}{2^{k+1}} \,, \qquad S_n = \sum_{k=0}^{n} (-1)^k \frac{\Delta^k a_0}{2^{k+1}} \,, \tag{1.61}$$

where

$$\Delta^k a_0 = (-1)^k \sum_{j=0}^{k} (-1)^j \binom{k}{j} a_j \,.$$

If there exist $N \in \mathbb{N}$ and $C > 0$ such that $|\Delta^n a_0| \leq C$ for all $n > N$, the series (1.61) converges faster than geometrically with

$$|S - S_n| \leq \frac{C}{2^{n+1}} \,, \qquad n > N \,.$$

In practical algorithms, we first form the partial sums

$$s_n^{(0)} = \sum_{k=0}^{n} (-1)^k a_k \,, \qquad n = 0, 1, \ldots, N-1 \,,$$

and recursively compute the *partial Euler transforms*

$$s_n^{(j+1)} = \frac{1}{2} \left(s_n^{(j)} + s_{n+1}^{(j)} \right) \,, \qquad j = 0, 1, \ldots \,. \tag{1.62}$$

The values $T_n \equiv s_0^{(n)}$ represent the improved (accelerated) approximations of the partial sums S_n (Fig. 1.12 (left)). The procedure is numerically demanding, since it requires $O(n^2)$ operations and $O(n)$ of memory for a complete transformation of a series with n terms. It turns out that the optimally precise results are obtained not by using the transform (1.62) with $j = N - 1$ and $n = 0$, but with $j = \lceil 2N/3 \rceil$ and $n = \lceil N/3 \rceil$. An efficient implementation is given in [51].

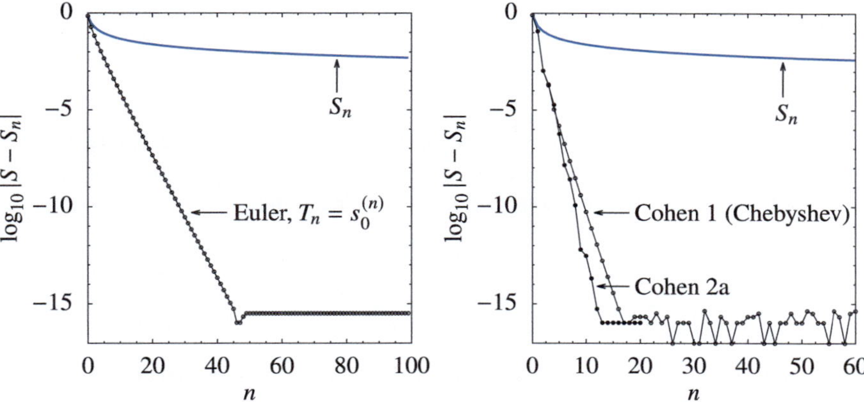

Fig. 1.12 Examples of acceleration of alternating series. Shown are the partial sums $S_n = \sum_{k=0}^{n}(-1)^k a_k$ without and with acceleration. [LEFT] Euler's method (1.62) for $a_k = 1/(k+1)$ (limit $S = \lim_{n\to\infty} S_n = \log 2$). [RIGHT] Cohen–Villegas–Zagier's algorithm 1 from page 45 by using Chebyshev polynomials (1.66) and algorithm 2a in [52] for $a_k = 1/(2k+1)$ (limit $S = \lim_{n\to\infty} S_n = \pi/4$). To compute the partial sum to ≈ 15 significant digits typically less than $\approx 10 - 20$ terms of the accelerated series are required

Generalizing the Euler's method Euler's transformation can be generalized by using the theory of measures. In this fresh approach to the summation of alternating series [52] we assume that for a series $\sum_{k=0}^{\infty}(-1)^k a_k$ there exists a positive function w such that the series terms a_k are its moments on the interval $[0, 1]$,

$$a_k = \int_0^1 x^k w(x)\, dx \,. \tag{1.63}$$

The sum of the series can then be written as

$$S = \sum_{k=0}^{\infty}(-1)^k a_k = \int_0^1 \left(\sum_{k=0}^{\infty}(-1)^k x^k w(x)\right) dx = \int_0^1 \frac{w(x)}{1+x}\, dx \,.$$

(In the final summation formula the weight function does not appear.) In the last step, we have used the identity

$$\sum_{k=0}^{n-1}(-1)^k x^k = \frac{1-(-x)^n}{1+x} \,, \qquad |x| < 1 \,, \tag{1.64}$$

in the limit $n \to \infty$. We now choose a sequence of polynomials $\{P_n\}$, where P_n has a degree n and $P_n(-1) \neq 0$. To the sequence $\{P_n\}$ we assign the numbers

$$S_n = \frac{1}{P_n(-1)} \int_0^1 \frac{P_n(-1) - P_n(x)}{1+x} \, w(x) \, dx \ .$$

The numbers S_n are linear combinations of the series terms a_k. This can be seen by inserting the expression for a general polynomial $P_n(x) = \sum_{k=0}^n p_k(-x)^k$ into the equation for S_n and observe (1.64) and a_k (1.63). We obtain

$$S_n = \frac{1}{d_n} \sum_{k=0}^{n-1} (-1)^k c_k^{(n)} a_k \ , \qquad d_n = \sum_{k=0}^n p_k \ , \qquad c_k^{(n)} = \sum_{j=k+1}^n p_j \ .$$

The S_n defined in this way represent the partial sums of S that converge to S when n is increased. The difference between the partial sum S_n and the sum S can be constrained as

$$\left| S - S_n \right| \leq \frac{1}{|P_n(-1)|} \int_0^1 \frac{|P_n(x)|}{1+x} \, w(x) \, dx \leq \frac{M_n}{|P_n(-1)|} |S| \ ,$$

where $M_n = \sup_{x \in [0,1]} |P_n(x)|$ is the maximum value of the polynomial P_n on $[0, 1]$. The sufficient condition for the convergence of the partial sums S_n to S is therefore $\lim_{n \to \infty} M_n / P_n(-1) = 0$. The authors of [52] recommend to choose a sequence of polynomials $\{P_n\}$ such that $M_n / P_n(-1)$ converges to zero as quickly as possible. The following three choices are the most fruitful.

The first type of the polynomials P_n that may cross one's mind, is

$$P_n(x) = (1-x)^n = \sum_{k=0}^n \binom{n}{k}(-x)^k \ , \qquad P_n(-1) = 2^n \ , \qquad M_n = 1 \ .$$

Namely, the corresponding partial sums are

$$S_n = \frac{1}{2^n} \sum_{k=0}^{n-1} (-1)^k c_k^{(n)} a_k \ , \qquad c_k^{(n)} = \sum_{j=k+1}^n \binom{n}{j} \ , \tag{1.65}$$

and they are identical to the partial sums of the Euler transform (1.61), except for a different subscripting (the sums (1.61) with subscript n are equal to the sums (1.65) with subscript $n+1$). By this choice we obtain $|S - S_n| \leq |S|/2^n$. Faster convergence, $|S - S_n| \leq |S|/3^n$, can be obtained by using the polynomials

$$P_n(x) = (1-2x)^n = \sum_{k=0}^n 2^k \binom{n}{k}(-x)^k \ , \qquad P_n(-1) = 3^n \ , \qquad M_n = 1 \ .$$

Here the partial sums have the form

$$S_n = \frac{1}{3^n} \sum_{k=0}^{n-1} (-1)^k c_k^{(n)} a_k \,, \qquad c_k^{(n)} = \sum_{j=k+1}^{n} 2^j \binom{n}{j} \,.$$

A third choice is a special family of Chebyshev polynomials, which have other beneficial algebraic properties and are orthogonal. We define these polynomials implicitly by $P_n(\sin^2 t) = \cos(2nt)$ or explicitly by

$$P_n(x) = T_n(1 - 2x) = \sum_{j=0}^{n} 4^j \frac{n}{n+j} \binom{n+j}{2j} (-x)^j \,,$$

where $T_n(x) = \cos(n \arccos x)$ are the Chebyshev polynomials of degree n on $[-1, 1]$. The polynomials of this sequence are computed by using the recurrence $P_{n+1}(x) = 2(1 - 2x)P_n(x) - P_{n-1}(x)$ initialized by $P_0(x) = 1$, $P_1(x) = 1 - 2x$, and one can show that $P_n(-1) = \frac{1}{2}[(3 + \sqrt{8})^n + (3 - \sqrt{8})^n]$ and $M_n = 1$. The partial sums

$$S_n = \frac{1}{P_n(-1)} \sum_{k=0}^{n-1} (-1)^k c_k^{(n)} a_k \,, \qquad c_k^{(n)} = \sum_{j=k+1}^{n} 4^j \frac{n}{n+j} \binom{n+j}{2j} \,, \qquad (1.66)$$

converge to the final sum as

$$\left| S - S_n \right| \le \frac{2|S|}{(3 + \sqrt{8})^n} < \frac{2|S|}{5.828^n} \,,$$

so we need to sum only $n \approx 1.31\, D$ terms for a precision of D significant digits! The coefficients $c_k^{(n)}$ and other constants can be computed iteratively and the whole computation of S_n can be implemented in a very compact algorithm [52]

Input: numbers $a_0, a_1, \ldots, a_{n-1}$ of an alternating series $\sum_{k=0}^{n-1}(-1)^k a_k$
$d = (3 + \sqrt{8})^n; d = (d + 1/d)/2;$
$b = -1; c = -d; s = 0;$
for $k = 0$ **step** 1 **to** $n - 1$ **do**
$\quad c = b - c;$
$\quad s = s + c\, a_k;$
$\quad b = (k + n)(k - n)b/((k + 1/2)(k + 1));$
end
Output: partial sum $S_n = s/d$

This algorithm requires $O(1)$ of memory and $O(n)$ of CPU. Similar results can be obtained by using other families of orthogonal polynomials; the paper [52] describes further algorithms in which the coefficients of the partial sums can not be generated as easily, but yield even faster convergence. For many types of sequences, these algorithms allow us to achieve convergence rates of $|S - S_n| \leq |S|/7.89^n$, in some cases even the breath-taking $|S - S_n| \leq |S|/17.93^n$. However, they require $O(n)$ of memory and $O(n^2)$ of CPU:

1.4.5 Levin's Transformations

Levin's transformations [53] are among the most generally useful, handy, and efficient methods to accelerate the convergence of series by semi-linear extrapolation. We implement them by using divided differences which are computed recursively:

$$\delta^k f_n = \frac{\delta^{k-1} f_{n+1} - \delta^{k-1} f_n}{t_{n+k} - t_n} \,, \qquad \delta^0 f_n = f_n \,,$$

where $t_n = (n + n_0)^{-1}$ and we usually take $n_0 = 0$ or $= 1$. To compute the extrapolated partial sums we need the partial sums S_n and auxiliary functions ψ which depend on the terms of the sequence and its character (monotonous or alternating). We use

$$S_{k,n} = \delta^k \left(\frac{S_n}{\psi(n)} \right) \left[\delta^k \left(\frac{1}{\psi(n)} \right) \right]^{-1} \,, \qquad k = 1, 2, \dots \,. \tag{1.67}$$

and take $S_{k,0}$ (with $n_0 = 1$) or $S_{k,1}$ (with $n_0 = 0$) as the extrapolated sum. Levin's transformations differ by the functional forms of ψ. The best known are

$$T : \ \psi(n) = a_n \,, \qquad U : \ \psi(n) = (n + n_0) a_n \,, \qquad W : \ \psi(n) = a_n^2/(a_{n+1} - a_n) \,.$$

The T-transformation is best for alternating series in which the partial sums behave as $S_n \sim r^n$. The U-transformation works well with monotonous sequences for which $S_n \sim n^{-r}$. The W-transformation can be used in either case, although it is more sensitive to rounding errors than the U- and T-methods. The U-method is recommended [47] as a reliable way to speed up the summation of any series. The U-transformation, including the extrapolation to the limit and providing the remainder estimates, is implemented in the GSL library [54] in the `gsl_sum_levin_u_accel()` function.

Example Let us sum the slowly converging series $S_n = \sum_{k=0}^n (-1)^k/(k + 1)$ with the limit $S = \lim_{n \to \infty} S_n = \log 2$. We choose the Levin's T-method and $n_0 = 0$, thus $t_n = n^{-1}$ and $\psi(n) = (-1)^n/(n + 1)$. By using (1.67) with $n = 1$ we obtain

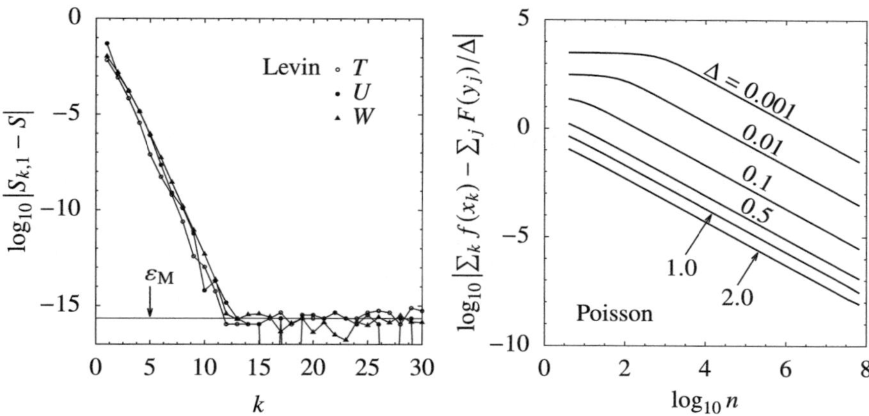

Fig. 1.13 [LEFT] Precision of Levin's methods T, U, and W in accelerating the convergence of the partial sums $S_n = \sum_{k=0}^{n}(-1)^k/(k+1)$ with the limit $S = \lim_{n\to\infty} S_n = \log 2$. [RIGHT] Precision of the Poisson's summation formula (1.68) for $f(x) = 1/(1+x^2)$ with different samplings $x_k = k\,\Delta$ on the real axis, where $-n \leq \{j,k\} \leq n$ and $n \gg 1$

$k = 1$	$S_k = 0.5$	$S_{k,1} = 0.7$
2	0.8333333333333334	0.69230769230769924
3	0.5833333333333334	0.6932153392330384
4	0.7833333333333333	0.6931436119116234
5	0.6166666666666667	0.6931472579962513
6	0.7595238095238096	0.6931471858853886
7	0.6345238095238096	0.6931471799439380
8	0.7456349206349208	0.6931471805844429
9	0.6456349206349208	0.6931471805603313
10	0.7365440115440117	0.6931471805598398

While the partial sums S_k merely hint at convergence, the accelerated sum $S_{k,1}$ at $k = 10$ is already precise to 12 digits. See also Fig. 1.13 (left). ◁

1.4.6 Poisson Summation

Often we encounter sums of function values f at equidistant points on the real axis,

$$S = \sum_{k\in\mathbb{Z}} f(x_k)\,, \qquad x_k = k\,\Delta\,, \qquad \Delta = x_{k+1} - x_k\,.$$

Assume that f is differentiable, that it decreases sufficiently fast at infinity, and that its Fourier transform

$$F(y) = \int\limits_{-\infty}^{\infty} f(x)\, e^{-i 2\pi xy}\, dx$$

exists. The sum of the values $f(x_k)$ and the sum of the transforms $F(y_j)$, computed at $y_j = j/\Delta$, $j \in \mathbb{Z}$, are linked by the Poisson's summation formula (see Fig. 1.13 (right))

$$\sum_{k=-\infty}^{\infty} f(x_k) = \frac{1}{\Delta} \sum_{j=-\infty}^{\infty} F(y_j) \,. \tag{1.68}$$

1.4.7 Borel Summation

Perturbative solutions in classical and quantum mechanics often appear as formally divergent series which, by appropriate means of summation, can yield a conditionally valid final result. Assume we have a sequence $\{a_k\}_{k\in\mathbb{N}_0}$ with the sum $\sum_{k=0}^{\infty} a_k$ that diverges. In the case when all a_k can be explicitly expressed as functions of the index k, the series can be summed by *Borel resummation*. The original sum is *resummable* in the form

$$S = \lim_{\xi\to\infty} e^{-\xi} \sum_{n=0}^{\infty} \frac{\xi^n}{n!}\, S_n \,, \qquad S_n = \sum_{k=0}^{n} a_k \,, \tag{1.69}$$

if the corresponding limit exists. In practice, the summation parameter ξ is not allowed to go to infinity; rather, we try to locate a range of its values in which the value of S stabilizes when ξ is being increased.

The resummation (1.69) is defined in its differential form. Even more often, we use the integral form, in which the series terms a_k (not the partial sums S_n) are used:

$$S = \int\limits_{0}^{\infty} e^{-\xi} \left(\sum_{k=0}^{\infty} \frac{a_k \xi^k}{k!} \right) d\xi \,.$$

This form is particularly useful when the function $f(\xi) = \sum_{k=0}^{\infty} a_k \xi^k / k!$ can be written in closed form or is well known in the region where it contributes most significantly to the integral $\int_0^{\infty} f(\xi)\, e^{-\xi}\, d\xi$. To do this, we can use the Padé approximation (see Sect. 1.2.2 and Problem 1.6.5).

1.4.8 Abel Summation

Assume that the series $S = \sum_{k=0}^{\infty} a_k$ formally diverges, but that the limit of the expression $S(x) = \sum_{k=0}^{\infty} x^k a_k$ still exists when $x \nearrow 1$. We can also introduce an auxiliary parameter ε such that $x = e^{-\varepsilon}$ and observe the limit $\varepsilon \searrow 0$. Then the value

$$S_A = \lim_{x \nearrow 1} S(x) = \lim_{x \nearrow 1} \sum_{k=0}^{\infty} x^k a_k = \lim_{\varepsilon \searrow 0} \sum_{k=0}^{\infty} e^{-\varepsilon k} a_k$$

is called the Abel's generalized sum of the series S. Like with the Borel summation, we introduce an intermediate parameter to regularize a divergent series and then try to sum it in the hope that the generalized limit is finite.

Example The divergent series

$$S = \sum_{k=0}^{\infty} k \cos kr = \operatorname{Re} \sum_{k=0}^{\infty} k\, e^{i kr} , \qquad 0 < r < 2\pi ,$$

has the Abel sum

$$S_A = \operatorname{Re} \lim_{x \nearrow 1} \sum_{k=0}^{\infty} x^k k\, e^{i kr} = \operatorname{Re} \lim_{x \nearrow 1} \sum_{k=0}^{\infty} k \left(x\, e^{ir} \right)^k = \operatorname{Re} \frac{e^{ir}}{(1 - e^{ir})^2} = -\frac{1}{4 \sin^2 r/2} .$$

As an exercise, change the parameter $0 < x < 1$ (approach $x \nearrow 1$) and the upper range of the sum, and watch what happens to the values S and S_A. ◁

An analogous method can be used with divergent integrals. If the integral $S = \int_a^{\infty} f(x)\, dx$ diverges, its generalized Abel sum is given by

$$S_A = \lim_{\varepsilon \searrow 0} \int_a^{\infty} e^{-\varepsilon x} f(x)\, dx .$$

Example The divergent integral

$$S = \int_0^{\infty} \sqrt{x} \cos \alpha x\, dx , \qquad \alpha > 0 .$$

has the generalized value

$$S_A = \lim_{\varepsilon \searrow 0} \int_0^{\infty} e^{-\varepsilon x} \sqrt{x} \cos \alpha x\, dx = \operatorname{Re} \lim_{\varepsilon \searrow 0} \int_0^{\infty} \sqrt{x}\, e^{-(\varepsilon + i\alpha)x}\, dx = -\frac{\sqrt{\pi}}{(2\alpha)^{3/2}} .$$

For exercise, take $\alpha = 1$. Change the upper integration limit in the expressions for S and S_A, and force the limiting parameter ε to approach zero from above. Compare the results to the analytic limit $-\sqrt{\pi}/(2\alpha)^{3/2}$.   ◁

1.5 Series Reversion

Let us study how the quantity y in the vicinity of y_0 depends on the variable x. For x near x_0 the dependence may be represented as a power series for some positive integer p,

$$y = f(x) = y_0 + a(x - x_0)^p h(x - x_0) , \tag{1.70}$$

where

$$h(x) = 1 + \alpha_1 x + \alpha_2 x^2 + \cdots .$$

Can we find the inverse function f^{-1}, i.e. the dependence of x on y? Indeed this can be done by solving the equation $f(x) = y$ for x at each given y separately by using general equation solvers. But a more convenient way to proceed is to calculate the reversed series

$$x = f^{-1}(y) = x_0 + \sum_{j=1}^{\infty} A_j (y - y_0)^{j/p} , \tag{1.71}$$

which is valid for y in the vicinity of y_0. Note that y must be larger than y_0 for even p. The $p = 1$ case is canonical and historically the best known. As we show later, the $p \neq 1$ case can be translated to $p = 1$, but one can also generalize the $p = 1$ formulas to $p \neq 1$. Namely, by plugging y from (1.70) into (1.71), expanding it in x around x_0 and matching the terms on the left and right side, we obtain a set of equations for the coefficients A_i with the solution

$$A_1 = a^{-1/p} , \qquad A_2 = -\frac{a^{-2/p}}{p}\alpha_1 , \qquad A_3 = \frac{a^{-3/p}}{2p^2}\left((3 + p)\alpha_1^2 - 2p\alpha_2\right) ,$$

$$A_4 = -\frac{a^{-4/p}}{3p^3}\left((p^2 + 6p + 8)\alpha_1^3 - 3p(p + 4)\alpha_1\alpha_2 + 3p^2\alpha_3\right) ,$$

and so on. The coefficients can be directly expressed by using an appropriately adapted Lagrange inverse formula pertaining to the $p = 1$ case [21]:

$$A_n = \frac{a^{-n/p}}{n!}\left[\frac{d^{n-1}}{dx^{n-1}}h^{-n/p}(x)\right]\Bigg|_{x=x_0} , \qquad n = 1, 2, \ldots .$$

There are many equivalent formulations of this inversion. One such convenient form typically used in computer algebra can be obtained in the following manner. Define the operator $[x_n]$ which extracts the coefficient of the x^n term in a series, i.e. $[x_n](\sum_k a_k x^k) = a_n$. Then the Lagrange inverse formula can be written as

$$A_n = [u_n] f^{-1}(u^p + y_0) = \frac{1}{n}[v_{n-1}]\left(\frac{v}{f(v + x_0)^{1/p}}\right)^n .$$

It is easy to see that the coefficients of the reversed series scale with the leading coefficient of the initial series a as

$$A_n = O\left(a^{-n/p}\right) .$$

This indicates that the series becomes problematic for vanishing a, but such a regime essentially means that y as a function of x is constant. For the convergence of the reversed series (1.71) it is important that the quantity $a^{-1/p}(y - y_0)$ is as small as possible.

By introducing new variables $u = x - x_0$ and $z = (y - y_0)^{1/p}$ we can transform the $p > 1$ case into the $p = 1$ case. The original series (1.70) then takes the form

$$z(u) = a^{1/p} u h^{1/p}(u) ,$$

while the reversed series is given by

$$u(z) = \sum_{j=1}^{\infty} A_j z^j .$$

Notice that the coefficients are identical to the ones appearing in (1.71). Consequently, it is sufficient in theory to discuss only $p = 1$ cases, but it is sometimes not very convenient. In the considered case the coefficients of the reversed series can be expressed as

$$A_n = \frac{1}{na^n} \sum_{s,t,u,\ldots} (-1)^{s+t+u+\cdots} \frac{n(n+1)\ldots(n-1+s+t+u+\cdots)}{s!\,t!\,u!\cdots} \alpha_1^s \alpha_2^t \alpha_3^u \cdots ,$$

where the sum runs over non-zero integers obeying $s + 2t + 3u + \ldots = n - 1$ (see pp. 411–413 of [34]). Explicit expressions for the coefficients in the form of determinants are given in [55].

Typically the reversed series are infinite, have complicated coefficients and small radii of convergence. We demonstrate the latter two properties in the Example below. In practice, therefore, the reversed series are used mainly for basic approximations of the inverse in the vicinity of specific points, which can then be refined by using more general non-linear equation solvers.

Example A simple series can have complicated reversed series. Consider the quadratic function $y = f(x) = x + bx^2$. The coefficients of the reversed series can be found analytically and are given by

$$A_n = \frac{b^{-n+1}}{n}\binom{2(n-1)}{n-1} \qquad n = 1, 2, \ldots ,$$

hence a finite series has generated an infinite reversed series with diverging coefficients. The convergence radius is equal to $\lim_{n\to\infty} A_n^{-1/n} = 1/(2b)$, hence the series converges absolutely for $x \in [-1/(2b), 1/(2b)]$.   ◁

The direct approach to computing the coefficients of the n-term reversed series by using Lagrange inverse formula in any form implies a numerical cost of $O(n^3)$ [8], although faster versions exist [56, 57]. The speed of calculating the coefficients is usually not a major obstacle as the reversed series considered in practice use only a few terms and they are typically calculated in advance for a known problem.

Example The Lagrange point L_1 in a binary system — the point on the line connecting the two stars where the acceleration is zero — is an important astrophysical parameter [58]. The potential of the primary star in the co-rotating reference frame with the primary star at the origin and the secondary star at $(d, 0, 0)$ is defined as

$$\Omega = \frac{1}{\sqrt{x^2 + y^2 + z^2}} + q \left(\frac{1}{\sqrt{(x - d)^2 + y^2 + z^2}} - \frac{x}{d^2} \right) + \frac{1}{2}(1 + q)(x^2 + y^2) \,,$$

where q is the mass ratio of the secondary versus the primary star. For simplicity, set $d = 1$. The Lagrange point L_1 is the hyperbolic point of Ω lying on the abscissa with $x \in [0, 1]$ satisfying

$$\frac{\partial}{\partial x}\Omega(x, 0, 0) = q \left(x + \frac{1}{(x - 1)^2} - 1 \right) - \frac{1}{x^2} + x = 0 \,.$$

We want to express the position of L_1 as a function of the reciprocal mass ratio q. The above equation can be rewritten as

$$\frac{1}{q} = \frac{x^3 \left(x^2 - 3x + 3\right)}{(1 - x)^3 \left(x^2 + x + 1\right)} := F(x) = 3x^3 + 3x^4 + 4x^5 + O(x^6) \,,$$

and we can empirically check that $F(1 - x)F(x) = 1$. We have thus obtained an additional relation between q and x, namely $q = F(1 - x)$, allowing us to conclude that for large q the coordinate of L_1 is small and vice-versa. Applying the Lagrange inversion formula to $1/q = F(x)$ yields the L_1 point approximation for large q,

$$x(L_1) = F^{-1}\left(q^{-1}\right) = \frac{q^{-1/3}}{3^{1/3}} - \frac{q^{-2/3}}{3^{5/3}} - \frac{q^{-1}}{27} + O\left(q^{-4/3}\right) \,,$$

while from $q = F(1 - x)$ we can extract the approximation

$$x(L_1) = 1 - F^{-1}(q) \,,$$

which is valid for small q.   ◁

1.6 Problems

1.6.1 Integral of the Normal Distribution

The probability density function $(1/\sqrt{2\pi})\,\exp(-x^2/2)$ of the standardized normal distribution pervades all branches of probability and statistics. Usually we need the probability over an interval, which can be computed by the integral

$$\operatorname{erf}(x) = \frac{2}{\sqrt{\pi}} \int_0^x \exp(-t^2)\,dt\,, \qquad \operatorname{erfc}(x) = 1 - \operatorname{erf}(x)\,. \tag{1.72}$$

Let us restrict the discussion to $x \geq 0$. The erf function monotonously increases and rapidly converges to 1 for large arguments. Its values are can be found in data tables, but for general purposes we wish to be able to compute them by exploiting different representations and approximations of the erf and erfc functions. If $\operatorname{erf}(x)$ is known, $\operatorname{erfc}(x)$ is also known (and vice-versa), but it is preferable to compute the function having a smaller argument because the error is easier to control. We can switch between calculations of erf and erfc at the point $x \approx 0.4769362762044695$ where $\operatorname{erf}(x) = \operatorname{erfc}(x) = \frac{1}{2}$.

For small x, we can use the power expansion

$$\operatorname{erf}(x) = \frac{2x}{\sqrt{\pi}} \sum_{k=0}^{\infty} (-1)^k \frac{x^{2k}}{k!\,(2k+1)}\,.$$

The convergence radius of this series is infinite, but its terms alternate and without acceleration it is not suitable for the computation of $\operatorname{erf}(x)$ at x much larger than 1. For large arguments, $x \gg 1$, the asymptotic expansion is suitable:

$$\operatorname{erfc}(x) \sim \frac{e^{-x^2}}{x\sqrt{\pi}} \left[1 + \sum_{k=1}^{\infty} (-1)^k \frac{(2k-1)!!}{(2x^2)^k} \right]\,.$$

In the range of x where both the power and the asymptotic expansions provide a poor description of erf, a rational approximation

$$\operatorname{erfc}(x) = e^{-x^2} \frac{\sum_{k=0}^{7} a_k x^k}{\sum_{k=0}^{7} b_k x^k} (1 + \varepsilon(x))$$

may be used. The parameters of this formula are listed in the following table.

k	a_k	b_k
0	1.000000000000013	1.000000000000000
1	1.474885681937094	2.603264849035166
2	1.089127207353042	3.026597029346489
3	0.481934851516365	2.046071816911715
4	0.133025422885837	0.873486411474986
5	0.021627200301105	0.237214006125950
6	0.001630015433745	0.038334123870994
7	-0.000000000566405	0.002889083295887

The relative precision of this parameterization is $\varepsilon(x) < 10^{-14}$ on $x \in [0, 5]$. An elegant, fast, but almost impenetrable implementation of the power expansion of erf and of the computation of erfc by means of tabulated values with a precision of 14 to 16 digits in C can be found in [59]. The algorithms in the GNU C library are also based on rational approximations with many parameters [60].

$\odot$ Examine the applicability and usefulness of different methods to compute $\mathrm{erf}(x)$. Watch the convergence of the power and asymptotic series. In the latter, sum the terms until the series appears to be converging. Does the convergence improve if Euler's method or Levin's U-transformation is used? By using all three ways of computation, write a program to calculate $\mathrm{erf}(x)$ in double precision on the whole real axis with an error less than 10^{-10}. Try to maximize the speed of the program. Show a comparison table of $\mathrm{erf}(x)$ for $x = 0\,(0.2)\,3$ and $\mathrm{erfc}(x)$ for $x = 3\,(0.5)\,8$.

The integral (1.72) can also be integrated numerically. What would be the required size of the subintervals you should use with the Simpson's formula in order to achieve a precision that would be comparable to the summation methods used above?

$\oplus$ Write a program to compute the inverse function erf^{-1} with an absolute precision of 10^{-9}. Now that we are in possession of an efficient procedure to compute $\mathrm{erf}(x)$, the inverse can be found by finding the root of the equation

$$F(x) = \mathrm{erf}(x) - y = 0$$

by using a method to solve non-linear equations. You can use bisection because erf is monotonous, but since the derivative $[\mathrm{erf}(x)]' = 2e^{-x^2}/\sqrt{\pi}$ is also known, the much faster Newton's method can be used. Compare the computed $\mathrm{erf}^{-1}(y)$ for small values of y to the power expansion [61]

$$\mathrm{erf}^{-1}(y) = \sum_{k=0}^{\infty} \frac{c_k}{2k+1} \left(\frac{\sqrt{\pi}}{2}y\right)^{2k+1} , \qquad c_0 = 1 , \quad c_{k\geq 1} = \sum_{m=0}^{k-1} \frac{c_m c_{k-1-m}}{(m+1)(2m+1)} .$$

1.6.2 Airy Functions

In physics, the Airy functions Ai and Bi (Fig. 1.14) appear in optics and quantum mechanics [62]. They are defined as the independent solutions of the equation

$$y''(x) - x y(x) = 0$$

and have the integral representations

$$\mathrm{Ai}(x) = \frac{1}{\pi} \int_0^\infty \cos(t^3/3 + xt)\, \mathrm{d}t\,, \quad \mathrm{Bi}(x) = \frac{1}{\pi} \int_0^\infty \left[\mathrm{e}^{-t^3/3 + xt} + \sin(t^3/3 + xt) \right] \mathrm{d}t\,.$$

For small x the functions Ai and Bi can be expressed by the Maclaurin series

$$\mathrm{Ai}(x) = \alpha f(x) - \beta g(x)\,, \qquad \mathrm{Bi}(x) = \sqrt{3}\left[\alpha f(x) + \beta g(x) \right],$$

where, at $x = 0$, we have $\alpha = \mathrm{Ai}(0) = \mathrm{Bi}(0)/\sqrt{3} \approx 0.355028053887817239$ and $\beta = -\mathrm{Ai}'(0) = \mathrm{Bi}'(0)/\sqrt{3} \approx 0.258819403792806798$. The series for f and g are

$$f(x) = \sum_{k=0}^\infty \left(\frac{1}{3}\right)_k \frac{3^k x^{3k}}{(3k)!}\,, \qquad g(x) = \sum_{k=0}^\infty \left(\frac{2}{3}\right)_k \frac{3^k x^{3k+1}}{(3k+1)!}\,,$$

where $(z)_n = \Gamma(z+n)/\Gamma(z)$ and $(z)_0 = 1$.

For large $|x|$ the Airy functions can be approximated by their asymptotic expansions. By substituting $\xi = \frac{2}{3}|x|^{3/2}$ and by using the asymptotic series

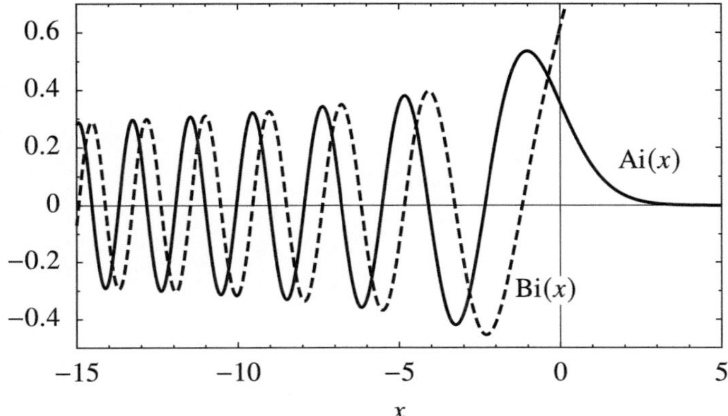

Fig. 1.14 Airy functions Ai and Bi for real arguments. Ai is bound everywhere, while Bi diverges at $x \to \infty$. The zeros of both functions occur only on the negative semi-axis

$$L(z) \sim \sum_{s=0}^{\infty} \frac{u_s}{z^s} \,, \qquad P(z) \sim \sum_{s=0}^{\infty} (-1)^s \frac{u_{2s}}{z^{2s}} \,, \qquad Q(z) \sim \sum_{s=0}^{\infty} (-1)^s \frac{u_{2s+1}}{z^{2s+1}} \,,$$

with coefficients

$$u_s = \frac{\Gamma(3s + \frac{1}{2})}{54^s s! \, \Gamma(s + \frac{1}{2})} \,,$$

we get, for large positive x,

$$\mathrm{Ai}(x) \sim \frac{e^{-\xi}}{2\sqrt{\pi} x^{1/4}} L(-\xi) \,, \qquad \mathrm{Bi}(x) \sim \frac{e^{\xi}}{\sqrt{\pi} x^{1/4}} L(\xi) \,,$$

while for large negative x we get

$$\mathrm{Ai}(x) \sim \frac{1}{\sqrt{\pi}(-x)^{1/4}} \Big[\, \sin(\xi - \pi/4)\, Q(\xi) + \cos(\xi - \pi/4)\, P(\xi) \Big] \,,$$

$$\mathrm{Bi}(x) \sim \frac{1}{\sqrt{\pi}(-x)^{1/4}} \Big[-\sin(\xi - \pi/4)\, P(\xi) + \cos(\xi - \pi/4)\, Q(\xi) \Big] \,.$$

$\odot$ Find an efficient procedure to compute the values of the Airy functions Ai and Bi on the whole real axis with a precision better than 10^{-10} by using a combination of the Maclaurin series and the asymptotic expansion. When estimating the errors, use programs that are capable of arbitrary-precision computations, e.g. MATHEMATICA.

$\oplus$ The zeros of Ai have an important role in mathematical analysis when one tries to determine the intervals containing zeros of other special functions and orthogonal polynomials [63], as well as in physics in computation of energy spectra of quantum systems [33]. Compute the first hundred zeros $\{a_s\}_{s=1}^{100}$ of the Airy function Ai and the first hundred zeros $\{b_s\}_{s=1}^{100}$ of Bi at $x < 0$, and compare the computed values to the formulas

$$a_s = -f\left(\frac{3\pi(4s - 1)}{8}\right) \,, \qquad b_s = -f\left(\frac{3\pi(4s - 3)}{8}\right) \,, \qquad s = 1, 2, \dots \,,$$

where f has the asymptotic expansion [21]

$$f(z) \sim z^{2/3} \left(1 + \frac{5}{48} z^{-2} - \frac{5}{36} z^{-4} + \frac{77125}{82944} z^{-6} - \frac{108056875}{6967296} z^{-8} + \dots \right) \,.$$

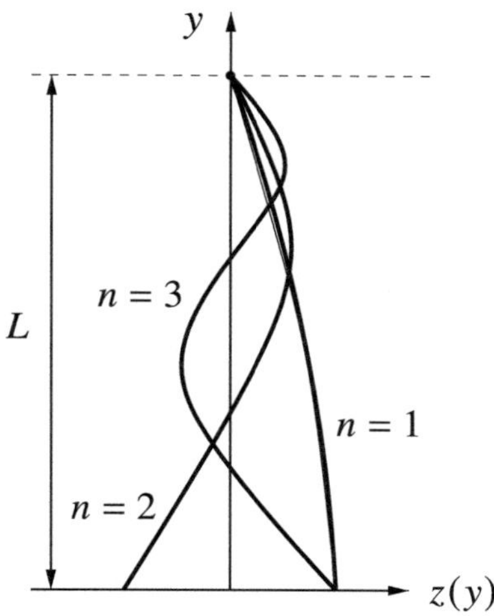

Fig. 1.15 Small transverse deflections of a rope in the gravitational field

1.6.3 Bessel Functions

Solving differential equations in circular or spherical geometry often leads to Bessel functions of the first kind J_ν and second kind Y_ν. A physical example which does not belong to this group but is related to it, is the problem of the oscillations of a freely hanging heavy rope in a constant gravitational field (Fig. 1.15).

A rope of length L is suspended from a ceiling and the origin of the y axis is placed at the free end of the rope. We study small deflections $z(y)$ of the rope in the field of the gravitational acceleration g. Three lowest eigenmodes are shown. The eigenmodes are described by the equation

$$\frac{\mathrm{d}}{\mathrm{d}y}\left(y\,\frac{\mathrm{d}z(y)}{\mathrm{d}y}\right) + \frac{\omega^2}{g}z(y) = 0\,,$$

where $z(y)$ is the deflection of the rope at y when oscillating with the angular frequency ω. The eigenfrequencies are determined by the equation and the boundary conditions $|z(0)| < \infty$ (deflection bounded at $y = 0$) and $z(L) = 0$ (fixed end at $y = L$). By the substitution $t = 2\omega\sqrt{y/g}$ the differential equation can be transformed to $\ddot{z}(t) + \dot{z}(t)/t + z(t) = 0$. The solution of this equation is a linear combination of the Bessel functions J_0 and Y_0. With $\alpha = \sqrt{L/g}$ the boundary conditions become $|z(t = 0)| < \infty$ and $z(t = 2\omega\alpha) = 0$. The latter eliminates Y_0 which is singular at the origin. The condition for the eigenfrequencies is then

$$\omega_n = \xi_{0,n}/(2\alpha)\,, \qquad n = 1, 2, \dots\,,$$

where $\xi_{0,n}$ is the nth zero of J_0. The values of $J_0(x)$ for small x can be computed by using the power expansion

$$J_0(x) = \sum_{k=0}^{\infty} \frac{(-1)^k (x/2)^{2k}}{(k!)^2} \, ,$$

while at large enough arguments, we use the formula

$$J_0(x) = \sqrt{\frac{2}{\pi x}} \Big[P(x) \cos(x - \pi/4) + Q(x) \sin(x - \pi/4) \Big], \qquad x \to \infty \, ,$$

where $P(x)$ and $Q(x)$ are known in terms of the asymptotic series [64]

$$P(x) \sim \sum_{k=0}^{\infty} (-1)^k \frac{a_{2k}^2}{(2k)!(8x)^{2k}} \, , \qquad Q(x) \sim \sum_{k=0}^{\infty} (-1)^k \frac{a_{2k+1}^2}{(2k+1)!(8x)^{2k+1}}$$

with coefficients $a_n = (2n - 1)!!$. (Note $(-1)!! = 0!! = 1$.) For intermediate x we compute $J_0(x)$ by using Bessel functions of higher orders J_n. In the limit $n \to \infty$ and constant x we have $J_n(x) \sim (x/2)^n/(n!)$, and for $x \sim 1$ and $n > 2x$ this is a very good approximation. Suppose we wish to calculate $J_0(x)$ at given x in arithmetic with precision ε_M. With the asymptotic approximation we determine an even $N \gg 2x$ such that $\varepsilon = J_N(x) \ll \varepsilon_M$. This value of $J_N(x)$ is not normalized. Let us denote J_n temporarily by C_n and, for given x, start the iteration

$$C_{N+1} = 0 \, , \qquad C_N = \varepsilon \, , \qquad C_{n-1}(x) = \frac{2n}{x} C_n(x) - C_{n+1}(x) \, .$$

We obtain $C_0(x)$ which differs by a factor from the true value, $J_0(x) = C_0(x)/A$. The factor A is determined by using the identity $J_0(x) + 2 \sum_{k=1}^{\infty} J_{2k}(x) = 1$:

$$A = C_0(x) + 2 \sum_{k=1}^{N/2} C_{2k}(x) \, .$$

$\odot$ Write a procedure to compute J_0 on the whole real axis to an absolute precision of 10^{-12} and compare it to the result of a tool of higher precision, like MATHEMATICA or MATLAB. Determine the first $N = 10000$ zeros $\{\xi_{0,n}\}_{n=1}^{N}$ of J_0 and compare them to the asymptotic formula [21, 29]

$$\xi_{0,n} \sim \beta + \frac{1}{8\beta} - \frac{31}{384\,\beta^3} + \frac{3779}{15360\,\beta^5} - \frac{6277237}{3440640\,\beta^7} + \dots , \qquad n \to \infty \, ,$$

where $\beta = (n - 1/4)\pi$. Draw the first five eigenmodes of the oscillating rope.

1.6.4 Alternating Series

Some alternating series are literally famous, for example

$$\frac{\pi}{4} = \sum_{k=0}^{\infty} \frac{(-1)^k}{2k+1} \, , \qquad \log 2 = \sum_{k=0}^{\infty} \frac{(-1)^k}{k+1} \, , \qquad \frac{\pi^2}{12} = \sum_{k=0}^{\infty} \frac{(-1)^k}{(k+1)^2} \, .$$

Just as well-known, the natural algorithm has the expansions

$$\log x = \sum_{k=1}^{\infty} \frac{(-1)^{k+1}}{k} (x-1)^k \qquad ; \ |x-1| \le 1 \, , \ x \ne 1 \, , \ \text{or}$$

$$\log x = 2 \sum_{k=0}^{\infty} \frac{1}{2k+1} \left(\frac{x-1}{x+1} \right)^{2k+1} \quad ; \ x > 0 \, ,$$

that enable us to compute $\log x$ on the whole positive semi-axis.

$\odot$ Compute the sums of these series to a precision of $\varepsilon = 10^{-7}$ by using different methods applicable to alternating series, and study their convergence. (The second series in (1.73) is an exception since it is not alternating.) Compare these methods to simple summation. Make a detailed analysis of the rounding errors first by using single-precision data types (type `float` in C or C++), and then by using double precision (type `double`).

$\oplus$ Sum the series given above by using a data type that allows for variable precision (for example, by using the GMP library mentioned in Appendix B.3). Draw a diagram of computational (CPU) times versus the required precision ε in the range $\log_{10} \varepsilon \in [-300, -4]$.

1.6.5 Coulomb Scattering Amplitude and Borel Resummation

An eloquent example [65] of trouble we may face by careless summation of divergent series is the Rutherford scattering amplitude for Coulomb scattering

$$f(\theta) = -\frac{\eta}{2k \sin^2(\theta/2)} \exp\left[\mathrm{i} \left(2\sigma_0 - \eta \log \sin^2(\theta/2) \right) \right] . \qquad (1.73)$$

We know that the asymptotic expansion of this amplitude is

$$f(\theta) \sim \frac{1}{2\mathrm{i}k} \sum_{l=0}^{\infty} (2l+1) \, P_l(\cos\theta) \left(\mathrm{e}^{2\mathrm{i}\sigma_l} - 1 \right) , \qquad (1.74)$$

where σ_l is the phase shift for Coulomb scattering in the partial wave with the orbital angular momentum quantum number l, and P_l is the Legendre polynomial of degree l (see Eq. (5.32)). The phase shift in the lth partial wave is

$$\sigma_l = \arg \Gamma(l + 1 + i\eta) \, .$$

In the limit $l \to \infty$ the phase shifts behave as $\sigma_l \sim \log l$, while the values $P_l(\cos\theta)$ at fixed θ fade out like $P_l(\cos\theta) \sim l^{-1/2}$. Therefore, for $l \to \infty$, the terms in the sum (1.74) oscillate within an envelope that is proportional to $l^{1/2}$, and the series diverges.

$\odot$ Sum the series (1.74) directly by summing the terms up to $l_{max} = 100$ by using $\eta = 1$ and $k = 1\,\mathrm{fm}^{-1}$. Compute the sum for angles $30° \le \theta \le 180°$ in steps of $10°$ and compare the results to (1.73). Do not calculate the phase shifts by using the gamma function. Rather, use backward recurrence: start with the value of σ_l at $l = l_{max}$, for which the Stirling approximation applies:

$$\sigma_l \sim \left(l + \frac{1}{2}\right)\beta + \eta\log\alpha - \eta - \frac{\sin\beta}{12\alpha} + \frac{\sin 3\beta}{360\alpha^3} - \frac{\sin 5\beta}{1260\alpha^5} + \frac{\sin 7\beta}{1680\alpha^7} - \frac{\sin 9\beta}{1188\alpha^9} + \cdots \, ,$$

where

$$\alpha = \sqrt{(l + 1)^2 + \eta^2} \, , \qquad \beta = \arctan\left(\frac{\eta}{l + 1}\right) \, .$$

Then use the recurrence to compute the phase shifts at lower l:

$$\sigma_l = \sigma_{l+1} - \arctan\left(\frac{\eta}{l + 1}\right) \, , \qquad l = l_{max} - 1, l_{max} - 2, \ldots, 0 \, .$$

Similarly, compute the Legendre polynomials by using the three-term recurrence formula $(l + 1)\, P_{l+1}(x) = (2l + 1)\, x\, P_l(x) - l\, P_{l-1}(x)$, which is initialized by $P_0(x) = 1$ and $P_1(1) = x$.

In addition, sum the series by Borel resummation in differential form (1.69). Compare the exact value (1.73) to the numerical one at angles $\theta = 10°, 60°, 120°,$ and $150°$ with the parameter ξ in the range $5 \le \xi \le 100$ in steps of 5.

$\oplus$ Compute the Rutherford sum by applying the Wynn algorithm (1.13). At scattering angles $\theta = 10°, 60°, 120°,$ and $150°$, calculate the diagonal Padé approximations $[n/n]$ for $0 \le n \le 20$. Stop the recurrence in the algorithm when the denominator of the fraction on the right side of (1.13) becomes equal to zero.

References

1. T.H. Cormen, C.E. Leiserson, R.L. Rivest, C. Stein, *Introduction to Algorithms*, 2nd edn. (MIT Press/McGraw-Hill, Cambridge/New York, 2001)
2. D.E. Knuth, *The Art of Computer Programming, Vol. 1: Fundamental Algorithms*, 3rd edn. (Addison-Wesley, Reading, 1997)
3. IEEE Standard 754–2008 for binary floating-point arithmetic. IEEE Standards Association (2008). Available at http://standards.ieee.org/ieee/754/4211/
4. D. Goldberg, What every computer scientist should know about floating-point arithmetic. ACM Comp. Surv. **23**, 5 (1991)
5. M.H. Holmes, *Introduction to Numerical Methods in Differential Equations* (Springer, New York, 2007); see Example in Appendix A.3.1
6. GNU Multi Precision (GMP), Free library for arbitrary precision arithmetic. Available at http://gmplib.org
7. H.J. Wilkinson, *Rounding Errors in Algebraic Processes* (Dover Publications, Mineola, 1994)
8. D.E. Knuth, *The Art of Computer Programming, Vol. 2: Seminumerical Algorithms*, 2nd edn. (Addison-Wesley, Reading, 1980)
9. J. Stoer, R. Bulirsch, *Introduction to Numerical Analysis, Texts in Applied Mathematics*, vol. 12, 3rd edn. (Springer, Berlin, 2002)
10. D. O'Connor, Floating Point Arithmetic, In *Dublin Area Mathematics Colloquium*, 5 Mar 2005
11. M.J.D. Powell, *Approximation Theory and Methods* (Cambridge University Press, Cambridge, 1981). See also C. Hastings, *Approximations for Digital Computers* (Princeton University Press, Princeton, 1955), which is a pedagogical jewel; a seemingly simplistic outward appearance hides a true treasure-trove of ideas yielding one insight followed by another
12. W. Fraser, A survey of methods of computing minimax and near-minimax polynomial approximations for functions of a single variable. J. Assoc. Comput. Mach. **12**, 295 (1965)
13. H.M. Antia, *Numerical Methods for Scientists and Engineers*, 2nd edn. (Birkhäuser, Basel, 2002); see Sects. 9.11 and 9.12
14. R. Pachón, L.N. Trefethen, *Barycentric-Remez Algorithms for Best Polynomial Approximation in the Chebfun System* (Oxford University Computing Laboratory, NAG Report No. 08/20)
15. G.A. Baker, P. Graves-Morris, *Padé approximants, Encyclopedia of Mathematics and its Applications*, No. 59, 2nd edn. (Cambridge University Press, Cambridge, 1996)
16. P. Gonnet, S. Güttel, L.N. Trefethen, Robust Padé approximation via SVD. SIAM Rev. **55**, 101 (2013)
17. P. Wynn, On the convergence and stability of the epsilon algorithm. SIAM J. Num. Anal. **3**, 91 (1966)
18. P.R. Graves-Morris, D.E. Roberts, A. Salam, The epsilon algorithm and related topics. J. Comput. Appl. Math. **12**, 51 (2000)
19. W. van Dijk, F.M. Toyama, Accurate numerical solutions of the time-dependent Schrödinger equation. Phys. Rev. E **75**, 036707 (2007)
20. K. Kormann, S. Holmgren, O. Karlsson, J. Chem. Phys. **128**, 184101 (2008); see also C. Lubich, Quantum simulations of complex many-body systems: from theory to algorithms, in J. Grotendorst, D. Marx, A. Muramatsu, editors. *John von Neumann Institute for Computing, Jülich, NIC Series*, vol. 10 (2002), p. 459
21. M. Abramowitz, I.A. Stegun, *Handbook of Mathematical Functions*, 10th edn. (Dover Publications, Mineola, 1972)
22. I.S. Gradshteyn, I.M. Ryzhik, *Tables of Integrals, Series and Products* (Academic Press, New York, 1980)
23. P. C. Abbott, Asymptotic expansion of the Keesom integral, J. Phys. A: Math. Theor. **40**, 8599 (2007); M. Battezzati, V. Magnasco, Asymptotic evaluation of the Keesom integral. J. Phys. A: Math. Theor. **37**, 9677 (2004)
24. T.M. Apostol, *Mathematical Analysis*, 2nd edn. (Addison-Wesley, Reading, 1974)
25. P.D. Miller, *Applied Asymptotic Analysis, Graduate Studies in Mathematics*, vol. 75 (AMS, Providence, 2006)

26. A. Erdélyi, *Asymptotic Expansions* (Dover, New York, 1987)
27. R. Wong, *Asymptotic Approximations of Integrals* (SIAM, Philadelphia, 2001)
28. J. Wojdylo, Computing the coefficients in Laplace's method. SIAM Rev. **48**, 76 (2006). While Ref. [27] in Section II.1 describes the classical way of computing the coefficients c_s by series inversion, this article discusses a modern explicit method
29. F.J.W. Olver, *Asymptotics and Special Functions* (A. K. Peters, Wellesley, 1997)
30. S. Wolfram, *Wolfram Mathematica*. http://www.wolfram.com
31. N. Bleistein, R.A. Handelsman, *Asymptotic Expansion of Integrals* (Holt, Reinhart and Winston, New York, 1975)
32. D.O. Gough, An elementary introduction to the JWKB approximation. Astron. Nachr. **328**, 273 (2007)
33. L.D. Landau, E.M. Lifshitz, *Course in Theoretical Physics, Vol. 3: Quantum Mechanics*, 3rd edn. (Pergamon Press, Oxford, 1991)
34. P.M. Morse, H. Feshbach, *Methods of Theoretical Physics*, vol. 1 (McGraw-Hill, Reading, New York, 1953)
35. A.D. Polyanin, V.F. Zaitsev, *Handbook of Exact Solutions for Ordinary Differential Equations*, 2nd edn. (Chapman & Hall/CRC, Boca Raton, 2003), p. 215
36. J. Boos, *Classical and Modern Methods in Summability* (Oxford University Press, Oxford, 2000). The collection of formulas and hints contained in the book T.J. I'a. Bromwich, in *An Introduction to the Theory of Infinite Series* (Macmillan, London, 1955), also remains indispensable
37. A. Sofo, *Computational Techniques for the Summation of Series* (Kluwer Academic/Plenum Publishing, New York, 2003)
38. M. Petkovšek, H. Wilf, D. Zeilberger, *A=B* (A. K. Peters Ltd., Wellesley, 1996)
39. D.M. Priest, On properties of floating-point arithmetics: Numerical stability and the cost of accurate computations. Ph.D thesis (University of California at Berkeley, 1992)
40. N.J. Higham, The accuracy of floating point summation. SIAM J. Sci. Comput. **14**, 783 (1993)
41. J. Demmel, Y. Hida, Accurate and efficient floating point summation. SIAM J. Sci. Comput. **25**, 1214 (2003)
42. W. Kahan, Further remarks on reducing truncation errors. Comm. ACM **8**, 40 (1965)
43. P. Linz, Accurate floating-point summation. Comm. ACM **13**, 361 (1970)
44. T.O. Espelid, On floating-point summation. SIAM Rev. **37**, 603 (1995)
45. I.J. Anderson, A distillation algorithm for floating-point summation. SIAM J. Sci. Comput. **20**, 1797 (1999)
46. G. Bohlender, Floating point computation of functions with maximum accuracy. IEEE Trans. Comput. **26**, 621 (1977)
47. D. Laurie, *Convergence Acceleration*, Appendix A, pp. 227–261 in the fascinating book F. Bornemann, D. Laurie, S. Wagon, J. Waldvögel, *The SIAM 100-Digit Challenge. A Study in High-Accuracy Numerical Computing* (SIAM, Philadelphia, 2004)
48. C. Brezinski, M.R. Zaglia, *Extrapolation Methods* (North-Holland, Amsterdam, 1991) A comprehensive historical review is offered by C. Brezinski, Convergence acceleration during the 20th century. J. Comput. Appl. Math. **122**, 1 (2000)
49. E.J. Weniger, Nonlinear sequence transformations for the acceleration of convergence and the summation of divergent series. Comp. Phys. Rep. **10**, 189 (1989)
50. K. Knopp, *Theory and Application of Infinite Series* (Blackie & Son, London, 1951)
51. The implementation described here is particularly attractive, as it automatically changes N and J such that the required maximum error is achieved most rapidly http://numerical.recipes/webnotes/nr3web5.pdf
52. H. Cohen, F.R. Villegas, D. Zagier, Convergence acceleration of alternating series. Exp. Math. **9**, 4 (2000)
53. H.H.H. Homeier, Scalar Levin-type sequence transformations. J. Comput. Appl. Math. **122**, 81 (2000)
54. GSL (GNU Scientific Library). http://www.gnu.org/software/gsl
55. E.T. Whittaker, On the reversion of series. Gaz. Mat. **12**(50), 1 (1951)

56. R.P. Brent, T.H. Kung, Fast algorithms for manipulating formal power series. J. Assoc. Comput. Mach. **25**, 581 (1978)
57. F. Johansson, A fast algorithm for reversion of power series. Math. Comp. **84**, 475 (2015)
58. D.A. Leahy, J.C. Leahy, A calculator for Roche lobe properties. Comput. Astrophys. Cosmol. **2**, 4 (2015)
59. G. Marsaglia, Evaluation of the normal distribution. J. Stat. Soft. **11**, 1 (2004)
60. J.F. Hart et al., *Computer Approximations* (Wiley, New York, 1968). See also W.J. Cody, Rational Chebyshev approximations for the error function. Math. Comp. **23**, 631 (1969)
61. J.R. Philip, The function inverfc θ. Austral. J. Phys. **13**, 13 (1960). See also L. Carlitz, The inverse of the error function. Pacific J. Math. **13**, 459 (1963)
62. O. Vallée, M. Soares, *Airy Functions and Applications to Physics* (Imperial College Press, London, 2004)
63. G. Szegö, *Orthogonal Polynomials* (AMS, Providence, 1939)
64. G.N. Watson, *Theory of Bessel Functions* (Cambridge University Press, Cambridge, 1922)
65. W.R. Gibbs, *Computation in Modern Physics* (World Scientific, Singapore, 1994); see Sects. 12.4 and 10.4

Chapter 2
Solving Non-linear Equations

Abstract Solving non-linear equations and systems of such equations is one of the daily chores of a physicist or engineer. In this Chapter basic techniques for scalar equations are explained: bisection, Newton-Raphson with optional convergence improvements, the secant and Müller's method. Vector non-linear equations are treated by Newton-Raphson and Broyden's method, illustrated by a robot engineering example. Solving polynomial equations of a single variable is described next, along with the efficient means of counting and locating the zeros. Special attention is given to the sensitivity of zeros to perturbations. After the presentation of the novel Hubbard-Schleicher-Sutherland method the Chapter ends with a discussion on how to solve algebraic equations of several variables (frequently occurring in automated assembly of mechanical systems, robot control, coding and cryptography), for which powerful algorithms based on Gröbner bases have been developed. The Examples and Problems include Kepler's equation, Wien's law of black-body radiation, Heisenberg's model in the mean-field approximation, energy levels of one-dimensional quantum-mechanical systems, fluid flow through systems of pipes, and automated assembly of three-dimensional structures.

In all areas of physics, mathematics, and engineering, we need to solve non-linear equations

$$f(x) = 0 , \qquad x, f(x) \in X , \tag{2.1}$$

where X is a vector space ($X = \mathbb{R}^n$ or $X = \mathbb{C}^n$). We assume that the number of unknowns in (2.1) is equal to the number of equations. In this Chapter the symbol $\boldsymbol{\xi}$ (ξ in scalar equations) denotes the exact solution of (2.1), so $f(\boldsymbol{\xi}) = \mathbf{0}$.

Some famous non-linear equations have exact solutions, like $x = a + b \sin x$ (Kepler's problem, see Example on page 70 and [1]) or $x e^x = a$ (solution given by the Lambert function, see [2]), but in general, non-linear equations need to be solved numerically. Efficient methods for solving non-linear equations are built into many numerical libraries, but to achieve greater flexibility, we may be inclined to write our own program.

There are two groups of methods for solving non-linear equations. The first group comprises the methods that can be written in the iterative form

$$x_{k+1} = \boldsymbol{\phi}(x_k) , \tag{2.2}$$

S. Širca and M. Horvat, *Computational Methods in Physics*, Graduate Texts in Physics, https://doi.org/10.1007/978-3-031-68566-8_2

where ϕ is the iteration function (mapping) and ξ is the fixed point of this mapping. The form of the function ϕ in general depends on the function f and its derivatives. The iteration yields successive approximations $x_1, x_2, \ldots$ of the exact solution ξ, and we terminate the sequence when, at chosen $\varepsilon > 0$, we reach

$$\|x_{k+1} - \phi(x_k)\| < \varepsilon .$$

These methods rest on the Banach contraction principle, so they are equally useful and efficient in all Banach spaces: scalar problems, $X = \mathbb{R}$, are easily generalized to vector problems, $X = \mathbb{R}^n$. The typical representatives of this class are the Newton's method (Sect. 2.1.2) and the Müller's method (Sect. 2.1.4). The second group contains the methods that can not be written in the form (2.2) and are hard to extend to multiple dimensions, e.g. the bisection (Sect. 2.1.1) or the secant method (Sect. 2.1.3).

Order of convergence In methods of the form (2.2) the convergence of the sequence $\{x_k\}$ to the fixed point ξ of ϕ has different speeds. We say that the sequence $\{x_k\}$ converges to ξ at least with order p if for any large enough k we have

$$\|e_{k+1}\| \le C \|e_k\|^p , \qquad e_k = x_k - \xi , \tag{2.3}$$

where $0 < C < 1$ if $p = 1$. Ideally, we always strive to find the largest *order of convergence p* and the smallest *quotient C*. The three lowest orders ($p = 1, 2$, and 3) are called linear, quadratic, and cubic, respectively. If the initial error e_0 is known, we may use (2.3) to determine the error of the kth approximation of the solution ξ:

$$\|e_k\| \le C^{(p^k-1)/(p-1)} \|e_0\|^{p^k} .$$

In linear convergence, the approach to the solution is exponential, and the speed of convergence depends exclusively on the magnitude of C. At higher orders, the approach is super-exponential and the speed also depends on p. The order and the quotient depend on the local properties of the mapping ϕ.

Order of the root To each root of the equation we assign its own *order*. The order r of a root of $f(x) = 0$ (i.e. of a zero of f) is defined as the largest r for which there exist real positive constants α, β, and ε, such that

$$\alpha \|e\|^r \le \|f(\xi + e)\| \le \beta \|e\|^r , \qquad \|e\| \le \varepsilon . \tag{2.4}$$

In general the orders of the roots are positive and real, while for analytic functions in the complex plane they are integer and named *root multiplicities*. If $r = 1$, the root is *simple*; otherwise, it is *multiple*. The root orders are important pieces of information in seeking possible or optimal ways to solve the equation.

Sensitivity of the roots to perturbations An important issue in solving the equation $f(x) = 0$ is the sensitivity of its roots to the perturbation of the function f by another function g. Let us consider only the scalar problem

$$f(x) + \varepsilon g(x) = 0 \,, \tag{2.5}$$

where $\varepsilon \in \mathbb{R}$ is the perturbation parameter. Let f be an analytic function with the zero ξ of order m, and let $g(x) \neq 0$ near ξ. In this vicinity of ξ we use the power expansion of f, in which only terms with powers higher than $m - 1$ survive:

$$f(\xi + h) = h^m \frac{f^{(m)}(\xi)}{m!} + \mathcal{O}(h^p) \,, \qquad p > m \,. \tag{2.6}$$

In the absence of the perturbation ($\varepsilon = 0$), the root $\xi(\varepsilon)$ of (2.5) is simply $\xi = \xi(0)$. When $\varepsilon \neq 0$, we insert the expansion (2.6) into (2.5) and obtain the dependence of the root upon the perturbation: in leading order,

$$\xi(\varepsilon) \doteq \xi + \varepsilon^{1/m} \left(-\frac{m!\, g(\xi)}{f^{(m)}(\xi)} \right)^{1/m} \,.$$

Since the mth root is a multi-valued complex function, multiple zeros of $f + \varepsilon g$ may shift far away (in the complex plane) from zeros of the unperturbed function f. We are interested in the effect of numerical errors in $f(x)$ on the precision of the computed zero ξ. If $f(x)$ is known to a precision of δf, it follows that

$$\delta\xi \approx \left| \frac{m!}{f^{(m)}(\xi)} \, \delta f \right|^{1/m} \,.$$

The error of the zero, $\delta\xi$, may therefore become very large if the multiplicity of the zero, m, is large. If f is a polynomial, special estimates for the sensitivities of its coefficients to perturbations can be derived: see Sect. 2.4.5.

2.1 Scalar Equations

Finding the roots of the equation $f(x) = 0$, where f is a scalar function and x is its scalar argument, is among the oldest mathematical tasks for which a rich assortment of methods exists; here we discuss only the most famous ones.

2.1.1 Bisection

Bisection is the simplest and the most reliable numerical procedure to find a root of odd order. It is also used as a building block of other methods. Initially we isolate an interval $[a, b]$ on which the function f definitely has a zero. This means that in one of the points from this interval, f changes its sign,

$$f(a)f(b) < 0 .$$

(Recall (1.2)!) In subsequent steps we gradually narrow the intervals containing the zero. In the kth step we use the point $c_k = (a_k + b_k)/2$ to divide the interval $[a_k, b_k]$, on which $f(a_k)f(b_k) < 0$, into subintervals $[a_k, c_k]$ and $[c_k, b_k]$. Only one of them contains the zero, and the one for which $f(a_{k+1})f(b_{k+1}) < 0$ applies, is used as the interval $[a_{k+1}, b_{k+1}]$ to be divided in the next step. This procedure is repeated until the step n at which the condition $|a_n - b_n| \leq \varepsilon$ is fulfilled, and the final approximation of the zero is $\xi \approx (a_n + b_n)/2$. The lengths of the intervals decrease as $e_{k+1} = e_k/2$: the method has linear convergence. The approximate number of steps N needed to determine the zero to precision ε, is

$$N \approx \frac{\log((b - a)/\varepsilon)}{\log 2} .$$

Bisection can not be used to find zeros of even orders, since $f(a_k)f(b_k) < 0$ does not hold true. Bisection can be generalized to multiple dimensions; see [3].

2.1.2 The Family of Newton's Methods and the Newton–Raphson Method

The Newton's family of methods contains methods for solving the equation $f(x) = 0$ based on power expansions of f and seeking its inverse f^{-1} around the point 0. Assume that ξ is the zero of f and S is an open neighborhood of ξ. Let f be at least $(n + 1)$-times continuously differentiable in S and at least n-times in its closure $\overline{S}$. The function f can be written as a Taylor series (1.22) where $x, x_0 \in S$ and $x^\star \in [x, x_0]$. If $f' \neq 0$ in S, the implicit-function theorem tells us that to f there corresponds an inverse function g on $\Omega = f(S)$, which is at least $(n + 1)$-times continuously differentiable on Ω and at least n-times on the closure $\overline{\Omega}$, and which has the form

$$g(u) = g(v) + g'(v)(u - v) + \cdots + \frac{g^{(n)}(v)}{n!}(u - v)^n + \frac{g^{(n+1)}(\delta)}{(n + 1)!}(u - v)^{n+1} .$$

Here $u, v \in \Omega$ and $\delta \in [u, v]$. The derivatives of the inverse function are

$$g^{(n)}(f(x)) = \lim_{x_0 \to x} \frac{\mathrm{d}^{n-1}}{\mathrm{d}x^{n-1}} \left(\frac{x - x_0}{f(x) - f(x_0)} \right)^n ;$$

the first derivative is $g'(f(x)) = 1/f'(x)$, while the second and the third are

$$g''(f(x)) = -\frac{f''(x)}{f'(x)^3} , \qquad g'''(f(x)) = \frac{3f''(x)^2 - f'(x)f'''(x)}{f'(x)^5} .$$

We expand g around $v = f(x)$, evaluate it at $u = f(\xi) = 0$, and end up with an expression for ξ which is valid for $f(x) \in \Omega$:

$$\xi = x + \underbrace{\sum_{k=1}^{n} \frac{g^{(k)}(f(x))(-f(x))^k}{k!}}_{\phi_n(x)} + \frac{g^{(n+1)}(\delta)}{(n+1)!}(-f(x))^{n+1}, \qquad \delta \in [0, f(x)].$$

The sum $\phi_n(x)$ can be used in the iteration of a method of order $n+1$ like

$$x_{k+1} = \phi_n(x_k),$$

as advertised at the outset (Eq. (2.2)). This iteration operates equally well on the real axis and in the complex plane, and can therefore be used to seek both real and complex roots. (If f is real and we wish to find a complex root, the initial approximation also needs to be complex.) The most important representatives of this family of methods are the Newton–Raphson's method ($n = 1$),

$$x_{k+1} = x_k - \frac{f(x_k)}{f'(x_k)} \tag{2.7}$$

(Fig. 2.1), and the modified Newton–Raphson's method ($n = 2$), with the iteration

$$x_{k+1} = x_k - \frac{f(x_k)}{f'(x_k)} - \frac{f''(x_k)f(x_k)^2}{2f'(x_k)^3}.$$

Methods of higher orders can be constructed by augmenting the iteration function, but their applicability is limited as the numerical computation, or even the existence of higher derivatives of f, are questionable. Methods of orders greater than two are rarely used.

The Newton–Raphson's method with the iteration (2.7) is the most simple, generally useful and explored method from the Newton's family. The search for the zero by using this formula is illustrated in Fig. 2.1. Due to this characteristic graphical interpretation this procedure is also known as the *tangent method*.

It is clear from this construction that the Newton–Raphson's iteration has a quadratic convergence if the derivative $f'(x)$ exists in some open neighborhood S of ξ and is continuous and non-zero ($f'(x) \neq 0 \; \forall S$). This can be confirmed by expanding the iteration formula around ξ in powers of the difference $e_k = x_k - \xi$. We get

$$e_{k+1} \approx \frac{f''(\zeta)}{2f'(x_k)} e_k^2, \qquad \zeta = \zeta(x_k) \in S.$$

The method therefore has quadratic convergence when $\max_{x \in S} |f''(x)| \neq 0$. The speed of convergence is essentially driven by the ratio of the second and the first

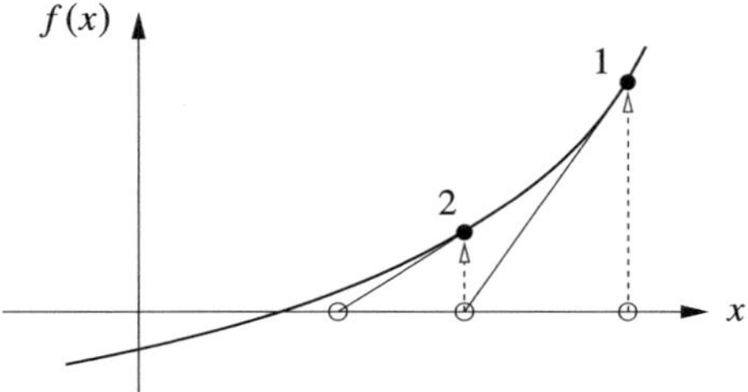

Fig. 2.1 Searching for the approximations of the zero by using the Newton–Raphson's method. At each point of the iteration $(x_k, f(x_k))$ we compute the tangent to the function. The intersection of the tangent with the abscissa is the new approximation of the zero x_{k+1}. See also Fig. 2.2 (right)

derivative evaluated at the zero of f. If the zero is multiple, the Newton–Raphson's method slows down and becomes only linearly convergent. In such cases, quadratic convergence can be restored by writing the iteration (2.7) as

$$x_{k+1} = x_k - m \frac{f(x_k)}{f'(x_k)} \, ,$$

where m is the order of the zero of f. If the order of ξ is unknown and there is a chance that the zero is multiple, one should not try to solve $f(x) = 0$ but rather find the zero of the function $u(x) = f(x)/f'(x)$. When u is inserted in (2.7), we obtain an iteration of the form

$$x_{k+1} = x_k - \frac{f(x_k)}{f'(x_k) - f(x_k) f''(x_k)/f'(x_k)} \, ,$$

which also accommodates multiple zeros.

The Newton–Raphson sequence is locally convergent, but it does not converge globally, that is, for arbitrary initial values; see Example below. Global convergence is ensured under the following conditions: if f is twice continuously differentiable on $[a, b]$, if $f(a)f(b) < 0$ (the interval contains the zero) and $f''(x) \neq 0$ for any $x \in [a, b]$, and if we choose the initial approximation such that $x_0 \in [a, b]$ and $f(x_0)f''(x_0) > 0$, the sequence from (2.7) converges monotonously (and quadratically) to ξ [4]. In plain words, the initial approximation should always be picked from that side of the interval on which the function and its second derivative have equal signs.

Example (Adapted from [5].) Kepler's equation

$$E = M + \varepsilon \sin E \tag{2.8}$$

connects the parameters of the planet's orbital motion around the Sun: eccentric anomaly E, average anomaly M, and eccentricity $\varepsilon = \sqrt{1 - b^2/a^2}$ (Fig. 2.2 (left)).

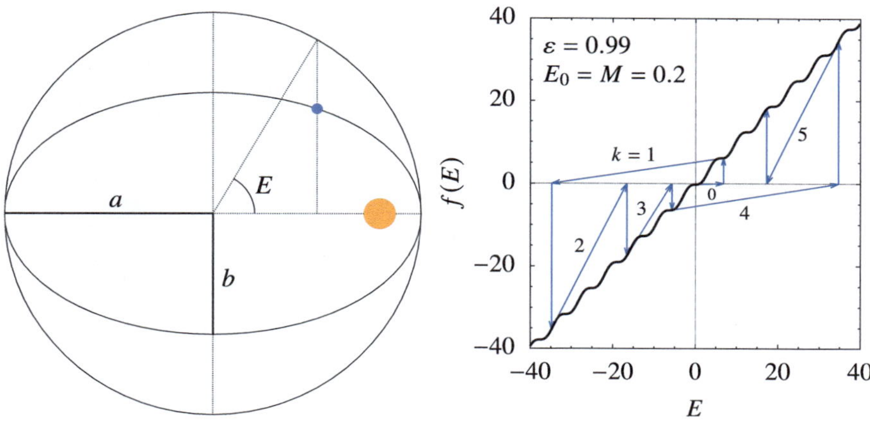

Fig. 2.2 Solving Kepler's equation by Newton's method. [LEFT] The definition of the anomaly E. [RIGHT] Poor convergence of the approximations E_k with a strongly eccentric orbit and bad initial approximation E_0. The function $f(E) = E - M - \varepsilon \sin E$ is drawn

The time dependence of M is given by $M(t) = 2\pi(t - t_p)/T$, where T is the period and t_p is the time of the passage through the perihelion (the point on the elliptic orbit closest to the Sun). Various forms of exact solutions exist [1].

Equation (2.8) can be solved by simple iteration, $E_{k+1} = M + \varepsilon \sin E_k$, which converges very slowly. It is therefore much better if the equation is rewritten in the Newton form (2.7) with the function $f(E) = E - M - \varepsilon \sin E$. We get

$$E_{k+1} = \frac{M - \varepsilon[E_k \cos E_k - \sin E_k]}{1 - \varepsilon \cos E_k}, \qquad k = 0, 1, 2, \ldots .$$

With a prudent choice of the initial approximation the sequence $\{E_k\}$ converges very rapidly: for $\varepsilon = 0.95$, $M = 245°$, and the initial approximation $E_0 = 215°$, we obtain the solution $E \approx 3.740501878977461$ to machine precision in just three steps; the simple iteration would require 131 steps for the same precision.

For small eccentricities, $\varepsilon \ll 1$, Newton's method is rather insensitive to the initial approximation, as in such cases $E = M + \varepsilon \sin E \approx M$, and the simple guess $E_0 = M$ does the job. For large ε, however, the sequence of the approximations $\{E_k\}$ with a poorly chosen initial value E_0 may diverge. Assume that we always set $E_0 = M$ initially. If $(\varepsilon, M) = (0.991, 0.13\pi)$ or $(0.993, 0.13\pi)$, the sequence $\{E_k\}$ converges, but for the values $(0.992, 0.13\pi)$ it diverges! Similar behavior can be observed near $(\varepsilon, M) = (0.99, 0.2)$ (Fig. 2.2 (right)). The reason for the divergence is the bad choice of E_0 combined with the tricky form of the function $f(E) = E - M - \varepsilon \sin E$ with almost horizontal parts where the tangent to f is "thrown" far away (see the jumps at $k = 1$ and $k = 4$). A good approximation for any ε, which can be used to avoid the pitfalls seen above, is $E_0 = \pi$ [5].   ◁

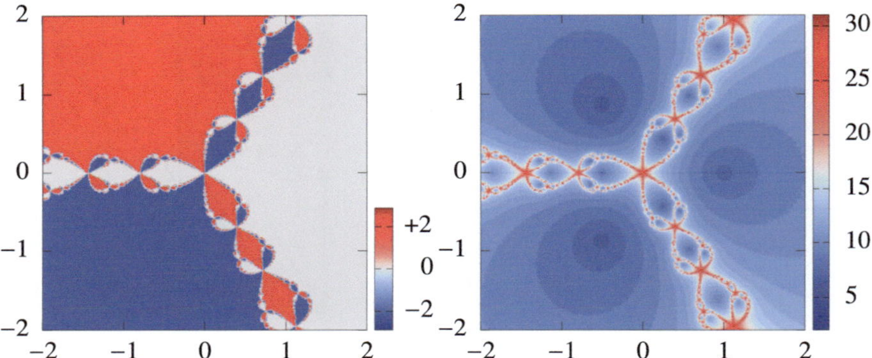

Fig. 2.3 Solving $z^3 - 1 = 0$ by Newton's method. [LEFT] The phase of the final value z as a function of the initial approximation specified by Re z and Im z on the abscissa and the ordinate. The three shaded areas are the *basins of attraction* corresponding to the phases 0, $2\pi/3 \approx 2.094395$, and $-2\pi/3 \approx -2.094395$ to which the individual initial approximation is "attracted" (see [6] and Sect. 2.4.7). [RIGHT] The number of iterations needed to achieve convergence, as a function of the initial approximation

Example We use Newton's method to compute the complex roots of $z^3 - 1 = 0$. The exact solutions are 1, $\mathrm{e}^{\mathrm{i}2\pi/3}$, and $\mathrm{e}^{-\mathrm{i}2\pi/3}$. We choose the initial approximation $z_0 = x_0 + \mathrm{i}\,y_0 \in [-2, 2] \times [-2, 2]$ and observe the value of z at the end of the iteration $z_{k+1} = z_k - f(z_k)/f'(z_k)$, where $f(z) = z^3 - 1$. Figure 2.3 (left) shows the phase of the final value of z, while Fig. 2.3 (right) shows the number of steps needed to achieve convergence to machine precision. ◁

2.1.3 The Secant Method and Its Relatives

Together with bisection, the secant method and its variants are the oldest known iterative procedures. From the pairs $(x_{k-1}, f(x_{k-1}))$ and $(x_k, f(x_k))$ with oppositely signed $f(x_{k-1})$ and $f(x_k)$ we determine the straight line intersecting the abscissa at x_{k+1}, and we compute the value of f at this very point, $f(x_{k+1})$. The process is repeated on the subinterval on which the function's values at its boundary points are oppositely signed, until convergence is achieved.

This method of finding the zero (regula falsi, see below) is just a special case in the family of methods based on the Newton–Raphson's iteration where the exact derivative of the function is replaced by the approximate one:

$$f'(x) \approx \frac{f(x_k) - f(x_{k-1})}{x_k - x_{k-1}}\,.$$

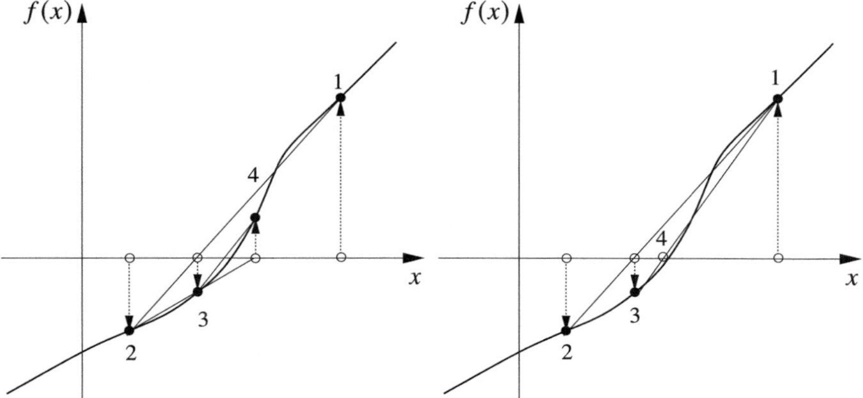

Fig. 2.4 Finding the approximations of the zeros of f by using the methods from the family defined by (2.9). [LEFT] Secant method. [RIGHT] Method of false position

When this approximation is inserted in (2.7), we obtain the iteration

$$x_{k+1} = x_k - \frac{f(x_k)(x_k - x_{k-1})}{f(x_k) - f(x_{k-1})} = \frac{f(x_k)x_{k-1} - f(x_{k-1})x_k}{f(x_k) - f(x_{k-1})} . \tag{2.9}$$

To compute x_{k+1} we need the information from *two* previous points, hence this method can not be written in the one-point form $x_{k+1} = \phi(x_k)$. The iteration (2.9) can be used in various ways which yield the secant method, the method of false position (regula falsi), and the chord method. Among these, only the secant method has super-linear convergence with the optimal order $(1 + \sqrt{5})/2 \approx 1.618$. All other methods from this family have approximately linear convergence.

Secant method This method does not require specific initial conditions (for example, it does not require the function to have opposite signs at the boundaries of the interval) and therefore does not always converge, in particular if the initial values are far from the zero. The secant method works best for functions with simple (not multiple) zeros. We use (2.9) directly; see Fig. 2.4 (left).

Regula falsi To use this method the function should have opposite signs at the boundary points of $[a, b]$, $f(a)f(b) < 0$. Then (2.9) is used in the form

$$c = \frac{f(b)a - f(a)b}{f(b) - f(a)} , \tag{2.10}$$

where c is the new approximation of the zero. According to the value $f(c)$ we choose the next interval on which the zero is located, and assign $a = c$ or $b = c$, just like with bisection (see Fig. 2.4 (right)). With minimal modifications of (2.10) we can make the regula falsi method achieve super-linear convergence: see [7].

Chord method This method is a simplified version of the regula falsi method. We also require $f(a)f(b) < 0$ (there is at least one zero on the interval) and that on $[a, b]$ the sign of the second derivative of f *does not* change. One of the ends of the interval may therefore be fixed and the iteration (2.9) can be refurbished according to the endpoint at which f satisfies $f(x)f''(x) > 0$. Two possible iteration procedures follow. If the point $x = a$ is fixed, we iterate

$$x_{k+1} = a - \frac{(x_k - a)f(a)}{f(x_k) - f(a)},$$

while if $x = b$ is fixed,

$$x_{k+1} = b - \frac{(b - x_k)f(b)}{f(b) - f(x_k)}.$$

2.1.4 Müller's Method

Like the secant methods, Müller's method is a one-point iteration method *with memory*: in computing the approximation of the zero in the current step, one uses the interpolation of the values $f(x_k)$ and (approximations of) derivatives $f^{(j)}(x_k)$ at points accessed in previous steps. Using only the values would be best, but then at most quadratic convergence can be achieved. If derivatives are used to improve the interpolation, arbitrary orders of convergence can be achieved in principle, but this is not practical. For details, see [8, 9].

The most well-known form of the Müller's method can be derived by generalizing the secant method: instead of the linear interpolation through the last two points, we use the quadratic interpolation through the last three points of the iteration $(x_k, f(x_k))$, $(x_{k-1}, f(x_{k-1}))$, and $(x_{k-2}, f(x_{k-2}))$. The next approximation of the zero ξ of f is one of the zeros $x_\pm$ of the interpolation polynomial

$$y(x) = a(x - x_k)^2 + 2b(x - x_k) + f(x_k)$$

with the coefficients

$$a = \left[\frac{f(x_k) - f(x_{k-1})}{x_k - x_{k-1}} - \frac{f(x_k) - f(x_{k-2})}{x_k - x_{k-2}} \right] \frac{1}{x_{k-1} - x_{k-2}},$$

$$b = \frac{1}{2} \left[\frac{f(x_k) - f(x_{k-1})}{x_k - x_{k-1}} + \frac{f(x_k) - f(x_{k-2})}{x_k - x_{k-2}} - \frac{f(x_{k-1}) - f(x_{k-2})}{x_{k-1} - x_{k-2}} \right].$$

The zeros of the interpolation polynomial are

$$x_\pm = x_k + \frac{1}{a} \left[-b \pm \sqrt{b^2 - af(x_k)} \right]$$

and are related by $x_+ x_- = (a x_k^2 - 2 b x_k + f(x_k))/a$ and $(x_+ - x_k)(x_- - x_k) = f(x_k)/a$. The iteration step of the Müller's method is then

$$x_{k+1} = x_\pm = x_k - \frac{f(x_k)}{b \pm \sqrt{b^2 - a f(x_k)}} \,.$$

We choose the sign which corresponds to the largest magnitude of the denominator since this minimizes the rounding errors. The iteration can be initialized with an arbitrary triplet of points x_0, x_1, and x_2 in the vicinity of ξ, but it is advisable to space them uniformly.

The Müller's method also allows us to search for complex zeros, with an important advantage over the secant and Newton's method: we enter the complex plane even if the initial approximation is on the real axis. On the other hand, this property may be a nuisance if we are searching only for strictly real zeros.

In the case that f is at least three times continuously differentiable in the vicinity of its simple zero ξ, and that the initial approximations x_0, x_1, and x_2 have been chosen close enough to ξ, one can show that for the sequence $\{x_k\}$, obtained in the Müller's iteration the following holds true:

$$\lim_{k \to \infty} \frac{|x_k - \xi|}{|x_{k-1} - \xi|^p} = \left| \frac{1}{6} \frac{f'''(\xi)}{f'(\xi)} \right|^{(p-1)/2} ,$$

where the order of convergence is $p \approx 1.83929$. Since the coefficient $2b$ is equal to the derivative of the interpolation polynomial at x_k, Müller's iteration can also be defined — in analogy to the Newton's method — as

$$x_{k+1} = x_k - \frac{f(x_k)}{2b} \,.$$

This simple iteration has the same order of convergence, but it becomes useless for $b \approx 0$. It can be put to good use, however, in the immediate vicinity of the real or complex zero or in cases where strictly real zeros are searched and we wish to avoid the jumps into the complex plane typical of the Müller's method.

Further reading In iterative approaches in this Section, we have devoted more attention to the discussion of one-point methods, in which the values of the function (and its derivatives) at a single point are needed to compute the next approximation of the zero. This is the recommended and the most widely used practice. Methods with memory and multi-step methods are discussed in [8]. In general these methods are only locally convergent, and locating a good initial approximation of the zero is imperative. In very specific cases (e.g. in polynomial equations) globally convergent "*sure-fire*" methods exist that withstand practically arbitrary initial approximations. For a deeper insight see [4, 8, 10, 11].

2.2 Vector Equations

In contrast to scalar equations, the solutions of vector equations are sought in n-dimensional ($n > 1$) vector spaces. Equation (2.1) should therefore be read as

$$f(x) = 0 \,, \qquad f : \mathbb{R}^n \to \mathbb{R}^n \ \ (\text{or } f : \mathbb{C}^n \to \mathbb{C}^n) \,, \qquad f : x \mapsto f(x) \,.$$

The argument x of the function f and the solution of the equation $f(x) = 0$, its root ξ, are vectors,

$$x = (x_1, x_2, \ldots, x_n)^{\mathrm{T}} \,, \quad \xi = (\xi_1, \xi_2, \ldots, \xi_n)^{\mathrm{T}} \,.$$

Similarly, a n-dimensional vector is used to represent the n scalar functions of x:

$$f(x) = \begin{pmatrix} f_1(x) \\ f_2(x) \\ \vdots \\ f_n(x) \end{pmatrix} = \begin{pmatrix} f_1(x_1, x_2, \ldots, x_n) \\ f_2(x_1, x_2, \ldots, x_n) \\ \vdots \\ f_n(x_1, x_2, \ldots, x_n) \end{pmatrix} \,.$$

In iterative solution of vector equations the approach to the root is no longer confined to a straight line. This introduces severe algorithmic complications and restricts the set of applicable methods. Here we follow [12] and introduce a generalization of the Newton–Raphson's and secant method to multiple dimensions.

2.2.1 Newton–Raphson's Method

In the vector variant of the Newton–Raphson's method we formally proceed in the same spirit as in the scalar case. In the vicinity of the root ξ the components of the function f have the Taylor expansions

$$0 = f_i(\xi) \approx f_i(x) + \sum_j \frac{\partial f_i(x)}{\partial x_j} (\xi_j - x_j) + \mathcal{O}(|\xi - x|^2) \,. \tag{2.11}$$

By using the Jacobi matrix of derivatives of f with respect to x,

$$J(x) = \begin{bmatrix} \dfrac{\partial f_1(x)}{\partial x_1} & \cdots & \dfrac{\partial f_1(x)}{\partial x_n} \\[2mm] \vdots & \ddots & \vdots \\[2mm] \dfrac{\partial f_n(x)}{\partial x_1} & \cdots & \dfrac{\partial f_n(x)}{\partial x_n} \end{bmatrix} \,,$$

Equation (2.11) can be rewritten as

$$\mathbf{0} = f(\boldsymbol{\xi}) \approx f(\boldsymbol{x}) + J(\boldsymbol{x})(\boldsymbol{\xi} - \boldsymbol{x}) + \cdots .$$

If $\boldsymbol{x}$ is close enough to $\boldsymbol{\xi}$, we may neglect the quadratic term in the expansion, and proclaim the solution $\boldsymbol{\zeta}$ of the equation $f(\boldsymbol{x}) + J(\boldsymbol{x})(\boldsymbol{\zeta} - \boldsymbol{x}) = \mathbf{0}$ as the improved approximation of the root, $\boldsymbol{\zeta} = \boldsymbol{x} - J^{-1}(\boldsymbol{x}) f(\boldsymbol{x})$. This leads us to the general iteration step

$$\boldsymbol{x}_{k+1} = \boldsymbol{x}_k - J^{-1}(\boldsymbol{x}_k) f(\boldsymbol{x}_k) , \qquad k = 0, 1, 2, \ldots , \tag{2.12}$$

where k is the iteration index.

It is sometimes very hard or even impossible to compute the Jacobi matrix analytically. In such cases, we resort to approximations of derivatives, like

$$\frac{\partial f_i(\boldsymbol{x})}{\partial x_j} \approx \frac{f_i(\boldsymbol{x} + h_j \boldsymbol{e} f_j) - f_i(\boldsymbol{x})}{h_j} ,$$

where $\boldsymbol{e}_j$ is the unit vector along the jth coordinate axis (a n-dimensional vector with the jth component set to 1). If the chosen h_j is too small, the errors in computing the values of $f(\boldsymbol{x})$ may become very large; if the h_j is too large, the derivative is locally poorly approximated. The optimal h_j can be estimated by

$$h_j^{(\mathrm{opt})} \left\| \frac{\partial f(\boldsymbol{x})}{\partial x_j} \right\| \approx \sqrt{\varepsilon} \, \| f(\boldsymbol{x}) \| \qquad \forall j ,$$

where ε is the relative precision of the computation of the values $f(\boldsymbol{x})$, for example, $\varepsilon = \varepsilon_{\mathrm{M}}$. By using this estimate, both $f(\boldsymbol{x})$ and $f(\boldsymbol{x} + h_j \boldsymbol{e}_j)$ are precise to about half of the digits allowed by the floating-point arithmetic being used.

The vector Newton's method (2.12) is exact for affine functions f. This means that the iterated values $\boldsymbol{x}$ are invariant with respect to a linear transformation $f \to A f$ with an arbitrary non-singular square matrix A, and that the iterated values of the method with the function $f(A\boldsymbol{x})$ can be obtained by simply applying the method with the function $f(\boldsymbol{x})$, and multiplying the vector of the iterated values by A^{-1}. The subsequent approximations converge to the root under the conditions of the Newton–Kantorovich theorem; see [4].

2.2.2 Broyden's (Secant) Method

In vector methods of the secant type we exploit the idea that at each point of the iteration $\boldsymbol{x}_k$ the function f can locally be approximated by a linear model,

$$f(\boldsymbol{x}) \approx f(\boldsymbol{x}_k) + B_k(\boldsymbol{x} - \boldsymbol{x}_k) .$$

From the viewpoint of the power expansion, the optimal choice is $B_k = J(x_k)$, which corresponds to the Newton–Raphson's method. But the matrix B_k may also be understood as an approximation of the Jacobi matrix which, at some step of the iteration, can be determined by the linear equation

$$B_k(x_k - x_{k-1}) = f(x_k) - f(x_{k-1}) \, .$$

If the system of equations is linear, B_k is constant, while in non-linear systems, B_k changes during the iteration $k = 0, 1, \dots$. The changes (or *updates*) of B_k may be very general or quite small. Various methods are characterized by different ways of constructing B_k. The most useful are known as *Broyden's methods* [13]; they are the generalizations of the secant method (2.9) to vector equations.

Broyden's methods belong to the class of methods in which the matrix B_k is only minimally changed during the steps of the iteration (*least-change secant update methods*). A nice, physically intuitive insight into the genesis of these methods is given by their author in [14]. Empirically, "good" and "bad" versions of the Broyden's methods were discovered. In the *good Broyden's method*, the approximation B_k of the Jacobi matrix in a single iteration step is calculated by

$$B_k = B_{k-1} + \frac{(y_{k-1} - B_{k-1}s_{k-1})s_{k-1}^{\mathrm{T}}}{s_{k-1}^{\mathrm{T}}s_{k-1}} \, ,$$

where

$$s_{k-1} = x_k - x_{k-1} \, , \qquad y_{k-1} = f(x_k) - f(x_{k-1}) \, .$$

The update of B_{k-1} at each step has rank one: this means that B_{k-1} does not change in any direction that is orthogonal to the vector s_{k-1}. By using the inverse of the updated matrix, B_k^{-1}, we then proceed with the next step of the iteration

$$x_{k+1} = x_k - B_k^{-1} f(x_k) \, . \tag{2.13}$$

In the program, this is implemented as a system of equations $B_k x_{k+1} = B_k x_k - f(x_k)$, which we solve for x_{k+1}. We start the iteration by specifying the initial approximation of the solution x_0, while B_0 is set to be the exact or the approximate Jacobi matrix $J(x_0)$. One can also use the updates of the inverse matrix

$$B_k^{-1} = B_{k-1}^{-1} + \frac{(s_{k-1} - B_{k-1}^{-1}y_{k-1})s_{k-1}^{\mathrm{T}}B_{k-1}^{-1}}{s_{k-1}^{\mathrm{T}}B_{k-1}^{-1}y_{k-1}} \, , \tag{2.14}$$

which is only allowed if $s_{k-1}^{\mathrm{T}}B_{k-1}^{-1}y_{k-1} \neq 0$. We need to calculate only one inverse of the matrix, B_0^{-1}, which we then update as in (2.14) and iterate (2.13).

If f is continuously differentiable, and we see that the approximations x_k converge to the root $\boldsymbol{\xi}$ (where s_k from some k on are linearly independent), we also find

$\lim_{k\to\infty} B_k = J(\boldsymbol{\xi})$: in the iteration B_k approaches the Jacobi matrix $J(\boldsymbol{\xi})$ (Fig. 2.6 (right)). For a precise formulation of the theorem and its proof see [15].

Example In this classic problem (Fig. 2.5), we seek the angles θ_1 and θ_2 at which the robot's arm with handles l_1 and l_2 reaches to the point (x, y). We are solving a system of non-linear equations $f_1(\theta_1, \theta_2) = 0$, $f_2(\theta_1, \theta_2) = 0$, with

$$f_1(\theta_1, \theta_2) = l_1 \cos \theta_1 + l_2 \cos(\theta_1 + \theta_2) - x \,,$$

$$f_2(\theta_1, \theta_2) = l_1 \sin \theta_1 + l_2 \sin(\theta_1 + \theta_2) - y \,.$$

The Jacobi matrix corresponding to this system is

$$J(\theta_1, \theta_2) = \begin{pmatrix} \partial f_1/\partial \theta_1 & \partial f_1/\partial \theta_2 \\ \partial f_2/\partial \theta_1 & \partial f_2/\partial \theta_2 \end{pmatrix} \,.$$

There may be n unique solution, multiple solutions, or it may not even exist. Let the lengths of the arms be $l_1 = l_2 = 1$ and the initial position of the arm be $(\theta_1, \theta_2) = (20°, 20°)$. We wish to reach the point $(x, y) = (0.7, 1.2)$.

The system can be solved by Broyden's updates (2.14) in the iteration (2.13) or its improved version (2.16) and (2.15). Since at each point $\boldsymbol{\theta} = (\theta_1, \theta_2)^{\mathrm{T}}$ we know the explicit form of the Jacobi matrix, the solution can also be computed by Newton's method (2.12). Figure 2.6 (left) shows the convergence of the sequence of approximations $\boldsymbol{\theta}_k$ for both iterations. The final position of the robotic arm is $(\theta_1, \theta_2) \approx (105.746°, 267.994°)$; if the initial approximation is $(20°, 100°)$ we obtain the second solution $\approx (13.741°, 92.006°)$. Figure 2.6 (right) shows the approach of the Broyden's matrix to the exact Jacobi matrix. $\triangleleft$

For a successful iteration it is imperative that the matrix B_k is non-singular, which can be controlled by monitoring its determinant [16]. During the iteration, the determinant changes according to

Fig. 2.5 A two-segment robot's arm attempting to reach a specific point in the plane

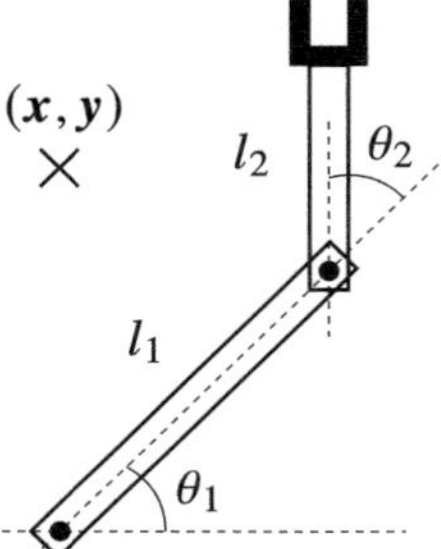

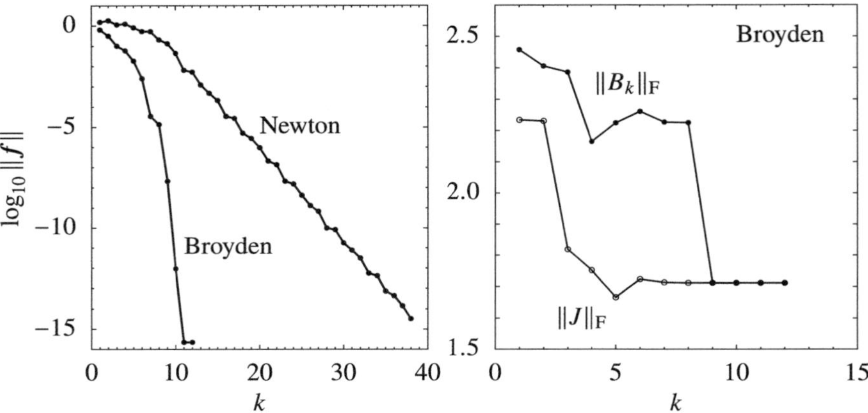

Fig. 2.6 The solution of the robotic arm problem. [LEFT] The deviation of $f(\theta_k)$ from the values $\mathbf{0}$ as a function of the iteration index in the Newton's and Broyden's method. (Other initial conditions may yield a rather different figure.) [RIGHT] The convergence of the Broyden's matrix to the exact Jacobi matrix as a function of the iteration index (measured in the Frobenius matrix norm)

$$\det(B_{k+1}) = \frac{s_k^{\mathrm{T}} B_k^{-1} y_k}{s_k^{\mathrm{T}} s_k} \det(B_k) \ .$$

In the kth step, B_k may become singular and remain singular in all subsequent steps if $s_k \perp B_k^{-1} y_k$. Even approximate orthogonality and near-vanishing of the determinant may have disastrous consequences. One obtains a much better control of the stability of the Broyden's method if the iteration is generalized as

$$x_{k+1} = x_k - \lambda_k B_k^{-1} f(x_k) \,, \tag{2.15}$$

$$B_k = B_{k-1} + \eta_{k-1} \frac{(y_{k-1} - B_{k-1} s_{k-1}) s_{k-1}^{\mathrm{T}}}{s_{k-1}^{\mathrm{T}} s_{k-1}} \ . \tag{2.16}$$

The control parameters η_k and λ_k can be adjusted during the iteration. Reasonable values are $\eta_k \in (0, 2)$ and $\lambda_k \in (0, 2)$. If η_k is calculated as

$$\eta_k = \begin{cases} 1 & ; \ |\gamma_k| \geq 1/10 \,, \\[2mm] \dfrac{1 - \mathrm{sign}(\gamma_k)/10}{1 - \gamma_k} & ; \ |\gamma_k| < 1/10 \,, \end{cases} \qquad \gamma_k = \frac{s_k^{\mathrm{T}} B_k^{-1} y_k}{s_k^{\mathrm{T}} s_k} \,,$$

which bounds $\eta_k \in [0.9, 1.1]$, it turns out [16] that the non-singularity of B_k is guaranteed; in addition, we establish $|\det(B_{k+1})| \geq |\det(B_k)|/10$. The sequence of approximations x_k or the deviations from the root $e_k = x_k - \xi$ obtained by the generalized Broyden's scheme satisfy

$$\frac{\|e_{k+1}\|_2}{\|e_k\|_2} \le \frac{\varepsilon_k + |\lambda_k - 1|}{1 - \varepsilon_k} ,$$

where $\{\varepsilon_k\}$ is a sequence for which $\lim_{k \to \infty} \varepsilon_k = 0$. In the basic Broyden's method ($\lambda_k = 1$) the right side of the inequality vanishes as $k \to \infty$, and the method is superlinear. This applies only near the solution. Farther away, it makes more sense to vary λ_k within a *line search*: We seek the largest λ_k such that

$$\|f(x_k - \lambda_k d_k)\|_2 < \|f(x_k)\|_2 , \qquad d_k = B_k^{-1} f(x_k) ,$$

or restrict the parameter λ_k to assume only the values of negative powers of 2,

$$\lambda_k = 2^{-j} , \qquad j = \min\{i \ge 0 : \|f(x_k - 2^{-i} d_k)\|_2 < \|f(x_k)\|_2\} .$$

We can also search for λ that minimizes the quantity

$$g(\lambda) = \|f(x_k - \lambda d_k)\|_2 ,$$

but if this is implemented carelessly, we may face large numerical costs and slow convergence. A very good estimate for the optimal value of the parameter λ can be obtained by approximating $g(\lambda)$ by a parabolic interpolant through the points $g_0 = g(0)$, $g_t = g(t)$, $t \in (0, 1)$, and $g_1 = g(1)$:

$$g(\lambda) \approx g_0 + \frac{t^2(g_1 - g_0) + g_0 - g_t}{t(t-1)} \lambda - \frac{t(g_1 - g_0) + g_0 - g_t}{t(t-1)} \lambda^2 .$$

Assume that λ at which $g(\lambda)$ has a minimum,

$$\lambda = \frac{1}{2} \frac{t^2(g_1 - g_0) + g_0 - g_t}{t(g_1 - g_0) + g_0 - g_t} ,$$

is a good approximation for the optimal value. If this approximation lies outside the interval $(0, 1)$, we discard it and make the next step by using $\lambda = 1$.

2.3 Convergence Acceleration ⋆

The convergence of iterative methods (and, more generally, all convergent sequences) can be accelerated by using *Aitken's procedures*. In the following, we present the most commonly known Aitken's methods to accelerate the iterations or sequences with linear or quadratic convergence.

Assume we are seeking the solution ξ of the equation $f(x) = 0$. If the sequence $\{x_k\}$ of approximations of ξ converges linearly, we know that the differences $e_k = x_k - \xi$ near the root satisfy $e_{k+1} = \alpha e_k$ and $e_{k+2} = \alpha e_{k+1}$. When the former of these equations is divided by the latter, we get $e_{k+1}^2 = e_k e_{k+2}$. We insert $e_k = x_k - \xi$, solve for ξ, and get an improved approximation of the root,

$$X_k = x_k - \frac{(x_{k+1} - x_k)^2}{x_{k+2} - 2x_{k+1} + x_k} \,, \tag{2.17}$$

which can represent the final result or the initial value for the next step. [Compare Eqs. (2.17) and (1.59)!] In the optimal case, the sequence $\{X_k\}$ converges *quadratically* to ξ. In general, however, one can guarantee only that the sequence $\{X_k\}$ converges *faster* than $\{x_k\}$. Due to its characteristic form of the numerator in (2.17) this method is known as the Aitken's Δ^2 process.

Typically we do not accelerate quadratically convergent sequences. Yet this does become sensible when a single iteration step implies excessive numerical costs and we wish to squeeze as much as possible from the numerical procedure. As in the linear case, we divide the convergence equations in two subsequent steps, $e_{k+1} = \alpha e_k^2$ and $e_{k+2} = \alpha e_{k+1}^2$, thereby eliminating α. We get $e_{k+1}^3 = e_{k+2} e_k^2$, into which we insert $e_k = x_k - \xi$, and end up with a quadratic equation for ξ,

$$\left(3x_{k+1} - x_{k+2} - 2x_k\right)\xi^2 + \left(-3x_{k+1}^2 + 2x_k x_{k+2} + x_k^2\right)\xi + x_{k+1}^2 - x_{k+2}x_k^2 = 0 \,.$$

To compute the improved approximation of the root, we take the average of its two solutions:

$$X_k = \frac{\xi_+ + \xi_-}{2} = \frac{x_k^2 + 2x_k x_{k+2} - 3x_{k+1}^2}{4x_k - 6x_{k+1} + 2x_{k+2}} \,.$$

2.4 Polynomial Equations of a Single Variable

Polynomials [17] are frequent guests in scientific and technical applications. In most cases, we are dealing with real polynomials of degree n, defined as

$$p(x) = a_n x^n + a_{n-1} x^{n-1} + \ldots + a_1 x + a_0 = \sum_{k=0}^{n} a_k x^k \,, \tag{2.18}$$

where a_k are real coefficients and $a_n \neq 0$. The zeros (roots) of polynomial equations $p(x) = 0$ are assigned orders (multiplicities), just as the roots of equations $f(x) = 0$ with general functions f—see Eq. (2.4). A real polynomial may have real or complex zeros. Complex roots always appear in complex conjugate pairs.

Division of polynomials One of the auxiliary tools for solving polynomial equations and simplifying polynomial expressions is the division of polynomials. When a polynomial u is divided by a non-zero polynomial v,

$$u(x) = u_n x^n + \ldots + u_1 x + u_0 \,, \qquad v(x) = v_m x^m + \ldots + v_1 x + v_0 \,,$$

we obtain the quotient q and the remainder r, $u(x) = q(x)v(x) + r(x)$, where

$$q(x) = q_{n-m}x^{n-m} + \ldots + q_1 x + q_0 , \qquad r(x) = r_{m-1}x^{m-1} + \ldots + r_1 x + r_0 ,$$

and where $\mathrm{degree}(r) < \mathrm{degree}(v)$. We divide the polynomials u and v by using the algorithm for *synthetic division* [18]:

> **Input**: coefficients $\{u_0, \ldots, u_n\}$ and $\{v_0, \ldots, v_m\}$ of polynomials u and v
> **for** $k = n - m$ **step** -1 **to** 0 **do**
> > $q_k = u_{m+k}/v_m$;
> > **for** $j = m + k - 1$ **step** -1 **to** k **do**
> > > $u_j = u_j - q_k v_{j-k}$
> >
> > **end**
>
> **end**
> **Output**: coefficients of the remainder, $\{r_0, \ldots, r_{m-1}\} = \{u_0, \ldots, u_{m-1}\}$, and
> of the quotient, $\{q_0, \ldots, q_{n-m}\}$

Computing the value of the polynomial The division of a polynomial by the linear function $x - \zeta$ leads to an elegant and economical method to compute the value of the polynomial at $x = \zeta$ by applying the rule $p(x) = (x - \zeta)q(x) + p(\zeta)$. We write $p(\zeta)$ in factorized form $p(\zeta) = (\ldots((a_n\zeta + a_{n-1})\zeta + a_{n-2})\zeta + \ldots)\zeta + a_0$. From here we infer that synthetic division simplifies to the *Horner's* or *Ruffini's algorithm*, which can be memorized in terms of the table

$$
\begin{array}{ccccccc}
a_n & a_{n-1} & \cdots & a_2 & a_1 & a_0 \\
 & b_n\zeta & \cdots & b_3\zeta & b_2\zeta & b_1\zeta \\
b_n & b_{n-1} & \cdots & b_2 & b_1 & b_0 = p(\zeta)
\end{array} \quad .
$$

The first row are the coefficients a_k of p. The third row contains the sums of the corresponding terms in the first and second rows. The second row contains the products of ζ with the coefficients b_k from the third row. The value $p(\zeta)$ is in the lower right corner. For example, compute the value of $p(x) = x^3 - 2x^2 + 3x - 4$ at $\zeta = 2$. With $a_3 = 1$, $a_2 = -2$, $a_1 = 3$, $a_0 = -4$, we get $b_3 = a_3 = 1$, $b_2 = a_2 + b_3\zeta = 0$, $b_1 = a_1 + b_2\zeta = 3$ and $b_0 = a_0 + b_1\zeta = p(\zeta) = 2$.

In a computer implementation we may use the following algorithm which supplies not only the value of the polynomial, but also its derivative at ζ:

> **Input**: coefficients $\{a_0, \ldots, a_n\}$ of the polynomial p, value of ζ
> $p = a_n$; $q = 0$;
> **for** $i = n - 1$ **step** -1 **until** 0 **do**
> > $q = p + q\zeta$;
> > $p = a_i + p\zeta$;
>
> **end**
> **Output**: $p(\zeta) = p$ and $p'(\zeta) = q$

Horner's procedure can also be used to compute the numbers in various bases, for example, the value of a natural number in base b:

$$(a_n a_{n-1} \ldots a_1 a_0)_b = \sum_{k=0}^{n} a_k b^k , \qquad a_k \in \{0, 1, \ldots, b-1\} ,$$

This method is much faster than computing the powers and requires only $\mathcal{O}(n)$ arithmetic operations. The procedure can also be parallelized by splitting the polynomial into a part with odd and a part with even powers:

$$p(x) = p_{\text{even}}(x^2) + x p_{\text{odd}}(x^2) , \quad p_{\text{even}}(t) = \sum_{k=0}^{\lfloor n/2 \rfloor} a_{2k} t^k , \quad p_{\text{odd}}(t) = \sum_{k=0}^{\lceil n/2 \rceil - 1} a_{2k+1} t^k .$$

The numerical cost of Horner's scheme is optimal. However, for a generic polynomial, there exists an algorithm which requires only $\lfloor n/2 \rfloor + 1$ multiplications and n summations, but it can not be found easily [18]. If floating-point arithmetic with precision ε_{M} is used (Table 1.1), the computed value $p_{\text{num}}(\zeta)$ differs from the exact value $p(\zeta)$, and the difference can be bounded by

$$|p_{\text{num}}(\zeta) - p(\zeta)| < \max_{0 \leq k \leq n} |a_k| \left[(1 + \varepsilon_{\text{M}})^{2n} - 1 \right] \frac{|\zeta|^{n+1} - 1}{|\zeta| - 1} .$$

The coefficients a_k and the value ζ are assumed to be exact; the differences appear due to rounding errors in multiplication and summation. Clearly the precision deteriorates when the degree of the polynomial or its coefficients increase.

2.4.1 *Locating the Regions Containing Zeros*

In general, zeros of polynomials lie in bounded regions of the complex plane. In the following, we describe some generally useful methods used to determine the extent of these regions [19]. The classical estimates are due to Cauchy and Fujiwara: the magnitudes of all zeros $\xi_j \in \mathbb{C}$ of a polynomial of the form (2.18) are smaller than the only real zero r of the Cauchy polynomial

$$g(x) = x^n - \sum_{k=0}^{n-1} \left| \frac{a_k}{a_n} \right| x^k .$$

Thus r can be understood as the most conservative estimate for the radius of the disc in the complex plane that contains all zeros ξ_j. Cauchy's estimate is

$$r \leq \max\left\{ \left|\frac{a_0}{a_n}\right|, \ 1 + \left|\frac{a_1}{a_n}\right|, \ \ldots, \ 1 + \left|\frac{a_{n-1}}{a_n}\right| \right\}. \tag{2.19}$$

Much later, Fujiwara [20] derived the estimate

$$r \leq 2\max\left\{ \left|\frac{a_0}{2a_n}\right|^{1/n}, \ \left|\frac{a_1}{a_n}\right|^{1/(n-1)}, \ \ldots, \ \left|\frac{a_{n-1}}{a_n}\right| \right\}, \tag{2.20}$$

which may result in a narrower bound. The estimates (2.19) and (2.20) are also applicable to complex polynomials and are thus also useful for complex zeros.

Determining the interval with real zeros It can be shown [21] that arbitrary real numbers $\{v_1, v_2, \ldots, v_n\}$, for which we define $\alpha = \sum_{k=1}^{n} v_k$ and $\beta = \sum_{k=1}^{n} v_k^2$, lie on a closed interval with the boundary points

$$\frac{1}{n}\left[\alpha \pm \sqrt{(n-1)(n\beta - \alpha^2)}\right].$$

This property can be used for the determination of the interval containing real zeros of a real polynomial. Let a real polynomial of degree n,

$$p(x) = a_n \prod_{k=1}^{n}(x - \xi_k),$$

which is just a factorized form of (2.18), possess real zeros $\{\xi_1, \xi_2, \ldots, \xi_n\}$. For the numbers v_k we choose $v_k(x) = (x - \xi_k)^{-1}$, where $x \neq \xi_k$ for $\forall k$. It is clear from the structure of the polynomial that α and β can be written as

$$\alpha = \sum_{k=1}^{n} v_k = \frac{p'(x)}{p(x)}, \qquad \beta = \sum_{k=1}^{n} v_k^2 = -\left(\frac{p'(x)}{p(x)}\right)' = \frac{p'(x)^2 - p(x)p''(x)}{p(x)^2}.$$

The values $v_k(x)$ for each x thus lie on the interval with the boundary points

$$u_\pm(x) = \frac{1}{np(x)}\left[p'(x) \pm \sqrt{[(n-1)p'(x)]^2 - n(n-1)p(x)p''(x)}\right],$$

thus

$$u_-(x) \leq \frac{1}{x - \xi_k} \leq u_+(x), \qquad k = 1, 2, \ldots, n. \tag{2.21}$$

To u_-, v_k, and u_+, we apply the functional $\Phi : f(x) \mapsto \lim_{x \to \infty} x(x f(x) - 1)$ which maps $u_- \mapsto \Phi(u_-)$, $v_k \mapsto \xi_k$ and $u_+ \mapsto \Phi(u_+)$. In this mapping, the sense of the relation (2.21) does not change. From the inequality $\Phi(u_-) \leq \xi_k \leq \Phi(u_+)$ we therefore infer that the zeros ξ_k lie on the interval $[A_-, A_+]$, where

$$A_{\pm} = \Phi(u_{\pm}) = \frac{1}{na_n}\left[-a_{n-1} \pm \sqrt{(n-1)^2 a_{n-1}^2 - 2n(n-1)a_n a_{n-2}}\right], \quad (2.22)$$

if $a_n > 0$, or on the interval $[A_+, A_-]$, if $a_n < 0$.

Determining the interval containing at least one zero For each point x we can determine the interval $[x - r, x + r]$ containing at least one zero of the polynomial. We differentiate the sum α from the previous paragraph m-times:

$$\alpha(x) = \frac{p'(x)}{p(x)} = \sum_{k=1}^{n} \frac{1}{x - \xi_k}, \qquad \alpha^{(m)}(x) = \sum_{k=1}^{n} \frac{(-1)^m (m+1)!}{(x - \xi_k)^{m+1}}.$$

This tells us that the derivative $\alpha^{(m)}$ is bound from above as $\left|\alpha^{(m)}(x)\right| \leq n(m+1)!/r^{m+1}$, where r measures the distance from x to the nearest zero. This inequality can be read as the estimate for the upper bound for r:

$$r \leq \min_{m \in \mathbb{N}_0} \left(\frac{n(m+1)!}{\left|\alpha(x)^{(m)}\right|}\right)^{1/(m+1)}. \quad (2.23)$$

A meaningful estimate for r can already be obtained at orders $m = 0$, 1, or 2.

Example The real polynomial of degree $n = 5$,

$$p(x) = 3\prod_{k=1}^{5}(x - k) = 3x^5 - 45x^4 + 255x^3 - 675x^2 + 822x - 360, \quad (2.24)$$

has five real zeros $\xi_k = k$, $1 \leq k \leq 5$. Cauchy's estimate (2.19) is almost useless, as it tells us that all zeros are contained in a very broad interval $[-275, 275]$. Fujiwara's estimate (2.20) succeeds in narrowing down this range significantly, to $[-30, 30]$. From (2.22) we get

$$[A_-, A_+] \approx [0.171573, 5.82843].$$

We also estimate the interval centered at $x = 10$ which contains at least one zero of p. We use Eq. (2.23) at $m = 0$, 1, 2, 3, and 100. At these orders, we calculate

$$\begin{aligned}
[x - r, x + r] \approx \;\; &[3.294, 16.706] \quad (m = 0)\,, \\
&[0.721, 19.279] \quad (m = 1)\,, \\
&[0.737, 19.264] \quad (m = 2)\,, \\
&[1.098, 18.902] \quad (m = 3)\,, \\
&[4.682, 15.318] \quad (m = 100)\,,
\end{aligned}$$

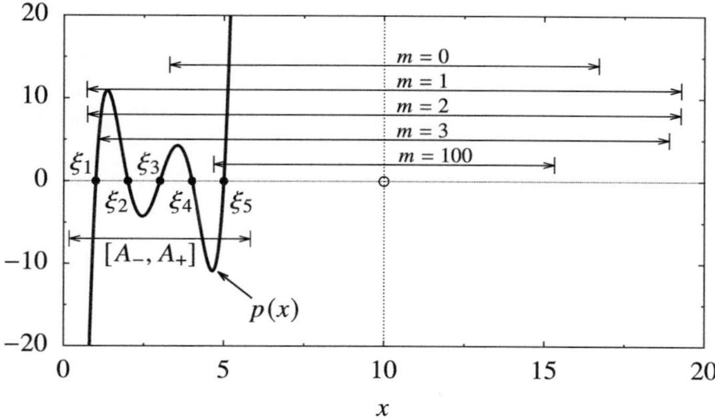

Fig. 2.7 Regions containing zeros of the polynomial (2.24). Arrows indicate the intervals centered at $x = 10$ which contain at least one zero according to Eq. (2.23). Also shown is the interval $[A_-, A_+]$ based on Eq. (2.22) which contains all zeros

shown in Fig. 2.7. With $m = 100$ we clearly get a good estimate for the location of ξ_5, but computing the hundredth derivative of p'/p is unfeasible; in a practical situation we might simply keep the $m = 0$ estimate. ◁

2.4.2 Descartes' Rule and the Sturm's Method

Descartes' rule [11] allows us to determine the maximum number of positive and negative zeros of the polynomial of the form (2.18). According to this rule, the number of positive zeros N_+ (where each zero is counted as many times as its multiplicity) is equal to or smaller by an even integer than the number of sign changes M_+ in the sequence of coefficients $\{a_n, a_{n-1}, \ldots, a_1, a_0\}$,

$$N_+ = M_+ - 2i , \qquad i \in \mathbb{N}_0 .$$

The zero coefficients should be disregarded in the sequence. By the same token, the number of negative zeros N_- does not exceed the number of sign changes M_- in the sequence of coefficients $\{(-1)^n a_n, (-1)^{n-1} a_{n-1}, \ldots, -a_1, a_0\}$,

$$N_- = M_+ - 2j , \qquad j \in \mathbb{N}_0 .$$

The maximum possible number of all real zeros is therefore $N = N_+ + N_-$.

To compute the number of real zeros of the polynomial lying on the chosen interval $[a, b]$, and to determine the intervals containing the individual zeros, we use the Sturm's method. Let us discuss polynomials with simple zeros only (multiple zeros can be eliminated, see Sect. 2.4.4). We first compute the derivative p'. The division of p by p' yields the remainder $-r_1$ which is used with the opposite sign in the following step. We then divide the derivative p' by r_1, yielding the remainder $-r_2$, which is again taken with the opposite sign. We repeat the procedure until the remainder is a constant:

$$
\begin{aligned}
p(x) &= p'(x)q_0(x) - r_1(x) \,, \\
p'(x) &= r_1(x)q_1(x) - r_2(x) \,, \\
r_1(x) &= r_2(x)q_2(x) - r_3(x) \,, \\
&\cdots \qquad \cdots \\
r_{k-1}(x) &= r_k(x)q_k(x) - r_{k+1} \,, \qquad r_{k+1} = \text{const} \,.
\end{aligned}
\tag{2.25}
$$

The sequence $S(x) = \{p(x), p'(x), r_1(x), r_2(x), \ldots\}$ is called the *Sturm's sequence*. At given x we count the number of sign changes of its terms and denote this number by $M(x)$. The number of zeros on $[a, b]$, each of which is counted with the corresponding multiplicity, is then $N = |M(a) - M(b)|$. By dividing the real axis and computing N for such divisions, individual zeros can be isolated.

Example Define the real polynomials

$$
p(x) = \prod_{k=1}^{7}(x - k) \,, \qquad p_+(x) = p(x) + x^3 \,.
\tag{2.26}
$$

The polynomial p contains the cubic term $-1960x^3$. What happens to the real zeros of p if the term $+1x^3$ is added to it? The polynomial p has seven zeros on the real axis: $\xi_k = k$, $1 \le k \le 7$ (Fig. 2.8 (left)). Let us choose the interval containing all zeros, e.g. $[a, b] = [0.5, 7.5]$. By Sturm's method we check that this interval indeed contains all seven zeros. Sturm's sequence, computed by using the procedure (2.25), has eight terms: $\{p(x), p'(x), r_1(x), \ldots, r_6(x)\}$. The values of the elements of this sequence at $x = a$ and $x = b$ are listed in the second and third columns of Table 2.1, respectively. In the second column we count $M(a) = 7$ sign changes, while there are no sign changes in the third, $M(b) = 0$. Therefore, p has $N = |M(a) - M(b)| = 7$ zeros on $[a, b]$.

A seemingly inconsequential perturbation $+1x^3$ distorts p to an extent that the perturbed polynomial p_+ has only three real zeros $\xi_1 \approx 0.999$, $\xi_2 \approx 2.085$, and $\xi_3 \approx 2.630$ (Fig. 2.8 (right)). Sturm's sequence for p_+ also contains eight terms, and their values at a and b are listed in the fourth and fifth columns of Table 2.1. From these columns, we now read off $M(a) = 5$ sign changes at $x = a$ and $M(b) = 2$ sign changes at $x = b$. On $[a, b]$ we therefore expect only $N = |M(a) - M(b)| = 3$ zeros of p_+, as confirmed by Fig. 2.8 (right).   ◁

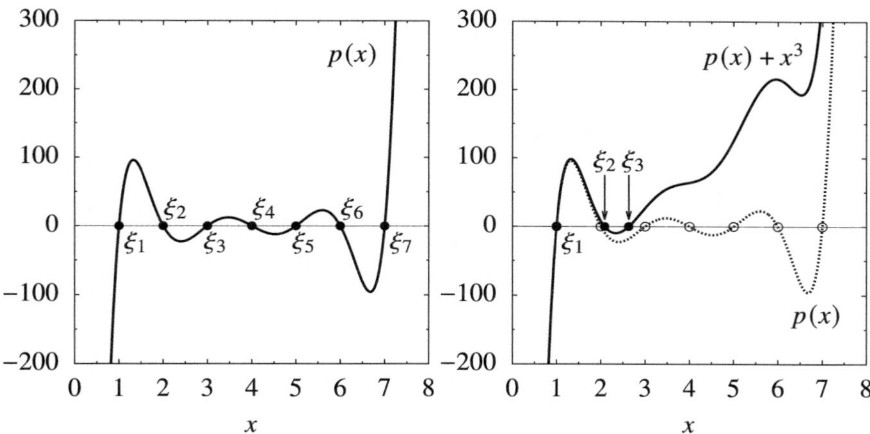

Fig. 2.8 Polynomials of degree seven defined by (2.26). [LEFT] Seven real zeros of the polynomial p. [RIGHT] Three real zeros of the polynomial p_+

Table 2.1 Values of the Sturm's sequence polynomials at $x = a$ and $x = b$, and the number of sign changes in this sequence for the polynomials p and p_+ defined by (2.26)

f	polynomial p		polynomial p_+	
	$f(a)$	$f(b)$	$f(a)$	$f(b)$
p	-1055.74	1055.74	-1055.62	1477.62
p'	4128.23	4128.23	4128.98	4296.98
r_1	-1008.38	1008.38	-1008.88	670.88
r_2	2048.06	2048.06	2050.37	-187.88
r_3	-357.00	357.00	-118.83	-830.57
r_4	503.00	503.00	-5404.29	-7278.65
r_5	-54.55	54.55	-14.92	45.58
r_6	36.00	36.00	283.82	283.82
M	7	0	5	2

2.4.3 Newton's Sums and Vièto's Formulas

Solving polynomial equations is an important part of many physical problems, and their solutions often reflect the symmetry properties of these problems. If the zeros in certain expressions appear on equal footing, it may be easier to compute these expressions by using Newton's sums and Vièto's formulas [22]. Both tools are also applicable as constraints in searching for zeros. For polynomials of the form (2.18) with zeros $\xi_1, \xi_2, \ldots, \xi_n$ Newton's sums are defined as

$$S_k = \sum_{l=1}^{n} \xi_l^k \, , \qquad k = 1, 2, \ldots, n \, .$$

The S_k are the sums of the powers of the zeros. For $1 \le k \le n$ they are related by linear equations $a_n S_k + a_{n-1} S_{k-1} + \cdots + a_{n-k+1} S_1 + k a_{n-k} = 0$ which can be solved recursively,

$$S_k = -\frac{1}{a_n} \left[a_{n-1} S_{k-1} + \cdots + a_{n-k+1} S_1 + k a_{n-k} \right] \, .$$

Vièto's formulas relate the sums of the products of the zeros,

$$\widetilde{S}_k = \sum_{1 \le i_1 < \cdots < i_k \le n} \xi_{i_1} \cdots \xi_{i_k} \, , \qquad (i_1, i_2, \ldots, i_k) \in \mathbb{N}_0^k \, ,$$

which can be expressed by the coefficients of the polynomial as

$$\widetilde{S}_k = (-1)^k \frac{a_{n-k}}{a_n} \, .$$

2.4.4 Eliminating Multiple Zeros of the Polynomial

Many methods to compute the zeros may exhibit a dramatic drop in convergence speed and become inefficient in the case of multiple zeros. Let us discuss a m-fold zero,

$$p(x) = (x - a)^m q_1(x) \, ,$$
$$p'(x) = (x - a)^{m-1} [\, m q_1(x) + (x - a) q_1'(x) \,] \equiv (x - a)^{m-1} q_2(x) \, ,$$

where q_1 and q_2 are non-constant polynomials that do not have a common divisor with the polynomials $p(x)$ and $(x - a)$. We see that the common divisor of p and p' contains the factor $(x - a)^{m-1}$, and an analogous situation can be observed for all multiple zeros. To eliminate multiple zeros, we compute the greatest common divisor of p and its derivative p', and divide it out from p. The polynomial obtained in this manner contains only simple zeros.

In principle, the *greatest common divisor* $\gcd(p, p')$ of the polynomials p and p' can be computed by using the Euclid's algorithm (2.25). The procedure is repeated until the remainder r_{k+1} is equal to zero, and then

$$\gcd(p, p')(x) = r_k(x) \, ;$$

on the other hand, if r_{k+1} is non-zero, the greatest common divisor of the polynomials is just a real number. This procedure may be numerically unstable [4] and we should avoid it if possible; if not, it should be accepted as necessary evil.

2.4.5 Conditioning of the Computation of Zeros

Computing the zeros of polynomials is an ill-conditioned problem. If the coefficient a_k of the polynomial (2.18) is changed by an infinitesimal amount δa_k, the zero ξ_j shifts by

$$\delta\xi_j = -(\delta a_k)\xi_j^k / p'(\xi_j) \,.$$

The sensitivity of the position of the zero with respect to the perturbations in the coefficients can be expressed by the ratio of the relative change of the zero ξ_j to the relative change of the coefficient a_k,

$$\kappa = \frac{|\delta\xi_j|}{|\xi_j|}\left[\frac{|\delta a_k|}{|a_k|}\right]^{-1} = \left|\frac{a_k\xi_j^{k-1}}{p'(\xi_j)}\right| \,,$$

which plays the role of a condition number [23]. An ill-conditioned problem exhibits large values of κ. The classical example of the Wilkinson's polynomial $p_{20}(x) = \prod_{k=1}^{20}(x - k) = a_0 + a_1 x + \cdots + a_{19}x^{19} + x^{20}$ is shown in Fig. 2.9.

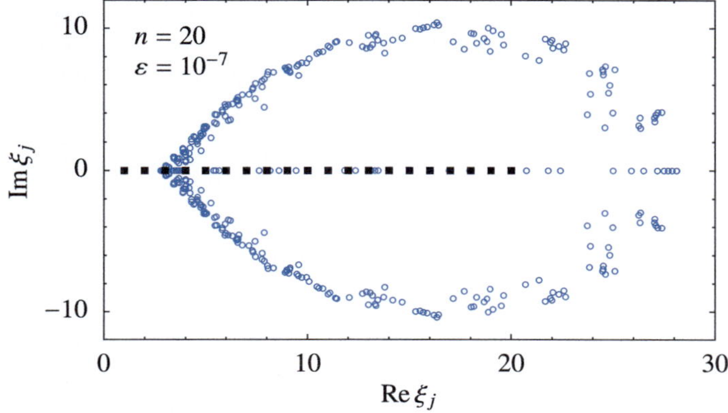

Fig. 2.9 Ill conditioning of the computation of zeros of the Wilkinson's polynomial p_{20} of degree 20. Full squares: zeros of the exact polynomial lying on the real axis, $\xi_j = j$ for $1 \le j \le 20$. Empty circles: zeros of twenty different polynomials in which the coefficients are perturbed as $a_k \to a_k(1 + \varepsilon R_k)$, where $\varepsilon = 10^{-7}$ and R_k is a normally distributed random number (zero average and unit variance). The most sensitive zero is ξ_{15} and has the strongest dependence on the coefficient $a_{15} \approx 1.67 \cdot 10^9$. The problem is extremely ill-conditioned, as $\kappa \approx 5.1 \cdot 10^{13}$. In spite of the tiny change in the coefficients, real zeros shift deeply into the complex plane

2.4.6 General Hints for the Computation of Zeros

Multiple polynomial zeros frequently stem from the symmetry properties of the physical problem. If this is the case, it is recommended to remove them "by hand"; otherwise, they can be removed by the procedure from Sect. 2.4.4. By using Sturm's method we first locate the intervals containing the individual zeros, which are then computed one by one either by using the general-purpose methods of Sect. 2.1 or more specific techniques tailored to polynomials, for instance, the Bernoulli's method, the Horner's linear method, the Bairstow's (Horner's quadratic) method, the Laguerre's method, or the Maehly–Newton-Raphson's method. Laguerre's method is among the fastest available, as it converges cubically in the case of simple zeros, and it can also be used to search for complex zeros. See [24, 25] for an in-depth review of these individual approaches.

When some zero ξ is found, it can be used to decrease the degree of the polynomial $p(x)/(x - \xi) \to p(x)$ and thereby reduce the complexity of further calculations, a process known as *deflation*. However, deflation may be numerically unstable and we might wish to avoid it and harness the entire polynomial instead. To stabilize the computation, the zeros already computed can be used in the Maehly–Newton's method.

If multiple zeros are anticipated, an alternative way is to divide the polynomial p by its derivative p', yielding a rational function $R(x) = p(x)/p'(x)$. The zeros of R can then be computed by using one of the general methods, for example, by the Newton–Raphson's method since the derivative of R is explicitly known, $R'(x) = 1 - p(x)p''(x)/p'(x)^2$.

The most efficient "sure-fire" and "black-box" method currently on the market, which is implemented in major software packages like MATHEMATICA, MATLAB or the IMSL library, is the three-stage Jenkins–Traub method [26]. It is well suited for the computation of zeros of real and complex polynomials and has more than quadratic order. Even though the method is globally convergent, it is reported to become unreliable for degrees higher than $n \approx 50$ [27].

2.4.7 The Hubbard–Schleicher–Sutherland Method

Recall the example in Fig. 2.3 where the sensitivity of the Newton's method to initial conditions (and lack of global convergence) has been observed. In an exciting new development [28] it was shown that it is possible to find *all* roots of a degree-n complex polynomial by Newton's method without recourse to deflation, starting from a set of $\mathcal{O}(n \log^2 n)$ initial points ($\mathcal{O}(n)$ if all roots are real) that can be constructed explicitly. The method works for degrees on the order of thousands or even millions [29], a task not unheard of in modern computer algebra systems and geometric modeling. In the following we quote the authors' main result in recipe form.

We construct a set of initial points I_n whose main feature is that at least one of them is in the basin of attraction of every root. It turns out that the points of this set lie along one or more concentric circles in the complex plane, the outermost of which has the radius $r(1 + \sqrt{2})$, where r is an estimate for the radius of the disc containing all roots, estimated one way or another or calculated, say, by using Eqs. (2.19) or (2.20). The number of required circles is

$$s = \lceil 0.26632 \log n \rceil \,,$$

and we need to place

$$N = \lceil 8.32547 \, n \log n \rceil$$

points on each circle. For $n \le 42$, a single circle is needed ($s = 1$). For $n \le 1825$, $n \le 78015$ and $n \le 3\,333\,550$ one needs 2, 3 and 4 circles, respectively. The points along the circles are evenly spaced in angle. We set

$$\rho(v) = r \left(1 + \sqrt{2}\right) \left(1 - \frac{1}{n}\right)^{(2v-1)/4s} \,, \qquad \theta_j = \frac{2\pi j}{N} \,,$$

for $v = 1, 2, \ldots, s$ and $j = 0, 1, \ldots, N - 1$. The initial grid then consists of sN points $\rho(v) \exp(\mathrm{i}\theta_j)$. An example of such a grid required to solve

$$p(x) = \prod_{k=1}^{5} (x - k) = -120 + 274\,x - 225\,x^2 + 85\,x^3 - 15\,x^4 + x^5 = 0 \quad (2.27)$$

($n = 5, s = 1, N = 67$) is shown in Fig. 2.10 (left).

We then start the Newton's iteration from each point $z_0 \in I_n$ and make it stop when $|z_i - z_{i-1}| < \varepsilon/n$ or after at most

$$K = \lceil n \log(r/\varepsilon) \rceil$$

iterations, where ε is the desired precision of the root. (Typically $\varepsilon \approx 10^{-6}$ is taken, with an option for later root refinement.) This condition guarantees that there is at least one root ξ_j for which $|z_i - \xi_j| < \varepsilon$. If the root ξ_j approximated by z_i is different from all previous roots, store it as a valid result, otherwise discard its "trajectory" entirely. If Newton's method has been applied K times to z_0 without converging to a root, save the value z_K in a set I_n^1 for possible future use. In addition, if $|z_i| > r(1 + \sqrt{2})$ for any $i > 1$ (a kind of "runaway" z), store z_i in I_n^1 as well. If all points in I_n have been tried and the number of located roots is less than n, repeat the whole procedure by taking I_n^1 as the new initial set and place any non-convergent points in the next auxiliary set I_n^2. Continue until all n roots are found.

The number of iterations required to find all roots with a precision of ε does not exceed $\mathcal{O}(n^2 \log^4 n + n \log |\log \varepsilon|)$. Note that the number of initial points can be reduced by increasing r, but starting from large radii implies that more

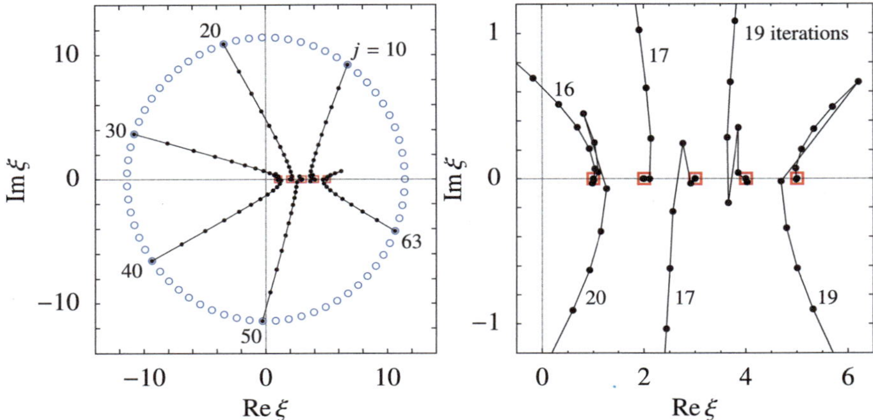

Fig. 2.10 Solving a fifth-degree polynomial equation (Eq. (2.27)) by using the Hubbard–Schleicher–Sutherland method. [LEFT] The initial grid I_5 consisting of $N = 67$ points along a single circle ($s = 1$) with a radius of ≈ 11.4 and some Newton "trajectories" starting from the points with angular indices $j = 10, 20, 30, 40, 50$ and 63, converging on the zeros at $\mathrm{Re}\,\xi = 4, 2, 1, 1, 3$ and 5, respectively. [RIGHT] Zoom-in of the region containing the zeros. The numbers indicate the required number of Newton iterations terminating at a tolerance below 10^{-6}

Newton steps need to be taken until convergence. For a comprehensive analysis of this computational trade-off and other details see [28].

2.5 Algebraic Equations of Several Variables ⋆

Polynomial or algebraic equations [30] and their systems, like, for example,

$$x_1^2 - x_2 = 0 \,, \qquad x_1^2 + x_2^2 = 1 \tag{2.28}$$

(the solution is the intersection of a parabola and a circle), or general systems

$$f_i(x_1, x_2, \ldots, x_n) = 0 \,, \qquad i = 1, 2, \ldots, m \,, \tag{2.29}$$

can be solved by algorithms from previous Sections, assuming that such systems possess a finite number of solutions. Yet in some instances, their algebraic properties can be exploited so that they can be solved more elegantly. One such approach takes us to *Gröbner bases* that are used in a variety of applications, e.g. in automated assembly of mechanical systems [31], in seeking intersections of volumes in three-dimensional space, and in robot control. They are also useful in coding and cryptography [32], as well as in logical and combinatorial problems [33]. Moreover, the theory of Gröbner bases provides tools to "solve" also systems with infinite sets of solutions [34].

Gröbner bases (invented by Buchberger) are a generalization of orthogonal bases known from linear algebra to algebraic expressions. This Section conveys only their background and the basic idea. Mathematically precise definitions and pedagogically thorough introductions can be found in [35–37].

To solve the equations by the Gröbner approach, we exploit the *elimination property* of Gröbner bases, whose consequence is that the original system of equations given in Eq. (2.29) can be rewritten as a different system involving Gröbner basis functions g_i:

$$g_i(x_1, x_2, \ldots, x_n) = 0 , \qquad i = 1, 2, \ldots, M . \tag{2.30}$$

In many cases the system with functions g_i turns out to be much simpler than that involving f_i, and this simplification is one of the main allures of Gröbner bases. Moreover, it can happen that the resulting system has the form

$$g_1(x_n) , \quad g_2(x_n, x_{n-1}) , \quad \ldots , \quad g_k(x_n, x_{n-1}, x_{n-2}) , \quad \ldots$$

The system of equations $g_i = 0$ for $i = 1, 2, \ldots, M$ can then be solved recursively: first one solves the equation $g_1 = 0$, then its solution is used in solving the second equation, $g_2 = 0$, and so on.

The Gröbner basis functions can be computed in several ways. The most widely known are the Buchberger's algorithm [36, 38] and its improved implementation, the Faugère's algorithm F_5 [39, 40]. These procedures are available in software packages like Singular (commands `groebner` or `slimgb`), Magna, MATHEMATICA (command `GroebnerBasis`) and Maple (command `gbasis`), as well as in the SymPy library of Python. In these environments, Gröbner bases are implemented for symbolic solution of algebraic systems. Yet when seeking specific solutions of complex systems of equations, it may be preferable to avoid these built-in capabilities and utilize Gröbner bases directly. This allows us to hand-pick and control the solutions when the equations $g_1 = 0$, $g_2 = 0$, and so on, are solved in sequence.

Example Let us see how Gröbner bases are used to solve Eq. (2.28) over the real field $\mathbb{R}[x_1, x_2]$. In this case the Gröbner basis consists of two functions,

$$g_1 = -1 + x_2 + x_2^2 , \qquad g_2 = x_1^2 - x_2 .$$

In MATHEMATICA, for example, they can be obtained from the functions

$$f_1(x_1, x_2) = x_1^2 - x_2 , \qquad f_2(x_1, x_2) = x_1^2 + x_2^2 - 1 ,$$

read off from Eq. (2.28) by using the command

```
GroebnerBasis[{x1^2 - x2, x1^2 + x2^2 - 1}, {x1, x2}]{.}
```

The default ordering in this case is lexicographic, $x_1 \succ x_2$, and the command returns the Gröbner basis

```
{g1, g2} = {-1 + x2 + x2^2, x1^2 - x2}{.}
```

The function g_1 depends solely on x_2, so $g_1(x_2) = 0$ can be solved for x_2 explicitly, yielding

$$x_2 \in \left\{ \frac{-1 - \sqrt{5}}{2}, \frac{-1 + \sqrt{5}}{2} \right\}.$$

For each of these two values of x_2, we then solve the equation $g_2(x_1, x_2) = 0$, which involves only the x_1 variable, and realize that only $x_2 = (-1 + \sqrt{5})/2$ leads to a real value of x_1. This brings us to the solutions of the second equation,

$$x_1 \in \left\{ \sqrt{x_2}, -\sqrt{x_2} \right\},$$

hence the two real solutions of Eq. (2.28) are approximately $(0.7862, 0.6180)$ and $(-0.7862, 0.6180)$.   ◁

In some instances it may happen that the number of equations M in (2.30) is exponentially larger than the number of equations m in (2.29). A typical reduced basis, however, has a dimension M that exceeds m at most by a few orders, and this represents only a minor problem in computations involving Gröbner bases. A much more severe issue is revealed by the following Example.

Example Let us discuss the system of equations $f_1 = f_2 = f_3 = f_4 = 0$ with

$$f_1(x_1, x_2, x_3) = 8x_1^2 x_2^2 + 5x_1 x_2^3 + 3x_1^3 x_3 + x_1^2 x_2 x_3,$$
$$f_2(x_1, x_2, x_3) = x_1^5 + 2x_2^3 x_3^2 + 13x_2^2 x_3^3 + 5x_2 x_3^4,$$
$$f_3(x_1, x_2, x_3) = 8x_1^3 + 12x_2^3 + x_1 x_3^2 + 3,$$
$$f_4(x_1, x_2, x_3) = 7x_1^2 x_2^4 + 18x_1 x_2^3 x_3^2 + x_2^3 x_3^3.$$

The reduced Gröbner basis for the ideal generated by f_1, f_2, f_3, and f_4 over the field of rational numbers $\mathbb{Q}[x_1, x_2, x_3]$ is exceedingly simple [41]:

$$g_1 = x_3^2, \qquad g_2 = x_2^3 + \frac{1}{4}, \qquad g_3 = x_1.$$

The system of equations $f_1 = f_2 = f_3 = f_4 = 0$ is equivalent to the system $g_1 = g_2 = g_3 = 0$ which is apparently trivial to solve. However, the size of the polynomial coefficients appearing in the algorithm that produces the Gröbner basis itself may grow exponentially. In the discussed case, the following polynomial appears in intermediate computations: $x_2^3 - 1735906504290451290764747182\ldots$. The integer in the second term of this polynomial contains roughly 80000 digits, and is actually the numerator of a fraction that contains about the same number of digits in the denominator. This demonstrates that Gröbner bases, in particular over the field of rational numbers, may in some cases be extremely difficult to compute. An interesting approach to overcoming this obstacle is described in [41].   ◁

2.6 Problems

2.6.1 Wien's Law and Lambert's Function

The distribution of the spectral energy density of black-body radiation [42] in terms of the wavelengths at temperature T is given by the Planck's formula

$$u(\lambda) = \frac{4\pi}{c}\frac{\mathrm{d}j}{\mathrm{d}\lambda} = \frac{8\pi hc}{\lambda^5}\frac{1}{\exp(hc/\lambda kT) - 1}\,,$$

while the distribution in terms of the frequencies is given by

$$u(\nu) = \frac{4\pi}{c}\frac{\mathrm{d}j}{\mathrm{d}\nu} = \frac{8\pi h\nu^3}{c^3}\frac{1}{\exp(h\nu/kT) - 1}\,.$$

Both distributions have maxima that shift with temperature (Fig. 2.11). Wien's law describes the temperature dependence of the position of these maxima, and can be derived by solving the appropriate equations for the local extrema. In the case of the distribution in terms of the wavelengths, this translates to

$$\frac{\mathrm{d}^2 j}{\mathrm{d}\lambda^2} = 0 \quad \Longrightarrow \quad (x_\lambda - 5)\mathrm{e}^{x_\lambda} + 5 = 0\,, \tag{2.31}$$

while in the case of the frequency distribution, one has

$$\frac{\mathrm{d}^2 j}{\mathrm{d}\nu^2} = 0 \quad \Longrightarrow \quad (x_\nu - 3)\mathrm{e}^{x_\nu} + 3 = 0\,, \tag{2.32}$$

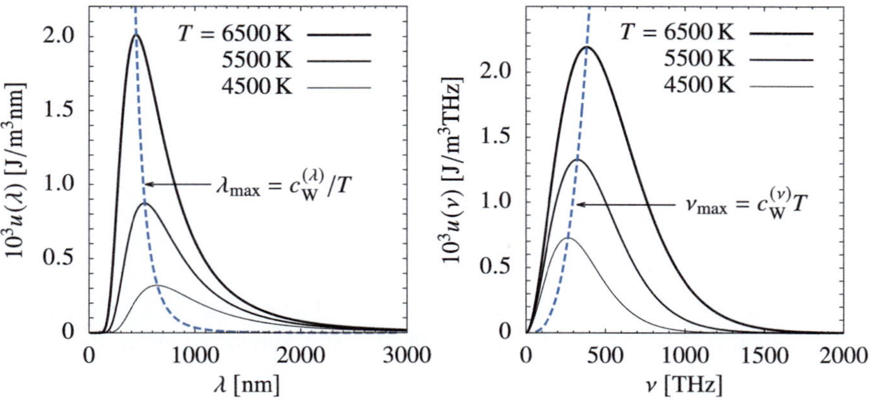

Fig. 2.11 Planck's law. [LEFT] Temperature dependence of the wavelength distribution of the spectral energy density. [RIGHT] Temperature dependence of the frequency distribution. Also shown are the Wien's curves connecting the distributions' maxima

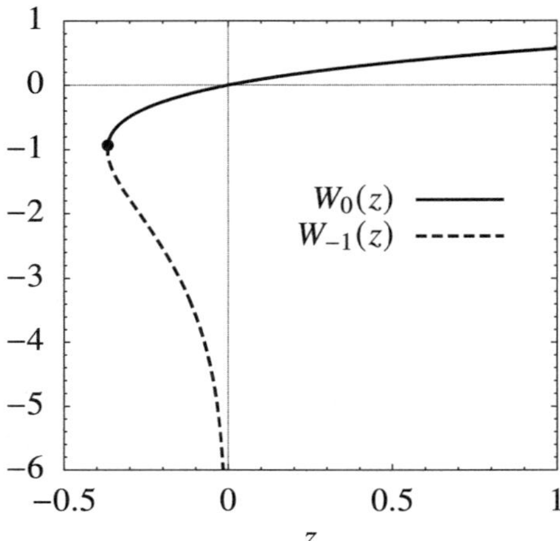

Fig. 2.12 Graphs of the Lambert's functions W_0 and W_{-1}. For details see [43]

where we have introduced $x_\lambda = hc/\lambda kT$ and $x_\nu = h\nu/kT$. These equations can be solved either analytically or numerically. The exact solution is related to the Lambert's function W_k, which represents different solutions of the equation

$$W_k e^{W_k} = z, \qquad z \in \mathbb{C}.$$

Lambert's functions are specified by the index k. We are concerned only about values $z \in \mathbb{R}$ and $W_k(z) \in \mathbb{R}$, for which two solutions are possible,

$$W_{-1}(z) \text{ at } z \in [-1/e, 0], \qquad W_0(z) \text{ at } z \in [-1/e, \infty],$$

shown in Fig. 2.12. The solutions of Eqs. (2.31) and (2.32) can be expressed as

$$x_\lambda = 5 + W_0(-5e^{-5}),$$
$$x_\nu = 3 + W_0(-3e^{-3}).$$

In the Wien's law, they appear as

$$\lambda_{\max} = \left(\frac{hc}{kx_\lambda}\right)\frac{1}{T} \equiv \frac{c_W^{(\lambda)}}{T}, \qquad \nu_{\max} = \left(\frac{k}{h}x_\nu\right)T \equiv c_W^{(\nu)}T. \tag{2.33}$$

⊙ Apply various numerical methods for the computation of zeros of non-linear scalar functions to determine the values of W_0. Use simple iteration, the secant method, and the Newton–Raphson's method. Optimize the computer program for speed and robustness. Draw the graph of W_0 for $x \in [-1/e, 2]$ and calculate the constants x_λ and x_ν appearing in Wien's laws (2.33).

⊕ Find the power expansion of the function W_0 in the vicinity of the point 0 by using the implicit-function theorem (page 68). According to this theorem, for each function f that is analytic near the point a, an inverse function g can be found, with the following power expansion near $f(a)$:

$$g(z) = a + \sum_{n=1}^{\infty} \lim_{x \to a} \left[\frac{d^{n-1}}{dx^{n-1}} \left(\frac{x - a}{f(x) - f(a)} \right)^n \right] \frac{(z - f(a))^n}{n!} .$$

For the Lambert's function, $f(x) = x e^x$ and $g(z) = W_0(z)$. Estimate the convergence radius of the power expansion of W_0 (use symbolic computation software).

2.6.2 Heisenberg's Model in the Mean-Field Approximation

Exact computations of thermo-dynamical equilibria of quantum spin systems are extremely time-consuming. One often resorts to simplified models in which knowledge from classical and quantum mechanics is combined. An instructive example is offered by the Heisenberg's system of electron spins in a magnetic field [44]. The electrons are arranged along a circle. The energy of the system is given by the Hamiltonian

$$H = -\sum_{ij} J_{ij}\, s_i \cdot s_j + \gamma B \cdot \sum_i s_i , \qquad \gamma = g\mu_B ,$$

where $\mu_B = e_0 \hbar / (2m_e)$ is the Bohr magneton, $g = 2$ is the gyro-magnetic ratio of the electron, and $s_i = (s_i^x, s_i^y, s_i^z)$ are the classical spin vectors with lengths $|s_i| = 1/2$. The coupling between the spins J_{ij} has the property $J_{ij} = J(|i - j|)$. Matter with $J_{ij} > 0$ is ferromagnetic; matter with $J_{ij} < 0$ is anti-ferromagnetic. In the *mean-field approximation* (MFA) we substitute

$$s_i \cdot s_j \longrightarrow \langle s \rangle \cdot s_j + \langle s \rangle \cdot s_i - | \langle s \rangle |^2 ,$$

where $\langle s \rangle$ is the statistical average of the spin vectors, which is a macroscopic observable of the system. We will determine it by requiring thermodynamic equilibrium. The z-axis is pointing along B so that $B = (0, 0, B)^{\mathrm{T}} \parallel (0, 0, \langle s^z \rangle)^{\mathrm{T}}$. In the MFA the Hamiltonian can be rewritten in the form

$$H_{\mathrm{MFA}} = -\gamma \sum_i B_{\mathrm{eff}}\, s_i^z , \qquad B_{\mathrm{eff}} = B - \frac{J_0}{\gamma} \langle s^z \rangle , \qquad J_0 = \frac{1}{N} \sum_{i,j=1}^{N} J_{ij} ,$$

describing a system of decoupled spins in the effective magnetic field B_{eff}. We quantize the system again and allow only $s_i^z \in \{-1/2, 1/2\}$. The statistical sum Z corresponding to a single spin in a chain of N independent spins, is

$$Z = \exp(-\beta\gamma B_{\text{eff}}/2) + \exp(\beta\gamma B_{\text{eff}}/2) = 2\cosh(\beta\gamma B_{\text{eff}}/2) \,,$$

where $\beta^{-1} = k_{\text{B}} T$. This sum can be used to compute the statistically averaged spin

$$\langle s^z \rangle = -\frac{1}{2} \tanh\left(\frac{1}{2}\beta(\gamma B - J_0 \langle s^z \rangle) \right) \,,$$

which is the self-consistent equation of the system. By appropriate substitutions, the equation can be cast in dimensionless form $z = \tanh(az - b)$, where we have introduced the dimensionless quantities $z \propto \langle s^z \rangle$, $a \propto \beta J_0$, and $b \propto \beta\gamma B$.

$\odot$ Find the solutions for the average dimensionless spin z in the range of parameters $(a, b) \in [-2, 2] \times [-5, 5]$. Display the solutions in a manner clearly indicating the transitions where the character of the solution changes. Draw the curves of $z(a, b)$ at $b = 0, 0.1$, and 0.5, by using $a \in [0, 5]$.

$\oplus$ Discuss in more detail the region in the vicinity of the transition between the ferromagnetic phase ($|z| > 0$) and para-magnetic phase ($z = 0$). A well-known transition point is $(a, b) = (1, 0)$. Increase the strength of the magnetic field b and locate the point a at which the transition occurs. This gives you the dependence of a on b: plot it.

2.6.3 Energy Levels of Simple One-Dimensional Quantum Systems

Often, numerical methods are the only way to compute the spectra (eigenenergies) of quantum systems. This problem deals with one-dimensional systems for which the equations for the eigenenergies E or the corresponding wave-numbers k are relatively simple, but their determination requires us to solve transcendental equations. The non-relativistic Hamiltonian operator

$$\hat{H} = -\frac{\partial^2}{\partial x^2} + U(x) \,,$$

contains the kinetic and the potential term. We are solving the stationary Schrödinger equation for the eigenenergies E and the eigenstates ψ_E,

$$\hat{H}\psi_E = E\psi_E \,, \qquad E = k^2 \,.$$

The spectrum may be discrete or continuous. Oscillation theorems tell us that energy degeneracies are impossible if the spectrum is discrete. Here we focus our attention to discrete spectra. The most simple systems involve piecewise constant potentials: three examples are shown in Fig. 2.13. These systems are described by the following eigenvalue equations: for a particle in the finite potential well with depth U_0 and width a (Fig. 2.13 a)):

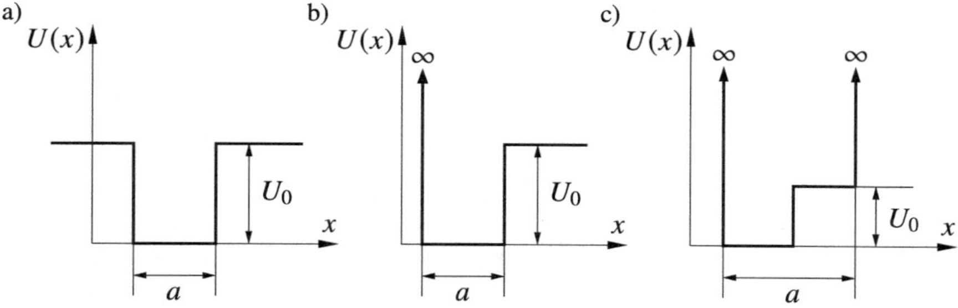

Fig. 2.13 Examples of one-dimensional potentials

$$\tan ka = \frac{2kk'}{k^2 - k'^2} \, , \qquad k' = \sqrt{U_0 - k^2} \, ;$$

for a particle in a semi-infinite well with a step of height U_0 on one side (Fig. 2.13 b)):

$$k' \tan ka = -k \, , \qquad k' = \sqrt{U_0 - k^2} \, ;$$

and for a particle in the infinite well with width a and a step of height U_0 reaching across one half of the well (Fig. 2.13 c):

$$k' \tan \tfrac{1}{2} ka = -k \tan \tfrac{1}{2} k'a \, , \qquad k' = \sqrt{k^2 - U_0} \, .$$

We are interested in the states with energies exceeding U_0 (so $k^2 - U_0 > 0$). In the cases (a) and (b) we obtain a finite number of discrete energy states, while there are infinitely many in the case (c).

⊙ Find the energy states of particles in at least one of the cases specified above by using the method of bisection, the secant method, and the Newton–Raphson's method. Set $U_0 = 1$. Compare the speed of convergence of all methods. Determine the order of convergence by plotting $\log |x_{k+1} - x_k|$ versus $\log |x_k - x_{k-1}|$, where x_k is the approximation of the root of the equation and k the iteration index. Consider the analytic structure of the equations when trying to figure out the initial approximations.

⊕ In the case (c) find all energy states starting from the lowest possible level and up to the level high enough that the asymptotic behavior of the solution can be ascertained. Confirm this behavior analytically.

2.6.4 Propane Combustion in Air

In concurrent chemical reactions proceeding in propane combustion in air,

$$C_3H_8 + \frac{R}{2}\,(4N_2 + O_2) \quad \longrightarrow \quad \text{products}\,,$$

different amounts x_i of the reaction products are formed, depending on the fraction R of air with respect to propane: CO_2 (x_1), H_2O (x_2), N_2 (x_3), CO (x_4), H_2 (x_5), H (x_6), OH (x_7), O (x_8), NO (x_9), and O_2 (x_{10}). The variables x_i correspond to the number of moles of the ith reaction product per mole of propane. A physical solution exists for $R > 3$. The equilibrium state at pressure 1 bar and temperature 2200 K is described by a system of non-linear equations [45]

$$
\begin{aligned}
x_1 + x_4 - 3 &= 0\,, \\
2x_1 + x_2 + x_4 + x_7 + x_8 + x_9 + 2x_{10} - R &= 0\,, \qquad R = 4.056734\,, \\
2x_2 + 2x_5 + x_6 + x_7 - 8 &= 0\,, \\
2x_3 + x_9 - 4R &= 0\,, \\
K_5 x_2 x_4 - x_1 x_5 &= 0\,, \qquad K_5 = 0.193\,, \\
K_6 \sqrt{x_2 x_4 S} - x_6 \sqrt{x_1} &= 0\,, \qquad K_6 = 0.002597\,, \\
K_7 \sqrt{x_1 x_2 S} - x_7 \sqrt{x_4} &= 0\,, \qquad K_7 = 0.003448\,, \\
K_8 x_1 S - x_4 x_8 &= 0\,, \qquad K_8 = 1.799 \cdot 10^{-5}\,, \\
K_9 x_1 \sqrt{x_3 S} - x_4 x_9 &= 0\,, \qquad K_9 = 2.155 \cdot 10^{-4}\,, \\
K_{10} x_1^2 S - x_4^2 x_{10} &= 0\,, \qquad K_{10} = 3.846 \cdot 10^{-5}\,,
\end{aligned}
$$

where $S = \sum_{i=1}^{10} x_i$. (This sum can be taken as exact and inserted in the system for x_i, $1 \le i \le 10$, or it can be treated as an independent, eleventh equation.)

  $\odot$ Solve the system of ten (or eleven) equations given above by using the Newton-Raphson's and Broyden's method for vector equations described in Sect. 2.2. Exploit the structure of the system in order to devise the best initial approximations for the iteration. How sensitive is the solution to the variations of the parameters R and K_i?

  $\oplus$ With poor initial approximations, negative arguments of the square roots may appear during the iteration. If you encounter such problems, replace

$$x_{i,\text{new}}^2 = x_{i,\text{old}}\,, \qquad i = 1, 2, 3, 4\,,$$

and repeat the calculation as before. You can also get rid of the negative arguments by taking their absolute values under all square roots, e.g. $\sqrt{|x_2 x_4 S|}$ instead of $\sqrt{x_2 x_4 S}$, or all equations are rewritten such that none of the x_i appears under the root sign. In all these approaches (see [45] for details), can you spot any differences in the convergence speed of Newton's or Broyden's methods?

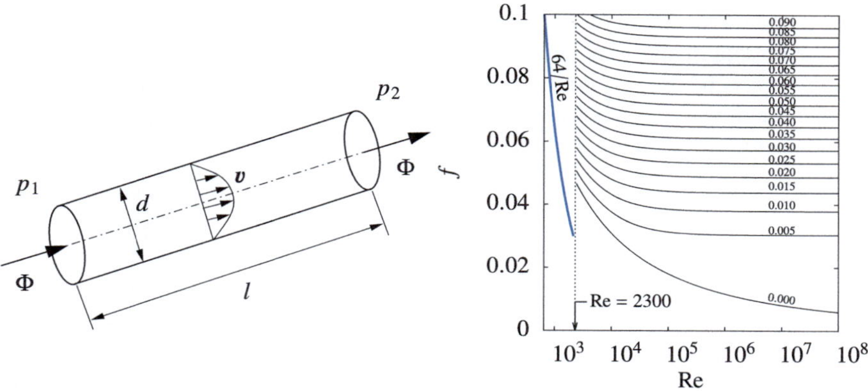

Fig. 2.14 [LEFT] Fluid flow in a pipe. The fluid enters at height z_1 and pressure p_1 with the average velocity $\bar{v}_1$, and exits at height z_2 and pressure p_2 with the velocity $\bar{v}_2$. The hydraulic diameter d for a pipe with circumference o is defined by the ratio $4S/o$. [RIGHT] Darcy's friction factor f as a function of the Reynolds number Re for different relative roughnesses e/d (simplified Moody diagram)

2.6.5 Fluid Flow Through Systems of Pipes

Flow through pipes is not just a technological problem: it also offers lovely mathematical physics insights. Assume that the flow is unidirectional and occurs due to pressure gradients, neglecting any possible complex dynamics [46, 47]. Curved pipes are approximated by sequences of smoothly joined straight segments of length l and hydraulic diameter d (Fig. 2.14 (left)), along which Bernoulli's equation applies:

$$p_1 + \frac{1}{2}\rho\bar{v}_1^2 + \rho g z_1 = p_2 + \frac{1}{2}\rho\bar{v}_2^2 + \rho g z_2 + \Delta p\,,$$

where Δp is the pressure drop. The average velocity of the fluid $\bar{v}$, the cross-sectional area of the pipe S and the density ρ determine the mass flux $\Phi = \rho S \bar{v}$ which is constant along the pipe.

Due to the viscosity of the fluid, the velocity at the wall pipe is zero and the velocity across the pipe diameter is not constant. The pressure drops Δp along the pipe appear because of the viscosity, the roughness of the pipe walls, and the dynamics in the fluid itself. Phenomenologically, the pressure drop is given by the Darcy–Weisbach equation [48]

$$\Delta p = f\frac{l}{2d}\rho\bar{v}|\bar{v}| = R\Phi|\Phi|\,, \qquad R = f\frac{l}{2d\rho S^2}\,,$$

where f is the Darcy friction factor and R the resistance coefficient. For long pipes R depends only on the relative roughness e/d and the Reynolds number $\mathrm{Re} = \rho\bar{v}d/\mu = 4\Phi/(\pi\mu d)$, where μ is the dynamical viscosity of the fluid. The roughness e is defined as the standard deviation of the pipe radius, averaged over its interior surface. In the laminar regime of the flow (Re < 2300) f is determined by the Hagen–Poiseuille equation, $f = 64/\text{Re}$, while in the intermediate ($2300 < \text{Re} < 4000$) and in the turbulent regime (Re > 4000) we can read it off from the Moody diagram (Fig. 2.14 (right)). In the turbulent regime, a good approximation for f is given by the Colebrook–White implicit equation

$$\frac{1}{\sqrt{f}} = -2\log_{10}\left(\frac{e/d}{3.7} + \frac{2.51}{\text{Re}\sqrt{f}}\right).$$

In pipe systems the dominant pressure drops (dominant energy losses) are caused by long pipes; relatively small losses are caused by the joining elements or sharp turns. At the known flux $\Phi = \rho S \overline{v}$, the pressure drop along some element is given by the equation

$$\Delta p = K \tfrac{1}{2}\rho \overline{v}|\overline{v}|,$$

where $K = fl/d$ is the loss coefficient (e.g. 0.35 for a 45° elbow or 0.75 for a 90° elbow).

An example of a network of pipes in the gravitational field is illustrated by the graph of vertices and connections in Fig. 2.15. We are interested in the stationary flow along the pipes when the pressures at the free ends of the graph are known. For the mass flux Φ_j in the individual segment, Darcy–Weisbach equation applies. For the jth pipe connecting vertices i and k, it reads

$$R_j|\Phi_j|\Phi_j = \tilde{p}_i - \tilde{p}_k, \qquad \tilde{p}_i = p_i + \rho g h_i,$$

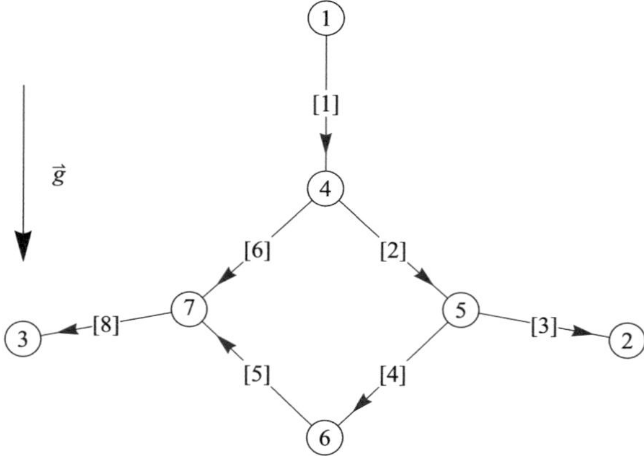

Fig. 2.15 The graph of a test pipe system. The vertices are denoted by indices in circles, and the connections are denoted by indices in brackets. We have N connections, M vertices, and m free ends. The remaining $M' = M - m$ vertices are at known heights h_i and the losses in them are assumed to be negligible. The pressures in the vertices are denoted by p_i where i is the index of the vertex

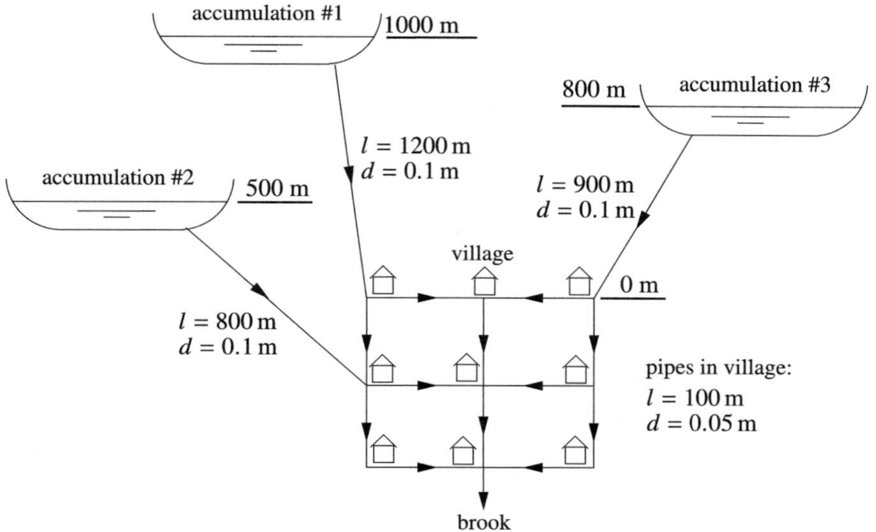

Fig. 2.16 A cartoon of a system of pipes in a mountain village

where $\tilde{p}_i$ are the hydrostatic pressures. (The resistance coefficients R_j depend on Re $\propto \Phi_j$.) At each vertex, the sum of the incoming mass fluxes should be equal to the sum of the outgoing mass fluxes. By defining the vector of fluxes $\boldsymbol{\Phi} = (\Phi_i)_{i=1}^N$ all continuity equations can be written in the matrix form $A\boldsymbol{\Phi} = \boldsymbol{0}$, where the matrix $A \in \mathbb{R}^{M' \times N}$ contains the information about the connectedness of the graph. The hydrostatic pressures in the internal vertices $\{i_1, \ldots, i_{M'}\}$ are arranged in the vector $\tilde{\boldsymbol{p}} = (\tilde{p}_{i_1}, \ldots, \tilde{p}_{i_{M'}})^{\mathrm{T}}$. In addition, we define the diagonal matrix $\Sigma(\boldsymbol{\Phi}) = \mathrm{diag}\,(R_j|\Phi_j|)_{j=1}^N$. All Darcy–Weisbach equations for the system can then be written in the matrix form $\Sigma\boldsymbol{\Phi} + A^{\mathrm{T}}\tilde{\boldsymbol{p}} = \boldsymbol{P}$, where the vector $\boldsymbol{P} = (P_1, \ldots, P_N)^{\mathrm{T}}$ contains only the hydrostatic pressures at the free ends ($\rho g h$ for points at height h, e.g. the lakes in Fig. 2.16). The Darcy–Weisbach and the continuity equations must be solved simultaneously and can be rewritten as a single system of non-linear equations $F(\boldsymbol{x})\boldsymbol{x} = \boldsymbol{b}$, where

$$F(\boldsymbol{x}) = \begin{bmatrix} \Sigma(\boldsymbol{\Phi}) & A^{\mathrm{T}} \\ A & 0 \end{bmatrix}, \qquad \boldsymbol{x} = \begin{bmatrix} \boldsymbol{\Phi} \\ \tilde{\boldsymbol{p}} \end{bmatrix}, \qquad \boldsymbol{b} = \begin{bmatrix} \boldsymbol{P} \\ \boldsymbol{0} \end{bmatrix}.$$

The equation can be solved by the iteration

$$\boldsymbol{x}_{n+1} = F_n^{-1}\boldsymbol{b}, \qquad F_n = F(\boldsymbol{x}_n).$$

We form the initial matrix F_0 in the laminar regime where, for the jth pipe, we have $R_j|\Phi_j| = 32\mu l_j/(\rho d_j^2 S_j)$.

⊙ A mountain village is supplied by water from three accumulations lying at the mountain slopes. The whole village is at constant altitude. The consumption of water in the village is negligible; the remainder of the water drains into the brook. The cartoon of the system of pipes is shown in Fig. 2.16. The relative roughness of the pipes is $e/d = 0.01$. Draw the graph of this system and compute the mass fluxes through the individual pipes if the Colebrook–White equation is used to compute the friction factor. Neglect all losses.

⊕ Discuss the fluid flow in a random two-dimensional pipeline on a Cartesian grid. On this grid, select the origin whence a random walker starts his walk $\mathcal{N}$-times. The speed of the walk is one unit of length on the mesh per unit time. An individual walk takes $\mathcal{M}$ units of time. We interpret all $\mathcal{N}$ walks as a random pipeline composed of straight pipes of equal lengths. At the origin, attach a water reservoir with a constant over-pressure p_0. The loss coefficient in the pipes K is constant. (Choose the units such that $K = p_0 = 1$.) We are interested in the distribution of the flows in the pipes in the cases $\mathcal{N} = \mathcal{M} = 10$, $\mathcal{N} = \mathcal{M} = 100$, and $\mathcal{N} = \mathcal{M} = 1000$. Can you infer a scaling law from these distributions?

2.6.6　*Automated Assembly of Structures*

Because so many building blocks are involved, the simulation of robots, motion and collisions of vehicles, proteins and other complex systems of bodies require the use of automated, *computer-aided assembly* [31]. The manner in which individual components are connected is embodied in the equations known as *constraints*. Based on these constraints, the program computes the structure of the body in three-dimensional space. In virtually all physically relevant types of connections, the constraints can be written as systems of algebraic equations.

Here we discuss the assembly of a simple linear structure in the (x, y) plane in the reference space $\mathbb{R}^3$ with the coordinate axes $\{x, y, z\}$. We would like to automatically assemble a chain of m connected rods. We denote the rods by their indices $1 \leq i \leq m$ and their lengths by l_i. The $(i + 1)$th rod should be rotated with respect to the ith rod by the angle θ_i. The center-of-mass of the ith rod is the origin of its local reference frame $\mathcal{S}^i$ with the axes $\{\xi_i, \eta_i, \zeta_i\}$ such that the ξ_i axis is along the rod, the ζ_i axis is aligned with the z axis of the global reference frame, and the edges of the rod are at $s_{\pm}^i = (\pm l_i/2, 0, 0)^{\mathrm{T}}$ (see Fig. 2.17 (left)). The position of the ith rod is described by the vector $r^i = (x^i, y^i, z^i)^{\mathrm{T}}$ from the origin of the frame $\mathcal{S}^i$, while its orientation is specified by the vector of Euler parameters $e^i = (e_0^i, e_1^i, e_2^i, e_3^i)^{\mathrm{T}}$ [49]. The Euler parameters $e = (e_0, e_1, e_2, e_3)^{\mathrm{T}}$ are used to define the rotation by an angle ϕ around an arbitrary axis $\hat{n}$ ($\hat{n}^{\mathrm{T}}\hat{n} = 1$) as

$$e_0 = \cos\frac{\phi}{2}, \qquad \begin{pmatrix} e_1 \\ e_2 \\ e_3 \end{pmatrix} = \hat{n}\sin\frac{\phi}{2}, \qquad \sum_{\rho=0}^{3} e_\rho^2 = 1.$$

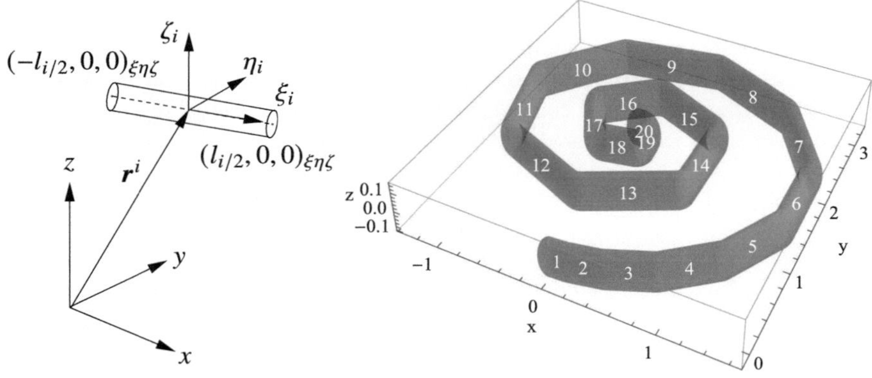

Fig. 2.17 [LEFT] A drawing of the ith rod with its local and global reference frames. [RIGHT] An example of a linear structure assembled from rods of diameter $d = 0.2$

The corresponding rotation matrix $R(e) = [R_{ij}(e)]^3_{i,j=1}$ is then

$$R_{ij}(e) = \delta_{i,j} \left(e_0^2 - e_k e_k\right) + 2e_i e_j + 2\varepsilon_{ijk} e_0 e_k \,,$$

where $\delta_{i,j}$ is the Kronecker delta, ε_{ijk} is the anti-symmetric tensor and Einstein's summation convention applies: at the right-hand side, we sum over all k.

All information on the ith rod are collected in the vector $q^i = (r^i, e^i)$ of dimension 7. The condition

$$R_{33}(e^i) - 1 = 0 \qquad \forall i$$

restricts the assembly of the chain to the (x, y) plane. The joining of the end of the ith rod by the beginning of the $(i + 1)$th rod is expressed by the constraint

$$R(e^i)s_+^i + r^i - R(e^{i+1})s_-^{i+1} - r^{i+1} = 0 \,.$$

In our case $\hat{n} = (0, 0, 1)^{\mathrm{T}}$, so the Euler parameters e^i for the ith rod depend only on the rotation angles of the individual rod, ϕ_i, and these are simply determined by the constraints $\phi_{i+1} - \phi_i - \theta_i = 0$. These constraints are not analytic in the parameters r^i and e^i; due to the periodicity of the angles, they are also not unique. We therefore choose slightly more complicated constraints

$$R_{11}(e^{i+1}) + \mathrm{i}\, R_{21}(e^{i+1}) - \exp\left(\mathrm{i}\sum_{j=1}^{i}\theta_j\right)\left(R_{11}(e^i) + \mathrm{i}\, R_{21}(e^i)\right) = 0 \,,$$

for $i = 1, 2, \ldots, m - 1$, which are analytic. We exploit the fact that the projections of the rod's orientation on the x and y axes are hidden in the components of the rotation matrices R_{11} and R_{21}. Without loss of generality, the first rod may be oriented along the x axis, which implies an additional constraint $R_{11}(e^1) - 1 = 0$.

$\odot$　Let the linear structure described above reside in the plane at $z = 0$ and let the beginning of the first rod coincide with the origin of the global reference frame. For some m, automatically assemble the structure corresponding to the lengths and angles

$$l_i = \sin \frac{\pi i}{m+1}, \qquad \theta_i = \frac{\pi i}{2(m+1)}, \qquad i = 1, 2, \ldots, m.$$

The system can be described by the vector of positions and Euler parameters of all rods

$$q = (q^i)_{i=1}^m$$

of dimension $M = 7m$. All constraints of the system can be collected in the vector equation

$$F(q) = \big(F_i(q)\big)_{i=1}^N = 0, \tag{2.34}$$

which is analytic in q with $N \geq M$. The solution of this system requires us to find the global minimum of the quantity

$$\Phi(q) = F(q)^{\mathrm{T}} F(q), \tag{2.35}$$

which, in general, is troublesome. Here, we are only seeking the local minimum, so we are solving $\nabla \Phi(q) = 0$. Since Φ is analytic in q, its minimum can be found by using the Newton's method of *unconstrained minimization* [12]. This method requires us to know the first derivative of (2.35),

$$D(q) = \big(\partial_i \Phi(q)\big)_{i=1}^M, \qquad \partial_i \Phi(q) = 2[\partial_i F_j(q)] F_j(q),$$

where $\partial_i \equiv \partial/\partial q_i$, as well as the second derivative, given by the Hessian matrix,

$$H(q) = [\partial_i \partial_j \Phi(q)]_{i,j=1}^M, \qquad \partial_i \partial_j \Phi(q) = 2[\partial_i \partial_j F_k(q)] F_k(q) + 2[\partial_j F_k(q)] \partial_i F_k(q).$$

(Einstein's summation convention applies throughout.) We choose a good enough initial approximation of the solution $q^{(0)}$ and iterate

$$q^{(k+1)} = q^{(k)} + \big[H\big(q^{(k)}\big)\big]^{-1} D(q^{(k)}). \tag{2.36}$$

This procedure is well suited for solving small or medium-sized systems where m is on the order of several 10 to several 100, and M on the order of 100 to 1000. We choose $m = 20$. The solution should appear as a structure shown in Fig. 2.17 (right). For larger systems, solving the system of equations

$$H\big(q^{(k)}\big)\big(q^{(k+1)} - q^{(k)}\big) = D\big(q^{(k)}\big)$$

at each iteration step (2.36) may be too time-consuming and numerically unstable. In such cases, other methods should be called to rescue: see [12].

$\oplus$ Another possibility to solve Eq. (2.34) leads through Gröbner bases (see Sect. 2.5). High numerical costs of this approach restrict m to about a few times 10; we choose $m = 3$. For a given set of functions $\{F_i(q)\}_{i=1}^N$ compute the reduced Gröbner basis $\mathcal{G} = \{g_i\}_{i=1}^M$ with lexicographic ordering. Solve the equations $g_i(q) = 0$ for $i = 1, 2, \ldots, M$ in this very same index sequence, i.e. make use of the solutions of equations with index $j \in [1, i - 1]$ in computing the solution of the equation with index i. Along the way, non-physical solutions can be eliminated. With such an approach, each equation $g_i(q) = 0$ depends only on a single variable, and the index of the variable increases with the index of the equation under consideration. Hence, standard methods for seeking zeros of polynomials can be applied. If multiple solutions appear, they all describe the same spatial structure, which one easily confirms by drawing it.

References

1. A.E. Dubinov, I.N. Galidakis, Explicit solution of the Kepler equation. Phys. Part. Nucl. Lett. **4**, 213 (2007)
2. R. Luck, J.W. Stevens, Explicit solutions for transcendental equations. SIAM Rev. **44**, 227 (2002)
3. G.R. Wood, The bisection method in higher dimensions. Math. Prog. **55**, 319 (1992); W. Baritompa, Multidimensional bisection: a dual viewpoint. Computers Math. Applic. **27**, 11 (1994)
4. J. Stoer, R. Bulirsch, *Introduction to numerical analysis*. In *Texts in Applied Mathematics* vol. 12, 3rd edn. (Springer, Berlin, 2002)
5. E.D. Charles, J.B.Tatum, The convergence of Newton–Raphson iteration with Kepler's equation. Celest. Mech. Dyn. Astr. **69**, 357 (1998); See also B.A. Conway, An improved algorithm due to Laguerre for the solution of Kepler's equation. Celest. Mech. **39**, 199 (1986)
6. H. Susanto, N. Karjanto, Newton's method's basins of attraction revisited. Appl. Math. Comp. **215**, 1084 (2009)
7. J.A. Ford, Improved algorithms of Illinois-type for the numerical solution of nonlinear equations. *Technical Report CSM-257* (University of Essex, 1995)
8. A. Ralston. H.S.Wilf, *Mathematical Methods of Digital Computers, vol. 2* (Wiley, New York, 1967) (Chap. 9)
9. J.F. Traub, *Iterative methods for the solution of equations* (Prentice-Hall, Englewood Cliffs, 1964)
10. W.H. Press, B.P. Flannery, S.A. Teukolsky, W.T. Vetterling, *Numerical Recipes: The Art of Scientific Computing*, 3rd edn. (Cambridge University Press, Cambridge, 2007). See also the equivalent handbooks in Fortran, Pascal and C, as well as http://numerical.recipes
11. G. Dahlquist, Å. Björck, *Numerical Methods in Scientific Computing*, vols. 1 and 2 (SIAM, Philadelphia, 2008)
12. J.E. Dennis Jr., R.B. Schnabel, *Numerical methods for unconstrained optimization and nonlinear equations* (SIAM, Philadelphia, 1996)
13. C.G. Broyden, A class of methods for solving nonlinear simultaneous equations. Math. Comp. **19**, 577 (1965); see also J.J. Moré, J.A. Trangenstein, On the global convergence of Broyden's method. Math. Comp. **30**, 523 (1976) and D.M. Gay, Some convergence properties of Broyden's method. SIAM J. Numer. Anal. **16**, 623 (1979)
14. C.G. Broyden, On the discovery of the "good Broyden" method. Math. Program. B **87**, 209 (2000)

15. D. Li, J. Zeng, S. Zhou, Convergence of Broyden-like matrix. Appl. Math. Lett. **11**, 35 (1998)
16. J.M. Martínez, Practical quasi-Newton methods for solving nonlinear systems. J. Comput. Appl. Math. **124**, 97 (2000)
17. P. Henrici, *Applied and Computational Complex Analysis*, vol. 1 (Wiley, New York, 1974)
18. D.E. Knuth, *The Art of Computer Programming, vol. 2: Seminumerical Algorithms*, 2nd edn. (Addison-Wesley, Reading, 1980)
19. M. Marden, *The Geometry of Zeros of a Polynomial in the Complex Variable* (AMS, New York, 1949)
20. M. Fujiwara, Über die obere Schranke des absoluten Betrages der Wurzeln einer algebraischen Gleichung. Tohoku Math. J. **10**, 167 (1916)
21. D. Kincaid, W. Cheney, *Numerical Analysis. Mathematics of Scientific Computing* (Brooks/Cole Publishing Company, Belmont, 1991)
22. E.B. Vinberg, *A Course in Algebra* (AMS, Providence, 2003)
23. L.N. Trefethen, D. Bau, *Numerical Linear Algebra* (SIAM, Philadelphia, 1997)
24. J.M. McNamee (ed.), *Numerical Methods for Roots of Polynomials—Part I, Studies in Computational Mathematics*, vol. 14 (Elsevier Science, 2007)
25. J.M. McNamee, V.Y. Pan (eds.), *Numerical Methods for Roots of Polynomials—Part II, Studies in Computational Mathematics* vol. 16 (Elsevier, 2013)
26. M.A. Jenkins, J.F. Traub, A three-stage variable-shift iteration for polynomial zeros and its relation to generalized Rayleigh iteration. Numer. Math. **14**, 252 (1970); M.A. Jenkins, J.F. Traub, A three-stage algorithm for real polynomials using quadratic iteration. SIAM J. Numer. Anal. **7**, 545 (1970)
27. V.Y. Pan, Solving a polynomial equation: some history and recent progress. SIAM Rev. **39**, 187 (1997)
28. J. Hubbard, D. Schleicher, S. Sutherland, How to find all roots of complex polynomials by Newton's method. Invent. Math. **146**, 1 (2001)
29. D. Schleicher, R. Stoll, Newton's method in practice: finding all roots of polynomials of degree one million efficiently. Theoret. Comput. Sci. **681**, 146 (2017)
30. H.J. Stetter, *Numerical Polynomial Algebra* (SIAM, Philadelphia, 2004)
31. E.J. Haug, *Computer aided kinematics and dynamics of mechanical systems*, vol. 1 (Allyn and Bacon, Boston, 1989)
32. M. Sala, T. Mora, L. Perret, S. Sakata, C. Traverso (eds.), *Gröbner Bases, Coding, and Cryptography* (Springer, Berlin, 2009)
33. J. Gago-Vargas et al., Sudokus and Gröbner bases: not only a divertimento, In V.G. Ganzha, E.W. Mayr, E.V. Vorozhtsov, editors. *CASC 2006, Lecture Notes in Computer Science*, vol. 4194 (Springer, Berlin, 2006), p. 155
34. V.R. Romanovski, D.S. Shafer, *The Center and Cyclicity Problems: A Computational Algebra Approach* (Birkhäuser, Boston, 2009)
35. W. Adams, P. Loustaunau, *An Introduction to Gröbner Bases, Graduate Studies in Mathematics*, vol. 3 (AMS, Providence, 1994)
36. D. Cox, J. Little, D. O'Shea, *Ideals, Varieties, and Algorithms*, 3rd edn. (Springer, Berlin, 2007)
37. E. Roanes-Lozano, E. Roanes-Macías, L.M. Laita, Some applications of Gröbner bases, in *Computing in Science and Engineering, May/Jun 2004*, p. 56; *The Geometry of Algebraic Systems and their Exact Solving Using Gröbner Bases*, ibid., Mar/Apr 2004, p. 76
38. B. Buchberger, An algorithm for finding the basis elements of the residue class ring of a zero dimensional polynomial ideal. J. Symbolic Comput. **41**, 475 (2006)
39. T. Stegers, *Faugère's F5 algorithm revisited*, diploma thesis, Technische Universität Darmstadt, 2005/2007
40. J.C. Faugère, *A new efficient algorithm for computing Gröbner bases without reduction to zero* (F_5), in *Proceedings of International Symposium on Symbolic and Algebraic Computation, Lille, France* (2002), p. 75
41. V.R. Romanovski, M. Prešern, An approach to solving systems of polynomials via modular arithmetics with applications. J. Comput. Appl. Math. **236**, 196 (2011)
42. L. Schiff, *Quantum Mechanics* (McGraw-Hill, New York, 1968)

43. F. Chapeau-Blondeau, A. Monir, Numerical evaluation of the Lambert W-function and application to generation of generalized Gaussian noise with exponent $1/2$. IEEE Trans. Signal Processing **50**, 2160 (2002)
44. H. Gould, J. Tobochnik, *Statistical and Thermal Physics: With Computer Applications* (Princeton University Press, Princeton, 2010)
45. K. Meintjes, A.P. Morgan, Chemical equilibrium systems as numerical test problems. ACM Trans. Math. Softw. **16**, 143 (1990)
46. O. Reynolds, An experimental investigation of the circumstances which determine whether the motion of water shall be direct or sinuous and of the law of resistance in parallel channels. Proc. R. Soc. Lond. **35**, 84 (1883)
47. H. Faisst, B. Eckhardt, Sensitive dependence on initial conditions in transition to turbulence in pipe flow. J. Fluid Mech. **504**, 343 (2004)
48. B.E. Larock, R.W. Jeppson, G.Z. Watters, *Hydraulics of Pipeline Systems* (CRC Press, Boca Raton, 2002)
49. H. Goldstein, *Classical Mechanics*, 2nd edn. (Addison-Wesley, Reading, 1980)

Chapter 3
Numerical Integration and Differentiation

Abstract This Chapter describes numerical integration for approximate computation of the values of definite integrals, as well as techniques for differentiation of functions in the presence of noise. Integration over finite intervals is introduced via the classic trapezoidal, Simpson's and Romberg's methods, followed by a presentation of Gaussian, Clenshaw–Curtis and Gauss–Kronrod quadrature, as well as contour integration. The trapezoidal method and Gaussian quadrature are discussed again in the context of integration over semi-infinite and infinite domains, where we also present double-exponential methods based on variable transformation. Computation of Cauchy principal values is discussed, and several methods for the integration of rapidly oscillating functions are illustrated. Techniques of stable numerical differentiation as a notoriously ill-posed problem are described, based on three characteristic approaches from the theory of inverse problems: regularization of finite differences, which yields low-noise Lanczos and related families of noise-robust differentiators; differentiation by using smoothing kernels; and variational regularization of the derivative, which can be translated into the problem of solving an integral equation for the derivative function.

Numerical integration or *quadrature* is any method for approximate computation of the values of definite integrals. The goal is to attain the desired accuracy with as few evaluations of the integrand as possible. The key factors one needs to consider are dimensionality (1, 2, 3 etc.), the integration domain (finite vs. (semi-)infinite intervals, surface areas, volumes) and the nature of the integrand (smooth, singular, oscillatory etc.). Numerical differentiation is the opposite operation and the most important question is how to do it in a stable manner in the presence of noise.

3.1 Integration Over Finite Intervals

Any quadrature method calls for the evaluation of the integrand at a finite set (the "grid" or "mesh") of points known as *abscissas* or *nodes*, then taking their weighted average and possibly exploiting further procedures—e.g. adaptively refining the grid or extrapolating the intermediate results—to yield the approximate value of the

© The Author(s), under exclusive license to Springer Nature Switzerland AG 2025
S. Širca and M. Horvat, *Computational Methods in Physics*, Graduate Texts in Physics,
https://doi.org/10.1007/978-3-031-68566-8_3

integral. The task at hand is to find the optimal quadrature nodes and weights that maximize the accuracy. Many quadrature rules have been devised over the centuries [1]; we shall discuss only the formulas relevant to the topics of this book.

3.1.1   Trapezoidal Rule

To calculate the numerical approximation of the integral

$$I[f] = \int_a^b f(x)\,dx \tag{3.1}$$

we partition the interval $[a, b]$ into subintervals $[x_j, x_{j+1}]$ defined by the $N+1$ evenly spaced nodes (including the endpoints)

$$x_j = a + jh\,, \quad h = (b-a)/N\,, \quad j = 0, 1, \ldots, N\,. \tag{3.2}$$

If on each subinterval the integrand is approximated by a linear function, we obtain the trapezoidal rule

$$I[f] \approx T_N[f] = h\left(\frac{f(a)}{2} + \sum_{j=1}^{N-1} f(x_j) + \frac{f(b)}{2}\right). \tag{3.3}$$

The partitioning of the interval $[a, b]$ need not be equidistant; if the nodes are chosen in a non-uniform manner, as in

$$\pi_N : a = x_0 < x_1 < \cdots < x_N = b\,, \tag{3.4}$$

the trapezoidal rule can be rewritten as

$$T[f; \pi_N] = \sum_{j=1}^{N} \frac{x_j - x_{j-1}}{2}\left(f(x_{j-1}) + f(x_j)\right).$$

In the case of evenly spaced nodes (3.2) the rule $T[f; \pi_N]$ reduces to $T_N[f]$. (Note that non-uniformity here is a different matter than in quadrature based on orthogonal polynomials where the nodes are non-equidistant by design.) Assuming that the integrand has a continuous second derivative, the error estimate (or the remainder, denoted by R_N) is given by

$$R_N[f] = I[f] - T_N[f] = -\frac{b-a}{12}h^2 f''(\xi)\,, \quad a < \xi < b\,, \tag{3.5}$$

i.e. the rule converges at least as fast as $h^2 \propto 1/N^2$. The functional form of the remainder (3.5) can be used to calculate the improved estimates for the value of the integral and the error. The integral is first computed by using N and subsequently $2N$ subintervals, so that we can write $I = T_N + R_N = T_{2N} + R_{2N}$, where

$$|R_N| = \frac{b-a}{12} h^2 |f''(\xi_1)| , \qquad |R_{2N}| = \frac{b-a}{12} \left(\frac{h}{2}\right)^2 |f''(\xi_2)| .$$

If we assume $|f''(\xi_1)| \approx |f''(\xi_2)|$, the ratio of the remainders is $|R_{2N}/R_N| \approx 1/4$, and it follows that

$$|T_{2N} - T_N| = |R_{2N} - R_N| \approx \frac{3}{4} |R_N| . \tag{3.6}$$

The sum T_{2N} is therefore an improved numerical estimate for the integral, while a naive (and conservative) estimate for the error is

$$\frac{4}{3} |T_{2N} - T_N| . \tag{3.7}$$

In tools and libraries providing numerical integration capabilities such interval subdivisions are performed repeatedly until the estimated error drops below the predefined tolerance. Error estimation is the critical component of any adaptive quadrature routine, and many schemes are available: see [2] and Chap. 7 of [3] for an extensive review.

The convergence rate of the trapezoidal rule can be increased if in addition to the values some other information about the integrand is accessible. For instance, if the first derivatives at the endpoints are known, the basic rule (3.3) can be promoted to

$$\widetilde{T}_N[f] = T_N[f] + \frac{h^2}{12}\left(f'(a) - f'(b)\right) , \tag{3.8}$$

resulting in the error estimate

$$\widetilde{R}_N = I[f] - \widetilde{T}_N[f] = \frac{b-a}{720} h^4 \max_{a \le x \le b} |f^{(4)}(\xi)| , \quad a < \xi < b ,$$

which drops off as $h^4 \propto 1/N^4$ [1]. In principle there is no need to stop at second order in h as in (3.8), as the infinite sequence of correction terms to $T_N[f]$ has been known since Euler and Maclaurin:

$$\widetilde{T}_N[f] = T_N[f] + \frac{h^2}{12}\left(f^{(1)}(a) - f^{(1)}(b)\right) - \frac{h^4}{720}\left(f^{(3)}(a) - f^{(3)}(b)\right)$$
$$+ \frac{h^6}{30240}\left(f^{(5)}(a) - f^{(5)}(b)\right) - \cdots \tag{3.9}$$

Thus the basic second-order (h^2) accuracy becomes fourth-order if $f'(a) = f'(b)$, sixth-order if $f'''(a) = f'''(b)$, and so on for all odd-order derivatives. For particular types of functions discussed below all corrections terms in the series vanish and the convergence becomes exponential (spectral). The lesson is that the accuracy of the trapezoidal scheme is driven by endpoint details, yet canceling the derivatives is not the only option to attain higher order: a different series of correction terms to $T_N[f]$ exists that, instead of requiring derivatives, manipulates the quadrature weights in the vicinity of the endpoints; see [4] for details.

3.1.2 Exponential Convergence of the Trapezoidal Rule

The first important class of functions for which the trapezoidal rule exhibits exponential convergence is the class of real or complex 2π-periodic functions on the real line. The trapezoidal rule applied to the integral

$$I[f] = \int_0^{2\pi} f(x)\,dx \tag{3.10}$$

reads

$$T_N[f] = \frac{2\pi}{N} \sum_{j=1}^{N} f(x_j), \quad x_j = \frac{2\pi j}{N}. \tag{3.11}$$

(Because f is periodic, $f(0) = f(2\pi)$, no factors $1/2$ are needed since $f(0)/2$ and $f(2\pi)/2$ merge into a single term in the abbreviated sum.) If f is analytic and satisfies $f(x) \le M$ in the half-plane $\mathrm{Im}(x) > -a$ for some $a > 0$, then for any $N \ge 1$ it holds that

$$|T_N[f] - I[f]| \le \frac{2\pi M}{e^{aN} - 1}, \tag{3.12}$$

and the constant 2π is as small as possible [5]. If the period of f is T instead of 2π, this error estimate becomes

$$|T_N[f] - I[f]| \le \frac{TM}{e^{2\pi aN/T} - 1}. \tag{3.13}$$

Similarly, if f is analytic and satisfies $f(x) \le M$ in the strip $-a < \mathrm{Im}(x) < a$ for some $a > 0$, then for any $N \ge 1$ the error estimate has the same form as (3.12), except that 2π in the numerator is replaced by 4π, with the obvious modification for periods $T \ne 2\pi$. See also Chap. 8 of [3].

Error estimates like (3.12) and (3.13) may be extracted from the general expression behind (3.9) tailored to 2π-periodic functions. If f is $(2m + 2)$ times continuously differentiable on $[0, 2\pi]$ for some $m \ge 0$, the error of the trapezoidal rule is

$$T_N[f] - I[f] = \sum_{k=1}^{m} \frac{B_{2k}}{(2k)!} h^{2k} \left(f^{(2k-1)}(2\pi) - f^{(2k-1)}(0) \right)$$

$$+ 2\pi h^{2m+2} \frac{B_{2m+2}}{(2m+2)!} f^{(2m+2)}(\xi) \tag{3.14}$$

for some $x \in [0, 2\pi]$, where B_k are the Bernoulli numbers [6]. This again makes it clear that the more odd-order derivatives at the endpoints one cancels, the higher order of accuracy is attained, as illustrated by the following Examples.

Examples The function $f_1(x) = |\sin(x/2)|$ is continuously differentiable to arbitrary order, yet its first derivative has a jump discontinuity at the endpoints, $f_1'(0) \neq f_1'(2\pi)$; the $k = 1$ term in the sum in (3.14) therefore does not cancel and trapezoidal quadrature will show the standard $h^2 \propto 1/N^2$ convergence rate. The function $f_2(x) = |\sin^3(x/2)|$, on the other hand, has $f_2'(0) = f_2'(2\pi)$, while $f_2'''(0) \neq f_2'''(2\pi)$, therefore the convergence rate improves to $h^4 \propto 1/N^4$. Genuine spectral convergence is attained, for instance, with the function $f_3(x) = 1/(a - \cos x)$, which is infinitely continuously differentiable and has no discontinuities in any of its derivatives, thus the error of the trapezoidal rule will drop off faster than any power of N; an easy calculation [6] shows that it is indeed exponential: it scales as $r^{N+1}/(1 - r^2)/(1 - r^N)$ where $r = a - \sqrt{a^2 - 1}$. For large N we have $1 \pm r^N \approx 1$ and the error is essentially of the order r^N; the smaller r (or the larger a), the faster the convergence. These findings are shown in Fig. 3.1.   ◁

Another case of quadrature problems for which the trapezoidal rule attains exponential convergence is the integration of functions over a circle in the complex plane. Let f be a real or complex function defined on the unit circle $|z| = 1$. We set $z = e^{i\theta}$ and by analogy to (3.10) define

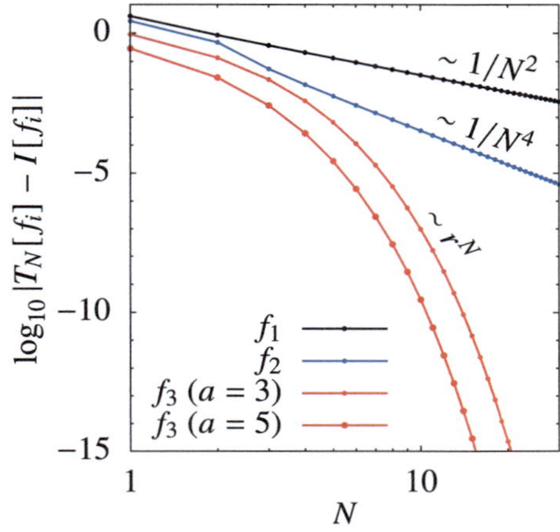

Fig. 3.1 The absolute error of the trapezoidal rule for the integration of the infinitely differentiable 2π-periodic functions $f_1(x) = |\sin(x/2)|$, $f_2(x) = |\sin^3(x/2)|$ and $f_3(x) = 1/(a - \cos x)$ for $a = 3$ and $a = 5$. Only the latter function has all derivatives periodic, which brings about the exponential convergence of the rule

$$I[f] = -\int\limits_{|z|=1} i\, z^{-1} f(z)\, \mathrm{d}z = \int\limits_{0}^{2\pi} f\!\left(e^{i\theta}\right) \mathrm{d}\theta\ .$$

Just as in (3.11) the trapezoidal approximation of this integral is

$$T_N[f] = \frac{2\pi}{N} \sum_{j=1}^{N} f(z_j)\ ,$$

where $\{z_j\}$ are the Nth roots of unity, i.e. $z_j = \exp(2\pi i j/N)$. If f is analytic and satisfies $|f(z)| \le M$ in the annulus $r^{-1} < |z| < r$ for some $r > 1$, then for any $N \ge 1$ the error of the trapezoidal rule is given by

$$|T_N[f] - I[f]| \le \frac{4\pi M}{r^N - 1}\ .$$

Several related variants of error estimates for integration over circles in the complex plane are discussed in [5].

3.1.3 Simpson's Rule

If a piecewise quadratic instead of linear function is used to approximate the integrand and equidistant nodes (3.2) are employed, we obtain the Simpson's rule

$$S_N[f] = \frac{h}{3}\left(f(x_0) + 4 \sum_{j=1}^{N/2} f(x_{2j-1}) + 2 \sum_{j=1}^{N/2-1} f(x_{2j}) + f(x_N) \right).$$

Note that N must be even. The leading term of the remainder is

$$R_N = -\frac{b-a}{180} h^4 f^{(4)}(\xi)\ , \qquad a < \xi < b\ , \tag{3.15}$$

and just as we have experienced with the trapezoidal rule in Eqs. (3.6) and (3.7) the functional form of this remainder can be used to obtain an improved estimate for the value of the integral and the error. We calculate the integrals on two grids with $(N+1)$ and $(2N+1)$ points and equate $I = S_N + R_N = S_{2N} + R_{2N}$, where

$$|R_N| = \frac{b-a}{180} h^4 \left| f^{(4)}(\xi_1) \right|\ , \qquad |R_{2N}| = \frac{b-a}{180} \left(\frac{h}{2}\right)^4 \left| f^{(4)}(\xi_2) \right|\ .$$

If we assume $|f^{(4)}(\xi_1)| \approx |f^{(4)}(\xi_2)|$, we obtain $|R_{2N}/R_N| \approx 1/16$, whence $|S_{2N} - S_N| = |R_{2N} - R_N| \approx \frac{15}{16}|R_N|$. The sum S_{2N} is therefore an improved numerical estimate for the integral, while a conservative estimate for the error is

$$\frac{16}{15}\,|S_{2N} - S_N|\;.$$

Given the default $h^4 \propto 1/N^4$ convergence rate of the Simpson's formula, its popularity is not surprising, and quips like "95 % of the work in numerical analysis boils down to applications of Simpson's rule and linear interpolation" (quoted on p. 57 of [1]) are easily understandable. Considering the exponential convergence of the trapezoidal rule for several classes of functions most relevant to a physicist, however, it is not always clear that the extra effort is "worth it".

Unevenly spaced abscissas If the values of the integrand are provided at unevenly spaced points (3.4), a case frequently encountered in practice, one can still devise a Simpson-like quadrature formula based on overlapping quadratic functions conjoined with a smoothing feature. Let

$$p(x_{j-1}, x_j, x_{j+1}) = a_j x^2 + b_j x + c$$

be a quadratic function that interpolates f at three neighboring points x_{j-1}, x_j, x_{j+1} $(j = 1, 2, \ldots, N - 1)$. Then for $j = 1, 2, \ldots, N - 2$ we may use the approximation

$$\int_{x_j}^{x_{j+1}} f(x)\,\mathrm{d}x = \frac{1}{2} \int_{x_j}^{x_{j+1}} \left(p(x_{j-1}, x_j, x_{j+1}) + p(x_j, x_{j+1}, x_{j+2}) \right)$$

$$= \frac{a_j + a_{j+1}}{2} \left(\frac{x_{j+1}^3 - x_j^3}{3} \right) + \frac{b_j + b_{j+1}}{2} \left(\frac{x_{j+1}^2 - x_j^2}{2} \right)$$

$$+ \frac{c_j + c_{j+1}}{2} (x_{j+1} - x_j)\,,$$

while the first and the last subinterval are treated separately:

$$\int_{x_0}^{x_1} f(x)\,\mathrm{d}x \approx \int_{x_0}^{x_1} p(x_0, x_1, x_2)\,\mathrm{d}x\,,$$

$$\int_{x_{N-1}}^{x_N} f(x)\,\mathrm{d}x \approx \int_{x_{N-1}}^{x_N} p(x_{N-2}, x_{N-1}, x_N)\,\mathrm{d}x\,.$$

An analogous technique using cubic interpolation polynomials (cubic splines) exists and is used both in interpolation of data and quadrature, but pushing the degrees of the interpolation polynomials upwards from three is generally not recommended in cases where the accuracy of the data is significantly poorer than the arithmetic accuracy.

Example Simpson's method requires many function evaluations to attain high precision (see Fig. 3.2 (left)). By recursive halving of the subintervals, which is repeated

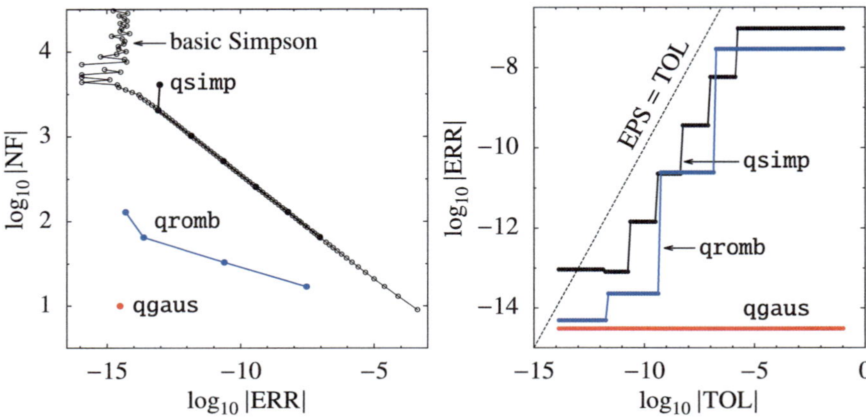

Fig. 3.2 Numerical integration of the function $f(x) = x^2(x^2 - 2)\sin x$ on $[0, \pi/2]$ by using the basic Simpson's method and by using routines qsimp, qromb and qgaus from the NUMERICAL RECIPES library [7]. [LEFT] The number of integrand evaluations, NF, as a function of the relative error ERR of the integral (with respect to the exact value). [RIGHT] The achieved relative error compared to the required tolerance TOL in subsequent halvings of the subinterval. Due to rounding errors, the value of ERR can be even larger than TOL if we require TOL to be very small: note the crossing of the EPS = TOL line at extremely low EPS and TOL when using the qsimp routine. For non-adaptive Gaussian quadrature with *fixed* number of nodes (qgaus) "tolerance" has no meaning

until the relative error drops below a specified value, the error of the final result can be reduced by many orders of magnitude: see Fig. 3.2 (right) corresponding to qsimp routine from [7]. Extrapolation methods (like the Romberg algorithm, see qromb in the Figure and Sect. 3.1.4) are even faster. But quadrature formulas (like the Gauss quadrature presented in Sect. 3.1.5 and denoted by qgaus in the Figure), are just as tempting: they require very few function evaluations and typically achieve spectacular precision for smooth functions.   ◁

3.1.4 Romberg Integration

Romberg's method is a generalization of a low-order (e.g. trapezodial) quadrature formula to a high-order scheme by resorting to extrapolation. Suppose that we wish to compute the value of (3.1) by a quadrature rule Q with node spacing h. The behavior of the remainder $I - Q(h)$ as a function of h is always of the form $I - Q(h) \propto h^p$ with $p > 0$. If the quadrature formula is evaluated with $h/2$ and subsequently with h, a better approximation for I may be obtained by Richardson's formula:

$$I \approx \frac{2^p Q(h/2) - Q(h)}{2^p - 1} = Q(h/2) + \frac{Q(h/2) - Q(h)}{2^p - 1}, \qquad (3.16)$$

see, for instance, Eq. (8.16). We have already made this step by exploiting the h-dependence of the remainders (3.5) or (3.15) to obtain improved estimates of I; with further grid refinements, however, consecutive approximations of the integral value can effectively be extrapolated to the $h = 0$ limit, yielding very precise results very quickly. The sequence of approximations is generated as follows. Let

$$Q_{k0} = \frac{b-a}{2^{k+1}}\left[f(a) + 2\sum_{j=1}^{2^k-1} f\left(j\frac{b-a}{2^k}\right) + f(b)\right], \quad k = 0, 1, \ldots .$$

Romberg's method amounts to calculating the entries of the table

$$
\begin{array}{llll}
Q_{00} & Q_{01} & \cdots & Q_{0,n-1}\ \ Q_{0n} \\
Q_{10} & Q_{11} & \cdots & Q_{1,n-1} \\
\cdots & \cdots & \cdots & \\
Q_{n-2,0} & Q_{n-2,1} & Q_{n-2,2} \\
Q_{n-1,0} & Q_{n-1,1} \\
Q_{n0}
\end{array}
$$

in which any element Q_{kl} is a particular approximation of I. The quantities Q_{k0} contained in the first column are the quadrature sums corresponding to the trapezoidal formula with diminishing h; once these quantities are calculated, the elements of subsequent columns are obtained by using the recurrence

$$Q_{k,l+1} = \frac{2^{2l+2}Q_{k+1,l} - Q_{kl}}{2^{2l+2} - 1}, \quad k = 0, 1, \ldots, n-l-1 . \tag{3.17}$$

The errors corresponding to the approximate values Q_{k0} from the first column are given by

$$I - Q_{k0} = -\frac{b-a}{12}h^2 f''(\xi), \quad h = \frac{b-a}{2^k}, \quad k = 0, 1, \ldots, n ,$$

as shown in Eq. (3.5), and by one application of Richardson's formula one obtains a better approximation to I,

$$Q_{k1} = \frac{4Q_{k+1,0} - Q_{k0}}{3}, \quad k = 0, 1, \ldots, n-1 .$$

Proceeding one step, we find that the quantities T_{k1} are nothing but Simpson's quadrature sums for ever smaller h,

$$I - Q_{k1} = -\frac{b-a}{180}h^4 f^{(4)}(\xi), \quad h = \frac{b-a}{2^{k+1}}, \quad k = 0, 1, \ldots, n-1 ,$$

to which (3.16) can be applied again, etc.; ultimately Q_{0n} provides the most precise estimate of I. The excellent performance of Romberg's integration is demonstrated in Fig. 3.2. For smooth functions on finite integration intervals it typically achieves machine precision with only a handful of iterations of (3.17) and only $\approx 10\text{–}100$ evaluations of f.

3.1.5 Gaussian Quadrature

Gaussian quadrature is a method for numerical estimation of definite integrals which also relies on the system of abscissas (nodes) and quadrature weights, but the idea is to pick *optimal* abscissas at which the function is to be evaluated. To this end we exploit orthogonal polynomials with the property

$$\int_a^b p_k(x) p_l(x) w(x)\, \mathrm{d}x = \delta_{k,l}\,,$$

and it turns out that the optimal abscissas $x_0, x_1, \ldots, x_N$ of an $(N+1)$-point Gaussian quadrature formula are precisely the roots of the corresponding orthogonal polynomial of degree $(N+1)$ on the same interval and involving the same weight function w, as shown e.g. in Sect. 2.7 of [1]. Gaussian quadrature formulas have the form

$$\int_a^b f(x) w(x)\, \mathrm{d}x = \sum_{j=0}^{N} w_j f(x_j) + R_N[f]\,, \tag{3.18}$$

which is exact (the remainder R_N becomes zero) when f is a polynomial of degree $2N + 1$ or less. If $p_k(x) = a_k x^k + \cdots$ is an orthogonal polynomial with the leading coefficient $a_k > 0$, the weights attached to the nodes x_j are given by

$$w_j = -\frac{a_{N+2}}{a_{N+1}} \frac{1}{p_{N+2}(x_j) p'_{N+1}(x_j)}\,,$$

but recurrence relations can be employed to convert this general expression to other equivalent formulas or to ensure that the computation of the weights is stable for large N. Nowadays finding the roots of orthogonal polynomials is far less daunting than it used to be, as they are accessible in most libraries and software packages. In MATHEMATICA, for instance, the N roots of the Legendre polynomial $P_N(x)$ can be found to arbitrary precision by using

```
R[k_, n_]  := Root[LegendreP[k, #] &, n];
x[j_]  := N[R[NN, j]];
```

where $j = 1, 2, \ldots, N$ and NN stands for N. In adaptive integration routines in which the basic Gaussian quadrature formula is applied on ever smaller subintervals, N usually does not exceed ≈ 10 and tabulated values can be used [9]; in spectral methods N is typically a few times 10; numerical solution of singular integral equations, for example, may call for hundreds of thousands of nodes [10]. In contrast to the classical methods with a numerical cost of $\mathcal{O}(N^2)$ this can now be accomplished by recently developed $\mathcal{O}(N)$ algorithms [11, 12] which allow for a fast and stable computation of the Legendre–Gauss nodes in fractions of a second even if N is on the order of millions.

In the following we describe Gaussian quadrature on the finite interval $[a, b] = [-1, 1]$ based on Legendre polynomials (weight function $w(x) = 1$) and Chebyshev polynomials ($w(x) = (1 - x^2)^{-1/2}$); we refer heavily to these two types of quadrature in Sect. 5.3 on transformations with orthogonal polynomials and in Chapter 12 on spectral methods for ODE and PDE. Note that both types of quadrature are also applicable to arbitrary $[a, b]$ intervals with finite a and b: one only needs to transform the independent variable to the $[-1, 1]$ domain and rescale the weights:

$$\int_a^b f(x)w(x)\,\mathrm{d}x = \frac{b-a}{2} \sum_{j=0}^{N} w_j f\left(\frac{b-a}{2} x_j + \frac{a+b}{2}\right) + R_N[f]. \tag{3.19}$$

A further distinction in nomenclature is needed depending on whether quadrature nodes are completely specified by zeros of the appropriate orthogonal polynomial or some of them are predefined, i.e. when the endpoints $x = -1$ and $x = 1$ are included explicitly. In the former ("standard") case mentioned in the introduction of this Subsection the quadrature is named Legendre–Gauss or Chebyshev–Gauss; in the latter case we label it Legendre–Gauss–Radau or Legendre–Gauss–Lobatto, respectively, and analogously for the Chebyshev variant. (See Table 3.1.) Gaussian quadrature on (semi-)infinite domains is discussed in Sect. 3.2.3.

3.1.5.1 Legendre–Gauss (LG)

For Legendre–Gauss quadrature the nodes x_j ($j = 0, 1, \ldots, N$) are the roots of the Legendre polynomial P_{N+1}, which all lie in the interior of $[-1, 1]$. The corresponding weights are

$$w_j = \frac{2}{(1 - x_j^2)[P'_{N+1}(x_j)]^2} = \frac{2(1 - x_j^2)}{(N + 2)^2 P_{N+2}^2(x_j)}.$$

With these nodes and weights the quadrature formula is exact for polynomials of degree at most $2N + 1$, while the remainder is

$$R_N[f] = \frac{2^{2N+3}((N + 1)!)^4}{(2N + 3)\big((2N + 2)!\big)^3} f^{(2N+2)}(\xi), \quad -1 < \xi < 1. \tag{3.20}$$

Table 3.1 Nodes $\{x_j\}$ and weights $\{w_j\}$ for Gaussian quadrature (3.18) with Legendre polynomials P_k ($x \in [-1, 1]$, weight function $w(x) = 1$), Chebyshev polynomials T_k ($x \in [-1, 1]$, $w(x) = (1 - x^2)^{-1/2}$), Laguerre polynomials L_k ($x \in [0, \infty]$, $w(x) = e^{-x}$) and Hermite polynomials H_k ($x \in [-\infty, \infty]$, $w(x) = e^{-x^2}$). The notation $\mathcal{R}(\cdot)$ means "roots of $(\cdot)$". The Radau and Lobatto versions listed here include the left or both endpoints of the domain, respectively, with the obvious modifications if the endpoint $+1$ is to be included instead of -1 in the Legendre and Chebyshev case. Two alternative ways to find *all* $N + 1$ Laguerre–Gauss–Radau nodes are to calculate either $\mathcal{R}(L_{N+1} - L_N)$ or $\mathcal{R}(x L'_{N+1})$. Unless explicitly specified, the span of j in any given expression is $j = 0, 1, \ldots, N$

Polynomial	Gauss		Gauss–Radau		Gauss–Lobatto	
	x_j	w_j	x_j	w_j	x_j	w_j
Legendre $P_k[-1, 1]$	$\mathcal{R}(P_{N+1})$	$\dfrac{2(1 - x_j^2)}{(N + 2)^2 P_{N+2}^2(x_j)}$	$\mathcal{R}(P_N + P_{N+1})$	$\dfrac{1 - x_j}{(N + 1)^2 P_N^2(x_j)}$	$x_0 = -1, x_N = 1$ and $\mathcal{R}(P'_N)$; $1 \leq j \leq N-1$	$\dfrac{2}{N(N + 1) P_N^2(x_j)}$
Chebyshev $T_k[-1, 1]$	$\cos \dfrac{(2j + 1)\pi}{2N + 2}$	$\dfrac{\pi}{N + 1}$	$\cos \dfrac{2j\pi}{2N + 1}$	$\dfrac{\pi}{2N + 1}; j = 0$ $\dfrac{2\pi}{2N + 1}; 1 \leq j \leq N$	$\cos \dfrac{j\pi}{N}$	$\dfrac{\pi}{2N}; j = 0, N$ $\dfrac{\pi}{N}; 1 \leq j \leq N-1$
Laguerre $L_k[0, \infty]$	$\mathcal{R}(L_{N+1})$	$\dfrac{x_j}{(N + 2)^2 L_{N+2}^2(x_j)}$	$x_0 = 0$ and $\mathcal{R}(L'_{N+1})$; $1 \leq j \leq N$	$\dfrac{1}{(N + 1) L_N^2(x_j)}$	/	/
Hermite $H_k[-\infty, \infty]$	$\mathcal{R}(H_{N+1})$	$\dfrac{2^{N+2}(N + 1)!\sqrt{\pi}}{H_{N+2}^2(x_j)}$	/	/	/	/

3.1.5.2 Legendre–Gauss–Radau (LGR)

The nodes for Legendre–Gauss–Radau quadrature with the left endpoint included are the roots of $P_N(x) + P_{N+1}$, that is, $x_0 = -1$, $x_1, x_2, \ldots, x_N \neq 1$. The weights are

$$w_j = \frac{1 - x_j}{(N + 1)^2 P_N^2(x_j)} \,, \qquad j = 0, 1, \ldots, N \,.$$

(If the $x = 1$ endpoint is to be included instead of $x = -1$, substitute $x \to -x$ and replace x_j by $-x_j$ in the numerator.) The quadrature formula with such nodes and weights is exact for polynomials of degree at most $2N$, and the error term is

$$R_N[f] = \frac{2^{2N+1}(N + 1)(N!)^4}{\big((2N + 1)!\big)^3} \, f^{(2N+1)}(\xi) \,, \quad -1 < \xi < 1 \,.$$

3.1.5.3 Legendre–Gauss–Lobatto (LGL)

The most frequently used Legendre–Gauss–Lobatto quadrature explicitly includes both endpoints of the interval, $x_0 = -1$ and $x_N = 1$, while the interior nodes x_j $(j = 1, 2, \ldots, N - 1)$ are the roots of P_N'. The corresponding weights are

$$w_j = \frac{2}{N(N + 1)} \frac{1}{P_N^2(x_j)} \,, \qquad j = 0, 1, \ldots, N \,.$$

The quadrature rule with these nodes and weights is exact for polynomials of degree at most $2N - 1$, while the general form of the remainder is

$$R_N[f] = -\frac{(N + 1)N^3 2^{2N+1}\big((N - 1)!\big)^4}{(2N + 1)\big((2N)!\big)^3} \, f^{(2N)}(\xi) \,, \quad -1 < \xi < 1 \,.$$

3.1.5.4 Chebyshev–Gauss (CG)

For Chebyshev–Gauss quadrature the nodes x_j $(j = 0, 1, \ldots, N)$—all of which lie in the interior of $[-1, 1]$—and the weights are

$$x_j = \cos \frac{(2j + 1)\pi}{2N + 2} \,, \qquad w_j = \frac{\pi}{N + 1} \,, \qquad j = 0, 1, \ldots, N \,.$$

The x_j's are the zeros of the Chebyshev polynomial T_{N+1} and can be written down explicitly because of their property (5.40); note that increasing j implies decreasing abscissas, i.e. $m < n \Leftrightarrow x_m > x_n$, as well as $x_0 = 1$ and $x_N = -1$. With these nodes and weights the quadrature formula is exact for polynomials of degree at most $2N + 1$, while the remainder is

$$R_N[f] = \frac{\pi}{2^{2N+1}(2N+2)!}\, f^{(2N+2)}(\xi)\,, \quad -1 < \xi < 1\,.$$

3.1.5.5   Chebyshev–Gauss–Radau (CGR)

The nodes for Chebyshev–Gauss–Radau quadrature are the roots of $T_N(a)T_{N+1}(x) - T_{N+1}(a)T_N(x)$, where $a = \pm 1$, depending on which of the two endpoints we wish to include. If the left endpoint, $x_N = -1$, is to be included, the nodes and the weights are

$$x_j = \cos\frac{(2j+1)\pi}{2N+1}\,, \quad w_j = \begin{cases} \dfrac{2\pi}{2N+1} & ;\ j = 0, 1, \ldots, N-1\,, \\[2ex] \dfrac{\pi}{2N+1} & ;\ j = N\,. \end{cases}$$

(For the opposite case with $x_0 = +1$ included see Table 3.1. The quadrature formula with such nodes and weights is exact for polynomials of degree at most $2N$.

3.1.5.6   Chebyshev–Gauss–Lobatto (CGL)

The most frequently used Chebyshev–Gauss–Lobatto quadrature explicitly includes both endpoints of the interval. The nodes are the roots of T_N' (i.e. extrema of T_N), augmented by the endpoints $x_0 = 1$ and $x_N = -1$, which again can be computed explicitly and, together with the weights, are given by

$$x_j = \cos\frac{j\pi}{N}\,, \quad w_j = \begin{cases} \dfrac{\pi}{2N} & ;\ j = 0, N\,, \\[2ex] \dfrac{\pi}{N} & ;\ j = 1, 2, \ldots, N-1\,. \end{cases} \tag{3.21}$$

The quadrature rule with these nodes and weights is exact for polynomials of degree at most $2N - 1$, while the general form of the remainder is

$$R_N[f] = -\frac{\pi}{2^{2N-1}(2N)!}\, f^{(2N)}(\xi)\,, \quad -1 < \xi < 1\,.$$

Example Let us demonstrate the power of Gaussian quadrature for the integral

$$\int_{-1}^{1} f(x)w(x)\,\mathrm{d}x\,, \quad f(x) = 1 - \mathrm{e}^{-x}\sin(3x^2)\,, \tag{3.22}$$

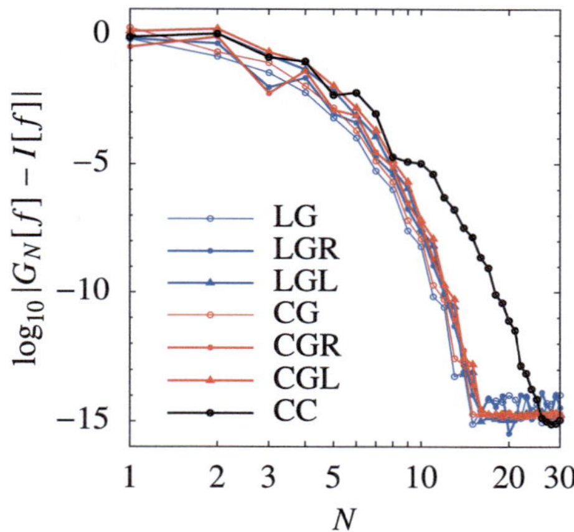

Fig. 3.3 Errors of Legendre–Gauss (LG, LGR, LGL) and Chebyshev–Gauss (CG, CGR, CGL) quadrature formulas for the integral (3.22) as a function of N. Note the exponential (spectral) convergence with the terminal plateau at the level of machine precision. Also shown is the error of Clenshaw–Curtis quadrature (CC)

with LG, LGR and LGL rules, where $w(x) = 1$. The blue curves in Fig. 3.3 show the absolute errors of the quadrature formulas (let us denote them by $G_N[f]$) with respect to the exact value as a function of N in log–log scale: note that the convergence is spectral (i.e. exponential), as can be inferred from the expressions for the remainder R_N in all three cases. The red curves show the errors of in Chebyshev quadrature with the same f as above, but with $w(x) = (1 - x^2)^{-1/2}$. The black curve shows the error of Clenshaw–Curtis quadrature (CC) discussed below.   ◁

A note on optimality   One often encounters the statement that Legendre–Gauss quadrature is *optimal*, but this is precisely true only in the context of polynomial integrands. Specifically, it can be shown to be optimal (by certain measures) for integrating functions that are analytic within the Bernstein ellipse $\mathcal{E}_\rho$, defined as the open set in the complex plane bounded by the image of the circle of radius ρ centered at the origin under the map $z = (u + u^{-1})/2$,

$$\mathcal{E}_\rho = \left\{ z \in \mathbb{C} : z = \tfrac{1}{2}(u + u^{-1}),\ u = \rho\, e^{i\phi},\ \rho \geq 1,\ 0 \leq \phi < 2\pi \right\} .$$

If f is analytic in $\mathcal{E}_\rho$, with $f(z) \leq M$ for some $\rho > 1$, then the Gaussian quadrature converges geometrically with the bound [1]

$$\left| G_N[f] - I[f] \right| \leq \frac{64M}{15(1 - \rho^{-2})\rho^{2N}} , \tag{3.23}$$

i.e. it exhibits geometric convergence with the rate $\mathcal{O}(\rho^{-2N})$ as $N \to \infty$. But, as pointed out in Sect. 4 of [13], "analyticity in an ellipse [...] is a skewed form of smoothness, requiring a function's Taylor coefficients for expansions in the middle of the interval to grow more slowly than those for expansions near the endpoints. Polynomials can resolve much faster wiggles near endpoints than in the interior, and Gauss quadrature [...] inherits this property"—as indeed can be inferred from the strong clustering of nodes in the vicinity of the endpoints. Of course this clustering has a purpose: it is there to mitigate the harmful Runge's oscillations (see Fig. 5.13). In practice, however, the integrands may be much more uniformly smooth over $[-1, 1]$, and the Bernstein ellipse, pertinent to the construction of quadrature weights from polynomial interpolation, can be abandoned in favor of some ε-neighborhood of this interval. This can be realized by transforming the quadrature formula by means of a conformal map which stretches the ellipse $\mathcal{E}_\rho$ along the real axis to form a more elongated analyticity region Ω_ρ, as shown in Fig. 3.4 [14].

The mapping function acting such that $g(\mathcal{E}_\rho) \subseteq \Omega_\rho$ must have the property $g(\pm 1) = \pm 1$; then $g([-1, 1])$ is an analytic curve in Ω_ρ, parameterized by $s \in [-1, 1]$, that connects the point -1 to the point $+1$. By Cauchy's theorem for analytic functions the integral of f over this curve is equal to the integral of f over $[-1, 1]$, thus

$$I[f] = \int_{-1}^{1} f(x)\, \mathrm{d}x = \int_{-1}^{1} f\big(g(s)\big) g'(s)\, \mathrm{d}s \, ,$$

and the corresponding transformed quadrature formula becomes

$$\widetilde{G}_N[f] = \sum_{j=0}^{N} \widetilde{w}_j f\big(\widetilde{x}_j\big) \, , \quad \widetilde{w}_j = w_j g'(x_j) \, , \quad \widetilde{x}_j = g(x_j) \, . \tag{3.24}$$

The net result for the error estimate is that instead of (3.23) we are left with

$$\big|\widetilde{G}_N[f] - I[f]\big| \leq \frac{64 M \gamma}{15(1 - \rho^{-2})\rho^{2N}} \, , \quad \gamma = \sup_{s \in \mathcal{E}_\rho} |g'(s)| \leq \infty \, ,$$

i.e. the transformed quadrature still converges geometrically. (If $\gamma = \infty$ in the above estimate, one can shrink ρ to make it finite.) For details on how the mapping functions g may be constructed see [14].

Figure 3.5 shows the effect of applying the transformation technique with the same mapping function as in Fig. 3.4 to the integrand $f(x) = \exp(-1/x^2)$. Shown is the absolute error as a function of N for standard LGL quadrature and for the transformed

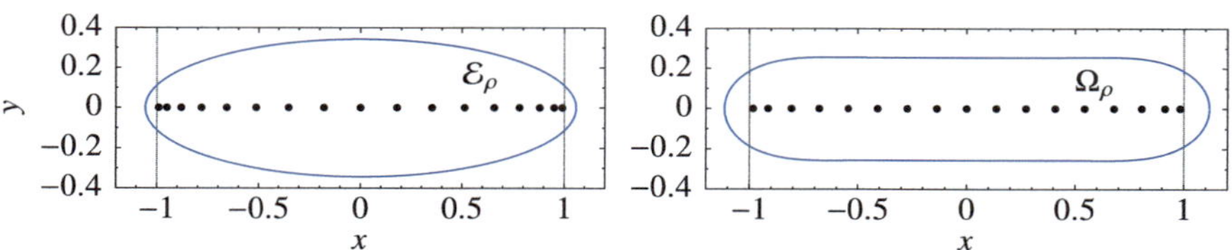

Fig. 3.4 Mapping the quadrature nodes ($N = 16$) and the surrounding Bernstein ellipse $\mathcal{E}_\rho$ with $\rho = 1.4$ (left) to an open set Ω_ρ (right), by using the function $g(s) = (40320s + 6720s^3 + 3024s^5 + 1800s^7 + 1225s^9)/53089$. Note the reduced density of the nodes near the endpoints in Ω_ρ

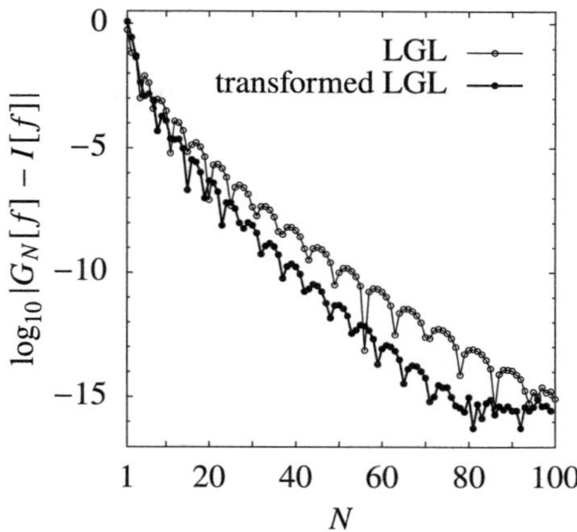

Fig. 3.5 Error of the standard LGL quadrature compared to the error of the transformed rule (3.24) for integrating the function $f(x) = \exp(-1/x^2)$. The comparison shows that the former is *not* optimal for this (non-polynomial) integrand

quadrature rule (3.24). Clearly the standard LGL rule is *not* optimal—which is not to say that it should be discarded. On the contrary, at least in one dimension and for smooth integrands it may safely remain the primary choice, except in the case of $(b - a)$-periodic functions which, as we have seen in Sect. 3.1.2, can be integrated very effectively by the trapezoidal rule. Transformed quadrature, on the other hand, becomes useful in multi-dimensional problems where a more uniform distribution of the nodes allows one to work with coarser grids and still maintain the same precision. For a more elaborate discussion see Sect. 6.9 ("Why Gaussian quadrature?") of [3].

3.1.6 Clenshaw–Curtis Quadrature

In Clenshaw–Curtis quadrature we rewrite the integral in (3.18) over $x \in [a, b] = [-1, 1]$—with $w(x) = 1$!—as an integral over $t \in [0, \pi]$,

$$\int_{-1}^{1} f(x)\, dx = \int_{0}^{\pi} f(\cos t)\, d(\cos t)\,,$$

and use the Chebyshev points

$$x_j = \cos t_j\,, \quad t_j = \frac{j\pi}{N}\,, \quad j = 0, 1, \ldots, N\,, \tag{3.25}$$

as in (3.21) instead of the nominally optimal Legendre–Gauss–Lobatto nodes. The weights are given by [15]

$$w_j = \frac{c_j}{N} \left(1 - \sum_{k=1}^{[N/2]} \frac{b_k}{4k^2 - 1} \cos(2k t_j) \right)\,, \tag{3.26}$$

where

$$c_j = \begin{cases} 1 \; ; \; j = 0, N \,, \\ 2 \; ; \; \text{otherwise} \,, \end{cases} \qquad b_k = \begin{cases} 1 \; ; \; k = N/2 \,, \\ 2 \; ; \; k < N/2 \,. \end{cases}$$

An example of use of this quadrature formula is given by the CC curve in Fig. 3.3.

The weights (3.26) can be computed by the (fast) discrete cosine transform at the computational cost of $\mathcal{O}(N \log N)$, as described in [16]. This, however, is not the main benefit of Clenshaw–Curtis over the standard Gauss quadrature in which, as we have seen, the state-of-the art computation of weights only takes $\mathcal{O}(N)$ as opposed to the classical $\mathcal{O}(N^2)$ algorithm [17]. The main advantage is rather its embedding (nesting) capability, so that higher-order approximations are able to re-use function evaluations at previous nodes, a feature that Gaussian formulas lack and that becomes important in adaptive quadrature, even more so in higher dimensions.

The error of the Clenshaw–Curtis quadrature formula with the nodes (3.25) and weights (3.26) can be estimated as

$$\frac{C_m}{m!} \max_{-1 \le x \le 1} \left| f^{(m)}(x) \right|, \quad C_m \sim \mathcal{O}(2^{-m} m^{-3}) \text{ as } m \to \infty \,,$$

where m is the smallest even integer $\ge N$ and we have assumed that $f \in C^m[-1, 1]$; see Sect. 2.5 of [1]. In practice Clenshaw–Curtis quadrature is known to attain accuracy and convergence rates comparable, and in some cases even better, to Gaussian quadrature: see [18] and Sect. 6.9 of [3] for a discussion. When Clenshaw–Curtis quadrature is applied to certain analytic function the convergence rate may exhibit a peculiar "kink phenomenon": for details see [19].

3.1.7 Gauss–Kronrod Quadrature

Let us revisit[1] the classical N-point Legendre–Gauss quadrature on $[-1, 1]$,

$$\int_{-1}^{1} f(x)\,\mathrm{d}x = \sum_{j=1}^{N} w_j f(x_j) + R_N = G_N + R_N \,. \tag{3.27}$$

The corresponding nodes and weights for $N = 7$ are listed in the upper part of Table 3.2. Turning to the question of the quadrature error, the estimate (3.20) for the remainder R_N is not very useful. It requires us to compute a high-order derivative of f and tends to grossly over-estimate the error. The error can be reduced in a controlled manner such that the quadrature weights are adaptively condensed or rearranged: new nodes are inserted between the nodes of the previous division, and the integration rules are used on these sub-grids. The integration precision is increased by covering

---

[1] In this Subsection we start the sums at $j = 1$ to remain in tune with the standard literature.

Table 3.2 Nodes and weights for Gauss–Kronrod quadrature (G_7, K_{15}) on $[-1, 1]$

Nodes for G_7	Weights for G_7
±0.949107912342759	0.129484966168870
±0.741531185599394	0.279705391489277
±0.405845151377397	0.381830050505119
0.000000000000000	0.417959183673469
Nodes for K_{15}	**Weights for K_{15}**
±0.991455371120813	0.022935322010529
±0.949107912342759	0.063092092629979
±0.864864423359769	0.104790010322250
±0.741531185599394	0.140653259715525
±0.586087235467691	0.169004726639267
±0.405845151377397	0.190350578064785
±0.207784955007898	0.204432940075298
0.000000000000000	0.209482141084728

the problematic regions of the integrand by ever denser quadrature formulas of lower orders. But such "adaptivity" greatly diminishes the symmetric beauty of Gauss formulas.

For a minimum number of evaluations of f we may consider the "optimal case": we make the existing grid denser by embedding one grid into another. Such nesting is realized in *Gauss–Kronrod quadrature*. First, we evaluate the Gauss quadrature G_N of order N by using (3.27). Second, $N + 1$ different nodes are added such that the resulting quadrature formula attains higher order. In addition to G_N in (3.27) we compute

$$K_{2N+1} = \sum_{j=1}^{N} a_j f(x_j) + \sum_{k=1}^{N+1} \alpha_k f(\xi_k) \,. \tag{3.28}$$

The quadratures G_N and K_{2N+1} share the N nodes $x_1, x_2, \ldots, x_N$, but differ in the weights: in the first sum of (3.28) we have a_j instead of w_j from (3.27). The second sum in (3.28) requires $N + 1$ additional nodes ξ_k and weights α_k. The nodes $\xi_1, \xi_2, \ldots, \xi_{N+1}$ are the zeros of the Stieltjes polynomial E_{N+1}, which satisfies

$$\int_{-1}^{1} P_N(x) E_{N+1}(x) x^k \, dx = 0 \,, \quad k = 0, 1, \ldots, N \,.$$

The method to compute the nodes and the weights is described in [21, 22]. In total, then, we need $2N + 1$ evaluations of the function f, but the resulting quadrature formula K_{2N+1} has order $3N + 1$ if N is even and $3N + 2$ if N is odd. The value

$$K_{2N+1}$$

may therefore be used as the—spectacularly!—improved estimate of the integral, while for the error estimate Ref. [20] recommends the formula

$$(200 \, |K_{2N+1} - G_N|)^{3/2} \, .$$

The nodes and weights for $2N + 1 = 15$ are listed in the lower part of Table 3.2; this is the standard Gauss–Kronrod pair (G_7, K_{15}) implemented in the QUADPACK library [23–25] and inherited by many other numerical packages; see also [26, 27].

3.1.8 Contour Integrals

Contour integrals are integrals of the form

$$\int_C f(x, y)\, dx \, , \qquad \int_C f(x, y)\, dy \, , \qquad \int_C f(x, y)\, ds \, , \tag{3.29}$$

where C is a contour in the xy plane. If C can be parameterized by $x = x(t), y = y(t)$, $t_0 \leq t \leq t_1$, the integrals (3.29) can be transformed to ordinary integrals of a single variable,

$$\int_{t_0}^{t_1} f\big(x(t), y(t)\big)\frac{dx}{dt}\, dt \, , \quad \int_{t_0}^{t_1} f\big(x(t), y(t)\big)\frac{dy}{dt}\, dt \, , \quad \int_{t_0}^{t_1} f\big(x(t), y(t)\big)\frac{ds}{dt}\, dt \, ,$$

and calculated by the methods described above.

If C is a *closed* contour and its parametric form can be expressed in terms of a central angle ϕ, then the integrand is automatically a 2π-periodic function of ϕ and can be very accurately integrated by the trapezoidal formula. For example, if the contour is an ellipse with $x = a \cos \phi$, $y = b \sin \phi$, $0 \leq \phi \leq 2\pi$, with $ds = ((dx)^2 + (dy)^2)^{1/2}$, we may set $\phi_j = 2\pi j / N$ and take

$$\int_C f(x, y)\, ds = \int_0^{2\pi} f(a \cos \phi, b \sin \phi)\big(a^2 \sin^2 \phi + b^2 \cos^2 \phi\big)^{1/2} d\phi$$

$$\approx \frac{2\pi}{N} \sum_{j=0}^{N-1} f\big(a \cos \phi_j, b \sin \phi_j\big)\big(a^2 \sin^2 \phi_j + b^2 \cos^2 \phi_j\big)^{1/2} \, .$$

If the integral of a function $f(z)$ with argument $z = x + iy$ needs to be evaluated over a contour in the complex plane, it can be split into its real and imaginary components, which are then handled separately: writing $f(z) = R(x, y) + iI(x, y)$ we have

$$\int_C f(z)\,\mathrm{d}z = \int_C (R + \mathrm{i}I)(\mathrm{d}x + \mathrm{i}\,\mathrm{d}y) = \int_C (R\,\mathrm{d}x - I\,\mathrm{d}y) + \mathrm{i}\int_C (I\,\mathrm{d}x + R\,\mathrm{d}y)\,.$$

When C is a circle, $|z| = \rho$, such integrals can also be very efficiently integrated by the trapezoidal rule. We write $z = \rho e^{\mathrm{i}\phi}$ and obtain

$$\int_C f(z)\,\mathrm{d}z = \mathrm{i}\rho \int_0^{2\pi} f(\rho e^{\mathrm{i}\phi})e^{\mathrm{i}\phi}\,\mathrm{d}\phi = \mathrm{i}2\pi\rho \int_0^1 f(\rho e^{\mathrm{i}2\pi t})e^{\mathrm{i}2\pi t}\,\mathrm{d}t$$

$$\approx \frac{\mathrm{i}2\pi\rho}{N} \sum_{j=0}^{N-1} f(\rho e^{\mathrm{i}2\pi j/N})e^{\mathrm{i}2\pi j/N}\,.$$

This simple rule is exact if f is a polynomial of degree at most $N - 1$.

3.2 Integrals Over (Semi-)infinite Intervals

In the following we describe quadrature methods for integrands defined on the intervals $[0, \infty]$ and $[-\infty, \infty]$. The computation of Cauchy principal values and integration of rapidly oscillating functions are discussed in Sects. 3.3 and 3.4.

3.2.1 Trapezoidal Rule

We have seen that the trapezoidal rule is an excellent choice for the integration of periodic integrands: see Eqs. (3.10) and (3.11). It turns out that it is extremely efficient also for real or complex functions on the real line. We wish to calculate the approximate value of the integral

$$I[f] = \int_{-\infty}^{\infty} f(x)\,\mathrm{d}x$$

by using the trapezoidal formula

$$I_h[f] = h \sum_{j=-\infty}^{\infty} f(x_j)\,, \quad x_j = jh\,, \tag{3.30}$$

where we assume that f is smooth and decays sufficiently fast at $\pm\infty$ so that $I[f]$ and $I_h[f]$ exist. We have the following theorems; see Sect. 5 of [5]. Suppose that f is analytic in the strip $-a_- < \mathrm{Im}(x) < a_+$ for some $a_- > 0$ and $a_+ > 0$, that $f(x) \to 0$ uniformly as $|x| \to \infty$ in the strip, and that for some M_- and M_+ it satisfies

$$\int_{-\infty}^{\infty} |f(x+ib_-)|\, dx \le M_- , \qquad \int_{-\infty}^{\infty} |f(x+ib_+)|\, dx \le M_+ ,$$

for all $b_- \in (-a_-, 0)$ and $b_+ \in (0, a_+)$. Then, for any $h > 0$, I_h defined by (3.30) exists and satisfies

$$|I_h[f] - I| \le \frac{M_-}{e^{2\pi a_-/h} - 1} + \frac{M_+}{e^{2\pi a_+/h} - 1} . \tag{3.31}$$

If the strip of analyticity is symmetric about the real axis, simply set $a_- = a_+ = a$ and $M_- = M_+ = M$ in the above formulas. A very similar theorem can be stated if f is analytic in the half-plane $\mathrm{Im}(x) > -a$ for some $a > 0$. On the other hand, if f is real on the real axis, the following holds true:

$$I_h[f] - I = -2\,\mathrm{Re}\left(\int_{-\infty+ia'}^{+\infty+ia'} \frac{f(x)}{1 - e^{-2\pi ix/h}}\, dx \right) .$$

In all cases, then, the trapezoidal formula with an infinite span in j will exhibit an exponential convergence rate. The question remains how to truncate the infinite series (3.30) to

$$I_{h,N}[f] = h \sum_{j=-N}^{N} f(x_j) , \quad x_j = jh , \tag{3.32}$$

provided that f decays rapidly enough as $x \to \pm\infty$, what is the optimal step size h, and what convergence rates one can expect. To determine the optimal N at which to truncate the sum, we write $|I_{h,N}[f] - I[f]| \le |I_h[f] - I[f]| + |I_h[f] - I_{h,N}[f]|$, i.e. we split the error into its discretization and truncation parts, and try to determine the optimal truncation condition by requiring the two contributions to be approximately equal. In Sect. 6 of [5] this exercise is performed for three functions:

$$
\begin{array}{l|cc}
f(x) & \text{optimal } h & \text{optimal rate} \\
\hline
f_1(x) = \dfrac{e^{-x\tanh x}}{1+x^2} & (2\pi/N)^{1/2} & \mathcal{O}\!\left(e^{-(2\pi N)^{1/2}}\right) \\[2ex]
f_2(x) = \dfrac{e^{-x^2}}{\sqrt{1+x^2}} & (2\pi/N^2)^{1/3} & \mathcal{O}\!\left(e^{-(2\pi N)^{2/3}}\right) \\[2ex]
f_3(x) = e^{-x^2} & (\pi/N)^{1/2} & \mathcal{O}\!\left(e^{-\pi N}\right)
\end{array}
\tag{3.33}
$$

These functions are typical representatives of the sort of integrands that we are most likely to encounter: f_1 is analytic in a strip with exponential decay, f_2 is analytic in a strip with Gaussian decay, so that in both cases the estimate (3.31) applies, and f_3 is an entire function with Gaussian decay. Figure 3.6 shows the error of the

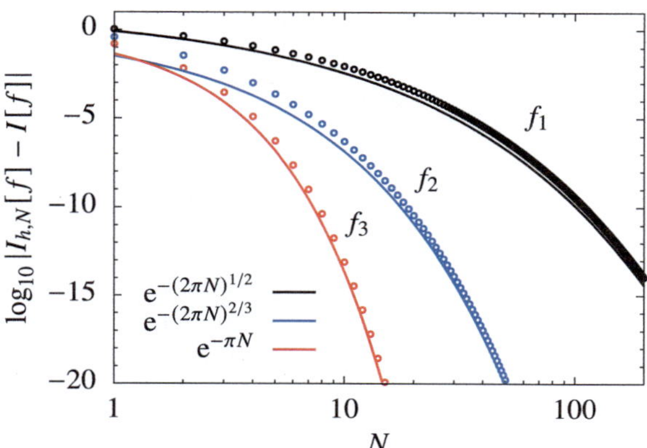

Fig. 3.6 Error of the trapezoidal formula (3.32) for the functions listed in (3.33), evaluated with the optimal step size h in each case, as a function of N (open circles). The curves indicate the expected asymptotic convergence rates

truncated trapezoidal rule for these three functions, each evaluated with its corresponding optimal h listed above, along with the anticipated asymptotic convergence rate.

For integrands that decay slower than exponentially as $x \to \pm\infty$ the trapezoidal rule is less efficient; the best approach to improve the accuracy and the convergence rate is to transform the slowly decaying integrand into a quickly decaying one by a change of variable: this is discussed in the next Subsection.

3.2.2 Variable Transformation: Double-Exponential Methods

Very efficient methods for numerical quadrature can be devised if we transform the integration variable such that the integral at hand in translated to an integral over the whole real axis, and use the trapezoidal rule to evaluate it. Let the given integral be

$$I[f] = \int_a^b f(x)\,\mathrm{d}x \ . \tag{3.34}$$

Here one or both limits may be infinite, but initially let us consider finite a and b. We wish to use a variable transformation $x = \phi(t)$ that will transform the interval $[a, b]$ to $[-\infty, \infty]$ such that $\phi(-\infty) = a$ and $\phi(\infty) = b$. The original integral then becomes

$$I[f] = \int_{-\infty}^{\infty} g(t)\,\mathrm{d}t \ , \quad g(t) = f\big(\phi(t)\big)\phi'(t) \ ,$$

and the trapezoidal formula with step size h applied to it takes the form

$$I_h[f] = h \sum_{j=-\infty}^{\infty} w_j f(x_j)\,, \quad x_j = \phi(jh)\,, \quad w_j = \phi'(jh)\,. \tag{3.35}$$

The sum has to be truncated at $j = \pm N$ large enough that $|w_j f(x_j)| < \varepsilon$ for $|j| > N$; one usually sets $\varepsilon = 10^{-p}$, where p is the desired numeric precision level in digits. If the transformation function ϕ is chosen such that ϕ' and higher derivatives tend to zero very rapidly for large t, the transformed integrand $f(\phi(t))\phi'(t)$ typically is a well-behaved function, even if f is oscillatory near or singular at one or both endpoints, and we may expect the truncated trapezoidal estimate of the integral to be very accurate. (The singularity must be integrable, of course.)

Many such transformation functions have been invented and analyzed, yet the one that seems to enjoy by far the greatest popularity for integrals of the form (3.34) is the "tanh–sinh" transformation [28, 29]:

$$\phi(t) = \tfrac{1}{2}(b + a) + \tfrac{1}{2}(b - a)\tanh(C \sinh t)\,. \tag{3.36}$$

The parameter C is usually taken to be 1 or (much more often) $\pi/2$, and we set $a = -1$, $b = 1$ without loss of generality to study integrals over a finite domain. The key property of (3.36) is that its derivative has a *double-exponential* decay,

$$\phi'(t) = \frac{b - a}{2} \frac{C \cosh t}{\cosh^2(C \sinh t)} \sim \mathcal{O}\left(e^{-C \exp(|t|)}\right) \quad \text{as } |t| \to \infty\,, \tag{3.37}$$

a landmark feature of the technique whereby we achieve the desired rapid decay of the transformed integrand at infinity. The error of the truncated sum

$$I[f] \approx I_{h,N}[f] = h \sum_{j=-N}^{N} \phi'(jh) f\big(\phi(jh)\big) \tag{3.38}$$

can be estimated as [7]

$$\big|I_{h,N}[f] - I[f]\big| \sim \mathcal{O}\left(e^{-kN/\log N}\right) \quad \text{as } N \to \infty\,, \tag{3.39}$$

where k is a positive constant. In practice tanh–sinh quadrature very often achieves the so-called "quadratic convergence" because reducing h by half and doubling N approximately doubles the number of correct digits in the result [30]. The optimal h is

$$h \approx \frac{\log(2\pi N d / C)}{N}\,,$$

where d is the distance from the real axis to the nearest singularity of the integrand. The function ϕ' in (3.37) has a pole when $\cosh^2 t = 0$, i.e. when $t = \pm i\,\pi/2$. If there are no poles in f closer to the real axis, then $d = \pi/2$.

The trapezoidal rule incorporating a double-exponential ϕ' is optimal in the sense that there is no quadrature formula obtained by variable transformation whose error decays faster than $\mathcal{O}(e^{-kN/\log N})$; see [31] for details. However, while (3.39) applies asymptotically, a much more useful and highly accurate estimate of the error in tanh–sinh quadrature is [32]

$$\left| I_{h,N}[f] - I[f] \right| \approx \frac{h^3}{4\pi^2} \sum_{j=-N}^{N} \left[f(\phi(t))\phi'(t) \right]''(jh) \ .$$

The idea of variable transformation can also be applied to integrals over the semi-infinite and infinite domains, where in each case the mapping function ϕ is chosen such that $\phi'(t)$ attains double-exponential asymptotics: see Table 3.3. Integrals of the form

$$\int_0^\infty f(x)\,dx \ , \tag{3.40}$$

for instance, can be evaluated by the above technique by using

$$\phi(t) = \exp\left(C \sinh t\right) \ . \tag{3.41}$$

(Integrals over $[0, \infty]$ can also be treated by the tanh–sinh quadrature employing (3.36), although it is primarily suitable for finite intervals: a substitution $s = 2\alpha/(x + 1) - \alpha$ yields an integral over $[-1, 1]$.) Often the integrand already contains an explicit exponential decay factor, so that $f(x) = f_1(x)e^{-x}$ in Eq. (3.40). In such cases only a single-exponential transform is needed to achieve double-exponential asymptotics of ϕ', and a standard choice for the mapping function is

$$\phi(t) = \exp\left(t - e^{-t}\right) \ .$$

Table 3.3 Mapping functions $\phi(t)$ for the truncated trapezoidal approximation (3.38) that result in double-exponential asymptotics of $\phi'(t)$ and ensure an exponentially decaying error (3.39), for various types of integrals; most often one takes $C = \pi/2$

Integral	$\phi(t)$	$\phi'(t)$
$\int_{-1}^{1} f(x)\,dx$	$\tanh(C \sinh t)$	$C \operatorname{sech}^2(C \sinh t)\cosh t$
$\int_0^\infty f(x)\,dx$, general $f(x)$	$\exp(C \sinh t)$	$C \exp(C \sinh t)\cosh t$
$\int_0^\infty f(x)\,dx$, with $f(x) = f_1(x)e^{-x}$	$\exp(t - e^{-t})$	$(1 + e^{-t})\exp(t - e^{-t})$
$\int_{-\infty}^\infty f(x)\,dx$	$\sinh(C \sinh t)$	$C \cosh(C \sinh t)\cosh t$

(This just adds one more exponential decay as $t \to \infty$, together with the double-exponential decay as $t \to -\infty$.) For integrals of the form

$$\int_{-\infty}^{\infty} f(x)\, \mathrm{d}x \tag{3.42}$$

the recommended transformation to use is

$$\phi(t) = \sinh(C \sinh t) \, . \tag{3.43}$$

Even if f in (3.42) decays slowly, the truncated trapezoidal rule (3.38) employing the mapping (3.43) will turn a poorly converging integral into a rapidly converging estimate. If the integrand already has the double-exponential nature (3.37) as in, say,

$$\int_{-\infty}^{\infty} \mathrm{e}^{-\cosh x + \mathrm{i}\,ax}\, \mathrm{d}x \, ,$$

no transformation is needed, and a direct application of the trapezoidal rule gives the best result. If, on the other hand, we would choose a simpler transformation $\phi(t) = \sinh t$ that yields only single-exponential asymptotics of ϕ' instead of (3.43) that provides double-exponential behavior, the trapezoidal formula would still perform better than the variable-mapped method. This has been proven for a class of entire functions of rapid decrease, i.e. functions f which satisfy

$$|f(x)| \le M \exp\!\left(-\alpha|x|^{\beta}\right), \quad x \in \mathbb{R},$$

and whose singularities are sufficiently removed from the real line. In [33] they established that the total errors incurred in approximating the integral by the plain trapezoidal rule and by the sinh-transformed formula, respectively, satisfy the inequalities

$$\left| \int_{-\infty}^{\infty} f(x)\, \mathrm{d}x - h \sum_{j=-N}^{N} f(jh) \right| \le C(N)\mathrm{e}^{-\gamma_N} \, ,$$

$$\left| \int_{-\infty}^{\infty} f(x)\, \mathrm{d}x - h \sum_{j=-N}^{N} f\!\left(\sinh(jh)\right)\cosh(jh) \right| \le C(N^{(\mathrm{s})})\mathrm{e}^{-\gamma_N^{(\mathrm{s})}} \, ,$$

where $C(N)$ and $C(N^{(\mathrm{s})})$ grow at most algebraically with N, and showed that for $\alpha > 0$, $\beta \ge 2$ the limit $\lim_{N \to \infty} \gamma_N / \gamma_N^{(\mathrm{s})}$ is *infinite*, i.e. the trapezoidal rule will be superior to the sinh-transformed quadrature. See also Section 14 of [5].

Example Let us illustrate the capabilities of the variable-transformation method on three characteristic integrals from the benchmark set listed in [28] whose exact values are known:

$$\int_{-1}^{1} f_1(x)\,dx = \int_{-1}^{1} \frac{\cos(\pi x)}{\sqrt{1+x^2}}\,dx = -\sqrt{2}\,C(2) \approx -0.690495\,,$$

$$\int_{0}^{\infty} f_2(x)\,dx = \int_{0}^{\infty} \frac{\exp(-1-x)}{1+x}\,dx = E_1(1) \approx 0.219384\,, \tag{3.44}$$

$$\int_{-\infty}^{\infty} f_3(x)\,dx = \int_{0}^{\infty} \frac{1}{1+x^4}\,dx = \pi/\sqrt{2}\,.$$

Here C is the Fresnel's cosine-integral and Ei is the exponential integral function. Figure 3.7 shows the absolute errors for the three quadratures employing the mapping functions ϕ listed in Table 3.3. ◁

Implementation details There are two implementation obstacles that may have caused tanh–sinh quadrature to enjoy less popularity than it deserves, both of which are easy to circumvent. The first one is the overflow of the denominator in (3.37) for even moderate t. With $C = \pi/2$ we find $\cosh^2(C\sinh(4.735)) \approx 3.4 \cdot 10^{38}$ and $\cosh^2(C\sinh(6.8075)) \approx 1.8 \cdot 10^{308}$, which correspond to the largest normal numbers in `float` and `double` precision, respectively; see Table 1.1. This can be avoided by introducing a new variable $q = e^{-2C\sinh t}$ so that

$$\phi'(t) = 2(b-a)\frac{Cq}{(1+q)^2}\cosh t\,.$$

For large t, q just underflows to zero, and so does the fraction $q/(1+q)^2$.

The second danger lurks in the loss of significant digits. If f has a singularity at the endpoint like $(1+x)^{-1+\mu}$ or $(1-x)^{-1+\mu}$, where μ is a small positive constant, this almost certainly implies large errors due to the loss of significant digits when x is very close to -1 or 1. This problem can be solved by computing and storing not the nodes $x_j = \tanh\big(C\sinh(jh)\big)$ but rather

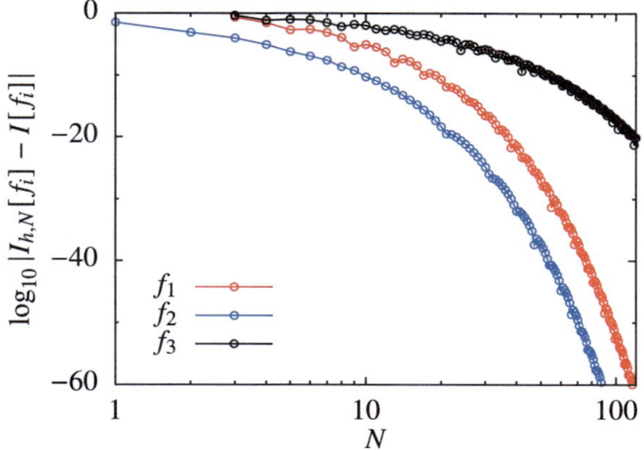

Fig. 3.7 Errors of the finite, semi-infinite and infinite interval quadrature based on double-exponential variable transformation, for the integrands of the set (3.44). Note the log–log scale and the large range on the ordinate axis: see Implementation details below to see how this level of precision can be attained

$$\xi = 1 + x = \exp(C \sinh t)/\cosh(C \sinh t) \,,$$

$$\eta = 1 - x = \exp(-C \sinh t)/\cosh(C \sinh t) \,,$$

were the required subtractions and additions must be performed in adequate floating-point precision. Then, instead of writing a routine for the evaluation of $f(x)$, we write a routine for $g(x, \xi, \eta)$. For example, if we wish to integrate

$$f(x) = \frac{1}{(x-2)(1+x)^{3/4}(1-x)^{1/4}}$$

over $[-1, 1]$, we should provide a routine for

$$g(x, \xi, \eta) = \frac{1}{(x-2)\xi^{3/4}\eta^{1/4}} \,.$$

Regardless of the integration domain, when a mapping function ϕ has been chosen, the abscissas x_j and the weights w_j can be computed once for a given step size h, and then used in a variety of problems. Typically the grid is refined by downsizing h in m steps like $h = 2^{-m}$. High-precision quadrature benchmarks indicate that $m = 12$ is more than sufficient to evaluate most integrals to 1000-digit accuracy.

Oscillatory integrands If the integrand in (3.40) is oscillatory and slowly decaying at infinity, a direct application of the trapezoidal rule employing the variable transformation (3.41) is typically rather ineffective. Yet recently specialized double-exponential methods have also been developed for calculation of such integrals, i.e. of continuous sine and cosine transforms. They are presented in Sect. 3.4.3.

3.2.3 Gaussian Quadrature

Gaussian quadrature on semi-infinite and infinite domains is devised along the same principles as on finite intervals (Sect. 3.1.5). The integral is approximated by a quadrature sum, while a remainder term is provided as an error estimate. On $[0, \infty]$ the most frequently used quadrature is based on Laguerre polynomials (weight function $w(x) = e^{-x}$, see Sect. 5.3.3). For integrands defined on $[-\infty, \infty]$ the most common quadrature is based on Hermite polynomials (weight function $w(x) = e^{-x^2}$, see Sect. 5.3.5).

Laguerre–Gauss In the Laguerre–Gauss case the origin ($x = 0$) is not included, and the nodes x_j are the zeros of the Laguerre polynomial L_{N+1}. The quadrature formula is

$$\int_0^\infty f(x) e^{-x}\, dx = \sum_{j=0}^N w_j f(x_j) + \frac{((N+1)!)^2}{(2N+2)!} f^{(2N+2)}(\xi) \,, \quad 0 < \xi < \infty \,,$$

where the weights are given by

$$w_j = \frac{x_j}{(N+2)^2 L_{N+2}^2(x_j)} \,.$$

Laguerre–Gauss–Radau In this case the node at the origin is explicitly included, and the nodes x_j can be calculated by any of the following three methods: predefined $x_0 = 0$ plus $x_1, x_2, \ldots, x_N$, the zeros of L'_{N+1}; roots of $x L'_{N+1}$; or roots of $L_{N+1} - L_N$. The quadrature formula has the form

$$\int_0^\infty f(x)e^{-x}\,\mathrm{d}x = w_0 f(0) + \sum_{j=1}^{N} w_j f(x_j) + \frac{N!(N+1)!}{(2N+1)!}\, f^{(2N+1)}(\xi)\,, \quad 0 < \xi < \infty\,,$$

and the weights are given by

$$w_j = \frac{1}{(N+1)L_N^2(x_j)} \,.$$

Note that $w_0 = 1/(N+1)$ since $L_N(0) = 1$ for any N.

It often happens that the weighting factor $\exp(-x)$ in the integral is implicit, so instead of $f(x)\exp(-x)$ we wish to approximate the integral with an exponentially decaying function $g(x)$. Formally we can do this by merely relocating the exponential factor, yielding a quadrature formula with scaled weights on its right hand-side:

$$\int_0^\infty g(x)\,\mathrm{d}x \approx \sum_{j=0}^{N} w_j e^{x_j} g(x_j)\,.$$

Note, however, that asymptotically the unscaled weights decay exponentially,

$$w_j \sim \pi \sqrt{\frac{x_j}{N}} e^{-x_j} \quad \text{as } N \to \infty\,, \tag{3.45}$$

and this brings about a piece of advice considered in the Example below. The classical algorithm for the computation of Laguerre–Gauss nodes and weights [17] based on the recurrence relation for the Laguerre polynomials has a numerical cost of $\mathcal{O}(N^2)$. Fast methods with a cost of $\mathcal{O}(N)$ to compute the nodes and (scaled) weights are described in [34, 35]; see also Appendix A of [36].

Example (Adapted from Sect. 6 of [13].) The "Laguerre–G" curve in Fig. 3.8 (left) shows the error of Laguerre–Gauss quadrature rule applied to $f(x) = \cos(x^2)$, i.e. to the integral

$$\int_0^\infty \cos(x^2)e^{-x}\,\mathrm{d}x\,, \tag{3.46}$$

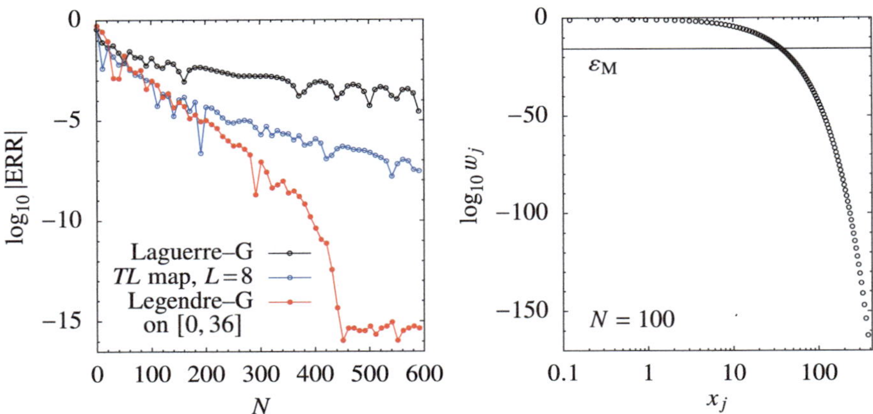

Fig. 3.8 [LEFT] The errors of direct Laguerre–Gauss quadrature on $[0, \infty]$, the quadrature with coordinate mapping that produces the $\{TL_k\}$ basis set, with scaling parameter $L = 8$, as well as Legendre–Gauss quadrature on the finite interval $[0, 36]$ for the integral (3.46), as a function of the quadrature order N. [RIGHT] Laguerre–Gauss weights plotted against their corresponding nodes. (Note the log–log scale and the tremendous range of the values w_j)

as a function of N. The convergence rate is poor, the rule attaining no more than about four-digit accuracy at $N \approx 500$. This kind of inefficiency can be traced back to the asymptotic behavior of the weights (3.45): for $N = 100$, for instance, about two thirds of the weights are smaller than ε_M, and the last weight is $w_{100} \approx 10^{-163}$, as shown in Fig. 3.8 (right); moreover, this fraction will increase with N as $\mathcal{O}(N^{1/2})$. If f is of order unity, the values $w_j f(x_j)$ at these nodes will make a negligible contribution to the sum.

A possible alternative would be to transform the integral by exploiting the same map that produces the basis of rational Chebyshev functions TL (see Sect. 5.3.4). With the map $x(t) = L \cot^2(t/2)$, where L is a scaling parameter, the integral of f becomes

$$I[f] = \int_0^\infty f(x)\,\mathrm{d}x = \int_0^\pi f\big(L \cot^2(t/2)\big) \frac{2L \sin t}{(1 - \cos t)^2}\,\mathrm{d}t \ .$$

The corresponding quadrature formula is

$$I_N[f] = \sum_{j=1}^N w_j f\big(L \cot^2(t_j/2)\big) ,$$

for which the nodes are given by $x_j = L \cot^2(t_j/2)$, with the evenly spaced Chebyshev points $t_j = j\pi/(N + 1)$, and the weights by [37]

$$w_j = \frac{2L \sin t_j}{(1 - \cos t_j)^2} \frac{2}{N+1} \sum_{m=1}^{N} \sin(mt_j) \frac{1 - \cos(m\pi)}{m} .$$

The error of such quadrature applied to the integral (3.46) with $L = 8$ is shown by the "*TL* map" curve in Fig. 3.8 (left). The convergence rate is faster, but still unsatisfactory: the reason is that the transformation $t_j \to x_j$ stretches the nodes farther and farther, which is the opposite of what we need. In such cases we are therefore much better served by simply truncating the interval and resorting to an unweighted formula like Legendre–Gauss. The "Legendre–G" curve in Fig. 3.8 (left) shows the result by using Eq. (3.19) with $f(x) = \cos(x^2)e^{-x}$ and $w(x) = 1$ on the interval $[a, b] = [0, 36]$, indicating that machine precision is reached at $N \approx 450$. Clenshaw–Curtis quadrature—using unmapped Chebyshev nodes on the *finite* interval—produces a very similar result.   ◁

Hermite–Gauss In Hermite–Gauss quadrature the nodes x_j are the zeros of H_{N+1}, and the quadrature rule has the form

$$\int_{-\infty}^{\infty} f(x)e^{-x^2}\, \mathrm{d}x = \sum_{j=0}^{N} w_j f(x_j) + \frac{(N+1)!\sqrt{\pi}}{2^{N+1}(2N+2)!} f^{(2N+2)}(\xi) , \quad -\infty < \xi < \infty .$$

There are several equivalent ways to compute the weights, for instance,

$$w_j = \frac{2^N (N+1)!\sqrt{\pi}}{(N+1)^2 H_N^2(x_j)} = \frac{2^{N+2}(N+1)!\sqrt{\pi}}{H_{N+2}^2(x_j)} .$$

The asymptotic decay of the weights is very fast,

$$w_j \sim \frac{\pi}{\sqrt{2N}} e^{-x_j^2} \quad \text{as } N \to \infty , \tag{3.47}$$

with similar consequences as illustrated in Fig. 3.8; see the following Example. Apart from the standard $\mathcal{O}(N^2)$ methods to compute the Hermite–Gauss nodes and weights for small N, fast methods with a numerical cost of $\mathcal{O}(N)$ exist to tackle large N: consult [35] and references therein.

Example (Adapted from Sect. 5 of [13].) Just as we have witnessed in the Laguerre case, direct application of the Hermite–Gauss quadrature rule on $[-\infty, \infty]$ may lead to major disappointment. The "Hermite–G" curve in Figure 3.9 (left) shows the error of the Hermite–Gauss quadrature rule applied to $f(x) = \cos(x^3)$, i.e. to the integral

$$\int_{-\infty}^{\infty} \cos(x^3)e^{-x^2}\, \mathrm{d}x , \tag{3.48}$$

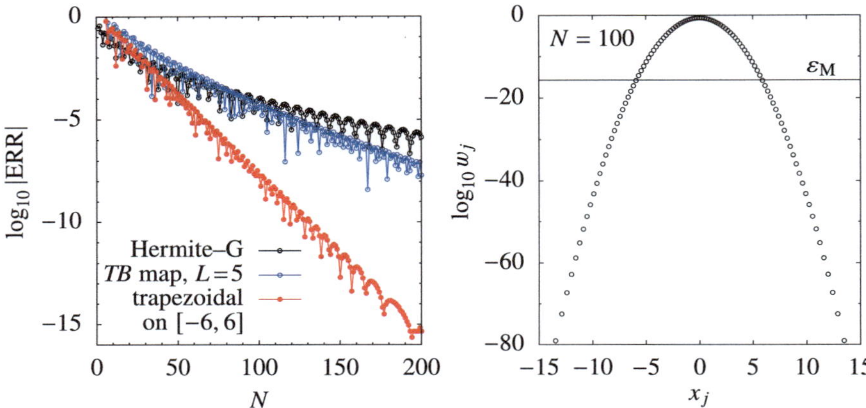

Fig. 3.9 The errors of direct Hermite–Gauss quadrature on $[-\infty, \infty]$, the quadrature with coordinate mapping that produces the $\{TB_k\}$ basis set, with scaling parameter $L = 5$, as well as trapezoidal quadrature on the finite interval $[-6, 6]$ for the integral (3.48), as a function of the quadrature order N. [RIGHT] Hermite–Gauss weights plotted against their corresponding nodes. Compare to Fig. 3.8

as a function of N. The convergence is poor, the rule attaining about six-digit accuracy at $N \approx 200$. Again, the slow convergence rate has its root cause in the asymptotic behavior of the weights (3.47): many of the weights are smaller than ε_M, as shown in Fig. 3.9 (right), and the fraction of such weights, which contribute next to nothing in the quadrature sum, tends to $100\,\%$ as $N \to \infty$.

An alternative approach, by analogy to the Laguerre $\leftrightarrow$ TL case, would be to transform the integral by exploiting the map that produces the basis of rational Chebyshev functions TB (see Sect. 5.3.6). With the map $x(t) = L \cot t$, where L is the scaling parameter, the integral of arbitrary f becomes

$$I[f] = \int_{-\infty}^{\infty} f(x)\,\mathrm{d}x = \int_{0}^{\pi} f\big(L \cot t\big) \frac{2L}{\sin^2 t}\,\mathrm{d}t \, ,$$

see e.g. Sect. 19.8.3 of [37]. The corresponding quadrature formula is

$$I_N[f] = \sum_{j=1}^{N} w_j f(L \cot t_j) \, ,$$

with the nodes $x_j = L \cot t_j$, where $t_j = j\pi/(N + 1)$. The weights are

$$w_j = \frac{L\pi}{(N + 1)\sin^2 t_j} \, .$$

The error of such quadrature applied to (3.48) with $L = 5$ is shown by the "*TB* map" curve in Fig. 3.9 (left) — not much of an improvement. In such cases it is better to use the trapezoidal rule with $f(x) = \cos(x^3)\exp(-x^2)$ and $w(x) = 1$ on the truncated interval, as demonstrated by the fastest decaying curve in Fig. 3.9 (left) obtained by the restriction on $[-6, 6]$. Machine precision is reached at $N \approx 200$. $\triangleleft$

3.3 Computation of Principal Values

In this Section we discuss only a specific topic related to singular integrands, the computation of Cauchy principal values. For a pedagogical treatment of integrable (removable) singularities—in particular how to eliminate them in order to produce a tractable regular integrand—see, for example, Chap. 15 of [38].

3.3.1 Finite Interval

The principal value of an integral over $[a, b]$ with a singularity at $x = \lambda$ is defined as

$$I(\lambda) = P \int_a^b \frac{f(x)}{x - \lambda}\, dx = \lim_{\varepsilon \to 0^+} \left[\int_a^{\lambda-\varepsilon} \frac{f(x)}{x - \lambda}\, dx + \int_{\lambda+\varepsilon}^b \frac{f(x)}{x - \lambda}\, dx \right].$$

(Anticipating the next Subsection, a and b may be infinite, provided that the integrand is regular and decays sufficiently fast.) We split such integrals into three parts,

$$I(\lambda) = \int_a^{\lambda-\Delta} \frac{f(x)}{x - \lambda}\, dx + \underbrace{P \int_{\lambda-\Delta}^{\lambda+\Delta} \frac{f(x)}{x - \lambda}\, dx}_{I_\Delta(\lambda)} + \int_{\lambda+\Delta}^b \frac{f(x)}{x - \lambda}\, dx ,$$

so that in the problematic integrand the singularity is framed *symmetrically* by the interval $[\lambda - \Delta, \lambda + \Delta]$. By substituting $x - \lambda = X$ and $x = X/\Delta$, the integral $I_\Delta(\lambda)$ can be approximated by a quadrature formula on $[-1, 1]$:

$$I_\Delta(\lambda) = P \int_{-\Delta}^{\Delta} \frac{f(X + \lambda)}{X}\, dX = P \int_{-1}^{1} \frac{f(x\Delta + \lambda)}{x}\, dx \approx \sum_{j=0}^{N} \frac{w_j}{x_j} f(x_j \Delta + \lambda) ,$$

where x_j and w_j are the nodes and weights, respectively. A natural quadrature for this domain is Legendre–Gauss; it is recommended to take a quadrature formula with a high order and with an even number of points (odd N) in order to avoid the

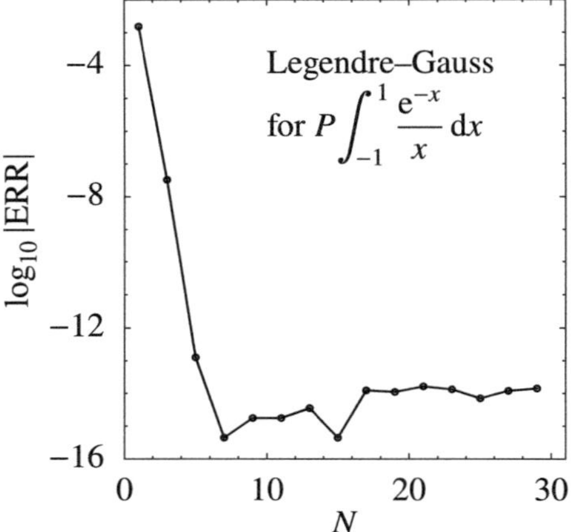

Fig. 3.10 Error of Legendre–Gauss quadrature on $[-1, 1]$ for the principal-value integral $P \int_{-1}^{1} (\exp(-x)/x)\mathrm{d}x \approx 2.114501750751457$, as a function of N. The flattening of the curve at ≈ -14 above $N = 7$ is caused by the slightly inaccurate Legendre–Gauss nodes and weights, calculated to approximately double precision

calculation of the derivative of f [39]. An example is shown in Fig. 3.10. See also [40] and Sect. 5.5. A robust principal-value integrator employing Clenshaw–Curtis quadrature, qawc, is implemented in QUADPACK [23–25], GSL etc.

3.3.2   Semi-infinite and Infinite Domains

The Cauchy principal-value integrals over the semi-infinite and infinite domain,

$$I[f; \lambda] = P \int_{0}^{\infty} \frac{f(x)}{x - \lambda} \mathrm{e}^{-x} \, \mathrm{d}x , \quad 0 < \lambda < \infty ,$$

$$I[f; \mu] = P \int_{-\infty}^{\infty} \frac{f(x)}{x - \mu} \mathrm{e}^{-x^2} \, \mathrm{d}x , \quad -\infty < \mu < \infty ,$$

can be converted to regular integrals by using the respective identities [1]

$$I[f; \lambda] = \int_{0}^{\infty} \frac{f(x) - f(\lambda)}{x - \lambda} \mathrm{e}^{-x} \, \mathrm{d}x + f(\lambda)I[1; \lambda] , \tag{3.49}$$

$$I[f; \mu] = \int_{-\infty}^{\infty} \frac{f(x) - f(\mu)}{x - \mu} \mathrm{e}^{-x^2} \, \mathrm{d}x + f(\mu)I[1; \mu] , \tag{3.50}$$

where

$$I[1; \lambda] = P \int_0^\infty \frac{e^{-x}}{x - \lambda} \, dx = -e^{-\lambda} \left[\gamma + \log \lambda + \sum_{n=1}^\infty \frac{\lambda^n}{n!n} \right] ,$$

$$I[1; \mu] = P \int_{-\infty}^\infty \frac{e^{-x^2}}{x - \mu} \, dx = -e^{-\mu^2} \sum_{n=0}^\infty \frac{(2\mu)^{2n+1} \Gamma(n + 1/2)}{(2n + 1)!}$$

$$= -\sqrt{\pi} \mu \, e^{-\mu^2} \int_{-1}^1 e^{\mu^2 x^2} \, dx ,$$

and γ is the Euler's constant. Since the integrals in (3.49) and (3.50) are already furnished with the appropriate $\exp(-x)$ and $\exp(-x^2)$ weights, they can be evaluated by Laguerre–Gauss and Hermite–Gauss quadrature formulas, respectively, or by the variable-transformation techniques presented in Sect. 3.2.2.

3.4 Integration of Rapidly Oscillating Functions

Introductory textbooks on numerical methods rarely discuss recipes for integration of rapidly oscillating functions although they are widely used in physics, chemistry, and engineering. Even in well-known libraries, such algorithms are scarce. Here, we present some basic approaches. We are mostly interested in the integrals of the form

$$I[f] = \int_a^b f(x) \, e^{i\omega g(x)} \, dx , \qquad -\infty < a < b < \infty , \tag{3.51}$$

where f and g are smooth functions and $|\omega|$ is large. A well-known example is the Fourier integral, in which the oscillation function or the *oscillator* is $g(x) = x$. For such integrals, the quadrature and extrapolation methods of the previous Section are woefully inadequate: when $1/|\omega|$ becomes much smaller than the number of nodes, quadrature formulas fail. To compute the integrals of the form (3.51) we use the asymptotic method, the Filon method, or their numerous improvements and generalizations. See also Sect. 1.3.

3.4.1 Asymptotic Method

Let us first discuss the case where $g'(x) \neq 0$ for $x \in [a, b]$. Then one can demonstrate [41] that asymptotically the integral behaves as $I[f] = \mathcal{O}(1/\omega)$ when $|\omega| \to \infty$, and that the value of the integral can be expanded in powers of $1/\omega$ as

$$I[f] \sim -\sum_{m=0}^{\infty} \frac{1}{(-i\omega)^{m+1}} \left[\frac{e^{i\omega g(x)}}{g'(x)} f_m(x) \right]_a^b , \qquad \omega \to \infty , \tag{3.52}$$

where for the functions f_m we have the recurrence

$$f_0(x) = f(x) ,$$
$$f_{m+1}(x) = \frac{d}{dx}\left(\frac{f_m(x)}{g'(x)} \right) , \qquad m = 0, 1, \ldots .$$

Even though this recurrence is analytically simple, it generates a clumsy sequence of functions which is hard to implement in computer code for non-trivial functions g, unless one exploits symbolic computation. In the special case of the Fourier oscillator, $g(x) = x$, i.e. in computing the integrals of the form

$$\int_a^b f(x) e^{i\omega x} \, dx , \tag{3.53}$$

this recurrence simplifies considerably. We have $g'(x) = 1$ and $g^{(r)}(x) = 0$ for $r > 1$, thus $f_m(x) = f^{(m)}(x)$, and the integral has the asymptotic expansion

$$I[f] \sim -\sum_{m=0}^{\infty} \frac{\left[e^{i\omega x} f^{(m)}(x) \right]_a^b}{(-i\omega)^{m+1}} = \sum_{m=0}^{\infty} \frac{e^{i\omega a} f^{(m)}(a) - e^{i\omega b} f^{(m)}(b)}{(-i\omega)^{m+1}} .$$

We should stress again that this expansion in general does not converge when $m \to \infty$, while it certainly converges when $|\omega| \to \infty$ (see Sect. 1.3.2).

We generate an asymptotic quadrature formula of order s if we keep only s terms in (3.52). For a general function g this means

$$I[f] \approx Q_{\mathrm{A}}^{(s)}[f] = \frac{e^{i\omega g(a)}}{g'(a)} \sum_{m=0}^{s-1} \frac{f_m(a)}{(-i\omega)^{m+1}} - \frac{e^{i\omega g(b)}}{g'(b)} \sum_{m=0}^{s-1} \frac{f_m(b)}{(-i\omega)^{m+1}} . \tag{3.54}$$

For the Fourier oscillator $g(x)=x$ we set $f_m(a) = f^{(m)}(a)$, $f_m(b) = f^{(m)}(b)$ and $g'(a) = g'(b) = 1$. The error of the asymptotic quadrature is

$$E_{\mathrm{A}}^{(s)}[f] = Q_{\mathrm{A}}^{(s)}[f] - I[f] = -\frac{I[f_s]}{(-i\omega)^s} \sim \omega^{-s-1} . \tag{3.55}$$

Figure 3.11 shows the numerical calculation of the integral $\int_0^1 \cos^2 x \, e^{i\omega x} \, dx$ by the asymptotic method (3.54). The left side shows the exact value of the integral as a function of ω, while the right side shows the dependence of the error with the characteristic $\sim \omega^{-s-1}$ trend. The reader might be particularly impressed by two

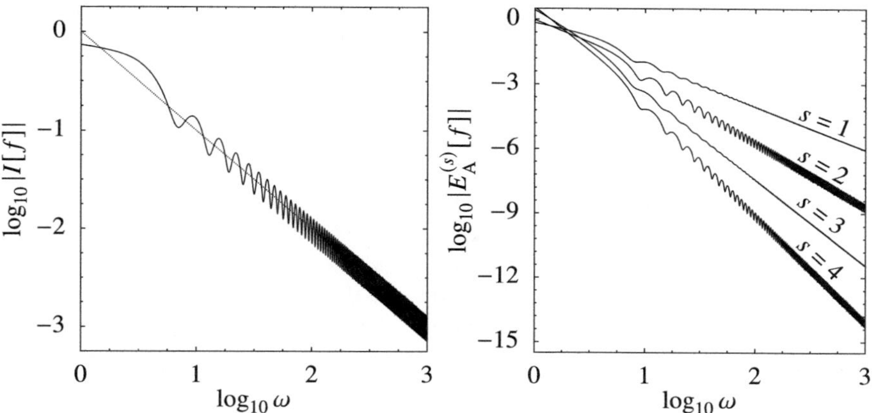

Fig. 3.11 Numerical calculation of the integral $I[f] = \int_0^1 \cos^2 x \, \mathrm{e}^{\mathrm{i}\omega x} \, \mathrm{d}x$ by the asymptotic method. [LEFT] Exact value of $I[f]$ with the asymptotics $|I[f]| \sim \mathcal{O}(1/\omega)$. [RIGHT] Error of the asymptotic method for $s = 1, 2, 3, 4$ with the behavior $|E_{\mathrm{A}}^{(s)}[f]| \sim \mathcal{O}(\omega^{-s-1})$

observations. The precision of the quadrature *improves* when ω in increased, which is counter-intuitive. The second surprise is the decent precision we obtain, for large enough ω, already at first order ($s = 1$): for $\omega \approx 1000$, the trivial formula

$$Q_{\mathrm{A}}^{(1)}[f] = \frac{\mathrm{i}}{\omega} \left[\mathrm{e}^{\mathrm{i}\omega a} f(a) - \mathrm{e}^{\mathrm{i}\omega b} f(b) \right]$$

is precise to about six digits!

The asymptotic method becomes only moderately more complicated when $g'(x) = 0$ on $x \in (a, b)$, or when $g'(a) = 0$ and/or $g'(b) = 0$: the formulas for such cases can be found in [42]. The largest obstacle in these methods is the annoying recursive computation of the functions f_m which is hard to automate for non-trivial g. The approach due to Filon [43] described in the following offers an appealing alternative.

3.4.2 Filon's Method

The basic idea of the Filon's method is to remove the oscillatory part of the integral $\mathrm{e}^{\mathrm{i}\omega g(x)}$ from the integrand in (3.51), and to translate the problem to simpler integrals that play the role of weights in a quadrature formula. The function f is replaced by a polynomial approximation $p(x) = \sum_{k=0}^{n} c_k x^k$ interpolating f at the points $x_0, x_1, \ldots, x_n$. The quadrature formula then becomes

$$I[f] \approx Q_{\mathrm{F}}[f] = I[p] = \int_a^b \left(\sum_{k=0}^{n} c_k x^k \right) \mathrm{e}^{\mathrm{i}\omega g(x)} \, \mathrm{d}x = \sum_{k=0}^{n} c_k \, \mu_k(\omega) \, .$$

To evaluate it, we need the coefficients c_k and $(n + 1)$ values of the integrals of the form

$$\mu_k(\omega) = \int\limits_a^b x^k \, e^{i\omega g(x)} \, dx \, , \tag{3.56}$$

which are called *moments*. We compute the coefficients c_k by solving the corresponding Vandermonde system: see Sect. 4.2.5 and [44]. (In MATHEMATICA this is accomplished by `InterpolatingPolynomial` and `CoefficientList`.)

The error of the Filon's formula $E_F[f]$ can be determined by using the difference function $E(x) = f(x) - p(x)$ that measures the deviation of f from its interpolation polynomial p. We have $E_F[f] = I[f] - Q_F[f] = I[f] - I[p] = I[f - p] = I[E]$. By analogy with (3.52), one immediately sees that

$$E_F[f] = I[E] \sim - \sum_{m=0}^{\infty} \frac{1}{(-i\omega)^{m+1}} \left[\frac{e^{i\omega g(x)}}{g'(x)} \, E(x) \right]_a^b \, ,$$

where we have assumed $g'(x) \neq 0$ for $x \in [a, b]$. Different asymptotic orders s can be achieved by using the Filon quadrature. If we use an interpolation polynomial p for which

$$p^{(m)}(a) = f^{(m)}(a) \, , \qquad p^{(m)}(b) = f^{(m)}(b) \, , \qquad m = 0, 1, \ldots, s - 1 \, , \tag{3.57}$$

i.e. if p interpolates all function's values and derivatives up to order $(s - 1)$ at the boundary points of the interval $[a, b]$, the error is $E_F[f] = \mathcal{O}(\omega^{-s-1})$, just like the error (3.55) of the asymptotic quadrature (see Fig. 3.13 (right)). In this case, Filon's quadrature $Q_F^{(s)}$ of order s is exact for all polynomials of degree $\leq 2s - 1$.

The integrals μ_k contain rapidly oscillating functions and are hard to compute for general g; to compute them, we can use the asymptotic method described earlier. In the following, we again restrict the discussion to the Fourier integrals (3.53) with $g(x) = x$. In this case, the moments (3.56) can be easily computed, and they are related by the recurrence

$$\mu_0 = \frac{e^{i\omega b} - e^{i\omega a}}{i\omega} \, , \qquad \mu_k = \frac{e^{i\omega b} b^k - e^{i\omega a} a^k - k\mu_{k-1}}{i\omega} \, , \qquad k = 1, 2, \ldots, n \, .$$

Figure 3.13 shows the error of the numerical calculation of the integral $\int_0^1 e^x \, e^{i\omega x} \, dx$ by using the Filon's method. Higher quadrature orders can be attained by trying to fulfill the requirements (3.57) at a chosen s as closely as possible. We add further ω-dependent points (Fig. 3.12, cases c-e) in the vicinity of the boundary points. By doing this, not only do we capture the correct function's values $f(a)$ and $f(b)$, but also the derivatives $f'(a)$, $f'(b)$, $f''(a)$, $f''(b)$, and so on.

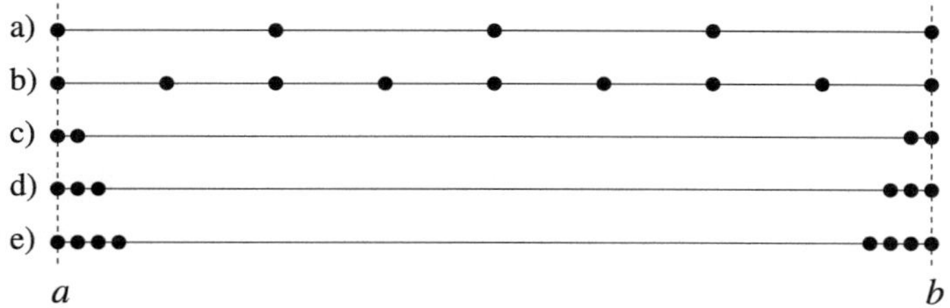

Fig. 3.12 The points used to interpolate f in the Filon's method. a) Equidistant mesh with 5 points. b) Mesh with 9 points. c) Inversely spaced mesh $\{a, a + 1/\omega, b - 1/\omega, b\}$. d) Inverse mesh $\{a, a + 1/\omega, a + 2/\omega, b - 2/\omega, b - 1/\omega, b\}$. e) Inverse mesh $\{a, a + 1/\omega, a + 2/\omega, a + 3/\omega, b - 3/\omega, b - 2/\omega, b - 1/\omega, b\}$. See also Fig. 3.13

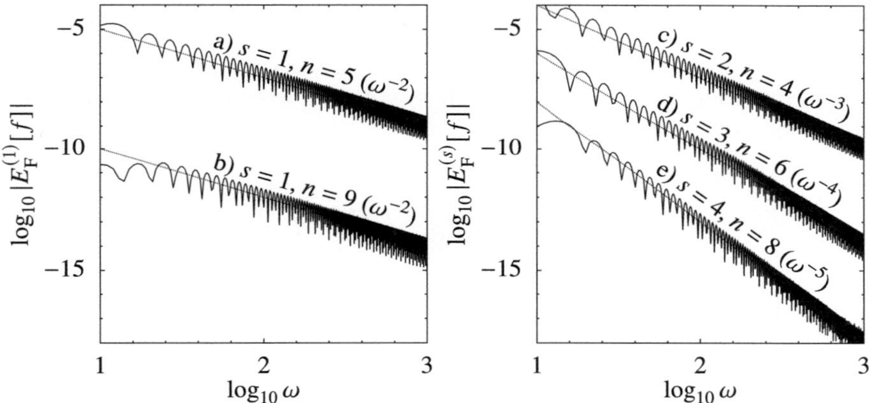

Fig. 3.13 Numerical calculation of the integral $I[f] = \int_0^1 e^x\, e^{i\omega x}\, \mathrm{d}x$ by the Filon's method. The quadrature nodes are defined in Fig. 3.12. [LEFT] Error of the first-order quadrature $|E_F^{(1)}[f]|$ with 5 and 9 equidistant points on $[a, b]$. By adding further interpolation points for f we can reduce the leading constant of the error, but the order of convergence remains the same. [RIGHT] Error of the quadrature $|E_F^{(s)}[f]|$ with three inverse quadrature meshes. By adding nodes in the vicinity of the boundaries ($x = a$ and $x = b$) we approximate higher derivatives of f, thus the order of convergence increases

Typically, we encounter only a limited set of oscillator functions $g(x)$, like x, x^2, $s_0 + s_1 x + s_2 x^2$, x^p, or $\sum_{j=0}^p s_j x^j$. Extensive derivations, a collection of formulas for the moments μ_k, and several numerical tricks to speed up the quadrature with such sets of functions can be found in [42]. A slightly different quadrature method to compute integrals of the form (3.51), utilizing collocation and the usual two-point Legendre–Gauss formula, is presented in [45]. The integrals containing Bessel functions of the type

$$\int\limits_a^b f(x)\, J_m(sx)\, \mathrm{d}x \quad \text{and} \quad \int\limits_a^b f(x)\, \cos(rx)\, J_m(sx)\, \mathrm{d}x$$

are described in [46]. An efficient approach to compute integrals of the form (3.51) with $b = \infty$ is discussed in the papers [47]. Robust subroutines for the numerical integration of oscillating functions on finite or infinite intervals can be found in QUADPACK [23–25] which is also incorporated in the GSL and SciPy libraries.

3.4.3 Double-Exponential Methods for Fourier-Type Integrals

Variable transformations described in Sect. 3.2.2 that resulted in double-exponential quadrature methods can also be applied to oscillating integrands, i.e. Fourier-type (sine and cosine) integrals

$$I^{(\mathrm{s})}[f] = \int\limits_0^\infty f(x)\,\sin(\omega x)\,\mathrm{d}x \quad \text{and} \quad I^{(\mathrm{c})}[f] = \int\limits_0^\infty f(x)\,\sin(\omega x)\,\mathrm{d}x\ ,$$

where f is a slowly (algebraically) decaying function. The variable transformation of choice is [48, 49]

$$x = M\phi(t)/\omega\ , \quad \phi(t) = \frac{t}{1 - \exp\!\big(-2t - \alpha(1 - \mathrm{e}^{-t}) - \beta(\mathrm{e}^t - 1)\big)}\ , \tag{3.58}$$

where M is a large positive constant, while the parameters α and β are taken to be

$$\beta = \frac{1}{4}\ , \quad \alpha = \beta/\sqrt{1 + M\log(1 + M)/(4\pi)}\ .$$

By imposing the relation $M = h\pi$ between the step size h and scaling parameter M, the mapping of x to t pertinent to the sine integral is indeed $x = M\phi(t)/\omega$, which leads to the trapezoidal rule

$$I^{(\mathrm{s})}_{h,N}[f] = \frac{\pi}{\omega} \sum_{j=-N}^N f\!\left(\frac{A_j}{\omega}\right) B_j\ , \quad A_j = M\phi(jh)\ , \quad B_j = \phi'(jh)\sin A_j\ .$$

For the cosine integral, a better mapping is $x = M\phi\big(t - \pi/(2M)\big)/\omega$, yielding

$$I^{(\mathrm{c})}_{h,N}[f] = \frac{\pi}{\omega} \sum_{j=-N}^N f\!\left(\frac{A_j}{\omega}\right) B_j\ , \quad A_j = M\phi\!\left(jh - \frac{\pi}{2M}\right)\ , \quad B_j = \phi'\!\left(jh - \frac{\pi}{2M}\right)\cos A_j\ .$$

Evaluation of sines and cosines of large arguments may lead to substantial errors: this can be avoided by calculating the quadrature weights for large positive j in the following manner [49]:

$$B_j = \begin{cases} (-1)^j \phi'(jh) \sin(d_j A_j) & ; \quad \text{in } I_{h,N}^{(s)}[f], \\ (-1)^j \phi'\big(jh - \pi/(2M)\big) \sin(d_j A_j) ; & \text{in } I_{h,N}^{(c)}[f], \end{cases}$$

where

$$d_j = \exp\big(-2t - \alpha(1 - e^{-t}) - \beta(e^t - 1)\big),$$

to be evaluated with $t = jh$ for the sine integral and with $t = jh - \pi/(2M)$ for the cosine integral.

Example The left panel of Fig. 3.14 shows the error of the variable-transformed trapezoidal quadrature for a sine-type integral with $f(x) = 1/x$,

$$I[f] = \int_0^\infty \frac{\sin x}{x}\, \mathrm{d}x = \frac{\pi}{2}, \tag{3.59}$$

while the right panel shows the error for a cosine-type integral with $f(x) = 1/(1 + x^2)$,

$$I[f] = \int_0^\infty \frac{\cos x}{1 + x^2}\, \mathrm{d}x = \frac{\pi}{2e}. \tag{3.60}$$

The errors are shown for various M as a function of N. In a practical implementation one may take for granted the phenomenologically established fact that the relative

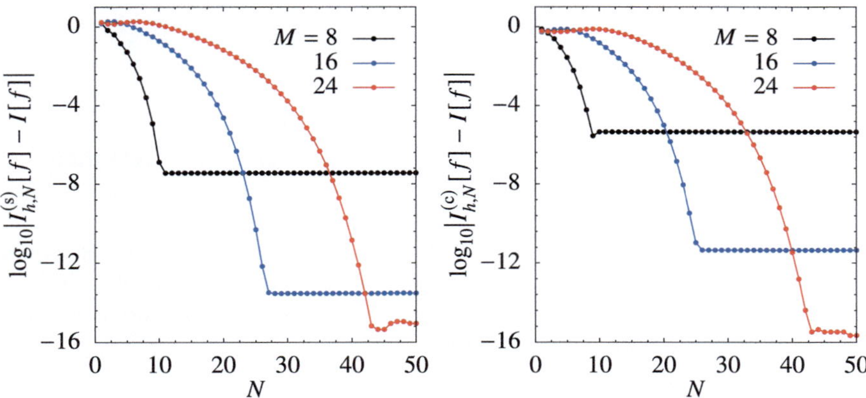

Fig. 3.14 Errors of the trapezoidal rule based on the variable transformation (3.58) for the [LEFT] Fourier-sine integral (3.59) and the [RIGHT] Fourier-cosine integral (3.60) as a function of N, for typical values of the scaling parameter M

error $\varepsilon = |I_{h,N}[f] - I[f]|/|I[f]|$ scales approximately as $\varepsilon \sim \mathcal{O}(\exp(-5M/\pi))$ [49]. Once M has been specified, the summation over j can be truncated at the point where two consecutive function values satisfy

$$\left| f(A_j/\omega) B_j \right| < \omega\varepsilon \left| I_{h,j} \right| .$$

For further details see [50, 51].   ◁

3.4.4 A Short Tale of Long Tail Integration

Taking the title cue from Ref. [52] we describe a simple endpoint correction to Fourier-type integrals that accounts for the domain omitted by the truncation. Suppose that the integral $\int_a^\infty f(x) \sin x \, dx$ has been truncated to an interval $[a, b]$, where $b = N\pi$ and N is some large integer: the truncation point then always coincides with a zero of the sine. Assuming that f is piecewise linear in the domain $[b, \infty]$ it can be shown that the unaccounted for contribution is given by

$$I_{\text{tail}} = \int_{N\pi}^\infty f(x) \sin x \, dx \approx I_{\text{tail}}^{(0)} = (-1)^N f(N\pi) , \tag{3.61}$$

that is, to obtain an estimate of the tail integral we only need to evaluate f at a single point, $x = N\pi$! Further assuming that the second derivative is also piecewise continuous, the next-to-leading-order estimate becomes

$$I_{\text{tail}}^{(2)} = (-1)^N \left(f(N\pi) - f^{(2)}(N\pi) \right) , \tag{3.62}$$

and this can be generalized to

$$I_{\text{tail}} = (-1)^N f(N\pi) + \sum_{i=1}^{k-1} (-1)^{N+i} f^{(2i)}(N\pi) + (-1)^k \int_{N\pi}^\infty f^{(2k)}(x) \sin x \, dx$$

for any $k \geq 1$. (Note that the assumption of piecewise linearity does not require monotonicity, i.e. f can be oscillating as long as its frequency is small compared to the frequency of the principal sine factor.) If the oscillating function is a cosine instead of a sine, Eq. (3.61) is replaced by

$$\int_{(N-1/2)\pi}^\infty f(x) \cos x \, dx \approx (-1)^N f\left((N-1/2)\pi \right) ,$$

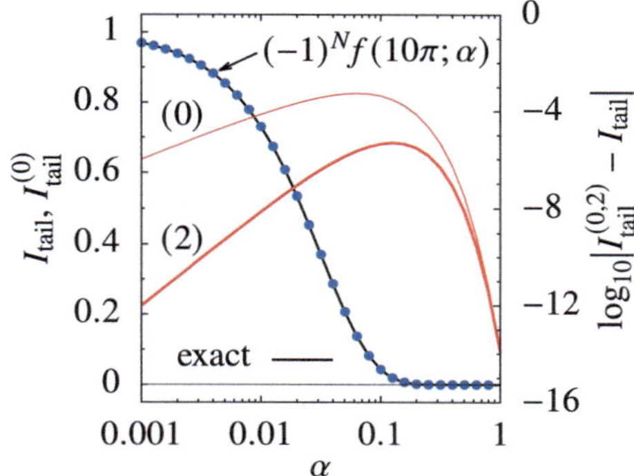

Fig. 3.15 Estimate for the residual tail contribution of the integral I truncated at $b = 10\pi$ (blue circles, left ordinate) and the errors of the leading-order and next-to-leading-order approximations (3.61) and (3.62), indicated by the red curves denoted by (0) and (2), respectively (right ordinate; note the log scale)

with analogous expressions for higher-order terms. Similar tail estimates can be obtained if the truncation point is chosen to be not a distant zero of the sine/cosine factor but one of its maxima: see [52] for details.

Example Figure 3.15 demonstrates the quality of these simple yet effective approximations for $f(x) = \exp(-\alpha x)$, with $\alpha > 0$. In this case the integrals over the whole domain and the truncated domain can be calculated exactly,

$$
I \equiv \int_0^\infty e^{-\alpha x} \sin x \, dx = \frac{1}{1 + \alpha^2} , \quad \widetilde{I}(N\pi) \equiv \int_0^{N\pi} e^{-\alpha x} \sin x \, dx = \frac{1 - e^{-\alpha N\pi}}{1 + \alpha^2} .
$$

The exact value of the tail contribution is $I_{\text{tail}} = I - \widetilde{I}(N\pi)$ and is shown as a function of α by the black curve (left ordinate). This it to be compared to the points suggested by the estimate (3.61), shown for $N = 10$ by the blue circles. The red curves (right ordinate) show the errors of the approximations (3.61) and (3.62), respectively. ◁

3.5 Quadrature in Two or More Dimensions

Quadrature in two or more dimensions is called *cubature*. A generalization of univariate rules in the manner of (3.18) to multivariate integrands is not trivial, and within the scope of this book it would be impossible to deal with all the subtleties of the subject at the same level as we have tried to attain in discussing univariate integrands; this section provides only the bare essentials and a road-map to select literature.

We wish to approximate an integral

$$
I[f] = \int_\Omega f(\boldsymbol{x}) w(\boldsymbol{x}) \, d\boldsymbol{x} ,
$$

where $\Omega \subset \mathbb{R}^n$, $I[1] < \infty$ and $w(\boldsymbol{x}) \geq 0$ for all $\boldsymbol{x} \in \mathbb{R}^n$, $n \geq 2$, by cubature rules of the form

$$Q[f] = \sum_{j=1}^{N} w_j f(\boldsymbol{x}_j) \,,$$

where $w_j \in \mathbb{R}$ and $\boldsymbol{x}_j \in \mathbb{R}^n$. The goal is to find such nodes $\boldsymbol{x}_j$ and weights w_j that cubature is exact — i.e. the remainder $I[f] - Q[f]$ vanishes—for products of monomials $x_1^{p_1} x_2^{p_2} \cdots$ with their cumulative order $d = p_1 + p_2 + \cdots + p_n$ as high as possible. One strives for rules with as few nodes as possible (small N) for d as large as possible. The main problem is that the minimal number of nodes needed by a cubature formula to attain a certain degree d does not depend only on this degree and the dimension n, but also on the weight function w and the shape of the region Ω — and the functional form of this dependence in general is not known.

3.5.1 *Integration Over a Square*

The one region onto which one-dimensional quadrature formulas are relatively easy to transplant is the unit square. In square geometry with $(x, y) \in [-1, 1] \times [-1, 1]$ a Gauss quadrature, say, of equal orders in x and y has the form

$$\int_{-1}^{1} \int_{-1}^{1} f(x, y) \, \mathrm{d}x \, \mathrm{d}y \approx \sum_{j=1}^{N} w_j \, f(x_j, y_j) \,,$$

where the nodes and the weights satisfy a system of non-linear equations

$$\sum_{j=1}^{N} w_j x_j^p y_j^q = \frac{1 + (-1)^p}{p + 1} \cdot \frac{1 + (-1)^q}{q + 1} \,, \tag{3.63}$$

for $p = 0, 1, \ldots, M$, $q = 0, 1, \ldots, M - p$, and $N = \frac{1}{6}(M^2 + 3M + 2)$. Some popular low-order sets of nodes and weights are collected in [9] (pp. 891–895), but those formulas are known to be sub-optimal. The system (3.63) is hard to solve, in particular for high orders, and one often resorts instead to a simple Cartesian product of one-dimensional formulas. For example, the approximate value of the integral over the square $(x, y) \in [-1, 1] \times [-1, 1]$ can be computed by using

$$\int_{-1}^{1} \int_{-1}^{1} f(x, y) \, \mathrm{d}x \, \mathrm{d}y \approx \sum_{j=1}^{N} \sum_{k=1}^{M} w_j^{(N)} w_k^{(M)} \, f(x_j, y_k) \,,$$

where quadratures in x and y coordinates may be of different orders. This simple product formula has MN nodes and requires just as many evaluations of $f(x, y)$. The weight corresponding to the node (x_j, y_k) is simply a product of the weights from individual univariate quadratures. Optionally, one can even apply different quadrature rules in different directions, for example, if functions are tame in one independent variable but wild in another. For select recent research see e.g. [53–57].

3.5.2 Software Resources

The most exhaustive online resource for multivariate integration is the online Encyclopedia of Cubature Formulas [58]. It provides tables of cubature rules for the hypercube (i.e. square, cube, four-cube and n-cube), for the sphere (i.e. circle, sphere, four-sphere, n-sphere), for the space with weight functions e^{-r} and e^{-r^2} (i.e. plane, space, four-space, n-space) and for the simplex (i.e. triangle, tetrahedron, four-simplex, five-simplex and n-simplex). The papers [59–61] provide the main reference, but see also [62–64].

The classical adaptive quadratures for hyper-rectangles have been developed by [65] and [66], and various implementations derived from them have been incorporated into modern numerical libraries and tools, e.g. into the `NIntegrate` function of MATHEMATICA. The cubature routines implemented in MATLAB are based on algorithms presented in [67, 68]. A simple package written in C (and a wrapper for Python) for adaptive integration of vector-valued integrands over hypercubes is provided at [69]. It contains two main modules. The first one, based on [65, 66], is an h-adaptive cubature employing a recursive partitioning of the domain into smaller subdomains, applying the same rule to each, until convergence is achieved; it is best suited for dimensions up to ≈ 7 beyond which Monte-Carlo methods prevail. The second one is p-adaptive integration based on a tensor product of Clenshaw-Curtis quadrature rules, repeatedly doubling the degree of the rules until convergence is attained. According to the author this algorithm is often superior to h-adaptive integration for smooth integrands for low dimensions, but is a poor choice in higher dimensions or for non-smooth integrands. Another library for multivariate quadrature, also developed from [65, 66] and featuring C/C++, Fortran and MATHEMATICA interfaces is Cuba [70].

3.6 Stable Numerical Differentiation

Some solution methods for inverse problems in Chapter 13 call for numerical differentiation of perturbed (or "noisy") functions given either analytically or in discretized form: one way or another, we wish to estimate the derivative of a noise-free function

f, given its noisy realization f^δ, where $\|f^\delta - f\| \le \delta$. Differentiation of noise-contaminated data is an ill-posed problem [71], but at least three types of approaches exist which, if pursued with caution, provide stable (or "noise-robust") differentiation: (1) regularized finite-difference methods with a step size h that is made to be a function of δ, and thereby turned into a regularization parameter; (2) "mollification" methods where one tries to tame f^δ by smoothing it with an appropriate (e.g. Gaussian) kernel or describing it with splines, and then to differentiate the resulting approximation [72–74]; and (3) variational methods [75–80] in which the problem of differentiation is translated into a problem of solving a regularized integral equation. We describe these three types of methods in the following—without opening the closely related digital filtering front [81]— and restrict the initial discussion to $x \in I = [0, 1]$. See also [82].

3.6.1 Regularized Finite Differences

It is worthwhile to investigate under what kind of conditions stable differentiation is possible *at all*. In practice one does not know f but only f^δ, perhaps with some additional prior knowledge about f: one may know, for instance, that f belongs to the class of functions

$$\mathcal{K}(\delta, m) = \left\{ f : f \in W^{m,p}(0, 1)\,, \ \|f^{(m)}\|_p \le M_{m,p} < \infty\,, \ \|f^\delta - f\|_p \le \delta \right\},$$

where $m = 0$ or $m = 1$. Here $W^{m,p}(0, 1)$ is the Sobolev space of functions whose mth derivative belongs to $L^p(0, 1)$. The range $1 < m \le 2$ is also of interest, but for $1 < m < 2$ the norms of the derivatives should be understood as

$$\|f^{(m)}\|_p = \sup_{x,y \in I, x \ne y} \frac{\|f'(x) - f'(y)\|_p}{|x - y|^{m_0}}\,, \quad m = 1 + m_0\,, \quad 0 < m_0 < 1\,.$$

The task at hand is to find an operator $\mathcal{D}^{(1)}_{h(\delta)} : L^p(0, 1) \to L^p(0, 1)$ such that

$$\sup_{f \in K(\delta, m)} \left\| \mathcal{D}^{(1)}_{h(\delta)} f^\delta - f' \right\|_p \le \eta(\delta) \to 0 \quad \text{as} \quad \delta \to 0\,, \tag{3.64}$$

where $\eta(\delta)$ is some positive continuous function of $\delta \in (0, \delta_0)$ and the step size $h(\delta)$ of a finite-difference formula to be devised acts as a regularization parameter whose magnitude is driven by the level of noise, δ. It was shown in [83] that for $p = 0$ or $p = \infty$ an operator satisfying (3.64) for $m \le 1$ does not exist; however, it does exist if $m > 1$ and $p \ge 1$. An example of such an operator for $m = 2$ is

$$\mathcal{D}_{h(\delta)}^{(1)} f^{\delta}(x) = \begin{cases} \dfrac{1}{h}\left[f^{\delta}(x+h) - f^{\delta}(x)\right], & 0 < x < h, \\[2mm] \dfrac{1}{2h}\left[f^{\delta}(x+h) - f^{\delta}(x-h)\right], & h \le x \le 1-h, \\[2mm] \dfrac{1}{h}\left[f^{\delta}(x) - f^{\delta}(x-h)\right], & 1-h < x < 1, \end{cases}$$

where $h = h(\delta) > 0$. If $f^{\delta} \in L^{\infty}(0, 1)$ and $f \in W^{2,p}(0, 1)$, the corresponding error estimate is

$$\left\| D_{h(\delta)}^{(1)} f^{\delta}(x) - f' \right\|_{p} \le \frac{2\delta}{h} + \frac{M_{2,p} h}{2}, \tag{3.65}$$

where $M_{2,p}$ is the upper bound on the second derivative, i.e. $\|f''\|_p \le M_{2,p}$. (If only the narrower domain $[h, 1-h]$ is considered, the error estimate is slightly more restrictive, amounting to $\delta/h + M_{2,p} h/2$.) The right-hand side of (3.65) reaches the absolute minimum $2(\delta M_{2,p})^{1/2}$ at the optimal value of h given by

$$h(\delta) = 2(\delta/M_{2,p})^{1/2}.$$

(If one uses the error estimate for the central subinterval $[h, 1-h]$, the optimal h becomes $h(\delta) = (2\delta/M_{2,p})^{1/2}$.) One can further show that if $f \in W^{3,p}(0, 1)$ the differentiation formula can be refined at the endpoints so it attains order $\mathcal{O}(h^2)$ of the error estimate as $h \to 0$. For example, one can use

$$\mathcal{D}_{h(\delta)}^{(1)} f^{\delta}(x) = \begin{cases} \dfrac{1}{2h}\left[4f^{\delta}(x+h) - f^{\delta}(x+2h) - 3f^{\delta}(x)\right], & 0 < x < 2h, \\[2mm] \dfrac{1}{2h}\left[f^{\delta}(x+h) - f^{\delta}(x-h)\right], & 2h \le x \le 1-2h, \\[2mm] \dfrac{1}{2h}\left[3f^{\delta}(x) + f^{\delta}(x-2h) - 4f^{\delta}(x-h)\right]. & 1-2h < x < 1. \end{cases}$$

It is straightforward to extend the above formalism to a more general class of functions $\mathcal{K}(\delta, m) = \left\{f : f \in C^1(\mathbb{R}), \|f^{(m)}\| \le M_m < \infty, \|f^{\delta} - f\| \le \delta\right\}$, with the norm $\|f\| = \operatorname{ess\,sup}_{x\in\mathbb{R}} |f(x)|$. Consider the functions $f \in \mathcal{K}(\delta, m)$ that possess derivatives up to order m, with $m > 2$ and odd, $m = 2Q + 1$, and let $f^{\delta} \in L^{\infty}(\mathbb{R})$ be given. The differentiator of such a function can be written in the form

$$\mathcal{D}_{h}^{(1)} f^{\delta} = \frac{1}{h} \sum_{j=-Q}^{Q} A_{j}^{Q} f^{\delta}\left(x + j\frac{h}{Q}\right), \tag{3.66}$$

where the sets of coefficients A_{j}^{Q} for $Q \in \{1, 2, 3, 4\}$ are given by [84, 85]

$$A_0^1 = 0 \,, \quad A_{\pm 1}^1 = \pm 1/2 \,,$$

$$A_0^2 = 0 \,, \quad A_{\pm 1}^2 = \pm 4/3 \,, \quad A_{\pm 2}^2 = \mp 1/6 \,,$$

$$A_0^3 = 0 \,, \quad A_{\pm 1}^3 = \pm 9/4 \,, \quad A_{\pm 2}^3 = \mp 9/20 \,, \quad A_{\pm 2}^3 = \pm 1/20 \,,$$

$$A_0^4 = 0 \,, \quad A_{\pm 1}^4 = \pm 16/5 \,, \quad A_{\pm 2}^4 = \mp 4/5 \,, \quad A_{\pm 2}^4 = \pm 16/105 \,, \quad A_{\pm 2}^4 = \mp 1/70 \,,$$

and so on. The differentiator (3.66) embodies nothing but the standard central differences (with coefficients multiplied by Q). For $Q = 1$ ($m = 3$) we have

$$\mathcal{D}_{h_3(\delta)}^{(1)} f^\delta = \frac{f^\delta \big(x + h_3(\delta)\big) - f^\delta \big(x - h_3(\delta)\big)}{2h_3(\delta)} \,, \tag{3.67}$$

and for $Q = 2$ ($m = 5$), omitting the indication of δ-dependence for clarity, we get

$$\mathcal{D}_{h_5(\delta)}^{(1)} f^\delta = \frac{-f^\delta \big(x + h_5\big) + 8 f^\delta \big(x + h_5/2\big) - 8 f^\delta \big(x - h_5/2\big) + f^\delta \big(x - h_5\big)}{6h_5(\delta)} \,.$$

The error of the differentiator can be estimated as [86]

$$\varepsilon_m(\delta) = \big|\mathcal{D}_h^{(1)} f^\delta - f'\big| \le \underbrace{\frac{\delta}{h} \sum_{j=-Q}^{Q} \big|A_j^Q\big|}_{\alpha_m} + h^{m-1} M_m \underbrace{\frac{1}{m!} \sum_{j=-Q}^{Q} \left|\left(\frac{j}{Q}\right)^m A_j^Q\right|}_{\beta_m} \,,$$

and setting the derivative of this expression with respect to h to zero we find out that the optimal step size for given m is

$$h = h_m(\delta) = \left(\frac{\delta}{M_m} \frac{\alpha_m}{(m-1)\beta_m}\right)^{1/m} \,,$$

corresponding to the minimum error

$$\varepsilon_m(\delta) = m \left(\delta^{m-1} M_m \frac{\alpha_m^{m-1} \beta_m}{(m-1)^{m-1}}\right)^{1/m} \,.$$

The formula (3.67), therefore, is best evaluated with the optimal step size

$$h_3(\delta) = (3\delta/M_3)^{1/3} \,. \tag{3.68}$$

Moreover, it can be shown rigorously that among all linear or non-linear operators $T : L^\infty(\mathbb{R}) \to L^\infty(\mathbb{R})$ the operator (3.67) used with (3.68) gives the best possible estimate of f' in the class of functions $f \in \mathcal{K}(\delta, M_3)$. A similar result can be proven for the second derivative: the operator

$$\mathcal{D}^{(2)}_{h_3(\delta)} f^\delta = \frac{f^\delta\big(x + h_3(\delta)\big) - 2f^\delta(x) + f^\delta\big(x - h_3(\delta)\big)}{h_3^2(\delta)} \,, \quad h_3(\delta) = 2(3\delta/M_3)^{1/3} \,,$$

gives the best possible estimate of f'' in the class $f \in \mathcal{K}(\delta, M_3)$.

3.6.1.1 Low-Noise Lanczos Differentiators

The above formalism can be extended to the differentiation of functions of the form

$$f^\delta(x) = f(x) + e^\delta(x) \,, \tag{3.69}$$

where the error $e^\delta(x)$ is independent and identically distributed with zero mean and variance δ^2, and where we assume that $f \in C^2(\mathbb{R})$ is twice differentiable, with the bound $\| f'' \| \le M_2$. The structure of the differentiator remains as in (3.66), only the coefficients A_j^Q and, consequently, the estimates for the optimal h and the corresponding minimum error change [86]. The differentiator can be formulated as

$$\mathcal{D}^{(1)}_h f^\delta = \frac{1}{h} \frac{3}{(Q + 1)(2Q + 1)} \sum_{j=-Q}^{Q} j f^\delta\left(x + j\frac{h}{Q}\right) \,. \tag{3.70}$$

The optimal step size is

$$h(\delta) = \left(\frac{\delta}{M_2}\right)^{1/2} \left(\frac{a_Q}{b_Q}\right)^{1/4} \,, \tag{3.71}$$

where

$$a_Q = \frac{3Q}{(Q + 1)(2Q + 1)} \,, \quad b_Q = \frac{9(Q + 1)^2}{16(2Q + 1)^2} \,,$$

and the standard deviation of the optimal estimate is

$$\left(\mathrm{E}\left[\big(\mathcal{D}^{(1)}_h f^\delta - f'\big)^2\right]\right)^{1/2} \le (a_Q b_Q)^{1/4} \sqrt{2\delta M_2} \,. \tag{3.72}$$

Asymptotically $(a_Q b_Q)^{1/4} \sim Q^{-1/4}$ as $Q \gg 1$ and at least in principle it is possible to attain an arbitrary accuracy of the approximation by taking Q sufficiently large. The usual practitioner's problem, however, is the ignorance about δ and M_2. If these two quantities are unknown, Ref. [86] suggest the following simple consideration: ignore (3.71), take h sufficiently small and choose Q in such a way that $h^2 Q \to \infty$ as $h \to 0$. With this strategy, alas, the error estimate (3.72) can not be used.

With the substitution $h \leftarrow h/Q$ the differentiator (3.70) can be rewritten as

$$\mathcal{D}_h^{(1)} f^\delta = \frac{1}{h} \frac{3}{Q(Q+1)(2Q+1)} \sum_{j=1}^{Q} j\left(f^\delta(x + jh) - f^\delta(x - jh)\right), \qquad (3.73)$$

which gives rise to the more familiar *low-noise Lanczos differentiation formulas*. For $Q \in \{1, 2, 3, 4\}$ they are given by

$$Q = 1: \quad f'_\mathrm{L}(x_0) \approx \frac{f_1 - f_{-1}}{2h},$$

$$Q = 2: \quad f'_\mathrm{L}(x_0) \approx \frac{(f_1 - f_{-1}) + 2(f_2 - f_{-2})}{10h},$$

$$Q = 3: \quad f'_\mathrm{L}(x_0) \approx \frac{(f_1 - f_{-1}) + 2(f_2 - f_{-2}) + 3(f_3 - f_{-3})}{28h},$$

$$Q = 4: \quad f'_\mathrm{L}(x_0) \approx \frac{(f_1 - f_{-1}) + 2(f_2 - f_{-2}) + 3(f_3 - f_{-3}) + 4(f_4 - f_{-4})}{60h},$$

where we have denoted $f_j = f(x + jh)$. Their "low-noise" attribute stems from the low-pass property of the differentiators in the frequency domain: the higher the Q, the stronger the suppression of high-frequency components.

Example We use (3.73) to differentiate the function (3.69) with

$$f(x) = \mathrm{e}^{-x} \sin(\pi x), \quad x \in [0, 1],$$

sampled at 100 equidistant points on $[0, 1]$ whose values at each of these points are contaminated by Gaussian noise with zero mean and variance $\delta^2 = (0.01)^2$. Figure 3.16 (top) shows the results obtained by low-noise Lanczos differentiators with $Q = 4$ (9-point formula), $Q = 6$ (13 points) and $Q = 9$ (19 points), along with the derivative obtained by the conventional 9-point central difference formula

$$f'_\mathrm{cd}(x_0) \approx \frac{672(f_1 - f_{-1}) - 168(f_2 - f_{-2}) + 32(f_3 - f_{-3}) - 3(f_4 - f_{-4})}{840h}.$$

$$(3.74)$$

Figure 3.16 (bottom) shows the results obtained by *smooth noise-robust differentiators* of Ref. [87] which were developed in order to cure the main deficiency of Lanczos differentiators, namely that their suppression of high-frequency components is not monotonous, resulting in a residual "wavy" nature of the derivative. These differentiators have the form

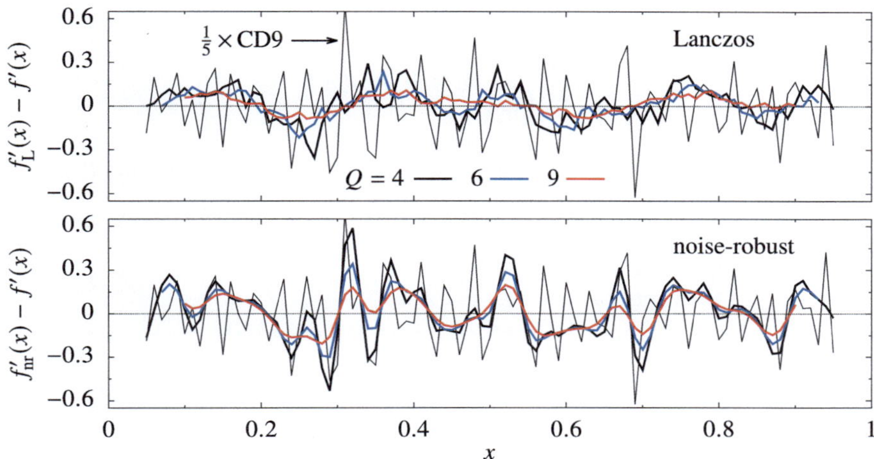

Fig. 3.16 Errors of the differentiation of the noisy function $e^{-x}\sin(\pi x) + \delta$, $\delta \sim N(0, 0.01)$, by using [TOP] low-noise Lanczos differentiators and [BOTTOM] noise-robust differentiators of Ref. [87]. Note that the error of the standard (and extremely noise-prone) 9-point central difference formula, denoted by CD9, is shown at only 20 % of its true magnitude

$$f'_{\mathrm{nr}}(x_0) \approx \frac{1}{h}\sum_{j=1}^{Q} c_j^Q (f_j - f_{-j}), \quad c_j^Q = \frac{1}{2^{2Q-1}}\left[\binom{2(Q-1)}{Q-j} - \binom{2(Q-1)}{Q-j-2}\right].$$

The explicit formulas for $Q \in \{1, 2, 3, 4\}$ are

$$Q = 1: \quad f'_{\mathrm{nr}}(x_0) \approx \frac{f_1 - f_{-1}}{2h},$$

$$Q = 2: \quad f'_{\mathrm{nr}}(x_0) \approx \frac{2(f_1 - f_{-1}) + (f_2 - f_{-2})}{8h},$$

$$Q = 3: \quad f'_{\mathrm{nr}}(x_0) \approx \frac{5(f_1 - f_{-1}) + 4(f_2 - f_{-2}) + (f_3 - f_{-3})}{32h},$$

$$Q = 4: \quad f'_{\mathrm{nr}}(x_0) \approx \frac{14(f_1 - f_{-1}) + 14(f_2 - f_{-2}) + 6(f_3 - f_{-3}) + (f_4 - f_{-4})}{128h}.$$

Smooth noise-robust *second-order* differentiators (to compute f'') have also been developed. The general centered formula spanning $2Q + 1$ equidistant points (the central x_0 and up to Q points to the left and to the right), with $Q \geq 2$, is

$$f_{\mathrm{nr}}''(x_0) \approx \frac{1}{2^{2Q-2}h^2}\left(d_0 f_0 + \sum_{j=1}^{Q} d_j\left(f_j + f_{-j}\right)\right).$$

The coefficients d_j can be computed by the following recursive algorithm [87]:

Algorithm: $\mathrm{d}(j) \equiv d_j$

Input: Q, j

if $(j > Q)$ **then return** 0;

if $(j = Q)$ **then return** 1;

return $\big((2Q-4)\,\mathrm{d}(j+1) - (Q+j+2)\,\mathrm{d}(j+2)\big)/(Q-j)$;

Output: d_j

The explicit formulas for $Q \in \{2, 3, 4\}$ are

$$Q = 2: \quad f_{\mathrm{nr}}''(x_0) \approx \frac{-2f_0 + (f_2 + f_{-2})}{4h^2},$$

$$Q = 3: \quad f_{\mathrm{nr}}''(x_0) \approx \frac{-4f_0 - (f_1 + f_{-1}) + 2(f_2 + f_{-2}) + (f_3 + f_{-3})}{16h^2},$$

$$Q = 4: \quad f_{\mathrm{nr}}''(x_0) \approx \frac{-10f_0 - 4(f_1 + f_{-1}) + 4(f_2 + f_{-2}) + 4(f_3 + f_{-3}) + (f_4 + f_{-4})}{64h^2}.$$

An illustration is given in Fig. 3.17. In addition to central-difference formulas one can also construct one-sided differences in order to handle the vicinity of the endpoints where ever wider stencils run out of space. Similar machinery has been employed to design noise-robust smoothing filters. For details see [87]. ◁

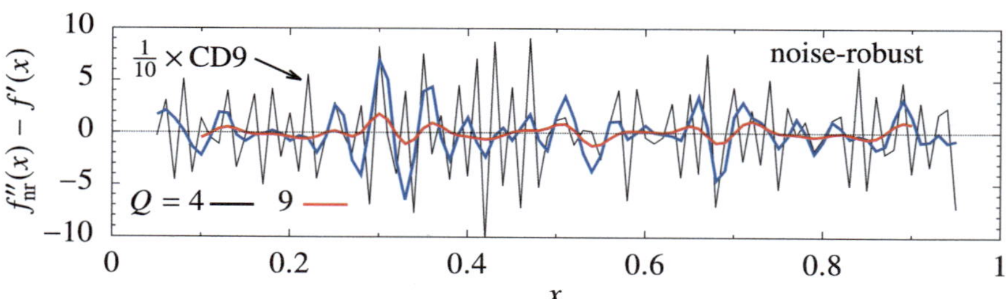

Fig. 3.17 Errors of the second derivative of the noisy function $e^{-x}\sin(\pi x) + \delta$, $\delta \sim N(0, 0.001)$, by using 9-point ($Q = 4$) and 19-point ($Q = 9$) noise-robust differentiators. Note that the error of the standard 9-point central difference formula, denoted by CD9, is shown scaled by $1/10$

3.6.2 Differentiation by Using Smoothing Kernels

An alternative approach is to make the noisy data smoother by using splines [72–74]; or to convolute the data with a Gaussian $g_h(x) = (1/(h\sqrt{\pi}))\exp(-x^2/h^2)$ and then differentiate the resulting smooth approximation:

$$\left(G_{h(\delta)}\right)' f^\delta(x) = \left(g_h' * f^\delta\right)(x) = \int_0^1 g_h'(x - t) f^\delta(t)\, dt ,$$

where $f^\delta \in L^2(0, 1)$ and $\|f^\delta - f\|_2 \le \delta$. For $f \in H^1(0, 1)$ with $f(0) = f(1) = 0$ and the second derivative bounded by $\|f''\|_2 \le M_2$, one can show that [71]

$$\varepsilon \equiv \left\|\left(G_{h(\delta)}\right)' f^\delta(x) - f'\right\|_2 \le \frac{2\delta}{h\sqrt{\pi}} + hM_2 .$$

Choosing the optimal step size $h = h(\delta) = (2\delta/(M_2\sqrt{\pi}))^{1/2}$ corresponds to the error estimate $\varepsilon \le (2\delta M_2/\sqrt{\pi})^{1/2}$.

3.6.3 Variational Regularization of the Derivative

The third approach leads by the minimization of the functionals

$$F_\alpha(u) = \frac{1}{2}\left\|Au - f^\delta\right\|_2^2 + \frac{\alpha}{2}\left\|u^{(m)}\right\|_2^2 , \qquad m = 0 \text{ or } 1 \text{ or } \cdots , \tag{3.75}$$

where

$$Au(x) = \int_0^x u(t)\, dt$$

and $\alpha > 0$ is the regularization parameter. Given $\|f^\delta - f\|_2 \le \delta$ and with a suitable choice of $\alpha = \alpha(\delta)$, the functionals of the above form possess unique minimizers $u_{\alpha(\delta)}^\delta$ such that $\|u_{\alpha(\delta)}^\delta - f'\|_2 \to 0$ as $\delta \to 0$. (For the derivation of these error estimates and further details see Sect. 2.5.1 of [71].) Minimizing (3.75) for $m = 0$ with respect to u gives

$$u_\alpha^\delta = (A^*A + \alpha)^{-1} A^* f^\delta , \tag{3.76}$$

a well-known result from the theory of inverse problems: see Eq. (13.19). A value must be chosen for the regularization parameter; by resorting to the discrepancy principle (see Sect. 13.3.7), α is given as the root of the equation

$$\|Au_\alpha^\delta - f^\delta\| = C\delta \,, \quad C = \text{const.} \geq 1 \,, \tag{3.77}$$

but there is no specific choice of C that is theoretically justified and optimal in some sense. Finding the derivative of a noisy function, then, implies solving (3.76) along with (3.77). However, one can also depart from the realization that if f is the function to be differentiated and u is its unknown derivative, u satisfies the Volterra equation

$$f(x) = f(0) + Au(x) = f(0) + \int_0^x u(t)\, dt \,.$$

Staying with the domain $x \in [0, 1]$ for simplicity and assuming that $f(0) = 0$ without loss of generality the appropriate regularized equation involving the noisy source function f^δ and its derivative u^δ reads

$$Au^\delta + \alpha u^\delta = f^\delta \,. \tag{3.78}$$

This problem can be discretized and solved as a system of linear equations for the values of u at the grid points. Alternatively, it lends itself to a solution in simple closed form [88]:

$$u^\delta(x) = -\frac{1}{\alpha} \int_{-x/\alpha}^{0} f^\delta(x + \alpha t) e^t\, dt + \frac{f^\delta(x)}{\alpha} \,. \tag{3.79}$$

Example (Adapted from Examples 5.1 and 5.2.1 in [88].) We wish to compute the derivative of the function $f(x) = \sin(\pi x)$, $x \in [0, 1]$, assuming that only the noisy values of f are given, and the noise function is $e(x) = \delta \cos(3\pi x)$. We have

$$f^\delta(x) = f(x) + e(x) = \sin(\pi x) + \delta \cos(3\pi x) \,, \tag{3.80}$$

and we find $u^\delta(x)$ by using (3.79). The results for $\delta = 0.02$ and $\delta = 0.05$ are shown in Fig. 3.18 along with the derivative of the exact f, $u(x) = f'(x) = \pi \cos(\pi x)$. The optimal regularization parameters α that minimize the norm $\|u^\delta - u\|_2$ are:

$$\delta = 0.02 : \quad \alpha \approx 0.0059 : \quad \|u^\delta - u\|_2 \approx 0.140 \,,$$
$$\delta = 0.05 : \quad \alpha \approx 0.0135 : \quad \|u^\delta - u\|_2 \approx 0.346 \,.$$

The parameter α should satisfy $\lim_{\delta \to 0} \alpha(\delta) = 0$ and $\lim_{\delta \to 0}\big(\delta/\alpha(\delta)\big) = 0$, e.g. $\alpha(\delta) = \delta^p/c$, where $p \in (0, 1)$ and $c = \text{const.}$ A phenomenological formula found in [88] for this example is $\alpha(\delta) = \delta^{0.9}/5$ but there is no known general algorithm for choosing p and c. Solving (3.78) for u^δ yields practically the same results if a sufficiently fine grid is chosen. See also Problem 3.7.3. $\triangleleft$

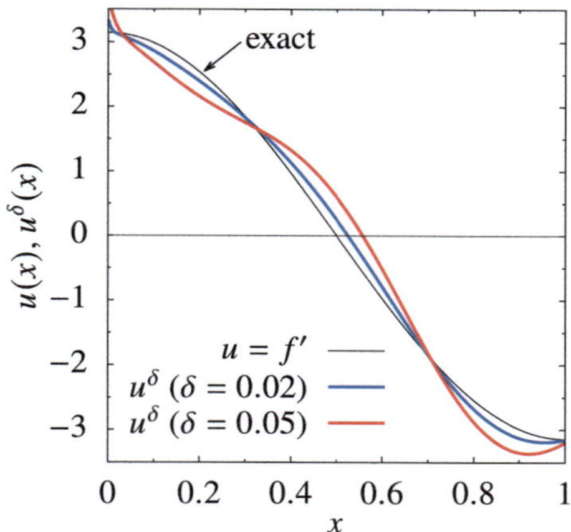

Fig. 3.18 Differentiating the function (3.80) containing an admixture of a noise function with amplitudes $\delta = 0.02$ and 0.05, by solving the regularized problem (3.78). Each δ requires its own regularization parameter α: the optimal values that minimize the norm $\|u^\delta - u\|_2$ are 0.0059 and 0.0135, respectively

3.7 Problems

3.7.1 Hyperbolic Volume

The integral

$$\frac{24}{7\sqrt{7}} \int_{\pi/3}^{\pi/2} \log\left|\frac{\tan t + \sqrt{7}}{\tan t - \sqrt{7}}\right|$$

appears in analyses of volumes of ideal tetrahedra in hyperbolic space [89].

⊙ By using any of the quadrature methods discussed in this Chapter (and any method for the summation of series from Sect. 1.4) show to the maximum level of precision you are able to achieve that the value of this integral is equal to

$$\sum_{n=0}^{\infty} \left(\frac{1}{(7n+1)^2} + \frac{1}{(7n+2)^2} - \frac{1}{(7n+3)^2} + \frac{1}{(7n+4)^2} - \frac{1}{(7n+5)^2} - \frac{1}{(7n+6)^2}\right).$$

This "identity" has been verified numerically to a precision of 20.000 digits, but no proof is known.

3.7.2 A Parametric Integral

Let

$$I(a) = \int_0^1 \frac{\arctan(\sqrt{x^2 + a^2})}{\sqrt{x^2 + a^2}(x^2 + 1)}\, dx \; .$$

It can be shown [32] that

$$
\begin{aligned}
I(0) &= (\pi \log 2)/8 + G/2 \; , \\
I(1) &= \pi/4 - \pi/\sqrt{2}/2 + 3\sqrt{2}\arctan(\sqrt{2})/2 \; , \\
I(\sqrt{2}) &= 5\pi^2/96 \; ,
\end{aligned}
\tag{3.81}
$$

where $G = \sum_{k \geq 0}(-1)^k/(2k+1)^2 \approx 0.91596559417\ldots$ is the Catalan's constant.

⊙ By any quadrature rule at your disposal verify as precisely as possible the three particular values (3.81) and show that the above integral for a general a equals

$$I(a) = \frac{\pi}{2\sqrt{a^2 - 1}}\left(2\arctan(\sqrt{a^2 - 1}) - \arctan(\sqrt{a^4 - 1})\right) \; .$$

(This has been proven rigorously.)

3.7.3 Differentiation of a Noisy Function

We wish to compute the derivative of the function

$$f(x) = \sin\big((\pi x)^4\big) \; , \quad x \in [0, 1] \; ,
\tag{3.82}$$

in the presence of noise, that is, when only the values of the noise-contaminated function

$$f^\delta(x) = f(x) + \delta\big(\cos(2x) + \cos(3x^2)\big) \; , \quad \delta = 0.1 \; ,$$

are available. The exact derivative is $f'(x) = u(x) = 4\pi^4 x^3 \cos\big((\pi x)^4\big)$ and is shown in Fig. 3.19 (left).

⊙ Compute the derivative of f^δ by using the method proposed in Eq. (3.79). Choose the regularization parameter according to the formula

$$\alpha(\delta) = \delta^p/c \; , \quad p \in (0, 1) \; , \quad c = \text{const.}
\tag{3.83}$$

Investigate the quality of the approximate derivative by calculating the overall error $\|u^\delta - u\|_2$ and the error of the value at the origin, $|u^\delta(0) - u(0)|$, for different values of the parameters p and c: try to find an optimal combination of the two. The local

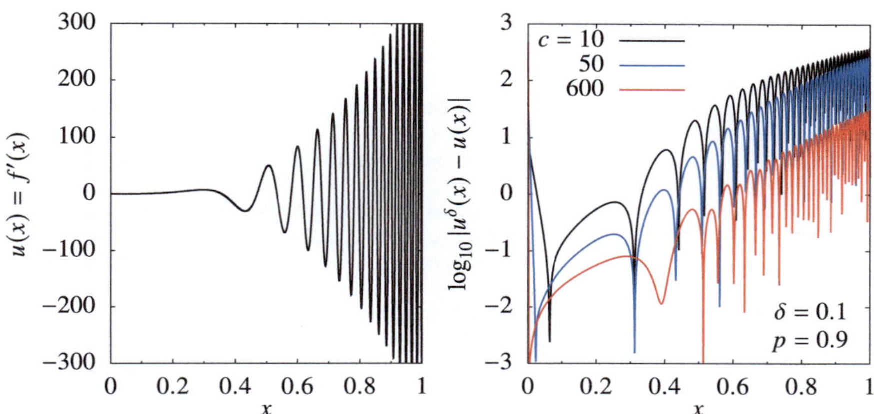

Fig. 3.19 [LEFT] The exact derivative of (3.82). [RIGHT] The point-wise errors of the approximate numerical derivatives $u^\delta(x)$ for $\delta = 0.1$ and three different values of the scaling constant c appearing in the phenomenological formula (3.83) for the optimal regularization parameter

(point-wise) errors for $p = 0.9$ and $c = 10, 50$ and 600—resulting in the global errors $\|u^\delta - u\|_2 \approx 0.01259, 0.00252$ and 0.00021, respectively—are shown in Fig. 3.19 (right).

References

1. P.J. Davis, P. Rabinowitz, *Methods of Numerical Integration*, 2nd ed. (Academic Press, San Diego, 1984)
2. P. Gonnet, A review of error estimation in adaptive quadrature. ACM Comput. Surv. **44**(Art. 22) (2012)
3. H. Brass, K. Petras, *Quadrature Theory. Providence, The Theory of Numerical Integration on a Compact Interval* (AMS, 2011)
4. B. Fornberg, Improving the accuracy of the trapezoidal rule. SIAM Rev. **63**, 167 (2021)
5. L.N. Trefethen, J.A.C. Weideman, The exponentially convergent trapezoidal rule. SIAM Rev. **56**, 385 (2014)
6. J.A.C. Weideman, Numerical integration of periodic functions: a few examples. Amer. Math. Monthly **109**, 21 (2002)
7. W.H. Press, B.P. Flannery, S.A. Teukolsky, W.T. Vetterling, *Numerical Recipes: The Art of Scientific Computing*, 3rd edn. (Cambridge University Press, Cambridge, 2007). See also the equivalent handbooks in Fortran, Pascal and C, as well as http://numerical.recipes
8. W. Romberg, Vereinfachte numerische Integration. Norske Vid. Sels. Forh. **28**, 30 (1955)
9. M. Abramowitz, I.A. Stegun, *Handbook of Mathematical Functions*, 10th edn. (Dover Publications, Mineola, 1972)
10. S. Olver, R.M. Slevinsky, A. Townsend, Fast algorithms using orthogonal polynomials. Acta Numerica **29**, 573 (2020)
11. N. Hale, A. Townsend, Fast and accurate computation of Gauss-Legendre and Gauss-Jacobi quadrature nodes and weights. SIAM J. Sci. Comput. **35**, A652 (2013)
12. I. Bogaert, Iteration-free computation of Gauss-Legendre quadrature nodes and weights. SIAM J. Sci. Comput. **36**, A1008 (2014)

13. L.N. Trefethen, Exactness of quadrature formulas. SIAM Rev. **64**, 132 (2022)
14. N. Hale, L.N. Trefethen, New quadrature formulas from conformal maps. SIAM J. Numer. Anal. **46**, 930 (2008)
15. W.M. Gentleman, Implementing Clenshaw-Curtis quadrature, I: methodology and experience. Commun. ACM **15**, 337 (1972)
16. J. Waldvogel, Fast construction of the Fejér and Clenshaw-Curtis quadrature rules. BIT Numer. Math. **46**, 195 (2006)
17. G.H. Golub, J.H. Welsch, Calculation of Gauss quadrature rules. Math. Comput. **23**, 221 (1969)
18. L.N. Trefethen, Is Gauss quadrature better than Clenshaw-Curtis? SIAM Review **50**, 67 (2008)
19. J.A.C. Weideman, L.N. Trefethen, The kink phenomenon in Fejér and Clenshaw-Curtis quadrature. Numer. Math. **107**, 707 (2007)
20. D. Kahaner, C. Moler, S. Nash, *Numerical Methods and Software* (Prentice-Hall, Englewood Cliffs, 1989)
21. D. Calvetti, G.H. Golub, W.B. Gragg, L. Reichel, Computation of Gauss-Kronrod quadrature rules. Math. Comput. **69**, 1035 (2000)
22. D.P. Laurie, Calculation of Gauss-Kronrod quadrature rules. Math. Comput. **66**, 1133 (1997)
23. R. Piessens, E. De Doncker-Kapenga, C.W.Überhuber, *QUADPACK: A Subroutine Package for Automatic Integration* (Springer, Berlin, 1983). The original Fortran77 routines are available at http://www.netlib.org/quadpack/; a new version converted to modern free-form syntax is available at https://github.com/jacobwilliams/quadpack
24. P. Favati, G. Lotti, F. Romani, Algorithm 691: improving QUADPACK automatic integration routines. ACM Trans. Math. Softw. **17**, 218 (1991)
25. W. Gander, W. Gautschi, Adaptive quadrature—revisited. BIT Numer. Math. **40**, 94 (2000)
26. S. Ehrich, Error bounds for Gauss-Kronrod quadrature formulae. Math. Comput. **62**, 295 (1994)
27. S.E. Notaris, Gauss-Kronrod quadrature formulae—A survey of fifty years of research. Electron. Trans. Numer. Anal. **45**, 371 (2016)
28. H. Takahasi, M. Mori, Double exponential formulas for numerical integration. Publ. RIMS Kyoto Univ. **9**, 721 (1974)
29. M. Mori, Quadrature formulas obtained by variable transformation and the DE–Rule. J. Comput. Appl. Math. **12 & 13**, 119 (1985)
30. D.H. Bailey, K. Jeyabalan, X.S. Li, A comparison of three high-precision quadrature schemes. Exp. Math. **14**, 317 (2005)
31. M. Mori, Discovery of the double exponential transformation and its developments. Publ. RIMS Kyoto Univ. **41**, 897 (2005)
32. D.H. Bailey, J.M. Borwein, High-precision numerical integration: progress and challenges. J. Symb. Comput. **46**, 741 (2011)
33. N. Eggert, J. Lund, The trapezoidal rule for analytic functions of rapid decrease. J. Comput. Appl. Math. **27**, 389 (1989)
34. D. Huybrechs, P. Opsomer, Construction and implementation of asymptotic expansions for Laguerre-type orthogonal polynomials. IMA J. Numer. Anal. **38**, 1085 (2018)
35. A. Gil, J. Segura, N.M. Temme, Fast, reliable and unrestricted iterative computation of Gauss-Hermite and Gauss-Laguerre quadratures. Numer. Math. **143**, 649 (2019)
36. J. Shen, Stable and efficient spectral methods in unbounded domains using Laguerre functions. SIAM J. Numer. Anal. **38**, 1113 (2000)
37. J.P. Boyd, *Chebyshev and Fourier Spectral Methods*, 2nd edn. (Dover Publications, Mineola, 2000). Available at http://www-personal.engin.umich.edu/~jpboyd
38. F.S. Acton, *Numerical Methods that Work* (The Mathematical Association of America, Washington, D.C., 1997)
39. J.V. Noble, Gauss–Legendre principal value integration. Comput. Sci. Eng. 92 (2000); see also W.J.Thompson, Principal-value integrals by a simple and accurate finite-interval method. Comp. Phys. **12**, 94 (1998)
40. P. Rabinowitz, Gauss-Kronrod integration rules for Cauchy principal value integrals. Math. Comput. **41**, 63 (1983)

41. A. Iserles, S.P.Nørsett, On quadrature methods for highly oscillatory integrals and their implementation. BIT Numer. Math. **44**, 755 (2004); A. Iserles, S.P. ørsett, Efficient quadrature of highly oscillatory integrals using derivatives. Proc. R. Soc. A **461**, 1383 (2005)
42. J. Vanbiervliet, Numerical methods for highly oscillatory integrals and integral equations. MSc Thesis (KU Leuven, Leuven, 2005)
43. A. Iserles, On the numerical quadrature of highly-oscillating integrals I: Fourier transforms. IMA J. Num. Anal. **24**, 365 (2004); A. Iserles, On the numerical quadrature of highly-oscillating integrals II: Irregular oscillators. IMA J. Num. Anal. **25**, 25 (2005)
44. N.J. Higham, Fast solution of Vandermonde-like systems involving orthogonal polynomials. IMA J. Num. Anal. **8**, 473 (1988); Å. Björck, V. Pereyra, Solution of Vandermonde systems of equations. Math. Comput. **24**, 893 (1970); D. Calvetti, L. Reichel, Fast inversion of Vandermonde-like matrices involving orthogonal polynomials. BIT Numer. Math. **33**, 473 (1993)
45. S. Xiang, Efficient quadrature for highly oscillatory integrals involving critical points. J. Comput. Appl. Math. **206**, 688 (2007)
46. S. Xiang, Numerical analysis of a fast integration method for highly oscillatory functions. BIT Numer. Math. **47**, 469 (2007)
47. T. Sauter, Computation of irregularly oscillating integrals. Appl. Num. Math. **35**, 245 (2000); T. Sauter, *Integration of highly oscillatory functions*. Comp. Phys. Comm. **125**, 119 (2000)
48. T. Ooura, M. Mori, The double exponential formula for oscillatory functions over the half infinite interval. J. Comput. Appl. Math. **38**, 353 (1991)
49. T. Ooura, M. Mori, A robust double exponential formula for Fourier-type integrals. J. Comput. Appl. Math. **112**, 229 (1999)
50. T. Ooura, A double exponential formula for the Fourier transforms. Publ. RIMS Kyoto Univ. **41**, 971 (2005)
51. E. Denich, P. Novati, Some notes on the trapezoidal rule for Fourier type integrals. Appl. Numer. Math. **198**, 160 (2024)
52. X. Luo, P.V. Shevchenko, A short tale of long tail integration. Numer. Algor. **56**, 577 (2011)
53. I.P. Omelyan, V.B. Solovyan, Improved cubature formulas of high degree of exactness for the square. J. Comput. Appl. Math. **188**, 190 (2006)
54. L.N. Trefethen, Cubature, approximation, and isotropy in the hypercube. SIAM Rev. **59**, 469 (2017)
55. J.R. Van Zandt, Efficient cubature rules. Electron. Trans. Numer. Anal. **51**, 219 (2019)
56. R. Orive, J.C. Santos-León, M.M. Spalević, Cubature formulas for the Gaussian weight. Some old and new rules. Electron. Trans. Numer. Anal. **53**, 426 (2020)
57. J. Glaubitz, Stable high-order cubature formulas for experimental data. J. Comput. Phys. **447**, 110693 (2021)
58. https://nines.cs.kuleuven.be/research/ecf/ecf.html
59. R. Cools, P. Rabinowitz, Monomial cubature rules since "Stroud": a compilation. J. Comput. Appl. Math. **48**, 309 (1993)
60. R. Cools, Monomial cubature rules since "Stroud": a compilation—Part 2. J. Comput. Appl. Math. **112**, 21 (1999)
61. R. Cools, Constructing cubature formulae: the science behind the art. Acta Numerica **6**, 1 (1997)
62. CUBPACK, https://nines.cs.kuleuven.be/software/CUBPACK
63. R. Cools, A. Haegemans, Algorithm 824: CUBPACK: a package for automatic cubature; framework description. ACM Trans. Math. Softw. **29**, 287 (2003)
64. A.R. Krommer, C.W. Überhuber, Construction of cubature formulas, in *Computational Integration* (SIAM, Philadelphia, 1998), pp. 155–165
65. A.C. Genz, A.A. Malik, Remarks on algorithm 006: an adaptive algorithm for numerical integration over an N-dimensional rectangular region. J. Comput. Appl. Math. **6**, 295 (1980)
66. J. Berntsen, T.O. Espelid, A. Genz, An adaptive algorithm for the approximate calculation of multiple integrals. ACM. Trans. Math. Softw. **17**, 437 (1991)

67. L.F. Shampine, Vectorized adaptive quadrature in MATLAB. J. Comput. Appl. Math. **211**, 131 (2008)
68. L.F. Shampine, MATLAB program for quadrature in 2D. Appl. Math. Comput. **202**, 266 (2008)
69. https://github.com/stevengj/cubature
70. Cuba—A library for multidimensional numerical integration. https://feynarts.de/cuba/
71. A.G. Ramm, *Inverse Problems. Mathematical and Analytical Techniques with Applications to Engineering* (Springer, New York, 2005)
72. R. Qu, A new approach to numerical differentiation and integration. Math. Comput. Modell. **24**, 55 (1996)
73. M. Hanke, O. Scherzer, Inverse problems light: numerical differentiation. AMS Monthly **108**, 512 (2001)
74. F. Jauberteau, J.L. Jauberteau, Numerical differentiation with noisy signal. Appl. Math. Comput. **215**, 2283 (2009)
75. J. Cullum, Numerical differentiation and regularization. SIAM J. Numer. Anal. **8**, 254 (1971)
76. R.S. Anderssen, P. Bloomfield, Numerical differentiation procedures for non-exact data. Numer. Math. **22**, 157 (1974)
77. R.S. Anderssen, F.R. de Hoog, Finite difference methods for the numerical differentiation of non-exact data. Computing **33**, 259 (1984)
78. I. Knowles, R. Wallace, A variational method for numerical differentiation. Numer. Math. **70**, 91 (1995)
79. Z. Zhao, Z. Meng, G. He, A new approach to numerical differentiation. J. Comput. Appl. Math. **232**, 227 (2009)
80. H. Xu, J. Liu, Stable numerical differentiation for the second order derivatives. Adv. Comput. Math. **33**, 431 (2010)
81. M. Schmid, D. Rath, U. Diebold, Why and how Savitzky-Golay filters should be replaced. ACS Meas. Sci. Au **2**, 185 (2022)
82. F. van Breugel, J.N. Kutz, B.W. Brunton, Numerical differentiation of noisy data: a unifying multi-objective optimization framework. IEEE Access **8**, 196865 (2022)
83. A.G. Ramm, A. Smirnova, Stable numerical differentiation: when is it possible? J. Korean Soc. Indl. Appl. **7**, 1 (2003)
84. A.G. Ramm, Estimates of derivatives of random functions, I. J. Math. Anal. Appl. **102**, 244 (1984)
85. T.L. Miller, A.G. Ramm, Estimates of derivatives of random functions, II. J. Math. Anal. Appl. **110**, 429 (1985)
86. A.G. Ramm, A. Smirnova, On stable numerical differentiation. Math. Comput. **70**, 1131 (2001)
87. http://www.holoborodko.com/pavel/numerical-methods/
88. S. Ahn, U.J. Choi, A.G. Ramm, A scheme for stable numerical differentiation. J. Comput. Appl. Math. **186**, 325 (2006)
89. D.H. Bailey, J.M. Borwein, V. Kapoor, E.W. Weisstein, Ten problems in experimental mathematics. Am. Math. Monthly **113**, 481 (2006)

Chapter 4
Matrix Methods

Abstract Solving systems of linear equations, linear least-square problems and matrix eigenvalue problems is handled by a myriad of freely available and commercial tools. Yet even the most basic operations like the multiplication of two matrices or computing a determinant can be excessively time-consuming and inaccurate if performed recklessly. This Chapter elucidates the crucial aspects of matrix manipulation and standard algorithms for the classes of matrices most frequently encountered by a physicist, with particular attention to the analysis of errors and condition estimates, and to treating algebraic and eigenvalue problems involving sparse matrices. Singular value decomposition is illustrated in the framework of image compression. A separate Section is devoted to pseudospectra, and another to random matrices, with emphasis on Gaussian orthogonal and unitary ensembles and their circular counterparts. The Examples and Problems include the calculation of energy states of particles in one-dimensional and two-dimensional potentials, percolations in random lattices, electric circuits of linear elements, Anderson localization, and spectra of symmetric random matrices.

For physicist's needs, numerical linear algebra is so comprehensively covered by classic textbooks [1, 2] that another detailed description of the algorithms here would be pointless. To a larger extent than in other chapters we wish to merely set up road-signs between approaches to solving systems of linear equations, least-squares problems, and eigenvalue problems. Only Sect. 4.9 on random matrices conveys a somewhat different tone. Above all, we shall pay attention to classes of matrices for which analytic and numerical tools are particularly thoroughly developed and efficient.

To numerically solve problems in linear algebra, never write your own routines! This advice applies even to seemingly straightforward operations like matrix multiplication, and even more so to solving systems of equations $Ax = b$ or eigenvalue problems $Ax = \lambda x$. For these tasks we almost invariably use specialized libraries (see, for example, [3] and Appendix J).

S. Širca and M. Horvat, *Computational Methods in Physics*, Graduate Texts in Physics,
https://doi.org/10.1007/978-3-031-68566-8_4

4.1   Basic Operations

4.1.1   Matrix Multiplication

Classical multiplication of $n \times n$ matrices in the form

$$C_{ij} = \sum_{k=1}^{n} A_{ik} B_{kj} , \quad i, j = 1, 2, \ldots, n , \tag{4.1}$$

where the $A_{ik} B_{kj}$ multiplication is nested into three loops over i, j, and k, requires $F = 2n^3$ multiplications or additions and $M = n^3 + 3n^2$ memory accesses. The optimal ratio is $F/M \approx n/2$ [1], so this method of multiplication is not optimal. (Note that M may depend on processor architecture, peculiarities of the programming language, implementation, and compiler.) In standard libraries BLAS3 or LAPACK (see Appendix J) multiplication is realized in block form that also requires $F = 2n^3$ arithmetic operations, but has a better asymptotic ratio F/M for large n. For all forms of direct multiplication in floating-point arithmetic we have the estimate

$$\mathrm{fl}(AB) = AB + E , \quad |E| \le n \frac{\varepsilon_{\mathrm{M}}}{2} |A| \, |B| + \mathcal{O}(\varepsilon_{\mathrm{M}}^2) , \tag{4.2}$$

where $|A|$ means $(|A|)_{ij} = |A_{ij}|$. We also have

$$\|\mathrm{fl}(AB) - AB\|_1 \le n \frac{\varepsilon_{\mathrm{M}}}{2} \|A\|_1 \|B\|_1 + \mathcal{O}(\varepsilon_{\mathrm{M}}^2) . \tag{4.3}$$

Faster algorithms exist, e.g. Strassen's [4] that splits the matrices A and B into smaller blocks which are then recursively multiplied and summed. The asymptotic cost of the algorithm is $4.70 \, n^{2.81}$, so it really starts to soar when used with matrices of dimensions n in the hundreds or thousands. In some cases, Strassen's method may exhibit instabilities, but robust versions are contained in the fast version BLAS3 [5] and we should use them if speed is our absolute priority. For optimally fast multiplication we need to understand the connection between hardware and software; see Fig. 4.1 and [6]. We can see in the anatomically detailed paper [7] just how deep this knowledge should be in order to design the best algorithms. For Strassen's multiplication the estimate (4.3) still applies while (4.2) is *almost always* true (see [2] and Chap. 23 in [8]).

The currently lowest asymptotic cost is claimed by the Coppersmith–Winograd algorithm [9] which is of order $\mathcal{O}(n^{2.38})$, but the error constant in front of $n^{2.38}$ is so large that the algorithm becomes useful only at very large n. A standard version is described in [10] and the best adaptive implementation in [11]. The theoretical lower limit of the numerical cost for any matrix multiplication algorithm is, of course, $2n^2$, since each element of A and B needs to be accessed at least once. See also [12–14].

Fig. 4.1 Strassen's multiplication of 8×8 matrices. The diagrams show the elements of A and B which are accessed in memory in order to form the elements of C. For example, to get the C_{18} element (upper right corners) we need A_{1k} ($1 \le k \le 8$) and B_{k8} ($1 \le k \le 8$) as in classical multiplication (4.1), but a different pattern for other elements

4.1.2 Computing the Determinant

The determinant of a matrix A or its permutation PA should never be computed with the school-book method via sub-determinants. Rather, we use LU decomposition (4.7) and read off the determinant from the diagonal elements of U:

$$\det(PA) = (-1)^{\det(P)} \prod_{i=1}^{n} U_{ii} \, , \quad \det(P) = \text{order of permutation} \, .$$

4.2 Systems of Linear Equations

This Section deals with methods to solve systems of linear equations $Ax = b$, where A is a $n \times n$ matrix and b and x are vectors of dimension n. The routines from good linear algebra libraries (e.g. LAPACK) allow us to solve the system for r right-hand sides simultaneously (we are then solving $AX = B$ where B and X are $n \times r$ matrices). Solving $Ax = b$ is equivalent to "computing the inverse of A" although A^{-1} almost never needs to be explicitly known or computed. Table 4.1 summarizes some of the most widely used routines.

Table 4.1 A collection of double-precision routines to solve systems of linear equations $Ax = b$ from the LAPACK, GSL, NUMERICAL RECIPES (third edition for C++) libraries and the `scipy.linalg` module for Python, for different types of $n \times n$ matrices A. The routines from the LAPACK library and Python allow the system to be solved simultaneously for r right-hand sides, while in GSL and NR3E one is limited to $r = 1$. The k_l and k_u denote the number of sub-diagonals and super-diagonals in banded matrices ($k_\mathrm{l} = k_\mathrm{u} = k$ for symmetric matrices). Many routines from LAPACK are also available in single precision (first letter S instead of D) for real variables, as well as in double and single precision for complex variables (initial letters Z or C). For details, see [3]. The GSL library and SciPy also contain complex versions of routines

Matrix type	Numerical cost	LAPACK	GSL	NR3E	Python
General	$\frac{2}{3}n^3 + 2n^2 r$	DGESV DGESVX	gsl_linalg_LU_decomp gsl_linalg_LU_solve	LUdcmp	.solve
General banded	$\mathcal{O}(nk_\mathrm{l}(k_\mathrm{l} + k_\mathrm{u})$ $+n(2k_\mathrm{l} + k_\mathrm{u})r)$ if $n \gg k_\mathrm{l}, k_\mathrm{u}$	DGBSV DGBSVX	/	Bandec	.solve_banded
General tridiagonal	$\mathcal{O}(nr)$	DGTSV DGTSVX	/	tridag	/
Symmetric indefinite	$\frac{1}{3}n^3 + \mathcal{O}(n^2 r)$	DSYSV DSYSVX	/	/	/
Symmetric pos. definite	$\frac{1}{3}n^3 + \mathcal{O}(n^2 r)$	DPOSV DPOSVX	gsl_linalg_cholesky_decomp gsl_linalg_cholesky_solve	Cholesky	.cho_factor .cho_solve
Symm. banded pos. definite	$\mathcal{O}(n(k + 1)^2$ $+nkr)$ if $n \gg k$	DPBSV DPBSVX	/	/	/
Symm. tridiag. pos. definite	$\mathcal{O}(nr)$	DPTSV DPTSVX	gsl_linalg_solve_symm_tridiag	/	/
Toeplitz	$\mathcal{O}(n^2)$ $\mathcal{O}(n \log^2 n)$ [18]	/	/	toeplz	.solve_toeplitz
Vandermonde	$\mathcal{O}(n^2)$ $\mathcal{O}(n \log^2 n)$ [21–23]	/	/	vander	/

4.2.1 Analysis of Errors

Numerical errors in solving the problem $Ax = b$ can be analyzed in two ways. One way is to observe the sensitivity of the solution x to small perturbations of A or b. If the vector $\widehat{x}$ solves the perturbed equation $(A + \delta A)\widehat{x} = b + \delta b$ where $\|\delta A\| \leq \varepsilon \|A\|$, $\|\delta b\| \leq \varepsilon \|b\|$, and $\varepsilon \|A^{-1}\| \|A\| < 1$, the deviation $x - \widehat{x}$ from the exact solution can be bounded as

$$\frac{\|x - \widehat{x}\|}{\|\widehat{x}\|} \leq \frac{\varepsilon}{1 - \varepsilon \|A^{-1}\| \|A\|} \left\{ \|A^{-1}\| \|A\| + \frac{\|A^{-1}\| \|b\|}{\|x\|} \right\} \leq \frac{2\varepsilon \kappa(A)}{1 - \varepsilon \kappa(A)}, \quad (4.4)$$

where $\kappa(A) \equiv \|A^{-1}\| \|A\|$ is the *condition number* of A with respect to matrix inversion (see Sect. 4.2.6). Matrices with large $\kappa(A)$ are *ill-conditioned*, those with small $\kappa(A)$ are *well-conditioned*. The value of the condition number depends on the norm in which it is measured. For example, in the Euclidean norm we have $\kappa_2(A) = \|A^{-1}\|_2 \|A\|_2 = \sigma_{\max}(A)/\sigma_{\min}(A)$ where $\sigma_{\max}(A)$ is the largest and $\sigma_{\min}(A)$ the smallest singular value of A (see Sect. 4.5.2). For non-singular A the inverse value of the condition number $1/\kappa_p(A)$ measures the distance (in p-norm) to the nearest singular problem [15],

$$\frac{1}{\kappa_p(A)} = \min_{A + \Delta A \text{ singular}} \frac{\|\Delta A\|_p}{\|A\|_p}.$$

Example A singular matrix has $\det(A) = 0$ but $\det(A) \approx 0$ does not necessarily mean that the problem $Ax = b$ is "almost singular". Let us elucidate this contrast by two examples ([2], p. 82). The upper-triangular $n \times n$ matrix U_n with values of 1 along the main diagonal and -1 on all super-diagonals, has $\det(U_n) = 1$ but $\kappa_\infty(U_n) = n2^{n-1}$. On the other hand, the $n \times n$ matrix $D_n = \mathrm{diag}(10^{-1}, 10^{-1}, \ldots, 10^{-1})$ has a condition number of $\kappa_p(D_n) = 1$ while $\det(D_n) = 10^{-n}$. Ill conditioning of the matrix (large κ) and the determinant being close to zero are therefore, in general, weakly correlated.   ◁

The bound (4.4) may over-estimate the actual error by many orders of magnitude, which does not hurt, but there is also no benefit. The estimate [8]

$$\frac{\|x - \widehat{x}\|_\infty}{\|\widehat{x}\|_\infty} \leq \frac{\| |A^{-1}| \left(|r| + \gamma_{n+1} \left(|A||\widehat{x}| + |b|\right)\right) \|_\infty}{\|\widehat{x}\|_\infty}, \quad (4.5)$$

where

$$r = A\widehat{x} - b, \quad \gamma_n = \frac{n\varepsilon_M}{1 - n\varepsilon_M},$$

is much more useful. To evaluate (4.5) we need to compute the remainder r and the combination $|A||\widehat{x}| + |b|$ (trivial), as well as $|A^{-1}||r|$ (non-trivial, but there is no need to compute A^{-1} explicitly; see Sect. 4.2.6).

In the second way of analyzing errors we think backwards. The smallest ω for which such $|\delta A| \leq \omega |A|$ and $|\delta b| \leq \omega |b|$ exist that $(A + \delta A)\widehat{x} = b + \delta b$, is called the *relative backward error*. In other words, ω is the smallest relative change in any element of A or b such that $\widehat{x}$ represents the exact solution of the perturbed problem. The *lower* limit of the relative backward error is

$$\omega = \max_i \left\{ \frac{|r_i|}{(A\,|\widehat{x}| + b)_i} \right\} . \tag{4.6}$$

If we encounter a division of the type $|r_i|/0$ in (4.6) we assign a value of zero to the fraction if $|r_i| = 0$ or ∞ otherwise. Such cases occur in problems with sparse matrices when for some part of the matrix $A_{ij}x_j = 0$ for $\forall j$ and the denominator in (4.6) becomes zero or small enough to cause an overflow when computing the fraction. Error estimates built into library routines avoid this problem in several ways; see Sect. 7.7 in [8] for details.

4.2.2 Gauss Elimination

The most broadly useful algorithm to solve systems of equations $Ax = b$ with square matrices A without particular structure is the Gauss elimination with partial pivoting (GEPP). The core of the algorithm is the decomposition of the row-permuted matrix PA to a lower-triangular matrix L (values of 1 on the diagonal and non-zero elements at most below the diagonal) and an upper-triangular matrix U (non-zero elements on the diagonal and above it) [2]:

$$PA = LU . \tag{4.7}$$

Partial pivoting enables the algorithm to proceed and prevents the absolute values of the elements of L to exceed unity. The condition number of A (Sect. 4.2.6) therefore remains comparable to those of L and U. In other words, by using partial pivoting we *almost always* [1] avoid numerical instabilities which might render the norm $\|A - LU\|$ comparable to $\|A\|$. The measure for the stability of GEPP is the *pivot growth factor* $g_{pp} \equiv \|U\|_{max}/\|A\|_{max}$. For GEPP we have

$$\|\delta A\|_\infty \leq \tfrac{3}{2} g_{pp} n^3 \varepsilon_M \|A\|_\infty . \tag{4.8}$$

Figure 4.2 (left) shows the upper limits for the relative backward error $\tfrac{3}{2} g_{pp} n^3 \varepsilon_M$ and $\tfrac{3}{2} n \varepsilon_M \| |L| \cdot |U| \|_\infty / \|A\|_\infty$ compared to the true error $\|Ax - b\|_\infty / (\|A\|_\infty \|x\|_\infty)$.

GEPP in double precision is implemented in the routines DGESVX (ZGESVX in complex) from LAPACK, gsl_linalg_LU_decomp/_solve from GSL, LUdcmp from NR3E and scipy.linalg.solve in Python (Table 4.1). In rare cases in which GEPP fails, we use elimination with complete pivoting (GECP) which is slower than GEPP: while GEPP requires $\tfrac{2}{3} n^3 + 2n^2 r$ operations (r is the number of right-hand sides), finding the pivots in GECP implies additional $\mathcal{O}(n^3)$ operations ($\mathcal{O}(n^2)$ per step).

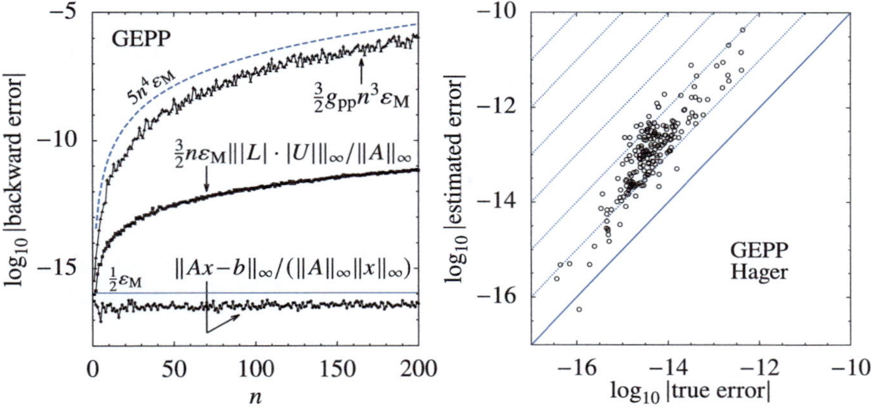

Fig. 4.2 Error estimates in solving the system of equations $Ax = b$ with random matrices by Gauss elimination with partial pivoting. [LEFT] Backward error versus matrix dimension. The pivot growth factor g_{pp} Eq. (4.8) typically does not increase faster than $\mathcal{O}(n)$. [RIGHT] Hager's estimate compared to the true error. The error is over-estimated by about one order of magnitude; rare exceptions to this rough rule are mentioned in [1]

If A has special properties, the numerical cost may be reduced. For general or positive definite symmetric matrices the decomposition is accomplished by one of the variants of the Cholesky method which requires $n^3/3 + \mathcal{O}(n^2 r)$ operations. We can use the `DSYSVX` and `DPOSVX` routines from LAPACK, the `gsl_linalg_cholesky_decomp/_solve` from GSL, `Cholesky` from NR3E or `scipy.linalg.cho_factor/_solve` in Python.

4.2.3 Systems with Banded Matrices

Banded matrices have k_l sub-diagonals and k_u super-diagonals. For general banded matrices with a small bandwidth, $(n \gg k_l, k_u)$ fast methods to solve $Ax = b$ exist, at a cost of $\mathcal{O}(nk_l(k_l + k_u) + n(2k_l + k_u))$ (see Table 4.1). In double precision we recommend the use of the `DGBSVX` routine from LAPACK, `Bandec` from NR3E or `scipy.linalg.solve_banded` for Python.

Tridiagonal matrices are a special case of banded matrices. In physics, they can often be traced to the difference approximation of spatial or temporal differential operators on a mesh, like in Eqs. (10.6) or (10.7). The three values at the neighboring mesh points appear in characteristic triplets in each matrix row. (If the difference involves more points, as in (10.12), we get a five- or more-diagonal matrix.) The algorithms for tridiagonal matrices (`DGTSV` and `DGTSVX` from LAPACK or `tridag` from NR3E) have the numerical cost of $\mathcal{O}(nr)$. Algorithms for symmetric tridiagonal positive-definite matrices (`DPTSV` and `DPTSVX` from LAPACK and `gsl_linalg_solve_symm_tridiag` from GSL) require $8nr$ operations. Such small costs motivate us to try to translate the physics problem such that tridiagonal or banded matrices appear instead of general ones.

4.2.4 Toeplitz Systems

Toeplitz matrices play a key role in problems of the theory of systems control, signal analysis, image processing, and in many related areas [16]. Two typical cases of Toeplitz matrices are

$$
T = \begin{pmatrix}
t_0 & t_1 & t_2 & t_3 & \cdots \\
t_{-1} & t_0 & t_1 & t_2 & \cdots \\
t_{-2} & t_{-1} & t_0 & t_1 & \cdots \\
t_{-3} & t_{-2} & t_{-1} & t_0 & \cdots \\
\vdots & \vdots & \vdots & \vdots & \ddots
\end{pmatrix} , \quad
T_k = \begin{pmatrix}
1 & t_1 & \cdots & t_{k-2} & t_{k-1} \\
t_1 & 1 & & & t_{k-2} \\
\vdots & & 1 & & \vdots \\
t_{k-2} & & & 1 & t_1 \\
t_{k-1} & t_{k-2} & \cdots & t_1 & 1
\end{pmatrix} ,
$$

where T_k is positive definite. The common property of Toeplitz matrices is their insensitivity to a shift along a symmetry axis of the matrix. The physics background of this invariance are temporal or spatial shifts or displacements. Due to the characteristic appearance of these matrices we say that they possess *displacement structure* [17]. The most relevant problems involving Toeplitz matrices are the Yule–Walker equation $T_n x = -(t_1, t_2, \ldots, t_n)^\mathrm{T}$ in the theory of linear prediction (Sect. 7.7) and the solution of a general system $Tx = b$. Spatial derivatives in spectral methods for partial differential equations can also be represented by Toeplitz matrices (look carefully at the structure of the matrix (12.11) in Sect. 12.1).

Toeplitz systems can be solved by methods with a numerical cost of $\mathcal{O}(n^2)$. For general purposes, we recommend the use of the `toeplz` routine from NR3E or `scipy.linalg.solve_toeplitz` for Python. Super-fast $\mathcal{O}(n \log^2 n)$ Toeplitz methods exist [18]. See also [19].

4.2.5 Vandermonde Systems

A Vandermonde matrix $V \in \mathbb{R}^{(n+1)\times(m+1)}$ is defined by the elements $V_{ij} = (x_j)^i$, where $0 \le i \le m, 0 \le j \le n$. We encounter such matrices in two types of problems: interpolation and quadrature. They are illustrated by the following two examples.

Example Assume we have a set of measurements $\{(x_j, y_j)\}_{j=0}^n$, the physics background of which is the function $y(x) = \sin(2\pi x) + 0.002 \cos(100x)$ (a sine-wave with a tiny modulation or noise). We would like to interpolate the data by a polynomial $p(x) = a_n x^n + a_{n-1} x^{n-1} + \cdots + a_0$ such that $p(x_j) = y_j$. We calculate the vector of coefficients $a = (a_0, a_1, \ldots, a_n)^\mathrm{T}$ by solving the system

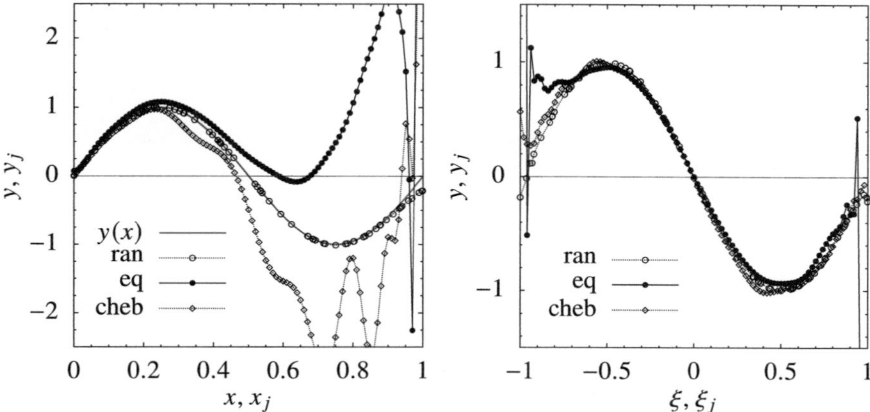

Fig. 4.3 The quality of the polynomial interpolation for different symmetries and choices of x_j. [LEFT] Interpolation through $\{(x_j, y(x_j))\}_{j=0}^{80}$ with randomly (ran) or uniformly (eq) distributed $x_j \in [0, 1]$. [RIGHT] Interpolation through $\{(\xi_j, y((\xi_j + 1)/2))\}_{j=0}^{80}$ with points ξ_j on a symmetric interval $[-1, 1]$. Also shown are the interpolation polynomials in the case when ξ_j are the roots of the Chebyshev polynomial of degree 80 (cheb)

$$V^{\mathrm{T}} a = y, \quad V = \begin{pmatrix} 1 & 1 & \cdots & 1 \\ x_0 & x_1 & \cdots & x_n \\ \vdots & \vdots & \ddots & \vdots \\ x_0^n & x_1^n & \cdots & x_n^n \end{pmatrix}, \quad y = \begin{pmatrix} y_0 \\ y_1 \\ \vdots \\ y_n \end{pmatrix}, \tag{4.9}$$

where V is a square matrix. We sample the function y at $n + 1$ randomly chosen points x_j from the interval $[0, 1]$ and solve the system (4.9) by Gauss elimination. The points are shown by the $\circ$ symbols and the corresponding polynomial by the (ran) curve in Fig. 4.3 (left). If we use equidistant points ($\bullet$) we get the (eq) curve, and with the Chebyshev nodes ($\diamond$) we get the (cheb) curve in the same Figure.

On the non-symmetric interval $[0, 1]$ the interpolation works well close to the origin, but fails elsewhere. The reason for this is the imprecise calculation of the polynomial coefficients due to the large condition number $\kappa_2(V)$. By choosing the Chebyshev points x_j, $\kappa_2(V)$ can be greatly reduced, but interpolation is still poor. Its reliability can be enhanced by mapping the points $x_j \in [0, 1]$ to the symmetric interval $[-1, 1]$ by using $\xi_j = 2x_j - 1$ (Fig. 4.3 (right)). The randomly distributed points also work well, but this should not be a taken as a rule. ◁

Example The second class of problems with Vandermonde matrices involves quadrature. If we know the sums $\{s_i\}_{i=0}^{n}$ of a quadrature formula $s_i = \sum_j w_j x_j^i$, the weights w_j are determined by solving $V w = s$. Let us choose the points $\{x_j\}_{j=0}^{n}$ from $[a, b]$ such that $a \le x_0 < x_1 < \ldots < x_{n-1} < x_n \le b$. We wish to determine such weights in the equation

$$\int_a^b f(x)\,\mathrm{d}x \approx \sum_{j=0}^{n} w_j f(x_j)\,,$$

that the formula will be exact for polynomials of degree $n - 1$ at least, so

$$
\begin{aligned}
f(x) &= 1 &&: & w_0 \quad &+ w_1 \quad &+ \cdots + w_n \quad &= \int_a^b \mathrm{d}x \quad &= s_0\,, \\
f(x) &= x &&: & w_0 x_0 \quad &+ w_1 x_1 \quad &+ \cdots + w_n x_n \quad &= \int_a^b x\,\mathrm{d}x \quad &= s_1\,, \\
&\vdots && & \vdots \quad & \vdots \quad & \vdots \quad & \vdots \quad & \vdots \\
f(x) &= x^{n-1} &&: & w_0 x_0^{n-1} \quad &+ w_1 x_1^{n-1} \quad &+ \cdots + w_n x_n^{n-1} \quad &= \int_a^b x^{n-1}\,\mathrm{d}x \quad &= s_n\,.
\end{aligned}
$$

The system is already in the form $Vw = s$ and can be solved, for example, by using the `vander` routine from the NR3E library.   ◁

Vandermonde systems are extremely ill-conditioned; the condition number $\kappa_2(V)$ grows exponentially with the matrix size. Even the lower limit of κ is known [20]. The algorithms to solve systems of the form $V^{\mathrm{T}}a = y$ and $Vw = s$ differ technically, but they have the same numerical cost. Systems of both types are best solved by Gauss elimination with complete pivoting which requires $\mathcal{O}(n^3)$ operations. Some Vandermonde systems can be solved by methods requiring only $\mathcal{O}(n^2)$ or even $\mathcal{O}(n \log^2 n)$ operations [21–23].

4.2.6   Condition Estimates for Matrix Inversion

The condition number of A corresponding to the sensitivity of $Ax = b$ to perturbations $(A + \delta A)\widehat{x} = b + \delta b$ depends on the type of these perturbations and on the matrices and vectors used in comparisons to δA and δb [8]. The error in the $\|\cdot\|_\infty$ norm, as in (4.4), corresponds to the condition number

$$\kappa_\infty(A) \equiv \|A^{-1}\|_\infty \|A\|_\infty\,.$$

For the error (4.6) with the $\|\cdot\|_\infty$ norm, the appropriate condition numbers are

$$\mathrm{cond}(A, x) = \| |A^{-1}||A||x| \|_\infty / \|x\|_\infty \quad \text{or} \quad \mathrm{cond}(A) = \| |A^{-1}||A| \|_\infty \le \kappa_\infty(A)\,.$$

Regardless of the definition we must obtain $\|A^{-1}\|$ or $\| |A^{-1}|\xi \|$ for $\xi \in \mathbb{R}^n$ in order to compute the condition number, and we would like to do this as precisely as possible without explicitly calculating the inverse A^{-1} which in general requires $\mathcal{O}(n^3)$ operations. Several *condition estimators* exist with a numerical cost of $\mathcal{O}(n^2)$.

Among the most popular is the Hager's estimator [24] which is simple to code and suffices for almost all uses. Its more refined form [25] is used in the LAPACK routines xyyCON and in the routines ending in the letter X from Table 4.1. This estimator returns the inverse value of the condition number RCOND. The comparison of the value of the estimator and the true error in solving a multitude of systems involving random matrices is shown in Fig. 4.2 (right). In Python the condition number can be calculated by using `numpy.linalg.cond`. See also [26] and Chap. 15 in [8].

Balancing If the problem $Ax = b$ is ill-conditioned, a more precise solution can be obtained by first solving the *scaled* or *balanced* problem $(D_1^{-1}AD_2)y = D_1^{-1}b$ where D_1 and D_2 are diagonal, and then calculate $x = D_2y$. Balancing may do more harm than good (a nice example is given in [27]), and there is no general recipe. In addition, balancing may be used to precondition matrices in linear and eigenvalue solvers [28]. Practical routines can be found in LAPACK [3].

4.3   Solving $Ax = b$ with sparse matrices

A matrix is considered to be sparse as opposed to "full" or "dense" if it contains enough zeros that this very fact can be—or, rather, must be—exploited as the key circumstance allowing us to store the matrix and solve the system $Ax = b$ as economically as possible. For example, the matrices displayed in Fig. 11.2 stemming from discrete formulations of problems involving elliptic partial differential equations are sparse, with the added benefit of being structured (in these prototypical cases, tridiagonal and banded). The matrix shown in Fig. 7.6 (right) is also sparse: as it lacks any symmetry or structure, one would be forced to use a dense-matrix method to solve the corresponding linear system even though it will be mostly crunching zeros.

For large sparse matrices, the dense methods of Sect. 4.2 become too costly in terms of storage ($\mathcal{O}(n^2)$) and time (in general $\mathcal{O}(n^3)$). "Large" stands for, say, $n \approx 10^6$, a size one easily cooks up in multi-dimensional problems. (The scope of "large" tends to move up by a factor 10 to 100 in n about every 10 years.) By using sparse methods one strives to solve $Ax = b$ with storage and time proportional to $\mathcal{O}(n) + \mathcal{O}(\tau)$, where τ is the number of non-zeros in A. To solve such systems, two classes of methods stand at our disposal: direct and iterative.

4.3.1   Direct Methods

In direct methods [29] a sparse matrix A is factorized into a product of other sparse matrices through which the system becomes easier to solve. The factorizations discussed below are all based on classical Gaussian elimination but the $\mathcal{O}(n) + \mathcal{O}(\tau)$ goal adds an immense layer of complexity. Direct sparse methods typically consist of four distinct phases: a *pre-ordering* phase in which anything from simple permutations to sophisticated graph-theory tools is used to rearrange the original matrix to a more condensed or more symmetric form; a *symbolic analysis* phase in which the matrix structure is analyzed in order to predict and reduce the storage and run-time cost of the subsequent factorization, as well as to provide it with the appropriate sparse data structures; the *factorization* phase in which the actual numerical factorization is performed; and a *solution* phase in which the computed matrix factors are used to solve the system.

For a general (non-symmetric) $n \times n$ sparse matrix A one usually computes its LU decomposition $A = LU$, where L is unit lower-triangular and U is upper-triangular. If A is symmetric indefinite, it is typically factorized as $A = LDL^{\mathrm{T}}$, where L is unit lower-triangular and D is block-diagonal. If A is symmetric and positive-definite, the standard approach is the Cholesky factorization $A = LL^{\mathrm{T}}$, where L is lower-triangular. If A is a $m \times n$ rectangular matrix, with $m \geq n$, we compute its QR decomposition $A = QR$, where Q is a $m \times m$ orthogonal matrix and R is a $m \times n$ upper-trapezoidal matrix (upper triangular if $m = n$).

The main concern is that the factor matrices in any decomposition may no longer be sparse: factorization should engender as few additional non-zeros (the so-called *fill-in* values) as possible. Consider the following symmetric positive-definite sparse matrix A and its Cholesky factor L:

$$
A = \begin{pmatrix} 4 & -1 & -1 & -1 \\ -1 & 3 & 0 & 0 \\ -1 & 0 & 2 & 0 \\ -1 & 0 & 0 & 1 \end{pmatrix}, \quad
L = \begin{pmatrix} 2 & 0 & 0 & 0 \\ -0.5 & 1.6583 & 0 & 0 \\ -0.5 & -0.1508 & 1.3143 & 0 \\ -0.5 & -0.1508 & -0.2075 & 0.8272 \end{pmatrix}.
$$

Obviously L is dense, having picked up three fill-ins L_{32}, L_{42} and L_{43}. An efficient way to avoid excessive fill-in is permutation: if A is symmetric, one can use a permutation matrix P to construct PAP^{T} instead of A and solve the system

$$
PAP^{\mathrm{T}}y = Pb \,,
$$

followed by $x = P^{\mathrm{T}}y$. (If A is non-symmetric, use $A \rightarrow PAQ$.) In the above example, a 4×4 permutation matrix P with $P_{12} = P_{23} = P_{34} = P_{41} = 1$ and all other P_{ij} equal to zero does the job, as it produces the Cholesky factor which preserves the number of zero entries:

$$
L = \begin{pmatrix} 1.7321 & 0 & 0 & 0 \\ 0 & 1.4142 & 0 & 0 \\ 0 & 0 & 1 & 0 \\ -0.5774 & -0.7071 & -1 & 1.4720 \end{pmatrix},
$$

Permutation is just one of the basic tools that sparse packages use to produce a good fill-reducing ordering. Finding a permutation matrix that leads to the sparsest factorization and minimal time to compute it is very difficult, hence various heuristic approaches are used; see Chap. 8 of [29].

A discussion of specific properties of individual direct sparse-matrix algorithms and their customizations is beyond the scope of this book. A comprehensive list of available packages and libraries is given in Sect. 13 of [29]; consult also Chap. 6 in [30]. Benchmark comparisons of tools for large sparse symmetric systems have been compiled by [31, 32]. See also [33, 34].

4.3.2 Iterative Methods Based on Krylov Subspaces

The second choice for solving sparse systems are iterative methods. Their classical representatives are the Jacobi, Gauss–Seidel and SOR (SSOR) methods presented separately in Sect. 11.2 in the context of solving elliptic PDEs. Here we discuss a class of iterative methods based on Krylov subspaces [35].

In Krylov methods we attempt to solve the system $Ax = b$ with $A \in \mathbb{R}^{n \times n}$ by extracting its approximate solution $\widetilde{x}$ from a specially devised subspace of $\mathbb{R}^n$. We achieve this goal by constructing two m-dimensional subspaces of $\mathbb{R}^n$, the *subspace of candidate approximants* $\mathcal{K}_m$ and the *subspace of [orthogonality] constraints* $\mathcal{L}_m$. Let x_0 be the initial approximation of the exact solution x_* and let $r_0 = b - Ax_0$ be the corresponding initial residual. We then seek the approximate solution by iteratively picking trial vectors δ from $\mathcal{K}_m$ and ensuring that the obtained residual is orthogonal to all vectors from $\mathcal{L}_m$:

$$\widetilde{x} = x_0 + \delta , \quad \delta \in \mathcal{K}_m ,$$
$$b - A\widetilde{x} = r_0 - A\delta \perp w , \quad \forall w \in \mathcal{L}_m .$$

One usually considers two options: either $\mathcal{L}_m = \mathcal{K}_m$ (for symmetric positive definite matrices A) or $\mathcal{L}_m = A\mathcal{K}_m$ (for general A). Krylov methods of the first type, also known as *error projection methods*, operate by calculating a series of approximate vectors $\widetilde{x}$ such that the error $\widetilde{e} = x_* - \widetilde{x}$ of the new approximant diminishes with respect to the initial error $e_0 = x_* - x_0$ as $\|\widetilde{e}\|_A \leq \|e_0\|_A$, where $\|z\|_A = (z^{\mathrm{T}} Az)^{1/2}$. The methods of the second type—the *residual projection methods*—ensure that the new residual $\widetilde{r} = b - A\widetilde{x}$ is reduced with respect to the initial one, $\|\widetilde{r}\|_2 \leq \|r_0\|_2$.

The two most versatile representatives of these approaches are the conjugate gradients (CG) method and the generalized minimum-residual (GMRES) method, which are described in the following.

Conjugate Gradients (CG) Method

The conjugate gradients method [36, 37] is a Krylov subspaces projection method suitable for sparse symmetric positive definite matrices A, making use of

$$\mathcal{L}_m = \mathcal{K}_m = \mathrm{span}\{r_0, Ar_0, A^2 r_0, \ldots, A^{m-1} r_0\} .$$

In essence, the method seeks an approximate solution x_m from the affine space $x_0 + \mathcal{K}_m$ by imposing the orthogonality condition $b - Ax_m \perp \mathcal{K}_m$ at each step. By doing this, the iterative projection process effectively reduces the value of the function $f(x) = \frac{1}{2}x^{\mathrm{T}}Ax - bx$ which is minimal when $\nabla f = Ax - b = 0$, i.e. when $Ax = b$. The minimum of f is approached by a sequence of search directions and corrections to x_m along them such that x_{m+1} minimizes f over all previously determined directions. The convergence rate estimate of the CG method is given by the relation

$$\|x_* - x_m\|_A \leq 2 \left(\frac{\sqrt{\kappa} - 1}{\sqrt{\kappa} + 1} \right)^m \|x_* - x_0\|_A \, ,$$

where x_m is the approximate solution obtained at the mth step of the iteration and $\kappa = \lambda_{\max}/\lambda_{\min}$ is the ratio of the largest and the smallest eigenvalue of A. A typical stopping condition for the iteration may be $\|Ax - b\|/\|b\| \leq \mathrm{TOL}$ or $\|M^{-1}(Ax - b)\|/\|M^{-1}b\| \leq \mathrm{TOL}$ if a preconditioner M is used (see Sect. 4.3.3). Consult Sects. 6.7 and 6.11 of [35] for details of implementation and possible runtime issues. See also Sect. 13.3.4.

An "upgrade" of the CG method is the biconjugate gradients (BCG) method [38, 39] which solves not only the original system $Ax = b$ but also its dual involving A^{T}. The matrix A may be general (non-symmetric) and need not be positive definite. The BCG method is implemented in the `linbcg` routine of [40].

Generalized Minimum-Residual (GMRES) Method

The generalized minimum-residual method [41] is a projection method devised for arbitrary square matrices A. It operates in the Krylov subspaces

$$\mathcal{K}_m = \mathrm{span}\{v_1, Av_1, A^2v_1, \ldots, A^{m-1}v_1\} \tag{4.10}$$

and $\mathcal{L}_m = A\mathcal{K}_m$ with $v_1 = r_0/\|r_0\|_2$. Similarly to the CG method, the successive iterates x_m keep on reducing the norm of the residual r over all vectors in the subspace $x_0 + \mathcal{K}_m$, effectively minimizing the function $g(x) = r^{\mathrm{T}}r = \|Ax - b\|_2^2$ and thus solving $Ax = b$ to required precision. Assume that A is diagonalizable such that $A = X\Lambda X^{-1}$, where $\Lambda = \mathrm{diag}\,(\lambda_1, \lambda_2, \ldots, \lambda_n)$, and define

$$\varepsilon^{(m)} = \min_{p \in \mathbb{P}_n, p(0)=1} \max_{i=1,2,\ldots,n} |p(\lambda_i)| \, .$$

Then the norm of the residual attained in the mth step of the GMRES iteration satisfies the inequality

$$\|r_m\|_2 \leq \kappa_2(X)\varepsilon^{(m)}\|r_0\|_2 \, ,$$

where $\kappa_2(X) = \|X\|_2\|X^{-1}\|_2$ and $\varepsilon^{(m)}$ is a scalar quantity that can be estimated from the spectrum of the matrix A. If the spectrum of A in the complex plane is contained in the ellipse that *excludes the origin* and has the center c, focal distance d and major semi-axis a—see Fig. 4.4 for two possible cases—an upper bound for $\varepsilon^{(m)}$ can be found (see [35], p. 218):

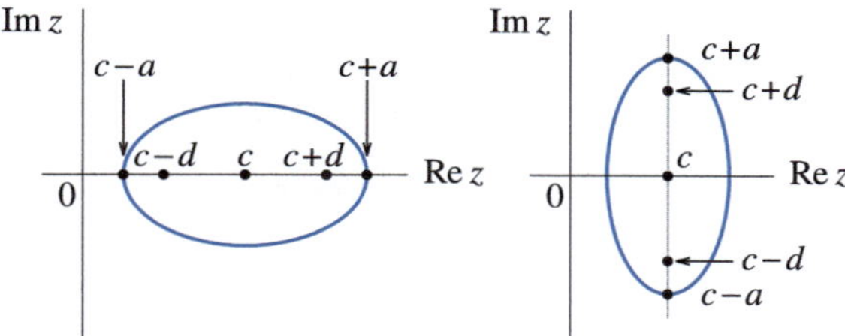

Fig. 4.4 Ellipses containing the spectrum of the matrix A used in estimating the convergence rate of the GMRES method

$$\varepsilon^{(m)} \leq \frac{T_m(a/d)}{T_m(c/d)} \approx \left(\frac{a + \sqrt{a^2 - d^2}}{c + \sqrt{c^2 - d^2}}\right)^m ,$$

where T_k are the Chebyshev polynomials (see p. 267). See Sects. 6.5, 6.8, 6.9 and 6.11 of [35] for implementation details.

4.3.3 Preconditioning in Projection Methods

In terms of convergence speed and robustness, naive formulations of Krylov methods are not significantly better than direct or classical relaxation approaches. The crucial remedy that allows iterative solvers to truly excel is *preconditioning*. (The balancing procedure mentioned on p. 183 is also a kind of preconditioning.) The basic idea is to transform the system $Ax = b$ to a form which might be easier to solve by writing it as

$$M^{-1}Ax = M^{-1}b \tag{4.11}$$

or, by a change of variables $y = Mx$, as

$$AM^{-1}y = b , \quad x = M^{-1}y , \tag{4.12}$$

or even in factorized form as

$$M_L^{-1}AM_R^{-1}y = M_L^{-1}b , \quad x = M_R^{-1}y .$$

Determining the appropriate form and placement of the preconditioning matrix or matrices usually involves some black magic in addition to computational science. The first guide, however, is that M in (4.11) or (4.12), for instance, should be "similar" to A—in the hope that these systems become better conditioned and diagonally dominant so that an iterative method will digest them faster. This suggests at least a trivial possibility: letting M to be the diagonal part of A. The second requirement for M is that the auxiliary system $Mx = b$ should be relatively easy to solve, as this system usually constitutes a part of the preconditioned algorithm. The third desirable property of M is that it preserves the symmetry of the original problem. For details of preconditioning in iterative methods for sparse linear systems refer to Chaps. 9 and 10 of [35] or Chaps. 8 and 9 of [30].

4.4 Solving Matrix Equations

Signal and image processing, certain problems in statistics and system control, studies of stability of linear dynamical systems and many other application areas call for a solution of equations of the form

$$AXD + CXB = W , \tag{4.13}$$

where all symbols denote matrices. Assume that A, B, C and D are square, while the dimensionality of W conforms to that of X (which need not be square). If $C = D = I$, (4.13) is called the *Sylvester equation.* If in addition $B = A^\dagger$, so that

$$AX + XA^\dagger = W ,$$

it is known as the *Lyapunov equation.* These two special cases are by far the most common in practice.

Note that if C and D are both non-singular, (4.13) can be left-multiplied by C^{-1} and right-multiplied by D^{-1} and thus recast in Sylvester form, but this temptation should be resisted: if either C or D are ill-conditioned, such a transformation may lead to instabilities. Instead, one should use dedicated solution methods for each of the equation types. In the following we describe three solution methods for equations involving small- or medium-sized matrices (typically up to $n \approx 10^4$). For computations beyond $n \approx 10^6$ refer to [42].

4.4.1 Sylvester Equations

The solution of the Sylvester equation

$$AX + XB = W , \tag{4.14}$$

where $A \in \mathbb{C}^{n \times n}$, $B \in \mathbb{C}^{m \times m}$ and $W \in \mathbb{C}^{n \times m}$, exists and is unique if and only if the spectra of A and $-B$ are disjunct, i.e. if their intersection is an empty set. Such equations with matrices of small and moderate sizes can be solved robustly and efficiently by the Bartels–Stewart method [43] used, with some modifications, in the LAPACK and SLICOT [44] libraries. The algorithm is:

1. Compute the Schur factorizations $A^\dagger = URU^\dagger$ and $B = VSV^\dagger$ of $A^\dagger$ and B, where R and S are upper-triangular. This is usually accomplished by QR iteration (see Sect. 4.6.1).
2. Solve

$$R^\dagger Y + YS = U^\dagger WV$$

   for Y. Since $R^\dagger$ and S are lower- and upper-triangular, respectively, this matrix system can be solved explicitly element by element, line by line. For example, Y_{11} is given by $R_{11}^\dagger Y_{11} + Y_{11}S_{11} = (U^\dagger WV)_{11}$, then the known value of Y_{11} can be plugged into $R_{11}^\dagger Y_{12} + Y_{11}S_{12} + Y_{12}S_{22} = (U^\dagger WY)_{12}$ to yield Y_{12}, and so on.
3. Calculate $X = UYV^\dagger$.

Large A **or** B If either A or B are large, the Schur decompositions may become too costly in terms of computational time or storage requirements, and a different approach is warranted. Assume that B is "small" (m on the order of 1000) so that its spectral decomposition $B = U\Lambda U^{-1}$ with $\Lambda = \mathrm{diag}\,(\lambda_1, \lambda_2, \ldots, \lambda_m)$ can be computed in a cheap and stable manner. In contrast, let A be "large", such that $n \gg m$. (The opposite case is handled analogously.) Then (4.14) reads

$$AY + Y\Lambda = Z, \quad Y = XU, \quad Z = WU.$$

Each column of Y is obtained by solving a "shifted" linear system

$$(A + \lambda_i I)(Y)_i = (Z)_i, \quad i = 1, 2, \ldots, m,$$

where $(Y)_i$ and $(Z)_i$ denote the ith columns of Y and Z, respectively, by using any chosen method of Sect. 4.2. Finally,

$$X = YU^{-1}.$$

4.4.2 Lyapunov Equations

Lyapunov matrix equations

$$AX + XA^\dagger = W$$

with small- or moderate-sized matrices A are best solved by the Bartels–Stewart method mentioned above, while large-scale problems of this type are better served by a Newton-like procedure derived in [45] and given below. *If all eigenvalues of A have negative real parts,* the iteration

$$\begin{aligned}
A_{i+1} &= \tfrac{1}{2}\left(A_i + A_i^{-1}\right), & A_0 &= A, \\
W_{i+1} &= \tfrac{1}{2}\left(W_i + A_i^{-1}W_i A_i^{-\mathrm{T}}\right), & W_0 &= W,
\end{aligned} \quad i = 0, 1, 2, \ldots$$

converges quadratically to $A_\infty = -I$ and $W_\infty = -2X$, hence

$$X = -\frac{1}{2} W_\infty \, .$$

The method is globally convergent, but may be slow in the initial stages of the iteration. Various techniques of its acceleration are discussed in [42] and in Sect. 5.5 of [46].

4.5 Linear Least-Squares Problem and Orthogonalization

In the previous Section we have discussed systems of the form $Ax = b$ where $A \in \mathbb{R}^{n \times n}$ and $x, b \in \mathbb{R}^n$. In this Section, we are dealing with over-determined systems, in which there are more equations than unknowns, that is

$$Ax = b \, , \quad A \in \mathbb{R}^{m \times n} \, , \quad b \in \mathbb{R}^m \, , \quad x \in \mathbb{R}^n \, , \quad m \geq n \, .$$

Solving such systems can be understood as seeking a vector x that minimizes the remainder $r = Ax - b$ in some vector norm, e.g. $\|r\|_1$, $\|r\|_2$, or $\|r\|_\infty$. Due to its analytic properties (differentiability with respect to x) the $\|\cdot\|_2$ norm is most widely used. The problem of minimizing $\|Ax - b\|_2$ is also called the *linear least-squares (LS) problem*, as we attempt to minimize $[\sum_i r_i^2]^{1/2}$.

In the following we assume that the matrix $A \in \mathbb{R}^{m \times n}$ for $m \geq n$ has full rank, so it has n linearly independent columns. For rank-deficient or nearly rank-deficient matrices the matters become more complicated. For a matrix A with an exactly deficient rank the solution of the least-squares problems is not unique: if $\mathrm{rank}(A) = r < n$, there exists a $(n - r)$-dimensional set of vectors x that minimize $\|Ax - b\|_2$. From this set we have to choose the vector with the smallest norm which may be the only one to possess physical meaning (see Sect. 3.5 in [1], Sect. 5.5 in [2] and Sect. 4.5.3 in the following).

The basic method to solve the linear least-squares problem is to transform it to the *normal system*. We seek a vector x for which the gradient of $f(x) = \|Ax - b\|_2^2 = (Ax - b)^\mathrm{T}(Ax - b)$ is equal to zero, that is, a vector x such that $\partial f / \partial x_i = 0$ for $i = 1, 2, \ldots, n$. A short calculation brings us to the $n \times n$ normal system $A^\mathrm{T} A x = A^\mathrm{T} b$ which has the solution

$$x_{\mathrm{LS}} = (A^\mathrm{T} A)^{-1} A^\mathrm{T} b \, . \tag{4.15}$$

For matrices $A \in \mathbb{R}^{m \times n}$ ($m \geq n$) with full rank the condition number for the least-squares problem is defined as the ratio of the largest and the smallest singular value, $\kappa_2(A) = \sigma_{\max}(A) / \sigma_{\min}(A)$ (see Sect. 4.5.2). The relative error of the vector $\widehat{x}_{\mathrm{LS}}$ computed by solving the normal system with respect to the true vector x_{LS} can be estimated as

$$\frac{\|\widehat{x}_{\text{LS}} - x_{\text{LS}}\|_2}{\|x_{\text{LS}}\|_2} \approx \varepsilon_{\text{M}}\kappa_2(A)^2 . \tag{4.16}$$

Note that this estimate involves the *square* of the condition number. An additional concern is that solving the normal system may be unstable: this means that the computed $\widehat{x}_{\text{LS}}$ does not necessarily minimize the "nearby" least-squares problem $\min_x \|(A + \delta A)x - (b + \delta b)\|_2$ where δA and δb are small perturbations. Still, the normal-system method is recommendable for well-conditioned matrices A as it comes at a numerical cost of just $(m + n/3)n^2$. In the NR3E library it is implemented in the `Fitlin` structure; see also Algorithm 5.3.1 in [2]. When higher precision and better stability are required, we resort to QR decomposition with column pivoting or singular value decomposition (SVD).

4.5.1 The QR decomposition

The QR decomposition is a method to orthogonalize a matrix $A \in \mathbb{R}^{m \times n}$ ($m \geq n$) where A is decomposed to an orthogonal matrix $Q \in \mathbb{R}^{m \times m}$ ($Q^{\text{T}}Q = QQ^{\text{T}} = I$) and an upper-triangular matrix $R \in \mathbb{R}^{m \times n}$ such that $A = QR$. The method with $Q \in \mathbb{R}^{m \times n}$ and $R \in \mathbb{R}^{n \times n}$ [1] is known as the "*thin*" variant of QR [2]. By using the QR decomposition, we hit two birds with one stone. If A has full rank n, its columns span a n-dimensional vector space $\mathcal{R}_n(A)$, and the first n columns of Q represent an orthonormal basis for this space; the decomposition and the computation of the basis vectors in double precision is accomplished by the `DGEQRF` and `DORGQR` routines from LAPACK. In addition, the QR decomposition also yields the solution of the least-squares problem. We see that if the solution of the normal system (4.15) is rewritten as

$$x_{\text{LS}} = (A^{\text{T}}A)^{-1}A^{\text{T}}b = (R^{\text{T}}Q^{\text{T}}QR)^{-1}R^{\text{T}}Q^{\text{T}}b = R^{-1}(R^{\text{T}})^{-1}R^{\text{T}}Q^{\text{T}}b = R^{-1}Q^{\text{T}}b .$$

When A is QR-decomposed, we obtain the solution x_{LS} by solving the upper-triangular system $Rx = Q^{\text{T}}b$ by back-substitution. The whole process is implemented in the LAPACK's routine `DGELS` with improvements [47], the routines `gsl_linalg_QR_decomp/_lssolve` from GSL as well as `scipy.linalg.lstsq` for Python (see Table 4.2). The lion's share of numerical work is hidden in the decomposition itself; the typical cost of these algorithms is $2n^2(m - n/3)$ to solve the least-squares problem, and an equal number of operations to construct the basis vectors for $\mathcal{R}_n(A)$ from Q.

There are other ways to orthogonalize a matrix $A \in \mathbb{R}^{m \times n}$. One of them is the classical Gram–Schmidt (GS) procedure, but it is unstable if the columns of A are only weakly dependent. In such cases, the GS procedure generates columns of Q which are not orthonormal, so that $\|I - Q^{\text{T}}Q\| \gg \varepsilon$, and should not be used! The modified Gram–Schmidt procedure (mGS, Algorithm 5.2.5 in [2]) is much better: it can be used to generate columns of the matrix $Q_1 = [q_1, q_2, \ldots, q_n]$ for which

Table 4.2 A collection of double-precision routines to solve the least-squares problem and its generalizations from the LAPACK, GSL, NUMERICAL RECIPES (third edition for C++) libraries and the `scipy.linalg` module for Python. In all cases we have $A \in \mathbb{R}^{m \times n}$ where $m \geq n$ (over-determined systems). The solution (4.15) by the normal-system methods is implemented in the `Fitlin` routine of NR3E. The solution based on the QR decomposition is implemented in `DGELS` (LAPACK), `gsl_linalg_QR_decomp/_solve` (GSL) and `scipy.linalg.lstsq` (Python). The numerical cost of the LAPACK QR routines for complex variables is about four times higher than the cost of the corresponding routines for real variables. The methods exploiting singular value decomposition (last line in the lower Table) may also be applied to full-rank matrices. The numerical cost of these routines strongly depends on the implementation details. See also caption to Table 4.1

Full-rank matrices, $\mathrm{rank}(A) = \min(m, n)$

Minimize	Constraint	Numerical cost	LAPACK	GSL	NR3E	Python
$\min_x \|Ax - b\|_2$	/	$n^2\left(m + \frac{1}{3}n\right)$	/	/	Fitlin	/
$\min_x \|Ax - b\|_2$	/	$2n^2\left(m - \frac{1}{3}n\right)$	DGELS	gsl_linalg_QR_decomp gsl_linalg_QR_lssolve	/	.lstsq
$\min_x \|Ax - b\|_2$	$Bx = c$	$\leq 2n^2\left(2m + \frac{1}{3}n\right)$	DGGLSE	/	/	/
$\min_x \|y\|_2$	$Ax + By = c$	$2\left(\frac{2}{3}m^3 - \frac{1}{3}n^3\right) + 4nm^2$	DGGGLM	/	/	/
$\min_x \|B^{-1}(c - Ax)\|_2$	B square	$\frac{14}{3}n^3$	DGGGLM	/	/	/

Rank-deficient matrices, $\mathrm{rank}(A) < \min(m, n)$

Minimize	Constraint	Numerical cost	LAPACK	GSL	NR3E	Python
$\min_x \|Ax - b\|_2$	$\min_x \|x\|_2$	$\approx \mathcal{O}(mn^2) + \mathcal{O}(n^3)$ ([2], Sect. 5.4)	DGELSY	gsl_linalg_QRPT_decomp gsl_linalg_QRPT_solve	/	.lstsq
$\min_x \|Ax - b\|_2$	$\min_x \|x\|_2$	$\approx \mathcal{O}(mn^2) + \mathcal{O}(n^3)$ ([2], Sect. 5.5)	DGELSS DGELSD	gsl_linalg_SV_decomp gsl_linalg_SV_solve	SVD Fitsvd	.svd

$$Q_1^\mathsf{T} Q_1 = I_n + E_\mathrm{mGS}\,, \quad \|E_\mathrm{mGS}\|_2 \approx \varepsilon_\mathrm{M} \kappa_2(A)\,. \tag{4.17}$$

If our requirements on orthonormality are very strict, we may use the mGS procedure only in cases when the condition number $\kappa_2(A)$ is small (columns of A well linearly independent). On the other hand, the mGS method is attractive due to its small cost of $\approx 2mn^2$.

The routines mentioned above are based on Householder and Givens transformations which automatically ensure stability. In analogy to solving systems of equations $Ax = b$, the stability of least-squares methods is gauged by the sensitivity of the solution x of the problem $\min_x \|Ax - b\|_2$ with the remainder $r = b - Ax$ to small perturbations of A and b. Let $\widehat{x}$ minimize the perturbed problem, $\widehat{x} = \min_x \|(A + \delta A)x - (b + \delta b)\|_2$, and let $\widehat{r} = (b + \delta b) - (A + \delta A)\widehat{x}$ and

$$\varepsilon = \max\left\{ \frac{\|\delta A\|_2}{\|A\|_2}\,, \frac{\|\delta b\|_2}{\|b\|_2} \right\}\,.$$

(To get a feeling, we may simply set $\varepsilon = \varepsilon_\mathrm{M}$.) Then the following holds [2]:

$$\frac{\|\widehat{x} - x\|_2}{\|x\|_2} < \varepsilon \left[\frac{2\|b\|_2}{\sqrt{\|b\|_2^2 - \rho_\mathrm{LS}^2}} \kappa_2(A) + \frac{\rho_\mathrm{LS}}{\sqrt{\|b\|_2^2 - \rho_\mathrm{LS}^2}} \kappa_2(A)^2 \right] + \mathcal{O}(\varepsilon^2)\,,$$

where $\rho_\mathrm{LS} = \|Ax_\mathrm{LS} - b\|_2$ and we have assumed $\rho_\mathrm{LS} \neq \|b\|_2$. We see that the condition number $\kappa_2(A)$ appears linearly and quadratically in the relative error of the solution of the least-squares problem. If the problem is well-conditioned and the remainders ρ_LS are small, the linear term dominates and the QR decomposition gives more precise results than the normal-system solution with the error (4.16) determined by the *square* of $\kappa_2(A)$. The QR decomposition based on Householder or Givens transformations also relieves us of the annoying presence of $\kappa_2(A)$ in the orthogonalization precision. Instead of (4.17)—see Fig. 4.5—we get

$$Q_1^\mathsf{T} Q_1 = I_n + E_\mathrm{HQR}\,, \quad \|E_\mathrm{HQR}\|_2 \approx \varepsilon_\mathrm{M}\,. \tag{4.18}$$

4.5.2 *Singular Value Decomposition (SVD)*

Orthogonalization of a matrix by singular value decomposition (SVD) [48] is one of the most remarkable methods of linear algebra with a limitless field of applications like data compression, linear least-squares problems, and principal component analysis. If A is a real $m \times n$ matrix, there exist orthogonal matrices

$$U = [u_1, u_2, \ldots, u_m] \in \mathbb{R}^{m \times m} \quad \text{and} \quad V = [v_1, v_2, \ldots, v_n] \in \mathbb{R}^{n \times n}\,,$$

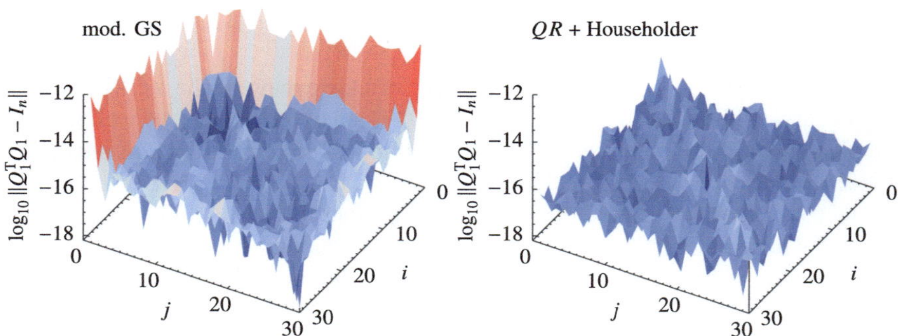

Fig. 4.5 The degree of orthonormality of the basis calculated by orthogonalizing a 30×30 matrix with the elements $A_{ij} = 1 + 0.01R$ where $R \in (-0.5, 0.5)$ is a uniformly distributed random number. The columns of A are weakly linearly dependent, so its condition number is large, $\kappa_2(A) = \sigma_{\max}(A)/\sigma_{\min}(A) \approx 6 \cdot 10^4$. [LEFT] Modified Gram–Schmidt procedure. Due to ill conditioning the error $\|Q_1^{\mathrm{T}} Q_1 - I_n\|$ increases (Eq. (4.17)). [RIGHT] QR decomposition with Householder transformations. The error is essentially constrained by the machine precision (Eq. (4.18))

such that A can be decomposed as

$$A = U\Sigma V^{\mathrm{T}}, \quad \Sigma = \mathrm{diag}(\sigma_1, \sigma_2, \ldots, \sigma_p) \in \mathbb{R}^{m \times n}, \quad p = \min\{m, n\},$$

where $\sigma_1 \geq \sigma_2 \geq \cdots \geq \sigma_p \geq 0$. Like in the QR decomposition, SVD is also known in its "thin" variety in which $A \in \mathbb{R}^{m \times n}$ for $m \geq n$ can be decomposed as $A = U_1 \Sigma_1 V_1^{\mathrm{T}}$ where $U_1 \in \mathbb{R}^{m \times n}$ and $\Sigma_1 = \mathrm{diag}(\sigma_1, \sigma_2, \ldots, \sigma_n) \in \mathbb{R}^{n \times n}$. The vectors u_k (or v_k) are known as left (or right) singular vectors of A, and the σ_k are known as singular values of A. We have $A v_i = \sigma_i u_i$ and $A^{\mathrm{T}} u_i = \sigma_i v_i$ for $1 \leq i \leq \min\{m, n\}$. The ratio of the largest and the smallest singular value determines the condition number of A, measured in the 2-norm,

$$\kappa_2(A) = \|A\|_2 \|A^{-1}\|_2 = \frac{\sigma_{\max}(A)}{\sigma_{\min}(A)}. \tag{4.19}$$

The singular values are equal to the lengths of the semi-axes of the hyper-ellipsoid $\mathcal{E} = \{Ax : \|x\|_2 = 1\}$, and the condition number (4.19) is the ratio of the longest and shortest semi-axis (the elongation) of $\mathcal{E}$. We see this [49] if we represent an element of the unit sphere in $\mathbb{R}^n$ by the expansion $x = x_1 v_1 + \cdots + x_n v_n$ where $\sum_{i=1}^{n} x_i^2 = 1$, and observe its image $Ax = \sigma_1 x_1 u_1 + \cdots + \sigma_k x_k u_k$. The vector x is mapped to $y_1 u_1 + \cdots + y_k u_k$, where $y_i = \sigma_i x_i$, and

$$\frac{y_1^2}{\sigma_1^2} + \frac{y_2^2}{\sigma_2^2} + \cdots + \frac{y_k^2}{\sigma_k^2} = \sum_{i=1}^{k} x_i^2 \leq 1. \tag{4.20}$$

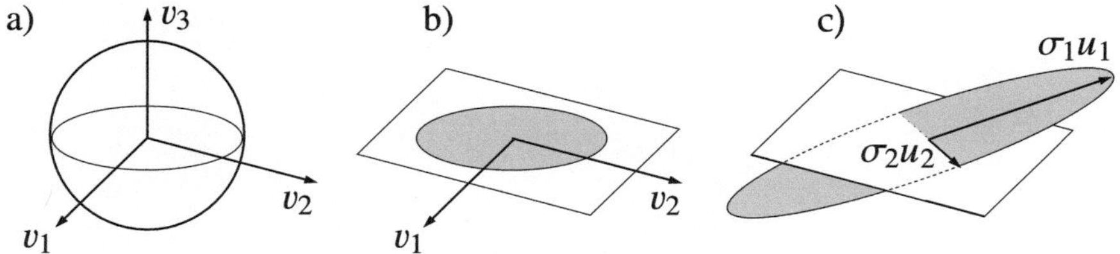

Fig. 4.6 Geometric interpretation of SVD [49] for a matrix $A \in \mathbb{R}^{m \times n}$ with $m = n = 3$ and $k = 2$. The mapping A collapses the $(n - k)$ dimensions of the space $\mathbb{R}^3$ (a). The unit sphere in the remaining k dimensions (b) is deformed into an ellipsoid by stretching and shrinking. This ellipsoid is finally embedded into $\mathbb{R}^m$ (c)

In other words, A maps the unit sphere in $\mathbb{R}^n$ into a k-dimensional hyper-ellipsoid with semi-axes of length σ_i in the directions u_i (Fig. 4.6). If A has full rank ($k = n$), the inequality in (4.20) becomes a strict equality and Ax resides on the surface of the hyper-ellipsoid.

The singular value decomposition unravels many important properties of a matrix related to its rank, condition, and norm. The singular values of A with rank $r < n$ are equal to zero from σ_r onwards, $\sigma_1 \geq \sigma_2 \geq \cdots \geq \sigma_r > \sigma_{r+1} = \cdots = \sigma_p = 0$, and the range of the mapping corresponding to this matrix is determined by the vectors $\{u_1, u_2, \ldots, u_r\}$. This brings us to the *SVD expansion*

$$A = \sum_{i=1}^{r} \sigma_i u_i v_i^{\mathrm{T}} , \tag{4.21}$$

which can be used to represent the matrix A by a limited set of singular values and vectors. The SVD expansion up to some rank $r < n$ is at the core of classical data reduction algorithms like, for instance, those used in image compression. An example is shown in Fig. 4.7; see also Problem 4.10.4.

LAPACK contains two double-precision routines for SVD, `DGESVD` and `DGESDD`. In the GSL library we have `gsl_linalg_SV_decomp`, in NR3E one resorts to `SVD`, while the `scipy` module for Python offers the `linalg.svd` routine: see Table 4.2. Typical numerical costs are $\mathcal{O}(mn^2)$ for $m > n$ and $\mathcal{O}(m^2 n)$ for $m = n$. If A has full rank, SVD also yields the solution of the linear least-squares problem $\min_x \|Ax - b\|_2$: its solution is $x = V \Sigma^{-1} U^{\mathrm{T}} b$. The numerical cost of SVD is comparable to the cost of the QR decomposition.

Even though this is far too costly from the numerical standpoint, we may also use SVD to solve systems of equations $Ax = b$ with $n \times n$ matrices A, and study the sensitivity of the solutions to perturbations in A and b. If $A \in \mathbb{R}^{n \times n}$ is non-singular, $A = U \Sigma V^{\mathrm{T}}$ is its singular value decomposition, and $b \in \mathbb{R}^n$, we have

$$x = A^{-1} b = \left(U \Sigma V^{\mathrm{T}} \right)^{-1} b = \sum_{i=1}^{n} \frac{u_i^{\mathrm{T}} b}{\sigma_i} v_i .$$

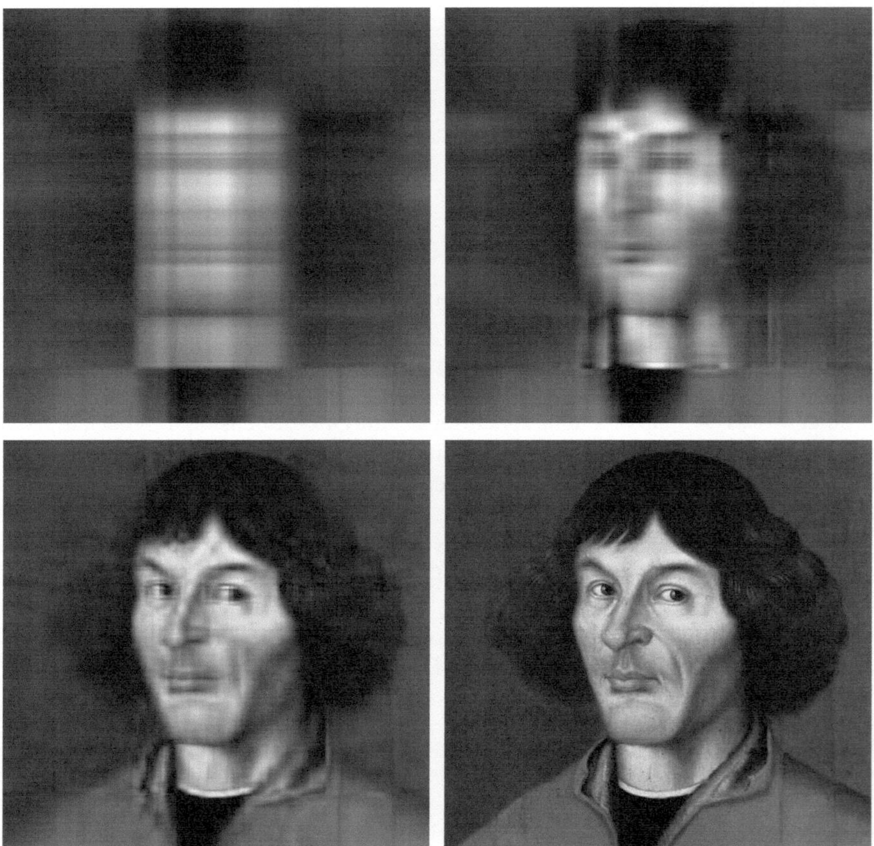

Fig. 4.7 Image compression by SVD. The original image is a 512×512 matrix of pixels with 256 gray levels. Shown are the approximations of rank 2 (99.6% compression), rank 5 (99.0%), rank 15 (97.1%), and rank 100 (80.4%). See Problem 4.10.4

Small perturbations in A and b may therefore cause large errors in x if σ_n and nearby singular values are small. With respect to conditioning of certain types of matrices, singular values tend to have a larger "predictive power" than the usual eigenvalues λ_i (see Sect. 4.6). In general, it holds that

$$\sigma_{\min}(A) \leq \min_i |\lambda_i| \leq \max_i |\lambda_i| \leq \sigma_{\max}(A) .$$

In particular with non-normal matrices, for which $A^\mathrm{T} A \neq A A^\mathrm{T}$ or $A^\dagger A \neq A A^\dagger$, it can happen that $\max_{i,j} |\lambda_i|/|\lambda_j| \ll \kappa_2(A)$.

A large gap between the largest and the smallest singular value of a matrix is usually an indication of its ill-conditioning. For some matrices, it may occur that from an index r upwards, the singular values are very small compared to the largest one (but not exactly zero), or that the spectrum exhibits two clearly separated sets of "large" and "small" singular values. Such a matrix has a *numerically defective rank*. Details can be found in [2].

The singular values of a matrix A are also intimately connected to its norms. If $A \in \mathbb{R}^{m \times n}$ has the decomposition $A = U \Sigma V^{\mathrm{T}}$, the following holds:

$$\|A\|_2 = \sigma_1 , \quad \|A\|_{\mathrm{F}}^2 = \sum_{i=1}^{p} \sigma_i^2 \quad (p = \min\{m, n\}) , \quad \min_{x \neq 0} \frac{\|Ax\|_2}{\|x\|_2} = \sigma_n \quad (m \geq n) .$$

Small perturbations δA of a matrix $A \in \mathbb{R}^{m \times n}$ $(m \geq n)$ induce changes in the singular values which are bounded by

$$\sum_{k=1}^{n} \Big(\sigma_k(A + \delta A) - \sigma_k(A) \Big)^2 \leq \|\delta A\|_{\mathrm{F}}^2 .$$

4.5.3 *The Minimum-Norm Solution of the Least-Squares Problem*

If the matrix A is rank-deficient, the solution of the least-squares problem may become extremely sensitive to small perturbations $b \to b + \delta b$. The condition of such a problem can be improved by imposing an additional constraint on the solution, which is known as *regularization*. We require that the solution has a minimal norm. The vector x that solves the regularized problem $\min_x \|Ax - b\|_2$ is then known as the *minimal solution of the least-squares problem*. The minimum-norm solution for a singular matrix A can be found by SVD. We decompose the matrix $A \in \mathbb{R}^{m \times n}$ with rank $r < n$ and $m \geq n$ as

$$A = U \Sigma V^{\mathrm{T}} = \begin{pmatrix} U_1 & U_2 \end{pmatrix} \begin{pmatrix} \Sigma_1 & 0 \\ 0 & 0 \end{pmatrix} \begin{pmatrix} V_1 & V_2 \end{pmatrix}^{\mathrm{T}} = U_1 \Sigma_1 V_1^{\mathrm{T}} , \tag{4.22}$$

where $\Sigma_1 \in \mathbb{R}^{r \times r}$ is non-singular, $U_1 \in \mathbb{R}^{m \times r}$, $U_2 \in \mathbb{R}^{m \times (n-r)}$, $V_1 \in \mathbb{R}^{n \times r}$, and $V_2 \in \mathbb{R}^{n \times (n-r)}$. All solutions x may then be written in the form

$$x = V_1 \Sigma_1^{-1} U_1^{\mathrm{T}} b + V_2 z , \tag{4.23}$$

where $z \in \mathbb{R}^{n-r}$ is an arbitrary vector. The solution x has a minimal norm $\|x\|_2$ precisely when $z = 0$. Its norm is bounded by $\|x\|_2 \leq \|b\|_2 / \sigma_{\min}(A)$. Hence the computation of the minimum-norm solution is well-conditioned when the smallest positive singular value of A is not too small.

By using (4.23) the solution of the least-squares problem may be written as

$$x = A^+b \,, \quad A^+ = V\Sigma^+U^{\mathrm{T}} \,, \quad \Sigma^+ = \begin{pmatrix} \Sigma_1 & 0 \\ 0 & 0 \end{pmatrix}^+ = \begin{pmatrix} \Sigma_1^{-1} & 0 \\ 0 & 0 \end{pmatrix} \,, \tag{4.24}$$

where A^+ is the Moore–Penrose *generalized inverse* or *pseudoinverse* of the matrix $A \in R^{m \times n}$ with $m \geq n$ (see Appendix A.7 for an alternative construction and example). If A is rank-deficient, the solution $x = A^+b$ is unique and has the minimum norm. Details can be found in Sects. 3.5 of [1] and 5.5 of [2]. The generalized inverse is exploited in Sect. 6.7 and in Chapter 13.

4.6 Eigenvalue Problems

For a physicist, solving the linear eigenvalue problem

$$Ax = \lambda x \,, \quad x \neq 0 \,, \tag{4.25}$$

especially in the context of real symmetric matrices, amounts to *diagonalizing a matrix*. In science and engineering, the eigenproblem (4.25) is ubiquitous: for example, solving systems of differential equations $\dot{x}(t) = Ax(t)$ or $\ddot{x}(t) = Ax(t)$ with suitable initial conditions can be translated to problems of the form (4.25) by the ansatz $x_i(t) = \exp(\lambda_i t)x_i(0)$. Equation (4.25) defines the *right eigenvectors* x and the corresponding eigenvalues. We may consider x as the direction which is insensitive to multiplication by A, and the corresponding eigenvalue λ is a representation of A in a subspace spanned by x. In this subspace, multiplying a vector by a matrix amounts to a rescaling along the eigenvector, thus

$$A^k x = \lambda^k x \,.$$

Similar reasoning applies to the *left eigenvectors* y which are solutions of the problem

$$y^\dagger A = \lambda y^\dagger \,, \quad y \neq 0 \,.$$

Eigenvalues of the matrix $A \in \mathbb{C}^{n \times n}$ are the roots of the characteristic polynomial $p(\lambda) = \det(\lambda I - A)$. In general, a real matrix may therefore possess complex eigenvalues. Due to potential numerical instabilities, individual λ_i should never be computed by seeking the roots of $p(\lambda) = 0$. Diagonalization algorithms convert the matrix A to a special form suitable for the computation of the eigenvalues and eigenvectors. The nature of this conversion depends on the symmetry properties of A.

We should not race ahead with diagonalization as some matrices are not diagonalizable at all. If $A \in \mathbb{C}^{n \times n}$ there exists a non-singular matrix X such that

$$X^{-1}AX = \mathrm{diag}(\lambda_1 I + N_1,\ \lambda_2 I + N_2,\ \ldots,\lambda_q I + N_q)\,,\quad N_i \in \mathbb{C}^{n_i \times n_i}\,,\quad (4.26)$$

where $\lambda_1, \lambda_2, \ldots, \lambda_q$ are distinct and all matrices N_i are strictly upper-triangular. The numbers n_i satisfy $n_1 + n_2 + \cdots + n_q = n$ and express the *algebraic multiplicity* of the individual λ_i. The *geometric multiplicity* of the eigenvalue λ_i is equal to the number of linearly independent eigenvectors belonging to λ_i. If the algebraic and the geometric multiplicities of λ_i are equal, the matrix A is diagonalizable or *non-defective*, and we can find a matrix $X \in \mathbb{C}^{n \times n}$ such that

$$X^{-1}AX = \mathrm{diag}(\lambda_1, \lambda_2, \ldots, \lambda_n)\,. \tag{4.27}$$

This can be done precisely when there exist scalars $\lambda_1, \lambda_2, \ldots, \lambda_n$ and linearly independent vectors $x_1, x_2, \ldots, x_n$ such that $Ax_i = \lambda_i x_i$ for $1 \leq i \leq n$. Then

$$\det(A) = \prod_{i=1}^{n} \lambda_i\,,\quad \mathrm{tr}(A) = \sum_{i=1}^{n} A_{ii} = \sum_{i=1}^{n} \lambda_i\,.$$

The set of eigenvalues $\lambda(A) = \{\lambda_1, \lambda_2 \ldots, \lambda_n\}$ is called the *spectrum*. When some algebraic multiplicity λ_i exceeds the geometric multiplicity, the eigenvalue is said to be *defective*, and the matrix is also defective (or non-diagonalizable).

Example Check that the matrix $A = ((2, 0, 0), (0, 2, 0), (0, 0, 2))$ and the matrix $B = ((2, 1, 0), (0, 2, 1), (0, 0, 2))$ have the same characteristic polynomial $p(\lambda) = (\lambda - 2)^3$ which yields the eigenvalue $\lambda = 2$ of algebraic multiplicity 3. For the matrix A, the geometric multiplicity is also 3, as three independent eigenvectors, for example e_1, e_2, and e_3, may be assigned to this λ. On the other hand, only one independent eigenvector (e_1) can be found for B, so the geometric multiplicity of λ in this case is only 1. Hence B is deficient. $\triangleleft$

Normal matrices $(AA^{\mathrm{T}} = A^{\mathrm{T}}A$ or $AA^{\dagger} = A^{\dagger}A)$ are diagonalizable (the opposite is not necessarily true), and tests of normality are frequently used as test of diagonalizability. This property can also be checked by using *minimal polynomials* [50]. Hermitian matrices $(A^{\dagger} = A)$ and real symmetric matrices $(A^{\mathrm{T}} = A)$ are normal, hence also diagonalizable, and have real eigenvalues.

4.6.1 Non-symmetric Problems

The basic tool for the diagonalization of dense non-symmetric matrices $A \in \mathbb{R}^{n \times n}$ or $A \in \mathbb{C}^{n \times n}$ is the iterative QR algorithm with implicit shifts [51]. (The QR iteration is not the same as the QR decomposition discussed in Sect. 4.5.1. The QR decomposition is just its ingredient.) The QR algorithm is based on the conversion of A to the upper Hessenberg form and on the iterative computation of the real Schur form which ultimately yields the eigenvalues. The eigenvectors are finally computed by inverse iteration [2]. The QR algorithm for real non-symmetric matrices is implemented in the DGEEVX routine from LAPACK (the complex variant is ZGEEVX), in gsl_eigen_nonsymmv from GSL, in Unsymmeig from NR3E and in the scipy.linalg.eig routine for Python (real and complex). The numerical cost of the QR algorithm is $\mathcal{O}(n^3)$ for each iteration step. See Table 4.3.

The eigenvalues and eigenvectors of $A \in \mathbb{C}^{n \times n}$ are sensitive to small perturbations in the matrix. Let λ be a simple eigenvalue of A, and let x and y be its right and left eigenvectors normalized to unity, $\|x\|_2 = \|y\|_2 = 1$. Let $\lambda + \delta\lambda$ be the eigenvalue of the perturbed matrix $A + \delta A$. Then we have [1]

$$|\delta\lambda| \leq \frac{\|\delta A\|}{|y^\dagger x|} + \mathcal{O}\left(\|\delta A\|^2\right) ,$$

where $\kappa = 1/|y^\dagger x| = 1/\cos\theta(x, y)$ is the condition number for the eigenproblem with this eigenvalue. In general, therefore, perturbations $\|\delta A\|$ on the order of $\approx \varepsilon_\mathrm{M}$ cause errors on the order of $\approx \kappa\varepsilon_\mathrm{M}$ in simple eigenvalues. For multiple eigenvalues, the sensitivity to perturbations δA may be much more pronounced. Namely, a large condition number $\kappa \gg 1$ for a simple eigenvalue λ implies that A is "near to" $A + \delta A$ which has a multiple eigenvalue, such that

$$\frac{\|\delta A\|_2}{\|A\|_2} \leq \frac{1}{\sqrt{\kappa^2 - 1}} \approx \frac{1}{\kappa} .$$

In A is non-normal, the value of κ *may be* very large, while for symmetric and Hermitian matrices, we have $|y^\dagger x| = 1$ and hence always $\kappa = 1$.

Example (Modified from [2]) The eigenvectors are particularly sensitive to perturbations in A if its simple eigenvalues lie close to each other or if there are multiple eigenvalues. Consider the matrices

$$A = \begin{pmatrix} 1.001 & 0.001 \\ 0.000 & 0.999 \end{pmatrix}, \quad A + \delta A = \begin{pmatrix} 1.001 & 0.001 \\ 0.000 & 1.000 \end{pmatrix} .$$

The matrix A has eigenvalues 1.001 and 0.999. The eigenvector corresponding to the smaller eigenvalue is $(-0.44721, 0.89443)^\mathrm{T}$. If we slightly perturb A to $A + \delta A$ (10^{-3} change in lower right corner), the perturbed matrix has eigenvalues 1.001 and 1.000. Now, the eigenvector corresponding to the smaller of the eigenvalues is $(-0.70711, 0.70711)^\mathrm{T}$! For details, see [52]. ◁

4.6.2 The Power Method and Inverse Iteration

Frequently we are interested in certain portions of the matrix spectrum, or perhaps even in just a specific eigenvalue. The simplest method to locate *only the largest*

Table 4.3 A collection of double-precision routines to solve real eigenproblems from the LAPACK, GSL, and NUMERICAL RECIPES (third edition for C++) libraries and the `scipy.linalg` module for Python. (The basic Jacobi method is described on page 202.) The DSYEVD, DSTEVD, DSYGVD, and DSBGVD routines based on divide-and-conquer algorithms [53] and the DSYEVR routine based on relatively robust representations [54] have the same asymptotic cost as the corresponding standard routines (last letter 'X'), but their leading constant is much smaller, and they are therefore much faster. The complex variants of the routines are mentioned in the main text. See also caption to Table 4.1

Problem $Ax = \lambda x$ $(A \in \mathbb{R}^{n \times n})$

Matrix A	Numerical cost	LAPACK	GSL	NR3E	Python
General	$\mathcal{O}(n^3)$	DGEESX/DGEEVX	gsl_eigen_nonsymmv	Unsymmeig	.eig
Symmetric	$\mathcal{O}(n^3)$	DSYEVX	gsl_eigen_symmv	Symmeig	.eigh
		DSYEVD			
		DSYEVR			
Symmetric banded	$\mathcal{O}(k_A n^2)$ for λ, $\mathcal{O}(n^3)$ for x	DSBEVX	/	/	.eig_banded
		DSBEVD			
Symmetric tridiagonal	$\mathcal{O}(n^2)$ for λ, $\mathcal{O}(n^3)$ for x	DSTEVX	/	/	.eigh_tridiagonal
	$\mathcal{O}(n^2)$ for λ, $\mathcal{O}(n^3)$ for x	DSTEVD			
	$\mathcal{O}(n^2)$ for all λ and x	DSTEGR/DSTEVR			

Problem $Ax = \lambda Bx$ $(A, B \in \mathbb{R}^{n \times n}, \det(A - \lambda B) \neq 0)$

Matrices A and B	Numerical cost	LAPACK	GSL	NR3E	Python
General A, B	$\mathcal{O}(n^3)$	DGGESX	gsl_eigen_genv	/	.eig
		DGGEVX			
Symmetric A, BB pos. definite	$\mathcal{O}(n^3)$	DSYGVX	/	/	.eigh
		DSYGVD			
Symmetric A, B Abanded, Bpos.definite	$\mathcal{O}(k_A n^2)$ for λ, $\mathcal{O}(n^3)$ for x	DSBGVX	/	/	.eigh
		DSBGVD			

eigenvalue (in magnitude) of an asymmetric matrix A and the corresponding eigenvector is the power method, embodied in the classic algorithm:

> **Input**: initial vector x
> **repeat**
> > $y = Ax$;
> > $x = y/\|y\|_2$;
> > $\lambda = x^{\mathrm{T}}Ax$;
>
> **until** *convergence*;
> **Output**: approximate largest λ and corresponding eigenvector x

The initial x is best chosen to be as random as possible. The convergence rate depends on the ratios $|\lambda_n/\lambda_1|$ for $n > 1$: if they happen to be close to 1, convergence may be very slow.

To find other specific eigenvalues and eigenvectors, the power method is applied to $(A - \sigma I)^{-1}$ where σ is the *shift* parameter. The resulting method is called *inverse iteration*: all we need to modify in the above code is replace A by $(A - \sigma I)^{-1}$. The algorithm then converges to the eigenvalue *closest to* σ and the corresponding eigenvector. For details, see [1], pp. 153–156.

4.6.3 Symmetric Problems

The $Ax = \lambda x$ eigenproblems with real symmetric matrices $(A = A^{\mathrm{T}})$ possess numerous particular properties which allow for specially tailored, more efficient solution methods. The symmetry of A commonly originates in the natural symmetries of the underlying physics problem or its mathematical formulation. All eigenvalues of a symmetric matrix are real, and we can always find an orthogonal matrix Q such that

$$Q^{\mathrm{T}}AQ = \mathrm{diag}(\lambda_1, \lambda_2, \ldots, \lambda_n)\,, \quad \lambda_1 \geq \lambda_2 \geq \cdots \geq \lambda_n\,,$$

and $\|A\|_2 = \max\{|\lambda_1|, |\lambda_n|\}$. Symmetric matrices are never defective, and their left eigenvectors are equal to their transposed right eigenvectors. Similarly, all eigenvalues of Hermitian matrices $(A^{\dagger} = A)$ are real, and their left eigenvectors are the Hermitian conjugates of their right eigenvectors. Symmetric tridiagonal matrices—omnipresent in physics—with non-zero sub-diagonals can only have simple eigenvalues.

The oldest procedure to compute all eigenvalues and eigenvectors of dense symmetric matrices is the Jacobi method, in which a sequence of orthogonal transformations is used to gradually (to a required precision) extinguish all off-diagonal elements. The method has a cost of $\mathcal{O}(n^3)$ with a large leading constant, but it is extremely robust and may compute small eigenvalues to a better relative precision than other, faster methods. Moreover, since the mentioned transformations are decoupled, it is well suited for parallelization. The basic version is implemented as `Jacobi` in the NR3E library and is recommended for matrices with sizes up to $n \approx 10$. A parallel implementation is given in [2].

By a sequence of Householder transformations, any symmetric matrix can be transformed to tridiagonal form from which eigenvalues and eigenvectors can be extracted. Fast methods therefore attempt to convert dense symmetric matrices to a tridiagonal form, which is then diagonalized. This conversion usually requires $\frac{4}{3}n^3 + \mathcal{O}(n^2)$ operations.

The basic method to compute all eigenvalues and (if needed) all eigenvectors of a symmetric tridiagonal matrix is the QR iteration which is a variant of the QR iteration for non-symmetric matrices. The computation of all eigenvalues requires $\approx 36n^2 + \mathcal{O}(n)$ operations, while the computation of all eigenvectors takes $\approx 9n^3 + \mathcal{O}(n^2)$ operations. The total cost of computing all eigenvalues of a dense symmetric matrix by tridiagonalization and QR iteration is $\frac{4}{3}n^3 + \mathcal{O}(n^2)$, and the total cost of computing all eigenvectors is $9n^3 + \mathcal{O}(n^2)$. Based on the anticipated rounding errors, QL iteration can be used instead of QR; consult [40] to see when this is appropriate.

To compute double-precision eigenvalues and eigenvectors of dense symmetric matrices by tridiagonalization and the QR algorithm, LAPACK offers the DSYEVX routine. The corresponding Hermitian routine is ZHEEVX (see Table 4.3 which also lists the numerical costs). The GSL library has `gsl_eigen_symmv` for real symmetric matrices and `gsl_eigen_hermv` for Hermitian matrices. The NR3E offers only a real implementation Symmeig but any $n \times n$ Hermitian eigenvalue problem can be transformed to $2n \times 2n$ real symmetric eigenvalue problem. Namely, the eigenproblem $Hz = \lambda z$ with a Hermitian matrix $H = A + \mathrm{i}\,B$ means $(A + \mathrm{i}\,B)(x + \mathrm{i}\,y) = \lambda(x + \mathrm{i}\,y)$ where A, B, x, and y are real, which can be written as

$$\begin{pmatrix} A & -B \\ B & A \end{pmatrix} \begin{pmatrix} x \\ y \end{pmatrix} = \lambda \begin{pmatrix} x \\ y \end{pmatrix} .$$

The matrix of this system is symmetric since $A^{\mathrm{T}} = A$ and $B^{\mathrm{T}} = -B$ if H is Hermitian ($H^{\dagger} = H$), and the eigenvalues and eigenvectors are duplicated. In Python the eigenproblem for real symmetric or complex Hermitian matrices can be solved by using the `scipy.linalg.eigh` routine.

A faster way to compute the eigenvalues and eigenvectors of dense real symmetric or Hermitian matrices leads through the "divide-and-conquer" algorithms [53]. In LAPACK they can be found in the DSYEVD routine (real symmetric) and the ZHEEVD (Hermitian). Methods based on *relatively robust representations*, RRR [54] are even more efficient. LAPACK contains the DSYEVR routine (real symmetric) and the ZHEEVR routine (Hermitian).

For real symmetric tridiagonal matrices, LAPACK offers DSTEVX (QR iteration), DSTEVD ("divide-and-conquer") and DSTEVR (RRR). For symmetric banded matrices, there are DSBEVX (QR) and DSBEVD ("divide-and-conquer") for real matrices and the corresponding pair ZHBEVX and ZHBEVD for Hermitian matrices. In Python, one can use `scipy.linalg.eigh_tridiagonal` for real symmetric tridiagonal matrices and `scipy.linalg.eig_banded` for real symmetric or complex Hermitian banded matrices.

The sensitivity of eigenvalues and eigenvectors to small perturbations δA of the matrix A can be measured in analogy to the non-symmetric case. Let A be symmetric, with eigenvalues $\lambda_1 \geq \lambda_2 \geq \cdots \geq \lambda_n$ and eigenvectors $x_1, x_2, \ldots, x_n$, and let δA be a symmetric perturbation matrix, so that the eigenvalues of $A + \delta A$ are $\widehat{\lambda}_1 \geq \widehat{\lambda}_2 \geq \cdots \geq \widehat{\lambda}_n$ and the corresponding eigenvectors are $\widehat{x}_1, \widehat{x}_2, \ldots, \widehat{x}_n$. If the perturbation is bounded by $\|\delta A\|_2 = \mathcal{O}(\varepsilon)\|A\|_2$, we have

$$|\widehat{\lambda}_i - \lambda_i| \leq \|\delta A\|_2 = \mathcal{O}(\varepsilon)\|A\|_2 , \quad 1 \leq i \leq n . \tag{4.28}$$

If A is normal, $\|A\|_2 = \max_j |\lambda_j|$ also applies. The sensitivity to perturbations can also be measured by the angle θ_i between the eigenvector x_i of the unperturbed problem and the eigenvector $\widehat{x}_i$ of the perturbed problem. It can be shown that

$$\frac{1}{2} \sin 2\theta_i \leq \|\delta A\|_2 \left[\min_{j \neq i} |\widehat{\lambda}_j - \widehat{\lambda}_i| \right]^{-1} . \tag{4.29}$$

When the largest and the smallest eigenvalues differ by orders of magnitude, the bounds (4.28) and (4.29) tend to strongly over-estimate the errors. See [1] for stricter bounds.

Example We would like to compute the energy states of electrons in a two-dimensional periodic potential $V(x, y)$ with the period d in x- and y-directions in the presence of a constant magnetic field defined in terms of the vector potential $A = B(0, -x, 0)^{\mathrm{T}}$. The corresponding Hamiltonian is

$$\hat{H} = \frac{1}{2m_{\mathrm{e}}} (p - e A)^2 + V(x, y) .$$

The ansatz for the wave-function of the electron is [55]

$$\psi(x, y) = \sum_{n,m \in \mathbb{Z}} a_{nm} e^{-i2\pi m\alpha x/d} \psi_0(x - nd, y - md) ,$$

where ψ_0 are the single-atom wave-functions and $\alpha = eBd^2/h$. The parameter α corresponds to the number of quanta of the magnetic flux per unit cell of the mesh. Because the Hamiltonian is periodic, the Bloch theorem [56] permits us to write the expansion coefficients a_{nm} in the form $a_{nm} = e^{im\phi} a_n$. The coupling between the neighboring mesh points is given by the overlap integrals

$$W_{nm} = \iint \psi_0(x - nd, y - md) \, e^{i2\pi m\alpha x/d} \, \hat{H}\big|_{B=0} \, \psi_0(x, y) \, \mathrm{d}x\mathrm{d}y .$$

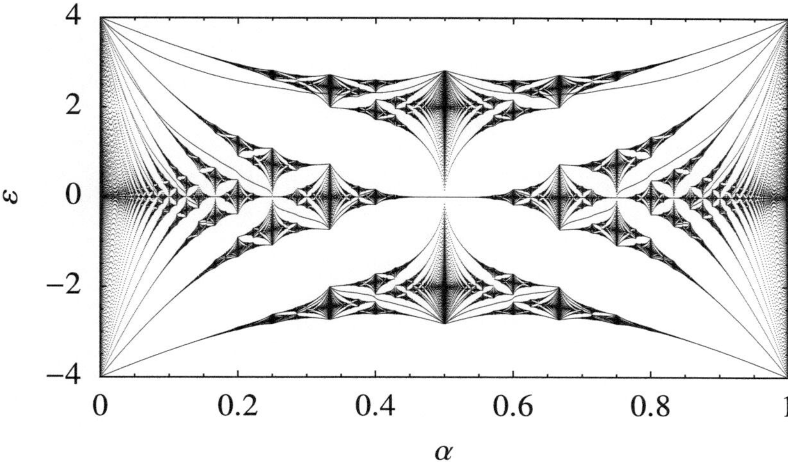

Fig. 4.8 Dimensionless eigenenergies ε of electrons in a two-dimensional periodic potential in the presence of a magnetic field B, as a function of $\alpha = eBd^2/h = p/q$, using Dirichlet boundary conditions for ψ and parameters $\lambda = 2, \phi = 0, q = 301$. In a crystal with $d \approx 0.2$ nm such patterns are not expected to appear until the field densities reach $B \approx 10^5$ T. But an almost identical image was seen in studies of microwave propagation through wave-guides possessing periodic scattering centers [55]

By using the ansatz ψ in the Schrödinger equation $\hat{H}\psi = E\psi$ and considering only the nearest-neighbor couplings ($W_{nm} = 0$ for $|n| > 1$ or $|m| > 1$) we obtain a tridiagonal system of equations for the expansion coefficients,

$$a_{n+1} + \lambda \, \cos(2\pi n\alpha - \phi)\, a_n + a_{n-1} = \varepsilon a_n \, , \quad 0 \le n \le q - 1 \, . \tag{4.30}$$

We introduce the dimensionless energy $\varepsilon = E/W_{10}$ and the ratio of coupling strengths in x- and y-directions, $\lambda = 2W_{10}/W_{01}$. The solutions of (4.30) are qualitatively different for rational or irrational values of α. For rational values $\alpha = p/q$ the sequence of dimensionless energies $\{\varepsilon_n\}$ has a period q, while for irrational α the sequence $\{\varepsilon_n\}$ constitutes a fractal. In the special case $\lambda = 2$ it is known as Hofstadter's butterfly (see Fig. 4.8 and [57]). ◁

4.6.4 Generalized Eigenvalue Problems

The textbook [1] describes the familiar physical example of masses connected by coupled springs with damping that can be described by the system of equations

$$M\ddot{x}(t) = -B\dot{x}(t) - Kx(t) \, , \quad x(0) = x_0 \, , \quad \dot{x}(0) = \dot{x}_0 \, , \tag{4.31}$$

where M is the mass matrix, B is the damping coefficient matrix and K is the string constant matrix. By using the ansatz $x_i(t) = \exp(\lambda_i t)x_i(0)$ Eq. (4.31) becomes a quadratic eigenvalue problem $(\lambda^2 M + \lambda B + K)x = 0$, but by introducing the vector $z = (x, y)^{\mathrm{T}}$ it can be transformed to the usual linear problem

$$\begin{pmatrix} 0 & I \\ -M^{-1}K & -M^{-1}B \end{pmatrix} \begin{pmatrix} x \\ y \end{pmatrix} = \lambda \begin{pmatrix} x \\ y \end{pmatrix} .$$

If M is ill-conditioned, it is advisable to rewrite this as

$$\begin{pmatrix} -K & -B \\ 0 & I \end{pmatrix} \begin{pmatrix} x \\ y \end{pmatrix} = \lambda \begin{pmatrix} 0 & M \\ I & 0 \end{pmatrix} \begin{pmatrix} x \\ y \end{pmatrix} ,$$

which has the form of a *generalized eigenvalue problem*

$$Az = \lambda Bz , \quad z \neq 0 . \tag{4.32}$$

Such eigenproblems can be solved by using the QZ algorithm which is a generalization of the QR algorithm from Sect. 4.6.1. In the case of general real matrices A and B we can use the DGGEVX from LAPACK or the gsl_eigen_genv from GSL. For general Hermitian matrices A and B where B is positive definite, we may use ZGGEVX from LAPACK or gsl_eigen_genhermv from GSL.

   To solve (4.32) with symmetric A and B, where B is positive definite, LAPACK offers us the DSYGVX and DSYGVD routines. For banded A and positive definite B, we may take DSBGVX and DSBGVD. The corresponding Hermitian quartet of routines is ZHEGVX, ZHEGVD, ZHBGVX, and ZHBGVD. The numerical cost of these algorithms is $\mathcal{O}(n^3)$ (see Table 4.3). Implementations in the LAPACK library also allow us to solve related problems of the form $ABx = \lambda x$ and $BAx = \lambda x$. In the scipy.linalg module for Python, generalized eigenvalue problems are accounted for in the .eig and .eigh routines with the appropriate choice of LAPACK drivers. Further reading can be found in [1] (Sect. 4.5) and [2] (Sects. 7.7 and 8.7).

Example Generalized eigenproblems are commonly found in looking for stationary values of quadratic forms $x^{\mathrm{T}}Ax$ with symmetric matrices $A \in \mathbb{R}^{n \times n}$ and constraints

$$c^{\mathrm{T}}x = 0 , \quad \|x\|_2 = \|c\|_2 = 1 , \quad x, c \in \mathbb{R}^n . \tag{4.33}$$

We use the function $\phi(x; \lambda, \mu) = x^{\mathrm{T}}Ax - \lambda(x^{\mathrm{T}}x - 1) + 2\mu\, x^{\mathrm{T}}c$ where λ and μ are the Lagrange multipliers [58]. When ϕ is differentiated with respect to x, we obtain

$$Ax - \lambda x + \mu c = 0 . \tag{4.34}$$

This is an inhomogeneous eigenproblem [59] but it can be transformed to the homogeneous one. We determine μ by multiplying Eq. (4.34) from the left by c^{T}, and by using (4.33), we get $\mu = -c^{\mathrm{T}} A x$. Hence

$$P A x = \lambda x \,, \quad P = I - c c^{\mathrm{T}} \,, \quad P^2 = I \,.$$

For eigenvalues of square matrices A and P it holds that $\lambda(PA) = \lambda(P^2 A) = \lambda(PAP)$, hence by solving $PAPz = \lambda z$, where $x = Pz$, we simultaneously solve (4.34). In general, the matrix PA is non-symmetric, but PAP is symmetric, and to solve $PAPz = \lambda z$, algorithms for symmetric eigenvalue problems can be used. $\lhd$

4.6.5 *The Quadratic Eigenvalue Problem*

The quadratic eigenvalue problem is to find the scalars λ and non-zero vectors x and y satisfying

$$(\lambda^2 M + \lambda C + K)x = 0 \,, \quad y^*(\lambda^2 M + \lambda C + K) = 0 \,,$$

where M, C and K ("mass", "damping" and "stiffness") are $n \times n$ complex matrices while x and y are the right and left eigenvectors, respectively, corresponding to the eigenvalue λ. As the matrix names suggest, such problems occur widely in dynamical analysis of mechanical systems and in study of flow stability in fluid mechanics. In contrast to the standard eigenvalue problem $Ax = \lambda x$ and its generalized partner $Ax = \lambda Bx$ the quadratic eigenvalue problem has $2n$ eigenvalues with up to $2n$ right and $2n$ left eigenvectors, of which up to n are linearly independent. The solution techniques depend on the symmetries of the matrices: see [60] for details.

4.6.6 *Converting a Matrix to Its Jordan Form*

Matrix diagonalization finds one of the most common uses in solving systems of ordinary differential equations with constant coefficients

$$\frac{\mathrm{d}u}{\mathrm{d}t} = Au \,, \quad A \in \mathbb{C}^{n \times n} \,, \quad u \in \mathbb{C}^n \,. \tag{4.35}$$

By using a linear transformation $u = Xv$ the system can be rewritten as

$$X \frac{\mathrm{d}v}{\mathrm{d}t} = AXv \quad \text{or} \quad \frac{\mathrm{d}v}{\mathrm{d}t} = X^{-1} AXv = Jv \,. \tag{4.36}$$

If A is diagonalizable, i.e. if it can be decomposed as (4.27), the system (4.36) decouples to $\mathrm{d}v_i/\mathrm{d}t = \lambda_i v_i$. The general solution of this system is

$$v_i(t) = v_i(0)\, e^{\lambda_i t} \quad \Longrightarrow \quad u(t) = \sum_i v_i(0) x_i\, e^{\lambda_i t} \ .$$

If A is defective, a complete decoupling by such a linear transformation can not be achieved. But we can still simplify (4.35) significantly if the Jordan decomposition of A is known. Jordan decomposition is a special form of (4.26),

$$X^{-1}AX = \text{diag}\left(J_{n_1}(\lambda_1), J_{n_2}(\lambda_2), \ldots, J_{n_r}(\lambda_r)\right) = J \ ,$$

where

$$J_{n_i}(\lambda_i) = \begin{pmatrix} \lambda_i & 1 & & 0 \\ & \ddots & \ddots & \\ & & \ddots & 1 \\ 0 & & & \lambda_i \end{pmatrix}^{n_i \times n_i} \ , \qquad \sum_{i=1}^{r} n_i = n \ .$$

The matrix J is called the *Jordan canonical form*, and each sub-matrix $J_{n_i}(\lambda_i)$ is its Jordan block with the eigenvalue λ_i of algebraic multiplicity n_i. The decomposition is unique in the sense that only a permutation of the blocks J_{n_i} along the main diagonal of J is permissible. The general solution of (4.35) is

$$u(t) = X\, \exp(tJ)\, X^{-1} u(0) \ .$$

The matrix $\exp(tJ)$ is block-diagonal, $\exp(tJ) = \text{diag}\left(\exp(tJ_{n_i}(\lambda_i))\right)_{i=1}^{r}$, where

$$\exp(tJ_{n_i}(\lambda_i)) = e^{\lambda_i t} \begin{pmatrix} 1 & t/1! & t^2/2! & \cdots & t^{n_i-1}/(n_i-1)! \\ 0 & 1 & t/1! & \cdots & t^{n_i-2}/(n_i-2)! \\ \vdots & \vdots & \vdots & \ddots & \vdots \\ 0 & 0 & 0 & 1 & t/1! \\ 0 & 0 & 0 & 0 & 1 \end{pmatrix} \ .$$

Only equations belonging to the same block remain coupled, and the eigenvalue multiplicities appear naturally. This is the main charm of the Jordan decomposition. However, in the real world—in floating-point arithmetic—the definition of the decomposition itself needs great care: we expect multiple eigenvalues λ_i on the main diagonal of a block, exact values of 1 on its first super-diagonal, and pure zeros elsewhere! Moreover, the numerical computation of the Jordan form is extremely ill-conditioned [61]. The standard algorithm is described in [62]. Details on the computation of matrix exponents $\exp(A)$ are discussed in Appendix A.9, while further reading on general functions of matrices $f(A)$ can be found in [46] and Chap. 11 of [2].

4.7 Eigenvalue Problems with Sparse Matrices

Just like in solving systems of linear equations, direct methods to compute eigenvalues and eigenvectors become too time-demanding and memory-consuming when matrices become too large [63]. This remains true in the case of large sparse matrices unless their special structure is harnessed properly. Again, iterative methods offer a way out. Modern solution techniques for eigenproblems with large matrices exploit the projection methods based on Krylov subspaces introduced in Sect. 4.3. Two of the most efficient iterative approaches are the Arnoldi's method [64] and its derivative, the Lanczos method [65].

4.7.1 Arnoldi's Method

Arnoldi's method is suitable for general (possibly non-symmetric or non-Hermitian) matrices, in particular if one is interested only in the top portion of their spectrum. The algorithm pursued to its mth step constructs an orthogonal basis consisting of n-dimensional vectors $v_1, v_2, \ldots, v_m$ for the Krylov subspace of the form (4.10). The $n \times m$ matrix V_m composed of these basis vectors embodies a partial orthogonal transformation of A to its upper Hessenberg form,

$$V_m^\dagger A V_m = H_m \ . \tag{4.37}$$

Due to the particular structure of H_m its m eigenvalues —known as Ritz eigenvalues— are fairly easy to compute, and some of them (typically a small fraction of m) will converge to the original eigenvalues λ_i of A. Since H_m is a $m \times m$ matrix, it has at most m distinct eigenvalues, so the idea is to increase m until as many as possible eigenvalues of A—or at least the dominant ones—are found. We list the algorithm (Algorithm 6.9 from [1]) as it is so simple to code yet very efficient:

> **Input**: (non-symmetric) $n \times n$ matrix A, random vector $b \in \mathbb{R}^n$
> $v_1 = b / \|b\|_2$
> **for** $j = 1$ **step** 1 **to** m **do**
> $\quad$ $z = A v_j$
> $\quad$ **for** $i = 1$ **step** 1 **to** j **do**
> $\quad\quad$ $H_{i,j} = v_i^\mathrm{T} z$
> $\quad\quad$ $z = z - H_{i,j} v_i$
> $\quad$ **end**
> $\quad$ $H_{j+1,j} = \|z\|_2$
> $\quad$ **if** $(H_{j+1,j} = 0)$ **then break** $v_{j+1} = z / H_{j+1,j}$
> **end**
> **Output**: $m \times m$ matrix H_m

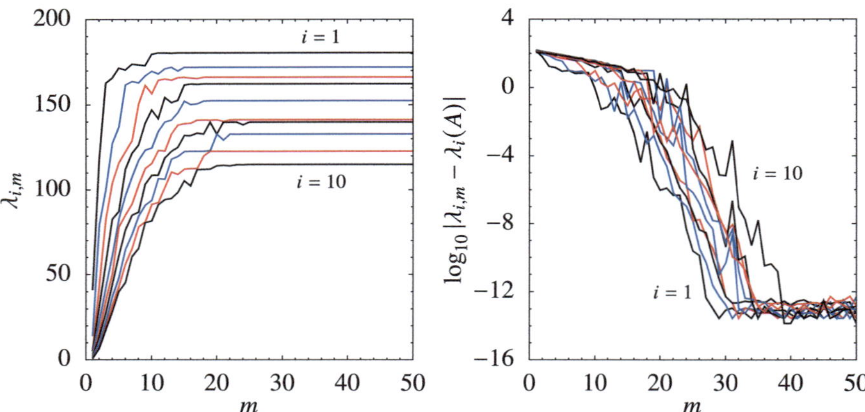

Fig. 4.9 Arnoldi iteration applied to a 100×100 non-symmetric matrix. [LEFT] Ten largest Arnoldi eigenvalues as a function of the iteration index m. [RIGHT] The deviations of the Arnoldi eigenvalues obtained by diagonalizing H_m from the corresponding eigenvalues calculated directly from A

The resulting matrix H_m must then be diagonalized, but because of its particular form the required computational cost is just $6m^2 + \mathcal{O}(m)$ instead of the standard $\mathcal{O}(m^3)$. Figure 4.9 shows a typical outcome of Arnoldi iteration applied to a 100×100 dense non-symmetric random matrix A with all eigenvalues real.

Even though the convergence properties of the Arnoldi's method are not yet fully understood, the ARPACK package [66, 67] based on it is presently the golden standard for Krylov-type solvers of eigenvalue problems, and is also included e.g. in the Eigenvalues function of MATHEMATICA, the eigs function of MATLAB and in the scipy.sparse.linalg.eigs function in SciPy. Further reading can be found in [68] and in Sects. 6.2 and 6.7 of [69].

4.7.2 Hermitian and Non-Hermitian Lanczos Algorithm

The Hermitian version of the Lanczos algorithm (see Sects. 6.3 and 6.6 of [69]) is a simplification of Arnoldi's method in the case when A is Hermitian. The most important advantage over the general case is that the Hessenberg matrix generated in the projection process turns out to be tridiagonal, symmetric and real, allowing for a much faster computation of its eigenvalues. Several varieties of the Lanczos iteration exist, but we quote the all-purpose Algorithm 7.2 from [1]:

Input: real symmetric or Hermitian $n \times n$ matrix A, random vector $b \in \mathbb{R}^n$
$v_0 = 0$, $\beta_0 = 0$, $v_1 = b/\|b\|_2$
for $j = 1$ **step** 1 **to** m **do**
$\quad z = Av_j$
$\quad \alpha_j = v_j^{\mathrm{T}} z$
$\quad z = z - \sum_{i=1}^{j}\left(z^{\mathrm{T}} v_i\right) v_i$ // full reorthogonalization
$\quad z = z - \sum_{i=1}^{j}\left(z^{\mathrm{T}} v_i\right) v_i$ // yes, twice!
$\quad \beta_j = \|z\|_2$
$\quad$ **if** $(\beta_j = 0)$ **then break** $v_{j+1} = z/\beta_j$
end
Output: tridiagonal matrix T_m with $\{\alpha_1, \alpha_2, \ldots, \alpha_m\}$ on the diagonal and
$\qquad\quad \{\beta_1, \beta_2, \ldots, \beta_{m-1}\}$ on the sub- and super-diagonal

The m Lanczos eigenvalues are obtained by diagonalizing T_m, which is a $\mathcal{O}(m^2)$ matter. Figure 4.10 shows a typical outcome of Lanczos iteration applied to a 100×100 real symmetric random matrix. The black curves denote the results of the algorithm in the above form. The (double) full reorthogonalization step is time-consuming but it is crucial in ensuring stable convergence. Indeed the relevant two lines can be replaced by $z = z - \alpha_j v_j - \beta_{j-1} v_{j-1}$, which is faster but orthogonalizes z only to a single preceding v_j; this generates round-off errors and may cause convergence failure, as indicated by the blue curves in the Figure. Note that methods exist through which orthogonality can be maintained without sacrificing too much of the computational speed: this can be accomplished, among other, by *selective (re)orthogonalization* described e.g. in Sect. 7.3 of [1].

The simpler structure of the transformed problem also allows us to assess the convergence of the algorithm in a more quantitative manner. Assume that the eigenvalues λ_i of A are ordered as $\lambda_1 \geq \lambda_2 \geq \cdots \geq \lambda_n$ and analogously for the approximate values $\lambda_{i,m}$ calculated by the algorithm. As in the Arnoldi algorithm the superscript m denotes the mth step of the iteration, i.e. the dimension of the corresponding Krylov subspace. Then the difference between the exact and approximate ith eigenvalue satisfies the relation

$$0 \leq \lambda_i - \lambda_{i,m} \leq (\lambda_1 - \lambda_n) \left(\frac{\kappa_{i,m} \tan \theta(v_1, u_i)}{T_{m-i}(1 + 2\gamma_i)} \right)^2 ,$$

where $\gamma_i = (\lambda_i - \lambda_{i+1})/(\lambda_{i+1} - \lambda_n)$, T_k are the Chebyshev polynomials, u_i is the exact eigenvector of A corresponding to λ_i, v_1 is the first Krylov basis vector and

$$\kappa_{1,m} = 1 , \quad \kappa_{i,m} = \prod_{j=1}^{i-1} \frac{\lambda_{j,m} - \lambda_n}{\lambda_{j,m} - \lambda_i} , \quad i > 1 . \tag{4.38}$$

The angle between the ith exact eigenvector u_i and the ith approximate eigenvector $u_{i,m}$ calculated at step m is bounded as

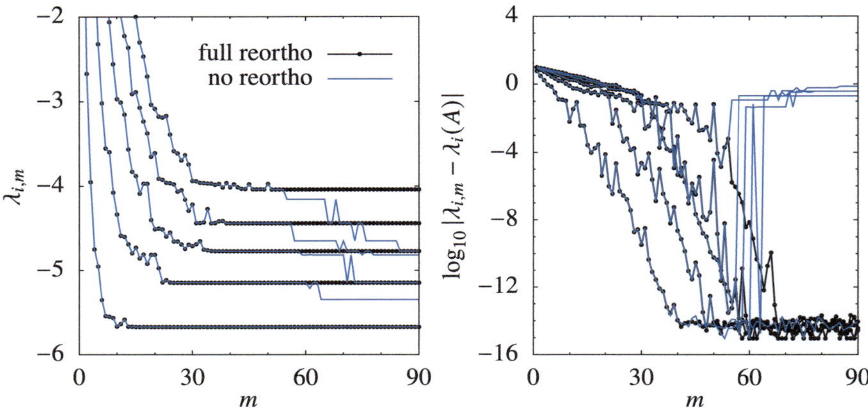

Fig. 4.10 Lanczos iteration applied to a 100×100 real symmetric matrix. [LEFT] Five smallest Lanczos eigenvalues as a function of the iteration index m. [RIGHT] The deviations of the Lanczos eigenvalues obtained by diagonalizing T_m from the eigenvalues calculated directly from A. In both panels the black curves and symbols denote the eigenvalues obtained by the algorithm with (double) full reorthogonalization, while the blue curves denote the eigenvalues if no reorthogonalization is performed

$$\sin \theta \left(u_i, u_{i,m}\right) \le \frac{\kappa_i \sqrt{1 + \beta_{m+1}^2/\delta_i^2}}{T_{m-1}(1 + 2\gamma_i)} \tan \theta(v_1, u_i),$$

where the κ_i are defined as in (4.38) but without the additional superscript m, β_k are scalar quantities computed by the algorithm, and δ_i is the distance between λ_i and the set of approximate eigenvalues other than $\lambda_{i,m}$.

The non-Hermitian Lanczos algorithm (see Sect. 6.4 of [69]) is a generalization of the previous method to non-Hermitian matrices. However, instead of constructing a single orthogonal basis for $\mathcal{K}_m$ as in the Arnoldi case, it builds a pair of bases $\{v_1, v_2, \ldots, v_m\}$ and $\{w_1, w_2, \ldots, w_m\}$ for the Krylov subspaces $\mathcal{K}_m(A, v_1) = \text{span}\{v_1, Av_1, \ldots, A^{m-1}v_1\}$ and $\mathcal{K}_m(A^\dagger, w_1) = \text{span}\{w_1, A^\dagger w_1, \ldots, (A^\dagger)^{m-1}w_1\}$. The resulting basis vectors are biorthogonal, $w_i^{\mathrm{T}} v_j = \delta_{i,j}$ $(1 \le i, j \le m)$, while Eq. (4.37) is replaced by an asymmetric transformation of the form $W_m^\dagger A V_m = T_m$, where V_m and W_m are constructed from the v_i's and w_i's, respectively. The matrix T_m remains tridiagonal but in general is no longer real and symmetric.

4.7.3 Preconditioning and Filtering

Just as in the case of sparse linear solvers, the efficiency of iterative projection algorithms for solving large eigenvalue problems can be greatly improved by preconditioning. In addition, their convergence can be accelerated by a technique called polynomial filtering: the idea is to pre-process the initial vectors or the initial Krylov subspace in order to reduce their components in the unwanted parts of the spectrum relative to those in the wanted parts. For details see Chaps. 7 and 8 of [69].

4.8 Pseudospectra of Matrices ⋆

The concept of pseudospectra is useful in the analysis of specific mathematical problems and physical systems in which the standard eigenvalue methods may fall short of completely explaining their known or presumed properties and behavior. Most often this can be witnessed in systems represented by matrices which are highly *non-normal*. Normal matrices are diagonalizable, i.e. they can be decomposed by (4.27) and hence possess a complete set of orthogonal eigenvectors such that $X^{-1} = X^{\mathrm{T}}$ or $X^{-1} = X^{\dagger}$, and the corresponding set of eigenvalues, the spectrum $\sigma(A)$. *Non-normality*, on the other hand, refers to the fact that the eigenvectors of A, if they exist, are *far from orthogonal*: this translates into the statement that A is far from normal when $\|X\| \|X^{-1}\| \gg 1$ in some matrix norm. In this Section we assume $\| \cdot \| = \| \cdot \|_2$.

4.8.1 *Definition of Pseudospectrum*

For arbitrary (usually small) $\varepsilon > 0$, the ε-pseudospectrum $\sigma_\varepsilon(A)$ of $A \in \mathbb{C}^{n \times n}$ is the set of points $z \in \mathbb{C}$ for which

$$\|(A - zI)^{-1}\| > \varepsilon^{-1}, \tag{4.39}$$

i.e. the norm of the resolvent of A at z exceeds some (usually large) value. Note that the proper spectrum implies $z \in \sigma(A) \Leftrightarrow \|(A - zI)^{-1}\| = \infty$. Indeed, for normal matrices $\|(A - zI)^{-1}\|$ is always large when z is near an eigenvalue. What makes non-normal matrices so special is the fact that this norm can be large even when z is far away from the spectrum, with important consequences. For instance, the transient (non-asymptotic) behavior of a non-normal system may be driven by the pseudospectrum, not the genuine eigenvalues alone. An example will be shown in the discussion of stability of stiff differential equations in Sect. 8.10, see also Fig. 12.10 (left). Contrary to common experience, the eigenvalues may also fail to predict the correct asymptotic and resonance behavior of such systems when variable coefficients, non-linearities or complicated force terms (right-hand sides) are present. A convincing motivation for pseudospectra and the proof of their all-pervading importance in wide areas of physics is offered by the landmark (and almost inexhaustible) monograph [70].

Since the eigenvalues may be extremely sensitive to perturbations (see Sect. 4.6.1), an equivalent definition of the pseudospectrum can be based on what happens to the eigenvalues of A when its matrix elements are changed by a small amount: the pseudospectrum $\sigma_\varepsilon(A)$ of A is the set of $z \in \mathbb{C}$ such that

$$z \in \sigma(A + E) \,, \tag{4.40}$$

where $E \in \mathbb{C}^{n \times n}$ is a small perturbation with $\|E\| < \varepsilon$. In other words, if one could generate all perturbations of A with $\|E\| < \varepsilon$, they would fill the area enclosed by the ε-boundary of the pseudospectrum.

Example (Adapted from [70].) Consider the tridiagonal non-symmetric Toeplitz matrix

$$A = \begin{pmatrix} 0 & 1 & & & \\ \frac{1}{4} & 0 & 1 & & \\ & \ddots & \ddots & \ddots & \\ & & \frac{1}{4} & 0 & 1 \\ & & & \frac{1}{4} & 0 \end{pmatrix} . \tag{4.41}$$

It can be symmetrized by $B = DAD^{-1}$, where $D = \mathrm{diag}\,(2, 4, \ldots, 2^n)$, yielding

$$B = \begin{pmatrix} 0 & \frac{1}{2} & & & \\ \frac{1}{2} & 0 & \frac{1}{2} & & \\ & \ddots & \ddots & \ddots & \\ & & \frac{1}{2} & 0 & \frac{1}{2} \\ & & & \frac{1}{2} & 0 \end{pmatrix} . \tag{4.42}$$

The similarity transformation preserves the eigenvalues, hence the spectra of A and B are identical, consisting of n distinct numbers on the real axis:

$$\lambda_i(A) = \lambda_i(B) = \cos \frac{i\pi}{n+1} \,, \quad 1 \le i \le n \tag{4.43}$$

(see Eq. (A.9)). The pseudospectra of A, however, reside far from the real axis. Figure 4.11 (left) shows the boundaries of $\sigma_\varepsilon(A)$ with $n = 64$ for several ε. We see that even at distances from the spectrum on the order of 1 the norm of the resolvent remains very large.

Figure 4.11 (right) illustrates the pseudospectrum of A according to the alternative definition (4.40). The plot shows a superposition of 6400 pseudoeigenvalues at $\varepsilon = 0.001$ generated by calculating the eigenvalues of 100 matrices $A + E$, where E is a random matrix whose elements are independent normal random deviates. The matrix E has been normalized such that $\|E\| = 0.001$.

Figure 4.11 is just a visualization of the pseudospectrum. But what does it *mean* for the behavior of A and B? Suppose that we wish to predict the norms of the powers A^k and B^k for various k. The classical approach to this problem is to resort to eigenvalues: first diagonalize A via $X^{-1}AX = \Lambda$, then, since taking the power of a diagonal matrix is trivial, calculate

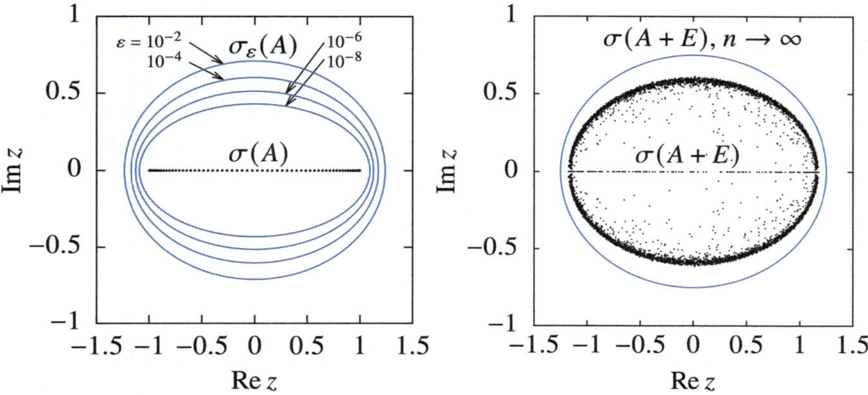

Fig. 4.11 [LEFT] The spectrum $\sigma(A)$ (dots on the real axis) and the pseudospectra $\sigma_\varepsilon(A)$ (ellipses) of the matrix (4.41) of dimension $n = 64$. [RIGHT] Superposition of eigenvalues of 100 matrices $A + E$, where E is a random matrix with $\|E\| = 0.001$

$$\Lambda^k = \left(X^{-1}AX\right)^k = X^{-1}AAA\cdots AX = X^{-1}A^kX\,,$$

and, finally, "un-diagonalize" Λ^k to obtain $A^k = X\Lambda^kX^{-1}$, and analogously for B^k. Furthermore, by (4.43), the spectral radii (the maximum absolute eigenvalues) of A and B are equal, $\rho(A) = \rho(B) = \cos(\pi/(n+1))$. Since these quantities are smaller than 1, by the procedure outlined above both A and B are power-bounded as $\|A^k\| \le C_A$ and $\|B^k\| \le C_B$ for all $k \ge 0$, with $\|A^k\| \to 0$ and $\|B^k\| \to 0$ as $k \to \infty$. Figure 4.12 confirms these expectations, yet it is obvious that the behavior of the powers of the non-normal matrix A and of the symmetric (and therefore normal) matrix B generated from it are completely different—even though their spectra are identical.

Fig. 4.12 The norms of the powers A^k and B^k of matrices (4.41) and (4.42) as functions of k, for $n = 32$ and $n = 64$

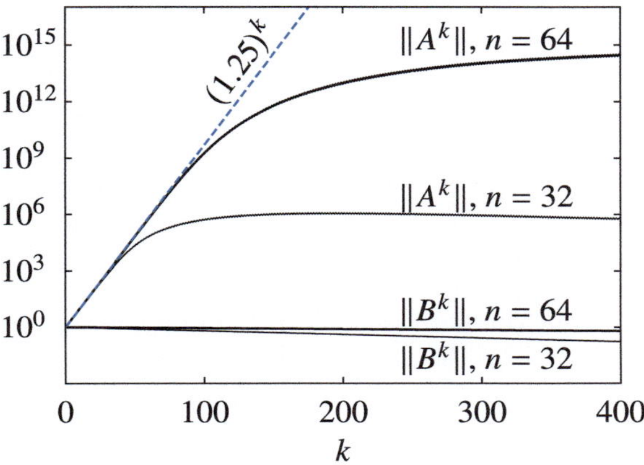

The pseudospectra help to elucidate the most interesting "transient" region where $k < n$. Here $\|A^k\|$ grows exponentially, with the slope in the log-linear plot given by $(1.25)^k$. The value 1.25 corresponds to the rightmost point of the dashed ellipse in Fig. 4.11 (right), i.e. to the absolutely maximum pseudoeigenvalue. It is precisely this pseudospectral radius—rather than the eigenvalues—that drives the behavior of $\|A^k\|$ for moderate k.   ◁

4.8.2 Pseudospectra of Linear Operators

The concept of matrix pseudospectra can be generalized to linear operators acting in infinitely dimensional spaces, for instance, in Banach spaces. Assuming that the inverse of such an operator A is bounded, its ε-pseudospectrum $\sigma_\varepsilon(A)$ can still be defined as the set of points $z \in \mathbb{C}$ such that

$$\|(A - z)^{-1}\| > \varepsilon^{-1}$$

by analogy to (4.39), but $\| \cdot \|$ is now an operator norm (see Appendix A.4). The most common operators of this kind are differential or integral operators. The simplest example is the derivative operator $A = \mathrm{d}/\mathrm{d}x$ acting in the space of functions $u \in L^2(0, 1)$,

$$Au = u' = \frac{\mathrm{d}u}{\mathrm{d}x}, \tag{4.44}$$

and subject to the boundary condition $u(1) = 0$. Clearly this differential equation could only be satisfied by functions of the form $u(x) = \mathrm{e}^{zx}$ for some $z \in \mathbb{C}$, but the boundary condition forbids any solution of this kind. Hence the spectrum of A is empty, $\sigma(A) = \{\}$. The pseudospectrum, however, exists: for $\mathrm{Re}\, z < 0$ it is determined by the resolvent norm

$$\|(A - z)^{-1}\| = \frac{\mathrm{e}^{|\mathrm{Re}\, z|}}{2\,|\mathrm{Re}\, z|} + \mathcal{O}\left(\frac{1}{|\mathrm{Re}\, z|}\right),$$

while for $\mathrm{Re}\, z > 0$ we have $\|(A - z)^{-1}\| \le 1/\mathrm{Re}\, z$ (see p. 37 of [70] for the derivation). The plot of the pseudospectrum is shown in Fig. 4.13.

How can this be of any use? Think of $A = (\mathrm{d}/\mathrm{d}x)$ as the coordinate representation of the momentum operator in quantum mechanics, $\widehat{p} = -\mathrm{i}\,\hbar(\mathrm{d}/\mathrm{d}x)$ or $\widehat{p} = -\mathrm{i}\,\hbar\nabla$, which is the generator of infinitesimal translations. The pseudospectrum of A, quantum or not, therefore gives insight into the behavior of the translation operator e^{tA}, which generates the solutions of the advection equation $u_t = u_x = Au$ (see (10.33) with $c = -1$). The solution of this problem with the boundary condition $u(1) = 0$ is

$$\mathrm{e}^{tA}u(x) = \begin{cases} u(x + t) \; ; \; x + t < 1, \\ 0 \qquad\quad ; \; x + t \ge 1. \end{cases}$$

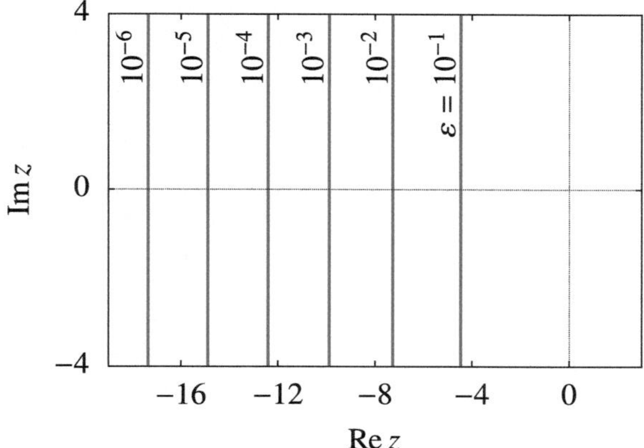

Fig. 4.13 Pseudospectra $\sigma_\varepsilon(A)$ of the differentiation operator $A = \mathrm{d}/\mathrm{d}x$ subject to the boundary condition $u(1) = 0$. The vertical lines are the right-hand boundaries of $\sigma_\varepsilon(A)$; there is no dependence on $\mathrm{Im}\, z$

By looking at the operator norm of e^{tA} as a function of t, we see that $\|\mathrm{e}^{tA}\| = 1$ for $0 \le t < 1$ and $\|\mathrm{e}^{tA}\| = 0$ for $t \ge 1$. The second property can be inferred from the pseudospectrum of A in the limit $\varepsilon \to 0$. The fact that $\|\mathrm{e}^{tA}\| \le 1$ for all $t \ge 0$ (the evolution implies a "contraction" or "dissipation") can be inferred from the behavior of $\sigma_\varepsilon(A)$ in the limit $\varepsilon \to \infty$ (see p. 39 of [70] for details).

Pseudospectral Properties of the Advection-Diffusion Operator

Differential operators employed in partial differential equations involving a combination of advection and diffusion terms (a sum of first and second spatial derivatives) are also non-normal in the case of variable coefficients or if boundaries are introduced, as in finite geometries. A typical representative (see Eq. (10.65)) is the advection-diffusion equation

$$u_t = Du_{xx} + u_x \equiv Au\,, \quad x \in (0, 1)\,, \tag{4.45}$$

where we impose the boundary conditions $u(0) = u(1) = 0$ and some initial condition, and where D is the diffusion constant. The eigenvalues and eigenfunctions of the operator $A = D(\mathrm{d}^2/\mathrm{d}x^2) + (\mathrm{d}/\mathrm{d}x)$ are

$$\lambda_n = -\frac{1}{4D} - Dn^2\pi^2\,, \quad u_n(x) = \mathrm{e}^{-x/2D}\sin(n\pi x)\,, \tag{4.46}$$

where $n = 1, 2, \ldots$ The easiest way to generate the pseudospectrum of A (or any other differential operator) is to discretize the problem on a grid, turning A into a n-dimensional matrix, and compute the pseudospectrum by (4.39) while pushing n as high as possible. (This approach avoids the nuisance of computing operator norms.) Let us choose the Chebyshev spectral derivatives defined by (12.17) and (12.18), embodied in the corresponding first and second derivative matrices (12.20) and (12.21). The boundary conditions are enforced by removing the first and last rows and columns from these matrices.

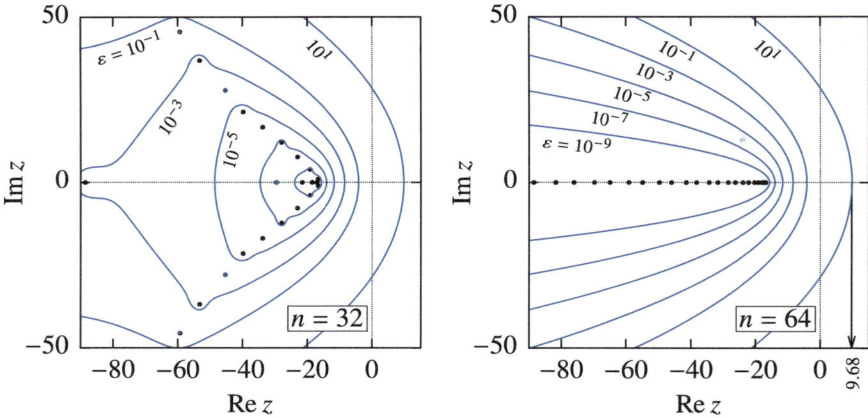

Fig. 4.14 Spectra (small circles) and pseudospectra (contours) of the discretized operator $A = D(\mathrm{d}^2/\mathrm{d}x^2) + (\mathrm{d}/\mathrm{d}x)$ for the advection-diffusion problem (4.45) with $D = 0.015$ and matrix dimension [LEFT] $n = 32$ and [RIGHT] $n = 64$

Sample pseudospectra for $D = 0.015$ are shown in Fig. 4.14. With increasing n, the eigenvalues of the finite matrix A indeed start to line up along the real axis as dictated by the continuum formula (4.46). One can also confirm—try this as an exercise!—that in the limit $D \to 0$ the operator A loses its diffusion character and becomes just (4.44), so its pseudospectra morph into straight vertical lines in the complex plane precisely as in Fig. 4.13.

Figure 4.15 shows the norm of the evolution operator e^{tA} as a function of t. Theorem 15.3 in [70] states that this norm is bounded as

$$\|\mathrm{e}^{tA}\| \geq \mathrm{e}^{t\alpha(A)}, \quad t \geq 0, \tag{4.47}$$

where $\alpha(A)$ is the spectral abscissa, i.e. the supremum of the real part of the spectrum of A, and that $\lim_{t \to \infty} t^{-1} \log \|\mathrm{e}^{tA}\| = \alpha(A)$. A similar bound (Theorem 15.4) can be formulated in terms of the *pseudospectral* abscissa, $\alpha_\varepsilon(A)$:

$$\|\mathrm{e}^{tA}\| \geq \mathrm{e}^{ta} - \frac{\mathrm{e}^{ta} - 1}{\alpha_\varepsilon(A)/M\varepsilon}, \quad t \geq 0, \tag{4.48}$$

where $a = \mathrm{Re}\, z$ and $M = \sup_{t \geq 0} \|\mathrm{e}^{tA}\|$. The large-$t$ behavior of $\|\mathrm{e}^{tA}\|$ is therefore entirely driven by the eigenvalues, as illustrated by the rapid falloffs at $t \gg 1$. The features at small t, however, are governed by the pseudospectrum. We can see that by comparing the two constraints given above. Evaluating (4.47) with $D = 0.015$ and the true spectral abscissa $\alpha(A) = \lambda_1 \approx -16.81$ at, say, $t = 0.1$, yields $\|\mathrm{e}^{tA}\| \geq \mathrm{e}^{0.1(-16.81)} \approx 0.1861$, way below the correct value $\|\mathrm{e}^{tA}\| \approx 0.9839$. Alternatively, we may exploit (4.48), noting that $M = 1$. Choosing, for instance, $\varepsilon = 10$, corresponding

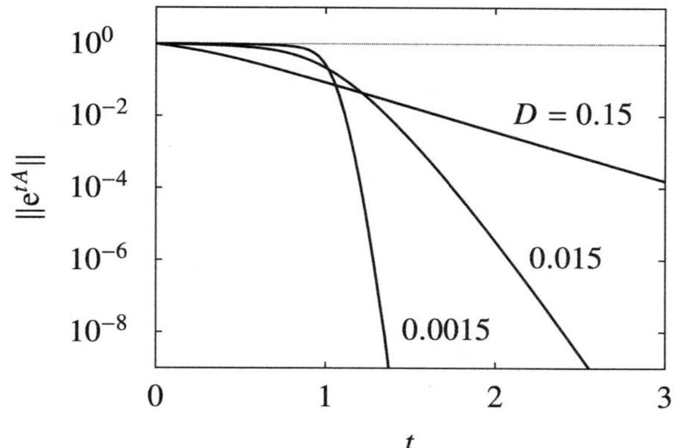

Fig. 4.15 The norm of the evolution operator e^{tA} with $A = D(d^2/dx^2) + (d/dx)$ acting as in (4.45), for three values of the diffusion constant D

to the pseudospectral abscissa $\alpha_\varepsilon(A) = a \approx 9.68$ denoted by the arrow in Fig. 4.14 (right), we get a much better estimate $\|e^{tA}\| \geq 0.9460$.

4.9 Random Matrices ⋆

Random matrices [71] are matrices with elements that are chosen randomly. They find their use in hypothesis confirmation in theoretical physics, as in the theory of quantum chaos [72], information theory [73], finance [74], in numerical algorithms [75] and elsewhere [76]. The theory focuses on specific classes of such matrices [77]. For a physicist, Gaussian and circular (orthogonal or unitary) ensembles of random matrices are the most relevant. In this Section we show how such matrices are generated and discuss their fundamental properties.

4.9.1 General Random Matrices

In certain applications we encounter square matrices with completely independent or weakly correlated random elements. Let the $n \times n$ matrix M_n have real or complex elements which are independent and identically distributed random numbers. Assume that the distribution of these numbers has zero mean and unit variance, and that all its higher moments are bounded. In addition, let $\{\lambda_i\}_{i=1}^n$ be the spectrum (the set of eigenvalues) of M_n and

$$\frac{d^2 P_{M_n}}{dx\,dy} \equiv \frac{d P_{M_n}}{dz} = p_{M_n}(z) = \frac{1}{n} \sum_{i=1}^n \delta(z - \lambda_i), \quad z \in \mathbb{C},$$

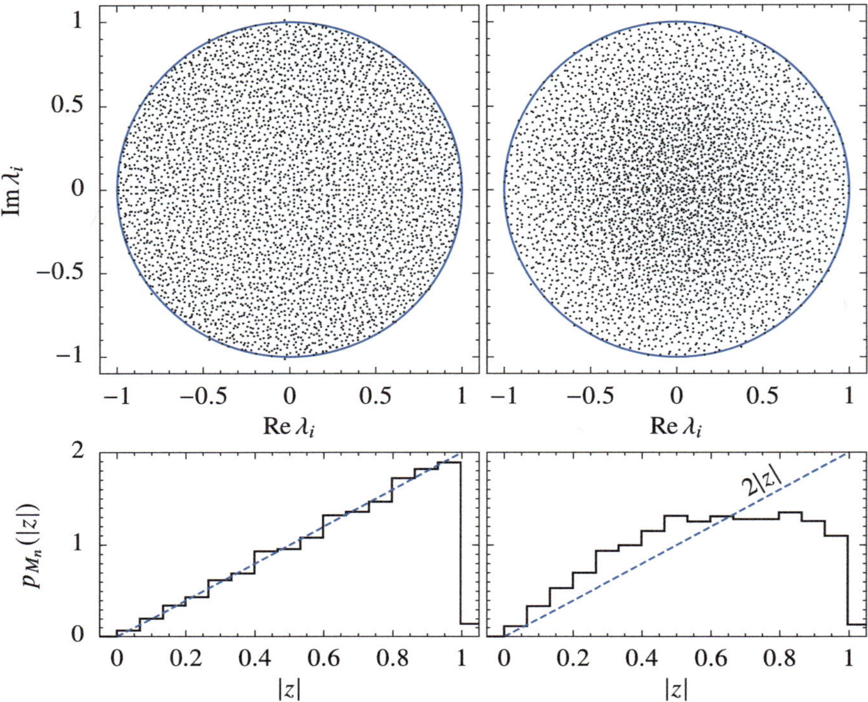

Fig. 4.16 Distribution of eigenvalues λ_i in the complex plane and the probability density of the distribution of $|\lambda_i|$ for real matrices M_n of dimension $n = 4000$. [LEFT] Completely independent matrix elements $(M_n)_{ij}$. The eigenvalues are uniformly distributed within the unit circle. [RIGHT] If the elements $(M_n)_{ij}$ are weakly correlated, the density of the eigenvalues may increase close to the origin and decreases near the unit circle compared to the uniform density. In both cases $\langle (M_n)_{ij} \rangle = 0$ and $\langle (M_n)_{ij}^2 \rangle = 1$

the probability density of eigenvalues in the complex plane. Then $p_{M_n}(z)$, in the sense of the probability that a chosen eigenvalue is found near some specific point, converges to the uniform density on the unit circle when n is increased [78],

$$\lim_{n \to \infty} p_{M_n}(z) = \begin{cases} \pi^{-1} & ; \ |z| \leq 1 \,, \\ 0 & ; \ \text{otherwise} \,, \end{cases} \tag{4.49}$$

which is known as the *circular law* or Girko's law [79]. (There is convincing numerical evidence that the circular law also applies to other classes of random matrices, for instance, to Markov matrices [80].) The validity of the law (4.49)—note that $\mathrm{d}P_{M_n}/\mathrm{d}r = 2\pi r (\mathrm{d}P_{M_n}/\mathrm{d}z) = 2r$ and $r \equiv |z|$—is illustrated in Fig. 4.16.

Marčenko–Pastur theorem Stating this important theorem calls for some preliminaries. Assume that the probability density function f of a random variable is known. The Stieltjes transformation [81] of f is

$$S_f(z) = \int\limits_{-\infty}^{\infty} \frac{f(x)}{x-z}\,\mathrm{d}x\,, \quad \mathrm{Im}\,z \neq 0\,,$$

while its inverse is

$$f(x) = \frac{1}{\pi} \lim_{\xi\to 0} \mathrm{Im}\,S_f(x+\mathrm{i}\xi)\,. \tag{4.50}$$

Let us define a $n \times n$ Hermitian matrix of the form

$$B_n = A_n + \frac{1}{n}\,X^\dagger T X\,,$$

where A_n is a $n \times n$ Hermitian matrix. When n is increased, let the distribution of its eigenvalues p_{A_n} *almost surely* converge to the distribution p_A with the corresponding Stieltjes transformation S_A. ("Almost sure convergence" means that the probability for the convergence of a sequence is equal to one [82].) Let X be a $m \times n$ complex matrix with independent and identically distributed elements X_{ij}, with the average $\langle X_{ij}\rangle = 0$ and variance $\langle |X_{ij}|^2\rangle = 1$. The diagonal matrix $T = \mathrm{diag}(\tau_1, \tau_2, \ldots, \tau_m)$ should contain elements for which almost surely a limit probability density $H(\tau)$ exists when $m \to \infty$.

With the assumptions listed above, the Marčenko–Pastur theorem [83] tells us that the probability density p_{B_n}, when n is increased, almost surely converges to the limit distribution p_B, and the Stieltjes transformation S_B of this distribution satisfies the self-consistency relation

$$S_B(z) = S_A\left(z - c\int\limits_{-\infty}^{\infty} \frac{\tau H(\tau)}{1+\tau S_B(z)}\,\mathrm{d}\tau\right)\,, \quad c = \lim_{n\to\infty} \frac{m(n)}{n} > 0\,. \tag{4.51}$$

Symmetric real or Hermitian complex matrices defined in terms of products $X^\dagger X$ are frequently seen in solutions of over-determined systems, in work with covariance matrices, or in applications of the singular value decomposition. In certain cases (see [84]) we may assume that the elements X_{ij} of X are independent and identically distributed with zero mean and unit variance. Then the Marčenko–Pastur theorem applies with $A_n = 0$, $S_A(z) = \int \delta(x)(x-z)^{-1}\,\mathrm{d}x = -z^{-1}$, $T = 1_m$, $H(\tau) = \delta(\tau - 1)$, and $B_n = \frac{1}{n}X^\dagger X$. When these quantities are inserted in (4.51), we obtain a quadratic equation $zS_B(z)^2 + S_B(z)(z - c + 1) + 1 = 0$. We solve it for S_B, and use the inverse Stieltjes transformation (4.50) to calculate

$$p_B(x) = \max\{0, 1 - c\}\delta(x) + \frac{\sqrt{(x - b_-)(b_+ - x)}}{2\pi x}\,I_{[b_-, b_+]}(x)\,,$$

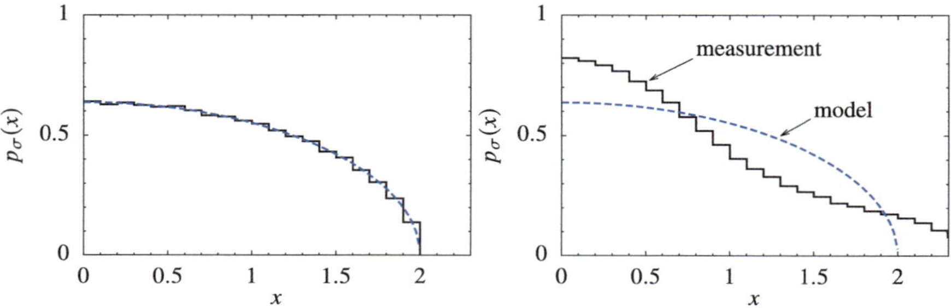

Fig. 4.17 Singular values of real matrices $X/\sqrt{n}$ of dimension $n = 4000$ with uniformly distributed elements X_{ij} having zero mean ($\langle X_{ij}\rangle = 0$) and unit variance ($\langle X_{ij}^2\rangle = 1$). [LEFT] Distribution for independent elements X_{ij}. [RIGHT] Distribution for weakly row-correlated X_{ij}. Due to the redundancy of the data the concentration of singular values increases near the origin and some of them may actually exceed the value 2

where $b_{\pm} = (1 \pm \sqrt{c})^2$, and $I_{[a,b]}(x) = 1$ for $x \in [a, b]$ or 0 otherwise. If X is a $n \times n$ (square) matrix, then $c = 1$, and the formula above can be simplified as

$$p_B(x) = \frac{1}{2\pi x}\sqrt{x(4 - x)}\, I_{[0,2]}(x) \,.$$

The Marčenko–Pastur theorem assists us in many ways. In singular value decomposition of X, we find unitary matrices U and V such that $X = U\Sigma V^{\dagger}$ and $\Sigma = \mathrm{diag}(\sigma_1, \sigma_2, \ldots, \sigma_m)$, where $\{\sigma_i \geq 0\}_{i=1}^{m}$ is the set of singular values of X. From the decomposition formula we see that $X^{\dagger}X = V\Sigma^2 V^{\dagger}$, so the eigenvalues of $X^{\dagger}X$ are simply the squares of the singular values of X (see also Appendix A.6). One of the consequences is the *quarter-circular law*: it states that the probability density of the distribution of singular values of $X/\sqrt{n}$ converges, in the sense probability, to

$$p_\sigma(x) = \frac{2}{\pi}\sqrt{1 - \left(\frac{x}{2}\right)^2}\, I_{[0,2]}(x) \,, \qquad (4.52)$$

when n is increased. In plain words, the singular values of the random matrix $X/\sqrt{n}$ do not exceed the value 2. Equation (4.52) is illustrated by Fig. 4.17.

4.9.2   Gaussian Orthogonal or Unitary Ensemble

An ensemble of matrices is defined by the set of its elements (matrices) and by their distribution in the set. In the following, we are dealing with $n \times n$ matrices. The *Gaussian orthogonal ensemble* (GOE) consists of real symmetric matrices H with $n(n + 1)/2$ free real parameters. The matrices from this set are distributed according to the probability density p which is invariant to orthogonal transformations O in the sense

$$p(H) = p(O^{\mathrm{T}} H O)\,.$$

The *Gaussian unitary ensemble* (GUE) consists of complex Hermitian matrices with n^2 free real parameters and a probability density p which is invariant with respect to unitary transformations U:

$$p(H) = p(U^{\dagger} H U)\,.$$

The probability density p of the matrix H should be understood as the probability density of its independent elements. We can use the invariance condition to prove that the most general form of the density is $p(H) = \exp(a\,\mathrm{tr}(H^2) + b\,\mathrm{tr}(H) + c)$ where a and b are adjustable parameters and c is a normalization constant [72]. The ratio $b/(2a)$ represents the center of mass of the eigenvalues of H, which we are allowed to shift arbitrarily, so we set $b = 0$. The probability density of matrices belonging to either of the two ensembles is then

$$p(H) = \exp(a\,\mathrm{tr}(H^2) + c)\,.$$

For simplicity, let us choose $a = 1/2$. The elements of matrices H from the GOE or the GUE can be randomly generated by using the formulae

$$
\begin{array}{lll}
 & \mathrm{GOE} & \mathrm{GUE} \\
i < j : & H_{ij} = 2^{-\frac{1}{2}} x_{ij}\,, & H_{ij} = 2^{-\frac{1}{2}} (x_{ij} + \mathrm{i}\, y_{ij})\,, \\
i = j : & H_{ii} = x_{ii}\,, & H_{ii} = x_{ii}\,,
\end{array}
$$

where x_{ij} and y_{ij} are distributed normally according to $N(0, 1)$.

In the limit of large n, the spectra of matrices from GOE and GUE possess certain well-defined statistical properties. Let H_n be a $n \times n$ matrix from GOE or GUE with the spectrum $\Lambda = \{\lambda_i\}_{i=1}^n$, and let us rescale the spectrum by using $\Lambda' = (a/n)^{1/2}\Lambda$ in the case of GOE, or by using $\Lambda' = (a/2n)^{1/2}\Lambda$ in the case of GUE. Then, if n is increased, the probability density of the distribution of the rescaled eigenvalues Λ' converges, in the sense of probability, to

$$
p_{\mathrm{sc}}(x) = \begin{cases} \frac{2}{\pi}\sqrt{1 - x^2} \; ; & x \in [-1, 1]\,, \\ 0 & ; \text{ otherwise}\,. \end{cases}
\tag{4.53}
$$

Equation (4.53) establishes *Wigner's semi-circular law* [77], which tells us that we should expect the magnitudes of the rescaled eigenvalues not to exceed unity (see Fig. 4.18).

For random matrices, the fluctuations of the splittings between consecutive eigenvalues with respect to their local averages are very instructive. It turns out that the statistics of these fluctuations is equal to the statistics of the spectra of Hamiltoni-

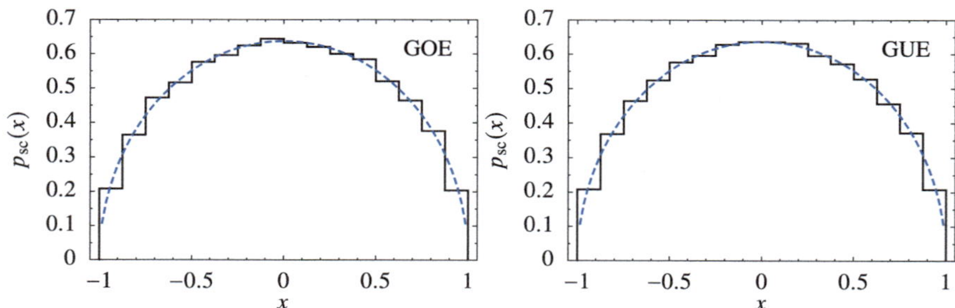

Fig. 4.18 Probability densities of the rescaled eigenvalues Λ' of a single realization of a matrix of dimension $n = 2000$ [LEFT] from the GOE and [RIGHT] from the GUE. The dashed lines represent the Wigner semi-circle (4.53)

ans corresponding to classically chaotic quantum systems [85]. The nature of the fluctuations becomes apparent when the spectrum is *unfolded*.

Unfolding of matrix spectra Assume that the spectrum Λ of a real symmetric or a Hermitian matrix is already known, and that the eigenvalues are sorted as $\{\lambda_i \in \mathbb{R} : \lambda_{i+1} \geq \lambda_i\}_{i=1}^n$. We use the eigenvalues to construct the density

$$d(\lambda) = \sum_{i=1}^{n} \delta(\lambda - \lambda_i)$$

and compute the cumulative distribution function

$$N(\lambda) = \int_{-\infty}^{\lambda} d(\lambda')\, d\lambda' = \#\{\lambda_i \leq \lambda\}\,,$$

which measures the numbers of eigenvalues not exceeding λ. This is a staircase-like, monotonously increasing function of λ. By using some method of local averaging, we split N into a smooth part $\overline{N}$ and an oscillatory part N_{osc},

$$N(\lambda) = \overline{N}(\lambda) + N_{\mathrm{osc}}(\lambda)\,,$$

so that the local average of N_{osc} over a few (say, 10) neighboring eigenvalues is equal to zero or is negligible compared to the increase in N. We use the average increase of the number of eigenvalues $\overline{N}$ to define the average density of eigenvalues, $\overline{d}(\lambda) = d\overline{N}(\lambda)/d\lambda$. In practice, this connection is often used in reverse to compute $\overline{N}$ if $\overline{d}$ is known. For a large enough Δ we can compute the average density

$$\overline{d}(\lambda) = \frac{1}{2\Delta} \int_{-\Delta}^{\Delta} d(\lambda + t)\, dt$$

and define

$$\overline{N}(\lambda) = \int\limits_{-\infty}^{\lambda} \overline{d}(\lambda')\,\mathrm{d}\lambda' \, .$$

The separation of the smooth and oscillatory parts is not unique, but in the limit of large matrices it does not influence significantly the conclusions about the statistics of the fluctuations. By using the function $\overline{N}$ the spectrum $\{\lambda_i\}_{i=1}^n$ is *unfolded* into

$$\widetilde{\lambda}_i = \overline{N}(\lambda_i)\,, \quad i = 1, 2, \ldots, n\,,$$

so that the average distance between the neighboring eigenvalues $s_i = \widetilde{\lambda}_{i+1} - \widetilde{\lambda}_i$ locally (over a couple of consecutive eigenvalues) and globally equals 1. Finally, we are interested in the probability density of this average distance after the unfolding, which is

$$p_{\mathrm{s}}(x) = \frac{1}{n-1} \sum_{i=1}^{n-1} \delta(x - s_i) \, .$$

For large matrices from GOE ($a = 1/(2n)$) or GUE ($a = 1/(4n)$) we may approximate $\overline{d}$ by the limit density (4.53), so $\overline{d}(x) = n p_{\mathrm{sc}}(x)$ with $\Delta \to 0$, and we get

$$\overline{N}(\lambda) \sim n \int\limits_{-1}^{\lambda} p_{\mathrm{sc}}(\lambda')\,\mathrm{d}\lambda' = n\left(1 + \frac{1}{\pi}\left[\arcsin\lambda - \lambda\sqrt{1-\lambda^2}\right]\right)\,, \quad \lambda \in [-1, 1]\,.$$

With increasing n, the probability density p_{s} of the distances between neighboring unfolded eigenvalues becomes very close, in the sense of probability, to the density

$$p_{\mathrm{s}}^{\mathrm{GOE}}(x) = \frac{\pi}{2}\,x\,\mathrm{e}^{-(\pi/4)x^2} \tag{4.54}$$

for matrices from GOE, or to the density

$$p_{\mathrm{s}}^{\mathrm{GUE}}(x) = \frac{32}{\pi^2}\,x^2\,\mathrm{e}^{-(4/\pi)x^2} \tag{4.55}$$

for matrices from GUE. Formulae (4.54) and (4.55) are known as the Wigner distributions [77, 86]. Both distributions vanish at $x = 0$, which is a sign of the *avoidance of level crossing* or *level repulsion* (repulsion between neighboring eigenvalues), a mechanism which is statistically relatively stronger for matrices from GUE. Fig. 4.19 shows the probability densities p_{s} for large matrices from GOE and GUE, as well as the comparison to the forms (4.54) and (4.55). The connections between the Wigner distributions and the spectra of quantum-mechanical systems are discussed in Problem 4.10.7.

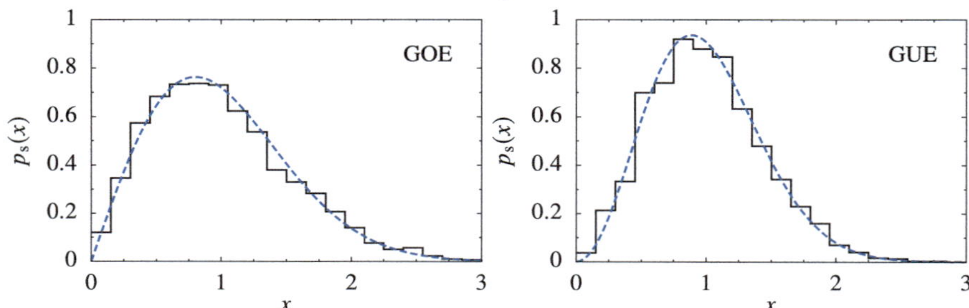

Fig. 4.19 Probability densities of the distribution of the splittings between the subsequent eigenvalues of a single realization of a matrix with $n = 2000$ after unfolding. [LEFT] A matrix from GOE. [RIGHT] A matrix from GUE. The dashed curves correspond to the Wigner distributions (4.54) and (4.55)

4.9.3 Circular Ensembles

In certain physical problems, we need to comb the space in all orthogonal directions, and the procedure needs to be repeated for randomly chosen sets of directions. Such sets can be generated by a special class of random matrices discussed in this Subsection. Again we restrict ourselves to $n \times n$ matrices.

The *circular orthogonal ensemble* (CRE) consists of real orthogonal matrices $O(n)$, the *circular unitary ensemble* (CUE) consists of unitary matrices $U(n)$, and the *circular orthogonal ensemble* (COE) consists of symmetric unitary (not necessarily real) matrices $V(n)$ whose measure is the normalized Haar measure μ_H [77].

Haar measure For a set of matrices Γ and the corresponding set of all its open subsets Σ, the Haar measure $\mu_H : \Sigma \to \mathbb{R}$ is defined as

$$\mu_H(\Gamma) = 1 \, ,$$
$$\mu_H(\gamma S) = \mu_H(S) \, , \quad \forall \gamma \in \Gamma \, , \quad \forall S \in \Sigma \, .$$

If the parameterization of the matrices in Γ is known, we prefer to specify the differential of the measure at some point, $\gamma \in \Gamma$, hence $\mathrm{d}\mu_H(\gamma) = p(\gamma) \, \mathrm{d}\gamma$, where $p(\gamma)$ is the probability density of the matrices over the set Γ. For example, the Haar measure for a set of one-dimensional unitary matrices is

$$U(1) = \left\{ \mathrm{e}^{\mathrm{i}\theta} : \theta \in [0, 2\pi) \right\} .$$

For 3×3 special orthogonal matrices, the Haar measure is

$$SO(3) = \left\{ R_z(\phi_2) R_x(\theta) R_z(\phi_1) : \theta \in [0, \pi) \, , \quad \phi_1, \phi_2 \in [0, 2\pi) \right\} ,$$

where $R_o(\alpha) \in \mathbb{R}^{3 \times 3}$ is the rotation matrix for a rotation around the axis o by the angle α. Both sets are uniquely generated, so the differential of the Haar measure in the case of $U(1)$ matrices is equal to

$$d\mu_{\mathrm{H}}(\gamma) = \frac{1}{2\pi}\, d\theta \,, \qquad \gamma \in U(1)\,,$$

while for the $SO(3)$ matrices it is

$$d\mu_{\mathrm{H}}(\gamma) = \frac{1}{8\pi^2}\, \sin\theta\, d\theta\, d\phi_1\, d\phi_2\,, \qquad \gamma \in SO(3)\,.$$

Generating matrices from CUE To generate random matrices from CUE we first generate a Hermitian matrix A from GUE and perform its spectral decomposition,

$$A = UDU^\dagger \,, \qquad U = [\,U_{ij}\,]_{i,j=1}^n \,.$$

By decomposing A we obtain a random unitary matrix $U \in U(n)$ and the diagonal matrix D containing the eigenvalues of A. Assume that we have used a diagonalization routine that sets the phase of the leading coefficient of individual eigenvectors to zero. To eliminate this regularity, we randomly pick the phases $\{\phi_i\}_{i=1}^n$ that are uniformly distributed on $[0, 2\pi)$, and use them to modify the phases of the columns of U by using

$$U_{\mathrm{rnd}} = \left[e^{i\,\phi_j}\, U_{ij} \right]_{i,j=1}^n \,.$$

In this manner, we have generated a random unitary matrix from CUE which is chosen randomly with respect to the Haar measure.

Generating matrices from COE A random matrix V from COE is obtained by taking a matrix U from CUE (see above) and forming $V = U^{\mathrm{T}}U$.

Generating matrices from CRE There are many methods to generate a random orthogonal matrix from CRE that preserves the Haar measure. The simplest way is to form a symmetric matrix A from GOE and diagonalize it,

$$A = ODO^{\mathrm{T}} \,, \qquad O = [O_{ij}]_{i,j=1}^n \,.$$

By doing this, we obtain a random orthogonal matrix $O \in O(n)$. Assume that the diagonalization has set the leading coefficient of each column of O to a positive value, as it is done by LAPACK routines [3]. This violates the condition that the matrices are chosen randomly with respect to the Haar measure. In this case, we randomly choose the signs $s_i \in \{-1, 1\}$ of the columns of O with equal probabilities, $\mathrm{Prob}(s = \pm 1) = 1/2$, and finally generate the desired random orthogonal matrix from CRE,

$$O_{\mathrm{rnd}} = \left[s_j\, O_{ij} \right]_{i,j=1}^n \,,$$

which is chosen randomly with respect to the Haar measure.

The above procedures for generation of circular matrices require $\mathcal{O}(n^3)$ operations with standard diagonalization algorithms. In spite of their simplicity, they are just by a constant factor slower than the optimal methods based on Householder reflections (for GOE, [87]) or QR decomposition (for GUE, [88]). When a random orthogonal or unitary matrix X is not needed explicitly (for example, if already the products AX or XA are sufficient), the generation requires only $\mathcal{O}(n^2)$ operations.

Eigenvalues of unitary matrices reside on the unit circle. It should therefore not surprise us that the eigenvalues of matrices from CUE in the limit of large dimensions are also uniformly distributed on the unit circle. Interesting mathematical details can be found in [72] and [77].

4.10 Problems

4.10.1 Percolation in a Random-Lattice Model

Percolation is the passage of substances from one region to another in the process of filtering. In this Problem it is discussed in the context of random lattices which appear frequently in the physical descriptions of gels, polymers, glasses, and other disordered matter [89]. Mathematically, random lattices are graphs [90]. Here, we are dealing only with *discrete random lattices* in fixed geometries and possessing a certain degree of statistical disorder in their connectedness. One should distinguish the *lattice of connections*, where connections are randomly chosen, from the *lattice of sites*, in which the occupation of sites is random [91]. We shall study the electric conductivity of both types of lattices.

A lattice made of conductive material is placed between electrodes to which a constant voltage is applied. We randomly drop a fraction p of the connections from the lattice and remove the loose ends which can not conduct. The lattice used in the calculations is shown in Fig. 4.20 (right). We would like to compute the current I flowing through the lattice, and determine how I changes (on average) with p.

The lattice is in fact a circuit of N resistors $\{R_i\}_{i=1}^N$ and voltage sources $\{U_j\}_{j=1}^N$ connected at M nodes. Each connection can be oriented (positive direction from its beginning to its end). A voltage source with subscript j represents a point at the end of a free connection with a known electric potential U_j. Through a resistor R_i, a current I_i is flowing, and the sums of the incoming and outgoing currents should be equal at each node, which can be written as the matrix equation $AI = 0$ for the vector of currents $I = (I_i)_{i=1}^N$. The matrix $A \in \mathbb{R}^{M \times N}$ with the elements $A_{ij} \in \{-1, 1\}$ contains all information on the connectedness of the circuit (it is the *node-edge adjacency matrix* in graph theory).

For the ith connection between the potential e at the beginning of the connection and e' at its end, Ohm's law applies: $e + R_i I_i = e'$. At the edges of the lattice, the potential is given by the external voltage source and then e or e' are kept in the vector U; inside the lattice, e and e' are collected in the vector E. For all connections, Ohm's law can be written in the matrix form $RI + A^{\mathrm{T}}E = U$ where $R = \mathrm{diag}(R_i)_{i=1}^N$ is the resistance matrix, so all equations can be merged into one:

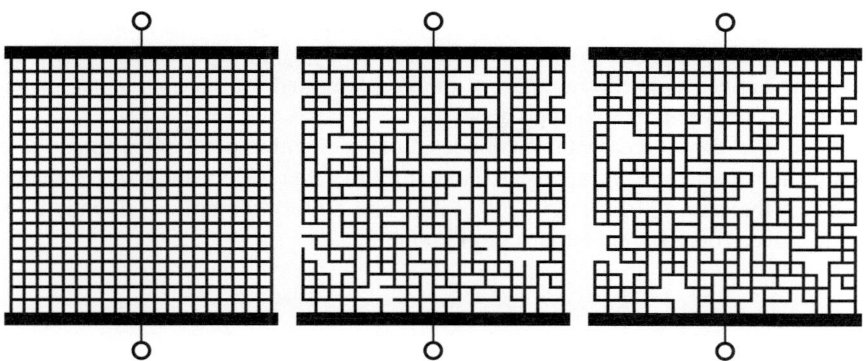

Fig. 4.20 Cartesian lattices of connections between the electrodes with an applied constant voltage. In the full lattice with 20×20 sites (left) we randomly drop $p = 20\%$ of connections (center) and finally remove the unconnected ends (right)

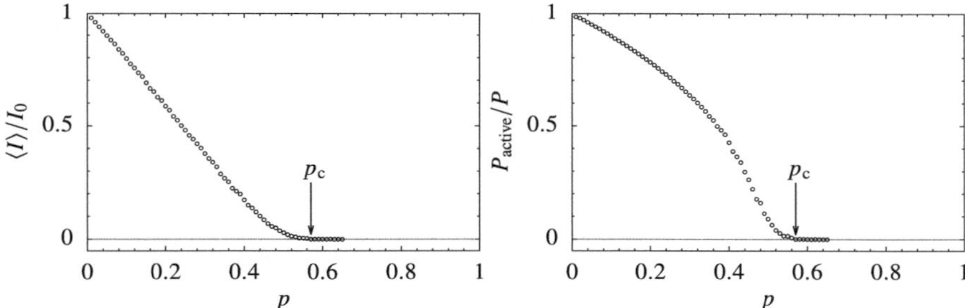

Fig. 4.21 [LEFT] The current flowing from the lattice of resistors as a function of the fraction p of dropped connections. [RIGHT] The fraction of the connections actually carrying current. At the critical value $p_c \approx 0.57$ the lattice no longer conducts

$$\begin{pmatrix} R & A^{\mathrm{T}} \\ A & 0 \end{pmatrix} \begin{pmatrix} I \\ E \end{pmatrix} = \begin{pmatrix} U \\ 0 \end{pmatrix}.$$

⊙ Calculate the average current $\langle I \rangle$ through 10×10, 20×20, and 50×50 lattices as a function of the fraction p of dropped connections. All connections have equal conductivities. To compute the total current, sum all components of I. Compute the average over 100 random lattices of equal size and normalize $\langle I \rangle$ to the current I_0 flowing through the complete lattice. The result should look approximately as in Fig. 4.21 (left). Draw the ratio of the number of connections P_{active} actually carrying current to the number of original connections P (Fig. 4.21 (right)). Determine the critical fraction of the dropped connections p_c at which the lattice, on average, becomes globally unconnected and stops conducting.

⊕ To implement the lattice, we can also use nodes with attached connections spanning half of the inter-node distance. We randomly fill the sites of an empty lattice with nodes with attached connections. Eventually, an agglomeration of nodes builds up and the lattice is filled to a degree r. At some critical degree r_c the lattice becomes globally connected and starts to conduct. Compute $\langle I \rangle / I_0$ in dependence of r for the same lattices as in the first part of the Problem, and estimate r_c. What are the differences between the two approaches?

At some r the lattice contains connected clusters with typical extensions of $l = \max |r_i - r_j|$ where $(i, j) \in$ cluster. How does the average cluster size $\langle l \rangle$ change with r in the case of 100×100, 200×200, and 500×500 lattices? Compute the average over 100 different lattices of the same size.

4.10.2   Electric Circuits of Linear Elements

Electric circuits are systems of resistors, coils, capacitors, diodes, transistors, and other components. By Kirchhoff's laws, the sum of the currents at each node is zero, and the sum of the voltage drops in any closed loop of the circuit is zero. The time evolution of circuits is hard to predict if its elements are non-linear, and may even be chaotic [92]. If the circuit contains only linear elements, the usual relations $U = RI$, $U = -L\dot{I}$, and $U = e/C$ apply, and the system is analytically solvable.

⊙ Devise a connected electric circuit containing resistors, coils, and capacitors. The number of the elements n should be larger than 5. By using Kirchhoff's laws, obtain a homogeneous system of linear differential equations for the voltage drops $\{U_i\}_{i=1}^{n}$ on individual elements. Introduce new variables such that the problem can be reformulated as a single differential equation of order one. Let us denote all variables by $\{X_i\}_{i=1}^{m \geq n}$. For the solution of the equation, we use the ansatz $X_i = \exp(i\omega t)Y_i$ where $Y_i \in \mathbb{C}$ are constants, so we obtain a non-homogeneous eigensystem. Find the spectrum of frequencies ω for this system by using any of the methods described in the text. Observe the changes in the spectrum when the values of the resistances (energy losses) are modified.

⊕ Another option is to use the ansatz $U_i(t) = \exp(i\omega t)V_i$ in the original equation, where $V_i \in \mathbb{C}$ are constants. This yields a linear homogeneous system of equations for the vector $v = \{V_i\}_{i=1}^{n}$ of the form $A(\omega)v = 0$. We expect non-trivial solutions at the frequencies ω which are zeros of the determinant of the matrix $A(\omega)$: $\det(A(\omega)) = 0$. In general, the polynomial $\det(A(\omega))$ is complex, and its roots can be found by general programs (see Appendix J).

4.10.3   Systems of Oscillators

Imagine a one-dimensional chain of n small spheres with masses m connected by springs with elastic coefficients k. At its ends, the chain is attached to the wall by springs with coefficients K (Fig. 4.22).

⊙ Analyze the spectrum of oscillations of this system when the ratio k/K is changed, and try to identify its thermodynamic limit $n \to \infty$. Discuss the spectrum

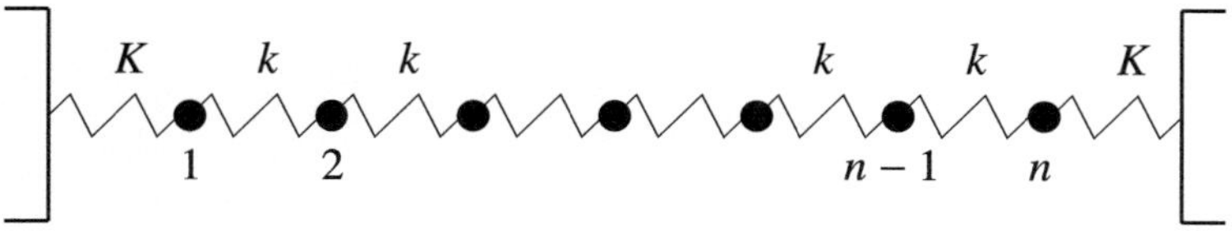

Fig. 4.22 A chain of n small masses connected by springs with elastic coefficients k. The chain is attached to the walls by springs with different coefficients K

in the case when one of the masses is heavier than the others. What happens if all but one sphere move without friction, while a single one of them is submerged in a liquid so that the linear drag law $F_\mathrm{u} = -6\pi\eta r v$ applies to it? Explore the dynamics of the spectrum of oscillations when the damping or the ratio η/k are increased!

⊕ Submerge the whole system in a liquid with viscosity η. Assume linear drag for all spheres and watch how the spectrum changes when η is increased. Set all elastic coefficients to be the same, $k = K$, and repeat the analysis.

4.10.4 *Image Compression by Singular Value Decomposition*

Singular value decomposition (SVD, Sect. 4.5.2) represents a particular choice of the orthonormal basis for the domain and range of the mapping $A : x \mapsto Ax$ such that the matrix in it becomes diagonal. In one form or another, SVD lies at the heart of many classical algorithms for data reduction by which we attempt to represent the data with as few parameters as possible. This Problem acquaints us with using SVD for image compression.

⊙ Perform the singular value decomposition of the 512×512 matrix representing the black-and-white image of Nicolas Copernicus (Fig. 4.7). The pixel map (in shades of gray) can be found at the website of the book in ASCII format. Calculate the Frobenius norm of the matrix. Generate the SVD expansions (4.21) of rank 1, 2, 4, 8, 16, 32, 64, and 128. Convert the matrices of the lower ranks to the format of the original image and see if what you see is identifiable as a face or whether it is pleasant to the eye. How does the norm of the approximate matrix and its relative norm with respect to the original norm change with the rank of the approximation? Draw the magnitudes of the singular values as a function of their index. At any given rank of the expansion (4.21), compare the amount of memory needed to store the approximate image to the amount of memory needed to store the original image (the compression ratio). At what rank your naked eye perceives the approximate image as indistinguishable from the original? What does the image look like if *all but the first two* components are included in the singular expansion?

⊕ Perform the singular value decomposition of the Jackson Pollock's painting *Lavender Mist* (Fig. 4.23) which is represented by the 1024×749 matrix (see website of the book). Again draw the magnitudes of the singular values as a function of

Fig. 4.23 Jackson Pollock (1912–1956), *Number 1, 1950 (Lavender Mist)*; oil, enamel and aluminum on canvas, 221 cm × 300 cm

their index, and calculate the relative error of the norm of the approximate matrix. Determine the rank needed to achieve a 10 % relative error between the approximate and the original matrix, as measured in the Frobenius norm. Compare your results to those obtained from the image of Copernicus. Do your conclusions change if you take a completely random matrix?

4.10.5 *Eigenstates of Particles in the Anharmonic Potential*

In this Problem (adapted from [93]) we study the one-dimensional linear harmonic oscillator (particle of mass m with kinetic energy $T(p) = p^2/(2m)$ in the quadratic potential $V(q) = m\omega^2 q^2/2$). The corresponding Hamiltonian is

$$H_0 = \frac{1}{2}\left(p^2 + q^2\right) ,$$

where energy is in units of $\hbar\omega$, momentum in units of $(\hbar m\omega)^{1/2}$, and linear dimensions in units of $(\hbar/m\omega)^{1/2}$. The eigenstates $|n\rangle$ of the unperturbed Hamiltonian H_0 are known from the introductory quantum mechanics courses: in coordinate representation, they are $|n\rangle = (2^n n!\sqrt{\pi})^{-1/2} e^{-q^2/2}\,\mathcal{H}_n(q)$, where $\mathcal{H}_n$ are the Hermite polynomials. The eigenfunctions satisfy the stationary Schrödinger equation

$$H_0|n^0\rangle = E_n^0|n^0\rangle$$

with non-degenerate eigenenergies $E_n^0 = n + 1/2$ for $n = 0, 1, 2, \ldots$. The matrix $\langle i|H_0|j\rangle$ is obviously diagonal, with values $\delta_{i,j}(i + 1/2)$ on the diagonal. Now let us add the anharmonic term to the unperturbed Hamiltonian,

$$H = H_0 + \lambda q^4 .$$

How does this perturbation modify the eigenenergies? We are looking for the matrix elements $\langle i|H|j\rangle$ of the *perturbed* Hamiltonian in the basis of the *unperturbed* wave-functions $|n^0\rangle$. In the calculation, we make use of the expectation value of the transition matrix element

$$q_{ij} = \langle i|q|j\rangle = \frac{1}{2}\sqrt{i+j+1}\,\delta_{|i-j|,1}\,,$$

which embodies the selection rule for electric dipole transitions between the levels of the harmonic oscillator. In the practical calculation the matrices q_{ij} and $\langle i|H|j\rangle$ of course need to be limited to finite sizes $N \times N$.

⊙ Use diagonalization methods to find the lowest eigenvalues (energies) and eigenfunctions (wave-functions) of the perturbed Hamiltonian $H = H_0 + \lambda q^4$ with the parameter $0 \le \lambda \le 1$. You are solving the eigenvalue problem

$$H|n\rangle = E_n|n\rangle\,.$$

The new (corrected) wave-functions $|n\rangle$ are, of course, linear combinations of the old (unperturbed) wave-functions $|n^0\rangle$. Make sure that $E_n \to E_n^0$ when $\lambda \to 0$. Examine the dependence of the results on the matrix dimension N and observe the convergence of eigenvalues at large N.

Instead of computing the matrix elements q_{ij} and treating $[q_{ij}]^4$ as the perturbation matrix, we might wish to compute the transition matrix elements of the *square* of the coordinate, $q_{ij}^{(2)} = \langle i|q^2|j\rangle$ and treat the perturbation λq^4 as the square of the corresponding matrix, or simply compute the matrix elements of the fourth power of the coordinate, $q_{ij}^{(4)} = \langle i|q^4|j\rangle$ and take *this* matrix as the perturbation:

$$\lambda q^4 \to \lambda\left[q_{ij}\right]^4 \quad \text{or} \quad \lambda q^4 \to \lambda\left[q_{ij}^{(2)}\right]^2 \quad \text{or} \quad \lambda q^4 \to \lambda\left[q_{ij}^{(4)}\right]\,.$$

Identify the differences between these three methods! Use the equations

$$\langle i|q^2|j\rangle = \frac{1}{2}\left[\sqrt{j(j-1)}\,\delta_{i,j-2} + (2j+1)\,\delta_{i,j} + \sqrt{(j+1)(j+2)}\,\delta_{i,j+2}\right],$$

$$\langle i|q^4|j\rangle = \frac{1}{2^4}\sqrt{\frac{2^i\,i!}{2^j\,j!}}\left[\delta_{i,j+4} + 4\,(2j+3)\,\delta_{i,j+2} + 12\,(2j^2+2j+1)\,\delta_{i,j}\right.$$

$$\left. + 16j\,(2j^2-3j+1)\,\delta_{i,j-2} + 16j\,(j^3-6j^2+11j-6)\,\delta_{i,j-4}\right],$$

which are easy to derive from the recurrence relations for Hermite polynomials.

⊕ Compute the lowest eigenenergies and eigenfunctions for the problem in the "Mexican-hat" potential with two minima,

$$H = \frac{p^2}{2} - 2q^2 + \frac{q^4}{10}\,.$$

4.10.6 Anderson Localization

One of the important successes of twentieth-century physics was the discovery of
the localization of quantum states at impurities in magnets and disordered systems,
which restrict the diffusion through such systems [94]. In this Problem we examine
the localization in a chain of N harmonic oscillators with random masses $M_i > 0$,
which is attached at its ends [95]. Let q_i be the deflection of the ith mass from
equilibrium, $p_i = M_i \dot{q}_i$ its momentum, and K the elastic constant of the springs
connecting the masses. The Hamiltonian of the system is

$$H = \frac{1}{2} \sum_{i=0}^{N-1} \frac{p_i^2}{2M_i} + \frac{K}{2} \sum_{i=0}^{N-1} (q_{i+1} - q_i)^2 .$$

The boundary condition requires $q_{-1} = q_N = 0$. The equation of motion for the ith
oscillator is $M_i \ddot{q}_i = K(q_{i+1} - 2q_i + q_{i-1})$. With the effective couplings $w_i = K/M_i$
and substitution $x_i = q_i/\sqrt{w_i}$ the system can be rewritten as

$$\frac{\mathrm{d}^2 x_i}{\mathrm{d}t^2} = \sqrt{w_i w_{i+1}}\, x_{i+1} - 2w_i x_i + \sqrt{w_i w_{i-1}}\, x_{i-1} , \quad i = 0, 1, \ldots, N-1 ,$$

or, in matrix form,

$$\ddot{z} = -Wz , \quad W = \begin{pmatrix} 2w_0 & -\sqrt{w_0 w_1} & 0 & 0 & \ldots \\ -\sqrt{w_0 w_1} & 2w_1 & -\sqrt{w_1 w_2} & 0 & \ldots \\ 0 & -\sqrt{w_1 w_2} & 2w_2 & -\sqrt{w_1 w_3} & \ldots \\ \ldots & \ldots & \ldots & \ldots & \ldots \end{pmatrix} , \quad (4.56)$$

where $z = (x_0, x_1, \ldots, x_{N-1})^{\mathrm{T}}$. By solving the eigenvalue problem

$$Wz = \omega^2 z , \quad z \neq 0 ,$$

we obtain the eigenvalues (angular frequencies) $\{\omega_i\}_{i=0}^{N-1}$ and the corresponding
eigenvectors (eigenmodes) $\{z_i\}_{i=0}^{N-1}$ which are normalized as $\|z_i\|_2 = 1$. By using
the known pairs $\{(\omega_i, z_i)\}_{i=0}^{N-1}$ we can rewrite any solution of the differential equation (4.56) as the sum

$$x(t) = \sum_{i=0}^{N-1} \left[a_i^{\mathrm{T}} z_i \, e^{+\mathrm{i}\omega_i t} + b_i^{\mathrm{T}} z_i \, e^{-\mathrm{i}\omega_i t} \right] ,$$

where the vectors $a_i, b_i \in \mathbb{C}^N$ are determined from the initial conditions $x(0)$ and
$\dot{x}(0)$. We are interested in the eigenmodes. The energies ω_i^2 of the eigenmodes
are homogeneous functions of the reciprocal masses w_i, so $(w_i \to \lambda w_i) \Leftrightarrow (\omega_i^2 \to \lambda \omega_i^2)$. This enables us to fix one moment of the distribution of the values w_i, for

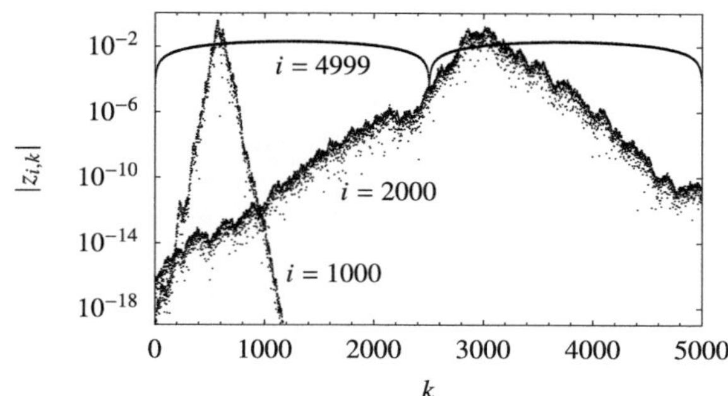

Fig. 4.24 Examples of eigenmodes in a slightly disordered harmonic chain with $N = 5000$ in the case when the weights w_i are uniformly distributed on $[0.9, 1.1]$

example, its average $\langle w \rangle = 1$. The high-energy eigenmodes typically have a limited range, which is known as the *localization of modes*. Some examples are shown in Fig. 4.24.

A physically relevant quantity is the localization length. It can be defined by the average of any quantity f with respect to the ith eigenmode $z_i = (z_{i,k})_{k=0}^{N-1}$,

$$\langle f \rangle_{\alpha,i} = \frac{\sum_k f(k)|z_{i,k}|^\alpha}{\sum_k |z_{i,k}|^\alpha} ,$$

where $z_{i,k}$ denotes the kth component of the ith mode, and $\alpha > 0$. Assuming that the eigenmode has an exponential localization $|z_{i,k}| \propto \exp(-|k - a|/\xi_i)$, the localization length ξ_i may be approximated by

$$\xi_i \approx \sqrt{\langle k^2 \rangle_{\alpha,i} - \langle k \rangle_{\alpha,i}^2} .$$

⊙ Analyze the chains of harmonic oscillators with different distributions of weights w_i for dimensions $N = 10^3$ and 10^4. First use the binomial distribution: to the weight w_i assign the value W_0 with probability p and the value W_1 with probability $1 - p$, so $\mathrm{Prob}(w_i = W_0) = p$ and $\mathrm{Prob}(w_i = W_1) = 1 - p$. Then use the exponential distribution: the weights w_i are distributed according to the cumulative distribution function $P(w_i \le w) = 1 - \exp(-\lambda w)$ at constant $\lambda > 0$. Check that for different λ the system of eigenmodes does not change significantly, and that similar conclusions apply for the distribution of $\lambda \omega_i^2$.

For each of these cases, draw the shapes of the eigenmodes. Plot z_i, x_i, and the distribution of the energies $\{\omega_i^2\}_{i=0}^{N-1}$, of the average position $\{\langle k \rangle_{\alpha=1,i}\}_{i=0}^{N-1}$, and of the localization length $\{\xi_i\}_{i=0}^{N-1}$. Is there a statistically meaningful correlation between these quantities? To diagonalize the matrix W, use programs for tridiagonal matrices (see Table 4.3).

⊕ Study the spectral properties of a more abstract system of first-order differential equations

$$\dot{x}_i = -x_{i+1} + w_i x_i - x_{i-1} , \quad i = 0, 1, \ldots, N - 1 ,$$

where $x_{-1} = x_N = 0$. Distribute the weights w_i around the value zero ($\langle w_i \rangle = 0$) with a standard deviation of $\langle w_i^2 \rangle^{1/2} = 1$. Find the eigenmodes by using the ansatz $x_i(t) = y_i \exp(\lambda t)$. Due to the symmetry of the matrix elements we have $\lambda \in \mathbb{R}$. Calculate the distribution of the localization lengths and eigenenergies in the case of uniformly and normally distributed weights w_i for $N = 10^2$, 10^3, and 10^4. Make sure that all eigenmodes are localized and that their centers-of-gravity are uniformly distributed over the modes.

4.10.7   Spectra of Random Symmetric Matrices

In quantum mechanics the state of the system is given by the vector $\psi \in \mathbb{C}^n$. Its time evolution is described by the Schrödinger equation

$$ i\hbar \frac{\mathrm{d}\psi}{\mathrm{d}t} = \widehat{H}\psi \,, $$

in which the Hamiltonian $\widehat{H}$ is represented by the Hermitian matrix H. In time-reversal symmetric systems a basis in the space of states exists such that H is real and symmetric. Without this symmetry, H is general Hermitian.

Quantum systems whose classical analogues are chaotic behave statistically in some aspects [72]. The paradigmatic classes of such systems are the particle in a multi-dimensional potential well of irregular shape or the system of strongly correlated particles like nucleons in heavy nuclei. In such systems, at sufficiently long times, the wave-function becomes statistically similar to a *random wave* [96]. Moreover, the unfolded energy spectra E_i or the fluctuations of the splittings between the neighboring levels ($E_{i+1} - E_i$) possess universal statistics (see Sect. 4.9.2 and [85]). In the case when the system is time-reversal symmetric, the splittings obey the statistics of the GOE matrix ensemble (Eq. (4.54)), while if there is no such invariance, they obey the statistics for GUE (Eq. (4.55)). This finding established the connection between the theory of random matrices and quantum physics.

  ⊙   Let us restrict ourselves to matrices from the GOE. Check (analytically and numerically) that the differences between the eigenenergies of 2×2 matrices from the GOE of the form $((a, b), (b, c))$ are distributed according to (4.54), where a, $b/\sqrt{2}$, and c are distributed randomly according to $N(0, 1)$. (Randomly generate a large number of such matrices, compute the eigenvalues for each of them, and plot their distribution.)

  Randomly generate the matrix $H \in \mathbb{R}^{n \times n}$ from the GOE with a size as large as possible ($n \approx 5000$), and find its eigenenergies by any method adapted to real symmetric matrices. Compute the distribution of the energies as a function of $E/\sqrt{2n}$, and compare it to the Wigner's semi-circular distribution

$$ p(E) \propto \sqrt{1 - E^2/(2n)} \,. $$

Unfold the energy spectrum $\{E_i\}_{i=1}^n$ by following the procedure outlined in Sect. 4.9.2: form the cumulative distribution $N(E) = \#\{E_i \leq E\}$, separate N to its smooth and oscillatory part, $N(E) = \overline{N}(E) + N_{\mathrm{osc}}(E)$, and then unfold the spectrum into $\{\tilde{E}_i = \overline{N}(E_i)\}_{i=1}^n$ such that the average distance between the neighboring levels $S = \tilde{E}_{i+1} - \tilde{E}_i$ becomes locally (over a few levels) and globally equal to 1. Finally, compute the distribution of the splittings (as a function of S) between neighboring levels, and compare it to the distribution (4.54).

$\oplus$ Let A and B be $n \times n$ random matrices from the GOE. We use their linear combination to define the Hamiltonian

$$H(t) = \sqrt{\frac{n}{4\mathrm{tr}(\mathcal{H}(t)^2)}}\, \mathcal{H}(t)\,, \quad \mathcal{H}(t) = (1-t)\,A + t\,B\,, \quad t \in [0,1]\,.$$

Compute the spectrum of H for $n = 10, 50$, and 100 as a function of the parameter t. You should observe the typical behavior of the energy levels similar to the one shown in Fig. 4.25. At some places, the levels appear to approach each other and then recede, the phenomenon known as the *avoidance of crossing* or *level repulsion*. This is a characteristic feature of classically chaotic systems and can already be inferred from the distributions (4.54) and (4.55) since $p_s(0) = 0$. Make sure that for each t and large enough n (say, $n = 2000$) the distributions of energies and energy splittings are as they should be for matrices from the GOE.

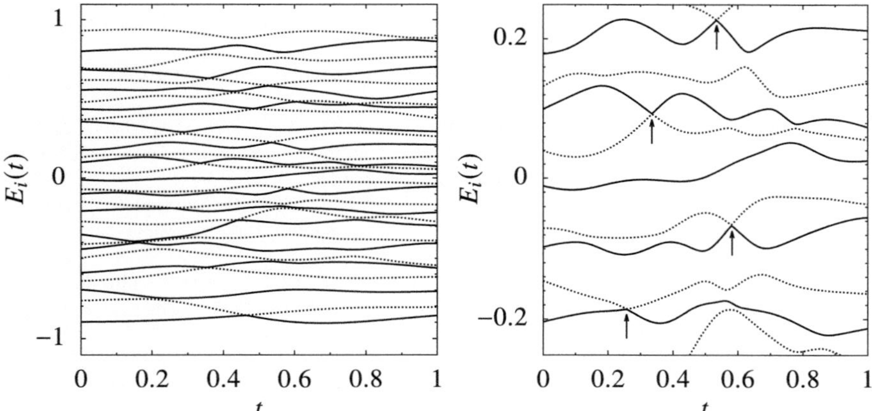

Fig. 4.25 [LEFT] Energy levels corresponding to the 30×30 Hamiltonian matrix in dependence of the parameter t. [RIGHT] A blow-up of the energy axis near its origin. The arrows indicate the most prominent examples of the avoidance of crossing

References

1. J.W. Demmel, *Applied Numerical Linear Algebra* (SIAM, Philadelphia, 1997)
2. G.H. Golub, C.F. Van Loan, *Matrix Computations*, 3rd edn (Johns Hopkins University Press, Baltimore, 1996)
3. *LAPACK, The Linear Algebra PACKage*, http://netlib.org/lapack (version for Fortran 77), `../lapack95` (version for Fortran 95), `../clapack` (version for C), `../lapack++` (version for C++). For the Java interface see https://icl.utk.edu/f2j. The reference handbook is E. Anderson et al., *LAPACK Users' Guide*, 3rd edn (SIAM, Philadelphia, 1999)
4. V. Strassen, Gaussian elimination is not optimal. Num. Math. **13**, 354 (1969)
5. J. Demmel, N.J. Higham, Stability of block algorithms with fast level 3 BLAS. ACM Trans. Math. Softw. **18**, 274 (1992). See also N.J. Higham, Exploiting fast matrix multiplication within the level 3 BLAS. ACM Trans. Math. Softw. **16**, 352 (1990)
6. S. Chatterjee, A.R. Lebeck, P.K. Patnala, M. Thottethodi, Recursive array layouts and fast matrix multiplication. IEEE Trans. Parallel Distrib. Syst. **13**, 1105 (2002)
7. K. Goto, R.A. van de Geijn, Anatomy of high-performance matrix multiplication. ACM Trans. Math. Softw. **34**, art. 12 (2008)
8. N.J. Higham, *Accuracy and Stability of Numerical Algorithms*, 2nd edn (SIAM, Philadelphia, 2002)
9. D. Coppersmith, S. Winograd, Matrix multiplication via arithmetic progressions. J. Symbolic Comput. **9**, 251 (1990)
10. C.C. Douglas, M.A. Heroux, G. Slishman, R.M. Smith, GEMMW: a portable level-3 BLAS Winograd variant of Strassen's matrix-matrix multiply algorithm. J. Comp. Phys. **110**, 1 (1994)
11. P. D'Alberto, A. Nicolau, Adaptive Winograd's matrix multiplications. ACM Trans. Math. Softw. **36**, art. 3 (2009)
12. J. Demmel, I. Dumitriu, O. Holtz, R. Kleinberg, Fast matrix multiplication is stable. Numer. Math. **106**, 199 (2007)
13. J.W. Demmel, N.J. Higham, Stability of block algorithms with fast level-3 BLAS. ACM Trans. Math. Softw. **18**, 274 (1992)
14. D.H. Bailey, K. Lee, H.D. Simon, Using Strassen's algorithm to accelerate the solution of linear systems. J. Supercomputing **4**, 357 (1990)
15. J. Demmel, The componentwise distance to the nearest singular matrix. SIAM J. Matrix Anal. Appl. **13**, 10 (1992). See also J.W. Demmel, On condition numbers and the distance to the nearest ill-posed problem. Numer. Math. **51**, 251 (1987)
16. R.M. Gray, Toeplitz and circulant matrices: a review. Found. Trends Commun. Inf. Theory **2**, 155 (2006). See also J.R. Bunch, Stability of methods for solving Toeplitz systems of equations. SIAM J. Sci. Stat. Comput. **6**, 349 (1985)
17. T. Kailath, J. Chun, Generalized displacement structure for block-Toeplitz, Toeplitz-block, and Toeplitz-derived matrices. SIAM J. Matrix Anal. Appl. **15**, 114 (1994). See also T. Kailath, A.H. Sayed, Displacement structure: theory and applications. SIAM Rev. **37**, 297 (1995)
18. G.S. Anmar, W.B. Gragg, Superfast solution of real positive definite Toeplitz systems. SIAM J. Matrix Anal. Appl. **9**, 61 (1988). See also T.F. Chan, P.C. Hansen, A look-ahead Levinson algorithm for indefinite Toeplitz systems. SIAM J. Matrix Anal. Appl. **13**, 490 (1992)
19. M.K. Ng, *Iterative Methods for Toeplitz Systems* (Oxford University Press, Oxford, 2004)
20. W. Gautschi, G. Inglese, Lower bound for the conditional number of Vandermonde matrices. Numer. Math. **52**, 241 (1988)
21. Å. Björck, V. Pereyra, Solution of Vandermonde systems of equations. Math. Comput. **24**, 893 (1970). See also Å. Björck, T. Elfving, Algorithms for confluent Vandermonde systems. Numer. Math. **21**, 130 (1973)
22. N.J. Higham, Fast solution of Vandermonde-like systems involving orthogonal polynomials. IMA J. Num. Anal. **8**, 473 (1988). See also D. Calvetti, L. Reichel, Fast inversion of Vandermonde-like matrices involving orthogonal polynomials. BIT Numer. Math. **33**, 473 (1993)

23. H. Lu, Fast solution of confluent Vandermonde linear systems. SIAM J. Matrix Anal. Appl. **15**, 1277 (1994); H. Lu, Solution of Vandermonde-like systems and confluent Vandermonde-like systems. SIAM J. Matrix Anal. Appl. **17**, 127 (1996); see also Sect. 2.4 in D. Bini, V.Y. Pan, *Polynomial and Matrix Computations, Vol. 1: Fundamental Algorithms* (Birkhäuser, Boston, 1994)

24. W.W. Hager, Condition estimates. SIAM J. Sci. Stat. Comput. **5**, 311 (1984). See also Algorithm 2.5 in [1]

25. N.J. Higham, FORTRAN codes for estimating the one-norm of a real or complex matrix, with applications to condition estimation. ACM Trans. Math. Softw. **14**, 381 (1988) and N.J. Higham, F. Tisseur, A block algorithm for matrix 1-norm estimation, with an application to 1-norm pseudospectra. SIAM J. Matrix Anal. Appl. **21**, 1185 (2000)

26. N.J. Higham, Experience with a matrix norm estimator. SIAM J. Sci. Stat. Comput. **11**, 804 (1990) and N.J. Higham, A survey of condition number estimation for triangular matrices. SIAM Rev. **29**, 575 (1987)

27. D.S. Watkins, A case where balancing is harmful. ETNA **23**, 1 (2006)

28. E.E. Osborne, On pre-conditioning of matrices. J. Assoc. Comput. Mach. **7**, 338 (1960) and B.N. Parlett, C. Reinsch, Balancing a matrix for calculation of eigenvalues and eigenvectors. Numer. Math. **13**, 293 (1969). Balancing of sparse matrices is described by T.-Y. Chen, J.W. Demmel, Balancing sparse matrices for computing eigenvalues. Lin. Alg. Appl. **309**, 261 (2000)

29. T.A. Davis, S. Rajamanickam, W.M. Sid-Lakhdar, A survey of direct methods for sparse linear systems. Acta Numerica **25**, 383 (2016)

30. J.J. Dongarra, I.S. Duff, D.C. Sorensen, H.A. van der Vorst, *Numerical Linear Algebra for High-Performance Computers* (SIAM, Philadelphia, 1998)

31. N.I.M. Gould, J.A. Scott, Y. Hu, A numerical evaluation of sparse direct solvers for the solution of large sparse symmetric linear systems of equations. ACM Trans. Math. Softw. **33**, art. 10 (2007)

32. J.A. Scott, Y. Hu, Experiences of sparse direct symmetric solvers. ACM Trans. Math. Softw. **33**, art. 18 (2007)

33. E.J. Haunschmid, C.W. Ueberhuber, *Direct Solvers for Sparse Systems*, SFB F011 "AURORA" Report (TU Wien, 1999)

34. I.S. Duff, M.A. Heroux, R. Pozo, An overview of the sparse basic linear algebra subprograms: the new standard from the BLAS technical forum. ACM Trans. Math. Softw. **28**, 239 (2002)

35. Y. Saad, *Iterative Methods for Sparse Linear Systems*, 2nd edn (SIAM, Philadelphia, 2003)

36. M.R. Hestenes, E.L. Stiefel, Methods of conjugate gradients for solving linear systems. J. Res. Natl. Bur. Stand. **49**, 409 (1952)

37. J.R. Shewchuk, *An Introduction to the Conjugate Gradient Method Without the Agonizing Pain* (Carnegie Mellon University, Pittsburgh, 1994). https://dl.acm.org/doi/10.5555/865018

38. C. Lanczos, Solution of systems of linear equations by minimized iterations. J. Res. Natl. Bur. Stand. **49**, 33 (1952)

39. R. Fletcher, Conjugate gradient methods for indefinite systems, in Proceedings of the Dundee Biennal Conference on Numerical Analysis 1974, ed. by G.A. Watson (Springer, New York, 1975), p. 73

40. W.H. Press, B.P. Flannery, S.A. Teukolsky, W.T. Vetterling, *Numerical Recipes: The Art of Scientific Computing*, 3rd edn (Cambridge University Press, Cambridge, 2007). See also the equivalent handbooks in Fortran, Pascal and C, as well as http://numerical.recipes

41. Y. Saad, M.H. Schulz, GMRES: a generalized minimal residual algorithm for solving nonsymmetric linear systems. SIAM J. Sci. Stat. Comp. **7**, 856 (1986)

42. V. Simoncini, Computational methods for linear matrix equations. SIAM Rev. **58**, 377 (2016)

43. R.H. Bartels, G.W. Stewart, Algorithm 432: solution of the matrix equation $AX + XB = C$. Comm. ACM **15**, 820 (1972)

44. M. Slowik, P. Benner, V. Sima, Evaluation of the linear matrix equation solvers in SLICOT. J. Numer. Anal. Ind. Appl. Math. **2**, 11 (2007). See also http://www.slicot.org

45. Z. Bai, W. Gao, Y. Su (eds.), *Matrix Functions and Matrix Equations*, Series in Contemporary Applied Mathematics, vol. 19 (World Scientific, Singapore, 2015), p. 67
46. N.J. Higham, *Functions of Matrices. Theory and Computation* (SIAM, Philadelphia, 2008)
47. The routines xGELS offer only the basic solution without iterative improvement. This novelty is now offered by the xGELS_X and xGELS_RFSX routines; see J. Demmel, Y. Hida, E.J. Riedy, X.S. Li, Extra-precise iterative refinement for overdetermined least squares problems. ACM Trans. Math. Softw. **35**, art. 28 (2009)
48. G. Golub, W. Kahan, Calculating the singular values and pseudo-inverse of a matrix. SIAM J. Numer. Anal. B **2**, 205 (1965)
49. D. Kalman, A singularly valuable decomposition: the SVD of a matrix. Coll. Math. J. **27**, 2 (1996)
50. M. Abate, When is a linear operator diagonalizable? Am. Math. Mon. **104**, 824 (1997)
51. D.S. Watkins, Understanding the QR algorithm. SIAM Rev. **24**, 427 (1982); D.S. Watkins, The QR algorithm revisited. SIAM Rev. **50**, 133 (2008)
52. C. Van Loan, On estimating the condition of eigenvalues and eigenvectors. Lin. Alg. Applic. **88/89**, 715 (1987) and J.H. Wilkinson, Sensitivity of eigenvalues. Utilitas Math. **25**, 5 (1984)
53. J.J.M. Cuppen, A divide and conquer method for the symmetric tridiagonal eigenproblem. Numer. Math. **36**, 177 (1981)
54. I.S. Dhillon, B.N. Parlett, Multiple representations to compute orthogonal eigenvectors of symmetric tridiagonal matrices. Lin. Alg. Applic. **387**, 1 (2004); I.S. Dhillon, B.N. Parlett, C. Vömel, The design and implementation of the MRRR algorithm. ACM Trans. Math. Softw. **32**, 533 (2006)
55. H.-J. Stöckmann, *Quantum Chaos. An Introduction* (Cambridge University Press, Cambridge, 2006)
56. N.W. Ashcroft, N.D. Mermin, *Solid State Physics* (Harcourt College Publishers, Fort Worth, 1976)
57. D.R. Hofstadter, Energy levels and wave functions of Bloch electrons in rational and irrational magnetic fields. Phys. Rev. B **14**, 2239 (1976)
58. G.H. Golub, Some modified matrix eigenvalue problems. SIAM Rev. **15**, 318 (1973)
59. R.M.M. Mattheij, G. Söderlind, On inhomogeneous eigenvalue problems. Lin. Alg. Appl. **88/89**, 507 (1987)
60. F. Tisseur, K. Meerbergen, The quadratic eigenvalue problem. SIAM Rev. **43**, 235 (2001)
61. G.H. Golub, J.H. Wilkinson, Ill-conditioned eigensystems and the computation of the Jordan canonical form. SIAM Rev. **18**, 578 (1976). See also A. Bujosa, R. Criado, C. Vega, Jordan normal form via elementary transformations. SIAM Rev. **40**, 947 (1998)
62. B. Kågström, A. Ruhe, An algorithm for numerical computation of the Jordan normal form of a complex matrix. ACM Trans. Math. Softw. **6**, 398 (1980)
63. W. Kerner, Large-scale complex eigenvalue problem. J. Comput. Phys. **85**, 1 (1989)
64. W.E. Arnoldi, The principle of minimized iteration in the solution of the matrix eigenvalue problem. Quart. Appl. Math. **9**, 17 (1951)
65. C. Lanczos, An iteration method for the solution of the eigenvalue problem of linear differential and integral operators. J. Res. Natl. Bur. Stand. **45**, 491 (1950)
66. R.B. Lehoucq, D.C. Sorensen, C. Yang, *ARPACK Users Guide: Solution of Large Scale Eigenvalue Problems by Implicitly Restarted Arnoldi's Methods* (SIAM, Philadelphia, 1998)
67. https://github.com/opencollab/arpack-âŁ‹ng
68. J.A. Scott, An Arnoldi code for computing selected eigenvalues of sparse, real, unsymmetric matrices. ACM Trans. Math. Softw. **21**, 432 (1995)
69. Y. Saad, *Numerical Methods for Large Eigenvalue Problems*, 2nd edn (SIAM, Philadelphia, 2011)
70. L.N. Trefethen, M. Embree, *Spectra and Pseudospectra. The Behavior of Nonnormal Matrices and Operators* (Princeton University Press, New Jersey, 2005)
71. G. Livan, M. Novaes, P. Vivo, *Introduction to Random Matrices. Theory and Practice* (Springer, Berlin, 2018)
72. F. Haake, *Quantum Signatures of Chaos*, 3rd edn (Springer, Berlin, 2006)

73. A.M. Tulino, S. Verdú, *Random Matrix Theory and Wireless Communications*, Foundations and Trends in Communications and Information Theory (Now Publishers, Hanover, 2004); full text available at: http://dx.doi.org/10.1561/0100000001
74. V. Plerou et al., Random matrix approach to cross correlations in financial data. Phys. Rev. E **65**, 066126 (2002)
75. A. Edelman, N.R. Rao, Random matrix theory. Acta Numerica **14**, 233 (2005)
76. P. Bleher, A. Its, *Random Matrix Models and Their Applications* (Cambridge University Press, Cambridge, 2001)
77. M.L. Mehta, *Random Matrices*, 2nd edn (Academic Press, San Diego, 1990)
78. T. Tao, V. Vu, From the Littlewood-Offord problem to the circular law: universality of the spectral distribution of random matrices. Bull. Amer. Math. Soc. **46**, 377 (2009)
79. V.L. Girko, Circular law. Theory Probab. Appl. **29**, 694 (1985)
80. M. Horvat, The ensemble of random Markov matrices. J. Stat. Mech. **2009**, P07005 (2009)
81. D.V. Widder, The Stieltjes transform. Trans. Amer. Math. Soc. **43**, 7 (1938)
82. R.M. Dudley, *Real Analysis and Probability* (Cambridge University Press, Cambridge, 2002)
83. V.A. Marčenko, L.A. Pastur, Distribution of eigenvalues for some sets of random matrices. Math. USSR Sbornik **1**, 457 (1967)
84. R.R. Nadakuditi, *Applied Stochastic Eigen-Analysis*, doctoral dissertation (Massachusetts Institute of Technology, Cambridge, 1999). Accessible at: http://hdl.handle.net/1912/1647
85. O. Bohigas, M.J. Giannoni, C. Schmit, Characterization of chaotic quantum spectra and universality of level fluctuation laws. Phys. Rev. Lett. **52**, 1 (1984)
86. E.P. Wigner, On the distribution of the roots of certain symmetric matrices. Ann. Math. **67**, 325 (1958)
87. G.W. Stewart, The efficient generation of random orthogonal matrices with an application to condition estimators. SIAM J. Num. Anal. **17**, 403 (1980)
88. F. Mezzadri, How to generate random matrices from the classical compact groups. Notes AMS **54**, 592 (2007)
89. S. Bunde, S. Havlin, *Fractals and Disordered Systems* (Springer, Berlin, 1991)
90. R.J. Wilson, J.J. Watkins, *Graphs: An Introductory Approach* (John Wiley & Sons, New York, 1990)
91. D. Stauffer, A. Aharony, *Introduction to Percolation Theory*, 2nd edn (Taylor & Francis, London, 1992)
92. G. Chen, T. Ueta, *Chaos in Circuits and Systems* (World Scientific, Singapore, 2002)
93. W. Kinzel, G. Reents, *Physics by Computer: Programming of Physical Problems Using Mathematica and C* (Springer, Berlin, 1997)
94. P.W. Anderson, Absence of diffusion in certain random lattices. Phys. Rev. **109**, 1492 (1958)
95. F.J. Dyson, The dynamics of a disordered linear chain. Phys. Rev. **92**, 1331 (1953)
96. M.V. Berry, Regular and irregular semiclassical wavefunctions. J. Phys. A **10**, 2083 (1977)

Chapter 5
Transformations of Functions and Signals

Abstract The discrete Fourier transformation is one of the most important tools in the analysis of functions and signals, but its detailed aspects are often disregarded. Fourier and sampling theorems, Parseval's equality, and power spectral densities are introduced, and the concepts of signal uncertainty, aliasing, and leakage are discussed. Sparse and non-uniform Fourier transformations are introduced. Transformations with orthogonal polynomials (Legendre, Chebyshev, Laguerre, Hermite), as well as the Chebyshev TL and TB functions are described. Their relation to quadrature formulas is explained. The numerical computation of the Laplace transformation and its inverse is introduced in the context of differential equations for transfer functions of physical systems. The continuous and discrete Hilbert transformations are discussed next, introducing the concept of the analytic signal, and their relation to the Kramers–Krönig dispersion relations is given. Finally, we describe the continuous and discrete wavelet transformations. The Examples and Problems include multiplication of polynomials by FFT, Fourier analysis of realistic acoustic signals and the Doppler effect, and the analysis of brain potentials by the Hilbert transform.

In this Chapter we describe the most important continuous and discrete transformations of functions (signals), which can be used to represent these functions in a different space (for instance, in terms of frequency instead of time, or momentum instead of coordinate)—a key tool in spectral methods discussed in Chap. 12. A quick roadmap for Sects. 5.1–5.3 is given in Table 5.1.

5.1 Fourier Transformation

The continuous Fourier transformation $\mathcal{F}$ of the function f ("the signal") for $x \in \mathbb{R}$ is defined as

$$F(\omega) = \mathcal{F}[f](\omega) = \int_{-\infty}^{\infty} f(x)\, e^{-i\omega x}\, dx \, . \tag{5.1}$$

© The Author(s), under exclusive license to Springer Nature Switzerland AG 2025   243
S. Širca and M. Horvat, *Computational Methods in Physics*, Graduate Texts in Physics,
https://doi.org/10.1007/978-3-031-68566-8_5

Table 5.1 A roadmap for the choice of appropriate basis needed in transforms of functions f with specific properties on various domains. (Adapted from [1].)

Property of f	Suitable basis set
Periodic	Fourier ($\exp(\pm ikx)$, sin, cos)
Periodic, symmetric (even)	Fourier (cos)
Periodic, antisymmetric (odd)	Fourier (sin)
$x \in [a, b]$, f non-periodic	Chebyshev or Legendre polynomials
$x \in [0, \infty]$, f decays exponentially as $x \to \infty$	Chebyshev TL or Laguerre functions
$x \in [0, \infty]$, f has asymptotic series in $1/x^p$	Chebyshev TL functions
$x \in [-\infty, \infty]$, f decays exponentially as $\lvert x \rvert \to \infty$	Chebyshev TB or Hermite functions
$x \in [-\infty, \infty]$, f has asymptotic series in $1/x^p$	Chebyshev TB functions

The sufficient conditions for the existence of F are that f is absolutely integrable, i.e. $\int_{-\infty}^{\infty} \lvert f(x) \rvert \, \mathrm{d}x < \infty$, and that f is piecewise continuous or has a finite number of discontinuities. The inverse transformation is

$$f(x) = \mathcal{F}^{-1}[F](x) = \frac{1}{2\pi} \int_{-\infty}^{\infty} F(\omega) \, \mathrm{e}^{\mathrm{i}\omega x} \, \mathrm{d}\omega \,. \tag{5.2}$$

Changing the sign of the argument x of f (flipping the space or time coordinate) implies a change of the sign of the frequency ω in the transform:

$$g(x) = f(-x) \quad \Longleftrightarrow \quad G(\omega) = F(-\omega) \,.$$

Fourier theorem If the function f is piecewise continuous on the interval $[-\pi, \pi]$, one can form the Fourier series

$$\frac{1}{2\pi} \sum_{n \in \mathbb{Z}} \int_{-\pi}^{\pi} f(x) \, \mathrm{e}^{\mathrm{i}n(\xi - x)} \, \mathrm{d}x = \frac{1}{2}\left[f(\xi + 0) + f(\xi - 0) \right], \quad \xi \in [-\pi, \pi] \,.$$

It follows from this theorem that the function f which is continuous on $\mathbb{R}$ and its Fourier transform F for any $a \in \mathbb{R}$, $a > 0$, are related by the *Poisson sum*

$$\sum_{n \in \mathbb{Z}} F(n\,a) = \frac{2\pi}{a} \sum_{m \in \mathbb{Z}} f\left(\frac{2\pi m}{a} \right)$$

(see also (Eq. 1.68)).

Parseval's equality The integral of the square of the function f over x and the integral of the square of its transform F over ω are related by the Parseval's equality

$$\int_{-\infty}^{\infty} |f(x)|^2 \, dx = \frac{1}{2\pi} \int_{-\infty}^{\infty} |F(\omega)|^2 \, d\omega \, . \tag{5.3}$$

This invariance of the norm means that power is preserved in the transition from the temporal to the frequency representation of a signal. Parseval's equality is valid for all function expansions in Hilbert spaces; see e.g. [2].

Fourier uncertainty Let f be normalized as $\int_{-\infty}^{\infty} |f(x)|^2 \, dx = 1$. Then $|f|^2$ is non-negative and can be understood as the probability density of the signal with respect to x, with the first and second moments (mean and variance)

$$\langle x \rangle = \int_{-\infty}^{\infty} x |f(x)|^2 \, dx \, , \quad \sigma_x^2 = \int_{-\infty}^{\infty} [x - \langle x \rangle]^2 \, |f(x)|^2 \, dx \, .$$

By Eq. (5.3) the Fourier transform of f is also normalized, $\int_{-\infty}^{\infty} |F(\omega)|^2 \, d\omega = 2\pi$. Hence $|F(\omega)|^2/(2\pi)$ is the probability density with respect to frequencies (the *power spectral density*), with the first and second moments

$$\langle \omega \rangle = \frac{1}{2\pi} \int_{-\infty}^{\infty} \omega |F(\omega)|^2 \, d\omega \, , \quad \sigma_\omega^2 = \frac{1}{2\pi} \int_{-\infty}^{\infty} [\omega - \langle \omega \rangle]^2 \, |F(\omega)|^2 \, d\omega \, .$$

The product of the variances in both representations of f is

$$\sigma_x^2 \sigma_\omega^2 \geq \tfrac{1}{4} \, . \tag{5.4}$$

Inequality (5.4) expresses the *Fourier uncertainty*: spatial or temporal variances of the signal are not independent of frequency variances. If variances decrease in the configuration space, they increase in the transform space, and vice-versa. The lower bound of the uncertainty is achieved by Gaussian signals,

$$f(x) = (2 s/\pi)^{1/4} \, e^{-s(x-a)^2} \, , \quad s > 0 \, , \quad a \in \mathbb{R} \, . \tag{5.5}$$

We say that such signals have a *minimum possible width*.

Example We compute the Fourier uncertainty of two "bell"-shaped functions. The first one is

$$f(x) = \sqrt{\frac{2}{\pi}} \frac{1}{1 + x^2} \, , \quad F(\omega) = \sqrt{2\pi} \, e^{-|\omega|} \, .$$

We get $\langle x \rangle = 0$, $\sigma_x^2 = 1$, $\langle \omega \rangle = 0$, and $\sigma_\omega^2 = \tfrac{1}{2}$, thus $\sigma_x^2 \sigma_\omega^2 = \tfrac{1}{2}$. The second one is a narrow Gaussian shifted from the origin (Eq. (5.5) with $s = 2$ and $a = 5$):

$$f(x) = (4/\pi)^{1/4} e^{-2(x-5)^2} \, , \quad F(\omega) = e^{-\frac{1}{8}\omega(\omega + 40\,i)} \, .$$

We obtain $\langle x \rangle = 5$, $\sigma_x^2 = \frac{1}{8}$, $\langle \omega \rangle = 0$, and $\sigma_\omega^2 = 2$, hence $\sigma_x^2 \sigma_\omega^2 = \frac{1}{4}$, which indeed is the smallest possible product of variances, in accordance with Eq. (5.4). $\triangleleft$

Sampling theorem Assume that the function (signal) f is continuous on the whole real axis and that (5.1) is its Fourier transform. The signal is uniformly sampled at the points spaced Δx apart, like $f_n = f(n\,\Delta x)$ for $n \in \mathbb{Z}$. If the range of the frequencies corresponding to the signal f is bounded, i.e. if

$$F(\omega) = 0 \quad \forall |\omega| \geq \omega_{\mathrm{c}} = \frac{\pi}{\Delta x} \,, \tag{5.6}$$

where ω_{c} is the Nyquist (critical) frequency, the signal f can be reconstructed *with no loss of information* from the discrete sample $\{f_n : n \in \mathbb{Z}\}$ by using [3]

$$f(x) = \sum_{n \in \mathbb{Z}} f_n \frac{\sin \omega_{\mathrm{c}}(x - n\,\Delta x)}{\omega_{\mathrm{c}}(x - n\,\Delta x)} \,.$$

5.2   Fourier Series

5.2.1   Continuous Fourier Expansion

Following the notational conventions of [4], the Fourier transform of the function f which is bounded and piecewise continuous on the interval $[0, 2\pi]$, is

$$\widehat{f_k} = \frac{1}{2\pi} \int_0^{2\pi} f(x)\, \mathrm{e}^{-ikx}\, \mathrm{d}x \,, \quad k = 0, \pm 1, \pm 2, \ldots . \tag{5.7}$$

The functions $\phi_k(x) = \mathrm{e}^{ikx}$ are orthogonal on $[0, 2\pi]$,

$$\int_0^{2\pi} \phi_j(x)\phi_k^*(x)\, \mathrm{d}x = 2\pi \delta_{j,k} \,,$$

and form the basis of the space $\mathcal{T}_N$ of trigonometric polynomials of degree $\leq N/2$. The cosine and sine transforms are the real and imaginary parts of the transform (5.7),

$$a_k = \frac{1}{2\pi} \int_0^{2\pi} f(x) \cos kx\, \mathrm{d}x \,, \quad b_k = \frac{1}{2\pi} \int_0^{2\pi} f(x) \sin kx\, \mathrm{d}x \,,$$

where all three forms are related by $\widehat{f}_k = a_k - ib_k$. In general, the functions f and their Fourier coefficients $\widehat{f}_k$ are complex. Specific symmetry properties of the signals correspond to specific properties of their transforms. The most relevant cases are:

$$
\begin{aligned}
&\text{if} \qquad\quad f \text{ is real}, &\qquad \widehat{f}_{-k} &= \widehat{f}_k^* \,; \\
&\qquad\qquad f \text{ is real and even}, &\qquad \widehat{f}_k &\text{ is real and even}\,; \\
&\qquad\qquad f \text{ is real and odd}, &\qquad \widehat{f}_k &\text{ is imaginary and odd}\,.
\end{aligned}
$$

If f is real, the coefficients of its cosine and sine transform a_k and b_k are also real, and $\widehat{f}_{-k} = \widehat{f}_k^*$. The function f can be represented by the Fourier series

$$
Sf(x) = \sum_{k=-\infty}^{\infty} \widehat{f}_k \phi_k(x)\,, \tag{5.8}
$$

with the coefficients given by (5.7), or by its truncated form,

$$
S_N f(x) = \sum_{k=-N/2}^{N/2-1} \widehat{f}_k \phi_k(x)\,. \tag{5.9}
$$

Remark on notation The normalization factor $(2\pi)^{-1}$ in Eqs. (5.7) and (5.8) has changed its place relative to Eqs. (5.1) and (5.2). In both cases the inverse transformation of the transform restores the original function. Other variants are widely used, e.g. swapping the roles of $\phi_k(x) = \mathrm{e}^{ikx}$ and $\phi_k^*(x) = \mathrm{e}^{-ikx}$: the opposite phases cause only a sign change in the imaginary components. These conventions partly originate in different interpretations of the sign of the frequencies: the Schrödinger equation $i\hbar\partial\psi/\partial t = E\psi$ forces the physicist to associate $\mathrm{e}^{i\omega t}$ with a quantum state with negative energy $E = -\hbar\omega$, while an electronics engineer will see it as a signal with positive frequency.

If the function f is continuous, periodic, and has a bounded variation on $[0, 2\pi]$, i.e. $\int_0^{2\pi} |f'(x)|\, \mathrm{d}x < \infty$, the series $Sf(x)$ uniformly converges to $f(x)$, so $\lim_{N\to\infty} \max_x |f(x) - S_N f(x)| = 0$. If f has a bounded variation on $[0, 2\pi]$, then $S_N f(x)$ converges to $(f(x+0) + f(x-0))/2$ for each $x \in [0, 2\pi]$. If f is continuous and periodic, the Fourier series does not necessarily converge at each $x \in [0, 2\pi]$. The Fourier series of $f \in L^2(0, 2\pi)$ converges to f in the sense

$$
\lim_{N\to\infty} \int_0^{2\pi} |f(x) - S_N f(x)|^2\, \mathrm{d}x = 0\,.
$$

The truncation of $Sf(x)$ to $S_N f(x)$ containing a finite number of terms implies an error. Its magnitude is determined by the asymptotics of the Fourier coefficients,

$$
\max_{0 \le x \le 2\pi} |f(x) - S_N f(x)| \le \sum_{k < -N/2} |\widehat{f}_k| + \sum_{k > N/2-1} |\widehat{f}_k|\,.
$$

If f is m-times ($m \geq 1$) continuously differentiable on $[0, 2\pi]$ and its jth derivative is periodic for all $j \leq m - 2$, we have

$$\widehat{f_k} = \mathcal{O}(k^{-m}) \, .$$

There are several ways to estimate the truncation error [4]. In the norm in $L^2(0, 2\pi)$, $S_N f$ is the best approximation for f among all functions from $\mathcal{T}_N$. For any m-times differentiable function f it holds that $\|f - S_N f\|_{L^2(0,2\pi)} \leq C N^{-m} \|f^{(m)}\|_{L^2(0,2\pi)}$.

5.2.2 Discrete Fourier Expansion

The Fourier expansion of a function f for which the values on $[0, 2\pi]$ are known at an even number of points (nodes)

$$x_j = 2\pi j/N \, , \quad j = 0, 1, \ldots, N - 1 \, , \quad N \text{ even} \, , \tag{5.10}$$

is represented by the coefficients of the *discrete Fourier transform* (DFT)

$$\widetilde{f_k} = \frac{1}{N} \sum_{j=0}^{N-1} f(x_j)\, e^{-ikx_j} \, , \quad -N/2 \leq k \leq N/2 - 1 \, , \tag{5.11}$$

see Fig. 5.1). For real functions f it holds that $\widetilde{f}_{N-k} = \widetilde{f}_k^*$ and $\widetilde{f}_0 \in \mathbb{R}$.

Another remark on notation Since N is even, $\widetilde{f}_{-N/2} = \widetilde{f}_{N/2}$, and thus an expansion with $|k| \leq N/2$ is also sensible: for the coefficient at $k = N/2$ we then use the normalization $1/(2N)$ instead of $1/N$. The index k representing the frequency ω may also run from 0 to $N - 1$. The lower portion of the spectrum ($1 \leq k \leq N/2 - 1$) then corresponds to the positive frequencies, $0 < \omega < \omega_c$, while the upper portion ($N/2 + 1 \leq k \leq N - 1$) corresponds to the negative ones, $-\omega_c < \omega < 0$. The transform with index $k = N/2$ maps to both ω_c and $-\omega_c$, while the value at $k = 0$ belongs to frequency zero (the average of the signal or its "DC component").

The function value at the point x_j is obtained by the inverse DFT, that is, by the sum

$$f(x_j) = \sum_{k=-N/2}^{N/2-1} \widetilde{f}_k \, e^{ikx_j} \, , \quad j = 0, 1, \ldots, N - 1 \, . \tag{5.12}$$

Let $I_N f$ denote the interpolant of the function f at N points. The discrete Fourier series

$$I_N f(x) = \sum_{k=-N/2}^{N/2-1} \widetilde{f}_k \, e^{ikx} \tag{5.13}$$

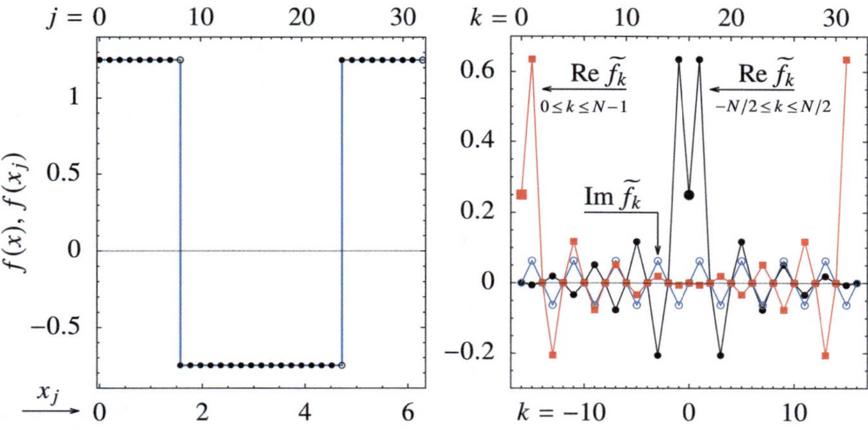

Fig. 5.1 Discrete Fourier transformation. [LEFT] The sum of the constant value 0.25 and the unit square wave on $[0, 2\pi]$ is sampled at $N = 32$ points $x_j = 2\pi j/N, 0 \le j \le N - 1$ (the • symbols). At the points denoted by ∘ the function is not sampled! [RIGHT] The real and imaginary parts of the Fourier coefficients for two different ways of indexing ($-N/2 \le k \le N/2$ or $0 \le k \le N - 1$). The thick symbols at $k = 0$ denote the zero-frequency component, which is the average of the signal—precisely the 0.25 vertical shift on the left figure. (If we subtract the average of the signal, the zeroth component of its transform is zero)

interpolates f, hence $I_N f(x_j) = f(x_j)$. The points x_j at which the value of f exactly coincides with the value of the interpolant $I_N f$ are known as the *Fourier collocation points*. The interpolant can also be written as

$$I_N f(x) = \sum_{j=0}^{N-1} f(x_j) g_j(x), \tag{5.14}$$

where

$$g_i(x) = \frac{1}{N} \sin \frac{N(x - x_j)}{2} \bigg/ \frac{N(x - x_j)}{2} \tan \frac{x - x_j}{2}, \quad N \text{ even} \tag{5.15}$$

are the characteristic Lagrange trigonometric polynomials with the property $g_j(x_i) = \delta_{i,j}$. If N is odd, the characteristic polynomials have the form

$$g_j(x) = \frac{1}{N} \sin \frac{N(x - x_j)}{2} \bigg/ \tan \frac{x - x_j}{2}, \quad N \text{ odd}.$$

The first three polynomials for $N = 8$ are shown in Fig. 5.2. An example of the trigonometric interpolation of a periodic function is shown in Fig. 5.3.

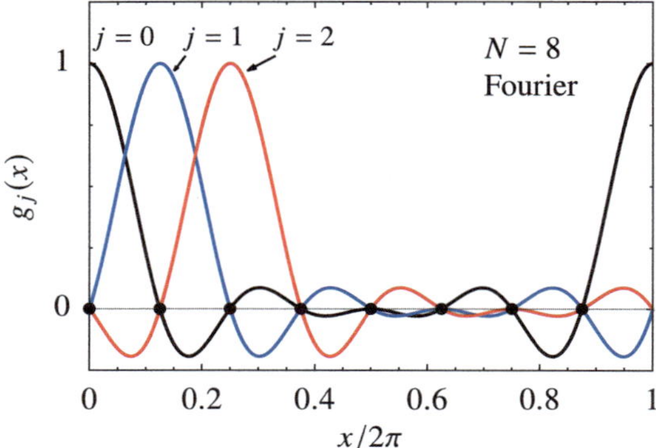

Fig. 5.2 The first three Lagrange trigonometric polynomials $g_j(x)$ in the case $N = 8$. The maxima are at the Fourier collocation points $x_j = 2\pi j / N$ ($j = 0, 1, \ldots, N - 1$)

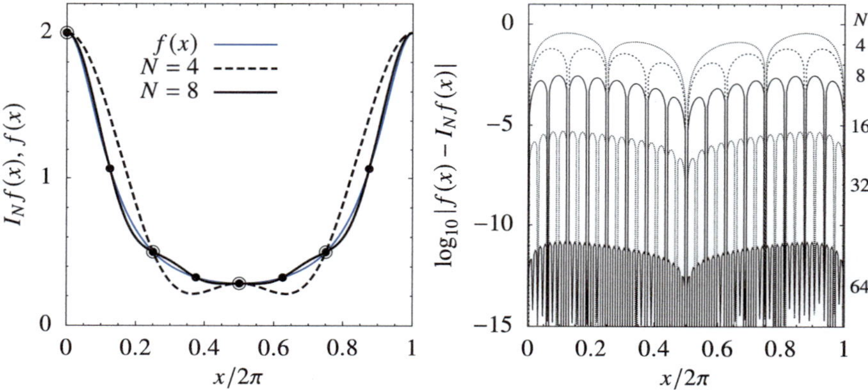

Fig. 5.3 Trigonometric interpolation of the function $f(x) = 2/(4 - 3\cos x)$. [LEFT] The interpolants $I_4 f$ (matching the function f at four points, symbols ○) and $I_8 f$ (matching at eight points, symbols ●). [RIGHT] The error of the Fourier series for $N = 4, 8, 16, 32$, and 64. Note the exceptionally rapid drop-off of the error with increasing N (right vertical axis): it is known as spectral convergence and attains its full relevance in the spectral methods of Chap. 12. The interpolation property $I_N f(x_j) = f(x_j)$ is seen in the characteristic downward spikes of vanishing error

The quadrature

$$\frac{1}{2\pi} \int_0^{2\pi} f(x)\, \mathrm{d}x = \frac{1}{N} \sum_{j=0}^{N-1} f(x_j)$$

with the Fourier points (5.10) is exact for any function e^{-ikx} for $k \in \mathbb{Z}$, $|k| < N$ or any trigonometric polynomial of degree less than N which is a linear combination of such functions. The DFT can therefore also be seen as an approximation of the continuous Fourier transform of a periodic function f on $[0, 2\pi]$:

$$\frac{1}{2\pi} \int\limits_{0}^{2\pi} f(x)\,\mathrm{e}^{-\mathrm{i}kx}\,\mathrm{d}x = \frac{1}{N}\sum_{j=0}^{N-1} f(x_j)\,\mathrm{e}^{-\mathrm{i}2\pi jk/N} + \mathcal{R}_N\,,$$

where $\mathcal{R}_N = -f''(\xi)(2\pi)^2/(12N^2)$ and $\xi \in [0, 2\pi]$.

5.2.3 Aliasing

The coefficients of the discrete Fourier transform (5.11) and the coefficients of the exact expansion (5.7) are related by

$$\widetilde{f}_k = \widehat{f}_k + \sum_{\substack{m=-\infty \\ m\neq 0}}^{\infty} \widehat{f}_{k+Nm}\,, \quad k = -N/2, \ldots, N/2 - 1\,. \tag{5.16}$$

By using (5.9) and (5.13) this can be written as $I_N f = S_N f + R_N f$. The remainder

$$R_N f = I_N f - S_N f = \sum_{k=-N/2}^{N/2-1} \left(\sum_{\substack{m=-\infty \\ m\neq 0}}^{\infty} \widehat{f}_{k+Nm} \right) \phi_k \tag{5.17}$$

is the *aliasing error* and it measures the difference between the interpolation polynomial and the truncated series. Aliasing implies that the component with the wavenumber $(k + Nm)$ behaves like the one with the wave-number k. Because the basis functions are periodic, $\phi_{k+Nm}(x_j) = \phi_k(x_j)$, such components are indistinguishable (Fig. 5.4): on a discrete mesh the kth component of $I_N f$ depends not only on the kth component of f, but also on the higher components mimicking the kth. The aliasing error is orthogonal to the truncation error, thus $\|f - I_N f\|^2 = \|f - S_N f\|^2 + \|R_N f\|^2$; the interpolation error is always larger than the truncation error.

Example Knowing how to control aliasing has practical importance. The signals on audio compact discs are sampled at the frequency of $\nu_s = 44.1$ kHz, corresponding to the critical frequency (5.6) of $\nu_c = \nu_s/2 = 22.05$ kHz. Such fine sampling prevents aliasing in the audible part of the spectrum and there is no distortion of the signal. Had we wished, however, to compress the signal, we should not do this by simply decreasing the sampling frequency, since this would map the high-frequency components into the low-frequency part of the spectrum, causing distortion. Instead, we should use a filter to remove the high-frequency part of the spectrum and only then down-sample. Figure 5.5 shows the appearance of aliasing in the frequency spectrum of a signal sampled at 44.1 kHz and 882 Hz without filtering. ◁

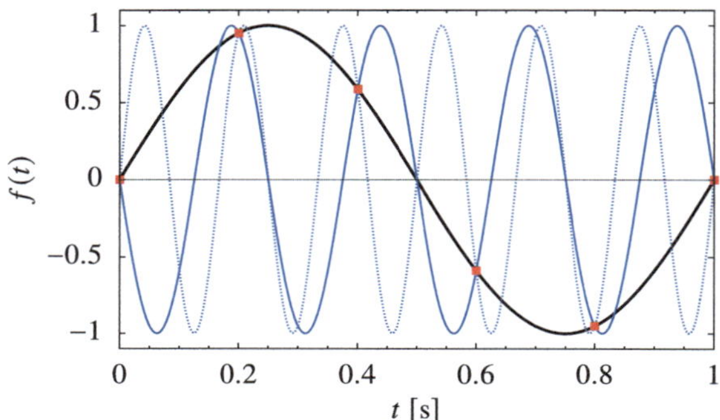

Fig. 5.4 The signal $\sin(2\pi\nu t)$ with $\nu = 1\,\mathrm{s}^{-1}$ (full line) is sampled at $\nu_s = 5\,\mathrm{s}^{-1}$ (squares): the critical frequency (5.6) is $\nu_c = \omega_c/(2\pi) = 2.5\,\mathrm{s}^{-1}$. We obtain exactly the same values by sampling the functions $\sin(2\pi(\nu - \nu_s)t)$ (thin full line) or $\sin(2\pi(\nu + \nu_s)t)$ (dotted line). At this sampling rate, aliasing causes all three signals to be represented with the frequency ν in the spectrum

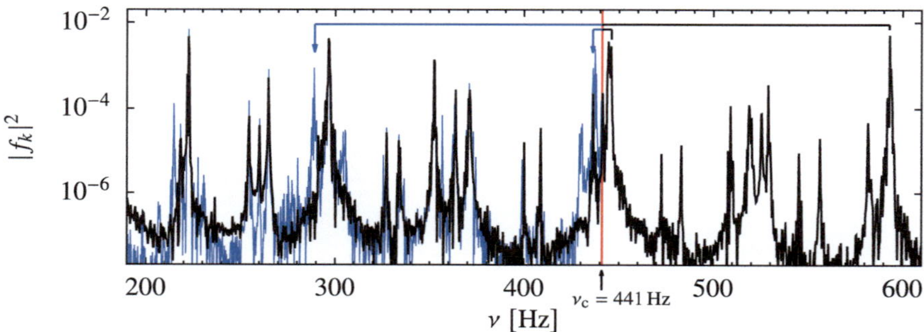

Fig. 5.5 Frequency spectrum of the concluding chord of the Toccata and Fugue for organ in *d*-minor, BWV 565, of J. S. Bach. Black: sampling at 44.1 kHz. Blue: sampling at a smaller frequency 882 Hz causes aliasing. The arrows indicate the peak at 593 Hz which is mirrored across the critical frequency $\nu_c = 441$ Hz onto the frequency $(441 - (593 - 441))\,\mathrm{Hz} = 289\,\mathrm{Hz}$, and the peak at 446 Hz which maps onto 436 Hz

5.2.4 Leakage

The discrete Fourier transformation of realistic signals of course involves only finitely many values. From an infinite sequence we pick (multiply by one) only a sample of length N, whereas the remaining values are dropped (multiplied by zero). Due to this restriction, known as *windowing*, the frequency spectrum exhibits the *leakage* phenomenon shown in Fig. 5.6 (adapted from [5]). To some extent leakage can be controlled by using *window (or weight) functions* that engage a larger portion of the signal and smoothly fade out instead of crudely multiplying the signal by one and the remainder by zero. Instead of Eq. (5.11), then, one has

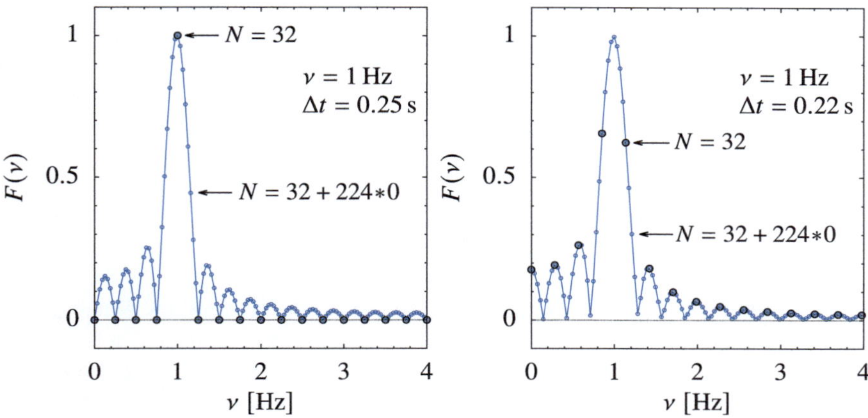

Fig. 5.6 Leakage in the frequency spectrum in the discrete Fourier transform of a sine wave with the frequency 1 Hz. [LEFT] Sampling at $N = 32$ points 0.25 s apart encompasses precisely four complete waves. The only non-zero component of the transform is the one corresponding to the frequency of 1 Hz. [RIGHT] Sampling at $N = 32$ points 0.22 s apart covers only 3.52 waves. Many non-zero frequency components appear. The curves connect the discrete transform of the same signals, except that the 32 original samples of the signal are followed by 224 zeros (total of 256 points). Adding zeros in the temporal domain is known as *zero padding* and improves the resolution in the frequency domain. In the limit $N \to \infty$ we approach the continuous Fourier transform

$$\widetilde{f}_k = \frac{1}{N} \sum_{j=0}^{N-1} w(x_j) f(x_j) \, e^{-ikx_j} \,, \quad -N/2 \le k \le N/2 - 1 \,,$$

where $w(x_j)$ are the weights corresponding to the chosen window function. The advantages and weaknesses of some popular window functions are discussed in [6].

5.2.5 *Fast Discrete Fourier Transformation (FFT)*

The discrete Fourier transformation (5.11) is a mapping between the vector spaces of dimensions N, $\mathcal{F}_N : \mathbb{C}^N \to \mathbb{C}^N$. Let us rewrite it in a more transparent form,

$$F = \mathcal{F}_N[f] \,, \quad F_k = \frac{1}{N} \sum_{j=0}^{N-1} f_j \, e^{-2\pi i jk/N} \,, \tag{5.18}$$

where we have denoted $f = \{f_j\}_{j=0}^{N-1}$ and $F = \{F_k\}_{k=0}^{N-1}$. Note that the indices j and k run symmetrically, both from 0 to $N - 1$. The inverse transformation is

$$f = \mathcal{F}_N^{-1}[F] , \quad f_j = \sum_{k=0}^{N-1} F_k \, e^{2\pi i j k/N} .$$

Then Eq. (5.18) can be written as

$$F_k = \frac{1}{N} \sum_{j=0}^{N-1} W_N^{kj} f_j , \quad W_N = e^{-2\pi i/N} . \tag{5.19}$$

To evaluate the DFT by computing this sum we need $\mathcal{O}(N^2)$ operations. But precisely the same result can be achieved with far fewer operations by using the Cooley-Tukey algorithm. Let N be divisible by m. Then the sum can be split into m partial sums, and each of them runs over the elements f_j of the array f with the same modulus of the index $j \bmod m$:

$$F_k = \frac{1}{N} \sum_{l=0}^{m-1} W_N^{kl} \sum_{j=0}^{N/m-1} W_{N/m}^{kj} f_{mj+l} .$$

Let us denote by $f^{(l)} = \{f_{mj+l}\}_{j=0}^{N/m-1}$ the components of the array f which have the same modulus of the index with respect to m. We have thus recast the transform of the original array f of dimension N as a sum of m transforms of the shorter arrays $f^{(l)}$ of dimension N/m. This can be written symbolically as

$$\mathcal{F}_N[f]_k = \frac{1}{N} \sum_{l=0}^{m-1} W_N^{kl} \left(\frac{N}{m} \mathcal{F}_{N/m}[f^{(l)}] \right)_k .$$

This is a recursive computation of the DFT for the array f that follows the idea of *divide-and-conquer* algorithms. The array f for which the DFT should be computed is gradually broken down into sub-arrays, thus reducing the amount of necessary work. The optimal factorization is $N = 2^p$ ($p \in \mathbb{N}$) in which at each step the array is split into two sub-arrays containing elements of the original array with even and odd indices, respectively. This method requires $\mathcal{O}(N \log_2 N)$ operations for the full DFT instead of $\mathcal{O}(N^2)$ by direct summation, lending it the name *Fast Fourier Transformation*. Similar speeds are attainable if N is factorizable to primes, e.g.

$$N = 2^{p_1} 3^{p_2} 5^{p_3} 7^{p_4} , \quad p_i \in \mathbb{N} ,$$

which is supported by all modern FFT libraries. Good implementations of the FFT are complicated, as the factorization should be carefully matched to the addressing of the arrays. The most famous library is the multiple-award winning FFTW3 (Fastest Fourier Transform in the West) [7]; see also [8, 9]. A comparison of the CPU cost of the standard DFT and FFT is illustrated in Fig. 5.7 (left).

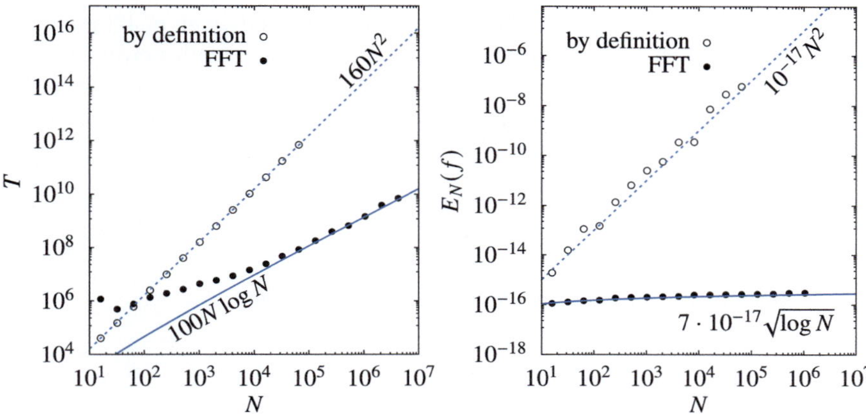

Fig. 5.7 [LEFT] The numerical cost (number of CPU cycles T) of the computation of the DFT by the basic definition (5.19) and by using the FFT, as a function of the sample size N. [RIGHT] The average measure of deviation $E_N(f)$ for the computation of the DFT by definition and by using the FFT

Example Due to the fewer operations, FFT is not only essentially faster than the naive DFT; it is also more precise, as can be confirmed by a numerical experiment. We form the array $f = \{f_0, f_1, \ldots, f_{N-1}\}$ of random complex numbers. We apply the DFT to compute the transform of f, to which we apply the inverse DFT. Finally, we compute the deviation of the resulting array from the original array:

$$\Delta f = (\mathcal{F}_N^{-1} \circ \mathcal{F}_N) f - f \,.$$

In arithmetic with precision ε, we get $\Delta f \neq 0$. We define the average deviation as $E_N(f) = \langle \|\Delta f\|_2 / \|f\|_2 \rangle$, where the average $\langle \cdot \rangle$ is over a large set of random arrays. The results are shown in Fig. 5.7 (right). When the DFT is computed by Eq. (5.18), we get $E_N(f) \sim \mathcal{O}(\varepsilon N^2)$, while the FFT gives $E_N(f) \sim \mathcal{O}(\varepsilon \sqrt{\log N})$ [10]. In short, the FFT algorithm is unbeatable! All decent numerical libraries support the computation of the DFT by FFT algorithms (see Appendix J).   ◁

5.2.6 Multiplication of Polynomials by Using the FFT

Multiplication of polynomials is one of the tasks in computing with power bases, e.g. in expansions in powers of the perturbation parameters in classical and quantum mechanics. The multiplication of $p(x) = \sum_{i=0}^{n} a_i x^i$ and $q(x) = \sum_{i=0}^{m} b_i x^i$ in the form

$$q(x)p(x) = \sum_{i=0}^{n+m} c_i x^i \,, \quad c_i = \sum_{k=0}^{i} a_k b_{i-k} \,,$$

requires $\mathcal{O}((n+1)(m+1))$ operations to determine the coefficients c_i. If the number of terms is large ($n, m \gg 1$) this process is slow and prone to rounding errors. A faster and a more precise way is offered by the FFT. We form two arrays of length $N = m + n + 1$. The coefficients of the polynomials p and q are stored at the beginning of these arrays while the remaining elements are set to zero:

$$A = \{A_i\}_{i=0}^{N-1} = \{a_0, \ldots, a_n, \underbrace{0, \ldots, 0}_{m}\}, \quad B = \{B_i\}_{i=0}^{N-1} = \{b_0, \ldots, b_m, \underbrace{0, \ldots, 0}_{n}\}.$$

The coefficients of the product are given by the convolution

$$c_i = \sum_{k=0}^{N-1} A_k B_{i-k}, \quad i = 0, 1, \ldots, N-1,$$

where we assume periodic boundary conditions, $A_k = A_{N+k}$, $B_k = B_{N+k}$. The convolution is then evaluated by first computing the FFT of the arrays A and B,

$$\hat{A} = \{\hat{A}_i\}_{i=0}^{N-1} = \mathcal{F}_N[A], \quad \hat{B} = \{\hat{B}_i\}_{i=0}^{N-1} = \mathcal{F}_N[B],$$

multiplying the transforms component-wise into a new array $\hat{C} = \{\hat{A}_i \hat{B}_i\}_{i=0}^{N-1}$, and computing its inverse FFT,

$$C = \{C_i\}_{i=0}^{N-1} = N\mathcal{F}_N^{-1}[\hat{C}].$$

This procedure has a numerical cost of $\mathcal{O}(N \log_2 N)$ which, for $n, m \gg 1$, is much smaller than the cost of directly computing the sums of the products.

5.2.7 Power Spectral Density

The Fourier transformation is a decomposition of a signal into a linear combination of the functions $A_\omega e^{i\omega x}$, and $|A_\omega|^2$ is the *signal power* at frequency ω. In dealing with real signals we are mostly interested in the total power at the absolute value of the frequency, $|A_\omega|^2 + |A_{-\omega}|^2$ for $\omega \geq 0$. If the signal f is continuous and has the Fourier transform (5.1), we define the *double-sided power spectral density* (PSD) as

$$S(\omega) = |F(\omega)|^2, \quad \omega \in \mathbb{R},$$

while the *single-sided power spectral density* is

$$S(\omega) = |F(-\omega)|^2 + |F(\omega)|^2, \quad \omega \in \mathbb{R}_+.$$

Often, the double-sided spectral density of a signal is defined via the single-sided density in which the power of a component with the frequency ω is equal to the power of the component with the frequency $-\omega$, and then $S(\omega) = 2|F(\omega)|^2$.

For discrete data $\{f_j\}$ with the transform (5.18) the discrete double-sided power spectral distribution $\{S_k\}$ is defined analogously to the continuous case,

$$S_k = |F_k|^2 , \quad k = 0, 1, \ldots, N - 1 .$$

Now S_k measures the power at the frequency $2\pi k/N$, while S_{N-k} corresponds to the frequency $-2\pi k/N$, where $k = 0, 1, \ldots, N/2 - 1$. For the single-sided distribution we sum over the negative and positive frequencies; for odd N we get

$$
\begin{aligned}
S_0 &= |F_0|^2 , \\
2S_k &= |F_k|^2 + |F_{N-k}|^2, \quad k = 1, 2, \ldots, (N - 1)/2 ,
\end{aligned}
$$

while for even N the distribution $\{S_k\}$ is given by

$$
\begin{aligned}
S_0 &= |F_0|^2 , \\
2S_k &= |F_k|^2 + |F_{N-k}|^2 , \quad k = 1, 2, \ldots, N/2 , \\
S_{N/2} &= |F_{N/2}|^2 .
\end{aligned}
\tag{5.20}
$$

In the discrete case, Parseval's equality applies in the form

$$\frac{1}{N} \sum_{j=0}^{N-1} |f_j|^2 = \sum_{k=0}^{N-1} |F_k|^2 .$$

If windowing is used (weighting the signal components by weights w_j), resulting in the transforms $\widetilde{F}_k$ instead of F_k, the above formulas are replaced by

$$
\begin{aligned}
S_0 &= \frac{1}{W} |\widetilde{F}_0|^2 , \\
2S_k &= \frac{1}{W} \left[|\widetilde{F}_k|^2 + |\widetilde{F}_{N-k}|^2 \right] , \quad k = 1, 2, \ldots, N/2 , \\
S_{N/2} &= \frac{1}{W} |\widetilde{F}_{N/2}|^2 ,
\end{aligned}
$$

where

$$W = \sum_{j=0}^{N-1} w_j^2 ,$$

and similarly in the case of odd N.

5.2.8 Sparse FFT

In many applications including audio and video compression, magnetic resonance imaging, radio astronomy and global positioning systems the overwhelming majority of the Fourier coefficients of a signal are small or equal to zero. While the standard FFT is insensitive to this sparseness of the spectrum and grinds away at its immutable $\mathcal{O}(N \log N)$, a "sparse FFT" should be able to exploit precisely this feature. In other words, given a N-dimensional vector (set of complex values) f with the Fourier transform F, a sparse FFT algorithm should produce an approximation $\widehat{F}$ to F that satisfies

$$\|F - \widehat{F}\|_2 \le C \min_{k-\text{sparse } G} \|F - G\|_2 \,,$$

where C is a constant and the minimization runs over k-sparse transforms G, i.e. transforms with k non-zero frequency components. Transforms possessing at most k non-zero components are called *exactly sparse*; if the other $N - k$ components still represent a non-zero, yet relatively much smaller fraction of the signal power, the transform is said to be *approximately sparse*—the latter being the usual case due to the presence of noise.

Sparse Fourier transformations have a long history, but in an interesting recent development, the authors have shown that such transforms can be obtained at a computational cost of $\mathcal{O}(\log N \sqrt{Nk \log N})$ or $\mathcal{O}(k \log N \log(N/k))$ in the approximately sparse case, depending on the details of the implementation (see [11–13], sFFT2 library), and as low as $\mathcal{O}(k \log N)$ in the exactly sparse case [13]. Taking architecture-specific details into account, the performance can be further improved (see [14, 15], sFFT3 library, exactly sparse only). Fig. 5.8 shows the average computation times for the evaluations of the sparse FFT by using the FFTW3, sFFT2 and sFFT3 libraries.

5.2.9 Non-uniform (Non-equispaced) FFT

Unevenly spaced time series are ubiquitous in science, in particular in astrophysics and seismology: there the signals come at times which are beyond our control, as in earthquakes. It may also happen that the detectors are inoperational for a fraction of the time, or that certain data values are flagged or proclaimed as suspicious and must be discarded. An example of such a signal is given in Fig. 5.9. Simplistic approaches to dealing with the missing data problem—illustrated in the bottom panel—do not work well. Large gaps in the data (case (a), for instance) tend to amplify the low-frequency portion of the power spectrum, corresponding to wavelengths on the order of the gap size.

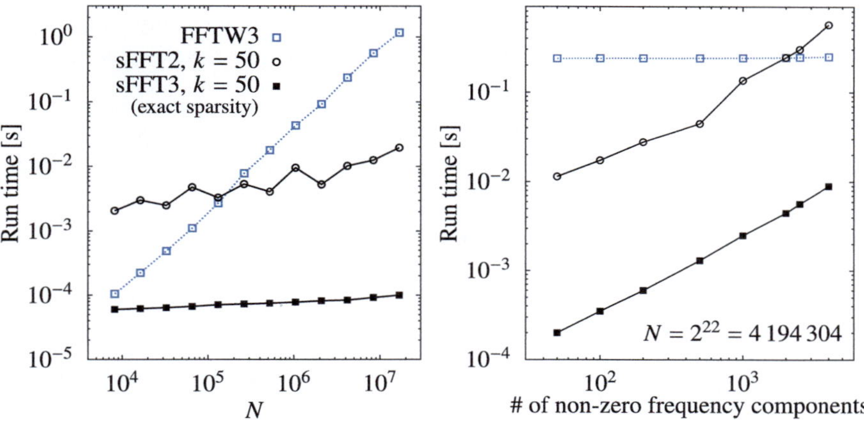

Fig. 5.8 [LEFT] Average times needed to compute the discrete Fourier transforms of length-N signals with sparse ($k = 50$) spectra by using FFTW3 [7], sFFT2 and sFFT3. [RIGHT] Average computation times as functions of the number of non-zero frequency components at fixed signal size N. (Notation as in the left panel)

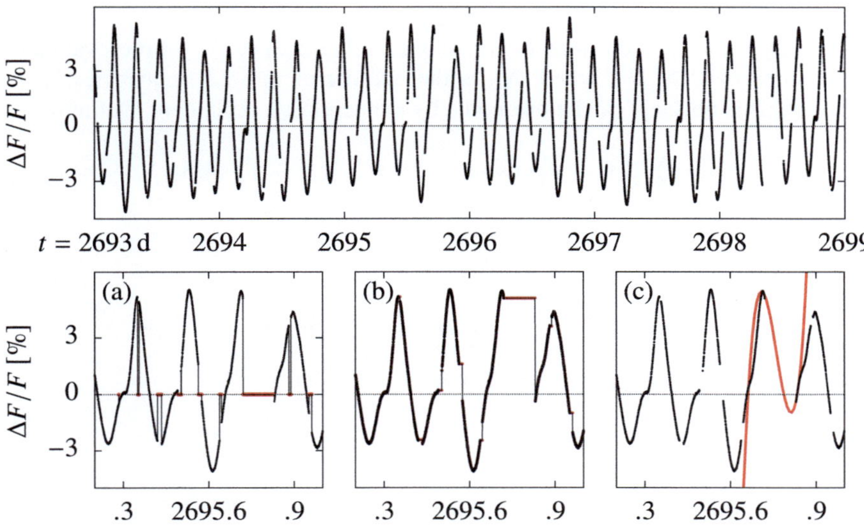

Fig. 5.9 [TOP] A part of the measured light curve (normalized flux F) of the variable star HD 180642 acquired by the CoRoT satellite [16]. [BOTTOM] Naive (and inappropriate) strategies to cope with the missing data: (a) setting them to zero, (b) "clamping" to the last known value, (c) a dangerous step into the dark by cubic interpolation

The most popular and reliable spectral analysis method for unevenly sampled data $\{f_i\}_{i=0}^{N-1}$ acquired at arbitrary times $\{t_i\}_{i=0}^{N-1}$ is to construct the Lomb–Scargle power spectrum (periodogram) [17]

$$P_{\mathrm{LS}}(\omega) = \frac{1}{2\sigma^2} \left\{ \frac{\left[\sum_i (f_i - \overline{f}) \cos \omega(t_i - \tau)\right]^2}{\sum_i \cos^2 \omega(t_i - \tau)} + \frac{\left[\sum_i (f_i - \overline{f}) \sin \omega(t_i - \tau)\right]^2}{\sum_i \sin^2 \omega(t_i - \tau)} \right\}, \qquad (5.21)$$

where

$$\overline{f} = \frac{1}{N} \sum_{i=0}^{N-1} f_i, \quad \sigma^2 = \frac{1}{N-1} \sum_{i=0}^{N-1} \left(f_i - \overline{f}\right)^2,$$

are the signal mean and variance, respectively, and the parameter τ is defined by

$$\tan 2\omega\tau = \frac{\sum_i \sin 2\omega t_i}{\sum_i \cos 2\omega t_i}.$$

The offset τ makes $P_{\mathrm{LS}}(\omega)$ independent of shifting all the t_i's by a constant. Moreover, introducing τ in this manner causes the linear least-squares fit of the data to the model $f(t) = A \cos \omega t + B \sin \omega t$ to yield precisely (5.21).

As in the naive DFT, the numerical cost of calculating $P_{\mathrm{LS}}(\omega)$ by direct evaluation of the sums is $\mathcal{O}(N_\omega N)$, where N_ω is the number of desired frequencies in the spectrum. This obstacle can be overcome by a much faster procedure. Defining

$$\begin{aligned} C_f &= \sum_i (f_i - \overline{f}) \cos \omega t_i, \quad & S_f &= \sum_i (f_i - \overline{f}) \sin \omega t_i, \\ C_2 &= \sum_i \cos 2\omega t_i, \quad & S_2 &= \sum_i \sin 2\omega t_i, \end{aligned} \qquad (5.22)$$

it holds that

$$\begin{aligned} \sum_i (f_i - \overline{f}) \cos \omega(t_i - \tau) &= C_f \cos \omega\tau + S_f \sin \omega\tau, \\ \sum_i (f_i - \overline{f}) \sin \omega(t_i - \tau) &= S_f \cos \omega\tau - C_f \sin \omega\tau, \\ \sum_i \cos^2 \omega(t_i - \tau) &= \frac{N}{2} + \frac{C_2}{2} \cos 2\omega\tau + \frac{S_2}{2} \sin 2\omega\tau, \\ \sum_i \sin^2 \omega(t_i - \tau) &= \frac{N}{2} - \frac{C_2}{2} \cos 2\omega\tau - \frac{S_2}{2} \sin 2\omega\tau. \end{aligned} \qquad (5.23)$$

Note that if the t_i's were evenly spaced, one could calculate C_f, S_f, C_2 and S_2 by two complex FFTs, plug the results into (5.23) and use these to compute (5.21). The key question, therefore, is how to evaluate the sums of (5.22) for arbitrarily spaced data. This is accomplished by fast non-uniform FFT algorithms (NUFFT) with typical numerical costs of $\mathcal{O}(N \log N)$ or $\mathcal{O}(N \log N + |\log \varepsilon| N)$, where ε is the desired numerical precision.

Several types of NUFFT are available, depending on whether the sampling in the configuration space or in the transform space—or both—are non-uniform, since the forward and the inverse transformation are not simply the two sides of the same coin as in the standard FFT. An introduction to various classes of algorithms is given in [18, 19], see also [20–22]. Most methods rely on some sort of local rearrangement of the data from an irregular grid onto a regular one, in conjunction with the classic FFT to evaluate the intermediate sums. In the `fasper` routine of the NR3E library based on [23] this rearrangement is accomplished by inverse interpolation. In an alternative approach, trigonometric polynomials $p(x) = \sum_k p_k \exp(-2\pi i k x)$ are approximated by linear combinations of translated window functions φ that are well localized in both configuration and transform spaces; the core of the algorithm is again the transformation of Fourier integrals of φ (calculated analytically) by means of the FFT. This is at the heart of NFFT3 [24], one of the most comprehensive and fastest NUFFT packages available based on the FFTW3 library and elaborated in [25–27]. For NFFT3 at work in the astrophysical example mentioned above, see [28]. The state-of-the-art parallel implementation of NUFFT of all types (non-uniform to uniform, uniform to non-uniform, or non-uniform to non-uniform) in 1, 2, or 3 dimensions and with an interface to several languages is FINUFFT [29].

5.3 Transformations with Orthogonal Polynomials

Functions can also be expanded in terms of orthogonal polynomials [30]. Let us work in $\mathcal{P}_N$, the space of polynomials of degree $\leq N$, and restrict the presentation to the polynomials most relevant to a physicist: Legendre and Chebyshev polynomials, orthogonal on $[a, b] = [-1, 1]$, as well as Laguerre and Hermite polynomials (and the corresponding Laguerre and Hermite basis functions), which are orthogonal on $[0, \infty]$ and $[-\infty, \infty]$, respectively.

In all these cases, orthogonality is ensured by applying the appropriate weight function w which depends on the type of the polynomials. In the function space $L_w^2(a, b)$ we define the scalar product

$$\langle u, v \rangle_w = \int_a^b u(x)v(x)w(x)\, \mathrm{d}x \tag{5.24}$$

and the norm $\|u\|_w = \sqrt{\langle u, u \rangle_w}$. Orthogonality means that

$$\int_a^b p_i(x)p_j(x)w(x)\, \mathrm{d}x = \begin{cases} \text{const.} & ; \ i = j \,, \\ 0 & ; \ i \neq j \,. \end{cases} \tag{5.25}$$

In particular, we have $w(x) = 1$ for Legendre, $w(x) = (1 - x^2)^{-1/2}$ for Chebyshev, $w(x) = \mathrm{e}^{-x}$ for Laguerre and $w(x) = \mathrm{e}^{-x^2}$ for Hermite polynomials. The expansion of a function f in terms of orthogonal polynomials has the form

$$f(x) \equiv Sf = \sum_{k=0}^{\infty} \widehat{f_k} \, p_k(x) \,, \quad \widehat{f_k} = \frac{1}{\|p_k\|_w^2} \int_a^b f(x) p_k(x) w(x) \, dx \,.$$

Here $\widehat{f_k}$ are the expansion coefficients which may be understood as the transforms of f, by analogy to (5.8). For $N \in \mathbb{N}$ we have a finite series

$$S_N f \equiv \sum_{k=0}^{N} \widehat{f_k} \, p_k(x) \,.$$

Relation to quadrature formulas Orthogonal polynomials are closely related to quadrature formulas (see Sect. 3.1.5). A quadrature requires *collocation points* or *nodes* $x_j \in [a, b]$ as well as *quadrature weights* $w_j > 0$ for $j = 0, 1, \ldots, N$. With these quantities the integral of a continuous function can be approximated by the sum of the products of the weights and the function values at the nodes,

$$\int_a^b f(x) w(x) \, dx = \sum_{j=0}^{N} f(x_j) w_j + R_N \,. \tag{5.26}$$

(Note the indexing starting at $j = 0$.) The remainder R_N may be zero if f is a polynomial of a certain maximum degree. These degrees depend on the choice of the nodes.

Discrete transformation Following the notational conventions of [4], we define the discrete transformation with orthogonal polynomials as

$$f(x_j) = \sum_{k=0}^{N} \widetilde{f_k} \, p_k(x_j) \,, \tag{5.27}$$

where

$$\widetilde{f_k} = \frac{1}{\gamma_k} \sum_{j=0}^{N} f(x_j) p_k(x_j) w_j \,, \quad \gamma_k = \sum_{j=0}^{N} p_k^2(x_j) w_j \,. \tag{5.28}$$

Equations (5.27) and (5.28) for polynomial transforms are analogous to the discrete Fourier transforms with trigonometric polynomials.

In collocation methods (for example, for the solution of partial differential equations in Chap. 12) we can represent a smooth function f on the given interval by its discrete values at the collocation points, while the derivatives of f are approximated by the derivatives of its interpolation polynomial. The interpolation polynomial is an element of $\mathcal{P}_N$, and its values at the collocation points are equal to the function values, $I_N f(x_j) = f(x_j)$ for $0 \le j \le N$. It is formed by the sum

$$I_N f(x) = \sum_{k=0}^{N} \widetilde{f}_k p_k(x) \, . \tag{5.29}$$

It can be shown—just like in Fourier expansions—that the interpolant $I_N f$ is the projection of f on the space $\mathcal{P}_N$ with respect to the scalar product

$$\langle u, v \rangle_N = \sum_{j=0}^{N} u(x_j) v(x_j) w_j \, .$$

This means that $\langle I_N f, v \rangle_N = \langle f, v \rangle_N$ for any continuous function v. Expression (5.28) can therefore also be read as $\gamma_k \widetilde{f}_k = \langle f, p_k \rangle_N$, and the orthogonality relation as $\langle p_j, p_k \rangle_N = \gamma_k \delta_{j,k}$ for $0 \le k \le N$. The relation between the discrete polynomial coefficients and the coefficients of the continuous expansion is

$$\widetilde{f}_k = \widehat{f}_k + \frac{1}{\gamma_k} \sum_{j > N} \langle p_j, p_k \rangle_N \widehat{f}_j \, . \tag{5.30}$$

In general $\langle p_j, p_k \rangle_N \neq 0$ for $j > N$, so the kth component of the interpolant $I_N f$ depends on the kth component of f and *all* components with indices $k > N$. We can again rewrite Eq. (5.30) in the form $I_N f = S_N f + R_N f$, where in

$$R_N f = I_N f - S_N f = \sum_{k=0}^{N} \left(\frac{1}{\gamma_k} \sum_{j > N} \langle p_j, p_k \rangle_N \widehat{f}_j \right) p_k \tag{5.31}$$

we again recognize the aliasing error (compare Eqs. (5.30) and (5.31) to the corresponding Eqs. (5.16) and (5.17) for the DFT). The aliasing error is orthogonal to the series truncation error $f - S_N f$, hence

$$\| f - I_N f \|_w^2 = \| f - S_N f \|_w^2 + \| R_N f \|_w^2 \, .$$

5.3.1 *Legendre Polynomials*

Legendre polynomials P_k for $k = 0, 1, \ldots$ are the eigenfunctions of the singular Sturm–Liouville problem

$$\left((1 - x^2) P_k'(x) \right)' + k(k + 1) P_k(x) = 0 \, , \quad x \in [-1, 1] \, ,$$

which is of the form (9.84) with $p(x) = 1 - x^2$, $q(x) = 0$, and $w(x) = 1$. The polynomials P_k for even (odd) k are even (odd). With the normalization $P_k(1) = 1$ they can be generally written as

$$P_k(x) = \frac{1}{2^k} \sum_{l=0}^{\lfloor k/2 \rfloor} (-1)^l \binom{k}{l} \binom{2k - 2l}{k} x^{k-2l} , \qquad (5.32)$$

where $\lfloor k/2 \rfloor$ is the integer part of $k/2$. Legendre polynomials possess a three-term recurrence relation

$$(k + 1) P_{k+1}(x) = (2k + 1)x P_k(x) - k P_{k-1}(x) , \quad P_0(x) = 1 , \quad P_1(x) = x ,$$

which is a great tool to actually compute the polynomial values. Some typical polynomials P_k are shown in Fig. 5.10 (left).

By using the polynomials P_k any function $f \in L^2(-1, 1)$ can be expanded in a series

$$f(x) = \sum_{k=0}^{\infty} \widehat{f_k} P_k(x) , \quad \widehat{f_k} = \frac{2k + 1}{2} \int_{-1}^{1} f(x) P_k(x) \, \mathrm{d}x . \qquad (5.33)$$

The discrete expansion in terms of Legendre polynomials is defined by Eqs. (5.27) and (5.28) by inserting $p_k(x) = P_k(x)$, while the collocation points x_j, the weights w_j, and the normalization factors γ_k depend on the type of the quadrature (5.26). For quadrature with Legendre polynomials there are no explicit formulas for the collocation points x_j; we must compute the zeros of the corresponding polynomials. There are three ways to accomplish this.

Gauss For Legendre–Gauss quadrature the collocation points x_j $(j = 0, 1, \ldots, N)$ are the roots of the equation $P_{N+1}(x) = 0$, which all lie in the interior of the interval $[-1, 1]$. The corresponding weights and normalization factors are

$$w_j = \frac{2}{(1 - x_j^2)[P'_{N+1}(x_j)]^2} = \frac{2(1 - x_j^2)}{(N + 2)^2 P_{N+2}^2(x_j)} , \quad \gamma_k = \frac{2}{2k + 1} . \qquad (5.34)$$

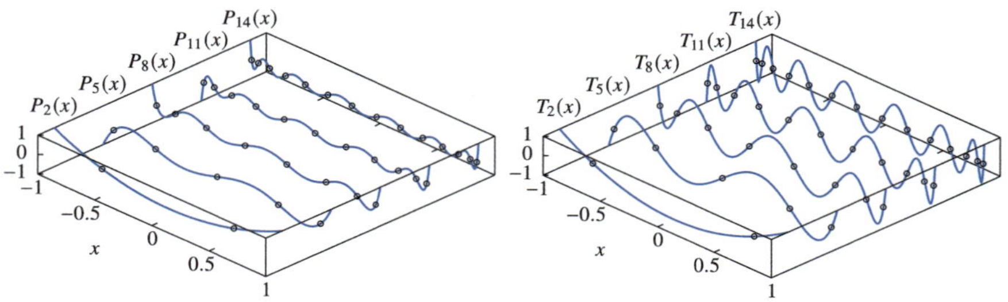

Fig. 5.10 Orthogonal polynomials used in the transformations of functions, collocation methods, and spectral methods for partial differential equations (Chap. 12). [LEFT] Legendre polynomials P_k. [RIGHT] Chebyshev polynomials T_k. Shown are the polynomials of degree $k \in \{2, 5, 8, 11, 14\}$. There is a characteristic clustering of the zeros in the vicinity of the boundary points $x = -1$ and $x = 1$

With these nodes and weights the quadrature formula is exact for polynomials of degree at most $2N + 1$.

Gauss–Radau The collocation points for Legendre–Gauss–Radau quadrature are the roots $x_0 = -1, x_1, x_2, \ldots, x_N \neq 1$ of the equation $P_N(x) + P_{N+1}(x) = 0$, and the weights are

$$w_0 = \frac{2}{(N+1)^2} \,, \quad w_j = \frac{1 - x_j}{(N+1)^2 P_N^2(x_j)} \,, \quad j = 1, 2, \ldots, N \,. \tag{5.35}$$

The normalization factors are $\gamma_k = 2/(2k + 1)$. The quadrature formula with such nodes and weights is exact for polynomials of degree at most $2N$.

Gauss–Lobatto The most frequently used Legendre–Gauss–Lobatto quadrature explicitly includes both endpoints of the interval, $x_0 = -1$ and $x_N = 1$, while the interior points x_j are the roots of the equation $P_N'(x) = 0$:

$$x_j = \begin{cases} -1 & ; \ j = 0 \,, \\ \text{solutions of } P_N'(x) = 0 & ; \ 1 \le j \le N - 1 \,, \\ +1 & ; \ j = N \,. \end{cases} \tag{5.36}$$

In this case the weights are

$$w_j = \frac{2}{N(N+1)} \frac{1}{P_N^2(x_j)} \,, \quad j = 0, 1, \ldots, N \,. \tag{5.37}$$

The normalization factors are $\gamma_k = 2/(2k + 1)$ for $k < N$ and $\gamma_N = 2/N$. The corresponding quadrature formula is exact for polynomials of degree at most $2N - 1$.

Example Let us determine the discrete Legendre–Gauss expansion of the function $f(x) = x \cos\big((\pi x)^2\big)$ by using ten points ($N = 9$). The collocation points x_j are the roots of the equation $P_{N+1}(x) = 0$ (all values rounded to four digits):

$$-x_0 = x_9 = 0.9739 \,, \quad -x_1 = x_8 = 0.8651 \,, \quad -x_2 = x_7 = 0.6794 \,,$$
$$-x_3 = x_6 = 0.4334 \,, \quad -x_4 = x_5 = 0.1489 \,.$$

The weights w_j are given by Eq. (5.34):

$$w_0 = w_9 = 0.0667 \,, \quad w_1 = w_8 = 0.1495 \,, \quad w_2 = w_7 = 0.2191 \,,$$
$$w_3 = w_6 = 0.2693 \,, \quad w_4 = w_5 = 0.2955 \,.$$

The coefficients (5.28) of the discrete expansion (5.27) are then

$$\tilde{f}_1 = -0.1084 \,, \quad \tilde{f}_3 = -0.1784 \,, \quad \tilde{f}_5 = -0.4807 \,, \quad \tilde{f}_7 = -1.0136 \,, \quad \tilde{f}_9 = -0.0923 \,,$$

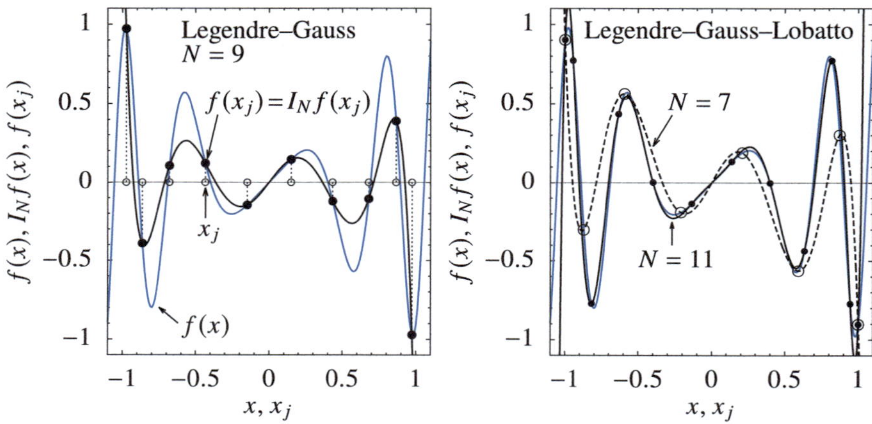

Fig. 5.11 Interpolation polynomials corresponding to the transformation of functions by Legendre polynomials. [LEFT] Computation at the Legendre–Gauss nodes with $N = 9$. [RIGHT] Computation at the Legendre–Gauss–Lobatto nodes with $N = 7$ and $N = 11$

while $\widetilde{f}_0 = \widetilde{f}_2 = \widetilde{f}_4 = \widetilde{f}_6 = \widetilde{f}_8 = 0$: since the function f is odd, its expansion also involves only odd Legendre polynomials. Figure 5.11 (left) shows the points x_j, the function f, and its interpolant $I_N f$ (5.29).

We follow the same path in Legendre–Gauss–Lobatto collocation, where the endpoints $-x_0 = x_N = 1$ are always included. Let us take eight points ($N = 7$). The nodes are given by Eq. (5.36),

$$-x_0 = x_7 = 1.0000 \,, \quad -x_1 = x_6 = 0.8717 \,,$$
$$-x_2 = x_5 = 0.5917 \,, \quad -x_3 = x_4 = 0.2093 \,,$$

while the weights are given by Eq. (5.37),

$$-w_0 = w_7 = 0.0357 \,, \quad -w_1 = w_6 = 0.2107 \,,$$
$$-w_2 = w_5 = 0.3411 \,, \quad -w_3 = w_4 = 0.4125 \,.$$

The seventh-degree interpolation polynomial $I_7 f$ of f is shown in Fig. 5.11 (right). The same Figure also shows the eleventh-degree interpolant $I_{11} f$.   ◁

Note that in the Gauss–Lobatto case, the polynomial $I_N f$ that interpolates f and matches it at the points x_j can be written in two ways: as an expansion in $P_k(x)$ with the coefficients $\widetilde{f}_k$, or in the form of the expansion in Lagrange interpolation polynomials where the coefficients are given by the values of f at x_j:

$$I_N f(x) = \sum_{k=0}^{N} \widetilde{f}_k P_k(x) = \sum_{j=0}^{N} f(x_j) l_j(x) \,, \tag{5.38}$$

where

$$l_j(x) = \frac{1}{N(N+1)} \frac{x^2 - 1}{x - x_j} \frac{P'_N(x)}{P_N(x_j)} \tag{5.39}$$

is the Lagrange interpolation polynomial. (At $x = x_j$ both its numerator and denominator are zero and l'Hôspital is called to rescue.) In the discrete Fourier transformation we have expressed the interpolant $I_N f$ by the sum (5.14) over trigonometric polynomials (5.15). Expressions (5.38) and (5.39) are their equivalents for the discrete Legendre transformation.

5.3.2 Chebyshev Polynomials

Chebyshev polynomials T_k for $k = 0, 1, \dots$ [31] are the eigenfunctions of the singular Sturm–Liouville problem

$$\left(\sqrt{1 - x^2}\, T'_k(x) \right)' + \frac{k^2}{\sqrt{1 - x^2}}\, T_k(x) = 0, \quad x \in [-1, 1],$$

which is of the form (9.84) with $p(x) = (1 - x^2)^{1/2}$, $q(x) = 0$, and $w(x) = (1 - x^2)^{-1/2}$. The polynomials T_k for even (odd) k are even (odd). With the normalization $T_k(1) = 1$ they are defined as

$$T_k(x) = \cos(k \arccos x) = \sum_{i=0}^{\lfloor k/2 \rfloor} \binom{k}{2i} (x^2 - 1)^i x^{k-2i}. \tag{5.40}$$

Some typical examples are shown in Fig. 5.10 (right). The polynomials are related by the three-term recurrence

$$T_{k+1}(x) = 2x T_k(x) - T_{k-1}(x), \quad T_0(x) = 1, \quad T_1(x) = x, \tag{5.41}$$

which allows us to compute the values $T_n(x)$ in just $\mathcal{O}(n)$ operations. Chebyshev polynomials are orthogonal on $[-1, 1]$ with respect to the weight $(1 - x^2)^{-1/2}$,

$$\int_{-1}^{1} T_k(x) T_l(x) \frac{dx}{\sqrt{1 - x^2}} = \begin{cases} 0 & ; \; k \neq l, \\ \pi/2 & ; \; k = l \neq 0, \\ \pi & ; \; k = l = 0, \end{cases}$$

and are also "point-wise orthogonal": on the discrete set of points

$$x_j = \cos \frac{(2j + 1)\pi}{2N}, \quad j = 0, 1, \dots, N - 1,$$

corresponding to the N zeros of the polynomial T_N, Chebyshev polynomials are orthogonal in the sense of the sum

$$\sum_{j=0}^{N-1} T_k(x_j) T_l(x_j) = \begin{cases} 0 & ; \ k \neq l \,, \\ N/2 & ; \ k = l \neq 0 \,, \\ N & ; \ k = l = 0 \,. \end{cases} \tag{5.42}$$

(More general formulas can be found in Eqs. (1.141) and (1.144) in [31].) Any function $f \in L^2(-1, 1)$ can be expanded in terms of Chebyshev polynomials T_k as

$$f(x) = \sum_{k=0}^{\infty} \widehat{f_k} T_k(x) \,, \quad \widehat{f_k} = \frac{2}{\pi c_k} \int_{-1}^{1} f(x) T_k(x) w(x) \, dx \,, \tag{5.43}$$

where $c_0 = 2$ and $c_k = 1$ for $k \geq 1$. In contrast to quadrature involving Legendre polynomials, the nodes and the weights for Gauss, Gauss–Radau and Gauss–Lobatto quadrature involving Chebyshev polynomials are given by simple *explicit* formulas which are listed in the following. Note that the nodes are indexed such that the values of x_j decrease when j increases.

Gauss For Chebyshev–Gauss collocation the nodes and the weights are

$$x_j = \cos \frac{(2j + 1)\pi}{2N + 2} \,, \quad w_j = \frac{\pi}{N + 1} \,, \quad j = 0, 1, \ldots, N \,,$$

while the factors γ_k in (5.28) are equal to $\gamma_k = \pi c_k/2$ for $0 \leq k < N$ and $\gamma_N = \pi/2$.

Gauss–Radau For Chebyshev–Gauss–Radau collocation with the node $x_0 = +1$ included we have

$$x_j = \cos \frac{2j\pi}{2N + 1} \,, \quad w_j = \begin{cases} \dfrac{\pi}{2N + 1} & ; \ j = 0 \,, \\ \dfrac{2\pi}{2N + 1} & ; \ j = 1, 2, \ldots, N \,, \end{cases} \tag{5.44}$$

while $\gamma_k = \pi c_k/2$ for $0 \leq k < N$ and $\gamma_N = \pi/2$.

Gauss–Lobatto The most frequently used Gauss–Lobatto collocation includes the endpoints $x_0 = 1$ and $x_N = -1$. Here, the nodes and the weights are given by

$$x_j = \cos \frac{j\pi}{N} \,, \quad w_j = \begin{cases} \dfrac{\pi}{2N} & ; \ j = 0, N \,, \\ \dfrac{\pi}{N} & ; \ j = 1, 2, \ldots, N - 1 \,. \end{cases} \tag{5.45}$$

The normalization factors are $\gamma_k = \pi c_k/2$ for $0 \leq k < N$ and $\gamma_N = \pi$.

Discrete Chebyshev transformation on the interval $x \in [-1, 1]$ is most often associated with Gauss–Lobatto collocation, and this is the only case we discuss henceforth. The expansion of the function f in a finite series has the form

$$I_N f(x) = \sum_{k=0}^{N} \widetilde{f}_k T_k(x) , \quad \widetilde{f}_k = \frac{2}{N\overline{c}_k} \sum_{j=0}^{N} \frac{1}{\overline{c}_j} f(x_j) T_k(x_j) , \tag{5.46}$$

where $\overline{c}_0 = \overline{c}_N = 2$ and $\overline{c}_j = 1$ for $j = 1, 2, \ldots, N - 1$. This is a discrete equivalent of the continuous expansion defined in Eqs. (5.43). At the Gauss–Lobatto quadrature nodes (5.45) the expansion can be rewritten as

$$I_N f(x_j) = \sum_{k=0}^{N} \widetilde{f}_k \cos \frac{\pi j k}{N} , \quad \widetilde{f}_k = \frac{2}{N\overline{c}_k} \sum_{j=0}^{N} \frac{1}{\overline{c}_j} f(x_j) \cos \frac{\pi j k}{N} ,$$

which is nothing but the cosine transform. The transliteration of the Chebyshev transformation to the Fourier transformation makes a great impact in spectral methods (Sect. 12.1.3).

Example As in the Legendre case (5.38), the transformation with Chebyshev polynomials allows us to write the interpolation polynomial in two ways: by using Eqs. (5.46) or by expanding it in terms of different interpolation polynomials,

$$I_N f(x) = \sum_{k=0}^{N} \widetilde{f}_k T_k(x) = \sum_{j=0}^{N} f(x_j) l_j(x) ,$$

where

$$l_j(x) = \frac{1}{\overline{c}_j} \frac{(-1)^j}{N^2} \frac{x^2 - 1}{x - x_j} T'_N(x) \tag{5.47}$$

is the interpolation polynomial for Gauss–Lobatto collocation (5.45). The polynomials l_0, l_1, and l_2 for $N = 8$ are shown in Fig. 5.12.

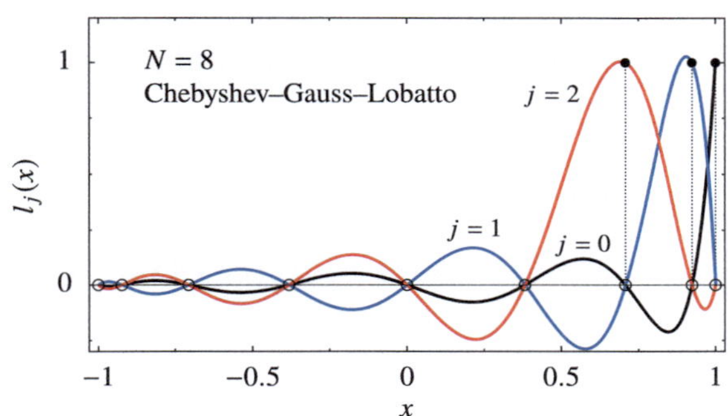

Fig. 5.12 The first three Lagrange interpolation polynomials $l_j(x)$ for Chebyshev–Gauss–Lobatto collocation with $N = 8$. The collocation points $x_j = \cos(\pi j/N)$, at which $l_j(x_k) = \delta_{j,k}$ holds, follow from right to left for indices $j = 0, 1, \ldots, N$

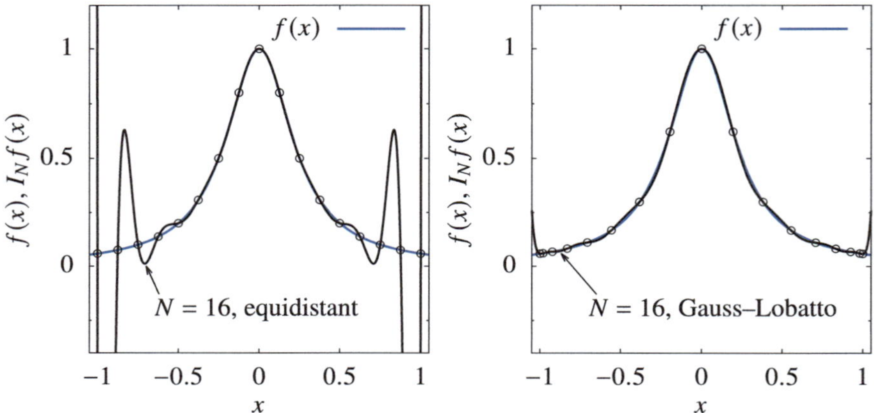

Fig. 5.13 Lagrange interpolation of the function $f(x) = 1/(1 + 16x^2)$ on the interval $[-1, 1]$. [LEFT] The interpolant $I_{16}f$ through 17 equidistant points $x_j = 2j/N - 1$ $(j = 0, 1, \ldots, N)$. When the number of points N is increased, the error $\|f - I_N f\|$ exponentially increases, the annoyance known as the Runge phenomenon. [RIGHT] The interpolant $I_{16}f$ through 17 Gauss–Lobatto points (5.45). By increasing N the error $\|f - I_N f\|$ exponentially decreases, which can be exploited in spectral methods (Chap. 12)

A correct choice of the collocation points is of key importance. Namely, on the interval $[-1, 1]$, a different Lagrange interpolation polynomial with the general form

$$l_j(x) = \frac{q_N(x)}{(x - x_j)q_N'(x_j)} \, , \quad q_N(x) = \prod_{j=0}^{N}(x - x_j) \, , \tag{5.48}$$

can be spanned through any set of function values at the $N + 1$ nodes. But for the chosen class of orthogonal polynomials, only one set of nodes is optimal: for Chebyshev polynomials with Gauss–Lobatto collocation these are precisely the points (5.45). Figure 5.13 shows the classic comparison between the interpolation of the function at equidistant points according to (5.48) and the interpolation at the Gauss–Lobatto points (5.45), which prevents the wild oscillations of the interpolant.   ◁

5.3.3 *Laguerre Polynomials and Basis Functions*

Laguerre polynomials L_k for $k = 0, 1, \ldots$ are the eigenfunctions of the singular Sturm–Liouville problem

$$x L_k''(x) + (1 - x)L_k'(x) + k L_k(x) = 0 \, , \quad x \in [0, \infty] \, ,$$

and are suitable (with an appropriate weight) for problems defined on the semi-infinite domain. They can be expressed as

$$L_k(x) = \frac{e^x}{k!}\frac{d^k}{dx^k}\left(e^{-x}x^k\right) = \frac{1}{k!}\left(\frac{d}{dx} - 1\right)^k x^k$$

and satisfy the recurrence relation

$$(k+1)L_{k+1}(x) = (2k+1-x)L_k(x) - kL_{k-1}(x)\,, \quad L_0(x) = 1\,, \quad L_1(x) = 1 - x\,.$$

Laguerre polynomials are orthogonal with respect to the weight $w(x) = e^{-x}$,

$$\int_0^\infty L_k(x)L_l(x)e^{-x}dx = \delta_{k,l}\,.$$

The nodes x_j $(j = 0, 1, \ldots, N)$ for Gauss quadrature (collocation) are the roots of the polynomial L_{N+1}, while the weights are given by

$$w_j = \frac{x_j}{(N+2)^2 L_{N+2}^2(x_j)}\,. \tag{5.49}$$

Many modern numerical libraries include generators of nodes and weights, based on the algorithms presented in [33] which allow for a stable and precise calculation in double precision; see also Sect. 3.2.3 for the most recent developments. Up to $N = 15$ they are listed on p. 923 of [34].

In spectral methods on the semi-infinite domain $[0, \infty]$ discussed in Chap. 12 it is convenient to use basis functions which differ from the Laguerre polynomials by an additional exponential:

$$\widehat{L}_k(x) = L_k(x)\,e^{-x/2}\,. \tag{5.50}$$

This makes them orthonormal, $\int_0^\infty \widehat{L}_k(x)\widehat{L}_l(x)dx = \delta_{k,l}$, and applicable to cases in which the function to be expanded—or the solution of the differential equation to be found—decays exponentially at infinity; in particular, $\widehat{L}_k(x)$ are the exact radial parts of the energy eigenfunctions of the hydrogen atom. The five lowest-order Laguerre functions are shown in Fig. 5.14.

5.3.4   *Rational Chebyshev Functions TL*

With a specific coordinate transformation the semi-infinite domain can be mapped into a finite domain, allowing us to devise other basis sets for $y \in [0, \infty]$ that are the images of Chebyshev polynomials (with $x \in [-1, 1]$) or trigonometric functions (with $t \in [0, \pi]$). The most commonly used mappings are

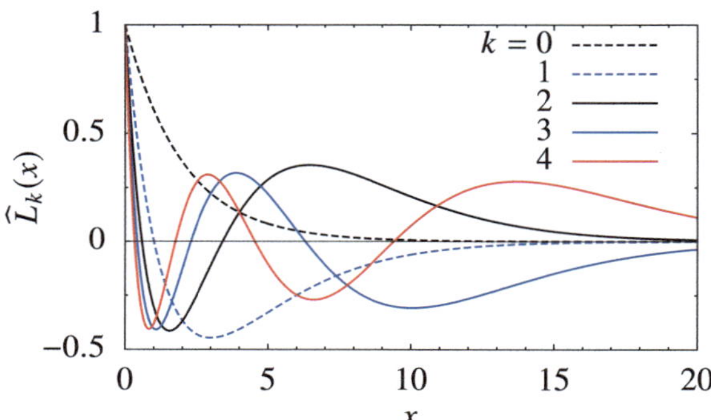

Fig. 5.14 The five lowest-order basis functions (5.50) suitable for the expansion of functions on $[0, \infty]$ that decay sufficiently fast as $x \to \infty$

$$y(x) = L\frac{1+x}{1-x} \quad \Leftrightarrow \quad x(y) = \frac{y-L}{y+L},$$

$$y(t) = L \cot^2\left(\frac{t}{2}\right) \quad \Leftrightarrow \quad t(y) = 2\operatorname{arccot}\left(\sqrt{\frac{y}{L}}\right),$$

$$(5.51)$$

and they produce the *rational Chebyshev functions* (denoted by TL) with the property

$$TL_k(y) = T_k(x) = \cos(kt), \quad x = \cos(t), \tag{5.52}$$

where T_k are the Chebyshev polynomials. Apart from $TL_0(y) = 1$ the first few are

$$TL_1(y; L) = (y - L)/(y + L),$$
$$TL_2(y; L) = \left(y^2 - 6yL + L^2\right)/(y + L)^2,$$
$$TL_3(y; L) = (y - L)\left(y^2 - 14yL + L^2\right)/(y + L)^3,$$
$$TL_4(y; L) = \left(y^4 - 28y^3L + 70y^2L^2 - 28yL^3 + L^4\right)/(y + L)^4,$$

and they are shown in Fig. 5.15. Here L is a parameter that defines how much he scale shrinks when mapped from y to x or t (or expands in the opposite direction). A rule-of-thumb is to choose L roughly commensurate with the scale of variation of the solution; see [1] for details on how L may be optimized. In spite of the change of variables the rational Chebyshev functions remain orthogonal:

$$\int_0^\infty TL_k(y; L)TL_l(y; L)\frac{\sqrt{L}}{(y + L)\sqrt{y}}\, \mathrm{d}y = \begin{cases} 0 & ; k \neq l, \\ \pi & ; k = l = 0, \\ \pi/2 & ; k = l > 0. \end{cases}$$

If the function $f(y)$ to be expanded is rational and non-singular at infinity, its series in terms of $TL_k(y)$ will have geometric spectral convergence, i.e. its coefficients f_k will asymptotically ($k \gg 1$) behave like $f_k \sim \mathcal{O}\left(\exp(-ck^p)\right)$ with $p = 1$. If the function (or the solution of a differential equation) is singular at infinity, the convergence will be subgeometric ($p < 1$). If $f(y)$ only decays algebraically as

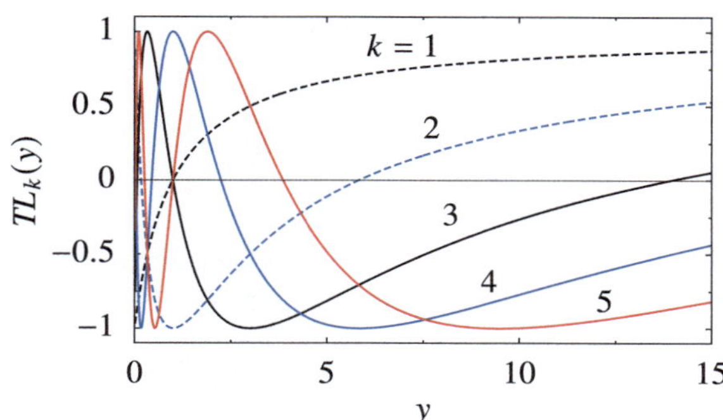

Fig. 5.15 Rational Chebyshev functions $TL_k(y)$ on $[0, \infty]$, for $1 \leq k \leq 5$ and $L = 1$. The shapes of the curves are universal, i.e. the graphs are valid for any L if we replace y by y/L. Note that $\lim_{y \to \infty} TL_k(y) = 1$ for all k regardless of L

$y \to \infty$, its series in terms of TL_k may still retain spectral convergence—and this is a good reason to use TL_k as the basis set instead of the "raw" Laguerre polynomials in such cases: an expansion in terms of the unweighted L_k becomes inefficient unless $f(y)$ decays exponentially as $y \to \infty$ [1].

As shown in Sect. 12.7 it is precisely the mapping to t that enables us to compute the spectral derivatives quickly and efficiently. The corresponding collocation points (pseudospectral points in the parlance of spectral methods) are given by

$$y_j = L \cot^2 \left(\frac{t_j}{2} \right) , \quad t_j = \frac{(2j+1)\pi}{2N+2} , \quad j = 0, 1, \ldots, N .$$

5.3.5 Hermite Polynomials and Basis Functions

Hermite polynomials H_k for $k = 0, 1, \ldots$ are the eigenfunctions of the singular Sturm–Liouville problem

$$H_k''(x) - 2x H_k'(x) + 2n H_k(x) = 0 , \quad n \in \mathbb{N}_0 , \quad x \in [-\infty, \infty] ,$$

and are suitable (with an appropriate weight) for problems defined on the whole real axis. They can be expressed as

$$H_k(x) = (-1)^k e^{x^2} \frac{d^k}{dx^k} e^{-x^2}$$

and satisfy the recurrence relation

$$H_{k+1}(x) = 2x H_k(x) - 2k H_{k-1}(x) , \quad H_0(x) = 1 , \quad H_1(x) = x .$$

Hermite polynomials are orthogonal with respect to the weight $w(x) = \mathrm{e}^{-x^2}$,

$$\int_{-\infty}^{\infty} H_k(x) H_l(x) \mathrm{e}^{-x^2} \mathrm{d}x = \delta_{k,l} \ .$$

The nodes x_j $(j = 0, 1, \ldots, N)$ for Gauss collocation are the roots of the polynomial H_{N+1}, while the weights are given by

$$w_j = \frac{2^N (N+1)! \sqrt{\pi}}{(N+1)^2 H_N^2(x_j)} = \frac{2^{N+2}(N+1)! \sqrt{\pi}}{H_{N+2}^2(x_j)} \ . \tag{5.53}$$

Generators of Hermite quadrature nodes and weights are included in many numerical libraries; see Sect. 3.2.3. Up to $N = 15$ they are listed on p. 924 of [34].

In spectral methods on the infinite domain $[-\infty, \infty]$ discussed in Chap. 12 it is convenient to furnish the basic polynomials by an additional exponential and a normalization factor:

$$\widehat{H}_k(x) = \left(2^k k! \sqrt{\pi}\right)^{-1/2} H_k(x)\, \mathrm{e}^{-x^2/2} \ . \tag{5.54}$$

This makes them orthonormal, $\int_{-\infty}^{\infty} \widehat{H}_k(x) \widehat{H}_l(x) \mathrm{d}x = \delta_{k,l}$, and applicable to cases in which the function to be expanded—or the solution of the differential equation to be found—decays exponentially at $\pm\infty$. The five lowest-order Hermite functions are shown in Fig. 5.16.

5.3.6 *Rational Chebyshev Functions TB*

In an analogous fashion as in the mapping (5.51) applicable to the positive real semi-axis, we can map the infinite domain $[-\infty, \infty]$ into $[-1, 1]$ or $[0, \pi]$ so that

Fig. 5.16 The five lowest-order basis functions (5.54) suitable for the expansion of functions on $[-\infty, \infty]$ that decay sufficiently fast as $x \to \pm\infty$

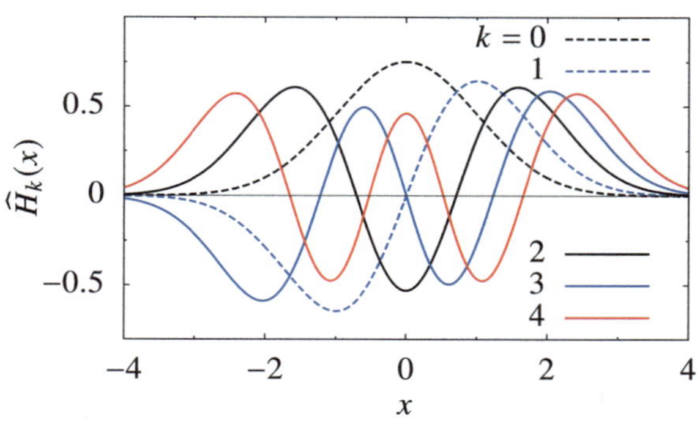

Chebyshev polynomials or trigonometric functions can be used in the expansions and calculations of the derivatives. The most frequently used mappings are

$$
y(x) = L\frac{x}{\sqrt{1-x^2}} \Leftrightarrow x(y) = \frac{y}{\sqrt{y^2+L^2}} ,
$$
$$
y(t) = L\cot(t) \quad \Leftrightarrow t(y) = \mathrm{arccot}(y/L) ,
$$

(5.55)

and this *algebraic mapping* of Chebyshev polynomials produces Chebyshev basis functions in the infinite interval (denoted by TB) with the property

$$
TB_k(y) = T_k(x) = \cos(kt) , \quad x = \cos(t) ,
$$

(5.56)

where T_k are the Chebyshev polynomials. Apart from $TB_0(y) = 1$ the first few are

$$
TB_1(y; L) = y\big/\big(y^2 + L^2\big)^{1/2} ,
$$
$$
TB_2(y; L) = \big(y^2 - L^2\big)\big/\big(y^2 + L^2\big) ,
$$
$$
TB_3(y; L) = \big(y^3 - 3yL^2\big)\big/\big(y^2 + L^2\big)^{3/2} ,
$$
$$
TB_4(y; L) = \big(y^4 - 6y^2L^2 + L^4\big)\big/\big(y^2 + L^2\big)^{2} ,
$$

and they are shown in Fig. 5.17.

The basis functions of even orders are symmetric about $y = 0$ and are rational functions of y, while the odd-order functions are anti-symmetric about $y = 0$ and have an additional non-rational (square-root) factor in the denominator; in spite of this distinction all TB's, just as the TL's, are somewhat confusingly also known as rational Chebyshev functions [on the infinite domain]. The scaling parameter L has the same role as in the case of TL functions. The TB functions are orthogonal:

$$
\int_{-\infty}^{\infty} TB_k(y; L)TB_l(y; L)\frac{L}{y^2+L^2}\,\mathrm{d}y =
\begin{cases}
0 & ; k \neq l , \\
\pi & ; k = l = 0 , \\
\pi/2 & ; k = l > 0 .
\end{cases}
$$

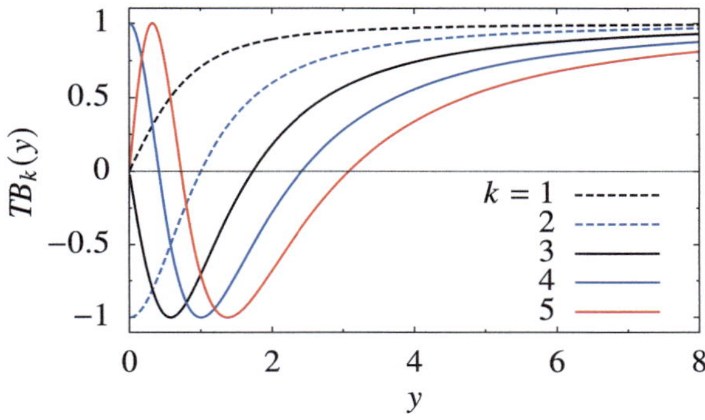

Fig. 5.17 Algebraically mapped Chebyshev polynomials of orders 1 to 5, for $L = 1$. The plot shows only the positive real semi-axis: the functions TB_k of even (odd) orders are symmetric (anti-symmetric) with respect to $y = 0$. For $L \neq 1$, replace y by y/L

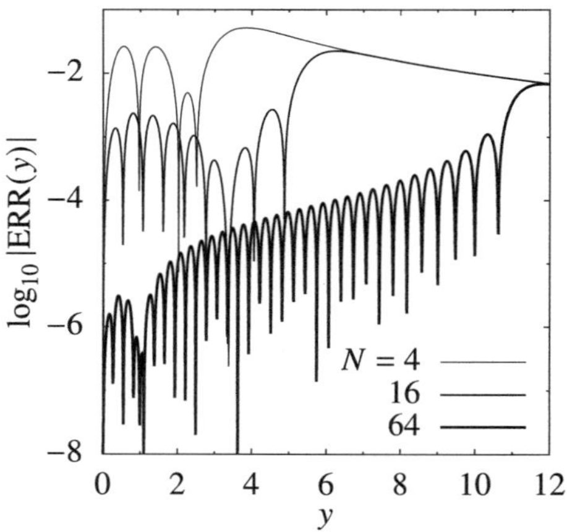

Fig. 5.18 Error of the expansion of $f(y) = 1/(1 + y^2)$ in terms of Hermite basis functions (5.54) up to orders $N = 4$, 16 and 64. (Only the graphs at $y \geq 0$ are shown.) Note the rapid improvement of the accuracy at small y with increasing N, in contrast to a persistently large error in the tails

The corresponding collocation points are

$$y_j = L \cot(t_j) , \quad t_j = \frac{(2j + 1)\pi}{2N + 2} , \quad j = 0, 1, \ldots, N .$$

Illustration By the same reasoning as in the *TL* case an expansion of a function $f(y)$ in terms of Hermite functions will converge exponentially only if $f(y)$ decays rapidly enough as $y \to \pm\infty$. If $f(y)$ tends to a constant or decays slowly as $y \to \pm\infty$, such an expansion will require a huge number of terms because the tails of all basis functions are suppressed by the $e^{-y^2/2}$ factor. Let us demonstrate this by expanding the function

$$f(y) = \frac{1}{1 + y^2} , \tag{5.57}$$

which falls off only algebraically as $y \to \pm\infty$, in terms of Hermite basis functions (5.54). Figure 5.18 shows the error of the approximation for $N = 4$, 16 and 64.

At $N = 64$ the maximum error still hovers around 10^{-2}; in order to improve it by two more orders of magnitude (accuracy of about four decimal places) one needs at least $N \approx 5000$—a preposterously large number. On the other hand, the *exact* $TB_k(y)$ series for (5.57) contains precisely two terms:

$$f(y) = \frac{1}{2}\big(TB_0(y) - TB_2(y)\big) !$$

Although our specific choice of $f(y)$ may seem contrived, it is clear that for functions that decay algebraically rather than exponentially, and for functions that asymptotically tend to a constant, the recommended basis functions must be the TB_k's.   ◁

5.4 Laplace Transformation

The continuous Laplace transformation of the function f is defined as

$$F(s) = \mathcal{L}[f](s) = \int_0^\infty e^{-st} f(t)\, \mathrm{d}t\,, \quad s \in \mathbb{C}\,.$$

The sufficient conditions for the existence of the transform $\mathcal{L}[f]$ are that f is piecewise continuous on $\mathbb{R}_+$ and that f is of exponential order in the limit $t \to \infty$: this means that real constants $C > 0$, a, and $T > 0$ exist such that $|f(t)| \le Ce^{at}$ for $\forall t > T$. The transformation is linear, $\mathcal{L}[c_1 f_1 + c_2 f_2] = c_1 \mathcal{L}[f_1] + c_2 \mathcal{L}[f_2]$. The transforms of some typical functions are enumerated in Table 5.2. The algorithms for the fast discrete Laplace transformation are described in [32].

Laplace transforms of real functions for which we do not know the suitable elementary integral, can be computed by using the Gauss–Laguerre quadrature of high order (see Fig. 5.19 (left)). Assuming that f can be sufficiently well described by a polynomial, the transform with $s > 0$ can be computed as

$$\mathcal{L}[f](s) = \frac{1}{s} \sum_{j=0}^{N} w_j f(x_j/s) + R_N\,, \quad R_N = \frac{(N!)^2}{(2N)!} f^{(2N)}(\xi)\,, \tag{5.58}$$

where $\xi \in \mathbb{R}$, $\{x_j\}_{j=0}^{N}$ are the zeros of the Laguerre polynomial $L_{N+1}(x)$, and the weights are given by Eq. (5.49). The quadrature (5.58) may fail at small values of s, in particular with oscillatory functions f, because the values of $f(x_j/s)$ change too quickly; in such cases it is preferable to use the formula

$$\mathcal{L}[f](s) \approx \sum_{j=0}^{N} w_j\, e^{(1-s)x_j} f(x_j)\,.$$

Table 5.2 Laplace transforms of some common functions. The transforms of the products $\mathcal{L}[\theta(t-c)f(t-c)] = e^{-cs}F(s)$, $\mathcal{L}[e^{ct}f(t)] = F(s-c)$, and $\mathcal{L}[(-t)^n f(t)] = F^{(n)}(s)$ are also very useful. For a much more complete list see Sect. 29.3 in [34]

$f(t) = \mathcal{L}^{-1}[F]$	$F(s) = \mathcal{L}[f]$	Assumptions
t^p	$\Gamma(p+1)/s^{p+1}$	$p > -1$, $\mathrm{Re}\{s\} > 0$
e^{at}	$1/(s-a)$	$\mathrm{Re}\{s\} > a$
$\sin at$	$a/(s^2 + a^2)$	$\mathrm{Re}\{s\} > 0$
$\cos at$	$s/(s^2 + a^2)$	$\mathrm{Re}\{s\} > 0$
$\delta(t-c)$	e^{-cs}	$\mathrm{Re}\{s\} > 0$
$\theta(t-c)$	e^{-cs}/s	$\mathrm{Re}\{s\} > 0$

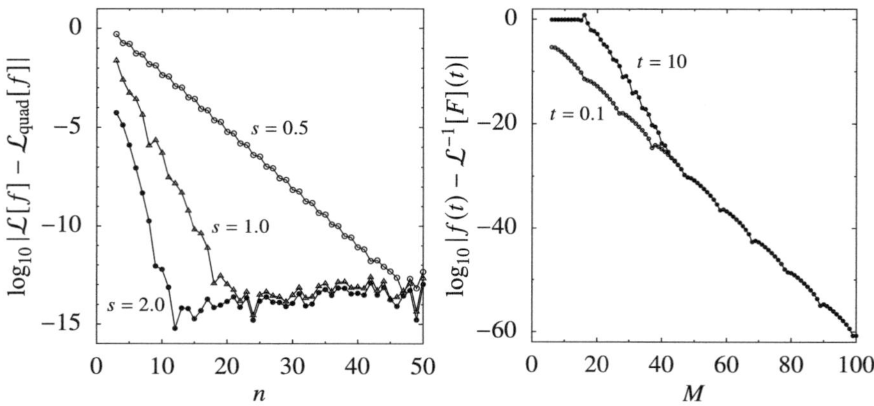

Fig. 5.19 [LEFT] The precision of the quadrature formula (5.58) for the Laplace transformation of the function $f(t) = \cos t$ (the exact transform is $F(s) = s/(s^2 + 1)$). [RIGHT] The precision of the FT algorithm from [36] to compute the inverse Laplace transform $F(s) = s/(s^2 + 1)$ to arbitrary precision (the exact solution is $f(t) = \cos t$). Shown is the error $\log_{10}|f(t) - \mathcal{L}^{-1}[F](t)|$ as a function of the number of terms in the quadrature sum for two different t

5.4.1 Use of Laplace Transformation with Differential Equations

One of the most fruitful application areas of the Laplace transformation is the study of electric circuits and mechanical systems where we wish to understand the solutions of linear differential equations with discontinuous or impulse forcing terms. The generic example is the equation $\ddot{x} + \beta \dot{x} + \omega_0^2 x = f(t)$ with the Heaviside (step) function $f(t) = \theta(t)$ or with the impulse $f(t) = \delta(t)$. The kernel of the Laplace transformation e^{-st} determines the natural scale of the physical process.

Laplace transformation draws its true strength from the relations between the transforms of the function f itself and the transforms of its derivatives. For a piecewise continuous f' it holds that

$$\mathcal{L}[f'] = s\mathcal{L}[f] - f(0)$$

or, with similar assumptions for the higher derivatives [35],

$$\mathcal{L}[f^{(n)}] = s^n \mathcal{L}[f] - s^{n-1} f(0) - \cdots - s f^{(n-2)}(0) - f^{(n-1)}(0) . \tag{5.59}$$

By using the relation (5.59) in the initial-value problem with constant coefficients

$$a\ddot{x} + b\dot{x} + cx = f(t) ,$$

with given initial conditions $x(0)$ and $\dot{x}(0)$, we obtain

$$a\left[s^2 X(s) - sx(0) - \dot{x}(0)\right] + b\left[s X(s) - x(0)\right] + c X(s) = F(s) \,. \qquad (5.60)$$

Instead of solving the differential equation for $x(t)$ we have succeeded in rephrasing the problem in terms of an algebraic equation for $X(s)$ into which the initial conditions have already been "built in". From the function $X(s)$ we then obtain the solution $x(t)$ by computing the inverse Laplace transform.

Formally, the inverse Laplace transform is given by the formula

$$\mathcal{L}^{-1}[F](t) = \frac{1}{2\pi i} \int_C e^{st} F(s)\,\mathrm{d}s \,, \qquad (5.61)$$

where C is a vertical line in the complex plane, defined by $C = \xi + i\eta$, and ξ is chosen such that all singularities of the transform $F(s)$ lie to the left of C. In other words, F is analytic on the half-plane $\mathrm{Re}\{s\} > \xi$. The sufficient conditions for the existence of the inverse are

$$\lim_{s\to\infty} F(s) = 0 \,, \quad \lim_{s\to\infty} |s F(s)| < \infty \,.$$

The inverse Laplace transformation is linear.

The world of the inverse Laplace transformation is not rosy: the expression for $X(s)$, which we read off from (5.60), first has to be reshuffled such that in its various parts of the functions $F(s)$ from Table 5.2 are identified, and these are then associated with the corresponding parts of the solution $x(t)$. For example,

$$X(s) = \frac{s-1}{s^2 - s - 2} = \frac{1}{3}\frac{1}{(s-2)} + \frac{2}{3}\frac{1}{(s+1)} \,,$$

which corresponds to

$$x(t) = \frac{1}{3}e^{2t} + \frac{2}{3}e^{-t}$$

(see the second row of Table 5.2). This was an easy example, as the available set of pairs $F(s)$ and $f(t)$ for which the table of known elementary functions and their transforms is read from right to left, is quickly exhausted. The inverse transform then needs to be computed numerically (see Problem 5.8.3).

The numerical computation of the inverse Laplace transform is described in an immense body of papers. Excellent insight is offered by [37, 38], and [39]. One of the best approaches is a deformation of the curve C in the integral (5.61) such that the integral converges as quickly as possible, and the use of the corresponding quadrature formulas [40]. Another good way is to rewrite the integral as a Fourier series which can be computed by the FFT algorithm [41]. A very robust and simple algorithm for the inverse transformation at arbitrary arithmetic precision can be found in [36]; see also the example in Fig. 5.19 (right).

5.5 Hilbert Transformation ★

The Hilbert transformation $\mathcal{H}$ of the function $s : \mathbb{R} \to \mathbb{R}$ is defined as the convolution of the function s with the function $h(t) = 1/(\pi t)$:

$$\hat{s}(t) = \mathcal{H}[s](t) = -(s * h)(t) = \frac{1}{\pi} P \int_{-\infty}^{\infty} \frac{s(\tau)}{\tau - t}\, d\tau \,, \tag{5.62}$$

where $*$ denotes the convolution. The integral should be understood as the Cauchy principal value (symbol P). The principal value integral of the function f over the interval $[a, b]$, on which f has a singularity at $c \in [a, b]$, is defined as

$$P \int_{a}^{b} f(t)\, dt = \lim_{\varepsilon \searrow 0} \left(\int_{a}^{c-\varepsilon} f(t)\, dt + \int_{c+\varepsilon}^{b} f(t)\, dt \right) .$$

The value of the transform does not change if an arbitrary constant is added to the function, $\mathcal{H}[s + \text{const}] = \mathcal{H}[s]$. The transform (5.62) of the function $s \in L^p(\mathbb{R})$ for $1 < p < \infty$ exists only in the case that s tends to zero at positive and negative infinity [42], $\lim_{t \to \pm\infty} s(t) = 0$. The Hilbert transformation is linear and bounded [43]:

$$\|\mathcal{H}[s]\|_p \le C_p \|s\|_p \,, \quad C_p = \begin{cases} \tan(\pi/2p) & ;\ p = 1, 2\,, \\ \tan(\pi/2p)^{-1} & ;\ p > 2\,. \end{cases}$$

The inverse Hilbert transformation is

$$s(t) = \mathcal{H}^{-1}[\hat{s}](t) = (\hat{s} * h)(t) = -\frac{1}{\pi} P \int_{-\infty}^{\infty} \frac{\hat{s}(\tau)}{\tau - t}\, d\tau \,,$$

hence $\mathcal{H}^{-1} = -\mathcal{H}$ and $\mathcal{H}^2 = -\text{id}$.

Relation to the Fourier transform The Fourier transforms (5.1) of the signal s, $S(\omega) = \mathcal{F}[s](\omega)$, and of the function $h(t) = 1/(\pi t)$,

$$H(\omega) = \mathcal{F}[h](\omega) = -i\,\text{sign}(\omega) \,, \quad \text{sign}(\omega) = \begin{cases} 1 & ;\ \omega > 0\,, \\ 0 & ;\ \omega = 0\,, \\ -1 & ;\ \omega < 0\,, \end{cases}$$

are related to the Hilbert transform of s by

$$\hat{s} = -\mathcal{F}^{-1}[SH] \,, \tag{5.63}$$

where $\mathcal{F}^{-1}$ is the inverse Fourier transformation (5.2). The norm of the function $s \in L^2(\mathbb{R})$ with the Fourier transform $S \in L^2(\mathbb{R})$ is the same in all three representations,

$$
\int_{-\infty}^{\infty} |\mathcal{H}[s](t)|^2 \, dt = \frac{1}{2\pi} \int_{-\infty}^{\infty} |\mathrm{sign}(\omega) S(\omega)|^2 \, d\omega = \lim_{\varepsilon \searrow 0} \left(\int_{-\infty}^{-\varepsilon} + \int_{\varepsilon}^{\infty} \right) |s(t)|^2 \, dt \, ,
$$

where we have used (5.63) and Parseval's equality (5.3) for the Fourier transform. The Hilbert transformation therefore preserves the total power of the signal.

A real signal s and its Hilbert transform $\hat{s} = \mathcal{H}[s]$ are orthogonal, i.e.

$$
\int_{-\infty}^{\infty} s(t) \hat{s}(t) \, dt = \frac{i}{2\pi} \int_{-\infty}^{\infty} \mathrm{sign}(\omega) |S(\omega)|^2 \, d\omega = 0 \, ,
$$

if s, $\hat{s}$, and the Fourier transform $S = \mathcal{F}[s]$ are in $L^1(\mathbb{R})$ or $L^2(\mathbb{R})$ (we assumed $S(\omega)^* = S(-\omega)$). A similar conclusion can be made by examining (5.63) for the Fourier modes, since for the functions $\cos \omega t$ and $\sin \omega t$, which are orthogonal for all $\omega \neq 0$, we have

$$
\mathcal{H}[\cos \omega t] = -\mathrm{sign}(\omega) \sin \omega t \, , \quad \mathcal{H}[\sin \omega t] = \mathrm{sign}(\omega) \cos \omega t \, . \tag{5.64}
$$

Warning The numerical computation of the Hilbert transform of the function f on the whole real axis may be very problematic (see Fig. 5.20). Even small perturbations $\varepsilon(t)$ that are not in the space $L^p(\mathbb{R})$, $p > 1$ may cause a singularity in $\mathcal{H}[f + \varepsilon]$. The problems of this type are not unique to the Hilbert transformation, but because its kernel $h(t) = 1/(\pi t)$ is singular, the instabilities become more pronounced. We therefore often resort to simplifications; a few guidelines can be found in [44]. The Hilbert transform $\mathcal{H}[f](t)$ at large parameters t can be elegantly computed by its asymptotic expansion presented in [45] (in this case the asymptotics of f has to be known). We discuss the methods for numerical computation of the continuous Hilbert transform in Sect. 5.5.3, while the discrete transform is discussed in Sect. 5.5.4.

5.5.1 Analytic Signal

Hilbert transformation is a commonly used tool in the analysis of signals where it is used for their "complexification". This means that a real signal (function) $s(t)$ is assigned a complex function

$$
s^{\#}(t) = s(t) - i \, \mathcal{H}[s](t) \, , \tag{5.65}
$$

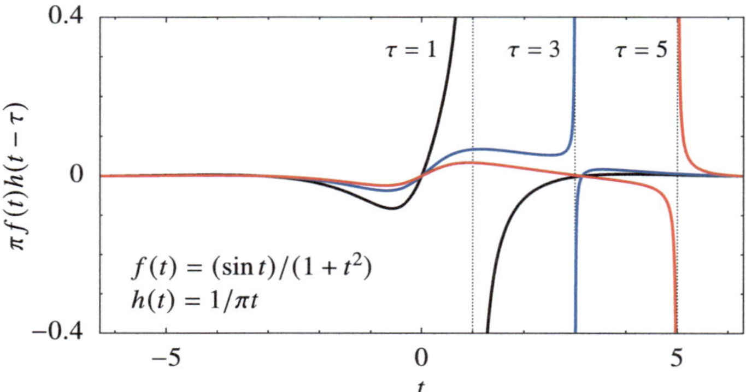

Fig. 5.20 The integrands for the computation of the Hilbert transform of $f(t) = \sin t/(1 + t^2)$ for three values $\tau = 1$, $\tau = 3$, and $\tau = 5$. We integrate over the poles migrating along the real axis and "sample" f in a very sensitive manner. See also Sect. 3.3

known as the *analytic signal* (s and $\hat{s} = \mathcal{H}[s]$ are real functions). If the signal is $s(t) = \cos \omega t$, $\omega > 0$, the analytic signal is $s^{\#}(t) = \cos \omega t + \mathrm{i} \sin \omega t = \exp(\mathrm{i}\omega t)$, as can be inferred from (5.64). The Fourier transform of the analytic signal is

$$S^{\#}(\omega) = S(\omega) \left[1 + \mathrm{sign}\,(\omega) \right] ,$$

where $S(\omega) = \mathcal{F}[s](\omega)$ is the Fourier transform (5.1) of the signal s. The function $S^{\#}$ is zero for negative frequencies $\omega < 0$: this is a fundamental property of the analytic signals. If the function $s^{\#}$ is analytic on the upper complex half-plane and satisfies the Cauchy integral equation

$$P \int_{-\infty}^{\infty} \frac{s^{\#}(\xi)}{\xi - x} \, \mathrm{d}\xi = \mathrm{i}\pi s^{\#}(x) , \tag{5.66}$$

we are referring to a *strongly analytic signal.* In addition to analyticity, the sufficient condition for the validity of the integral equation is that the function falls off quickly enough in the upper complex half-plane. If the signal $s^{\#}$ is strongly analytic, we may insert $s^{\#} = s - \mathrm{i}\,\hat{s}$ into (5.66) and obtain the relations between the real and imaginary parts of the analytic signal,

$$\mathcal{H}[s] = \hat{s} , \quad \mathcal{H}[\hat{s}] = -s .$$

The analytic signal can be represented in the complex plane as

$$s^{\#}(t) = A(t)\,\mathrm{e}^{\mathrm{i}\phi(t)} ,$$

where

$$A(t) = |s^{\#}(t)| = \sqrt{(s(t))^2 + (\mathcal{H}[s](t))^2} \tag{5.67}$$

is the analytic amplitude, and

$$\phi(t) = \arg(s^{\#}(t)) = \operatorname{atan}\left(\frac{\operatorname{Im} s^{\#}(t)}{\operatorname{Re} s^{\#}(t)}\right) \tag{5.68}$$

is the analytic phase. In this representation we define various *instantaneous* quantities [46], like the *instantaneous signal power* $E_i(t) = |A(t)|^2$ and the *instantaneous complex phase* $\Phi_i(t) = \log s^{\#}(t) = \log A(t) + i\phi(t)$. By using the time derivative of the complex phase we also define the *instantaneous complex frequency* $\Omega_i(t) = \dot{\Phi}_i(t) = \alpha_i(t) + i\beta_i(t)$, where $\alpha_i(t) = \dot{A}(t)/A(t)$ is the *instantaneous radial frequency* and $\beta_i(t) = \dot{\phi}(t)$ is the *instantaneous angular frequency*. These quantities help us in unraveling the characteristics of the signal s which are not apparent from observing just the time dependence of $s(t)$.

Example Hilbert transformation is used as a tool in the analysis of brain potentials for the study of brain states and functions. The signals used in the analysis are obtained by digitizing the electro-encephalograph (EEG) recordings. An example of the EEG recording during an epileptic seizure is shown in Fig. 5.21 (top left). We use the sampled signal s_n to form the Hilbert transform $\mathcal{H}[s]_n$ (Fig. 5.21 (top right)). By using (5.67) and (5.68) we then compute the analytic amplitude and phase (Fig. 5.21 (bottom left and right)).

The actual (physiological) changes in the brain states can then be inferred from the sudden increases or abrupt drops in the analytic amplitude, jumps in the analytic phase or from the behavior of other observables formed from the Fourier and Hilbert transform of the original signal, for example, the instantaneous frequencies from (5.78). (However, similar effect may be caused by time-dependent interferences between different frequency components of the signal.) A good guide to work in this field can be found in the papers [47]. $\quad\triangleleft$

5.5.2 *Kramers–Kronig Relations*

Relations between real and imaginary parts of the Fourier transforms play prominent roles in many areas of physics. These relations have the form of Hilbert transforms. Let $s(t)$ be a real signal with the complex Fourier transform $S(\omega) = \mathcal{F}[s](\omega)$ which is an analytic function on the upper half-plane of the complex variable ω. If S satisfies (5.66), its real and imaginary parts are related by

$$\operatorname{Re} S(\omega) = \frac{1}{\pi} P \int_{-\infty}^{\infty} \frac{\operatorname{Im} S(\omega')}{\omega' - \omega}\, d\omega' , \quad \operatorname{Im} S(\omega) = -\frac{1}{\pi} P \int_{-\infty}^{\infty} \frac{\operatorname{Re} S(\omega')}{\omega' - \omega}\, d\omega' ,$$

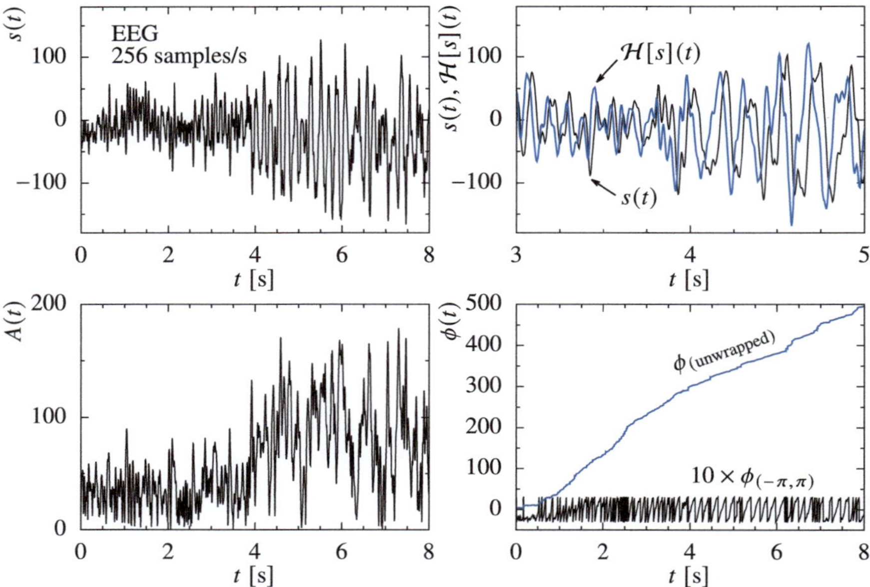

Fig. 5.21 Analysis of electro-encephalograms by using Hilbert transformation. [TOP LEFT] The raw signal s (2048 values, 256 samples per second) [48]. [TOP RIGHT] The signal s and its Hilbert transform $\mathcal{H}[s]$ at $t \approx 4\,$s where the nature of the signal changes. [BOTTOM LEFT] The analytic amplitude A (Eq. (5.67)). [BOTTOM RIGHT] The analytic phase ϕ (Eq. (5.68)). The lower part of the figure shows the phase on the interval $(-\pi, \pi)$, while there is no such constraint in the upper part (at each crossing of the branch cut in the complex plane for the arctan function we add 2π to the phase)

in short, $\operatorname{Re} S(\omega) = \mathcal{H}[\operatorname{Im} S](\omega)$ and $\operatorname{Im} S(\omega) = -\mathcal{H}[\operatorname{Re} S](\omega)$. These equations are known as the *Kramers–Kronig (dispersion) relations*. Relations of this type can be formulated for all Fourier transforms of signals which appear in the descriptions of the system's response to an external perturbation. They are very general: their derivation requires only that the response of the system is causal. This means that the signal s at times $t > 0$ is a consequence of the events occurring at $t \le 0$.

Example The most famous Kramers–Kronig relations connect the real and imaginary parts of the electric susceptibility $\chi(\omega) = \operatorname{Re} \chi(\omega) + \mathrm{i} \operatorname{Im} \chi(\omega)$ or the refraction index $N(\omega) = n(\omega) + \mathrm{i}\,\kappa(\omega)$ [49]. The polarization $\boldsymbol{P}$ represents the linear response of the matter to the electric field $\boldsymbol{E}$,

$$\boldsymbol{P}(t) = \varepsilon_0 \int\limits_{-\infty}^{\infty} \chi(t - t')\boldsymbol{E}(t')\,\mathrm{d}t' , \quad \chi(\tau) = 0 \ \text{ at } \ \tau < 0 .$$

In the frequency representation this means $\boldsymbol{P}(\omega) = \varepsilon_0 \chi(\omega)\boldsymbol{E}(\omega)$, where $\chi(\omega)$ is the susceptibility which depends on the frequency of the external perturbation $\boldsymbol{E}(\omega)$.

Susceptibility χ satisfies the above Kramers–Kronig relations, which can be further simplified by using symmetry properties. We have $\chi(\omega)^* = \chi(-\omega)$, or Re $\chi(-\omega) =$ Re $\chi(\omega)$ and Im $\chi(-\omega) = -$Im $\chi(\omega)$, so the integration range can be narrowed down to positive frequencies only,

$$\text{Re } \chi(\omega) = \frac{2}{\pi} P \int_0^\infty \frac{\omega' \text{ Im } \chi(\omega')}{\omega'^2 - \omega^2} \, d\omega' \,, \tag{5.69}$$

$$\text{Im } \chi(\omega) = -\frac{2}{\pi} P \int_0^\infty \frac{\omega \text{ Re } \chi(\omega')}{\omega'^2 - \omega^2} \, d\omega' \,.$$

We need only one more step to find the relations between the real (dispersive) and the imaginary (absorption) part of the complex refraction index. In the limit of small susceptibilities we have

$$N(\omega) = \sqrt{1 + \text{Re } \chi(\omega) + i \text{ Im } \chi(\omega)} \approx 1 + \underbrace{\frac{1}{2} \text{ Re } \chi(\omega)}_{n(\omega)} + \underbrace{\frac{i}{2} \text{ Im } \chi(\omega)}_{i\kappa(\omega)} \,.$$

From Eq. (5.69) it then follows that

$$n(\omega) = 1 + \frac{2}{\pi} P \int_0^\infty \frac{\omega' \kappa(\omega')}{\omega'^2 - \omega^2} \, d\omega' \,. \tag{5.70}$$

It is relatively hard to measure the frequency dependence of the real part of the refraction index n. But by using the derived relations it can be computed from the imaginary part κ which can be determined easily, just by measuring the ratio of the incident and transmitted wave intensities at different frequencies. As an example, Fig. 5.22 shows $n(\omega)$ and $\kappa(\omega)$ for infrared light in water.　　　　　◁

5.5.3 Numerical Computation of the Continuous Hilbert Transform

The numerical computation of the Hilbert transform is almost always a tough nut to crack, as the transformation involves the integral of a singular function. Here we mention a few strategies for a quick and stable computation. More information can be found in the review article [50].

Transforming the integrand to a sum of orthogonal polynomials Let us first generalize the definition of the Hilbert transformation to integrals over the interval $I \subset \mathbb{R}$,

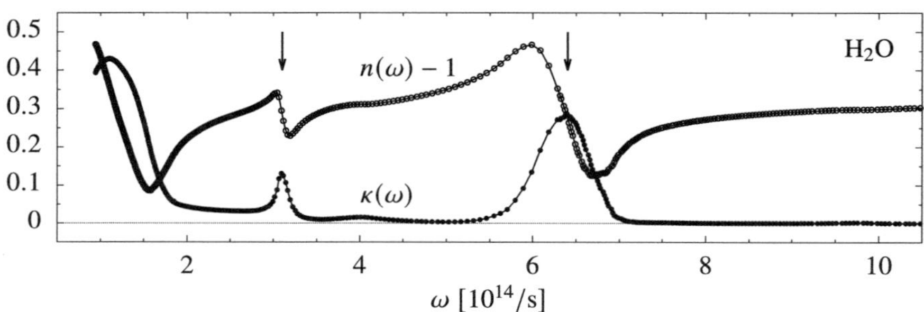

Fig. 5.22 Frequency dependence of the real and imaginary part of the complex refraction index of water in the range of micrometer wavelengths. They are related by Eq. (5.70). The arrows indicate the regions of anomalous dispersion (a sudden drop of the real part) and absorption (a rapid increase of the imaginary part)

$$\mathcal{H}[f](y) = \frac{1}{\pi} \, P \int_I \frac{f(x)}{x - y} \, dx \; .$$

If the interval $I = [a, b]$ is finite, we can rewrite the integral as

$$\mathcal{H}[f](y) = \frac{1}{\pi} \, f(y) \log \left| \frac{b - y}{a - y} \right| + \frac{1}{\pi} \int_a^b \frac{f(x) - f(y)}{x - y} \, dx \; ,$$

which eliminates the singularity. We write the integrand as the product of a positive weight function w and the remaining factor, $f(x) - f(y) = w(x)g(x; y)$. With foresight, we would like to find a particular pair (interval I, weight function w) that corresponds to the definition domain and the weight function of some system of orthogonal polynomials [30]. Then we could approximate the function g by the expansion

$$g(x; y) \approx \sum_{j=0}^n d_j(y) q_j(x) \; ,$$

where $\{q_j\}_{j \in \mathbb{N}_0}$ are orthogonal polynomials, and compute the Hilbert transform as the weighted sum of the transforms of the individual terms in this expansion:

$$\mathcal{H}[wg](y) \approx \sum_{j=0}^n d_j(y) \psi_j(y) \; , \quad \psi_j(y) = \mathcal{H}[wq_j](y) \; .$$

All systems of orthogonal polynomials q_j possess three-term recurrence relations which also apply to the functions ψ_j, e.g. $\psi_{j+1}(x) = (A_j x + B_j)\psi_j(x) + C_j \psi_{j-i}(x)$, where A_j, B_j, and C_j are constants that do not depend on x. This relation between q_j and ψ_j is based on the property of the Hilbert transformation

$$\mathcal{H}[xf](y) = y\mathcal{H}[f](y) + \frac{1}{\pi} \int\limits_{-\infty}^{\infty} f(x)\,\mathrm{d}x$$

and the orthogonality of the polynomials q_j to a constant, $\int_{-\infty}^{\infty} q_j(x)w(x)\,\mathrm{d}x = 0$ for $j > 0$. Using the recurrence to compute ψ_j speeds up tremendously the computation of the transform $\mathcal{H}[wf](y)$. For the computation with the Chebyshev polynomials (weight $w(x) = (1 - x^2)^{-1/2}$ and $I = [-1, 1]$) see [51]; for the computation with the Hermite polynomials ($w(x) = \mathrm{e}^{-x^2}$ and $I = \mathbb{R}$) and the generalized Laguerre polynomials ($w(x) = x^\alpha \mathrm{e}^{-x}$ and $I = \mathbb{R}_+$) see [52].

Quadrature formulas If our knowledge about the behavior of the integrand is very limited, the Hilbert transform can be computed by using quadrature formulas. A simple one is

$$\mathcal{H}[f](y) \approx \frac{2}{\pi} \sum_{n \in \mathbb{Z}} \frac{f(y + (2n + 1)h)}{2n + 1}, \tag{5.71}$$

which has been suggested and analyzed in [53]. Of course, in a practical implementation, we restrict $|n| \le N$, see Fig. 5.23 (left).

Collocation method A slightly different approach to the Hilbert transformation has been charted by [55] and it exploits its spectral properties. (Note that in this part of the text we define $\mathrm{sign}(0) = 1$.) The authors use the functions

$$\rho_n(x) = \frac{(1 + \mathrm{i}x)^n}{(1 - \mathrm{i}x)^{n+1}}, \quad n \in \mathbb{Z}, \tag{5.72}$$

which satisfy $\mathcal{H}[\rho_n](x) = \mathrm{i}\,\mathrm{sign}(n)\rho_n(x)$ and are therefore the eigenfunctions of the Hilbert transformation. The functions ρ_n on $L^2(\mathbb{R})$ form a complete orthogonal basis and fulfill $\int_{-\infty}^{\infty} \rho_m^*(x)\rho_n(x)\,\mathrm{d}x = \pi\delta_{m,n}$. An arbitrary function $f \in L^2(\mathbb{R})$ can be expanded in terms of these basis functions as

$$f(x) = \sum_{n \in \mathbb{Z}} a_n \rho_n(x), \quad a_n = \frac{1}{\pi} \int\limits_{-\infty}^{\infty} \rho_n(x)^* f(x)\,\mathrm{d}x. \tag{5.73}$$

The Hilbert transform of the function f is then

$$\mathcal{H}[f](x) = \mathrm{i} \sum_{n \in \mathbb{Z}} \mathrm{sign}(n)a_n \rho_n(x) \approx \mathrm{i} \sum_{n=-N}^{N-1} \mathrm{sign}(n)a_n \rho_n(x). \tag{5.74}$$

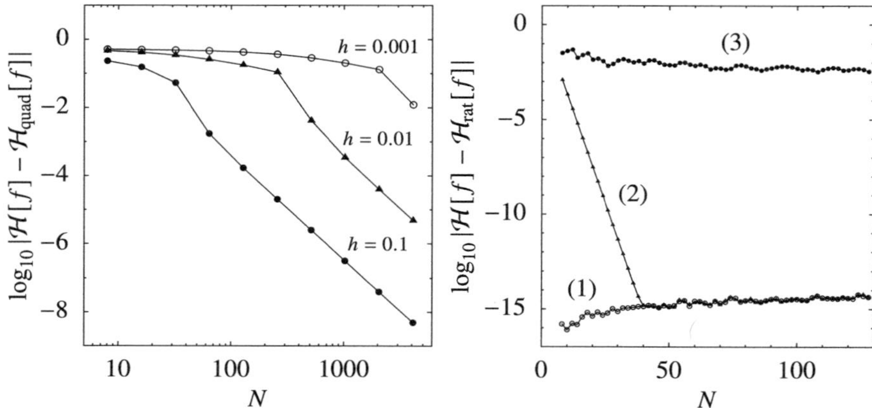

Fig. 5.23 The numerical computation of the Hilbert transform. [LEFT] The precision of the formula (5.71) in dependence of N and the width of the interval h, for the function $f(x) = (\sin x)/(1 + x^2)$. [RIGHT] The precision of the algorithm (5.74) with the rational approximation (5.72) in dependence of N for three different functions: (1) $f(x) = 1/(1 + x^2)$, (2) $f(x) = 1/(1 + x^4)$, and (3) $f(x) = (\sin x)/(1 + x^2)$. Since the function (3) oscillates, we should obviously try a bit harder: see [54]

If the function f is real, the expansion coefficients satisfy $a_n = a^*_{-n-1}$, so one needs to compute only half of the coefficients a_n for $n = 0, 1, \ldots, N - 1$. The coefficients a_n are given by (5.73). Alternatively, one can use the substitution $x = \tan(\phi/2)$ to rewrite the very same expression as

$$a_n = \frac{1}{2\pi} \int_{-\pi}^{\pi} g_n(\phi)\, \mathrm{d}\phi\,, \quad g_n(\phi) = \left(1 - \mathrm{i}\tan\frac{\phi}{2}\right) f\left(\tan\frac{\phi}{2}\right) \mathrm{e}^{-in\phi}\,.$$

The function g is periodic on $[-\pi, \pi]$ and $\lim_{\phi \to \pm\pi} g(\phi) = 0$ is assumed. The approximation of the integral in the above expression can be obtained by various integration methods, for example, by using the trapezoidal rule with the points $\phi_j = \pi j/N$ for $|j| < N$. This results in the approximation

$$a_n \approx \frac{1}{2N} \sum_{j=-N+1}^{N-1} g_n(\phi_j)\,.$$

By evaluating this formula with the fast Fourier transform we can compute the coefficients a_n for all n at once, requiring only $\mathcal{O}(N \log N)$ operations. This approach is reliable in cases where the coefficients a_n rapidly decrease as n increases. This behavior is typical for functions f that themselves fall off rapidly at infinity (see Fig. 5.23 (right)). Details can be found in [55].

5.5.4 *Discrete Hilbert Transformation*

We sample a continuous signal s equidistantly in t in steps of Δ and obtain the discrete signal $\{s_k\}_{k\in\mathbb{Z}}$ to which we assign a discrete Hilbert transform $\{\hat{s}_k\}_{k\in\mathbb{Z}}$. While there are several ways to do this, all of them reproduce the continuous transform (5.62) in the limit $\Delta \to 0$ if s is smooth enough. The discrete Hilbert transform is an arbitrarily good approximation of the continuous one.

Time-domain approach The discrete variant of the Hilbert transform for a finitely long signal $\{s_k\}$, specified at times $t_k = k\Delta$, can be cast in the form

$$\hat{s}_n = \mathcal{H}_N[s]_n = \frac{1}{\pi} \sum_{m=1}^{N} \frac{s_{n+m} - s_{n-m}}{m} \, ,$$

where N is the number of points in the sample to the left and right of t_n which are included in the sum. The discrete transform reverts to the continuous one when $N \to \infty$ and $\Delta \to 0$. "Infinitely long" signals occur when there is an incessant flow of data from the measuring apparatus and we wish to process them in real time, or if the amount of data is large compared to the available memory. We should therefore always estimate the extent of the signal with respect to some reference point that we still wish to use in the computation of the transform.

It is possible to compute the discrete Hilbert transform by numerically evaluating the integral in the continuous transform (5.62). We either use the interpolant of the signal s or a chosen model function is fitted to the signal: both can be simply and stably convoluted with the function $h(t) = 1/(\pi t)$. By linear interpolation of the signal values [56] we obtain the discrete transform

$$\mathcal{H}_N[s]_n = -\frac{1}{\pi}\Bigg\{ \sum_{k=0}^{n-2} s_k\, \phi(k-n+1) + (s_k - s_{k+1})\Big[1 + (n-k)\phi(n-k)\Big]$$
$$+ s_{n-1} - s_{n+1} + \sum_{k=n+2}^{N-1} s_k\, \phi(k-n) + (s_{k-1} - s_k)\Big[1 + (k-n)\phi(k-n)\Big] \Bigg\} \, ,$$

where we sample the signal s_k at $k\Delta t$ ($k = 0, 1, \ldots, N-1$) and $\phi(k) = \log(1 - 1/k)$. This computation requires $\mathcal{O}(N^2)$ operations. Boche's approach [57] which is tailored to signals of limited bandwidth, is also found in practice.

Fourier-domain approach The Fourier approach is fruitful for periodic signals $\{s_k\}_{k=0}^{N-1}$ (for which $s_{k+N} = s_k$). The discrete Fourier transform of the signal is

$$S_n = \mathcal{F}_N[s]_n = \frac{1}{N} \sum_{k=0}^{N-1} s_k\, e^{-i2\pi nk/N} \, .$$

We define the discrete Hilbert transform for such a signal as the convolution

$$\hat{s}_n = \mathcal{H}_N[s]_n = \sum_{k=0}^{N-1} h_{n-k} s_k = \sum_{k=0}^{N-1} h_k s_{n-k} \, , \qquad (5.75)$$

where $\{h_k\}_{k=0}^{N-1}$ is the convolution kernel and we assume $h_{k+N} = h_k$, $s_{k+N} = s_k$. We determine the kernel by translating the property (5.64) of the continuous Hilbert transform to discrete language, namely $\mathcal{H}[e](t) = \mathrm{i}\,\mathrm{sign}(\omega) e(t)$, where $e(t) = \exp(\mathrm{i}\,\omega t)$. This means

$$\mathcal{H}_N[e^{(k)}]_n = \mathrm{i}\,\mathrm{sign}(N - 2k)\, e_n^{(k)} \, , \quad e_n^{(k)} = \exp(\mathrm{i}\, 2\pi k n / N) \, . \qquad (5.76)$$

We have already taken into account the definitions of the negative and positive frequencies in the discrete Fourier transform. By using the relation between the discrete Fourier transform and the convolution, we can express the Hilbert transform (5.75) as

$$\mathcal{H}_N[s]_n = -N \mathcal{F}_N^{-1}[F]_n \, , \quad F = \{H_i S_i\}_{i=0}^{N-1} \, ,$$

where we have denoted $S = \mathcal{F}_N[s]$ and $H = \mathcal{F}_N[h]$. Condition (5.76) is satisfied if $H_k = -\mathrm{i}\,\mathrm{sign}(N - 2k)/N$. After using Eq. (5.76) the path to the discrete Hilbert transform becomes clear: we compute the components of the Fourier transform S_k of the signal s, multiply them by $\mathrm{i}\,\mathrm{sign}(N - 2k)$, and compute the inverse Fourier transform of the product:

$$\mathcal{H}_N[s]_n = \mathrm{i} \sum_{k=0}^{N-1} S_k \,\mathrm{sign}(N - 2k)\, \mathrm{e}^{\mathrm{i}\,2\pi n k / N} \, . \qquad (5.77)$$

By analogy with the continuous case (5.65) we define the complex analytic signal corresponding to the discrete Hilbert transform:

$$s_n^{\#} = s_n - \mathrm{i}\,\hat{s}_n \, , \quad \hat{s}_n = \mathcal{H}_N[s]_n \, .$$

Its real and imaginary parts can again be used to elegantly compute the instantaneous quantities (see page 281). For example, the instantaneous radial frequency α_n and the instantaneous angular frequency β_n become

$$\alpha_n = \frac{s_n' s_n + \hat{s}_n' \hat{s}_n}{s_n^2 + \hat{s}_n^2} \, , \quad \beta_n = \frac{\hat{s}_n' s_n - s_n' \hat{s}_n}{s_n^2 + \hat{s}_n^2} \, . \qquad (5.78)$$

where $'$ denotes the time derivative. Note that even these derivatives can be computed by using the discrete Fourier transformation. If the array $\{f_k\}_{k=0}^{N-1}$ corresponds to the discrete Fourier transform $\{F_n\}_{n=0}^{N-1}$, the time derivative of the component f_k is given by

$$f'_k = \sum_{n=0}^{N-1} \omega_n F_n \, \mathrm{e}^{\mathrm{i} 2\pi nk/N} \, , \quad \omega_n = \frac{2\pi}{N} \left[\left(n - \frac{N}{2} \right) \operatorname{sign}(N - 2n) + \frac{N}{2} \right] ,$$

where the physical angular frequencies ω_k belong to the individual Fourier modes. Only $\mathcal{O}(N \log N)$ operations are needed if FFT is used to compute the discrete Hilbert transform (5.77).

The Fourier-domain approach is the fastest among the possibilities described here, but it does have its deficiencies. The signal is periodic but rapid changes in the signal may be aliased in the frequency spectrum and therefore misinterpreted when the convolution is performed. It is not wise to use the Fourier approach if the Hilbert transform falls off too slowly at infinity as the finite Fourier series is not optimal for the description of the long tails. A detailed discussion of the variants of the Hilbert transform with emphasis on signal processing applications can be found in [46, 58, 59] and in the monumental work [60].

5.6 Continuous Wavelet Transformation ⋆

We may think of the *wavelet transformation* of a signal [61–64] as an extension of its Fourier analysis, through which not only the strengths of the signal's frequency components are determined, but also the times at which these components occur. The classic example in Fig. 5.24 illustrates this basic idea for the signal $\sin(t^2)$ whose frequency linearly increases with time. By using the wavelet transformation we can also locate changes in the signal that are not immediately apparent from its temporal behavior alone (Fig. 5.25).

The *continuous wavelet transform* (CWT) of the function f is defined as

$$L_\psi[f](s, t) = \frac{1}{\sqrt{c_\psi s}} \int_{-\infty}^{\infty} f(\tau) \, \psi^* \left(\frac{\tau - t}{s} \right) \mathrm{d}\tau \, , \quad t \in \mathbb{R}, \quad s \neq 0 \, ,$$

where t is the time at which a feature of scale s is observed in the function f, and c_ψ is the normalization constant. The function ψ, whose properties are given in the following, should allow us to change the parameter s (the typical scale of a structure in the signal f) as well as its shift t with respect to the signal f. By denoting

$$\psi_s(t) = \psi^*(-t/s)$$

we can rewrite the definition in the form of a convolution

$$L_\psi[f](s, t) = \frac{1}{\sqrt{c_\psi s}} \int_{-\infty}^{\infty} f(\tau) \, \psi_s(t - \tau) \, \mathrm{d}\tau \, . \tag{5.79}$$

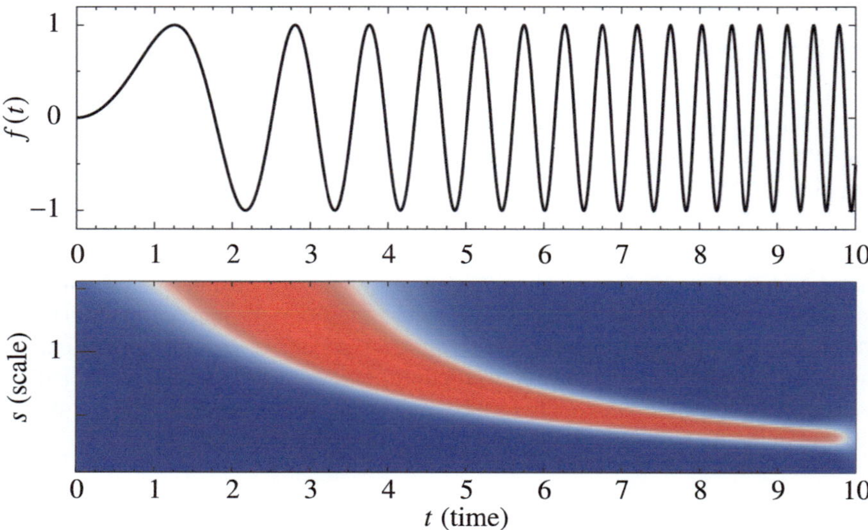

Fig. 5.24 The basic idea of the continuous wavelet transform. [Top] The signal $f(t) = \sin \omega t$, $\omega \propto t$. [Bottom] In this portion of the signal, the continuous wavelet transformation detects large structures at short times (where the waves have a typical scale $s \approx 1.5$) and small structures at long times (scale $s \approx 0.3$). The frequency and the scale of the oscillations are inversely proportional, which generates the typical curvature of the transform ($s \propto \omega^{-1} \propto t^{-1}$)

The function ψ should satisfy certain conditions. Its "energy" should be bounded, which means $\int_{-\infty}^{\infty} |f(t)|^2 \, \mathrm{d}t < \infty$, and the weighted integral of its spectral density (the square of the Fourier transform $\hat{\psi}$) should be finite:

$$c_\psi = 2\pi \int_{-\infty}^{\infty} \frac{1}{|\omega|} \left| \hat{\psi}(\omega) \right|^2 \, \mathrm{d}\omega < \infty \, .$$

The functions ψ found in the literature usually fulfill this *admissibility condition* by design. Moreover, we require the functions ψ to fulfill

$$\int_{-\infty}^{\infty} \psi(\eta) \, \mathrm{d}\eta = 0 \, . \tag{5.80}$$

The functions ψ therefore oscillate around the abscissa and fall off rapidly at large distances from the origin, giving them the appearance of small waves and the nickname *wavelets*. The simplest wavelet is the Haar function

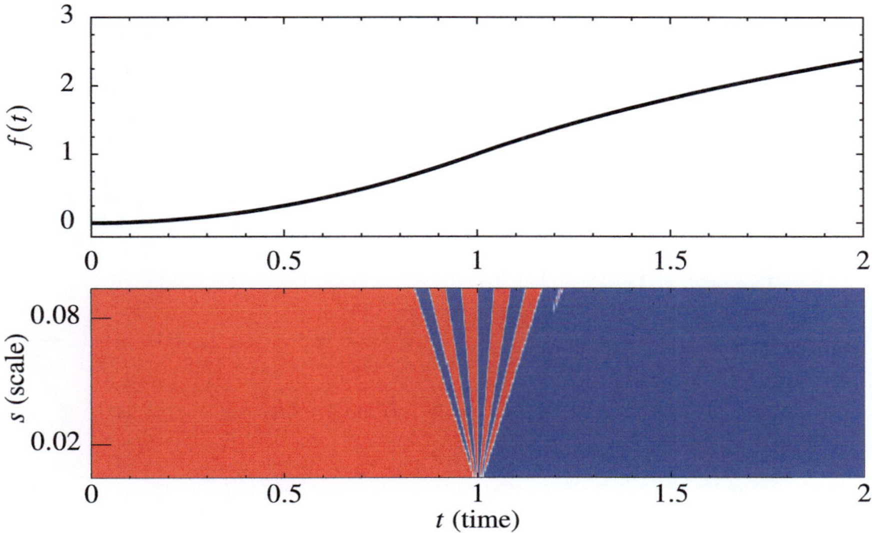

Fig. 5.25 Continuous transform of a real signal with a complex Morlet wavelet (5.82) [TOP] The signal $f(t) = t^2$ $(t < 1)$ or $f(t) = 1 + 2\log t$ $(t \geq 1)$ is continuous and has a continuous first derivative at $t = 1$ while its second derivative is discontinuous. [BOTTOM] The phase of the wavelet transform in the vicinity of $t = 1$ oscillates a couple of times, and reveals the location of the critical point when the scale s is decreased

$$\psi_{\text{Haar}}(\eta) = \begin{cases} 1 \; ; \; 0 \leq \eta < 1/2 \, , \\ -1 \; ; \; 1/2 \leq \eta < 1 \, , \\ 0 \; ; \; \text{otherwise} \, . \end{cases}$$

The *derivative of Gaussian* wavelets DOG(m) are also very simple to use. We obtain them by successive derivatives of the Gauss function,

$$\psi_{\text{DOG}(m)}(\eta) = \frac{(-1)^{m+1}}{\sqrt{\Gamma(m + 1/2)}} \frac{\mathrm{d}^m}{\mathrm{d}\eta^m} \left(\mathrm{e}^{-\eta^2/2} \right) . \tag{5.81}$$

Another useful wavelet is the complex Morlet wavelet

$$\psi_{\text{Morlet}}(\eta) = \pi^{-1/4} \, \mathrm{e}^{\mathrm{i}\omega_0 \eta} \, \mathrm{e}^{-\eta^2/2} \, , \quad \omega_0 \in [5, 6] \, . \tag{5.82}$$

(For the Morlet wavelet Eq. (5.80) is not exactly fulfilled; the absolute precision to which the equality is valid improves when ω_0 is increased, and amounts to at least $\approx 10^{-5}$ for $\omega_0 > 5$.) When complex wavelets are used, the corresponding transforms should be specified in terms of their magnitudes and phases (see Fig. 5.25). Some typical wavelets are shown in Fig. 5.26.

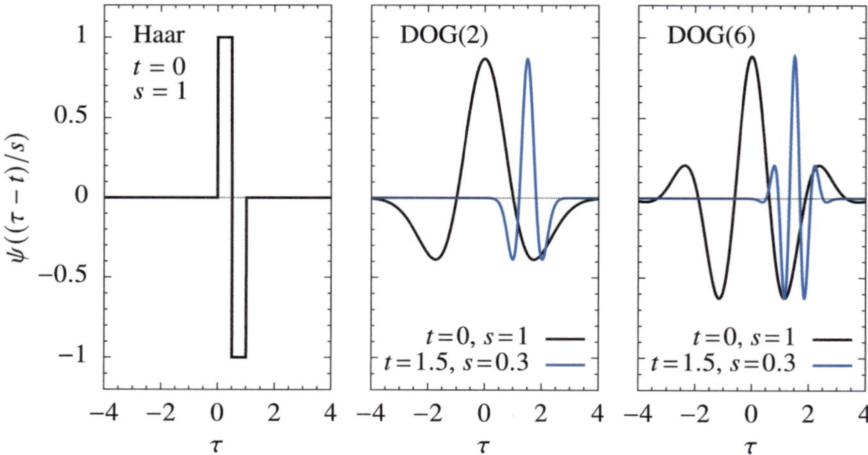

Fig. 5.26 Examples of wavelets used in the continuous wavelet transformation. By horizontal shifts and changes of scale the wavelet probes the features of the investigated signal and the times at which these features appear. [LEFT] Haar wavelet. [CENTER] The DOG(2) wavelet known as the "Mexican hat". [RIGHT] The DOG(6) wavelet

5.6.1 Numerical Computation of the Wavelet Transform

The continuous wavelet transform (5.79) of the discrete values of the signal f_k with the chosen scale parameter s is evaluated by computing the convolution sum [65]

$$L_\psi[f](s, t_n) = \frac{1}{\sqrt{c_\psi s}} \sum_{k=0}^{N-1} f_k \, \psi^* \left(\frac{(k-n)\Delta t}{s} \right) , \quad n = 0, 1, \ldots, N-1 . \quad (5.83)$$

We take the values of s from an arbitrary set $\{s_m\}_{m=0}^{M-1}$ with some $M < N$. The most simple choice is $s_m = (m+1)\Delta t$. The computation of the sum becomes unacceptably slow for large N and M as the time cost increases as $\mathcal{O}(MN^2)$. Since the convolution of two functions in configuration space is equivalent to the multiplication of their Fourier transforms in the Fourier space, the CWT can be computed by using the fast Fourier transformation (FFT).

Wavelets given as continuous functions The procedure is simple when wavelet functions exist in closed forms and for which the analytic form of their Fourier transforms in known. First we use the FFT to compute the Fourier transform F of the signal f which has been sampled at N points with uniform spacings Δt:

$$F_k = \mathcal{F}_N[f]_k = \frac{1}{N} \sum_{n=0}^{N-1} f_n \, e^{-i \, 2\pi kn/N} . \quad (5.84)$$

If the wavelet is given by the function $\psi(t/s)$ in configuration space, its correlate in Fourier space (in the continuous limit) is the function $\hat{\psi}(s\omega)$. For example, the family of wavelets (5.81) corresponds to the family of transforms

$$\hat{\psi}_{\mathrm{DOG}(m)}(s\omega) = \frac{-\mathrm{i}^m}{\sqrt{\Gamma(m+1/2)}}\,(s\omega)^m\,\mathrm{e}^{-(s\omega)^2/2}\,.$$

The wavelet transform is then evaluated by using the inverse FFT to compute the sum

$$L_\psi[f](s,t_n) = \frac{1}{\sqrt{c_\psi s}}\sum_{k=0}^{N-1} F_k\,\hat{\psi}^*(s\omega_k)\,\mathrm{e}^{\mathrm{i}\omega_k n\Delta t}\,,$$

where

$$\omega_k = \begin{cases} 2\pi k/(N\Delta t)\;;\; k \le N/2\,, \\ -2\pi k/(N\Delta t)\;;\; k > N/2\,. \end{cases}$$

The numerical cost of this procedure is $\mathcal{O}(MN\log N)$.

Wavelets given at discrete points If the wavelet is not specified as an analytic function or if its Fourier representation is not known, we need to compute both the discrete Fourier transform of the sampled signal (5.84) and the discrete Fourier transform of the wavelet at scale s, i.e.

$$Y_k = \mathcal{F}_N[\psi_s]_k = \frac{1}{N}\sum_{n=0}^{N-1}\psi^*(-n\Delta t/s)\,\mathrm{e}^{-\mathrm{i}2\pi kn/N}\,.$$

The convolution is at the heart of this procedure and the way the sampling is done requires some attention. For even wavelets it makes sense to adopt the sampling which preserves the symmetry of the wavelet about its origin. The wavelet and the signal are sampled as shown in Fig. 5.27.

We obtain the wavelet transform by multiplying the arrays $F_k \equiv \mathcal{F}_N[f]_k$ and $Y_k \equiv \mathcal{F}_N[\psi_s]_k$ component-wise, diving the result by $\sqrt{c_\psi s}$, and computing the inverse Fourier transform of the product array W_k:

$$L_\psi[f](s) = N\mathcal{F}_N^{-1}[W]\,,\quad W_k \equiv \frac{1}{\sqrt{c_\psi s}}\,F_k Y_k\,.$$

The inverse procedure is at hand: we reconstruct the original signal from the continuous wavelet transform by deconvolution, i.e. by dividing the Fourier representation of the transform by the Fourier representation of the wavelet. We form the arrays $L_k \equiv \mathcal{F}_N[L_\psi[f](s)]_k$ and $Y_k \equiv \mathcal{F}_N[\psi_s]_k$, divide them, multiply them by $\sqrt{c_\psi s}$, and compute the inverse Fourier transform of the quotient array:

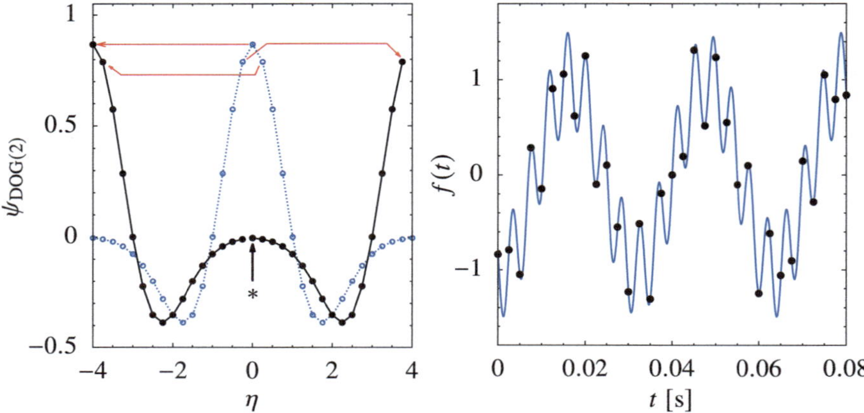

Fig. 5.27 Sampling at $N = 32$ points for the computation of the continuous wavelet transform. [LEFT] Periodic sampling of the wavelet in its natural (dimensionless) scale at the maximum scaling parameter s. The wavelet is sampled at N points on $[\eta_{\min}, \eta_{\max}] = [-4, 4]$. When we wish to reduce the parameter s, we insert zeros at the location marked by the thick arrow and the symbol $*$. [RIGHT] Sampling of the signal at N points of the physical (time) scale or the interval $[0, N\Delta t]$. The relation between the dimensionless variable η and the physical time t is $\eta = t(\eta_{\max} - \eta_{\min})/(N\Delta t)$. For details see [66]

$$f = N^{-1}\mathcal{F}_N^{-1}[F], \quad F_k \equiv \sqrt{c_\psi s}\,(L_k/Y_k).$$

(The parameter s obviously has to be chosen such that $Y_k \neq 0$ for all k.)

5.7 Discrete Wavelet Transformation ★

In Sect. 5.6 we have sampled the signal $f(t)$ and the wavelet $\psi(t)$ only at discrete points, but the described transformation can still be considered continuous, since the scale parameter s and the time axis span all $M \times N$ values. The continuous wavelet transform of a function sampled at N points has $M \times N$ values and is therefore highly redundant. In contrast, the wavelet transformation that represents the signal uniquely at just N points is known as *discrete wavelet transformation* (DWT).

The basic idea of the DWT [67, 68] is to break down a signal into two partial, half-length subsignals called the *trend* and the *detail*, respectively. The former is a coarser version of its predecessor, while the latter encodes the details that were lost in downsampling the original. The fact that this process can be repeated recursively leads to a fast version of the algorithm.

5.7.1 One-Dimensional DWT

A single step of the discrete wavelet transformation of a signal $f(t)$ sampled at times $t_n = n\Delta t, n = 0, 1, \ldots, N - 1$, and represented by a N-dimensional vector

$$f = (f_0, f_1, \ldots, f_{N-1})^{\mathrm{T}}$$

amounts to applying digital filters to f [67, 69]. These filters have the form

$$\widetilde{f}_i = \sum_{k=0}^{K-1} c_k f_{2i+k}, \quad i = 0, 1, 2, \ldots,$$

where c_k are the filter coefficients and K is the filter length. Note the $2i$: the signals are being downsampled. Two types of filters are used: "low-pass" (denoted by $\mathcal{H}$ and specified by the coefficients h_k) and "high-pass" ($\mathcal{G}$ and g_k).

The simplest variety of the DWT with $K = 2$ is the discrete version of the continuous transformation with Haar wavelets. The first filter with the coefficients $h_0 = h_1 = 1/\sqrt{2}$ (see Table 5.3) is used to compute the running averages of the signal, $f_0^1 = (f_0 + f_1)/\sqrt{2}$, $f_1^1 = (f_2 + f_3)/\sqrt{2}, \ldots$, resulting in the trend subsignal of length $N/2$,

$$f^1 = \left(f_0^1, f_1^1, \ldots, f_{N/2-1}^1\right)^{\mathrm{T}}. \tag{5.85}$$

Clearly f^1 still represents the dominant features of the original signal, but at twice poorer resolution: the filter sorts out the highest frequencies, i.e. doubles the scale. The second filter with the coefficients $g_0 = -g_1 = 1/\sqrt{2}$ is used to compute the corresponding differences, $(f_0 - f_1)/\sqrt{2}$, $(f_2 - f_3)/\sqrt{2}, \ldots$, yielding the detail signal

$$d^1 = \left(d_0^1, d_1^1, \ldots, d_{N/2-1}^1\right)^{\mathrm{T}}$$

describing the *fluctuations* around the trend (5.85). These filter actions are equivalent to computing the components of f^1 and d^1 by the scalar products

$$f_n^1 = f^{\mathrm{T}} H_n^1, \quad d_n^1 = f^{\mathrm{T}} G_n^1, \quad n = 0, 1, \ldots, N/2 - 1, \tag{5.86}$$

where $H_0^1 = \left(1/\sqrt{2}, 1/\sqrt{2}, 0, \ldots\right)^{\mathrm{T}}$, $H_1^1 = \left(0, 0, 1/\sqrt{2}, 1/\sqrt{2}, 0, \ldots\right)^{\mathrm{T}}, \ldots$ and $G_0^1 = \left(1/\sqrt{2}, -1/\sqrt{2}, 0, \ldots\right)^{\mathrm{T}}$, $G_1^1 = \left(0, 0, 1/\sqrt{2}, -1/\sqrt{2}, 0, \ldots\right)^{\mathrm{T}}, \ldots$ We denote the effect of this single-step (level-1) DWT by

$$f \mapsto \left(f^1 \big| d^1\right).$$

The process can now be repeated as many times as the signal can be halved: see Fig. 5.30 (top). In general, a signal containing $N = 2^L$ values can be subject to a maximum of L consecutive transformations, each halving the dimension of the

Table 5.3 The $\mathcal{H}$-filter coefficients belonging to the Daubechies (db$m$) wavelets with m ranging from 1 (Haar wavelet) to 5

	$m = 1$	$m = 2$	$m = 3$	$m = 4$	$m = 5$
h_0	$1/\sqrt{2}$	$(1 + \sqrt{3})/4\sqrt{2}$	0.332671	0.230378	0.160102
h_1	$1/\sqrt{2}$	$(3 + \sqrt{3})/4\sqrt{2}$	0.806892	0.714847	0.603829
h_2		$(3 - \sqrt{3})/4\sqrt{2}$	0.459878	0.630881	0.724309
h_3		$(1 - \sqrt{3})/4\sqrt{2}$	-0.135011	-0.0279838	0.138428
h_4			-0.0854413	-0.187035	-0.242295
h_5			0.0352263	0.0308414	-0.0322449
h_6				0.032883	0.0775715
h_7				-0.0105974	-0.00624149
h_8					-0.0125808
h_9					0.00333573

preceding trend and detail vectors. The level-2 DWT results in the $N/4$-dimensional trend and detail signals $\boldsymbol{f}^2$ and $\boldsymbol{d}^2$ derived from $\boldsymbol{f}^1$, as well as the previous $N/2$-dimensional detail vector $\boldsymbol{d}^1$,

$$\boldsymbol{f} \mapsto \left(\boldsymbol{f}^2 \big| \boldsymbol{d}^2, \boldsymbol{d}^1\right),$$

the level-3 DWT produces the $N/8$-dimensional trend and detail signals $\boldsymbol{f}^3$ and $\boldsymbol{d}^3$ derived from $\boldsymbol{f}^2$ as well as the two previous detail signals,

$$\boldsymbol{f} \mapsto \left(\boldsymbol{f}^3 \big| \boldsymbol{d}^3, \boldsymbol{d}^2, \boldsymbol{d}^1\right),$$

and so on. The components of $\boldsymbol{f}^2$ are given by the scalar products

$$f_n^2 = \boldsymbol{f}^{\mathrm{T}} \boldsymbol{H}_n^2, \quad d_n^2 = \boldsymbol{f}^{\mathrm{T}} \boldsymbol{G}_n^2, \quad n = 0, 1, \ldots, N/4 - 1, \tag{5.87}$$

where $\boldsymbol{H}_n^2 = h_0 \boldsymbol{H}_{2n}^1 + h_1 \boldsymbol{H}_{2n+1}^1$ and $\boldsymbol{G}_n^2 = g_0 \boldsymbol{H}_{2n}^1 + g_1 \boldsymbol{H}_{2n+1}^1$, the components of $\boldsymbol{f}^3$ are given by

$$f_n^3 = \boldsymbol{f}^{\mathrm{T}} \boldsymbol{H}_n^3, \quad d_n^3 = \boldsymbol{f}^{\mathrm{T}} \boldsymbol{G}_n^3, \quad n = 0, 1, \ldots, N/8 - 1, \tag{5.88}$$

where $\boldsymbol{H}_n^3 = h_0 \boldsymbol{H}_{2n}^2 + h_1 \boldsymbol{H}_{2n+1}^2$ and $\boldsymbol{G}_n^3 = g_0 \boldsymbol{H}_{2n}^2 + g_1 \boldsymbol{H}_{2n+1}^2$, and so on: compare (5.87) and (5.88) to (5.86) and note the diminishing range of n! An example of a three-level DWT of a 64-dimensional signal is shown in Fig. 5.28.

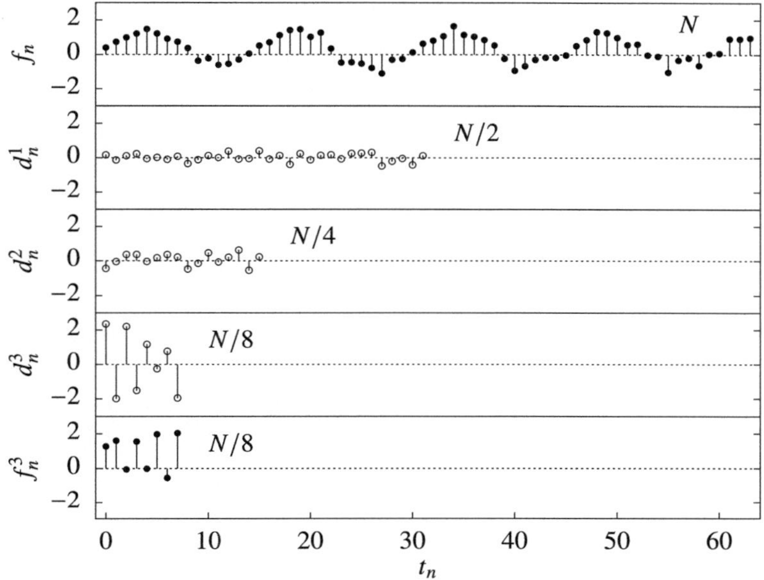

Fig. 5.28 A three-level discrete wavelet transform of signal f of length $N = 64$, resulting in the trend f^3 and three detail subsignals d^1, d^2 and d^3

The recursive division of the trend signals to their subordinate trends and details can be tracked on a time versus scale mesh which is a discrete representation of the landscape shown, for instance, in Fig. 5.24 (bottom). Assuming that the original signal was continuous, given by $f(t)$, the detail subsignals $d^1, d^2, \ldots, d^k$ and the final f^k contain the DWT of f at very specific points dictated by this recurrence:

$$L_\psi[f]\left(s = 2^l, t = 2^l n\right) = \frac{1}{\sqrt{c_\psi}}\, d_n^l\,, \quad l = 1, 2, \ldots, L\,,$$

which may be compared to the CWT given by formula (5.83). An example of such a mesh is shown in Fig. 5.29. The redundancy of values plaguing the continuous transform is gone: the DWT encodes all information in just N points.

The Inverse Transformation

The f^1 and d^1 taken together (or f^2, d^2 and d^1 taken together, and so on) encode all information contained in f. This is ensured by requiring that the filter coefficients satisfy $\sum_k h_k^2 = \sum_k g_k^2 = 1$ (energy conservation) as well as $\sum_k h_k = \sqrt{2}$ and $\sum_k g_k = 0$ [61]. In addition, the transformation should be invertible, i.e. the signal f should be recoverable from the transforms $(f^1|d^1)$, $(f^2|d^2, d^1)$, and so on. The inverse DWT is realized by applying the so-called *dual filters* $\mathcal{H}^*$ and $\mathcal{G}^*$ to subsequent trend and detail signals in reverse order, thereby reconstructing the original signal: see Fig. 5.30 (bottom).

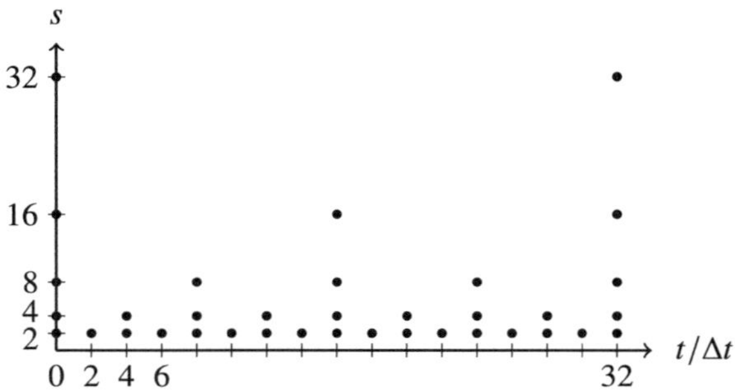

Fig. 5.29 The discrete time versus scale mesh on which the DWT stores its trend and detail subsignals

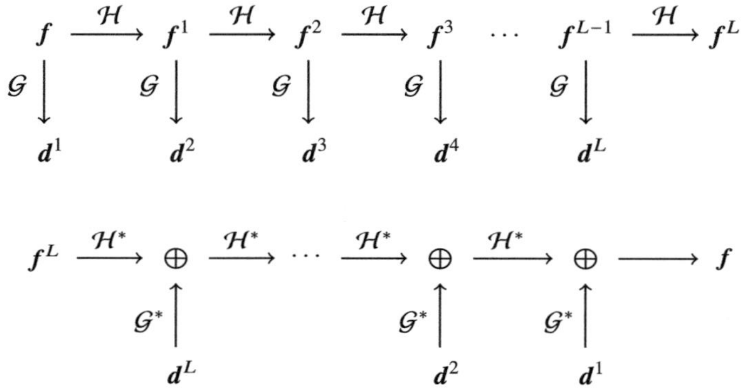

$$f \xrightarrow{\mathcal{H}} f^1 \xrightarrow{\mathcal{H}} f^2 \xrightarrow{\mathcal{H}} f^3 \quad \cdots \quad f^{L-1} \xrightarrow{\mathcal{H}} f^L$$

$$\mathcal{G}\Big\downarrow \qquad \mathcal{G}\Big\downarrow \qquad \mathcal{G}\Big\downarrow \qquad \mathcal{G}\Big\downarrow \qquad \mathcal{G}\Big\downarrow$$

$$\boldsymbol{d}^1 \qquad \boldsymbol{d}^2 \qquad \boldsymbol{d}^3 \qquad \boldsymbol{d}^4 \qquad \boldsymbol{d}^L$$

$$f^L \xrightarrow{\mathcal{H}^*} \oplus \xrightarrow{\mathcal{H}^*} \cdots \xrightarrow{\mathcal{H}^*} \oplus \xrightarrow{\mathcal{H}^*} \oplus \longrightarrow f$$

$$\mathcal{G}^*\Big\uparrow \qquad \mathcal{G}^*\Big\uparrow \qquad \mathcal{G}^*\Big\uparrow$$

$$\boldsymbol{d}^L \qquad \qquad \boldsymbol{d}^2 \qquad \boldsymbol{d}^1$$

Fig. 5.30 The logical flow of the [TOP] L-level discrete wavelet transformation by using the $\mathcal{H}$ and $\mathcal{G}$ filters and [BOTTOM] its inverse by using their respective duals

For example, the three-level DWT given by (5.86), (5.87) and (5.88) can be inverted by using the same filter coefficients as the "forward" DWT, by computing

$$\boldsymbol{F}^3 = \sum_{n=0}^{N/8-1} f_n^3 \boldsymbol{H}_n^3 \tag{5.89}$$

as well as

$$\boldsymbol{D}^3 = \sum_{n=0}^{N/8-1} d_n^3 \boldsymbol{G}_n^3\,, \quad \boldsymbol{D}^2 = \sum_{n=0}^{N/4-1} d_n^2 \boldsymbol{G}_n^2\,, \quad \boldsymbol{D}^1 = \sum_{n=0}^{N/2-1} d_n^1 \boldsymbol{G}_n^1\,, \tag{5.90}$$

and finally summing the parts to yield $f = F^3 + D^3 + D^2 + D^1$. For a L-level transformation, this obviously generalizes to

$$f = F^L + D^L + D^{L-1} + \cdots + D^2 + D^1 .$$

Daubechies Wavelets

Many invertible filters can be designed that fulfill the above mentioned requirements for h_k and g_k. Each admissible set of filter coefficients corresponds to a specific family of wavelet functions themselves. The $h_0 = h_1 = 1/\sqrt{2}$ filter corresponds to Haar wavelets, which are the simplest members of a broader class of widely used Daubechies wavelets [70, 71] denoted by db$m$. The m stands for the number of vanishing moments of the wavelet function $\psi_{\text{db}m}(x)$. The requirement that $\int_{-\infty}^{\infty} t^p \psi(t)\, dt = 0$ for p as high as possible is driven by the desire to maximize the sensitivity of the wavelet transformation: if $\psi(t)$ has enough vanishing moments, the smooth parts of the signal will yield close-to-zero transform strength, while the transform will be concentrated around the instances of rapid signal changes.

A DWT based on db$m$ wavelets has $2m$ filter coefficients for each of the $\mathcal{H}$ and $\mathcal{G}$ filters. Table 5.3 lists the first five $\mathcal{H}$-sets. The corresponding $\mathcal{G}$-coefficients are given by

$$g_k = (-1)^k h_{2m-k-1}, \quad k = 0, 1, \ldots, 2m - 1 .$$

Fig. 5.31 shows the db$m$ wavelet functions for $m = 2, 4, 6, 10, 15$ and 20. Note that these functions need not be coded explicitly—and actually can not be as they do not possess closed-form expressions. It is the filters that embody the transformation as if it were done with precisely these functions.

Example The three-level DWT based on the db2 wavelets is calculated by using the formulas (5.86), (5.87) and (5.88), but now the H_n^1 vectors are given by

$$\begin{aligned}
H_0^1 &= (h_0, h_1, h_2, h_3, 0, 0, \ldots, 0)^{\mathrm{T}}, \\
H_1^1 &= (0, 0, h_0, h_1, h_2, h_3, 0, 0, \ldots, 0)^{\mathrm{T}}, \\
&\;\vdots \\
H_{N/2-2}^1 &= (0, 0, \ldots, 0, h_0, h_1, h_2, h_3)^{\mathrm{T}}, \\
H_{N/2-1}^1 &= (h_2, h_3, 0, 0, \ldots, 0, h_0, h_1)^{\mathrm{T}},
\end{aligned} \tag{5.91}$$

with $\{h_k\}_{k=0}^{3}$ listed in the $m = 2$ column of Table 5.3. The G_n^1 vectors have the same structure with h_k replaced by g_k. Since the filter is specified by four coefficients, one has to decide what to do with the overflowing indices $N/2$ and $N/2 + 1$. The usual solution is to wrap around the coefficients as shown in (5.91): this corresponds to the assumption that the signal is periodic. By the same token, the H_n^2 and H_n^3 vectors are constructed as

$$\begin{aligned}
H_n^2 &= h_0 H_{2n}^1 + h_1 H_{2n+1}^1 + h_2 H_{2n+2}^1 + h_3 H_{2n+3}^1 , & n = 0, 1, \ldots, N/4 - 2 , \\
H_n^3 &= h_0 H_{2n}^2 + h_1 H_{2n+1}^2 + h_2 H_{2n+2}^2 + h_3 H_{2n+3}^2 , & n = 0, 1, \ldots, N/8 - 2 ,
\end{aligned}$$

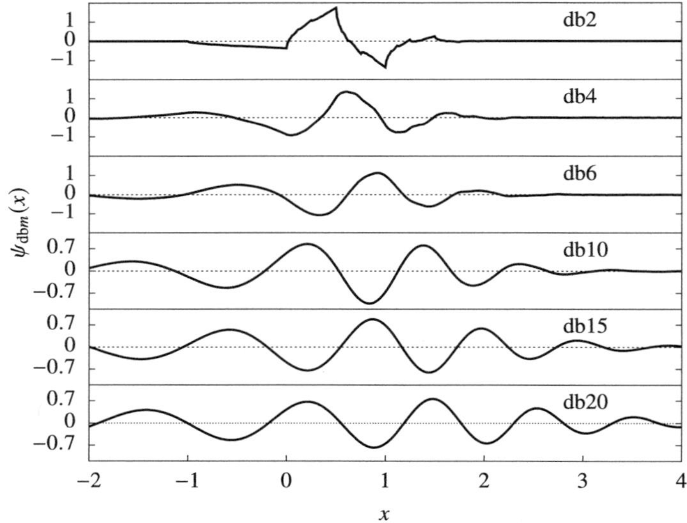

Fig. 5.31 Daubechies wavelets for $m = 2, 4, 6, 10, 15$ and 20

while computing the stranded $H^1_{N/4-1}$ and $H^2_{N/8-1}$ again involves wrap-around:

$$H^2_{N/4-1} = h_2 H^1_0 + h_3 H^1_1 + h_0 H^1_{N/2-2} + h_1 H^1_{N/2-1} \,,$$
$$H^3_{N/8-1} = h_2 H^2_0 + h_3 H^2_1 + h_0 H^2_{N/4-2} + h_1 H^2_{N/4-1} \,.$$

The calculation of the G^2_n and G^3_n vectors proceeds along the same lines, with h_k replaced by g_k. The inverse transformation is performed by using (5.89) and (5.90), followed by $f = F^3 + D^3 + D^2 + D^1$. The DWT involving db$m$ with larger m require correspondingly more coefficients to be wrapped around: four for $m = 3$, six for $m = 4$, and so on.   ◁

5.7.2 Two-Dimensional DWT

The generalization of the discrete wavelet transformation to two-dimensional signals (e.g. images) is straightforward. Assume that we are dealing with a square ($N \times N$) grayscale image f represented by a matrix of values

$$f_{ij} \,, \quad i, j = 0, 1, \ldots, N - 1 \,.$$

A single-step DWT now also involves processing the image with the $\mathcal{H}$ and $\mathcal{G}$ filters, but this proceeds in four different ways. Scanning with $\mathcal{H}$ along both the columns

(c) and rows (r) produces the approximate image $f^1 = \mathcal{H}_r\mathcal{H}_c[f]$. Operating with $\mathcal{H}$ along the columns and $\mathcal{G}$ along the rows results in the "vertical detail" image $d^{1v} = \mathcal{G}_r\mathcal{H}_c[f]$, while the opposite procedure gives the "horizontal detail" image $d^{1h} = \mathcal{H}_r\mathcal{G}_c[f]$. Applying $\mathcal{G}$ to both the columns and rows encodes the "diagonal detail" image $d^{1d} = \mathcal{G}_r\mathcal{G}_c[f]$. The first-level DWT of a $N \times N$ image f therefore produces four $N/2 \times N/2$ subimages usually arranged as

$$f \mapsto \begin{pmatrix} f^1 & d^{1h} \\ d^{1v} & d^{1d} \end{pmatrix} .$$

The procedure can then be recursively repeated by exact analogy to the one-dimensional case. The approximate ("trend") $N/4 \times N/4$ and $N/8 \times N/8$ images f^2 and f^3, for instance, are generated by calculating

$$f^1 \mapsto \begin{pmatrix} f^2 & d^{2h} \\ d^{2v} & d^{2d} \end{pmatrix} , \quad f^2 \mapsto \begin{pmatrix} f^3 & d^{3h} \\ d^{3v} & d^{3d} \end{pmatrix} , \quad \cdots$$

An example of a three-level two-dimensional DWT is shown in Fig. 5.32.

The inverse transformation proceeds analogously to the flowchart of Fig. 5.30 (bottom), where $\mathcal{H}^*$ and $\mathcal{G}^*$ are replaced by the appropriate products of $\mathcal{H}_r^*$, $\mathcal{H}_c^*$, $\mathcal{G}_r^*$ and $\mathcal{G}_c^*$, and the $\oplus$ operation stands for the summation of matrices.

Image compression and denoising

Wavelet transformations lie at the heart of modern data compression algorithms. The simplest way one-dimensional data can be compacted by means of the *continuous* wavelet transformation is illustrated in Problem 5.8.4. Filtering and compressing two-dimensional data like images, however, calls for fast two-dimensional discrete transformations.

The basic trick of image compression consists of transforming the image matrix by DWT, manipulating the transform, and calculating the inverse transform. The most obvious strategy of this "manipulation" is to truncate the transform, i.e. to strip it of its insignificant coefficients and keep only the prominent ones. This "cut-off" can be realized in many ways, the simplest of which are taking only the coefficients with absolute values above a certain threshold, or keeping a specified number of coefficients. The latter approach is demonstrated in Fig. 5.33. The original image (see caption of Fig. 5.32) has been transformed by a 4-level DWT based on db4 wavelets, the transform truncated, and inverted.

The image format standard JPEG2000 (also known as J2K) [72] exploits a special variety of the DWT relying on *biorthogonal filters*. The biorthogonal DWT utilizes *two* sets of low-pass filters, *two* sets of high-pass filters (and their inverses), thus implying *two* types of wavelets. For lossy J2K compression one uses the 9/7 Cohen–Daubechies–Feauveau (CDF) wavelets, while the lossless compression relies on 5/3 CDF wavelets [73, 74]. Their coefficients and further application details can be found in Sects. 3.7 and 3.8 of [69].

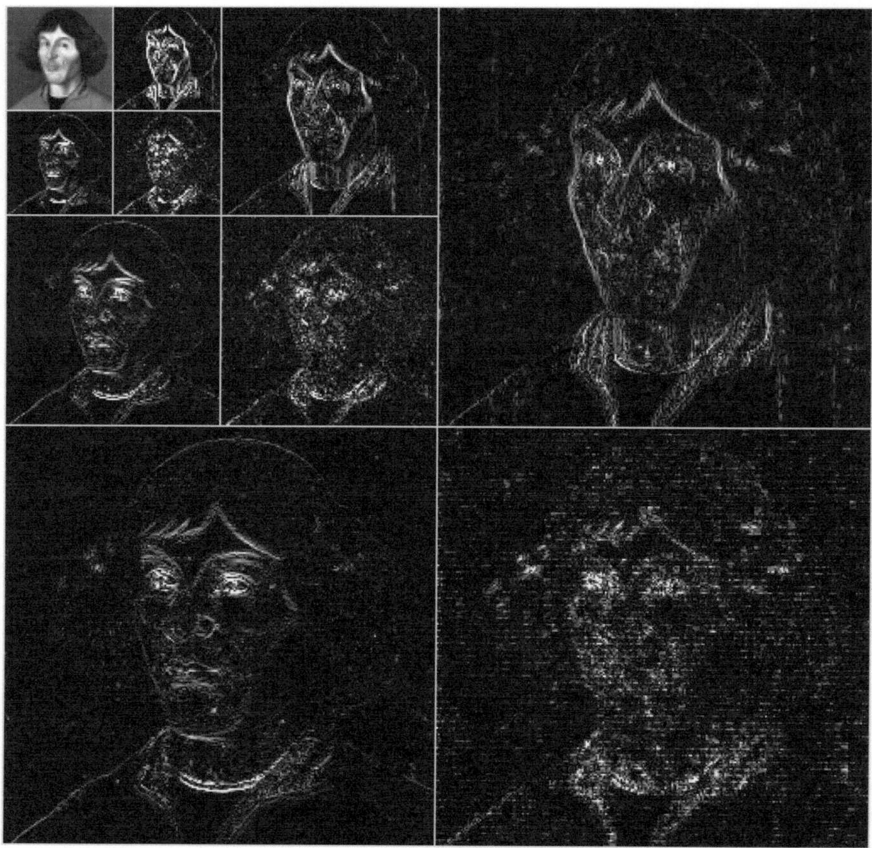

Fig. 5.32 A two-dimensional DWT of a 512×512 image. (The original image can be found on the website of the book, the "almost" original is in the bottom right panel of Fig. 4.7.) The 64×64 level-3 approximate image f^3 is in the top left corner, encircled (outward and clockwise) by the 64×64 detail images d^{3h}, d^{3d}, d^{3v}, followed by their 128×128 predecessors d^{2h}, d^{2d}, d^{2v} and their 256×256 parents d^{1h}, d^{1d} and d^{1v}

Software Implementations

For C/C++ environments, the DWT with Daubechies and CDF wavelets is implemented in the GSL library, while the LIBDWT project [75] focuses on DWT and CWT with biorthogonal wavelets optimized for image processing. PYWAVELETS [76] is a free open-source Python-based library offering several varieties of the wavelet transform that also supports custom wavelets. The one-dimensional and two-dimensional DWT are implemented in both MATHEMATICA and the Wavelet Toolbox of MATLAB. For freely accessible MATLAB/OCTAVE routines see [77].

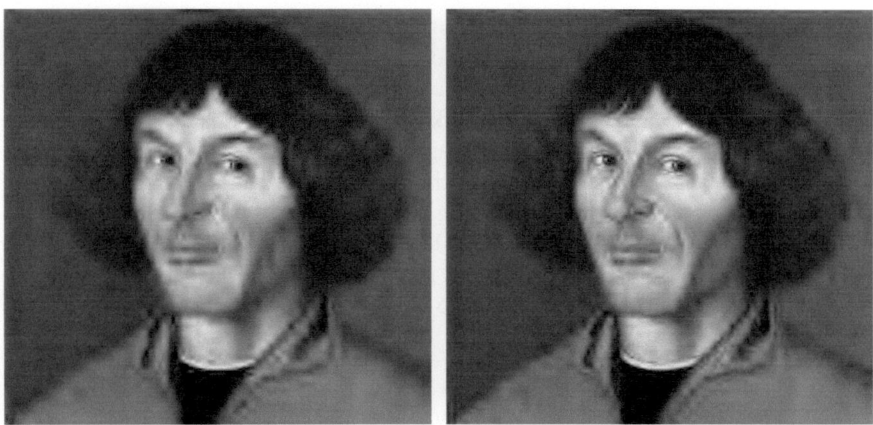

Fig. 5.33 Image compression by DWT. The original image is a 512×512 matrix of pixels with 256 gray levels. Shown are the DWT reconstructions by using the db4 wavelets up to refinement level $L = 4$ and keeping the largest [LEFT] 262 transform coefficients (99.9% compression) and [RIGHT] largest 655 coefficients (99.8% compression). Compare to the top two panels of Fig. 4.7 as well as Problems 4.10.4 and 5.8.4

5.8 Problems

5.8.1 Fourier Spectrum of Signals

The discrete Fourier transformation (DFT, Sect. 5.2.2) is the fundamental tool of signal analysis. First we confirm the formulas for known pairs of periodic signals $f = \{f_j\}_{j=0}^{N-1}$ and their transforms $F = \{F_k\}_{k=0}^{N-1} = \mathcal{F}_N[f]$, for example,

$$f_j = e^{iaj/N} \Leftrightarrow F_k = \frac{1}{N} \left\{ \left(e^{ia} - 1 \right) \Big/ \left(e^{i(a-2k\pi)/N} - 1 \right) \right\},$$

$$\mathrm{Re}\, f_j \,,\ \mathrm{Im}\, f_j \sim \mathcal{N}(0, 1) \Leftrightarrow \mathrm{Re}\, F_k \,,\ \mathrm{Im}\, F_k \sim \mathcal{N}(0, N) \,,$$

where $\mathcal{N}(\mu, \sigma^2)$ is the normal distribution with mean μ and variance σ^2.

⊙ Compute the Fourier transforms of simple signals, in which several frequencies are represented. Compare the results in the case when the sample is periodic on the interval (the chosen frequencies are integer multiples of the fundamental frequency), to those cases when it is not. Observe the effect of aliasing for a sample that includes frequencies above the critical value. Show that for real signals

$$F_{N-k} = F_k^*$$

and that $\mathcal{F}_N$ and $\mathcal{F}_N^{-1}$ are truly inverse operations.

Perform a Fourier analysis of the sound generated by tapping on a rectangular wood-box resonator, creating transient acoustic waves in its interior. Plot the power spectral density and identify the eigenmodes of the resonator from its most prominent peaks. Analyze the sound of the tuning fork attached to the resonator of a guitar. (The signals can be found at the book's web-page.)

$\oplus$ Fourier-analyze the concluding chord of the Toccata and Fugue for organ in d-minor, BWV 565, of J. S. Bach (Fig. 5.5). The signal from the audio CD has been sampled at 44100 Hz, 11025 Hz, 5512 Hz, 2756 Hz, 1378 Hz, and 882 Hz. By listening to the recordings in the .mp3 format find out what happens when the sampling frequency is reduced, then try to confirm this by Fourier analysis.

5.8.2   *Fourier Analysis of the Doppler Effect*

Here we discuss the experiment set up in Salzburg, the birthplace of C. Doppler (1805-1853). A speaker is attached to a rotating wheel with radius R, and there is a microphone in the plane of rotation at a distance L from the center of the wheel, as shown in Fig. 5.34. The rotation angular frequency is $\Omega = v/R$, where v is the tangential velocity. The radius vector $\boldsymbol{r}(t)$ describes the relative position of the microphone with respect to the speaker. The speaker periodically approaches and recedes from the microphone (period $2\pi/\Omega$).

The microphone is used to measure the pressure differences $\delta p(t)$ given by the formula

$$\delta p(t) = A \, \frac{\cos[\omega(t - x(\Omega t)/c)]}{x(\Omega t)} \, .$$

Here A is the amplitude, ω is the angular frequency of the emitted sound and c is its velocity. The function $x(t)$ is defined implicitly as the positive solution of the equation

$$l^2 + r^2 + 2lr \sin(t - x) = x^2 \, ,$$

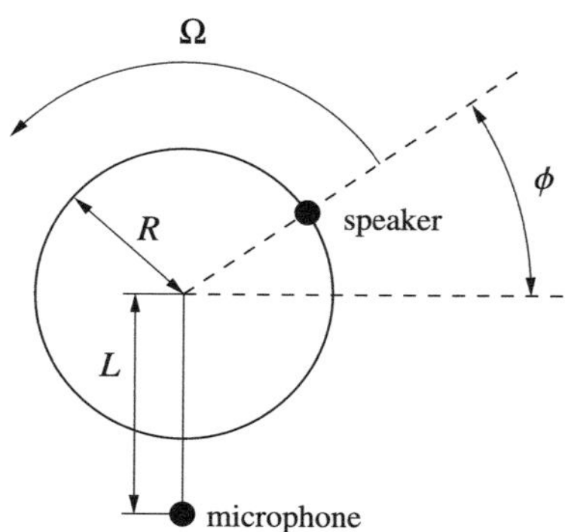

Fig. 5.34 A speaker-microphone setup used for the Fourier analysis of the Doppler effect

where $l = \Omega L/c$ and $r = \Omega R/c$.

⊙ Let $v = 100\,\text{m/s}$ and $A = 1$. Sample the signal $\delta p(t)$ at time intervals $\Delta t = 1/(\Omega N)$ over two periods. This gives you the discrete signal $f_j = \delta p(j\,\Delta t)$ at $j = 0, 1, \ldots, N-1$. Compute the DFT and the Hilbert transform of the discrete signal in the limits $L \ll R$, $L \gg R$, and $L \approx R$, and analyze its single-sided power spectral density (PSD), its instantaneous amplitude, and instantaneous frequency. An example of the signal and its spectral density is shown in Fig. 5.35. Compare the results to the classical Doppler prediction.

⊕ Analyze the PSD of the signal at smaller time windows which open just slightly ahead of the time when the speaker is closest to the microphone, and close just after that.

5.8.3 *Use of Laplace Transformation and Its Inverse*

The time dependence of the charge on a capacitor connected in series to a electric generator, a resistor, and a coil, is described by the differential equation

$$L\,\frac{\mathrm{d}^2 Q}{\mathrm{d}t^2} + R\,\frac{\mathrm{d}Q}{\mathrm{d}t} + \frac{1}{C}\,Q = U_\mathrm{g}(t)$$

or, in dimensionless form, at given R, C and L,

$$\ddot{x} + \dot{x} + x = g(t)\,.$$

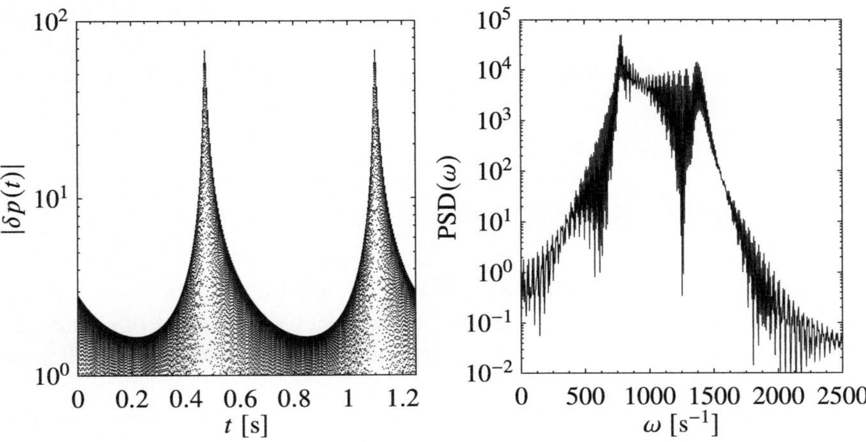

Fig. 5.35 [LEFT] An example of the signal at $v = 100\,\text{m/s}$, $R = 10\,\text{m}$, $L = 10.5\,\text{m}$, $\omega = 1000/\text{s}$, and $c = 340\,\text{m/s}$. [RIGHT] The Fourier power spectral density. The peaks in the spectrum occur approximately at $\omega/(1 + v/c) = 773\,\text{Hz}$ and $\omega/(1 - v/c) = 1417\,\text{Hz}$

With initial conditions $x(0) = \dot{x}(0) = 0$ the source generates a voltage pulse of the form

$$g(t) = 1 - \theta(t - \pi) = \begin{cases} 1 \; ; \; 0 \leq t < \pi \, , \\ 0 \; ; \; t \geq \pi \, . \end{cases}$$

By using the properties (5.59) and Table 5.2, and by taking the initial conditions into account, the Laplace transform of the differential equation is

$$s^2 X(s) + s X(s) + X(s) = \mathcal{L}[1] - \mathcal{L}[\theta(t - \pi)] = (1 - e^{-\pi s})/s$$

or

$$X(s) \equiv \left(1 - e^{-\pi s}\right) H(s) \, , \quad H(s) = \frac{1}{s \left(s^2 + s + 1\right)} = \frac{1}{s} - \frac{\left(s + \frac{1}{2}\right) + \frac{1}{2}}{\left(s + \frac{1}{2}\right)^2 + \frac{3}{4}} \, .$$

Let us denote $h(t) = \mathcal{L}^{-1}[H]$ and remember that the inverse Laplace transformation is linear. Then by looking at Table 5.2 once more and considering the property $\mathcal{L}[e^{ct} f(t)] = F(s - c)$, we find the exact solution

$$x(t) = \mathcal{L}^{-1}[X] = \mathcal{L}^{-1}[H] - e^{-\pi s} \mathcal{L}^{-1}[H] = h(t) - h(t - \pi)\theta(t - \pi) \, ,$$

where

$$h(t) = \mathcal{L}^{-1}[H] = 1 - \frac{1}{3} e^{-t/2} \left(3 \cos \frac{\sqrt{3}}{2} t + \sqrt{3} \sin \frac{\sqrt{3}}{2} t \right) \, .$$

⊙   Use the inverse Laplace transformation to find the solution $x(t)$ from the function $X(s) = (1 - e^{-\pi s}) H(s)$. Since a discontinuous function $g(t)$ appears in the time domain, the numerical inverse should be computed by dedicated algorithms specially tailored to such functions. You may use the methods from the den Iseger's paper in [41].

5.8.4 Use of the Wavelet Transformation

The continuous wavelet transformation is one of the basic tools in signal processing and analysis. We use it to determine the frequency components of signals and the instances in time when individual components actually occur: this is its main advantage over the usual Fourier transformation.

⊙   Use the continuous wavelet transformation on a simple periodic signal to which components with higher frequencies have been admixed at certain subintervals, for example,

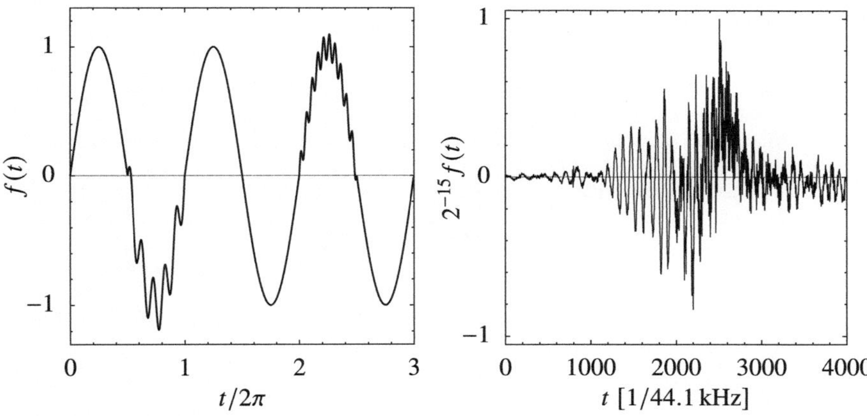

Fig. 5.36 [LEFT] The signal (5.92) with three frequency components, two of which appear only occasionally. [RIGHT] A recording of the intensity of sound caused by quickly waving a bamboo-stick through air

$$
f(t) = \begin{cases}
\sin t + 0.2 \, \sin 10t & ; \ \pi \le t \le 2\pi \ , \\
\sin t + 0.1 \, \sin 20t & ; \ 4\pi \le t \le 5\pi \ , \\
\sin t & ; \ \text{otherwise}
\end{cases}
\tag{5.92}
$$

(see Fig. 5.36 (left)). You can additionally perturb the signal by mixing it with noise of various amplitudes. How do such admixtures influence the quality (the resolution capability) of the wavelet transform? Use different types of wavelet functions. What differences can you observe?

$\oplus$ The continuous wavelet transformation also gives us some basic feeling for the properties of compression and decompression of data. Figure 5.36 (right) shows a typical acoustic signal. (The file can be found at the book's web-page.) By using the procedure described in Sect. 5.6.1 compute the continuous wavelet transform for this signal. Set to zero all components in the transform that are below a certain chosen threshold (for example, keep only 20%, 10%, or 5% of the strongest components). Then perform the inverse wavelet transformation and compare the result to the original signal. How has it changed? Do you reach the same conclusion if you repeat the exercise with the Fourier transform? (Use the FFT to map the signal into Fourier space, keep 20%, 10%, or 5% of the strongest components, and compute the inverse FFT.)

$\oplus$ Repeat the exercise by using the discrete variety of the wavelet transformation discussed in Sect. 5.7.

References

1. J.P. Boyd, *Chebyshev and Fourier Spectral Methods*, 2nd edn (Dover Publications, Mineola, 2000). Available at: http://www-personal.engin.umich.edu/jpboyd
2. B.D. MacCluer, *Elementary Functional Analysis* (Springer Science+Business Media, New York, 2010)
3. C. Shannon, Communication in the presence of noise. Proc. IEEE **86**, 447 (1998); M. Unser, Sampling—50 years after Shannon. Proc. IEEE **88**, 569 (2000); H. Nyquist, Certain topics in telegraph transmission theory. Proc. IEEE **90**, 280 (2002)
4. C. Canuto, M.Y. Hussaini, A. Quarteroni, T.A. Zang, *Spectral Methods* (Fundamentals in Single Domains (Springer, Berlin, 2006)
5. D. Donnelly, B. Rust, The fast Fourier transform for experimentalists, Part I: concepts. Comput. Sci. Eng, p. 80 (2005)
6. F.J. Harris, On the use of windows for harmonic analysis with the discrete Fourier transform. Proc. IEEE **66**, 51 (1978)
7. M. Frigo, S.G. Johnson, The design and implementation of FFTW3. Proc. IEEE **93**, 216 (2005); documentation can be found at: http://www.fftw.org
8. W.H. Press, B.P. Flannery, S.A. Teukolsky, W.T. Vetterling, *Numerical Recipes: The Art of Scientific Computing*, 3rd edn (Cambridge University Press, Cambridge, 2007). See also the equivalent handbooks in Fortran, Pascal and C, as well as http://numerical.recipes
9. P. Duhamel, M. Vetterli, Fast Fourier transforms: a tutorial review and a state of the art. Signal Process. **19**, 259 (1990)
10. J.C. Schatzman, Accuracy of the discrete Fourier transform and the fast Fourier transform. SIAM J. Sci. Comput. **17**, 1150 (1996)
11. H. Hassanieh, P. Indyk, D. Katabi, E. Price, *Simple and practical algorithm for sparse Fourier transform*, in Proceedings of the 23rd Annual ACM-SIAM Symposium on Discrete Algorithms, ed. by Y. Rabani (SIAM, Philadelphia, 2012), p. 1183
12. https://groups.csail.mit.edu/netmit/sFFT
13. H. Hassanieh, P. Indyk, D. Katabi, E. Price, *Nearly optimal sparse Fourier transform*, in Proceedings of the 44th Annual ACM Symposium on Theory of Computing (ACM, New York, 2012), p. 563
14. J. Schumacher, M. Püschel, *High-performance sparse fast Fourier transforms*, in Proceedings IEEE Workshop on Signal Processing Systems (IEEE, 2014), p. 1
15. http://spiral.net/software/sfft.html
16. P. Degroote et al., Evidence for nonlinear resonant mode coupling in the β Cephei star HD 180642 (V1449 Aquilae) from CoRoT photometry. Astron. Astrophys. **506**, 111 (2009)
17. J.T. VanderPlas, Understanding the Lomb-Scargle periodogram. Astrophys. J. Suppl. Ser. **236**, 16 (2018)
18. A. Dutt, V. Rokhlin, Fast Fourier transforms for nonequispaced data. SIAM J. Sci. Comput. **14**, 1368 (1993)
19. A. Dutt, V. Rokhlin, Fast Fourier transforms for nonequispaced data. II. Appl. Comput. Harm. Anal. **2**, 85 (1995)
20. L. Greengard, J.-Y. Lee, Accelerating the nonuniform fast Fourier transform. SIAM Rev. **46**, 443 (2004)
21. M. Lyon, J. Picard, The Fourier approximation of smooth but non-periodic functions from unevenly spaced data. Adv. Comput. Math. **40**, 1073 (2014)
22. N. Seghouani, High-resolution spectral analysis of unevenly spaced data using a regularization approach. Mon. Not. R. Astron. Soc. **468**, 3312 (2017)
23. W.H. Press, G.B. Rybicki, Fast algorithm for spectral analysis of unevenly sampled data. Astrophys. J. **338**, 277 (1989)
24. J. Keiner, S. Kunis, D. Potts, Using NFFT3–a software library for various nonequispaced Fourier transforms. ACM Trans. Math. Softw. **36**, 1 (2009)
25. D. Potts, G. Steidl, Fast summation at nonequispaced knots by NFFTS. SIAM J. Sci. Comput. **24**, 2013 (2003)

26. D. Potts, G. Steidl, M. Tasche, Fast Fourier transforms for nonequispaced data: a tutorial, in *Modern Sampling Theory: Mathematics and Applications*, ed. by J.J. Benedetto, P.J.S.G. Ferreira (Birkhäuser, Boston, 2001), p. 246

27. G. Steidl, A note on the fast Fourier transforms for nonequispaced grids. Adv. Comput. Math. **9**, 337 (1998)

28. B. Leroy, Fast calculation of the Lomb-Scargle periodogram using nonequispaced fast Fourier transforms. Astron. Astrophys. **545**, A50 (2012)

29. A.H. Barnett, J. Magland, L. af Klintenberg, A parallel non-uniform fast Fourier transform library based on an "exponential of semicircle" kernel. SIAM J. Sci. Comput. **41**, C479 (2019)

30. G. Szegö, *Orthogonal Polynomials* (AMS, Providence, 1939)

31. T.J. Rivlin, *The Chebyshev Polynomials* (John Wiley & Sons, New York, 1974)

32. V. Rokhlin, A fast algorithm for discrete Laplace transformation. J. Complexity **4**, 12 (1988) and J. Strain, A fast Laplace transform based on Laguerre functions. Math. Comp. **58**, 275 (1992)

33. S. Zhang, J. Jin, *Computation of Special Functions* (Wiley-Interscience, 1996)

34. M. Abramowitz, I.A. Stegun, *Handbook of mathematical functions*, 10th edn. (Dover Publications, Mineola, 1972)

35. W.E. Boyce, R.C. DiPrima, *Elementary Differential Equations and Boundary Value Problems*, 5th edn. (John Wiley & Sons, New York, 1992)

36. J. Abate, P.P. Valkó, Multi-precision Laplace transform inversion. Int. J. Numer. Meth. Eng. **60**, 979 (2004)

37. A.M. Cohen, *Numerical Methods for Laplace Transform Inversion* (Springer Science+Business Media, New York, 2007)

38. B. Davies, B. Martin, Numerical inversion of the Laplace transform: a survey and comparison of methods. J. Comp. Phys. **33**, 1 (1979); see also D.G. Duffy, On the numerical inversion of Laplace transforms: comparison of three new methods on characteristic problems from applications. ACM Trans. Math. Softw. **19**, 333 (1993)

39. C.L. Epstein, J. Schotland, The bad truth about Laplace's transform. SIAM Rev. **50**, 504 (2008)

40. R. Piessens, Gaussian quadrature formulas for the numerical integration of Bromwich's integral and the inversion of the Laplace transform. J. Eng. Math. **5**, 1 (1971); see also L.N. Trefethen, J.A.C. Weideman, T. Schmelzer, Talbot quadratures and rational approximations. BIT Numer. Math. **46**, 653 (2006)

41. J. Abate, W. Whitt, Queueing systems **10**, 5 (1992); see also P. den Iseger. Prob. Eng. Inform. Sci. **20**, 1 (2006) and A. Yonemoto, T. Hisikado, K. Okumura, IEEE Proc. Circuits Devices Syst. **150**, 399 (2003). Among the most efficient are also the algorithms based on the expansion of the function f in terms of Laguerre polynomials and the corresponding computation of $\mathcal{L}[f]$ and $\mathcal{L}^{-1}[F]$; see J. Abate, G.L. Choudhury, Informs. J. Comput. **8**, 413 (1996)

42. E. Titchmarsh, *Introduction to the Theory of Fourier Integrals*, 2nd edn. (Clarendon Press, Oxford, 1948)

43. L. Grafakos, *Classical and Modern Fourier Analysis* (Prentice Hall, New Jersey, 2003)

44. M.L. Glasser, Some useful properties of the Hilbert transform. SIAM J. Math. Anal. **15**, 1228 (1984)

45. R. Wong, Asymptotic expansion of the Hilbert transform. SIAM J. Math. Anal. **11**, 92 (1980)

46. S.L. Hahn, *Hilbert Transform in Signal Processing* (Artech House, Boston, 1996)

47. W.J. Freeman, Origin, structure, and role of background EEG activity, Part 1. Analytic amplitude. Clin. Neurophys. **115**, 2077 (2004); Part 2. Analytic phase, Clin. Neurophys. **115**, 2089 (2004); Part 3. Neural frame classification. Clin. Neurophys. **116**, 1118 (2005); Part 4. Neural frame simulation. Clin. Neurophys. **117**, 572 (2006)

48. EEG Recorded by M. West and A. Krystal, Duke University

49. C. Kittel, *Introduction to Solid State Physics*, 8th edn. (John Wiley & Sons, New York, 2005)

50. M. Nandagopal, N. Arunajadai, On finite evaluation of finite Hilbert transform. Comput. Sci. Eng. **9**, 90 (2007)

51. T. Hasegawa, T. Torii, Hilbert and Hadamard transforms by generalized Chebyshev expansion. J. Comput. Appl. Math. **51**, 71 (1994)

52. W. Gautschi, J. Waldvogel, Computing the Hilbert transform of the generalized Laguerre and Hermite weight functions. BIT Num. Math. **41**, 490 (2001)
53. V.R. Kress, E. Martensen, Anwendung der Rechteckregel auf die reelle Hilberttransformation mit unendlichem Intervall. Z. Angew. Math. Mech. **50**, 61 (1970)
54. F.W. King, G.J. Smethells, G.T. Helleloid, P.J. Pelzl, Numerical evaluation of Hilbert transforms for oscillatory functions: a convergence accelerator approach. Comput. Phys. Commun. **145**, 256 (2002)
55. J.A.C. Weideman, Computing the Hilbert transform on the real line. Math. Comp. **64**, 745 (1995). Attention: the authors use a non-standard definition sign(0) = 1; in Eq. (22) correct $1/N \to 1/(2N)$
56. X. Wang, *Numerical implementation of the Hilbert transform*, Ph.D. thesis (University of Saskatchewan, Saskatoon, 2006)
57. H. Boche, M. Protzmann, A new algorithm for the reconstruction of the band-limited functions and their Hilbert transform. IEEE Trans. Instr. Meas. **46**, 442 (1997)
58. A.V. Oppenheim, R.W. Schafer, *Discrete-Time Signal Processing*, 2nd edn. (Prentice Hall, New Jersey, 1989)
59. R. Bracewell, *The Fourier Transform and Its Applications*, 2nd edn. (McGraw-Hill, Reading/New York, 1986)
60. F.W. King, *Hilbert transforms*, Vols. 1 & 2, *Encyclopedia of Mathematics and its Applications*, Vols. 124 & 125 (Cambridge University Press, Cambridge, 2009)
61. H.-G. Stark, *Wavelets and Signal Processing. An Application-Based Introduction.* (Springer, Berlin, 2005)
62. P.S. Addison, *The Illustrated Wavelet Transform Handbook* (Institute of Physics, Bristol, 2002)
63. G. Kaiser, *A Friendly Guide to Wavelets* (Birkhäuser, Boston, 1994). A physicist might find particular pleasure in physical (acoustic and electro-magnetic) wavelets: see G. Kaiser, J. Phys. A: Math. Gen. **36**, R291 (2003)
64. J.F. Kirby, Which wavelet best reproduces the Fourier power spectrum? Comput. Geosci. **31**, 846 (2005)
65. C. Torrence, G.P. Compo, A practical guide to wavelet analysis. Bull. Am. Meteorol. Soc. **79**, 61 (1998)
66. D. Jordan, R.W. Miksad, E.J. Powers, Implementation of the continuous wavelet transform for digital time series analysis. Rev. Sci. Instrum. **68**, 1484 (1997)
67. G. Strang, T. Nguyen, *Wavelets and Filter Banks*, 2nd edn. (Wellesley-Cambridge Press, Cambridge, 1996)
68. D.B. Percival, A.T. Walden, *Wavelet Methods for Time Series Analysis* (Cambridge University Press, Cambridge, 2000)
69. J.S. Walker, *A Primer on Wavelets and Their Scientific Applications*, 2nd edn. (Chapman & Hall/CRC, Boca Raton, 2008)
70. I. Daubechies, The wavelet transform time-frequency localization and signal analysis. IEEE Trans. Inform. Theory **36**, 961 (1990)
71. I. Daubechies, *Ten Lectures on Wavelets* (SIAM, Philadelphia, 1992)
72. D.S. Taubman, M.W. Marcellin, *JPEG2000: Image Compression Fundamentals, Standards and Practice* (Kluwer, Boston, 2002)
73. A. Cohen, I. Daubechies, J.-C. Feauveau, Biorthogonal bases of compactly supported wavelets. Comm. Pure Appl. Math. **45**, 485 (1992)
74. A.R. Calderbank et al., Wavelet transforms that map integers to integers. Appl. Comput. Harmonic Anal. **5**, 332 (1998)
75. http://sourceforge.net/projects/libdwt
76. G.R. Lee, R. Gommers, F. Waselewski, K. Wohlfahrt, A. O'Leary, PyWavelets: a Python package for wavelet analysis. J. Open Source Softw. **4**(36), 1237 (2019). See also https://github.com/PyWavelets/pywt
77. http://ltfat.org

Chapter 6
Statistical Analysis and Modeling of Data

Abstract Outcomes of physical measurements are most frequently recorded as signals that need to be processed statistically in order to infer their overall properties and features, and later compared to theoretical or phenomenological models. This Chapter starts with an introduction to the basic statistical techniques of computing the averages and moments of distributions and their uncertainties. Particular emphasis is given on ways of identifying outliers and robust estimates of location (averages) and scale (dispersion). We introduce commonly used methods of computing confidence intervals for the sample means and variances, of comparing the means of samples with equal or different variances, of comparing two distributions, and computing correlations. Simple linear and multiple linear, as well as non-linear regression techniques are explained, again with attention to robust measures. Powerful multi-variate methods of principal component analysis, cluster analysis, linear discriminant analysis, and factor analysis are discussed in a separate Section each. The illustrations in the Problems include the study of Raman spectra in fabric yarns, the analysis of geyser eruptions, radar reflections in the ionosphere, and the correlation analysis of astrophysical objects.

We record the outcomes of physical measurements as signals (sequences of values), where we are not interested in each value in particular but the characteristics of the signal as a whole. Signals can be analyzed in the statistical sense, where the time ordering of the data is irrelevant, or in the functional sense, where it becomes essential: then we imagine that the signal (the measurement of a quantity) originates in the source (the dynamical system), and we attempt to infer the properties of that system from the properties of the signal. In this Chapter we introduce the basic methods of both approaches [1].

© The Author(s), under exclusive license to Springer Nature Switzerland AG 2025   313
S. Širca and M. Horvat, *Computational Methods in Physics*, Graduate Texts in Physics,
https://doi.org/10.1007/978-3-031-68566-8_6

6.1 Basic Data Analysis

6.1.1 Probability Distributions

Figure 6.1 (left) shows the probability density function

$$p(x) = \sqrt{\frac{2}{\pi}}\, \frac{1}{\sigma^3}\, x^2 \exp\left(-x^2/2\sigma^2\right)\,, \qquad \sigma = 2\,, \tag{6.1}$$

describing the classical Maxwell's velocity distribution, while Fig. 6.1 (right) shows the corresponding cumulative distribution function $P(x)$. The value $P(x)$ is the probability that a random variable X has a value not exceeding x,

$$P(x) = \mathrm{Prob}(X \le x)\,, \quad 1 - P(x) = \mathrm{Prob}(X > x)\,, \quad P(0) = 0\,, \quad P(\infty) = 1\,,$$

while the probability density $p(x) = \mathrm{d}P(x)/\mathrm{d}x$ is a measure of probability for X assuming a value on the interval $[x, x + \mathrm{d}x]$,

$$\mathrm{Prob}(a \le X \le b) = \int_a^b p(x)\,\mathrm{d}x = P(b) - P(a)\,.$$

A physicist dealing with the analysis of finite data samples typically encounters just a handful of probability distributions. But even within this limited set she sometimes finds it difficult to judge whether the measurements are in agreement with an assumed distribution, or if perhaps there is an unexpected deviation from it.

In addition to the analytic form of the probability density, Fig. 6.1 (left) also illustrates a typical situation with two sets of measurements (full and empty circles with uncertainties) generating the usual every-day questions. Are the two sets of data mutually consistent? How can they be compared at all? Is the arithmetic mean (ave), the median (med), or the maximum of the distribution (max) the best measure of consistency? Could one say that the two scattered data sets are "similar to" or "compatible with" each other? Is either of the sets consistent with the distribution that is expected (in this case, Maxwell's)? What can we do if a couple of data values appear to lie very far from the rest?

6.1.2 Moments of Distributions

The basic quantities that can be computed from the data set (the sample) $x = \{x_i\}_{i=0}^{n-1}$ gathered from a very large population, are the arithmetic mean $\bar{x}$ and the variance (dispersion) s_x^2 or the standard deviation s_x,

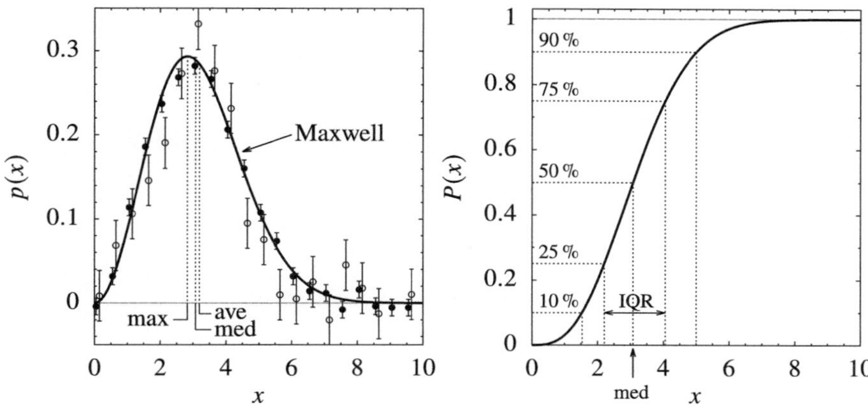

Fig. 6.1 Maxwell's distribution. [LEFT] Probability density (6.1) with indicated maximum (max), median (med), and average (ave). [RIGHT] Cumulative distribution function $P(x) = \int_0^x p(t)\,dt$. Shown are the first and last decile, the first and last quartile with the inter-quartile range (IQR), and the median

$$\overline{x} = \frac{1}{n} \sum_{i=0}^{n-1} x_i \,, \qquad s_x^2 = \frac{1}{n-1} \sum_{i=0}^{n-1} (x_i - \overline{x})^2 \,, \qquad s_x = \sqrt{s_x^2} \,. \tag{6.2}$$

The sample mean $\overline{x}$ is a non-biased estimate for the true average value of the whole population, $\langle x \rangle$, and the sample variance s_x^2 is a non-biased estimate for the true variance of the population, σ_x^2.

The average of an observable f with respect to the probability density p is

$$\langle f \rangle = \int f(x) p(x) \, dx \,. \tag{6.3}$$

If the data x_i come from the population in which the elements are distributed according to the probability density p, the measures (6.2) are the approximations of their first two central moments

$$\langle x \rangle = \lim_{n \to \infty} \overline{x} \,, \qquad \sigma_x^2 = \langle (x - \langle x \rangle)^2 \rangle = \lim_{n \to \infty} s_x^2 \,.$$

The convergence of the statistical measures $\overline{x}$ and s_x with increasing n depends on the properties of the data distribution. In certain cases these measures do not even exist. Assume that the data is spread throughout the whole real axis and that their probability density function p falls off as $\mathcal{O}(|x|^{-\rho})$, where $\rho > 1$. Then the average does not exist for $\rho \leq 2$ and the deviation does not exist for $\rho \leq 3$. If the first and the second central moments of a random variable are bounded and $n \gg 1$, the exact values can be related to the measured ones within some statistical confidence interval (see Sect. 6.1.3).

When we are dealing with histogrammed and non-normalized data, as in Fig. 6.1 (left), the estimates $\overline{x}$ and s_x^2 in the sense of the definition (6.3) can be computed by the weighted averages

$$\overline{x} = \frac{\sum_{i=0}^{n-1} x_i\, p(x_i)}{\sum_{i=0}^{n-1} p(x_i)}\ , \qquad s_x^2 = \frac{n}{n-1}\,\frac{\sum_{i=0}^{n-1} (x_i - \overline{x})^2\, p(x_i)}{\sum_{i=0}^{n-1} p(x_i)}\ . \tag{6.4}$$

6.1.3 Uncertainties of Moments of Distributions

In most cases we do not know the statistical nature of the individual measurements' uncertainties. We do assume that the uncertainties (the "errors") behave randomly, so that each measurement consists of a true value $\langle x \rangle$ and a random error Δx_i,

$$x_i = \langle x \rangle + \Delta x_i\ , \tag{6.5}$$

but we do not know the random process that generates the errors Δx_i. Physicists usually assume that Δx_i are normally distributed ("Gaussian"), so that they possess the probability density

$$p(x) = \frac{1}{\sqrt{2\pi\sigma^2}}\exp\left(-\frac{(x - \langle x \rangle)^2}{2\sigma^2}\right) \tag{6.6}$$

defined by its moments $\langle x \rangle$ and $\sigma^2 = \langle (x - \langle x \rangle)^2 \rangle$. Then, for $n \gg 1$, $\overline{x}$ is also normally distributed, and for the estimates (6.2) we have the uncertainties

$$\Delta \overline{x} = \frac{s_x}{\sqrt{n}}\ , \qquad \Delta s_x^2 = s_x^2\sqrt{\frac{2}{n-1}}\ , \qquad \Delta s_x = \frac{s_x}{\sqrt{2(n-1)}}\ . \tag{6.7}$$

We give the measurement results in the form $\langle x \rangle = \overline{x} \pm \Delta\overline{x}$, $\sigma^2 = s_x^2 \pm \Delta s_x^2$, $\sigma = s_x \pm \Delta s_x$, where the uncertainties correspond to confidence intervals (more will be said in Sect. 6.3 at Fig. 6.5). Note, however, that any sufficiently smooth observable f of the quantity x is distributed approximately normally, with the confidence interval

$$f(\langle x \rangle) \pm \left| f'(\langle x \rangle) \right| \Delta x\ .$$

Only the local behavior of f near $\langle x \rangle$ is relevant for this estimate. The case of two noisy and statistically independent quantities $x \pm \Delta x$ and $y \pm \Delta y$ can be described by a function of two variables $f(x, y)$, and the variance computed as

$$(\Delta f)^2 = \left(\frac{\partial f(x, y)}{\partial x}\right)^2 (\Delta x)^2 + \left(\frac{\partial f(x, y)}{\partial y}\right)^2 (\Delta y)^2\ .$$

The estimates (6.2) and their uncertainties (6.7) apply to n independent measurements with equal measurement errors, from which we wish to infer the unknown quantities $\langle x \rangle$ and σ_x. If the values x_i have different but known errors σ_i, the weighted averaging (6.4) results in the non-biased estimate for $\langle x \rangle$ and its uncertainty

$$\overline{x} = \frac{1}{w} \sum_{i=0}^{n-1} w_i x_i \,, \qquad w_i = \frac{1}{\sigma_i^2} \,, \qquad w = \sum_{i=0}^{n-1} w_i \,, \qquad \Delta \overline{x} = \frac{1}{\sqrt{w}} \,. \tag{6.8}$$

6.2 Robust Statistics

Sometimes the population sample contains values that clearly deviate from the majority of other values. They are known as *outliers* [2], and may represent measurement errors or even genuine data that indeed strongly deviate from the rest. The data from the Nimbus 7 satellite, for instance, indicated a hole in the ozone layer above Antarctica, but the researchers discarded them as instrument errors [3]. Robust statistics is a methodology to determine the parameters that characterize well the samples with relatively few outliers [4]. "Robustness" implies a small sensitivity of the estimated mean and variance to the inclusion or exclusion of outliers from the sample.

Example The classical case leading us to consider the use of robust methods are the 24 measurements of copper content in bread flour [5],

$$2.20\ 2.20\ 2.40\ 2.40\ 2.50\ 2.70\ 2.80\ 2.90\ 3.03\ 3.03\ 3.10\ \ 3.37$$
$$3.40\ 3.40\ 3.40\ 3.50\ 3.60\ 3.70\ 3.70\ 3.70\ 3.70\ 3.77\ 5.28\ 28.95\ \text{'}$$

shown in Fig. 6.2. The arithmetic mean of the whole sample is $\overline{x} = 4.28$ and the standard deviation is $s_x = 5.30$. If the value 28.95 is excluded from the sample, we get $\overline{x} = 3.21$ and $s_x = 0.69$. A single outlier therefore substantially modifies both $\overline{x}$ and s_x, so clearly these two quantities do not give us robust estimates for the properties of the complete population. ◁

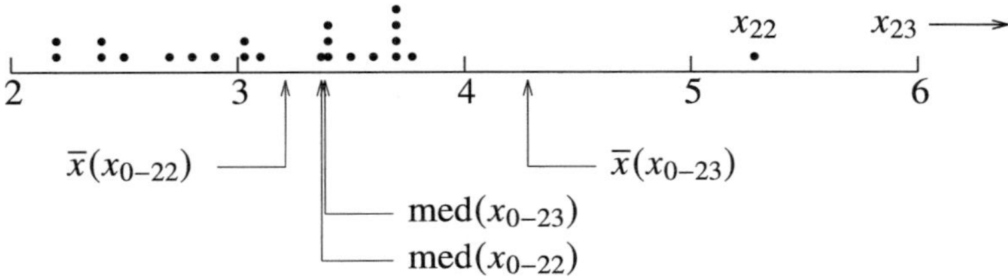

Fig. 6.2 The sample (24 values) of copper content in flour in μg/g. The value $x_{23} = 28.95$ (and potentially $x_{22} = 5.28$) represent outliers. The median (med) is much less sensitive to the exclusion of the extreme outlier than the arithmetic mean (ave)

A much more robust measure for the "center" of the sample $x = \{x_i\}_{i=0}^{n-1}$ is the *median*. It is defined as the value that splits the ordered sample ($x_i \leq x_{i+1}$) in two, such that its lower and upper portions contain half of the data each,

$$\mathrm{med}(x) = \begin{cases} x_{(n-1)/2} & ;\ n\ \text{odd}, \\ \frac{1}{2}(x_{n/2-1} + x_{n/2}) & ;\ n\ \text{even}. \end{cases}$$

If the probability density has the form $p(x - \mu)$, where $\mu = \bar{x} = \mathrm{med}(x)$, the variance of the sample median tends to $1/[4np(0)^2]$ if $p(0) > 0$ and $n \gg 1$. The spread of the data around the median can be estimated by the *median absolute deviation* (MAD)

$$\mathrm{MAD}(x) = \mathrm{med}(|x - \mathbf{1}_n \mathrm{med}(x)|), \tag{6.9}$$

where $\mathbf{1}_n = \{1, 1, \ldots, 1\}$ is the sequence of values 1 with length n. It makes sense to define a quantity that can be directly compared to the standard deviation,

$$\mathrm{MADN}(x) = 1.4826\,\mathrm{MAD}(x).$$

The factor 1.4826 is determined such that for the normal distribution $N(\mu, \sigma^2)$, $\mathrm{MADN} = \sigma$ holds. The values for all data in the flour sample are med $= 3.385$ and MADN $= 0.526$, while they are 3.37 and 0.504, respectively, if the outlier at 28.95 is excluded. Both values change only insignificantly (see Fig. 6.2).

It is clear from the uncertainties (6.7) that the arithmetic mean for $n \gg 1$ is distributed normally as $N(\mu, \sigma^2/n)$ if the measurement errors in (6.5) are distributed normally as $N(0, \sigma^2)$. With the same assumption for the errors, the median is distributed normally as $N(\mu, (\pi/2)\sigma^2/n)$, so the variance of the median for the normal distribution is $\pi/2 \approx 1.571$-times larger than the variance of the arithmetic mean. We say that the median in the case of the normal distribution has *low asymptotic efficiency*, that is, eff $\approx 1/1.571 \approx 64\,\%$.

Estimating the "center" and "spread" of data by using the median and MADN only weakly depends on the values in the distribution tails, but the computation of both estimates is numerically more costly than the computation of $\bar{x}$ and s_x^2, since sorting the data requires $\mathcal{O}(n \log n)$ operations and $\mathcal{O}(n)$ of memory.

6.2.1 Hunting for Outliers

Chasing outliers in one dimension is but a tiny subgroup of activities related to the search for subsets of data that do not behave according to expectations or deviate from the bulk of the data set. In literature, these activities are known under the names *anomaly detection, outlier detection, novelty detection,* or *exception mining* (see the review articles [6–9]), and represent one of the tools of *data mining.*

The school-book way of eliminating outliers is the "3σ-rule". It compels us to eliminate all data deviating from the sample mean $\bar{x}$ by more than $\pm 3s_x$, or to assign all such data the values $\bar{x} \pm 3s_x$. This method has numerous deficiencies. For example, since the interval $[\bar{x} - 3s_x, \bar{x} + 3s_x]$ for large n includes 99.7 % of data, this recipe forces us to needlessly remove approximately three data points from an impeccable normally distributed sample of size $n = 1000$, Moreover, the computation of the mean and the variance themselves is sensitive to outliers. Instead, it is preferable to use the criterion

$$x_i \text{ outlier} \iff \left| \frac{x_i - \text{med}(x)}{\text{MADN}(x)} \right| > 3.5 . \tag{6.10}$$

A simple way to visually identify the outlier candidates is to draw a *box diagram*. First we compute the median of the sample and the first and third quartiles Q_1 and Q_3: this splits the sample into four compartments with a fourth of the data in each of them. Then the *inter-quartile range* (IQR) is formed, along with the boundaries X_- and X_+ beyond which outliers can be expected,

$$\text{IQR} = Q_3 - Q_1 , \qquad X_- = Q_1 - \tfrac{3}{2}\text{IQR} , \qquad X_+ = Q_3 + \tfrac{3}{2}\text{IQR} .$$

This method identifies the values x_{22} and x_{23} from the flour sample as outliers (see Fig. 6.3). The interval $[X_-, X_+]$ for large n contains 99.3 % of all data, and for the normal distribution, the method is approximately equivalent to the "3σ-rule".

In general, all statistical methods for the identification of outliers in the sample x already assume a certain probability distribution for the whole population (in most cases, normal). The outliers are sought in the range

$$\{x : |x - \widehat{\mu}_n| > \widehat{\sigma}_n\, g(n, \alpha_n)\} , \tag{6.11}$$

where $\widehat{\mu}_n$ is an estimate for the population average, $\widehat{\sigma}_n$ is the estimate for its standard deviation, and the function $g(n, \alpha_n)$ depends on the sample size n. With certain approximations, all values x lying within (6.11) are considered as outliers with a probability of $1 - \alpha_n$. The criterion (6.10) is just Eq. (6.11) with $\widehat{\mu}_n = \text{med}(x)$, $\widehat{\sigma}_n = \text{MADN}(x)$, and $g(n, \alpha_n) = 3.5$, and is known as the Hampel identifier. A more precise definition of these quantities is given in [10] from which we take (with minor simplifications) the formulas valid for $n \geq 10$:

Fig. 6.3 The box diagram for a visual identification of outliers. The outliers can be expected outside of the interval $[X_-, X_+] = [Q_1 - \tfrac{3}{2}\text{IQR}, Q_3 + \tfrac{3}{2}\text{IQR}]$, where $\text{IQR} = Q_3 - Q_1$'

$$g(n, 0.05) \approx 2.9 + 12.5(n - 5.5)^{-0.572} , \qquad g(n, 0.01) \approx 3.8 + 25.3(n - 7.0)^{-0.601} .$$

The criterion (6.11) also performs well if $\widehat{\mu}_n$ is set to the center of the data, while $\widehat{\sigma}_n$ is set to the length of the *shortest half-sample* of the size $[n/2] + 1$ (Rousseeuw identifier). The older methods which are still in broad use—partly due to ignorance of better options—are described in [11, 12].

6.2.2 M-Estimates of Location

Eliminating outliers is not the best idea since this inevitably implies a certain loss of information [13]. In a softer approach one tries to preserve the outliers but take them into account with a smaller weight. One such approach is pursued by *M*-estimates [14]; these methods attempt to estimate the properties of data sets $\{x_i\}_{i=0}^{n-1}$ by characterizing the majority of the data in the sample in a stable manner, regardless of the presence or absence of outliers. In general, the *M-estimate of location* $\widehat{\mu}$, telling us something about the location of the bulk of the data, is defined as

$$\widehat{\mu} = \arg\min_{\mu} \sum_{i=0}^{n-1} \rho(x_i - \mu) . \tag{6.12}$$

This notation means that the parameter $\widehat{\mu}$ is equal to μ which minimizes the sum on the right-hand side. Here $\rho(x) = -\log p(x)$, $p(x) = P'(x)$, and $P(x)$ is the cumulative distribution function, according to which the error Δx_i in Eq. (6.5) is distributed. The definition (6.12) follows from the requirement that the estimate $\widehat{\mu}$ maximizes the *likelihood function*; for details see [14]. If ρ is differentiable and $\psi(x) = \rho'(x)$, we can differentiate (6.12) and obtain

$$0 = \sum_{i=0}^{n-1} \psi(x_i - \widehat{\mu}) . \tag{6.13}$$

Example In the measurement errors Δx_i are normally distributed, as it is usually assumed, then $p(x) = \exp(-x^2/2)/\sqrt{2\pi}$, $\rho(x) = x^2/2$, and $\psi(x) = x$, so Eqs. (6.12) and (6.13) simplify to

$$\widehat{\mu} = \arg\min_{\mu} \sum_{i=0}^{n-1} (x_i - \mu)^2 , \qquad \sum_{i=0}^{n-1} (x_i - \widehat{\mu}) = 0 .$$

Both equations relate well-known stories: the latter is obviously solved by the mean, $\widehat{\mu} = \overline{x}$, while the former tells us that this $\overline{x}$ also minimizes the sum of the squares of the differences between the measured data and $\overline{x}$. Similarly, one can show that the median minimizes the sum $\sum_{i=0}^{n-1} |x_i - \mu|$. The arithmetic mean and the median are just special cases of M-estimates. $\triangleleft$

M-estimates are devised as weighted averages and as such they are also computed [14]. For most realistic distributions $\psi(0) = 0$ and $\psi'(0)$ exists, so that ψ is approximately linear near the origin. Equation (6.13) can then be written as

$$\sum_{i=0}^{n-1} (x_i - \widehat{\mu}) \, W(x_i - \widehat{\mu}) = 0$$

or

$$\widehat{\mu} = \frac{\sum_{i=0}^{n-1} w_i x_i}{\sum_{i=0}^{n-1} w_i} \, , \qquad w_i = W(x_i - \widehat{\mu}) \, , \tag{6.14}$$

where W is the weight function. The essence of a good M-estimate is the choice of a weight function that sufficiently damps the outliers at large $|x|$. Many functions are in use, based on the presumed distributions of the measurement errors. In terms of robustness, good representatives are the Tukey function

$$\psi(t) = \begin{cases} t \, (1 - t^2/\tau^2)^2 & ; \ |t| \le \tau \\ 0 & ; \ |t| > \tau \end{cases} , \qquad W(t) = \begin{cases} \psi(t)/t & ; \ |t| \le \tau \\ 0 & ; \ |t| > \tau \end{cases} \tag{6.15}$$

and the Cauchy function

$$\psi(t) = \frac{t}{1 + \frac{1}{2} t^2} \, , \qquad W(t) = \psi(t)/t \, , \tag{6.16}$$

shown in Fig. 6.4. When τ is increased in the Tukey function, more distant outliers are admitted while the robustness of the M-estimate is reduced; on the other hand, its asymptotic efficiency increases (and vice-versa). The values $\tau = 3.44$, 3.88, and 4.68 correspond to asymptotic efficiencies eff $= 85\,\%$, $90\,\%$, and $95\,\%$ (asymptotic efficiency has been defined on page 318).

Equation (6.14) is implicit since the desired parameter $\widehat{\mu}$ occurs at left and in the weights w_i, but it can be solved iteratively [14]. The convergence is guaranteed by favorable properties of the function W. If we wish to compute the M-estimate of location in the case when the dispersion of the data $\widehat{\sigma}$ is already known, we define the weights

$$w_i^{(k)} = W\left(\frac{x_i - \widehat{\mu}^{(k)}}{\widehat{\sigma}}\right) \, , \qquad i = 0, 1, \ldots, n - 1 \, , \tag{6.17}$$

and rewrite Eq. (6.14) as an iteration

$$\widehat{\mu}^{(k+1)} = \frac{\sum_{i=0}^{n-1} w_i^{(k)} x_i}{\sum_{i=0}^{n-1} w_i^{(k)}} \, , \tag{6.18}$$

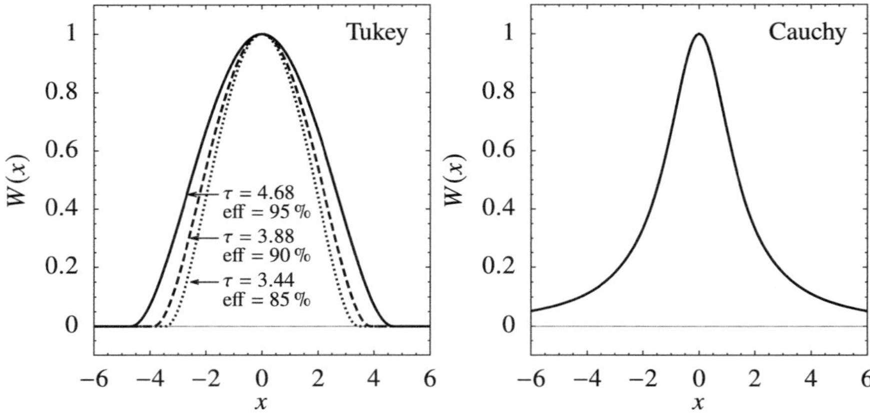

Fig. 6.4 Weight functions for M-estimates of location. [LEFT] Tukey function (6.15). The parameter τ determines the asymptotic efficiency. [RIGHT] Cauchy function

where k is the iteration index. The algorithm to compute the M-estimate of location is then

> **Input**: Values $x = \{x_i\}_{i=0}^{n-1}$ and the relative precision ε of the M-estimate
> $k = 0$; $\widehat{\mu}^{(k)} = \mathrm{med}(x)$; $\widehat{\sigma} = \mathrm{MADN}(x)$;
> **while** $(\,|\widehat{\mu}^{(k+1)} - \widehat{\mu}^{(k)}| > \varepsilon\,\widehat{\sigma}|\,)$ **do**
> | Compute the weights (6.17) with W from (6.15) or (6.16);
> | Compute the estimate (6.18);
> | $k = k + 1$;
> **end**
> **Output**: M-estimate of location, $\widehat{\mu}$

For the data in Fig. 6.2 and the Tukey function (parameter $\tau = 4.68$, asymptotic efficiency eff $= 95\,\%$) we get $\widehat{\mu} = 3.144$ for all data, $\widehat{\mu} = 3.143$ by omitting the extreme value 28.95, and $\widehat{\mu} = 3.127$ by omitting the value 5.28. (Check these numbers and monitor the usual ("naive") mean $\overline{x}$ during this procedure!)

6.2.3   *M-Estimates of Scale*

Robust estimates of data dispersion are called *M-estimates of scale*. Analogously to M-estimates of location they are introduced through maximum likelihood functions [14]. In contrast to Eq. (6.5) one imagines that the measurements x_i are distributed around zero with some error $\sigma\,\Delta x_i$ and that Δx_i are distributed according to some probability density p, while σ is an unknown parameter measuring the scale of this error. The M-estimate of scale $\widehat{\sigma}$ is defined as

$$\widehat{\sigma} = \arg\max_{\sigma} \frac{1}{\sigma^n} \prod_{i=0}^{n-1} p\left(\frac{x_i}{\sigma}\right) . \qquad (6.19)$$

The parameter $\widehat{\sigma}$ is equal to σ maximizing the product at the right, while $\rho(x) = x\psi(x)$ and $\psi(x) = -p'(x)/p(x)$. If the error is normally distributed (as (6.6) with $\langle x \rangle = 0$ and $\sigma = 1$), we have $\rho(x) = x^2$ and Eq. (6.19) is solved by the *root mean square* $\widehat{\sigma} = \text{RMS}(x) = (\langle x^2 \rangle)^{1/2}$. The measure of dispersion $\text{RMS}(x)$ is therefore just a special case of M-estimates of scale.

In practice, we usually do not know the true probability density p for the distribution of errors. Still, we would like to ensure that the estimate of dispersion is robust with respect to potential outliers in the sample. M-estimates of scale are devised analogously to M-estimates of location, as weighted averages,

$$\widehat{\sigma} = \sqrt{\frac{c}{n} \sum_{i=0}^{n-1} w_i\, x_i^2} , \qquad w_i = W\left(\frac{x_i}{\widehat{\sigma}}\right) , \qquad (6.20)$$

where the function W and the constant c depend on the model that we use to understand the source or the probability density of the errors Δx_i. If little is known about it, the following empirical formulas can be recommended [14]:

$$W(t) = \min\left\{1/t^2,\ t^4 - 3t^2 + 3\right\} , \qquad c = 2 . \qquad (6.21)$$

The parameter $\widehat{\sigma}$ in Eq. (6.20) again occurs at the left and in the weights w_i, so the equation is solved iteratively,

$$\widehat{\sigma}^{(k+1)} = \sqrt{\frac{c}{n} \sum_{i=0}^{n-1} w_i^{(k)} x_i^2} , \qquad w_i^{(k)} = W\left(\frac{x_i}{\widehat{\sigma}^{(k)}}\right) ,$$

where k is the iteration index. We start the iteration with $\widehat{\sigma}^{(0)} = \text{MADN}(x)$ and terminate it when

$$|\widehat{\sigma}^{(k+1)} - \widehat{\sigma}^{(k)}| < \varepsilon\widehat{\sigma}^{(k)} .$$

For normally distributed errors Δx_i, the estimate $\widehat{\sigma}$ becomes the standard deviation if we divide it by 1.56 [14].

6.3 Statistical Tests

In this Section we describe basic statistical tests useful in quantifying the properties of various data samples or comparing the samples to each other. Some of the questions posed in Fig. 6.1 (left) will find their answers here. These standard procedures are founded and explained in greater detail in all good statistics textbooks; see e.g. [15].

6.3.1   Computing the Confidence Interval
for the Population Mean

For a set of uncorrelated data $\{x_i\}_{i=0}^{n-1}$ we have used Eq. (6.2) to compute the mean $\overline{x}$ and variance s_x^2. Assume that the measured data are normally distributed. We are seeking a quantitative measure to determine the quality of the approximation $\overline{x}$ for the true average $\langle x \rangle$. To do this, we use $\overline{x}$ and s_x^2 to form the statistic

$$t = \frac{\overline{x} - \langle x \rangle}{\sqrt{s_x^2}} \sqrt{n} \, .$$

If x_i are normally distributed according to $N(\langle x \rangle, \sigma^2)$, the statistic t is distributed according to $S(t; n - 1)$ [16], where

$$S(t; \nu) = \frac{\mathrm{d}P}{\mathrm{d}t}(t; \nu) = \frac{1}{\sqrt{\nu}\, B\left(\frac{\nu}{2}, \frac{1}{2}\right)} \left(1 + \frac{t^2}{\nu}\right)^{-(\nu+1)/2} \tag{6.22}$$

is the Student's distribution for ν *degrees of freedom* and $B(a, b)$ is the usual beta function. The number ν is not necessarily integer and it depends on the number of measurements and of the type of the statistical test. Examples are given later on. The integral of the probability density (6.22) enables us to determine the intervals on which, with some probability $1 - \alpha$ chosen in advance, we may expect to find the values of t or $\langle x \rangle$, while with probability α the value $\langle x \rangle$ will be outside the established interval. The Student's probability density function is even in t, so we define symmetric limits t_- and t_+ such that

$$\int_{t_-}^{t_+} \frac{\mathrm{d}P}{\mathrm{d}t}\, \mathrm{d}t = 1 - \alpha \, , \qquad -t_- = t_+ = t_* \, . \tag{6.23}$$

This means that we may believe, at *confidence level* $1 - \alpha$, that $|t| \leq t_*$, or that we may expect $|t| > t_*$ with probability (*risk level*) α. The statistic t has therefore been bounded by the conditions

$$t_- \leq \frac{\overline{x} - \langle x \rangle}{s_x} \sqrt{n} \leq t_+ \, .$$

In other words, the true average of a large population, from which the sample $\{x_i\}_{i=0}^{n-1}$ has been acquired, can be estimated by $\langle x \rangle = \overline{x}$, and $\langle x \rangle$ will be found with probability $1 - \alpha$ on the confidence interval

$$\left[\overline{x} - \frac{t_* s_x}{\sqrt{n}} , \, \overline{x} + \frac{t_* s_x}{\sqrt{n}} \right] \, . \tag{6.24}$$

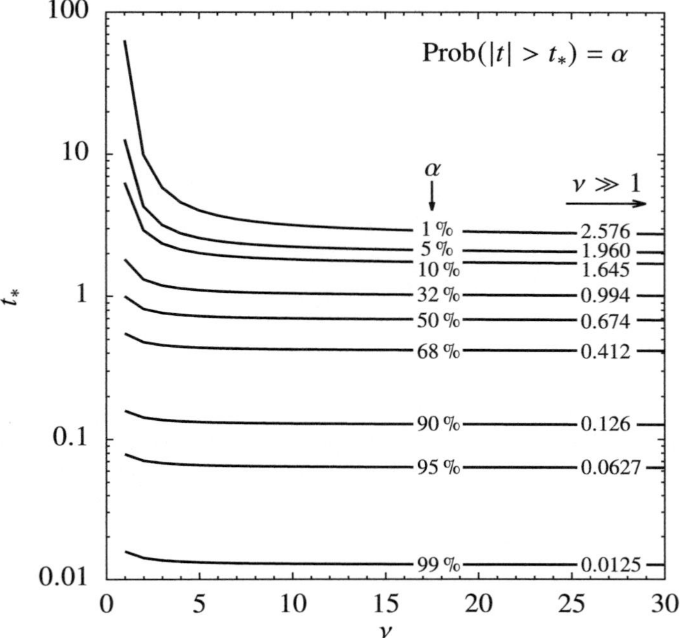

Fig. 6.5 Determining the confidence interval. The value t_* in (6.24) as a function of the number of degrees of freedom ν for various α. The values t_* in the limit $\nu \gg 1$ are shown at the extreme right. The risk level 31.7 % (confidence level 68.3 %) corresponds to $t_* \approx 1$ and the confidence interval (6.24) is $[\overline{x} - s_x/\sqrt{n},\ \overline{x} + s_x/\sqrt{n}]$, as told by (6.7)

Figure 6.5 shows t_* in dependence of the number of degrees of freedom ν, for various values of α.

6.3.2   Comparing the Means of Two Samples with Equal Variances

The Student's t-test frequently appears in a different disguise. Assume that we have two sets of data with equal variances, as would perhaps occur when using the same apparatus to measure a quantity which, we suspect, has changed between the two sets of measurements. From the samples

$$\boldsymbol{x} = \{x_i\}_{i=0}^{n_x-1}\,, \qquad \boldsymbol{y} = \{y_i\}_{i=0}^{n_y-1}\,,$$

we first form the means $\overline{x}$ and $\overline{y}$, then the statistics s_d and t_d,

$$s_d = \left[\frac{\sum_{i=0}^{n_x-1}(x_i - \overline{x})^2 + \sum_{i=0}^{n_y-1}(y_i - \overline{y})^2}{n_x + n_y - 2} \left(\frac{1}{n_x} + \frac{1}{n_y} \right) \right]^{1/2}, \qquad t_d = \frac{\overline{x} - \overline{y}}{s_d}.$$

With t_d we compute the value $0 \leq \alpha \leq 1$ by using the formula

$$\alpha = 1 - \int_{-|t_d|}^{|t_d|} \frac{dP}{dt}(t; \nu)\, dt = \frac{B_x(\nu/2, 1/2)}{B(\nu/2, 1/2)}, \qquad x = \frac{\nu}{\nu + t_d^2}, \qquad (6.25)$$

where $B(a, b)$ is the beta and $B_x(a, b)$ the incomplete beta function (accessible in standard numerical libraries) and we set $\nu = n_x + n_y - 2$. The value α ($0 \leq \alpha \leq 1$) measures the probability that for samples x and y, the situation $|t| \geq |t_d|$ occurs by chance. The smaller the value of α, the more statistically significant is the estimated difference of the means $\overline{x} - \overline{y}$. Computing α by Eq. (6.25) is simpler than the inverse task (6.23) of computing the integration limits t_* for a given α.

6.3.3 Comparing the Means of Two Samples with Different Variances

The Student's t-test can also be used on two samples with different variances. We face this situation when a quantity is measured by different devices (for example, one more and one less precise) and we wish to know whether the average of this quantity has changed between the two experiments. From the samples

$$x = \{x_i\}_{i=0}^{n_x-1}, \qquad y = \{y_i\}_{i=0}^{n_y-1},$$

we form the means $\overline{x}$ and $\overline{y}$, the variances s_x^2 and s_y^2, and the statistic

$$t = \frac{\overline{x} - \overline{y}}{\sqrt{s_x^2/n_x + s_y^2/n_y}}.$$

The statistic t is distributed approximately according to the Student's distribution (6.22) with the number of degrees of freedom

$$\nu = \left[\frac{s_x^2}{n_x} + \frac{s_y^2}{n_y} \right]^2 \left[\frac{1}{n_x - 1}\left(\frac{s_x^2}{n_x} \right)^2 + \frac{1}{n_y - 1}\left(\frac{s_y^2}{n_y} \right)^2 \right]^{-1}.$$

6.3.4 Determining the Confidence Interval for the Population Variance

This unit complements Sect. 6.3.1 where we have determined the confidence interval for the mean of presumably normally distributed uncorrelated data $\{x_i\}_{i=0}^{n-1}$. Again we compute the sample mean and variance, $\overline{x}$ and s_x^2, and form the statistic

$$\chi^2 = \frac{(n-1)s_x^2}{\Sigma^2} \, .$$

Here Σ^2 is the desired parameter for which the confidence interval should be found, and thus allow us to gauge the reliability of the dispersion estimate s_x^2. The statistic χ^2 is distributed with the density $dP/d\chi^2(\chi^2; n-1)$, where

$$\frac{dP}{d\chi^2}(\chi^2; v) = \frac{1}{2^{v/2}\,\Gamma(v/2)} \left(\chi^2\right)^{v/2-1} e^{-\chi^2/2}\,, \qquad \chi^2 \geq 0\,, \tag{6.26}$$

and where v is the number of degrees of freedom. (Note that for large v the χ^2 distribution resembles the normal distribution with the mean v and variance $2v$.) We define the limits χ_-^2 and χ_+^2 such that the integrals of the extreme lower and the extreme upper tail of the distribution are equal,

$$\int_0^{\chi_-^2} \frac{dP}{d\chi^2}\, d\chi^2 = \int_{\chi_+^2}^{\infty} \frac{dP}{d\chi^2}\, d\chi^2 = \frac{\alpha}{2}\, . \tag{6.27}$$

Here α is the chosen risk level quantifying the expectation that χ^2 will by chance lie outside of the confidence interval, that is, below χ_-^2 or above χ_+^2. (Compare Eqs. (6.27)–(6.23).) The variance estimate is $\Sigma^2 = s_x^2$ and we anticipate, at $1-\alpha$ confidence level, that the true variance σ_x^2 is within the confidence interval

$$\frac{(n-1)s_x^2}{\chi_+^2} \leq \sigma_x^2 \leq \frac{(n-1)s_x^2}{\chi_-^2}\, .$$

The values χ_-^2 and χ_+^2 satisfying (6.27) can be computed by standard numerical packages, while for everyday use they can be read off from Figs. 6.6 or 6.13.

Example Assume that we have used (6.2) to estimate the variance s_x^2 of a sample $\{x_i\}$ of $n = 11$ values. Let us determine the interval to which, at 90 % confidence level, the true variance σ_x^2 can be constrained. We use Fig. 6.6 to locate the curve corresponding to $v = n - 1 = 10$. The χ_-^2 and χ_+^2 are found as values of χ^2 on the abscissa where the horizontal lines at $\alpha = 0.05$ and $\alpha = 0.95$ intersect this curve: in this case ≈ 18.3 and ≈ 3.94. The risk level of 10 % thus corresponds to the confidence interval

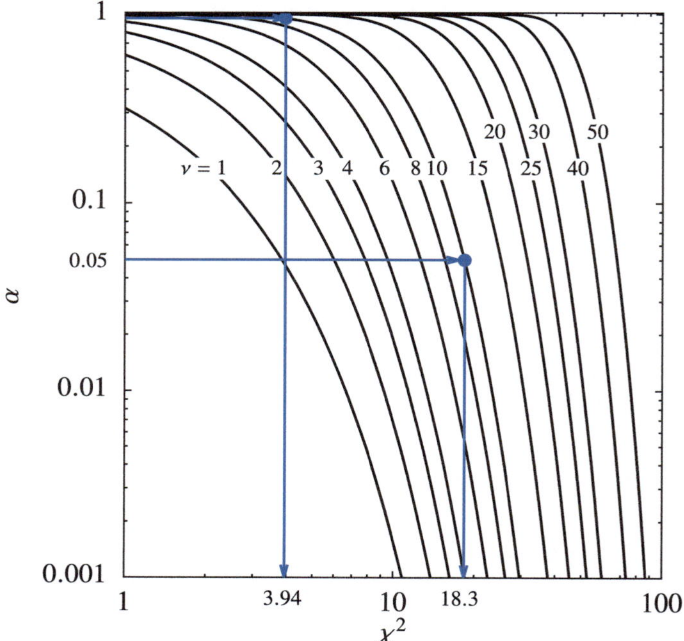

Fig. 6.6 The solution of $\mathrm{Prob}(\chi^2 > \chi_+^2) = \alpha$ and $\mathrm{Prob}(\chi^2 < \chi_-^2) = \alpha$. The symbols ● denote the points $(\chi_-^2, 1 - \alpha) = (3.94, 0.95)$ and $(\chi_+^2, \alpha) = (18.3, 0.05)$ in Example (6.28)

$$\frac{10s_x^2}{18.3} \leq \sigma_x^2 \leq \frac{10s_x^2}{3.94} \, . \tag{6.28}$$

See also Fig. 6.13.   ◁

6.3.5   *Comparing Two Sample Variances*

Another standard task is the comparison of variances of two data samples

$$\boldsymbol{x} = \{x_i\}_{i=0}^{n_x-1} \, , \qquad \boldsymbol{y} = \{y_i\}_{i=0}^{n_y-1} \, .$$

We compute their sample variances s_x^2 and s_y^2, and form the ratio

$$F = s_x^2/s_y^2 \, . \tag{6.29}$$

The larger variance should be put in the numerator and the smaller in the denominator. (If needed, rename the data $\boldsymbol{x} \leftrightarrow \boldsymbol{y}$.) The ratio F is distributed according to $\mathrm{d}P/\mathrm{d}F(n_x - 1, n_y - 1)$, where

$$\frac{\mathrm{d}P}{\mathrm{d}F}(\nu_1, \nu_2) = \left(\frac{\nu_1}{\nu_2}\right)^{\nu_1/2} \frac{\Gamma((\nu_1 + \nu_2)/2)}{\Gamma(\nu_1/2)\Gamma(\nu_2/2)} F^{\nu_1/2-1} \left(1 + \frac{\nu_1}{\nu_2} F\right)^{-(\nu_1+\nu_2)/2} .$$

Similar to the previous tests, we define the limit F_* for which

$$\int_0^{F_*} \frac{\mathrm{d}P}{\mathrm{d}F}(\nu_1, \nu_2)\,\mathrm{d}F = 1 - \frac{\alpha}{2}, \tag{6.30}$$

and where α is the confidence level or *significance*. The values F_* at $\alpha = 10\%$ in dependence of ν_1 and ν_2 are shown in Fig. 6.7. The assumption that the true variances σ_x^2 and σ_y^2 are equal should be rejected at confidence level α if the value F from Eq. (6.29) is larger than F_*.

Example (Taken from [16], p. 217.) We have two samples of sizes $n_x = 10$ and $n_y = 7$,

$$\begin{aligned}
x &= \{100, 101, 103, 98, 97, 98, 102, 101, 99, 101\}, \\
y &= \{97, 102, 103, 96, 100, 101, 100\},
\end{aligned} \tag{6.31}$$

with the means $\bar{x} = 100.0$ and $\bar{y} = 99.8$, and the sample variances $s_x^2 = 3.78$ and $s_y^2 = 6.50$, respectively. Is the deviation of the variance ratio $F = s_y^2/s_x^2 = 1.72$ from unity statistically significant, at significance $\alpha = 10\%$? From Fig. 6.7 we read off the value F_* along the curve $\nu_2 = \nu_x = n_x - 1 = 9$ at the abscissa $\nu_1 = \nu_y = n_y - 1 = 6$. We get $F_* = 3.37$. Since $F > F_*$ does *not* hold, the hypothesis $\sigma_x^2 = \sigma_y^2$ can not be discarded.								◁

6.3.6 Comparing Histogrammed Data to a Known Distribution

Assume that we can measure the distribution of a physical quantity with respect to x that may have values from $[-2.75, 2.75]$. Suppose that the experimental apparatus restricts us in a way that only allows us to acquire data on a narrower interval $[-1.35, 1.35]$. We display the results on this interval as a histogram with $\mathcal{N} = 27$ bins, as shown in Fig. 6.8. Is the measured distribution consistent with the theoretically predicted uniform distribution?

We use N_i to denote the number of measured events in the ith bin, while n_i denotes the corresponding theoretically anticipated number. We form the statistic

$$\chi^2 = \sum_{i=1}^{\mathcal{N}} \frac{(N_i - n_i)^2}{n_i},$$

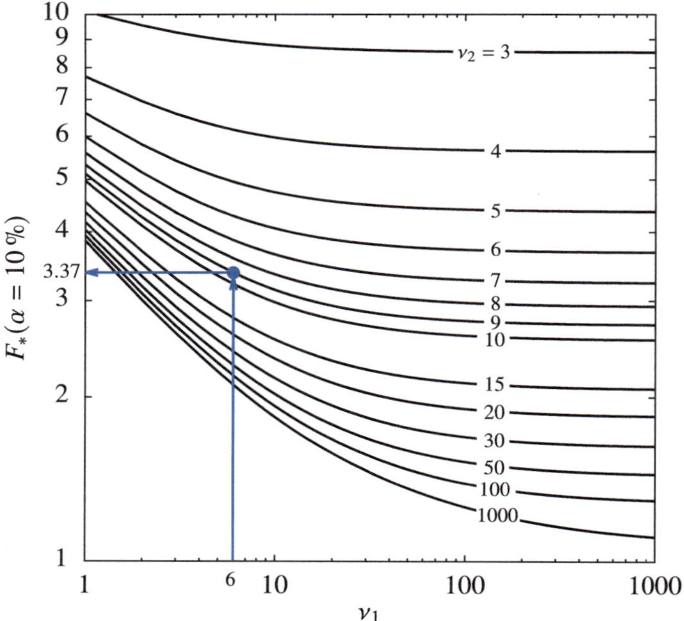

Fig. 6.7 The solutions of Eq. (6.30) for $\alpha = 10\,\%$ for various degrees of freedom v_1 and v_2. The symbol $\bullet$ denotes the value $F_* = 3.37$ ($v_1 = 6$, $v_2 = 9$) from Example (6.31)

where the sum runs over all bins $\mathcal{N}$. The number of all events is $N = \sum_{i=1}^{\mathcal{N}} N_i$. In the limit $N \gg 1$, χ^2 is distributed according to $\mathrm{d}P/\mathrm{d}\chi^2(\mathcal{N} - 1)$ (see Eq. (6.26)). We choose a risk level α at which the assumed distribution is discarded even though it is correct. With the chosen α we determine χ_+^2 in the cumulative distribution

$$\int_{\chi_+^2}^{\infty} \frac{\mathrm{d}P}{\mathrm{d}\chi^2}(\chi^2; \mathcal{N} - 1)\,\mathrm{d}\chi^2 = \alpha \,.$$

If the assumed distribution with the values n_i is consistent with the measured data N_i, we may expect $\chi^2 < \chi_+^2$ at confidence level $1 - \alpha$, while the outcome $\chi^2 > \chi_+^2$ leads us to the conclusion that the theoretical distribution is inconsistent with the experimental one. The *reduced values* $\chi_+^2/(\mathcal{N} - 1)$ for some typical α can be read off from Fig. 6.13.

Example Consider the data in Fig. 6.8. The total number of events in all bins is $N = 838$. Is the measured distribution consistent with the uniform distribution, according to which we would expect $n_i = N/\mathcal{N} = 31.04$? We compute $\chi^2 = 59.38$ or, for $\mathcal{N} - 1 = 26$ degrees of freedom, $\chi^2/(\mathcal{N} - 1) = 2.28$. At $\alpha = 5\,\%$ we read off $\chi_+^2/(\mathcal{N} - 1) \approx 1.5$ from Fig. 6.13, while for $\alpha = 1\,\%$ we find $\chi_+^2/(\mathcal{N} - 1) \approx 1.8$. In any case

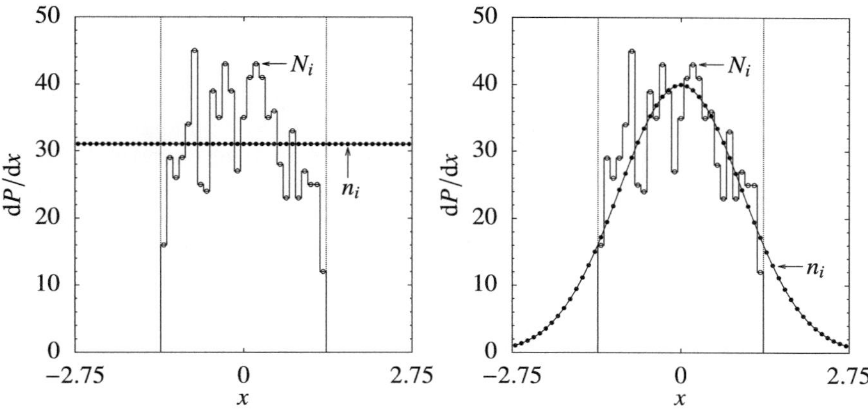

Fig. 6.8 Consistency of histogrammed data and a known distribution. [LEFT] Comparison to the uniform distribution. [RIGHT] Comparison to the normal distribution

we obtain $\chi^2 > \chi_+^2$ which points to the conclusion that the uniform distribution is inconsistent with the measured distribution.

We have narrowed the range in Fig. 6.8 on purpose: this is what one often finds in practice, forcing us to see the data as nothing but "a constant". What do we get with the values n_i corresponding to the normal distribution $N(0, 1)$? Now we compute $\chi^2 = 35.66$ or $\chi^2/(\mathcal{N} - 1) = 1.37$, which is less than $\chi_+^2(\alpha = 5\,\%)$ and $\chi_+^2(\alpha = 1\,\%)$. At confidence level of at least $99\,\%$ we may therefore claim that the measured and the assumed theoretical distribution are consistent.   ◁

The value of χ^2 depends on the number of bins $\mathcal{N}$. The classic choice [17] for the number of bins $\mathcal{N}_{\mathrm{opt}}$ that makes the χ^2-test optimally sensitive is $\mathcal{N}_{\mathrm{opt}} \approx 4\,N^{2/5}$ (at $\alpha = 1\,\%$), but see [18] for modern alternatives.

6.3.7 *Comparing Two Sets of Histogrammed Data*

The χ^2-test described above can also be used to question the mutual consistency of two sets of histogrammed data N_i and M_i in bins $i = 1, 2, \ldots, \mathcal{N}$. In this case, the statistic χ^2 should be formed as

$$\chi^2 = \sum_{i=1}^{\mathcal{N}} \frac{(\sqrt{M/N}\, N_i - \sqrt{N/M}\, M_i)^2}{N_i + M_i}\,, \qquad N = \sum_{i=1}^{\mathcal{N}} N_i\,, \qquad M = \sum_{i=1}^{\mathcal{N}} M_i\,.$$

In general $N \neq M$. The χ^2-test with the chosen risk level α is performed as before, with the number of degrees of freedom $\mathcal{N} - 1$. The values $\chi^2 > \chi_+^2$ indicate that the measured data N_i and M_i originate in different distributions.

6.3.8 Kolmogorov–Smirnov Test

If we wish to compare non-histogrammed data to a continuous distribution on the
domain that is identical to the data range, we resort to the Kolmogorov–Smirnov
(KS) test [19, 20]. Of course the data as well as the model distribution can be
grouped in bins and compared by using the previously described χ^2-test, but a direct
comparison has several advantages. The test statistic of the KS test is the maximum
distance between the cumulative distribution functions of the data and the assumed
continuous distribution.

We rearrange the sample $\{x_i\}_{i=0}^{n-1}$ so that $x_0 \le x_1 \le \cdots \le x_{n-1}$. For this sorted set
we define the *empirical distribution function*

$$\widetilde{P}_n(x) = \begin{cases} 0 & ;\, x < x_0 \,, \\ (i+1)/n & ;\, x_i \le x < x_{i+1} \,, \quad i = 0, 1, \ldots, n-2 \,, \\ 1 & ;\, x \ge x_{n-1} \,. \end{cases}$$

This is a monotonously increasing function that jumps upwards by $1/n$ at each point
x_i. For the data

$$x = \{0.22,\ -0.87,\ -2.39,\ -1.79,\ 0.37,\ -1.54,\ 1.28,\ -0.31,\ -0.74,\ 1.72, \\ 0.38,\ -0.17,\ -0.62,\ -1.10,\ 0.30,\ 0.15,\ 2.30,\ 0.19,\ -0.50,\ -0.09\}\,, \quad (6.32)$$

for instance, it is shown by the "staircase" graph in Fig. 6.9 (left).

In the basic version of the KS test the empirical distribution function $\widetilde{P}_n$ is compared to the model distribution function P, shown by the smooth curve in the same
Figure. The null hypothesis is $H_0 : \widetilde{P}_n(x) = P(x)$. The test statistic is the maximum
distance between these distributions,

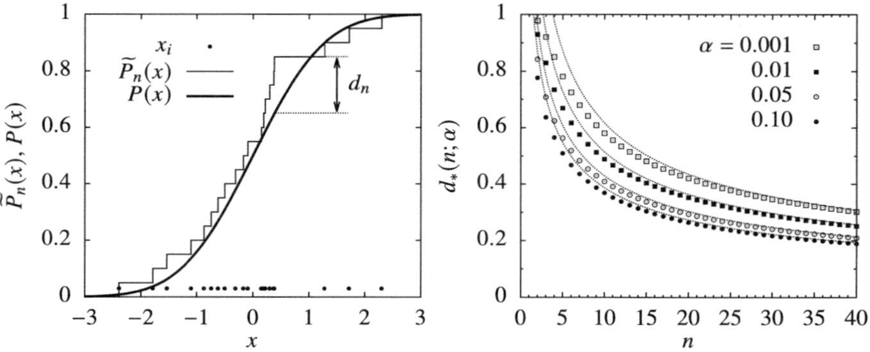

Fig. 6.9 Kolmogorov–Smirnov test. [LEFT] Data sample (•), the corresponding empirical distribution function $\widetilde{P}_n$, model distribution function P and the greatest distance between them, d_n. [RIGHT]
Critical values d_* as a function of n for various statistical significances α. The curves correspond
to the asymptotic formulas (6.34)

$$D_n = \sup_x \left| \widetilde{P}_n(x) - P(x) \right| = \max_{0 \le i \le n-1} \left\{ \frac{i+1}{n} - P(x_i),\ P(x_i) - \frac{i}{n} \right\} . \qquad (6.33)$$

The smaller the distance d_n (value of D_n), the better the agreement between $\widetilde{P}_n$ and P, pointing to the acceptance of H_0. If, however, the calculated d_n is larger than the critical value $d_*(n; \alpha) \equiv d(\alpha)/\sqrt{n}$ at chosen α, H_0 may be rejected. The critical values are tabulated; a method (and a MATLAB code) to compute them can be found in [21]; see also [22]. The symbols in Fig. 6.9 (right) represent $d_*(n; \alpha)$ for a few typical α. The Figure also contains the asymptotic curves

$$d_*(n; \alpha) \approx \frac{1}{\sqrt{n}} \sqrt{-0.5 \log(\alpha/2)} , \qquad n \gtrsim 35 . \qquad (6.34)$$

It is an astonishing property of the KS test that the distribution of D_n is known and, moreover, *does not depend on the distribution function P*. In the $n \gg 1$ limit it holds that

$$P_{\mathrm{KS}}(z) = \lim_{n \to \infty} \mathrm{Prob}\left(D_n \le \frac{z}{\sqrt{n}} \right) = 1 - 2 \sum_{k=1}^{\infty} (-1)^{k-1} e^{-2k^2 z^2} , \qquad (6.35)$$

which is suitable for the calculation at large z, while the form

$$P_{\mathrm{KS}}(z) = \frac{\sqrt{2\pi}}{z} \sum_{k=1}^{\infty} \exp\left(-\frac{(2k-1)^2 \pi^2}{8z^2} \right)$$

is preferable for small z. Either way

$$\mathrm{Prob}\left(D_n > z \right) = 1 - \mathrm{Prob}\left(D_n \le z \right) = 1 - P_{\mathrm{KS}}\left(\sqrt{n}\, z \right) .$$

Everyday work is made simpler by the approximation

$$\mathrm{Prob}\left(D_n > z \right) \approx 1 - P_{\mathrm{KS}}\left[\sqrt{n_{\mathrm{eff}}}\, z \right] ,$$

where

$$\sqrt{n_{\mathrm{eff}}} = \sqrt{n} + 0.12 + 0.11/\sqrt{n} .$$

This approximation works well already for $n \gtrsim 5$ and possesses the correct asymptotics. This can be exploited for the calculation of the critical $d_*(n; \alpha)$ for arbitrary, even small n, if tables are not at hand. Namely, we can insert $z = d_*$ in Eq. (6.35) to obtain

$$\alpha \approx 2 \sum_{k=1}^{\infty} (-1)^{k-1} e^{-2 n_{\mathrm{eff}} k^2 d_*^2} .$$

With given n we must then figure out d_* such that the sum on the right equals the chosen α on the left. When we succeed, we have found $d_* = d_*(n_{\text{eff}}; \alpha)$. For further details consult [23].

Example At $\alpha = 0.05$ we wish to test the null hypothesis that the sample (6.32) stems from a standard normal population with the corresponding distribution function P. By using (6.33) we obtain $d_n = 0.202$ indicated in Fig. 6.9. The sample size is $n = 20$, and the exact critical value is $d_*(20; 0.05) = 0.294$. By formula (6.34), not expected to apply at such low n, one gets a very similar value, $d_*(20; 0.05) = 0.304$. In either case $d_n < d_*$, the hypothesis can not be rejected. The data are consistent with the normal distribution. ◁

6.4 Correlation

Here we discuss measures of correlation between data sets. The correlation strength is measured by correlation coefficients, while we use suitable statistics to confirm whether the observed correlation is statistically significant or not.

6.4.1 Linear Correlation

The basic measure for the degree of correlation between two data sets is the linear correlation coefficient ρ. The corelatedness of two-dimensional data sometimes simply "pops out": typical images in the (x_i, y_i) plane for correlation coefficients $\rho \approx 1$ (almost complete positive correlation), $\rho \approx -1$ (almost complete anti-correlation), or $\rho \approx 0$ (uncorrelated data) are shown in Fig. 6.10.

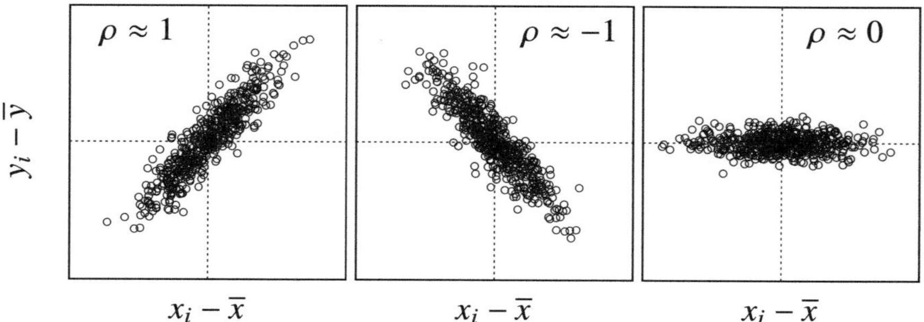

Fig. 6.10 Typical images of almost complete correlation ($\rho \approx 1$), almost complete anti-correlation ($\rho \approx -1$), and almost uncorrelated data ($\rho \approx 0$) in the (x_i, y_i) plane

The estimate for the linear correlation between the data $\{x_i\}_{i=0}^{n-1}$ and $\{y_i\}_{i=0}^{n-1}$ is

$$\widehat{\rho} = \frac{\sum_{i=0}^{n-1}(x_i - \overline{x})(y_i - \overline{y})}{\sqrt{\sum_{i=0}^{n-1}(x_i - \overline{x})^2}\,\sqrt{\sum_{i=0}^{n-1}(y_i - \overline{y})^2}}\,, \qquad -1 \le \widehat{\rho} \le 1\,. \tag{6.36}$$

The coefficient (6.36) is appropriate for the estimate of the degree of correlation once we have already confirmed that the correlation exists and that it has a certain statistical significance. Just like for the sample mean and variance, we can determine the confidence interval for the sample correlation coefficient $\widehat{\rho}$. In order to do this, we use the statistic

$$z = \frac{1}{2}\log\frac{1 + \widehat{\rho}}{1 - \widehat{\rho}} = \mathrm{Atanh}\,\widehat{\rho}$$

and assume that the measured data x_i and y_i are distributed according to the binormal (two-dimensional normal) distribution. For sample sizes n of more than a few times ten, the statistic z is then approximately normally distributed, with

$$N\left(\overline{z}, \sigma_z^2\right) = N\left(\frac{1}{2}\left[\log\frac{1 + \rho}{1 - \rho} + \frac{\rho}{n - 1}\right], \frac{1}{n - 3}\right)\,,$$

where ρ is the true correlation coefficient. The best estimate for the correlation coefficient is nothing but $\rho = \widehat{\rho}$, while the significance level at which we may claim that the measured $\widehat{\rho}$ differs from ρ, is given by

$$\alpha = 1 - \mathrm{erf}\left(\frac{|z - \overline{z}|\sqrt{n - 3}}{\sqrt{2}}\right)\,.$$

In determining whether the measurements of the quantities x_i and y_i from two time periods ("1" and "2") are correlated differently, we compare the correlation coefficients $\widehat{\rho}_1$ and $\widehat{\rho}_2$. The statistical significance of the difference between $\widehat{\rho}_1$ and $\widehat{\rho}_2$ is

$$\alpha = 1 - \mathrm{erf}\left(\frac{|z_1 - z_2|}{\sqrt{2}}\sqrt{\frac{(n_1 - 3)(n_2 - 3)}{n_1 + n_2 - 6}}\right)\,.$$

We may also ask the inverse question: to what confidence interval $[\rho_-, \rho_+]$ will the correlation coefficient be restricted at confidence level $1 - \alpha$? For $1 - \alpha \approx 96\,\%$, which is appropriate for everyday use, the values ρ_- and ρ_+ can be computed as

$$\rho_- = \tanh\left(\mathrm{Atanh}\,\widehat{\rho} - \frac{2}{\sqrt{n}}\right)\,, \qquad \rho_+ = \tanh\left(\mathrm{Atanh}\,\widehat{\rho} + \frac{2}{\sqrt{n}}\right)\,.$$

6.4.2 Non-parametric Correlation

The formula for the linear correlation coefficient (6.36) involves the sample means $\overline{x}$ and $\overline{y}$ which are strongly sensitive to the presence of outliers (see Sect. 6.2). We need a more robust tool. One option is to define the correlation by referring only to the positions (ranks) r_i and s_i that x_i and y_i occupy in the ordered samples x and y. When more (for instance, m) equal values share m positions, we assign to all these values an average rank that they would have, had they been only slightly different. In addition, we compute the mean ranks $\overline{r} = (\sum_{i=1}^{n} r_i)/n$ and $\overline{s} = (\sum_{i=1}^{n} s_i)/n$ (the ranks are indexed from 1 upwards).

Example Determine the mean rank of the sample $\{x_i\}_{i=0}^{7} = \{2, 3, 9, 3, 4, 9, 7, 3\}$! We first order the sample and obtain the array $\{x_0, x_1, x_3, x_7, x_4, x_6, x_2, x_5\}$. The values $x_1 = x_3 = x_7 = 3$ share the ranks 2 to 4, so their mean rank is $(2 + 3 + 4)/3 = 3$. The values $x_2 = x_5 = 9$ share the ranks 7 and 8, so their rank is 7.5. Finally we obtain $\{r_i\}_{i=1}^{8} = \{1, 3, 3, 3, 5, 6, 7.5, 7.5\}$ and $\overline{r} = 4.5$.   ◁

We use the ranks r_i and s_i as well as the mean ranks $\overline{r}$ and $\overline{s}$ to define the *rank correlation coefficient*

$$\widehat{\rho}_{\mathrm{p}} = \frac{\sum_{i=1}^{n} (r_i - \overline{r})(s_i - \overline{s})}{\sqrt{\sum_{i=1}^{n} (r_i - \overline{r})^2} \sqrt{\sum_{i=1}^{n} (s_i - \overline{s})^2}} . \tag{6.37}$$

In computing $\widehat{\rho}_{\mathrm{p}}$ we are only referring to the mutual placement of the data, hence this type of correlation estimate is called non-parametric. The distribution of the ranked data is uniform, and if the sample contains relatively few repeat values, the estimate (6.37) is much more robust than (6.36). The statistical significance of a measured correlation coefficient $\widehat{\rho}_{\mathrm{p}} \neq 0$ can be established by the t-test. We form the statistic

$$t_{\mathrm{p}} = \widehat{\rho}_{\mathrm{p}} \sqrt{\frac{n - 2}{1 - \widehat{\rho}_{\mathrm{p}}^2}} ,$$

which is distributed approximately according to the Student's distribution (6.22) with $n - 2$ degrees of freedom. The confidence level at which the assumption that the measured correlation coefficient $\widehat{\rho}_{\mathrm{p}}$ is equal to the true coefficient ρ_{p} can be discarded, is computed by Eq. (6.25) in which we use t_{p} instead of t_{d} and set $\nu = n - 2$.

6.5 Linear Regression

On an almost daily basis, we encounter the problem of fitting a smooth curve to a set of values

$$(x_i, y_i \pm \sigma_i) , \qquad i = 0, 1, \ldots, n - 1 . \tag{6.38}$$

The curve fitted to the data is specified by its analytic form that contains a certain set of model parameters. The fitting procedure as well as the final values of these parameters should somehow reflect the precision (the uncertainties) of the data. The search for the appropriate model curve is known as *regression*. According to the linear or non-linear dependence of the fitting curve with respect to the model parameters, we distinguish linear and non-linear regression.

6.5.1 *Fitting by a Polynomial, Straight Line, or Constant*

In linear regression, the fitting curve (model) linearly depends on the parameters characterizing it. We use this type of regression when we wish to find, for instance, a polynomial

$$f(x) = a_m x^m + a_{m-1} x^{m-1} + \cdots + a_1 x + a_0 \,,$$

that fits the data (6.38) as well as possible. One assumes that the points y_i are realizations of a random variable that is distributed around the unknown exact value with an absolute error σ_i. We wish to perform the fitting such that the sum of the squares of the errors $(y_i - f(x_i))^2$ with respect to the uncertainties σ_i will be as small as possible. If the fitting function is linear ($a_i = 0$ for $i \geq 2$), we are dealing with "straight-line" linear regression, while if it is a general polynomial, we are referring to polynomial (or general linear) regression, as the dependence on the model parameters is still linear.

The posed problem does not have a unique solution, since many measures of deviation can be devised. However, the *least-squares method* mentioned above is by far the most popular. In this case, we are seeking the parameters a_j ($j = 0, 1, \ldots, m$) that minimize the weighted sum of the squares of differences between the polynomial $f(x_i)$ and the values y_i,

$$\chi^2 = \sum_{i=0}^{n-1} \frac{(y_i - f(x_i))^2}{\sigma_i^2} \,. \tag{6.39}$$

The deviation is "punished" inversely proportional to the squared absolute error in y_i. The measure of deviation χ^2 is minimized by fulfilling the requirements

$$\frac{\partial \chi^2}{\partial a_j} = 0 \,, \qquad j = 0, 1, \ldots, m \,. \tag{6.40}$$

This translates into a system of linear equations for the parameters a_j,

$$\sum_{j=0}^{m} \Sigma_{kj} a_j = b_k \,, \qquad \Sigma_{kj} = \sum_{i=0}^{n-1} \frac{x_i^{k+j}}{\sigma_i^2} \,, \qquad b_k = \sum_{i=0}^{n-1} \frac{x_i^k y_i}{\sigma_i^2} \,.$$

Introducing the vectors of parameters $\boldsymbol{a} = (a_j)_{j=0}^{m}$ and coefficients $\boldsymbol{b} = (b_j)_{j=0}^{m}$, as well as the matrix $\Sigma = [\Sigma_{kj}]_{k,j=0}^{m}$, the system can be written in matrix form,

$$\Sigma \boldsymbol{a} = \boldsymbol{b} \quad \text{or} \quad \boldsymbol{a} = \Sigma^{-1} \boldsymbol{b} . \tag{6.41}$$

By using the Vandermonde matrix $V = [V_{ij}]$, where $V_{ij} = x_i^{j}$ ($0 \leq i \leq n-1, 0 \leq j \leq m$), and the weight matrix $D = \text{diag}(\sigma_i^{-2})_{i=0}^{n-1}$, the system becomes

$$V^{\mathrm{T}} D V \boldsymbol{a} = V^{\mathrm{T}} D \boldsymbol{y} , \qquad \Sigma = V^{\mathrm{T}} D V , \qquad \boldsymbol{b} = V^{\mathrm{T}} D \boldsymbol{y} ,$$

where $\boldsymbol{y} = (y_i)_{i=0}^{n-1}$. The parameters a_j depend on the values y_i which are statistically distributed around the exact values. Assuming that y_i are good approximations of the exact values, the variances and covariances of a_j can be estimated as

$$\text{var}(a_j) = \sigma^2(a_j) = \left(\Sigma^{-1}\right)_{jj} , \qquad \text{cov}(a_j, a_k) = \left(\Sigma^{-1}\right)_{jk} . \tag{6.42}$$

The matrix Σ tends to be very poorly conditioned, since its condition number grows exponentially with its dimension, $\kappa(\Sigma) = C \exp(\mathcal{O}(n))$ (Sect. 4.2.5). If we assume $\kappa(\Sigma) = C \exp(\alpha n)$, the desired relative precision of the coefficients a_j is ε, and the arithmetic precision is ε_{M}, we may include polynomials of degrees $n \leq (1/\alpha) \log(\varepsilon_{\mathrm{M}}/C\varepsilon)$. In double-precision floating-point arithmetic this typically means $n \leq 10$.

Regression with Orthogonal Polynomials

At least some stability problems can be avoided if the points $\{x_i\}_{i=0}^{n-1}$ coincide with the definition domain of some system of orthogonal polynomials. The most useful in regression are *orthogonal polynomials of a discrete variable*. These are polynomials $\{p_k(x) : k = \text{degree}(p_k)\}_{k=0}^{m}$ that are linearly independent and orthogonal on a discrete set of points $\{x_i\}$ in the sense

$$\sum_{i=0}^{n-1} \frac{1}{\sigma_i^2} p_k(x_i) p_l(x_i) = A_k \delta_{k,l}$$

with some weight $1/\sigma_i^2$. The model function fitted to the data can be designed as the linear combination $f(x) = \sum_{k=0}^{m} a_k p_k(x)$. We determine the expansion coefficients a_k by minimizing the measure of deviation (6.39). We obtain

$$\chi^2 = \sum_{k=0}^{m} a_k^2 A_k - 2 \sum_{k=0}^{m} a_k B_k + C , \qquad B_k = \sum_{i=0}^{n-1} \frac{p_k(x_i) y_i}{\sigma_i^2} , \qquad C = \sum_{i=0}^{n-1} \frac{y_i^2}{\sigma_i^2} .$$

From the condition for the minimum $\partial \chi^2/\partial a_k = 0$ we get $2a_k A_k - 2B_k = 0$ or

$$a_k = \frac{B_k}{A_k} \quad \text{and} \quad \chi^2 = C - \sum_{k=0}^{m} \frac{B_k^2}{A_k} = \min .$$

Example A most popular system of orthogonal polynomials of a discrete variable are the Chebyshev polynomials (5.40) that exhibit orthogonality by points (see Eq. (5.42)). A linear combination $f(x) = \sum_{k=0}^{m} a_k T_k(x)$ of these polynomials minimizes the measure of deviation $\chi^2 = \sum_{i=0}^{n-1}(y_i - f(x_i))^2$ with the coefficients

$$a_0 = \frac{1}{n} \sum_{i=0}^{n-1} y(x_i) , \qquad a_k = \frac{2}{n} \sum_{i=0}^{n-1} y(x_i) T_k(x_i) , \quad 1 \le k \le m ,$$

which is known as the Chebyshev approximation formula. (The problem is well defined for $m + 1 \le n$. When $m + 1 = n$ the function f interpolates the data y_i and therefore $\chi^2 = 0$.) Chebyshev polynomials have good approximation properties (Sect. 1.2.1) and can be efficiently computed by (5.41), and are therefore frequently used for the solution of least-squares problems [24].   ◁

Fitting by a Straight Line

In the case of linear regression we are seeking the straight line $f(x) = a_1 x + a_0$ that best fits the data (x_i, y_i) with known uncertainties (errors) σ_i. When χ^2 (see Eq. (6.39)) is minimized with respect to the parameters a_0 and a_1, we obtain an analytically solvable system of equations

$$a_0 S + a_1 S_x = S_y ,$$
$$a_0 S_x + a_1 S_{xx} = S_{xy} ,$$

where we have denoted

$$S = \sum_{i=0}^{n-1} \frac{1}{\sigma_i^2} , \quad S_x = \sum_{i=0}^{n-1} \frac{x_i}{\sigma_i^2} , \quad S_{xx} = \sum_{i=0}^{n-1} \frac{x_i^2}{\sigma_i^2} , \quad S_{xy} = \sum_{i=0}^{n-1} \frac{x_i y_i}{\sigma_i^2} , \quad S_y = \sum_{i=0}^{n-1} \frac{y_i}{\sigma_i^2} .$$

In this system, we immediately recognize the matrix Σ from (6.41) and its inverse,

$$\Sigma = \begin{pmatrix} S & S_x \\ S_x & S_{xx} \end{pmatrix} , \qquad \Sigma^{-1} = \frac{1}{\det(\Sigma)} \begin{pmatrix} S_{xx} & -S_x \\ -S_x & S \end{pmatrix} ,$$

where we have assumed $\det(\Sigma) = S_{xx} S - S_x^2 \ne 0$. From (6.41) it follows that the coefficients a_0 and a_1 minimizing the measure of deviation χ^2 are

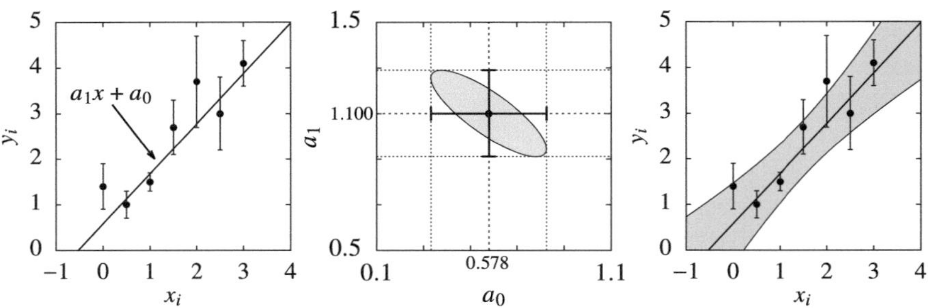

Fig. 6.11 Fitting a straight line to the data values $y_i = (1.4, 1.0, 1.5, 2.7, 3.7, 3.0, 4.1)$ with the uncertainties $\sigma_i = (0.5, 0.3, 0.2, 0.6, 1.0, 0.8, 0.5)$ at $x_i = (0, 0.5, 1, 1.5, 2, 2.5, 3)$ by using the least-squares method (example adapted from [16]). [LEFT] The function $f(x) = a_1 x + a_0$ minimizing χ^2 ($a_0 = 0.578$, $a_1 = 1.100$). [CENTER] The covariance ellipse centered at (a_0, a_1) with the uncertainties $\pm\sigma(a_0) = \pm 0.247$ and $\pm\sigma(a_1) = \pm 0.190$. Any (a_0, a_1) parameter pair within this ellipse corresponds to a straight line contained in the shaded area of the [RIGHT] panel. The straight line corresponding to the true a_0 and a_1 has a probability of $1 - e^{-1/2}$ of lying within this area

$$a_0 = \frac{S_{xx} S_y - S_x S_{xy}}{S_{xx} S - S_x^2}, \qquad a_1 = \frac{S S_{xy} - S_x S_y}{S_{xx} S - S_x^2}. \tag{6.43}$$

An example is shown in Fig. 6.11 (left).

If the errors are constant ($\sigma_i = \sigma$), the parameters of the straight line are $a_1 = \mathrm{cov}(x, y)/s_x^2$, $a_0 = \overline{y} - a_1 \overline{x}$, where $\overline{x} = (\sum_{i=0}^{n-1} x_i)/n$ and $\overline{y} = (\sum_{i=0}^{n-1} y_i)/n$ are the arithmetic means of the data, while

$$s_x^2 = \frac{1}{n-1} \sum_{i=0}^{n-1} (x_i - \overline{x})^2 \quad \text{and} \quad \mathrm{cov}(x, y) = \frac{1}{n-1} \sum_{i=0}^{n-1} (x_i - \overline{x})(y_i - \overline{y})$$

are their variance and covariance. We see that the straight line runs through the "center-of-mass" of the data $(\overline{x}, \overline{y})$.

From (6.42) we obtain the variances of a_0 and a_1 (the diagonal elements of Σ^{-1}) that do not depend on the position of the points along the y-axis:

$$\sigma^2(a_0) = \left(\Sigma^{-1}\right)_{00} = \frac{S_{xx}}{S_{xx} S - S_x^2}, \qquad \sigma^2(a_1) = \left(\Sigma^{-1}\right)_{11} = \frac{S}{S_{xx} S - S_x^2}. \tag{6.44}$$

The off-diagonal elements of the covariance matrix Σ^{-1} are

$$\mathrm{cov}(a_0, a_1) = \left(\Sigma^{-1}\right)_{01} = \left(\Sigma^{-1}\right)_{10} = \frac{-S_x}{S_{xx} S - S_x^2},$$

thus the coefficient of linear correlation between a_0 and a_1 is

$$\rho = \frac{\text{cov}(a_0, a_1)}{\sigma(a_0)\sigma(a_1)} \,.$$

The parameters a_0 and a_1, their uncertainties $\sigma(a_0) \equiv \sigma_0$ and $\sigma(a_1) \equiv \sigma_1$, and the coefficient ρ define the *covariance ellipse* centered at (a_0, a_1) with the semi-axes r_0 and r_1, rotated by α in the (a_0, a_1)-plane [16]. The parameters are

$$\tan 2\alpha = 2\rho\sigma_0\sigma_1 \,/\, (\sigma_0^2 - \sigma_1^2) \,,$$
$$r_0^2 = \sigma_0^2\sigma_1^2(1 - \rho^2) \,\Big/\, \left[\sigma_1^2 \cos^2\alpha - \rho\sigma_0\sigma_1 \sin 2\alpha + \sigma_0^2 \sin^2\alpha\right] \,, \qquad (6.45)$$
$$r_1^2 = \sigma_0^2\sigma_1^2(1 - \rho^2) \,\Big/\, \left[\sigma_1^2 \sin^2\alpha + \rho\sigma_0\sigma_1 \sin 2\alpha + \sigma_0^2 \cos^2\alpha\right] \,.$$

An example is shown in Fig. 6.11 (center). All points on the ellipse are equally probable and represent the "1σ" confidence interval for the parameters a_0 and a_1. This interval corresponds to an infinite set of possible straight lines. For the indicated points on the ellipse they are shown in Fig. 6.11 (right).

How do the uncertainties of the model parameters, $\sigma^2(a_0)$ and $\sigma^2(a_1)$, change when the number of points n is increased? Assume that the points $\{x_i\}_{i=0}^{n-1}$ on some interval $[\alpha, \beta]$ are distributed uniformly, so $x_i = \alpha + i\Delta x$ with $\Delta x = (\beta - \alpha)/(n - 1)$, and that the measurement error is $\sigma_i = \sigma$. We compute $S = n/\sigma^2$, $S_x = n(\alpha + \beta)/(2\sigma^2)$ and $S_{xx} = n[(\alpha^2 + \alpha\beta + \beta^2)(2n - 1) - 3\alpha\beta]/[6(n - 1)\sigma^2]$. From Eqs. (6.44) we read off the asymptotic behavior in the limit $n \to \infty$,

$$\sigma^2(a_0) \sim \frac{4(\alpha^2 + \alpha\beta + \beta^2)\sigma^2}{n(\alpha - \beta)^2} + \mathcal{O}\left(\frac{1}{n^2}\right) \,, \qquad \sigma^2(a_1) \sim \frac{12\sigma^2}{n(\alpha - \beta)^2} + \mathcal{O}\left(\frac{1}{n^2}\right) \,.$$

With increasing n, the regression coefficients a_0 and a_1 indeed gain in precision at the same rate as any statistical averages, $\sigma(a_0), \sigma(a_1) \sim \mathcal{O}(n^{-1/2})$.

Unknown errors If the errors σ_i of individual y_i are unknown, the uncertainties of a_0 and a_1 can be estimated by using the procedure fully elaborated in [25]. We first set $\sigma_i = 1$ for $\forall i$ and compute a_0 and a_1 and their variances $\sigma^2(a_0)$ and $\sigma^2(a_1)$ by using Eqs. (6.44). We use these a_0 and a_1 to compute the value of χ^2 from (6.39), and multiply both variances by $\chi^2/(n - 2)$. The estimates for the uncertainties of the parameters a_0 and a_1 are therefore rescaled as

$$\sigma(a_0) \to \sigma(a_0)\sqrt{\chi^2/(n - 2)} \,, \qquad \sigma(a_1) \to \sigma(a_1)\sqrt{\chi^2/(n - 2)} \,.$$

Fitting a Straight Line to Data with Errors in Both Coordinates

In straight-line regression with the function $f(x) = a_1 x + a_0$, where the errors occur in both variables (y_i as well as x_i), we try to minimize

$$\chi^2 = \sum_{i=0}^{n-1} \frac{(y_i - a_1 x_i - a_0)^2}{a_1^2 \sigma_{xi}^2 + \sigma_{yi}^2} \, .$$

To determine a_0 and a_1 we have to fulfill $\partial \chi^2 / \partial a_0 = 0$ and $\partial \chi^2 / \partial a_1 = 0$, but a_1 appears non-linearly, so the problem is non-trivial; for a review, see [26]. Reference [27] provides a reliable algorithm to compute the parameters a_0 and a_1 and their uncertainties, $\sigma(a_0)$ and $\sigma(a_1)$. It typically converges in ≈ 10 iterations. The code is available at the book's web site.

Fitting by a Constant

In zeroth-order regression a constant function is fitted to the values y_i. We minimize (6.39) with $f(x_i) = a$. From the condition $d\chi^2 / da = 0$ we get

$$a = \frac{1}{S} \sum_{i=0}^{n-1} \frac{y_i}{\sigma_i^2} \, , \qquad S = \sum_{i=0}^{n-1} \frac{1}{\sigma_i^2} \, , \qquad \sigma(a) = \frac{1}{\sqrt{S}} \, , \tag{6.46}$$

which is precisely the weighted average (6.8). We use this method instead of the arithmetic mean whenever the measured values of the same quantity have different errors. For $n \to \infty$ the uncertainty of a scales as $\sigma(a) \sim \mathcal{O}(n^{-1/2})$.

After obtaining a, we may compute $\chi^2 = \sum_{i=0}^{n-1} (y_i - a)^2 / \sigma_i^2$ to ascertain whether the premise of the normally distributed errors σ_i is justified or not. At the chosen risk level α (e.g. $\alpha = 5\,\%$) we determine χ_+^2 from the equation

$$\int_{\chi_+^2}^{\infty} \frac{dP}{d\chi^2}(\chi^2; n - 1)\, d\chi^2 = \alpha \, ,$$

where $dP/d\chi^2$ is given in Eq. (6.26), and compare χ_+^2 to the χ^2 computed from the data. If $\chi^2 > \chi_+^2$, the premise is questionable; in the opposite case, we may fit the constant a and its uncertainty is considered as consistent with the data.

Example Numerous dangers lurk behind all this simplicity. Figure 6.12 (left) shows $n = 10$ data points (x_i, y_i) to which we fit a constant a. Following the procedure described above we get $a = 2.73 \pm 0.10$ and $\chi^2 / (n - 1) = 36.5$. At $\alpha = 5\,\%$ for $v = n - 1 = 9$ we read off $\chi_+^2(\alpha = 5\,\%)/9 = 1.88$ from Fig. 6.13. Since $\chi^2 / (n - 1) > \chi_+^2 / (n - 1)$, the assumption about the normal distribution of errors should be discarded. (In other words, with a probability much greater than α, the constant probability density is inconsistent with the data sample.) But if we omit the outliers y_3 and y_6, we get $a = 1.71 \pm 0.12$ and $\chi^2 / 7 = 1.10$, while $\chi_+^2(\alpha = 5\,\%)/7 = 2.01$. Now the fit is consistent with the data. Here we again run into the problems of robustness. More on this will be told in Sect. 6.5.3.

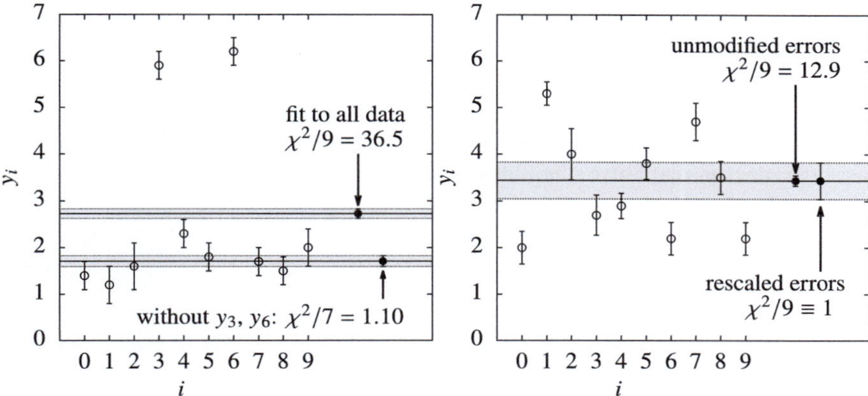

Fig. 6.12 Fitting a constant to the data $(y_i \pm \sigma_i)_{i=0}^9$ (adapted from [16]). [LEFT] The weighted average in the presence of two outliers and without them. [RIGHT] The weighted average of the data with an unexplained systematic error. The fit procedure yields the parameter a with a very small uncertainty inconsistent with the actual scatter of data. Rescaling of the errors according to (6.47) gives a more sensible result

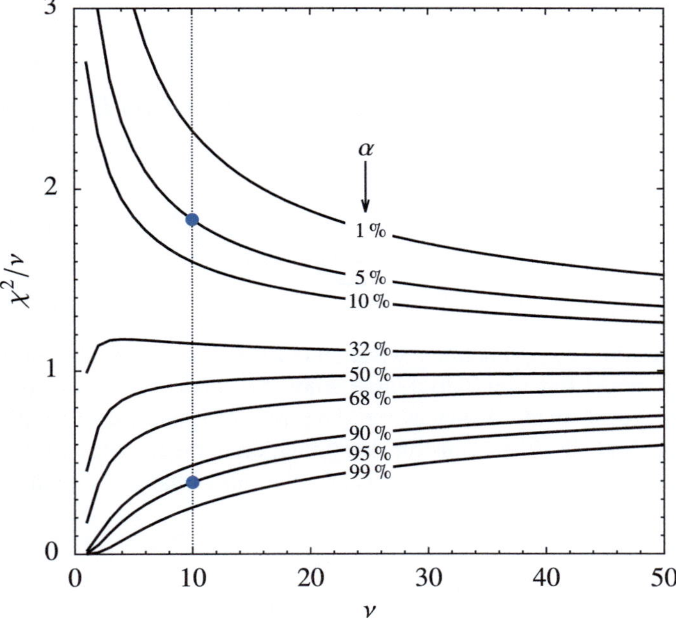

Fig. 6.13 The reduced value χ^2/ν used in the χ^2-test as a function of the number of degrees of freedom, at some typical risk levels α. The symbols ● denote the points $(\nu, \chi_-^2/\nu) = (10, 0.394)$ and $(\nu, \chi_+^2/\nu) = (10, 1.83)$ from Example (6.28)

Another pitfall is illustrated by Fig. 6.12 (right). Applying the method from Sect. 6.3.1 we determine that some data points fall outside of the confidence interval for the sample mean. By using the χ^2-test we obtain $\chi^2/9 = 12.9$ and again $\chi_+^2(\alpha = 5\,\%)/9 = 1.88$. But individual outliers should not be blamed for the large value of χ^2, as obviously the data have an underestimated systematic error. In such cases the measurement error should be *rescaled* [28]

$$\sigma_i' = \sigma_i \sqrt{\frac{\chi^2}{n-1}}\,, \qquad i = 0, 1, \ldots, n-1\,. \tag{6.47}$$

We still form the weighted average for the computation of the fit parameter a by using Eq. (6.46), but its uncertainty becomes

$$\sigma'(a) = \sqrt{\frac{\chi^2}{n-1}} \left(\sum_{i=0}^{n-1} \frac{1}{\sigma_i^2}\right)^{-1/2}\,.$$

Thus we get a more sensible result corresponding to $\chi^2/(n-1) = 1$ (this example is shown in Fig. 6.12 (right)). $\lhd$

6.5.2   Generalized Linear Regression by Using SVD

In generalized linear regression

$$f(x) = \sum_{j=0}^{m-1} a_j \phi_j(x)\,,$$

where ϕ_j are some basis functions, we attempt to use a small number of parameters a_j to fit the model function f to a large amount of data. Such problems are over-determined. Yet in many instances the data are not rich enough to allow for an unambiguous and physically sensible determination of the linear combination of the basis functions: often we can obtain several solutions of the problem that minimize

$$\chi^2 = \sum_{i=0}^{n-1} \left[\frac{y_i - \sum_{j=0}^{m-1} a_j \phi_j(x_i)}{\sigma_i}\right]^2$$

almost equally well. (From the strict mathematical point of view, the solution of the over-determined system by the least-squares methods is, of course, unique.)

We can obtain a much better handle over the meaningfulness of the parameters $a = (a_0, a_1, \ldots, a_{m-1})^{\mathrm{T}}$ by the use of singular value decomposition (SVD, see Sect. 4.5.2 and 4.5.3). Solving the least-squares problem by the SVD is appealing in the case of over-determined systems where we are trying to describe the measured data with a set of model parameters that is unnecessarily large. The singular values obtained by the SVD help us to eliminate the superfluous combinations of the basis functions. We denote

$$A_{ij} = \frac{\phi_j(x_i)}{\sigma_i} , \qquad b_i = \frac{y_i}{\sigma_i} ,$$

and perform the "thin" SVD of the matrix $A = U \Lambda V^{\mathrm{T}} \in \mathbb{R}^{n \times m}$ (page 194) where $n \geq m$. The matrix $U = (\boldsymbol{u}_0, \boldsymbol{u}_1, \ldots, \boldsymbol{u}_{m-1})$ has columns $\boldsymbol{u}_i$ of dimension n, the matrix $V = (\boldsymbol{v}_0, \boldsymbol{v}_1, \ldots, \boldsymbol{v}_{m-1})$ has columns $\boldsymbol{v}_i$ of dimension m, and the diagonal matrix $\Lambda = \mathrm{diag}(\lambda_0, \lambda_1, \ldots, \lambda_{m-1})$ contains the singular values λ_i. We obtain the vector of parameters $\boldsymbol{a}$ by summing

$$\boldsymbol{a} = \sum_{i=0}^{m-1} \frac{\boldsymbol{u}_i^{\mathrm{T}} \boldsymbol{b}}{\lambda_i} \boldsymbol{v}_i .$$

The covariance matrix of the parameters $\boldsymbol{a}$ is $\mathrm{cov}(a_j, a_k) = \sum_{i=0}^{m-1} V_{ji} V_{ki} / \lambda_i^2$ and its diagonal elements represent the individual variances

$$\mathrm{var}(a_j) = \sigma^2(a_j) = \sum_{i=0}^{m-1} \frac{V_{ji}^2}{\lambda_i^2} . \tag{6.48}$$

We should pay attention to all singular values λ_i for which the ratio $\lambda_i / \lambda_{\max}$ is smaller than $\approx n \varepsilon_{\mathrm{M}}$. Such values increase the error (6.48) and indicate that the inclusion of new model parameters is meaningless. Moreover, they do not contribute significantly to the minimization of χ^2, so we exclude them. We do this by setting $1/\lambda_i = 0$ (for a detailed explanation, see the comment to the `Fitsvd` algorithm in [25]). We may exclude also those singular values for which the ratio $\lambda_i / \lambda_{\max}$ is *larger* than $\approx n \varepsilon_{\mathrm{M}}$, until χ^2 starts to increase visibly.

6.5.3 Robust Methods for One-Dimensional Regression

Like all estimates of location and dispersion, regression methods are sensitive to outliers (Sect. 6.2.1). The straight line LS in Fig. 6.14 (left) corresponds to the standard regression on the data $\{(x_i, y_i)\}_{i=0}^{15}$ with unknown errors: here we minimize the sum of the squares of the residuals $r_i = y_i - (a_1 x_i + a_0)$, so the straight line LS fails to describe the bulk of the data because of the outliers at x_0, x_1, and x_3. If we remove these three outliers, we obtain the straight line denoted by LS_{-013}. Robust regression methods [29] yield a good description of the majority of the data without the need to remove individual outliers by hand.

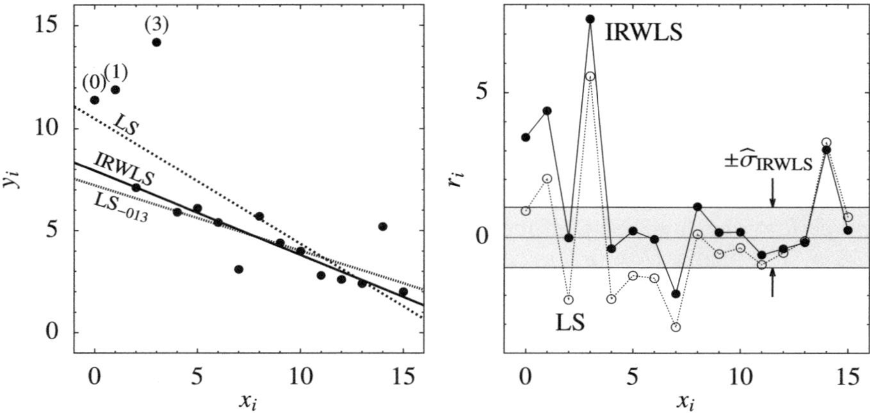

Fig. 6.14 Robust linear regression. (Example adapted from [14].) [LEFT] The regression straight line by the method of least squares (LS), by omitting three outliers (LS$_{-013}$), and by the iterative reweighting method (IRWLS). [RIGHT] The residuals r_i and the estimate of dispersion by the IRWLS method

The literature on robust regression techniques is abundant: for the identification of outliers (*regression diagnostics*) see [30]; for a review of methods see [31]. Numerous methods exist, all having some advantages and deficiencies; many of them are awkward to implement or entail high numerical costs. One of such methods is the regression in which it is not $\sum_i r_i^2$ that is minimized, but rather $\sum_i |r_i|$—an L_1-type estimator; see e.g. [32, 33]. Here we describe two procedures with proven good properties for everyday use.

IRWLS The *iterative reweighted least squares method* (IRWLS) closely follows the logic of M-estimates of location (Sect. 6.2.2 and algorithm on page 322). The iteration to the final values of the parameters a_0 and a_1 of the regression straight line typically converges in ≈ 30 steps. The determination of their uncertainties, the estimate for the dispersion, and other details are discussed in [14], p. 105. An example is shown in Fig. 6.14 (left, IRWLS line). Figure 6.14 (right) shows the residuals r_i in the LS and IRWLS methods. A sample IRWLS code is available at the book's web site.

LMS The second method resembles standard linear regression, except that the median of the squares of the residuals $r_i = y_i - (a_1 x_i + a_0)$ is minimized, hence its name, *least median of squares* (LMS). We seek a_0 and a_1 such that

$$\mathrm{med}(y_i - a_1 x_i - a_0)^2 = \min . \tag{6.49}$$

The LMS method [34] is very robust and behaves well even in the rare circumstances in which IRWLS fails. An example is shown in Fig. 6.15 (left). We have a sample of $n_1 = 30$ data $y_i = a x_i + b + u_i$, where $a = 1$, $b = 2$, $x_i \sim U(1, 4)$, and $u_i \sim N(0, 0.2)$, and a set of $n_2 = 20$ points assumed to be outliers, $x_i \sim N(7, 0.5)$, $y_i \sim$

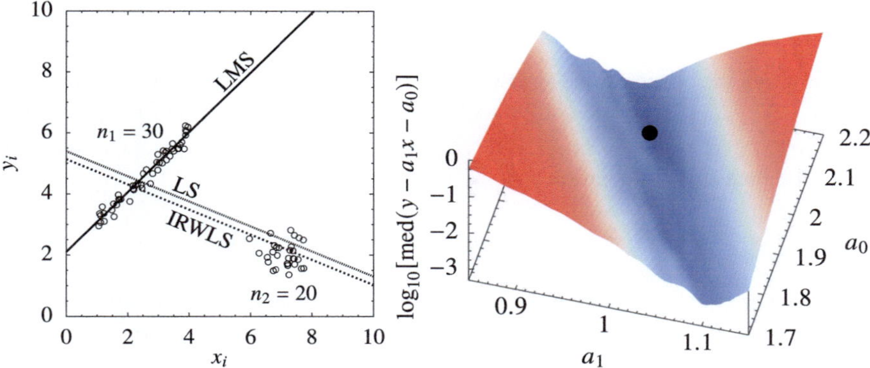

Fig. 6.15 Robust linear regression on the data with many outliers. [LEFT] Only the LMS method correctly describes the majority of the data. [RIGHT] The function being minimized in the LMS method. The symbol • denotes the global minimum

$N(2, 0.5)$. We would like to fit a straight line to the data such that the result will be oblivious to the outlier portion of this compound set. Neither the LS nor the IRWLS method yield meaningful results: both simply run the line through both data portions. In contrast, the LMS line fits the non-outlier group well.

The main nuisance of the LMS method is precisely the numerical minimization (6.49). The function we wish to minimize with respect to the parameters a_j has $\mathcal{O}(n^{m+1})$ local minima, where n is the number of data points (x_i, y_i) and m is the degree of the regression polynomial. In the example from the Figure we have $n = n_1 + n_2 = 50$ and $m + 1 = 2$ (straight line), so there are ≈ 2500 local minima, among which the global minimum needs to be located, as shown in Fig. 6.15 (right). This is best accomplished—there are dangers since the function is continuous, but not continuously differentiable—by forming the function

$$M(a_1) = \min_{a_0} \left\{ \mathrm{med} \, (y_i - a_1 x_i - a_0)^2 \right\} ,$$

where for each a_1 we determine a_0, and then minimize $M(a_1)$ with respect to a_1. This implementation becomes very costly at large n. Fast computations of the regression parameters by LMS methods are non-trivial: see [35, 36] as well as a sample LMS code provided at the book's web site.

6.6 Non-linear Regression

In non-linear regression the dependence of the model function on regression parameters is non-linear, for example

$$f(x; \boldsymbol{a}) = a_0 + a_1 \, \mathrm{e}^{a_2 x} + a_3 \sin(x + a_4) \,.$$

If we arrange the observations y_i at x_i into vectors $\boldsymbol{x} = (x_0, x_1, \ldots, x_{n-1})^{\mathrm{T}}$ and $\boldsymbol{y} = (y_0, y_1, \ldots, y_{n-1})^{\mathrm{T}}$, and the components of the regression function into $\boldsymbol{f}(\boldsymbol{x}; \boldsymbol{a}) = \big(f(x_0; \boldsymbol{a}), f(x_1; \boldsymbol{a}), \ldots, f(x_{n-1}; \boldsymbol{a})\big)^{\mathrm{T}}$, where $\boldsymbol{a} = (a_0, a_1, \ldots, a_{p-1})^{\mathrm{T}}$, we can define the measure of deviation as

$$\chi^2 = \big(\boldsymbol{y} - \boldsymbol{f}(\boldsymbol{x}; \boldsymbol{a})\big)^{\mathrm{T}} \Sigma_{\boldsymbol{y}}^{-1} \big(\boldsymbol{y} - \boldsymbol{f}(\boldsymbol{x}; \boldsymbol{a})\big) \,. \tag{6.50}$$

If the measurement errors are uncorrelated, the covariance matrix is diagonal, $\Sigma_{\boldsymbol{y}} = \mathrm{diag}\,(\sigma_0^2, \sigma_1^2, \ldots, \sigma_{n-1}^2)$, and the above expression can be simplified to

$$\chi^2 = \sum_{i=0}^{n-1} \frac{\big(y_i - f(x_i; \boldsymbol{a})\big)^2}{\sigma_i^2} \,. \tag{6.51}$$

The minimization of χ^2 now implies solving- a system of p (in general non-linear) equations $\partial \chi^2 / \partial a_j = 0$ $(j = 0, 1, \ldots, p - 1)$. Such problems can be solved iteratively: we ride on the sequence

$$\boldsymbol{a}^{(\nu+1)} = \boldsymbol{a}^{(\nu)} + \Delta \boldsymbol{a}^{(\nu)} \,, \qquad \nu = 0, 1, 2, \ldots \,, \tag{6.52}$$

where $\Delta \boldsymbol{a}^{(\nu)}$ is obtained by solving a *linear* problem, to approach the optimal parameter set. In the νth step the update can be calculated by minimizing

$$\chi^2(\Delta \boldsymbol{a}) = \Big[\boldsymbol{y} - \boldsymbol{f}\big(\boldsymbol{a}^{(\nu)} + \Delta \boldsymbol{a}\big)\Big]^{\mathrm{T}} \Sigma_{\boldsymbol{y}}^{-1} \Big[\boldsymbol{y} - \boldsymbol{f}\big(\boldsymbol{a}^{(\nu)} + \Delta \boldsymbol{a}\big)\Big] \,,$$

where we have suppressed the $\boldsymbol{x}$-dependence of $\boldsymbol{f}$. If $\Delta \boldsymbol{a}$ is small, $\boldsymbol{f}$ can be expanded as $\boldsymbol{f}\big(\boldsymbol{a}^{(\nu)} + \Delta \boldsymbol{a}\big) \approx \boldsymbol{f}\big(\boldsymbol{a}^{(\nu)}\big) + J\big(\boldsymbol{a}^{(\nu)}\big) \Delta \boldsymbol{a}$, where J is the Jacobi matrix with the elements

$$J_{ij} = \left(\frac{\partial f_i}{\partial a_j}\right)_{\boldsymbol{a} = \boldsymbol{a}^{(\nu)}} \,, \qquad i = 0, 1, \ldots, n - 1 \,, \quad j = 0, 1, \ldots, p - 1 \,.$$

Let us denote $\boldsymbol{f}_\nu = \boldsymbol{f}\big(\boldsymbol{a}^{(\nu)}\big)$ and $J_\nu = J\big(\boldsymbol{a}^{(\nu)}\big)$. We must minimize the expression

$$\chi^2(\Delta \boldsymbol{a}) = \Big[(\boldsymbol{y} - \boldsymbol{f}_\nu) - J_\nu \Delta \boldsymbol{a}\Big]^{\mathrm{T}} \Sigma_{\boldsymbol{y}}^{-1} \Big[(\boldsymbol{y} - \boldsymbol{f}_\nu) - J_\nu \Delta \boldsymbol{a}\Big] \,.$$

From the requirement $\partial \chi^2 / \partial (\Delta \boldsymbol{a}) = \boldsymbol{0}$ it follows that

$$\big[J_\nu^{\mathrm{T}} \Sigma_{\boldsymbol{y}}^{-1} J_\nu\big] \Delta \boldsymbol{a} = J_\nu^{\mathrm{T}} \Sigma_{\boldsymbol{y}}^{-1} (\boldsymbol{y} - \boldsymbol{f}_\nu) \,.$$

This system of *linear* equations must be solved in order to obtain the update $\Delta a^{(\nu)}$ in the νth iteration (6.52). If the data uncertainties are uncorrelated, the procedure can be summarized as

$$G\big(a^{(\nu)}\big)\,\Delta a^{(\nu)} = g\big(a^{(\nu)}\big)\,,\tag{6.53}$$

$$a^{(\nu+1)} = a^{(\nu)} + \Delta a^{(\nu)}\,,\qquad \nu = 0,1,2,\ldots\,,\tag{6.54}$$

where

$$G_{jk}\big(a^{(\nu)}\big) = \sum_{i=0}^{n-1}\frac{1}{\sigma_i^2}\left(\frac{\partial f_i}{\partial a_j}\frac{\partial f_i}{\partial a_k}\right)_{a=a^{(\nu)}}\,,\qquad g_k\big(a^{(\nu)}\big) = \sum_{i=0}^{n-1}\frac{y_i - f_i}{\sigma_i^2}\left(\frac{\partial f_i}{\partial a_k}\right)_{a=a^{(\nu)}}.$$

We are still missing the uncertainties of the resulting $\widehat{a}$. Near the optimum χ^2 is approximately parabolic, so the covariance matrix of the regression parameters can be computed by

$$\Sigma_{\widehat{a}} = \big[J^{\mathrm{T}}(\widehat{a})\,\Sigma_y^{-1}J(\widehat{a})\big]^{-1}$$

or simply $\Sigma_{\widehat{a}} = \big[G(\widehat{a})\big]^{-1}$ if σ_i are uncorrelated. The diagonal elements of this matrix are the parameter variances, and the off-diagonal are the covariances. If the uncertainties of y are unknown, the parameter covariance matrix can be calculated by using

$$\Sigma_{\widehat{a}} = \frac{\chi^2(\widehat{a})}{n-p}\big[J^{\mathrm{T}}(\widehat{a})J(\widehat{a})\big]^{-1}\,.\tag{6.55}$$

Example (Adopted from [23].) The circles in Fig. 6.16 (left) show $n = 132$ values of annual global release of CO_2 from fossil fuels over 1880–2011, measured in millions of tons [37]. We would like to model the data by a two-parameter ($p = 2$) function of the form

$$f(x;a) = a_0\,\mathrm{e}^{a_1 x}\,,\tag{6.56}$$

where $x = t - 1880$ (origin shift). The data errors are unknown, which implies $\sigma_i = 1$ for all i. The iterative method requires the derivatives

$$\frac{\partial f_i}{\partial a_0} = \mathrm{e}^{a_1 x_i}\,,\qquad \frac{\partial f_i}{\partial a_1} = a_0 x_i\,\mathrm{e}^{a_1 x_i}\,,$$

to form the 2×2 matrix G,

$$G_{00}(a) = \sum_{i=0}^{n-1}\mathrm{e}^{2a_1 x_i}\,,\qquad G_{01}(a) = a_0\sum_{i=0}^{n-1}x_i\,\mathrm{e}^{2a_1 x_i}\,,\qquad G_{11}(a) = a_0^2\sum_{i=0}^{n-1}x_i^2\,\mathrm{e}^{2a_1 x_i}\,,$$

where $G_{10} = G_{01}$, as well as the right-hand side of Eq. (6.53):

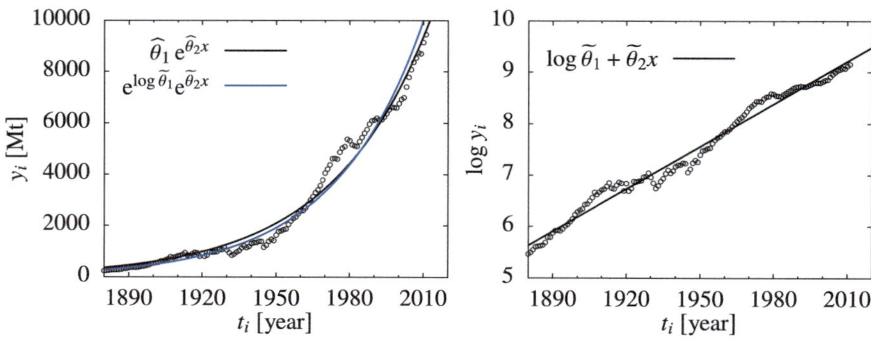

Fig. 6.16 Global release of CO_2. [LEFT] Exponential model, calculated by non-linear regression (thick curve), and the fit obtained by transforming the straight-line regression on logarithmic data (thin curve). [RIGHT] Linear regression on logarithmic data

$$g_0(\boldsymbol{a}) = \sum_{i=0}^{n-1} \left(y_i - a_0 e^{a_1 x_i}\right) e^{a_1 x_i} , \qquad g_1(\boldsymbol{a}) = \sum_{i=0}^{n-1} \left(y_i - a_0 e^{a_1 x_i}\right) a_0 x_i \, e^{a_1 x_i} .$$

Iteration (6.54) is started by the initial approximation $(a_0, a_1)^{(0)} = (200, 0.02)$. After less than ten iterations we obtain the final values

$$\widehat{a}_0 = (363.8 \pm 18.3)\text{Mt} , \qquad \widehat{a}_1 = (2.499 \pm 0.045) \cdot 10^{-2}/\text{yr} ,$$

where the uncertainties have been computed by formula (6.55). The result of non-linear regression is shown by the thick curve in Fig. 6.16 (left). However, do think about it: a standard polynomial regression of such strongly scattered data would yield an equally likable result.

There is another way to proceed. If we take the logarithm of the values y_i, non-linear regression becomes linear, since

$$\log f(x; \boldsymbol{a}) = \log a_0 + a_1 x . \tag{6.57}$$

Now the regression parameters are $\log a_0$ (not a_0) and a_1. By formula (6.43) we compute their optimal values

$$\log \widetilde{a}_0 = (5.630 \pm 0.026) \log(\text{Mt}) , \qquad \widetilde{a}_1 = (2.755 \pm 0.034) \cdot 10^{-2}/\text{yr} ,$$

corresponding to the straight line in Fig. 6.16 (right). But if the obtained $\widetilde{a}_0$ and $\widetilde{a}_1$ are reinserted in (6.56), one gets a different description of the non-logarithmic data, namely the thin curve in Fig. 6.16 (left). The reason for this disagreement is clear. Even though the models (6.56) and (6.57) are mathematically identical, the statistical models $y_i = a_0 \, e^{a_1 x_i} + \sigma_i$ and $\log y_i = \log a_0 + a_1 x_i + \varepsilon_i$ are *not* equivalent because the uncertainties σ_i and ε_i in each case have different distributions. Besides, the method of least squares by itself is not "invariant" with respect to such transformations. $\triangleleft$

In more involved cases the minimization of (6.50) or (6.51) requires a multi-dimensional minimum search in many, sometimes hundreds of dimensions. The well-tested tool for this purpose is the Levenberg–Marquardt method, which is implemented in standard libraries. Several benchmark examples are given in Sect. 9.7 of [23], see also [38].

6.7 Multiple Linear Regression

The remaining Sections of this Chapter, heavily indebted to the wonderfully clear exposition and carefully chosen examples of [39], introduce some of the most appealing methods of *multivariate analysis*, which is the simultaneous statistical analysis of a collection of variables. For example, we would like to understand the influence of several independent variables to a single dependent variable, or to study the mutual dependence of several variables. Such methods can be applied in innumerable areas of natural and social sciences.

6.7.1 The Basic Method

Section 6.5 was devoted to simple one-dimensional linear regression where each independent (input) quantity x_i corresponded to a single dependent (output) quantity y_i. An obvious generalization is multiple linear regression, by which we identify the influence of many input variables on a single output variable.

Assume that we have n measurements or other sources of a set of m independent variables x. We would like to ascertain whether a linear relation can be established between the values of x and the n dependent variables y. In other words, can the whole data set be described by the model

$$y_i = f(x_i) + \Delta y_i = a_0 + \sum_{j=1}^{m} a_j X_{ij} + \Delta y_i , \qquad i = 0, 1, \ldots, n - 1 \qquad (6.58)$$

(see Problem 6.13.1)? Here X_{ij} stands for the ith measurement of the jth independent variable, Δy_i are the unknown errors, and a_j are the regression parameters to be determined. We collect a single (ith) measurement of m independent variables into the vector $x_i = (x_{i1}, x_{i2}, \ldots , x_{im})^{\mathrm{T}}$, so that all n measurements can be assembled in the matrix $X = (x_0, x_1, \ldots , x_{n-1})^{\mathrm{T}}$. (Note the index ranges $i = 0, 1, 2, \ldots$ and $j = 1, 2, 3, \ldots$) We compute the mean vector (of dimension m)

$$\overline{x} = \frac{1}{n} \sum_{i=0}^{n-1} x_i \, , \tag{6.59}$$

and form the matrix of data from which their average has been subtracted,

$$\overline{X} = (\overline{x}, \overline{x}, \dots , \overline{x})^{\mathrm{T}} \, , \qquad X_{\mathrm{c}} = X - \overline{X} \, . \tag{6.60}$$

The matrices X, $\overline{X}$, and X_{c} have size $n \times m$. The dependent variables are handled similarly. We compute the mean of n measured values

$$\overline{y} = \frac{1}{n} \sum_{i=0}^{n-1} y_i \, ,$$

and use it to form the vector of dependent variables from which the mean has been subtracted,

$$\overline{y} = (\overline{y}, \overline{y}, \dots , \overline{y})^{\mathrm{T}} \, , \qquad y_{\mathrm{c}} = y - \overline{y} \, .$$

The vectors y, $\overline{y}$, and y_{c} have dimension n. Again, our basic requirement is to minimize the squares of the deviations (residuals) $y_i - f(x_i)$ with respect to the parameters a_j. The measure of deviation is given in Eq. (6.39) which we generalize to more than one dimension and set $\sigma_i = 1$. For a more compact notation, we add a column of values 1 to the data matrix, yielding a $n \times (m + 1)$ matrix, while the regression coefficients (including the zeroth one) are arranged in a vector of dimension $(m + 1)$,

$$\Xi = \left(\mathbf{1}_n, X \right) , \qquad \widehat{a} = \left(\widehat{a}_0, \widehat{a}_{(m)}^{\mathrm{T}} \right)^{\mathrm{T}} \, .$$

We obtain the vector of regression parameters $\widehat{a}$ by solving the normal system (4.15),

$$\widehat{a} = \left(\Xi^{\mathrm{T}} \Xi \right)^{-1} \Xi^{\mathrm{T}} y \, .$$

The first component of the vector $\widehat{a}$ (the zeroth regression parameter or the *intercept*) and the m *slopes* of the regression "straight lines" are then given by

$$\widehat{a}_0 = \overline{y} - \overline{x}^{\mathrm{T}} \widehat{a}_{(m)} \, , \qquad \widehat{a}_{(m)} = (a_1, a_2, \dots , a_m)^{\mathrm{T}} = \left(X_{\mathrm{c}}^{\mathrm{T}} X_{\mathrm{c}} \right)^{-1} X_{\mathrm{c}}^{\mathrm{T}} y_{\mathrm{c}} \tag{6.61}$$

(see Sect. 5.2 in [39]). Of course we would also like to know the uncertainties of the estimated parameters and the quality of the model description (6.58). The measure for the dispersion of the dependent variables is the *total sum of squares*

$$S_y = y_{\mathrm{c}}^{\mathrm{T}} y_{\mathrm{c}} \, .$$

The total dispersion has two contributions. The first contribution can be explained by linear regression in the independent variables x, and is equal to

$$S_{\text{reg}} = \widehat{\boldsymbol{a}}^{\mathrm{T}} (\Xi^{\mathrm{T}} \Xi) \widehat{\boldsymbol{a}} \ .$$

The remaining portion of the dispersion that is not encompassed by the presumed linear model, is called the *residual sum of squares* and amounts to

$$S_{\text{res}} = S_y - S_{\text{reg}} = (\boldsymbol{y} - \Xi\,\widehat{\boldsymbol{a}})^{\mathrm{T}} (\boldsymbol{y} - \Xi\,\widehat{\boldsymbol{a}}) \ .$$

Finally, we compute the *residual variance* $\widehat{\sigma}^2$ which also determines the variance of the individual regression parameters:

$$\widehat{\sigma}^2 = \frac{S_{\text{res}}}{n - m - 1} \ , \qquad \text{var}(\widehat{\boldsymbol{a}}) = \widehat{\sigma}^2 \left(\Xi^{\mathrm{T}} \Xi \right)^{-1} \ . \tag{6.62}$$

The fraction of the total dispersion in $\boldsymbol{y}$ that can be explained by linear regression in m independent variables x, is given by the correlation coefficient

$$R^2 = \frac{S_{\text{reg}}}{S_y} = \frac{S_y - S_{\text{res}}}{S_y} = 1 - \frac{S_{\text{res}}}{S_y} \ .$$

The value R^2 lies between 0 and 1. Values close to 1 roughly indicate that the linear model (6.58) is an adequate description of data.

If we wish to judge the relevance of the individual parameters a_j, we must assume that some distribution law governs the errors Δy_i in the model (6.58). In most cases $n \gg 1$ and the errors are normally distributed, $\Delta y_i \sim N(0, \sigma^2)$. For each estimated regression parameter $\widehat{a}_j$ we form the statistic

$$t_j = \frac{\widehat{a}_j}{\widehat{\sigma}\sqrt{\xi_j}} \ , \qquad \xi_j = \left[(\Xi^{\mathrm{T}} \Xi)^{-1} \right]_{jj} \ ,$$

where $\widehat{\sigma}$ is given by Eq. (6.62). A value $|t_j| > 2$ indicates that the corresponding parameter $a_j \neq 0$. A value $|t_j| \leq 2$ (and even more so $|t_j| \approx 0$) hints at $a_j \approx 0$.

6.7.2 *Principal Component Multiple Regression*

The computation of the coefficients $\boldsymbol{a}$ and their variances in multiple regression involves the matrix $X_{\mathrm{c}}^{\mathrm{T}} X_{\mathrm{c}}$, where X_{c} is the centered data matrix (6.60). If X_{c} is ill-conditioned or contains weakly linearly dependent columns, or when the number of measurements in smaller than the number of independent variables ($n < m$ or even $n \ll m$), $X_{\mathrm{c}}^{\mathrm{T}} X_{\mathrm{c}}$ becomes singular or nearly singular (see Fig. 6.17 and the following text). In such cases the regression parameters $\boldsymbol{a}$ can not be uniquely determined or may have unphysical values and variances.

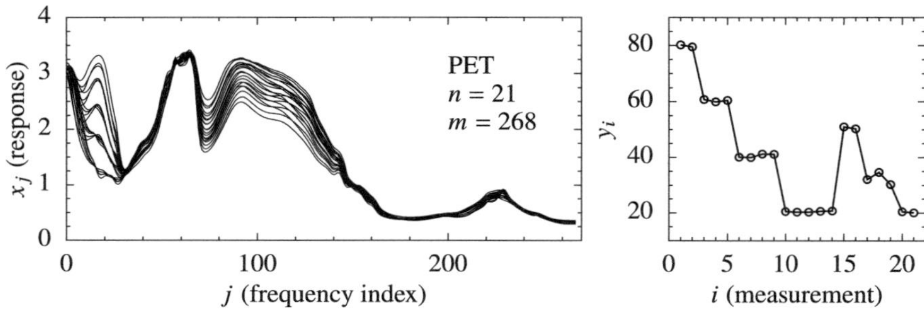

Fig. 6.17 Raman absorption spectrum of PET yarns as a typical experiment in which the number of measurements n is much smaller than the number of independent variables m in the individual measurement. (Example from [39] based on original data from [40].) [LEFT] The $n = 21$ measurements at $m = 268$ frequencies. [RIGHT] Values of the dependent variables (yarn density) in n measurements

If the matrix $X_c^T X_c$ is singular or nearly singular, we can use its pseudo-inverse (see Sect. 4.5.3). Assume that $X_c^T X_c$ has known rank r $(1 \le r \le m)$. It can be diagonalized as

$$X_c^T X_c = V \Lambda V^T , \qquad \Lambda = \mathrm{diag}\,(\lambda_1, \lambda_2, \ldots, \lambda_r) , \qquad V^T V = I_r ,$$

while the smallest $m - r$ eigenvalues are zero. The columns of the $m \times r$ matrix V are eigenvectors $\boldsymbol{v}_j$ corresponding to the individual eigenvalues λ_j. The pseudo-inverse of $X_c^T X_c$ is then

$$\left(X_c^T X_c\right)^+ = V \Lambda^{-1} V^T = \sum_{j=1}^{r} \frac{\boldsymbol{v}_j \boldsymbol{v}_j^T}{\lambda_j} .$$

Instead of the true inverse $\left(X_c^T X_c\right)^{-1}$, which does not exist in this case, we use the pseudo-inverse in Eq. (6.61). Thus we obtain a unique vector of m regression coefficients by the method of *generalized* or *pseudo-inverse regression* (denoted by the superscript "pi" in the following):

$$\widehat{\boldsymbol{a}}_{(r)}^{\mathrm{pi}} = \left(X_c^T X_c\right)^+ X_c^T \boldsymbol{y}_c = \sum_{j=1}^{r} \frac{(\boldsymbol{v}_j \boldsymbol{v}_j^T) X_c^T \boldsymbol{y}_c}{\lambda_j} . \qquad (6.63)$$

The values of dependent variables $\boldsymbol{y}$ (vector of dimension n) corresponding to the regression parameters (6.63) are

$$\widehat{\boldsymbol{y}}^{\mathrm{pi}} = \overline{\boldsymbol{y}} + X_c \widehat{\boldsymbol{a}}_{(r)}^{\mathrm{pi}} .$$

The estimate for the regression coefficients can also be obtained by decomposing the matrix X_c or the product $X_c^T X_c$ to principal components (see Sect. 6.8). The basic idea is to form the matrices $Z_r = X_c V$ that contain the first r principal components of X_c and correspond to those directions in r-dimensional space along which the data

X_c have the largest variance, while the remaining $m - r$ components are discarded. As input variables in the regression we use the matrix Z_r, while the output variable is y_c. The regression coefficients for the principal components Z_r (not for the variables X_c) are [39]

$$\boldsymbol{\zeta}_{(r)} = (\zeta_1, \zeta_2, \ldots, \zeta_r)^{\mathrm{T}} = \left(Z_r^{\mathrm{T}} Z_r\right)^{-1} Z_r^{\mathrm{T}} y_c = \Lambda^{-1} V^{\mathrm{T}} X_c^{\mathrm{T}} y_c \,.$$

We compute the regression coefficients for the original independent variables (the data in the matrix X_c) by first forming the m-dimensional vectors

$$\widetilde{\boldsymbol{a}}_j = \zeta_j \boldsymbol{v}_j \,, \qquad j = 1, 2, \ldots, r \,,$$

as $V \in \mathbb{R}^{m \times r}$. Then we compute the partial sums (the superscript "pc" denotes principal components)

$$\boldsymbol{a}_{(\rho)}^{\mathrm{pc}} = \sum_{j=1}^{\rho} \widetilde{\boldsymbol{a}}_j = \sum_{j=1}^{\rho} \zeta_j \boldsymbol{v}_j \,, \qquad \rho = 1, 2, \ldots, r \,. \tag{6.64}$$

The vector of regression coefficients $\boldsymbol{a}_{(\rho)}^{\mathrm{pc}}$ has m components (equal to the number of independent variables in a single measurement). The ρth partial sum gathers ρ principal components in it. The values of the dependent variables y corresponding to the regression parameters (6.63) are the same as in the pseudo-inverse method,

$$\widehat{\boldsymbol{y}}^{\mathrm{pc}} = \overline{\boldsymbol{y}} + X_c V \widehat{\boldsymbol{\zeta}}_{(r)} = \overline{\boldsymbol{y}} + X_c \, \boldsymbol{a}_{(r)}^{\mathrm{pc}} = \widehat{\boldsymbol{y}}^{\mathrm{pi}} \,.$$

The procedure outlined here is known as *principal-component regression* (PCR). Above all, it is applicable in the cases where the number of variables m is much larger than the number of measurements n. In the PCR method, it is precisely the restriction to a limited number of principal components that helps us to avoid the problems due to the near-singularity of the data matrix.

Example Principal component regression is a good tool to seek the connections between the infrared Raman spectrum of the polyethylene-terephthalate (PET) yarns and their density. Figure 6.17 (left) shows the $n = 21$ measurements of the spectrum at $m = 268$ frequencies, which are stored in the data matrix X (or X_c after centering). In each of the n measurements we also measure the density of the yarn, and store the results in the vector of dependent variables y (or y_c after centering). The dependent variables are shown in Fig. 6.17 (right).

The final result of the regression is the coefficient vector (6.64) which we gradually augment by additional principal components. Figure 6.18 (left) shows the partial sums up to the fourth. Typically we include only a few principal components; further admixtures merely tend to increase the noise. Figure 6.18 (right) clearly shows how well the measured dependent variables can be described by taking the first one, two, or three principal components. ◁

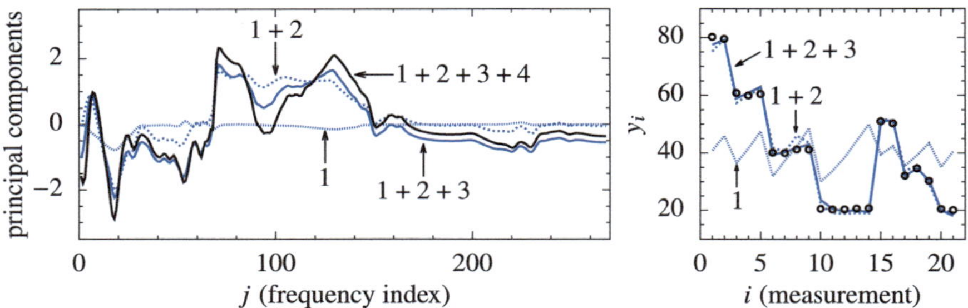

Fig. 6.18 Principal component regression. [LEFT] Vectors of coefficients (6.64). The first four partial sums are shown. [RIGHT] Values of dependent variables (empty circles). Already by including just three major principal components (three curves for 1, 2, and 3 components) one obtains a good description of the data. See also caption to Fig. 6.17

6.8 Principal Component Analysis

Principal component analysis (PCA) is one of the most important tools to reduce the dimensionality of correlated multivariate data. The basic idea can be illustrated by the example of a data set $x_i = (x_{i1}, x_{i2})$ distributed according to the two-dimensional normal distribution $N(x; \mu, \Sigma)$ (see (6.73)) with the average $\mu = (0, 0)$ and covariance matrix $\Sigma = ((1, \rho), (\rho, 1))$, where $\rho = 0.85$ is the correlation coefficient. The data points are shown in Fig. 6.19 (left). The eigenvectors of the covariance matrix Σ are $v_1 = (1, 1)/\sqrt{2}$ and $v_2 = (-1, 1)/\sqrt{2}$, corresponding to the eigenvalues $\lambda_1 = 1 + \rho = 1.85$ and $\lambda_2 = 1 - \rho = 0.15$.

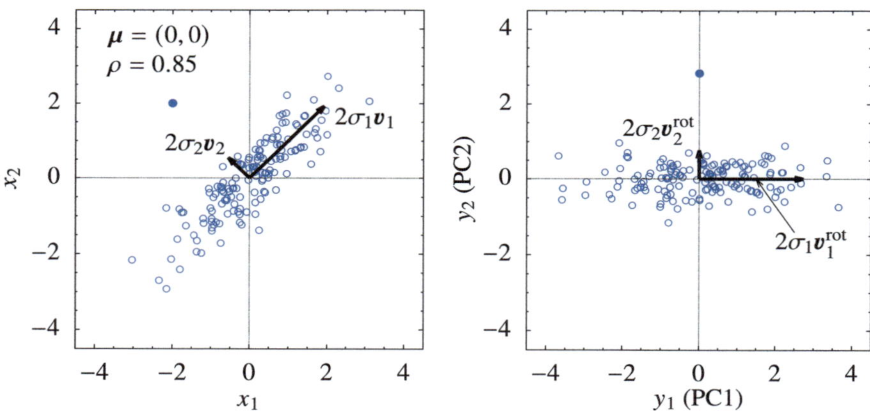

Fig. 6.19 [LEFT] Bivariate data with the rescaled eigenvectors of the covariance matrix $2\sigma_1 v_1$ and $2\sigma_2 v_2$ (the lengths of the vectors are twice the standard deviations along their directions). [RIGHT] Transformed data. The symbol ● represents the data point slightly detached from the central group, which can not be identified from the analysis of the distribution of the variables x_1 or x_2 alone, yet it is easily caught as an outlier in the y_2 variable

The components x_{i1} and x_{i2} are strongly correlated, so it is obvious that in order to describe the data in Fig. 6.19 (left), a single dimension is almost sufficient: that along the eigenvector $\boldsymbol{v}_1$. It carries most of the dispersion while the variance along the vector $\boldsymbol{v}_2$ is relatively small. The basic idea of PCA is to determine the orthogonal linear transformation which projects the data $\boldsymbol{x}$ to a space with smaller dimension. For the example discussed above, we can use the transformation

$$
\boldsymbol{y}_i = V^{\mathrm{T}}\boldsymbol{x}_i \,, \qquad V = (\boldsymbol{v}_1, \boldsymbol{v}_2) = \frac{1}{\sqrt{2}} \begin{pmatrix} 1 & -1 \\ 1 & 1 \end{pmatrix} \,,
$$

to map all data to the vicinity of the y_1 axis, while just a minor part of their dispersion remains along the y_2 axis. Namely, the variances of the transformed variables y_1 and y_2 are

$$
\mathrm{var}(y_{1,2}) = \mathrm{var}\left(\frac{x_2 \pm x_1}{\sqrt{2}}\right) = \frac{1}{2}\,\mathrm{var}(x_2 \pm x_1)
$$

$$
= \frac{1}{2}\left(\mathrm{var}(x_1) + \mathrm{var}(x_2) \pm 2\,\mathrm{cov}(x_1, x_2)\right) = \frac{1}{2}\,(1 + 1 \pm 2\rho) = \lambda_{1,2} \,.
$$

The eigenvalues of the covariance matrix, λ_1 and λ_2, are therefore equal to the variances of the transformed variables y_1 and y_2.

These two-dimensional thoughts can be generalized to the reduction of dimensionality of a set of n *correlated* multivariate data $\boldsymbol{x}_i = (x_{i1}, x_{i2}, \dots, x_{im})^{\mathrm{T}}$, where $i = 0, 1, \dots, n-1$. The leading example in this Section are the data obtained in measuring radar reflections in the ionosphere (Problem 6.13.3, Fig. 6.30), where $n = 351$ measurements of $m = 33$ variables have been acquired. We wish to find such a linear combination of the data components $\boldsymbol{x}_i$,

$$
y_{ji} = v_{j1}x_{i1} + v_{j2}x_{i2} + \dots + v_{jm}x_{im} = \boldsymbol{v}_j^{\mathrm{T}}\boldsymbol{x}_i \,, \qquad j = 1, 2, \dots, r \,, \qquad r \leq m \,,
$$

that this transformation preserves the information about the dispersion of the original data as truthfully as possible, and that the variables y_{ji} become *uncorrelated.* We call the product $\boldsymbol{v}_j^{\mathrm{T}}\boldsymbol{x}$ the jth *principal component*, and the value of this product y_{ji} for a specific data component $\boldsymbol{x}_i$ the *principal component score.* In summary, we are seeking a coefficient vector $\boldsymbol{v}_1$ that maximizes the variance

$$
\frac{1}{n-1}\sum_{i=0}^{n-1}(y_{1i} - \overline{y}_1)^2 \,,
$$

then the vector $\boldsymbol{v}_2$ that maximizes $\sum_i (y_{2i} - \overline{y}_2)^2$ and simultaneously ensures that y_{1i} is uncorrelated to y_{2i}, then the vector $\boldsymbol{v}_3$ that maximizes $\sum_i (y_{3i} - \overline{y}_3)^2$ and ensures that y_{3i} is uncorrelated to both y_{2i} and y_{1i}, and so on. Due to this chain construction with intermediate requirements on the uncorrelatedness of the scores y_{ji}, PCA is also known as *decorrelation of variables*. This goal can be achieved by following either of the two paths: by diagonalizing the covariance matrix, or by the singular value decomposition of the data matrix [41].

6.8.1 Principal Components by Diagonalizing the Covariance Matrix

To obtain the principal components from the covariance matrix, we first subtract the average (6.59) from each x_i, and construct the $n \times m$ centered data matrix X_c according to Eq. (6.60). The estimate for the covariance matrix is

$$\widehat{\Sigma}_{xx} = \frac{1}{n-1} \sum_{i=0}^{n-1} (x_i - \overline{x})(x_i - \overline{x})^{\mathrm{T}} = \frac{1}{n-1} X_c^{\mathrm{T}} X_c . \tag{6.65}$$

Then we compute the eigenvalues and eigenvectors of $\widehat{\Sigma}_{xx}$. This matrix is symmetric and its singular value decomposition has the form

$$\widehat{\Sigma}_{xx} = V \Lambda V^{\mathrm{T}} , \qquad \Lambda = \mathrm{diag}\,(\lambda_1, \lambda_2, \ldots, \lambda_m) , \qquad V^{\mathrm{T}} V = I_m , \tag{6.66}$$

where λ_j are the eigenvalues of $\widehat{\Sigma}_{xx}$ ordered in decreasing magnitude, and the columns of V are its eigenvectors. The first r principal components of x are

$$y_j = v_j^{\mathrm{T}} x , \qquad j = 1, 2, \ldots, r , \qquad r \le m .$$

The eigenvalues λ_j are the estimates for the variance of the jth principal component. What we wish to achieve by PCA is to use just a couple of the lowest principal components to account for nearly all variation of the original data.

The characteristic drop-off of the eigenvalues is shown in Fig. 6.20 (left, left y-axis). We define the percentage of the explained variance as

$$\eta_r = \frac{\sum_{j=1}^{r} \lambda_j}{\sum_{j=1}^{m} \lambda_j} = 1 - \frac{\sum_{j=r+1}^{m} \lambda_j}{\sum_{j=1}^{m} \lambda_j} . \tag{6.67}$$

With increasing rank r of the principal component approximation, η_r also increases (Fig. 6.20 (left, right y-axis)). For a crude description of the data at least those principal components should be retained that correspond to the eigenvalues from the steepest portion of the λ_j curve. By this guideline, we may accept eigenvalues from λ_1 to λ_6. Typically, we retain enough components that $\approx 90\,\%$ of the data variance is accounted for (up to including λ_{18} in Fig. 6.20).

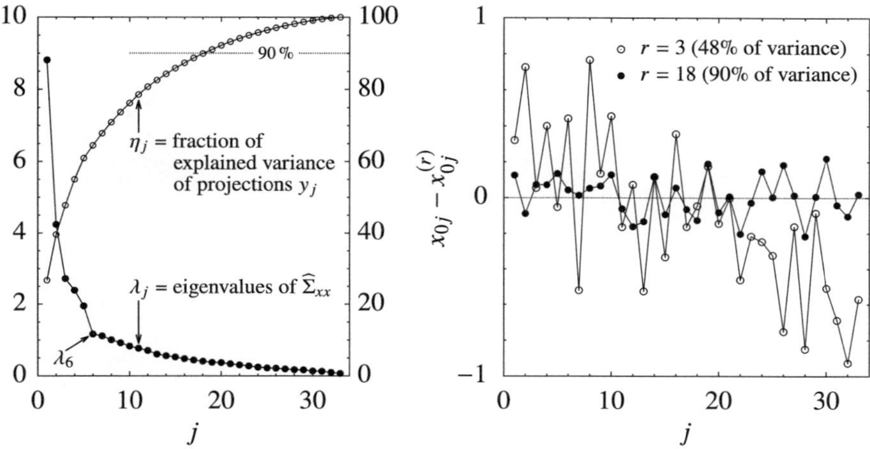

Fig. 6.20 Data from Problem 6.13.3 studied by PCA. [LEFT] Eigenvalues of the covariance matrix (left y-axis) and the percentage of the data variance (6.67) (right y-axis). The first eigenvalue λ_1 accounts for $\approx 27\,\%$ of the variance, while the sum up to including λ_{18} covers $\approx 90\,\%$. [RIGHT] Reconstruction of the data x_0 ($i = 0$) by the sum of principal components up to ranks 3 and 18

The rank-r approximation for each original data point x_i is

$$x_i^{(r)} = \overline{x} + \left(\sum_{j=1}^{r} v_j v_j^{\mathrm{T}} \right) (x_i - \overline{x}) , \qquad i = 0, 1, \ldots, n - 1 , \qquad r \le m .$$

With increasing rank r, $x_i^{(r)}$ becomes an increasingly good approximation of x_i; for $r = m$ we should, of course, retrieve $x_i^{(m)} \equiv x_i$. Figure 6.20 (right) shows the difference between the rank-r approximation and the actual data x_0 for $r = 3$ and $r = 18$. Indeed, the approximation of a higher rank describes the data better (a smaller difference and a larger percentage of the described variance).

PCA is not just a data reduction method. By evaluating the scores of the jth principal component for an individual (ith) data point,

$$y_{ji} = v_j^{\mathrm{T}} x_i , \qquad i = 0, 1, \ldots, n - 1 , \tag{6.68}$$

we use the two-dimensional distribution of y_{ji} versus $y_{j'i}$ ($j' \neq j$) to reveal peculiarities in the structure of the data, e.g. locate outliers (see Fig. 6.19) or identify clustering. Figure 6.21 shows the scores y_{3i} versus y_{1i} and y_{3i} versus y_{2i} for the data of Problem 6.13.3. In this case, two-dimensional images of the first few principal components help us to clearly separate the good reflections from the bad ones.

Fig. 6.21 The PCA scores for the data of Problem 6.13.3. [LEFT] Component 3 versus component 1. [RIGHT] Component 3 versus component 2

6.8.2   Standardization of Data for PCA

When the estimated variances of the individual data elements x_i are comparable, PCA should be initialized with the data averaged by (6.59) and (6.60). But if the variances differ widely, or if the data elements are incomparable, the data should be standardized. This is a crucial step if quantities are given in different measurement units. Otherwise, the first few principal components will be totally dominated by the influence of the data components x_i with the largest numerical spread. The data should therefore always be centered and divided by the estimates of their standard deviations:

$$x_i \longleftarrow \frac{x_i - \overline{x}}{s_{xi}} \tag{6.69}$$

(the division of vectors is meant to be component-wise). The standardization of data means that the PCA procedure, *de facto*, is carried out with the correlation (and not the covariance) matrix. Further arguments in favor of the standardization or against it can be found in [41], Chap. 2.3.

6.8.3   Principal Components from the SVD of the Data Matrix

The second option to compute the principal components of a set of multivariate data leads through the singular value decomposition (SVD) of the data matrix X_c. The SVD process has been described in Sect. 4.5.2. We use SVD to decompose the $n \times m$ matrix X_c as

$$X_c = U \Lambda V^T , \qquad U \in \mathbb{R}^{n \times n} , \quad \Lambda \in \mathbb{R}^{n \times m} , \quad V \in \mathbb{R}^{m \times m} , \tag{6.70}$$

where the columns of V are the eigenvectors v_j of Σ_{xx}. In full SVD the matrix Λ has size $n \times m$ and the square $m \times m$ submatrix at its upper edge contains the *square roots* of the eigenvalues of the covariance matrix, multiplied by $(n - 1)$. In other words, the eigenvalues λ_j from the decomposition (6.66) are equal to the values $\lambda_j^2/(n - 1)$ from the decomposition (6.70). In addition, SVD rewards us with bonus principal component scores: in scaled form they are hidden in the columns of U. We compute the scores (6.68) for the jth principal component as

$$y_{ji} = U_{ij}\sqrt{\lambda_j} , \qquad i = 0, 1, \ldots, n - 1 .$$

PCA is usually carried out with the SVD of the centered data matrix X_c, not by diagonalizing the covariance matrix Σ_{xx}. The terms PCA and SVD are therefore often considered to be synonymous.

6.8.4 *Improvements of PCA: Non-linearity, Robustness*

PCA is a linear method that we can use to determine whether the data from $\mathbb{R}^m$, to a good approximation, actually "reside" in a smaller space $\mathbb{R}^r$, $r < m$. Yet the method fails, for example, in the simple case when the data in $\mathbb{R}^2$ is distributed approximately uniformly along the perimeter of the unit circle. Obviously such data has only one degree of freedom—the angle $\phi \in [0, 2\pi)$—while linear PCA will allow us to span these data on two practically arbitrary orthogonal directions (any two orthogonal unit vectors can describe the variance of such data). Linear PCA is also sensitive to outliers: even though the method is supposed to help us to find the directions with the largest variation, it is unacceptable that a single outlier should radically alter the direction of any of the eigenvectors. Non-linear improvements to the PCA are described in [39] in Sects. 16.2 and 16.5. The initial reading on robust PCA can be found in Chap. 10 of [41] and Sect. 6.10 in [14]; see also [42].

6.9 Cluster Analysis ⋆

Cluster analysis is one of the basic methods of *unsupervised learning*. Its primary task is to classify sets of multivariate data into clusters, groups, or classes (all synonyms in common use). The number of classes is not necessarily known in advance. This is the key distinction with respect to discriminant analysis (discussed in Sect. 6.10) where we initially declare the number of classes and where previously processed data may be used to classify new data for which class membership is yet to be determined. A famous case is shown in Fig. 6.22 (left). We would like to know whether

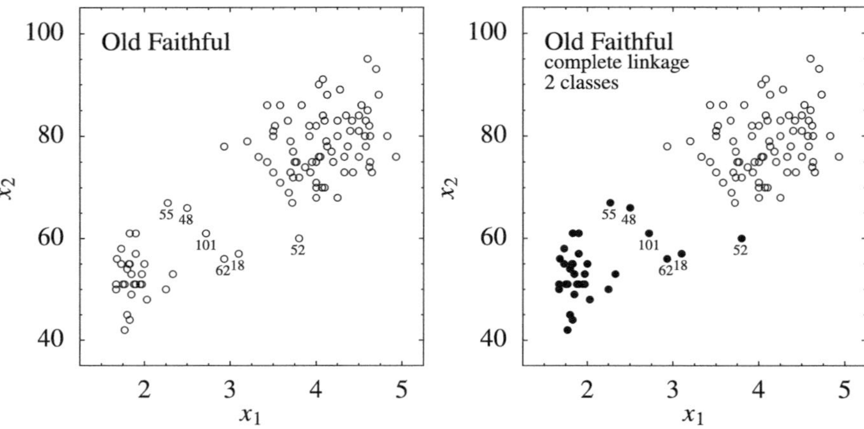

Fig. 6.22 Eruptions of the Old Faithful geyser (Yellowstone National Park). We have 107 measurements of the eruption duration (x_1) and the times between two consecutive eruptions (x_2), both in minutes. [LEFT] Unclassified data. [RIGHT] Data classified in two clusters by using agglomerative clustering with complete linkage. See also Fig. 6.23. (Data indices are offset by 1)

in the distribution of geyser eruptions in terms of their duration and pauses between the eruptions any natural ordering or classification can be established, reflecting an underlying physical mechanism [43].

A cluster can loosely be defined as a group of objects (points in the plane or in space) in which the objects are "close" to each other, while the objects within one group are "far away" from the objects in other groups. The way the distance between the objects and between the clusters is measured depends on the individual method. Often, the results can be in conflict with our subjective judgment: in Fig. 6.22 (left) we may recognize two or three clusters, depending on how we interpret the data 55, 48, 101, 62, 18, and 52.

Data clustering methods come in two broad varieties: hierarchical and non-hierarchical (or partitioning). Here we discuss the most typical representatives of each type. Further reading can be found in [44], in Chap. 11 of [45], in Chap. 12 of [39], and in specialized monographs [46–49].

6.9.1 Hierarchical Clustering

In hierarchical clustering the data are split in clusters of different sizes, among which some hierarchy is established. We initiate the process at the bottom of the hierarchy and interpret each data point as a cluster of its own. We locate the nearest two clusters, merge them, and repeat this until the last two clusters merge to a single cluster containing all data. This is the basic idea of *agglomerative* clustering. In the second approach, known as *divisive* clustering, all data are understood as a single cluster which is gradually broken down into smaller clusters until we are left with as many clusters as there are data entries. Agglomerative clustering has more adherents, and we discuss it in the following.

We represent each object (multivariate data item with m values) by the vector $\boldsymbol{x}_i = (x_{i1}, x_{i2}, \ldots, x_{im})^{\mathrm{T}}$, where $i = 0, 1, \ldots, n - 1$. When the ranges of numerical values of the individual components differ widely, they should be standardized, $x_{ij} \leftarrow (x_{ij} - \overline{x}_j)/s_{x_j}$, where $\overline{x}_j = n^{-1} \sum_i x_{ij}$ and $s_{x_j} = (n - 1)^{-1} \sum_i (x_{ij} - \overline{x}_j)^2$. To measure the distance between the objects we use

$$
d(\boldsymbol{x}_i, \boldsymbol{x}_j) = \left[\sum_{k=1}^{m} | x_{ik} - x_{jk} |^p \right]^{1/p} ,
$$

in particular the cases $p = 2$ (Euclidean distance) and $p = 1$ (so-called Manhattan street distance). We use the n data entries to construct the symmetric $n \times n$ *proximity matrix* with the elements

$$
D_{i,j} = \begin{cases} d(\boldsymbol{x}_i, \boldsymbol{x}_j) \; ; \; i \neq j , \\ \quad 0 \qquad ; \; i = j , \end{cases} \tag{6.71}
$$

where $i, j = 0, 1, \ldots, n - 1$, and follow the algorithm [39]

Input: Multivariate data $\{\boldsymbol{x}_i\}_{i=0}^{n-1}$, each one is its own cluster
Compute the matrix $D^{(0)}$ by using (6.71).
for $s = 0$ **to** $n - 1$ **do**
> Find the shortest distance $D_{I,J}$ in the matrix $D^{(s)}$.
> Merge clusters I and J into cluster IJ. // (*)
> Compute the distance $D_{IJ,K}$ between the new cluster IJ and all remaining clusters $K \neq IJ$ by using
>
> $$ D_{IJ,K} = \max\{ D_{I,K}, D_{J,K} \} \quad \text{(complete linkage)} $$
>
> Construct a new $(n - s + 1) \times (n - s + 1)$ matrix $D^{(s+1)}$ by deleting the rows and columns I and J from $D^{(s)}$ and adding the row and column IJ with distances $D_{IJ,K}$. At the end of the loop $D^{(n-1)} = 0$.

end
Output: The list of clusters merged at step (*); the list of distances $D_{IJ,K}$ at the individual merge

Example Let us revisit the Old Faithful (Fig. 6.22). The final results of the algorithm above are the list of clusters that were merged into a larger cluster, and the list of distances $D_{IJ,K}$ at this merge. We illustrate this hierarchy graphically by a *dendrogram* (Fig. 6.23) in which the distances $D_{IJ,K}$ are represented by the different spacings between the sites where two clusters merge. A larger distance implies that the corresponding clusters are more disjunct. Vertical cuts through the branches of the dendrogram reveal the number of clusters. According to the first possible cut (to

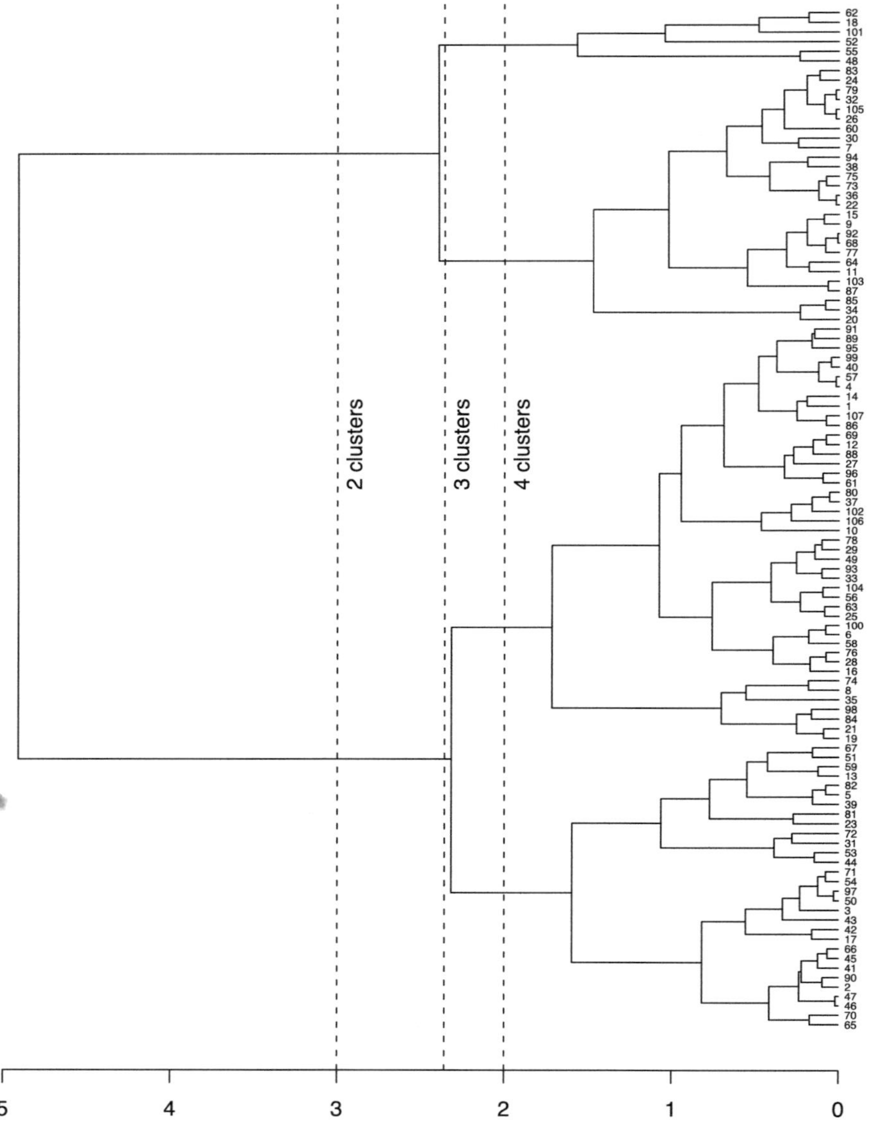

Fig. 6.23 Dendrogram for hierarchical clustering of the data from Fig. 6.22 by complete linkage. The intersections of vertical lines with the branches of the dendrogram define the number of classes (clusters). All elements to the right of the cut belong to the corresponding sub-cluster. The element 52 is misclassified in two-class clustering. (Data indices are offset by 1)

the right of the value $D \approx 3$, two clusters) the elements from 62 to 20 belong to one cluster, while the elements from 91 to 65 belong to the other. According to the next possible cut (to the right of $D \approx 2.36$, three clusters) the elements from 62 to 48 fit into one, the elements from 83 to 20 into another, and those from 91 to 65 into the third cluster.   ◁

The presented algorithm has allowed us to perform clustering by *complete linkage* where we have determined the distance $D_{IJ,K}$ by finding the *maximum* value $\max\{\,D_{I,K},\,D_{J,K}\,\}$. A typical outcome of such linkage is a large number of small, compact clusters. Another option could be to use *single linkage* where $D_{IJ,K}$ would be determined by finding the *minimum* values $\min\{\,D_{I,K},\,D_{J,K}\,\}$. This tends to result in long chains of clusters made of just single data elements. The details on different types of linkage can be found in [47].

To compute and draw the dendrogram, we may resort to dedicated software. For example, in MATHEMATICA, we store the data x_{ij} in the matrix x and type

```
Needs["HierarchicalClustering'"]
x={{x_11,x_12,...,x_1m},...,{x_n1,x_n2,...,x_nm}};
DendrogramPlot[x,DistanceFunction->EuclideanDistance,LeafLabels->(#&)];
```

(Note that this command does not just draw the dendrogram but hides the complete clustering algorithm!) The same can be accomplished in the R environment [50] which is a powerful suite of tools for statistical data analysis. In this environment the data should first be converted to the proximity matrix, then the clusters are formed, and finally the dendrogram is plotted:

```
require(graphics)
require(utils)
DATA <- matrix(scan("data.dat",0),ncol=2,byrow=TRUE)
hc <- hclust(dist(DATA),"complete")
dend <- as.dendrogram(hc)
plot(dend,horiz=TRUE)
```

If the input data already form relatively well delineated clusters, the final outcomes of hierarchical methods do not depend strongly on the specifics of linkage and the details of the algorithms. The main deficiency of the hierarchical methods is their unpredictability. Since the clustering proceeds by a stiff recipe which can not be modified or adapted at any of the intermediate steps, the final distribution of the data into clusters can be arbitrarily bad. Moreover, hierarchical algorithms are numerically expensive: in naive implementations they require $\mathcal{O}(n^2)$ of memory and $\mathcal{O}(n^3)$ of processor time.

6.9.2 Partitioning Methods: k-Means

Many deficiencies of hierarchical methods can be avoided by using *partitioning methods*. Their main characteristic is that clustering is initiated with a predefined number of clusters K, and that the classification into K clusters is not necessarily hierarchically related to a classification into another number of clusters: partitioning methods are therefore *non-hierarchical*. The number of clusters K is an input parameter that in many instances may be readily available. For example, in the bivariate data in Fig. 6.22 we will obviously attempt either $K = 2$ or $K = 3$; with multivariate data, the decision might be much harder.

A very popular approach is the method of k-*means*. In the standard implementation we randomly distribute the multivariate data set $\{x_i\}_{i=0}^{n-1}$ over the chosen number of clusters K, and use the data within each group to determine their centers-of-mass (centroids). In the next step, each element is assigned to the centroid nearest to it, and we use this reordered configuration to compute the new centroids. We repeat this until the configuration no longer changes:

> Standardize the data as $x_{ij} \leftarrow (x_{ij} - \overline{x}_j)/s_{x_j}$, choose number of clusters K.
> Randomly distribute the data x_i into K clusters and compute the centroids (means) $\overline{x}_k$ of each cluster $k = 1, 2, \ldots, K$.
> Compute the sum of squares of distances of each data point to its centroid,

$$\mathcal{D} = \sum_{k=1}^{K} \sum_{c(i)=k} \| x_i - \overline{x}_k \|_2^2 , \tag{6.72}$$

where $c(i)$ is the cluster containing the element x_i.
> Locate the nearest centroid $\overline{x}_k$ of the element x_i and reassign x_i to cluster k. After the reassignments, compute the new centroids $\overline{x}_k$.
> Repeat from step (6.72) until no further readjustments of the elements are needed. Then $\mathcal{D}$ reaches its minimum. The final result is the list of elements within the clusters $c(i)$.

There is no need to randomize the initial assignments of the elements if we know better. For each presumed cluster the centroids can be computed in advance; for example, by looking at Fig. 6.22 and assuming $K = 2$ we might decide for $\overline{x}_1 = (2, 50)$ and $\overline{x}_2 = (4, 80)$, while with the assumption $K = 3$ we might start with $(2, 50)$, $(3, 60)$, and $(4, 80)$ (or corresponding standardized values).

Regardless of the initial configuration the algorithm converges very quickly: the numerical cost is just $\mathcal{O}(nKv)$, where v is the number of iterations needed for convergence (typically $v \approx 3$). Because of its speed—and because it is so wonderfully simple to program—it is a popular tool to classify large data sets. It can also be used to roughly locate the cluster centroids, thereby providing the initial step for processing with other, more sophisticated algorithms. Figure 6.24 shows the initial configuration and the situation after the first iteration of the k-means algorithm for the geyser data from Fig. 6.22.

The method of k-means fails in classifying clusters with very different characters (e.g. with $K = 2$, in the case of one large and one small cluster, or of one compact and one elongated cluster, or of two approximately parallel elongated clusters as shown in Fig. 6.25). The method fails because it relies solely on the sum of distances (6.72)

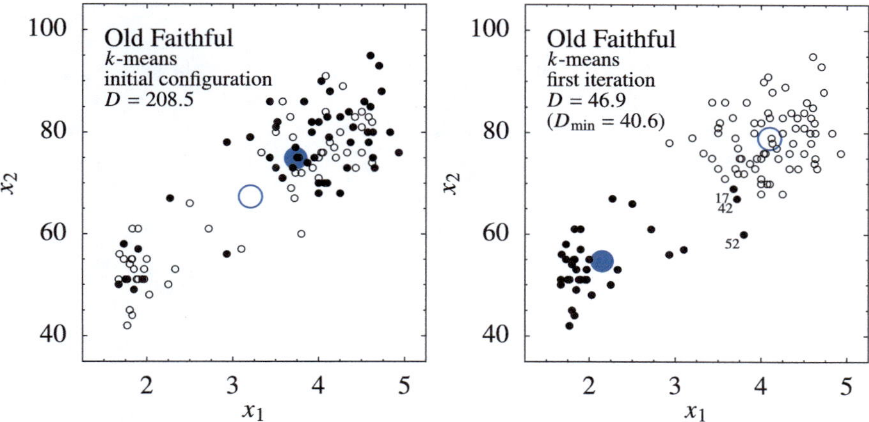

Fig. 6.24 Clustering bivariate data from Fig. 6.22 by using the method of k-means. [LEFT] At the beginning of the iteration the data are randomly distributed among both clusters. The centroids (large symbols • and ○) are therefore at completely wrong locations and the sum of distances $\mathcal{D} = 208.5$ (Eq. (6.72)) is large. [RIGHT] Situation after the first iteration. The data are already almost correctly classified, the centroids are at almost their final positions, and $\mathcal{D} = 46.9$ is near its final value 40.6. The iteration stops in the next step, in which the elements 17, 42, and 52 become rearranged. (Data indices are offset by 1)

between the elements and the centroids, and is therefore insensitive to the relative sizes or shapes of the clusters. One may therefore run the algorithm with several random initial configurations and ultimately accept the solution for which $\mathcal{D}$ reaches its global minimum (Fig. 6.25 (right)).

6.9.3 *Gaussian Mixture Clustering and the EM Algorithm*

The principle of the Gaussian mixture method is the search for a limited set of multivariate normal distributions that best reproduce the whole data set as a conglomerate of clusters (which may or may not overlap). In other words, we would like to frame the data $x_i = (x_{i1}, x_{i2}, \dots, x_{im})^{\mathrm{T}}$, where $i = 0, 1, \dots, n - 1$, into K ($k = 1, 2, \dots, K$) multi-dimensional Gaussian distributions

$$N(\boldsymbol{x}; \boldsymbol{\mu}_k, \Sigma_k) = (2\pi)^{-m/2} (\det \Sigma_k)^{-1/2} \exp\left\{-\frac{1}{2}(\boldsymbol{x} - \boldsymbol{\mu}_k)^{\mathrm{T}} \Sigma_k^{-1} (\boldsymbol{x} - \boldsymbol{\mu}_k)\right\} . \tag{6.73}$$

Here $\boldsymbol{\mu}_k$ is the distribution mean (a vector of dimension m) and Σ_k the corresponding $m \times m$ covariance matrix. We determine the parameters $\boldsymbol{\mu}_k$ and Σ_k by using an iterative procedure called the EM (*expectation-maximization*) algorithm. With the EM algorithm we maximize the quantity

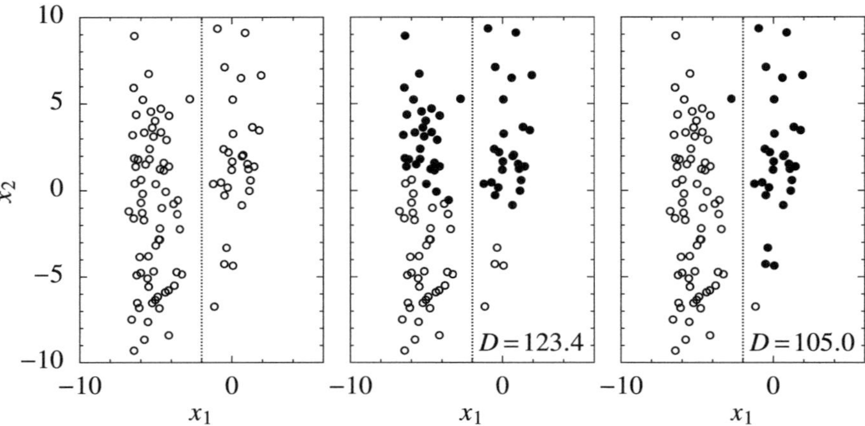

Fig. 6.25 The method of k-means sometimes fails. In such cases the algorithm typically locates just the local minimum of the sum of the squares of distances $\mathcal{D}$ (see Eq. (6.72)). [LEFT] Unclassified data. [CENTER] A wrong final configuration in a classification into two clusters (local minimum $\mathcal{D} = 123.4$). [RIGHT] The correct final configuration (global minimum $\mathcal{D} = 105.0$)

$$\mathcal{L} = \prod_{i=0}^{n-1} \mathcal{P}(\boldsymbol{x}_i) , \tag{6.74}$$

where $\mathcal{P}(\boldsymbol{x}_i)$ is the probability that some point is found at $\boldsymbol{x}_i$, i.e.

$$\mathcal{P}(\boldsymbol{x}_i) = \sum_{k=1}^{K} \mathcal{P}(k)\, N(\boldsymbol{x}_i; \boldsymbol{\mu}_k, \Sigma_k) ,$$

where $\mathcal{P}(k)$ is the probability that a randomly chosen data point belongs to the cluster k. At the same time, $\mathcal{P}(k)$ measures the fraction of all n data that belong to the cluster k. The corresponding probability for the kth component of the ith data point is then

$$\mathcal{P}_{ik} = \frac{\mathcal{P}(k)\, N(\boldsymbol{x}_i; \boldsymbol{\mu}_k, \Sigma_k)}{\sum_k \mathcal{P}(k)\, N(\boldsymbol{x}_i; \boldsymbol{\mu}_k, \Sigma_k)} = \frac{\mathcal{P}(k)\, N(\boldsymbol{x}_i; \boldsymbol{\mu}_k, \Sigma_k)}{\mathcal{P}(\boldsymbol{x}_i)} .$$

The computation of $\mathcal{L}$ and $\mathcal{P}_{ik}$ with the estimates $\widehat{\boldsymbol{\mu}}_k$, $\widehat{\Sigma}_k$, and $\widehat{\mathcal{P}}(k)$, represents the *expectation step* (E) of the EM algorithm. We estimate the averages (centroids) of the K distributions, their covariance matrices, and probabilities $\mathcal{P}(k)$, by

$$\widehat{\boldsymbol{\mu}}_k = \frac{\sum_i \mathcal{P}_{ik} \boldsymbol{x}_i}{\sum_i \mathcal{P}_{ik}} , \qquad \widehat{\Sigma}_k = \frac{\sum_i \mathcal{P}_{ik}(\boldsymbol{x}_i - \widehat{\boldsymbol{\mu}}_k)(\boldsymbol{x}_i - \widehat{\boldsymbol{\mu}}_k)^{\mathrm{T}}}{\sum_i \mathcal{P}_{ik}} , \qquad \widehat{\mathcal{P}}(k) = \frac{1}{n} \sum_i \mathcal{P}_{ik} .$$

These three estimates represent the second step (*maximization step*, M).

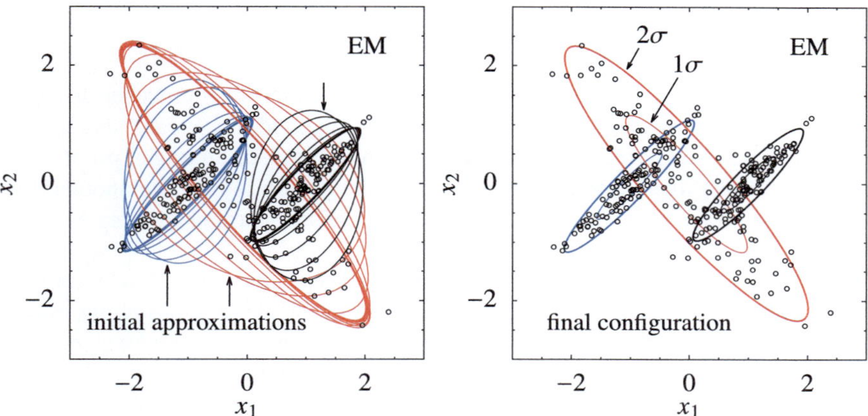

Fig. 6.26 Classification of $n = 300$ bivariate data into $K = 3$ clusters by the EM algorithm. [LEFT] Translation and rotation of the "2σ" covariance ellipses during the iteration. The semi-axes of the ellipses are $2r_0$ and $2r_1$ long (see Eq. (6.45)). [RIGHT] The situation at the end of the iteration

Example The algorithm is implemented iteratively. At the beginning we approximately determine the parameters $\widehat{\boldsymbol{\mu}}_k$, $\widehat{\boldsymbol{\Sigma}}_k$, and $\widehat{\mathcal{P}}(k)$. Figure 6.26 (left) shows $n = 300$ bivariate data ($m = 2$) which we would like to classify into three clusters ($K = 3$). We initialize the iteration with the estimates for the three centroids and three covariance matrices, for example,

$$\widehat{\boldsymbol{\mu}}_1 = (-1, 0)\,, \qquad \widehat{\boldsymbol{\mu}}_2 = (0, 0)\,, \qquad \widehat{\boldsymbol{\mu}}_3 = (1, 0)\,, \qquad \widehat{\boldsymbol{\Sigma}}_{1,2,3} = \begin{pmatrix} 0.25\ 0.00 \\ 0.00\ 0.05 \end{pmatrix}$$

and $\widehat{\mathcal{P}}(1) = \widehat{\mathcal{P}}(2) = \widehat{\mathcal{P}}(3) = 1/3$, and use them to obtain $\mathcal{L}$ and $\mathcal{P}_{ik}$ (step E). Then we compute the new estimates $\widehat{\boldsymbol{\mu}}_k$, $\widehat{\boldsymbol{\Sigma}}_k$, and $\widehat{\mathcal{P}}(k)$ (step M). We repeat this procedure until $\mathcal{L}$ stops increasing. (Multiple tries with different initial configurations may convince us that we have indeed reached the global, not a local, maximum.) During the iteration the covariance ellipses entangle the individual data clusters ever more closely, and after typically ≈ 10 rounds the procedure converges to the final Gaussian mixture that optimally fits the data and delineates K clusters. The ultimate configuration for this case is shown in Fig. 6.26 (right).   ◁

For large n, many small probabilities are multiplied in (6.74); as a result, range underflows in finite-precision arithmetic may occur. The whole procedure should therefore be implemented solely by taking the logarithms of all quantities: instead of $\mathcal{L} = \prod_i \mathcal{P}(\boldsymbol{x}_i)$ we should use $\log \mathcal{L} = \sum_i \log \mathcal{P}(\boldsymbol{x}_i)$ and

$$\log N(\boldsymbol{x}; \widehat{\boldsymbol{\mu}}_k, \widehat{\boldsymbol{\Sigma}}_k) = -\frac{1}{2}(\boldsymbol{x} - \widehat{\boldsymbol{\mu}}_k)^{\mathrm{T}} \widehat{\boldsymbol{\Sigma}}_k^{-1} (\boldsymbol{x} - \widehat{\boldsymbol{\mu}}_k) - \frac{m}{2}\log 2\pi - \frac{1}{2}\log \det \widehat{\boldsymbol{\Sigma}}_k\,.$$

Additional numerical details can be found in [25]; see also [42].

6.9.4 Spectral Methods

It is not hard to imagine a layout of bivariate data for which all approaches discussed so far (the hierarchical agglomerative method, the method of k-means, and the EM algorithm) will fail. Typical examples are shown in Fig. 6.27 hinting at problematic types of clusters like concentric regions or clusters that bite into one another. One of the most powerful modern methods of unsupervised learning from such data is *spectral clustering.*

The foundation of the method is, again, a kind of proximity matrix. But by applying a few simple transformations of this matrix the problematic data regions can be converted into more compact groups which are easier to cluster by simpler algorithms. A good introduction to these spectacularly effective methods is offered by the paper [51]. From [52] we adopt the algorithm for spectral clustering of multivariate data x_i $(i = 0, 1, \ldots, n - 1)$ into K clusters:

▷ Use the data $\{x_i\}_{i=0}^{n-1}$ to construct the *affinity matrix*

$$
A_{ij} = \begin{cases} \exp\left[-\|x_i - x_j\|^2/(2\sigma^2)\right] & ; \ i \neq j \,, \\ 0 & ; \ i = j \,. \end{cases}
$$

The way to determine the parameter σ is described below. The matrix A contains all information about the distances between the data. By using A, compute

$$
L = D^{-1/2} A \, D^{-1/2} \,, \qquad D = \mathrm{diag}(d_0, d_1, \ldots, d_{n-1}) \,, \qquad d_i = \sum_{j=0}^{n-1} A_{ij} \,.
$$

▷ Find the K eigenvectors v_k corresponding to the largest eigenvalues of L, and arrange them in the columns of the matrix $V = (v_1, v_2, \ldots, v_K) \in \mathbb{R}^{n \times K}$. Normalize the rows of V to unit length,

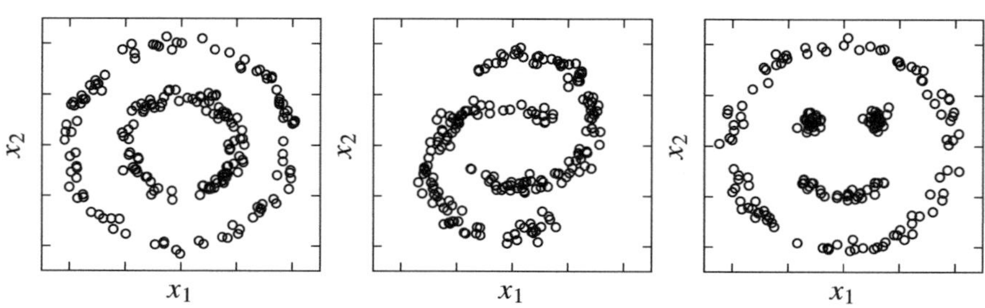

Fig. 6.27 Typical examples of bivariate data for which standard clustering methods fail. They can be handled effortlessly by spectral clustering

$$W_{ij} = V_{ij} \left[\sum_{k=1}^{K} V_{ik}^2 \right]^{-1/2} .$$

▷ Each row of the matrix $W \in \mathbb{R}^{n \times K}$ is a new point $\boldsymbol{w}_i$ (multivariate data entry) in space $\mathbb{R}^K$. By the method of k-means (Sect. 6.9.2) classify the points $\boldsymbol{w}_i$ into K clusters. Choose the parameter σ such that the points accumulate in compact, well separated groups (sometimes easily discernible). These groups are easy to handle by the k-means algorithm.

▷ Assign the original data points $\boldsymbol{x}_i$ to the cluster j precisely in the case that the ith row of the matrix W has been assigned to cluster j.

6.10 Linear Discriminant Analysis ⋆

Assume that an experiment gives us n measurements of m variables. Suppose we anticipate the results of each of the n measurements will originate in one of the two possible classes R_r, $r \in \{1, 2\}$. A well-known example is the analysis of n microscopic samples of needle biopsies of potentially cancerous tissues, where each sample allows us to determine m properties of the tissue cells [53]. For example, one can monitor the shape, the symmetry, the size, and other cell properties, and arrange the ith measurement of all m variables in the vector

$$\boldsymbol{x}_{ri} = \left(x_1^{(r)}, x_2^{(r)}, \ldots, x_m^{(r)} \right)_i^{\mathrm{T}} , \qquad i = 0, 1, \ldots, n-1 , \qquad r \in \{1, 2\} . \qquad (6.75)$$

If, by surgical methods, we realize that a microscopic sample corresponds to a benign tissue, we assign it to the class $r = 1$; if it is malignant, we assign it to the class $r = 2$.

Discriminant analysis allows us to solve two sequential problems. Firstly, we wish to devise a tool for classification into classes R_1 and R_2 on the basis of unambiguously identified n_1 benign and $n_2 = n - n_1$ malignant tissue samples. Secondly, we would like to apply this very classification tool to determine the class of a new, as yet unclassified observation (sample). This is one of the aspects distinguishing discriminant analysis from clustering analysis (Sect. 6.9): in the former the number of classes is known in advance, while in the latter both the number of classes (clusters) and the class assignments are unknown beforehand.

6.10.1 Binary Classification

Here we discuss the simplest case of *binary classification*, where only two classes are involved. We refer to the Bayes theorem of conditional probabilities. We assume that the prior probabilities

$$\mathcal{P}(\boldsymbol{\xi} \in R_r) = \mathcal{P}_r , \qquad r = 1, 2 ,$$

are already known, i.e. we know the probabilities that a randomly chosen measurement $\boldsymbol{\xi} = \boldsymbol{x}$ belongs to either class R_1 or class R_2. If we do not know them, we can estimate them from the population sample at hand: in the cancerous tissue example above, we can simply set $\mathcal{P}_1 = n_1/n$ and $\mathcal{P}_2 = n_2/n$. Suppose that the conditional probability density for class r is

$$p(\boldsymbol{\xi} = \boldsymbol{x} \mid \boldsymbol{\xi} \in R_r) = p_r(\boldsymbol{x}) . \tag{6.76}$$

By using Bayes theorem $\mathcal{P}(A|B) = p(B|A)\mathcal{P}(A)/p(B)$, where $\mathcal{P}$'s denote the probabilities and p's the probability densities, we obtain the posterior probabilities

$$\widetilde{\mathcal{P}}_r(\boldsymbol{x}) = \mathcal{P}(\boldsymbol{\xi} \in R_r \mid \boldsymbol{\xi} = \boldsymbol{x}) = \frac{p_r(\boldsymbol{x})\mathcal{P}_r}{p_1(\boldsymbol{x})\mathcal{P}_1 + p_2(\boldsymbol{x})\mathcal{P}_2} , \tag{6.77}$$

which are the probabilities that the observed $\boldsymbol{x}$ belongs to the class R_1 or R_2. In binary classification we choose R_1 to be the class corresponding to the larger prior probability $\mathcal{P}_1$. This means that the observed $\boldsymbol{\xi}$ should be assigned to R_1 if $\widetilde{\mathcal{P}}_1(\boldsymbol{x})/\widetilde{\mathcal{P}}_2(\boldsymbol{x}) > 1$, or to R_2 otherwise. We can use Eq. (6.77) to assign $\boldsymbol{x}$ to R_1 if $p_1(\boldsymbol{x})/p_2(\boldsymbol{x}) > \mathcal{P}_2/\mathcal{P}_1$, or to R_2 otherwise. For practical use we prefer to define

$$L(\boldsymbol{x}) = \log \frac{p_1(\boldsymbol{x})\mathcal{P}_1}{p_2(\boldsymbol{x})\mathcal{P}_2} \tag{6.78}$$

and classify

$$L(\boldsymbol{x}) > 0 \iff \boldsymbol{x} \in R_1 , \qquad L(\boldsymbol{x}) < 0 \iff \boldsymbol{x} \in R_2 . \tag{6.79}$$

Usually we assume that the probability densities $p_1(\boldsymbol{x})$ and $p_2(\boldsymbol{x})$ correspond to the normal distributions, and that in general they have different means $\boldsymbol{\mu}_1$ and $\boldsymbol{\mu}_2$, yet equal covariance matrices $\Sigma_1 = \Sigma_2 = \Sigma_{xx}$, thus

$$p_r(\boldsymbol{x}) = (2\pi)^{-m/2} \, |\, \Sigma_{xx} \,|^{-1/2} \, \exp\left[-\frac{1}{2} \, (\boldsymbol{x} - \boldsymbol{\mu}_r)^{\mathrm{T}} \, \Sigma_{xx}^{-1} \, (\boldsymbol{x} - \boldsymbol{\mu}_r) \right] .$$

(The procedure for $\Sigma_1 \neq \Sigma_2$ is described in [39], Sect. 8.3.7.) When this is plugged in Eq. (6.78), L can be cast in the form $L(\boldsymbol{x}) = a_0 + \boldsymbol{a}^{\mathrm{T}}\boldsymbol{x}$, where

$$a_0 = -\frac{1}{2} \left\{ \boldsymbol{\mu}_1^{\mathrm{T}} \Sigma_{xx}^{-1} \boldsymbol{\mu}_1 - \boldsymbol{\mu}_2^{\mathrm{T}} \Sigma_{xx}^{-1} \boldsymbol{\mu}_2 \right\} - \log \frac{\mathcal{P}_2}{\mathcal{P}_1} , \qquad \boldsymbol{a} = \Sigma_{xx}^{-1} (\boldsymbol{\mu}_1 - \boldsymbol{\mu}_2) .$$

In practice we are only dealing with finite population samples, so we do not know the exact $\boldsymbol{\mu}_1$, $\boldsymbol{\mu}_2$ and Σ_{xx}. But we can estimate them: we compute the vectors $\widehat{\boldsymbol{\mu}}_1$ in $\widehat{\boldsymbol{\mu}}_2$ (both of dimension m) by averaging over all n_1 and n_2 measurements,

$$\widehat{\boldsymbol{\mu}}_r = \overline{\boldsymbol{x}}_r = \frac{1}{n_r} \sum_{i=0}^{n_r-1} \boldsymbol{x}_{ri} \,, \qquad r = 1, 2$$

(see definition (6.75)). We estimate the $m \times m$ covariance matrix by

$$\widehat{\Sigma}_{xx} = \frac{1}{n_1 + n_2} \left[S_{xx}^{(1)} + S_{xx}^{(2)} \right] \,,$$

where

$$S_{xx}^{(r)} = \sum_{i=0}^{n_r-1} (\boldsymbol{x}_{ri} - \overline{\boldsymbol{x}}_r)(\boldsymbol{x}_{ri} - \overline{\boldsymbol{x}}_r)^{\mathrm{T}} \,, \qquad r = 1, 2 \,.$$

The discriminant function is then

$$\widehat{L}(\boldsymbol{x}) = \widehat{a}_0 + \widehat{\boldsymbol{a}}^{\mathrm{T}} \boldsymbol{x} \,, \tag{6.80}$$

where

$$\widehat{a}_0 = \frac{1}{2} \left\{ \overline{\boldsymbol{x}}_2^{\mathrm{T}} \widehat{\Sigma}_{xx}^{-1} \overline{\boldsymbol{x}}_2 - \overline{\boldsymbol{x}}_1^{\mathrm{T}} \widehat{\Sigma}_{xx}^{-1} \overline{\boldsymbol{x}}_1 \right\} + \log \frac{n_1}{n_2} \,, \qquad \widehat{\boldsymbol{a}} = \widehat{\Sigma}_{xx}^{-1} (\overline{\boldsymbol{x}}_1 - \overline{\boldsymbol{x}}_2) \,.$$

We have described the most basic procedure of binary classification with the discrimination function (6.80). This function is linear in $\boldsymbol{x}$ and we have therefore performed a *linear discriminant analysis* (LDA). When some sample from a two-class population (n measurements of m variables belonging either to class R_1 or to class R_2) has been used to determine the $m + 1$ parameters a_0 and $\boldsymbol{a}$ of the function L, the criterion (6.79) allows us to classify all further measurements $\boldsymbol{x}$. We simply compute the value of $L(\boldsymbol{x})$ for each newly acquired $\boldsymbol{x}$ and check its sign: if it is negative, $\boldsymbol{x}$ belongs to R_1, otherwise to R_2.

The quality of the classification can be tested as follows. From the complete set of n measurements we omit a single measurement (one vector of dimension m) and estimate the parameters a_0 and $\boldsymbol{a}$ of L from the remaining $n - 1$ data entries. Then the omitted entry is classified according to this modified estimate for L. We repeat this procedure for all n measurements and obtain the values

n_{11} — data that belongto R_1 and were indeed assigned to R_1 ;
n_{12} — data that belong to R_2 but were assigned to R_1 ;
n_{21} — data that belong to R_1 but were assigned to R_2 ;
n_{22} — data that belong to R_2 and were indeed assigned to R_2 .

Of course $n_{11} + n_{12} + n_{21} + n_{22} = n_1 + n_2 = n$. Ideally we would like to achieve $n_{11} = n_1$ and $n_{22} = n_2$, but in practice we almost invariably find $n_{12} \neq 0$ and/or $n_{21} \neq 0$. The measure for the inefficiency of binary classification (the *misclassification rate*) is the ratio $(n_{12} + n_{21})/n$. In the histogrammed measurements the classification inefficiency is apparent in the areas where the distribution for $L > 0$ leaks across the $L = 0$ boundary to the side with $L < 0$ and vice-versa (see Fig. 6.28).

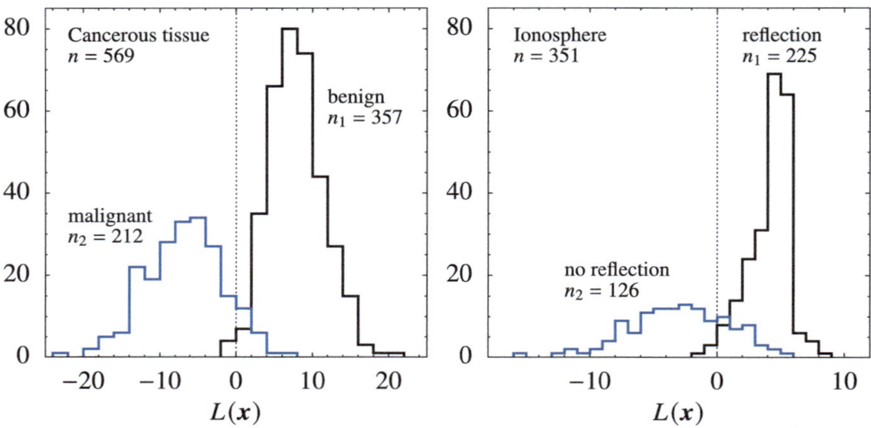

Fig. 6.28 Distribution of the measurements with respect to the linear discriminant function $L(x)$. [LEFT] Cancerous tissue (misclassification rate 4.2 %). [RIGHT] Reflections of radar signals in the ionosphere (Problem 6.13.3; misclassification rate 13.7 %)

6.10.2 Logistic Discriminant Analysis

The usual linear discrimination analysis relies on a linear discriminant function (6.80) and is therefore strongly sensitive to outliers. Obviously, we may encounter problems of the same type by analyzing data that are not distributed normally or data from two classes with strongly differing covariance matrices. To some extent, these obstacles can be overcome by *logistic discriminant analysis*. The assignment of an individual measurement x to one of the classes is recorded by keeping track of the dependent variable

$$y = \begin{cases} 1 \; ; \; x \in R_1 \,, \\ 0 \; ; \; x \in R_2 \,. \end{cases}$$

Then we collect the data into pairs $(x, y)_i$ and finally compute the parameters $\widehat{a_0}$ and $\widehat{a}$ directly by maximizing the conditional probability

$$\mathcal{L}(a_0, a) = \prod_{i=0}^{n-1} [p_1(x_i; a_0, a)]^{y_i} \, [1 - p_1(x_i; a_0, a)]^{1-y_i} \,,$$

where $p_1(x) = p(\xi = x \mid \xi \in R_1)$ (see Eq. (6.76)). Of course, the method has its deficiencies. Among others, it requires substantially larger data samples than linear discrimination for the same asymptotic efficiency. An efficient iterative scheme for the maximization of $\mathcal{L}$ with respect to a_0 and a and other details can be found in [39], Sect. 8.3.5.

6.10.3 Assignment to Multiple Classes

Multi-class discriminant analysis is a generalization of two-class analysis to multiple classes. Initial reading is offered by [39], Sect. 8.5, and [45], Chap. 12.

6.11 Canonical Correlation Analysis ⋆

In addition to the principal component analysis (Sect. 6.8), one of the most widespread and generally useful methods of multivariate analysis is the *canonical correlation analysis* (CCA). The CCA method allows us to handle sets of data of the form

$$x_i = (x_{i1}, x_{i2}, \ldots, x_{ip})^{\mathrm{T}}, \qquad y_i = (y_{i1}, y_{i2}, \ldots, y_{iq})^{\mathrm{T}}, \qquad i = 0, 1, \ldots, n-1,$$

where $p \neq q$ in general. The basic premise is that the correlation between x_i and y_i is the most relevant carrier of information for this data set. By CCA we wish to reduce the dimensionality of the data by projecting x_i and y_i to a smaller set of canonical variables, among which we attempt to impose maximum correlation. From this viewpoint, the task of CCA is precisely opposite to the task of PCA, where data should become *decorrelated*.

An example are the $n = 3462$ multivariate data on measured properties of galaxies (Problem 6.13.4, adapted from [39] based on data [54, 55]). Each of the n measurements represents a data point x_i ($0 \leq i \leq n-1$) with $p = 23$ values describing the brightnesses and luminosities of galaxies in various spectral ranges, while each y_i stores $q = 6$ variables describing other physical properties. We are interested in correlations between the individual components of these data.

From the data we estimate the means $\overline{x} = (\sum_{i=0}^{n-1} x_i)/n$ and $\overline{y} = (\sum_{i=0}^{n-1} y_i)/n$, and use them to estimate the covariance matrices

$$\Sigma_{xx} = \frac{1}{n-1} \sum_{i=0}^{n-1} (x_i - \overline{x})(x_i - \overline{x})^{\mathrm{T}}, \qquad \Sigma_{yy} = \frac{1}{n-1} \sum_{i=0}^{n-1} (y_i - \overline{y})(y_i - \overline{y})^{\mathrm{T}},$$

and

$$\Sigma_{xy} = \Sigma_{yx}^{\mathrm{T}} = \frac{1}{n-1} \sum_{i=0}^{n-1} (x_i - \overline{x})(y_i - \overline{y})^{\mathrm{T}}.$$

For general x and y we construct the linear combinations

$$\begin{aligned}
\xi_j &= a_{1j}x_1 + a_{2j}x_2 + \cdots + a_{pj}x_p = a_j^{\mathrm{T}}x, \\
\zeta_j &= b_{1j}y_1 + b_{2j}y_2 + \cdots + b_{qj}y_q = b_j^{\mathrm{T}}y,
\end{aligned} \tag{6.81}$$

where $j = 1, 2, \ldots, r$ and $r \leq \min(p, q)$. The coefficient vectors

$$\boldsymbol{a}_j = (a_{1j}, a_{2j}, \ldots, a_{pj})^\mathrm{T}, \qquad \boldsymbol{b}_j = (b_{1j}, b_{2j}, \ldots, b_{qj})^\mathrm{T},$$

are determined—this is the key step of CCA—such that the pairs (ξ_j, ζ_j) are arranged in decreasing strength of linear correlation

$$\rho_j = \mathrm{corr}(\xi_j, \zeta_j) = \frac{\boldsymbol{a}_j^\mathrm{T} \Sigma_{xy} \boldsymbol{b}_j}{\sqrt{\boldsymbol{a}_j^\mathrm{T} \Sigma_{xx} \boldsymbol{a}_j} \sqrt{\boldsymbol{b}_j^\mathrm{T} \Sigma_{yy} \boldsymbol{b}_j}}, \qquad j = 1, 2, \ldots, r, \qquad (6.82)$$

so that $\rho_1 \geq \rho_2 \geq \cdots \geq \rho_r$. Moreover, ξ_j should be uncorrelated to all ξ_k with lower indices,

$$\mathrm{cov}(\xi_j, \xi_k) = \boldsymbol{a}_j^\mathrm{T} \Sigma_{xx} \boldsymbol{a}_k = 0, \qquad k < j,$$

and analogously for all ζ_j,

$$\mathrm{cov}(\zeta_j, \zeta_k) = \boldsymbol{b}_j^\mathrm{T} \Sigma_{yy} \boldsymbol{b}_k = 0, \qquad k < j.$$

In plain words, arbitrary correlations between the original variables $\boldsymbol{x}$ and $\boldsymbol{y}$ are thus systematically relocated to an ordered sequence of decreasing correlations between the pairs (ξ_j, ζ_j) which are known as *canonical variates*, while the corresponding mutual correlations (6.82) are the *canonical correlation coefficients*.

The coefficients vectors $\boldsymbol{a}_j$ and $\boldsymbol{b}_j$ are obtained after a short calculation in which $\mathrm{corr}(\xi, \zeta) = \boldsymbol{a}^\mathrm{T} \Sigma_{xy} \boldsymbol{b}$ is maximized. The full derivation can be found in [39], here we just quote the final result. We compute the products

$$\Omega_a = \Sigma_{xx}^{-1/2} \Sigma_{xy} \Sigma_{yy}^{-1} \Sigma_{yx} \Sigma_{xx}^{-1/2}, \qquad \Omega_b = \Sigma_{yy}^{-1/2} \Sigma_{yx} \Sigma_{xx}^{-1} \Sigma_{xy} \Sigma_{yy}^{-1/2},$$

and determine the eigenvectors and eigenvalues λ_j of the symmetric matrices $\Omega_a \in \mathbb{R}^{p \times p}$ and $\Omega_b \in \mathbb{R}^{q \times q}$. We order the eigenvalues in decreasing order and arrange the corresponding eigenvectors accordingly. We get

$$\boldsymbol{a}_j^\mathrm{T} = \sqrt{\lambda_j}\, \boldsymbol{u}_j^\mathrm{T} \Sigma_{xx}^{-1/2}, \qquad \boldsymbol{b}_j^\mathrm{T} = \boldsymbol{v}_j^\mathrm{T} \Sigma_{yy}^{-1/2}, \qquad (6.83)$$

where $\boldsymbol{u}_j$ is the eigenvector of the matrix Ω_a corresponding to the eigenvalue λ_j, and $\boldsymbol{v}_j$ is the eigenvector of Ω_b (the first $r = \min(p, q)$ eigenvalues of Ω_a and Ω_b are identical). In the above expressions we need the "square roots" of the matrices Σ_{xx} and Σ_{yy}. Appendix A.8 teaches you how to do this.

The coefficient vectors (6.83) are all we need in the last step. We use them to compute (6.81) for an arbitrary pair of data $\boldsymbol{x}_i$ and $\boldsymbol{y}_i$ ($i = 0, 1, \ldots, n - 1$). This gives us the *canonical variate scores*

$$\xi_{ij} = \boldsymbol{a}_j^\mathrm{T} \boldsymbol{x}_i, \qquad \zeta_{ij} = \boldsymbol{b}_j^\mathrm{T} \boldsymbol{y}_i, \qquad j = 1, 2, \ldots, r, \qquad (6.84)$$

where for each j we also compute the corresponding correlation coefficient (6.82). If some significant correlation is present in the original data, it should express itself most acutely in a two-dimensional rendering of the set of pairs (ξ_{i1}, ζ_{i1}). The next set of pairs (ξ_{i2}, ζ_{i2}) will most likely exhibit a smaller correlation, the third set (ξ_{i3}, ζ_{i3}) an even smaller one, and so on.

The images of canonical variate scores often reveal unexpected structures in the data or highlight outliers which would remain hidden in the original variables. A basic feel for the method is conveyed by Problem 6.13.4. Further reading on CCA can be found in Chap. 14 of [45] and Sect. 7.3 of [39].

6.12 Factor Analysis ⋆

Factor analysis is a multivariate statistical method in which we strive to explain a set of measured data by a much smaller number of variables known as *factors*. The factors do not necessarily correspond to physically measurable quantities and should therefore be considered as *latent* variables used only for a formal parameterization of the data. Classical factor analysis therefore belongs to the class of *latent variable models*; see Sects. 15.4 and 15.5 in [39]. In the context of time series we encounter one such model in Sect. 7.9. Numerous fields of application of factor analysis in natural sciences are discussed in [56].

Depending on how the data and the factors are related, several types of factor analysis exist. By far the most common is a linear relation,

$$X = FA^{\mathrm{T}} + E \,. \tag{6.85}$$

The $n \times m$ matrix X contains n multivariate data $\boldsymbol{x}_i = (x_{i1}, x_{i2}, \ldots, x_{im})^{\mathrm{T}}$, one per row. The $n \times k$ matrix F contains the *factor scores* for the ith data. The number of factors k which, as a rule, is much smaller than the number of data components m, is chosen in advance. The $m \times k$ matrix A contains the *factor loadings* which determine the role of a given factor in the data $\boldsymbol{x}_i$ (see Fig. 6.29). The $n \times m$ matrix E contains the remainders which represent data errors.

In Eq. (6.85) only the data matrix X is known, while we are completely ignorant of all matrices at the right-hand side of the equation. A unique solution without additional assumptions does not exist. From many possible ways [56] we follow, as an illustration, the path of *principal-component factor analysis*. In this approach the factors are obtained by the additional requirement that the factors should describe the maximum variance of all measured data X.

The data should first be centered as in (6.60). The factors and the weights are then determined by fitting the matrix product FA^{T} to the data X_{c} in the least-squares sense. This means that for the chosen number of factors $k < m$ we must determine F and A^{T} such that the sum of squares of the elements of

$$E = X_{\mathrm{c}} - FA^{\mathrm{T}}$$

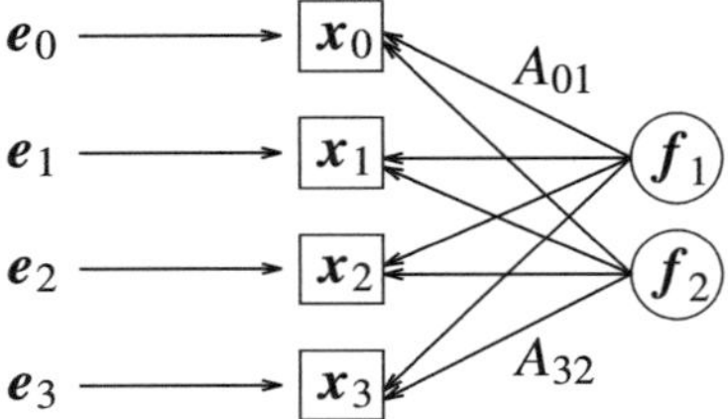

Fig. 6.29 A schematic representation of factor analysis. We construct a linear combination of a small number of factors (two in this example, f_1 and f_2) and remainders e_i in order to represent a larger body of data x_i. The weights are stored in the matrix A such that for each data entry we have $x^{\mathrm{T}} = A f^{\mathrm{T}} + e^{\mathrm{T}}$, where x, f, and e are the corresponding rows of the matrices X, F, and E

will be minimal. We know the solution of this problem from Sect. 4.5. The optimal approximation of the rank-r matrix $X \in \mathbb{R}^{n \times m}$ by the matrix $W \in \mathbb{R}^{n \times m}$ of a lower rank $k < r$ in the least-squares sense is given by the singular-value decomposition $X = U \Gamma V^{\mathrm{T}}$, such that

$$F A^{\mathrm{T}} = \gamma_1 u_1 v_1^{\mathrm{T}} + \gamma_2 u_2 v_2^{\mathrm{T}} + \cdots + \gamma_k u_k v_k^{\mathrm{T}} = U_k \Gamma_k V_k^{\mathrm{T}} , \tag{6.86}$$

where u_i $(1 \le i \le k)$ are the first k columns of U, and v_i $(1 \le i \le k)$ are the first k columns of V. The vectors u_i (dimension n) and v_i (dimension m) are arranged in decreasing order of the corresponding singular values in the matrix $\Gamma_k = \mathrm{diag}\,(\gamma_1, \gamma_2, \ldots, \gamma_k)$. In this approach we neglect the measurement errors (remainders) e, that is, we assume $\|E\| \ll \|X_c\|$ in some matrix norm.

6.12.1 Determining the Factors and Weights from the Covariance Matrix

Since the singular values of X are closely related to the eigenvalues of the product $X^{\mathrm{T}} X$ (see Appendix A.6), we prefer to use the covariance matrix

$$\widehat{\Sigma}_{xx} = \frac{1}{n-1} X_c^{\mathrm{T}} X_c \tag{6.87}$$

instead of the centered data matrix X_c. At first sight, the decomposition (6.86) offers two solutions. The first possibility is

$$F = U_k , \qquad A^{\mathrm{T}} = \Gamma_k V_k^{\mathrm{T}} . \tag{6.88}$$

The singular values γ_i from the decomposition of the data matrix X_c are the square roots of the singular values λ_i from the decomposition of the covariance matrix $\widehat{\Sigma}_{xx} = U_k \Lambda_k V_k^{\mathrm{T}}$, thus we obtain

$$A = \left(\Gamma_k V_k^{\mathrm{T}}\right)^{\mathrm{T}} = V_k \Gamma_k = V_k \Lambda_k^{1/2} \tag{6.89}$$

and

$$F = U_k = X_{\mathrm{c}} V_k \Gamma_k^{-1} = X_{\mathrm{c}} A \Gamma_k^{-1} \Gamma_k^{-1} = X_{\mathrm{c}} A \Lambda_k^{-1} . \tag{6.90}$$

By following this procedure we obtain factors that are uncorrelated and have unit variance. The columns of the weight matrix A are directly comparable to each other. The matrix of correlations *between the data and the factors* is

$$C = S^{-1} A = S^{-1} V_k \Lambda_k^{1/2} , \tag{6.91}$$

where $S \in \mathbb{R}^{m \times m}$ is diagonal: its diagonal elements are the standard deviations of the data components,

$$S = \mathrm{diag} \left(\sqrt{\widehat{\Sigma}_{11}}, \sqrt{\widehat{\Sigma}_{22}}, \ldots, \sqrt{\widehat{\Sigma}_{mm}} \right) .$$

The first method of determining the factors is then

1. Compute the covariance matrix (6.87) and its singular-value decomposition $U_k \Lambda_k V_k^{\mathrm{T}}$.
2. Compute the weight matrix A by (6.89) and the factor matrix F by (6.90).
3. The data-factors correlation is then given by (6.91).

The second possibility that can be guessed from the decomposition (6.86) is

$$F = U_k \Gamma_k , \qquad A^{\mathrm{T}} = V_k^{\mathrm{T}} . \tag{6.92}$$

Again we compute the singular values of the covariance matrix instead of the singular values of the data matrix, and use $U_k = X_{\mathrm{c}} V_k \Gamma_k^{-1}$. We obtain

$$A = V_k , \qquad F = X_{\mathrm{c}} V_k \Gamma_k^{-1} \Gamma_k = X_{\mathrm{c}} V_k = X_{\mathrm{c}} A . \tag{6.93}$$

This option also gives us mutually uncorrelated factors, but they have different variances, and the columns of the weight matrix A are therefore not directly comparable. Nonetheless, the matrix of correlations between the data and the factors is still given by (6.91). We follow the procedure

1. Compute the covariance matrix (6.87) and its singular-value decomposition $U_k \Lambda_k V_k^{\mathrm{T}}$.
2. Compute the weight matrix A and the factor matrix F by (6.93).
3. The data-factors correlation is given by (6.91).

Example (Adapted from [56].) In some measurement we have acquired a set of $n = 10$ multivariate data with $m = 5$ components each, $x_i = (x_{i1}, x_{i2}, \ldots, x_{i5})^{\mathrm{T}}$. We assume that the whole data set can be described by only two factors, which we would like to determine such that individual data components are well correlated to either one of the factors. We arrange the data in the matrix

$$
X = \begin{pmatrix}
5.0 & 25.0 & 15.0 & 5.0 & 5.0 \\
10.0 & 30.0 & 17.0 & 17.0 & 8.0 \\
3.0 & 6.0 & 10.0 & 13.0 & 25.0 \\
7.5 & 27.2 & 16.0 & 11.0 & 6.5 \\
4.6 & 21.9 & 14.0 & 6.6 & 9.0 \\
3.8 & 13.6 & 12.0 & 9.8 & 17.0 \\
8.3 & 26.6 & 15.9 & 14.2 & 9.1 \\
6.1 & 22.7 & 14.6 & 10.2 & 9.9 \\
7.6 & 24.2 & 15.2 & 13.8 & 10.8 \\
3.9 & 10.3 & 11.2 & 12.6 & 21.3
\end{pmatrix} .
\tag{6.94}
$$

From each row of X we subtract the mean vector

$$
x = \frac{1}{n} \sum_{i=0}^{n-1} x_i = (5.98, 20.75, 14.09, 11.32, 12.16)^{\mathrm{T}} ,
$$

and get the centered data matrix X_{c} by which we compute the covariance matrix

$$
\widehat{\Sigma}_{xx} = \begin{pmatrix}
5.257 & 15.768 & 4.739 & 4.585 & -10.150 \\
15.768 & 63.818 & 18.209 & 1.569 & -50.618 \\
4.739 & 18.209 & 5.241 & 1.177 & -14.044 \\
4.585 & 1.569 & 1.177 & 13.006 & 6.068 \\
-10.150 & -50.618 & -14.044 & 6.068 & 44.305
\end{pmatrix} .
$$

The eigenvalues of $\widehat{\Sigma}_{xx}$ are $\lambda_1 = 114.039$, $\lambda_2 = 17.5673$, $\lambda_3 = 0.0216273$, and $\lambda_4 \approx \lambda_5 \approx 0$. We wish to describe the data by using just two factors ($k = 2$), so we keep only the two largest eigenvalues λ_1 and λ_2 which are stored in $\Lambda_2 = \mathrm{diag}(\lambda_1, \lambda_2)$. They correspond to the first two columns of V,

$$
V_2 = \begin{pmatrix}
-0.173 & 0.323 \\
-0.744 & 0.187 \\
-0.211 & 0.101 \\
0.015 & 0.860 \\
0.609 & 0.334
\end{pmatrix} .
$$

We choose the first variant of determining the factor weights and factor scores. By using Eqs. (6.89) and (6.90) we get

$$A = V_2 \Lambda_2^{1/2} = \begin{pmatrix} -1.851 & 1.353 \\ -7.949 & 0.783 \\ -2.249 & 0.424 \\ 0.157 & 3.603 \\ 6.506 & 1.401 \end{pmatrix}, \qquad F = X_c A \Lambda_2^{-1} = \begin{pmatrix} -0.716 & -1.731 \\ -0.997 & 1.625 \\ 1.892 & 0.383 \\ -0.835 & -0.066 \\ -0.243 & -1.277 \\ 0.849 & -0.463 \\ -0.652 & 0.830 \\ -0.278 & -0.301 \\ -0.363 & 0.705 \\ 1.342 & 0.296 \end{pmatrix}.$$

By using Eq. (6.91) and the matrix $S = \mathrm{diag}(2.293, 7.989, 2.289, 3.606, 6.656)$ we finally obtain the data-factor correlation matrix:

$$C = S^{-1} V_k \Lambda_k^{1/2} = \begin{pmatrix} \underline{-0.807} & 0.590 \\ \underline{-0.995} & 0.098 \\ \underline{-0.983} & 0.185 \\ 0.044 & \underline{0.999} \\ \underline{0.977} & 0.211 \end{pmatrix}. \tag{6.95}$$

The first column of (6.95) establishes a correlation of the first factor to the first, second, third, and fifth component of x_i. The second column reveals a strong correlation between the second factor and the fourth data component. ◁

The solution for the matrices F and A^{T} by using Eq. (6.86) is obviously not unique, as we may perform the transformations $F' = FT$ and $A' = AT^{-1}$ with any non-singular matrix T and still fulfill the equation. The solutions (6.88) and (6.92) presented here are just the most obvious ones: no transformation is applied to the factors. Still, even though the matrix (6.95) distinctly reveals the basic connections between the data and the factors, we would ideally wish to obtain a matrix with an even simpler structure,

$$C \approx \begin{pmatrix} \underline{-1} & 0 \\ \underline{-1} & 0 \\ \underline{-1} & 0 \\ 0 & \underline{1} \\ \underline{1} & 0 \end{pmatrix},$$

which would pronounce the correlation yet more clearly. This can be achieved by carefully chosen transformations of the factors; the most commonly used in practice are the so-called *varimax rotations*. Details can be found in [56].

6.12.2 Standardization of Data and Robust Factor Analysis

In Sect. 6.12.1 and in the Example on page 379 we have used the centered data matrix X_c. Principal component factor analysis is not suitable for data with exceedingly different variances, for example, for data with components in different measurement units. (This problem is common to all approaches exploiting the principal component method; see Sect. 6.8.2.)

If the variances of the data components are very different, the data should be standardized: we should subtract the means and divide out the appropriate standard deviations as in (6.69). This turns the covariance matrix into the correlation matrix. To determine the factors and weights we still follow the procedure of Sect. 6.12.1, but slightly different matrices A, F, and C are obtained. For the data (6.94) from the Example on page 379 we get

$$
C = \begin{pmatrix}
\underline{-0.917} & 0.398 \\
\underline{-0.992} & -0.124 \\
\underline{-0.999} & -0.035 \\
-0.178 & \underline{0.984} \\
\underline{0.907} & 0.421
\end{pmatrix} ,
$$

which we may compare to the result (6.95).

The approaches to factor analysis outlined here are, again, very sensitive to outliers. A single runaway data component can strongly modify the covariance matrix and may result in completely different factor and weight matrices F and A, as well as the correlation matrix C. Robust factor analysis which is not hampered by these obstacles is discussed in the papers [57, 58]; see also Sect. 15.4 in [39] and the often-quoted geo-chemical research work [59].

6.13 Problems

6.13.1 Multiple Regression

The lean body mass (the difference between the mass of the whole body and the mass of the fatty tissue) is an indicator of an individual's predisposition for certain types of diseases. It can be measured by underwater weighting [60] but that procedure is cumbersome and expensive. In this Problem (adapted from [39]) we try to establish whether by much simpler measurements of the body mass, the circumference of certain body parts, and by knowing the person's age, multiple regression may be a tool to reach the comparable estimate of the lean body mass.

We collect n measurements of $m = 13$ independent variables x_1 (age), x_2 (body mass), x_3 (body height) and the circumferences x_4 (neck), x_5 (chest), x_6 (abdomen),

x_7 (hip), x_8 (thigh), x_9 (knee), x_{10} (ankle), x_{11} (biceps), x_{12} (forearm), x_{13} (wrist). Assume that the dependent variable, the mass of the fatty tissue, can be described by the model

$$y = a_0 + \sum_{j=1}^{m} a_j x_j + \Delta y \, ,$$

which applies for each of the n measurements (see Eq. (6.58)) and where Δy is the unknown error. The a_j are the regression parameters to be determined.

$\odot$ On the website [60] one can find the data for $n = 252$ measurements of $m = 13$ input variables x_j and one output variable y. Use the method of multiple regression described in Sect. 6.7.1 to determine these regression parameters a_j (and establish the corresponding independent variables) that are statistically significant for the determination of the lean body mass.

$\oplus$ If there are fewer measurements than there are independent variables in each individual measurement ($n \ll m$), the data matrix $X_c^{\mathsf{T}} X_c$ becomes non-invertible. In such cases we may apply the principal component multiple regression (see, for example, the absorption spectrum for the PET yarns in Sect. 6.7.2). Try to reproduce Fig. 6.18. The data can be found at the book's web page.

6.13.2   Nutritional Value of Food

We are used to specifying the nutritional value of individual food items by quoting their protein content (x_1), carbohydrates (x_2), fats (x_3) and saturated fats (x_4), as well as cholesterol (x_5), the total energy value (x_6), and the mass (x_7). Suppose we measure these $m = 7$ variables for each of the n food items, where $n \gg m$. Can we reduce the dimensionality of this set of data by using the principal component analysis (PCA)?

$\odot$ At the book's web page you can find the data for $n = 691$ different food items. The values were measured for items with very different masses x_7, so let us first normalize all data as

$$x_j \leftarrow x_j / x_7 \, , \qquad 1 \le j \le 6 \, .$$

Since the measured values are expressed in different measurement units, you should also standardize the data by using (6.69), and construct the standardized data matrix X_c. Apply the PCA method from Sect. 6.8 in either of its two forms: by diagonalizing the covariance (or the correlation) matrix, or by singular value decomposition of the data matrix.

Regardless of the way you compute the principal components, show how the percentage of the variance explained by inclusion of r principal components changes with increasing r. For a few chosen x_i, check how well r principal components reproduce the original data point. Plot the scores (6.68) for all n data at the same time, for a few of the leading principal components PC (in particular, PC1 vs. PC2, PC1 vs. PC3, and PC2 vs. PC3). Use these plots to determine whether some food items tend to form clusters and whether there are any outliers.

6.13.3 Discrimination of Radar Signals from Ionospheric Reflections

A system of sixteen antennas in Goose Bay (Labrador) has been used to measure the reflection of radar waves from free electrons in the ionosphere. (Example from [39] based on original data of [61].) Those reflected signals that hinted at a structure in the ionosphere, have been assigned to the class of "good" signals (class R_1). The remaining reflections were labeled as "bad" (class R_2). They have determined the character of $n = 351$ signals, each of which was represented by $m = 34$ independent variables $x_i = (\text{Re}\, x_i, \text{Im}\, x_i)$, $1 \leq i \leq 17$, corresponding to 17 pairs of complex values (the imaginary part of the first signal is always zero, so the actual number of independent variables is just $m = 33$).

⊙ Fig. 6.30 makes it abundantly clear why a simple bivariate (for example, correlation) analysis does not suffice. By comparing the pairs of individual components of the signal alone, the classes can not be cleanly separated, as "good" reflections are interspersed with the "bad" ones. Perform a linear discriminant analysis of the radar signals from these measurements by using the procedure described in Sect. 6.10. Try to reproduce Fig. 6.28 (right) and compute the misclassification rate.

⊕ Classify the same radar reflection data by the method of logistic discriminant analysis (Sect. 6.10.2). Compute the misclassification rate.

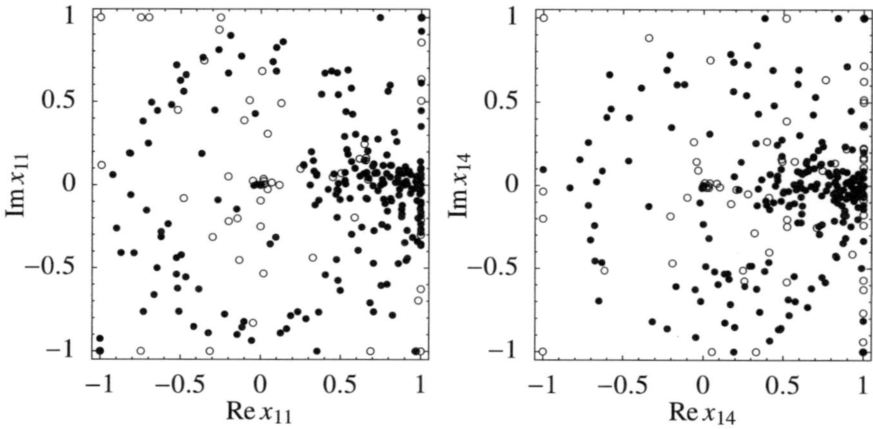

Fig. 6.30 Reflection of radar signals in the ionosphere. Shown are the "good" (symbols •) and "bad" reflections (symbols ○). [LEFT] Variable x_{11}. [RIGHT] Variable x_{14}

6.13.4 Canonical Correlation Analysis of Objects in the CDFS Area

Chandra Deep Field South (CDFS) is one of the most frequently studied areas in the sky. In a comprehensive study [54] the astronomers have collected astrometric, photometric, and morphological data on 63501 astrophysical objects (for example, galaxies, stars, and quasars). The measurements of the spectro-photometric data have been performed in 17 wavelength bands spanning the range between 350 and 930 nm. The whole COMBO-17 (*Classifying Objects by Medium-Band Observations—A spectrophotometric 17-filter survey*) data set is accessible at [55].

This Problem deals only with the data which the researchers believe belongs to galaxies. We have $n = 3462$ multivariate data entries. From the complete set of variables in each measurement we select $p = 23$ variables of the data x_i and $q = 6$ variables of the data y_i (see Sect. 6.11). The first ten elements of x_i are the absolute magnitudes of individual galaxies in ten different spectral bands:

$$\texttt{UjMag, BjMag, VjMag, usMag, gsMag, rsMag,}$$
$$\texttt{UbMag, BbMag, VbMag, S280Mag.}$$

The remaining elements are the observed brightnesses in thirteen spectral bands:

$$\texttt{W420F\_E, W462F\_E, W485F\_D, W518F\_E, W571F\_S, W604F\_E, W646F\_D,}$$
$$\texttt{W696F\_E, W753F\_E, W815F\_S, W856F\_D, W914F\_D, W914F\_E.}$$

The elements of y_i are

$$\texttt{Rmag, ApD\_Rmag, mu\_max, MC\_z, MC\_z\_ml, chi2red}$$

(total R-band magnitude, aperture difference of $\texttt{Rmag}$, central surface brightness in $\texttt{Rmag}$, mean shift in the distribution of the observed sources with respect to the red shift z, the peak of the red shift distribution, and the quality of the fit).

⊙ Use the canonical correlation analysis to compute the pairs of canonical variables (ξ_j, ζ_j) for $1 \le j \le 6$ with the coefficients (6.83) and their scores (6.84) for the data from the COMBO-17 database. Keep only those data from the database in which none of the 23 elements of x_i and none of the 6 elements of y_i are missing. For each j plot the pairs of scores and compute the corresponding canonical correlation coefficient.

References

1. J.E. Gentle, W. Härdle, Y. Mori (eds.), *Handbook of Computational Statistics (Concepts and Methods* (Springer-Verlag, Berlin, 2004)
2. V. Barnett, T. Lewis, *Outliers in Statistical Data*, 3rd edn. (John Wiley & Sons, New York, 1994)

3. R. Kandel, *Our Changing Climate*, p. 110 (McGraw-Hill, New York, 1991)
4. L. Davies, U. Gather, *Robust Statistics*, Chapter III.9 in III, pp. 655–695
5. Analytical Methods Committee, Robust statistics—how not to reject outliers, Part 1: basic concepts. Analyst **114**, 1693 (1989); Part 2: Inter-laboratory trials. Analyst **114**, 1699 (1989)
6. V. Chandola, A. Banerjee, V. Kumar, Anomaly detection: a survey. ACM Comput. Surv. **41**, art. 15 (2009)
7. A. Patcha, J.-M. Park, An overview of anomaly detection techniques: existing solutions and latest technological trends. Comp. Networks **51**, 3448 (2007)
8. M. Agyemang, K. Barker, R. Alhajj, A comprehensive survey of numeric and symbolic outlier mining techniques. Intell. Data Anal. **10**, 521 (2006)
9. V.J. Hodge, J. Austin, A survey of outlier detection methodologies. Artif. Intell. Rev. **22**, 85 (2004)
10. L. Davies, U. Gather, The identification of multiple outliers. J. Amer. Statistical Assoc. **88**, 782 (1993); See also B. Iglewicz, J. Martinez, Outlier detection using robust measures of scale. J. Statist. Comput. Simul. **15**, 285 (1982)
11. F.E. Grubbs, Procedures for detecting outlying observations in samples. Technometrics **11**, 1 (1969)
12. W. J. Dixon, Ratios involving extreme values. Ann. Math. Stat. **22**, 68 (1951); W.J. Dixon, Analysis of extreme values. Ann. Math. Stat. **21**, 488 (1950)
13. R.J. Beckman, R.D. Cook, Outlier..........s. Technometrics **25**, 119 (1983)
14. R.A. Maronna, R.D. Martin, V.J. Yohai, *Robust Statistics. Theory and Methods* (John Wiley & Sons, Chichester, 2006)
15. M.R. Spiegel, *Schaum's Outline of Theory and Problems of Probability and Statistics* (McGraw-Hill, New York, 1975)
16. S. Brandt, *Data Analysis*, 3rd edn. (Springer-Verlag, New York, 1999)
17. H.B. Mann, A. Wald, On the choice of the number of class intervals in the application of the chi square test. Ann. Math. Stat. **13**, 306 (1942)
18. W.C.M. Kallenberg, J. Oosterhoff, B.F. Schriever, The number of classes in chi-squared goodness-of-fit tests. J. Amer. Stat. Assoc. **80**, 959 (1985) and references therein. See also W.C. Kallenberg, On moderate and large deviations in multinomial distributions. Ann. Stat. **13**, 1554 (1985)
19. A. Kolmogorov, Sulla determinazione empirica di una legge di distribuzione. Giornalo dell'Istituto Italiano degli Attuari **4**, 461 (1933), Translated in A.N. Shiryayev (ed.), *Selected Works of A.N. Kolmogorov*, vol. II, p. 139 (Springer Science+Business Media, Dordrecht, 1992)
20. N. Smirnov, Sur les écarts de la courbe de distribution empirique. Rec. Math. **6**, 3 (1939)
21. S. Facchinetti, A procedure to find exact critical values of Kolmogorov-Smirnov test. Statistica Applicata-Ital. J. Appl. Stat. **21**, 337 (2009)
22. M.A. Stephens, Use of the Kolmogorov-Smirnov, Cramer-von Mises and related statistics without extensive tables. J. Royal Stat. Soc. B **32**, 115 (1970)
23. S. Širca, *Probability for Physicists* (Springer International Publishing AG Switzerland, 2016)
24. A.F. Nikiforov, S.K. Suslov, V.B. Uvarov, *Classical Orthogonal Polynomials of a Discrete Variable* (Springer-Verlag, Berlin, 1991)
25. W.H. Press, B.P. Flannery, S.A. Teukolsky, W.T. Vetterling, *Numerical Recipes: The art of Scientific Computing*, 3rd edn (Cambridge University Press, Cambridge, 2007). See also the equivalent handbooks in Fortran, Pascal and C, as well as http://numerical.recipes
26. C.A. Cantrell, Technical note: review of methods for linear least-squares fitting of data and application to atmospheric chemistry problems. Atmos. Chem. Phys. **8**, 5477 (2008)
27. D. York et al., Unified equations for the slope, intercept, and standard errors of the best straight line. Am. J. Phys. **72**, 367 (2004)
28. R.L. Workman et al. (Particle Data Group), Review of Particle Physics. Prog. Theor. Exp. Phys. 2022, 083C01 (2022); see Section 5 of the Introduction
29. M.C. Ortiz, L.A. Sarabia, A. Herrero, Robust regression techniques. A useful alternative for the detection of outlier data in chemical analysis. Talanta **70**, 499 (2006)

30. J. Ferré, *Regression Diagnostics*, Section 3.20 in the Encyclopedia, in *Comprehensive chemometrics: chemical and biochemical data analysis*, 2nd edn, vol. 3, p. 431, S.D. Brown, R. Tauler, B. Walczak ed. by (Elsevier, 2020)

31. P.J. Rousseeuw, A.M. Leroy, *Robust Regression and Outlier Detection* (John Wiley & Sons, Hoboken, 2003)

32. I. Barrodale, F.D.K. Roberts, An improved algorithm for discrete l_1 linear approximation. SIAM J. Numer. Anal. **10**, 839 (1973)

33. S. Portnoy, R. Koenker, The Gaussian hare and the Laplacian tortoise: computability of squared-error versus absolute-error estimators. Stat. Sci. **12**, 279 (1997)

34. P.J. Rousseeuw, Least median of squares regression. J. Amer. Stat. Assoc. **79**, 871 (1984)

35. T. Bernholt, *Computing the Least Median of Squares Estimator in Time $O(n^d)$*, in *Lecture Notes in Computer Science* vol. 3480, p. 697, ed. by O. Gervasi et al. (Springer-Verlag, Berlin, 2005)

36. A. Stromberg, Computing the exact least median of squares estimate and stability diagnostics in multiple linear regression. SIAM J. Sci. Comp. **14**, 1289 (1993)

37. T.A. Boden, G. Marland, R.J. Andres, *Global, Regional, and National Fossil-Fuel CO_2 Emissions*, Carbon Dioxide Information Analysis Center, Oak Ridge National Laboratory, https://doi.org/10.3334/CDIAC/00001_V2015

38. B.W. Rust, *Fitting nature's basic functions. Part I: Polynomials and linear least squares*, Computers in Science and Engineering Sep/Oct 2001, p. 84; *Part II: Estimating uncertainties and testing hypotheses*, CISE Nov/Dec 2001, p. 60; *Part III: Exponentials, sinusoids, and nonlinear least squares*, CISE Jul/Aug 2002, p. 72; *Part IV: The variable projection algorithm*, CISE Mar/Apr 2003, p. 74

39. A.J. Izenman, *Modern Multivariate Statistical Techniques* (Springer-Verlag, Berlin, 2008)

40. H. Swierenga, A.P. de Weijer, R.J. van Wijk, L.M.C. Buydens, Strategy for constructing robust multivariate calibration models. Chemom. Intell. Lab. Syst. **49**, 1 (1999)

41. I.T. Jolliffe, *Principal Component Analysis*, 2nd edn. (Springer-Verlag, Berlin, 2002)

42. S. Roweis, Z. Ghahramani, A unifying review of linear Gaussian models. Neural Comput. **11**, 305 (1999)

43. A. Azzalini, A.W. Bowman, A look at some data on the Old Faithful geyser. J. Royal Stat. Soc. C **39**, 357 (1990)

44. A.K. Jain, M.N. Murty, Data clustering: a review. ACM Comput. Surv. **31**, 264 (1999)

45. W. Härdle, L. Simar, *Applied Multivariate Statistical Analysis* (Springer-Verlag, Berlin, 2007)

46. R. Xu, D.C. Wunsch II., *Clustering* (John Wiley & Sons, Hoboken, 2009)

47. G. Gan, C. Ma, J. Wu, *Data Clustering. Theory, Algorithms, and Applications* (SIAM, Philadelphia, 2007)

48. J. Kogan, *Introduction to Clustering Large and High-Dimensional Data* (Cambridge University Press, Cambridge, 2007)

49. J. Valente de Oliveira, W. Pedrycz (eds.), *Advances in Fuzzy Clustering and Its Applications* (John Wiley & Sons, Chichester, 2007)

50. The R Project for Statistical Computing, see http://www.r-project.org Attention: the R reference manual has approximately 3000 pages! A good introductory text for R is J. Maindonald, J. Braun, *Data Analysis and Graphics Using R*, 2nd edn (Cambridge University Press, Cambridge, 2006). R is an open-source alternative to the S/S+ systems ("R is to S what OCTAVE is to MATLAB")

51. U. von Luxburg, *A tutorial on Spectral Clustering*, Max-Planck-Institut für biologische Kybernetik, Technical Report No. Tr-149 (2006)

52. A.Y. Ng, M.I. Jordan, Y. Weiss, On spectral clustering: analysis and an algorithm. Adv. Neural Inf. Process. Syst. **14**, 849 (2001). See also Ref. [11] in this paper

53. O.L. Mangasarian, W.N. Street, W.H. Wolberg, Breast cancer diagnosis and prognosis via linear programming. Oper. Res. **43**, 570 (1995)

54. C. Wolf et al., A catalogue of the Chandra deep field south with multi-colour classification and photometric redshifts from COMBO-17. Astron. Astrophys. **421**, 913 (2004). See also the update C. Wolf et al., Calibration update of the COMBO-17 CDFS catalogue. Astron. Astrophys. **492**, 933 (2008)

55. https://www2.mpia-hd.mpg.de/COMBO/combo_CDFSpublic.html. The data are available at https://astrostatistics.psu.edu/datasets/COMBO17.html
56. R.A. Reyment, K.G. Jöreskog, L.F. Marcus, *Applied Factor Analysis in the Natural Sciences* (Cambridge University Press, Cambridge, 1993)
57. G. Pison, P.J. Rousseeuw, P. Filzmoser, C. Croux, Robust factor analysis. J. Multivariate Anal. **84**, 145 (2003)
58. P. Filzmoser, K. Hron, C. Reimann, R. Garrett, Robust factor analysis for compositional data. Comput. Geosci. **35**, 1854 (2009)
59. C. Reimann, P. Filzmoser, R.G. Garrett, Factor analysis applied to regional geochemical data: problems and possibilities. Appl. Geochem. **17**, 185 (2002)
60. See http://lib.stat.cmu.edu/datasets/bodyfat where all data are collected and the corresponding original literature is cited
61. V.G. Sigillito, S.P. Wing, L.V. Hutton, K.B. Baker, Classification of radar returns from the ionosphere using neural networks. Johns Hopkins APL Tech. Digest **10**, 262 (1989). The data can be found at http://archive.ics.uci.edu/ml or at the book's web page

Chapter 7
Modeling and Analysis of Time Series

Abstract This Chapter is devoted to the analysis of sequential measurements of physical quantities, in which the "time" variable may be either continuous or discrete. Formal definitions and classifications of random variables and random processes precede the topics of the central limit theorem and stable distributions, both in the context of random walks and their asymptotic regimes. We discuss discrete-time and continuous-time Markov chains, and describe methods of generating noise with arbitrary spectral properties. Correlation and auto-correlation of signals and the auto-regression analysis of discrete-time signals are explained along with the concepts of resonance spectra, Fourier spectra, and maximum entropy estimators. We present optimal (Kalman) filtering and independent component analysis. Finally, we outline the methods of state-space reconstruction, a key tool of non-linear time-series analysis based on the Takens embedding theorem. The Examples and Problems range from the computation of auto-correlations in the logistic and standard (Chirikov) maps to phase transitions in the two-dimensional Ising model, the analysis of spectra of acoustic resonators, and the study of electro-cardiograms in the presence of noise.

A *time series* or *signal* $s(t) \in \Sigma$ should be understood as a sequential measurement of some quantity. The time variable and the corresponding signal may be discrete or continuous. The signals may be real, complex, or integer. In the analysis of time series we use mathematical tools to extract their basic characteristics and learn about the properties of their sources. The sources of signals are also called *processes* and are commonly identified with the signals that these processes generate. We divide the processes according to their statistical properties, as shown in Fig. 7.1. At the fundamental level, processes may be considered to be either deterministic or random.

Deterministic processes In deterministic processes, each value of the signal is precisely determined by a mathematical or physical law, a rule, or a table of values. If we understand the process dynamics, such processes are known as *dynamical systems* [1, 2]. The state of a dynamical system in space Ξ can be uniquely described by the mapping of the system from state $x(t_0)$ at time t_0 to state $x(t)$ at time t, i.e.

$$x(t) = \phi(t, t_0, x(t_0)).$$

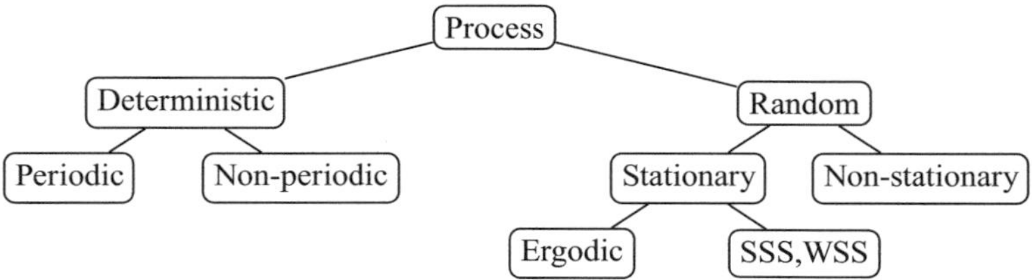

Fig. 7.1 Classification of processes. In deterministic processes the values of the signals are determined by rules or laws. In random processes the signals are formed based on chance. The SSS and WSS notation is defined on page 394

The observed signal s is the result of a function $F : \Xi \to \Sigma$ acting on the system evolving in time, $s(t) = F(x(t))$. Signals from deterministic processes are simply reproducible and we know how the signal behaved in the past and what it is going to look like in the future.

Random processes In random processes the generated signal is not determined in advance by the process parameters; it comes about by chance. Before the measurement it is impossible to precisely foresee its value. The signal originating in a random process is known as a *random signal* and represents *one possible realization* of that process. In spite of the misleading nomenclature, the random signal is not necessarily random in time; just the choice of its realization is random (see Example on page 395). Details are discussed in Sect. 7.2.

7.1   Random Variables

A random or stochastic variable is a variable whose value is not determined in advance and which, at each measurement of an observable, assumes a value with a certain probability. The measurement of a quantity is known as the *realization of a random variable* or *drawing*. The values of a random variable are drawn randomly and independently. We denote random variables by Roman capitals, e.g. X, and the corresponding values with small letters, e.g. x. In this Section we use the conventional notation of theories of measure [3, 4] and probability [5, 6].

7.1.1   Basic Definitions

A random variable X is characterized by its *sample space* Ω and its *probability measure* $P : A \to [0, 1]$ defining the probability that X is in some subset $A \subset \Omega$, which we write as $P(A) = \mathrm{Prob}(X \in A)$. If the sample space is continuous, we may use the differential volume element $\mathrm{d}V(x)$ at $x \in \Omega$ to write the differential of the probability measure as

$$\mathrm{d}P(x) = p(x)\,\mathrm{d}V(x)\,.$$

We use it to define the probability density p of the variable X. If Ω is countable, $P(x)$ represents the probability that the value x is drawn.

The most important operation on random variables is statistical averaging. We define the statistical average of a function (observable) f of a random variable X according to the type of the sample space Ω:

$$\text{discrete } \Omega: \quad \langle f \rangle = \sum_{x \in \Omega} f(x) P(x)\,, \qquad \text{continuous } \Omega: \quad \langle f \rangle = \int_{\Omega} f(x)\,\mathrm{d}P(x)\,.$$

In the literature we also find the notation $E[f] = \langle f \rangle$, helping us to define the variance or dispersion $V[f] = \langle (f - E[f])^2 \rangle$, as well as the standard deviation $\sigma[f] = (V[f])^{1/2}$.

An important quantity for a random variable X with the sample space Ω that is a subset of $\mathbb{R}$ or $\mathbb{Z}$, is the characteristic function

$$\phi_X(t) = \langle e^{itX} \rangle\,, \qquad \phi_X(0) = 1\,, \qquad |\phi_X(t)| < 1 \quad \forall t \neq 0\,. \tag{7.1}$$

The characteristic function allows us to construct the statistical moments $\langle X^n \rangle, n = 0, 1, 2, \ldots$ since if ϕ_X is n-times continuously differentiable at the origin,

$$\lim_{t \to 0} \frac{\mathrm{d}^n}{\mathrm{d}t^n} \phi_X(t) = i^n \langle X^n \rangle\,.$$

The characteristic function and the probability density are related by the Fourier transformation,

$$p(x) = \frac{1}{2\pi} \int_{-\infty}^{\infty} \phi_X(t)\, e^{-itx}\,\mathrm{d}t$$

in the continuous case ($\Omega \subset \mathbb{R}$), while we have

$$P(x) = \frac{1}{N} \sum_{k=0}^{N-1} \phi_X(k)\, e^{-i\,(2\pi k/N)x}$$

in the discrete case ($\Omega \subset \mathbb{Z}_N$). The probability density and the characteristic function are equivalent descriptions of statistical properties of a random variable.

7.1.2 Generation of Random Numbers

In numerical calculations of statistical averages and in simulations involving random numbers we use random number generators (Appendix C). These are algorithms that *deterministically* produce numbers with desired probability distributions giving an impression of randomness. Such generators may therefore be seen as sources of realizations of random variables. For each generated sequence of numbers $\{x_i\}$ and arbitrary smooth observable f, a good random number generator should ensure *ergodicity*,

$$\langle f(x_t)\rangle_t = \lim_{T\to\infty} \langle f(x_t)\rangle_{t,T} = \langle f \rangle \;, \qquad \langle f(x_t)\rangle_{t,T} = \frac{1}{T}\sum_{t'=0}^{T-1} f(x_{t'}) \;,$$

that is, the equality of statistical and time averages, as well as the property of *mixing*, for which we require that the quantity

$$\lim_{T\to\infty} \left\langle \Big(f(x_t) - \langle f\rangle\Big)\Big(f(x_{t+j}) - \langle f\rangle\Big)\right\rangle_{t,T} \tag{7.2}$$

vanishes when $j \to \infty$. (In Sect. 7.6 we shall see that (7.2) represents the auto-correlation of f at shift j.) Ergodicity and mixing are concepts from the theory of dynamical systems [7], where the latter implies the former, but not the other way around. Mixing is equivalent to the requirement that the drawn numbers are as independent as possible. For smooth enough f, the exponential convergence of the auto-correlation (7.2) ensures the statistically fastest possible convergence of the average $\langle f(x_t)\rangle_{t,T}$ to the limit $\langle f\rangle$, so $\langle f(x_t)\rangle_{t,T} - \langle f\rangle = \mathcal{O}(T^{-1/2})$. Fast mixing is a desirable (but not essential) property of random number generators.

7.2 Random Processes

Random processes are omnipresent in physics. Quantum mechanics and statistical physics, for example, are founded on assumptions about the randomness of Nature at certain distance or energy scales [8]. In the following four Sections we present the fundamentals of the theory of random processes and describe three typical representatives: random walks, Markov chains, and noise.

7.2.1 Basic Definitions

The random process is a generalization of the concept of random variables where, instead of a value, we randomly choose a time signal which we name the *realization of a random process* or the *sample path*. Random processes can be defined in two ways illustrated in Fig. 7.2.

In the first approach we imagine a random variable S with values s distributed over the set Λ. We define an ensemble of functions $\{x(t; s) \in \Omega\}_{s\in\Lambda}$ with sample

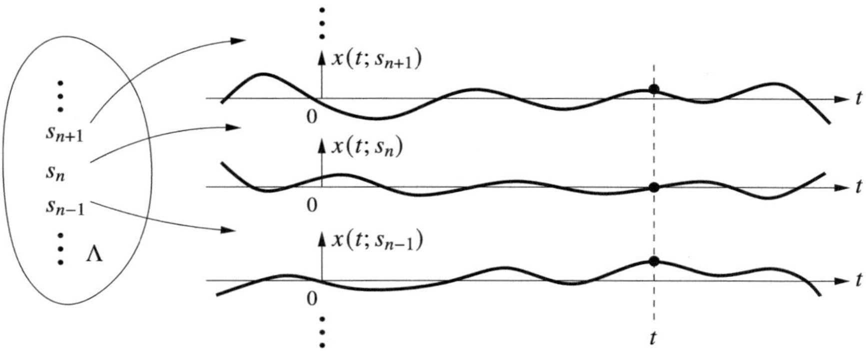

Fig. 7.2 A random process $X(t)$ and its three possible realizationsA random process $X(t)$ and its three possible realizations

space Ω, where the time t runs over the set $\mathbb{T}$. The random process $X(t)$ represents a random function from the ensemble which we select by drawing the value s. At a given particular t we can understand $X(t)$ as a random variable $X(t) = x(t; S)$. In other words, if we make a vertical cut through Fig. 7.2 at a chosen t, the obtained values behave as a random variable. On the other hand, we can interpret the random process as drawing a set of random variables $\{X(t) \in \Omega\}_{t \in \mathbb{T}}$ which may be correlated.

If $\mathbb{T} = \mathbb{R}$, we are referring to continuous-time processes, while with $\mathbb{T} = \mathbb{N}$ or $\mathbb{Z}$ we are dealing with discrete-time processes. In analogy to random variables, random processes are also denoted by capital Roman letters, e.g. X, while we use small ones, like x, for their realizations. Sample spaces of random processes are commonly labeled by capital Greek letters, e.g. Ω. The statistical average over the realizations of the random process is denoted by $\langle \cdot \rangle$ or $E[\cdot]$.

Random processes are almost never defined by the probability distributions of the elements of their signals; rather, we use the time dependence of the statistical moments of the signals, for example, the average $\mu_X(t) = \langle X(t) \rangle$, the two-point auto-correlation $R_{XX}(t_1, t_2) = \langle X(t_1) X(t_2) \rangle$, and so on. A rare exception is the discrete real Gaussian random process. For n points of the signal at times $t_1, t_2, \ldots, t_n$ it can be described by the probability density

$$p_{X(t_1), X(t_2), \ldots, X(t_n)}(\boldsymbol{x}) = \frac{1}{\sqrt{(2\pi)^n \det(R)}} \exp\left[-\frac{1}{2}(\boldsymbol{x} - \boldsymbol{\mu})^\mathrm{T} R^{-1}(\boldsymbol{x} - \boldsymbol{\mu}) \right] ,$$

where $\boldsymbol{x} = (x_1, x_2, \ldots, x_n)^\mathrm{T}$. The probability density $p_{X(t_1), X(t_2), \ldots, X(t_n)}(\boldsymbol{x})$ is defined by the vector of averages $\boldsymbol{\mu} = (\mu_X(t_1), \mu_X(t_2), \ldots, \mu_X(t_n))^\mathrm{T}$ and the auto-correlation matrix $R = [R_{XX}(t_i, t_j)]_{ij}$, which we will get to know in an instant. Most often we encounter Gaussian processes with the property $R_{XX}(t_1, t_2) = R_{XX}(t_1 - t_2)$.

Non-stationarity and stationarity Random processes are distinguished by their statistical properties (we follow [9]). The process may be stationary or non-stationary. Roughly speaking, the statistics of stationary processes do not change with time; a more precise definition of stationarity of a random process X with sample space Ω requires us to use the joint probability measure for n points of the process at times $t_1, t_2, \ldots, t_n$,

$$P_{X(t_1), X(t_2), \ldots, X(t_n)}(A) = \mathrm{Prob}(X(t_1), X(t_2), \ldots, X(t_n) \in A) \,,$$

where $A \subset \Omega^n$. A random process X is *stationary of order n* if this measure for arbitrary n points is invariant with respect to translations along the time axis,

$$P_{X(t_1), X(t_2), \ldots, X(t_n)} = P_{X(t_1+t), X(t_2+t), \ldots, X(t_n+t)} \qquad \forall t \,.$$

Note that a process that is stationary of order $n + 1$ is also stationary of order n, while the reverse is not necessarily true. In the following we assume that the sample space of the random process is $\Omega \subset \mathbb{R}$. In the first-order stationary process the measure $P_{X(t)}$ is constant, and therefore also all single-point averages are constant, like the moments: $\langle X(t)^m \rangle = \mathrm{const}$ for $m = 1, 2, \ldots$ For the second-order stationary process, the two-point measure $P_{X(t_1), X(t_2)}$ is translationally invariant. It follows that all two-point averages depend only on time *differences*. One such quantity is the two-point correlation of the observables f and g of the random process X:

$$\langle f(X(t_1))g(X(t_2)) \rangle = \langle f(X(0))g(X(t_2 - t_1)) \rangle = \langle f(X(t_1 - t_2))g(X(0)) \rangle \,.$$

A very interesting average is the auto-correlation

$$R_{XX}(t_1, t_2) = \langle X(t_1)X(t_2) \rangle = R_{XX}(t_1 - t_2) \,,$$

for which

$$|R_{XX}(\tau)| \leq R_{XX}(0) = \langle X(t)^2 \rangle \,, \qquad R_{XX}(-\tau) = R_{XX}(\tau) \,. \tag{7.3}$$

If the process is stationary of arbitrary order, it is said to be *strict-sense stationary* (SSS). In contrast, a random process X is *wide-sense stationary* (WSS) if the following holds: that the average is constant, $\langle X(t) \rangle = \mathrm{const}$; that the auto-correlation depends on time difference only, thus $R_{XX}(t_1, t_2) = R_{XX}(t_1 - t_2)$; and that the variance is bounded for all times, $\langle (X(t) - \langle X(t) \rangle)^2 \rangle < \infty$. Every second-order stationary process in wide-sense stationary, while the reverse is not necessarily true.

Ergodicity Define the time average of the function f in the continuous case,

$$\langle f(t) \rangle_t = \lim_{T \to \infty} \frac{1}{2T} \int_{-T}^{T} f(t') \, dt' \,, \tag{7.4}$$

where $\mathbb{T} = \mathbb{R}$, or in the discrete case,

$$\langle f(t)\rangle_t = \lim_{T\to\infty} \frac{1}{2T+1} \sum_{t'=-T}^{T} f(t') , \qquad (7.5)$$

where $\mathbb{T} = \mathbb{Z}$ (be alert to the notation $\langle\cdot\rangle_t$). Assume that almost every realization x of the process X fulfills the conditions that the time average is equal to the statistical average,

$$\langle x(t)\rangle_t = \langle X(t)\rangle ,$$

and that the time auto-correlation is equal to the statistical one, i.e.

$$\langle x(t)x(t+\tau)\rangle_t = R_{XX}(\tau) .$$

Then we say that the random process X is *ergodic*. We commonly refer to the first condition as *ergodicity on average*, while the second condition is known as *ergodicity in auto-correlation*. The first condition is fulfilled precisely when the average of the auto-covariance

$$C_{XX}(\tau) = \left\langle \left(X(t) - \mu_X\right)\left(X(t+\tau) - \mu_X\right)\right\rangle , \qquad \mu_X = \langle X(t)\rangle ,$$

is equal to zero, so $\langle C_{XX}(t)\rangle_t = 0$. The sufficient condition for ergodicity on average is $\lim_{t\to\infty} C_{XX}(t) = 0$ [10].

In practice we are dealing only with *realizations* of random processes. To compute the statistical properties of processes from which these realizations originate, we commonly assume ergodicity and wide-sense stationarity: this allows us to use the signal to make inferences about the underlying process.

Assume that in a discrete-time random process $X(t)$, $t \in \mathbb{Z}$, we obtained a finite sample $\{x(t)\}_{t=0}^{N-1}$ from its infinitely long realization $x(t)$. If the process is ergodic, the average and the auto-correlation can be approximated by the sums

$$\langle X(t)\rangle \approx \frac{1}{T} \sum_{t'=0}^{N-1} x(t') , \qquad R_{XX}(\tau) \approx \frac{1}{N-\tau} \sum_{t'=0}^{N-1-\tau} x(t')x(t'+\tau) ,$$

where the symmetry $R_{XX}(-\tau) = R_{XX}(\tau)$ has been exploited (see Eq. (7.3)).

Example Imagine a random process represented by the sum

$$X(t) = \sum_{i=1}^{n} w_i \cos(\omega_i t + Y_i) , \qquad (7.6)$$

where the weights w_i and the frequencies ω_i are constant, and the independent random variables Y_i are distributed normally on the interval $[0, 2\pi)$. We immediately realize that $\langle X(t)\rangle = 0$, while the auto-correlation is

$$R_{XX}(t_1, t_2) = \langle X(t_1) X(t_2) \rangle = \frac{1}{2} \sum_{i=1}^{n} w_i^2 \cos(\omega_i(t_1 - t_2)) = R_{XX}(t_1 - t_2) \,.$$

The process X is therefore wide-sense stationary. It is easy to show that the time average of an individual *realization* is also zero, $\langle x(t) \rangle_t = 0$, while its time autocorrelation is equal to the statistical one, $\langle x(t)x(t + \tau) \rangle_t = R_{XX}(\tau)$. We conclude that the random process X is also ergodic.

Equation (7.6) also elucidates the initial definitions of random processes and their realizations (page 390), claiming that "*the individual random signal is not necessarily random along the time axis; what is random is just the choice of its realization*". Namely, the dependence of the signal originating in the process (7.6) on the time variable t is known explicitly! It is by drawing the variable Y_i that randomness or unpredictability is built into the process.   ◁

7.3 Stable Distributions and Random Walks

7.3.1 Central Limit Theorem

Let $X_1, X_2, \ldots, X_n$ be real independent and identically distributed random variables with the probability density p_X, whose average $\mu_X = \langle X_i \rangle$ and variance $\sigma_X^2 = \langle (X_i - \mu_X)^2 \rangle$ are bounded. Define the sum of random variables $Y_n = \sum_{i=1}^{n} X_i$ with the average $\langle Y_n \rangle = n\mu_X$ and variance $\sigma_{Y_n}^2 = n\sigma_X^2$. The probability density p_Y of the summed variable Y_n is given by the n-fold convolution of the densities of the variables X_i, i.e.

$$p_{Y_n} = \underbrace{p_X * p_X * \cdots * p_X}_{n} \,.$$

Let us define the rescaled variable $Z_n = (Y_n - n\mu_X)/(\sqrt{n}\sigma_X)$. Then, in the limit $n \to \infty$, the cumulative distribution function of the variable Z_n converges to the cumulative distribution function of the normal distribution $N(0, 1)$,

$$\lim_{n \to \infty} \mathrm{Prob}(Z_n < z) = \Phi(z) = \frac{1}{\sqrt{2\pi}} \int_{-\infty}^{z} e^{-\frac{1}{2}t^2} \mathrm{d}t \,.$$

In other words, the probability density p_{Y_n} in the limit $n \to \infty$ converges to the normal (Gaussian) probability density $N(\mu_{Y_n}, \sigma_{Y_n}^2)$, the statement known as the *central limit theorem* (CLT). The speed of convergence to the normal distribution is the substance of the Berry–Esséen theorem [6]. If the third moment of $|X - \mu_X|$ is bounded, i.e. $\rho = \langle |X - \mu_X|^3 \rangle < \infty$, we have

$$| \operatorname{Prob}(Z_n < z) - \Phi(z) | \leq \frac{C\rho}{\sqrt{n}\sigma_X^3} \, ,$$

where $C \geq 0.4097$ [11]. The central limit theorem and the mathematical form of the estimate remain valid even when we sum the variables X_i distributed according to different probability distributions, but only if the dispersion of the values is not excessive (Lindeberg criterion, see [6]).

7.3.2 Stable Distributions

The normal distribution as the limit distribution in summation of independent random variables can be generalized by introducing the concept of *stable distributions* [12–14]. Assume that we have independent random variables X_1, X_2, and X_3 with the same distribution with respect to the sample space Ω. We say that this distribution is *stable* if for each pair of numbers a and b, a pair c and d exists such that the distribution of the linear combination $aX_1 + bX_2$ is the same as the distribution of $cX_3 + d$, so if it holds that

$$\operatorname{Prob}(aX_1 + bX_2 \in A) = \operatorname{Prob}(cX_3 + d \in A) \quad \forall A \subset \Omega \, .$$

Such random numbers are also said to be "stable"; a superposition of stable random numbers is a linear function of the stable random number with the same distribution.

In general, stable distributions are most easily described by their characteristic functions. Among the many possible notations we follow that of [12]. We say that a random variable X has a stable distribution $p_{\text{stab}}(x; \alpha, \beta, \gamma, \delta)$ if the logarithm of its characteristic function (7.1) has the form

$$\log \phi_X(t) = \mathrm{i}\delta t - \gamma^\alpha |t|^\alpha [1 - \mathrm{i}\beta \Phi_\alpha(t)] \, ,$$

where

$$\Phi_\alpha(t) = \begin{cases} \operatorname{sign}(t)\,\tan(\pi\alpha/2) \; ; \; \alpha \neq 1 \, , \\ -\frac{2}{\pi}\operatorname{sign}(t)\,\log|t| \; ; \; \alpha = 1 \, . \end{cases}$$

The parameter $\alpha \in (0, 2]$ is the *stability index* or the *characteristic exponent*, while the parameter $\beta \in [-1, 1]$ defines the *skewness* of the distribution. There is also the scaling parameter $\gamma > 0$ and the position parameter $\delta \in \mathbb{R}$. For $\alpha \in (1, 2]$ the average exists and is equal to $\langle X \rangle = \delta$. For general $\alpha \in (0, 2]$ the statistical moments $\langle |X|^p \rangle$ exist, where $p \in [0, \alpha)$.

For practical computations it is useful to replace the random variable X by another random variable Z,

$$X = \begin{cases} \gamma Z + \delta & ; \; \alpha \neq 1 \, , \\ \gamma \left(Z + \frac{2}{\pi}\beta \log\gamma \right) + \delta & ; \; \alpha = 1 \, , \end{cases}$$

because the characteristic function for Z is slightly simpler,

$$\log \phi_Z(t) = -|t|^\alpha [1 - i\beta \Phi_\alpha(t)] \,,$$

as it depends only on two parameters, α and β. The probability density p_Z of the variable Z is computed by the inverse Fourier transformation of the characteristic function ϕ_Z. We obtain

$$p_Z(z; \alpha, \beta) = \frac{1}{\pi} \int_0^\infty \exp(-t^\alpha) \cos\left(zt - t^\alpha \beta \Phi_\alpha(t)\right) \, dt \,.$$

It holds that $p_Z(-z; \alpha, \beta) = p_Z(z; \alpha, -\beta)$. The values of the functions p_Z and p_X can be computed by using integrators specially adapted to rapidly oscillating functions (see Appendix 3.4; a modest programming support for stable distributions can also be found in [15]). With respect to α and β, the function p_Z with the argument z has the definition ranges

$$z \in \begin{cases} [-\infty, 0] \,; \ \alpha < 1 \,, \ \beta = -1 \,, \\ [0, \infty] \quad ; \ \alpha < 1 \,, \ \beta = 1 \,, \\ \mathbb{R} \qquad \ \ ; \ \text{otherwise} \,. \end{cases}$$

The dependence of the stable distribution p_{stab} (p_X or p_Z with appropriate scaling) on the parameter α is shown in Fig. 7.3 (top left and right), while the dependence on β is shown in the same Figure at bottom left and right.

A formulation as flexible as this allows us to generate all possible stable distributions. The most familiar ones are

$$\begin{aligned} \text{normal}: \ & \alpha = 2 \,, \ \beta = 0 \,, \ \ p_X(t) = \frac{1}{\sqrt{2\pi}} \exp\left(-\frac{1}{2}t^2\right) \,, & t \in \mathbb{R} \,; \\ \text{Cauchy}: \ & \alpha = 1 \,, \ \beta = 0 \,, \ \ p_X(t) = \frac{1}{\pi} \frac{1}{1+t^2} \,, & t \in \mathbb{R} \,; \\ \text{Lévy}: \ & \alpha = \tfrac{1}{2} \,, \ \beta = 1 \,, \ \ p_X(t) = \frac{1}{\sqrt{2\pi}} \exp\left(-\frac{1}{2t}\right) t^{-\frac{3}{2}} \,, & t \in \mathbb{R}_+ \,. \end{aligned}$$

A well-known property of stable distributions with $\alpha \in (0, 2)$ is the characteristic asymptotic behavior of their probability densities, known as *power* or *fat tails*. For the cumulative probabilities the following holds:

$$\begin{aligned} \beta \in (-1, 1] : \ & \int_x^\infty p_Z(z; \alpha, \beta) \, dz \sim \frac{1}{2} c_\alpha (1 + \beta) x^{-\alpha} \,, & x \to \infty \,, \\[2mm] \beta \in [-1, 1) : \ & \int_{-\infty}^x p_Z(z; \alpha, \beta) \, dz \sim \frac{1}{2} c_\alpha (1 - \beta)(-x)^{-\alpha} \,, & x \to -\infty \,, \end{aligned}$$

$$(7.7)$$

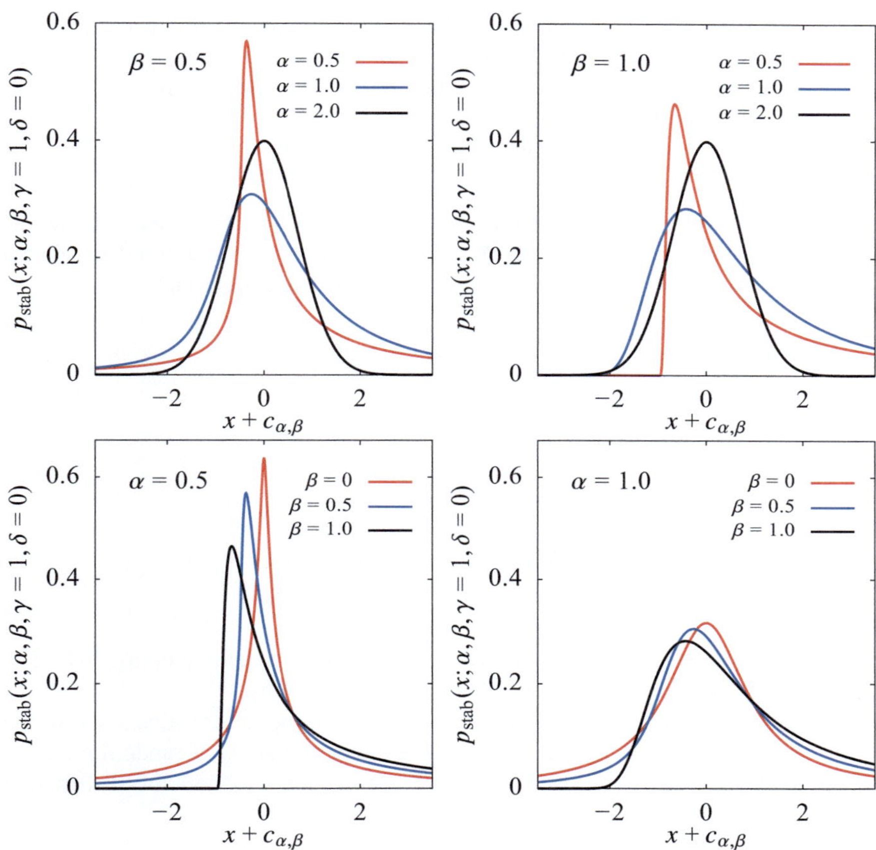

Fig. 7.3 Stable distributions $p_{\text{stab}}(x; \alpha, \beta, \gamma, \delta)$. [TOP LEFT AND RIGHT] Dependence on α for $\beta = 0.5$ and 1.0. [BOTTOM LEFT AND RIGHT] Dependence on β for $\alpha = 0.5$ and 1.0. At $\alpha \neq 1$ the independent variable is shifted by $c_{\alpha,\beta} = \beta \tan(\pi\alpha/2)$

where $c_\alpha = 2\sin(\pi\alpha/2)\Gamma(\alpha)/\pi$. For $\beta \in (-1, 1)$ such asymptotic behavior can be established in both limits, $x \to \pm\infty$. Recall that a probability density has the asymptotics $\mathcal{O}(|x|^{-\alpha-1})$ if the corresponding cumulative probability has the asymptotics $\mathcal{O}(|x|^{-\alpha})$.

7.3.3   *Generalized Central Limit Theorem*

By using our knowledge of the stable distributions (Sect. 7.3.2) we can write a generalized central limit theorem (or Lévy's limit theorem) elaborated in detail by [6, 14]. Here we convey only its essence.

Assume that we have a sequence of independent and identically distributed random numbers $\{X_i\}_{i\in\mathbb{N}}$ from which we form the partial sum $Y_n = \sum_{i=1}^n X_i$. Assume that their distribution has power tails, so that for $\alpha \in (0, 2]$ the limits

$$\lim_{x\to\pm\infty} |x|^\alpha \, \mathrm{Prob}(\pm X > x) = d_\pm$$

exist and $d = d_+ + d_- > 0$. Then real coefficients $a_n > 0$ and b_n exist such that the rescaled partial sum $Z_n = (Y_n - nb_n)/a_n$ in the limit $n \to \infty$ is stable and distributed according to $p_{\mathrm{stab}}(x; \alpha, \beta, 1, 0)$. The skewness of the stable distribution is

$$\beta = (d_+ - d_-)/(d_+ + d_-) \,,$$

and the coefficients a_n and b_n are [12, 16]

$$a_n = \begin{cases} (d\,n/c_\alpha)^{1/\alpha} & ; \alpha \in (0, 2) \,, \\ \sqrt{(d\,n\log n)/2} & ; \alpha = 2 \,, \end{cases}$$

$$b_n = \begin{cases} \langle X_i \rangle & ; \alpha \in (1, 2] \,, \\ \langle X_i\,\theta(|X_i| - a_n) \rangle & ; \text{otherwise} \,, \end{cases}$$

where θ is the Heaviside (step) function. The constant c_α is defined in Eq. (7.7). The coefficient a_n for $\alpha < 2$ diverges as $\mathcal{O}(n^{1/\alpha})$ when n increases.

The generalized CLT is applicable to random walk processes (discussed in the following), which are analogous to extending the partial sums of random numbers Y_n. From practical experience we know that the convergence to the stable distribution at $n \to \infty$ becomes ever slower and "fussier" when α decreases.

7.3.4 Discrete-Time Random Walks

Random walks are non-stationary random processes used in modeling of numerous physical and engineering phenomena. In this section, we present discrete-time random walks [6, 17, 18], while continuous-time random walks [17–20] are discussed in the next. Note that the meaning of α and β in the following is different from that in Sects. 7.3.2 and 7.3.3.

Imagine a discrete-time random process X that we observe as a sequence of random variables $\{X(t)\}_{t\in\mathbb{N}}$. The partial sums of this sequence are

$$Y(t) = Y(0) + \sum_{i=1}^t X(i) = Y(t-1) + X(t) \,. \tag{7.8}$$

They represent a new discrete-time random process Y and a sequence of random variables $\{Y(t)\}_{n\in\mathbb{N}_0}$. The process Y is a *random walk*, a single step of which is the process $X(t)$. Let the sample space Ω of the processes X and Y be continuous. We are interested in the time evolution of the probability density $p_{Y(t)}$ of the random variable Y with a known initial density $p_{Y(0)}$.

If we assume that Y is a process in which the state of each point depends only on the state of the previous point, the time evolution of $p_{Y(t)}$ is given by

$$p_{Y(t)}(y) = \int_\Omega p\big(Y(t) = y \mid Y(t-1) = x\big)\, p_{Y(t-1)}(x)\, \mathrm{d}x \, ,$$

where $p\big(Y(t) = y \mid Y(t-1) = x\big)$ is the conditional probability density that Y evolves from the value x at time $t-1$ to the value y at time t. We also assume that the process X is completely independent of its previous states, so that

$$p\big(X(t) = x \mid Y(t-1) = y - x\big) = p_{X(t)}(x)$$

holds. It then follows that

$$p_{Y(t)}(y) = \int_\Omega p_{X(t)}(x) p_{Y(t-1)}(y-x)\, \mathrm{d}x = \big(p_{X(t)} * p_{Y(t-1)}\big)(y) \, ,$$

where $*$ denotes the convolution. By using this formula, we can express $p_{Y(t)}$ as a convolution of the initial distribution $p_{Y(0)}$ with the distribution of the sum of the steps until t, p_X^{*t}:

$$p_{Y(t)} = p_{Y(0)} * p_X^{*t} \, , \qquad p_X^{*t} = p_{X(1)} * p_{X(2)} * \cdots * p_{X(t)} \, . \tag{7.9}$$

The evolution of $p_{Y(t)}(y)$ is best computed in Fourier space, as it is given by the product of the Fourier transforms $\mathcal{F}$ of the probability densities,

$$\mathcal{F}\big[p_{Y(t)}\big] = \mathcal{F}\big[p_{Y(0)}\big] \prod_{i=1}^{t} \mathcal{F}\big[p_{X(i)}\big] \, .$$

One frequently assumes that the value of the process Y at time zero is known and that $p_{Y(0)}(y) = \delta(y)$. This assumption is particularly useful when we are interested in the qualitative behavior of $p_{Y(t)}$ after long times.

Asymptotics The time asymptotics of the distributions $p_{Y(t)}$ is most easily established in the case of one-dimensional real random walks. Assume that all steps have the same distribution with the probability density $p_{X(t)} = p_X$, and that $Y(0) = 0$. The distribution that corresponds to the process Y is given by

$$p_{Y(t)} = \mathcal{F}^{-1}\big[(\mathcal{F}[p_X])^t\big]$$

for all times t. The behavior of $p_{Y(t)}$ in the limit $t \to \infty$ is determined by the central limit theorem (Sect. 7.3.1) and its generalization (Sect. 7.3.3). These theorems tell us that for increasing t, $p_{Y(t)}$ converges to the limit distribution which is expressible by one of the stable distributions p_{stab}, such that

$$p_{Y(t)}(y) \sim L(t)\, p_{\text{stab}}\big(L(t)y + t\mu(t)\big)$$

for suitably chosen functions L and μ. The function L represents the effective width of the central part of the distribution $p_{Y(t)}$, where the majority of the probability is located, and is known as the *characteristic spatial scale*. The function μ has the role of the distribution average.

For the distribution of the steps p_X with a bounded variance ($\sigma_X^2 < \infty$) we also know, based on the CLT, that $p_{Y(t)}$ tends to the Gaussian with the width $L = \sigma_{Y(t)} \sim t^{1/2}$. This particular asymptotic dependence of the spatial scale on time is typical for *normal diffusion* and the same nomenclature pertains to the corresponding regime of the random walk (Fig. 7.4 (left)).

If the distribution p_X has the asymptotic behavior

$$p_X(x) \sim \frac{C_\pm}{|x|^{\alpha+1}}\,, \qquad x \to \pm\infty\,,$$

where $C_\pm$ are constants, the distribution is said to have a *power* or *fat tail* (see Sect. 7.3.2). For $\alpha \in (0, 2)$ the second moment of the distribution no longer exists and $p_{Y(t)}$ at long times tends to the distribution with $L \sim t^{1/\alpha}$. Because in this case the characteristic scale changes faster than in normal diffusion, we are referring to *super-diffusion*. The dynamics of the process Y in this regime are known as *Lévy flights*. The diffusion with $\alpha = 1$ is said to be *ballistic*: particles propagate without constraints with given velocities, so $L \sim t$. Near $\alpha = 2$ we have $L(t) \sim (t \log t)^{1/2}$ and name the corresponding regime *log-normal diffusion*.

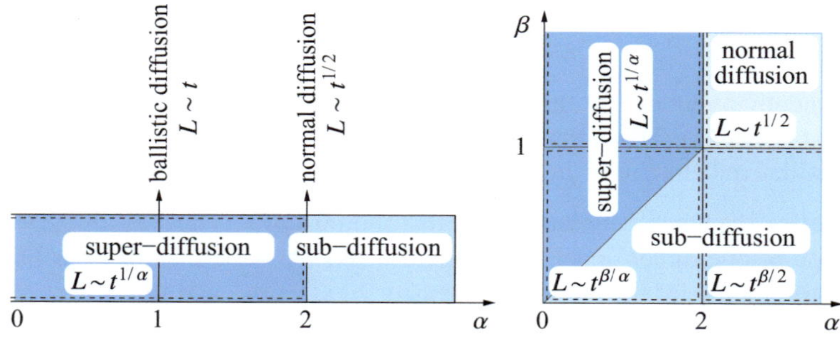

Fig. 7.4 The dependence of the characteristic spatial scale L on time t. [LEFT] Discrete-time random walk. [RIGHT] Continuous-time random walk

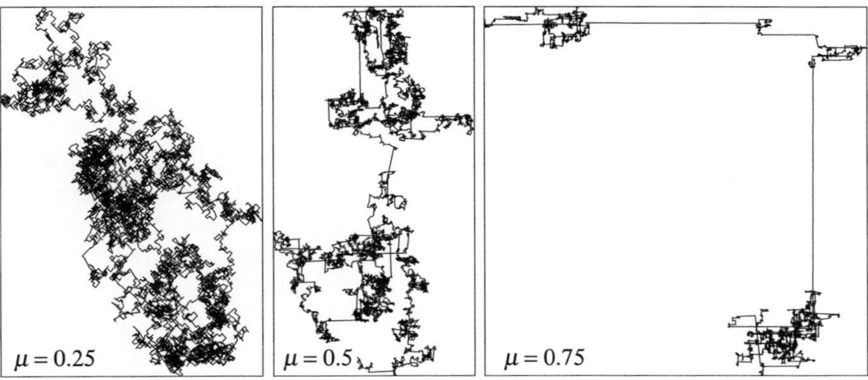

Fig. 7.5 Examples of random walks (x_t, y_t) with 10^4 steps, generated by the formulas $x_{t+1} = x_t + \text{sign}(X)|X|^{-\mu}$ and $y_{t+1} = y_t + \text{sign}(Y)|Y|^{-\mu}$, where X and Y are independent random variables distributed uniformly on the interval $[-1, 1]$

Properties of one-dimensional random walks described above are easily generalized to more dimensions. We observe the projection of a multi-dimensional walk $\hat{n}^T Y(t)$ along the unit vector $\hat{n}$, and its probability density $p_{\hat{n}^T Y(t)}$. For an individual $\hat{n}$ we apply the CLT or its generalization and determine the scale $L_{\hat{n}}$. In such a random walk, a particular direction $\hat{n}^*$ exists along which the scale is maximum or increases most rapidly with t: the characteristic scale of the distribution is then $L_{\hat{n}^*}$. Examples of two-dimensional random walks where the distributions in x and y directions are independent, are shown in Fig. 7.5.

If the densities $p_{X(t)}$ have power tails, the $p_{Y(t)}$ also has power tails. This applies regardless of the CLT or its generalization. Assume that in the limit $t \to \infty$ we have $p_{X(t)}(x) \sim C_{\pm,t}|x|^{-\alpha-1}$. During the process of forming the distribution $p_{Y(t)}$, the amplitudes of the tails add up, so that $p_{Y(t)}(x) \sim \left(\sum_{i=1}^{t} C_{\pm,i}\right)|x|^{-\alpha-1}$ in the limit $x \to \pm\infty$. This means that with increasing time, chances will increase that an observation of $Y(t)$ will yield an extreme event, since

$$\text{Prob}\big(|Y(t)| > y\big) \sim \sum_{i=1}^{t} \text{Prob}\big(|X(i)| > y\big), \qquad y \to \infty .$$

The dispersion of the values generated in such processes can be estimated by robust methods (Sect. 6.2). For example, in sub-diffusive random walks we prefer to compute the MAD (6.9) instead of the standard deviation $\sigma_{Y(t)}$.

7.3.5 Continuous-Time Random Walks

In continuous-time random walks [17–20] the number of steps $N(t)$ performed until time t becomes a random variable. We reformulate the definition for the discrete-time random walk (7.8) as

$$Y(t) = Y(0) + \sum_{i=1}^{N(t)} X(i) .$$

Obviously $Y(t)$ can not be written in the iterative form $Y(t) = Y(t-1) + \cdots$ as in (7.8). The number of steps $N(t)$ has the probability distribution $P_{N(t)}$. Assume that $N(t)$ and $X(i)$ are independent processes, which is not always true, as in a given time it is not possible to perform arbitrarily many steps [21, 22]. If $X(i)$ at different times are independent and correspond to probability densities $p_{X(i)}$, the probability density of the random variable $Y(t)$ is

$$p_{Y(t)}(y) = \sum_{n=0}^{\infty} P_{N(t)}(n)\left(p_{Y(0)} * p_X^{*n}\right)(y) ,$$

where p_X^{*n} is defined in Eq. (7.9).

We adopt the manner of interpretation of such a random walk and the choice of the distribution $P_{N(t)}$ from [19]. A random walk can be envisioned as a sequence of steps with randomly chosen lengths $X(i)$ and *waiting times* $T(i)$ between the steps. After N steps from the departure at the origin, the walk has brought us to the point $\mathcal{X}(N)$. Until then, time $\mathcal{T}(N)$ has elapsed, so that

$$\mathcal{X}(N) = \sum_{i=1}^{N} X(i) , \qquad \mathcal{T}(N) = \sum_{i=1}^{N} T(i) , \qquad \mathcal{X}(0) = \mathcal{T}(0) = 0 .$$

Within the allotted time, a certain point can be reached in various numbers of steps N. If the step lengths $X(i)$ and waiting times $T(i)$ are independent, the number of steps $N(t)$ until time t is determined by the process of drawing the waiting times. We introduce the probability that the ith step does not occur before time t,

$$P_{T(i)}(t) = \int_{t}^{\infty} p_{T(i)}(t) \, \mathrm{d}t ,$$

where $p_{T(i)}$ is the probability density of the waiting times. The probability that n steps are performed in a time frame $[0, t]$ is then

$$P_{N(t)}(n) = \int_{0}^{t} p_T^{*n}(t') P_{T(n+1)}(t - t') \, \mathrm{d}t' = \left(p_T^{*n} * P_{T(n+1)}\right)(t) ,$$

where $p_T^{*n} = p_{T(1)} * p_{T(2)} * \cdots * p_{T(n)}$.

We compute $P_{N(t)}$ by using the Laplace transformation in the time variable and the Fourier transformation in the spatial variable. This allows us to work comfortably with function products in transform spaces instead of with convolutions. The procedure leads to the Montroll–Weiss equation (see [19] for details). This equation allows us to identify four regions of parameters defining the distributions p_X and p_T with different dependencies of the scale L on time t, which in turn determine the diffusion properties of the random walk. These regions are specified in the following and are shown in Fig. 7.4 (right). We assume that the distributions of steps and waiting times do not change during the walk, so that $p_{X(i)} = p_X$ and $p_{T(i)} = p_T$.

Normal diffusion with the scale $L \sim t^{1/2}$ occurs when $\langle T \rangle < \infty, \sigma_X < \infty$.

Sub-diffusion with the scale $L \sim t^{\beta/2}$ occurs with $\langle T \rangle = \infty$, $\sigma_X < \infty$, and the distribution of waiting times

$$p_T(t) \sim \frac{1}{t^{1+\beta}} \,, \qquad \beta \in (0, 1) \,.$$

Super-diffusion with the scale $L \sim t^{1/\alpha}$ occurs for $\langle T \rangle < \infty$, $\sigma_X = \infty$, and the distribution of the steps

$$p_X(t) \sim \frac{1}{t^{1+\alpha}} \,, \qquad \alpha \in (0, 2) \,.$$

When $\langle T \rangle = \infty$ and $\sigma_X = \infty$, and

$$p_X(t) \sim \frac{1}{t^{1+\alpha}} \,, \quad p_T(t) \sim \frac{1}{t^{1+\beta}} \,, \qquad \alpha \in (0, 2) \,, \quad \beta \in (0, 1) \,,$$

applies, the scale is $L \sim t^{\beta/\alpha}$. The walks are super-diffusive if $2\beta > \alpha$, and sub-diffusive otherwise. Processes with $\langle T \rangle = \infty$ are deeply non-Markovian: this means that the values of the process at some time depend on its complete history, not solely on the state just before that time. Further reading can be found in [18, 20].

7.4 Markov Chains ⋆

In seeking the solutions of the dynamics of complex systems one is often forced to resort to probabilistic description. The true dynamics is simplified to jumping between states with probabilities that, at each jump, depend only on the initial state (before the jump) and the final state (immediately afterwards). In this Section we discuss one type of such random processes, the Markov chain [5, 23, 24].

7.4.1 Discrete-Time or Classical Markov Chains

Think of a random process X in the countable sample space Ω and discrete time $t = 0, 1, \ldots$ Define the conditional probability for a transition from the state x at time t to the state y at time $t + 1$,

$$\text{Prob}\big(X(t + 1) = y \mid X(t) = x\big) = [P(t)]_{x,y} \, .$$

We arrange these probabilities in a *Markov* or *stochastic matrix* $P(t) = [P(t)]_{x,y}$, where $x, y \in \Omega$. Conservation of probability requires $\sum_{y \in \Omega} [P(t)]_{x,y} = 1$. A random process X is called a *Markov chain* with the initial probability distribution μ in Ω and the transition matrix $P(t)$, if the following conditions are fulfilled:

$$\text{Prob}(X(0) = x_0) = \mu(x_0) \, ,$$

$$\text{Prob}\big(X(t + 1) = x_{t+1} \mid X(0) = x_0, \ldots, X(t) = x_t\big) = [P(t)]_{x_t,x_{t+1}} \, .$$

The second condition says that the probability for the present state to occur depends only on the state immediately preceding it. In other words, the probability to reach the point x_{t+1} at time $t + 1$ depends on none of the previous points except x_t. We say that such a Markov chains describes a random process *without memory*. For a system *with memory* consisting of m steps we could imagine a larger sample space Ω^m and define the random process

$$Y(t) = \big(X(t), X(t - 1), X(t - 2), \ldots, X(t - m + 1)\big) \in \Omega^m \, ,$$

which again is a Markov chain. If the Markov matrix P does not depend on time, we are referring to a time-homogeneous Markov chain.

A Markov chain represents a description of the time evolution of a probability distribution in the sample space. If $\boldsymbol{p}(t) = \{p_x(t) = \text{Prob}(X(t) = x)\}_{x \in \Omega}$ is the probability distribution at time t, the distribution at later times $t' > t$ is given by

$$\boldsymbol{p}^{\mathrm{T}}(t) P(t) P(t + 1) \cdots P(t') = \boldsymbol{p}^{\mathrm{T}}(t') \, .$$

Markov chains are linear discrete dynamical systems of time evolution of the probability distribution [25]. The probability for the transition from the state x_0 at time t_0 to the state x_1 at time t_1 is given by the matrix element (x_0, x_1) of the product of Markov matrices $P(t)$ with the times t on the interval $[t_0, t_1]$,

$$\text{Prob}\big(X(t_1) = x_1 \mid X(t_0) = x_0\big) = [P(t_0)P(t_0 + 1) \cdots P(t_1)]_{x_0,x_1} \, , \qquad (7.10)$$

which is one of the forms of the Chapman–Kolmogorov equation [26]. If the Markov chain is time-homogeneous ($P(t) = P$), the expression above simplifies to

$$\text{Prob}\big(X(t_1) = x_1 \mid X(t_0) = x_0\big) = [P^{t_1 - t_0}]_{x_0,x_1} \, .$$

This relation is often used to test whether a process is a Markov chain or not.

Reducibility If a non-zero probability exists that from any state in the chain we can arrive at any other state, we say that the states *communicate* and the corresponding chain is said to be *irreducible*. In the opposite case, we can form subsets of states with respect to which the chain is irreducible.

Periodicity, reproducibility, ergodicity In Markov chains it is important whether we are able to return to the original state from the sample space Ω and how. A state is *periodic* if we can return to it along the paths with the number of steps whose greatest common divisors are larger than 1. In the opposite case the state is non-periodic. A state is *reproducible* if we can return to it in finite time. If the Markov chain is irreducible on the whole sample space Ω and all states are non-periodic and reproducible, the chain is said to be *ergodic*.

Stationary distributions If the Markov chain is time-homogeneous, the Perron–Frobenius theorem [27] guarantees that for each irreducible part of the chain at least one set of points (vector) $\boldsymbol{\pi} = \{\pi_x > 0\}_{x \in \Omega}$ exists such that

$$\boldsymbol{\pi}^{\mathrm{T}} P = \boldsymbol{\pi}^{\mathrm{T}} ,$$

with the normalization $\sum_x \pi_x = 1$. The vector $\boldsymbol{\pi}$ is a left eigenvector of the matrix P with the maximum possible eigenvalue of 1. We call such vectors *stationary* (or *invariant*, or *equilibrium*) distributions of the Markov chain. Stationary distributions play the role of probability distributions that are preserved within the dynamics of the Markov chain. If a Markov chain is ergodic, exactly one stationary distribution $\boldsymbol{\pi}$ exists to which any initial probability distribution $\boldsymbol{p}$ converges as

$$\boldsymbol{\pi}^{\mathrm{T}} - \boldsymbol{p}^{\mathrm{T}} P^t = \mathcal{O}(|\nu|^t) ,$$

where ν is the second largest eigenvalue of P. A stationary distribution $\boldsymbol{\pi}$ may also exist in time-inhomogeneous chains, where we require

$$\boldsymbol{\pi}^{\mathrm{T}} P(t) = \boldsymbol{\pi}^{\mathrm{T}} \qquad \forall t ,$$

but its existence is not guaranteed.

Detailed balance In physically motivated Markov chains one frequently encounters the condition of *detailed balance* which states that in equilibrium the occupation probability of all states is equal, so

$$\pi_x \, \mathrm{Prob}\big(X(t+1) = y \,\big|\, X(t) = x\big) = \pi_y \, \mathrm{Prob}\big(X(t+1) = x \,\big|\, X(t) = y\big) ,$$

or

$$\pi_x [P(t)]_{x,y} = \pi_y [P(t)]_{y,x} ,$$

where $\boldsymbol{\pi} = \{\pi_x\}_{x\in\Omega}$ is a stationary distribution. If a distribution $\boldsymbol{\pi}$ exists for which the condition of detailed balance applies, the corresponding Markov chain is called *reversible*.

Entropy One way to think about a Markov chain is to establish a probabilistic description of a random motion of a particle or a more general system between the states in Ω. Assume that the Markov chain is time-homogeneous and that the transition probabilities are given by the matrix P. We are interested in the probability that, given some initial discrete probability distribution μ, the particle follows the trajectory $\{x_i\}_{i=0}^t$ beginning at the state x_0, from whence it jumps to x_1, to $x_2, \ldots$ and arrives at x_t after time t. This discrete probability distribution is

$$p(x_0, x_1, \ldots, x_t) = \mu(x_0) P_{x_0, x_1} P_{x_1, x_2} \cdots P_{x_{t-1}, x_t} \, .$$

The entropy of such a random process until time t is defined as

$$H(t) = - \sum_{(x_0, x_1, \ldots, x_t) \in \Omega^{t+1}} p(x_0, x_1, \ldots, x_t) \log p(x_0, x_1, \ldots, x_t)$$

and tells us something about the richness (complexity) of the possible trajectories of the process in the sample space [28]. We expect physics-inspired Markov chains to be ergodic, and in such chains the entropy increases linearly with time. Of course, this does not apply to periodic states where the complexity of the dynamics is strongly diminished. The rate of entropy growth in an ergodic chain with the Markov matrix P and stationary distribution $\boldsymbol{\pi}$ is given by

$$h = \lim_{t\to\infty} \frac{1}{t} H(t) = - \sum_{x,y \in \Omega} \pi_x P_{x,y} \log P_{x,y} \, .$$

If a Markov chain describes the dynamics of a sufficiently complex system, h is an approximation of the information entropy for that system, and is proportional to its statistical (Boltzmann) entropy [29]. The connection between the rate of entropy growth and eigenvalues of Markov matrices is discussed in [30].

Example A tourist travels within the pre-brexit EU ($n = 27$ states). She decides about the entry into the next state upon looking at its relative size with respect to the sizes of all states accessible at that moment. The transitions between the states are a random process that can be described as a discrete-time Markov chain on the set of indices $\{i\}_{i=1}^n$ denoting individual states. The possibilities for the transitions are shown by the graph in Fig. 7.6 (left) and are summarized by the function

$$c(i, j) = \begin{cases} 1 \; ; \; \text{state } i \text{ connected to state } j \text{ by road or rail} \, , \\ 0 \; ; \; \text{otherwise} \, . \end{cases}$$

Clearly, each state is connected to itself, hence $c(i, i) = 1$. The tourist departs the ith state heading for the jth state with the conditional probability

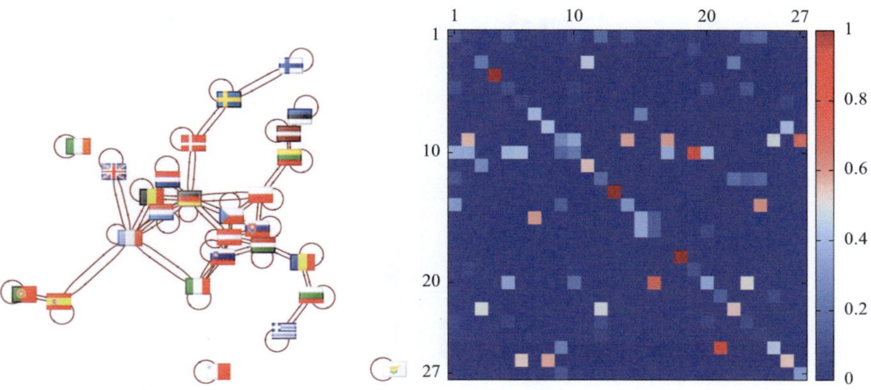

Fig. 7.6 Traveling within the (pre-brexit) EU as a discrete-time Markov chain. [LEFT] The graph of road and rail connections. [RIGHT] The matrix elements of the Markov matrix P

$$\text{Prob}(j|i) = \frac{c(i,j)\,S_j}{\sum_k S_k}\,, \qquad c(k,i) \neq 0\,,$$

where S_i is the size of the ith state. We arrange all conditional probabilities in the Markov matrix $P = [\text{Prob}(j|i)]_{i,j=1}^n$ shown in Fig. 7.6 (right).

The probability distribution of the tourist is represented by the array $\boldsymbol{p}(t) = \{p_i(t)\}_{i=1}^n$. This array can be treated as a vector of dimension n that evolves as

$$\boldsymbol{p}^{\mathrm{T}}(t) = \boldsymbol{p}^{\mathrm{T}}(0)\,P^t\,,$$

where $t \in \mathbb{N}_0$ counts the decisions on further travel ("time"). We are interested in the probability that the tourist will appear in the state i after time t if she started her travel in the state s, so $p_i(0) = \delta_{i,s}$. An example of a trip commenced in Austria is shown in Fig. 7.7. Gradually she wanders everywhere except Cyprus, Malta and Ireland, which are not connected to the remaining states. All n states can thus be split in four connected subsets on which the Markov chain is irreducible. On each of these subsets we could devise a chain with reproducible states, but only on the largest subset of 24 states the chain is non-periodic and therefore also ergodic. Any initial position of the tourist ends up in a unique stationary distribution, the approximation of which is shown in Fig. 7.7 at $t = 100$.   ◁

7.4.2 Continuous-Time Markov Chains

It is not hard to generalize discrete-time Markov chains to continuous-time chains. Assume that a random process X with a countable sample space Ω is observed in continuous time $t \in \mathbb{R}$. In such a process, the probability for a transition from state x at time t to state y at a later time $t + \Delta t$ is

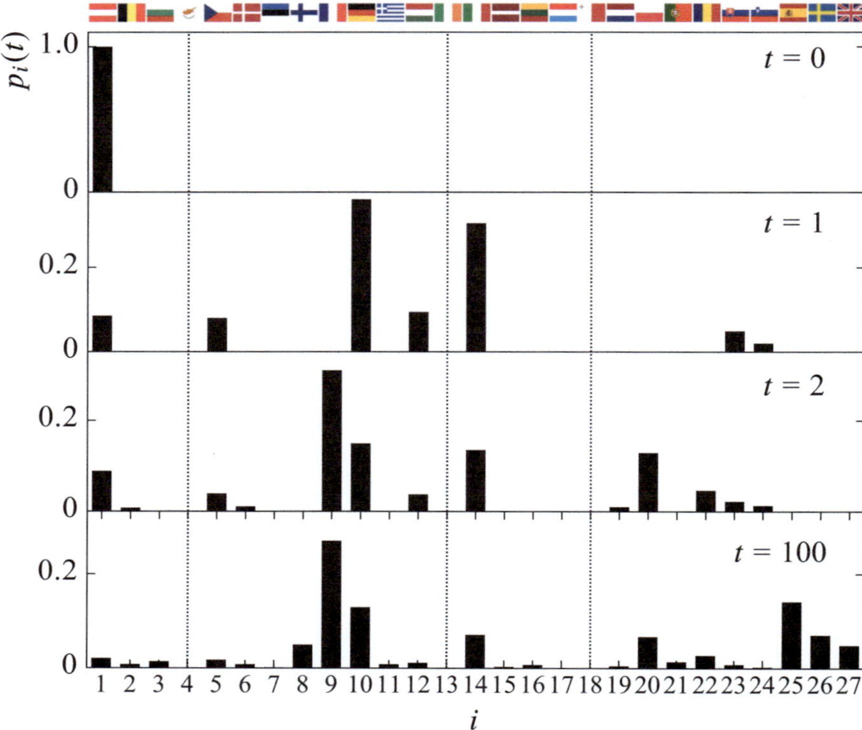

Fig. 7.7 The time evolution of the tourist's probability distribution for a trip started in Austria ($i = 1$). The distribution at $t = 100$ is an excellent approximation for the stationary distribution. Note that the shown bar sizes are *not* proportional to the sizes of the states! A good example is Finland ($i = 8$) that is reachable only through Sweden. The average probability of meeting the tourist in Finland is therefore only the seventh largest (4.9 %) although Finland is the fifth largest state. Three dotted vertical lines correspond to Cyprus, Malta and Ireland, all of which the tourist never enters

$$\mathrm{Prob}\big(X(t + \Delta t) = y \,\big|\, X(t) = x\big) \,.$$

If the conditional probabilities satisfy the continuous analogue of Eq. (7.10), such a process represents a continuous-time Markov chain. In the limit $\Delta t \to 0$ it then holds that

$$\mathrm{Prob}\big(X(t + \Delta t) = y \,\big|\, X(t) = x\big) = \delta_{x,y} + [Q(t)]_{x,y}\Delta t + \mathcal{O}(\Delta t) \,,$$

where $Q(t)$ is the *transition rate matrix*. Its matrix elements are $[Q(t)]_{x,y} \geq 0$ for $x \neq y$ and $[Q(t)]_{x,x} \leq 0$. From conservation of probability $\sum_{y \in \Omega} \mathrm{Prob}\big(X(t + \Delta t) = y \,|\, X(t) = x\big) = 1$ we obtain $\sum_y [Q(t)]_{x,y} = 0$ for each $x \in \Omega$, hence the decay rate of the individual state is

$$[Q(t)]_{x,x} = -\sum_{y \neq x} [Q(t)]_{x,y}\,.$$

Just like in the case of discrete chains (Sect. 7.4.1) the continuous-time Markov chains may be seen as a probabilistic description of the dynamics of some system in the set of states Ω. Let $\boldsymbol{p}(t) = \{p_x(t) = \mathrm{Prob}(X(t) = x)\}_{x \in \Omega}$ be the probability distribution of the system with respect to the states at time t. The time evolution of this distribution (as a vector) is then given by a system of ordinary differential equations

$$\frac{\mathrm{d}}{\mathrm{d}t} \boldsymbol{p}^{\mathrm{T}}(t) = \boldsymbol{p}^{\mathrm{T}}(t) Q(t)\,. \tag{7.11}$$

The solution of this system is

$$\boldsymbol{p}^{\mathrm{T}}(t') = \boldsymbol{p}^{\mathrm{T}}(t)\, \mathcal{P}(t, t')\,,$$

where the form of the matrix $\mathcal{P}$ depends on whether the Markov matrix Q is time-dependent or not. For a homogeneous chain ($Q(t) = Q$) we get

$$\mathcal{P}(t, t') = \exp((t' - t)Q)\,.$$

In the case of a non-homogeneous chain (Q depends on time) the formal solution is

$$\mathcal{P}(t, t') = \hat{\mathcal{T}} \exp\left(\int_{t}^{t'} Q(\tau)\,\mathrm{d}\tau\right),\tag{7.12}$$

where the time-ordering operator $\hat{\mathcal{T}}$ [31] creates the correct time ordering of the integration limits once the exponential function of matrices is power-expanded. Due to its complicated structure the solution (7.12) is generally useful at short times only; in other cases, we need to tackle the system (7.11) directly.

A continuous-time Markov chain is converted to a discrete-time chain by choosing the step size Δt and setting $P(t) = \mathcal{P}(t, t + \Delta t)$. If one assumes that the random process X at time t is in the state x, the probability that the process remains in this state until time $t + \Delta t$ is

$$\mathrm{Prob}\big(X(r) = x \,\big|\, X(t) = x\big) = \exp\big(\Delta t[Q]_{x,x}\big) \qquad \forall r \in (t, t + \Delta t]$$

in the case of a homogeneous chain, while it is

$$\mathrm{Prob}\big(X(r) = x \,\big|\, X(t) = x\big) = \hat{\mathcal{T}} \exp\left(\int_{0}^{\Delta t} [Q(t + \tau)]_{x,x}\,\mathrm{d}\tau\right) \qquad \forall r \in (t, t + \Delta t]$$

in the case of a non-homogeneous one. Reducibility, periodicity, reproducibility, and ergodicity of states for continuous-time Markov chains are introduced by complete analogy to discrete-time chains (Sect. 7.4.1).

Example Imagine an array of n sites with labels from 1 to n. Particles may jump between neighboring sites. We describe the dynamics of the particles by a continuous-time homogeneous Markov chain, where the sample space are the sites on the chain, $\Omega = \{i\}_{i=1}^{n}$. The particle at site i can jump to the neighboring site to the left (label $i - 1$) or to the right (label $i + 1$) with the rate r, if this site exists. This rule can be embodied in the transition rate matrix

$$
Q = \begin{pmatrix}
-r & r & & & 0 \\
r & -2r & r & & \\
 & \ddots & \ddots & \ddots & \\
 & & r & -2r & r \\
0 & & & r & -r
\end{pmatrix} .
$$

The distribution of the particles along the chain at time t is given by the array $\boldsymbol{p}(t) = \{p_i(t)\}_{i=1}^{n}$, which we interpret as a vector $(p_1(t), p_2(t), \ldots, p_n(t))^{\mathrm{T}}$. Its time evolution is given by the system of differential equations

$$
\frac{\mathrm{d}}{\mathrm{d}t} \boldsymbol{p}^{\mathrm{T}}(t) = \boldsymbol{p}^{\mathrm{T}}(t)\, Q .
$$

At $t = 0$ all particles are at the chosen site s, thus $p_i(0) = \delta_{i,s}$. The time evolution of the distribution for $r = 0.1$ and $s = 11$ in the case of a chain of length $n = 21$ is shown in Fig. 7.8 (left). Early on, the distribution acquires a Gauss-like shape, but after long times it reaches the boundaries and flattens out. Ultimately we reach the limit $\lim_{t \to \infty} p_i(t) = 1/n$ representing the stationary distribution, $p_i = \pi_i$.

The average of an observable f with respect to the distribution $\boldsymbol{p}$ is $\langle f(i) \rangle_{i,t} = \sum_{i=1}^{n} f(i) p_i(t)$. The dynamics of the system is such that the distribution average does not change with time, $\langle i \rangle_{i,t} = s$. What does change is the variance,

$$
\sigma^2(t) = \left\langle (i - s)^2 \right\rangle_{i,t} ,
$$

shown in Fig. 7.8 (right). We see that $\sigma^2(t) \approx 2rt$ until the distribution spreads out to the boundaries of the chain which it is unable to cross. The direct proportionality of the variance and time indicates that the diffusion of particles in this system is normal, with the diffusion constant r.   ◁

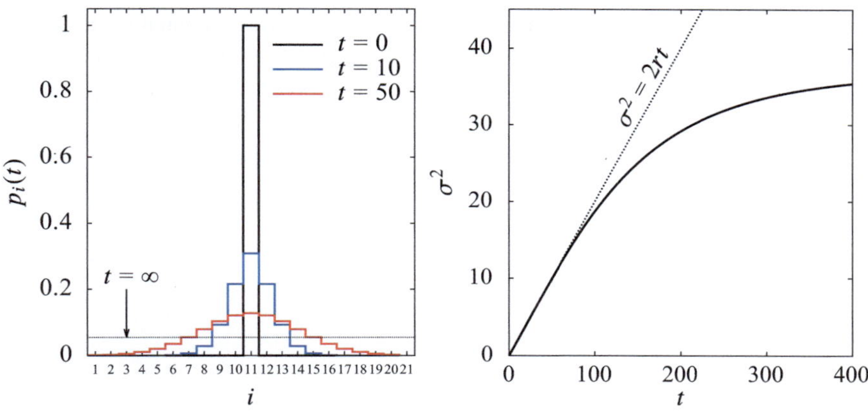

Fig. 7.8 Motion of particles between the states of a one-dimensional array of length $n = 21$ as a continuous-time Markov chain. [LEFT] Time evolution of the distribution $\boldsymbol{p}$ with particles initially at $s = 11$, with $r = 0.1$. After long times we obtain a constant distribution $p_i(t) = 1/21 \approx 0.04762$. [RIGHT] Time evolution of the distribution variance

7.5 Noise

Noise (for continuous or discrete time) is defined as a real random process Z which is wide-sense stationary and ergodic (see page 394), and has zero average,

$$\langle Z(t) \rangle = 0 \,,$$

a rapidly decreasing temporal auto-correlation R_{ZZ},

$$R_{ZZ}(t_1, t_2) = \langle Z(t_1) Z(t_2) \rangle = R_{ZZ}(t_1 - t_2) \,,$$

and probability density

$$p_Z(x) = \langle \delta(x - Z(t)) \rangle \,.$$

Here $\langle \cdot \rangle$ denotes the statistical average over various realizations of the random process. Because noise is ergodic, the statistical average is equal to the time average. Noise as an abstract entity and, above all, in its individual realizations, is frequently used in stochastic simulations in physics, electric engineering, and acoustics.

7.5.1 Types of Noise

Noise is classified according to the functional forms of its probability density p_Z and auto-correlation R_{ZZ}. Auto-correlation is the most important property of noise defining its characteristics. It is usually presented in Fourier space by computing the *average power spectral density* (PSD). The power density S_Z is equal to the square of the Fourier transform of the signal's auto-correlation (see Eqs. (5.1) and (5.18), as well as Sect. 5.2.7). For continuous-time noise we obtain (up to a factor)

$$S_Z(\omega) \propto \mathcal{F}[R_{ZZ}](\omega) \, ,$$

while for discrete-time noise we get

$$S_{Z,k} \propto (\mathcal{F}_N[R_{ZZ}])_k \, .$$

Noises with a spectral density of the form

$$S_Z(\omega) \propto \omega^{-a} \, , \qquad a > 0 \, ,$$

are said to possess *colors*, which are assigned according to the values of a. The most well-known noises are white ($a = 0$), pink ($a = 1$), and red ($a = 2$). Examples are shown in the left panels of Fig. 7.9. The corresponding distributions of the values of the noise are shown in the right panels.

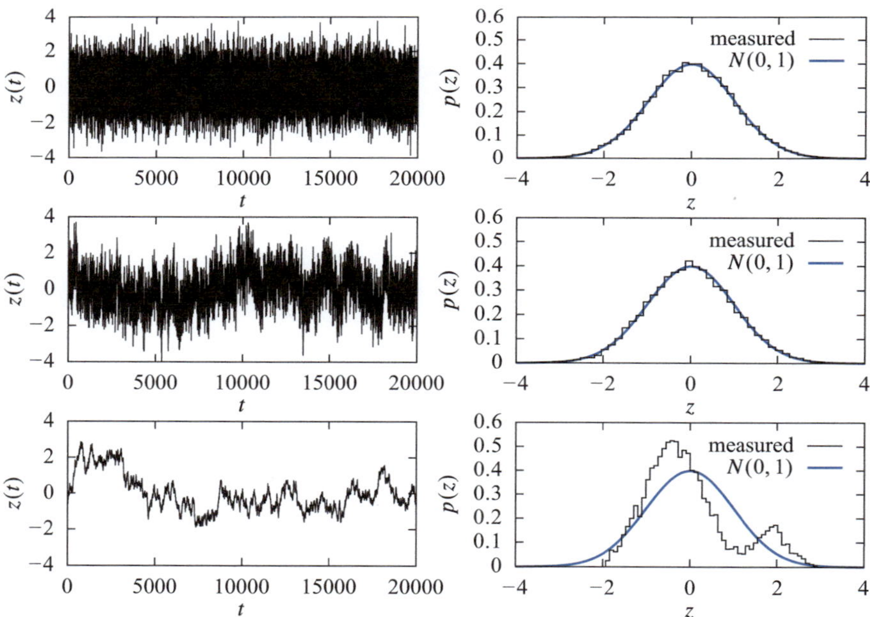

Fig. 7.9 Examples of realizations of various types of noise (left panels) and the corresponding distributions of values in these realizations (right panels). The color of the noise is defined by the parameter a in the power spectral density $S_Z(\omega) \propto \omega^{-a}$. [TOP] White noise ($a = 0$). [CENTER] Pink noise ($a = 1$). [BOTTOM] Red (or Brownian) noise ($a = 2$). The figures at the right also show the standard normal distribution $N(0, 1)$

In the strict sense, colored noises with parameters $a \geq 1$ are non-stationary random processes [32], since $S_Z \propto \omega^{-a}$ has a singularity at the origin. In practical calculations we require that S_Z differs from zero only for frequencies ω larger than some minimal frequency ω_0, e.g. $\omega_0 \approx \mathcal{O}(N^{-1})$, where N is the sample size.

White noise The Fourier spectrum of *white noise* is "flat", $S_Z(\omega) = \text{const}$. The noise signal is therefore δ-auto-correlated in the statistical sense. For continuous-time noise this implies

$$R_{ZZ}(\tau) = \langle Z(t)^2 \rangle \delta(\tau) ,$$

while for discrete-time noise

$$R_{ZZ}(\tau) = \langle Z(t)^2 \rangle \delta_{\tau,0} .$$

White noise is strict-sense stationary and resembles the noise caused by thermal fluctuations of charge carriers at finite temperatures. We are commonly referring to Gaussian white noise since its probability density p_Z is the Gauss function.

Pink noise *Pink* or *flicker noise* has the power spectral density $S_Z(\omega) \propto \omega^{-1}$. It occurs, for example, in circuits with semiconductor elements.

Red (Brownian) noise *Red* or *Brownian noise* has the power spectral density $S_Z(\omega) \propto \omega^{-2}$ and is typically encountered in random-walk processes. Realizations $z(t)$ of red noise can be obtained from white noise $w(t)$ by means of the Wiener process [33] defined by

$$z(t) = z(0) + \int_0^t w(\tau)\, d\tau .$$

Once this equation is rewritten as $dz(t)/dt = w(t)$, the Fourier transformation can be applied to both sides, and we get

$$S_Z(\omega) = \frac{\langle w(t)^2 \rangle}{\omega^2} .$$

7.5.2 Generation of Noise

Generation of noise with appropriate statistical and spectral properties is of crucial importance in certain applications. A system with pronounced sensitivity to high-frequency content of signals will respond differently if white or red noise is introduced at its input. Two classes of noise generation methods exist: *online* and *offline*. In online mode we generate the signal values $\{z_i\}$ of noise Z sequentially, for example, while the simulation is running; the values should therefore be generated as quickly as possible. In offline mode, the complete realization of the noise $\{z_i\}_{i=0}^{N-1}$ with an arbitrary spectral density S_Z is generated in a single shot. Both methods are described in the following.

Online mode White noise with a desired probability density p_Z can be generated directly by using a generator of random numbers that are distributed according to p_Z (see Appendix C).

A review of pink-noise generators can be found at [34]. The improved correlated generator [35] based on random weighted summation of n white-noise signals is particularly simple and effective. To generate a signal with the spectrum $S_Z(\omega) \propto \omega^{-1}$ we have to choose the appropriate weights $\boldsymbol{w} = \{w_i\}_{i=1}^n$ for the contributions of white noise, the bounding probabilities $\boldsymbol{p} = \{p_i\}_{i=1}^n$ for random summation of individual contributions, and the vector describing the state of the generator, $\boldsymbol{c} = \{c_i\}_{i=1}^n$. The components of this vector should be initialized with random numbers distributed according to $U(-1, 1)$ (uniformly on $[-1, 1]$). The algorithm to obtain the next value of the signal of pink noise is:

> **Data**: arrays of weights $\boldsymbol{w} = \{w_i\}_{i=1}^n$ and probabilities $\boldsymbol{p} = \{p_i\}_{i=1}^n$.
> **Input**: state vector of the generator $\boldsymbol{c} = \{c_i\}_{i=1}^n$.
> Draw a random x from $U(0, 1)$;
> **for** $i = 1, 2, \ldots, n$ **do**
> > **if** ($x \le p_i$) **then**
> > > Draw a random y from $U(-1, 1)$;
> > > $c_i = w_i y$;
> > > Break the loop;
> >
> > **end**
>
> **end**
> $z = \sum_{i=1}^n c_i$;
> **Output**: next value of the signal z and the state vector $\boldsymbol{c}$.

The uniform generator used to generate x and y in this algorithm may be very simple, as its properties do not influence significantly the quality of the generated pink noise. Ref. [35] recommends $n = 5$ and

$$\boldsymbol{w} = \{0.95060, \ 0.74235, \ 0.64925, \ 0.77175, \ 0.85015\},$$
$$\boldsymbol{p} = \{0.00198, \ 0.01478, \ 0.06378, \ 0.23378, \ 0.91578\}.$$

The generated signals have values on the interval $[-W, W]$, where $W = \sum_{i=1}^n w_i$. The values are distributed approximately according to $N(0, 1)$.

Red or Brownian noise can be generated by using a one-dimensional random walk from which the average of the last n steps is continuously subtracted. Imagine a random walk defined by the recurrence

$$a_{t+1} = a_t + \varepsilon w_t,$$

where w_t is a random number distributed according to $N(0, 1)$, and ε is the length of the step. The sequence of values of approximately red noise $\boldsymbol{z} = \{z_t\}_{t=0}^\infty$ is obtained by applying a linear filter on the sequence $\boldsymbol{a} = \{a_t\}_{t=0}^\infty$,

$$z_t = a_t - \frac{1}{m} \sum_{i=0}^{m-1} a_{t-i} \, .$$

The shape of the spectrum is controlled by the filter length m. Assume that we have computed the signal $z = \{z_i\}_{i=0}^{N-1}$ of length N and its single-sided spectrum $S = \{S_i\}_{i=0}^{N/2} = \text{PSD}[z]$. For $i > K = N/(2\,m)$ we obtain $S \sim \mathcal{O}(i^{-2})$, while for $i < K$ the spectrum S does not change significantly. To generate a signal resembling red noise, we therefore need $N = \mathcal{O}(m)$. If we choose $\varepsilon = 2/\sqrt{m}$ and $K \gg 1$, the generated values are distributed according to $N(0, 1)$.

Offline mode The simplest algorithm to generate a complete sample of noise with an arbitrary spectral density S_Z is based on the Fourier transformation. Here we adopt, with minor modifications, the algorithm from [36]. Generate the array $c = \{c_i\}_{i=0}^{N-1}$ (with N odd) by using

$$c_0 = 0 \, , \qquad c_{N-j}^* = c_j = \sqrt{S_Z\left(\frac{j\omega'}{N}\right)} \, (a_j + i\,b_j) \, , \qquad j = 1, 2, \ldots, (N-1)/2 \, ,$$

where a_j and b_j are random numbers distributed according to $N(0, 1)$, and ω' is a characteristic frequency that we set according to the desired frequency range of the generated spectrum. By computing the inverse Fourier transform of c,

$$z_k = \frac{1}{\|c\|_2} \sum_{j=0}^{N-1} e^{i\,2\pi jk/N} c_j \, ,$$

we obtain the desired realization of noise. If S_Z is a bounded function of ω, the average probability density of the values z_k converges to $N(0, 1)$ with increasing N. The same applies to colored noises with the parameter $a \in [0, 1]$. In contrast, for a larger than 1 the probability density of the values in individual signals (and on average) deviates far from the normal distribution. Figure 7.9 shows the signals of white ($a = 0$), pink ($a = 1$), and red noise ($a = 2$), as well as the probability densities of values in these signals obtained by the above method.

7.6 Time Correlation and Auto-Correlation

Correlations measure statistical dependence or similarity of quantities. The mathematical formulation of correlation varies according to the type of the quantities, but its meaning remains the same. Consider two continuous-time random processes F and G. We define their time correlation as the statistical average of the product of the processes, one of which is shifted with respect to the other by τ [9]:

$$R_{FG}(\tau) = \left\langle F(t)^* G(t + \tau) \right\rangle .$$

The time shift τ is the parameter of the correlation. If the processes F and G are ergodic and their correlation is equal to the time correlation of their realizations f and g, such processes are known as *jointly ergodic*, and we have

$$R_{FG}(\tau) = R_{fg}(\tau) = \left\langle f(t)^* g(t + \tau) \right\rangle_t = \lim_{T \to \infty} \frac{1}{2T} \int_{-T}^{T} f(t')^* g(t' + \tau) \, dt' ,$$

where the average $\langle \cdot \rangle_t$ is defined in Eq. (7.4). For discrete times the integral in the expression above should be rewritten as a sum, in accordance with the discrete definition of the time average, Eq. (7.5).

More formally, correlation can be understood as a binary operation that assigns the correlation function R_{fg} to the functions f and g. The precise form of the assignment depends on the definition domains X of the functions, e.g.

$$R_{fg}(\tau) = \int_X f(t)^* g(t + \tau) \, d^n t , \qquad X = \mathbb{R}^n, \mathbb{C}^n ,$$

or

$$R_{fg}(\tau) = \sum_{t \in X} f(t)^* g(t + \tau) , \qquad X = \mathbb{Z}^n, \mathbb{Z}_N^n .$$

Correlation therefore represents an overlap integral in the domain of the shifted functions, where the shift is the argument of the correlation. The correlation of functions in the domain $\mathbb{Z}_N^2$, where $N = 21$, is illustrated by the Example below (see Fig. 7.10). An overview of use of time correlations in theoretical physics is given by [37].

Basic properties and relation to convolution Regardless of the character of the objects f and g (realizations of random processes, discrete or continuous functions), their correlation has some common properties. A correlation is *symmetric*,

$$R_{gf}(-\tau) = R_{fg}(\tau)^* , \tag{7.13}$$

and bounded in the sense

$$\left| R_{fg}(\tau) \right|^2 \leq R_{ff}(0) R_{gg}(0) , \qquad \left| R_{fg}(\tau) \right| \leq \tfrac{1}{2} \left(R_{ff}(0) + R_{gg}(0) \right) .$$

A correlation between different random processes or their realizations (functions) f and g is also called *cross-correlation*, while for $f = g$ we are referring to an *auto-correlation function* (ACF). Auto-correlation is the sum of auto-correlations of non-correlated parts: let $f = \sum_i s_i$ and $R_{s_i s_j} = 0$ for $i \neq j$. Then

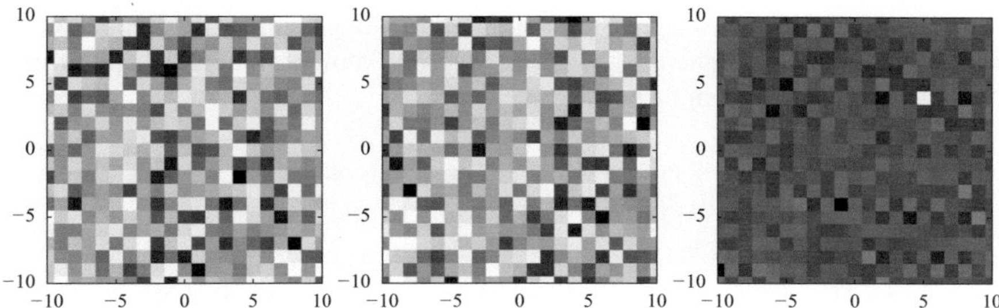

Fig. 7.10 The basic operational idea of an optical mouse. [LEFT] First acquired image: a two-dimensional sample (scalar field) F of size 20×20 with values $F_{i,j} \in [0, 1]$ and periodic boundary conditions. [CENTER] The shifted version of the image $\widetilde{F}_{i,j} = F_{i-5,j-4}$. [RIGHT] The correlation of the original and shifted fields $C_{i,j} = (1/N^2) \sum_{\alpha,\beta} \widetilde{F}_{i+\alpha,j+\beta} F_{\alpha,\beta}$ indicates the shift of the mouse from the current position at $(0, 0)$ to $(5, 4)$

$$R_{ff} = \sum_i R_{s_i s_i} \,.$$

The correlation R_{fg} can be expressed by the convolution, $* : (f, g) \mapsto f * g$, as

$$R_{fg}(\tau) = \left[f^*(-t) * g(t) \right] (\tau) \,.$$

When the existence of the correlation or of the convolution itself is questioned, we rely on the property that $f, g \in L^p(X)$ implies $R_{fg} \in L^p(X)$. In computations of correlations and convolutions where Fourier transformations are involved, we frequently assume that f and g are rapidly decreasing smooth functions [38]. For details see [39, 40].

Example An optical mouse is a standard external device for modern computers. By moving the mouse we communicate to a certain program our desire to move the graphical pointer. The mouse senses the motion by a small CCD camera capturing the reflection of light from the surface illuminated by a light-emitting diode. (The image acquisition rate is usually about 1 kHz.) The program compares the subsequent images and uses correlations between them to determine the direction and length of the move. An example of such a computation is shown in Fig. 7.10. The more random the acquired samples are, the more precise the determination of the shift can be. This is why optical mice perform poorly on polished reflective surfaces or surfaces with fine periodic textures.   ◁

7.6.1 Sample Correlations of Signals

In practice we are not dealing with infinitely long signals but with their finite chunks or *samples*. We would like to use the sample to learn as much as possible about the whole signal, for example, by computing the approximation of its time average or a correlation between two signals. Let an infinitely long signal f be defined on the whole real or discrete time axis. Assume we only know its sample $\hat{f}$ in the time interval $[t_0, t_0 + T)$ of length T, thus

$$
\hat{f}(t) = \begin{cases} f(t_0 + t) \; ; \; t \in [0, T) \subset \mathbb{R} & \text{(continuous signal)}, \\ f(t_0 + t) \; ; \; t \in [0, T - 1] \subset \mathbb{Z} & \text{(discrete signal)}. \end{cases}
$$

For continuous time, a correlation of two such samples $\hat{f}$ and $\hat{g}$ is defined as

$$
\mathcal{R}_{\hat{f}\hat{g}}(\tau) = w(\tau, T) \int\limits_0^{T-|\tau|} dt \begin{cases} \hat{f}(t)^*\hat{g}(t + \tau) \; ; \; \tau \in [0, T), \\ \hat{f}(t + |\tau|)^*\hat{g}(t) \; ; \; \tau \in (-T, 0), \end{cases}
$$

while for discrete time we use

$$
\mathcal{R}_{\hat{f}\hat{g}}(\tau) = w(\tau, T) \sum_{t=0}^{T-|\tau|-1} \begin{cases} \hat{f}(t)^*\hat{g}(t + \tau) \; ; \; \tau \in [0, T - 1], \\ \hat{f}(t + |\tau|)^*\hat{g}(t) \; ; \; \tau \in [-T + 1, 0]. \end{cases}
$$

The weight w will be determined later; for the moment we require that it behaves asymptotically as $w(\tau, T) \sim T^{-1}$ when $T \to \infty$, so that we have the limit

$$
\lim_{\substack{T \to \infty \\ t_0 \to -\infty}} \mathcal{R}_{\hat{f}\hat{g}} = R_{fg} .
$$

Sample correlations defined in this manner have the symmetry $\mathcal{R}_{\hat{f}\hat{g}}(-\tau) = \mathcal{R}_{\hat{g}\hat{f}}(\tau)^*$, just as in Eq. (7.13), but they do not necessarily possess other previously enumerated properties of correlations. Which properties are shared among the infinite-signal and sample correlations strongly depends on the choice of the weight.

Choice of the weight In choosing the weight to compute the sample correlation we are usually forced to make a compromise between mathematical correctness and technical usefulness. We exploit the following facts. If the signals f and g are realizations of jointly ergodic random processes F and G that are wide-sense stationary (WSS, page 394), it holds that

$$
\left\langle R_{\hat{f}\hat{g}}(\tau) \right\rangle = (T - |\tau|)\, w(\tau, T)\, R_{FG}(\tau) . \tag{7.14}
$$

If F is a discrete-time Gaussian random process and f is its realization, the variance of the auto-correlation is

$$
\text{var}\left[\mathcal{R}_{\hat{f}\hat{f}}(\tau)\right] \approx T\, w(\tau, T)^2 \sum_{t \in Z} \left\{ R_{FF}(t)^2 + R_{FF}(t + \tau)R_{FF}(t - \tau) \right\} , \tag{7.15}
$$

and this dependence on T and τ roughly applies also to other processes from the WSS class [41]. We conclude that it is sensible to use the weight

$$w(\tau, T) = \frac{1}{T - |\tau|} \, ,$$

since by (7.14) the sample correlation then yields the correct estimate of the correlation of random processes $\langle \mathcal{R}_{\hat{f}\hat{g}}(\tau)\rangle = R_{FG}(\tau)$. Such sample correlations are *unbiased*. Their main weakness is that for fixed T and increasing τ the amount of the signal used in the computation of the sample correlation decreases, causing the statistical error to grow as $\mathcal{O}((T - |\tau|)^{-1/2})$ according to (7.15). Therefore, for $\tau = \mathcal{O}(T)$ the sample correlation becomes completely unpredictable. Besides, the sample correlation does not necessarily fulfill the inequality

$$|\mathcal{R}_{\hat{f}\hat{f}}(\tau)| \leq |\mathcal{R}_{\hat{f}\hat{f}}(0)| \, , \tag{7.16}$$

which is crucial in some applications. Therefore, one often prefers the weight

$$w(\tau, T) = \frac{1}{T} \, ,$$

which generates *biased* correlation estimates. In this case the sample correlation does not accurately estimate the correlation of the processes

$$\left\langle \mathcal{R}_{\hat{f}\hat{g}}(\tau)\right\rangle = \left(1 - \frac{|\tau|}{T}\right) R_{FG}(\tau) \, ,$$

but the statistical error is reduced: it becomes almost independent of the shift τ and is of the order $\mathcal{O}(T^{-1/2})$; at the same time inequality (7.16) holds true.

7.6.2 Representation of Time Correlations

Assume that for signals f and g we computed the time averages $\mu = \langle f(t)\rangle_t$ and $\nu = \langle g(t)\rangle_t$. In terms of these averages and the remainders we can write

$$f(t) = \mu + x(t) \, , \qquad g(t) = \nu + y(t) \, .$$

The time correlation of f and g is then

$$R_{\mu+x, \, \nu+y}(\tau) = \mu\nu + R_{xy}(\tau) \, .$$

Therefore, the correlation with the product of the averages subtracted is

$$R^{(1)}_{fg}(\tau) = R_{fg}(\tau) - \mu\nu = R_{f-\mu,\,g-\nu}(\tau)\,.$$

Alternatively, the averages are subtracted from the signal prior to the calculation of the correlation. We see that at shift $\tau = 0$, the correlation $R^{(1)}_{fg}$ is equal to the covariance with respect to the time average,

$$R^{(1)}_{fg}(0) = \langle (f(t) - \mu)(g(t) - \nu)\rangle_t\,.$$

In auto-correlation we have $f = g$ and the covariance is equal to the variance of the signal with respect to the time average, $R^{(1)}_{ff}(0) = \langle (f(t) - \mu)^2\rangle_t$. Because the variance of f is known, we divide this auto-correlation by the variance, and display

$$R^{(2)}_{ff}(\tau) = \frac{R^{(1)}_{ff}(\tau)}{R^{(1)}_{ff}(0)}\,.$$

7.6.3   *Fast Computation of Discrete Sample Correlations*

The correlation of finite sequences $a = \{a_i\}_{i=0}^{N-1}$ and $b = \{b_i\}_{i=0}^{N-1}$ with periodic boundary conditions $a_{i+N} = a_i$ and $b_{i+N} = b_i$ is a new periodic sequence $c = \{c_i\}_{i=0}^{N}$ with the terms $c_i = \sum_{j=0}^{N-1} a_j^* b_{j+i}$. This direct summation requires $\mathcal{O}(N^2)$ operations. A faster way of computing the correlation is possible by using the discrete Fourier transformation (DFT, see Eqs. (5.11) and [42]). Namely, the Fourier transforms of the arrays a, b, and c are related by

$$(\mathcal{F}_N[c])_i = N\,(\mathcal{F}_N[a])_i^*(\mathcal{F}_N[b])_i\,, \qquad i = 0, 1, \ldots, N - 1\,,$$

so that the inverse DFT gives us the correlation of periodic arrays

$$c = \mathcal{F}_N^{-1}[d]\,, \qquad d = N\left((\mathcal{F}_N[a])_i^*(\mathcal{F}_N[b])_i\right)_{i=0}^{N-1}\,. \tag{7.17}$$

By using the fast DFT (FFT, Sect. 5.2.5) the correlation c according to the formula above can be computed in $\mathcal{O}(N \log_2 N)$ operations.

In computing the sample correlation of samples $\{a_i\}$ and $\{b_i\}$ we wish to effectively compute the sum of the form

$$c_i = \sum_{j=0}^{N-i-1} a_j^* b_{j+i}\,, \qquad i = 0, 1, \ldots, N - 1\,.$$

This is done by extending the arrays $\{a_i\}$ and $\{b_i\}$ to double their length and filling the attached part with zeros. Thus we obtain two new arrays,

$$\tilde{a} = \{a_0, a_1, \ldots, a_{N-1}, 0, \ldots, 0\}, \qquad \tilde{b} = \{b_0, b_1, \ldots, b_{N-1}, 0, \ldots, 0\},$$

for which we assume periodic boundary conditions $\tilde{a}_{i+2N} = \tilde{a}_i$ and $\tilde{b}_{i+2N} = \tilde{b}_i$. We rewrite the sample correlation as a correlation of periodic arrays

$$c_i = \sum_{j=0}^{2N-1} \tilde{a}_j^* \tilde{b}_{j+i}, \tag{7.18}$$

where only the first N terms of c are used in the computation of the sample correlation. A sum of the form (7.18) can be computed via Eq. (7.17) by using the FFT, except that the transforms appearing in it act on arrays of length $2N$. (The procedure described here closely resembles the multiplication of polynomials by using the FFT; see Sect. 5.2.6.)

In the case of two *non-periodic* signals whose correlation length is much smaller than the size of the samples picked from these signals, we may assume, without major loss of precision, that the samples are periodic. This allows us to again use Eq. (7.17) to compute the sample correlation.

Example We often perform measurements by perturbing the system by a real periodic signal

$$s(t) = S \cos(\omega t)$$

with known amplitude S and angular frequency ω, and observe the response of the system $r(t)$. An example of such analysis is the determination of the resonance spectrum of an acoustic resonator shown in Fig. 7.11. On the resonator wall we use a loudspeaker to generate the perturbation $s(t)$ with the frequency $v = \omega/(2\pi)$, and measure the response $r(t)$ by a microphone. This spectroscopic method of testing objects or materials is known as the *continuous wave technique*. In all systems with a linear response to external perturbations the response can be written in the form

$$r(t) = R \cos(\omega t + \phi),$$

where the amplitude of the response $R \propto S$ and the phase ϕ can be functions of ω. In complex notation, the response $\tilde{r}(t) = R\, \mathrm{e}^{\mathrm{i}(\omega t + \phi)}$ is just the perturbation $\tilde{s}(t) = S\, \mathrm{e}^{\mathrm{i}\omega t}$ multiplied by the factor $Z = (R/S)\, \mathrm{e}^{\mathrm{i}\phi}$, which we wish to determine. Because we

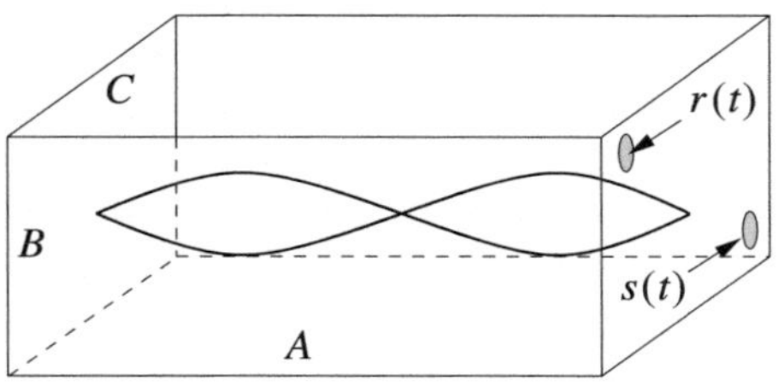

Fig. 7.11 Determination of the resonance spectrum of an acoustic resonator

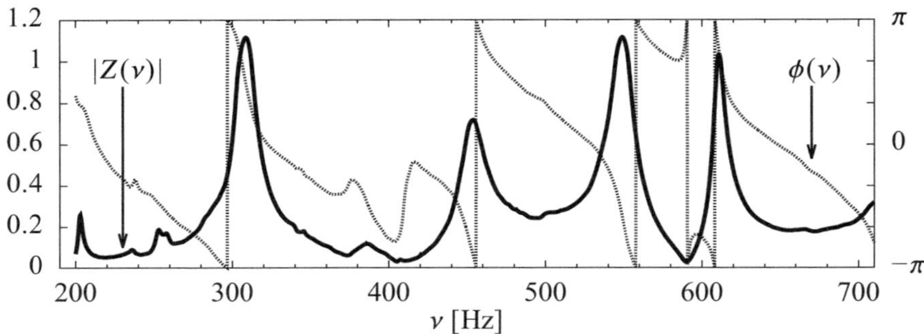

Fig. 7.12 The frequency dependence of the amplitude $|Z|$ and phase $\phi = \arg(Z)$ of the ratio between the complex response and complex perturbation during the excitation of a shoe-box acoustic resonator with sides $A = 56.7\,\mathrm{cm}$, $B = 38.5\,\mathrm{cm}$, and $C = 24\,\mathrm{cm}$. Near the resonances the phase changes rapidly and is approximately $\pm\pi/2$

generate the perturbation and know it exactly, Z can be computed by correlating the response and the complex perturbation,

$$\langle \widetilde{s}(t)^* r(t) \rangle_t = \tfrac{1}{2} S R \, \mathrm{e}^{\mathrm{i}\phi} = \tfrac{1}{2} Z S^2 \,.$$

An example of the determination of the amplitude $|Z|$ and the phase $\phi = \arg(Z)$ as functions of the frequency ν is shown in Fig. 7.12.   ◁

7.7   Auto-Regressive Analysis of Discrete-Time Signals ★

In the field of digital signal analysis [10, 32] the *auto-regressive model* (ARM) is a method that is also known as the *infinite impulse response filter* (IIR) or the *all-pole filter*, while in physical applications it goes under the name of *maximum-entropy model* (MEM). Here we discuss a very specific use of the theory, the prediction of signals and the estimate of their spectra.

Auto-regressive modeling is closely related to linear prediction, and often in literature no distinction is made between them. A concise overview of linear prediction can be found in the classic paper [43], and a broader description of well-established methods and technicalities in [41]. General theory of linear prediction is presented in [44, 45], and the connection to linear models in [46, 47].

Preparing the data for AR analysis Let us discuss only discrete-time signals defined at times $t \in \mathbb{Z}$. If the signal $s(\tau)$ is continuous, we discretize it as $s_{\mathrm{discrete}}(t) = s_{\mathrm{continuous}}(\tau_0 + t\,\Delta t)$, where Δt is the time step and τ_0 the chosen origin. By discretizing the signal its spectrum is artificially constrained to the frequency range $[-\nu_{\mathrm{c}}, \nu_{\mathrm{c}}]$, where $\nu_{\mathrm{c}} = 1/(2\Delta t)$ is the Nyquist frequency (see Eq. (5.6)). From the signal we then try to remove all constant contributions or obvious dependencies, for example,

$$x(t) = s(t) - \text{power dependency of the signal} .$$

The signal x obtained in this way becomes approximately wide-sense stationary (WSS) and comparable to a realization of some ergodic random process, and its average is practically zero. Occasionally we attempt to use power transformations to modify the distribution of data so that it becomes more similar to one of the standardized distributions (for example, normal distribution) [48].

7.7.1   Auto-Regressive (AR) Model

An auto-regressive model for a discrete-time signal x is defined as the recurrence

$$x(t) = -\sum_{i=1}^{p} a_i \, x(t-i) + \varepsilon(t) , \tag{7.19}$$

where $\varepsilon(t)$ is the *remainder* (or *deviation*, or *error*), while p is the *order* of the AR model. We wish to find the order p and coefficients $\{a_i\}_{i=1}^{p}$ such that the remainder ε is minimal and possibly similar to a realization of white noise.

Assume that x and ε are realizations of independent random processes X and E that are wide-sense stationary. Equation (7.19) then represents a relation between these processes. We multiply it by $X(t')^*$ from the left and right, and average over time. We obtain a recurrence relation for the auto-correlation

$$R_{XX}(t'-t) = -\sum_{i=1}^{p} a_i \, R_{XX}(t'-t-i) + \delta_{t,t'} \left\langle E(t)^2 \right\rangle . \tag{7.20}$$

By introducing the vector of coefficients $\boldsymbol{a} = (a_1, a_2, \ldots, a_p)^{\mathrm{T}}$ and denoting $\rho = \langle E(t)^2 \rangle$, the equation above can be rewritten in the Yule–Walker matrix form

$$R \begin{pmatrix} 1 \\ \boldsymbol{a} \end{pmatrix} = \begin{pmatrix} \rho \\ \boldsymbol{0} \end{pmatrix} , \tag{7.21}$$

where R is the auto-correlation matrix

$$R = \begin{pmatrix} R_{XX}(0) & R_{XX}(1)^* & R_{XX}(2)^* & \cdots & R_{XX}(p)^* \\ R_{XX}(1) & R_{XX}(0) & R_{XX}(1)^* & \cdots & R_{XX}(p-1)^* \\ \vdots & \vdots & \vdots & & \vdots \\ R_{XX}(p) & R_{XX}(p-1) & R_{XX}(p-2) & \cdots & R_{XX}(0) \end{pmatrix}$$

The properties of auto-correlation (page 418) ensure that R is positive definite.

The Yule–Walker system (7.21) relates the coefficients a_i of the AR model, the average square of the remainder, ρ, and the auto-correlation of the process, R_{XX}. If we know R_{XX}, we can determine a_i and ρ, or vice-versa. The matrix R has a Toeplitz structure and the system can be effectively solved by the Levinson–Durbin method [49] in $\mathcal{O}(p^2)$ steps or one of the fast methods (see Sect. 4.2.4). With the solution we also obtain the coefficients a_i. Moreover, the solution at order p can be recursively used to find the solution at order $p + 1$. Levinson–Durbin recurrence lies at the heart of digital signal analysis [41, 45].

If the processes X and E are ergodic, statistical averages in (7.21) may be replaced by time averages, $R_{XX} = R_{xx} = \langle x(t')^* x(t) \rangle_t$ and $\langle E(t)^2 \rangle = \langle \varepsilon(t)^2 \rangle_t$. Because the model (7.19) is so general and the constraints on the remainder ε so loose, the coefficients $\{a_i\}_{i=1}^p$ at chosen p can be calculated in several ways [41]. The most well-known are the method of least squares in its auto-correlation or covariance version, and the Burg's algorithm.

Determining the parameters by the auto-correlation method Let us select a sample of length T from the signal x (signal x different from zero on the interval $t \in [0, T - 1]$ and vanishing outside). The remainder at time t is

$$\varepsilon(t) = x(t) + \sum_{i=1}^{p} a_i\, x(t - i)\,. \tag{7.22}$$

Define the vector of remainders $\boldsymbol{\varepsilon}_{[\alpha,\beta]} = (\varepsilon(\alpha), \varepsilon(\alpha + 1), \ldots, \varepsilon(\beta))^{\mathrm{T}}$ for an arbitrary time interval $[\alpha, \beta]$. Each equation of this type can be rewritten as

$$\boldsymbol{\varepsilon}_{[\alpha,\beta]} = X_{[\alpha,\beta]} \begin{pmatrix} 1 \\ \boldsymbol{a} \end{pmatrix}\,,$$

where

$$X_{[\alpha,\beta]} = \begin{pmatrix} x(\alpha) & x(\alpha - 1) & \cdots & x(\alpha - p) \\ x(\alpha + 1) & x(\alpha) & \cdots & x(\alpha - p + 1) \\ \vdots & \vdots & & \vdots \\ x(\beta - 1) & x(\beta - 2) & \cdots & x(\beta - p - 1) \\ x(\beta) & x(\beta - 1) & \cdots & x(\beta - p) \end{pmatrix}\,.$$

We determine the coefficients a_i by minimizing the sum of the squares of the remainders $\rho^{(\mathrm{a})}$ for the given sample, which can be written as

$$\rho^{(\mathrm{a})} = \sum_{t=0}^{T+p-1} |\varepsilon(t)|^2 = \left(1,\, \boldsymbol{a}^\dagger\right) R^{(\mathrm{a})} \begin{pmatrix} 1 \\ \boldsymbol{a} \end{pmatrix}\,, \qquad R^{(\mathrm{a})} = X_{[0,T+p-1]}^\dagger X_{[0,T+p-1]}\,.$$

The sum also contains the values of the signal outside the interval $[0, T - 1]$ according to the assumptions. The $(p + 1) \times (p + 1)$ matrix $R^{(\mathrm{a})}$ is Hermitian, with the elements

$$R_{ij}^{(a)} = \sum_{t=0}^{T-(i-j)-1} x^*\big(t + (i - j)\big)x(t) \tag{7.23}$$

that depend only on the differences of indices i and j. The sum of the squares of the remainders is minimized by $\partial \rho^{(a)}/\partial a = 0$, which can be written as

$$R^{(a)} \begin{pmatrix} 1 \\ a \end{pmatrix} = \begin{pmatrix} \rho^{(a)} \\ 0 \end{pmatrix}. \tag{7.24}$$

Here $R^{(a)}$ is a Toeplitz matrix, so the system of equations is of the Yule–Walker type (7.21) and can be solved by the methods described on page 426. Note, however, that this approach may fail if the matrix is ill-conditioned: see [50].

For short signal samples the biased auto-correlation estimate is not necessarily a good approximation of the auto-correlation of the process. In such cases we divide Eq. (7.24) on the left and on the right by T and replace the matrix elements of $R^{(a)}/T$ by the unbiased auto-correlation estimates.

Determining the parameters by the covariance method Instead of minimizing the remainders computed by Eq. (7.22) we can also opt to minimize the corresponding sum that involves only the values of the signal within the sample,

$$\rho^{(c)} = \sum_{t=p}^{T-1} |\varepsilon(t)|^2 = \big(1, \, a^\dagger\big) R^{(c)} \begin{pmatrix} 1 \\ a \end{pmatrix}, \qquad R^{(c)} = X_{[p,T-1]}^\dagger X_{[p,T-1]} \, .$$

The equation for the minimum, $\partial \rho^{(c)}/\partial a = 0$, can again be cast in the form (7.24),

$$R^{(c)} \begin{pmatrix} 1 \\ a \end{pmatrix} = \begin{pmatrix} \rho^{(c)} \\ 0 \end{pmatrix}, \qquad R_{ij}^{(c)} = \sum_{t=p}^{T-1} x^*(t - i)x(t - j) \, . \tag{7.25}$$

The $(p + 1) \times (p + 1)$ matrix $R^{(c)}$ is still Hermitian, but its elements are not functions of differences of indices and it does not have a Toeplitz structure. The system (7.25) is therefore not of the Yule–Walker type, but it can still be solved in $\mathcal{O}(p^2)$ steps by using the CORVAR method described in [41]. A necessary (but not sufficient) condition for the non-singularity of $R^{(c)}$ is $p < T/2$. For details and other approaches to the minimization of remainders see [41, 43, 51].

Burg's algorithm Burg's algorithm, known also as the maximum-entropy algorithm, can be used to compute the coefficients of a stable AR model, and is therefore primarily used in signal prediction (Sect. 7.7.2). If implemented correctly, it requires $\mathcal{O}(3Tp) + \mathcal{O}(p^2)$ operations and $\mathcal{O}(3T + p)$ of memory [41, 44]. It can be found in the NUMERICAL RECIPES library [52] and in MATLAB [53].

Optimal order There is no general rule to choose the order p at which the data in a signal sample would be described optimally. When p is increased, the size of the remainder, quantified by $\langle \varepsilon^2 \rangle_t = \rho/(T + \mathcal{O}(1))$ or by the corresponding normalized quantity $\langle \varepsilon^2 \rangle_t / R_{XX}(0)$, decreases. In general, $\langle \varepsilon^2 \rangle_t$ initially drops rapidly, but at some p its decrease slows down. The value of p at which the transition between these two regimes occurs can be taken as optimal. For details see [43].

7.7.2   Application of AR Models

Auto-regressive models are used in many ways, but the most well-known are *modeling* and *prediction*. In *modeling*, we use a known signal to compute the coefficients $\{a_i\}$ from which we infer the properties of the process that generated this signal. An example of modeling is the search for an approximation of the spectral density of a signal. We describe this in Sect. 7.7.3. In *linear prediction*, we determine the coefficients from the history of the signal $\{x(t - p), x(t - p + 1), \ldots, x(t - 1)\}$ and use them to predict its current value $x(t)$ and the forthcoming values $x(t + 1), x(t + 2), \ldots$ by using (7.19).

The resonance spectrum of the AR model The dynamics of the auto-correlation R_{XX} defined by Eq. (7.20) can be calculated analytically. Let us define the characteristic polynomial of the AR model,

$$p_{\mathrm{AR}}(x) = x^p + \sum_{i=1}^{p} a_i \, x^{p-i} = \prod_{i=1}^{p} (x - z_i) \,. \tag{7.26}$$

If we know the zeros $\{z_i\}_{i=1}^{p}$ of the polynomial p_{AR}, which we call the *resonances of the AR model*, the time evolution of the auto-correlation can be written as $R_{xx}(t) = \sum_{i=1}^{p} A_i z_i^t$, where A_i are constants. We determine the values of A_i from the initial conditions given by the derivatives $R_{XX}^{(i)}(0)$ with $i = 0, 1, \ldots, p - 1$, or from p points of the auto-correlation at different times. If all zeros lie on the unit circle, we are referring to a *line spectral process* in which auto-correlations merely oscillate in time [45]. Similarly, the time evolution of the signal in the noiseless limit can be written as

$$x(t) = \sum_{i=1}^{p} B_i \, z_i^t \,,$$

where B_i are constants. This equation represents an auto-regressive approximation of the signal. When we use AR models, we often try to find a decomposition of precisely this kind, hoping that the signal-to-noise ratio is large.

If the coefficients a_i are real, the resonances z_i of the AR model occur in complex-conjugate pairs if they reside away from the real axis in the complex plane. It happens that the zeros z_i fall outside of the unit circle (see Example below and Fig. 7.14 (right)). The signal that was generated by such an AR model then diverges as $\mathcal{O}((\max_i |z_i|)^t)$. Quite often, the zeros lying outside of the unit circle are just artifacts due to the presence of noise, which we strive to eliminate anyway, in particular when we wish to generate or predict long signal samples. In such cases we "fix" the problematic zeros by replacing z_i by $\hat{z}_i = z_i/|z_i|$ (which does not alter the phase) and computing the coefficients $\hat{a}_i$ of a slightly modified AR model with the characteristic polynomial

$$\prod_{i=1}^{p}(x - \hat{z}_i) = x^p + \sum_{i=1}^{p}\hat{a}_i x^{p-i} \, .$$

Example Assume that a process is described by the AR model with coefficients

$$\{a_i\}_{i=1}^{p} = \{0.1,\ 0.2,\ -0.3,\ 0.4,\ 0.5,\ 0.1,\ 0.2,\ -0.2,\ 0.5,\ -0.1\} \tag{7.27}$$

and that $\varepsilon(t)$ is white noise with $\sigma_\varepsilon = 0.1$. We use Eq. (7.19) to generate a signal sample $\{x(t)\}_{t=0}^{T-1}$ of length $T = 1000$, where the first p values are used as the initial condition of the iteration. A possible realization of the signal is shown in Fig. 7.13. The average square of the remainder in the model is shown in Fig. 7.14 (left). The

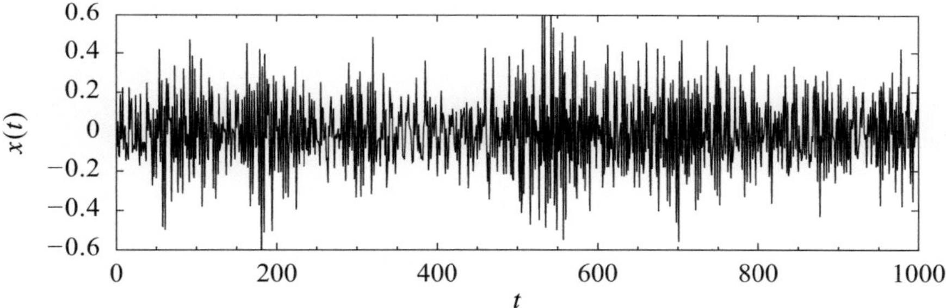

Fig. 7.13   The signal generated by the AR model (7.19) with the coefficients (7.27) and an admixture of Gaussian white noise ε with standard deviation $\sigma_\varepsilon = 0.1$

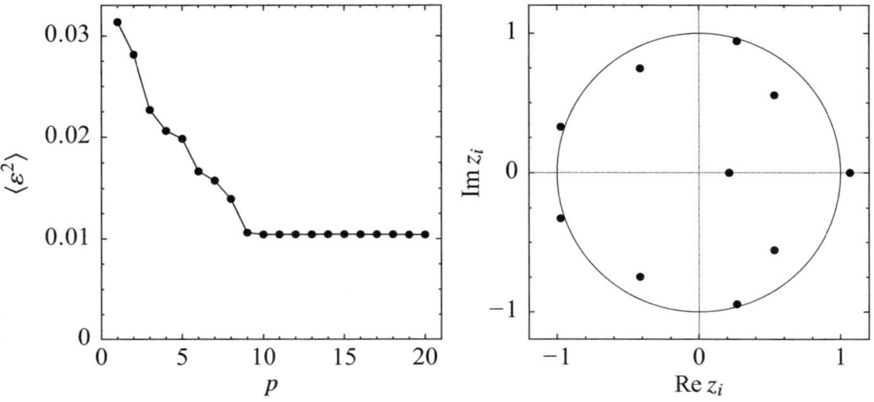

Fig. 7.14   [LEFT] The average square of the remainder in the AR model with respect to the sample values, as a function of order p. [RIGHT] The resonance spectrum of the model used to generate the signal in Fig. 7.13. Three resonances lie outside of the unit circle, thus the model is unstable

spectrum of the resonances shown in Fig. 7.14 (right) reveals that this AR model is unstable, as three out of ten resonances lie outside of the unit circle.

As an exercise, let us study the generated signal by the AR model with the coefficients $\{\tilde{a}_i\}_{i=1}^{\tilde{p}}$ for different orders $\tilde{p}$ and by applying the covariance method. With increasing $\tilde{p}$ the average square of the remainder $\langle\varepsilon^2\rangle_t$ decreases until it stabilizes at $\mathcal{O}(\sigma_\varepsilon^2)$. This occurs at $\tilde{p} = 10$, which is the order of the AR model used. For $\tilde{p} > p$, ε resembles the noise that was used in the generation of the signal. With increasing $\tilde{p}$ the coefficients $\tilde{a}_i$ approach a_i of the original model:

$\tilde{p}$	$\tilde{a}_1$	$\tilde{a}_2$	$\tilde{a}_3$	$\tilde{a}_4$	$\tilde{a}_5$	$\tilde{a}_6$	$\tilde{a}_7$	$\tilde{a}_8$	$\tilde{a}_9$	$\tilde{a}_{10}$
1	0.343									
2	0.453	0.321								
3	0.310	0.120	−0.442							
6	0.312	0.225	−0.443	0.374	0.288	0.400				
8	0.326	0.132	−0.472	0.179	0.462	0.374	0.095	−0.340		
10	0.095	0.202	−0.310	0.386	0.477	0.089	0.199	−0.202	0.470	−0.131

The approach of $\tilde{a}_i$ to the original coefficients a_i depends on the sample size and the noise intensity. Convergence is guaranteed only for infinite WSS signals. The coefficients $\tilde{a}_i$ for $i \geq \tilde{p} > p$ are not zero (for $\tilde{p} = 11$ we get $\{\tilde{a}_i\}_{i=1}^{10} = \{0.097, 0.194, -0.306, 0.383, 0.475, 0.080, 0.192, -0.197, 0.467, -0.133\}$, then $\tilde{a}_{11} = -0.018$): rather, they are random and on the order of $\mathcal{O}(\sigma_\varepsilon)$.   ◁

7.7.3   Estimate of the Fourier Spectrum

The power spectral density of a discrete-time signal $x(t)$, $t \in \mathbb{Z}$, is defined as

$$P_x(\omega) = \lim_{T \to \infty} \frac{1}{2T+1} \left| \sum_{t=-T}^{T} x(t)\, \mathrm{e}^{-\mathrm{i}\omega t} \right|^2 , \qquad -\pi \leq \omega < \pi .$$

(See [51, 54].) If x was obtained by discretizing a continuous signal with the step Δt, ω is related to frequency ν as $\omega = 2\pi\nu\Delta t$. The relation between the time autocorrelation of the signal R_{xx} and the spectral density is

$$P_x(\omega) = \sum_{t=-\infty}^{\infty} R_{xx}(t)\, \mathrm{e}^{-\mathrm{i}\omega t} .$$

For samples of length T we only know the sample auto-correlation $\mathcal{R}_{\hat{x}\hat{x}}(t)$ and then the formula above can be rewritten as a *correlogram*

$$P_{\hat{x}}(\omega) = \sum_{t=-L}^{L} \mathcal{R}_{\hat{x}\hat{x}}(t)\, \mathrm{e}^{-\mathrm{i}\omega t} = \mathcal{R}_{\hat{x}\hat{x}}(0) + 2\sum_{t=1}^{L} \mathcal{R}_{\hat{x}\hat{x}}(t)\cos\omega t , \qquad (7.28)$$

where the sample auto-correlation is used up to the maximum shift $L \leq T - 1$. A correlogram is an approximation of the spectral density of the complete signal. If we assume that the signal is a realization of a WSS process, we can find the averages of the sample spectral densities for both defined approximations [41]:

$$\text{unbiased} : \langle P_{\hat{x}}(\omega) \rangle = P_x(\omega) * D_L(\omega) \,, \tag{7.29}$$

$$\text{biased} : \langle P_{\hat{x}}(\omega) \rangle = P_x(\omega) * \frac{2}{L} D_L^2(\omega/2) \,, \tag{7.30}$$

where $*$ denotes convolution and $D_L(\omega) = \mathrm{e}^{-\mathrm{i}\,2\pi\omega(L-1)} \sin(\pi\omega L)/\sin(\pi\omega)$ is the Dirichlet kernel. The sample spectral densities are biased approximations of the true spectral density $P_x(\omega)$, since the averages $\langle P_{\hat{x}}(\omega) \rangle$ are its convolutions with the Dirichlet kernel. With increasing L the Dirichlet kernel widens and it is sensible to use $L \ll T$ (at most $L \approx T/10$). For $L = T - 1$ the correlogram is equal to the power spectral density of the discrete signal as computed by the DFT,

$$\tilde{P}_x(\omega) = \frac{1}{T} \left| \sum_{t=0}^{T-1} x(t)\, \mathrm{e}^{-\mathrm{i}\,\omega t} \right|^2 \,, \tag{7.31}$$

but this is statistically unstable. The correlogram with the maximum shift of $L < T/2$ for a sample of size T can be computed at the points $\{\omega_k = 2\pi k/T\}_{k=0}^{T-1}$ by using the DFT. We arrange the values of the sample auto-correlation $\mathcal{R}_{\hat{x}\hat{x}}$ in an array of length T, $\mathcal{R} = \{\mathcal{R}_{\hat{x}\hat{x}}(0), \ldots, \mathcal{R}_{\hat{x}\hat{x}}(L), 0, \ldots, 0, \mathcal{R}_{\hat{x}\hat{x}}(-L), \ldots, \mathcal{R}_{\hat{x}\hat{x}}(-1)\}$. When we apply DFT to it, we obtain the spectral density

$$\left\{ P_{\hat{x}}(\omega_k) \right\}_{k=0}^{T-1} = T \mathcal{F}_T[\mathcal{R}] \,.$$

For the computation we should chose L such that $P_{\hat{x}}(\omega) \geq 0$, which is fulfilled in the case of auto-correlations, but does not necessarily apply to *sample* auto-correlations. A comparison of the sample spectral density (computed by DFT) and the correlogram is shown in Fig. 7.15. We see that the correlogram is less sensitive to the presence of noise. Details on estimates of spectral densities of signals are given in [41].

Assume that we know the AR model of order p with the coefficients $\{a_i\}_{i=1}^{p}$ for a signal x. Let us look at this computation in the reverse direction: we use a linear transformation of the signal to generate the remainder $\varepsilon(t)$,

$$\varepsilon(t) = x(t) + \sum_{i=1}^{p} a_i\, x(t-i) \,.$$

We apply the Fourier transformation $\mathcal{F}[f](\omega) = \sum_{t \in \mathbb{Z}} f(t)\, \mathrm{e}^{-\mathrm{i}\omega t}$ to both sides of the equation and get $\mathcal{F}[\varepsilon](\omega) = \mathcal{F}[x](\omega) \exp(-\mathrm{i}\omega p)\, p_{\mathrm{AR}}(\exp(\mathrm{i}\omega))$, where p_{AR} is the characteristic polynomial (7.26). The power spectral density of the signal is

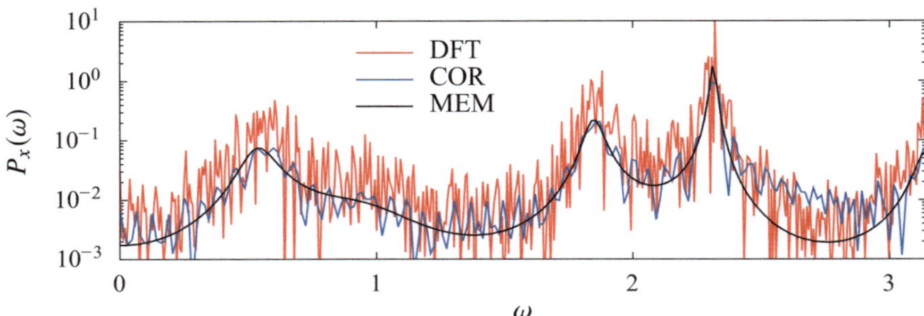

Fig. 7.15 Three approximations of the spectral density of the signal x from the Example on page 429. The sample has a length of $T = 10^4$. Shown are the curves of the sample spectral density (DFT, Eq. (7.31)), the correlogram (COR, Eq. (7.28)) with $L = T/10$, and the maximum-entropy approximation at order $p = 10$ (MEM, Eq. (7.32))

$$P_x(\omega) = \frac{P_\varepsilon(\omega)}{|p_{AR}(\exp(i\omega))|^2} \, , \tag{7.32}$$

where $P_\varepsilon(\omega)$ is the power spectral density of the remainder ε. We almost invariably assume that the remainder is white noise with variance σ_ε^2, so that $P_\varepsilon(\omega) = \sigma_\varepsilon^2$. This assumption implies that this process has the maximum possible entropy among all processes with the same auto-correlation (the *maximum entropy principle*). The estimate (7.32) is therefore known as the *maximum entropy spectrum estimator* (MESE), while the method to find this approximation is called the *maximum entropy method* (MEM) [45, 55].

Figure 7.15 makes it clear that the MEM estimate for orders $p \ll T$ is smoother than either the sample spectral density or the correlogram. The MEM estimate is also resilient to noise until p is much smaller than T and below the limit at which the remainder starts to resemble noise. When p is increased further, aliasing effects sneak into the MEM estimate, recognizable as contributions oscillating with a period of $2\pi/p$ around the expected spectral density (Fig. 7.16). This Figure shows that increasing p reveals more and more resonances in the signal. When all essential resonances have been accounted for, the remainder in the AR model resembles noise. If we know that the spectrum of the signal should contain k prominent frequencies, it makes sense to choose the order $p = 2k$, since resonances occur in complex-conjugate pairs.

We recommend the use of the MEM spectral approximation when the signal can be described by an AR model of order p that is much smaller than the size of the sample. Moreover, p should be small enough that the methods we use to compute the coefficients remain stable. In most cases $p < 40$ suffices.

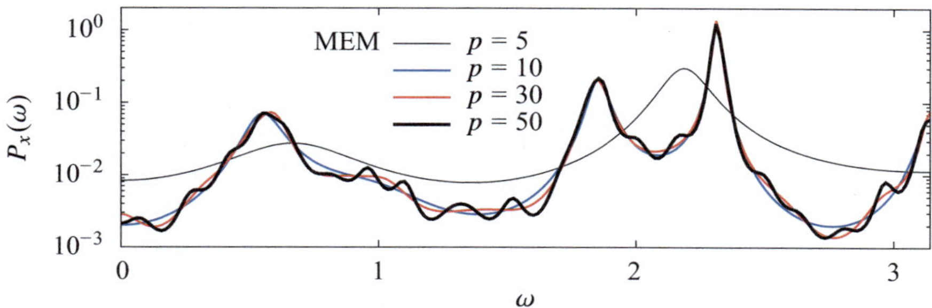

Fig. 7.16 Maximum-entropy spectrum approximations for different orders p of the AR models. The coefficients of the models were computed by the least-squares method for the data from Fig. 7.13. The curve of the exact spectral density (which corresponds to the model used to generate the sample) is almost identical to the curve at $p = 10$

7.8 Optimal Filtering

Optimal filtering [56] is a powerful and versatile tool to combine uncertain information about a dynamical system with (noisy) measurements obtained from it into a less uncertain estimation of its past, present or future states. In modern parlance, optimal filtering in discrete-time systems is almost synonymous with applications of the Kalman filter [57, 58], which is discussed in this Section in its original linear formulation as well as in its non-linear variants.

7.8.1 Linear Kalman Filter

The main task is to estimate time-dependent random state vectors x_n from the corresponding measurements z_n obtained sequentially at times t_n ($n = 1, 2, \ldots$). We assume that the dynamics of the system is represented by the linear *model* (or *process*) *equation*

$$x_{n+1} = F_n x_n + C_n u_n + q_n \,, \tag{7.33}$$

where F_n is the (known) transition matrix describing the transition from the state at t_n to the state at t_{n+1}, while q_n is the vector of (Gaussian) *driving noise* with zero mean and covariance matrix $Q_n = \langle q_n q_n^{\mathrm{T}} \rangle$. If applicable, a control term $C_n u_n$ based on some auxiliary input vector u_n can be added. We further assume that the measurements z_n are related to the states x_n by the linear *measurement equation*

$$z_n = H_n x_n + r_n \,, \tag{7.34}$$

where H_n is the measurement or ("sensor") matrix mapping the state x_n into the corresponding measurement z_n, and r_n is the vector of (Gaussian) measurement noise with zero mean and covariance matrix $R_n = \langle r_n r_n^{\mathrm{T}} \rangle$.

Let $\hat{\boldsymbol{x}}_n^-$ and $\hat{\boldsymbol{x}}_n^+$ denote the prior and posterior optimal linear estimators of $\boldsymbol{x}_n$ based on measured $z_1, z_2, \ldots, z_n$. Similarly, let $\boldsymbol{\xi}_n^- = \boldsymbol{x}_n - \hat{\boldsymbol{x}}_n^-$ and $\boldsymbol{\xi}_n^+ = \boldsymbol{x}_n - \hat{\boldsymbol{x}}_n^+$ denote the errors of these estimators, and $P_n^- = \langle \boldsymbol{\xi}_n^- \boldsymbol{\xi}_n^{-\mathrm{T}} \rangle$ and $P_n^+ = \langle \boldsymbol{\xi}_n^+ \boldsymbol{\xi}_n^{+\mathrm{T}} \rangle$ the corresponding covariance matrices. If the noises $\boldsymbol{q}_n$ and $\boldsymbol{r}_n$ are uncorrelated, the optimal linear estimator of $\boldsymbol{x}_n$ from the measurements $z_1, z_2, \ldots, z_n$ up to time t_n is obtained by the following recurrence ($n = 1, 2, \ldots$):

$$
\begin{aligned}
\text{Propagation}: \quad \hat{\boldsymbol{x}}_n^- &= F_{n-1}\hat{\boldsymbol{x}}_{n-1}^+ + C_{n-1}\boldsymbol{u}_{n-1}\,, & \text{(P1)} \\
P_n^- &= F_{n-1} P_{n-1}^+ F_{n-1}^{\mathrm{T}} + Q_{n-1}\,, & \text{(P2)} \\
\text{Update}: \quad K_n &= P_n^- H_n^{\mathrm{T}} \left(H_n P_n^- H_n^{\mathrm{T}} + R_n \right)^{-1}\,, & \text{(U1)} \\
\hat{\boldsymbol{x}}_n^+ &= \hat{\boldsymbol{x}}_n^- + K_n (z_n - H_n \hat{\boldsymbol{x}}_n^-)\,, & \text{(U2)} \\
P_n^+ &= (I - K_n H_n) P_n^-\,. & \text{(U3)}
\end{aligned}
\tag{7.35}
$$

Implementation details This scheme should be initialized with $\hat{\boldsymbol{x}}_0^+ = \langle \boldsymbol{x}_0 \rangle$ or, if the latter value is unknown, with an appropriate educated guess. Setting the prior state estimation uncertainty (the "initial ignorance") P_0^+ demands more attention. Ideally, the initial condition should be $P_0^+ = \langle (\boldsymbol{x}_0 - \langle \boldsymbol{x}_0 \rangle)(\boldsymbol{x}_0 - \langle \boldsymbol{x}_0 \rangle)^{\mathrm{T}} \rangle$: this guarantees that the state estimates $\hat{\boldsymbol{x}}_n^+$ are *unbiased*. Although the filter is usually very efficient in readjusting the P_n's during the iteration, these covariances should remain reasonably commensurate with the measurement covariances given by R_n in order to avoid computational artifacts. In the extreme situation in the scalar case, if P_0 is much larger than R_0, such that adding R_0 to P_0 results in the same value P_0 due to roundoff in floating-point arithmetic, the scheme leads to $P_n = 0$; see Sect. 6.3.2 of [59] for the anatomy of this failure. Similarly, if the numerical accuracy is insufficient, the covariance matrix produced by step (U3) may fail to be positive-definite, which is unacceptable for a covariance matrix. The solution is to propagate it in its Cholesky-factorized form $P = P^{1/2}(P^{1/2})^{\mathrm{T}}$, where the "square root" of the matrix, $P^{1/2}$, is lower-triangular.

Also note that steps (P2), (U1) and (U3) are data-independent, so they may be calculated in advance in order to save computation time if applicable. If, however, the Kalman gain matrix K_n must be calculated for each n, the matrix inversion required in step (U1) is the most time-consuming operation and becomes the bottle-neck in real-time applications. Two ways in which matrix inversion can be circumvented are described in Chap. 7 of [60].

Relation to Auto-Regressive Processes

Devising the Kalman filter for the estimation of discrete-time signals originating in a stochastic process contaminated by additive noise is a straightforward exercise in transcribing the discrete-time model into the appropriate process and measurement equations. For example, the auto-regressive (AR) model of Eq. (7.19) can be represented by the $p \times 1$ state vector $\boldsymbol{x}_n = \big(x(n-p+1), x(n-p+2), \ldots, x(n-1), x(n) \big)^{\mathrm{T}}$ and the n-independent (time-invariant) $p \times p$ transition matrix

$$
F = \begin{pmatrix}
0 & 1 & 0 & \cdots & 0 \\
\vdots & \ddots & \ddots & \ddots & \vdots \\
0 & \cdots & 0 & 1 & 0 \\
0 & \cdots & \cdots & 0 & 1 \\
-a_p & -a_{p-1} & \cdots & -a_2 & -a_1
\end{pmatrix},
$$

while the error $\varepsilon(t)$ of the model translates to the driving noise $q(n) = \varepsilon(n)$ as the last component of the $p \times 1$ vector $\boldsymbol{q}_n = (0, 0, \ldots, q(n))^{\mathrm{T}}$, with the corresponding covariance matrix $Q_n = \boldsymbol{q}_n \boldsymbol{q}_n^{\mathrm{T}} = Q = \mathrm{diag}(0, 0, \ldots, 0, \sigma_q^2)$. The $1 \times p$ measurement matrix is $H = (0, 0, \ldots, 0, 1)$, while the measurement covariance "matrix" is just a scalar, $R_n = R = \sigma_r^2$. The model and measurement equations are then simply $\boldsymbol{x}_{n+1} = F\boldsymbol{x}_n + \boldsymbol{q}_n$ and $z(n) = H\boldsymbol{x}_n + r(n)$, respectively, and the filter algorithm can be applied directly.

Dealing with Correlated Noise Sources

The standard Kalman filter is based on the assumption that the driving noise, $\boldsymbol{q}_n$, and the measurement noise, $\boldsymbol{r}_n$, are uncorrelated, such that $\langle \boldsymbol{q}_m \boldsymbol{r}_n^{\mathrm{T}} \rangle = 0$ for all m, n. If the two noises are mutually correlated as

$$
\langle \boldsymbol{q}_m \boldsymbol{r}_n^{\mathrm{T}} \rangle = S_m \delta_{m,n} ,
$$

all one needs to modify in the filter algorithm is the expression for the Kalman gain and the update equation for the state covariance [59]:

$$
\begin{aligned}
K_n &= \left(P_n^- H_n^{\mathrm{T}} + S_n \right) \left(H_n P_n^- H_n^{\mathrm{T}} + R_n + H_n S_n + S_n^{\mathrm{T}} H_n^{\mathrm{T}} \right)^{-1} , \quad &\text{(U1)} \\
P_n^+ &= P_n^- - K_n \left(H_n P_n^- + S_n^{\mathrm{T}} \right) . &\text{(U3)}
\end{aligned}
$$

If, on the other hand, only the measurement noise itself is temporally correlated, such that

$$
\langle \boldsymbol{r}_m \boldsymbol{r}_n^{\mathrm{T}} \rangle \neq 0
$$

for at least one $m \neq n$, we proceed as follows. We assume that the measurement noise can also be represented by a model of the form

$$
\boldsymbol{x}_{n+1}^* = F_n^* \boldsymbol{x}_n^* + \boldsymbol{q}_n^* ,
$$

where $\boldsymbol{x}_n^*$ is the corresponding noise state vector, F_n^* is a known matrix, and $\boldsymbol{q}_n^*$ is a temporally correlated white (Gaussian) driving noise—contaminating the *noise component of the signal*—with zero mean and covariance matrix $\langle \boldsymbol{q}_n^* \boldsymbol{q}_n^{*\mathrm{T}} \rangle = Q_n^*$. The measurement noise $\boldsymbol{r}_n$ is assumed to be given by

$$
\boldsymbol{r}_n = H_n^* \boldsymbol{x}_n^* ,
$$

where H_n^* is also known (it may well be the identity matrix). The signal can still be represented in the form (7.33), while the measurement model (7.34) becomes

$$z_n = H_n x_n + r_n = H_n x_n + H_n^* x_n^* .$$

One can then concatenate the usual signal state vector, x_n, and the noise state vector, x_n^*, as well as the corresponding driving terms, q_n and q_n^*, into

$$\overline{x}_n \equiv \left(x_n^{\mathrm{T}}, x_n^{*\mathrm{T}}\right)^{\mathrm{T}} , \quad \overline{q}_n \equiv \left(q_n^{\mathrm{T}}, q_n^{*\mathrm{T}}\right)^{\mathrm{T}} ,$$

and use the standard filter algorithm with these augmented vectors. The model equation (7.33) becomes

$$\overline{x}_{n+1} = \overline{F}_n \overline{x}_n + \overline{q}_n , \quad \overline{F}_n = \begin{pmatrix} F_n & 0 \\ 0 & F_n^* \end{pmatrix} , \quad \overline{Q}_n = \begin{pmatrix} Q_n & 0 \\ 0 & Q_n^* \end{pmatrix} ,$$

while the measurement equation (7.34) becomes

$$\overline{z}_n = \overline{H}_n \overline{x}_n , \quad \overline{H}_n = \left(H_n, H_n^*\right) , \quad \overline{R}_n = 0 .$$

For a general strategy of dealing with colored (non-white) settings, for instance, when the process noise is white but the measurement noise is colored (and combinations thereof), see Chap. 5 of [60] and Chap. 11 of [56].

Dealing with Noisy Control Input

The case when the control input itself is noisy, that is, when the components of u_n in the model (7.33) are contaminated by Gaussian noise, can be accommodated by slightly modifying the formulas for the prior state covariance and the Kalman gain matrix. If the noise has zero mean and variance σ_u^2, the two items in the Kalman filter that need to be modified take the form [61]:

$$
\begin{aligned}
P_n^- &= F_{n-1} P_{n-1}^+ F_{n-1}^{\mathrm{T}} + \sigma_u^2 C_{n-1} C_{n-1}^{\mathrm{T}} + Q_{n-1} , \quad &\text{(P2)} \\
K_n &= P_n^- H_n^{\mathrm{T}} \left(H_n P_n^- H_n^{\mathrm{T}} + R_n - \sigma_u^2 I\right)^{-1} . \quad &\text{(U1)}
\end{aligned}
\tag{7.36}
$$

An example of filtering a two-component signal with noisy control input is illustrated in Fig. 7.17. (Note the huge reduction of the error of the dominant component.)

The *fixed-interval* smoother described above is suitable for offline (not real-time) use because the re-estimations take place after all measurements have been acquired. Real-time applications, on the other hand, call for either *fixed-lag* smoothers, which use measurements up to the current time, but generate improved estimates only after a certain delay (lag), or *fixed-point* smoothers, which generate improved estimates at a fixed time before, after or at the current time: in these cases they

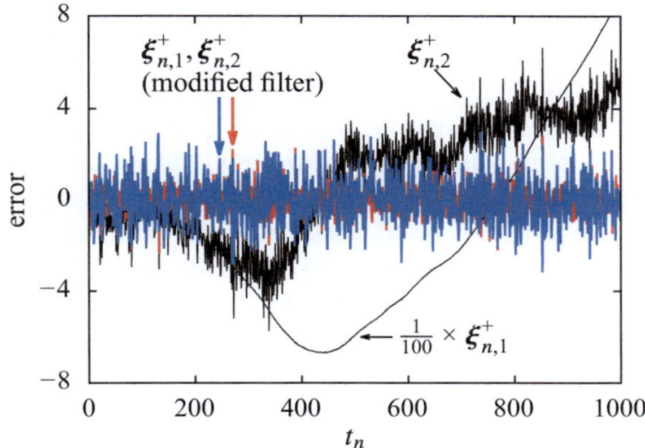

Fig. 7.17 Errors of the state estimates vs. the true states, by using the standard Kalman filter and Eq. (7.36) suitable for noisy control input, for a two-component model with $F = ((1, 1), (0, 1))$, $C = 0.5$, $Q = 0$, $H = (1, 0)$, $R = 1$, $\sigma_u^2 = 0.05$, initial conditions $\hat{x}_0^+ = (95, 1)^T$ and $P_0^+ = \mathrm{diag}(10, 1)$. Adapted from Example 1 of Ref. [61]

operate as smoothers, filters or predictors, respectively. For details of these two types of smoothers, which require more intricate algebraic implementations than the fixed-interval version, consult, for instance, Chap. 5 of [59].

Stability

The statements of stability of the linear Kalman filter rest upon the concepts of *controllability* and *observability*. (Complete) controllability of a dynamical system described by (7.33) means that it can be brought from the initial state x_0 to any possible state in a finite number of time steps with suitable inputs u_n. In turn, a system with the measurement part (7.34) is said to be (completely) observable if the states, x_n, can be inferred from the measurements, z_n. Defining the transition matrix from the state at "time" instant i to instant $j \geq i$ as

$$\Phi_{j,i} = \prod_{n=i}^{j-1} F_n \,,$$

the system (7.33) is *completely controllable* [62] if the matrix

$$\mathcal{C}_{n,m} = \sum_{i=m}^{n-1} \Phi_{n,i+1} Q_i \Phi_{n,i+1}^T \,,$$

defined for $0 < m < n$, is positive-definite for some such m and n. In a more restrictive sense, such a system is *uniformly (completely) controllable* if there exist $\alpha_1, \alpha_2 > 0$ such that $0 \leq \alpha_1 I \leq \mathcal{C}_{n,n-N} \leq \alpha_2 I$ for all $n \geq N$.[1] Similarly, the system (7.33) is *completely observable* through (7.34) if the matrix

---

[1] Here the notation $A > B$ ($A \geq B$) means that $A - B$ is positive-definite (positive-semidefinite).

$$\mathcal{O}_{n,m} = \sum_{i=m}^{n} \Phi_{i,m}^{\mathrm{T}} H_i^{\mathrm{T}} R_i^{-1} H_i \Phi_{i,m} \,,$$

defined for $0 < m < n$, is positive-definite for some such m and n. Furthermore, the system is *uniformly (completely) observable* if there exist $\beta_1, \beta_2 > 0$ such that $0 \le \beta_1 I \le \mathcal{O}_{n,n-N} \le \beta_2 I$ for all $n \ge N$.

These definitions allow us to state the stability conditions: if the dynamical system specified by (7.33) and (7.34) is uniformly controllable and uniformly observable, and if P_0^+ is positive-definite, the corresponding Kalman filter (7.35) is exponentially stable. This means that there exist $c_1, c_2 > 0$ (independent of n) such that

$$\|\Phi_{n,0}\| \le c_1 \mathrm{e}^{-c_2 n} \,, \quad \forall n \ge 0 \,.$$

If in addition the matrix C_n in the control term is uniformly bounded, such a system is also stable in the sense that bounded inputs produce bounded outputs, that is,

$$\sup_{n \ge 0} \|\boldsymbol{u}_n\| < \infty \;\Rightarrow\; \sup_{n \ge 1} \|\boldsymbol{x}_n\| < \infty \,.$$

The circumstance that some modes of the system may not be controllable or observable—for instance, when dealing with a three-component state injectively mapped into a two-component measurement—can be accommodated by the generalization of these two concepts to the ones of *stabilizability* and *detectability*, respectively. Even though we lack the knowledge about specific components or are unable to control them, the corresponding modes should remain stable; it turns out that if the system given by (7.33) and (7.34) is uniformly stabilizable and uniformly detectable, and P_0^+ is at least positive-*semi*definite, the algorithm (7.35) remains exponentially stable; see [62] for details.

Robust Filtering

A common nuisance in the application of the Kalman filter is its sensitivity to (too many) outliers in the measured data. While a single strongly deviating value should not derail the filter, a series of propagating outliers as encountered, for instance, in tracking problems, might result in very prominent fluctuations. Dealing with outliers typically relies on robust estimation techniques similar to those presented in Sect. 6.2. See [63, 64] for a concise discussion of modern approaches, and [65] for a comprehensive background review.

7.8.2 Kalman Predictor

Many applications call for a prediction of a future state, x_k, from measurements $z_1, z_2, \ldots, z_n$ available up to some earlier time n ($n < k$). Assuming that the control inputs are also known up to time k, the optimal linear predictions of the states and the corresponding covariance matrices are given by the recurrence

$$\hat{x}_k^+ = F_{k-1}\hat{x}_{k-1}^+ + C_{k-1}u_{k-1} , \tag{7.37}$$
$$P_k^+ = F_{k-1}P_{k-1}^+ F_{k-1}^{\mathrm{T}} + Q_{k-1} .$$

(Effectively, prediction is enforced by setting the Kalman gain K_n in step (U2) of the filter to zero, hence equating the prior and posterior quantities: the model equation is then used to just propagate the states from time n to time k.) If there are no control inputs to deal with, Eq. (7.37) reduces to $\hat{x}_k^+ = F_{k-1}F_{k-2}\cdots F_n\hat{x}_n^+$; in this case the matrix products can be computed in advance, so that only a single matrix-vector multiplication per prediction is required in real time.

7.8.3 (Fixed-Interval) Kalman Smoother

In contrast to the Kalman predictor, the Kalman *smoother* is a tool to obtain improved estimates of the states x_n for which optimal values have already been obtained by the "forward" pass of the Kalman filter. So given the outputs $\hat{x}_n^+$ computed by the Kalman filter for all $n = 1, 2, \ldots, N$, we wish to go back and compute, for each n, the improved ("smoothed") estimates, denoted by $\tilde{x}_n$, exploiting *all* available measurements. This is accomplished by setting $\tilde{x}_N$ to the last known optimal estimate, $\tilde{x}_N = \hat{x}_N^+$, and following the backward recurrence due to Rauch, Tung and Striebel [66]:

$$\tilde{x}_n = \hat{x}_n^+ + A_n(\tilde{x}_{n+1} - \hat{x}_{n+1}^-) ,$$
$$A_n = P_n^+ F_n^{\mathrm{T}}(P_{n+1}^-)^{-1} ,$$

down to $n = N - 1, N - 2, \ldots, 1$. To be able to use the backward formulas, the relevant quantities ($\hat{x}_n^-$, $\hat{x}_n^+$, P_n^- and P_n^+ for all n) must be saved from the forward pass. If needed, the covariances $\tilde{P}_n$ corresponding to the smoothed estimates can be calculated by the recurrence

$$\tilde{P}_n = P_n^+ + A_n(\tilde{P}_{n+1} - P_{n+1}^-)A_n^{\mathrm{T}}$$

starting with $\tilde{P}_N = P_N^+$ and $A_N = 0$. An illustration of this kind of Kalman smoother at work is given in Fig. 7.18. See also Problem 7.11.4.

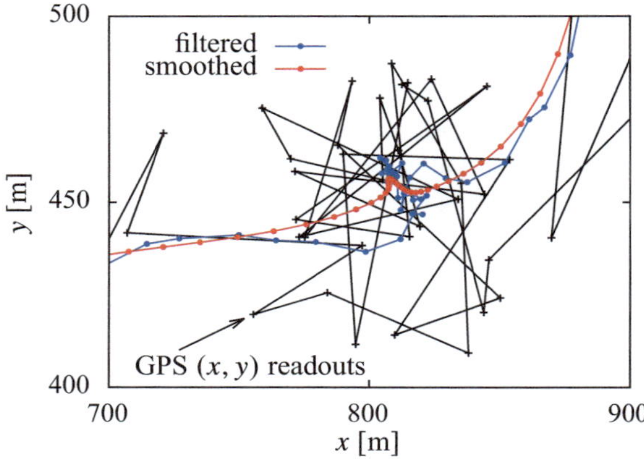

Fig. 7.18 Reconstruction of a vehicle trajectory based on GPS readouts providing *very* noisy position and velocity information. The trajectory obtained by the backward Rauch–Tung–Striebel recurrence [66] is much smoother than the trajectory obtained in the forward pass of the Kalman filter

7.8.4   *Extended Kalman Filter*

If the process and/or the measurement equations are nonlinear, Kalman filtering is commonly implemented by their linearization, resulting in the so-called *extended Kalman filter*. Consider the nonlinear process equation

$$x_{n+1} = f_n(x_n, u_n, q_n) \tag{7.38}$$

and the nonlinear measurement equation

$$z_n = h_n(x_n, r_n) , \tag{7.39}$$

where, as before, q_n and r_n represent temporally uncorrelated Gaussian driving and measurement noises with means $\overline{q}$ and $\overline{r}$ (usually assumed to be zero), and covariance matrices Q and R, respectively. The filtering scheme (compare to (7.35)) becomes

$$
\begin{aligned}
\text{Propagation}: \quad \hat{x}_n^- &= f_{n-1}(\hat{x}_{n-1}^+, u_{n-1}, \overline{q}) , & \text{(P1)} \\
P_n^- &= F_{n-1} P_{n-1}^+ F_{n-1}^T + G_{n-1} Q_{n-1} G_{n-1}^T , & \text{(P2)} \\
\text{Update}: \quad K_n &= P_n^- H_n^T \big(H_n P_n^- H_n^T + D_n R_n D_n^T\big)^{-1} , & \text{(U1)} \\
\hat{x}_n^+ &= \hat{x}_n^- + K_n\big(z_n - h_n(\hat{x}_n^-, \overline{r})\big) , & \text{(U2)} \\
P_n^+ &= (I - K_n H_n) P_n^- , & \text{(U3)}
\end{aligned}
$$

and is initialized by appropriate $\hat{x}_0^+$ and P_0^+. The matrices embodying the Taylor expansions of the functions f and h to first order are given by

$$
F_n = \left. \frac{\partial f_n(x, u_n, \overline{q})}{\partial x} \right|_{x=\hat{x}_n^+} , \quad
G_n = \left. \frac{\partial f_n(\hat{x}_n^+, u_n, q)}{\partial q} \right|_{q=\overline{q}}
$$

and

$$H_n = \left.\frac{\partial \boldsymbol{h}_n(\boldsymbol{x}, \overline{\boldsymbol{r}})}{\partial \boldsymbol{x}}\right|_{\boldsymbol{x}=\hat{\boldsymbol{x}}_n^-} \quad , \quad D_n = \left.\frac{\partial \boldsymbol{h}_n(\hat{\boldsymbol{x}}_n^-, \boldsymbol{r})}{\partial \boldsymbol{r}}\right|_{\boldsymbol{r}=\overline{\boldsymbol{r}}} .$$

Example Consider the one-dimensional problem with the nonlinear model equation

$$x_{n+1} = f_n(x_n, u_n, q_n) = \sqrt{5 + x_n} + q_n \tag{7.40}$$

and the nonlinear measurement equation

$$z_n = h_n(x_n, r_n) = x_n^3 + r_n \tag{7.41}$$

with $\sigma_q = 1$ ($Q = 1$) and $\sigma_r = \sqrt{2}$ ($R = 2$). We need $F_n = \frac{1}{2}/\sqrt{5 + \hat{x}_n^+}$, $G_n = 1$, $H_n = 3(\hat{x}_n^-)^2$ and $D_n = 1$. The result with the initial condition $\hat{x}_0^+ = 2$ and initial covariance $P_0^+ = 1$ is shown in Fig. 7.19 by the curves labeled by EKF.   ◁

The extended filter can be further improved by iteration: see [67]. Note, however, that any brand of "extended" (linearized) approach, although very popular, is just a heuristic extension of the linear filter that fails to account for a proper nonlinear propagation of Gaussian variables through the dynamic equations, and therefore the estimate optimality inherent to the linear case is no longer guaranteed.

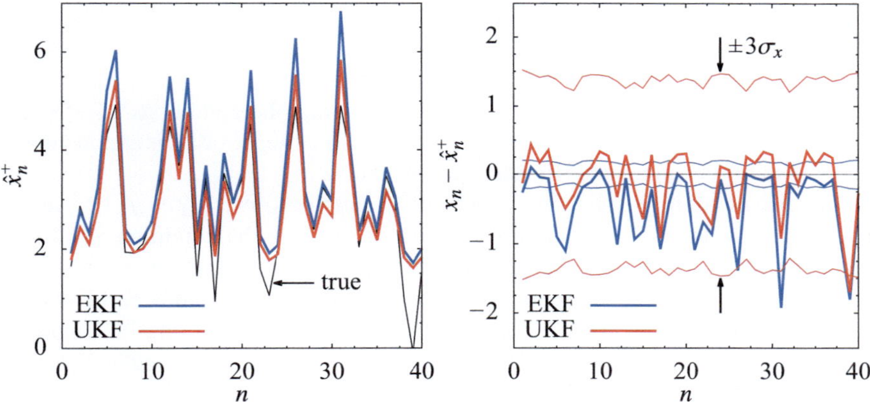

Fig. 7.19 [LEFT] The state estimates for the model defined by Eqs. (7.40) and (7.41), obtained by the extended (EKF) and the unscented (UKF) Kalman filter. [RIGHT] The errors of the estimates with respect to the true signal. The $\pm 3\sqrt{P} = \pm 3\sigma_x$ bands are also shown. Note the superior quality of the UKF estimates and their improved error bounds

Stability

By Taylor-expanding f_n and h_n near $(\hat{x}_n^+, 0)$ and $(\hat{x}_n^-, 0)$, respectively—and suppressing the u_n variable which is irrelevant to the stability statements—we get

$$f_n(x_n, q_n) = f_n(\hat{x}_n^+, 0) + F_n \xi_n^+ + G_n q_n + \phi_n(x_n, \hat{x}_n^+, q_n) ,$$
$$h_n(x_n, q_n) = h_n(\hat{x}_n^-, 0) + H_n \xi_n^- + D_n r_n + \chi_n(x_n, \hat{x}_n^-, r_n) ,$$

where $\xi_n^- = x_n - \hat{x}_n^-$ and $\xi_n^+ = x_n - \hat{x}_n^+$ denote the prior and posterior estimation errors of the filter, respectively, while ϕ_n and χ_n are the remainder functions. As outlined below, this allows us to formulate the stability theorem for the extended Kalman filter with most general (non-additive) noise [62].

(1) Suppose that for some $f, g, h, d, q_1, q_2, r_1, r_2, p_1, p_2 > 0$ the following bounds can be established for all $n \geq 0$:

$$\|F_n\| \leq f , \quad \|G_n\| \leq g , \quad \|H_n\| \leq h , \quad \|D_n\| \leq d , \tag{7.42}$$
$$q_1 I \leq Q_n \leq q_2 I , \quad r_1 I \leq R_n \leq r_2 I ,$$
$$p_1 I \leq P_n^+ \leq P_n^- \leq p_2 I . \tag{7.43}$$

(2) Suppose that for all $\varepsilon_\phi, \varepsilon_\chi > 0$ there exist $\kappa_\phi, \kappa_\chi > 0$ such that if $\|x_n - \hat{x}_n^+\|_2 < \varepsilon_\phi$ and $\|x_n - \hat{x}_n^-\|_2 < \varepsilon_\chi$, then

$$\|\phi_n(x_n, \hat{x}_n^+, q_n)\|_2 \leq \kappa_\phi \|x_n - \hat{x}_n^+\|_2^2 ,$$
$$\|\chi_n(x_n, \hat{x}_n^-, r_n)\|_2 \leq \kappa_\chi \|x_n - \hat{x}_n^-\|_2^2 .$$

Then (given (1) and (2)) there exist $\varepsilon, \delta > 0$ such that the conditions $\xi_1^- \leq \varepsilon$ and $Q_n, R_n \leq \delta I$ guarantee that the predicted (prior) estimation error ξ_n^- is exponentially bounded in mean square and stochastically bounded.

The case of additive noise, i.e. models of the simpler form $x_{n+1} = f_n(x_n, u_n) + q_n$ and $z_n = h_n(x_n) + r_n$ is slightly less restrictive: here the condition $\|\xi_0^+\| \leq \varepsilon$, together with $Q_n, R_n \leq \delta I$, guarantees that the (posterior) estimation errors ξ_n^+ are bounded. Also note that in general the upper limits of the bounds (7.42) and in particular (7.43) are unknown before the filter has actually been run; in a heuristic attitude, the state estimates may be considered trustworthy if the calculated F_n, H_n and P_n^+ remain bounded throughout the filtering process.

7.8.5　*Unscented Kalman Filter*

In the extended Kalman filter described above the distribution of the state components is assumed to be represented by Gaussian random variables, and these are then propagated through a linear approximation of the nonlinear system. This may result in false posterior estimates of the distribution moments (mean, covariance and so on) at each step, leading to poor filter performance or even divergence. The *unscented Kalman filter* [68] offers a remedy of this drawback by representing the state distribution by a set of sample points (called "sigma points") which, upon propagation through the full (true) nonlinear system, accurately capture the distribution moments at least to second order—and even to third order for Gaussian inputs—regardless of the functional forms of f and h. The modest price one has to pay for this is the need to operate with an augmented (M-dimensional) state vector,

$$x^{\mathrm{a}} = \left(x^{\mathrm{T}}, q^{\mathrm{T}}, r^{\mathrm{T}}\right)^{\mathrm{T}},$$

which is a concatenation of the original state vector and the usual noise vectors, and with the augmented block-diagonal covariance matrix,

$$P^{\mathrm{a}} = \mathrm{diag}(P, Q, R),$$

which encapsulates the corresponding covariance matrices. (For instance, if x and z and their respective noises q and r all have dimension 2, the augmented state x^{a} has dimension $M = 6$ and P^{a} is an $M \times M = 6 \times 6$ matrix.) Initializing the filter with

$$\hat{x}_0^{\mathrm{a}+} = \left(\hat{x}_0^{+\mathrm{T}}, \mathbf{0}^{\mathrm{T}}, \mathbf{0}^{\mathrm{T}}\right)^{\mathrm{T}}, \qquad P_0^{\mathrm{a}+} = \mathrm{diag}(P_0^+, Q_0, R_0),$$

one then proceeds (for $n = 1, 2, \ldots$) as follows. We form an $M \times (2M + 1)$ matrix $\mathcal{X}^{\mathrm{a}}$ whose $2M + 1$ columns are the M-dimensional sigma vectors $(\mathcal{X}^{\mathrm{a}})_i$:

$$\left(\mathcal{X}_{n-1}^{\mathrm{a}}\right)_i = \begin{cases} \hat{x}_{n-1}^{\mathrm{a}+} & ; \quad i = 0, \\[2mm] \hat{x}_{n-1}^{\mathrm{a}+} + \sqrt{M + \mu}\left(\sqrt{P_{n-1}^{\mathrm{a}+}}\right)_i & ; \quad i = 1, 2, \ldots, M, \\[2mm] \hat{x}_{n-1}^{\mathrm{a}+} - \sqrt{M + \mu}\left(\sqrt{P_{n-1}^{\mathrm{a}+}}\right)_{i-M} & ; \quad i = M+1, M+2, \ldots, 2M, \end{cases}$$

where $\left(\sqrt{P}\right)_i$ denotes the ith column of the "matrix square root" L as obtained from a lower-triangular Cholesky decomposition, such that $P = LL^{\mathrm{T}}$. Note that the $\mathcal{X}^{\mathrm{a}}$ matrix itself also consists of its state and noise parts, possessing the structure $\mathcal{X}^{\mathrm{a}} = \left((\mathcal{X}^x)^{\mathrm{T}}, (\mathcal{X}^q)^{\mathrm{T}}, (\mathcal{X}^r)^{\mathrm{T}}\right)^{\mathrm{T}}$. The scaling parameter μ is given by

$$\mu = \alpha^2(M + \kappa) - M. \tag{7.44}$$

The constant α influences the distance of the sigma points from the mean and is usually set to a small positive value, $10^{-4} \lesssim \alpha \leq 1$, while κ is a secondary scaling parameter that is set either to $3 - M$ (even if $M > 3$) or 0; another option is to define $\mu = \alpha^2(M + \kappa)^{-M}$ and use any $\kappa \geq 0$: see [69] for details.

The first of the two crucial steps follows at this point in the loop: instead of only the current (non-augmented) state estimate, $\hat{x}_n^+$, the $2M + 1$ sigma points constructed *around* the augmented state are transported through the nonlinearity:

$$\left(\mathcal{X}_n^{x-}\right)_i = \boldsymbol{f}_{n-1}\left(\left(\mathcal{X}_{n-1}^{x+}\right)_i, \boldsymbol{u}_{n-1}, \left(\mathcal{X}_{n-1}^{q}\right)_i\right) , \qquad i = 0, 1, \ldots, 2M . \qquad (7.45)$$

A weighted average of the x-components of the resulting vectors is then calculated to obtain the prior state and covariance estimates:

$$\hat{\boldsymbol{x}}_n^- = \sum_{i=0}^{2M} w_i^{(\text{m})} \left(\mathcal{X}_n^{x-}\right)_i ,$$

$$P_n^- = \sum_{i=0}^{2M} w_i^{(\text{c})} \left[\left(\mathcal{X}_n^{x-}\right)_i - \hat{\boldsymbol{x}}_n^-\right]\left[\left(\mathcal{X}_n^{x-}\right)_i - \hat{\boldsymbol{x}}_n^-\right]^{\text{T}} .$$

The weights for the means (superscript (m)) and the covariances (superscript (c)) are given by the formulas

$$w_0^{(\text{m})} = \frac{\mu}{M + \mu} , \qquad w_0^{(\text{c})} = \frac{\mu}{M + \mu} + 1 - \alpha^2 + \beta , \qquad (7.46)$$

$$w_i^{(\text{m})} = w_i^{(\text{c})} = \frac{1}{2(M + \mu)} , \qquad i = 1, 2, \ldots, 2M , \qquad (7.47)$$

where β is an additional parameter that can be used to incorporate prior knowledge on the distribution of $\boldsymbol{x}$ (for Gaussian distributions, $\beta = 2$ is optimal). Note that the placement of the sigma points is not unique, and the weights given above also represent only a specific popular choice [70, 71].

Next, by the same logic as in Eq. (7.45), not the state and the measurement noise vector are fed into the nonlinear model, but rather the corresponding sigma points:

$$\left(\mathcal{Z}_n\right)_i = \boldsymbol{h}_n\left(\left(\mathcal{X}_{n-1}^{x-}\right)_i, \left(\mathcal{X}_{n-1}^{r}\right)_i\right) , \qquad i = 0, 1, \ldots, 2M ,$$

whence we construct

$$\hat{\boldsymbol{z}}_n = \sum_{i=0}^{2M} w_i^{(\text{m})} \left(\mathcal{Z}_n\right)_i$$

to be used in the measurement update equations. Finally, calculating

$$P_{xz} = \sum_{i=0}^{2M} w_i^{(\text{c})} \left[\left(\mathcal{X}_n^{x-}\right)_i - \hat{\boldsymbol{x}}_n^-\right]\left[\left(\mathcal{Z}_n\right)_i - \hat{\boldsymbol{z}}_n\right]^{\text{T}} ,$$

$$P_{zz} = \sum_{i=0}^{2M} w_i^{(\text{c})} \left[\left(\mathcal{Z}_n\right)_i - \hat{\boldsymbol{z}}_n\right]\left[\left(\mathcal{Z}_n\right)_i - \hat{\boldsymbol{z}}_n\right]^{\text{T}} ,$$

allows us to evaluate the Kalman gain matrix,

$$K_n = P_{xz} P_{zz}^{-1} ,$$

and the posterior state and covariance estimates:

$$\hat{x}_n^+ = \hat{x}_n^- + K_n \left(z_n - \hat{z}_n \right) ,$$
$$P_n^+ = P_n^- - K_n P_{zz} K_n^{\mathrm{T}} .$$

This closes the $n = 1, 2, \ldots$ loop.

Example Let us revisit the nonlinear problem specified by Eqs. (7.40) and (7.41) which we have already solved by using the extended Kalman filter (Fig. 7.19). The curves labeled by UKF in both panels of that same Figure show the solution by using the unscented variant of the filter, with $M = 1$, $\alpha = 0.1$, $\beta = 2$ and $\kappa = 0$. Note the superior quality of the UKF estimates and improvement of the error bounds over their EKF counterparts. $\triangleleft$

Technically the stability theorems for the unscented filter are not more involved than those pertaining to the extended filter, but their statement requires more space than could be afforded in this book. See Chap. 5 of [62] for details.

7.8.6 Adaptive Non-linear Estimation

It often happens that in addition to the state, control and noise variables, the nonlinear process and nonlinear measurement equations also depend on a set of parameters θ, so that Eqs. (7.38) and (7.39) become

$$x_{n+1} = f_n(x_n, u_n, q_{x,n}; \theta_n) \tag{7.48}$$

and

$$z_n = h_n(x_n, r_{x,n}; \theta_n) , \tag{7.49}$$

respectively, where $q_x \sim N(\overline{q}_x, Q_x)$ and $r_x \sim N(\overline{r}_x, R_x)$. Suppose that the functional forms of f and h are known, but the parameters θ themselves are unknown or poorly known. Is it possible to provide optimal estimates for both the system states *and* its parameters? Indeed it is, and there are two popular ways to accomplish this task, known as *adaptive estimation*. The first one is *joint estimation* [72]: we may combine the states and the parameters into a joint vector

$$\tilde{x}_n = \left(x_n^{\mathrm{T}}, \theta_n^{\mathrm{T}} \right)^{\mathrm{T}} ,$$

then proceed with either the extended or the unscented Kalman filter operating on this joint state, taking into account all corresponding covariances and starting the loop with suitable initial conditions. This method works "in principle", but in practice the convergence tends to be slow. What is worse, the state covariances may seep into the parameter covariances and the other way around, causing hard-to-control drifts and uncertainties.

A better way to approach the problem is by *dual estimation* [73] employing *two* filters: one for the states and one for the parameters. The updated parameters are fed into the model and measurement equations (7.48) and (7.49), then the updated states are used in the dynamical equations for the parameters. These are simply

$$\boldsymbol{\theta}_{n+1} = \boldsymbol{\theta}_n + \boldsymbol{q}_{\theta,n} , \tag{7.50}$$

"instructing" the filter not to change the parameters $\boldsymbol{\theta}$, and

$$\boldsymbol{d}_{n+1} = \boldsymbol{h}_n \left[\boldsymbol{f}_n(\boldsymbol{x}_n, \boldsymbol{u}_n, \boldsymbol{q}_{x,n}; \boldsymbol{\theta}_n), \boldsymbol{r}_{x,n}; \boldsymbol{\theta}_n \right] + \boldsymbol{r}_{\theta,n} , \tag{7.51}$$

which acts as a "measurement on $\boldsymbol{\theta}_n$" at fixed $\boldsymbol{x}_n$. Here $\boldsymbol{q}_\theta \sim N(\overline{\boldsymbol{q}}_\theta, Q_\theta)$ and $\boldsymbol{r}_\theta \sim N(\overline{\boldsymbol{r}}_\theta, R_\theta)$. These *four* models—Eqs. (7.48)–(7.51)—can then be plugged into either the extended or the unscented filters running in parallel. An unscented implementation is provided in Appendix D.

There is, however, a fundamental obstacle to be considered: the filter may be unable to "decide" whether the states or the parameters should be adjusted, possibly resulting in failed or poor convergence. For instance, in the case of a model as simple as $x_{n+1} = f_n(x_n, q_n; \theta) = \sqrt{x_n + \theta} + q_n$, both x_n and θ appear additively under the square root sign, and there is almost complete arbitrariness in the filter's "decision" about which component to modify. In other words, there is no inherent guarantee that the procedure will converge to anything physically meaningful. One possible trick to stabilize the filtering process in such cases is to introduce an additional synthetic signal generated with a known underlying model, which is as similar as possible to the supposed true model, then incorporating this auxiliary signal into the filter along all the others. This obviously introduces some model dependence, but in some cases it may be the only means to achieve convergence. An example is discussed in Ref. [74].

And, finally, what if one has *no clue* about the analytical form of the underlying model? An exciting method exists [75] that addresses this problem by merging Kalman filtering with the Takens' technique of state-space reconstruction (Sect. 7.10).

7.9 Independent Component Analysis ⋆

Independent component analysis (ICA) is a multivariate analysis technique from the class of latent-variable methods (as is the factor analysis discussed in Sect. 6.12). The classical problem that can be solved by ICA is the decomposition of an unknown mixture of signals to its independent components. For example, we record the signals $x_1(t)$, $x_2(t)$, and $x_3(t)$ of three microphones picking up the sounds from two sources $s_1(t)$ and $s_2(t)$. We assume that the signals are linear combinations of the sources,

$$\begin{pmatrix} x_1(t) \\ x_2(t) \\ x_3(t) \end{pmatrix} = \begin{pmatrix} A_{11} & A_{12} \\ A_{21} & A_{22} \\ A_{31} & A_{32} \end{pmatrix} \begin{pmatrix} s_1(t) \\ s_2(t) \end{pmatrix} ,$$

where neither the sources s_i nor the matrix elements A_{ij} are known. A generalization of this example is the famous *cocktail-party problem*, in which we wish to reconstruct the signals of r speakers based on n measurements of signals from m microphones arranged in a room. The components of the vector representing the sources are the latent variables mentioned above.

A very similar problem is illustrated in Fig. 7.20 (see also Problem 7.11.5 and Fig. 7.32). The Figure shows eight traces of an electro-cardiogram of a pregnant woman recorded at various places on the thorax and the abdomen. In some signals (x_3, x_2 and, above all, x_1) we can also see the child's heartbeat which has a higher frequency than the mother's heartbeat, but its amplitudes are smaller and partly masked by noise. The task of ICA is to extract the original signals *sources* and to explain the mixing of these sources into the measured signals. The procedure should be done such that the computed sources are mutually as independent as possible.

Here we discuss only *static linear independent component analysis*, for which we assume that each vector of correlated measurements $x = (x_1, x_2, \ldots, x_m)^{\mathrm{T}}$ is generated by *linear* mixing of *independent sources* $s = (s_1, s_2, \ldots, s_r)^{\mathrm{T}}$, thus

$$x = As , \tag{7.52}$$

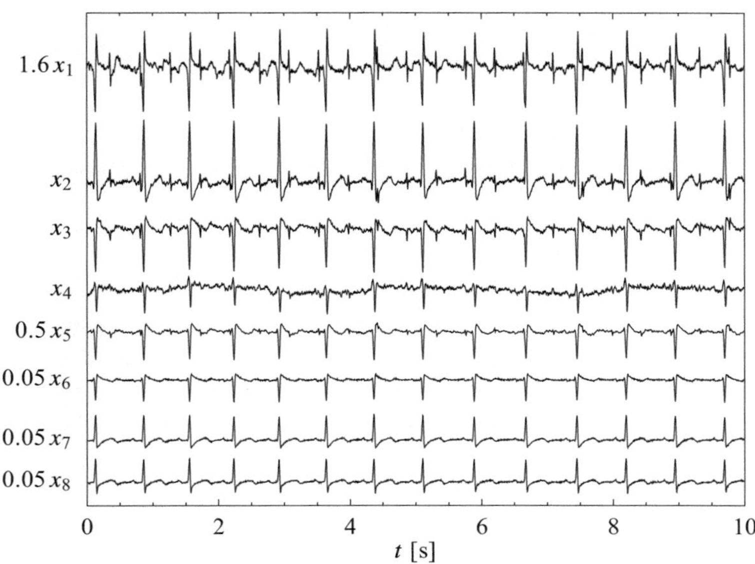

Fig. 7.20 Electro-cardiogram of a pregnant woman. The relatively weak and noisy signal of the child's heartbeat, visible in signals x_1, x_2, and x_3, mixes with the signal of the mother's heartbeat. (Example adapted from [76] based on data from [77])

where $A \in \mathbb{R}^{m \times r}$ is a time-independent *mixing matrix*. Usually the number of sources is smaller than the number of measured signals, $r \leq m$ (as above, $r = 2$ hearts and $m = 8$ electrodes). We neglect measurement errors that, in principle, could be added to the right side of Eq. (7.52). We assume that noise is already contained in the sources s: we are performing a *noiseless ICA*.

The independence of the sources is expressed by the statement that their joint probability density factorizes as $p(s_1, s_2, \ldots, s_r) = p_1(s_1)p_2(s_2) \cdots p_r(s_r)$ and therefore also $\langle f(s_i)g(s_j) \rangle = \langle f(s_i) \rangle \langle g(s_j) \rangle$ for arbitrary functions f and g. A pair of statistically independent variables $(s_i, s_{j \neq i})$ is uncorrelated and has the covariance $\mathrm{cov}(s_i, s_j) = 0$; the inverse is not necessarily true. This distinguishes ICA from the decorrelation of variables by PCA (Sect. 6.8), in which principal components of multivariate data are determined by maximizing their variances and minimizing their mutual correlations.

In the following we assume that the average of an individual vector of sources is zero, $\langle s \rangle = \mathbf{0}$, and that the corresponding covariance matrix is diagonal, $\Sigma_{ss} = \mathrm{cov}(s, s) = ss^{\mathrm{T}} = I$. Other than that, the components s_i may have any probability distribution except Gaussian. The requirement for a non-Gaussian character of the distributions is essential, otherwise the ICA method allows for the determination of the independent components only up to an orthogonal transformation. At most one component of the signal is allowed to be Gaussian: for a detailed explanation see Sect. 15.3 in [76, 78, 79].

7.9.1 Estimate of the Separation Matrix and the FastICA Algorithm

According to the model (7.52) with the matrix A of full rank, a *separation* or *unmixing matrix* W exists by which the sources s can be exactly reconstructed from the measured signals x:

$$s = Wx , \qquad W = (A^{\mathrm{T}}A)^{-1}A^{\mathrm{T}} = (w_1, w_2, \ldots, w_r)^{\mathrm{T}} \qquad (7.53)$$

(see Eq. (4.15)). But in practice, neither the matrix A nor the vector of the sources s in the model (7.52) are known. In order to reconstruct s, we therefore attempt to find an estimate of the separation matrix W. We resort to several additional requirements that should be fulfilled by the one-dimensional projections of the measured signals,

$$y_i = w_i^{\mathrm{T}}x , \qquad i = 1, 2, \ldots, r .$$

If the vector w_i is equal to some row of the generalized inverse W of A (see Eq. (7.53)), the projection y_i is already one of the independent components, $y_i = s_i$. Since A (or W) is unknown, we try to find the vectors w_i such that the probability distribution of the projection y_i will be *as non-Gaussian as possible*. Seeking such vectors is the key part of the independent component analysis.

Table 7.1 Typical choices for the function G and its derivatives appearing in the approximation of negentropy (7.54) and in the FastICA algorithm to determine the independent components. For general use we recommend the functions (1) with the parameter $\alpha = 1$. For signals with pronounced outliers or super-Gaussian probability distributions, we recommend the functions (2)

	$G(y)$	$G'(y)$	$G''(y)$	Remark
(1)	$\frac{1}{\alpha}\log\cosh\alpha y$	$\tanh\alpha y$	$\alpha(1 - \tanh^2\alpha y)$	$1 \leq \alpha \leq 2$
(2)	$-e^{-y^2/2}$	$y\,e^{-y^2/2}$	$(1 - y^2)\,e^{-y^2/2}$	

The deviation of the distribution of a continuous random variable Y from the normal ("Gaussian") distribution can be measured by the entropy

$$H(Y) = -\int p(y)\log p(y)\,\mathrm{d}y \,,$$

where $p(y)$ is the probability density of Y. Of all random variables with the same variance the normal random variable has the maximum entropy [26]. The value of the entropy can therefore be used as a tool to gauge the similarity of some unknown distribution to the normal distribution. Computationally it is more convenient to work with *negentropy*

$$I(Y) = H(Y_{\text{Gauss}}) - H(Y) \,, \qquad Y_{\text{Gauss}} \sim N(0, 1) \,,$$

which is a non-negative quantity and is equal to zero only in the case that both Y_{Gauss} and Y are distributed normally with zero average and unit variance.

For an exact calculation of $I(Y)$ we need the probability density $p(y)$, which we almost never know in practice. We therefore use approximations of the negentropy, most often [78]

$$I(Y) \approx \left[\langle G(Y)\rangle - \langle G(Y_{\text{Gauss}})\rangle\right]^2 \,, \tag{7.54}$$

where $G(y)$ is a non-quadratic function of y (two popular choices are given in Table 7.1). We compute the estimate y_i for a single independent component s_i by finding a vector $\boldsymbol{w}_i$ such that the projection $y_i = \boldsymbol{w}_i^{\mathrm{T}}\boldsymbol{x}$ will have maximum negentropy $I(y_i)$ (then its probability density will least resemble the Gaussian). We achieve the decomposition to independent components when we ultimately find all vectors $\boldsymbol{w}_i$ ($1 \leq i \leq r$) corresponding to local maxima of $I(y_i)$. Since the independent components s_i (or the estimates y_i for them) are uncorrelated, they can be computed individually. Finally, the vectors $\boldsymbol{w}_i$ are arranged in the matrix W and we use Eq. (7.53) to compute the independent components $\boldsymbol{s}$.

The FastICA Algorithm

The independent components of the signals $\boldsymbol{x}_i = (x_1, x_2, \ldots, x_m)_i^{\mathrm{T}}$, which are measured at n consecutive times $(0 \leq i \leq n - 1)$, can be determined by the FastICA algorithm described in the following. The measured signals are first arranged in the matrix

$$X = (\boldsymbol{x}_0, \boldsymbol{x}_1, \ldots, \boldsymbol{x}_{n-1})^{\mathrm{T}} \in \mathbb{R}^{n \times m} \,,$$

in which each of the n rows represents one m-variate data entry (for example, voltages on m microphones at time $t = n\Delta t$). We compute the mean $\overline{\boldsymbol{x}}$ by Eq. (6.59) and the covariance matrix Σ_{xx} by Eq. (6.65). We diagonalize Σ_{xx} as

$$\Sigma_{xx} = U \Lambda U^{\mathrm{T}}$$

and thus obtain the orthogonal matrix U and the diagonal matrix Λ. The data is then transformed as

$$\boldsymbol{x}_i \longleftarrow \Lambda^{-1/2} U^{\mathrm{T}} (\boldsymbol{x}_i - \overline{\boldsymbol{x}}) \,, \qquad i = 0, 1, \ldots, n - 1 \,. \tag{7.55}$$

The transformation (7.55) is known as *data whitening* and is equivalent to a decomposition of the standardized data to their principal components (see Sects. 6.8.2 and 6.8.3). Whitening removes any trace of scale or correlation from the data. We then follow the algorithm [80]:

1. Choose the number of independent components r you wish to determine.
2. Randomly initialize the vectors $\boldsymbol{w}_1, \boldsymbol{w}_2, \ldots, \boldsymbol{w}_r \in \mathbb{R}^m$ normalized to $\|\boldsymbol{w}_k\|_2 = 1$, and arrange them in the matrix $W = (\boldsymbol{w}_1, \boldsymbol{w}_2, \ldots, \boldsymbol{w}_r)^{\mathrm{T}} \in \mathbb{R}^{r \times m}$.
3. Perform a symmetric orthogonalization of W,

$$W \longleftarrow \left(W W^{\mathrm{T}}\right)^{-1/2} W \,.$$

This step ensures that all previously found $\boldsymbol{w}_k$ are mutually orthogonal. The square root of the matrix is computed as described in Appendix A.8.
4. For each $k = 1, 2, \ldots, r$ compute new vectors

$$\boldsymbol{w}_k \longleftarrow \frac{1}{n} \left\{ \sum_{i=0}^{n-1} \boldsymbol{x}_i \, G'\left(\boldsymbol{w}_k^{\mathrm{T}} \boldsymbol{x}_i\right) - \boldsymbol{w}_k \sum_{i=0}^{n-1} G''\left(\boldsymbol{w}_k^{\mathrm{T}} \boldsymbol{x}_i\right) \right\} \,, \tag{7.56}$$

where the function pair $G'(y)$ and $G''(y)$ can be chosen from Table 7.1. This step is the crucial part of the algorithm ensuring that the iteration leads to the fixed point corresponding to the maximum of negentropy (7.54). Normalize the obtained vectors to unit length,

$$\boldsymbol{w}_k \longleftarrow \boldsymbol{w}_k / \|\boldsymbol{w}_k\|_2 \,.$$

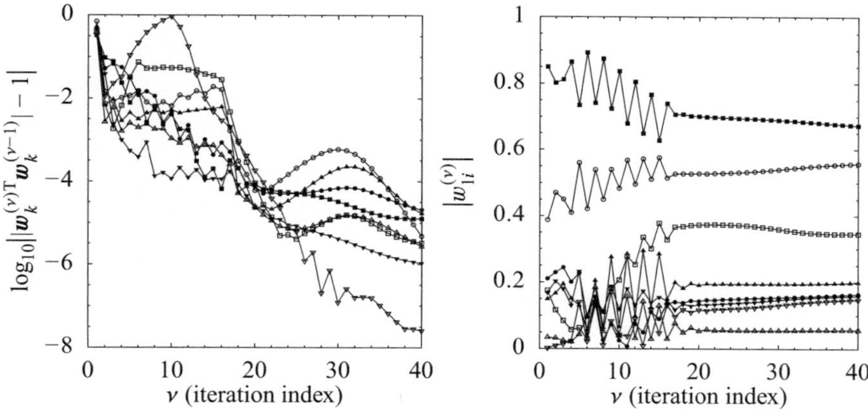

Fig. 7.21 Typical convergence in the FastICA algorithm applied to the relatively complex signals from Fig. 7.20. [LEFT] The difference of the scalar product $|w_k^{(v)\mathrm{T}} w_k^{(v-1)}|$ from unity in subsequent iterations (v) in the case eight independent components ($k = 1, 2, \ldots, 8$). [RIGHT] The convergence of eight components of the vector w_1 in consecutive iterations (v)

5. Repeat items 3 and 4 until convergence is achieved in all components of the vectors w_k. A good measure of convergence is the configuration in which the direction of the vector w_k in consecutive iterations (v) and ($v - 1$) no longer changes, i.e. when the absolute value of the scalar product $|w_k^{(v)\mathrm{T}} w_k^{(v-1)}|$ is close to unity. Typically a few times ten iterations are needed (see Fig. 7.21).
6. The independent components are the rows of the matrix $S = W X^{\mathrm{T}} \in \mathbb{R}^{r \times n}$.

The FastICA algorithm allows us to compute the matrix of independent components (sources) S in which the sources appear to be arranged arbitrarily. Namely, any permutation P of the source components s in (7.52) implies just a different mixing matrix, $x = (A P^{-1}) P s$. From the viewpoint of the ICA method, the vectors s and $P s$ are indistinguishable.

Moreover, the signals S determined by ICA are given up to a multiplicative constant: we may choose any constant c_j to divide the source s_j and multiply the jth row of the mixing matrix A without modifying the product $A s$. Even vectors of opposite directions (w_k and $-w_k$ as the rows of W) that frequently occur during the iterations of the FastICA algorithm, are therefore equivalent.

Stabilization of the FastICA Algorithm

The iteration step (7.56) follows from the Newton's method to search for the maximum of negentropy, so occasionally we may encounter convergence problems. In such cases the authors [80] recommend a stabilized version of the step (7.56), which is

$$
\boldsymbol{w}_k \longleftarrow \boldsymbol{w}_k - \mu \left[\frac{1}{n} \sum_{i=0}^{n-1} \boldsymbol{x}_i \, G'\!\left(\boldsymbol{w}_k^{\mathrm{T}} \boldsymbol{x}_i\right) - \beta \, \boldsymbol{w}_k \right] \left[\frac{1}{n} \sum_{i=0}^{n-1} G''\!\left(\boldsymbol{w}_k^{\mathrm{T}} \boldsymbol{x}_i\right) - \beta \right]^{-1} ,
$$

where

$$
\beta = \frac{1}{n} \sum_{i=0}^{n-1} \left(\boldsymbol{w}_k^{\mathrm{T}} \boldsymbol{x}_i\right) G''\!\left(\boldsymbol{w}_k^{\mathrm{T}} \boldsymbol{x}_i\right) ,
$$

and μ is a parameter used to control the stability of the iteration, with typical values much smaller than one, for example, $\mu \approx 0.1$ or $\mu \approx 0.01$. Details of other possible improvements of the algorithm can be found in [80].

7.10　State-Space Reconstruction ⋆

Imagine a dynamical system whose evolution law does not change with time—an *autonomous system*. It can be assigned a phase space X and furnished with a map $\phi^t : X \to X$ that represents its dynamics: if the system is initially in state $\boldsymbol{x}(0)$ then its trajectory is given by $\boldsymbol{x}(t) = \phi^t(\boldsymbol{x}(0))$. We will assume that X is bounded and that the map ϕ^t is continuously differentiable, which are reasonable assumptions for most physical systems.

Although the dynamics is defined on the whole phase space the system will not necessarily visit all of its parts equally frequently and perhaps certain parts will never be revisited at all. Here we are interested only in the subset of the phase space called the *attractor* [25, 81]. This subset is characterized by the fact that the system, once started in its vicinity, will be attracted to it, converge to it and stay there arbitrarily long; essentially this is the set in which the system is most likely to be found after long enough times. Due to dissipation, for instance, a damped oscillator will end up in the minimum of the potential, which is its attractor.

An interesting example of a dynamical system with an attractor residing in less than the full space is the three-dimensional Rössler system [82] defined by

$$
\begin{aligned}
\dot{x} &= -y - z , \\
\dot{y} &= x + ay , \\
\dot{z} &= b + z(x - c) ,
\end{aligned}
$$

with $a = 0.2$, $b = 0.2$ and $c = 5.7$. The state of the system at time t is given by $\boldsymbol{x}(t) = (x(t), y(t), z(t))$. A sample trajectory that we use for the demonstration of methods throughout this Section is presented in Fig. 7.22. The attractor of the Rössler system is a fractal of dimension approximately equal to two.

Consider a trajectory $\boldsymbol{x}(t)$ that densely covers the attractor and view it by using a real continuously differentiable observable $f : X \to \mathbb{R}$. This gives us the *signal* $f(\boldsymbol{x}(t))$, representing the outcome of an experiment where we observe the dynamics

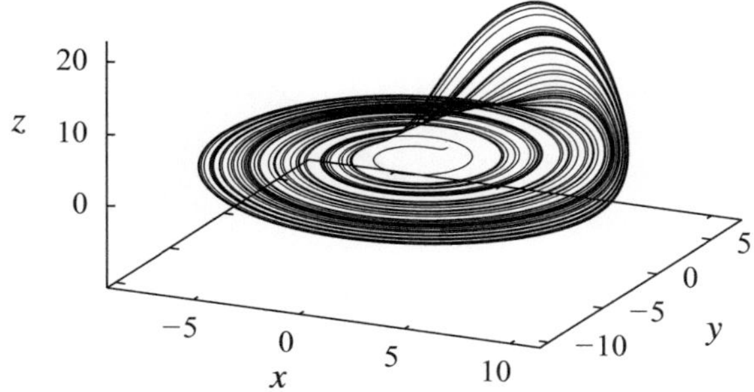

Fig. 7.22 A Rössler system trajectory for $t \in [0, 1000]$ starting at $(1, 1, 1)$

of a potentially very complex system through a single real measurement. In our illustrative case, we take the observable f to be the first component of the dynamical system, the variable x itself.

From the signal we construct d-tuples of observations called *reconstruction-space vectors* or *delay coordinates*

$$\boldsymbol{R}_d(t) = \left(f(\boldsymbol{x}(t)), f(\boldsymbol{x}(t + \tau)), \ldots, f(\boldsymbol{x}(t + (d - 1)\tau)) \right),$$

which reside in the reconstruction space $\mathbb{R}^d$. By introducing the *reconstruction map* or *delay map*

$$\Phi_{d,\tau}(\boldsymbol{x}) = \left(f(\boldsymbol{x}), f(\phi^\tau(\boldsymbol{x})), \ldots, f(\phi^{(d-1)\tau}(\boldsymbol{x})) \right)$$

each reconstruction-space vector obtained from the trajectory can be visualized as an image of the reconstruction map $\boldsymbol{R}_d(t) = \Phi_{d,\tau}(\boldsymbol{x}(t))$. The reconstruction map essentially associates a given phase-space point with consecutive measurements.

Suppose that a dynamical system has an attractor of dimension d_A. It was rigorously shown by Takens [83] and Mañé [84] that for $d > 2d_A$ and $\tau > 0$ the reconstruction map $\Phi_{d,\tau}$ is an *embedding* [85]: there is a one-to-one correspondence between the dynamics on the attractor and the dynamics in the reconstruction space using vectors $\boldsymbol{R}_d$. This means that by following the dynamics of a system in reconstruction space we should not encounter any self-crossing of trajectories, which is an indicator of non-deterministic dynamics. Such crossings can still be seen in Fig. 7.23 for the Rössler system, using the embedding dimension $d = 2$. In other words, embedding is an *unfolding of the attractor in reconstruction space*: as we gradually increase d, such self-crossings disappear. In principle τ can be chosen almost arbitrarily, but this choice is driven by practical considerations discussed below.

The Takens–Mañé theorem states only the sufficient condition for embedding and there may exist a smaller dimension d to achieve it. (For instance, the famous Lorenz "butterfly" system [86] has an attractor of dimension $d_A \approx 2.06$ that can be embedded in $d = 3$.) Let the minimal dimension enabling the embedding be denoted by d_E. Working with a larger reconstruction space has many drawbacks. Beside the obvious information redundancy and excessive memory usage, one may face problems with noise and round-off contaminating the values of the reconstruction-

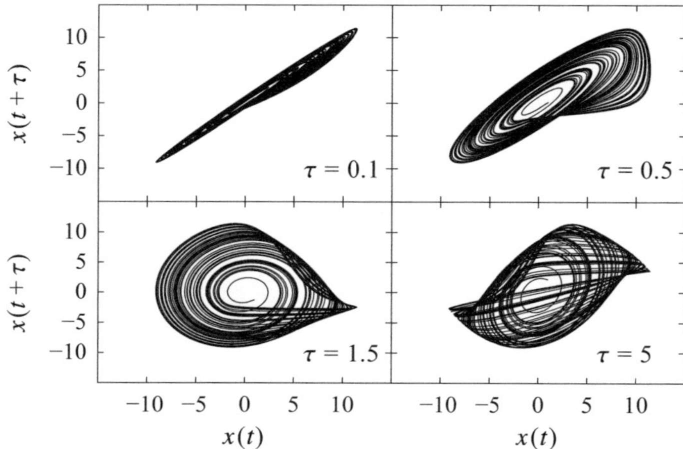

Fig. 7.23 Two-dimensional delay-coordinate embeddings of the Rössler system for different time delay parameters τ. Self-crossings indicate that the dimension of the embedding is too low

state vector in unnecessary dimensions. In chaotic systems, which are inherently sensitive to small changes in the initial conditions, larger than necessary dimensions can produce artificial separations between almost identical points.

If τ is too small, the points will lie near the diagonal of the reconstruction space—see Fig. 7.23—and in the presence of experimental or machine-precision noise some subsequent points will be indistinguishable. If, however, τ is much larger than the decay time of the correlation, the reconstruction-space vectors start to resemble random points in finite arithmetic or at given measurement precision. To avoid this, τ is best chosen such that it minimizes the statistical dependence of the components of the reconstruction-space vector, thereby maximizing its descriptive property.

Ideally we would like to work with a reconstruction space at minimal embedding dimension d_{E} and optimal time delay τ. After establishing the embedding, calculations with trajectories represented by the reconstruction-space vectors can be performed in the same manner as with phase-space trajectories, since both possess analogous dynamical properties. Herein is the true power of the method: in real-life experiments in which we may be completely ignorant of the nature of the dynamical system and from which we typically obtain only a simple scalar quantity—the signal—embedding helps us to determine whether the underlying system is deterministic or not, based only on this very signal. This makes embedding one of the key approaches of non-linear time-series analysis [87, 88], i.e. the analysis of signals by using the methods developed for (chaotic) dynamical systems.

In the following we present the most commonly used methods to determine the embedding parameters. Usually we first try to establish the optimal time delay and then use this knowledge to determine the minimal embedding dimension. For an overview of different techniques see [89]. A unified approach in which both parameters are determined simultaneously is discussed in [90].

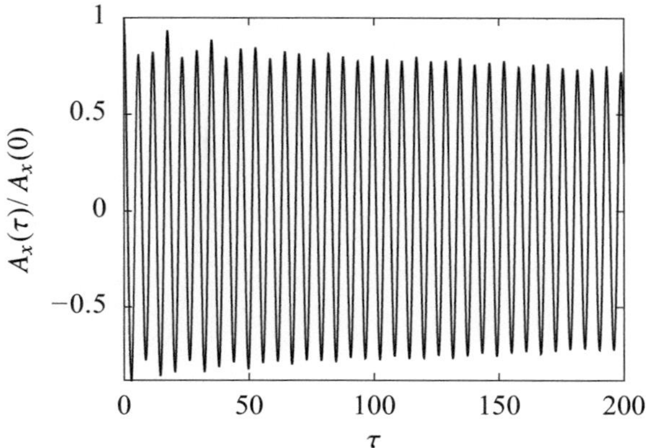

Fig. 7.24 The autocorrelation of the signal from the Rössler system. The optimal embedding time delay can be read off from its first zero-crossing at $\tau \approx 1.45$

7.10.1   Establishing the Optimal Time Delay

The optimal time delay can be estimated by using the autocorrelation function of the sample $\{x(t)\}_{t=0}^{N-1}$, here again taken to be the first component of the Rössler system. We use the unbiased autocorrelation given by

$$A(\tau) = \frac{1}{N - \tau} \sum_{t=0}^{N-1-\tau} \big(x(t) - \overline{x}\big)\big(x(t + \tau) - \overline{x}\big),$$

where $\overline{x}$ is the mean value of the sample. In systems with an exponential decay of correlations the optimal time delay τ in reconstruction space can be estimated by the instant at which the autocorrelation drops to $1/e$ of its initial value. In other systems τ can be chosen by the first zero or a significant local minimum of the autocorrelation [91]. In this way the linear statistical independence of the components of the reconstruction-space vector is maximized.

The Rössler system is chaotic [92], but has a very slow decay of correlations [93]. For practical purposes we can say that the autocorrelation oscillates almost periodically, as can be seen in Fig. 7.24, hence the optimal time delay can be estimated from the first zero of the autocorrelation function: $\tau \approx 1.45$.

A more efficient method for determining the time delay is to use the first minimum of the mutual information $I(\tau)$ between the signal $x(t)$ and its delayed version $x(t + \tau)$ as the optimal embedding delay [94]. This again makes the reconstruction-space vector components maximally independent in the probabilistic sense.

Assume that the signal $x(t)$ spans the range $\mathcal{I} = [x_{\min}, x_{\max}]$. Partition the range into M uniform bins $\{\mathcal{I}_n\}_{n=1}^{M}$. Introduce the probability $P_n = \mathrm{Prob}\big(x(t) \in \mathcal{I}_n\big)$ that the value of the signal is in the nth bin, and the probability $P_{n,m}(\tau) = \mathrm{Prob}\big(x(t) \in \mathcal{I}_n \wedge x(t + \tau) \in \mathcal{I}_m\big)$ that the values of the signal at times t and $t + \tau$ are in the nth and mth bin, respectively. The mutual information of the binned values is then given by

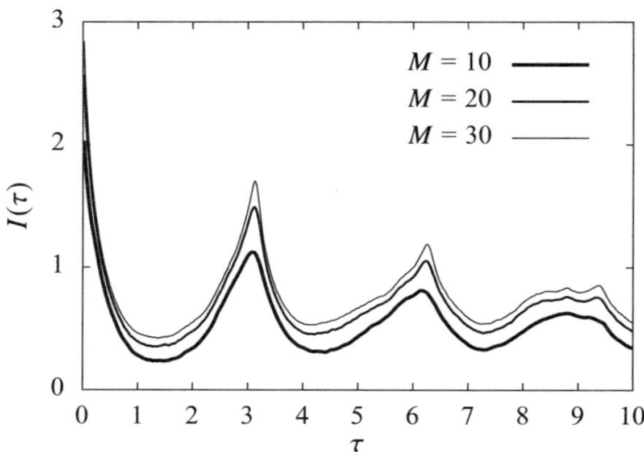

Fig. 7.25 Mutual information calculated for the signal from the Rössler system, for different range partition sizes M

$$I(\tau) = \sum_{n,m=1}^{M} P_{n,m}(\tau) \log P_{n,m}(\tau) - 2 \sum_{n=1}^{M} P_n \log P_n \, .$$

Ideally, the number of bins M should be such that the number of signal values falling into each bin for the calculation of P_n and into pairs of bins (n, m) for the calculation of $P_{n,m}$ is statistically significant. Consequently, the limit $M \to \infty$ can be approached only for infinitely long signals.

Figure 7.25 shows the mutual information for the signal from the Rössler system. The first minimum is found at $\tau = 1.41 \pm 0.05$, which agrees nicely with the autocorrelation estimate. These two approaches apparently give quite similar results in our test case; nevertheless, determining τ through mutual information is usually much more robust.

7.10.2　Determining the Embedding Dimension

An efficient algorithm for determining the minimal embedding dimension at given time delay was proposed in [95]. It is based on the fact that in a proper embedding for $d \geq d_{\mathrm{E}}$ a point in phase space uniquely determines the state of the system and that the points do not move dramatically relative to each other with increasing d.

Assume that $\boldsymbol{R}_d(t')$ for $t' \neq t$ is the nearest point to $\boldsymbol{R}_d(t)$, its immediate neighbor in d-dimensional space. We say that the point $\boldsymbol{R}_d(t)$ has a *false nearest neighbor* (FNN) if by incrementing d by unity the point $\boldsymbol{R}_{d+1}(t')$ dramatically increases its distance to $\boldsymbol{R}_{d+1}(t)$. This can be stated by the condition

$$\left| x(t + \tau d) - x(t' + \tau d) \right| > \max\left\{ r_{\mathrm{tol}} \| \boldsymbol{R}_d(t) - \boldsymbol{R}_d(t') \|, \, a_{\mathrm{tol}} \sigma_x \right\},$$

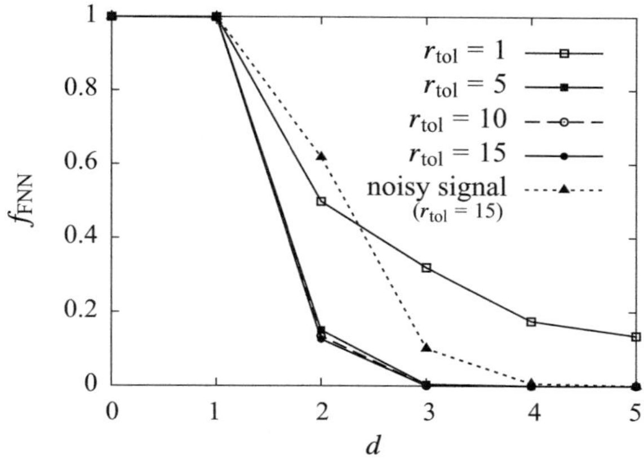

Fig. 7.26 Fraction of false nearest neighbors for a signal from the Rössler system, as a function of dimension d. The value of d at which f_{FNN} drops to zero may be identified with the minimal embedding dimension

where σ_x is the standard deviation of $x(t)$ while r_{tol} and a_{tol} are tunable tolerances depending on the considered system. Usually, a good choice is $r_{\mathrm{tol}} \geq 10$ and $a_{\mathrm{tol}} = 1$. The fraction of points $\boldsymbol{R}_d(t)$ that have a false nearest neighbor, f_{FNN}, decreases with increasing d and drops to zero for $d \geq d_{\mathrm{E}}$. This fact leads to an estimate for the embedding dimension d_{E}.

In Fig. 7.26 we plot f_{FNN} corresponding to the Rössler system for several tolerances r_{tol} and $a_{\mathrm{tol}} = 1$. We see that f_{FNN} monotonically decreases with d and that the curves are almost identical for $r_{\mathrm{tol}} \geq 5$, dropping to zero for $d \geq 3$. We conclude that for all practical purposes the minimal embedding dimension is $d_{\mathrm{E}} = 3$. In practice we determine the optimal r_{tol} by gradually increasing it to the value beyond which further decay of f_{FNN} with increasing d is no longer significant, i.e. yields the same minimal embedding dimension.

If we perturb the signal, say, by adding $x'(t) = x(t) + 0.5\xi$, where ξ is white noise distributed uniformly over $[-1, 1]$, we observe a much slower decay of f_{FNN} (dotted curve in the Figure), indicating that the embedding dimension might be larger. This is not surprising as noise, from the perspective of finite arithmetic, can in fact be seen as a dynamical system of very high dimension. This tells us that in realistic systems, where noise is certainly present, we can not determine the precise embedding dimension, but rather establish only an effective embedding dimension which, nevertheless, will be sufficient to perform calculations on the trajectories expressed by reconstruction-space vectors.

We can check whether an appropriate embedding dimension d_{E} has been found by comparing it to the fractal dimension d_{A} of the attractor, as we know that $d_{\mathrm{E}} = 2d_{\mathrm{A}} + 1$ is already sufficient for embedding. In most practical cases $d_{\mathrm{E}} \geq d_{\mathrm{A}}$, but this is not a rigorous statement. Here we present a simple method to estimate the fractal dimension of the attractor from the trajectory of the dynamical system [96].

Consider a signal $x(t)$ from which we construct a sample of d-dimensional reconstruction-state vectors $(\boldsymbol{R}_d(t))_{t=0}^{N-1}$ for a known τ. Then we can write the correlation sum

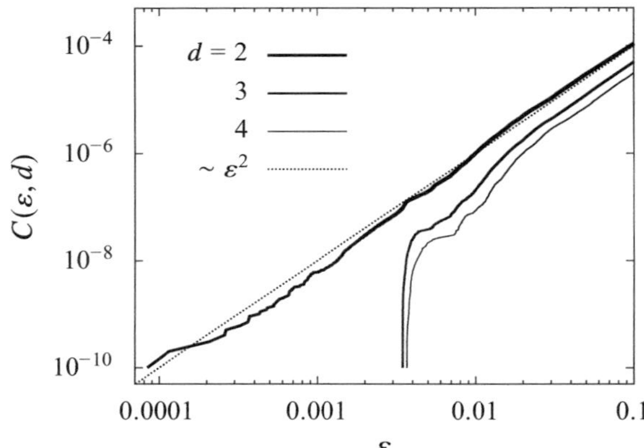

Fig. 7.27 Correlation sum (with $T = 0$) for the signal of the Rössler system, with equal sample sizes ($N = 10^5$) for each tested d

$$C(\varepsilon, d) = \frac{1}{2N(N - T)} \sum_{t=0}^{N-1} \sum_{t' < t - T} \Theta\big(\varepsilon - \|\mathbf{R}_d(t) - \mathbf{R}_d(t')\|\big) \,,$$

where Θ is the Heaviside function. This sum essentially measures the cumulative probability that pairs of points in reconstruction space are separated by less than ε. In general, the correlation between the points makes averaging over them strongly dependent on the sample size N, which can lead to spurious results. To counteract this an additional parameter T may be introduced as shown in the above sum [97], increasing independence between the sampled points.

It can be shown that for dimensions $d \geq d_{\mathrm{E}}$ the correlation sum behaves as $C(\varepsilon, d) \sim \varepsilon^{\nu}$ for $r \to 0$, where ν is the *correlation dimension* (or Renyi dimension D_2) [91]. It is a good estimate for the box-counting fractal dimension of the attractor, d_{A}. For a review of different methods see [98].

Figure 7.27 shows the correlation sum in the case of the Rössler system for different d. For $d \geq 2$ we get almost the same power dependence for small, but still statistically well represented, separations ε. (The rapid falloff of $C(\varepsilon, d)$ at $\varepsilon \approx 0.004$ for $d = 3$ and $d = 4$ is due to the finite sample size resulting in a lack of points which are extremely close to each other.) This indicates that $d_{\mathrm{A}} \approx 2$. (In fact d_{A} is almost exactly 2 [99].) We conclude that $d_{\mathrm{E}} \geq 2$.

Note that identifying the minimum embedding dimension is but one partial goal of a realistic embedding study. Its purpose, again, is not the recovery of the number of "degrees of freedom" in familiar dynamical systems—like 2 for the Rössler system based on Fig. 7.27, "more than 2" judging from Fig. 7.23, 3 according to Fig. 7.26; or 3 for the Lorenz system. The key advantage of embedding is to be able to exploit the reconstruction-space variables corresponding to realistic systems and perform all calculations *as if* we were dealing with original phase-space variables.

7.11 Problems

7.11.1 Logistic Map

In autonomous dynamical systems the time evolution of the points in phase space is defined by the map $\phi : \mathbb{R} \times \mathbb{R} \to \mathbb{R}$, where the function $x(t) = \phi(t - t_0, x(t_0))$ describes the trajectory of the system and represents its time evolution from a known initial state $x(0)$. The paradigmatic example of a discrete system is the logistic map [25], for which one time step is given by the recurrence

$$x(t + 1) = ax(t)(1 - x(t)), \qquad a \in [1, 4], \tag{7.57}$$

and phase space $[0, 1]$. A periodic orbit with period T is a set of points with the property $x(t + T) = x(t)$. By observing the trajectories in phase space we can only find stable periodic orbits, as these attract the trajectories from their neighborhood. The occurrence of stable periodic orbits in the logistic map has a rich dependence on the parameter a, as demonstrated by the diagram in Fig. 7.28 (left). At given a we take an arbitrary initial point in phase space and observe the trajectory after a sufficiently long transition time. If one period is destroyed and another is formed when the value of a changes, we are referring to a *bifurcation* of the dynamics. The sequences of distances between the points at which bifurcations occur possess intriguing universal properties [100].

Dynamical systems whose trajectories with *almost equal* initial points exponentially diverge with time, are called *chaotic*. The chaoticity of the system or its sensitivity to the precision of initial conditions can be measured by the Lyapunov exponent λ defined as

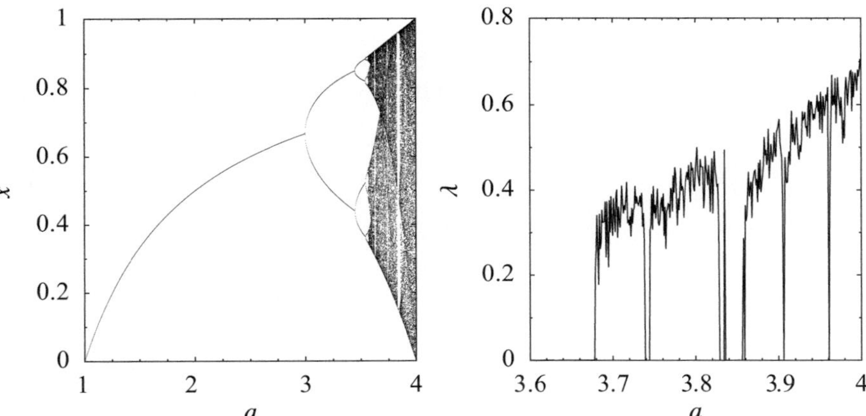

Fig. 7.28 Logistic map. [LEFT] Bifurcation diagram (dependence of the value from recurrence (7.57) on the parameter a after sufficient transition time). [RIGHT] Lyapunov exponent in dependence of a

$$\lambda = \lim_{\substack{t \to \infty \\ \varepsilon \to 0}} \frac{1}{t} \log \frac{|\phi(t, x(0) + \varepsilon) - \phi(t, x(0))|}{|\varepsilon|} \, .$$

The Lyapunov exponent is closely related to entropy formation in the dynamical system. The chaoticity of the logistic map rapidly changes with the parameter a and the coefficient λ is equal to zero when stable periodic orbits occur, as shown in Fig. 7.28 (right). At $a = 4$ even the exact value is known: $\lambda = \log 2$.

   $\odot$ For different values of a in the vicinity of 3.702, 3.74, 3.961, and 3.9778 compute $N = 1024$ values of the sequence $\{x(t)\}_{t=0}^{N-1}$ from the logistic map (7.57) with the chosen initial point $x(0) \in [0, 1]$. This sequence is your signal. Compare the single-sided power spectral densities (PSD) of this signal computed by the Fourier transformation and by the method of maximum entropy. In the ranges of a corresponding to chaotic dynamics the spectrum has no prominent peaks and is practically continuous, which is one of the landmarks of chaos.

For individual values of a, compute the auto-correlations

$$R_{xx}(\tau) = \langle x(t)x(\tau + t) \rangle_t - (\langle x(t) \rangle_t)^2$$

of the sequences $\{x(t)\}_{i=0}^{N-1}$ and determine an approximation of the correlation length ξ as a function of a. Choose N large enough that $R_{xx}(\tau)/R_{xx}(0)$ will be precise to three digits. Assume that the auto-correlation has the form

$$R_{xx}(\tau) = A \, e^{-|\tau|/\xi} \, .$$

It can be shown that

$$\xi = \log \left(\frac{|S_1^2 - S_2|}{S_1^2 + S_2} \right)^{-1} , \qquad S_n = \sum_{t=0}^{\infty} R_{xx}(t)^n \, .$$

Systems in which all correlations between quantities converge to zero at long times, are said to *mix the space*, which is one of the key properties permitting a statistical analysis of dynamics.

7.11.2 *Diffusion and Chaos in the Standard Map*

In some Hamiltonian systems, continuous dynamics can be translated to discrete dynamics. This applies in particular to systems on which we act with a periodic external force in the form of short pulses. The point (p, q) of the trajectory of the system (position and momentum) at time $t + 1$ is related to the point at time t when the system receives the pulse, by the equations

$$p(t+1) = p(t) + F(q(t), p(t)),$$
$$q(t+1) = q(t) + G(p(t+1)),$$

where F is the force pulse and G represents the dynamics between the pulses. In such systems a very clear connection between diffusion and auto-correlation exists. The diffusion coefficient is defined as

$$D = \lim_{t \to \infty} \frac{1}{2t} \langle [q(t) - q(0)]^2 \rangle,$$

where $\langle \cdot \rangle$ denotes the phase average (average over initial points $(q(0), p(0))$ distributed uniformly in the region of phase space which is invariant with respect to the dynamics). The time auto-correlation function G is defined as

$$R_{GG}(\tau) = \langle G(p(\tau))G(p(0)) \rangle - \langle G(p(0)) \rangle^2 .$$

By assuming that this system sufficiently quickly mixes the phase space, a relation between the diffusion coefficient and auto-correlation can be derived:

$$D = R_{GG}(0) + 2 \sum_{k=1}^{\infty} R_{GG}(k) . \tag{7.58}$$

⊙ Analyze the dynamical system of the standard (Chirikov's) map

$$p(t+1) = p(t) + \frac{K}{2\pi} \sin(2\pi q(t)) \mod 1,$$
$$q(t+1) = q(t) + p(t+1) \mod 1,$$

which is defined on the torus $(q(t), p(t)) \in [0, 1)^2$. This map played a major role in the development of the theory of classical and quantum chaos [101]. Examples of phase portraits obtained by propagating a number of uniformly distributed points in the case of $K = 0.5$ and $K = 1.0$ are shown in Fig. 7.29. By increasing K above ≈ 0.97 the chaotic region of the phase space, where the points are distributed uniformly, becomes ever larger, until diffusion occurs throughout.

Compute the time auto-correlation $R_{pp}(\tau) = \langle p(\tau + t)p(t) \rangle_t - \langle p(t) \rangle_t^2$ of the momentum samples $\{p(t)\}_{t=0}^{N-1}$ for the values $K = 1, 2, 5,$ and 10, with the initial point $(q(0), p(0))$ in the chaotic region of the phase space. The expected absolute error of $R_{pp}(\tau)$ should be less than 10^{-3}. If the auto-correlation decays rapidly, show it in the rescaled form $R_{pp}(\tau)/R_{pp}(0)$ in logarithmic scale. If you observe an exponential fall-off of the auto-correlation,

$$|R_{pp}(\tau)| \sim e^{-\tau/\xi},$$

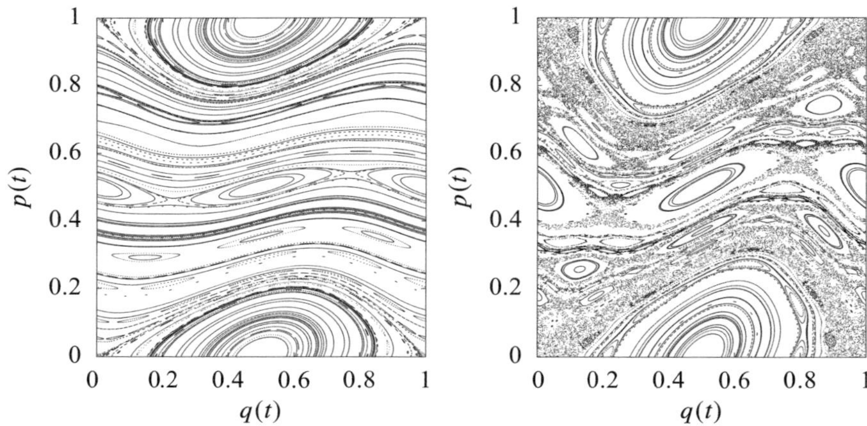

Fig. 7.29 Phase portraits of the standard map for different values of the parameter K. [LEFT] $K = 0.5$. [RIGHT] $K = 1.0$

estimate the correlation length ξ which represents the time after which the trajectories become statistically independent. To compute the auto-correlation use the FFT algorithm (Sect. 5.2.5).

$\oplus$ Observe the dependence of the diffusion coefficient D on the parameter K from the interval $[2, 20]$. Compute D by summing the auto-correlations of momentum (i.e. by using Eq. (7.58) in which R_{GG} is replaced by R_{pp}). The initial points of the map should be located in the chaotic region of the phase space. An analytic approximation for the diffusion coefficient can be found in [101].

7.11.3 Phase Transitions in the Two-Dimensional Ising Model

Some properties of systems consisting of magnetic dipoles with local interactions can be nicely explained by the *Ising model*. This model assumes that the dipoles are arranged in the nodes of a two-dimensional mesh and that the spins s_i are either "up" ($s_i = 1$) or "down" ($s_i = -1$). The Hamiltonian (the energy) of such a system in the presence of an external magnetic field H can be written in the form

$$E = -J \sum_{\langle i,j \rangle} s_i s_j - H \sum_i s_i \, , \tag{7.59}$$

where the indices i and j represent the coordinates of the spins on the mesh (in two dimensions $i \in \mathbb{N}_0^2$). The first sum in (7.59) runs over neighboring points only. The parameter J determines how the spins interact. For $J > 0$ the adjacent dipoles tend to point in the same direction, while for $J < 0$ they tend in the opposite directions. In the following we set $J > 0$. The total magnetization of the system is

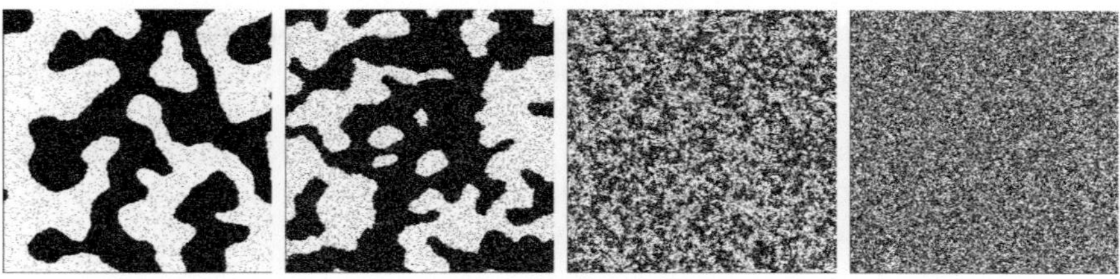

Fig. 7.30 A two-dimensional spin array of size 4096×4096 in the Ising model with $J = 1$ at equilibrium at various temperatures T. Black and white dots denote spins $s_i = 1$ and $s_i = -1$, respectively. Shown from left to right are images of the ferromagnetic phase at temperatures $T = 1.9$ and 2.1, followed by the paramagnetic phase at $T = 2.3$ and 2.4, all in units of J/k_B

$$M = \sum_i s_i \,.$$

At different temperatures the system may reside in ferromagnetic or paramagnetic phase. The temperature T_c of the phase transition between these phases in the absence of external field ($H = 0$) is given by the equation $\sinh(2J/(k_B T_c)) = 1$, which has the solution $T_c \approx 2.269185 \, J/k_B$.

⊙ At the book's website you can find the data for the thermodynamic equilibrium state of spins in the two-dimensional Ising model at various temperatures. A few examples are shown in Fig. 7.30. Compute the auto-correlation function of the orientation of spins on a $N \times N$ mesh,

$$R_{ss}(i, j) = \frac{1}{N^2} \sum_{k,l} s_{k+i,l+j} s_{k,l} \,,$$

where we assume periodic boundary conditions. Average the auto-correlation over the neighborhood of the points with approximately the same radius r. Assume that the auto-correlation has the form

$$|R_{ss}(i, j)| \sim C\, e^{-r/L} \,, \qquad r = \sqrt{i^2 + j^2} \,,$$

and estimate the correlation length L. Show L as a function of temperature: you should observe a strong increase in L near the critical temperature T_c.

7.11.4 Reconstruction of Vehicle Position and Velocity

A typical application of (linear) optimal filtering described in Sect. 7.8 is the reconstruction of the vehicle trajectory and velocity based on (noisy) GPS location data, information on instantaneous velocity from the vehicle and acceleration data provided by an accelerometer.

$\odot$ The file `kalman_cartesian_data.dat` (see book's website) contains the times $t_n = n\Delta t$ and the noisy GPS data on vehicle position, velocity and acceleration, i.e. $(t_n, x_n, y_n, v_{x,n}, v_{y,n}, a_{x,n}, a_{y,n})$ in SI units, where the sampling interval is $\Delta t = 1.783$ s. The state vector is $\boldsymbol{x}_n = (x, y, v_x, v_y)_n^{\mathrm{T}}$ while the control input is determined by the accelerations, $\boldsymbol{u}_n = (0, 0, a_x\Delta t, a_y\Delta t)_n^{\mathrm{T}}$. Based on simple kinematic considerations the transition matrices are constant,

$$F_n = \begin{pmatrix} I_{2\times2} & I_{2\times2}\Delta t \\ 0_{2\times2} & I_{2\times2} \end{pmatrix}, \quad C_n = H_n = I_{4\times4},$$

while the covariance matrices may be taken as $Q_n = \mathrm{diag}\big(0, 0, \sigma_a^2(\Delta t)^2, \sigma_a^2(\Delta t)^2\big)$ and $R_n = \mathrm{diag}\big(\sigma_{xy}^2, \sigma_{xy}^2, \sigma_{v,n}^2, \sigma_{v,n}^2\big)$, respectively, with $\sigma_a = 0.05\,\mathrm{m/s^2}$, $\sigma_{xy} = 25\,\mathrm{m}$ and $\sigma_{v,n} = \min(1\,\mathrm{km/h}, 0.01\|\boldsymbol{v}_n\|)$. Compute the filtered trajectory and the time dependence of the velocity components! What happens if only every fifth velocity readout and only every tenth position readout are available? How well can the trajectory (and the velocities) be determined if no velocity readouts are available at all? What if only the velocities and the accelerations are known? The reference (exact) data are in `kalman_cartesian_control.dat`.

$\oplus$ In truth the accelerometer is only able to provide the tangential and radial accelerations with respect to the current orientation of the vehicle, $\boldsymbol{a} = (a_{\mathrm{t}}, a_{\mathrm{r}})^{\mathrm{T}}$, which implies a non-constant transition matrix C_n incorporating the orthogonal transformation defined by the current velocity:

$$C_n = \begin{pmatrix} I_{2\times2} & 0_{2\times2} \\ 0_{2\times2} & C_n^{vv} \end{pmatrix}, \quad C_n^{vv} = \frac{1}{\|\boldsymbol{v}_n\|} \begin{pmatrix} v_x & -v_y \\ v_y & v_x \end{pmatrix}_n.$$

This makes the covariance of the acceleration measurement slightly more complicated, since one must account for the uncertainty of the current velocity. Therefore

$$Q_n^{vv} = (\Delta t)^2 \left[\sigma_a^2 I_{2\times2} + \frac{(\boldsymbol{v}_n^\perp)^{\mathrm{T}} P_n^{vv} \boldsymbol{v}_n^\perp}{\|\boldsymbol{v}_n\|^4} \big((C_n^{vv}\boldsymbol{a}_n^\perp)(C_n^{vv}\boldsymbol{a}_n^\perp)^{\mathrm{T}}\big) \right], \tag{7.60}$$

where Q_n^{vv} and P_n^{vv} are the velocity-velocity blocks of the corresponding 4×4 covariance matrices. The vectors $\boldsymbol{v}_n^\perp$ and $\boldsymbol{a}_n^\perp$ are the current velocity and acceleration, respectively, rotated by $\pi/2$ in the positive sense. The second term in (7.60) takes into account that the uncertainty of the velocity perpendicular to the direction of motion implies an uncertainty in the angle, diminishing the reliability of the acceleration data. Reconstruct the vehicle trajectory from the $(t_n, x_n, y_n, a_{\mathrm{t},n}, a_{\mathrm{r},n})$ data contained in `kalman_relative_data.dat`!

7.11.5 Independent Component Analysis

Independent component analysis (ICA) is a method that can be applied to a set of measured signals $x_j(t)$ in order to determine independent components (sources) $s_j(t)$ that caused these signals, as well as the nature of the mixing of sources. We assume that the measured signals and their sources are linearly related,

$$x(t) = As(t),$$

where A is the mixing matrix (see Sect. 7.9). At each instant t the sources s (for example, speakers in a room) and the measured signals x (for example, voltages on the microphones) have several components.

$\odot$ In the first part of the Problem we assume that the mixing matrix is known. Suppose that we have four sources:

$$
\begin{aligned}
s_{1i} &= \sin(10\,\pi i/n), \\
s_{2i} &= \exp[-10(i - n/2)^2/n^2] + 0.2[\mathcal{R}(0,1) - 0.5], \\
s_{3i} &= \sin^3(40\,\pi i/n)\exp(-1.5i/n), \\
s_{4i} &= \sin(100\,i^2/n^2),
\end{aligned}
\tag{7.61}
$$

shown in Fig. 7.31 (left). The index i measures the time, $t = i\,\Delta t$ ($i = 0, 1, \ldots, n - 1$) and $n = 1000$. We use $\mathcal{R}(0,1)$ to denote a uniformly distributed random number from the interval $[0, 1]$. The matrix A mixes the sources into the measured signals,

$$
\begin{pmatrix} x_1(t) \\ x_2(t) \\ x_3(t) \\ x_4(t) \end{pmatrix}
= \frac{1}{5}
\underbrace{
\begin{pmatrix}
-1 & 2 & -1 & 1 \\
2 & 1 & 3 & -2 \\
-2 & 2 & -1 & 1 \\
1 & -2 & 3 & -3
\end{pmatrix}}_{A}
\begin{pmatrix} s_1(t) \\ s_2(t) \\ s_3(t) \\ s_4(t) \end{pmatrix},
\tag{7.62}
$$

which are shown in Fig. 7.31 (right). Use the FastICA algorithm described in Sect. 7.9.1 to determine the independent components (sources) s from the signals x. Compare the computed sources to the original ones (7.61).

In the case described here the number of sources is equal to the number of measured signals, so the model $x = As$ is exactly invertible, $s = Wx = A^{-1}x$. The computed and exact sources can therefore be accurately compared. Discuss the case with fewer sources than signals (invent your own mixing matrix).

$\oplus$ Perform the independent component analysis of eight electro-cardiogram (ECG) traces of a pregnant woman shown in Fig. 7.20 [77]. There are 2500 voltage readouts with a sampling frequency of 500 Hz. The final result of the analysis are the eight independent components which should closely resemble those of Fig. 7.32. Observe the convergence of the algorithm by monitoring the direction of the vectors w_k and the values of their components during the iteration (use Fig. 7.21 as an

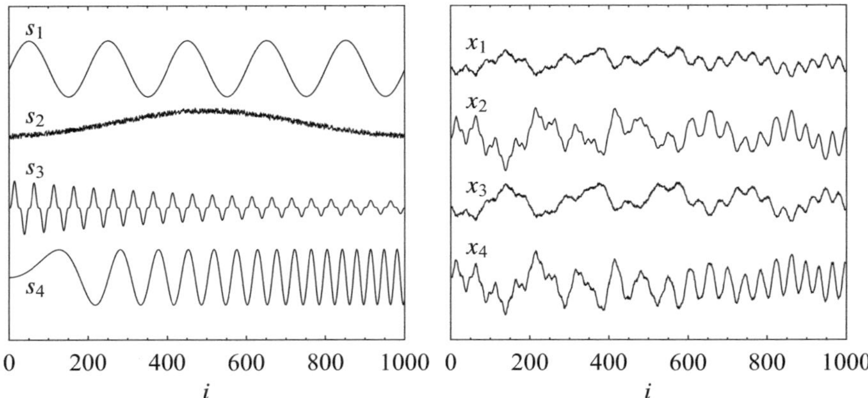

Fig. 7.31 [LEFT] The original signals (sources) from Eq. (7.61). [RIGHT] The measured signals which are known mixtures of the sources (Eq. (7.62))

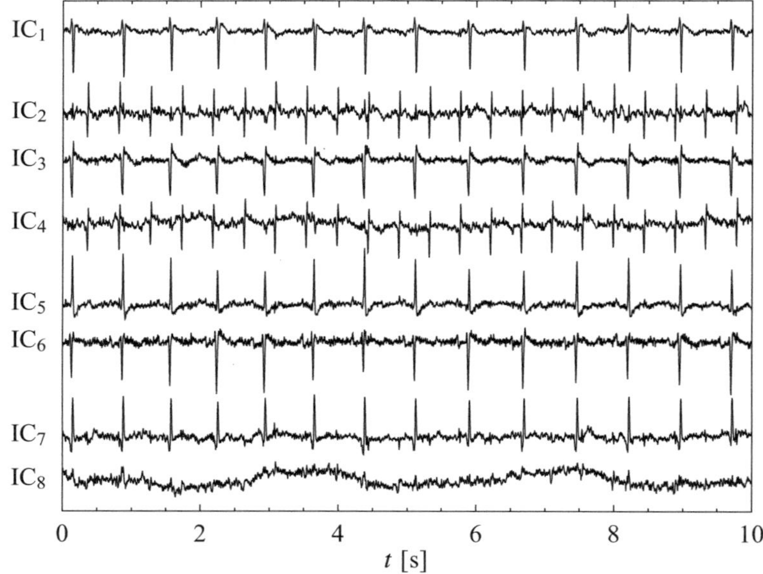

Fig. 7.32 Independent components (sources) of the signals shown in Fig. 7.20. The child's fast heartbeat is clearly identifiable in the independent components IC_2 and IC_4. Other components display the mother's heartbeat, except IC_8 which has probably been generated by breathing during the measurement

illustration). Use the stabilized version of the FastICA algorithm as described on page 452. How do your conclusions change if raw data are first processed by the principal component analysis (PCA) and only the scores of a few leading principal components are retained for the analysis with the FastICA algorithm?

References

1. C. Robinson, *Dynamical Systems*, 2nd edn. (CRC Press, Boca Raton, 1999)
2. S.H. Strogatz, *Nonlinear Dynamics and Chaos: With Applications to Physics, Biology, Chemistry, and Engineering* (Perseus Books, Reading, 1994)
3. P. Billingsley, *Probability and Measure*, 3rd edn. (Wiley-Interscience, New York, 1995)
4. R.M. Gray, *Probability, Random Processes and Ergodic Properties*, 2nd edn. (Springer-Verlag, New York, 2009)
5. W. Feller, *An Introduction to Probability Theory and Its Applications*, vol. 1, 3rd edn (John Wiley & Sons, New York, 1967)
6. W. Feller, *An Introduction to Probability Theory and Its Applications*, vol. 2, 2nd edn (John Wiley & Sons, New York, 1971)
7. V.I. Arnold, A. Avez, *Ergodic Problems of Classical Mechanics* (W. A. Benjamin, New York, 1968)
8. L. Schiff, *Quantum Mechanics* (McGraw-Hill, New York, 1968)
9. P.Z. Peebles, *Probability, Random Variables and Random Signal Principles* (McGraw-Hill, New York, 1980)
10. J.D. Hamilton, *Time Series Analysis* (Princeton University Press, Princeton, 1994)
11. I.S. Tyurin, On the accuracy of the Gaussian approximation. Doklady Math. **80**, 840 (2009)
12. J.P. Nolan, *Stable Distributions-Models for Heavy Tailed Data* (Birkhäuser, Boston, 2010)
13. See Chapter 1 in P. Čížek, W. Härdle, R. Weron, *Statistical Tools for Finance and Insurance* (Springer-Verlag, Berlin, 2005)
14. A. Janicki, A. Weron, Can one see α-stable variables and processes? Stat. Sci. **9**, 109 (1994)
15. GSL (GNU Scientific Library). http://www.gnu.org/software/gsl
16. R. LePage, M. Woodroofe, J. Zinn, Convergence to a stable distribution via order statistics. Ann. Prob. **9**, 624 (1981)
17. B.D. Hughes, *Random Walks and Random Environments: Vol. 1: Random Walks* (Oxford University Press, New York, 1995)
18. M. Bazant, Random walks and diffusion, MIT OpenCourseWare. Course 18.366. Available at: http://ocw.mit.edu
19. E.W. Montroll, G.H. Weiss, Random walks on lattices. II. J. Math. Phys. **6**, 167 (1965)
20. R. Metzler, J. Klafter, The random walk's guide to anomalous diffusion: a fractional dynamics approach. Phys. Reports **339**, 1 (2000)
21. M.F. Shlesinger, B.J. West, J. Klafter, Lévy dynamics of enhanced diffusion: application to turbulence. Phys. Rev. Lett. **58**, 1100 (1987)
22. V. Tejedor, R. Metzler, Anomalous diffusion in correlated continuous time random walks. J. Phys. A Math. Theor. **43**, 082002 (2010)
23. S. Meyn, R.L. Tweedie, *Markov Chains and Stochastic Stability* (Springer-Verlag, Berlin, 1995)
24. E. Parzer, *Stochastic Processes* (Holden-Day, San Francisco, 1962)
25. E. Ott, *Chaos in Dynamical Systems*, 2nd edn. (Cambridge University Press, Cambridge, 2002)
26. A. Papoulis, *Probability, Random Variables and Stochastic Processes*, 3rd edn. (McGraw-Hill, New York, 1991)
27. C. Meyer, *Matrix Analysis and Applied Linear Algebra* (SIAM, Philadelphia, 2000)
28. T. Cover, J. Thomas, *Elements of Information Theory*, 2nd edn (John Wiley & Sons, New York, 2006)
29. E.T. Jaynes, Information theory and statistical mechanics. Phys. Rev. **106**, 620 (1957); E.T. Jaynes, Information theory and statistical mechanics II. Phys. Rev. **108**, 171 (1957)
30. M. Horvat, The ensemble of random Markov matrices. J. Stat. Mech. **2009**, P07005 (2009). ((and references therein))
31. F. Schwabl, *Advanced Quantum Mechanics*, 4th edn. (Springer-Verlag, Berlin, 2008)
32. A.V. Oppenheim, R.W. Schafer, *Digital Signal Processing* (Prentice-Hall, Englewood Cliffs, 1975)

33. C.W. Gardiner, *Handbook of Stochastic Methods*, 2nd edn. (Springer-Verlag, Berlin, 1997)
34. R. Whittle, *DSP Generation of Pink (1/f) Noise* (2010). Accessible at the website. http://www.firstpr.com.au/dsp/pink-noise
35. L. Trammell, unpublished papers *Pink Noise Generator* and *Improvements in the Correlated Pink Noise Generator Evaluation*. Accessible at the book's web page
36. P. Bourke, *Generating Noise with Different Power Spectra Laws* (1998). Accessible at: http://paulbourke.net/fractals/noise
37. R. Zwanzig, Time-correlation functions and transport coefficients in statistical mechanics. Ann. Rev. Phys. Chem. **16**, 67 (1965)
38. M. Reed, B. Simon, *Methods of Modern Mathematical Physics, Vol. 1: Functional Analysis*, 2nd edn (Academic Press, San Diego, 1980)
39. J.B. Conway, *A Course in Functional Analysis* (Springer-Verlag, New York, 1985)
40. W. Rudin, *Functional Analysis* (McGraw-Hill, New York, 1973)
41. S.L. Marple Jr., *Digital Spectral Analysis* (Prentice-Hall, Englewood Cliffs, 1987)
42. T.H. Cormen, C.E. Leiserson, R.L. Rivest, C. Stein, *Introduction to Algorithms*, 2nd edn. (MIT Press, Cambridge, 2001)
43. J.M. Makhoul, Linear prediction: a tutorial review. Proc. IEEE **63**, 561 (1975)
44. S.V. Vaseghi, *Advanced Digital Signal Processing and Noise Reduction*, 4th edn. (John Wiley & Sons, New York, 2008)
45. P. Vaidyanathan, *The Theory of Linear Prediction. Synthesis Lectures on Signal Processing* (Morgan & Claypool Publishers, San Rafael, 2008)
46. P.J. Brockwell, R.A. Davis, *Introduction to Time Series and Forecasting* (Springer-Verlag, New York, 2002)
47. G.E.P. Box, G.M. Jenkins, G. Reinsel, *Time Series Analysis: Forecasting & Control*, 4th edn. (John Wiley & Sons, New York, 2008)
48. G.E.P. Box, D.R. Cox, An analysis of transformations. J. Roy. Statist. Soc. B **26**, 211 (1964)
49. G.H. Golub, C.F. van Loan, *Matrix Computations*, 3rd edn. (Johns Hopkins University Press, Baltimore, 1996)
50. M.J.L. de Hoon, T.H.J.J. van der Hagen, H. Schoonewelle, H. van Dam, Why Yule-Walker should not be used for autoregressive modelling. Ann. Nucl. Energy **23**, 1219 (1996)
51. S.M. Kay, S.L. Marple, Spectrum analysis–a modern perspective. Proc. IEEE **69**, 1380 (1981)
52. W.H. Press, B.P. Flannery, S.A. Teukolsky, W.T. Vetterling, *Numerical recipes: The Art of Scientific Computing*, 3rd edn (Cambridge University Press, Cambridge, 2007). See also the equivalent handbooks in Fortran, Pascal and C, as well as http://numerical.recipes
53. MATLAB, The MathWorks. http://www.mathworks.com
54. P. Stoica, R. Moses, *Spectral Analysis of Signals* (Prentice-Hall, Englewood Cliffs, 2005)
55. S. Haykin, *Adaptive Filter Theory*, 4th edn. (Prentice-Hall, Englewood Cliffs, 2001)
56. B.D.O. Anderson, J.B. Moore, *Optimal Filtering* (Prentice-Hall, Englewood Cliffs, 1979)
57. R.E. Kalman, A new approach to linear filtering and prediction problems. J. Basic Eng. (Series D) **82**, 35 (1960)
58. J. Humpherys, P. Redd, J. West, A fresh look at the Kalman filter. SIAM Rev. **54**, 801 (2012)
59. M.S. Grewal, A.P. Andrews, *Kalman Filtering: Theory and Practice Using Matlab*, 3rd edn. (John Wiley & Sons, Hoboken, 2008)
60. C.K. Chui, G. Chen, *Kalman Filtering with Real-Time Applications*, 5th edn. (Springer-Verlag, Berlin, 2017)
61. W. Ma, J. Qiu, J. Liang, B. Chen, Linear Kalman filtering algorithm with noisy control input variable. IEEE Trans. Circ. Syst. II Express Briefs **66**, 1282 (2019)
62. T. Karvonen, *Stability of linear and non-linear Kalman filters* (University of Helsinki, Helsinki, 2014)
63. P. Ruckdeschel, B. Spangl, D. Pupashenko, Robust Kalman tracking and smoothing with propagating and non-propagating outliers. Stat. Papers **55**, 93 (2014)
64. G. Agamennoni, J.I. Nieto, E.M. Nebot, An outlier-robust Kalman filter, in *2011 IEEE International Conference on Robotics and Automation* (Shanghai, China, 2011), p. 1551, https://doi.org/10.1109/ICRA.2011.5979605

65. I.R. Petersen, A.V. Savkin, *Robust Kalman Filtering For Signals and Systems with Large Uncertainties* (Birkhäuser, Boston, 1999)
66. H.E. Rauch, F. Tung, C.T. Striebel, Maximum likelihood estimate of linear dynamic systems. AIAA Journal **3**, 1445 (1965)
67. T. Lefebvre, H. Bruyninckx, J. de Schutter, Kalman filters for non-linear systems: a comparison of performance. Int. J. Control **77**, 639 (2004)
68. S. Julier, J. Uhlmann, H.F. Durant-Whyte, A new method for the nonlinear transformation of means and covariances in filters and estimators. IEEE Trans. Automat. Contr. **45**, 477 (2000)
69. S. Julier, J. Uhlmann, Unscented filtering and nonlinear estimation. Proc. IEEE **92**, 401 (2004)
70. E.A. Wan, R. van der Merwe, The unscented Kalman filter, in *Kalman Filtering and Neural Networks*, ed. by K. Haykin (John Wiley & Sons, New York, 2001), pp. 221–280
71. F. Gustafsson, G. Hendeby, Some relations between extended and unscented Kalman filters. IEEE Trans. Signal Process. **60**, 545 (2012)
72. E.A. Wan, A.T. Nelson, Dual extended Kalman filter methods, in *Kalman Filtering and Neural Networks*. ed. by S. Haykin (John Wiley & Sons, New York, 2001), pp.123–173
73. E.A. Wan, R. van der Merwe, A.T. Nelson, Dual estimation and the unscented transformation. Adv. Neural Inf. Process. Syst. **12**, 666 (1999)
74. G.L. Plett, Extended Kalman filtering for battery management systems of LiPB-based HEV battery packs. Part 3: state and parameter estimation. J. Power Sour. **134**, 277 (2004)
75. F. Hamilton, T. Berry, T. Sauer, Ensemble Kalman filtering without a model. Phys. Rev. X **6**, 011021 (2016)
76. A.J. Izenman, *Modern Multivariate Statistical Techniques* (Springer-Verlag, Berlin, 2008)
77. L. De Lathauwer, B. De Moor, J. Vandewalle, Fetal electrocardiogram extraction by blind source subspace separation. IEEE Trans. Biomed. Eng. **47**, 567 (2000). The signals are available at: http://www.esat.kuleuven.ac.be/~smc/daisy/daisydata.html
78. A. Hyvärinen, E. Oja, Independent component analysis: algorithms and applications. Neural Netw. **13**, 411 (2000)
79. A. Hyvärinen, Survey on independent component analysis. Neural Comput. Surv. **2**, 94 (1999)
80. A. Hyvärinen, Fast and robust fixed-point algorithms for independent component analysis. IEEE Trans. Neural Netw. **10**, 626 (1999)
81. B.R. Hunt, J.A. Kennedy, T.-Y. Li, H.E. Nusse (eds.), *The Theory of Chaotic Attractors* (Springer Science & Business Media, New York, 2004)
82. O.E. Rössler, An equation for continuous chaos. Phys. Lett. A **57**, 397 (1976)
83. F. Takens, Detecting strange attractors in turbulence, in *Dynamical systems and turbulence*, ed. by D.A. Rand, L.-S. Young. Lecture Notes in Mathematics, vol. 898 (Springer-Verlag, Berlin, 1981), pp.366–381
84. R. Mañé, On the dimension of the compact invariant sets of certain nonlinear maps, in: D.A. Rand, L.-S. Young, *Dynamical Systems and Turbulence*, Lecture Notes in Mathematics, vol. 898 (Springer-Verlag, Berlin, 1981), pp. 230–242
85. T. Sauer, J.A. Yorke, M. Casdagli, Embedology. J. Stat. Phys. **65**, 579 (1991)
86. E.N. Lorenz, Deterministic nonperiodic flow. J. Atmos. Sci. **20**, 130 (1963)
87. E. Ott, T. Sauer, J.A. York, *Coping with Chaos: Analysis of Chaotic Data and Exploitation of Chaotic Systems* (Wiley-Interscience, New York, 1994)
88. H. Kantz, T. Schreiber, *Nonlinear Time Series Analysis*, 2nd edn. (Cambridge University Press, Cambridge, 2003)
89. H.D.I. Abarbanel, *Analysis of Observed Chaotic Data* (Springer-Verlag, New York, 1996)
90. L.M. Pecora, L. Moniz, J. Nichols, T.L. Carroll, A unified approach to attractor reconstruction. Chaos **17**, 013110 (2007)
91. E. Bradley, H. Kantz, Nonlinear time-series analysis revisited. Chaos **25**, 097610 (2015)
92. C. Letellier, O.E. Rössler, Rössler attractor. Scholarpedia **1**(10), 1721 (2006)
93. M. Peifer, B. Schelter, M. Winterhalder, J. Timmer, Mixing properties of the Rössler system and consequences for coherence and synchronization analysis. Phys. Rev. E **72**, 026213 (2005)
94. A.M. Fraser, H.L. Swinney, Independent coordinates for strange attractors from mutual information. Phys. Rev. A **33**, 1134 (1986)

95. M.B. Kennel, R. Brown, H.D.I. Abarbanel, Determining embedding dimension for phase-space reconstruction using a geometrical construction. Phys. Rev. A **45**, 3403 (1992)
96. P. Grassberger, I. Procaccia, Characterization of strange attractors. Phys. Rev. Lett. **50**, 346 (1983)
97. J. Theiler, Spurious dimension from correlation algorithms applied to limited time-series data. Phys. Rev. A **34**, 2427 (1986)
98. J. Theiler, Estimating fractal dimension. J. Opt. Soc. Am. A **7**, 1055 (1990)
99. H.-O. Peitgen, H. Jürgens, D. Saupe, *Fractals for the Classroom, Part Two: Complex Systems and Mandelbrot Set* (Springer-Verlag, New York, 1992)
100. M. J. Feigenbaum, Quantitative universality for a class of non-linear transformations. J. Stat. Phys. **19**, 25 (1978); M.J. Feigenbaum. The universal metric properties of nonlinear transformations. J. Stat. Phys. **21**, 669 (1979)
101. J. Lichtenberg, M.A. Lieberman, *Regular and Stochastic Motion* (Springer-Verlag, New York, 1983)

Chapter 8
Initial-Value Problems for ODE

Abstract The ability to reliably solve initial-value problems for ordinary differential equations is essential in order to understand the evolution of dynamical systems. In this Chapter we deal with methods of advancing the given initial state of a system to later times, explaining clearly the role of stiffness, local discretization and round-off errors, and introducing the crucial concepts of consistency, convergence, and stability of difference schemes by which the differential equations are approximated. Single-step explicit and implicit methods are treated at length; automatic step-size control, embedding, and dense output are presented. Explicit and implicit multi-step (Adams, predictor-corrector, and backward-differentiation) methods are analyzed from the viewpoint of stability and long-term integration. Runge-Kutta-Nyström methods are introduced as a special means to solve conservative second-order equations, and basic concepts of solving stiff ODE are introduced. Geometric and Lie series integrators are discussed as most viable options for high-precision integration requiring the preservation of certain invariants. The Examples and Problems include the Lorentz system, synchronization of globally coupled oscillators, chaotic scattering in two dimensions, problems in chemical kinetics, and a variety of astrophysical problems involving planets, stars and galaxies.

Often we try to understand the evolution of a system from a known initial to an unknown later state by observing the temporal variation (dynamics) of quantities that we use to describe the system. There are endless examples. Physicists study oscillations of non-linear mechanical pendula and electric circuits, monitor the motion of charged particles in electro-magnetic fields, and predict orbits of satellites in the presence of other celestial bodies. Chemists investigate properties and reaction mechanisms of compounds. Biologists use models to describe the population dynamics of plant or animal species. Modern economics, among other, uses dynamical models to predict stock market exchange rates.

© The Author(s), under exclusive license to Springer Nature Switzerland AG 2025   471
S. Širca and M. Horvat, *Computational Methods in Physics*, Graduate Texts in Physics,
https://doi.org/10.1007/978-3-031-68566-8_8

8.1 Evolution Equations

In simpler cases the dynamics of the system is determined by ordinary (not necessarily linear) differential equations (ODE) with time derivatives and prescribed initial conditions. (Problems with specified boundary conditions, which are harder than initial-value problems, are discussed in Chap. 9. Partial differential equations that are even more difficult to solve due to the appearance of derivatives in variables other than time, are treated in Chaps. 10, 11, and 12.) Though the dynamical (evolution) equation for a physical problem is known, it is often not analytically solvable and we have to resort to numerical methods. We first discuss first-order ordinary differential equations of the form

$$y' = f(x, y), \qquad y(x_0) = y_0, \tag{8.1}$$

where f is a known function. In the language of dynamical analysis this equation defines a *non-autonomous system*. If the system does not depend on x,

$$y' = f(y),$$

it is called *autonomous*. The variable x most often means time.

Converting an Mth order equation to M first-order equations Problems with ordinary differential equations of higher order M can be rewritten as systems of M first-order differential equations by using auxiliary variables. We write the system as in Eq. (8.1), except that the solution y and the right-hand side of the equation f have more components and we treat them as vectors:

$$\mathbf{y}'(x) = \begin{pmatrix} y_1'(x) \\ y_2'(x) \\ \vdots \\ y_M'(x) \end{pmatrix} = \begin{pmatrix} f_1(x, y_1, y_2, \ldots, y_M) \\ f_2(x, y_1, y_2, \ldots, y_M) \\ \vdots \\ f_M(x, y_1, y_2, \ldots, y_M) \end{pmatrix} = \mathbf{f}(x, \mathbf{y}), \tag{8.2}$$

where $\mathbf{y} \in \mathbb{R}^M$ and $\{f_i\}_{i=1}^M$ are scalar functions, the components of $\mathbf{f} : \mathbb{R}^{M+1} \to \mathbb{R}^M$. We choose the new variables such that each intermediate variable is the derivative of the previous one. For a second-order equation $m\ddot{\mathbf{x}} = \mathbf{F}$ (Newton's law) this intermediate variable is the velocity $\mathbf{v}$, and the equation can be written as a system of two vector (six scalar) first-order equations

$$\begin{aligned} \dot{\mathbf{x}} &= \mathbf{v} \\ m\dot{\mathbf{v}} &= \mathbf{F}(t, \mathbf{x}, \mathbf{v}) \end{aligned} \quad \Longleftrightarrow \quad \begin{pmatrix} y_1' \\ y_2' \\ y_3' \\ y_4' \\ y_5' \\ y_6' \end{pmatrix} = \begin{pmatrix} y_4 \\ y_5 \\ y_6 \\ F_1(x, y_1, y_2, y_3, y_4, y_5, y_6)/m \\ F_2(x, y_1, y_2, y_3, y_4, y_5, y_6)/m \\ F_3(x, y_1, y_2, y_3, y_4, y_5, y_6)/m \end{pmatrix}.$$

If sufficiently many derivatives of the functions f_i with respect to x and y_j exist, f can be Taylor-expanded around any point $(x_*, \boldsymbol{y}_*)$:

$$f(x, y) = f(x_*, \boldsymbol{y}_*) + \left.\frac{\partial f}{\partial x}\right|_* (x - x_*) + \left.\frac{\partial f}{\partial \boldsymbol{y}}\right|_* \cdot (\boldsymbol{y} - \boldsymbol{y}_*) + \cdots .$$

We shall see in the discussion of the stability of numerical methods that the local behavior of the solution strongly depends on the last term which contains the Jacobi matrix of the partial derivatives of the functions f_i with respect to the variables y_j,

$$J = \frac{\partial f}{\partial \boldsymbol{y}} = \begin{pmatrix} \dfrac{\partial f_1}{\partial y_1} & \dfrac{\partial f_1}{\partial y_2} & \cdots & \dfrac{\partial f_1}{\partial y_M} \\[2ex] \dfrac{\partial f_2}{\partial y_1} & \dfrac{\partial f_2}{\partial y_2} & \cdots & \dfrac{\partial f_2}{\partial y_M} \\[1ex] \vdots & \vdots & & \vdots \\[1ex] \dfrac{\partial f_M}{\partial y_1} & \dfrac{\partial f_M}{\partial y_2} & \cdots & \dfrac{\partial f_M}{\partial y_M} \end{pmatrix} . \tag{8.3}$$

The stability analysis (explained in Sect. 8.5 and Appendix F) is closely related to the solution of the system of differential equations $\boldsymbol{y}' = J\boldsymbol{y}$ and the search for (in general complex) eigenvalues of the matrix J.

When does (8.1) have a unique solution? Suppose we are seeking the solution in the closed region $\mathcal{D} \equiv \{(x, y) : x_0 - \alpha \le x \le x_0 + \alpha,\ y_0 - \beta \le y \le y_0 + \beta\}$. If f is continuous in $\mathcal{D}$, bounded by $|f(x, y)| < B$, and if for each x, y_1, and y_2 from $\mathcal{D}$ we can find a constant Λ such that the Lipschitz inequality

$$|f(x, y_2) - f(x, y_1)| \le \Lambda\, |y_2 - y_1| \tag{8.4}$$

is fulfilled, then on the interval centered at x_0 with width $\Delta x = \min\{\alpha, \beta/B, \gamma/\Lambda\}$, where $0 < \gamma < 1$, precisely one solution of (8.1) exists [1]. The parameter Λ in (8.4) plays an important role in the error estimates of numerical methods for differential equations. The generalization of the existence theorem to systems of equations can be found in [1].

Road-signs To solve equations of the type (8.1), we use numerical methods in which the interval $[a, b]$ on the continuous x-axis is discretized by a mesh

$$x_0 = a < x_1 < x_2 < \cdots < x_N = b$$

of generally non-equidistant points, while the differential equation is transcribed as a difference scheme. The following Sections are devoted to the analysis of precision, stability, convergence properties, and computational efficiency of individual difference schemes. We use $\boldsymbol{y}_n$ to denote the approximate values of the solution at the points x_n and $\boldsymbol{y}(x_n)$ to denote the exact values.

Initial-value problems for ordinary differential equations can be solved by *single-step methods* in which we need just the previous values y_n in order to obtain the next ones, y_{n+1}, as well as *multi-step methods*, in which y_{n+1} is computed from several previous values y_n, y_{n-1}, ...

The methods also differ by how an individual time step is executed: in *explicit methods* the old values are used to compute the new ones by using a known explicit formula; in *implicit methods* (Sect. 8.8) the values in the next time step occur in two or more places at both sides of the equation and need to be computed at each step by solving a system of non-linear (and not necessarily algebraic) equations. In all methods we obtain discrete approximations for the values $y(x_n)$ determined by the equation that specifies the derivatives y': these methods are therefore commonly known as *integrators*.

8.2   Explicit Euler's Methods

Euler's methods are the simplest representatives of difference schemes in which the approximate solution of the differential equation at the point x_n is used to compute the approximate solution at the next point, x_{n+1}. We restrict the discussion to the equidistant mesh,

$$h = (b - a)/N = x_{n+1} - x_n \,, \qquad n = 0, 1, \ldots, N - 1 \,.$$

Explicit (basic) Euler's method is just a reshuffled difference approximation of the first derivative: $y' \approx (y(x + h) - y(x))/h$. The approximate value of the solution y_{n+1} at the next mesh point is obtained from the previous one, y_n, by using

$$y_{n+1} = y_n + h\, y_n' = y_n + h\, f(x_n, y_n) \,. \tag{8.5}$$

In other words, at each point x_n the curve of the exact solution is approximated by the tangent at that point, with the slope determined by the differential equation. We advance along this tangent to the next approximate solution value.

Local discretization error and the method order In one step from x_n to x_{n+1} we make a *discretization* or *truncation error*, which is defined as the difference between the numerical and exact solution in one step, assuming that these two solutions agree in all previous steps. (More on errors follows in Sect. 8.4.) By comparing the expression (8.5) for y_{n+1} with the Taylor expansion

$$y(x_n + h) = y(x_n) + h\, y'(x_n) + \tfrac{1}{2} h^2 y''(\xi_n) \,, \qquad \xi_n \in [x_n, x_{n+1}] \,,$$

we see that the local error of the Euler's method is $y(x_{n+1}) - y_{n+1} \sim O(h^2)$. A difference scheme with a local error of the order h^{p+1} is said to be a *method of order p*: the basic Euler's method is first order in h.

In the *improved Euler's method* the curve of the true solution is approximated by the tangent at the midpoint of the current interval $[x_n, x_{n+1}]$ and not at the initial point,

$$
\begin{aligned}
y_{n+\frac{1}{2}} &= y_n + \tfrac{1}{2} h\, y_n' , \\
y_{n+\frac{1}{2}}' &= f\left(x_{n+\frac{1}{2}},\, y_{n+\frac{1}{2}}\right) , \\
y_{n+1} &= y_n + h\, y_{n+\frac{1}{2}}' .
\end{aligned}
\tag{8.6}
$$

The local error of this method is $O\left(h^3\right)$ (i.e. the method is second order) if the third derivative of the solution is bounded in the vicinity of x_n, but the method may be unstable in certain cases.

There is also the *symmetrized Euler's method* which follows from the symmetric difference approximation of the derivative, $y' \approx (y(x+h) - y(x-h))/2h$. We use the difference scheme

$$
y_{n+1} = y_{n-1} + 2h y_n' ,
\tag{8.7}
$$

so we have to known *two* previous solution values at each step: we are dealing with a two-step method forcing us to use uniform mesh spacings. The initial condition $y_0 = y(x_0)$ is known, while the approximation of the solution y_1 can be, for example, computed by the basic Euler's method. The symmetrized Euler's method is also second order.

8.3 Explicit Methods of the Runge–Kutta Type

Methods of the Runge–Kutta (RK) type are the best known representatives of single-step methods. The solution is advanced in a sequence of short steps h like in the Euler's methods, but the right-hand side of the differential equation (8.1) is evaluated at particular intermediate points within each time step; by a suitable combination of these auxiliary quantities we wish the method to attain higher precision. The solution value at the next point of a step is computed as

$$
y_{n+1} = y_n + h\, F(h, x_n, y_n; f) ,
\tag{8.8}
$$

where

$$
F(h, x_n, y_n; f) = \sum_{i=1}^{m} b_i k_i .
$$

We express the quantities k_i by the right sides of (8.2),

$$k_i = f\left(x_n + c_i h,\ y_n + h\sum_{j=1}^{m} a_{ij}k_j\right), \qquad i = 1, 2, \ldots, m. \tag{8.9}$$

The procedure (8.8) is known as an *m-stage* Runge–Kutta (RK) method. We try to determine the parameters a_{ij}, b_i, and c_i such that the order of the method p is as high as possible. We Taylor-expand the numerical solution y_1 and the exact solution $y(x+h)$ around x and compare the coefficients in both expansions [2]. This results in a system of non-linear equations for the parameters that in general does not have a unique solution. Each solution corresponds to a different RK method. If we set $c_1 = 0$ and $a_{ij} = 0$ for $j \geq i$ (the sum in Eq. (8.9) then runs only over $1 \leq j \leq i - 1$), at each step $x_n \to x_{n+1}$ all k_i can be computed: the method is *explicit* since in the computation of k_i only the previous values $k_1, k_2, \ldots, k_{i-1}$ are used. In other cases the method is *implicit* and the system (8.9) needs to be solved for k_i at each step: such methods are discussed in Sect. 8.8.

In some cases the system for the parameters is easily solvable and leads to m-stage formulas requiring one evaluation of $f(x, y)$ per stage. The only solution for the coefficients in the case $p = m = 1$ is $b_1 = 1, b_i = 0\,(i > 1), c_i = 0, a_{ij} = 0$, and gives the basic Euler's method (8.5). One of the solutions for $p = m = 4$ corresponds to the popular fourth-order Runge–Kutta (RK4) method:

$$\begin{aligned}
k_1 &= f\left(x_n,\ y_n\right), \\
k_2 &= f\left(x_n + \tfrac{1}{2}h,\ y_n + \tfrac{h}{2}k_1\right), \\
k_3 &= f\left(x_n + \tfrac{1}{2}h,\ y_n + \tfrac{h}{2}k_2\right), \\
k_4 &= f\left(x_n + h,\ y_n + hk_3\right), \\
y_{n+1} &= y_n + \frac{h}{6}\left(k_1 + 2k_2 + 2k_3 + k_4\right).
\end{aligned} \tag{8.10}$$

We arrange the coefficients a_{ij}, b_i, and c_i of any RK method in a *Butcher tableau*. A general explicit scheme is shown at the left, the RK4 method at the right:

$$
\begin{array}{c|ccccc}
0 & & & & & \\
c_2 & a_{21} & & & & \\
c_3 & a_{31} & a_{32} & & & \\
\vdots & \vdots & \vdots & \ddots & & \\
c_m & a_{m1} & a_{m2} & \cdots & a_{m\,m-1} & \\
\hline
 & b_1 & b_2 & \cdots & b_{m-1} & b_m
\end{array}
\qquad\qquad
\begin{array}{c|cccc}
0 & & & & \\
1/2 & 1/2 & & & \\
1/2 & 0 & 1/2 & & \\
1 & 0 & 0 & 1 & \\
\hline
 & 1/6 & 2/6 & 2/6 & 1/6
\end{array}
$$

The order p can not be attained with less than p stages. The explicit RK4 method is of highest order at which we still have $p = m$, and this is one of the reasons for its popularity. For RK methods of orders $p > 4$ we need $m > p$ stages, and each additional stage implies another evaluation of $f(x, y)$, a consideration of prime importance if the functions f are computationally expensive. The world record in local precision is held by the 10th order methods requiring 18 [3] or 17 stages [4].

Dense output for the RK4 method Expressions (8.10) allow us to obtain the solution at the mesh points x_n. But often the point $x^\star$ at which the solution is sought is not known in advance, as it may implicitly depend on the computed solution and may be determined by some equation $g(x^\star, y(x^\star)) = 0$, for example, when seeking the Poincaré maps (i.e. surfaces of section) of the solution. The results at intermediate points $x^\star \in [x_n, x_{n+1}]$ can be obtained by using *dense output* [2] which requires further $(m^\star - m)$ stages to be added to the scheme. The approximate solution at an arbitrary point on the interval $[x_n, x_{n+1}]$ is given by

$$
y_{n+\theta} = y_n + h \sum_{i=1}^{m^\star} b_i(\theta) k_i , \qquad 0 \le \theta \le 1 . \tag{8.11}
$$

The polynomials b_i are determined by requiring $y_{n+\theta} - y(x_n + \theta h) = O(h^{p^\star+1})$. The dense output for the RK4 method can be formulated already with $m^\star = m$ (i.e. without computing additional k_i). We use

$$
b_1(\theta) = \theta - \frac{3\theta^2}{2} + \frac{2\theta^3}{3} , \quad b_2(\theta) = b_3(\theta) = \theta^2 - \frac{2\theta^3}{3} , \quad b_4(\theta) = -\frac{\theta^2}{2} + \frac{2\theta^3}{3} .
$$

A price has to paid for this convenient interpolation: while the basic method has order $p = 4$, the dense-output solution has only order $p^\star = 3$. An alternative (and cleaner) method to compute Poincaré maps is described in Appendix E.

Strong stability preserving Runge–Kutta methods A particular variety of Runge–Kutta methods known as *strong stability preserving* (SSP) has recently been developed to solve large systems of ordinary differential equations of the form (8.2), in particular those arising from the spatial discretization of hyperbolic partial differential equations whose solutions may exhibit discontinuities. SSP methods are introduced in Sect. 12.6.1.

8.4 Errors of Explicit Methods

When an explicit method is implemented in a program, two kinds of errors occur: *discretization* (also known as *truncation*) errors and *round-off* errors. The errors can be *local* (occurring in one step) or *global* (accumulated throughout the integration interval). In the following we also introduce the concepts of *consistency*, *convergence*, and *stability* of explicit methods.

8.4.1   Discretization and Round-Off Errors

The *discretization error* originates in the nature of the method and it appears because
the sum (8.8) or the Taylor expansion of the true solution to which we compare it, is
finite. This error occurs regardless of the arithmetic precision of the implementation.
It depends on the step size h, on the order of the method, and on the function f itself.
The local discretization error τ_{n+1} in the nth step of an explicit method is defined as

$$\tau_{n+1} = \| y(x_{n+1}) - y(x_n) - h\, F(h, x_n, y(x_n); f) \| \,. \tag{8.12}$$

Assuming that the numerical and exact solution match perfectly until the nth step,
we have $y(x_n) = y_n$ and $y(x_n) + h\, F(h, x_n, y(x_n); f) = y_{n+1}$, and then

$$\tau_{n+1} = \| y_{n+1} - y(x_{n+1}) \| \,.$$

In general, for an explicit method of order p it holds that

$$\tau_{n+1} = \Psi(y(x_n)) h^{p+1} + O(h^{p+2}) \,. \tag{8.13}$$

The first term at the right side of Eq. (8.12) is called the *principal local discretization
error* in which the function Ψ depends on the elementary differentials up to order
$p + 1$, evaluated at $y(x_n)$ [5].

Round-off errors occur because instead of the exact solution y_n of the *difference
scheme* we compute a numerically approximate value Y_n, with the difference $\rho_n \equiv
\| Y_n - y_n \|$. These errors arise due to the finite precision of floating-point arithmetic
and depend on the type of arithmetic operations performed and their quantity. The
round-off errors increase with the number of steps, and this is certainly one reason
not to decrease the step size h excessively.

The equation for τ_n and the analogous expression for ρ_n define the contributions to the local discretization and round-off errors at the level of the difference
scheme. But the success of an integrator is preferably judged by the *global error*
$E_N \equiv \| Y_N - y(x_N) \|$ that measures the cumulative difference between the true value
and its computed approximation at the last integration point ($n = N$). The error
that accumulates until some intermediate point x_n, $E_n \equiv \| Y_n - y(x_n) \|$, can also be
understood as global up to that point. The global error *is not* equal to the sum of the
local errors between $x_0 = a$ and $x_N = b$. It results from a propagation of local errors
along the whole integration interval.

Most often, we can determine only an upper limit for the global error. In general,
the order of the global error is one less than the order of the local error. If the local
error satisfies

$$\tau_{n+1} \le C h^{p+1}$$

(see Eq. (8.13)) and F is Lipschitz-continuous, $\| F(h, x, y; f) - F(h, x, z; f) \| \le
\Lambda \| y - z \|$, the global discretization error can be estimated as

$$\|E_N\| \le Ch^p \frac{1}{\Lambda} \left[e^{\Lambda(b-a)} - 1 \right] + e^{\Lambda(b-a)} \|y_0 - y(x_0)\| .$$

(The last term at the right vanishes if the initial condition is numerically exact.) In general the step h along $[a, b]$ can be non-uniform. In that case h in the above equation stands for $h = \max_n\{h_n\}$.

Example Useful global error estimates exist for simpler difference schemes. The local error can be propagated to the global error either by resorting to known or assumed properties of the exact solution, or by exploiting the properties of the difference scheme (see [2, 6], Chap. II). The global errors of the scalar basic and symmetrized Euler's methods with equidistant steps h (Eqs. (8.5) and (8.7)) can be estimated as

$$\text{Eq. (8.5)}: \quad |E_N| \le e^{\Lambda(b-a)} \left[|\rho_0| + \frac{1}{\Lambda}\left(\frac{hM_2}{2} + \frac{\rho}{h} \right) \right], \tag{8.14}$$

$$\text{Eq. (8.7)}: \quad |E_N| \le e^{\Lambda(b-a)} \left[\max\{|\rho_0|, |\rho_1|\} + \frac{1}{\Lambda}\left(\frac{h^2 M_3}{3} + \frac{\rho}{h} \right) \right], \tag{8.15}$$

where ρ_0 and ρ_1 are the round-off errors at $x_0 = a$ and x_1, while the largest round-off error along the whole integration interval is $\rho \equiv \max_{1 \le n \le N} |\rho_n|$. The Lipschitz constant Λ is equal to the upper limit for $|\partial f/\partial y|$. The estimates for the upper limits of $M_2 = \max |y''|$ and $M_3 = \max |y'''|$ (both on $a \le x \le b$) are harder to obtain, as at least something needs to be known about the true solution $y(x)$! Still, Eqs. (8.14) and (8.15) clearly reveal the interplay of the discretization and round-off errors. An optimal h exists at which the global error is minimal (Fig. 8.1 for the basic Euler's method.) ◁

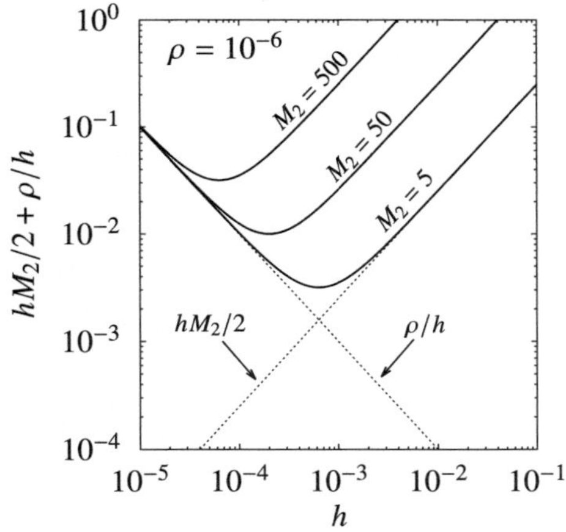

Fig. 8.1 The h-dependent part of the total global error (8.14) for the basic Euler's method on an equidistant mesh for $\rho = 10^{-6}$ and for three values $M_2 = 5, 50$, and 500. Separately shown for $M_2 = 5$ are the linearly increasing discretization component of the error and the inversely proportional round-off part

8.4.2   *Consistency, Convergence, Stability*

The differential equation and the difference scheme for it are *consistent* when

$$\lim_{h \to 0} F(h, x_n, y_n; f) = f(x_n, y_n)$$

(the difference scheme approximates the differential equation arbitrarily well in the limit $h \to 0$). A difference scheme is *convergent* if the approximate solution converges to the exact solution when the step size h is decreased,

$$\lim_{h \to 0} \|y_n - y(x_n)\| = 0 , \qquad nh = x_n - x_0 .$$

A difference scheme is *stable* if numerical errors do not accumulate, i.e. when for numerical solutions y_n and $\widetilde{y}_n$, with initial conditions y_0 and $\widetilde{y}_0$, we have

$$\|y_n - \widetilde{y}_n\| \le C \|y_0 - \widetilde{y}_0\| ,$$

where C does not depend on h. More on stability will follow in Sect. 8.5. For convergence of a difference scheme, both consistency and stability are needed:

$$\text{consistency} + \text{stability} \iff \text{convergence} .$$

All explicit RK methods are convergent within their stability regions [7]. The relation between the discretization error of the scheme, its stability, and the convergence of its solution to that of the differential equation also applies to difference schemes for partial differential equations (Sects. 10.3 and 10.5).

8.4.3   *Richardson Extrapolation*

Global error estimates usually provide crude, over-estimated upper bounds. This motivates us to control the step size h in explicit RK methods carefully, and use the error estimates to adjust that size accordingly. The local error in an explicit scheme of order p can be estimated if we assume that the factor of the h^{p+1} term in the remainder of the Taylor series does not change too quickly on the current interval of length h (or is constant, to order h^{p+1}). We rely on the fact that the principal discretization error in Eq. (8.12) is much larger than the remainder, which is not always the case (see Sect. 8.8). This is the idea behind *Richardson extrapolation* which can be used to improve the accuracy of the solution by one order at each step of the difference scheme.

We use the solution y_{n-1} to compute its next value y_n, once in a single step of h, then in two steps of $h/2$ each (Fig. 8.2). If the method is of order p, we have

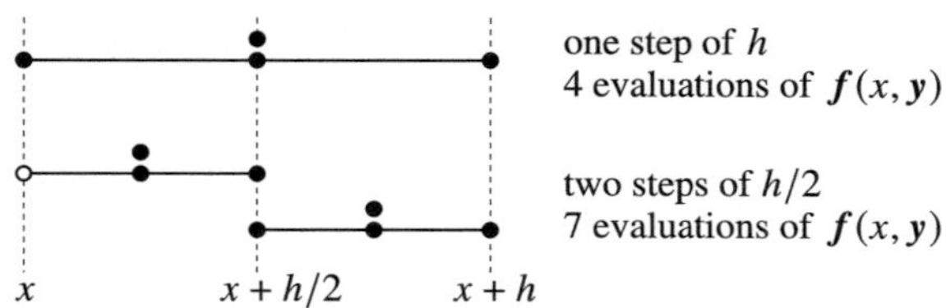

Fig. 8.2 Estimation of the local discretization error for the RK4 method by halving the step size h. The symbols • denote the evaluations of the function $f(x, y)$ (there is no need to repeat the evaluation at the point ○)

$$1 \cdot h : \quad y_n = y(x_n) + C h^{p+1} + O(h^{p+2}) ,$$
$$2 \cdot \tfrac{1}{2} h : \quad \widetilde{y}_n = y(x_n) + 2 C \left(\tfrac{1}{2} h\right)^{p+1} + O(h^{p+2}) .$$

We eliminate C from these equations and obtain

$$y(x_n) = \widetilde{y}_n + \frac{\widetilde{y}_n - y_n}{2^p - 1} + O(h^{p+2}) , \tag{8.16}$$

which is an improved approximation of the true solution $y(x_n)$.

The norm $\|\widetilde{y}_n - y_n\|$ is a good measure of the local discretization error that can be built into the computer program for any chosen method in order to assess whether the chosen step size h is too small or too large with respect to the prescribed precision. If the error in a given step is above the desired value, we decrease h and repeat the procedure.

8.4.4   Embedded Methods

To estimate the error with a single halving of the step, eleven evaluations of the function $f(x, y)$ are needed (Fig. 8.2), which may be too costly numerically. This served as a motivation for *embedded* Runge–Kutta methods [2] in which the local error can be estimated differently. They are m-stage RK methods which contain one linear combination of the intermediate quantities k_i to attain an approximation of order p,

$$y_{n+1} = y_n + h \left(b_1 k_1 + b_2 k_2 + \cdots + b_m k_m\right) + O(h^{p+1}) ,$$

and another combination yielding an approximation of order $\widehat{p} \neq p$,

$$\widehat{y}_{n+1} = y_n + h \left(\widehat{b}_1 k_1 + \widehat{b}_2 k_2 + \cdots + \widehat{b}_m k_m\right) + O(h^{\widehat{p}+1}) .$$

An embedded method is denoted by $p(\widehat{p})$; usually $\widehat{p} = p \pm 1$. The coefficients b_i and $\widehat{b}_i$ are determined as to minimize the leading error constant of either the low-order or the high-order solution approximation. This consideration dictates whether the estimate y_{n+1} or $\widehat{y}_{n+1}$ should be taken as the initial value in the next step of the integration. Among the most practical is the Dormand–Prince 5(4) method, in which the initial value for the next step is y_{n+1}, while the local error estimate is

$$\Delta = y_{n+1} - \widehat{y}_{n+1} = \sum_{i=1}^{m} \left(b_i - \widehat{b}_i \right) k_i \, .$$

By decreasing h we can bring the value $\|\Delta\|$ below the prescribed tolerance and thus control the magnitude of the local error. The coefficients of the Dormand–Prince 5(4) method are listed in this table:

0						
$\dfrac{1}{5}$	$\dfrac{1}{5}$					
$\dfrac{3}{10}$	$\dfrac{3}{40}$	$\dfrac{9}{40}$				
$\dfrac{4}{5}$	$\dfrac{44}{45}$	$-\dfrac{56}{15}$	$\dfrac{32}{9}$			
$\dfrac{8}{9}$	$\dfrac{19372}{6561}$	$-\dfrac{25360}{2187}$	$\dfrac{64448}{6561}$	$-\dfrac{212}{729}$		
1	$\dfrac{9017}{3168}$	$-\dfrac{355}{33}$	$\dfrac{46732}{5247}$	$\dfrac{49}{176}$	$-\dfrac{5103}{18656}$	
1	$\dfrac{35}{384}$	0	$\dfrac{500}{1113}$	$\dfrac{125}{192}$	$-\dfrac{2187}{6784}$	$\dfrac{11}{84}$
y_{n+1}	$\dfrac{35}{384}$	0	$\dfrac{500}{1113}$	$\dfrac{125}{192}$	$-\dfrac{2187}{6784}$	$\dfrac{11}{84}$ $\quad 0$
$\widehat{y}_{n+1}$	$\dfrac{5179}{57600}$	0	$\dfrac{7571}{16695}$	$\dfrac{393}{640}$	$-\dfrac{92097}{339200}$	$\dfrac{187}{2100}$ $\quad \dfrac{1}{40}$

An abundance of similar methods with such handy error estimates can be found in the literature, for example, the formerly popular fourth-order Fehlberg method with Cash–Karp parameters and 6 stages [8]. Until recently it was implemented in [9] where it has now been replaced by the Dormand–Prince 5(4) method [10]. The order 8(5,3) method [11] is also widely used.

Dense output for the Dormand–Prince 5(4) method　　Dense output of the form (8.11) for the Dormand–Prince 5(4) method, for which no additional evaluations of the function f are needed [2], is obtained by

$$b_1(\theta) = \theta^2(3 - 2\theta)b_1 + \theta(\theta - 1)^2$$
$$- 5\theta^2(\theta - 1)^2(2558722523 - 31403016\,\theta)/11282082432\,,$$
$$b_2(\theta) = 0\,,$$
$$b_3(\theta) = \theta^2(3 - 2\theta)b_3 + 100\,\theta^2(\theta - 1)^2(882725551 - 15701508\,\theta)/32700410799\,,$$
$$b_4(\theta) = \theta^2(3 - 2\theta)b_4 - 25\,\theta^2(\theta - 1)^2(443332067 - 31403016\,\theta)/1880347072\,,$$
$$b_5(\theta) = \theta^2(3 - 2\theta)b_5 + 32805\,\theta^2(\theta - 1)^2(23143187 - 3489224\,\theta)/199316789632\,,$$
$$b_6(\theta) = \theta^2(3 - 2\theta)b_6 - 55\,\theta^2(\theta - 1)^2(29972135 - 7076736\,\theta)/822651844\,,$$
$$b_7(\theta) = \theta^2(\theta - 1) + 10\,\theta^2(\theta - 1)^2(7414447 - 829305\,\theta)/29380423\,.$$

The expressions for the individual $b_i(\theta)$ follow from an interpolation of function values by Hermite polynomials, which yields a fourth-order dense output. We have written the coefficients carefully in terms of fractions: the strict precision requirements of Problem 8.13.6 teach us why.

8.4.5 Automatic Step Size Control

The spacings between the points $h = x_{n+1} - x_n$ used in difference schemes for ordinary differential equations need not be uniform and can be made smaller or larger according to the required local precision. To ensure their efficiency, difference schemes should be implemented by applying *adaptive step size control*: the algorithm should constantly "sense" when the right-hand side (8.2) dictates a wild change in the solution or when it is tracing a humble functional dependence, and adjust the step size accordingly. The usual criterion for a change in the step size is the local discretization error.

In extrapolation or embedded RK methods, two solution estimates are available at each step, y_{n+1} (of order p) and $\widehat{y}_{n+1}$ (of order $\widehat{p}$). We would like to constrain the local error $y_{n+1} - \widehat{y}_{n+1}$ by components, as

$$\left|y_{n+1,i} - \widehat{y}_{n+1,i}\right| \le S_i\,, \qquad S_i = A_i + R_i \max\left\{|y_{n,i}|, |y_{n+1,i}|\right\}\,,$$

where A_i are the prescribed absolute local errors of the ith solution component, and R_i are the relative local errors. As a joint measure of the computed error one may take

$$E = \sqrt{\frac{1}{M}\sum_{i=1}^{M}\left(\frac{y_{n+1,i} - \widehat{y}_{n+1,i}}{S_i}\right)^2}\,,$$

where M is the number of solution components. A well-established advice [2] for the choice of the optimal step size is

$$h_{\mathrm{opt}} = h\,(1/E)^{1/(q+1)}\,, \qquad q = \min(p, \widehat{p})\,,$$

but more polished programs include another "safety factor", which ensures that the error estimate in the next step is acceptable and that h neither increases nor decreases too quickly. For this safety factor V, various authors propose empirically determined values

$$V = 0.8\,,\ 0.9\,,\ (0.25)^{1/(q+1)}\ \text{or}\ (0.38)^{1/(q+1)}\,,$$

where V has a lower bound of $V_{\min} \approx 0.2$ and an upper bound of $V_{\max} \approx 5$. For the new value of the step size we ultimately take

$$h_{\text{new}} = h \cdot \min\left\{V_{\max},\ \max\left\{V_{\min},\ V \cdot (1/E)^{1/(q+1)}\right\}\right\}\,.$$

If at some step $E \le 1$, we accept the computed step size and continue the solution y_n with the new initial value y_{n+1} and step h_{new}. If $E > 1$, we reject the step size and try again with the initial value y_n and step h_{new}. If a step has been rejected, it is recommendable to set $V_{\max} = 1$. Details on the choice of the initial step size and step size control can be found in [2] as well as in [9, 12].

8.5   Stability of Explicit Single-Step Methods

Stability of explicit methods (for example, from the Runge–Kutta family) is judged by their *absolute* or *asymptotic stability* which is defined in Appendix F. The stability of a differential equation is not equivalent to the stability of the corresponding difference scheme, although they are closely related.

Stability of the differential equation The essence of stability can be elucidated with the one-dimensional linear initial-value problem $y' = \lambda y$, $y(x_0) = y_0$, $x \ge x_0$, $\lambda \in \mathbb{C}$. The exact solution is $y(x) = y_0 e^{\lambda(x-x_0)}$. The problem has a stable fixed point $y = 0$ if λ lies in the left complex half-plane ($\operatorname{Re}\lambda < 0$). A homogeneous system of linear differential equations is just a generalization of the above problem to

$$\boldsymbol{y}' = A\boldsymbol{y}\,, \qquad \boldsymbol{y}(x_0) = \boldsymbol{y}_0\,, \qquad x \ge x_0\,, \tag{8.17}$$

where A is a matrix with eigenvalues λ_i that all lie in the left complex half-plane. Namely, when A is diagonalized, we decouple the system (8.17) to independent one-dimensional problems of the form $y' = \lambda_i y$, where $\lambda_i \in \mathbb{C}$ are the eigenvalues of the Jacobi matrix $J = A$. This has a physical background: the local behavior of some component of the solution depends on the character of the corresponding eigenvalue (increasing solutions for positive real eigenvalues, decreasing for negative real eigenvalues, and oscillatory for imaginary eigenvalues).

The stability of the most general problem

$$\boldsymbol{y}' = \boldsymbol{f}(x, \boldsymbol{y})\,, \qquad \boldsymbol{y}(x_0) = \boldsymbol{y}_0\,, \qquad x \ge x_0\,, \tag{8.18}$$

depends on the eigenvalues λ_i of the Jacobi matrix J with the elements $J_{ij} = \partial f_i / \partial y_j$ that we compute *locally*, at current x and y. We use them to compute

$$r_{\max} = \max_i \{\operatorname{Re} \lambda_i(J)\} .$$

If $r_{\max} < 0$, the system of equations is locally stable, while for $r_{\max} > 0$ it is locally unstable. See also Example on page 975 in Appendix F.

Stability of the numerical method for a differential equation Consider Eq. (8.17) again. The individual components of the exact solution y tend to zero for $x \to \infty$ if the corresponding eigenvalues of A satisfy $\operatorname{Re} \lambda_i < 0$. For a numerical method this is not necessarily true. In a method of the form (8.8) the subsequent numerical solutions are computed as

$$y_{n+1} = S(h\lambda) y_n ,$$

where S is the *growth factor* or *stability function* [13]. We define the region of absolute stability of such a method as the set of complex numbers $h\lambda$ (h real positive number, $\lambda \in \mathbb{C}$), for which $\lim_{n \to \infty} \| y_n \| = 0$ (fixed point at origin is stable). The method is stable with values $h\lambda$ for which $|S(h\lambda)| \le 1$.

Example With the basic Euler's method (8.5) for the scalar problem $y' = \lambda y = f(y)$ we obtain $y_{n+1} = y_n + hf(y_n) = (1 + h\lambda)y_n$ or

$$S(h\lambda) = \frac{\mathrm{d}y_{n+1}}{\mathrm{d}y_n} = 1 + h\lambda .$$

The region with $|S| = |1 + h\lambda| \le 1$ defines a shifted circle in the complex plane $(\operatorname{Re}(h\lambda), \operatorname{Im}(h\lambda))$ (shaded area $p = 1$ in Fig. 8.3). The basic Euler's method remains stable only with a step size h for which $h\lambda$ lies within that area.

Similar calculations can be done for other methods. For an arbitrary scalar explicit method of order p with $q > p$ stages, the stability function becomes

$$S = \frac{\mathrm{d}y_{n+1}}{\mathrm{d}y_n} = 1 + h\lambda + \frac{(h\lambda)^2}{2!} + \cdots + \frac{(h\lambda)^p}{p!} + \sum_{i=p+1}^{q} \gamma_i \, (h\lambda)^i ,$$

where the coefficients γ_i depend on the given scheme, and the region of stability is defined by $|S| \le 1$. Figure 8.3 shows the region of stability in the complex plane $(\operatorname{Re}(h\lambda), \operatorname{Im}(h\lambda))$ for explicit Runge–Kutta methods of order p.

Stability in vector problems is related to the spectral radius of the Jacobi matrix. If we use the basic Euler's method to solve the system (8.17), we obtain $y_{n+1} = y_n + hA y_n = (I + hA)y_n$. In this case the spectral radius $\rho(I + hA)$ should not exceed unity. In solving (8.18) by the general method (8.8), the spectral radius $\rho(I + h[\partial F(h, x_n, y_n; f)/\partial y_n])$ should be less than 1. The procedures to compute the Jacobi matrix during the integration are described in [14].   ◁

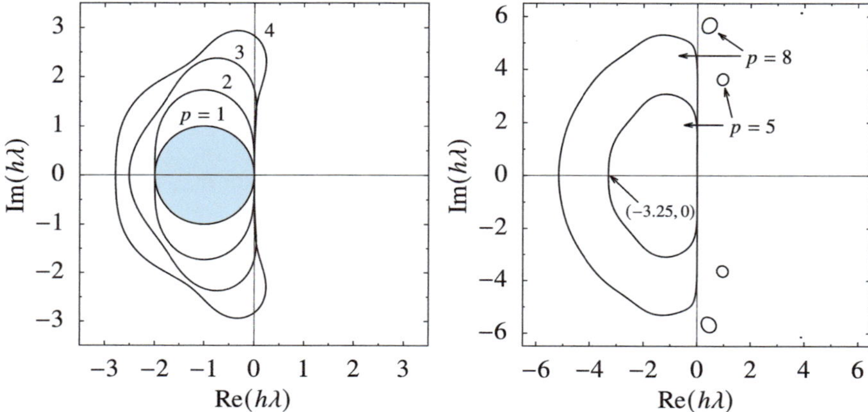

Fig. 8.3 [LEFT] Regions of stability of explicit pth order, p-stage methods $1 \leq p \leq 4$. The basic Euler's method corresponds to the interior of the circle denoted by $p = 1$, while the interior of the shape 4 corresponds to the RK4 method. [RIGHT] Regions of stability of Dormand–Prince methods of orders 5 and 8 (note the change of scales). The regions become larger with increasing order p, but they remain bounded for all explicit schemes

8.6 Extrapolation Methods ★

Extrapolation methods, first suggested by Bulirsch and Stoer [15], exploit a generalization of Richardson extrapolation to a sequence of crumblings of the integration interval. By using some pth order method, e.g. the basic Euler's method, we compute the solution on an interval $[x_n, x_n + H]$ with ever shorter steps $h_i = H/n_i$ ($i = 1, 2, \ldots, k$), where n_i are positive integers

$$n_1 < n_2 < \ldots < n_k \,.$$

For example, we compute the solution at $x_n + H$ in two steps of $H/2$, four steps of $H/4$, six steps of $H/6$, and so on. We obtain the quantities

$$\boldsymbol{T}_{i,1} = \boldsymbol{y}_{h_i}(x_0 + H) \,, \qquad i = 1, 2, \ldots, k \,.$$

The global error of the basic method of order p is of order p (page 478) and has the asymptotic expansion $\boldsymbol{y}(x) - \boldsymbol{y}_h(x) = \boldsymbol{e}_p(x)h^p + \boldsymbol{e}_{p+1}(x)h^{p+1} + \cdots + \boldsymbol{e}_N(x)h^N + O(h^{N+1})$. In the extrapolation method we eliminate the higher terms in this expansion by constructing an interpolation polynomial $\boldsymbol{P}(h) = \widehat{\boldsymbol{y}} - \boldsymbol{e}_p h^p - \boldsymbol{e}_{p+1}h^{p+1} - \cdots - \boldsymbol{e}_{p+k-2}h^{p+k-2}$, for which we require

$$\boldsymbol{P}(h_i) = \boldsymbol{T}_{i,1} \,, \qquad i = j, j-1, \ldots, j-k+1 \,. \tag{8.19}$$

We then "extrapolate to $h \to 0$" by using as the final result

$$P(0) = \widehat{y} = T_{j,k} \; .$$

Each value $T_{j,k}$ is an approximation of $y(x_n + H)$ of order $p + k - 1$. Equations (8.19) constitute a system of k equations for k unknowns $\widehat{y}, e_p, \ldots, e_{p+k-2}$ (which can even be solved analytically for $p = 1$). We generate the consecutive elements by

$$T_{j,k+1} = T_{j,k} + \frac{T_{j,k} - T_{j-1,k}}{n_j / n_{j-k} - 1} \; . \tag{8.20}$$

For example, with $n_1 = 1$ and $n_2 = 2$ we obtain $T_{22} = T_{21} + (T_{21} - T_{11})/(2^1 - 1)$, which is precisely Richardson's formula (8.16) for $p = 1$. By repeating the procedure for $j = 1, 2, \ldots$ and $k = 1, 2, \ldots, j$ we obtain a whole fan of estimates for the final value (each $T_{j,k}$ is labeled by the corresponding order of the result):

$$
\begin{aligned}
j \downarrow , k \to \; & T_{1,1}(p) \\
& T_{2,1}(p) \; T_{2,2}(p+1) \\
& T_{3,1}(p) \; T_{3,2}(p+1) \; T_{3,3}(p+2) \\
& \quad \cdots \qquad \cdots \qquad \cdots \qquad \cdots
\end{aligned}
$$

Various sequences of factors n_i by which the integration interval is fragmented, are available, e.g. the Romberg $1, 2, 4, 8, 16, 32, \ldots$ or the optimized Bulirsch sequence $1, 2, 3, 4, 6, 8, 12, 16, 24, 32, \ldots$ (powers 2^k and $1.5 \cdot 2^k$); the latter is more economical in higher orders of the method. Recently the harmonic sequence $1, 2, 3, 4, 5, 6, \ldots$ has gained popularity (see Fig. 8.4).

At low precisions the extrapolation integrators are just as efficient as Runge–Kutta methods. But they truly prosper in the regime of precision computations: by sequential construction of the elements $T_{j,k}$ we can, in principle, attain arbitrarily high orders, which means that they can also be arbitrarily faster than fixed-order methods. However, extrapolation methods are not suited for solving equations with singularities on the integration interval.

Extrapolation methods also permit the control of step size H. Moreover, they allow us to change the *order* p of the basic scheme along the integration interval. A simultaneous choice of the optimal step size H and of the order p is cumbersome, but it is doable [2]. Adaptive extrapolation methods may "jump" in very long steps, which calls for a reliable interpolation of the solution. If the basic method is the explicit or implicit Euler's scheme, one can derive closed formulas for dense output in analogy to Eq. (8.11); see [16]. As an impression, Fig. 8.5 shows the comparison of single-step integrators: a few equidistant explicit methods of the RK type, the adaptive embedded RK4 method (Cash–Karp), and the Bulirsch–Stoer extrapolation method.

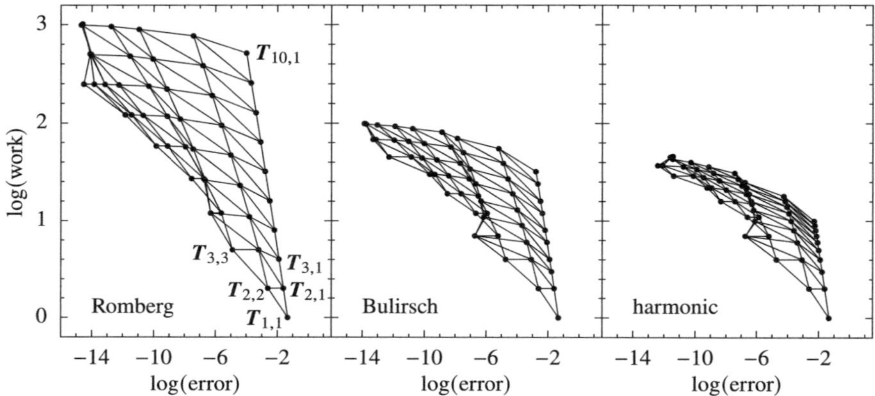

Fig. 8.4 Solving the differential equation $y' = (-y \sin x + 2 \tan x)y$ with the initial condition $y(\pi/6) = 2/\sqrt{3}$ by extrapolation methods on an interval of length $H = 0.2$. The basic method is the Euler's explicit scheme (8.5). Shown is the numerical cost ("work") versus the solution error for three different sequences $\{n_i\}$. The lines connecting the points $T_{j,k}$ overlap due to round-off errors appearing in the recurrence (8.20)

8.7   Conservative Second-Order Equations

Second-order vector differential equations of the form $\boldsymbol{y}'' = \boldsymbol{f}(x, \boldsymbol{y}, \boldsymbol{y}')$ are ubiquitous in physics: they embody every single instance of Newton's law $m\ddot{\boldsymbol{x}} = \boldsymbol{F}(t, \boldsymbol{x}, \dot{\boldsymbol{x}})$. Even the much simpler case $m\ddot{x} = F(x)$, where m is constant and F is an arbitrary integrable function, is interesting enough: such equations appear when forces are independent of the velocities, as in undamped oscillators ($F(x) = -kx$) or in the motion of bodies in the classical gravitational field. If m has units of mass and x units of length, the two terms of the first integral of this equation, $\frac{1}{2}m\dot{x}^2 - \int F(x)\mathrm{d}x = \mathrm{const}$, represent the kinetic energy $T(\dot{x})$ and the potential energy $U(x)$ (up to an additive constant), with the conservation law $T + U = E_0$ characteristic for conservative systems. It follows that $\dot{x} = \pm[2(E_0 - U(x))/m]^{1/2}$, and by integration we immediately obtain the dependence of the spatial coordinate on time,

$$t = t_0 + \int_{x_0}^{x} \frac{\mathrm{d}\xi}{\dot{x}(\xi)} = t_0 \pm \int_{x_0}^{x} \left[\frac{2}{m}\Big(E_0 - U(\xi)\Big)\right]^{-1/2} \mathrm{d}\xi \; .$$

The integral can be computed analytically if U is a polynomial of at most fourth degree (and even then we are dealing with elliptic integrals). Other cases of such

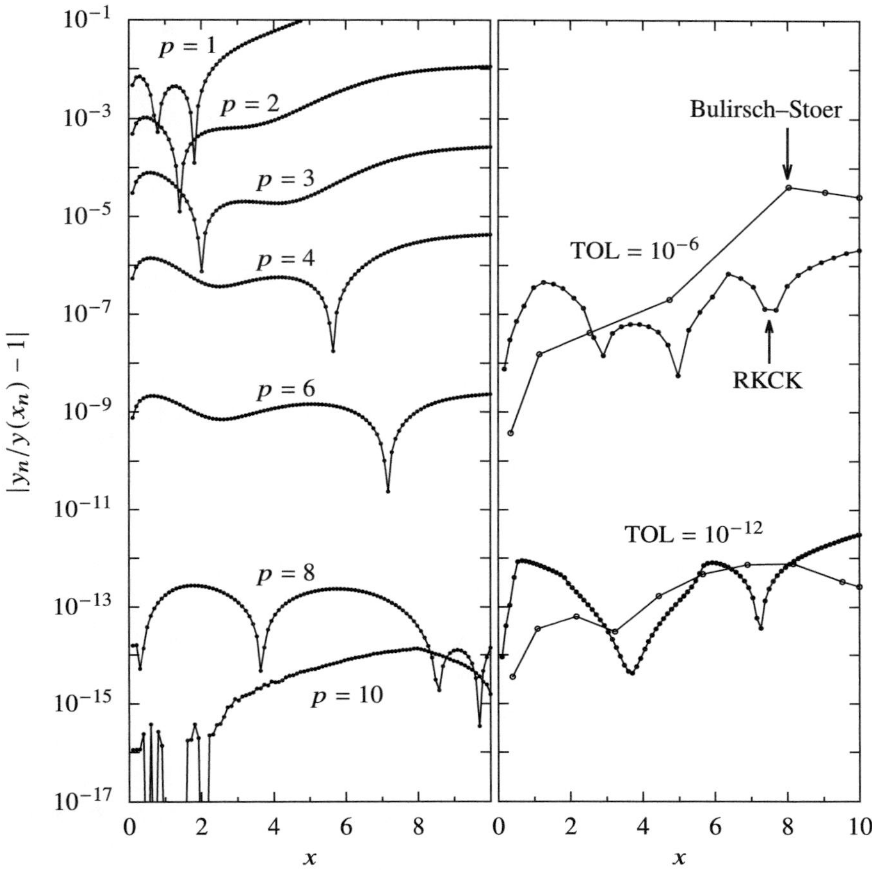

Fig. 8.5 Relative global error of integrators for $y'' = e^x \sin x - y - y' = f(x, y, y')$ with initial conditions $y(0) = 1$, $y'(0) = 0$. [LEFT] The error of the basic Euler ($p = 1$), improved Euler ($p = 2$), Kutta ($p = 3$), Runge–Kutta ($p = 4$), Butcher ($p = 6$), and Hairer explicit methods with $p = 8$ and $p = 10$, without step size control on a mesh of 100 points. The errors can be further reduced at all orders if we allow for more evaluations of the function f. [RIGHT] The error of the adaptive RK4 (Cash–Karp) method and the Bulirsch–Stoer extrapolation method (with the modified midpoint method as the underlying stepper) for two tolerances. Here comparable errors are attained with far fewer evaluations of f and at fewer mesh points

conservative problems must be approached numerically, and the classic method to integrate equations of the type $y'' = f(y)$ is the second-order leapfrog scheme:

$$
\begin{aligned}
y'_{n+1/2} &= y'_n + \tfrac{1}{2} h f(y_n) \,, \\
y_{n+1} &= y_n + h y'_{n+1/2} \,, \\
y'_{n+1} &= y'_{n+1/2} + \tfrac{1}{2} h f(y_{n+1}) \,.
\end{aligned}
\tag{8.21}
$$

8.7.1　Runge–Kutta–Nyström Methods

Equations of the form $y'' = f(x, y, y')$ can be rewritten as systems of two first-order equations

$$\begin{pmatrix} y \\ y' \end{pmatrix}' = \begin{pmatrix} y' \\ f(x, y, y') \end{pmatrix} ,$$

with initial conditions $y(x_0) = y_0$, $y'(x_0) = y_0'$. Such systems can be solved by using any method from previous Sections. But a substantial simplification can be achieved for equations without first derivatives,

$$y'' = f(x, y)$$

(Newton's law with velocity-independent forces, but allowing for explicit dependence on time). A special class of Runge–Kutta schemes exists for them [2], the Runge–Kutta–Nyström methods (RKN):

$$k_i' = f\left(x_n + c_i h, \ y_n + c_i h y_n' + h^2 \sum_{j=1}^{m} \overline{a}_{ij} k_j' \right) ,$$

$$y_{n+1} = y_n + h y_n' + h^2 \sum_{i=1}^{m} \overline{b}_i k_i' ,$$

$$y_{n+1}' = y_n' + h \sum_{i=1}^{m} b_i k_i' ,$$

where m is the number of stages. We determine the parameters $\overline{a}_{ij}$, b_i, $\overline{b}_i$, and c_i such that the order of the method is as high as possible, analogously to the methods of the core RK family. The very useful RKN method of fifth order,

c_i			$\overline{a}_{ij}$	
0				
$\dfrac{1}{5}$	$\dfrac{1}{50}$			
$\dfrac{2}{3}$	$-\dfrac{1}{27}$	$\dfrac{7}{27}$		
1	$\dfrac{3}{10}$	$-\dfrac{2}{35}$	$\dfrac{9}{35}$	
$\overline{b}_i$	$\dfrac{14}{336}$	$\dfrac{100}{366}$	$\dfrac{54}{336}$	0
$\overline{b}_i$	$\dfrac{14}{336}$	$\dfrac{125}{366}$	$\dfrac{162}{336}$	$\dfrac{35}{336}$

requires only *four* evaluations of the function f, while in the standard RK methods at least six are needed for the same order. The RKN methods also allow for local error control by adjusting the step size: this can be done either by extrapolation or within an embedded scheme (Sect. 8.4.4). The RKN method of order seven can be found in [17], and the methods of orders from 8 to 11 in [18].

8.8 Implicit Single-Step Methods

For stiff problems discussed in Sect. 8.10 we need difference schemes where scalar equations $y' = \lambda y$ ($\lambda \in \mathbb{C}$) would enjoy stability (bounded numerical solutions) at least in the whole left complex half-plane, i.e. for $\mathrm{Re}(h\lambda) \leq 0$ with $h > 0$. Such methods are called A-stable (Fig. 8.6).

The A-stability criterion is satisfied by the methods of the Runge–Kutta type. Implicit methods are given by Eqs. (8.8) and (8.9) in which $a_{ij} \neq 0$ for $j \geq i$. Let us replace the quantities k_i by the argument g_i such that $k_i = f(x_i, g_i)$ [19, 20]! The implicit Runge–Kutta methods can then be written in the form

$$g_i = y_n + h \sum_{j=1}^{m} a_{ij}\, f\left(x_n + c_j h,\ g_j\right), \tag{8.22}$$

$$y_{n+1} = y_n + h \sum_{j=1}^{m} b_j\, f\left(x_n + c_j h,\ g_j\right), \tag{8.23}$$

where m is the number of stages and $1 \leq i \leq m$. Equation (8.22) immediately discloses the price we have to pay for stability: at each step we need to solve a system of m (in general, non-linear, vector) equations for the quantities g_i.

The simplest single-step method is the *implicit Euler's method* of the first order. As its foundation we take the explicit Euler formula (8.5), but now we evaluate the right side of the equation at the next point, thus

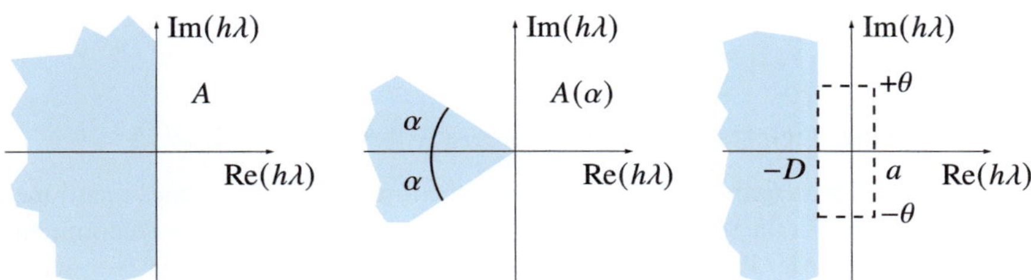

Fig. 8.6 Definitions of linear stability regions (scalar problem $y' = \lambda y$) for implicit methods. [LEFT] A-stability for single-step methods. [CENTER] $A(\alpha)$-stability for single-step methods. [RIGHT] "Stiff stability" for multi-step methods

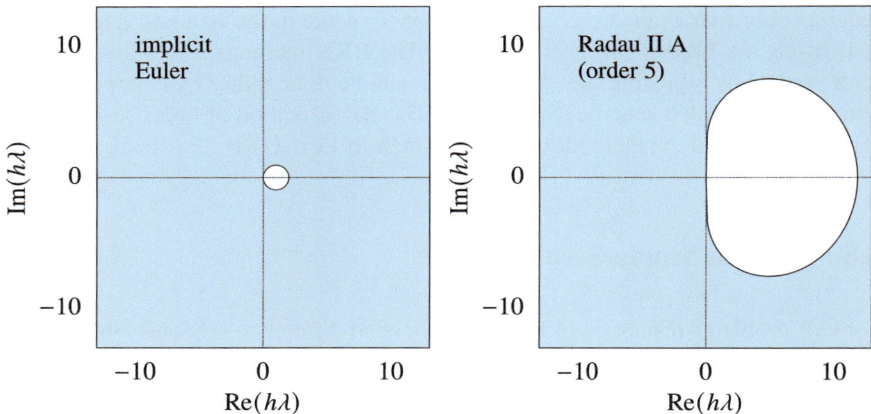

Fig. 8.7 Linear stability regions of implicit RK methods. [LEFT] Implicit Euler method corresponds to the whole exterior of the circle centered at $h\lambda = 1$ with radius 1 (compare to Fig. 8.3). [RIGHT] The shaded exterior of the oval is the stability region of the fifth-order "Radau 5" method from the Radau II A class [21]. The stability region of the implicit midpoint, trapezoidal, and the sixth-order "Gauss 6" method is the whole half-plane $\text{Re}(h\lambda) < 0$. All these methods are A-stable

$$\boldsymbol{y}_{n+1} = \boldsymbol{y}_n + h\boldsymbol{f}(x_{n+1}, \boldsymbol{y}_{n+1}) \,. \tag{8.24}$$

With the implicit method for the equation $y' = \lambda y$ we got $y_{n+1} = (1 + h\lambda)y_n$; with the implicit one we obtain $y_{n+1} = (1 - h\lambda)^{-1}y_n$. Now the region of absolute linear stability has become the *exterior* of the circle in the complex plane in Fig. 8.7 (left) and is outwardly *unbounded*: the implicit Euler method is therefore more than A-stable, since its stability region encompasses the whole left and almost the whole right complex half-plane. The second-order *implicit midpoint method*

$$\boldsymbol{y}_{n+1} = \boldsymbol{y}_n + h\boldsymbol{f}\left(\frac{x_n + x_{n+1}}{2}, \frac{\boldsymbol{y}_n + \boldsymbol{y}_{n+1}}{2}\right) \tag{8.25}$$

and the *implicit trapezoidal method*

$$\boldsymbol{y}_{n+1} = \boldsymbol{y}_n + \frac{h}{2}\big(\boldsymbol{f}(x_n, \boldsymbol{y}_n) + \boldsymbol{f}(x_{n+1}, \boldsymbol{y}_{n+1})\big)$$

are also A-stable: their stability region is precisely the half-plane $\text{Re}(h\lambda) < 0$.

Radau 5 and Gauss 6 methods The orders of the three implicit schemes mentioned above are too low for the integration of stiff problems. For serious use we recommend the implicit fifth-order Radau 5 method [21] from the Radau II A family (for details see [19]). Its coefficients are given by the Butcher tableau

$$
\begin{array}{c|ccc}
\dfrac{4-\sqrt{6}}{10} & \dfrac{88-7\sqrt{6}}{360} & \dfrac{296-169\sqrt{6}}{1800} & \dfrac{-2+3\sqrt{6}}{225} \\[2ex]
\dfrac{4+\sqrt{6}}{10} & \dfrac{296+169\sqrt{6}}{1800} & \dfrac{88+7\sqrt{6}}{360} & \dfrac{-2-3\sqrt{6}}{225} \\[2ex]
1 & \dfrac{16-\sqrt{6}}{36} & \dfrac{16+\sqrt{6}}{36} & \dfrac{1}{9} \\[2ex]
\hline
 & \dfrac{16-\sqrt{6}}{36} & \dfrac{16+\sqrt{6}}{36} & \dfrac{1}{9}
\end{array}
$$

This method is also A-stable (Fig. 8.7 (right)). We complete the collection by the sixth-order Gauss 6 method from the Gauss family (details in [19]). Apart from A-stability this method possesses other important properties relevant for geometric integration discussed in Sect. 8.11. Its coefficients are given in the tableau

$$
\begin{array}{c|ccc}
\dfrac{1}{2}-\dfrac{\sqrt{15}}{10} & \dfrac{5}{36} & \dfrac{2}{9}-\dfrac{\sqrt{15}}{15} & \dfrac{5}{36}-\dfrac{\sqrt{15}}{30} \\[2ex]
\dfrac{1}{2} & \dfrac{5}{36}+\dfrac{\sqrt{15}}{24} & \dfrac{2}{9} & \dfrac{5}{36}-\dfrac{\sqrt{15}}{24} \\[2ex]
\dfrac{1}{2}+\dfrac{\sqrt{15}}{10} & \dfrac{5}{36}+\dfrac{\sqrt{15}}{30} & \dfrac{2}{9}+\dfrac{\sqrt{15}}{15} & \dfrac{5}{36} \\[2ex]
\hline
 & \dfrac{5}{18} & \dfrac{4}{9} & \dfrac{5}{18}
\end{array}
$$

Extensive stability regions are one of the main attractions of implicit methods. Yet all implicit methods are not suitable for stiff problems, nor for geometric integration. The coefficients of other popular schemes are listed in [19]. For non-linear equations the theory of linear A-stability needs to be generalized to B-stability (see Appendix G).

8.8.1 Solution by Newton's Iteration

Implicit Runge–Kutta methods are more demanding than the explicit ones, as in general they call for solution of systems of non-linear equations. The main numerical obstacle is the system (8.22) that can be treated by Newton's iteration [19]. We introduce

$$
z_i = g_i - y_n ,
$$

which alleviates the influence of round-off errors. Then (8.22) becomes

$$z_i = h \sum_{j=1}^{m} a_{ij} f \left(x_n + c_j h , y_n + z_j \right) .$$

The Jacobi matrix used in the iteration is approximated by

$$J = \frac{\partial f}{\partial y}(x_n + c_j h , y_n + z_j) \approx \frac{\partial f}{\partial y}(x_n, y_n) ,$$

and assemble the solution in the kth step in the vector

$$\mathbf{Z}^{(k)} = (z_1^{(k)\mathrm{T}}, z_2^{(k)\mathrm{T}}, \ldots, z_m^{(k)\mathrm{T}})^{\mathrm{T}} .$$

Newton's iteration has the form

$$(I - hA \otimes J)\Delta \mathbf{Z}^{(k)} = - \mathbf{Z}^{(k)} + h(A \otimes I)\mathbf{F}(\mathbf{Z}^{(k)}) , \qquad (8.26)$$
$$\mathbf{Z}^{(k+1)} = \mathbf{Z}^{(k)} + \Delta \mathbf{Z}^{(k)} ,$$

where

$$I - hA \otimes J = \begin{pmatrix} I - ha_{11}J & \cdots & -ha_{1\,m}J \\ \vdots & & \vdots \\ -ha_{m1}J & \cdots & I - ha_{mm}J \end{pmatrix} , \quad A \otimes I = \begin{pmatrix} a_{11}I & \cdots & a_{1\,m}I \\ \vdots & & \vdots \\ a_{m1}I & \cdots & a_{mm}I \end{pmatrix} ,$$

and where we have denoted

$$\mathbf{F}(\mathbf{Z}^{(k)}) = \left(f^{\mathrm{T}}\left(x_n + c_1 h , y_n + z_1^{(k)} \right) , \ldots , f^{\mathrm{T}}\left(x_n + c_m h , y_n + z_m^{(k)} \right) \right)^{\mathrm{T}} .$$

In each iteration the function f needs to be computed m times, and a linear system with a $Mm \times Mm$ matrix solved, where M is the number of equations in the system. The simplest, but also the worst, initial approximation for Newton's iteration is $\mathbf{Z}^{(0)} = \mathbf{0}$. We control the convergence by monitoring the ratio

$$\zeta^{(k)} = \|\Delta \mathbf{Z}^{(k)}\| / \|\Delta \mathbf{Z}^{(k-1)}\| .$$

With the prescribed local error TOL we terminate the iteration when

$$\frac{\zeta^{(k)}}{1 - \zeta^{(k)}} \|\Delta \mathbf{Z}^{(k)}\| \leq \kappa \cdot \mathrm{TOL} ,$$

and accept the values $\mathbf{Z}^{(k+1)}$ as the final outcome. Experience shows that sensible values κ are approximately between 0.01 and 0.1. Numerous further instructions on effective solution of the system (8.26) can be found in [19].

When all z_i have been obtained by Newton's iteration, we need to compute the next solution $\mathbf{y}_{n+1}$ by using Eq. (8.23). At first sight it appears as if f needs to be evaluated yet m more times, but this can be avoided. Namely, Eq. (8.23) can be rewritten as

$$\mathbf{y}_{n+1} = \mathbf{y}_n + \sum_{i=1}^{m} d_i \mathbf{z}_i \, ,$$

where $(d_1, d_2, \ldots, d_m) = (b_1, b_2, \ldots, b_m) \, A^{-1}$. For example, for the Radau 5 method (page 492) we have simply $d_1 = d_2 = 0$ and $d_3 = 1$.

8.8.2 Rosenbrock Linearization

If an autonomous differential equation $\mathbf{y}' = f(\mathbf{y})$ is non-linear, even implicit schemes may become unstable, even if they are stable with respect to the linear problem $\mathbf{y}' = A\mathbf{y}$. Moreover, implicit equations can generally be solved only iteratively, so convergence problems may appear. With *Rosenbrock methods* we try to circumvent these problems by *linearization*. First we linearize the auxiliary quantities $\mathbf{k}_i$ in the m-stage implicit RK method,

$$\mathbf{k}_i = h f \underbrace{\left(\mathbf{y}_n + \sum_{j=1}^{i-1} a_{ij} \mathbf{k}_j + a_{ii} \mathbf{k}_i \right)}_{\mathbf{g}_i} \approx h f\left(\mathbf{g}_i\right) + h a_{ii} \underbrace{\frac{\partial f}{\partial \mathbf{y}}\left(\mathbf{g}_i\right)}_{J_y(\mathbf{g}_i)} \mathbf{k}_i \, ,$$

where $1 \leq i \leq m$. The computing cost at the right can be reduced if the true Jacobi matrix $J_y(\mathbf{g}_i)$ is replaced by the approximate, $J_y^{(n)} = J_y(\mathbf{y}_n)$, which needs to be computed only once at each step. On the other hand, the ansatz can be enriched by adding a few linear combinations $J_y(\mathbf{y}_n)\mathbf{k}_i$ at the right. With this small additional expense an m-stage Rosenbrock method takes the form

$$\left[I - h\gamma_{ii} J_y^{(n)}\right] \mathbf{k}_i = h f\left(\mathbf{y}_n + \sum_{j=1}^{i-1} \alpha_{ij} \mathbf{k}_j \right) + h J_y^{(n)} \sum_{j=1}^{i-1} \gamma_{ij} \mathbf{k}_j \, , \tag{8.27}$$

$$\mathbf{y}_{n+1} = \mathbf{y}_n + \sum_{j=1}^{s} b_j \mathbf{k}_j \, , \tag{8.28}$$

where $i = 1, 2, \ldots, m$. Again the coefficients α_{ij}, γ_{ij}, and c_i are tuned such that the maximum order of the method is attained. Just like in the non-linearized implicit RK methods, here too at each stage of the nth step we have to solve a system of m equations for the unknowns k_i, with the matrix $I - h\gamma_{ii} J_y^{(n)} = I - h\gamma_{ii}[\partial f/\partial y](y_n)$, except that now the system is linear.

Non-autonomous equations Rosenbrock methods were devised for autonomous problems $y' = f(y)$. Like in other approaches they can be generalized to non-autonomous problems $y' = f(x, y)$ by adding the equation $x' = 1$ to the system. In practical implementations (NUMERICAL RECIPES, NAG) this has already been taken care of. We can explicitly solve Eqs. (8.27) for x and thus obtain a system of linear equations for the quantities k_i:

$$k_i = h f \left(x_n + \alpha_i h , y_n + \sum_{j=1}^{i-1} \alpha_{ij} k_j \right) + h J_y^{(n)} \sum_{j=1}^{i} \gamma_{ij} k_j + h^2 \gamma_i J_x^{(n)} .$$

The concluding part of each time step—Eq. (8.28)—remains the same. Above we have used the abbreviations

$$\alpha_i = \sum_{j=1}^{i-1} \alpha_{ij} , \qquad \gamma_i = \sum_{j=1}^{i} \gamma_{ij} , \qquad J_x^{(n)} = \frac{\partial f}{\partial x}(x_n, y_n) , \qquad J_y^{(n)} = \frac{\partial f}{\partial y}(x_n, y_n) .$$

Implicit differential equations Implicit equations of the form $M y' = f(x, y)$ with constant non-singular matrices M can be rewritten in the equivalent form $y' = M^{-1} f(x, y)$, resulting in

$$M k_i = h f \left(x_n + \alpha_i h, y_n + \sum_{j=1}^{i-1} \alpha_{ij} k_j \right) + h J_y^{(n)} \sum_{j=1}^{i} \gamma_{ij} k_j + h^2 \gamma_i J_x^{(n)} .$$

By simple transformations among the variables and by exploiting the banded structure of the matrices M and J_y, the numerical efficiency of Rosenbrock methods can be strongly enhanced [19]. In similar ways, problems with non-constant matrices $M(x)$ can be harnessed [22].

8.9   Multi-step Methods ⋆

This Section is devoted to the basics of multi-step methods, in particular the predictor-corrector (PC) methods and the backward differentiation methods for equations of the type $y' = f(x, y)$. Together with symplectic integrators (Sect. 8.11) they play an important role in the integration of orbits in astronomy.

One characteristic of a multi-step method is that in order to initialize it, solutions $y_0, y_1, \ldots$ at $x_0, x_1, \ldots$ need to be known already. They can be computed by one of the single-step methods, but its order should at least match the order of the multi-step method they feed. We integrate Eq. (8.2) on $[x_n, x_{n+1}]$,

$$y_{n+1} = y_n + \int_{x_n}^{x_{n+1}} f(\xi, y(\xi)) \, d\xi \,, \tag{8.29}$$

and approximate f by the Newton interpolation polynomial through the points at which we already know the solution values, up to including x_n. We get [2]

$$y_{n+1} = y_n + h \sum_{j=0}^{k-1} \left[(-1)^j \int_0^1 \binom{-s}{j} \, ds \right] \nabla^j f_n \,. \tag{8.30}$$

We have used the usual backward differences $\nabla^{j+1} f_n = \nabla^j f_n - \nabla^j f_{n-1}$, where $\nabla^0 f_n = f_n$ and $f_n = f(x_n, y_n)$. It follows that

$$y_{n+1} = y_n + h \left[1 + \frac{1}{2} \nabla + \frac{5}{12} \nabla^2 + \frac{3}{8} \nabla^3 + \frac{251}{720} \nabla^4 + \frac{95}{288} \nabla^5 + \cdots \right] f_n \,.$$

If we express the backward differences by function values, we obtain explicit Adams methods of various orders. The order depends on the number of points spanned by the interpolation polynomial. From the third order upwards we have

$$y_{n+1} = y_n + \frac{h}{12} \left(23 f_n - 16 f_{n-1} + 5 f_{n-2} \right) ,$$

$$y_{n+1} = y_n + \frac{h}{24} \left(55 f_n - 59 f_{n-1} + 37 f_{n-2} - 9 f_{n-3} \right) , \tag{8.31}$$

$$y_{n+1} = y_n + \frac{h}{720} \left(1901 f_n - 2774 f_{n-1} + 2616 f_{n-2} - 1274 f_{n-3} + 251 f_{n-4} \right) ,$$

$$y_{n+1} = y_n + \frac{h}{1440} \left(4277 f_n - 7923 f_{n-1} + 9982 f_{n-2} - 7298 f_{n-3} + 2877 f_{n-4} \right.$$
$$\left. - 475 f_{n-5} \right) .$$

If the interpolation polynomial f also includes the point x_{n+1} at which the solution y_{n+1} is not yet known, we obtain

$$y_{n+1} = y_n + h \left[1 - \frac{1}{2} \nabla - \frac{1}{12} \nabla^2 - \frac{1}{24} \nabla^3 - \frac{19}{720} \nabla^4 - \frac{3}{160} \nabla^5 + \cdots \right] f_{n+1} \,.$$

This leads us to a class of implicit Adams methods, for example, from the third order (local error of fourth order) upwards,

$$y_{n+1} = y_n + \frac{h}{12}\left(5f_{n+1} + 8f_n - f_{n-1}\right)$$

$$y_{n+1} = y_n + \frac{h}{24}\left(9f_{n+1} + 19f_n - 5f_{n-1} + f_{n-2}\right),$$

$$y_{n+1} = y_n + \frac{h}{720}\left(251f_{n+1} + 646f_n - 264f_{n-1} + 106f_{n-2} - 19f_{n-3}\right), \tag{8.32}$$

$$y_{n+1} = y_n + \frac{h}{1440}\left(475f_{n+1} + 1427f_n - 798f_{n-1} + 482f_{n-2} - 173f_{n-3}\right.$$

$$\left. + 27f_{n-4}\right).$$

The equations are implicit since the unknown solution y_{n+1} occurs at the left as well as at the right, in $f_{n+1} = f(x_{n+1}, y_{n+1})$. We solve them iteratively.

8.9.1   *Predictor-Corrector Methods*

Explicit and implicit Adams methods can be merged into efficient procedures of various orders, known as *predictor-corrector methods* due to the characteristic sequence of approximations. In predicting y_{n+1} by one of the explicit formulas, e.g. (8.31), we rely on the extrapolation of the Newton polynomial to the point x_{n+1} which lies outside of the interpolation interval. We denote this prediction by

$$\text{P}: \qquad y_{n+1}^{(P)},$$

and call it the *predictor* (P). Correspondingly, the predicted derivative

$$\text{E}_1: \qquad y_{n+1}^{\prime(P)} = f\left(x_{n+1}, y_{n+1}^{(P)}\right). \tag{8.33}$$

is also questionable. We denote this step by E_1 (*evaluation*). In the next step we use one of the implicit methods, e.g. (8.32), which we solve by iteration. Especially for small h the iteration converges rapidly, so that instead of the unknown functions f_{n+1} we take the extrapolated derivatives (8.33). In this step we obtain the improved approximation for the solution, the *corrector* (C),

$$\text{C}: \qquad y_{n+1}^{(C)}.$$

Finally, we improve the derivative,

$$\text{E}_2: \qquad y_{n+1}^{\prime(C)} = f\left(x_{n+1}, y_{n+1}^{(C)}\right).$$

The usual operation cycle in one complete step from x_n to x_{n+1} is then PE_1CE_2 or $\text{PE}_1(\text{CE}_2)^m$ if the corrector step is executed after m iterations in the implicit part.

Local error estimate From the Taylor expansions of y_{n+1} around x_n and y_n around x_{n+1} simple local error estimates can be derived [23]. For the fourth-order pair of the Adams predictor and corrector (Eqs. (8.31) and (8.32)) we obtain

$$y(x_n) - y_n^{(P)} \approx \frac{251}{270}\left(y_n^{(C)} - y_n^{(P)}\right)\,, \qquad y(x_n) - y_n^{(C)} \approx \frac{19}{270}\left(y_n^{(P)} - y_n^{(C)}\right)\,,$$

if we assume that the fifth derivative of f is constant on $[x_n, x_{n+1}]$. This gives us the popular Adams–Bashforth–Moulton explicit method

$$y_{n+1}^{(P)} = y_n + \frac{h}{24}\left(55 f_n - 59 f_{n-1} + 37 f_{n-2} - 9 f_{n-3}\right)\,,$$

$$y_{n+1}^{*} = y_{n+1}^{(P)} + \frac{251}{270}\left(y_n^{(C)} - y_n^{(P)}\right)\,,$$

$$y_{n+1}^{(C)} = y_n + \frac{h}{24}\left(9 f_{n+1}^{*} + 19 f_n - 5 f_{n-1} + f_{n-2}\right)\,,$$

$$y_{n+1} = y_{n+1}^{(C)} + \frac{19}{270}\left(y_{n+1}^{(P)} - y_{n+1}^{(C)}\right)\,,$$

where we have denoted $f_{n+1}^{*} = f\left(x_{n+1}, y_{n+1}^{*}\right)$.

Such simple error estimates that depend only on the function values in individual steps, are one of the charms of predictor-corrector methods. Similar formulas can be derived in the case when x_{n-1} and y_{n-1} are used in Eq. (8.29) instead of x_n and y_n, which leads us to explicit Nyström predictors and implicit Milne–Simpson correctors [2].

Predictor-corrector methods perform best in tracing smooth solutions, when precision requirements are relatively severe, and when the evaluation of the functions f is numerically expensive. In one predictor or corrector step the function f needs to be evaluated only once!

8.9.2 Stability of Multi-step Methods

For simplicity we discuss here only scalar equations. A general k-step method has the form of a difference equation

$$\sum_{i=0}^{k} \alpha_i y_{n+i} = h \sum_{i=0}^{k} \beta_i f_{n+i}\,. \tag{8.34}$$

Stability is related to the generating polynomials for the coefficients of the method,

$$\rho(\zeta) = \sum_{i=0}^{k} \alpha_i \zeta^i\,, \qquad \sigma(\zeta) = \sum_{i=0}^{k} \beta_i \zeta^i\,.$$

When the differential equation $y' = 0$ is being solved and the solution of the corresponding difference equation (8.34) remains bounded, the multi-step method is said to be *zero-stable*. This is equivalent to the requirement that all zeros of ρ lie within the unit circle or on it; in the latter case the zeros should be simple. Zero stability therefore concerns only the coefficients α_i, not β_i.

Example Explicit and implicit k-step Adams methods have the generating polynomial $\rho(\zeta) = \zeta^k - \zeta^{k-1} = \zeta^{k-1}(\zeta - 1)$ with a simple zero $\zeta = 1$ and a zero $\zeta = 0$ of multiplicity $(k - 1)$, so they are zero-stable. The explicit Nyström and Milne-Simpson methods are also zero-stable, since for them $\rho(\zeta) = \zeta^k - \zeta^{k-2} = \zeta^{k-2}(\zeta - 1)(\zeta + 1)$. Nevertheless, the simple zero $\zeta = -1$ may cause problems with certain classes of differential equations [2]. $\qquad\qquad\triangleleft$

For stiff differential problems of the form $y' = f(y)$ (Sect. 8.10) stability can not be defined only in terms of the coefficients α_i, but β_i become relevant as well. In such problems we study the stability of a difference scheme for the problem $y' = \lambda y$ that can be understood as a linearization of $y' = f(y)$ in the vicinity of a fixed point. If the equation $y' = \lambda y$ is solved by (8.34), we get

$$\sum_{i=0}^{k} (\alpha_i - h\lambda\beta_i)\, y_{n+i} = 0 \,.$$

In this case, stability depends on the roots of the equation $\rho(\zeta) - \mu\sigma(\zeta) = 0$, where $\mu = h\lambda$. The stability region of the method (8.34) is the set of complex values μ for which all zeros of the polynomial $\rho - \mu\sigma$ lie within the unit circle or on it, and in the latter case the zeros should be simple.

Example For the explicit Adams method (Eq. (8.31), $k = 4$) we obtain

$$\rho(\zeta) - \mu\sigma(\zeta) = \zeta^4 - \zeta^3 - \mu \left(\frac{55}{24}\,\zeta^3 - \frac{59}{24}\,\zeta^2 + \frac{37}{24}\,\zeta - \frac{9}{24} \right) = 0\,,$$

from which we compute μ as a function of ζ. Let us write $\zeta = e^{i\phi}$ so that ζ goes around the whole unit circle for $0 \le \phi \le 2\pi$. When it does that, the variable μ outlines the edge of the stability region in the complex plane $(\mathrm{Re}(h\lambda), \mathrm{Im}(h\lambda))$ denoted by the index 4 in Fig. 8.8 (left). The Figure also shows the stability regions of explicit methods of orders 1, 2, and 3. Figure 8.8 (right) shows the stability regions of implicit Adams methods. $\qquad\qquad\triangleleft$

Example Note that a higher order of the local error of the scheme does not necessarily imply better stability. As an example [2] we solve $y' = y$, $y(0) = 1$ (exact solution $y(x) = e^x$) by using a third-order difference method

$$y_{n+2} + 4y_{n+1} - 5y_n = h\,[4f(x_{n+1}, y_{n+1}) + 2f(x_n, y_n)]\,,$$

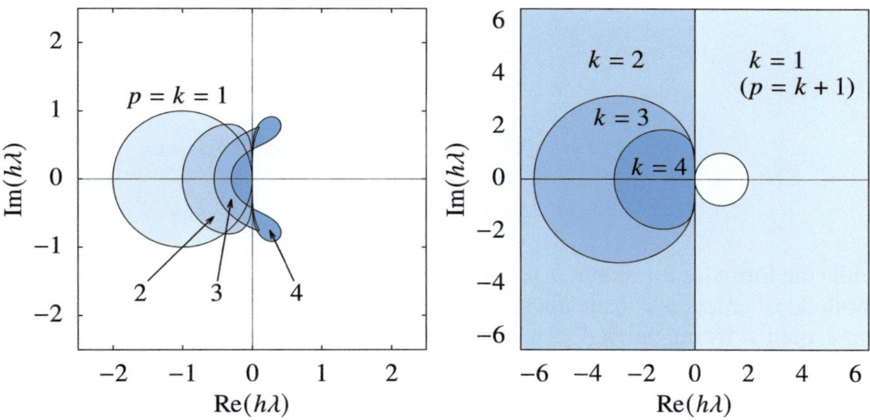

Fig. 8.8 Regions of linear stability of k-step explicit [LEFT] and implicit [RIGHT] Adams methods. (Note the different axis ranges.) The order of the explicit and implicit methods is $p = k$ and $p = k + 1$, respectively. With increasing order the regions of stability (of both explicit and implicit methods) shrink

which is in the form (8.34). The corresponding characteristic equation is

$$p(\zeta) - h\sigma(\zeta) = \zeta^2 + 4(1 - h)\zeta - (5 + 2h) = 0 \,,$$

with the solutions $\zeta_1 = 1 + h + O(h^2)$ and $\zeta_2 = -5 + O(h)$. With the values at the first two points, $y_0 = 1$ and $y_1 = e^h$, we obtain the solution $y_n = (1 + h)^n + (-5)^n$, which is obviously unstable at $n \to \infty$. As an exercise, plot the numerical solution computed with steps $h = 0.1, 0.05$, and 0.025!   ◁

8.9.3 *Backward Differentiation Methods*

Adams, Nyström, and Milne–Simpson methods are all based on the *integration of the right side* of the differential equation, as in (8.29). Instead, we could *differentiate the left side* of the equation and this is what we do in *backward differentiation formulas* (BDF), again by spanning an interpolation polynomial through a couple of known points. The most useful are implicit formulas generated by the series

$$\sum_{j=1}^{k} \frac{1}{j} \nabla^j y_{n+1} = h f_{n+1}$$

(compare to Eq. (8.30)). For an impression we list here the first-, second-, third- and fourth-order (local errors of $O(h^2), O(h^3), O(h^4)$, and $O(h^5)$, respectively) difference schemes:

$$y_{n+1} - y_n = h f_{n+1} \, ,$$

$$\frac{3}{2} y_{n+1} - 2 y_n + \frac{1}{2} y_{n-1} = h f_{n+1} \, ,$$

$$\frac{11}{6} y_{n+1} - 3 y_n + \frac{3}{2} y_{n-1} - \frac{1}{3} y_{n-2} = h f_{n+1} \, ,$$

$$\frac{25}{12} y_{n+1} - 4 y_n + 3 y_{n-1} - \frac{4}{3} y_{n-2} + \frac{1}{4} y_{n-3} = h f_{n+1} \, , \tag{8.35}$$

while the formulas for orders $p = 5$ and 6 are listed in [2]. Backward differentiation formulas of orders $p > 6$ are not stable, while the methods of order $1 \le p \le 6$ boast large, outwardly unbounded regions of linear stability (see Fig. 8.9 and compare the axis range to that in Fig. 8.8). The integrators of this section can also be implemented with a variable step size [2].

8.9.4 Multi-step Methods for Second-Order Conservative Equations

Multi-step methods can also be tailored to second-order conservative equations. We integrate the equation $y'' = f(x, y)$ twice [2] and obtain

$$y(x + h) - 2 y(x) + y(x - h)$$

$$= h^2 \int_0^1 (1 - s) \left[f\left(x + sh, y(x + sh)\right) + f\left(x - sh, y(x - sh)\right) \right] \mathrm{d}s \, .$$

Then we replace the function f by an interpolation polynomial. If (x_n, y_n) is the last point touched by the polynomial, we obtain the Störmer's family of explicit methods

$$y_{n+1} - 2 y_n + y_{n-1} = h^2 \left[1 + \frac{1}{12} \nabla^2 + \frac{1}{12} \nabla^3 + \frac{19}{240} \nabla^4 + \frac{3}{40} \nabla^5 + \cdots \right] f_n \, ,$$

while if also the point (x_{n+1}, y_{n+1}) is included, we obtain the Cowell's family

$$y_{n+1} - 2 y_n + y_{n-1} = h^2 \left[1 - \nabla + \frac{1}{12} \nabla^2 - \frac{1}{240} \nabla^4 - \frac{1}{240} \nabla^5 + \cdots \right] f_{n+1} \, .$$

Note that the third backward difference is absent in this expression, which results in a very useful fifth-order implicit Numerov's method

$$y_{n+1} - 2 y_n + y_{n-1} = \frac{h^2}{12} \left[f_{n+1} + 10 f_n + f_{n-1} \right] \, .$$

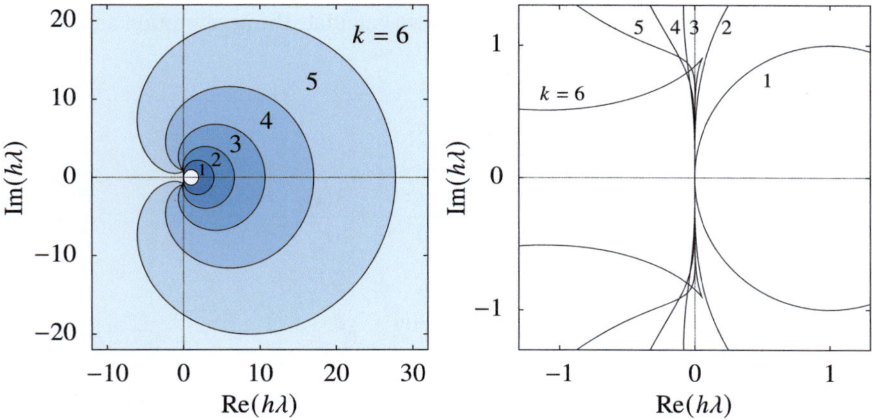

Fig. 8.9 [LEFT] Stability regions of implicit backward differentiation methods of order $p = k$. With increasing order the regions shrink: the stability region for the method of order p is the *exterior* of the geometric shape denoted by p. [RIGHT] Zoom-in of the left panel near the origin. Only the first- and second-order methods are stable throughout the left complex half-plane (Sect. 8.8)

The stability criteria for the methods mentioned above are described in [2].

8.9.5 *Integration of Gravitational Many-Body Problems*

The usual way to solve gravitational many-body initial-value problems like the evolution of planetary systems or calculation of orbits in stellar clusters or galaxies is by means of high-order integrators of the predictor-corrector type: provided that the number of bodies involved is sufficiently large, one may assume that the net force acting on any given object varies smoothly throughout its orbit, allowing us to use the past trajectory information to predict (and correct) its subsequent motion.

The standard method is the fourth-order scheme based on Hermite interpolation of all kinematic quantities: see [24]. (Higher-order schemes exist, but the fourth-order variant is the most popular.) In the *individual time-step scheme* [25] each particle i is furnished with its own time t_i, timestep h_i, as well as position, velocity, acceleration and jerk ($\boldsymbol{x}_i$, $\boldsymbol{v}_i$, $\boldsymbol{a}_i$ and $\dot{\boldsymbol{a}}_i$, respectively) at t_i. First, we select particle i with a minimum $t_i + h_i$, and set the global time t to be this minimum. We then predict the positions and velocities of all particles j (including particle i) by

$$\boldsymbol{x}_j^{(\mathrm{P})} = \boldsymbol{x}_{0,j} + \boldsymbol{v}_{0,j} h_j + \frac{1}{2}\boldsymbol{a}_{0,j} h_j^2 + \frac{1}{6}\dot{\boldsymbol{a}}_{0,j} h_j^3 \,,$$

$$\boldsymbol{v}_j^{(\mathrm{P})} = \boldsymbol{v}_{0,j} + \boldsymbol{a}_{0,j} h_j + \frac{1}{2}\dot{\boldsymbol{a}}_{0,j} h_j^2 \,.$$

From the predicted positions and velocities we calculate the acceleration and its time derivative for particle i by

$$a_{1,i} = G \sum_{j \neq i} m_j \frac{r_{ij}}{(r_{ij}^2 + \varepsilon^2)^{3/2}}, \tag{8.36}$$

$$\dot{a}_{1,i} = G \sum_{j \neq i} m_j \left[\frac{v_{ij}}{(r_{ij}^2 + \varepsilon^2)^{3/2}} - \frac{3(v_{ij} \cdot r_{ij}) r_{ij}}{(r_{ij}^2 + \varepsilon^2)^{5/2}} \right], \tag{8.37}$$

where

$$r_{ij} = x_j^{(P)} - x_i^{(P)}, \quad v_{ij} = v_j^{(P)} - v_i^{(P)}, \quad r_{ij} = |r_{ij}|.$$

(The same set of formulas can be used to evaluate the starting values of a_0 and $\dot{a}_0$.) Here ε is the *softening parameter*—a "poor man's solution" employed in order to alleviate the numerical consequences of r_{ij} occasionally becoming very small; a more rigorous and stable approach to harness such near-singularities in orbital mechanics is called *regularization* and is discussed in Appendix G. Next, the second and third derivative of the acceleration of particle i are estimated by

$$\ddot{a}_{0,i} = 2 \left[-3(a_{0,i} - a_{1,i}) - h_i(2\dot{a}_{0,i} + \dot{a}_{1,i}) \right] / h_i^2,$$
$$\dddot{a}_{0,i} = 6 \left[2(a_{0,i} - a_{1,i}) + h_i(\dot{a}_{0,i} + \dot{a}_{1,i}) \right] / h_i^3, \tag{8.38}$$

where $a_{0,i}$ and $\dot{a}_{0,i}$ are the acceleration and its derivative at time t_i. Finally, the corrected positions and velocities are calculated according to

$$x_i(t_i + h_i) = x_{1,i} = x_i^{(C)} = x_i^{(P)} + \frac{1}{24}\ddot{a}_{0,i} h_i^4 + \frac{1}{120}\dddot{a}_{0,i} h_i^5,$$
$$v_i(t_i + h_i) = v_{1,i} = v_i^{(C)} = v_i^{(P)} + \frac{1}{6}\ddot{a}_{0,i} h_i^3 + \frac{1}{24}\dddot{a}_{0,i} h_i^4,$$

and the cycle is repeated with a new assignment of h_i's and t.

What remains to be considered is a prudent estimation of the step sizes. In the individual time-stepping scheme, one usually applies the formula

$$h_i = \sqrt{ \eta \frac{|a_{1,i}||\ddot{a}_{1,i}| + |\dot{a}_{1,i}|^2}{|\dot{a}_{1,i}||\dddot{a}_{1,i}| + |\ddot{a}_{1,i}|^2} },$$

where all quantities are evaluated at the new time, $t_i + h_i$. The $a_{1,i}$ and $\dot{a}_{1,i}$ are known explicitly from (8.36) and (8.37); by the manner in which the integrator is constructed, $\ddot{a}_{1,i}$ is identical to $\ddot{a}_{0,i}$ from (8.38), while the updated jerk is given by

$$\dddot{a}_{1,i} = \dddot{a}_{0,i} + h_i \ddddot{a}_{1,i}.$$

The parameter η controls the precision. Since we are dealing with a fourth-order scheme, the local integration error is of the order $(\sqrt{\eta})^4 = \eta^2$.

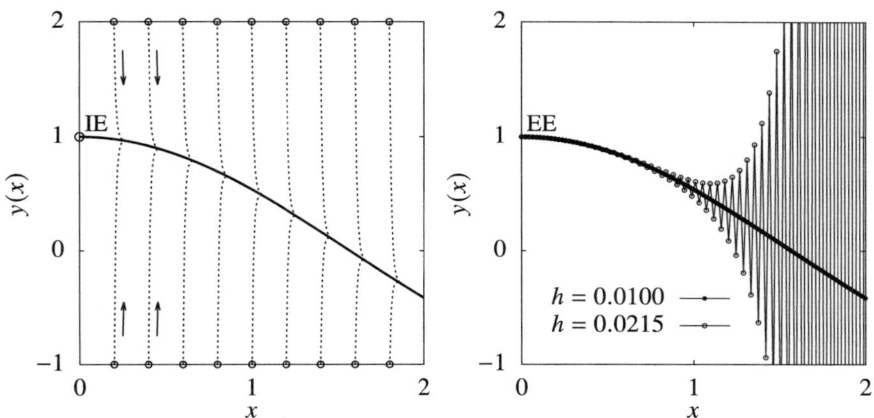

Fig. 8.10 Solving the stiff differential equation $y' = -100(y - \cos x) - \sin x$ with the initial condition $y(0) = 1$. [LEFT] Using the implicit Euler's method with step size $h = 0.0215$. Also shown are families of solutions with various initial conditions along the $y = 2$ and $y = -1$ axes. [RIGHT] Solutions by the explicit Euler's method with $h = 0.0215$ (unstable) and $h = 0.0100$ (stable)

When thousands or millions of bodies are being traced, individual time-stepping may become very costly due to excessive number of prediction steps (force evaluations). In sophisticated N-body codes such overheads are diminished by "quantizing" the time steps as $h_n = h_{\max}/2^{n-1}$ ($n = 1, 2, \ldots$). By doing so, and by appropriate synchronization at output time, whole groups of bodies can be advanced simultaneously with commensurate ("block") time steps, significantly speeding up the calculation. For details, see Chap. 2 in [26].

8.10 Stiff Problems

To obtain a feeling about the concept of stiffness, consider the problem

$$y'(x) = -100\big(y(x) - \cos x\big) - \sin x \, , \qquad y(0) = 1 \, , \tag{8.39}$$

with the exact solution $y(x) = \cos x$ (example adapted from [27]). Figure 8.10 shows its numerical solution by using the implicit (IE, left) and explicit (EE, right) Euler's method. The correct solution is the smooth curve in the center of the left panel. All other solutions with different initial conditions converge to this solution, but only after rapidly ramping up or down, following almost vertical trajectories. Such short-lived transient behavior that is dictated by the differential equation itself but is absent in the actual solution is one of the landmarks of a stiff problem.

Figure 8.10 also reveals that the implicit Euler's method nicely follows the solution while the explicit variant diverges unless the step size is sufficiently reduced. Provid-

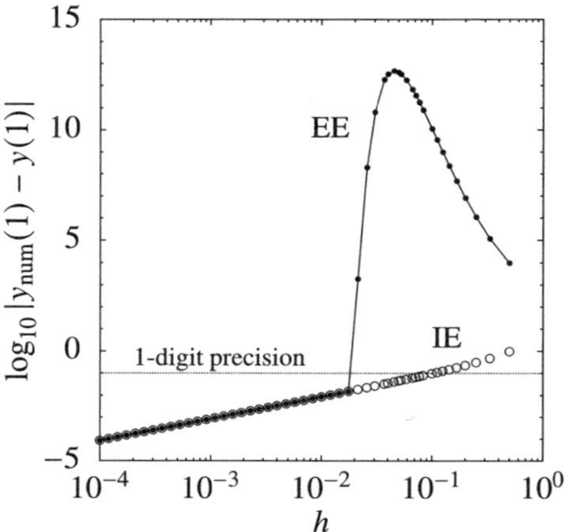

Fig. 8.11 The error of the numerical solution of (8.39) at $x = 1$ as a function of the step size h. The explicit Euler scheme (EE) is unstable unless a sufficiently small h is used, while the implicit method (IE) is stable for any h

ing stability in addition to accuracy is therefore another challenge in stiff problems. Figure 8.11 shows that for, say, single-digit precision of $y(1)$, the explicit method requires about five times shorter steps than the implicit one, and the disparity becomes much more pronounced if the leading coefficient -100 in (8.39) is increased. Thus, as a rule, stiff problems should be integrated by using implicit schemes: although in general the solution of implicit equations at each step may represent a complication, implicitness provides the needed stabilization.

This leads to the suspicion that initial-value problems with ordinary differential equations might be stiff when two or more very different scales are involved in the solution, for example, two characteristic times. Think of a robotic arm that moves with angular velocities of ≈ 1 Hz, which we hit by a small hammer, exciting oscillations with a frequency of ≈ 100 Hz. We wish to simulate the movement of the arm by numerical integration of the equations of motion. A variable step size integrator will be forced to reduce the step size in order to faithfully trace the physically totally irrelevant high-frequency component that dies out; at the same time, it will apply almost imperceptible corrections to the slowly changing low-frequency solution.

8.10.1 Eigenvalue Characterization of Stiffness

Let us promote the discussion to the most general, non-linear initial-value problem of the form (8.2). Suppose we are about to obtain its solution $\boldsymbol{y}_*(x)$ at some particular x_*. May we expect the computation of $\boldsymbol{y}(x)$ at $x \approx x_*$ to be stiff or not? Assume that the solution can be written as $\boldsymbol{y}(x) = \boldsymbol{y}_*(x) + \boldsymbol{w}(x)$, where $\boldsymbol{w}(x)$ is small in some norm. We linearize the problem by writing

$$\boldsymbol{y}'(x) = \boldsymbol{f}(x, \boldsymbol{y}_*) + J(x)\boldsymbol{w}(x) + o(\|\boldsymbol{w}\|),$$

where J is the $M \times M$ Jacobi matrix given by (8.3) and evaluated at $(x, \mathbf{y}_*)$. Subtracting $\mathbf{y}'_*(x) = \mathbf{f}(x, \mathbf{y}_*)$ from this equation we get

$$\mathbf{w}'(x) \approx J(x)\mathbf{w}(x) \,. \tag{8.40}$$

Linearization may be considered a good approximation to the original problem if $\mathbf{w}$ is small enough. Moreover, since stiffness and stability appear to be local, transient phenomena, it makes sense to *freeze* the coefficients of the system (8.40), so the Jacobi matrix becomes constant:

$$J(x_*) = J \,.$$

We are now treading the familiar terrain of Sect. 8.5 where stability of a difference scheme for a certain initial-value problem was related to the eigenvalues of the corresponding Jacobi matrix and their position within the method's stability region or outside of it. We diagonalize J to obtain M decoupled scalar, linear, constant-coefficient problems $y' = \lambda_i y$ ($i = 1, 2, \ldots, M$). A genuinely linear (or linearized) problem is said to be stiff when

$$\max_{1 \leq i \leq M} |\operatorname{Re} \lambda_i| \gg \min_{1 \leq i \leq M} |\operatorname{Re} \lambda_i| \,. \tag{8.41}$$

(Sometimes $|\lambda_i|$ are used instead of $\operatorname{Re} \lambda_i$.) The ratio of the left and right side of (8.41) is mostly a good measure of stiffness. If some eigenvalue is zero, the ratio is infinite: in such cases the problem is not stiff if other eigenvalues are small.

Example The famous Robertson chemical kinetics problem involves compounds A, B, and C linked by a chain of reactions with very different reaction rates,

$$
\begin{aligned}
A &\longrightarrow B & &\text{proceeds slowly (rate } 0.04) \,, \\
B + B &\longrightarrow B + C & &\text{very fast } (3 \cdot 10^7) \,, \\
B + C &\longrightarrow A + C & &\text{fast } (10^4) \,.
\end{aligned}
$$

The processes are described by a system of differential equations

$$
\begin{aligned}
A &: & y'_1 &= -0.04 y_1 + 10^4 y_2 y_3 & &, y_1(0) = 1 \,, \\
B &: & y'_2 &= 0.04 y_1 - 10^4 y_2 y_3 - 3 \cdot 10^7 y_2^2 & &, y_2(0) = 0 \,, \\
C &: & y'_3 &= 3 \cdot 10^7 y_2^2 & &, y_3(0) = 0 \,.
\end{aligned}
\tag{8.42}
$$

It turns out that the solution y_2 achieves its maximum $y_2(x) \approx 0.0000365$ at $x \approx 0.0045$, while y_3 still rapidly increases there, and $y_1 \approx 1$. The Jacobi matrix of partial derivatives (8.3) for the system (8.42) around that point is

$$
\begin{pmatrix}
-0.04 & 10^4 y_3 & 10^4 y_2 \\
0.04 & -10^4 y_3 - 6 \cdot 10^7 y_2 & -10^4 y_2 \\
0 & 6 \cdot 10^7 y_2 & 0
\end{pmatrix}
\approx
\begin{pmatrix}
-0.04 & 1.434 & 0.365 \\
0.04 & -2191 & -0.365 \\
0 & 2189 & 0
\end{pmatrix} \,,
$$

with eigenvalues $\lambda_1 \approx 0$, $\lambda_2 \approx -0.405$, $\lambda_3 \approx -2190.3$. Explicit methods fail here. Why? We find the answer in Fig. 8.3 (right) for the Dormand-Prince 5(4) method. Its stability region for $\mathrm{Re}(h\lambda) < 0$ reaches to $\mathrm{Re}(h\lambda) \approx -3.25$. Stability is guaranteed for $-3.25 \leq -2190\,h$ or $h \leq 0.0015$. The problem (8.42) is therefore stiff because of the eigenvalue λ_3 which is much larger than the rest. An explicit integrator will have to use a step this small in order to trace the solution that is very humble on almost the whole integration interval. $\qquad\qquad\qquad\qquad\qquad\qquad\qquad\qquad\qquad\quad \lhd$

8.10.2   Stiffness and Pseudospectra

The Jacobian of the problem (8.39) is just a constant, $J = \partial f / \partial y = -100 = \lambda$, whose modulus has no relation to the $O(1)$ scale of the solution $y(x) = \cos x$. On the other hand, solutions $\boldsymbol{y}$ of other initial-value problems are indeed often controlled by J—that is, by a subset of its eigenvalues—and then the ratio (8.41) should be a good measure of stiffness as long as the eigenvalues of J and the actual dynamics of the system driven by J are strongly correlated. In other words, stiffness is a transient, local phenomenon, whereas the eigenvalues are an asymptotic feature, and there may be no one-to-one correspondence between the two regimes. This fact is illustrated by the constant-coefficient system

$$ \boldsymbol{y}' = A\boldsymbol{y}\,, \qquad \boldsymbol{y}(0) = \boldsymbol{y}_0\,, \qquad \boldsymbol{y} \in \mathbb{R}^M\,, \quad A \in \mathbb{R}^{M \times M}\,, \qquad (8.43) $$

where $\boldsymbol{y}_0 = (1, 1, \ldots, 1)^{\mathrm{T}}$ and A is a triangular matrix with -1 on the main diagonal and -2 everywhere below it [28]. The matrix A possesses a single multiple eigenvalue, $\lambda = -1$. Judging by the ratio criterion (8.41)—A in this linear example is already the Jacobian—there is nothing stiff about this problem, yet the dependence of the error as a function of the step size h (Fig. 8.12 (left)) again indicates stiff behavior calling for an implicit solver.

Figure 8.12 (right) shows the ε-pseudospectrum of A (see Sect. 4.8). Obviously A "lives" in the region of size $O(M)$ in the left complex half-plane extending far away from the single eigenvalue at -1. This examples motivates an alternative definition of stiffness [28]: "An ordinary differential equation is stiff for the solution $\boldsymbol{y}_*(x)$ at $x \approx x_*$ if the pseudospectra of $A(x_*)$ extend far into the left half-plane as compared to the x-scale of the solution $\boldsymbol{y}_*(x)$ itself."

8.10.3   Automatic Stiffness Detection

Examples (8.39) and (8.42) are not very terrifying. For numerical modeling of the concentration of long-lived compounds in the Earth's atmosphere the integrators must handle time intervals on the order of centuries, while the time scales of certain chemical processes can be many orders of magnitude smaller. The integrator must

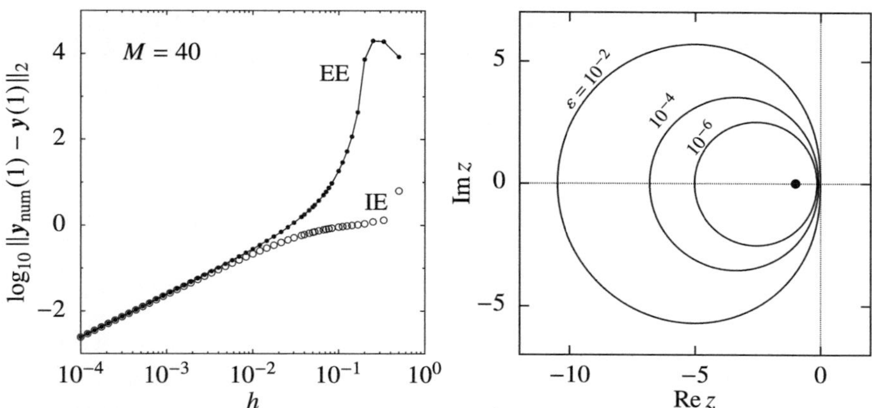

Fig. 8.12 [LEFT] The error of the numerical solution of (8.43) with $M = 40$ as a function of the step size h, for the explicit (EE) and implicit (IE) Euler methods. [RIGHT] The ε-pseudospectrum of the matrix A with respect to its only eigenvalue -1 (•)

therefore detect stiffness automatically and reliably, and the choice of the appropriate step size becomes critical.

When to check for stiffness Checking at each and every step of an adaptive step size method whether the problem is stiff or not is too costly. Instead, one possibility is to simply check "from time to time", for instance, after every 10 or 20 evaluations of the right-hand side f. Another option, which may be used in conjunction, is to check when more than ≈ 10 step rejections were encountered in the last ≈ 50 accepted steps and whether

$$\frac{1}{5}\,\overline{h} \le h_n \le 5\overline{h}\,.$$

Here h_n is the present step size and $\overline{h}$ is the running average of the step size since the beginning of the solution or since the last value of x at which stiffness has been checked.

How to check for stiffness The most commonly used method to check for the actual onset of stiffness in the spirit of (8.41) is based on estimating the leading eigenvalue of the current Jacobi matrix. Let v be an approximation of its corresponding eigenvector. Then

$$|\lambda| \approx \frac{\|f(x, y + v) - f(x, y)\|}{\|v\|}\,, \tag{8.44}$$

for $\|v\|$ sufficiently small, will be a good approximation of the dominant eigenvalue. In other words, we are estimating the Lipschitz constant from the inequality $\|f(x, y) - f(x, \tilde{y})\| \le \Lambda\|y - \tilde{y}\|$ while recalling $\Lambda = \sup\|J\|$ and the fact that the spectral radius of J satisfies the relation $\rho(J) \le \|J\| \le \Lambda$.

How (8.44) is evaluated in practice, depends on the particular integrator. Let us illustrate the procedure for the Dormand–Prince 5(4) method with its Butcher tableau given on p. 461. In this case $c_6 = c_7 = 1$ so both k_6 and k_7 are evaluated at $x + h$, yielding $k_6 = f(x + h, g_6)$ and $k_7 = f(x + h, g_7)$ with

$$g_6 = y_n + h\left[\frac{9017}{3168}\,k_1 - \frac{355}{33}\,k_2 + \frac{46732}{5247}\,k_3 + \cdots\right],$$
$$g_7 = y_n + h\left[\frac{35}{384}\,k_1 + 0\,k_2 + \frac{500}{1113}\,k_3 + \cdots\right].$$

Both g_6 and g_7 are approximations of the exact solution $y(x + h)$ and the Taylor expansion reveals that $\|g_7 - g_6\| = O(h^3)$, which is suitably small to be used as the denominator of (8.44). (It can be shown [19] that $g_7 - g_6$ is essentially the vector obtained by four iterations of the power method applied to the matrix hJ—see Sect. 4.6.2.) By using $\|k_7 - k_6\|$ as the numerator we obtain the estimate

$$\Lambda_{\mathrm{DP5(4)}} = \frac{\|k_7 - k_6\|}{\|g_7 - g_6\|} \approx \rho(J).$$

The equation is suspected to be stiff when $h\Lambda_{\mathrm{DP5(4)}}$ dwells near the edge of the method's stability region along the real axis: for the Dormand–Prince 5(4) method, this is at $\mathrm{Re}(h\lambda) \approx -3.25$ (see Fig. 8.3 (right)), i.e. when $h\Lambda_{\mathrm{DP5(4)}}/3.25 \approx 1$.

Example The coupled set of equations

$$\begin{aligned} y_1' &= y_2 \,, \\ y_2' &= \mu\left((1 - y_1^2)y_2 - y_1\right) \end{aligned} \tag{8.45}$$

with $y_1(0) = 2$, $y_2(0) = 0$, describes a popular model of a non-linear (Van der Pol) oscillator. This initial-value problem gradually becomes stiff when $\mu \gg 1$. Figure 8.13 (top) shows the first component of its solution, $y_1(x)$, calculated by the Dormand–Prince 5(4) method for $\mu = 300$.

The bottom panel shows $h\Lambda_{\mathrm{DP5(4)}}/3.25$ during integration using adaptive step size control of the form presented in Sect. 8.4.5. Values near unity indicate stiff behavior. Note that it is possible to stabilize the step size control to eliminate the wiggles: for details, consult [19], pp. 24–36.   ◁

8.10.4   *Implicit Multi-step Methods for Stiff Problems*★

Stiff problems can also be efficiently solved by using multi-step methods, whose stability was discussed in Sect. 8.9. Stability regions of multi-step methods are large, but any A-stable implicit multi-step method can have at most order $p = 2$ [29]. This constraint, known as the *Dahlquist barrier*, is clearly shown by Fig. 8.8 (right) for implicit Adams methods and Fig. 8.9 (right) for BDF methods.

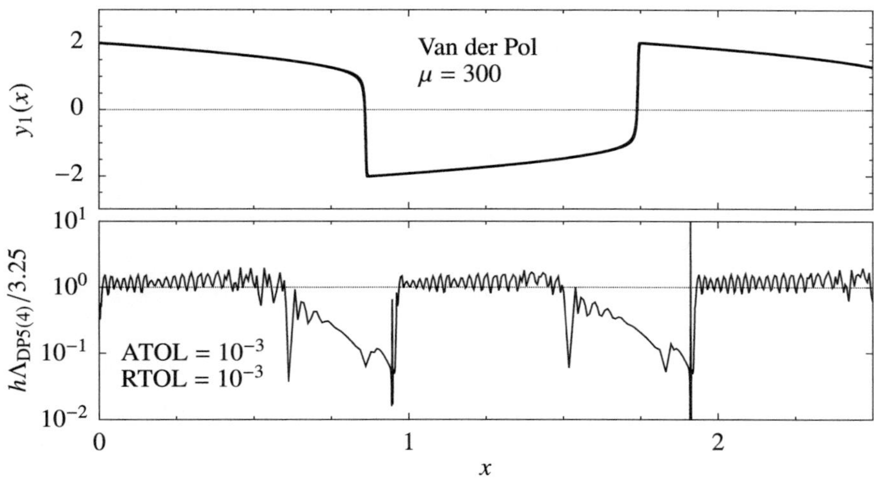

Fig. 8.13 [TOP] The first component of the system (8.45). [BOTTOM] Stiffness is diagnosed whenever the expression $h\Lambda_{\mathrm{DP5(4)}}/3.25$ approaches or exceeds unity

The low order of A-stable implicit multi-step methods diminishes their numerical precision. Yet not all stiff problems require stability on the whole half-plane $\mathrm{Re}(h\lambda) < 0$. Moreover, solutions of stiff equations with eigenvalues near the imaginary axis strongly oscillate and the step size needs to be reduced anyway in order to reproduce the high-frequency solution components. We therefore relax the requirements of strict A-stability and settle for $A(\alpha)$-*stable methods* which are stable at least on the slices defined by $|\arg(-z)| \le \alpha$ with $\alpha \le \pi/2$ (Fig. 8.6 (center)). A "completely" A-stable method is therefore $A(\pi/2)$-stable. We also define "stiff stability" for which we require the stability region $\mathrm{Re}\, z < -D$ for some $D > 0$ and a "sufficient precision" of the method within the rectangle $-D \le \mathrm{Re}\, z \le a$, $-\theta \le \mathrm{Im}\, z \le \theta$ for some $a > 0$ and $\theta \sim \pi/5$ (Fig. 8.6 (right)).

BDF methods of orders $3 \le p \le 6$ (Fig. 8.9) correspond to $\alpha = 86.03°$, $D = 0.083$ ($k = 3$), $\alpha = 73.35°$, $D = 0.667$ ($k = 4$), $\alpha = 51.84°$, $D = 2.327$ ($k = 5$), $\alpha = 17.84°$, $D = 6.075$ ($k = 6$). Figure 8.14 shows the region of $A(\alpha)$-stability and stiff stability for the fourth-order method (8.35). $A(\alpha)$-stable multi-step methods of higher orders with $\alpha \lesssim \pi/2$ do exist, but they possess large leading error constants and they are only of limited use. In order to cross the Dahlquist barrier more general multi-step methods can be devised: see [19].

8.11 Geometric Integration ⋆

In the following we use the variable pair (t, y) instead of (x, y), in the spirit of dynamical analysis pervading this Section. We are interested in the solutions of equations $\dot{y} = f(y)$ where $\dot{}$ denotes the time derivative, especially in the con-

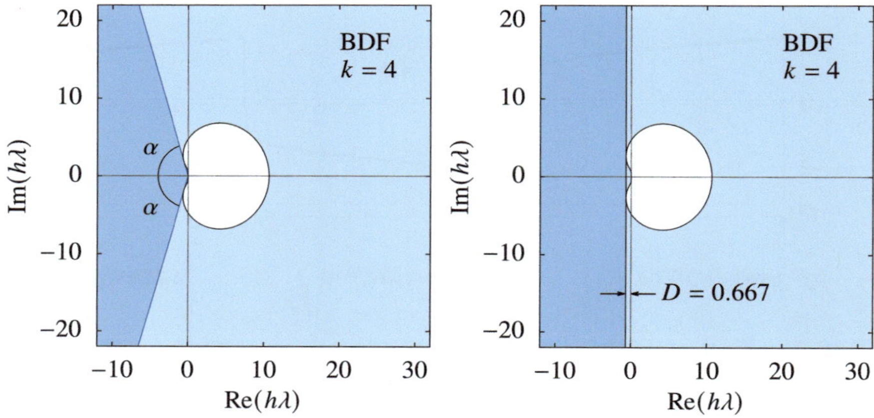

Fig. 8.14 Stability of the fourth-order BDF method (Eq. (8.35)). [LEFT] $A(\alpha)$-stability with angle $\alpha = 73.35°$. [RIGHT] Stiff stability with $D = 0.667$. See also Fig. 8.9

text of autonomous Hamiltonian systems [30]. Such systems are described by the Hamiltonians $H(\boldsymbol{p}, \boldsymbol{q})$, where

$$\dot{\boldsymbol{y}} = \boldsymbol{f}(\boldsymbol{y}) = J^{-1}\nabla H(\boldsymbol{y}) , \qquad \boldsymbol{y} = \begin{pmatrix} \boldsymbol{p} \\ \boldsymbol{q} \end{pmatrix} , \qquad J = \begin{pmatrix} 0 & 1 \\ -1 & 0 \end{pmatrix} , \qquad (8.46)$$

$\boldsymbol{p} \in \mathbb{R}^d$, $\boldsymbol{q} \in \mathbb{R}^d$ and $\nabla = (\nabla_p, \nabla_q)^{\mathrm{T}} = (\partial_{p_1}, \partial_{p_2}, \ldots, \partial_{p_d}, \partial_{q_1}, \partial_{q_2}, \ldots, \partial_{q_d})^{\mathrm{T}}$. The dynamical equations for the canonical variables $\boldsymbol{p}$ and $\boldsymbol{q}$ are

$$\dot{\boldsymbol{p}} = -\nabla_q H(\boldsymbol{p}, \boldsymbol{q}) , \qquad \dot{\boldsymbol{q}} = \nabla_p H(\boldsymbol{p}, \boldsymbol{q}) \qquad (8.47)$$

(see also Appendix H). How successful are the methods of previous Sections in such problems? We provide the answer from several viewpoints. All single-step methods discussed so far can be understood as mappings of the solution from "time" nh to "time" $(n + 1)h$,

$$\boldsymbol{y}_{n+1} = \phi_h(\boldsymbol{y}_n) . \qquad (8.48)$$

Does the numerical solution preserve the invariants of the continuous problem, for example, the Hamiltonian $H(\boldsymbol{p}, \boldsymbol{q})$? Does the numerical integration of the Hamiltonian system (the solution of the initial-value problem) preserve the symplectic structure of the phase space? We may also ask whether the integrator is symmetric and reversible. A numerical method that satisfies at least one of these requirements is known as a *geometric integrator.*

The motivation behind geometric methods is that preserving the phase-space geometry of the flow driven by the dynamical system is more relevant than reducing the local error. For instance, when hundreds or thousands of trajectories need to be traced (few-body encounters, planetary motion), geometric integrators are typically unnecessary; when this figure rises to millions and billions (charged particles in storage rings), the use of geometric integrators becomes imperative.

8.11.1 Preservation of Invariants

A non-constant function $I(y)$ is called the *first integral* (or *invariant*, or *constant of motion*) of Eq. (8.46) if

$$\nabla I(y) \cdot f(y) = 0 \qquad \forall y. \tag{8.49}$$

This means that at any point of the phase space (along any solution y) the gradient $\nabla I(y)$ is orthogonal to the vector field $f(y)$. A nice scalar example is the mathematical pendulum with the Hamiltonian $H(p, q) = \frac{1}{2} p^2 - \cos q$. The equations of motion (Newton's law) are $\dot{p} = -\sin q$ and $\dot{q} = p$, which can be written in the form (8.46) with $f(y) = (-\sin q, p)^{\mathsf{T}}$. Obviously $\nabla H(y) \cdot f(y) = (p, \sin q)(-\sin q, p)^{\mathsf{T}} = 0$. A further example are the three invariants of the classical Kepler problem, discussed in Problem 8.13.13.

Let us test some integrators on an even simpler case of the one-dimensional harmonic oscillator (linear pendulum) with the spring constant k^2, described by the Hamiltonian

$$H(p, q) = \frac{1}{2} \left(p^2 + k^2 q^2 \right), \tag{8.50}$$

with the exact solution

$$\begin{pmatrix} \widetilde{p}(h) \\ \widetilde{q}(h) \end{pmatrix} = \begin{pmatrix} \cos h & -k^2 \sin h \\ \sin h & \cos h \end{pmatrix} \begin{pmatrix} p(0) \\ q(0) \end{pmatrix} \tag{8.51}$$

at time $t = h$. The basic explicit Euler's method (8.5) is first order, so it also approximates the solution (8.51) only to first order. This implies

$$\begin{pmatrix} \widetilde{p}(h) \\ \widetilde{q}(h) \end{pmatrix}_{\text{Euler}} = \begin{pmatrix} 1 & -k^2 h \\ h & 1 \end{pmatrix} \begin{pmatrix} p(0) \\ q(0) \end{pmatrix}, \tag{8.52}$$

which can be seen if one explicit Euler's step is done for Eqs. (8.47). Instead of the correct value of the energy (8.50) we get

$$\tfrac{1}{2}(\widetilde{p}^2 + k^2 \widetilde{q}^2) = \tfrac{1}{2}(1 + k^2 h^2)(p^2 + k^2 q^2),$$

which is unbounded when the number of steps goes to infinity, regardless of the step size h. The RK4 method does not fare much better, as it results in damping

$$\tfrac{1}{2}(\widetilde{p}^2 + k^2 \widetilde{q}^2) = \tfrac{1}{2}\left(1 - k^6 h^6/72\right)(p^2 + k^2 q^2).$$

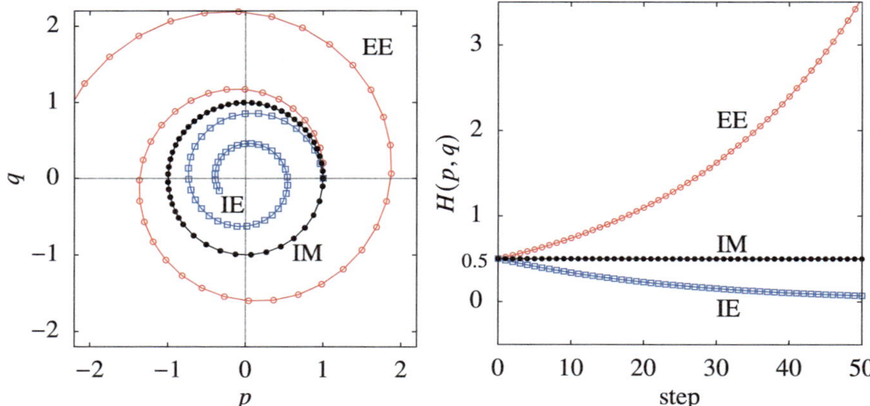

Fig. 8.15 [LEFT] The solution of the harmonic oscillator problem $\dot{p} = -k^2 q$, $\dot{q} = p$ with initial condition $(p, q) = (1, 0)$ (large symbol ●) and $k = 1$ by explicit Euler's (EE), implicit Euler's (IE), and implicit midpoint method (IM) in 50 steps of $h = 0.2$. [RIGHT] The values of $H(p, q) = \frac{1}{2}(p^2 + k^2 q^2)$ at current p and q for these three methods

We also obtain damping with the implicit Euler's method. Figure 8.15 (left) shows the solution (p, q) by the explicit Euler's (8.5), implicit Euler's (8.24), and implicit midpoint method (8.25) in the form for an autonomous Hamiltonian system:

$$y_{n+1} = y_n + h J^{-1} \nabla H \left(\tfrac{1}{2}(y_n + y_{n+1}) \right) . \tag{8.53}$$

In general, explicit and implicit RK methods preserve only linear invariants. As an example [19], consider the conservation of mass in the process (8.42): from the system of equations we see $\dot{y}_1 + \dot{y}_2 + \dot{y}_3 = 0$, so $I(y) = y_1 + y_2 + y_3$ is a linear invariant of the system. If the coefficients of an m-stage RK method (see Eq. (8.8) and Sect. 8.8) satisfy

$$b_i a_{ij} + b_j a_{ji} - b_i b_j = 0 , \qquad i, j = 1, 2, \ldots, m , \tag{8.54}$$

the method also preserves quadratic invariants of the form

$$Q(y) = y^{\mathrm{T}} C y , \tag{8.55}$$

where C is a symmetric square matrix. An example of such an invariant is the kinetic energy of a N-body system, $T(p) = \frac{1}{2} \sum_{i=1}^{N} p_i^{\mathrm{T}} p_i / m_i$. By definition (8.49), $Q(y)$ is the invariant of the system (8.46) when $y^{\mathrm{T}} C f(y) = 0$ for all y. The second-order implicit midpoint method (8.53), the two-stage scheme

$$k_1 = f\left(y_n + h\left[\qquad\quad \tfrac{1}{4}k_1 + \left(\tfrac{1}{4} - \tfrac{\sqrt{3}}{6}\right)k_2\right]\right),$$

$$k_2 = f\left(y_n + h\left[\left(\tfrac{1}{4} + \tfrac{\sqrt{3}}{6}\right)k_1 \qquad\quad + \tfrac{1}{4}k_2\right]\right),$$

$$y_{n+1} = y_n + \frac{h}{2}(k_1 + k_2)$$

of fourth-order, as well as the sixth-order, three-stage Gauss 6 method (defined on page 492) all satisfy the condition (8.54) and therefore preserve quadratic invariants of the form (8.55).

The energy of the linear harmonic oscillator $H(p, q)$ is a quadratic invariant in the variables p and q. Figure 8.15 (right) shows the dependence of $H(p, q)$ on the number of steps in the explicit Euler's, implicit Euler's, and the implicit midpoint method. Only the latter preserves the energy.

Partitioned RK methods When the equations of motion of dynamical systems can be written in separated form

$$\dot{y} = f(y, z), \qquad \dot{z} = g(y, z),$$

special methods are available for their solution that do not belong to the classical Runge–Kutta family. They are known as *partitioned Runge–Kutta methods* [19] and have the form

$$k_i = f\left(y_n + h\sum_{j=1}^{m} a_{ij}k_j,\; z_n + h\sum_{j=1}^{m} \widehat{a}_{ij}l_j\right),$$

$$l_i = g\left(y_n + h\sum_{j=1}^{m} a_{ij}k_j,\; z_n + h\sum_{j=1}^{m} \widehat{a}_{ij}l_j\right),$$

$$y_{n+1} = y_n + h\sum_{i=1}^{m} b_i k_i, \qquad z_{n+1} = z_n + h\sum_{i=1}^{m} \widehat{b}_i l_i.$$

In one step $n \to n + 1$, two different classical RK methods are merged: we seek the solutions y by a method with coefficients a_{ij} and b_i, and the solutions z by another method with coefficients $\widehat{a}_{ij}$ and $\widehat{b}_i$ (or vice-versa). The partitioned RK method of the lowest order joins two first-order Euler's methods: the implicit one in y and the explicit one in z (or vice-versa):

$$\begin{aligned} y_{n+1} &= y_n + h\,f\left(y_n, z_{n+1}\right), \\ z_{n+1} &= z_n + h\,g\left(y_n, z_{n+1}\right), \end{aligned} \quad \text{or} \quad \begin{aligned} y_{n+1} &= y_n + h\,f\left(y_{n+1}, z_n\right), \\ z_{n+1} &= z_n + h\,g\left(y_{n+1}, z_n\right). \end{aligned}$$

Another relevant partitioned method is the explicit second-order Störmer–Verlet scheme useful in solving the important class of problems $\ddot{q} = f(q)$ (that is, $\dot{q} = p$, $\dot{p} = f(q)$). This scheme is nothing but the leapfrog method (8.21) in disguise:

$$p_{n+1/2} = p_n + \tfrac{1}{2}hf(q_n) \,,$$

$$q_{n+1} = q_n + h\,p_{n+1/2} \,,$$ (8.56)

$$p_{n+1} = p_{n+1/2} + \tfrac{1}{2}hf(q_{n+1}) \,.$$

By interchanging the roles of p and q as in the Euler's pair, the adjoint Störmer–Verlet method can be derived (see next section).

In general the partitioned RK methods do not preserve quadratic invariants of the form $Q(y) = y^{\mathrm{T}} C y$. If the coefficients of the partitioned method satisfy

$$b_i \widehat{a}_{ij} + \widehat{b}_j a_{ji} - b_i \widehat{b}_j = 0 \,, \quad i, j = 1, 2, \ldots, m \,,$$
$$b_i - \widehat{b}_i = 0 \,, \quad i = 1, 2, \ldots, m \,,$$

the methods preserve the quadratic invariants

$$Q(y, z) = y^{\mathrm{T}} D z$$ (8.57)

(or $Q(p, q) = p^{\mathrm{T}} D q$), where D is a constant matrix. An example of such an invariant are the components of angular momentum of an N-body system, $L = \sum_{i=1}^{N} q_i \times p_i$. The partitioned Euler's method and the Störmer–Verlet method preserve the invariants of the form (8.57).

8.11.2 Preservation of the Symplectic Structure

In Hamiltonian systems the continuous mapping of (p, q) at time $t = 0$ to $(\widetilde{p}, \widetilde{q})$ at time $t = h$ along the solution preserves the symplectic structure,

$$\sum \mathrm{d}\widetilde{p} \wedge \mathrm{d}\widetilde{q} = \sum \mathrm{d}p \wedge \mathrm{d}q \,.$$

The numerical method (8.48) preserves this structure if it satisfies

$$\left[\phi_h(p, q)\right]^{\mathrm{T}} J \,\phi_h(p, q) = J \,,$$

which also implies $\det \phi_h(p, q) = 1$ or the preservation of volume in the phase space. Partitioned Euler's method from the previous section are symplectic. (Check this as an exercise.) For Hamiltonian systems they have the form

$$\begin{aligned} p_{n+1} &= p_n - h\nabla_q H(p_{n+1}, q_n) \,, \\ q_{n+1} &= q_n + h\nabla_p H(p_{n+1}, q_n) \,, \end{aligned} \quad \text{and} \quad \begin{aligned} p_{n+1} &= p_n - h\nabla_q H(p_n, q_{n+1}) \,, \\ q_{n+1} &= q_n + h\nabla_p H(p_n, q_{n+1}) \,. \end{aligned}$$ (8.58)

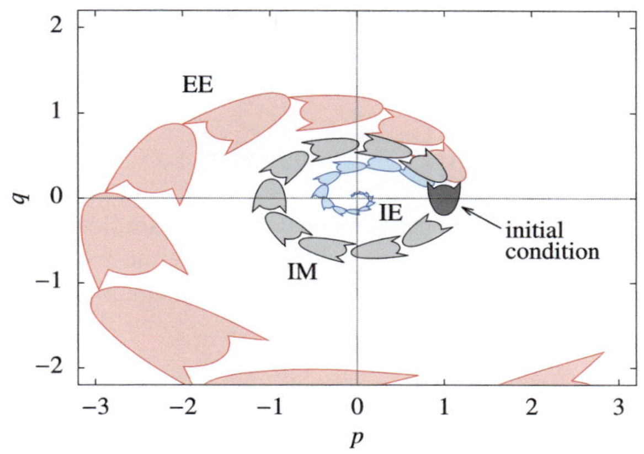

Fig. 8.16 Evolution of the Arnold's cat in the (p, q) plane for the harmonic oscillator problem with $k = 1.6$. The explicit Euler's method (EE) enlarges the area while the implicit (IE) reduces it. The implicit midpoint method (IM) preserves the area but not its shape

The Störmer-Verlet schemes,

$$
\boldsymbol{p}_{n+1/2} = \boldsymbol{p}_n - \frac{h}{2} \nabla_q H(\boldsymbol{p}_{n+1/2}, \boldsymbol{q}_n) \,,
$$

$$
\boldsymbol{q}_{n+1} = \boldsymbol{q}_n + \frac{h}{2} \left(\nabla_p H(\boldsymbol{p}_{n+1/2}, \boldsymbol{q}_n) + \nabla_p H(\boldsymbol{p}_{n+1/2}, \boldsymbol{q}_{n+1}) \right) \,, \tag{8.59}
$$

$$
\boldsymbol{p}_{n+1} = \boldsymbol{p}_{n+1/2} - \frac{h}{2} \nabla_q H(\boldsymbol{p}_{n+1/2}, \boldsymbol{q}_{n+1}) \,,
$$

or

$$
\boldsymbol{q}_{n+1/2} = \boldsymbol{q}_n + \frac{h}{2} \nabla_q H(\boldsymbol{p}_n, \boldsymbol{q}_{n+1/2}) \,,
$$

$$
\boldsymbol{p}_{n+1} = \boldsymbol{p}_n - \frac{h}{2} \left(\nabla_p H(\boldsymbol{p}_n, \boldsymbol{q}_{n+1/2}) + \nabla_p H(\boldsymbol{p}_{n+1}, \boldsymbol{q}_{n+1/2}) \right) \,, \tag{8.60}
$$

$$
\boldsymbol{q}_{n+1} = \boldsymbol{q}_{n+1/2} + \frac{h}{2} \nabla_q H(\boldsymbol{p}_{n+1}, \boldsymbol{q}_{n+1/2}) \,,
$$

are also symplectic, and so is the implicit midpoint method (8.53). Figure 8.16 shows the effect of three low-order integrators on the size and shape of Arnold's cat [31].

The symplectic Euler's schemes (8.58) and the Störmer–Verlet schemes (8.59) and (8.60) for general Hamiltonians $H(\boldsymbol{p}, \boldsymbol{q})$ are implicit. For separable Hamiltonians, $H(\boldsymbol{p}, \boldsymbol{q}) = T(\boldsymbol{p}) + V(\boldsymbol{q})$, all these methods become explicit. For partitioned RK methods the condition (8.54) also implies symplecticity (not only preservation of quadratic invariants). Classical explicit RK methods are not symplectic.

8.11.3 *Reversibility and Symmetry*

Conservative Hamiltonian systems are reversible. If at some point of the flow corresponding to the equation $\dot{\boldsymbol{y}} = \boldsymbol{f}(\boldsymbol{y})$, we change the direction of velocity, the form of the trajectory does not change; we just reverse the motion. An invertible linear transformation ρ helps us to define a more general notion of ρ-reversibility. A differential equation $\dot{\mathbf{y}} = \mathbf{f}(\mathbf{y})$ and the vector field $\mathbf{f}(\mathbf{y})$ are ρ-reversible if

$$\rho \mathbf{f}(\mathbf{y}) = -\mathbf{f}(\rho \mathbf{y}) \qquad \forall \mathbf{y} . \tag{8.61}$$

An analogous concept can be defined for single-step numerical methods (8.48). The method ϕ_h is symmetric (with respect to time reversal) if $\phi_h \circ \phi_{-h} = 1$. If the method ϕ_h, applied to a ρ-reversible differential equation, satisfies

$$\rho \circ \phi_h = \phi_{-h} \circ \rho , \tag{8.62}$$

then ϕ_h is a ρ-reversible mapping precisely when ϕ_h is symmetric. (Symmetry and ρ-reversibility of a discrete map are equivalent properties.) All explicit and implicit methods of the RK type satisfy Eq. (8.62) if (8.61) is valid. Partitioned RK methods satisfy the condition (8.62) if ρ can be written in the form $\rho(\mathbf{u}, \mathbf{v}) = (\rho_1(\mathbf{u}), \rho_2(\mathbf{v}))$, where ρ_1 and ρ_2 are invertible mappings. The symplectic Euler's methods (8.58) are not symmetric (or ρ-reversible). The Störmer–Verlet methods (8.59) and (8.60) are symmetric and therefore also ρ-reversible [32].

8.11.4   Modified Hamiltonians and Equations of Motion

Unfortunately it is difficult to find an integration method for general (also non-integrable) Hamiltonian systems that would preserve the symmetry properties of the system and its invariants, and at the same time fulfill the symplectic condition [33, 34]. We therefore often opt for methods satisfying at least one of these requirements: for the integration of astronomical orbits we might prefer to ensure the periodicity and preservation of energy, while in the study of charged particle motion in storage rings, where radiation losses can be compensated for, the behavior of the volume in phase space (spatial and momentum dispersion) may be more relevant. In the following we discuss only explicit methods for which we insist on symplecticity but sacrifice exact energy conservation. A surprise is in store.

In the basic Euler's method a small modification of the matrix in Eq. (8.52),

$$\begin{pmatrix} \widetilde{p}(h) \\ \widetilde{q}(h) \end{pmatrix}_{\text{symp. Euler}} = \begin{pmatrix} 1 - h^2 k^2 & -hk^2 \\ h & 1 \end{pmatrix} \begin{pmatrix} p(0) \\ q(0) \end{pmatrix} , \tag{8.63}$$

renders the mapping exactly symplectic ($\det \phi_h = 1$). The modified mapping does not preserve the energy, since after many repetitions of Eq. (8.63) we obtain $\widetilde{p}^2 + k^2 \widetilde{q}^2 = p^2 + k^2 q^2 + O(h^2)$: the value of the energy is only first-order accurate. On the example of the linear harmonic oscillator this can be explained by the conserved quantity (constant of motion)

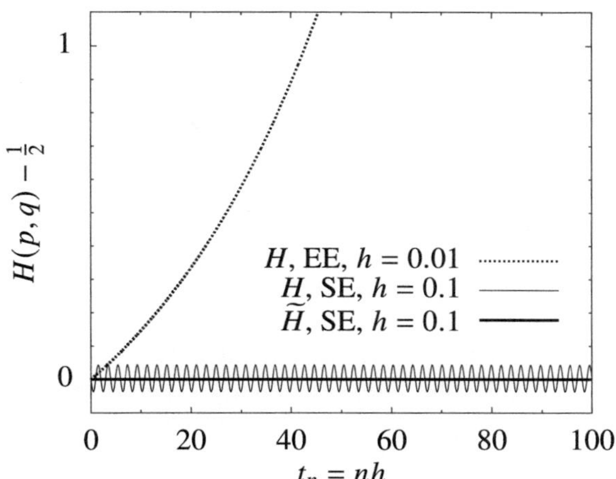

Fig. 8.17 Time evolution of the Hamiltonian for the linear harmonic oscillator with $k = 1.6$. Shown are the energy (8.50) for the explicit Euler's method (EE) with step $h = 0.01$ and the energies (8.50) and (8.64) for the symplectic Euler's method (SE) with step $h = 0.1$. This method *exactly* preserves the invariant (8.64)

$$\widetilde{H}(p, q) = \frac{1}{2}\left(p^2 + k^2 q^2\right) + \frac{h}{2} k^2 pq = \text{const.} \tag{8.64}$$

In other words, after many repetitions of (8.63) with initial conditions $(p, q) = (1, 0)$ and some small h, the points in phase space lie on the ellipse $p^2 + k^2 q^2 + h k^2 pq = 1$, which *always* differs from the exact solution $p^2 + k^2 q^2 = 1$ at most to order h. The error in the energy due to the local discretization error does not increase. The scheme (8.63) therefore preserves the symplectic *and* Hamiltonian structure, but the system evolves in accordance with a modified (or "perturbed") Hamiltonian (8.64). These observations are illustrated in Fig. 8.17.

This interesting realization is not confined to linear systems: an example is already the non-linearized pendulum with the Hamiltonian $H(p, q) = \frac{1}{2}p^2 - \cos q$. The true, "unperturbed" Hamiltonian equations of motion are $\dot{p} = -\sin q$, $\dot{q} = p$. By using the symplectic Euler's method (explicit in q and implicit in p) with step h we are in fact solving the equations

$$\begin{pmatrix} \dot{p} \\ \dot{q} \end{pmatrix} = \begin{pmatrix} -\sin q \\ p \end{pmatrix} + \frac{h}{2}\begin{pmatrix} p\cos q \\ -\sin q \end{pmatrix} + \frac{h^2}{12}\begin{pmatrix} (p^2 - 2\cos q)\sin q \\ 2p\cos q \end{pmatrix} + \cdots,$$

which correspond to the modified Hamiltonian

$$\widetilde{H}(p, q) = \frac{1}{2}p^2 - \cos q - \frac{h}{2}p\sin q + \frac{h^2}{12}\left(p^2 - \cos q\right)\cos q + \cdots.$$

The analysis of modified Hamiltonians and the corresponding equations of motion applies to a more general class of symplectic integrators. The sequence of mappings

$$q' = q + hc\left(\frac{\partial T}{\partial p}\right)_p, \qquad p' = p - hd\left(\frac{\partial V}{\partial q}\right)_{q=q'}, \tag{8.65}$$

each of which is symplectic by itself, faithfully describes the time evolution of the system in accordance with a modified Hamiltonian $\widetilde{H} = H + hH_1 + h^2H_2 + \cdots$. Higher-order symplectic integrators can be constructed by repeated applications of the tandem (8.65), where the parameters c and d are different at each step. An integrator of order p can be assigned the modified Hamiltonian

$$\widetilde{H}_p = H + h^p H_p + O(h^{p+1}) \, .$$

By reducing h we can therefore come (almost) arbitrarily close to exact energy conservation. However, we note that for non-linear systems the convergence in the term $h^p H_p$ has not yet been strictly proven.

The mathematical background and the construction of explicit high-order symplectic integrators are given in Appendix H. An excellent reference on geometric integration is offered by the monograph [32] and the review article [35]. Geometric integration is a crucial tool for the study of charged-particle trajectories in accelerators; a comprehensive discussion in this context is [36]. For partial differential equations (PDE), geometric methods are less well developed. The PDE solution approaches based on difference methods that ignore geometric aspects are discussed in Chaps. 10 and 11. An introduction to conservative discretizations of Hamiltonian PDEs can be found in [37]. See also [38, 39].

8.12 Lie-Series Integration ⋆

Lie-series integrators represent a class of explicit methods to solve ordinary differential equations. They are conceptually simple, but difficult to apply to general cases, as the equations of motion need to be differentiated explicitly. However, when the equations are given in terms of algebraic functions that can be manipulated efficiently, and when high precision is required, Lie-series integrators reveal their true power. Their typical area of use is dynamical astronomy [40–42]. Note that Lie-series integrators in their basic forms are *not* symplectic and their energy and angular momentum errors are unbounded [43].

Here we discuss autonomous dynamical systems $\phi : \mathbb{R} \times \mathbb{R}^n \to \mathbb{R}^n$ and denote the independent and dependent variables by $t \in \mathbb{R}$ and $x \in \mathbb{R}^n$, respectively. A trajectory starting at $x(0)$ is given by $x(t) = \phi(t, x(0))$ and is determined by the system of ODEs

$$\dot{x}(t) = F(x(t)) \, .$$

8.12.1 Taylor Expansion of the Trajectory

By assuming that the trajectories are smooth, they can be Taylor-expanded as

$$x(t + \Delta t) = \sum_{k=0}^{n} \frac{x^{(k)}(t)}{k!} (\Delta t)^k + R_n(t) \, ,$$

where $(\cdot)^{(k)}$ denotes the kth time derivative, and the remainder is

$$R_n(t) = \frac{1}{n!} \int_{t}^{t+\Delta t} x^{(n+1)}(\tau)(\tau - t)^n \, d\tau \, .$$

Defining the rescaled derivatives $F_k(x(t)) = x^{(k)}(t)/k!$ the Lie-series integrator of order n with time step $\Delta t > 0$ can be written explicitly as

$$x(t + \Delta t) = x(t) + \sum_{k=1}^{n} F_k(x(t)) \, (\Delta t)^k + R_n(t) \, .$$

The remainder $|R_n(t)|$ represents the local error of the integrator.

Note that for given equations of motion, a Lie-series integrator of order p can be constructed if F is continuously differentiable up to $p + 1$ times. The applicability of the integrator depends on whether the rescaled derivatives F_k can be efficiently generated from the right-hand side of the equations of motion, $F(x) = F_1(x)$. They can be obtained by using the recurrence relation

$$F_k(x) = \frac{1}{k} \, F'_{k-1}(x) F(x) \, . \tag{8.66}$$

For algebraic F_k this recurrence can be exploited efficiently to very high orders by using programs for symbolic computations.

Note that the local error has a very simple form, thus the global error can be well controlled. By the mean-value theorem, a $\xi \in [t, t + \Delta t]$ exists such that $R_n(t) = F_{n+1}(\xi)(\Delta t)^{n+1}$; if $F_{n+1}(x(t)) \neq 0$ and $\Delta t \ll 1$, the remainder becomes $R_n(t) \approx F_{n+1}(x(t)) \, (\Delta t)^{n+1}$. Hence $R_n(t)$ scales as $O((\Delta t)^{n+1})$, and for a given $x(t)$, it can be estimated by finding a convergence radius ρ of the series defined as $|F_k(x(t))| \asymp \rho^{-k}$ when $k \to \infty$. A fast convergence of the series is achieved if $\Delta t \ll \rho$. In this limit, $R_n(t)$ for large orders n can be approximated as

$$|R_n(t)| \asymp \sum_{k=n+1}^{\infty} \left(\frac{\Delta t}{\rho} \right)^k = \left(\frac{\Delta t}{\rho} \right)^{n+1} \left(1 - \frac{\Delta t}{\rho} \right)^{-1} \, .$$

Example   The Lorenz system (see [44, 45] and Problem 8.13.7) is a dynamical system with a three-dimensional phase space: the trajectory of the system is described by the vector $x(t) = (x_1(t), x_2(t), x_3(t))^{\mathrm{T}} \in \mathbb{R}^3$, and is determined by the system of equations (8.71); in the notation of this Section,

$$
\begin{aligned}
\dot{x}_1 &= \sigma(x_2 - x_1)   &&= F_1(x_1, x_2, x_3)\,, \\
\dot{x}_2 &= x_1(r - x_3) - x_2 &&= F_2(x_1, x_2, x_3)\,, \\
\dot{x}_3 &= x_1 x_2 - b x_3   &&= F_3(x_1, x_2, x_3)\,.
\end{aligned}
\tag{8.67}
$$

Here we use $\sigma = 10$, $r = 28$, and $b = 8/3$, for which the system is chaotic and possesses a finite strange attractor.

We would like to calculate the trajectory of the system over long times as precisely as possible by Lie-series integration. In order to minimize the round-off errors, we work in floating-point arithmetic with precision $\varepsilon_{\mathrm{M}} = 10^{-1024}$ by using the GMP library [46]. The Lie integrator of order p has the form

$$
x_i(t + \Delta t) = x_i(t) + \sum_{k=1}^{p} F_{i,k}(x(t))\,(\Delta t)^k\,,
$$

where $F_{i,k}$ denotes the rescaled derivative $x_i^{(k)}(t)/k!$. Any such derivative can be expressed as a polynomial of variables x_i in the form

$$
F_{i,k}(x) = \sum_{\alpha \in I_{i,k}} A_\alpha^{i,k} x_1^{\alpha_1} x_2^{\alpha_2} x_3^{\alpha_3}\,, \qquad \alpha = \{\alpha_1, \alpha_2, \alpha_3\}\,,
$$

where $I_{i,k}$ is the set of possible powers corresponding to $F_{i,k}(x)$. The explicit form of $F_{i,k}(x)$ can be obtained systematically by the recurrence (8.66), a task most swiftly performed by MATHEMATICA in exact arithmetic. (Note that since x and F are vectors, F' is a matrix.) The size of $I_{i,k}$ (the number of terms in $F_{i,k}$) for all i increases sub-exponentially with increasing k, as shown in Fig. 8.18 (left).

Because the attractor is finite and the dynamics is ergodic, a single trajectory visits the complete phase space on long time scales. Each trajectory experiences the same local errors, only at different times. Consequently, the maximum local error at given order p and time step Δt is $L_{\max}(p, \Delta t) = \max_{k \in \mathbb{N}} |R_p(k \Delta t)|$ and is equal for all coordinates. In finite floating-point arithmetic with precision ε_{M}, the maximum local error scales with ε_{M} and Δt as

$$
L_{\max}(p, \Delta t) = O\left(\varepsilon_{\mathrm{M}} \max_i \sum_{k=1}^{p} (\Delta t)^k S_{i,k}\, P_{i,k}\, \Lambda^{P_{i,k}}\right) + O\left((\Delta t)^{p+1}\right)\,,
$$

where $S_{i,k} = \sum_{\alpha \in I_{i,k}} \alpha$ is the sum of powers in $I_{i,k}$, $P_{i,k} = \max I_{i,k}$ is the maximum power in $I_{i,k}$, and Λ is the typical length scale of the attractor.

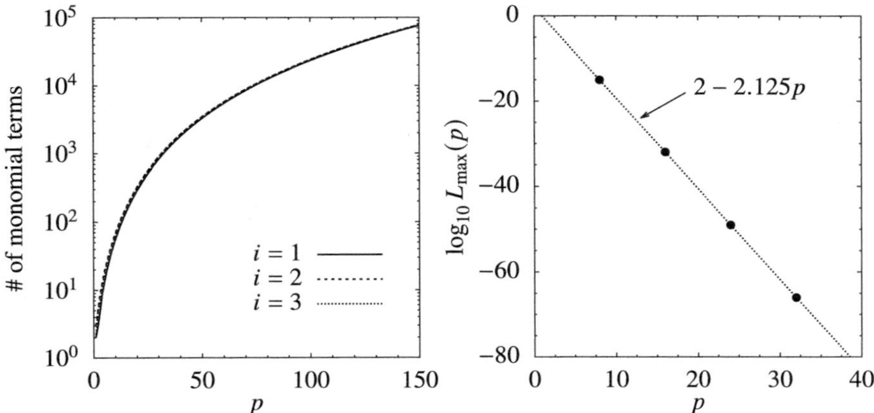

Fig. 8.18 [LEFT] The number of monomial terms in $F_{i,k}$ up to $k = p$ in the Taylor series expansion, as a function of order p. [RIGHT] The maximal local error $L_{\max}$ as a function of order p at time step $\Delta t = 10^{-3}$ and precision $\varepsilon_{\mathrm{M}} = 10^{-1024}$ (symbols ●)

The first term of $L_{\max}$ corresponds to round-off errors, while the second term appears because the integrator has a finite order. In our numerical calculation the arithmetic precision is so large that round-off errors have a negligible effect on the dynamics, hence the second term dominates. From Fig. 8.18 (right) we see that $L_{\max}$ depends exponentially on p like $L_{\max}(p, \Delta t) \asymp (\Delta t / \rho_{\mathrm{eff}})^p$, where ρ_{eff} is the effective Taylor series convergence radius for the attractor.

The deviation of the numerical trajectory $\hat{x}(t)$ obtained by the Lie-series integrator with step Δt, and the exact one, $x(t)$, which both start at the same point, $x(0) = \hat{x}(0)$, is represented by the global error $G_n = \|\hat{x}(n\Delta t) - x(n\Delta t)\|_2$. In the limit $G_n \ll 1$ the upper bound of the global error can be estimated as

$$G_n \leq G_{\max} \sim L_{\max}(p, \Delta t) \frac{\exp(\lambda_{\max} n \Delta t) - 1}{\exp(\lambda_{\max} \Delta t) - 1}, \qquad n \to \infty . \tag{8.68}$$

Here we have assumed that $G_{\max}$ grows *maximally* with n on average, i.e. it increases by a factor $\exp(\lambda_{\max} \Delta t)$ in one iteration, where $\lambda_{\max}$ is the maximum Lyapunov exponent. Consequently, $G_{\max}$ grows exponentially as $O(\exp(\lambda_{\max} n \Delta t))$. In the limit of small perturbations, $\varepsilon \to 0$, the maximum Lyapunov exponent $\lambda_{\max}$ can be determined by using the asymptotic relation

$$\|\phi(t, x + \varepsilon) - \phi(t, x)\|_2 \sim \|\varepsilon\|_2 \exp(\lambda_{\max} t) , \qquad t \to \infty .$$

In practice, this is approximately valid for times $t = n\Delta t$ and perturbations such that $\lambda_{\max} t \ll -\log \|\varepsilon\|_2$. From the results presented in Fig. 8.19 (left) we find $\lambda_{\max} = 0.90054 \pm 10^{-5}$, which agrees well with the value found in [47].

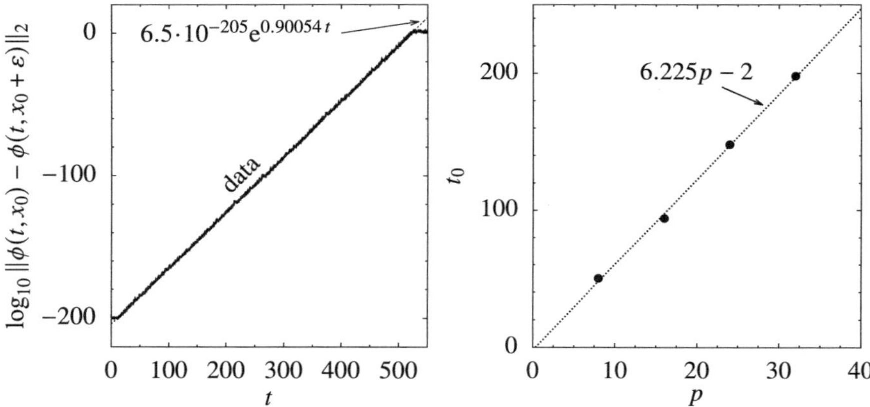

Fig. 8.19 [LEFT] The divergence of trajectories starting at neighboring points x_0 and $x_0 + \varepsilon$ for $x_0 = (1, 1, 1)^{\mathrm{T}}$ and $\varepsilon = (10^{-200}, 0, 0)^{\mathrm{T}}$. [RIGHT] The time t_0 until which the global error remains below a given threshold value ε, as a function of order p, for $\varepsilon = 0.1$ and a trajectory starting at $x(0) = (1, 1, 1)^{\mathrm{T}}$ (symbols •)

The upper bound of G_n given in Eq. (8.68) is quite loose, because the increase of the error is usually not maximal as assumed. A tighter bound can be found empirically by calculating the envelope of G_n along $n \in \mathbb{N}$, which is modeled by

$$G_{\max} \approx L_{\max} \exp(\mu n \Delta t) .$$

Here $\mu \leq \lambda_{\max}$ is the effective Lyapunov exponent that can be computed from the threshold time $t_0(\varepsilon)$ until which the global error remains below a specified threshold value ε. The threshold time as a function of the order of the integrator is shown in Fig. 8.19 (right), and $\mu = \log(\varepsilon/L_{\max})/t_0(\varepsilon)$. By applying this formula to the data in the Figure, we find $\mu \approx 0.79$: the global error increases significantly slower than is maximally possible on average.

For $\mu t \gg 1$ and $\mu \Delta t \ll 1$, these results can be condensed in an empirical formula $\log_{10} G_{\max} \approx -2.135p + 2 + 1.842t - \log_{10}(0.79\Delta t)$, which allows us to estimate the resources needed to compute the trajectory until $t = 10^3$ to an accuracy of one digit. By solving $G_{\max} \approx 1$ at $t = 10^3$ for p, we find the minimal order of the Lie-series integrator $p = 165$, while $G_{\max}$ at $p = 165$ and $t = 0$ gives the upper bound on the precision of the derivatives $F_{i,k}$, which is $\varepsilon_{\mathrm{der}} \approx 10^{-350}$. In order to compute the derivatives to precision $\varepsilon_{\mathrm{der}}$, we obviously need to work with arithmetic much finer than that, i.e. $\varepsilon_{\mathrm{M}} \ll \varepsilon_{\mathrm{der}}$.   ◁

8.13 Problems

8.13.1 Time Dependence of Filament Temperature

An example of a scalar first-order initial-value problem is the time dependence of the temperature of a filament (e.g. in a light bulb) [48]. At time zero we expose the filament with resistance R to a current pulse from a charged capacitor with capacitance C. The differential equation for the temperature is

$$mc_\mathrm{p}\frac{\mathrm{d}T}{\mathrm{d}t} = RI_0^2 \exp\left(-\frac{2t}{RC}\right) - \sigma S T^4.$$

By introducing dimensionless variables we obtain an equation with a single parameter a,

$$\frac{\mathrm{d}y}{\mathrm{d}x} = a\,\mathrm{e}^{-2x} - y^4 . \tag{8.69}$$

We neglect the radiation received by the filament from the environment, so the temperature will ultimately approach zero. Moreover, we assume that the current pulse is strong enough that the initial internal energy of the filament may be ignored as well. The initial condition for (8.69) is therefore $y(0) = 0$.

⊙ Solve this problem by using several Euler's methods discussed in Sect. 8.2, as well as by select methods of Sects. 8.3 and 8.4. Determine the required step size for the solution to remain stable. Choose the method (and the corresponding step sizes) to compute the families of solutions for $a = 2(4)18$! What would be your method of choice for a precise determination of the maximum temperature and the times at which these maxima occur? If the radiation from the environment is not neglected, Eq. (8.69) involves two parameters:

$$\frac{\mathrm{d}y}{\mathrm{d}x} = a\,\mathrm{e}^{-2x} - b\,(y^4 - 1) .$$

Find the families of solutions of these equations for $a = 5$ and $b = 0(0.5)2$! Explain the physical meaning of parameters a and b.

8.13.2 Oblique Projectile Motion with Drag Force and Wind

In this problem (adapted from [49]) we would like to compute the trajectory of an object launched under an angle with a specific initial velocity, and experiencing a quadratic drag force and the presence of wind. The equations of motion are

$$
\begin{aligned}
\dot{x} &= v \cos\theta\,, \\
\dot{z} &= v \sin\theta\,, \\
\dot{\theta} &= -\frac{g}{v}\cos\theta\,, \\
\dot{v} &= -\frac{F_{\mathrm{d}}}{m} - g\sin\theta\,,
\end{aligned}
$$

where the drag force is $F_{\mathrm{d}} = \frac{1}{2}c\rho S\left((\dot{x}-w(t))^2 + \dot{z}^2\right)$.

$\odot$ Solve this problem by using methods for the solution of systems of ordinary differential equations. We shoot the cannon-ball with mass $m = 15\,\mathrm{kg}$ and velocity $v_0 = 50\,\mathrm{m/s}$ from the origin $(x, z) = (0, 0)$ at an angle θ_0 with respect to the positive x-axis. Change the initial angle θ_0 in steps of $5°$. Other parameters are $c = 0.2$ (drag coefficient), $\rho = 1.29\,\mathrm{kg/m^3}$ (density of air), $S = 0.25\,\mathrm{m^2}$ (projectile cross-sectional area). The winds are blowing in the x-direction: no wind ($w(t) = 0$); constant-velocity headwind ($w(t) = -10\,\mathrm{m/s}$); non-uniform tailwind (blowing with $w(t) = +10\,\mathrm{m/s}$ every odd second and with $w(t) = 0$ every even second); and gusty wind (Gaussian distributed $w(t)$ with zero mean and $10\,\mathrm{m/s}$ standard deviation). Think of another aspect of the problem and discuss it accordingly.

$\oplus$ Solve the problem in the case that the winds blow in arbitrary directions in the (x, y) plane.

8.13.3 Influence of Fossil Fuels on Atmospheric CO_2 Content

The study of the influence of fossil fuel burning on the atmospheric CO_2 content is an important geochemical problem. The fraction of CO_2 in the present atmosphere is $\approx 3.5 \cdot 10^{-4}$, but a small change in this fraction may have strong repercussions on the global climate. A relatively simple model [50] can be used to simulate the interaction of carbon compounds in the atmosphere, shallow waters, and deep oceans. The main variables in the model are the partial pressure of CO_2 in the atmosphere, p, the concentrations of carbonates (dissolved C) in shallow and deep oceans, σ_s and σ_d, and the alkalinities in shallow and deep oceans, α_s and α_d. The auxiliary variables in shallow oceans, where chemical processes between the gaseous CO_2 and dissolved carbonates occur, are the concentrations of hydrogen carbonates and carbonates, h_s and c_s, and the partial pressure of CO_2 in water, p_s. The burning of fossil fuels from the beginning of the industrial age is the source of CO_2 entering as the function f. The corresponding system of differential equations is

$$\frac{\mathrm{d}p}{\mathrm{d}t} = \frac{p_\mathrm{s} - p}{d} + \frac{f(t)}{\mu_1} \,,$$

$$\frac{\mathrm{d}\sigma_\mathrm{s}}{\mathrm{d}t} = \frac{1}{v_\mathrm{s}} \left[w(\sigma_\mathrm{d} - \sigma_\mathrm{s}) - k_1 - \mu_2 \frac{p_\mathrm{s} - p}{d} \right] \,,$$

$$\frac{\mathrm{d}\sigma_\mathrm{d}}{\mathrm{d}t} = \frac{1}{v_\mathrm{d}} \left[k_1 - w(\sigma_\mathrm{d} - \sigma_\mathrm{s}) \right] \,,$$

$$\frac{\mathrm{d}\alpha_\mathrm{s}}{\mathrm{d}t} = \frac{1}{v_\mathrm{s}} \left[w(\alpha_\mathrm{d} - \alpha_\mathrm{s}) - k_2 \right] \,,$$

$$\frac{\mathrm{d}\alpha_\mathrm{d}}{\mathrm{d}t} = \frac{1}{v_\mathrm{d}} \left[k_2 - w(\alpha_\mathrm{d} - \alpha_\mathrm{s}) \right] \,.$$

It is complemented by the shallow-ocean equilibrium equations:

$$h_\mathrm{s} = \frac{\sigma_\mathrm{s} - (\sigma_\mathrm{s}^2 - k_3\alpha_\mathrm{s}(2\sigma_\mathrm{s} - \alpha_\mathrm{s}))^{1/2}}{k_3} \,, \qquad c_\mathrm{s} = \frac{\alpha_\mathrm{s} - h_\mathrm{s}}{2} \,, \qquad p_\mathrm{s} = k_4 \frac{h_\mathrm{s}^2}{c_\mathrm{s}} \,.$$

The numerical constants are $d = 8.64$, $\mu_1 = 4.95 \cdot 10^2$, $\mu_2 = 4.95 \cdot 10^{-2}$, $v_\mathrm{s} = 0.12$, $v_\mathrm{d} = 1.23$, $w = 0.001$, $k_1 = 2.19 \cdot 10^{-4}$, $k_2 = 6.12 \cdot 10^{-5}$, $k_3 = 0.997148$, $k_4 = 6.79 \cdot 10^{-2}$.

⊙ Solve the system of equations with the initial conditions $p = 1.00$, $\sigma_\mathrm{s} = 2.01$, $\sigma_\mathrm{d} = 2.23$, $\alpha_\mathrm{s} = 2.20$ and $\alpha_\mathrm{d} = 2.26$ in year $t = 1000$. The emission of CO_2 between the years 1000 and 5000 (Fig. 8.20 (left)) is approximated by the function

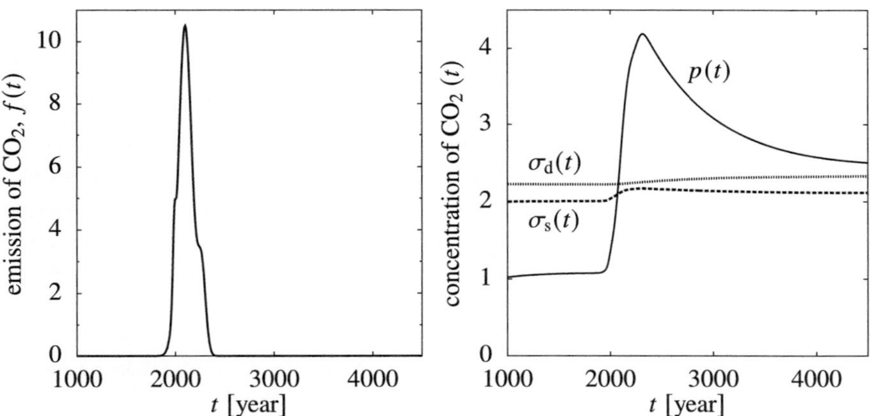

Fig. 8.20 [LEFT] The function f describing the emissions of CO_2 into the atmosphere from year 1000 to 5000. [RIGHT] Concentrations of CO_2 in the atmosphere (p), shallow (σ_s), and deep ocean (σ_d) as functions of time (solution by the adaptive RK4 method)

$$f(t) = \sum_{i=1}^{3} c_i \exp\left(-\frac{(t - t_i)^2}{s_i^2}\right) , \qquad \begin{aligned} c_1 &= 2.0 , \quad t_1 = 1988 , \quad s_1 = 21 , \\ c_2 &= 10.5 , \quad t_2 = 2100 , \quad s_2 = 96 , \\ c_3 &= 2.7 , \quad t_3 = 2265 , \quad s_3 = 57 . \end{aligned}$$

Determine the carbon concentrations in the atmosphere, shallow, and deep oceans from year 1000 to 5000 by select integrators and reproduce Fig. 8.20 (right). Which method is the most appropriate? Compare the initial and final concentrations. When does the atmospheric concentration of CO_2 reach its maximum? The chemical processes have quite different reaction rates, so the problem outlined above is mildly *stiff*: stiffness was introduced in Sect. 8.10.

8.13.4   *Synchronization of Globally Coupled Oscillators*

Kuramoto's model [51, 52] describes a large set of weakly (but non-linearly) coupled oscillators whose natural frequencies have a certain probability distribution. Due to the non-linearity of the coupling, a fraction of the oscillators may oscillate in phase ("collective synchronization") in specific conditions. This effect can be observed in networks of pacemaker cells in the brain and heart, in synchronous flashing of firefly swarms, with crickets chirping in unison, or even in electronic circuits based on Josephson junctions.

In this problem we discuss a finite number ($N \gg 1$) of coupled oscillators. Their dynamics is governed by the system of non-linear differential equations

$$\frac{\mathrm{d}\theta_i}{\mathrm{d}t} = \omega_i + \frac{K}{N} \sum_{j=1}^{N} \sin(\theta_j - \theta_i) , \qquad i = 1, 2, \ldots, N , \tag{8.70}$$

where θ_i is the phase of the ith oscillator, ω_i is its natural (proper, non-coupled) frequency, and K is the coupling constant. Let the natural frequencies ω_i be distributed according to some probability density $g(\omega)$ which is even and monotonously decreasing on either side of the mean value Ω, so $g(\Omega + \omega) = g(\Omega - \omega)$. Rotational symmetry allows us to set $\Omega = 0$ and redefine $\theta_i \to \theta_i + \Omega t$ for each t (we change to a system rotating uniformly with the frequency Ω): the equations of motion do not change, only the position of the peak of $g(\omega)$ shifts and we have $g(\omega) = g(-\omega)$.

The behavior of the set of oscillators is described by the complex order parameter that can be computed from the dynamical equations at any time t and measures the collective "rhythm" of the whole population of oscillators. We define it as

$$r(t)e^{i\psi(t)} = \frac{1}{N} \sum_{j} e^{i\theta_j(t)} .$$

The modulus of the parameter, $r(t)$, measures the level of coherence, while its phase, $\psi(t)$, measures the average phase; see Fig. 8.21. If K is smaller than the critical

Fig. 8.21 The phases of individual oscillators in Kuramoto's model, θ_j, and the corresponding average phase ψ

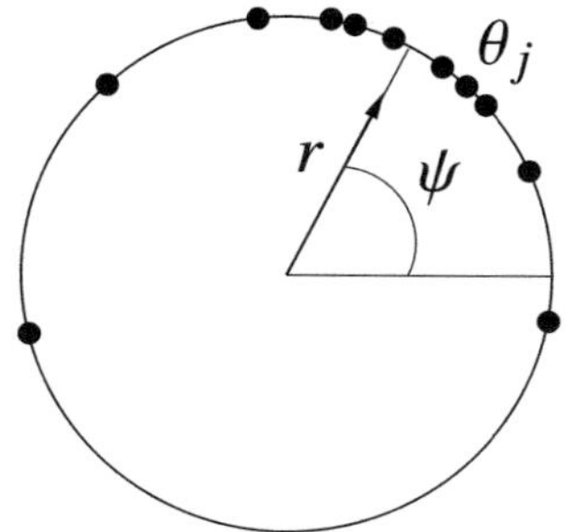

value $K_* = 2/\pi g(0)$, the oscillators oscillate incoherently (each with a different frequency and mutually uncorrelated in phase). After long times $r(t)$ oscillates out, with fluctuations on the order of $1/\sqrt{N}$, as anticipated for θ_j distributed uniformly on the interval $[0, 2\pi]$. When K is increased above K_*, an ever larger fraction of oscillators oscillate coherently: their phases tend to congregate around the average phase ψ, while the modulus of the order parameter asymptotically approaches a non-zero value $\lim_{t\to\infty} r(t) = r_\infty$. The fraction of the oscillators oscillating coherently after long times is

$$\int_{-Kr_\infty}^{+Kr_\infty} g(\omega)\, d\omega \ .$$

For certain distributions $g(\omega)$, both K_* and r_∞ can be calculated analytically [53]. For the Lorentz distribution of natural frequencies,

$$g(\omega) = \frac{\Delta}{\pi}\,\frac{1}{\omega^2 + \Delta^2} \ ,$$

we obtain $K_* = 2\Delta$ and the dependence

$$r_\infty = [1 - (K_*/K)^2]^{1/2} \ , \quad K > K_* \ .$$

⊙ Discuss the Kuramoto set of oscillators with Lorentz or Gauss distributions of natural frequencies. Map the chosen continuous distribution to a sufficiently large discrete set of oscillators (at least $N \approx 100$), and solve the system (8.70). At time zero the phases θ_j should be distributed uniformly on $[0, 2\pi]$. Use the solution $\theta_j(t)$ to compute the complex order parameter and monitor the behavior of $r(t)$ (Fig. 8.22 (left)) for the chosen value $K < K_*$ and for some value $K > K_*$. The critical value K_* is between 1.0 and 2.0. Compute $r(t)$ at long times (r_∞) for different values of the coupling parameter K in the vicinity of K_*. You ought to notice a phase transition with an approximately square-root dependence (Fig. 8.22 (right)). Draw the time

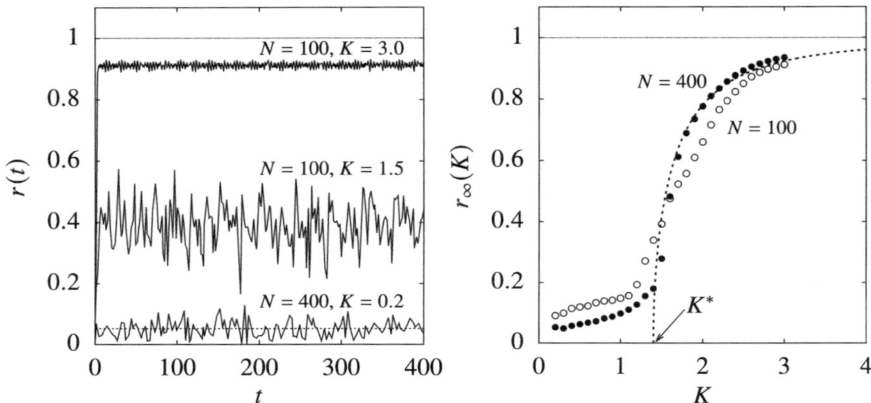

Fig. 8.22 [LEFT] The modulus of the order parameter as a function of t. The dotted line at the bottom is the average of $r(t)$ in $[0, 400]$, which is $0.052 \approx 1/\sqrt{400}$. [RIGHT] The value of r after long times as a function of K for $N = 100$ and $N = 400$

dependence of the phase $\psi(t)$. Increase the number of oscillators to a reasonable limit (in terms of CPU time) and compare the results to the previous ones. Do not run the integrator with too strict precision tolerances.

8.13.5   *Excitation of Muscle Fibers*

Non-linear evolution equations with highly interesting solutions appear in modeling of muscle fiber excitations by external voltage pulses [54]. The relevant quantities are the cell membrane potential $V(t)$ that measures the current deviation of the potential difference on two sides of the membrane from a reference value (for example, in the unexcited state), and the set of variables $w_i(t)$ describing the fraction of the ion channel that, at time t and specific potential difference, are in the conducting, non-conducting, or some other state. Already the relatively simple Morris–Lecar model [55] with two ion channels (Ca^{2+} and K^+) yields an extremely colorful physical picture. The dynamics of the voltage perturbation along the fiber is described by the coupled equations

$$C\frac{dV}{dt} = I - \overline{g}_{Ca} m_\infty(V)(V - V_{Ca}) - \overline{g}_K w(V - V_K) - \overline{g}_L(V - V_L) + s(t),$$

$$\frac{dw}{dt} = \phi \frac{w_\infty(V) - w}{\tau_w(V)},$$

where C is the membrane capacitance per unit surface, $\overline{g}_{Ca}$, $\overline{g}_K$, and $\overline{g}_L$ are the specific conductivities for the ion channels Ca^{2+}, K^+, and leakage, I is the static component of the surface current density through the fiber, and $s(t)$ is the external perturbation. The quantity w is the fraction of the open channels K^+. We have denoted

$$m_\infty(V) = \frac{1}{2}\left[1 + \tanh\left(\frac{V - V_1}{V_2}\right)\right],$$

$$w_\infty(V) = \frac{1}{2}\left[1 + \tanh\left(\frac{V - V_3}{V_4}\right)\right],$$

$$\tau_w(V) = \left[1 + \cosh\left(\frac{V - V_3}{2V_4}\right)\right]^{-1},$$

where $m_\infty(V)$ and $w_\infty(V)$ are the fractions of open channels Ca^{2+} and K^+ in the stationary state ($t \to \infty$). In the model, the characteristic time τ_w for the K^+ channel to open depends on the current value of V. (For the Ca^{2+} channel we assume instantaneous activation by V, otherwise we would have two equations, $dw_1/dt = \ldots$, $dw_2/dt = \ldots$ and two characteristic times $\tau_{w1} = \ldots, \tau_{w2} = \ldots$.)

⊙ Solve the Morris–Lecar system by two or three integration methods with various step sizes. Reproduce Fig. 8.23 (left) with the initial conditions listed in the first four rows of Table 8.1. Use the parameters $V_1 = -1.2$, $V_2 = 18.0$, $V_3 = 2.0$, $V_4 = 30.0$, $\overline{g}_{Ca} = 4.4$, $\overline{g}_K = 8.0$, $\overline{g}_L = 2.0$, $V_{Ca} = 120.0$, $V_K = -84.0$, $V_L = -60.0$, $\phi = 0.04$, and $C = 20.0$. (On purpose, we have not converted the equations to dimensionless form, as all quantities have a clear physical meaning. For units of all quantities see [55].)

Find the maximum and minimum values of the potential, V_{max} and V_{min}, at large times and determine the stationary-state curves (*nullclines*) that connect all points with the properties $dV/dt = 0$ or $dw/dt = 0$ in the phase diagram (V versus w), see Fig. 8.23 (right). The intersection of these curves is the asymptotic stationary state, if such a state exists.

Observe the time dependence of V on a wider interval $[0, 800]$ when you act on the system by an external pulse $s(t)$ of the form

$$s(t) = 30\, H(t - 100)H(105 - t) + 30\, H(t - 470)H(475 - t),$$

where $H(t)$ is the Heaviside (step) function. In an actual experiment this truly represents a 5 ms long step pulse with the peak current density $30\ \mu A/cm^2$ at time 100 ms and another such pulse at time 470 ms.

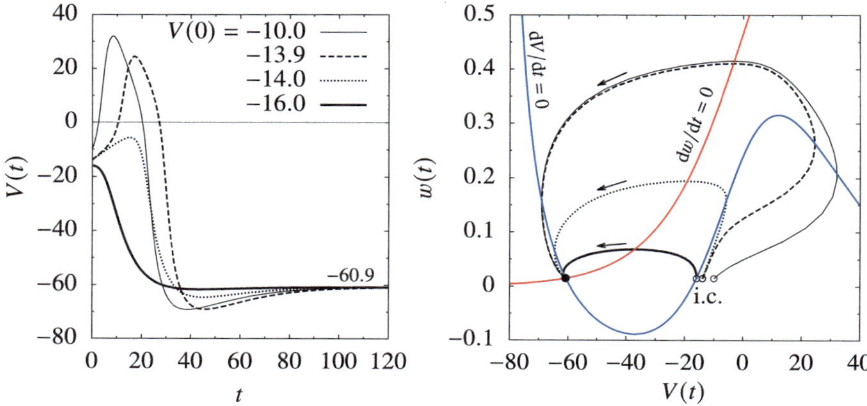

Fig. 8.23 Solutions of the Morris–Lecar model. [LEFT] The membrane potential V as a function of time for different initial conditions. [RIGHT] Phase diagram ($w(t)$ versus $V(t)$) for the same initial conditions (symbols ∘) as in the left figure. Also shown are the stationary-state curves connecting the points with $dV/dt = 0$ and $dw/dt = 0$. They intersect at $(V, w) \approx (-60.9, 0.015)$, the fixed point corresponding to the stationary state (symbol •)

Table 8.1 Initial conditions for V and w, the current density I, the integration interval length T, and the list of tasks for solving the Morris–Lecar model. The curves of the stationary state (s.s.) and the shape of the pulse $s(t)$ are defined in the text

$V(0)$	$w(0)$	I	T	Observe
-16.0	0.014915	0	150	$V(t)$, $w(t)$, $V(w)$, s.s. curves
-14.0	0.014915	0	150	
-13.9	0.014915	0	150	
-10.0	0.014915	0	150	
-10.0	0.014915	0(1)300	800	$V_{\max}(I)$, $V_{\min}(I)$
-26.59	0.129	90	800	$V(t)$ with double pulse $s(t)$

8.13.6 *Restricted Three-Body Problem (Arenstorf Orbits)*

A basic problem in astronomy is to find the trajectory of a light object (for example, a satellite) in the presence of two much heavier bodies whose motion is not influenced by the light object. The heavy bodies with a mass ratio $\mu : (1 - \mu)$ circle in the (x, y) plane with frequency 1 around their common center of gravity, which is at the origin: the ratio of their orbital radii is then $(1 - \mu) : \mu$. In general the light object may move outside the plane defined by the orbits of heavy bodies. The dimensionless equations of motion for the light body are

$$\ddot{x} = x + 2\dot{y} - \frac{(1 - \mu)(x + \mu)}{r^3} - \frac{\mu(x - 1 + \mu)}{s^3} \, ,$$

$$\ddot{y} = y - 2\dot{x} - \frac{(1 - \mu)y}{r^3} - \frac{\mu y}{s^3} \, ,$$

$$\ddot{z} = -\frac{(1 - \mu)z}{r^3} - \frac{\mu z}{s^3} \, ,$$

where

$$r = \sqrt{(x + \mu)^2 + y^2 + z^2} \, , \qquad s = \sqrt{(x - 1 + \mu)^2 + y^2 + z^2} \, .$$

A detailed analysis of planar solutions of the special case $\mu = 0.5$ (frequently used as a benchmark test for symplectic methods) can be found in [56].

⊙ Solve the restricted three-body problem corresponding to the Earth and Moon as the heavy objects, hence $\mu = M_{\text{Moon}}/M_{\text{Earth}} = 0.012277471$. Consider the planar case ($z = 0$) with the initial conditions (example parameters from [32])

$$(x(0), y(0)) = (0.994, \, 0) \, ,$$
$$(\dot{x}(0), \dot{y}(0)) = (0, \, -2.00158510637908252240533786224) \, ,$$

for which the solution is periodic with the period

$$T = 17.0652165601579625588917206249 \, .$$

Use the Euler's explicit method with step $h = T/24000$ and the RK4 method with $h = T/6000$. Plot the dependence of x, $\dot{x}$, y, and $\dot{y}$ on time, as well as the phase diagrams $(x, \dot{x})$ and $(y, \dot{y})$, as in Fig. 8.24 (left). Can you obtain a periodic solution by reducing the step size in either the Euler's or RK4 method? (Pretend that the solution is periodic if the deviation of $x(10T)$, $\dot{x}(10T)$, $y(10T)$, and $\dot{y}(10T)$ from the initial conditions after ten returns does not exceed 1 %.)

⊕ Use the Dormand–Prince method 5(4) with adaptive step size. Select a few tolerances on the local error and determine the number of steps needed for one cycle in the phase diagram at this precision (Fig. 8.24 (right)). Enrich the problem by allowing non-planar orbits of the third body ($z \neq 0$).

The problem outlined here is an example of motion of satellites along the "horseshoe" orbits in the gravitational field of the Sun–Earth system. Recently two nearby Earth asteroids have been discovered, 3753 Cruithne [57] and 2002 AA29 [58], whose orbits have similar properties. Even before that, the restricted three-body problem was analyzed theoretically for the Sun–Jupiter system [59]. You can find a rich set of initial conditions for it in [60] or [61].

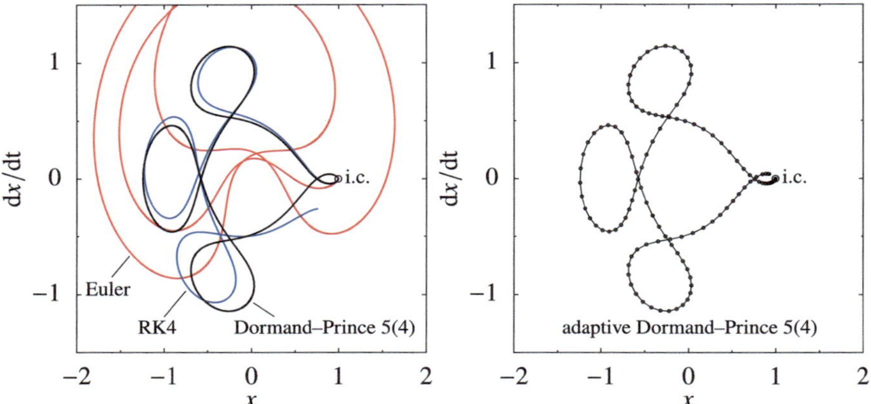

Fig. 8.24 Orbits of the light object in the restricted three-body problem. [LEFT] Solution by the explicit Euler's method (8.5) with step $h = T/24000$ and by the RK4 method (8.10) and Dormand–Prince 5(4) method (page 482) with steps $h = T/6000$. [RIGHT] Solution by the Dormand–Prince 5(4) method with adaptive step size. At 0.001 tolerance on the local error a mere 134 steps are needed

8.13.7 Lorenz System

A classical problem of atmospheric physics is the convection of a fluid with kinematic viscosity v, volume expansion coefficient β, and heat diffusion coefficient D, in a layer of thickness H in which a constant temperature difference $\Delta T_0 = T(0) - T(H) > 0$ persists between the top and bottom. Simpler cases were treated by Lord Rayleigh already in 1916, but in certain regimes of Prandtl ($\sigma = v/\beta$) and Rayleigh ($R = g\beta H^3 \Delta T_0 / Dv$) numbers the convection dynamics (velocity of motion, heat transfer) becomes very complex [62].

The Lorenz system [44]

$$\dot{X} = -\sigma X + \sigma Y \,,$$
$$\dot{Y} = -XZ + rX - Y \,, \qquad (8.71)$$
$$\dot{Z} = XY - bZ \,,$$

offers a simplified picture of such convection. The variable X represents the velocity of the convection current. The variable Y is proportional to the temperature difference between the ascending and descending convection currents (warmer fluid goes up, colder goes down), while Z is the deviation of the vertical temperature profile from the linear height dependence (a positive value implies strong gradients in the vicinity of the top or bottom edge). The derivatives in (8.71) are with respect to dimensionless time $\tau = \pi^2(1 + a^2)\kappa t / H^2$, where a is a parameter. The coefficient $r = R/R_c$ is the

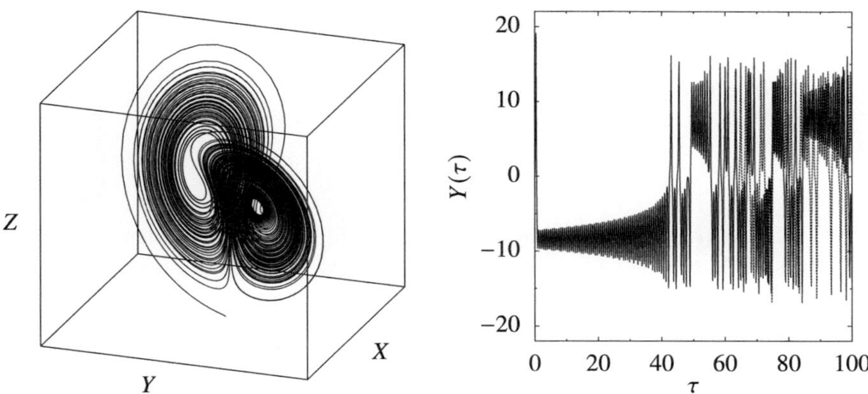

Fig. 8.25 The Lorenz problem. [LEFT] The solution on the time interval [0, 50] in the phase space of variables X, Y, and Z. [RIGHT] Dependence of the solution $Y(\tau)$ on the initial conditions. Shown are two solutions with the initial condition $Y(0) = 1$ (dashed curve) and with $Y(0) = 1 + 10^{-12}$ (full curve)

Rayleigh number in units of the critical value $R_c = \pi^4(1 + a^2)^3/a^2$ (with a minimum at $27\pi^4/4$), while $b = 4/(1 + a^2)$.

⊙ Solve the Lorenz system with parameters $\sigma = 10$, $b = 8/3$, $r = 28$, and initial conditions $X(0) = 0$, $Y(0) = 1$, $Z(0) = 0$, which are slightly "off" the convection-free state at $(0, 0, 0)$. Use the RK4 method. Plot the three-dimensional phase space (Fig. 8.25 (left)) and some typical two-dimensional slices through this space in all three planes. Quasi-stationary convection states correspond to the points $(6\sqrt{2}, 6\sqrt{2}, 27)$ and $(-6\sqrt{2}, -6\sqrt{2}, 27)$. Solve the system with slightly perturbed initial conditions. For example, take $Y(0) = 1 + \varepsilon$ instead of $Y(0) = 1$, where $|\varepsilon| \ll 1$ and observe the spread of the solutions as in Fig. 8.25 (right).

8.13.8 Sine Pendulum

In Newton's law $m\ddot{x} = F(x)$ for a non-linearized (sine) pendulum the restoring force is $F(x) = -\sin x$. Since F does not depend on the velocity, the system $m\dot{x} = p$, $\dot{p} = F(x)$, can be solved by a special trapezoidal-like scheme [48]

$$x(t + h) = x(t) + h\,u\left(t + \frac{1}{2}h\right),$$

$$u\left(t + \frac{1}{2}h\right) = u\left(t - \frac{1}{2}h\right) + h\,\frac{F(t, x(t))}{m}.$$

The auxiliary variable u best approximates the velocity $\dot{x}$ at the midpoints of $[t, t + h]$ and we assign it to those points.

⊙ Solve the differential equation

$$\frac{\mathrm{d}^2 x}{\mathrm{d}t^2} + \sin x = 0 \,, \qquad x(0) = 1 \,, \qquad \dot{x}(0) = 0 \,,$$

by the scheme given above and by the RK4 method, and compare the results. Find the step size h to achieve six-digit precision. Investigate the periodic stability of the methods: let the computation proceed over a large number (100, 1000, 10000) of oscillations and observe the change in the oscillation amplitudes. Compute the energy

$$E = V + T = 1 - \cos x + \frac{1}{2}\left(\frac{\mathrm{d}x}{\mathrm{d}t}\right)^2$$

and see if it remains constant. The exact solution can be expressed in terms of the Jacobi elliptic functions sn() and cn() implemented in the GSL library in the `gsl_sf_elljac_e()` routine. The period of oscillation of the exact solution is $4K(\sin\frac{1}{2}x(0))$, where $K(u)$ is the complete elliptic integral of the first kind: $K(\sin 0.5) \approx 1.67499$.

⊕ Study the resonance curve of the driven damped pendulum

$$\frac{\mathrm{d}^2 x}{\mathrm{d}t^2} + \beta\frac{\mathrm{d}x}{\mathrm{d}t} + \sin x = A \sin \omega_0 t \,,$$

where β is the damping coefficient, while A and ω_0 are the driving amplitude and frequency. When ω_0 is increased or decreased, the amplitudes at the same ω_0 differ: you can clearly observe hysteresis effects [63].

8.13.9 Charged Particles in Electric and Magnetic Fields

Motion of charged particles in complicated configurations of electric and magnetic fields abounds in experimental particle physics, for example, in electro-magnetic traps used to study the properties of individual particles or plasma, Wien filters with crossed electric and magnetic fields acting as velocity selectors in beam-lines, and in magnetic spectrometers. In all these cases the particle with mass m and charge e experiences the Lorentz force

$$\boldsymbol{F} = \frac{\mathrm{d}(\gamma m \boldsymbol{v})}{\mathrm{d}t} = e\left(\boldsymbol{E} + \boldsymbol{v} \times \boldsymbol{B}\right) . \tag{8.72}$$

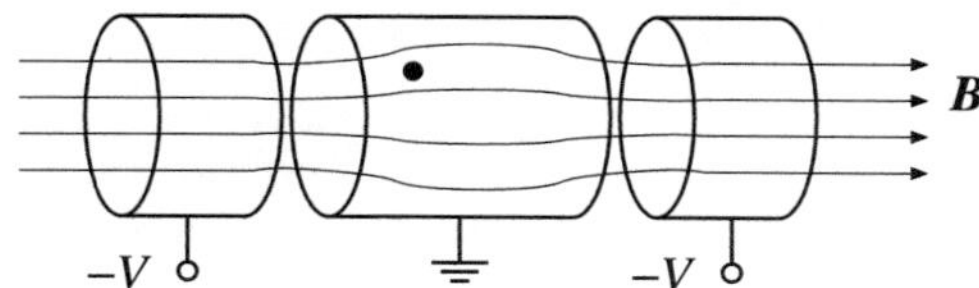

Fig. 8.26 Confinement of charged particles by a combination of electric and magnetic fields

In practical situations the hardest problem is the precise knowledge of the vector fields E and B at all places; the numerical integration of the equations of motion is the easier task.

⊙ Numerically integrate the equations of motion of a positively charged particle in the Penning trap shown in Fig. 8.26. Longitudinal confinement is provided by the electric potential, while the axially symmetric magnetic field confines the particles in the transverse directions. The system (8.72) can be solved as a system of six first-order equations,

$$\dot{r} = v \,, \qquad\qquad r(0) = r_0 \,,$$
$$\dot{v} = \kappa\,(E + v \times B) \,, \quad v(0) = v_0 \,,$$

where $\kappa = e/\gamma m$. (Note that the Lorentz factor γ depends on the magnitude of the velocity!) The electric and magnetic fields have the form

$$E_z(\pm z) = \pm\, E_0 \exp\!\left(-5(z \pm 3)^2\right) ,$$
$$B_z(\pm z) = \ \ B_0 \left(0.5 + 0.2\exp(-z^2)\right) ,$$

$E_0 = B_0 = 1$, and we neglect the transverse components of the magnetic field.

⊕ Study the precession of the average spin vector of a polarized beam of spin-1/2 particles in non-constant magnetic fields. The dynamics of the spatial part of the spin four-vector is determined by the Thomas equation

$$\frac{\mathrm{d}S}{\mathrm{d}t} = \kappa\, S \times \left[\frac{g}{2}\, B^{\parallel} + \left(1 + \frac{g-2}{2}\gamma\right) B^{\perp} \right], \qquad S(0) = S_0 \,,$$

where g is the gyro-magnetic ratio, and we split the magnetic field to its components parallel and perpendicular to the particle velocity vector, $B^{\parallel} \equiv (\hat{v} \cdot B)\,\hat{v}$ and $B^{\perp} \equiv B - B^{\parallel}$. Consider this vector equation as three new coupled scalar equations in the system, and run the integrator again. Observe the spin precession of protons with various momenta and incident angles in magnetic systems like a dipole magnet, a system of two quadrupole magnets rotated by $90°$, a sextupole magnet, or some other field configuration!

8.13.10 Chaotic Scattering

Classical planar scattering of a point particle with mass M on a potential $V(r)$ is described by the four-dimensional system of equations of motion

$$M\frac{\mathrm{d}v}{\mathrm{d}t} = -\nabla V(r) , \qquad \frac{\mathrm{d}r}{\mathrm{d}t} = v ,$$

where $r = (x, y)$ and $v = (v_x, v_y)$. Because the energy $E = \frac{1}{2}Mv^2 + V(r)$ is conserved, the phase space is only three-dimensional. Therefore the state of the particle is uniquely defined by the variables x, y, and the angle θ between the vector v and the positive x-axis, since the magnitude of the velocity can be calculated from the energy,

$$|v| = \sqrt{2(E - V(r))/M} .$$

⊙ Study the planar scattering of a particle of mass M on the potential

$$V(r) = x^2 y^2 \exp\left[-(x^2 + y^2)\right] ,$$

which has four maxima at $(x, y) = (1, \pm 1)$ and $(-1, \pm 1)$ with values $E_\mathrm{m} = 1/\mathrm{e}^2$ [64]. Let the particle with mass $M = 1$ and energy E impinge towards the central part of the potential from large distances with impact parameter b (parallel to the x-axis and at constant distance b from it).

Plot some typical particle trajectories in the (x, y) plane as a function of the parameter b (Fig. 8.27 (top right)). Compute the scattering angles ϕ for a large set of values of b, $-3 \le b \le 3$, at $E/E_\mathrm{m} = 1.626$ (regular scattering) and at $E/E_\mathrm{m} = 0.260$ (chaotic scattering). Look closer at the obtained results in the ever narrower regions of b, e.g. for $-0.6 < b < -0.1$, $-0.400 < b < -0.270$, and $-0.3920 < b < -0.3770$ (Fig. 8.27 (bottom left and right)). Compute the time delay (time spent by the particle in the central part of the potential) as a function of b. By using a constant step size integrator the delay is simply proportional to the number of steps taken.

8.13.11 Hydrogen Burning in the pp I Chain

Fusion of hydrogen nuclei occurs in stars with masses less than $\approx 1.3\,M_\odot$ in three branches of the pp chain. The first stage (pp I branch) involves the processes [65]

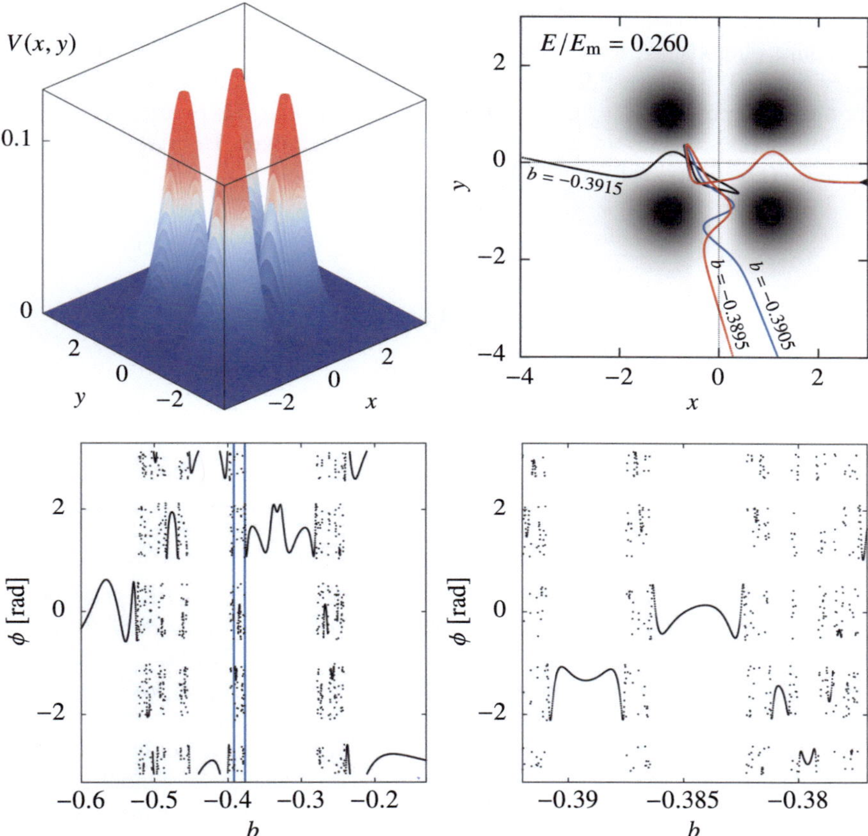

Fig. 8.27 Chaotic scattering on the potential $V(x, y) = x^2 y^2 \exp\left[-(x^2 + y^2)\right]$. [TOP LEFT] The potential $V(x, y)$. [TOP RIGHT] Dependence of the solution on initial conditions $x(0) = 3$, $y(0) = b$, $\dot{x}(0) = 0$, and $\dot{y}(0) = -\sqrt{2.0(E - V(x(0), y(0)))/M}$ with parameters $M = 1$, $E = 0.260 E_{\mathrm{m}}$. Shown are the solutions at three different impact parameters $b = -0.3915$, 0.3905, and 0.3895. [BOTTOM LEFT] Dependence of the scattering angle ϕ on b with $-0.60 \leq b \leq -0.13$. [BOTTOM RIGHT] The zoom-in in the region $-0.392 \leq b \leq -0.377$ (denoted by vertical lines in the left panel). We are witnessing classical chaos: we observe a large sensitivity of the system to the initial conditions, and self-similarity (equal or similar structures appear at different scales)

$$
\begin{aligned}
{}^{1}\mathrm{H} + {}^{1}\mathrm{H} &\longrightarrow {}^{2}\mathrm{H} + e^{+} + \nu_{e} && (\lambda_{\mathrm{pp}}), \\
{}^{1}\mathrm{H} + {}^{1}\mathrm{H} + e^{-} &\longrightarrow {}^{2}\mathrm{H} + \nu_{e} && (\lambda'_{\mathrm{pp}}), \\
{}^{2}\mathrm{H} + {}^{1}\mathrm{H} &\longrightarrow {}^{3}\mathrm{He} + \gamma && (\lambda_{\mathrm{pd}}), \\
{}^{3}\mathrm{He} + {}^{3}\mathrm{He} &\longrightarrow {}^{4}\mathrm{He} + {}^{1}\mathrm{H} + {}^{1}\mathrm{H} && (\lambda_{33}),
\end{aligned}
$$

where λ_i are the reaction rates. The pp I branch contributes 85 % to the solar luminosity, but the process rates in it are very different due to a large span of reaction cross-sections and Coulomb barriers [66, 67]. We express λ_i in units of reactions per unit time per $(\mathrm{mol/cm^3})^{N-1}$, where N is the number of interacting particles excluding photons. At temperatures in the solar interior ($\approx 15 \cdot 10^6\,\mathrm{K}$) they are $\lambda_{\mathrm{pp}} = 8.2 \cdot 10^{-20}$, $\lambda'_{\mathrm{pp}} = 2.9 \cdot 10^{-24}$, $\lambda_{\mathrm{pd}} = 1.3 \cdot 10^{-2}$, $\lambda_{33} = 2.3 \cdot 10^{-10}$.

For the listed processes $\lambda'_{\mathrm{pp}} \ll \lambda_{\mathrm{pp}} \ll \lambda_{33} \ll \lambda_{\mathrm{pd}}$. The first two processes are by far the slowest ones, as they are governed by the weak interaction. Here we approximate $\lambda'_{\mathrm{pp}} = 0$. The equations for the pp I branch then become

$$
\begin{aligned}
\frac{\mathrm{d}n_{\mathrm{p}}}{\mathrm{d}t} &= -\lambda_{\mathrm{pp}}n_{\mathrm{p}}^2 - \lambda_{\mathrm{pd}}n_{\mathrm{p}}n_{\mathrm{d}} + \lambda_{33}n_3^2 \,, \\
\frac{\mathrm{d}n_{\mathrm{d}}}{\mathrm{d}t} &= \lambda_{\mathrm{pp}}\frac{n_{\mathrm{p}}^2}{2} - \lambda_{\mathrm{pd}}n_{\mathrm{p}}n_{\mathrm{d}} \,, \\
\frac{\mathrm{d}n_3}{\mathrm{d}t} &= \lambda_{\mathrm{pd}}n_{\mathrm{p}}n_{\mathrm{d}} - \lambda_{33}n_3^2 \,, \\
\frac{\mathrm{d}n_4}{\mathrm{d}t} &= \lambda_{33}\frac{n_3^2}{2} \,,
\end{aligned}
\tag{8.73}
$$

where we have denoted the isotope concentrations by $n_{\mathrm{p}} = n(^1\mathrm{H})$, $n_{\mathrm{d}} = n(^2\mathrm{H})$, $n_3 = n(^3\mathrm{He})$, and $n_4 = n(^4\mathrm{He})$.

$\odot$ Solve the system (8.73) on the time interval $10^{10}\mathrm{s} \le t \le 10^{22}\mathrm{s}$. The initial condition for n_{p} can be computed from the estimate $(\lambda_{\mathrm{pp}}n_{\mathrm{p}})^{-1} \approx 10^{10}$ years which is valid at the conditions in the solar interior, and we set $n_{\mathrm{d}} = n_3 = n_4 = 0$. At first you may restrict the computation to large times, when $^2\mathrm{H}$ and $^3\mathrm{He}$ are in equilibrium $(\mathrm{d}n_{\mathrm{d}}/\mathrm{d}t = \mathrm{d}n_3/\mathrm{d}t = 0)$. In that case the stiff system of four equations reduces to a non-stiff system of two equations. To consider all times except the shortest, assume that $^2\mathrm{H}$ is in equilibrium: then the equation for $^3\mathrm{He}$ can be solved analytically by assuming that the concentrations of $^1\mathrm{H}$ and $^4\mathrm{He}$ do not change substantially, and by using the solution for n_3 in the equations for n_{p} and n_4. For finding the solution at arbitrary times, use an integrator tailored to stiff systems (see Fig. 8.28 (left)).

$\oplus$ Augment the basic system of equations for the pp I branch with contributions of heavier isotopes, and compare the results. Use the reactions given in [65–68], which list the reaction rates also for temperatures that are different from those in the solar interior.

8.13.12   Oregonator

Oregonator is a domesticated name for the chemical reaction of HBrO_2, Br^-, and $\mathrm{Ce(IV)}$. The dynamics is described by a stiff set of equations [69]

$$
\begin{aligned}
y_1' &= \kappa_1 \left(y_2 + y_1 \left(1 - \alpha y_1 - y_2\right)\right) \,, \\
y_2' &= \kappa_2 \left(y_3 - y_2 \left(1 + y_1\right)\right) \,, \\
y_3' &= \kappa_3 \left(y_1 - y_3\right) \,,
\end{aligned}
\tag{8.74}
$$

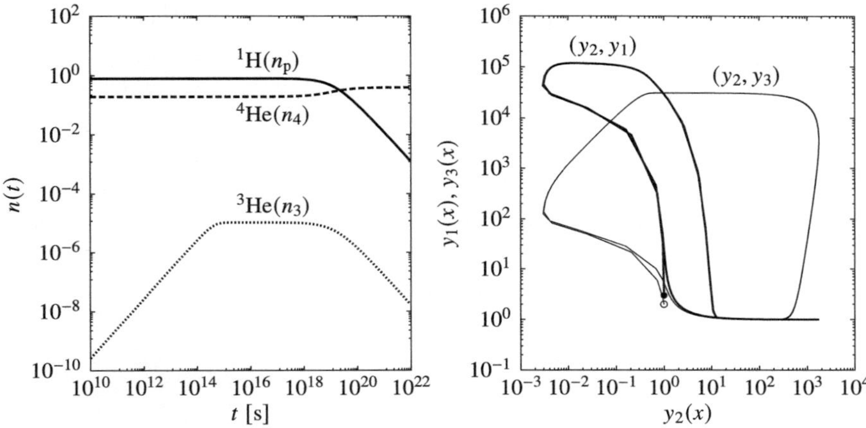

Fig. 8.28 Examples of stiff differential equations. [LEFT] Hydrogen burning in the pp I chain (Eqs. (8.73)). The solution for n_d lies further ≈ 8 orders of magnitude below the shown area. [RIGHT] Limit cycles of the Oregonator (Eqs. (8.74)). The symbols $\circ$ and $\bullet$ denote the initial conditions

where $\kappa_1 = 1/\kappa_2 = 77.27$, $\kappa_3 = 0.161$, and $\alpha = 8.375 \cdot 10^{-6}$. As expected from a stiff system, the solutions change over many orders of magnitude (Fig. 8.28 (left and right)).

$\odot$ Solve the system (8.74) with initial conditions $y_1(0) = 3$, $y_2(0) = 1$, $y_3(0) = 2$. Use an explicit integrator of your own choice and the implicit fifth-order integrator Radau 5 described on page 492. If possible, resort to adaptive step size control in both cases. Plot the solutions $y_1(x)$, $y_2(x)$, and $y_3(x)$ for $0 \leq x \leq 360$. The solutions on this interval form interesting three-dimensional limit cycles. Display them by plotting pairs of coordinates (y_2, y_1) and (y_2, y_3) as shown in Fig. 8.28 (right). What are the required step sizes in the computation with the explicit and implicit schemes, respectively?

8.13.13 Kepler's Problem

The solution of the Kepler's problem is one of the very first homeworks of any astronomer. We treat the planar motion of two gravitationally attracting bodies by placing one body at the origin and describing the other body by spatial coordinates (q_1, q_2) and momenta (p_1, p_2). The dimensionless equations of motion (Newton's law) are

$$\ddot{q}_1 = -\frac{q_1}{(q_1^2 + q_2^2)^{3/2}}, \qquad \ddot{q}_2 = -\frac{q_2}{(q_1^2 + q_2^2)^{3/2}}.$$

These equations describe the time evolution of a Hamiltonian system with three conserved quantities. The first one is the Hamiltonian (the total energy)

$$H(p_1, p_2, q_1, q_2) = \frac{1}{2}\left(p_1^2 + p_2^2\right) - \frac{1}{\sqrt{q_1^2 + q_2^2}}, \qquad p_i = \dot{q}_i .$$

The other quadratic invariant is the angular momentum $L(p_1, p_2, q_1, q_2) = q_1 p_2 - q_2 p_1$.

⊙ Solve Kepler's problem with eccentricity $e = 0.6$ and initial conditions

$$q_1(0) = 1 - e , \qquad \dot{q}_1(0) = 0 , \qquad q_2(0) = 0 , \qquad \dot{q}_2(0) = \sqrt{\frac{1+e}{1-e}} .$$

The orbital period is 2π, and the exact values of the energy and angular momentum are $H_0 = -1/2$ and $L_0 = \sqrt{1 - e^2}$. First use the explicit Euler's method (8.5) with a step size h that still ensures stability, say $h = 0.0005$, then use the same h in the implicit midpoint method (8.25). Observe the solution over at least 10 periods. Repeat the calculation by using the Störmer–Verlet method (8.56) with ten/hundred times larger h (and, accordingly, ten/hundred times fewer steps).

Plot the orbits ($q_2(t)$ versus $q_1(t)$), the energy, and the angular momentum. Check whether the integrators conserve the third invariant of Kepler's problem, the Laplace–Runge–Lenz vector $A = p \times L - r/r$, written by components as

$$\begin{pmatrix} A_1 \\ A_2 \\ 0 \end{pmatrix} = \begin{pmatrix} p_1 \\ p_2 \\ 0 \end{pmatrix} \times \begin{pmatrix} 0 \\ 0 \\ q_1 p_2 - q_2 p_1 \end{pmatrix} - \frac{1}{\sqrt{q_1^2 + q_2^2}} \begin{pmatrix} q_1 \\ q_2 \\ 0 \end{pmatrix} .$$

⊕ Solve the problem by using fourth- and sixth-order symplectic integrators with coefficients given in Appendix H.

8.13.14　Northern Lights

When charged particles emitted from the Sun (solar wind) encounter the Earth's magnetic field, they cause Northern lights. Assuming that the magnetic field is axially symmetric along the z-axis, the particle trajectories $(x(s), y(s), z(s))$ are determined by the equations (quoted in [2] based on work of [70]):

$$x'' = \frac{1}{r^5}\left(3yzz' - (3z^2 - r^2)y'\right) ,$$

$$y'' = \frac{1}{r^5}\left((3z^2 - r^2)x' - 3xzz'\right) ,$$

$$z'' = \frac{1}{r^5}\left(3xzy' - 3yzx'\right) ,$$

where $r^2 = x^2 + y^2 + z^2$ and $'$ denotes the derivative with respect to the parameter s. In polar coordinates, $x = R\cos\phi$ and $y = R\sin\phi$, we rewrite the system as

$$R'' = \left(\frac{2\gamma}{R} + \frac{R}{r^3}\right)\left(\frac{2\gamma}{R^2} + \frac{3R^2}{r^5} - \frac{1}{r^3}\right) , \tag{8.75}$$

$$z'' = \left(\frac{2\gamma}{R} + \frac{R}{r^3}\right)\frac{3Rz}{r^5} , \tag{8.76}$$

$$\phi' = \left(\frac{2\gamma}{R} + \frac{R}{r^3}\right)\frac{1}{R} , \tag{8.77}$$

where $r^2 = R^2 + z^2$, and the parameter γ is the integration constant when the equation for ϕ'' is integrated once.

⊙ Solve Eqs. (8.75) and (8.76) on the interval $0 \le s \le 0.3$ by one of the explicit Störmer methods for equations $y'' = f(x, y)$ from Sect. 8.7, e.g.

$$y_{n+1} - 2y_n + y_{n-1} = h^2 f_n ,$$

$$y_{n+1} - 2y_n + y_{n-1} = h^2\left(\frac{13}{12}f_n - \frac{1}{6}f_{n-1} + \frac{1}{12}f_{n-2}\right) ,$$

$$y_{n+1} - 2y_n + y_{n-1} = h^2\left(\frac{7}{6}f_n - \frac{5}{12}f_{n-1} + \frac{1}{3}f_{n-2} - \frac{1}{12}f_{n-3}\right) ,$$

and determine ϕ by integrating (8.77). The initial conditions are $R_0 = 0.257453$, $R_0' = \sqrt{Q_0}\cos u$, $z_0 = 0.314687$, and $z_0' = \sqrt{Q_0}\sin u$, where

$$r_0 = \sqrt{R_0^2 + z_0^2} , \qquad Q_0 = 1 - (2\gamma/R_0 + R_0/r_0^3)^2 , \qquad \gamma = -0.5 , \qquad u = 5\pi/4 .$$

8.13.15 Galactic Dynamics

Galaxies are assemblies of $N \gg 1$ stars in mutual gravitational attraction. One way of studying galactic dynamics is to compute the orbit of a single star in the gravitational potential generated by the remaining $N - 1$ stars. Assume that this potential is

$$V(x, y, z) = A\log\left(C + \frac{x^2}{a^2} + \frac{y^2}{b^2} + \frac{z^2}{c^2}\right),$$

and that it rotates around the galactic axis with uniform velocity, while its functional form does not change with time [71]. The Lagrangian for the system rotating with constant angular velocity Ω is

$$L = \frac{1}{2}\left((\dot{x} - \Omega y)^2 + (\dot{y} + \Omega x)^2 + \dot{z}^2\right) - V(x, y, z).$$

With the coordinates $q_1 = x$, $q_2 = y$, $q_3 = z$, and

$$p_1 = \frac{\partial L}{\partial \dot{x}} = \dot{x} - \Omega y, \qquad p_2 = \frac{\partial L}{\partial \dot{y}} = \dot{y} + \Omega x, \qquad p_3 = \frac{\partial L}{\partial \dot{z}} = \dot{z},$$

we obtain the Hamiltonian $H = p_1\dot{q}_1 + p_2\dot{q}_2 + p_3\dot{q}_3 - L$ or

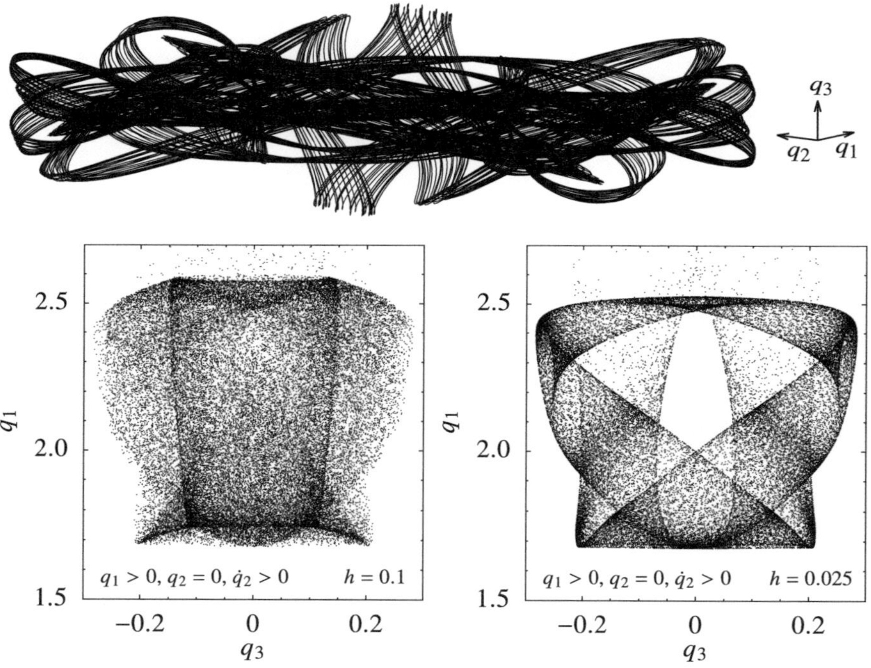

Fig. 8.29 Galactic dynamics. [TOP] Trajectory until $t = 10000$. [BOTTOM] Poincaré sections of the trajectory for $0 \le t \le 1000000$ (coordinates q_1 and q_3) with the half-plane $q_1 > 0$, $q_2 = 0$, $\dot{q}_2 > 0$. [BOTTOM LEFT] RK4 method, $h = 0.1$ (46230 intersections). [BOTTOM RIGHT] RK4 method, $h = 0.025$ (47022 intersections)

$$H = \frac{1}{2}\left(p_1^2 + p_2^2 + p_3^2\right) + \Omega\left(p_1 q_2 - p_2 q_1\right) + A\log\left(C + \frac{q_1^2}{a^2} + \frac{q_2^2}{b^2} + \frac{q_3^2}{c^2}\right).$$

⊙ From this Hamiltonian, derive the equations of motion, $\dot{q}_i = \partial H/\partial p_i$ and $\dot{p}_i = -\partial H/\partial q_i$. Solve them by using the initial values (taken from [2]):

$$q_1(0) = 2.5\,, \quad q_2(0) = 0\,, \quad q_3(0) = 0\,, \quad p_1(0) = 0\,, \quad p_3(0) = 0\,,$$

while $p_2(0)$ is the larger of the solutions of the equation $H = 2$. Use the parameters $A = C = 1$, $\Omega = 0.25$, $a = 1.25$, $b = 1$, and $c = 0.75$. Compute the solution on the interval $0 \le t \le 1000000$ by the RK4 method (8.10) with steps $h = 0.1$ and 0.025. Plot the orbits (connect the coordinates q_1, q_2, and q_3 as in Fig. 8.29 (top)). Plot the Poincaré sections of the solution (coordinates q_1 and q_3) with the half-plane $q_1 > 0$, $q_2 = 0$, $\dot{q}_2 > 0$ (Fig. 8.29 (bottom)). Plot the time evolution of the energy H.

⊕ Solve the problem by using the fifth-order implicit method Radau 5 and the sixth-order Gauss method defined in Sect. 8.8. Use the step sizes $h = 0.1$, 0.2, and 0.4. Again, plot the time evolution of the energy H.

References

1. E.A. Coddington, N. Levinson, *Theory of Ordinary Differential Equations* (Krieger Publishing, Malabar, 1984)
2. E. Hairer, S.P. Nørsett, G. Wanner, *Solving Ordinary Differential Equations I; Nonstiff Problems, Springer Series in Computational Mathematics*, vol. 8 (Springer-Verlag, Berlin, 2000)
3. A.R. Curtis, High-order explicit Runge Kutta formulae, their uses and limitations. J. Inst. Math. Appl. **16**, 35 (1975)
4. E. Hairer, A Runge-Kutta method of order 10. J. Inst. Math. Appl. **21**, 47 (1978)
5. J.C. Butcher, *Numerical Methods for Ordinary Differential Equations*, 2nd edn. (John Wiley & Sons, Chichester, 2008)
6. E. Isaacson, H.B. Keller, *Analysis of Numerical Methods* (John Wiley & Sons, New York, 1966)
7. P. Henrici, *Discrete Variable Methods in Ordinary Differential Equations* (John Wiley & Sons, New York, 1962)
8. J.R. Cash, A.H. Karp, A variable order Runge-Kutta method for initial value problems with rapidly varying right-hand sides. ACM Trans. Math. Softw. **16**, 201 (1990)
9. W.H. Press, B.P. Flannery, S.A. Teukolsky, W.T. Vetterling, *Numerical Recipes: The Art of Scientific Computing*, 3rd edn (Cambridge University Press, Cambridge, 2007). See also the equivalent handbooks in Fortran, Pascal and C, as well as http://numerical.recipes
10. J.R. Dormand, P.J. Prince, A family of embedded Runge-Kutta formulae. J. Comput. Appl. Math. **6**, 19 (1980)
11. P.J. Prince, J.R. Dormand, High order embedded Runge-Kutta formulae. J. Comput. Appl. Math. **7**, 67 (1981)
12. Numerical Algorithms Group. http://www.nag.com
13. J.H.E. Cartwright, O. Piro, The dynamics of Runge-Kutta methods. Int. J. Bifurc. Chaos **2**, 427 (1992)
14. J.D. Day, Run time estimation of the spectral radius of Jacobians. J. Comput. Appl. Math. **11**, 315 (1984)

15. R. Bulirsch, J. Stoer, Fehlerabschätzungen und Extrapolation mit rationalen Funktionen bei Verfahren vom Richardson-Typus. Numer. Math. **6**, 413 (1964)
16. E. Hairer, A. Ostermann, Dense output for extrapolation methods. Numer. Math. **58**, 419 (1990)
17. J.R. Dormand, P.J. Prince, New Runge-Kutta algorithms for numerical simulation in dynamical astronomy. Celest. Mech. **18**, 223 (1978)
18. S. Filippi, J. Gräf, New Runge-Kutta-Nyström formula-pairs of order 8(7), 9(8), 10(9) and 11(10) for differential equations of the form $y'' = f(t, y)$. J. Comp. Appl. Math. **14**, 361 (1986)
19. E. Hairer, G. Wanner, *Solving Ordinary Differential Equations II; Stiff and Differential-Algebraic Problems, Springer Series in Computational Mathematics*, vol. 14 (Springer-Verlag, Berlin, 2004)
20. J.C. Butcher, Implicit Runge-Kutta processes. Math. Comput. **18**, 50 (1964)
21. B.L. Ehle, High order A-stable methods for the numerical solution of systems of D.E.'s. BIT Numer. Math. **8**, 276 (1968)
22. L.F. Shampine, M.W. Reichelt, The Matlab ODE suite. SIAM J. Sci. Comput. **18**, 1 (1997)
23. J.D. Hoffman, *Numerical Methods for Engineers and Scientists*, 2nd edn. (Marcel Dekker, New York, 2001)
24. S.J. Aarseth, Direct N-body codes. Lect. Notes Phys. **760**, 1 (2008)
25. J. Makino, S.J. Aarseth, On a Hermite integrator with Ahmad-Cohen scheme for gravitational many-body problems. Publ. Astron. Soc. Japan **44**, 141 (1992)
26. S.J. Aarseth, *Gravitational N-Body Simulations* (Cambridge University Press, Cambridge, 2003)
27. C.F. Curtiss, J.O. Hirschfelder, Integration of stiff equations. Proc. Nat. Acad. Sci. **38**, 235 (1952)
28. L.N. Trefethen, M. Embree, *Spectra and Pseudospectra* (Princeton University Press, Princeton, 2005) (See Chapter 3)
29. G. Dahlquist, A special stability problem for linear multistep methods. BIT Numer. Math. **3**, 27 (1963)
30. H. Goldstein, *Classical Mechanics*, 2nd edn (Addison-Wesley, Reading, 1981)
31. V.I. Arnol'd, *Mathematische Methoden der klassischen Mechanik* (VEB Deutscher Verlag der Wissenschaften, Berlin, 1988)
32. E. Hairer, C. Lubich, G. Wanner, *Geometric Numerical Integration. Structure-Preserving Algorithms for Ordinary Differential Equations*, 2nd edn (Springer-Verlag, Berlin, 2006)
33. Z. Ge, J.E. Marsden, Lie-Poisson integrators and Lie-Poisson Hamilton-Jacobi theory. Phys. Lett. A **133**, 134 (1988)
34. J.M. Sanz-Serna, Runge-Kutta schemes for Hamiltonian systems. BIT Numer. Math. **28**, 877 (1988)
35. R.I. McLachlan, G.R.W. Quispel, Geometric integrators for ODEs. J. Phys. A: Math. Gen. **39**, 5251 (2006)
36. É. Forest, Geometric integration for particle accelerators. J. Phys. A Math. Gen. **39**, 5321 (2006)
37. T. Bridges, S. Reich, Numerical methods for Hamiltonian PDEs. J. Phys. A Math. Gen. **39**, 5287 (2006)
38. J.M. Sanz-Serna, Symplectic Runge-Kutta schemes for adjoint equations, automatic differentiation, optimal control, and more. SIAM Review **58**, 3 (2016)
39. E. Celledoni, H. Marthinsen, B. Owren, An introduction to Lie group integrators–basics, new developments and applications. J. Comp. Phys. **257**, 1040 (2014)
40. A. Hanslmeier, R. Dvorak, Numerical integration with Lie series. Astron. Astrophys. **132**, 203 (1984)
41. M. Delva, Integration of the elliptic restricted three-body problem with Lie series. Celest. Mech. **34**, 145 (1984)
42. H. Lichtenegger, The dynamics of bodies with variable masses. Celest. Mech. **34**, 357 (1984)
43. S. Eggl, R. Dvorak, *An Introduction to Common Numerical Integration Codes used in Dynamical Astronomy*, Lecture Notes in Physics vol. 790, p. 431 (Springer-Verlag, Berlin, 2010)

44. E.N. Lorenz, Deterministic nonperiodic flow. J. Atmos. Sci. **20**, 130 (1963)
45. A. Trevisan, F. Pancotti, Periodic orbits, Lyapunov vectors, and singular vectors in the Lorenz System. J. Atmos. Sci. **55**, 390 (1998)
46. GNU Multi Precision (GMP), free library for arbitrary precision arithmetic, Available at: http://gmplib.org
47. J.C. Sprott, *Chaos and Time-Series Analysis* (Oxford University Press, Oxford, 2003). See also http://sprott.physics.wisc.edu/chaostsa
48. I. Kuščer, A. Kodre, H. Neunzert, *Mathematik in Physik und Technik* (Springer-Verlag, Berlin, 1993)
49. C. Moler, *Numerical Computing with MATLAB* (SIAM, Philadelphia, 2008)
50. J.C.G. Walker, *Numerical Adventures with Geochemical Cycles* (Oxford University Press, Oxford, 1991)
51. Y. Kuramoto, *Chemical Oscillations, Waves, and Turbulence* (Springer-Verlag, Berlin, 1984)
52. S.H. Strogatz, From Kuramoto to Crawford: exploring the onset of synchronization in populations of coupled oscillators. Physica D **143**, 1 (2000)
53. E. Ott, *Chaos in Dynamical Systems*, 2nd edn (Cambridge University Press, Cambridge, 2002)
54. G.B. Ermentrout, J. Rinzel, Analysis of neural excitability and oscillations, in *Methods in Neuronal Modelling: From Synapses to Networks*, ed. by C. Koch, I. Segev (MIT Press, Cambridge, 1989)
55. C. Morris, H. Lecar, Voltage oscillations in the barnacle giant muscle fiber. Biophys. J. **35**, 193 (1981)
56. S.A. Chin, C.R. Chen, Forward symplectic integrators for solving gravitational few-body problems. Celest. Mech. Dyn. Astr. **91**, 301 (2005)
57. P.A. Wiegert, K.A. Innanen, An asteroidal companion to the Earth. Nature **387**, 685 (1997)
58. M. Connors, P. Chodas, S. Mikkola, Discovery of an asteroid and quasi-satellite in an Earth-like horseshoe orbit. Meteorit. Planet. Sci. **37**, 1435 (2002)
59. D.B. Taylor, Horseshoe periodic orbits in the restricted problem of three bodies for a Sun-Jupiter mass ratio. Astron. Astrophys. **103**, 288 (1981)
60. P.W. Sharp, Comparisons of integrators on a diverse collection of restricted three-body test problems. IMA J. Numer. Anal. **24**, 557 (2004)
61. E. Barrabés, S. Mikkola, Families of periodic horseshoe orbits in the restricted three-body problem. Astron. Astrophys. **432**, 1115 (2005)
62. B. Saltzman, Finite amplitude free convection as an initial value problem-I. J. Atmos. Sci. **19**, 329 (1962)
63. L.D. Landau, E.M. Lifshitz, *Course of Theoretical Physics, Vol. 1: Mechanics*, 3rd edn (Butterworth-Heinemann, Oxford, 2003)
64. S. Bleher, E. Ott, C. Grebogi, Routes to chaotic scattering. Phys. Rev. Lett. **63**, 919 (1990)
65. M. Stix, *The Sun. An introduction*, 2nd edn (Springer-Verlag, Berlin, 2002)
66. G.R. Caughlan, W.A. Fowler, Thermonuclear reaction rates V. At. Data Nucl. Data Tables **40**, 283 (1988)
67. Y. Xu, K. Takahashi, S. Goriely, M. Arnould, M. Ohta, H. Utsunomiya, NACRE II: an update of the NACRE compilation of charged-particle-induced thermonuclear reaction rates for nuclei with mass number A < 16. Nucl. Phys. A **918**, 61 (2013). The data are available at http://www.astro.ulb.ac.be/nacreii
68. E.L. Schatzman, F. Praderie, *The Stars* (Springer-Verlag, Berlin, 1993)
69. J.R. Field, R.M. Noyes, Oscillations in chemical systems. IV. Limit cycle behavior in a model of a real chemical reaction. J. Chem. Phys. **60**, 1877 (1974)
70. C. Störmer, Sur les trajectoires des corpuscules électrisés. Arch. Sci. Phys. Nat., Genève. **24**, 5–18, 113–158, 221–247 (1907)
71. J. Binney, S. Tremaine, *Galactic Dynamics* (Princeton University Press, Princeton, 1987)

Chapter 9
Boundary-Value Problems for ODE

Abstract This Chapter deals with various finite-difference methods for scalar boundary-value problems involving ordinary differential equations and for systems of such problems. The concepts of consistency, stability and convergence are introduced, as well as methods to increase the local solution precision by extrapolation. Shooting methods are offered as an alternative to difference methods in the case of non-linear equations and their systems. A separate Section is devoted to various types of discretizations that mirror the asymptotic physics regimes of the underlying differential equation. Collocation and weighted-residual methods are presented. Several approaches to boundary-value problems with eigenvalues are attempted: finite-difference methods, shooting methods involving the Prüfer transformation, and the Pruess method. We discuss problems with eigenvalues appearing in the boundary conditions and singular Sturm–Liouville problems. Examples and Problems include the non-linear Gelfand–Bratu equation, diffusion and reaction in a catalytic pellet, deflection of an inhomogeneous beam, the one-dimensional Schrödinger equation, and a boundary-layer problem.

In this Chapter we discuss methods for the solution of problems with ordinary differential equations, where we require the solution to satisfy the equation within the definition domain, and boundary conditions at its edges. We seek, for example, the function y that solves the equation

$$y'' = f(x, y, y')$$

for $x \in R = [a, b]$, with mixed boundary conditions

$$\alpha_0 y(a) - \alpha_1 y'(a) = \alpha \,, \quad |\alpha_0| + |\alpha_1| \neq 0 \,,$$
$$\beta_0 y(b) + \beta_1 y'(b) = \beta \,, \quad |\beta_0| + |\beta_1| \neq 0 \,,$$

where the coefficients satisfy $\alpha_0 \alpha_1 \geq 0$, $\beta_0 \beta_1 \geq 0$, and $|\alpha_0| + |\beta_0| \neq 0$. When does the corresponding solution exist at all? If f is continuous and continuously differentiable on R, and if it satisfies the Lipschitz condition (8.4) on R for both y and y', and if for some $M > 0$ we have $\partial f / \partial y > 0$ and $|\partial f / \partial y'| \leq M$, and, finally, if $\alpha_0 \alpha_1 \geq 0$ and $\beta_0 \beta_1 \geq 0$, such a boundary-value problem has a unique solution [1]. In the simplified case when f is linear in y and y',

© The Author(s), under exclusive license to Springer Nature Switzerland AG 2025   549
S. Širca and M. Horvat, *Computational Methods in Physics*, Graduate Texts in Physics,
https://doi.org/10.1007/978-3-031-68566-8_9

$$- y'' + p(x)y' + q(x)y = r(x) ,$$

where p, q, and r are continuous on R, for each α and β a unique solution exists precisely when $q(x) > 0$ for each $x \in R$. Sometimes we impose additional conditions that the solution should fulfill within the domain R. The corresponding existence theorems can be found in [2].

Uniqueness of solutions of non-linear problems is much harder to fathom than in linear problems: a seemingly simple problem $y'' = -\delta e^y$ with conditions $y(0) = y(1) = 0$ (Gelfand-Bratu equation of diffusion-reaction kinetics in a layer) has two solutions for $0 < \delta < \delta_c \approx 3.51$ while the solution does not exist for $\delta > \delta_c$. For the spherical diffusion-reaction problem $y'' + (2/x)y' = \phi^2 y + \exp[\gamma\beta(1 - y)/(1 + \beta(1 - \gamma))]$ with conditions $y'(0) = 0$, $y(1) = 1$ at least 15 solutions are known to exist. How do we numerically compute all of them?

9.1   Difference Methods for Scalar Boundary-Value Problems

In developing difference schemes for boundary-value problems we utilize our knowledge from the methods used for initial-value problems (Chap. 8). Our cornerstone will be the linear problem with Dirichlet boundary conditions,

$$Ly = -y'' + p(x)y' + q(x)y = r(x) , \qquad y(a) = \alpha , \quad y(b) = \beta , \qquad (9.1)$$

where L is a linear second-order differential operator and where p, q, and r are continuous on $[a, b]$. This is a classical two-point boundary-value problem in which the boundary conditions are expressed at two points ($x = a$ and $x = b$). More general multi-point problems with "boundary" conditions in the form $\sum_{j=1}^{N_b} \alpha_j y(\xi_j) = \alpha$ at $a = \xi_1 < \xi_2 < \cdots < \xi_{N_b} = b$, are discussed in [2]. To the interval $[a, b]$ we assign an equidistant mesh $a = x_0, x_1, \ldots, x_N = b$, where

$$x_j = a + jh , \qquad j = 0, 1, \ldots, N , \qquad h = (b - a)/N . \qquad (9.2)$$

Next, we discretize the derivatives and function values in the operator L, and denote $y(x_j) = y_j$. For example, the first derivative can be discretized by the non-symmetric (one-sided) differences with the discretization error $\mathcal{O}(h)$,

$$y_j' = \frac{y_{j+1} - y_j}{h} - \frac{h}{2} y''(\xi) , \qquad (9.3)$$

$$y_j' = \frac{y_j - y_{j-1}}{h} + \frac{h}{2} y''(\xi) , \qquad (9.4)$$

where ξ is some point positioned between the extreme points appearing in the formula. The order of the error can be improved by using symmetric (central) or non-symmetric differences that include more mesh points, for example:

$$y_j' = \frac{y_{j+1} - y_{j-1}}{2h} - \frac{h^2}{6} y'''(\xi) , \tag{9.5}$$

$$y_j' = \frac{-y_{j+2} + 8y_{j+1} - 8y_{j-1} + y_{j-2}}{12h} + \frac{h^4}{30} y^{(5)}(\xi) , \tag{9.6}$$

$$y_j' = \frac{-y_{j+2} + 4y_{j+1} - 3y_j}{2h} + \frac{h^2}{3} y'''(\xi) , \tag{9.7}$$

$$y_j' = \frac{3y_j - 4y_{j-1} + y_{j-2}}{2h} + \frac{h^2}{3} y'''(\xi) . \tag{9.8}$$

Expressions (9.5) and (9.6) are symmetric, but at the boundary points x_0 and x_N they require us to access the points x_{-1} and x_{N+1} (or even x_{-2} and x_{N+2}) beyond the mesh, and write additional difference equations for them. Expressions (9.7) and (9.8) are non-symmetric, but at the boundaries they can be applied directly. One-sided differences may have more natural physical backgrounds: we prefer to differentiate in the "upwind" direction from which the information is arriving, for example, facing the incoming wave. We shall encounter similar choices in the discretization of hyperbolic partial differential equations in Chap. 10.

The second derivative may be approximated by the central difference

$$y_j'' = \frac{y_{j+1} - 2y_j + y_{j-1}}{h^2} - \frac{h^2}{12} y^{(4)}(\xi) .$$

We use u_j to denote the approximate solution at x_j. By using the central difference for the first derivative we obtain the difference scheme

$$L_h u_j = -\frac{u_{j+1} - 2u_j + u_{j-1}}{h^2} + p(x_j)\frac{u_{j+1} - u_{j-1}}{2h} + q(x_j)u_j = r(x_j) , \tag{9.9}$$

where L_h is a second-order difference operator. The scheme is applicable for $j = 1, 2, \ldots, N - 1$, and the Dirichlet boundary conditions fix the remaining two values $u_0 = \alpha$ and $u_N = \beta$. We collect the solution components $u_1, u_2, \ldots, u_{N-1}$ in the vector $u = (u_1, u_2, \ldots, u_{N-1})^T$, and rewrite the difference equation in the matrix form $Au = r$, where

$$A = \begin{pmatrix} b_1 & c_1 & 0 & & & \\ a_2 & b_2 & c_2 & & 0 & \\ & \ddots & \ddots & \ddots & & \\ & 0 & & a_{N-2} & b_{N-2} & c_{N-2} \\ & & 0 & & a_{N-1} & b_{N-1} \end{pmatrix} , \qquad r = \begin{pmatrix} \frac{1}{2}h^2 r(x_1) - a_1\alpha \\ \frac{1}{2}h^2 r(x_2) \\ \vdots \\ \frac{1}{2}h^2 r(x_{N-2}) \\ \frac{1}{2}h^2 r(x_{N-1}) - c_{N-1}\beta \end{pmatrix} .$$

The matrix A is tridiagonal, with the matrix elements

$$a_j = -\frac{1}{2}\left[1 + \frac{h}{2}\,p(x_j)\right], \qquad b_j = 1 + \frac{h^2}{2}\,q(x_j), \qquad c_j = -\frac{1}{2}\left[1 - \frac{h}{2}\,p(x_j)\right].$$

In the case of mixed boundary conditions, $\alpha_0 y(a) - \alpha_1 y'(a) = \alpha$, we can take the one-sided difference (9.7) and use it to approximate the boundary condition as

$$\alpha_0 u_0 - \alpha_1 \frac{-u_2 + 4u_1 - 3u_0}{2h} = \alpha\,.$$

This gives us the expression

$$u_0 = \frac{\alpha_1}{2h\alpha_0 + 3\alpha_1}\,(4u_1 - u_2) + \frac{2h\alpha}{2h\alpha_0 + 3\alpha_1}\,,$$

which we insert in Eq. (9.9) at $j = 1$. When the coefficients in front of u_1 and u_2 are collected and the remainder is put to the right side of the equation, we again find a matrix equation for $\boldsymbol{u}$ with a tridiagonal matrix A, in which only the first row is modified:

$$b_1 \to b_1 + \frac{4\alpha_1 a_1}{2h\alpha_0 + 3\alpha_1}\,,$$

$$c_1 \to c_1 - \frac{\alpha_1 a_1}{2h\alpha_0 + 3\alpha_1}\,,$$

$$\tfrac{1}{2}h^2 r(x_1) - a_1\alpha \to \tfrac{1}{2}h^2 r(x_1) - \frac{2h a_1 \alpha}{2h\alpha_0 + 3\alpha_1}\,,$$

where the denominators should satisfy $2h\alpha_0 + 3\alpha_1 \neq 0$.

Consistency, Stability, and Convergence

The order of the numerical error of difference schemes for boundary-value problems is defined by analogy to initial-value problems (Sect. 8.4). The discretization error at x_j is $\tau_j = L_h y(x_j) - Ly(x_j)$. For the scheme (9.9) this means

$$\tau_j = -\left[\frac{u_{j+1} - 2u_j + u_{j-1}}{h^2} - y''(x_j)\right] + p(x_j)\left[\frac{u_{j+1} - u_{j-1}}{2h} - y'(x_j)\right]$$

$$= -\frac{h^2}{12}\left[y^{(4)}(\xi) - 2p(x_j)y'''(\eta)\right] = \mathcal{O}(h^2)\,, \tag{9.10}$$

where ξ and η are points from the interval $[x_{j-1}, x_{j+1}]$. The scheme (9.9) is *consistent* with the differential equation (9.1): for all functions y with a continuous second derivative y'' we observe $\tau_j \to 0$ when $h \to 0$. For functions y with a continuous fourth derivative on $[a, b]$ the difference operator L_h approximates the differential operator L to second order.

The notion of stability, however, needs to be introduced in a different manner than for initial-value problems. We say that the difference operator L_h is *stable* for some small enough h if a positive constant Λ can be found such that

$$|u_j| \leq \Lambda \left[\max\{|u_0|, |u_N|\} + \max_{1 \leq j \leq N-1} |L_h u_j| \right]. \tag{9.11}$$

If on the interval $[a, b]$ the functions p and q can be bounded by $|p(x)| \leq P_+$ and $0 < Q_- \leq q(x) \leq Q_+$, it can be shown that the scheme (9.9) is stable if we set $h \leq 2/P_+$. Then (9.11) holds with the constant $\Lambda = \max\{1, 1/Q_-\}$ [1]. If, in addition, y is four times continuously differentiable on $[a, b]$, the stability theorem also provides a useful error estimate ($0 \leq j \leq N$):

$$\left|u_j - y(x_j)\right| \leq \Lambda \frac{h^2}{12} \left[\max_{a \leq x \leq b} \left|y^{(4)}(x)\right| + 2P_+ \max_{a \leq x \leq b} \left|y'''(x)\right| \right]. \tag{9.12}$$

As in initial-value problems, we define convergence of difference schemes for boundary-value problems: a stable difference scheme that is consistent with the differential equation to order h^p, is convergent at the same order, thus

$$h \to 0 \quad \Rightarrow \quad \max_{0 \leq j \leq N} |u_j - y(x_j)| \to 0.$$

Increasing the order by extrapolation There is a lovely way to use the second-order difference scheme (9.9) and obtain the result which has fourth-order precision. We first solve the boundary-value problem on a mesh with spacings $h = (b - a)/N$, yielding the approximate solution $\boldsymbol{u}^{(h)}$. Then we use the scheme again, but on a mesh with spacings half as large, $h/2 = (b - a)/(2N)$, and get the values $\boldsymbol{u}^{(h/2)}$. The improved approximation for the values $y(x)$ on the mesh with spacings h is then

$$u_j^{(h/2,h)} = \frac{4}{3} u_{2j}^{(h/2)} - \frac{1}{3} u_j^{(h)}, \qquad j = 0, 1, \ldots, N. \tag{9.13}$$

The approximation (9.13) is of order four,

$$\| \boldsymbol{u}^{(h/2,h)} - \boldsymbol{y} \| = \mathcal{O}(h^4).$$

By further halvings even higher orders can be attained (see Sect. 9.2).

Non-linear Scalar Boundary-Value Problems

Solving non-linear boundary-value problems is more demanding than solving linear ones. The discretization of the differential equation itself is usually easy: the main work to be done is the solution of the system of non-linear equations that follows from this discretization. We discuss boundary-value problems of the form

$$Ly = -y'' + f(x, y, y') = 0 \tag{9.14}$$

on the interval $[a, b]$ with boundary conditions $y(a) = \alpha$ and $y(b) = \beta$. Assume that f is (once) continuously differentiable with respect to y and y', and that

$$0 < Q_- \leq \frac{\partial f}{\partial y} \leq Q_+ , \qquad \left| \frac{\partial f}{\partial y'} \right| \leq P_+ ,$$

for some positive constants Q_-, Q_+, and P_+. The difference approximation of the boundary-value problem (9.14) is

$$L_h u_j = -\frac{u_{j+1} - 2u_j + u_{j-1}}{h^2} + f\left(x_j, u_j, \frac{u_{j+1} - u_{j-1}}{2h}\right) = 0 , \tag{9.15}$$

and it applies for $j = 1, 2, \ldots, N - 1$, while the boundary conditions again dictate $u_0 = \alpha$ and $u_N = \beta$. With the above assumptions for $\partial f/\partial y$ and $\partial f/\partial y'$, and by choosing $h \leq 2/P_+$ the difference scheme (9.15) is stable, is consistent with the differential equation to order $\mathcal{O}(h^2)$, and has the error estimate (9.12) as for the linear scalar problem.

Equation (9.15) in matrix form represents a system of non-linear equations. We solve it by explicit iteration [1] or by using Newton's method in which at each step a matrix equation needs to be solved. The explicit iteration is

$$u_j^{(n+1)} = \frac{1}{1 + \omega} \left[\frac{1}{2}\left(u_{j+1}^{(n)} + u_{j-1}^{(n)}\right) + \omega u_j^{(n)} - \frac{h^2}{2} f\left(x_j, u_j^{(n)}, \frac{u_{j+1}^{(n)} - u_{j-1}^{(n)}}{2h}\right) \right] ,$$

where $j = 1, 2, \ldots, N - 1$, and the boundary values are $u_0 = \alpha$ and $u_N = \beta$. We start the iteration with some initial approximation $\boldsymbol{u}^{(0)}$, and the speed of convergence is determined by the parameter ω. By choosing $\omega \geq \frac{1}{2} h^2 Q_+$ the iteration converges for any initial approximation [1].

In order to use the Newton's method, we rewrite (9.15) as $\boldsymbol{F}(\boldsymbol{u}) = \boldsymbol{0}$, where $\boldsymbol{F} = (F_1(\boldsymbol{u}), F_2(\boldsymbol{u}), \ldots, F_{N-1}(\boldsymbol{u}))^{\mathrm{T}}$ with $F_j(\boldsymbol{u}) = \frac{1}{2}h^2(L_h u_j)$. For the system of equations $\boldsymbol{F}(\boldsymbol{u}) = \boldsymbol{0}$, Newton's method is an iteration to the fixed point

$$\boldsymbol{u}^{(n+1)} = \boldsymbol{G}\left(\boldsymbol{u}^{(n)}\right) ,$$

where

$$G(u) = u - [J(u)]^{-1} F(u)$$

is the iteration function. Here $J(u)$ is the Jacobi matrix which is tridiagonal,

$$J(u) = \frac{\partial F(u)}{\partial u} = \begin{pmatrix} B_1(u) & C_1(u) & & & \\ A_2(u) & B_2(u) & C_2(u) & & \\ & \ddots & \ddots & \ddots & \\ & & A_{N-2}(u) & B_{N-2}(u) & C_{N-2}(u) \\ & & & A_{N-1}(u) & B_{N-1}(u) \end{pmatrix},$$

with the matrix elements

$$A_j(u) = -\frac{1}{2}\left[1 + \frac{h}{2}\frac{\partial f}{\partial y'}\left(x_j, u_j, \frac{u_{j+1} - u_{j-1}}{2h}\right)\right],$$

$$B_j(u) = 1 + \frac{h^2}{2}\frac{\partial f}{\partial y}\left(x_j, u_j, \frac{u_{j+1} - u_{j-1}}{2h}\right),$$

$$C_j(u) = -\frac{1}{2}\left[1 - \frac{h}{2}\frac{\partial f}{\partial y'}\left(x_j, u_j, \frac{u_{j+1} - u_{j-1}}{2h}\right)\right].$$

(The boundary conditions $u_0 = \alpha$ and $u_N = \beta$ also enter through the quantities $B_1(u), C_1(u), A_{N-1}(u), B_{N-1}(u), F_1(u)$, and $F_{N-1}(u)$.) In each step of the Newton's iteration we are solving a system of linear equations

$$J(u^{(n)})\Delta u^{(n)} = -F(u^{(n)}), \tag{9.16}$$

and the solution $\Delta u^{(n)}$ of this system gives us the next approximation of the solution

$$u^{(n+1)} = u^{(n)} + \Delta u^{(n)}, \qquad n = 0, 1, 2, \ldots. \tag{9.17}$$

We repeat the iteration until the difference between subsequent approximations (as measured in some norm) drops below the specified tolerance.

Especially in the cases where the boundary-value problem is expected to have multiple solutions, one should choose the initial approximation $u^{(0)}$ carefully. We try to exploit a known symmetry property of the equation or boundary conditions, or perhaps the asymptotic behavior of a similar equation. The initial approximation should at least satisfy the boundary conditions. It also makes sense to precede the Newton's method by a few cycles of the explicit iteration.

Example A problem of the form (9.14) is the Gelfand–Bratu equation

$$y'' = -\delta\, e^y, \qquad 0 < x < 1, \qquad 0 < \delta < 3.51, \tag{9.18}$$

with boundary conditions $y(0) = y(1) = 0$. To solve it, we use Newton's method and the approximation (9.15), where $f(x, y, y') = -\delta\, e^y$. The function f depends only on y, so the off-diagonal elements of the Jacobi matrix are trivial:

$$A_j(\boldsymbol{u}) = -\frac{1}{2}\,, \qquad B_j(\boldsymbol{u}) = 1 + \frac{h^2}{2}\left(-\delta\, e^{u_j}\right), \qquad C_j(\boldsymbol{u}) = -\frac{1}{2}\,.$$

The components of the right side of (9.16) for $j = 1, 2, \ldots, N - 1$ are

$$F_j(\boldsymbol{u}) = \frac{1}{2}\, h^2 (L_h u_j) = \frac{h^2}{2}\left[-\frac{u_{j+1} - 2u_j + u_{j-1}}{h^2} - \delta\, e^{u_j}\right].$$

The boundary conditions, which do not change in (9.16–9.17), are $u_0 = u_N = 0$. We start the iteration with the initial approximation $y(x) = \mu x(1 - x)$,

$$u_j = \mu x_j(1 - x_j)\,, \qquad j = 0, 1, \ldots, N\,,$$

where μ is a positive real constant. For $0 < \delta < 3.51$ the Gelfand–Bratu problem has two solutions (Problem 9.9.1) and by using different values of μ Newton's iteration may bounce between these two solutions, or it may not converge at all (Fig. 9.1 (left)). The dependence of the norm of the last update in the iteration and the difference between the numerical and exact solution on the length of the subintervals is shown in Fig. 9.1 (right) in the case of $\delta = 1$.   ◁

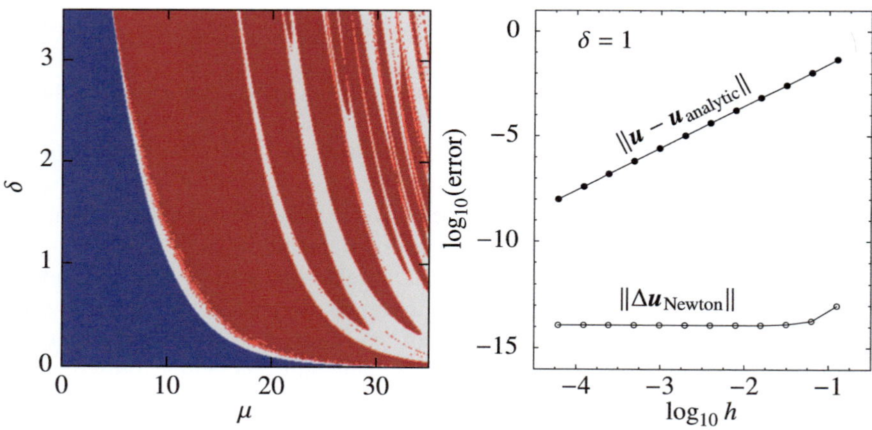

Fig. 9.1 Convergence of the Newton's method for the Gelfand–Bratu problem. [LEFT] The iteration for different values of μ in the initial approximation for $y(x)$ and δ converges to the first (red areas) or second solution (blue areas). In bright areas the iteration does not converge or the solution type can not be ascertained within the specified tolerance. [RIGHT] The norm of the last solution update $\Delta\boldsymbol{u}$ and the differences between the numerical and exact solution as a function of subinterval length h, for $\delta = 1$

When we solve non-linear boundary-value problems by simple iteration, Newton's method, or some other iterative method, we always obtain the numerical solution after a series of steps. During the iteration, the solution changes throughout the domain and *relaxes* to the true solution (the sequence of curves 1, 2, 3, 4 in Fig. 9.4 (left)); only the boundary conditions remain the same, e.g. as in the Dirichlet case $y(a) = \alpha$ and $y(b) = \beta$ in that Figure. Sect. 9.3 on *shooting methods* will show us how to obtain the solution in an altogether different manner.

9.2 Difference Methods for Systems of Boundary-Value Problems

A large class of boundary-value problems can be written in the form of a system of M (not necessarily linear) first-order differential equations

$$y' = f(x, y), \qquad a < x < b, \tag{9.19}$$

where $y \in \mathbb{R}^M$ and $\{f_i\}_{i=1}^M$ are scalar functions (components of $f : \mathbb{R}^{M+1} \to \mathbb{R}^M$) just as in Eq. (8.2). The general form of the boundary condition is

$$g(y(a), y(b)) = 0.$$

As in the scalar boundary-value problems we discretize the interval $[a, b]$ on a mesh with points $a = x_0, x_1, \ldots, x_N = b$. We collect the approximate solutions in the $M(N + 1)$-dimensional vector $u = (u_0^{\mathrm{T}}, u_1^{\mathrm{T}}, \ldots, u_N^{\mathrm{T}})^{\mathrm{T}}$, in which each component u_j $(j = 0, 1, \ldots, N)$ is a M-dimensional vector.

There are many ways [1] to discretize the derivatives and the arguments of the function f. The difference approximation of (9.19) with the trapezoidal formula for f is

$$L_h u_j = \frac{u_{j+1} - u_j}{h} - \frac{1}{2} \left[f\left(x_{j+1}, u_{j+1}\right) + f\left(x_j, u_j\right) \right] = 0,$$

where $j = 0, 1, \ldots, N - 1$, and the boundary condition has the form

$$g(u_0, u_N) = 0.$$

We write the discretized equation and boundary condition as a system of non-linear equations $F(u) = 0$, where $F = (F_0^{\mathrm{T}}(u), F_1^{\mathrm{T}}(u), \ldots, F_{N-1}^{\mathrm{T}}(u), g^{\mathrm{T}}(u_0, u_N))^{\mathrm{T}}$. The first N components of F are given by $F_j(u) = -h(L_h u_j)$. The system can again be solved by Newton's iteration. At each step of (9.16) we solve

$$
J\left(\boldsymbol{u}^{(n)}\right)\Delta\boldsymbol{u}^{(n)} =
\begin{pmatrix}
S_0 & R_0 & & & & \\
 & S_1 & R_1 & & & \\
 & & S_2 & R_2 & & \\
 & & & \ddots & \ddots & \\
 & & & & S_{N-1} & R_{N-1} \\
B_a & & & & 0 & B_b
\end{pmatrix}
\begin{pmatrix}
\Delta\boldsymbol{u}_0^{(n)} \\
\Delta\boldsymbol{u}_1^{(n)} \\
\Delta\boldsymbol{u}_2^{(n)} \\
\vdots \\
\Delta\boldsymbol{u}_{N-1}^{(n)} \\
\Delta\boldsymbol{u}_N^{(n)}
\end{pmatrix}
=
\begin{pmatrix}
\boldsymbol{q}_0^{(n)} \\
\boldsymbol{q}_1^{(n)} \\
\boldsymbol{q}_2^{(n)} \\
\vdots \\
\boldsymbol{q}_{N-1}^{(n)} \\
\boldsymbol{\beta}^{(n)}
\end{pmatrix}
$$

for $\Delta\boldsymbol{u}^{(n)}$ and use it to compute the next approximation $\boldsymbol{u}^{(n+1)}$ by Eq. (9.17). The Jacobi matrix has dimensions $M(N+1)\times M(N+1)$ and has a characteristic block structure. It consists of $M\times M$ matrices determined by the difference scheme and the boundary condition:

$$
S_j = \ I + \frac{h}{2}\frac{\partial\boldsymbol{f}}{\partial\boldsymbol{y}}\left(x_j,\boldsymbol{u}_j^{(n)}\right), \tag{9.20}
$$

$$
R_j = -I + \frac{h}{2}\frac{\partial\boldsymbol{f}}{\partial\boldsymbol{y}}\left(x_{j+1},\boldsymbol{u}_{j+1}^{(n)}\right), \tag{9.21}
$$

$$
B_a = \frac{\partial\boldsymbol{g}(\boldsymbol{u},\boldsymbol{v})}{\partial\boldsymbol{u}}, \quad B_b = \frac{\partial\boldsymbol{g}(\boldsymbol{u},\boldsymbol{v})}{\partial\boldsymbol{v}} \quad \text{with} \quad \boldsymbol{u}=\boldsymbol{u}_0^{(n)},\ \boldsymbol{v}=\boldsymbol{u}_N^{(n)}, \tag{9.22}
$$

while the right side of the equation is given by

$$
\boldsymbol{q}_j^{(n)} = \left(\boldsymbol{u}_{j+1}^{(n)}-\boldsymbol{u}_j^{(n)}\right) - \frac{h}{2}\left[\boldsymbol{f}\left(x_{j+1},\boldsymbol{u}_{j+1}^{(n)}\right)+\boldsymbol{f}\left(x_j,\boldsymbol{u}_j^{(n)}\right)\right], \tag{9.23}
$$

$$
\boldsymbol{\beta}^{(n)} = -\boldsymbol{g}(\boldsymbol{u}_0^{(n)},\boldsymbol{u}_N^{(n)}).
$$

(As an exercise, repeat the derivation above with a trapezoidal scheme in which $\boldsymbol{y}'$ and $\boldsymbol{f}$ are computed at x_{j-1} and x_j instead of x_j and x_{j+1}.)

Example The second-order Gelfand–Bratu problem (Eq. (9.18)) can be solved by the above method when it is written as a system of $M=2$ first-order equations:

$$
\boldsymbol{y}' = \boldsymbol{f}(x,\boldsymbol{y}), \qquad \boldsymbol{y} = \begin{pmatrix} u_0 \\ u_1 \end{pmatrix}, \qquad \boldsymbol{f}(x,\boldsymbol{y}) = \begin{pmatrix} u_1 \\ -\delta\,e^{u_0} \end{pmatrix}.
$$

We use $u_{0,j}$ and $u_{1,j}$ to denote the values of the first and second component, respectively, of the numerical solution $\boldsymbol{u}_j$ at x_j. The $M\times M = 2\times 2$ matrices (9.20) and (9.21) appearing in the Jacobi matrix are:

$$
S_j = \begin{pmatrix} 1 & \dfrac{h}{2} \\[2ex] -\dfrac{h}{2}\,\delta\exp\left(u_{0,j}^{(n)}\right) & 1 \end{pmatrix}, \qquad
R_j = \begin{pmatrix} -1 & \dfrac{h}{2} \\[2ex] -\dfrac{h}{2}\,\delta\exp\left(u_{0,j+1}^{(n)}\right) & -1 \end{pmatrix}.
$$

The right side of the Newton's system is given by (9.23),

$$q_j^{(n)} = \begin{pmatrix} u_{0,j+1} - u_{0,j} - \dfrac{h}{2}\left[u_{1,j+1} + u_{1,j}\right] \\[2ex] u_{1,j+1} - u_{1,j} + \dfrac{h}{2}\left[\delta\exp\left(u_{0,j+1}^{(n)}\right) + \delta\exp\left(u_{0,j}^{(n)}\right)\right] \end{pmatrix}.$$

Equation (9.22) tells us that the boundary condition $g(u_0, u_N) = 0$ is

$$B_a = \begin{pmatrix} 1 & 0 \\ 0 & 0 \end{pmatrix}, \qquad B_b = \begin{pmatrix} 0 & 0 \\ 1 & 0 \end{pmatrix}, \qquad \beta^{(n)} = -g\left(u_0^{(n)}, u_N^{(n)}\right) = \begin{pmatrix} 0 \\ 0 \end{pmatrix}.$$

We choose an initial approximation $u^{(0)}$ and run the iteration (9.16–9.17).   ◁

The Jacobi matrix has a deficiency which implies a higher numerical cost of solving the system of equations: the matrix B_a corresponding to the boundary condition at $x = a$ appears in the bottom left corner (Fig. 9.2a)). This form occurs because the boundary conditions are not separated, $g(u_0, u_N) = 0$ (or $B_a y(a) + B_b y(b) = \beta$ in the linear case). Problems with unseparated conditions can be rephrased as problems with separated conditions $g_1(y(a)) = 0$ and $g_2(y(b)) = 0$ (or $B_1 y(a) = \beta_1$ and $B_2 y(b) = \beta_2$ in the linear case), yielding a nicer form of the Jacobi matrix (Fig. 9.2b)). The price for this is the increase in the number of differential equations. Details on the separation of boundary conditions can be found in [2].

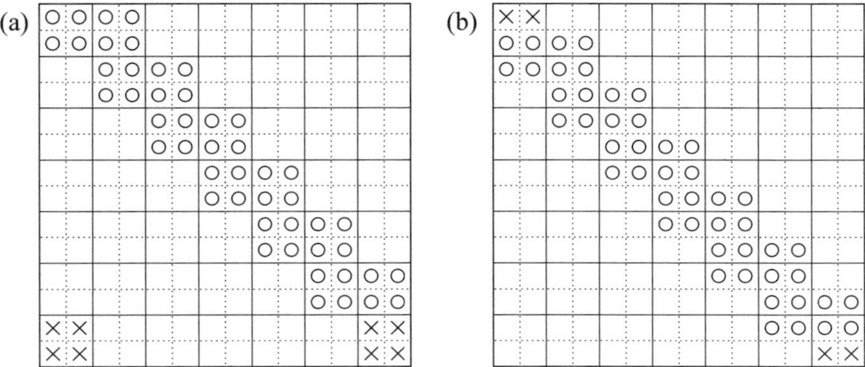

Fig. 9.2 The structure of the Jacobi matrix for systems of non-linear boundary-value problems for $M = 2$, $N = 6$. The symbols ○ denote the elements originating in the difference scheme, while the symbols × denote the elements fixed by the boundary conditions. Matrices of the form **a)** occur with unseparated boundary conditions $g(u_0, u_N) = 0$. For improved numerical efficiency, one could attempt to separate the boundary conditions as $g_1(y(a)) = 0$ and $g_2(y(b)) = 0$, leading to the banded structure **b)**

9.2.1 Linear Systems

We made an exception and treated the more general class of non-linear systems first.
Linear boundary-value problems with M first-order equations

$$y' = A(x)y + q(x), \qquad a < x < b, \tag{9.24}$$

and boundary conditions

$$B_a y(a) + B_b y(b) = \beta,$$

are just their special cases. Two simple difference schemes to solve linear systems
of the form (9.24) [2] are the trapezoidal scheme

$$u_{j+1} - u_j = \frac{h}{2}\left[A(x_{j+1})u_{j+1} + A(x_j)u_j\right] + \frac{h}{2}\left[q_{j+1} + q_j\right]$$

and the midpoint scheme

$$u_{j+1} - u_j = \frac{h}{2}A(x_{j+1/2})\left(u_{j+1} + u_j\right) + hq_{j+1/2},$$

applicable at $j = 1, 2, \ldots, N - 1$, while the boundary conditions in both cases are
$B_a u_0 + B_b u_N = \beta$. When the schemes are written in matrix form, we obtain a matrix
with the same structure as in the Jacobi matrix for the non-linear case,

$$\begin{pmatrix} S_0 & R_0 & & & & \\ & S_1 & R_1 & & & \\ & & S_2 & R_2 & & \\ & & & \ddots & \ddots & \\ & & & & S_{N-1} & R_{N-1} \\ B_a & & & & 0 & B_b \end{pmatrix} \begin{pmatrix} u_0 \\ u_1 \\ u_2 \\ \vdots \\ u_{N-1} \\ u_N \end{pmatrix} = \begin{pmatrix} q_0 \\ q_1 \\ q_2 \\ \vdots \\ q_{N-1} \\ \beta \end{pmatrix}, \tag{9.25}$$

where the $M \times M$ matrices R_j and S_j and vectors q_j (dimension M) are listed in
Table 9.1. The solution vector u and the vector $(q^{\mathrm{T}}, \beta^{\mathrm{T}})^{\mathrm{T}}$ have dimensions $M \times$
$(N + 1)$. Thus, the solutions of linear boundary-value problems can be found by
solving systems of linear equations: there is no need for iterative methods.

9.2.2 Schemes of Higher Orders

In order to improve the convergence for systems of boundary-value problems we try
to increase the order of the difference scheme. For systems of the form

Table 9.1 The elements of the matrix A and the vector at the right of Eq. (9.25) for the trapezoidal and midpoint scheme for the solution of the linear system $y' = Ay + q$. The $M \times M$ matrices B_a and B_b are given by the boundary condition $B_a y(a) + B_b y(b) = \beta$

Trapezoidal scheme	Midpoint scheme
$S_j = I + \frac{h}{2} A(x_j)$	$S_j = I + \frac{h}{2} A(x_{j+1/2})$
$R_j = -I + \dfrac{h}{2} A(x_{j+1})$	$R_j = -I + \dfrac{h}{2} A(x_{j+1/2})$
$q_j = -\dfrac{h}{2} \left[q(x_{j+1}) + q(x_j) \right]$	$q_j = -h\, q(x_{j+1/2})$

$$y' = f(x, y) ,$$

where f is easy to differentiate, a higher order scheme can be found by replacing f on each subinterval $[x_j, x_{j+1}]$ by its Hermite interpolant of order $p = 2k$, and integrating the system $y' = f(x, y)$ on this subinterval. We get a scheme of order p, for which the function f and its derivatives up to order $k - 1$ at x_j and x_{j+1} are needed (for details see [2]). For example, at $k = 2$ we obtain

$$\frac{u_{j+1} - u_j}{h} = \frac{1}{2} \left[f(x_j, u_j) + f(x_{j+1}, u_{j+1}) \right] + \frac{h}{12} \left[f'(x_j, u_j) - f'(x_{j+1}, u_{j+1}) \right] ,$$

which is of fourth-order and requires the computation of the total derivative

$$f'(x_j, u_j) = \left[\frac{\partial f}{\partial x} + \frac{\partial f}{\partial y} f \right]_{(x_j, u_j)} .$$

Higher orders can also be attained by extrapolating the mesh spacing h, in analogy to the scalar case (9.13). The approximation of a higher order is obtained by solving the boundary-value problem on a sequence of ever finer meshes (ever smaller spacings h) and constructing linear combinations of solutions from the individual meshes such that the discretization errors cancel out to an extent as large as possible. For example, we solve the boundary-value problem on two meshes with spacings h and $h/2$, and form the combination of the solutions

$$u_j^{(h/2,h)} = \frac{4}{3} u_{2j}^{(h/2)} - \frac{1}{3} u_j^{(h)} . \tag{9.26}$$

The finer mesh can also have a spacing of $h/3$, and we form

$$u_j^{(h/3,h)} = \frac{9}{8} u_{3j}^{(h/3)} - \frac{1}{8} u_j^{(h)} . \tag{9.27}$$

Either combination is fourth-order, $\| u^{(h/2,h)} - y \| \sim \| u^{(h/3,h)} - y \| = \mathcal{O}(h^4)$. Moreover, the approximations $u^{(h/2,h)}$ and $u^{(h/3,h)}$ can be merged in

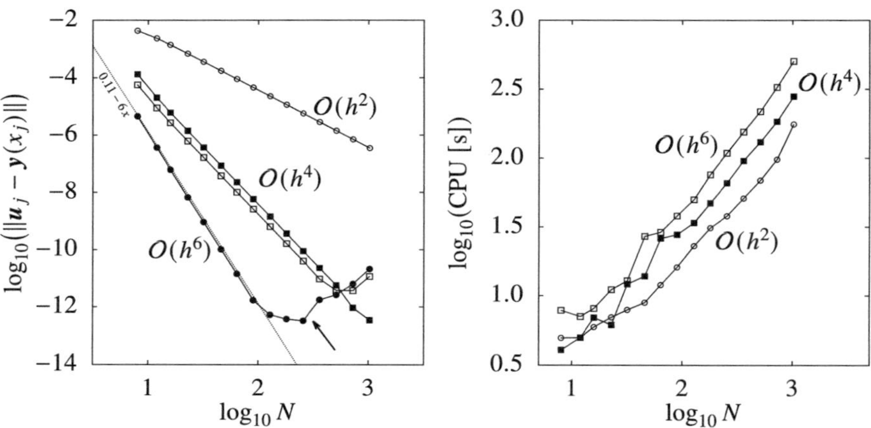

Fig. 9.3 [LEFT] Convergence of the numerical approximation to the exact solution of $y'' - \alpha(2x - 1)y' - 2\alpha y = 0$ with boundary conditions $y(0) = y(1) = 1$ when the mesh spacing is decreased and higher order methods are used. The basic method (9.9) is of order $\mathcal{O}(h^2)$, the approximations (9.26) and (9.27) are $\mathcal{O}(h^4)$, and (9.28) is $\mathcal{O}(h^6)$. We should not be oblivious to the possibility of round-off errors at large N (region to the right of the arrow)! [RIGHT] Numerical cost of the approximations

$$\boldsymbol{u}_j^{(h/3,h/2)} = \frac{9}{5}\,\boldsymbol{u}_j^{(h/3,h)} - \frac{4}{5}\,\boldsymbol{u}_j^{(h/2,h)} \,, \tag{9.28}$$

resulting in a *sixth-order* approximation, $\| \boldsymbol{u}^{(h/3,h/2)} - \boldsymbol{y} \| = \mathcal{O}(h^6)$. Of course, formulas (9.26), (9.27), and (9.28) are also applicable to scalar problems.

A gradual fragmentation of the interval in order to increase the numerical precision (Fig. 9.3) obviously requires additional computation time. But it also allows us to obtain reliable local error estimates which aid us in determining the convergence criteria or in choosing the mesh size in regions where the numerical error is expected to increase. Details can be found in [2].

9.3   Shooting Methods

The essence of shooting methods can be illustrated by the scalar linear problem

$$y'' + p(x)y' + q(x)y = r(x) \,, \qquad a \le x \le b \,,$$

with Dirichlet boundary conditions

$$y(a) = \alpha \,, \qquad y(b) = \beta \,.$$

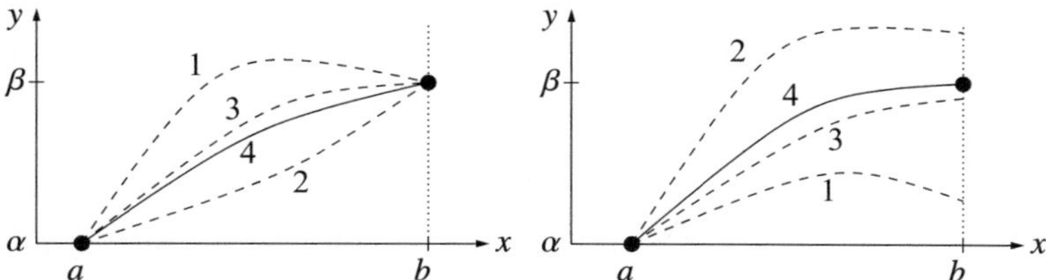

Fig. 9.4 [LEFT] The principle of a relaxation difference method. The numerical solution with an initial approximation satisfying the boundary conditions converges to the exact solution after some number of steps, maintaining $y(a) = \alpha$ and $y(b) = \beta$ at all times. [RIGHT] The basic idea of the shooting method. We integrate the differential equation with the initial value $y(a) = \alpha$ and slope $y'(a)$, and keep on changing this slope until the second boundary condition $y(b) = \beta$ is satisfied to the required precision

Assume that this problem has a unique solution $y(x)$, which we try to find by integrating the *initial-value problem* with some values $y(a)$ and $y'(a)$ towards the right boundary of the interval $[a, b]$. The value $y(a) = \alpha$ is given by the boundary condition, while the slope $y'(a)$ is unknown. Let us guess this slope, the "shooting angle", $y'(a) = s_1$, and compute the solution of the differential equation up to $x = b$. We obtain the solution $y_1(x)$ (for example, curve 1 in Fig. 9.4 (right)) which in general differs from the exact solution $y(x)$, since it does not satisfy the second boundary condition, $y_1(b) \neq y(b) = \beta$. Let us guess another angle, say, $y'(a) = s_2$, which corresponds to the solution $y_2(x)$ that in general also differs from the exact one, as $y_2(b) \neq y(b)$ and also $y_2(b) \neq y_1(b)$. Since the problem is *linear*, the true solution is the linear combination

$$y(x) = (1 - \theta)y_1(x) + \theta y_2(x) \,,$$

where

$$\theta = \frac{\beta - y_1(b)}{y_2(b) - y_1(b)} \,.$$

(Mathematicians call this a *convex combination of y_1 and y_2*.) A linear boundary-value problem can thus be solved in just two steps. Figure 9.4 (right) shows two approximate solutions "1" and "2" from which the final one ("4") is computed.

9.3.1 Second-Order Linear Equations

Let us approach the solution of linear boundary-value problems by shooting methods more generally. We follow [1] and discuss equations of the form

$$Ly = -y'' + p(x)y' + q(x)y = r(x) \,, \qquad a \leq x \leq b \,, \tag{9.29}$$

with the boundary conditions

$$a_0 y(a) - a_1 y'(a) = \alpha \,, \qquad a_0 a_1 \geq 0 \,, \qquad |a_0| + |a_1| \neq 0 \,,$$
$$b_0 y(b) + b_1 y'(b) = \beta \,, \qquad b_0 b_1 \geq 0 \,, \qquad |b_0| + |b_1| \neq 0 \,,$$

as well as $|a_0| + |b_0| \neq 0$. In addition to the continuity of the functions p, q, and r on $[a, b]$ we require that the corresponding *homogeneous problem* $Ly = 0$ with boundary conditions $a_0 y(a) - a_1 y'(a) = 0$ and $b_0 y(b) + b_1 y'(b) = 0$ has only a trivial solution, $y(x) = 0$. Then the problem (9.29) has a unique solution [1]. Define the auxiliary function $y^{(1)}$ that solves the initial-value problem

$$Ly^{(1)} = r(x) \,, \qquad y^{(1)}(a) = -\alpha c_1 \,, \qquad y^{(1)\prime}(a) = -\alpha c_0 \,,$$

and the function $y^{(2)}$ that solves the corresponding homogeneous problem

$$Ly^{(2)} = 0 \,, \qquad y^{(2)}(a) = a_1 \,, \qquad y^{(2)\prime}(a) = a_0 \,,$$

where c_0 and c_1 are *any such constants* that $a_1 c_0 - a_0 c_1 = 1$. The function

$$y(x) = y(x; s) = y^{(1)}(x) + s y^{(2)}(x) \tag{9.30}$$

obviously satisfies the left boundary condition $a_0 y(a) - a_1 y'(a) = \alpha(a_1 c_0 - a_0 c_1) = \alpha$ and also solves (9.29) if some s can be found—a sort of generalized "shooting angle"—such that the boundary condition at the right,

$$\phi(s) = b_0 y(b; s) + b_1 y'(b; s) - \beta = 0 \,,$$

is also satisfied. We insert (9.30) at $x = b$ in this expression and realize that the equation for $\phi(s)$ is linear in the parameter s, with the solution

$$s = \frac{\beta - b_0 y^{(1)}(b) - b_1 y^{(1)\prime}(b)}{b_0 y^{(2)}(b) + b_1 y^{(2)\prime}(b)} \,. \tag{9.31}$$

(This expression has a meaning if the denominator differs from zero. If it were zero, $y^{(2)}$ would be a non-trivial solution of the homogeneous problem $Ly = 0$, which is excluded by assumption.) Numerically we solve the initial-value problems $Ly^{(1)} = r(x)$ and $Ly^{(2)} = 0$ as two decoupled systems of two first-order differential equations,

$$\begin{pmatrix} y^{(1)} \\ v^{(1)} \end{pmatrix}' = \begin{pmatrix} v^{(1)} \\ pv^{(1)} + qy^{(1)} - r \end{pmatrix} \,, \qquad \begin{matrix} y^{(1)}(a) = -\alpha c_1 \\ v^{(1)}(a) = -\alpha c_0 \end{matrix} \,, \tag{9.32}$$

$$\begin{pmatrix} y^{(2)} \\ v^{(2)} \end{pmatrix}' = \begin{pmatrix} v^{(2)} \\ pv^{(2)} + qy^{(2)} \end{pmatrix} \,, \qquad \begin{matrix} y^{(2)}(a) = a_1 \\ v^{(2)}(a) = a_0 \end{matrix} \,. \tag{9.33}$$

These systems can be solved by any method for the solution of initial-value problems on the chosen mesh, e.g. (9.2). We denote the numerical solutions of (9.32) and (9.33) at x_j (mesh functions) by $\widetilde{y}_j^{(1)}$, $\widetilde{v}_j^{(1)}$, $\widetilde{y}_j^{(2)}$ in $\widetilde{v}_j^{(2)}$. At $x = a$ we apply the initial conditions

$$\widetilde{y}_0^{(1)} = -\alpha c_1 \,, \qquad \widetilde{v}_0^{(1)} = -\alpha c_0 \,, \qquad \widetilde{y}_0^{(2)} = a_1 \,, \qquad \widetilde{v}_0^{(2)} = a_0 \,,$$

and obtain the solution and its derivative at arbitrary x_j by the combinations

$$\widetilde{y}_j = \widetilde{y}_j^{(1)} + s\widetilde{y}_j^{(2)} \,, \tag{9.34}$$

$$\widetilde{v}_j = \widetilde{v}_j^{(1)} + s\widetilde{v}_j^{(2)} \,, \tag{9.35}$$

where

$$s = \frac{\beta - b_0 \widetilde{y}_N^{(1)} - b_1 \widetilde{v}_N^{(1)}}{b_0 \widetilde{y}_N^{(2)} + b_1 \widetilde{v}_N^{(2)}} \,. \tag{9.36}$$

Note that the approximate solution of the boundary-value problem is obtained only after both initial-value problems have been solved on the whole domain: in constructing (9.34) and (9.35) we need the parameter (9.36) that can be computed only when the mesh functions at $j = N$ become known.

For linear problems of the form (9.29) with smooth functions p, q, and r the procedure described above should work well: if a stable integration method with an error of order $\mathcal{O}(h^p)$ is used to solve the initial-value problem, the errors of the solution $\widetilde{y}$ and its derivative $\widetilde{v}$ are also of order $\mathcal{O}(h^p)$, thus

$$\left|\widetilde{y}_j - y(x_j)\right| = \mathcal{O}(h^p) \,, \qquad \left|\widetilde{v}_j - y'(x_j)\right| = \mathcal{O}(h^p) \,.$$

Still, this "naive" shooting method may lead to large numerical errors: why they occur and how we try to avoid them is discussed in Sect. 9.3.5.

9.3.2 Systems of Linear Second-Order Equations

Let us generalize this approach to systems of M linear second-order equations

$$L y = -y'' + P(x)y' + Q(x)y = r(x) \,, \qquad a \le x \le b \,, \tag{9.37}$$

where P and Q are $M \times M$ matrices whose elements are continuous in x on $[a, b]$, while y and r are M-dimensional vectors. Let the boundary conditions be

$$A_0 y(a) - A_1 y'(a) = \alpha \,, \qquad \det A_0 \ne 0 \,,$$
$$B_0 y(b) + B_1 y'(b) = \beta \,,$$

where A_0, A_1, B_0, and B_1 are constant matrices, and $\boldsymbol{\alpha}$ and $\boldsymbol{\beta}$ are constant vectors. We follow the steps from the scalar case (Sect. 9.3.1) and introduce a M-component vector function $\boldsymbol{y}^{(0)}$ that solves the initial-value problem

$$L\boldsymbol{y}^{(0)} = \boldsymbol{r}(x), \qquad A_0\boldsymbol{y}^{(0)}(a) - A_1\boldsymbol{y}^{(0)\prime}(a) = \boldsymbol{\alpha}, \qquad \boldsymbol{y}^{(0)\prime}(a) = \boldsymbol{0}. \qquad (9.38)$$

Define the M-component vector functions $\boldsymbol{y}^{(m)}(x)$, $m = 1, 2, \ldots, M$, that solve the system of M homogeneous initial-value problems

$$L\boldsymbol{y}^{(m)} = 0, \qquad A_0\boldsymbol{y}^{(m)}(a) - A_1\boldsymbol{y}^{(m)\prime}(a) = \boldsymbol{0}, \qquad \boldsymbol{y}^{(m)\prime}(a) = \boldsymbol{I}^{(m)}, \qquad (9.39)$$

where $\boldsymbol{I}^{(m)}$ is the unit vector with the value 1 at the mth position. We arrange the solutions of the system (9.39) in a $M \times M$ matrix,

$$Y(x) = \left(\boldsymbol{y}^{(1)}(x), \boldsymbol{y}^{(2)}(x), \ldots, \boldsymbol{y}^{(M)}(x)\right) .$$

The solution of the original boundary-value problem (9.37) is then

$$\boldsymbol{y}(x) = \boldsymbol{y}(x; \boldsymbol{s}) = \boldsymbol{y}^{(0)}(x) + Y(x)\boldsymbol{s} = \boldsymbol{y}^{(0)}(x) + \sum_{m=1}^{M} s_m \boldsymbol{y}^{(m)}(x) ,$$

where the vector $\boldsymbol{s} = (s_1, s_2, \ldots, s_M)^{\mathrm{T}}$ is the root of the equation

$$\boldsymbol{\phi}(\boldsymbol{s}) = B_0\boldsymbol{y}(b; \boldsymbol{s}) + B_1\boldsymbol{y}'(b; \boldsymbol{s}) - \boldsymbol{\beta} = \boldsymbol{0} .$$

This equation is linear in the parameter $\boldsymbol{s}$ that can be computed by solving

$$\left[B_0 Y(b) + B_1 Y'(b)\right] \boldsymbol{s} = \boldsymbol{\beta} - B_0\boldsymbol{y}^{(0)}(b) - B_1\boldsymbol{y}^{(0)\prime}(b)$$

(compare this to Eq. (9.31) and the corresponding commentary).

We compute the solution of (9.37) analogously to the scalar case. On the uniform mesh (9.2) we integrate the initial-value problems (9.38) and (9.39), and obtain the mesh functions $\widetilde{\boldsymbol{y}}_j^{(0)}$, $\widetilde{\boldsymbol{v}}_j^{(0)}$, $\widetilde{\boldsymbol{y}}_j^{(m)}$, and $\widetilde{\boldsymbol{v}}_j^{(m)}$, $m = 1, 2, \ldots, M$, at x_j. The approximations for the function $\boldsymbol{y}(x)$ and its derivative $\boldsymbol{y}'(x)$ are

$$\widetilde{\boldsymbol{y}}_j(\boldsymbol{s}) = \widetilde{\boldsymbol{y}}_j^{(0)} + \sum_{m=1}^{M} s_m \widetilde{\boldsymbol{y}}_j^{(m)} , \qquad\qquad (9.40)$$

$$\widetilde{\boldsymbol{v}}_j(\boldsymbol{s}) = \widetilde{\boldsymbol{v}}_j^{(0)} + \sum_{m=1}^{M} s_m \widetilde{\boldsymbol{v}}_j^{(m)} , \qquad\qquad (9.41)$$

where s is the root of the equation

$$\phi(s) = B_0 \widetilde{\mathbf{y}}_N(s) + B_1 \widetilde{\mathbf{v}}_N(s) - \boldsymbol{\beta} = \mathbf{0} .$$

We should stress again that one needs to solve the (vector) initial-value problem (9.38) *and* the system of M (vector) initial-value problems (9.39): this gives us the (vector) parameter s that we use to construct the final solution of the boundary-value problem (9.37) by using Eqs. (9.40) and (9.41). If the initial-value problems are integrated by a method with the error of order $\mathcal{O}(h^p)$, the error of the solution (or its derivative) of the boundary-value problem will be

$$\left\| \widetilde{\mathbf{y}}_j(s) - \mathbf{y}(x_j) \right\| = \mathcal{O}(h^p) , \qquad \left\| \widetilde{\mathbf{v}}_j(s) - \mathbf{y}'(x_j) \right\| = \mathcal{O}(h^p) .$$

9.3.3 *Non-linear Second-Order Equations*

Solving non-linear scalar boundary-value problems of the form

$$y'' = f(x, y, y') , \qquad a \le x \le b ,$$

with the boundary conditions

$$\begin{aligned}
a_0 y(a) - a_1 y'(a) &= \alpha , & a_0, a_1 &\ge 0 , \\
b_0 y(b) + b_1 y'(b) &= \beta , & b_0, b_1 &\ge 0 , & a_0 + b_0 &> 0 ,
\end{aligned}$$

allows us to apply the tools we learned in studying linear problems. With minor modifications, this section also closely follows [1]. Assume that the function $f(x, u, v)$ is continuous in x on the interval $[a, b]$ and is uniformly Lipschitz in u and v. If we further assume that

$$\frac{\partial f}{\partial u} > 0 , \qquad \left| \frac{\partial f}{\partial v} \right| \le C , \qquad C > 0 ,$$

such a boundary-value problem has a unique solution. As before, we seek its solution by solving the corresponding initial-value problem

$$\begin{aligned}
y'' &= f(x, y, y') , \qquad a \le x \le b , \\
y(a) &= a_1 s - c_1 \alpha , \\
y'(a) &= a_0 s - c_0 \alpha ,
\end{aligned}$$

where c_0 and c_1 are constants such that $a_1 c_0 - a_0 c_1 = 1$. The solution $u = u(x; s)$ of this initial-value problem is the solution of the original boundary-value problem precisely when the parameter s is the root of the equation

$$\phi(s) = b_0 y(b; s) + b_1 y'(b; s) - \beta = 0 .$$

Let us denote $y = u$ and $y' = v$, and transform the initial-value problem to the system of two first-order equations,

$$\begin{pmatrix} u \\ v \end{pmatrix}' = \begin{pmatrix} v \\ f(x, u, v) \end{pmatrix} , \qquad \begin{matrix} u(a) = a_1 s - c_1 \alpha \\ v(a) = a_0 s - c_0 \alpha \end{matrix} , \tag{9.42}$$

which can be numerically solved by some method discussed in Chap. 8. The solutions of the system at $x = b$ are $u(b; s)$ and $v(b; s)$, which we use to form

$$\phi(s) = b_0 u(b; s) + b_1 v(b; s) - \beta \tag{9.43}$$

and check whether at the present parameter s we attain $\phi(s) = 0$. We repeat this procedure until this requirement is met to specified precision, or until the parameter s settles to its final value $s^\star$. For the computation of $s^\star$, Newton's method can be used, as described in the following.

Solution by Newton's method We choose an initial approximation $s^{(0)}$ and iterate

$$s^{(\nu+1)} = s^{(\nu)} - \frac{\phi(s^{(\nu)})}{\phi'(s^{(\nu)})} , \qquad \nu = 0, 1, 2, \ldots , \tag{9.44}$$

where $'$ denotes the derivative with respect to s. The function $\phi(s)$ from Eq. (9.43) is already known; we obtain the function $\phi'(s)$ by first defining the functions

$$\xi(x) = \frac{\partial u(x; s)}{\partial s} , \qquad \eta(x) = \frac{\partial v(x; s)}{\partial s} ,$$

where $u = y$ and $v = y'$ are the solutions of (9.42). We differentiate the system (9.42) with respect to s and get an additional system of first-order equations

$$\begin{pmatrix} \xi \\ \eta \end{pmatrix}' = \begin{pmatrix} \eta \\ p(x; s)\eta + q(x; s)\xi \end{pmatrix} , \qquad \begin{matrix} \xi(a) = a_1 \\ \eta(a) = a_0 \end{matrix} , \tag{9.45}$$

where

$$p(x; s) = \frac{\partial f(x, u(x; s), v(x; s))}{\partial v} , \qquad q(x; s) = \frac{\partial f(x, u(x; s), v(x; s))}{\partial u} .$$

We use the solutions of (9.45) at $x = b$ to construct

$$\phi'(s) = b_0 \xi(b; s) + b_1 \eta(b; s) .$$

The initial-value problems (9.45) and (9.42) represent a system of four first-order differential equations which we solve for each s simultaneously on the same mesh: at the current s we use the solutions $u(b; s)$ and $v(b; s)$, as well as the corresponding value $\phi(s)$ at the extreme point of the interval, to compute the new value of $\xi(b; s)$ and $\eta(b; s)$, as well as the corresponding values $\phi'(s)$, and finally use $\phi(s)$ and $\phi'(s)$ in the iteration (9.44) until convergence is achieved. The additional time needed to solve (9.45) is compensated by the rapid convergence of the Newton's method.

9.3.4 Systems of Non-linear Equations

Boundary-value problems for systems of non-linear differential equations are solved by first transforming them to an equivalent system of M (in general non-linear) first-order equations

$$y' = f(x, y) , \qquad a < x < b , \tag{9.46}$$

where $y \in \mathbb{R}^M$, $f : \mathbb{R} \times \mathbb{R}^M \to \mathbb{R}^M$. We discuss boundary conditions of the form

$$A y(a) + B y(b) = \alpha . \tag{9.47}$$

Here A and B are constant $M \times M$ matrices and α is a constant M-dimensional vector. To the boundary-value problem (9.46)–(9.47) we assign the initial-value problem

$$u' = f(x, u) , \qquad u(a) = s . \tag{9.48}$$

If f is uniformly Lipschitz, and the Jacobi matrix $\partial f / \partial u$ exists, a unique solution $u(x; s)$ of this initial-value problem exists that smoothly depends on the components of s. We seek a vector of parameters $s^\star$ for which the solution of the initial-value problem $u(x; s)$ is also the solution of the boundary-value problem (9.46) and (9.47). This occurs precisely when for $s = s^\star$ we have

$$\phi(s) = As + Bu(b; s) - \alpha = 0 . \tag{9.49}$$

The initial-value problem is best solved on the usual uniform mesh (9.2) by using some stable method for the solution of initial-value problems. At the given s, each point x_j carries the numerical solution $\widetilde{u}_j$. If a method with the error of order $\mathcal{O}(h^p)$ is used, then $\|\widetilde{u}_j(s) - u(x_j; s)\| \leq \mathcal{O}(h^p)$. The solution $s^\star$ can be found by simple iteration or by using Newton's method.

Solution by simple iteration We are seeking the fixed point of the mapping

$$s = G(s) = s - (A + B)^{-1} [As + B\widetilde{u}_N(s) - \alpha] ,$$

where we have assumed that the matrix $Q = A + B$ is non-singular. (For more general boundary conditions (9.47) this is not necessarily true: then a different Q is needed to achieve convergence [1].) With an arbitrary initial approximation $s^{(0)}$ we iterate

$$s^{(\nu+1)} = G(s^{(\nu)}) , \qquad \nu = 0, 1, \dots .$$

The convergence in the parameters s is mirrored in the convergence of the solutions at the individual points x_j $(j = 0, 1, \dots, N)$, for which

$$\|\widetilde{u}_j(s^{(\nu)}) - u(x_j; s^\star)\| \le \mathcal{O}(h^p) + \omega^\nu \|s^\star - s^{(0)}\| , \qquad \nu = 0, 1, \dots ,$$

where $u(x; s^\star) \equiv y(x)$ is the desired solution of the boundary-value problem (9.46) and (9.47). The parameter ω $(0 < \omega < 1)$ satisfies the inequality

$$\int_a^b \mathcal{J}(x) \, dx \le \log\left(1 + \frac{\omega}{P}\right) ,$$

where $P = \|(A + B)^{-1} B\|_\infty$ and $\mathcal{J}$ is a function that satisfies $\mathcal{J}(x) \ge \left\| \partial f(x, u) / \partial u \right\|_\infty$.

Solution by Newton's method If the parameter ω in the simple iteration is close to one, the convergence is slow. A much faster approach is again offered by the Newton's method, which we start with the initial approximation $s^{(0)}$ and iterate

$$s^{(\nu+1)} = s^{(\nu)} + \Delta s^{(\nu)} , \qquad \nu = 0, 1, 2, \dots , \tag{9.50}$$

where $\Delta s^{(\nu)}$ is the solution of the system of linear equations

$$\frac{\partial \phi(s^{(\nu)})}{\partial s} \Delta s^{(\nu)} = -\phi(s^{(\nu)}) . \tag{9.51}$$

We decompose the Jacobi matrix $\partial \phi(s)/\partial s$, which is non-singular for all s [3], as

$$\frac{\partial \phi(s)}{\partial s} = A + B W(b; s) , \tag{9.52}$$

where the matrix $W(x; s) = \partial u(x; s)/\partial s$ is the solution of the system

$$W' = \frac{\partial f(x, u(x; s))}{\partial u} W , \qquad W(0) = I . \tag{9.53}$$

For Newton's method we repeat the following procedure (details are in [1]): in the νth iteration step (i.e. *at current values* $s^{(\nu)}$) on the mesh (9.2) we numerically solve the non-linear initial-value problem (9.48) with $s = s^{(\nu)}$, yielding the M components of the solution $\widetilde{u}(s^{(\nu)})$. Simultaneously, and on the same mesh, we solve M systems of linear differential equations (9.53), where we use the currently available $\widetilde{u}(s^{(\nu)})$ to compute the matrix $\partial f(x, u)/\partial u$. This gives us the approximate solution vector

$\widetilde{\boldsymbol{u}}(\boldsymbol{s}^{(\nu)})$ and the corresponding $M \times M$ matrix $W(x; \boldsymbol{s}^{(\nu)})$. We use them to compute the new vector $\boldsymbol{\phi}(\boldsymbol{s})$ according to (9.49) and the new matrix $\partial\boldsymbol{\phi}(\boldsymbol{s})/\partial\boldsymbol{s}$ according to (9.52). These are used to solve the matrix system for $\Delta\boldsymbol{s}^{(\nu)}$ (Eq. (9.51)) and, finally, we use Eq. (9.50) to compute the next approximate vector of the shooting parameters $\boldsymbol{s}^{(\nu+1)}$. We repeat the whole procedure until convergence to the final values $\boldsymbol{s}^{\star}$ is achieved.

9.3.5 Multiple (Parallel) Shooting

The basic version of scalar shooting from the beginning of Sect. 9.3 may be plagued by round-off errors. Such errors are generated, for example, when the combination (9.34) is formed and the contributions $\widetilde{y}_j^{(1)}$ and $s\widetilde{y}_j^{(2)}$ are almost equal in magnitude and oppositely signed, or when the denominator of (9.36) is very small. Such errors, due to which the solution of the corresponding initial-value problem blows up, are difficult to localize on an arbitrary mesh (9.2) and are almost unpreventable. We may face the same problems with systems of equations. Sometimes the most naive solution helps: increase arithmetic precision. If this does not work, one may opt for multiple shooting.

Here we describe multiple shooting for systems of non-linear equations (9.46) with boundary conditions (9.47), as approached by [2]. We rescale the problem with the variable $x \in [a, b]$ on the mesh (9.2) by introducing a new variable ξ that takes the values $\xi \in [0, 1]$ on each subinterval $[x_{j-1}, x_j]$,

$$\xi = \frac{x - x_j}{x_j - x_{j-1}} = \frac{x - x_j}{h}$$

(see Fig. 9.5). On these subintervals one solves decoupled initial-value problems, where initial conditions are imposed *for each subinterval* such that finally the main

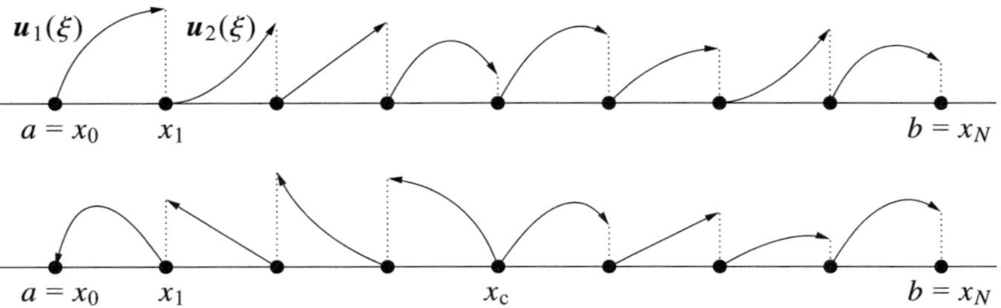

Fig. 9.5 Multiple (parallel) shooting. [Top] Shooting from $x = a$ towards $x = b$. We use the solutions $\boldsymbol{u}_j(\xi)$ from all subintervals $[x_{j-1}, x_j]$ to assemble the final solution on the whole interval $[a, b]$. [Bottom] Shooting from a critical point x_c (with a singularity, a known intermediate value, or another prescribed intermediate "boundary" condition) towards $x = a$ and $x = b$. We do not discuss such cases here; for further reading see [2]

boundary conditions at $x = a$ and $x = b$ are fulfilled. (If in the interior of $[a, b]$ further requirements need to be met, a non-uniform mesh may be utilized, so that certain conditions can be realized precisely at the required critical points. For details see [2].) We also transform the vectors y and f,

$$y_j(\xi) = y\left(x_{j-1} + \xi h\right) ,$$
$$f_j(\xi, z) = h\, f\left(x_{j-1} + \xi h, z\right) .$$

Thus, *on each (jth) subinterval* we are solving the system

$$\frac{\mathrm{d}y_j}{\mathrm{d}\xi} = f_j\left(\xi, y_j(\xi)\right) , \qquad 0 < \xi < 1 , \qquad j = 1, 2, \ldots, N ,$$

while the boundary conditions become

$$A y_1(0) + B y_N(1) = \alpha .$$

Moreover, we need to glue together continuously the solutions at each internal point of the mesh: we do this by requiring $y_j(0) - y_{j-1}(1) = 0$ (be warned again: the argument of these vector functions is the new variable ξ). The system of equations, the boundary conditions, and the local continuity conditions are condensed in the equations

$$\frac{\mathrm{d}Y}{\mathrm{d}\xi} = F(\xi, Y) , \qquad \widetilde{A}Y(0) + \widetilde{B}Y(1) = \widetilde{\alpha} , \tag{9.54}$$

where the vector

$$Y(\xi) = (y_1^{\mathrm{T}}(\xi), y_2^{\mathrm{T}}(\xi), \ldots, y_N^{\mathrm{T}}(\xi))^{\mathrm{T}} \in \mathbb{R}^{MN}$$

contains the complete solution, and the N vector functions $f_i : [0, 1] \times \mathbb{R}^M \to \mathbb{R}^M$ can be understood as a vector function $F : [0, 1] \times \mathbb{R}^{MN} \to \mathbb{R}^{MN}$,

$$F(\xi, Y) = (f_1^{\mathrm{T}}(\xi, y_1), f_2^{\mathrm{T}}(\xi, y_2), \ldots, f_N^{\mathrm{T}}(\xi, y_N))^{\mathrm{T}} .$$

The vector $\widehat{\alpha} = (\alpha^{\mathrm{T}}, 0^{\mathrm{T}}, \ldots, 0^{\mathrm{T}})^{\mathrm{T}} \in \mathbb{R}^{MN}$ contains the boundary and continuity conditions. The $MN \times MN$ matrices $\widetilde{A}$ and $\widetilde{B}$ have the block structure

$$\widetilde{A} = \begin{pmatrix} A & & & & \\ & I & & & \\ & & I & & \\ & & & \ddots & \\ & & & & I \end{pmatrix} , \qquad \widetilde{B} = \begin{pmatrix} 0 & & & & B \\ -I & 0 & & & \\ & -I & 0 & & \\ & & & \ddots & \ddots \\ & & & & -I & 0 \end{pmatrix} .$$

The matrices A and B from the underlying boundary-value problem, the identity I and the zero matrix 0 have size $M \times M$.

The system (9.54) corresponds to the vector initial-value problem consisting of MN first-order equations and the initial condition:

$$\frac{dU}{d\xi} = F(\xi, U), \qquad 0 < \xi \le 1, \qquad U(0) = s, \tag{9.55}$$

where the solution vector U has the same structure as the vector Y. We solve this problem on the interval $\xi \in [0, 1]$ by simple shooting. With respect to individual subintervals $[x_{j-1}, x_j]$, the system of equations (9.55) is decoupled and completely equivalent to N vector systems of dimension M,

$$\frac{du_j(\xi)}{d\xi} = f_j(\xi, u_j), \qquad u_j(0) = s_j, \qquad j = 1, 2, \ldots, N,$$

which can be solved independently (the vectors s_j of length M are the elements of the vector $s = (s_1^T, s_2^T, \ldots, s_N^T)^T$). As it has been explained in the previous two sections, we must solve, in addition to the system (9.55), the matrix initial-value problem

$$\frac{dW}{d\xi} = \frac{\partial F(\xi, U(\xi; s))}{\partial Y} W, \qquad W(0) = I. \tag{9.56}$$

The matrix W is block-diagonal, $W = \mathrm{diag}\big(W_1(\xi, s_1), W_2(\xi, s_2), \ldots, W_N(\xi, s_N)\big)$, so this system is decoupled as well, and can be written in the form

$$\frac{dW_j}{d\xi} = \frac{\partial f_j(\xi, u_j(\xi; s_j))}{\partial u_j} W_j, \qquad W_j(0) = I, \qquad j = 1, 2, \ldots, N.$$

Its solution can be again found by Newton's iteration with an appropriate initial approximation of the shooting parameters $s^{(0)}$. With the current approximation $s = s^{(v)}$ we solve the vector system (9.55), and use its solution to compute the matrix $\partial F / \partial Y$ at the right-hand side of (9.56). We use the resulting vector $U(1; s^{(v)})$ and matrix $W(1; s^{(v)})$ (at the maximum value $\xi = 1$) to compute

$$\phi(s^{(v)}) = \widetilde{A} s^{(v)} + \widetilde{B} U(1; s^{(v)}) - \widetilde{\alpha}$$

and solve the matrix system for the update to the vector of shooting parameters,

$$\left[\widetilde{A} + \widetilde{B} W(1; s^{(v)})\right] \Delta s^{(v)} = -\phi(s^{(v)}).$$

Finally, we compute the new shooting parameters,

$$s^{(v+1)} = s^{(v)} + \Delta s^{(v)},$$

and repeat the procedure until it converges.

Advantage of multiple shooting We had to wait until the end to learn why the strenuous solving of the system (9.48) on narrower subintervals with $\xi \in [0, 1]$ is preferable to solving the system (9.55) on the original interval with $x \in [a, b]$. If the systems are solved by using a stable method with the error of order $\mathcal{O}(h^p)$, the errors at the points on the subintervals can be estimated by

$$\|\boldsymbol{u}(x_j) - \boldsymbol{u}_j\| \le h^p M_1 \exp(\Lambda_1 |x_0 - x_j|) ,$$
$$\|\boldsymbol{U}(\xi_j) - \boldsymbol{U}_j\| \le h^p M_2 \exp(\Lambda_2 |\xi_0 - \xi_j|) ,$$

where the constants $M_{1,2}$ and $\Lambda_{1,2}$ can be bounded by the values of the functions $\boldsymbol{f}$ and $\boldsymbol{F}$, and their derivatives [4, 5]. Because the original mesh (9.2) is split into N subintervals, we have $\Lambda_2 \approx \Lambda_1 (b - a)/N$, so the upper bound for the error in the case of multiple shooting decreases exponentially or becomes proportional to $[\exp(\Lambda_1 |b - a|)]^{1/N}$ instead of to $[\exp(\Lambda_1 |b - a|)]$.

9.4   Asymptotic Discretization Schemes ⋆

In discussions of ordinary and partial differential equations we are often concerned about the limits in which some physical mechanism expressed by these equations dominates over other mechanisms. The limit (asymptotic) equation is an approximation of the complete physical picture, but it is usually much simpler than the full equation, and easier to use. We should be particularly alert in problems that allow one part of the domain to be treated asymptotically, while the rest requires a non-asymptotic approach. In such cases the continuous equation has to be discretized such that the solutions of the discrete equation converge to the true solution in both regimes when the mesh spacing is reduced.

Difference schemes with correct convergence properties in the asymptotic regime are called *asymptotic-preserving discretization schemes*. Inadequate discretization may yield an inefficient scheme, which actually leads to the correct asymptotic solution, but only by using mesh spacings that are much smaller than the spatial scale characteristic of the asymptotic solution.

Following closely [6, 7], we illustrate the idea of asymptotic discretization by the transport equation for particles in matter with absorption (absorption cross-section σ_a, absorptions per unit length) and scattering (scattering cross-section σ_s, scatterings per unit length). This equation originates in the stationary limit of the corresponding partial differential equation, but in its continuous and discrete form it is very instructive and possesses all the properties of the boundary-value problem, so we discuss it in this Chapter. The equation has the form

$$\mu \frac{\partial (v N(x, \mu))}{\partial x} + (\sigma_a + \sigma_s) v N(x, \mu) = \frac{\sigma_s}{2} \int_{-1}^{1} v N(x, \mu') \, d\mu' + Q(x, \mu) . \quad (9.57)$$

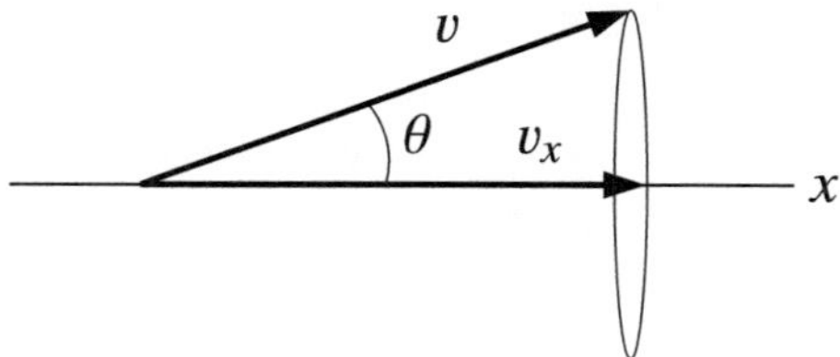

Fig. 9.6 Particle transport. All particles travel with equal magnitudes of velocity v in directions given by the cosines $\mu = \cos\theta = v_x/v$. Each parameter μ defines a cone of directions: even though the transport equation is one-dimensional, the density function in fact describes a three-dimensional region of an infinite layer

Its solution is the particle density function $N(x, \mu)$ depending on the coordinate (variable $x \in [0, 1]$) and angle ($\mu = \cos\theta = v_x/v \in [-1, 1]$), see Fig. 9.6. The number of particles on the interval $[x, x + dx]$ traveling in the directions $[\mu, \mu + d\mu]$ is $N(x, \mu)\,dx\,d\mu$. The particle absorption rate on these intervals is $\sigma_a v N(x, \mu)\,dx\,d\mu$, while the scattering rate is $\sigma_s v N(x, \mu)\,dx\,d\mu$. The average distance between particle interactions (the mean free path) is $\lambda = 1/\sigma_t$, where $\sigma_t = \sigma_a + \sigma_s$ is the total cross-section. The generation of new particles is described by the term $Q(x, \mu)$: the number of particles per unit time created at $[x, x + dx]$ with initial directions $[\mu, \mu + d\mu]$ is $Q(x, \mu)\,dx\,d\mu$. The solution of Eq. (9.57) is uniquely determined by the boundary conditions at $x = 0$ for $\mu > 0$ and at $x = 1$ for $\mu < 0$.

Equation (9.57) can be rewritten in a more compact form

$$\mu \frac{\partial \psi}{\partial x} + \sigma_t \psi = (\sigma_t - \sigma_a)\,\phi + Q\,,$$

where $\psi = v N$ is the *angular flux*, while its average over all directions,

$$\phi = \frac{1}{2} \int_{-1}^{1} \psi(x, \mu')\,d\mu'\,, \tag{9.58}$$

is the *scalar flux*. The key physics consideration follows now. We use a scaling parameter ε to extract the asymptotic behavior from the equation. Assume that we are studying a diffusion problem in a one-dimensional layer with thickness much larger than λ, in which scattering processes dominate (the absorption is weak). Suppose that the angular flux, the absorption and scattering cross-sections, and the source term Q are continuous and that they change insignificantly on distances comparable to the mean free path. This asymptotic diffusion limit is described by the equation

$$\mu \frac{\partial \psi}{\partial x} + \frac{\sigma_t}{\varepsilon}\,\psi = \left(\frac{\sigma_t}{\varepsilon} - \varepsilon\sigma_a\right)\phi + \varepsilon Q \tag{9.59}$$

in the limit $\varepsilon \to 0$, since then $\varepsilon\sigma_a \to 0$, $\sigma_t/\varepsilon \to \infty$ (or $\lambda/\varepsilon \to 0$) and $\varepsilon Q \to 0$. We obtain the *continuous* version of the transport equation valid in the asymptotic regime by expanding the solution in terms of the scaling parameter:

$$\psi = \sum_{n=0}^{\infty} \varepsilon^n \psi^{(n)} \, .$$

When this is inserted in Eq. (9.59) and the coefficients of the same powers of ε on both sides of the equation are matched, we get, to first three orders:

$$\psi^{(0)} = \phi^{(0)} \, ,$$
$$\psi^{(1)} = \phi^{(1)} - \frac{\mu}{\sigma_t} \frac{\partial \phi^{(0)}}{\partial x} \, ,$$
$$\psi^{(2)} = \phi^{(2)} - \frac{\sigma_a}{\sigma_t} \phi^{(0)} + \frac{\mu}{\sigma_t} \frac{\partial}{\partial x} \left[\phi^{(1)} - \frac{\mu}{\sigma_t} \frac{\partial \phi^{(0)}}{\partial x} \right] + \frac{Q}{\sigma_t} \, . \tag{9.60}$$

By integrating Eq. (9.60) over the directions μ, we obtain the diffusion equation

$$-\frac{\partial}{\partial x} \left[\frac{1}{3\sigma_t} \frac{\partial \phi^{(0)}}{\partial x} \right] + \sigma_a \phi^{(0)} = Q \, . \tag{9.61}$$

The asymptotic behavior of the transport equation with the assumptions specified above is therefore captured adequately by a simpler equation of the diffusion type, with a characteristic length $L = 1/\sqrt{3\sigma_t\sigma_a}$ that does not depend on ε.

9.4.1 Discretization

In the asymptotic limit, the solution of the full transport equation satisfies the diffusion equation. Does the analogous conclusion follow in the discrete case? The basic message of this Section is the warning that the discrete solution of the transport equation satisfies the discrete diffusion equation only in a discretization that preserves the physical content of such an asymptotic limit. In the following we show one asymptotic and one non-asymptotic discretization of the transport equation on a uniform mesh with cells (subintervals) $[x_{j-1/2}, x_{j+1/2}]$ of length $h = x_{j+1/2} - x_{j-1/2}$.

We discretize Eq. (9.59) in coordinates $(x_j, \; j = 1, 2, \ldots, N$, such that $x_{1/2} = 0$ and $x_{N+1/2} = 1)$ and angles $(\mu_m, \; m = 1, 2, \ldots, M)$. For clarity we assume constant σ_t, σ_a, and Q. We obtain

$$\mu_m \left(\psi_{m,j+\frac{1}{2}} - \psi_{m,j-\frac{1}{2}} \right) + \frac{h\sigma_t}{\varepsilon} \psi_{m,j} = h \left(\frac{\sigma_t}{\varepsilon} - \varepsilon\sigma_a \right) \phi_j + h\varepsilon Q \, .$$

The discrete solutions $\psi_{m,j\pm1/2}$ (computed at the cell edges) and $\phi_{m,j}$ (averaged over one cell) are functions of two indices. In a computer implementation we arrange the solution in a $M \times (N+1)$-dimensional array:

$$\psi = \left(\psi_{1,1/2}, \psi_{1,3/2}, \ldots, \psi_{1,N+1/2}, \quad \cdots \quad \psi_{M,1/2}, \psi_{M,3/2}, \ldots, \psi_{M,N+1/2}\right)^{\mathrm{T}}.$$

Let us assume Dirichlet boundary conditions. They are expressed as

$$\psi_{m,1/2} = f_m \quad (\mu_m > 0), \qquad \psi_{m,N+1/2} = g_m \quad (\mu_m < 0).$$

"Upwind" discretization The discrete angular flux has $N + 1$ spatial components. The discrete transport equation connects N pairs of cell boundaries. The one missing equation that relates the angular flux at the cell edge to the angular flux within the cell, will determine the asymptotic or non-asymptotic character of the whole scheme. We first attempt an *"upwind"* (U) discretization

$$\psi_{m,j} = \begin{cases} \psi_{m,j+1/2} \; ; \; \mu_m > 0 \, , \\ \psi_{m,j-1/2} \; ; \; \mu_m < 0 \, , \end{cases}$$

and apply the same asymptotic analysis in terms of powers of ε as in the continuous case. We find that the asymptotic solution satisfies the difference equation

$$\frac{1}{4h}\left[\phi_j^{(0)} - \phi_{j-1}^{(0)}\right] + \frac{1}{4h}\left[\phi_j^{(0)} - \phi_{j+1}^{(0)}\right] = 0 \, , \tag{9.62}$$

which is the discrete form of a "bare" diffusion equation $\partial^2\phi^{(0)}/\partial x^2 = 0$ from which scattering and absorption cross-sections have vanished! Therefore, the upwind discretization asymptotically generates a difference scheme which does not reflect the corresponding limit in the continuous equation.

"Diamond" discretization We arrive at a quite different asymptotic equation by the so-called *diamond* (D) discretization in the nomenclature of [6],

$$\psi_{m,j} = \frac{1}{2}\left(\psi_{m,j+1/2} + \psi_{m,j-1/2}\right) .$$

By applying the analysis in orders of the scaling parameter ε we realize that the diamond solution in the asymptotic diffusion limit satisfies the equation

$$-\frac{1}{3\sigma_{\mathrm{t}}}\frac{1}{h^2}\left[\phi_{j+3/2}^{(0)} - 2\phi_{j+1/2}^{(0)} + \phi_{j-1/2}^{(0)}\right]$$
$$+\frac{\sigma_{\mathrm{a}}}{4}\left[\phi_{j+3/2}^{(0)} + 2\phi_{j+1/2}^{(0)} + \phi_{j-1/2}^{(0)}\right] = \tfrac{1}{2}\left(Q_{j+1} + Q_j\right) , \tag{9.63}$$

which is a valid discretization of Eq. (9.61) and involves both cross-sections and the source term. The diamond discretization thus yields a difference scheme that preserves the character of the original equation in the asymptotic limit.

Example (Adapted from [9].) We check how the upwind and diamond schemes work in practice by computing the numerical solution on a mesh with $N = 10$ cells and parameters $Q = 1\,\mathrm{cm^{-1}s^{-1}}$, $\sigma_t = 10\,\mathrm{cm^{-1}}$, $\sigma_a = 0.1\,\mathrm{cm^{-1}}$, and $\varepsilon = 0.1$. We need to solve a system of linear equations with a $M(N + 1) \times M(N + 1)$ matrix, so excessive M and N may quickly exhaust our memory resources! The scalar flux ϕ_j, which we compute from the angular fluxes $\psi_{m,j}$ by integrating over angles in analogy to Eq. (9.58), is therefore best evaluated by Gaussian quadrature

$$\phi_j = \sum_{m=1}^{M} w_m \psi_{m,j} \,,$$

which is exceptionally precise even with few points ($M = 4, 8$, or 16). Here w_m are the quadrature weights and m is the index of the corresponding quadrature node μ_m (see Table 25.4 in [8] and Sect. 3.1.5). Figure 9.7 (left) shows the computed scalar flux for various parameters ε by using the upwind scheme. In the asymptotic limit $\varepsilon \to 0$ the solution converges to zero, as Eq. (9.62) does not contain the source term Q. Figure 9.7 (right) shows the flux computed by using the diamond scheme. In the limit $\varepsilon \to 0$ the numerical solution converges to the exact solution.

Figure 9.8 tells us that the upwind scheme does in fact converge to the exact solution, but only with a very large number of cells (a very fine spatial mesh). For a precise solution one needs to employ a mesh in which the cell length is several orders of magnitude smaller than the mean free path [7].   ◁

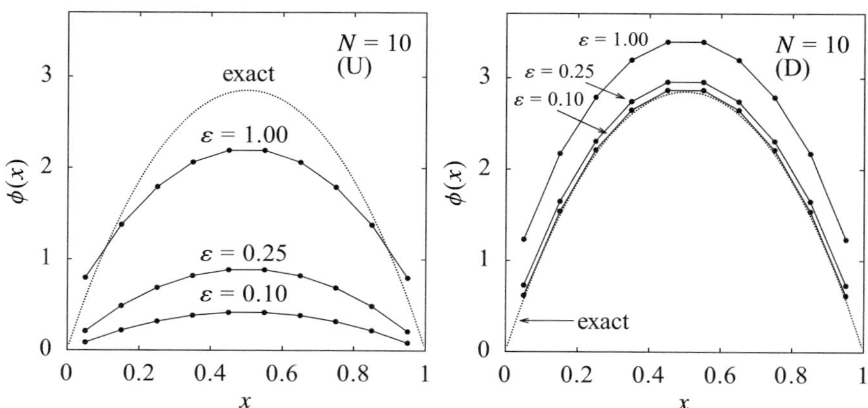

Fig. 9.7 The scalar flux from the solution of the discretized transport equation on a mesh of $N = 10$ cells for various scaling parameters ε. [LEFT] The upwind (U) scheme has no particle sources Q present in (9.62), so the solution converges to zero when $\varepsilon \to 0$. [RIGHT] The diamond (D) scheme (9.63) has the correct asymptotic behavior

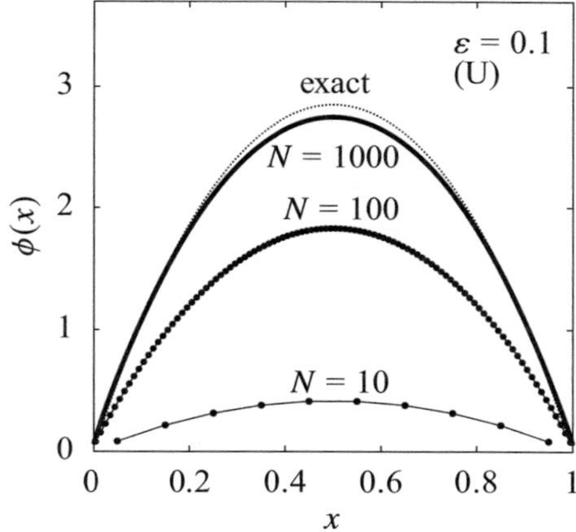

Fig. 9.8 Convergence of the scalar flux in the non-asymptotic difference scheme in upwind discretization with $N = 10$, 100, and 1000 cells, to the exact solution. The values of the parameters are $Q = 1\,\mathrm{cm^{-1}s^{-1}}$, $\sigma_\mathrm{t} = 10\,\mathrm{cm^{-1}}$, $\sigma_\mathrm{a} = 0.1\,\mathrm{cm^{-1}}$, and $\varepsilon = 0.1$. For a precise solution we need an extremely fine mesh: for $N = 1000$ the cell length is equal to $1/100$ of the mean free path

9.5 Collocation Methods ⋆

In Sect. 9.1 we discussed the solution of scalar boundary-value problems by difference methods. We obtained the solution by approximating the derivatives in the differential equations and boundary conditions, and solving the resulting difference equations. In collocation methods we expand the (as yet unknown) solution y as

$$y(x) \approx u(x) = \sum_n a_n \phi_n(x) \,, \qquad a \le x \le b \,,$$

and satisfy the requirements of the problem with appropriate coefficients a_n. The basis functions ϕ_n may be trigonometric functions, cubic B-splines, or polynomials. In the following we discuss the basic properties of collocation methods for three classes of scalar boundary-value problems: linear second-order problems by using B-spline collocation in Sect. 9.5.1, and linear and non-linear problems of higher orders by using Gauss collocation in Sects. 9.5.2 and 9.5.3, trailing closely the presentation in [2]. For greater clarity we use a uniform mesh, $x_{j+1} - x_j = h_j = h$, throughout this Section. Collocation methods for boundary-value problems on semi-infinite and infinite domains are discussed in the context of PDEs in Sect. 12.7.

9.5.1 Scalar Linear Second-Order Boundary-Value Problems

The basic idea of the collocation method for solving scalar boundary-value problems can be demonstrated by the linear problem of the form

$$y'' + p(x)y' + q(x)y = r(x) , \qquad 0 \le x \le 1 ,$$

with boundary conditions

$$y(0) = \alpha , \qquad y(1) = \beta .$$

The lowest degree of the polynomials for a continuous and continuously differentiable interpolation of the solution over all subintervals, is three. We can use piecewise cubic polynomials known as cubic B-splines. The basis B-spline B_j is defined as

$$B_j(x) = \begin{cases} 0 & ;\ x \le x_{j-2} , \\[2mm] \dfrac{(x - x_{j-2})^3}{6\Delta x^3} & ;\ x_{j-2} \le x \le x_{j-1} , \\[2mm] \dfrac{1}{6} + \dfrac{(x - x_{j-1})}{2\Delta x} + \dfrac{(x - x_{j-1})^2}{2\Delta x^2} - \dfrac{(x - x_{j-1})^3}{2\Delta x^3} & ;\ x_{j-1} \le x \le x_j , \\[2mm] \dfrac{1}{6} - \dfrac{(x - x_{j+1})}{2\Delta x} + \dfrac{(x - x_{j+1})^2}{2\Delta x^2} + \dfrac{(x - x_{j+1})^3}{2\Delta x^3} & ;\ x_j \le x \le x_{j+1} , \\[2mm] -\dfrac{(x - x_{j+2})^3}{6\Delta x^3} & ;\ x_{j+1} \le x \le x_{j+2} , \\[2mm] 0 & ;\ x_{j+2} \le x . \end{cases}$$

The function B_j is composed by "gluing" together cubic functions between x_{j-2} and x_{j-1}, between x_{j-1} and x_j, between x_j and x_{j+1}, and between x_{j+1} and x_{j+2}. The support of the spline B_j is the interval $[x_{j-2}, x_{j+2}]$, on which B_j is twice continuously differentiable. The resulting spline has values $B_j(x_j) = 2/3$ and $B_j(x_{j\pm1}) = 1/6$ at the central points, the first derivatives are $B'_j(x_j) = 0$ and $B'_j(x_{j\pm1}) = \mp 1/(2h)$, and the second derivatives are $B''_j(x_j) = -2/h^2$ and $B''_j(x_{j\pm1}) = 1/h^2$. At the remaining points $x_{j\pm k}$, $k \ge 2$, the values $B_j(x_j)$ are zero. Figure 9.9 shows the spline B_5 and its nearest neighbors B_4 and B_6 on a uniform mesh on $x \in [0, 1]$.

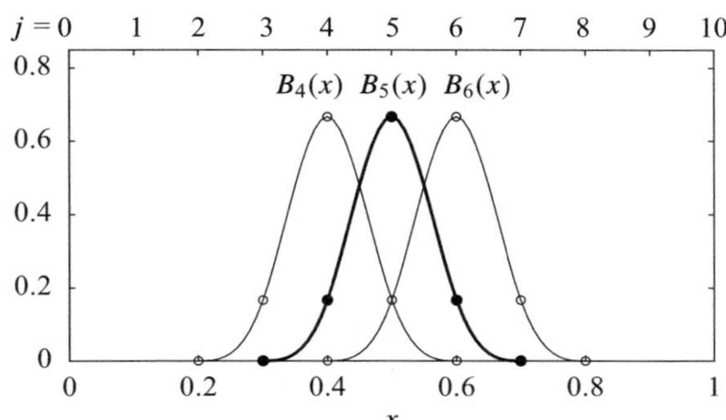

Fig. 9.9 The cubic spline B_5 and its nearest neighbors B_4 and B_6 on a uniform mesh $x \in [0, 1]$ with $(N + 1) = 11$ points (including the boundaries)

We write the solution of the boundary-value problem as

$$y(x) \approx u(x) = \sum_{j=-1}^{N+1} a_j B_j(x) \,. \tag{9.64}$$

The leftmost spline B_{-1} and the rightmost spline B_{N+1} are included in the sum because both have non-zero contributions on $[a, b]$: the former at $a = x_0$, the latter at $b = x_N$. The values of u and its first and second derivative at x_j are thus determined by the linear combinations of three adjacent coefficients a_j:

$$u(x_j) = \frac{1}{6} \left(a_{j+1} + 4a_j + a_{j-1} \right) \,,$$

$$u'(x_j) = \frac{1}{2h} \left(a_{j+1} - a_{j-1} \right) \,,$$

$$u''(x_j) = \frac{1}{h^2} \left(a_{j+1} - 2a_j + a_{j-1} \right) \,.$$

The key requirement of collocation is imposed at this stage. The remainder

$$y''(x) + p(x)y' + q(x)y - r(x)$$

should be *zero at the collocation points*. Here collocation and mesh points are the same, but in the following sections we also discuss "true" collocation where collocation points can be chosen differently. We denote the function values at x_j by $p(x_j) = p_j$, $q(x_j) = q_j$, and $r(x_j) = r_j$. When we account for boundary conditions, we obtain a system of equations for the coefficients a_j,

$$
\begin{pmatrix}
g_0 & h_0 & & & & \\
d_1 & g_1 & h_1 & & & \\
& d_2 & g_2 & h_2 & & \\
& & \ddots & \ddots & \ddots & \\
& & & d_{N-1} & g_{N-1} & h_{N-1} \\
& & & & d_N & g_N
\end{pmatrix}
\begin{pmatrix}
a_0 \\ a_1 \\ a_2 \\ \vdots \\ a_{N-1} \\ a_N
\end{pmatrix}
=
\begin{pmatrix}
R_0 \\ R_1 \\ R_2 \\ \vdots \\ R_{N-1} \\ R_N
\end{pmatrix} .
\tag{9.65}
$$

The matrix elements and the right side of the equation are given by

$$
g_j = \begin{cases}
-6 + 2hp_0 & ; \ j = 0 , \\
4(-3 + h^2 q_j) & ; \ j = 1, 2, \ldots, N-1 , \\
-6 - 2hp_N & ; \ j = N ,
\end{cases}
$$

$$
d_j = \begin{cases}
6 - 3hp_j + h^2 q_j & ; \ j = 1, 2, \ldots, N-1 , \\
-hp_N & ; \ j = N ,
\end{cases}
$$

$$
h_j = \begin{cases}
hp_0 & ; \ j = 0 , \\
6 + 3hp_j + h^2 q_j & ; \ j = 1, 2, \ldots, N-1 ,
\end{cases}
$$

$$
R_j = \begin{cases}
h^2 r_0 - \alpha(6 - 3hp_0 + h^2 q_0) & ; \ j = 0 , \\
6 h^2 r_j & ; \ j = 1, 2, \ldots, N-1 , \\
h^2 r_N - \beta(6 + 3hp_N + h^2 q_N) & ; \ j = N .
\end{cases}
$$

We use this method (and compare it to the difference method) in Problem 9.9.5.

9.5.2   Scalar Linear Boundary-Value Problems of Higher Orders

We discuss linear scalar boundary-value problems of the form

$$
Ly = y^{(M)} - \sum_{m=1}^{M} c_m(x) y^{(m-1)} = q(x) , \qquad a < x < b ,
\tag{9.66}
$$

where M is the order of the highest derivative. As the basic mesh we take (9.2) but pay attention to the indices: the subintervals $[x_j, x_{j+1}]$ are labeled by j, while j and the additional subscript k define the *collocation points* within these subintervals,

$$
x_{jk} = x_j + h\rho_k , \qquad 0 \le j \le N-1 , \qquad 1 \le k \le K .
\tag{9.67}
$$

The points on the subintervals are uniquely defined by the *canonical parameters*

$$
0 \le \rho_1 \le \rho_2 \le \ldots \le \rho_K \le 1
$$

that depend on the method. We usually choose them such that at some order K we obtain a quadrature formula of maximum possible precision on the subinterval. In Gauss collocation, ρ_k are the zeros of Legendre polynomials, where $\rho_1 > 0$ and $\rho_K < 1$, so none of the collocation points coincide with the mesh points. In Radau collocation we have $\rho_1 > 0$ and $\rho_K = 1$, so the last collocation point on the subinterval always coincides with a mesh point (except at $x = a$). In Lobatto collocation we have $\rho_1 = 0$ and $\rho_K = 1$. Figure 9.10 shows the basic uniform mesh x_j with the collocation points of the Gauss scheme of orders two and three. The canonical points for collocations of orders from $K = 2$ to $K = 5$ are listed in Table 9.2.

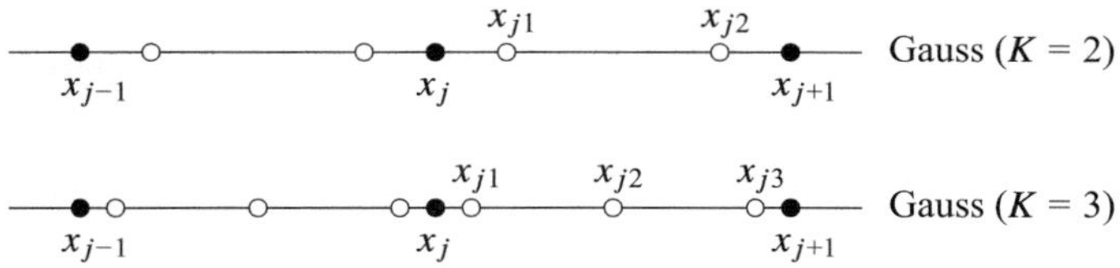

Fig. 9.10 Collocation points $x_{jk} = x_j + h\rho_k$ (symbols o) for Gauss collocation of order two with canonical points $\rho_{1,2} = 1/2 \mp \sqrt{3}/6$, and for Gauss collocation of order three with canonical points $\rho_{1,3} = 1/2 \mp \sqrt{15}/10$, and $\rho_2 = 1/2$ embedded in the basic uniform mesh x_j (symbols ●). In other collocations (Radau or Lobatto) some collocation points coincide with the mesh points

Table 9.2 Canonical parameters ρ_k of Gauss collocation points from order $K = 2$ to order $K = 5$, which are zeros of the Legendre polynomials $P_K(2x - 1)$ rescaled to the interval $[0, 1]$. The collocation points are given by Eq. (9.67)

ρ_k	$K = 2$	$K = 3$	$K = 4$	$K = 5$
ρ_1	$\dfrac{1}{2} - \dfrac{\sqrt{3}}{6}$	$\dfrac{1}{2} - \dfrac{\sqrt{15}}{10}$	$\dfrac{1}{2} - \dfrac{\sqrt{35(15 + 2\sqrt{30})}}{70}$	$\dfrac{1}{2} - \dfrac{\sqrt{7(35 + 2\sqrt{70})}}{42}$
ρ_2	$\dfrac{1}{2} + \dfrac{\sqrt{3}}{6}$	$\dfrac{1}{2}$	$\dfrac{1}{2} - \dfrac{\sqrt{35(15 - 2\sqrt{30})}}{70}$	$\dfrac{1}{2} - \dfrac{\sqrt{7(35 - 2\sqrt{70})}}{42}$
ρ_3		$\dfrac{1}{2} + \dfrac{\sqrt{15}}{10}$	$\dfrac{1}{2} + \dfrac{\sqrt{35(15 - 2\sqrt{30})}}{70}$	$\dfrac{1}{2}$
ρ_4			$\dfrac{1}{2} + \dfrac{\sqrt{35(15 + 2\sqrt{30})}}{70}$	$\dfrac{1}{2} + \dfrac{\sqrt{7(35 - 2\sqrt{70})}}{42}$
ρ_5				$\dfrac{1}{2} + \dfrac{\sqrt{7(35 + 2\sqrt{70})}}{42}$

In the collocation method we require the numerical solution u to satisfy the differential equation *at the collocation points*, so that *for each subinterval* indexed by $0 \leq j \leq N - 1$ and *for all collocation points* defined by the canonical parameters ρ_k ($1 \leq k \leq K$), it holds that

$$u^{(M)}(x_{jk}) - \sum_{m=1}^{M} c_m(x_{jk}) u^{(m-1)}(x_{jk}) - q(x_{jk}) = 0 . \tag{9.68}$$

The simple method with cubic splines from the preceding section corresponds to a kind of "degenerate" collocation, since the mesh and collocation points coincide. In the following we show the structure of "true" collocation [2] that is valid regardless of the type of the chosen basis functions. Let the degree of the polynomials on individual subintervals be $K + M$ for some $K \geq M$. On the subinterval $[x_j, x_{j+1}]$ we Taylor-expand the polynomial u around x_j,

$$
\begin{aligned}
u(x) &= \sum_{n=1}^{K+M} u^{(n-1)}(x_j) \frac{(x - x_j)^{n-1}}{(n-1)!} \\
&= \sum_{m=1}^{M} u_{jm} \frac{(x - x_j)^{m-1}}{(m-1)!} + h^M \sum_{k=1}^{K} z_{jk} \, \psi_k \left(\frac{x - x_j}{h} \right) ,
\end{aligned} \tag{9.69}
$$

where we have defined

$$z_{jk} = h^{k-1} u^{(M+k-1)}(x_j) , \qquad \psi_k(t) = \frac{t^{M+k-1}}{(M+k-1)!} , \qquad 0 \leq t \leq 1 ,$$

and where we have collected the components of the solution and its $(M - 1)$ derivatives *at the jth mesh point* into the vector $\boldsymbol{u}_j$ with components u_{jm}:

$$\boldsymbol{u}_j = \left(u_{j1}, u_{j2}, \ldots, u_{jM} \right)^{\mathrm{T}} = \left(u(x_j), u'(x_j), \ldots, u^{(M-1)}(x_j) \right)^{\mathrm{T}} . \tag{9.70}$$

When the collocation solution is used in Eq. (9.66), we obtain

$$Lu(x) = h^M \sum_{k=1}^{K} z_{jk} \, L \left[\psi_k \left(\frac{x - x_j}{h} \right) \right] - \sum_{m=1}^{M} c_m(x) \sum_{n=m}^{M} u_{jn} \frac{(x - x_j)^{n-m}}{(n-m)!} .$$

We further define $\boldsymbol{z}_j = (z_{j1}, z_{j2}, \ldots, z_{jK})^{\mathrm{T}}$ and $\boldsymbol{q}_j = (q(x_{j1}), q(x_{j2}), \ldots, q(x_{jK}))^{\mathrm{T}}$. The collocation requirement (9.68) results in a linear system

$$V_j \boldsymbol{u}_j + W_j \boldsymbol{z}_j = \boldsymbol{q}_j , \qquad 0 \leq j \leq N - 1 , \tag{9.71}$$

where the elements of the $K \times M$ matrix V_j and of the $K \times K$ matrix W_j are

$$(V_j)_{km} = -\sum_{l=1}^{m} c_l(x_{jk}) \frac{(h\rho_k)^{m-l}}{(m-l)!} \,, \qquad\qquad 1 \le k \le K \,, \ 1 \le m \le M \,,$$

$$(W_j)_{km} = \psi_m^{(M)}(\rho_k) - \sum_{l=1}^{M} c_l(x_{jk})\, h^{M-l+1} \psi_m^{(l-1)}(\rho_k) \,, \quad 1 \le k \le K \,, \ 1 \le m \le K \,.$$

Moreover, we require that the interpolation polynomial u and its $(M-1)$ derivatives are continuous at all internal boundary points of the subintervals. This introduces an additional system of equations

$$\boldsymbol{u}_{j+1} = C\boldsymbol{u}_j + D\boldsymbol{z}_j \,, \qquad 0 \le j \le N-1 \,, \tag{9.72}$$

with an upper-triangular $M \times M$ matrix C and a $M \times K$ matrix D, with the elements

$$C_{km} = \frac{h^{m-k}}{(m-k)!} \,, \qquad k \le m \,,$$

$$D_{km} = h^{M-k+1} \psi_m^{(k-1)}(1) \,, \quad 1 \le k \le M \,, \ 1 \le m \le K \,.$$

Finally, we consider the boundary conditions in the usual form $B_a\boldsymbol{u}_0 + B_b\boldsymbol{u}_N = \boldsymbol{\beta}$. The last step is the elimination of $\boldsymbol{z}_j$ from Eqs. (9.71) and (9.72). We obtain

$$\boldsymbol{u}_{j+1} = \Gamma_j \boldsymbol{u}_j + \boldsymbol{r}_j \,, \qquad 0 \le j \le N-1 \,,$$

where $\Gamma_j = C - DW_j^{-1}V_j$ and $\boldsymbol{r}_j = DW_j^{-1}\boldsymbol{q}_j$. In summary, the problem (9.66) has boiled down to solving the matrix system of the form

$$\begin{pmatrix} -\Gamma_0 & I & & & & \\ & -\Gamma_1 & I & & & \\ & & -\Gamma_2 & I & & \\ & & & \ddots & \ddots & \\ & & & & -\Gamma_{N-1} & I \\ B_a & & & & & B_b \end{pmatrix} \begin{pmatrix} \boldsymbol{u}_0 \\ \boldsymbol{u}_1 \\ \boldsymbol{u}_2 \\ \vdots \\ \boldsymbol{u}_{N-1} \\ \boldsymbol{u}_N \end{pmatrix} = \begin{pmatrix} \boldsymbol{r}_0 \\ \boldsymbol{r}_1 \\ \boldsymbol{r}_2 \\ \vdots \\ \boldsymbol{r}_{N-1} \\ \boldsymbol{\beta} \end{pmatrix} . \tag{9.73}$$

Note that each vector $\boldsymbol{u}_j$ has dimension M and contains the solution $u_j = u(x_j)$ at the mesh point x_j as well as its $(M-1)$ derivatives at this point (see definition (9.70)). The size of the matrix in (9.73) is $M(N+1) \times M(N+1)$. In the practical implementation we can compute in advance the quantities that do not depend on the choice of the mesh, like $\psi_m^{(l)}(\rho_k)$ and $\psi_m^{(l)}(1)$. We can also compute the vector of local variables $\boldsymbol{z}_j = W_j^{-1}(\boldsymbol{q}_j - V_j\boldsymbol{u}_j)$ that was eliminated in the derivation, and when we insert the values z_{jk} into (9.69), we obtain the complete collocation solution (not just at the mesh points). The method described here is illustrated by a linear fourth-order problem in the following Example (adapted from [2]), and in Problem 9.9.4.

Example We use the collocation method on the interval $[a, b] = [0, 1]$ to solve the linear second-order boundary-value problem

$$y'' = -\frac{1}{x} y' + \left(\frac{8}{8 - x^2}\right)^2, \qquad 0 < x < 1, \tag{9.74}$$

with the boundary conditions $y'(0) = y(1) = 0$, and the exact solution

$$y(x) = 2 \log\left(\frac{7}{8 - x^2}\right).$$

We rewrite the equation as

$$Ly = y'' + \frac{1}{x} y' = \left(\frac{8}{8 - x^2}\right)^2,$$

whence, by comparing it to (9.66) we infer $M = 2$ (a second-order problem) and

$$c_1(x) = 0, \qquad c_2(x) = -\frac{1}{x}, \qquad q(x) = \left(\frac{8}{8 - x^2}\right)^2.$$

We solve the problem on the uniform mesh $a = x_0, x_1, \ldots, x_N = b$ and opt for Gauss collocation with four canonical points (column $K = 4$ of Table 9.2) on each subinterval $[x_j, x_{j+1}]$. The collocation points are $x_{jk} = x_j + h\rho_k$, where $0 \leq j \leq N - 1$ and $1 \leq k \leq K$. We put the solution at each point x_j in the vector

$$\boldsymbol{u}_j = \left(u(x_j), u'(x_j)\right)^{\mathsf{T}}. \tag{9.75}$$

In computing the matrices V_j from Eq. (9.71) we use $c_1(x_{jk}) = 0$ and $c_2(x_{jk}) = -1/x_{jk}$, while to compute the matrices W_j we also need the derivatives of $\psi_k(t)$,

$$\psi_k(\rho_k) = \frac{\rho_k^{M+k-1}}{(M + k - 1)!} = \frac{\rho_k^{k+1}}{(k + 1)!}, \qquad \psi_k'(\rho_k) = \frac{\rho_k^{k}}{k!}, \qquad \psi_k''(\rho_k) = \frac{\rho_k^{k-1}}{(k - 1)!}.$$

The constant matrices C and D appearing in Eq. (9.72) are

$$C = \begin{pmatrix} 1 & h \\ 0 & 1 \end{pmatrix}, \qquad D = \begin{pmatrix} h^2\psi_1(1) & h^2\psi_2(1) & h^2\psi_3(1) & h^2\psi_4(1) \\ h\psi_1'(1) & h\psi_2'(1) & h\psi_3'(1) & h\psi_4'(1) \end{pmatrix}.$$

We write the boundary conditions $y'(0) = y(1) = 0$ in the form $B_a \boldsymbol{u}_0 + B_b \boldsymbol{u}_N = \boldsymbol{\beta}$, whence we read off

$$B_a = \begin{pmatrix} 0 & 1 \\ 0 & 0 \end{pmatrix}, \qquad B_b = \begin{pmatrix} 0 & 0 \\ 1 & 0 \end{pmatrix}, \qquad \boldsymbol{\beta} = \begin{pmatrix} 0 \\ 0 \end{pmatrix}.$$

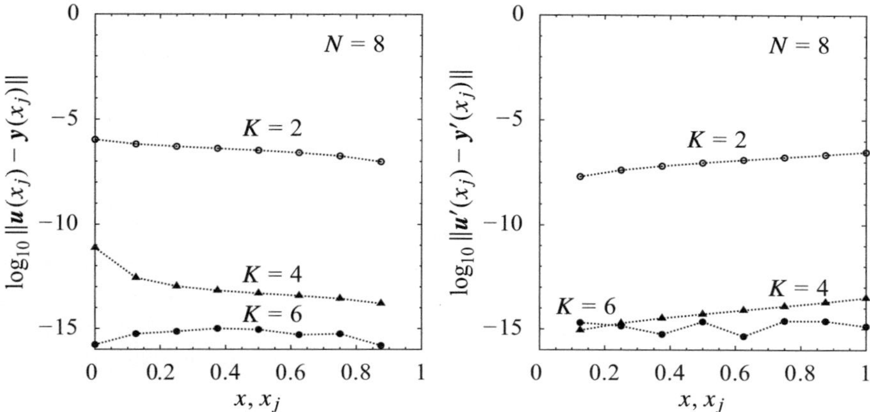

Fig. 9.11 Numerical errors in solving the problem (9.74) on a mesh with nine points ($N = 8$) by Gauss collocation of orders $K = 2$, $K = 4$, and $K = 6$ (two, four, and six canonical points on each subinterval). [LEFT] Error of the solution. Due to the boundary condition $y(1) = 0$ the logarithm of the error at x_N is undefined. [RIGHT] Error of the derivative. Due to the condition $y'(0) = 0$ the logarithm of the error at x_0 is undefined

From the right-hand side of the differential equation, described by the function q, we compute (at each collocation point) the vector

$$\boldsymbol{q}_j = \big(q(x_{j1}), q(x_{j2}), q(x_{j3}), q(x_{j4})\big)^{\mathrm{T}} .$$

Finally, we compute the matrices $\Gamma_j = C - DW_j^{-1}V_j$ and vectors $\boldsymbol{r}_j = DW_j^{-1}\boldsymbol{q}_j$, and solve the system (9.73) which gives us the values of the solution and its derivative (Eq. (9.75)) at each mesh point x_j. Figure 9.11 (left) shows the absolute error of the numerical solution, and Fig. 9.11 (right) the error of the derivative for $N = 8$ (nine points $x_0, x_1, \ldots, x_8$ or eight subintervals). ◁

9.5.3 *Scalar Non-linear Boundary-Value Problems of Higher Orders*

In the preceding section we learned how to deal with linear scalar problems of higher orders. Now let us discuss *non-linear* problems of the form

$$Ny = y^{(M)} - f\left(x, y, y', \ldots, y^{(M-1)}\right) = y^{(M)} - f(x, \boldsymbol{y}) = 0 , \qquad a < x < b ,$$

where the vector

$$\boldsymbol{y}(x) = \big(y(x), y'(x), y''(x), \ldots, y^{(M-1)}\big)^{\mathrm{T}}$$

again contains the values of the solution and its $(M - 1)$ derivatives at x (or the corresponding mesh or collocation points). The function f is non-linear in the arguments $\boldsymbol{y}$, and the boundary condition is

$$\boldsymbol{g}(\boldsymbol{y}(a), \boldsymbol{y}(b)) = \boldsymbol{0} \,.$$

One possible way towards the solution of such problems leads through *linearization.* The basic idea is to linearize the non-linear problem and use some iterative method (like Newton's) to solve the corresponding linear boundary-value problems. If linearization is performed correctly, we may hope that the sequence of solutions of the linearized problem will converge to the solution of the non-linear one. The basic form of the method can be obtained from the expansion around the exact solution,

$$y^{(M)}(x) = f\,(x, \boldsymbol{y}(x))$$

$$\approx f\,(x, \widetilde{\boldsymbol{y}}(x)) + \sum_{m=1}^{M} \frac{\partial f}{\partial y_m}(x, \widetilde{\boldsymbol{y}}(x)) \left(y^{(m-1)}(x) - \widetilde{y}^{(m-1)}(x)\right) \,,$$

where the ordering of the components of the vector $\widetilde{\boldsymbol{y}}$ is the same as in $\boldsymbol{y}$ (function, first derivative, second derivative, and so on). The difference between the current and the final numerical solution can be measured by the function

$$z(x) = y(x) - \widetilde{y}(x) \,,$$

and, by analogy to the vectors $\boldsymbol{y}$ and $\widetilde{\boldsymbol{y}}$, a similar relation applies to the whole vector $\boldsymbol{z}$, e.g. $z^{(M)}(x) = y^{(M)}(x) - \widetilde{y}^{(M)}(x)$. The expansion results in a *linear* differential equation

$$z^{(M)} - \sum_{m=1}^{M} \underbrace{\frac{\partial f(x, \widetilde{\boldsymbol{y}}(x))}{\partial y_m}}_{c_m(x, \widetilde{\boldsymbol{y}})} z^{(m-1)}(x) = -\left[\widetilde{y}^{(M)} - f(x, \widetilde{\boldsymbol{y}})\right] \tag{9.76}$$

for the correction of the solution, z. We solve Eq. (9.76) iteratively by some initial approximation $\widetilde{\boldsymbol{y}}$. The expression on its right-hand side plays the role of the source term $q(x)$ (compare Eqs. (9.76) and (9.66)) and we compute it in each iteration step by using the current solution $\widetilde{\boldsymbol{y}}$. The coefficients $c_m(x, \widetilde{\boldsymbol{y}})$ are also computed from the values of the current solution and its derivatives. We write the boundary condition $\boldsymbol{g}(\boldsymbol{y}(a), \boldsymbol{y}(b)) = \boldsymbol{0}$ in linearized form

$$B_a \boldsymbol{y}(a) + B_b \boldsymbol{y}(b) = \boldsymbol{0} \,,$$

where

$$B_a = \frac{\partial \boldsymbol{g}(\boldsymbol{u}, \boldsymbol{v})}{\partial \boldsymbol{u}} \,, \quad B_b = \frac{\partial \boldsymbol{g}(\boldsymbol{u}, \boldsymbol{v})}{\partial \boldsymbol{v}} \,, \quad \text{at } \boldsymbol{u} = \boldsymbol{y}(a) \,, \ \boldsymbol{v} = \boldsymbol{y}(b) \,.$$

In terms of the vector z this means

$$B_a z(a) + B_b z(b) = -g\left(\widetilde{y}(a), \widetilde{y}(b)\right) . \tag{9.77}$$

Solving non-linear scalar boundary-value problems of higher orders thus essentially translates to the use of Newton's method: the iteration is set off by some initial approximation for $\widetilde{y}$, which is used to solve the linear differential equation (9.76) with the boundary condition (9.77) by collocation. Then we update

$$\widetilde{y} \leftarrow \widetilde{y} + z ,$$

and repeat the procedure until $\|z\|$ drops below the desired precision. If the functions f and g have continuous second partial derivatives (plus a few additional mild assumptions), the method is quadratically convergent [2].

9.5.4 *Systems of Boundary-Value Problems*

Collocation methods are also widely used for systems of non-linear boundary-value problems of higher orders, but their discussion is beyond the scope of this book. Among the best-known collocation tools are COLSYS [10] and TWPBVP [11]. More recent is the (non-collocation) code MIRKDC [12], which is at the heart of the bvp4c routines built into the MATLAB environment. The latest development is the very fast BVP_SOLVER package [13] which is an enhanced version of the bvp4c package.

9.6 Weighted-Residual Methods ⋆

Weighted-residual methods are potent tools for the solution of boundary-value problems with ordinary and partial differential equations, and are at the core of numerous versions of the finite element method explained in Sect. 11.6. Their basic characteristic is the transformation of a boundary-value problem to a variational one. We restrict the discussion to boundary-value problems of the form

$$Lv = -\frac{d}{dx}\left(p(x)\frac{dv}{dx}\right) + q(x)v = f(x) \tag{9.78}$$

on $a < x < b$ with boundary conditions $x(a) = x(b) = 0$, where $p(x) > 0$ and $q(x) \geq 0$. Assume that the function p is continuously differentiable on $[a, b]$, while q and f are continuous on $[a, b]$. We multiply the differential equation (9.78) by some function w in the scalar-product sense (A.3), and obtain $\langle w, Lv - f \rangle = 0$. The solution of this problem is also the solution of the boundary-value problem (9.78) for all functions w for which this scalar product exists. By using

$$\langle w, Lv - f \rangle = 0 \qquad \forall w \in L^2(a, b) \tag{9.79}$$

we have thus turned the boundary-value problems in its variational form: we find the solution v of the boundary-value problem (9.78) precisely when (9.79) applies for *any* square-integrable function w. The name of the method comes from the manner in which the residual $Lv - f$ (measuring the deviation from the exact fulfillment of the differential equation) is weighted by the function w in the integral (9.79).

In a concrete method we can only find a numerical solution u which is an approximation of the true solution v. We write u as a linear combination of some simpler functions,

$$v(x) \approx u(x) = \sum_{j=1}^{J} c_j \phi_j(x) \,, \tag{9.80}$$

and call it the *trial function*. The functions ϕ_j are the basis functions of a finite-dimensional space of trial functions T^J. We usually think of $\phi_j \in T^J \subset L^2(a, b)$. The *weight* (or *test*) function is also set up as the sum

$$w(x) \approx \widetilde{w}(x) = \sum_{j=1}^{J} d_j \psi_j(x) \,, \tag{9.81}$$

where, in general, ψ_j are different from ϕ_j and span the space of weight functions, W^J. These functions are frequently taken from $\psi_j \in W^J \subset L^2(a, b)$, but not necessarily so: for example, the Dirac delta-"functions" may also act as weight functions.

The primary goal is to determine the unknown coefficients c_j with chosen ϕ_j and ψ_j such that u is a sufficiently good approximation of v. In their basic outline the weighted-residual methods are no different from, say, Fourier or collocation methods. What distinguishes all these methods from one another is the manner in which the residual is defined and its magnitude quantified. In weighted-residual methods we attempt to fulfill the equation $\langle \widetilde{w}, Lv - f \rangle = 0$ for each function $\widetilde{w}$: the residual $Lv - f$ should be orthogonal to all $\widetilde{w}$ from the space of weight functions. This means

$$\sum_{j} d_j \langle \psi_j, Lv - f \rangle = 0 \,.$$

Since this condition should be valid for *any* choice of the coefficients d_j, we must also have

$$\langle \psi_j, Lv - f \rangle = 0 \qquad \forall j \,.$$

Weighted-residual methods are classified according to the choice of function spaces for the trial functions ϕ_j and weight functions ψ_j. In the Galerkin method we take ϕ_j and ψ_j from the same space and $\phi_j = \psi_j$. If we choose ψ_j to be delta-

"functions" at the collocation points z_j, so that $\langle \psi_j, f \rangle = f(z_j)$, the residual satisfies the condition $\langle \psi_j, L u_c - f \rangle = 0$, where u_c is the collocation solution (see Sect. 9.5). When ϕ_j and ψ_j are taken from closely related but *different* function spaces (for example, from two spaces of polynomial splines of different orders), we are referring to Petrov–Galerkin methods.

Galerkin method In the following we discuss the Galerkin method which is the most common. We can achieve a more symmetric and numerically practicable form of the variational formulation if we integrate (9.79) by parts and thereby eliminate the terms involving second derivatives. (This turns out to be helpful in implementations of the two- and three-dimensional finite element method where it is difficult to construct continuously differentiable function approximations. This is why it makes sense to stick to very simple basis functions.) We get

$$\int_a^b w \left[-(pv')' + qv - f \right] \mathrm{d}x = -wpv' \Big|_a^b + \int_a^b \left[w'pv' + wqv - wf \right] \mathrm{d}x .$$

If w fulfills the same boundary conditions as v, the integrated part is zero, and the Galerkin condition assumes a nice symmetric form

$$A(w, v) - \langle w, f \rangle = 0 ,$$

where

$$A(w, v) = \int_a^b \left[w'pv' + wqv \right] \mathrm{d}x . \tag{9.82}$$

In boundary-value problems describing mechanical systems, the bilinear form $A(u, v)$ represents the internal or *strain energy*. The functions v and w must satisfy the condition [14]

$$\int_a^b \left[(z')^2 + z^2 \right] \mathrm{d}x < \infty , \qquad z(a) = z(b) = 0 , \qquad z \in \{v, w\} ,$$

and we are solving the variational problem $A(w, v) = \langle w, f \rangle$ for $\forall w$. In the actual implementation we replace v by u, and w by $\widetilde{w}$. According to Galerkin, we span u and $\widetilde{w}$ in the same function space ($u, \widetilde{w} \in T^J$) with the basis functions ϕ_j. We therefore compute the approximation u by solving the variational problem

$$A(\widetilde{w}, u) = \langle \widetilde{w}, f \rangle \qquad \forall \widetilde{w} \in T^J . \tag{9.83}$$

When we insert the expansions (9.80) and (9.81) for u and $\widetilde{w}$, we obtain a system of linear equations for the expansion coefficients c_k of the approximate solution,

$$\sum_{k=1}^{J} c_k A\left(\phi_j, \phi_k\right) = \langle \phi_j, f \rangle, \qquad j = 1, 2, \ldots, J.$$

How difficult it is to compute the integrals (9.82) depends on the choice of the basis functions ϕ_j (and, of course, on the functions p and q in the differential equation). If the basis functions are polynomials, it is sensible to choose low degrees. The fruits of the discussion in this Section are collected in Sect. 11.6 describing the finite element method in one and two dimensions. We will encounter weighted-residual methods again in numerical approaches to partial differential equations in Chap. 12.

9.7   Regular Boundary-Value Problems with Eigenvalues

In boundary-value problems with eigenvalues (in short, *eigenvalue problems*) we seek not only the function satisfying the differential equation and boundary conditions, but also the scalars appearing in the formulation of the problem. To a physicist, the most relevant are the one-dimensional Sturm–Liouville problems [15–17]

$$-\frac{\mathrm{d}}{\mathrm{d}x}\left(p(x)\frac{\mathrm{d}y}{\mathrm{d}x}\right) + q(x)y = \lambda w(x)y, \qquad a < x < b, \tag{9.84}$$

where we wish to compute the functions y and the scalars λ. The coefficient functions p, q, and w are known real functions. We assume that p and w are strictly positive on (a, b). Moreover, we assume that p, q, and w are defined on the closed interval $[a, b]$ and are at least piecewise continuous on it. In this Section we discuss regular Sturm–Liouville problems, where a and b are finite, and the boundary conditions at both endpoints can be expressed as

$$\begin{aligned}
a_1 y(a) - a_2 p(a) y'(a) &= 0, & a_1^2 + a_2^2 &> 0, \\
b_1 y(b) + b_2 p(b) y'(b) &= 0, & b_1^2 + b_2^2 &> 0.
\end{aligned} \tag{9.85}$$

The value of λ for which a non-trivial solution of Eq. (9.84) with boundary conditions (9.85) exists is called the *eigenvalue* and the corresponding solution y is the eigenfunction. Singular Sturm–Liouville problems are treated in Sect. 9.8.

9.7.1   Transformation to Liouville Normal Form

By using the substitutions

$$\xi(x) = \int_a^x \sqrt{\frac{w(s)}{p(s)}}\,\mathrm{d}s, \qquad Y\big(\xi(x)\big) = y(x)\,[w(x)p(x)]^{1/4}, \qquad B = \xi(b),$$

Eq. (9.84) can be turned into its *Liouville normal form* [18]:

$$-\frac{d^2 Y}{d\xi^2} + Q(\xi)Y = \lambda Y \,, \qquad \xi \in [0,\, B] \,, \tag{9.86}$$

where

$$Q(\xi) = \frac{q\big(x(\xi)\big)}{w\big(x(\xi)\big)} + \frac{1}{m(\xi)}\frac{d^2 m}{d\xi^2} \,, \qquad m(\xi) = \big[w\big(x(\xi)\big)\, p\big(x(\xi)\big)\big]^{1/4} \,.$$

In Eq. (9.86) we recognize, for instance, the one-dimensional Schrödinger equation for bound states of a quantum particle in the potential $Q(\xi)$. The parameter λ determines the value of the eigenenergy, and the solution y is the corresponding eigenfunction. Any regular Sturm–Liouville problem can be turned into an equation of this type by using the above transformation, assuming that p, p', q, w, w', $(wp)'$, and $(wp)''$ are continuous on $[0,\, B]$. The boundary conditions (9.85) change accordingly. They become

$$\begin{aligned}
\alpha_1 Y(0) - \alpha_2 Y'(0) &= 0 \,, \quad \alpha_1^2 + \alpha_2^2 > 0 \,, \\
\beta_1 Y(B) + \beta_2 Y'(B) &= 0 \,, \quad \beta_1^2 + \beta_2^2 > 0 \,,
\end{aligned} \tag{9.87}$$

where

$$\begin{aligned}
\alpha_1 &= \frac{1}{[w(a)p(a)]^{1/4}}\, a_1 + \frac{(wp)'(a)}{4[w(a)]^{5/4}[p(a)]^{1/4}}\, a_2 \,, \\
\alpha_2 &= [w(a)p(a)]^{1/4} a_2 \,, \\
\beta_1 &= \frac{1}{[w(b)p(b)]^{1/4}}\, b_1 - \frac{(wp)'(b)}{4[w(b)]^{5/4}[p(b)]^{1/4}}\, b_2 \,, \\
\beta_2 &= [w(b)p(b)]^{1/4} b_2 \,.
\end{aligned}$$

9.7.2 *Properties of Eigenvalues*

The eigenvalues λ_n ($n = 0, 1, \ldots$) of regular (non-singular) Sturm–Liouville problems are real and distinct (there are no pairs of linearly independent eigenfunctions with equal eigenvalues). The eigenvalues are invariant under the Liouville transformation. The asymptotics of the eigenvalues for $n \to \infty$ is well established. Assuming that the problem has been posed in Liouville normal form, there are four characteristic cases: $\alpha_2 \neq 0$, $\beta_2 \neq 0$ (mixed boundary conditions at both endpoints, but can represent pure Neumann conditions if $\alpha_1 = \beta_1 = 0$); $\alpha_2 \neq 0$, $\beta_2 = 0$ (mixed at $\xi = 0$, Dirichlet at $\xi = B$); $\alpha_2 = 0$, $\beta_2 \neq 0$ (vice-versa); and $\alpha_2 = \beta_2 = 0$ (homogeneous Dirichlet at both endpoints). We have:

$$\alpha_2 \neq 0 \,, \beta_2 \neq 0 : \ \sqrt{\lambda_n} = \frac{n\pi}{B} + \frac{C_1}{n\pi} + \mathcal{O}\left(\frac{1}{n^3}\right) , \tag{9.88}$$

$$C_1 = \frac{\alpha_1}{\alpha_2} + \frac{\beta_1}{\beta_2} + \frac{1}{2}\int_0^B Q(\xi)\,\mathrm{d}\xi \,,$$

$$\alpha_2 \neq 0 \,, \beta_2 = 0 : \ \sqrt{\lambda_n} = \frac{(n+1/2)\pi}{B} + \frac{C_1}{(n+1/2)\pi} + \mathcal{O}\left(\frac{1}{n^3}\right) , \tag{9.89}$$

$$C_1 = \frac{\alpha_1}{\alpha_2} + \frac{1}{2}\int_0^B Q(\xi)\,\mathrm{d}\xi \,,$$

$$\alpha_2 = 0 \,, \beta_2 \neq 0 : \ \sqrt{\lambda_n} = \frac{(n+1/2)\pi}{B} + \frac{C_1}{(n+1/2)\pi} + \mathcal{O}\left(\frac{1}{n^3}\right) , \tag{9.90}$$

$$C_1 = \frac{\beta_1}{\beta_2} + \frac{1}{2}\int_0^B Q(\xi)\,\mathrm{d}\xi \,,$$

$$\alpha_2 = \beta_2 = 0 : \ \sqrt{\lambda_n} = \frac{(n+1)\pi}{B} + \frac{C_1}{(n+1)\pi} + \mathcal{O}\left(\frac{1}{n^3}\right) , \tag{9.91}$$

$$C_1 = \frac{1}{2}\int_0^B Q(\xi)\,\mathrm{d}\xi \,.$$

The $1/n^3$ terms (and even the $1/n^5$ piece in the pure Dirichlet case) are also known, and contain derivatives of Q. The somewhat lengthy expressions for their corresponding coefficients C_3 and C_5 are given in [18]. For example, by including the $1/n^3$ term in (9.91) and squaring it we obtain the asymptotic expansion of λ_n itself,

$$\lambda_n = \left(\frac{(n+1)\pi}{B}\right)^2 + A_0 + \frac{A_2}{[(n+1)\pi]^2} + \mathcal{O}\left(\frac{1}{n^4}\right) , \tag{9.92}$$

$$A_0 = \frac{1}{B}\int_0^B Q(\xi)\,\mathrm{d}\xi \,,$$

$$A_2 = -\frac{B^2}{4}A_0^2 + \frac{B}{4}\left[\int_0^B Q^2(\xi)\,\mathrm{d}\xi - (Q'(B) - Q'(0))\right] ,$$

as $n \to \infty$. Further details on the dependence of the eigenvalues on the endpoints, the boundary conditions, and the coefficients of w can be found in [19].

9.7.3 Properties of Eigenfunctions

The eigenfunctions $y(x, \lambda_n)$ belonging to the individual eigenvalues are orthogonal with respect to the scalar product (A.1), i.e.

$$\int_a^b y(x, \lambda_m) y(x, \lambda_n) w(x)\, \mathrm{d}x = \delta_{m,n} \ .$$

The eigenfunctions form a complete system on (a, b). While the Liouville transformation changes the eigenfunctions themselves, it does preserve their *norms*,

$$\int_a^b |y(x, \lambda)|^2 w(x)\, \mathrm{d}x = \int_0^B |Y(\xi, \lambda)|^2\, \mathrm{d}\xi \ .$$

The asymptotic expansions of the normalized eigenfunctions corresponding to the Liouville standard form are:

$$\alpha_2 \neq 0\,,\ \beta_2 \neq 0 :\ \ y(x, \lambda_n) = \sqrt{\frac{2}{B}}\, \cos\left(\sqrt{\lambda_n}\, x\right) + \mathcal{O}\left(\frac{1}{n}\right) ,$$

$$\alpha_2 \neq 0\,,\ \beta_2 = 0 :\ \ y(x, \lambda_n) = \sqrt{\frac{2}{B}}\, \cos\left(\sqrt{\lambda_n}\, x\right) + \mathcal{O}\left(\frac{1}{n}\right) ,$$

$$\alpha_2 = 0\,,\ \beta_2 \neq 0 :\ \ y(x, \lambda_n) = \sqrt{\frac{2}{B}}\, \sin\left(\sqrt{\lambda_n}\, x\right) + \mathcal{O}\left(\frac{1}{n}\right) ,$$

$$\alpha_2 = \beta_2 = 0 :\ \ y(x, \lambda_n) = \sqrt{\frac{2}{B}}\, \sin\left(\sqrt{\lambda_n}\, x\right) + \mathcal{O}\left(\frac{1}{n}\right) ,$$

in which the respective expansions (9.88)–(9.91) should be inserted. The detailed structure of the $\mathcal{O}(1/n)$ terms and the asymptotic expansions in the untransformed case (Eqs. (9.84) and (9.85)) are listed in [18].

Two useful theorems The eigenfunctions $y(x, \lambda_n)$ have precisely n zeros on the open interval (a, b). The following theorems, stated without proof, are useful in practical work for the book-keeping of solution zeros.

The *Sturm comparison theorem* relates the positions of solution zeros corresponding to *different* boundary-value problems. Assume that the function y_1 on (a, b) is a non-trivial solution of the equation $(p_1(x)y')' + q_1(x)y = 0$, and y_2 a non-trivial solution of $(p_2(x)y')' + q_2(x)y = 0$, where $0 < p_2 \leq p_1$ and $q_1 \leq q_2$ for $x \in (a, b)$. Then between any two zeros of y_1 there is at least one zero of y_2, except when y_2 is just a constant multiple of y_1.

The consequence is the theorem on the *interlacing of zeros* which relates the positions of zeros of functions that solve the *same* differential equation. If y_1 and y_2 are linearly independent solutions of $(p(x)y')' + q(x)y = 0$ on (a, b), and $p > 0$

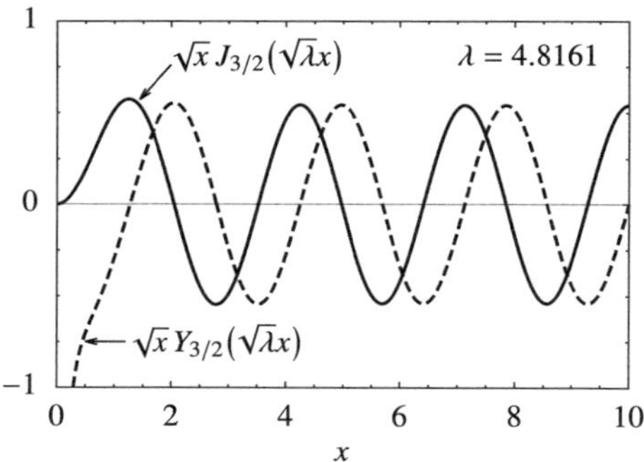

Fig. 9.12 Bessel's equation in Liouville form $-y''(x) + (v^2 - 1/4)x^{-2}y(x) = \lambda y(x)$ for $x \in [0, \infty]$ has two linearly independent solutions: the regular $\sqrt{x}\,J_v(\sqrt{\lambda}\,x)$ and $\sqrt{x}\,Y_v(\sqrt{\lambda}\,x)$ with a singularity at the origin. Shown are the solutions with $v = 3/2$ and $\lambda = 4.8161$ determined by the boundary conditions. The zeros of one eigenfunction interlace with the zeros of the other. See also Fig. 1.14

on (a, b), then between any two zeros of one function there is precisely one zero of the other function (see example in Fig. 9.12).

9.7.4 Solution by Difference Methods

The building blocks of the difference method for solving eigenvalue problems of the form (9.86) with Dirichlet boundary conditions $y(a) = 0$ and $y(b) = 0$ can be taken from the difference scheme (9.9) for problems (9.1) that do not involve eigenvalues. As usual we discretize all quantities on the uniform mesh (9.2). We denote the approximate value of the exact solution $y(x_j)$ by u_j, and the value $q(x_j)$ by q_j. By using the central difference for the second derivative we obtain a system of $(N - 1)$ equations,

$$-\frac{u_{j-1} - 2u_j + u_{j+1}}{h^2} + q_j u_j = \Lambda u_j, \qquad j = 1, 2, \ldots, N - 1, \qquad (9.93)$$

where Λ is an approximation of the exact eigenvalue λ. The boundary conditions translate to $u_0 = 0$ and $u_N = 0$. It is clear that at given N, solving this system will give us only $(N - 1)$ approximate eigenvalues $\{\Lambda_k\}_{k=1}^{N-1}$ (note the indexing starting from $k = 1$), while the continuous problem has infinitely many. The difference scheme can then be packaged as

$$A u = \Lambda u, \qquad (9.94)$$

where $u = (u_1, u_2, \ldots, u_{N-1})^{\mathrm{T}}$ is the vector of solution components, except for the trivial ones at the endpoints, and the matrix is

$$A = \frac{1}{h^2} D + Q ,$$

where

$$D = \begin{pmatrix} 2 & -1 & & & \\ -1 & 2 & -1 & & \\ & & \ddots & & \\ & & -1 & 2 & -1 \\ & & & -1 & 2 \end{pmatrix} , \qquad Q = \begin{pmatrix} q_1 & & & & \\ & q_2 & & & \\ & & \ddots & & \\ & & & q_{N-2} & \\ & & & & q_{N-1} \end{pmatrix} .$$

The error of the method is $\mathcal{O}(k^4 h^2)$, more precisely,

$$|\lambda_k - \Lambda_k| \le C k^4 h^2 , \qquad k = 1, 2, \ldots, N - 1 ,$$

where C depends only on the function q. The $\mathcal{O}(h^2)$ dependence should not come as a surprise once we recall the discretization error estimate (9.10). But in the context of eigenvalues, the $\mathcal{O}(k^4)$ dependence is much more relevant: we anticipate large errors of approximate eigenvalues with high indices k.

More general boundary conditions of the form (9.85) represent only a minor complication. Let us take just $p(a) = p(b) = 1$, and define $\alpha = a_2/a_1 \ne 0$ and $\beta = b_2/b_1 \ne 0$. The boundary conditions then access points beyond the mesh,

$$\frac{u_1 - u_{-1}}{2h} \approx \frac{u_0}{\alpha} , \qquad \frac{u_{N+1} - u_{N-1}}{2h} \approx \frac{u_N}{\beta} .$$

Equation (9.93) can also be considered at $x = a$ ($j = 0$) and $x = b$ ($j = N$), so u_{-1} and u_{N+1} can be eliminated. This results in a $(N + 1) \times (N + 1)$ matrix system of the form $A'u = \Lambda M u$, where M is a positive diagonal matrix.

Increasing the order of error By using a simple correction due to Numerov and Cowell, the discretization error can be improved by two orders: the scheme

$$-\frac{u_{j-1} - 2u_j + u_{j+1}}{h^2} + \left(q_j - \Lambda\right) u_j$$

$$-\frac{1}{12} \left[\left(q_{j-1} - \Lambda\right) u_{j-1} - 2\left(q_j - \Lambda\right) u_j + \left(q_{j+1} - \Lambda\right) u_{j+1}\right]$$

with boundary conditions $u_0 = 0$ and $u_N = 0$ leads to the matrix equation

$$A u = \Lambda M u , \tag{9.95}$$

where

$$A = \frac{1}{h^2} D + MQ \,, \qquad M = I - \frac{1}{12} D \,.$$

In general A, is not symmetric, but D and M commute, while Q is diagonal, so we can rewrite the generalized matrix eigenvalue problem (9.95) in the form $M^{-1} A u = \Lambda u$, where $M^{-1} A$ is symmetric (but sparse). The order of the Numerov–Cowell method is $\mathcal{O}(k^6 h^4)$.

The trick (9.13) that we used to extrapolate the solutions to an ever finer mesh can be applied to boundary-problems with eigenvalues as well. We first compute the eigenvalues on a mesh with spacing h, then do it again on a mesh with spacing $h/2$. The improved estimate for the kth eigenvalue is then

$$\Lambda_k^{(h/2,h)} = \frac{4}{3} \Lambda_k^{(h/2)} - \frac{1}{3} \Lambda_k^{(h)} \,.$$

If the "standard" difference scheme of order $\mathcal{O}(k^4 h^2)$ is used in the extrapolation, the order of the eigenvalue error becomes $\mathcal{O}(k^6 h^4)$, and if the Numerov–Cowell scheme of order $\mathcal{O}(k^6 h^4)$ is used, the resulting error of the eigenvalues is $\mathcal{O}(k^8 h^6)$. We realize that an improved spatial discretization implies an even larger error in the determination of high-lying eigenvalues. Simple difference schemes without further improvements are therefore suitable only for the computation of low-lying eigenvalues (λ_k at small k).

A very efficient way to overcome this obstacle is the correction [20] that reduces the error of the standard difference scheme to $\mathcal{O}(k^0 h^2)$ (the error becomes independent of k), and the error of the Numerov–Cowell scheme to $\mathcal{O}(k^2 h^4)$. The improved approximation for λ_k is given by

$$\widetilde{\Lambda}_k = \Lambda_k + \delta \Lambda_k \,, \tag{9.96}$$

where, for either method, the correction $\delta \Lambda_k$ is a simple function of the argument that depends only on the type of the boundary conditions:

$$\delta \Lambda_k = \begin{cases} \Delta(k) & ; & y(a) = y(b) = 0 \,, \\ \Delta\left(k - \frac{1}{2}\right) & ; & y(a) = y'(b) = 0 \,, \\ \Delta(k - 1) & ; & y'(a) = y'(b) = 0 \,. \end{cases}$$

In the case of the "standard" scheme (9.93) we use

$$\Delta(k) = \frac{\pi^2}{(b-a)^2} \left[k^2 - \frac{4N^2}{\pi^2} \sin^2\left(\frac{k\pi}{2N} \right) \right] \,,$$

while for the Numerov-Cowell scheme (9.95) we use

$$\Delta(k) = \frac{\pi^2}{(b-a)^2}\left[k^2 - \frac{4N^2}{\pi^2}\sin^2\left(\frac{k\pi}{2N}\right)\left(1 - \frac{1}{3}\sin^2\left(\frac{k\pi}{2N}\right)\right)^{-1}\right].$$

For Sturm–Liouville problems in the more general form (9.84) we can write a difference scheme

$$-\frac{1}{h}\left(p_{j+1/2}\frac{u_{j+1}-u_j}{h} - p_{j-1/2}\frac{u_j-u_{j-1}}{h}\right) + q_j u_j = \Lambda w_j u_j ,$$

which is of order $\mathcal{O}(k^4 h^2)$ if the functions p''', q'', and w'' are continuous. For this scheme a correction of the form (9.96) is not available.

9.7.5 Shooting Methods with Prüfer Transformation

Boundary-value problems with eigenvalues can also be solved by shooting (Sect. 9.3): we need to solve initial-value problems by integration from $x = a$ to $x = b$ (or "backwards" from $x = b$ to $x = a$, or "two-way" from $x = a$ to $x = c \in [a, b]$ from the left, and from $x = b$ to $x = c$ from the right). The role of the shooting parameter, unknown in advance, is played by the value of λ. If the boundary conditions are satisfied at the integration endpoint for some λ, that λ is the eigenvalue, and the corresponding solution is the eigenfunction. For example, when shooting from $x = a$ to $x = b$, we start with the values that satisfy the boundary conditions at $x = a$,

$$(pu')(a) = a_1 , \qquad u(a) = a_2 .$$

We solve the initial-value problem and denote the solution at $x = b$ by $u_L(x, \lambda)$. A meaningful function that measures the deviation from the exact fulfillment of the boundary conditions at $x = b$, is

$$D(\lambda) = b_1 u_L(b, \lambda) - b_2(pu_L')(b, \lambda) .$$

The sought eigenvalues λ are the zeros of $D(\lambda)$, which can be found by using the methods from Chap. 2. An analogous procedure can obviously be derived for "backward" shooting from $x = b$ to $x = a$.

Exponentially growing components of the solution in simple one-way shooting may cause instabilities that one can again try to fend off by shooting in two directions (Fig. 9.5 (bottom)). We shoot from $x = a$ with the initial conditions $(pu')(a) = a_1$ and $u(a) = a_2$, yielding the "left part" of the solution, $u_L(x, \lambda)$. By shooting backwards from $x = b$ with the initial conditions $(pu')(b) = b_1$ and $u(b) = b_2$, we obtain the "right part" of the solution, $u_R(x, \lambda)$. A suitable quantity that measures the difference between the solution parts at the point c, where $a < c < b$, is the Wronski's determinant,

$$D(\lambda) = \begin{vmatrix} (pu'_{\mathrm{L}})(c, \lambda) & (pu'_{\mathrm{R}})(c, \lambda) \\ (u_{\mathrm{L}})(c, \lambda) & (u_{\mathrm{R}})(c, \lambda) \end{vmatrix} . \tag{9.97}$$

This expression vanishes precisely when the solutions $u_{\mathrm{L}}(x, \lambda)$ and $u_{\mathrm{R}}(x, \lambda)$ are smoothly joined at $x = c$. Then u_{L} and u_{R}, taken together, represent the eigenfunction across the whole interval, and λ is the corresponding eigenvalue.

In problems where eigenvalues are found in clusters, the functions $D(\lambda)$ defined as in (9.97) may exhibit wild oscillations around zero, with large changes in the amplitudes and distances between the nodes. Such behavior can be alleviated by Prüfer transformation [20] that is used to transform the functions in the Sturm–Liouville problem to polar coordinates in the (pu', u) plane. The basic (unscaled) form of the transformation is

$$pu' = r \cos \theta \ ,$$
$$u = r \sin \theta \ .$$

The functions r and θ satisfy the differential equations

$$\frac{r'}{r} = \left[\frac{1}{p} - (\lambda w - q) \right] \sin \theta \ \cos \theta \ ,$$
$$\theta' = \frac{1}{p} \cos^2 \theta + (\lambda w - q) \sin^2 \theta \ . \tag{9.98}$$

In new variables, the boundary conditions (9.85) become

$$\theta(a) = \alpha \ , \quad \tan \alpha = a_2/a_1 \ ,$$
$$\theta(b) = \beta \ , \quad \tan \beta = b_2/b_1 \ .$$

The Prüfer transformation now reveals its two great charms. If we wish to determine just the eigenvalues (and not the eigenfunctions), we need to solve just one first-order equation (9.98). Besides, their determination is very direct: for a regular Sturm–Liouville problem, where the coefficient functions p, q, and w are piecewise continuous with $p, q > 0$, and where the parameters α and β are normalized such that

$$\alpha \in [0, \pi) \ , \quad \beta \in (0, \pi] \ ,$$

the eigenvalue λ_k is the value of λ that solves Eq. (9.98) and satisfies

$$\theta(a, \lambda) = \alpha \ , \quad \theta(b, \lambda) = \beta + k\pi \ .$$

In solving the eigenvalue problem with Dirichlet boundary conditions $y(a) = 0$ and $y(b) = 0$, we have to fulfill $\theta(a) = 0$ and $\theta(b) = \pi + k\pi$ (both determined up to an integer multiple of π). Thus, when shooting from $x = a$ to $x = b$, the eigenvalues λ_k are the roots of the equation $D(\lambda) = \theta(b, \lambda) - (k + 1)\pi = 0$.

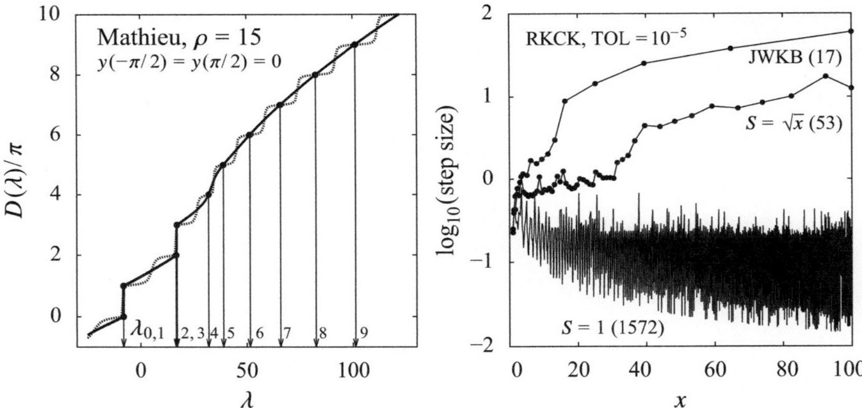

Fig. 9.13 [LEFT] The function $D(\lambda)/\pi$ in solving (9.99) with $\rho = 15$ and boundary conditions $y(-\pi/2) = y(\pi/2) = 0$ by Prüfer-transformed shooting from $x = -\pi/2$ to $x = \pi/2$. The dotted curve shows the unscaled version, $S(x, \lambda) = 1$, while the full curve shows the rescaled one, $S(x, \lambda) = \sqrt{\max\{1, \lambda - q(x)\}}$. Both curves go through the points (λ_k, k). [RIGHT] The step size in integration of the Prüfer equation (9.100) for the initial-value problem $u'' + xu = 0$, $u(1) = 0$, by the adaptive RK4 method (tolerance 10^{-5}): without scaling ($S = 1$), with scaling ($S = \sqrt{x}$), by solving the modified equation $\theta' = \sqrt{x}(1 + 5\sin^2 \theta/(16x^3))$. The numbers in brackets denote the number of integration steps for integration up to $x = 100$ with a tolerance of TOL $= 10^{-5}$

Example: rescaled Prüfer transformation (Adapted from [20].) There are pitfalls in seeking the eigenvalues by Prüfer-transformed shooting. The dotted curve in Fig. 9.13 (left) shows $D(\lambda)$ for solving the Mathieu equation

$$- y'' + (2\rho \cos 2x)y = \lambda y \,, \qquad y(-\pi/2) = y(\pi/2) = 0 \,, \qquad \rho = 15 \,, \quad (9.99)$$

by shooting from $a = -\pi/2$ to $b = \pi/2$. Note the staircase-like character of $D(\lambda)$, where the abscissas of the intercepts of the curve with the ordinates $k\pi$ determine the eigenvalues λ_k. In places where eigenvalues form clusters (e.g. the pairs $\lambda_{0,1}$ or $\lambda_{2,3}$), the problem of finding the intercept is poorly conditioned, because the curve there is almost vertical: no matter how hard we try, when locating the intercept, we tumble into one of the two eigenvalues—it is next to impossible to locate both, in particular if ρ is large. For higher eigenvalues, the problem is poorly conditioned because the curve there is almost horizontal.

One can resort to two tricks: rescaling the Prüfer transformation and changing the way we shoot. The rescaled Prüfer transformation has the form

$$pu' = \sqrt{S}\, r \cos \theta \,,$$

$$u = \frac{1}{\sqrt{S}}\, r \sin \theta \,,$$

where the function $S(x, \lambda)$ should be such that the staircase behavior of $D(\lambda)$ in a broad range of eigenvalues will become smoother. The differential equations for r and θ are

$$\frac{r'}{r} = \left[\frac{S}{p} - \frac{\lambda w - q}{S}\right] \sin\theta \, \cos\theta - \frac{1}{2}\frac{S'}{S} \cos 2\theta \, ,$$

$$\theta' = \frac{S}{p} \cos^2\theta + \frac{\lambda w - q}{S} \sin^2\theta + \frac{S'}{S} \sin\theta \, \cos\theta \, , \qquad (9.100)$$

and the boundary conditions become

$$\theta(a) = \alpha \, , \quad \tan\alpha = S(a)a_2/a_1 \, ,$$

$$\theta(b) = \beta \, , \quad \tan\beta = S(b)b_2/b_1 \, .$$

In "two-way" shooting with any parameter λ, we obtain the "left" solution $\theta_{\mathrm{L}}(x, \lambda)$ and the "right" solution $\theta_{\mathrm{R}}(x, \lambda)$ that both solve (9.100) in their respective domains, and satisfy

$$\theta_{\mathrm{L}}(a) = \alpha \, , \qquad \theta_{\mathrm{R}}(b) = \beta \, .$$

The departure from the boundary condition is best measured by the *Prüfer miss-distance function* $D(\lambda)$,

$$D(\lambda) = \theta_{\mathrm{L}}(c, \lambda) - \theta_{\mathrm{R}}(c, \lambda) \, .$$

The eigenvalues λ_k are then uniquely determined by the equation

$$D(\lambda_k) = k\pi \, , \qquad k = 0, 1, 2, \ldots \, .$$

The full curve in Fig. 9.13 (left) shows the function $D(\lambda)/\pi$ for the Prüfer shooting solution of the Mathieu equation (9.99) with the scaling function

$$S(x, \lambda) = \begin{cases} 1 & ; \lambda - 2\rho\cos 2x \leq 1 \, , \\ \sqrt{\lambda - 2\rho\cos 2x} & ; \lambda - 2\rho\cos 2x > 1 \, . \end{cases}$$

Above $\lambda \approx 20$ the scaling function has smoothened the jumpy $D(\lambda)$ to a humble curve, along which it is much easier to frame the intercepts (λ_k, k) and locate the eigenvalues λ_k. Two-way shooting requires additional care, as the continuity of the solution at the joining point $c \in [a, b]$ has to be ensured. Special attention in rescaling the solution is also called for if the function $S(x)$ has a discontinuity at $x = c$.

The choice of an efficient function $S(x, \lambda)$ and the optimal joining point c sometimes requires a stroke of luck, as illustrated in Fig. 9.13 (right). The figure shows the step size in solving the Prüfer equation (9.100) for the initial-value problem $u'' + xu = 0$ by adaptive RK4 integration without scaling, by using the scaling function $S = \sqrt{x}$, and by calculating with the modified equation (JWKB)

$\theta' = \sqrt{x}(1 + 5\sin^2\theta/(16x^3))$—see [20], where you can find further instructions on the choice of the joining point, the scaling function, and on ways of controlling the eigenvalue errors. $\lhd$

9.7.6 Pruess Method

The Pruess method is based on the approximation of the coefficient functions of the *differential equation* with simpler, piecewise continuous functions. We approximate the continuous problem (9.84) and (9.85) by the regular problem

$$-\frac{\mathrm{d}}{\mathrm{d}x}\left(P(x)\frac{\mathrm{d}Y}{\mathrm{d}x}\right) + Q(x)Y = \Lambda W(x)Y, \qquad a < x < b, \tag{9.101}$$

with boundary conditions

$$a_1 Y(a) - a_2 P(a)Y'(a) = 0,$$
$$b_1 Y(b) - b_2 P(b)Y'(b) = 0,$$

where P, Q, and W are approximations of p, q, and w [21]. One could, for example, choose them to be piecewise polynomial functions of degree m that are continuous at the joining points of the subintervals (x_{j-1}, x_j). Because we are approximating the equation rather than its solution, the resulting boundary-value problem preserves the set of eigenvalues (spectrum) of the original problem, which can also be infinite. Certainly we should not expect this property from difference methods with finite-dimensional matrices.

The convergence of the approximate eigenvalues Λ_k to the exact values λ_k of course depends on the polynomial degree m. If P, Q, and W are polynomials that interpolate p, q, and w *on each subinterval* at the Gauss collocation points (see Fig. 9.10 and Table 9.2), we have $|\lambda_k - \Lambda_k| \leq C(k)h^{2m+2}$. The simplest approximating functions are constant functions, $m = 0$, and they are used in most practical implementations of the Pruess approximation. We approximate the values $f(x)$ on the subintervals (x_{j-1}, x_j) by the values at the central points $f((x_{j-1} + x_j)/2)$, resulting in a *piecewise continuous midpoint approximation*. Figure 9.14 shows an example on a uniform mesh. For $m = 0$ the piecewise constant approximation is equivalent to Gauss interpolation, and we have

$$|\lambda_k - \Lambda_k| \leq Ck^3 h^2.$$

In [20] a more restrictive estimate $|\lambda_k - \Lambda_k| \leq Ckh^2 \max\{1, |\lambda_k|\}$ can be found that holds for large k. For Sturm–Liouville problems in normal form (9.86), we also have the estimate $|\lambda_k - \Lambda_k| \leq Ch^2\sqrt{\max\{1, \lambda_k\}}$.

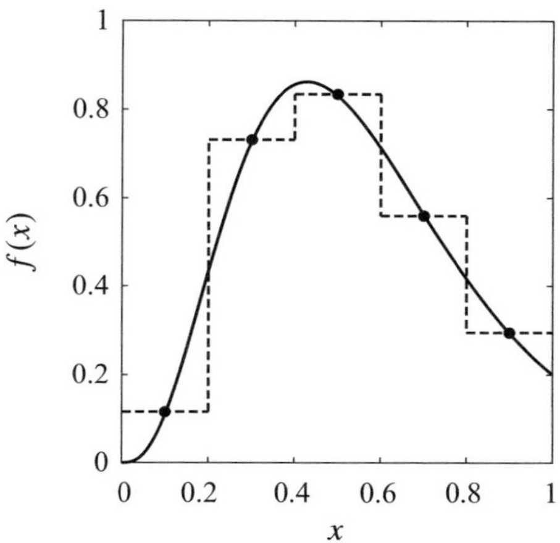

Fig. 9.14 Piecewise constant approximation of the function f, approximating the values $f(x)$ on subintervals (x_{j-1}, x_j) by the midpoint values $f((x_{j-1} + x_j)/2)$

Assume that we have approximated the functions p, q, and w on intervals (x_{j-1}, x_j) by the constant values $P = p_j$, $Q = q_j$, and $W = w_j$. The solution of Eq. (9.101) on the interval $[x_{j-1}, x_j]$ then has the general form [20]

$$Y(x) = f_j F_j(x) + g_j G_j(x) \,,$$

where F_j and G_j are independent solutions of $-Y'' = k_j Y$, and where

$$k_j = \frac{\Lambda w_j - q_j}{p_j} \,.$$

The explicit solution is

$$Y(x) = Y(x_{j-1}) \, \Psi_j \left(x - x_{j-1} \right) + (PY')(x_{j-1}) \, \Phi_j \left(x - x_{j-1} \right) \Big/ p_j \,, \qquad (9.102)$$

where

$$\Psi_j(s) = \begin{cases} \cos \omega_j s & ; \ k_j > 0 \,, \\ 1 & ; \ k_j = 0 \,, \\ \cosh \omega_j s & ; \ k_j < 0 \,, \end{cases} \qquad \Phi_j(s) = \begin{cases} \sin(\omega_j s)/\omega_j & ; \ k_j > 0 \,, \\ s & ; \ k_j = 0 \,, \\ \sinh(\omega_j s)/\omega_j & ; \ k_j < 0 \,, \end{cases}$$

and $\omega_j = \sqrt{|k_j|}$. By using Eq. (9.102) and its derivative it is easy to show (check it as an exercise) that the solution on one interval can be joined by the solution on the adjacent interval by the matrices

$$\begin{pmatrix} PY' \\ Y \end{pmatrix}_j = T_j \begin{pmatrix} PY' \\ Y \end{pmatrix}_{j-1} \,, \qquad T_j = \begin{cases} T_j^{(+)} & ; \ k_j > 0 \,, \\ T_j^{(0)} & ; \ k_j = 0 \,, \\ T_j^{(-)} & ; \ k_j < 0 \,, \end{cases} \qquad (9.103)$$

where

$$
T_j^{(+)} = \begin{pmatrix} \cos(\omega_j h_j) & -p_j \omega_j \sin(\omega_j h_j) \\ \sin(\omega_j h_j)/(p_j \omega_j) & \cos(\omega_j h_j) \end{pmatrix},
$$

$$
T_j^{(0)} = \begin{pmatrix} 1 & 0 \\ h_j/p_j & 1 \end{pmatrix},
$$

$$
T_j^{(-)} = \begin{pmatrix} \cosh(\omega_j h_j) & p_j \omega_j \sinh(\omega_j h_j) \\ \sinh(\omega_j h_j)/(p_j \omega_j) & \cosh(\omega_j h_j) \end{pmatrix}.
$$

We have denoted $h_j = x_j - x_{j-1}$ and have armed ourselves—an exception in this book—for the optional non-uniform mesh. Each inverse matrix T^{-1} can be obtained by simply replacing h by $(-h)$ in T. A good implementation of the Pruess approximation utilizes two-way shooting, in which one uses the mesh (9.2) to shoot from $x = a$ and $x = b$ towards the joining point $c \in [a, b]$. We let the joining point coincide with one of the mesh points, so that $c = x_s$, $0 < s < N$. The sketch of the algorithm [20] is

Input: Index s of point $c = x_s$, initial values $(PY')_0$ and Y_0 from the boundary
condition at $x_0 = a$, and initial values $(PY')_N$ and Y_N from the
boundary condition at $x_N = b$, tolerance ε

repeat

    Choose (or change) the value λ ;

    for $j = 1$ **step** 1 **to** s **do**

      | Compute $(PY', Y)_j$ from $(PY', Y)_{j-1}$ by Eq. (9.103)

    end

    $(PY')_{\mathrm{L}}(c) = (PY')_s$;

    $Y_{\mathrm{L}}(c) = Y_s$;

    for $j = N$ **step** -1 **to** $s + 1$ **do**

      | Compute $(PY', Y)_{j-1}$ from $(PY', Y)_j$ by Eq. (9.103)

      | and take into account that $T_j^{-1}(h_j) = T_j(-h_j)$

    end

    $(PY')_{\mathrm{R}}(c) = (PY')_s$;

    $Y_{\mathrm{R}}(c) = Y_s$;

    Form the Wronski's determinant $D(\lambda)$ by Eq. (9.97);

until $D(\lambda) \leq \varepsilon$;

Output: Eigenvalue λ

In state-of-the-art codes (see Appendix J) this algorithm is enhanced by Prüfer transformation and a prudent choice of the scaling function, by a generator of non-uniform meshes, and numerical error control. Further details can be found in [20].

9.7.7 *Eigenvalue-Dependent Boundary Conditions*

Now and then, one encounters Sturm–Liouville problems in which the eigenvalues appear as parameters in the boundary conditions, for example, as in

$$- \left(py'\right)' + q(x)y = \lambda w(x)y\,, \qquad x \in [a, b]\,,$$

with boundary conditions

$$\alpha_1 y(a) - \alpha_2 (py')(a) = \lambda \left[\alpha_1' y(a) - \alpha_2'(py')(a)\right]\,,$$
$$\beta_1 y(b) + \beta_2 (py')(b) = 0\,,$$

or

$$\alpha_1(\lambda)y(a) - \alpha_2(\lambda)p(a)y'(a) = 0\,,$$
$$\beta_1(\lambda)y(b) + \beta_2(\lambda)p(b)y'(b) = 0\,.$$

The standard difference schemes like (9.94) or (9.95) are inappropriate for such problems, since the corresponding sets of equations become too complicated. In general we obtain equations of the type $K(\Lambda)\boldsymbol{u} = \Lambda\boldsymbol{u}$ or $K(\Lambda)\boldsymbol{u} = \Lambda M(\Lambda)\boldsymbol{u}$. Problems of this type are therefore best attacked by shooting.

Boundary-value problems containing eigenvalues in the boundary conditions can be transformed such that various approaches or approximations can be used to determine the asymptotic behavior of eigenvalues and to compute the eigenfunctions (see [22, 23], and [24]). So far methods have been developed only for very specific analytic forms of the boundary conditions. Here we mention just a few of the most physically interesting ones, for boundary-value problems in Sturm–Liouville normal form

$$- y'' + qy = \lambda y\,, \qquad x \in [0, 1]\,.$$

Ref. [25] discusses transformations between the problems in which both boundary conditions are of "constant" (C) form,

$$\frac{y'}{y}(0) = \alpha\,, \qquad \frac{y'}{y}(1) = \beta\,,$$

and problems in which the eigenvalue appears in an "affine" (A),

$$\frac{y'}{y}(1) = \alpha\lambda + \beta\,,$$

or "bilinear" (B) manner,

$$\frac{y'}{y}(1) = \frac{\alpha\lambda + \beta}{\gamma\lambda + \delta}\,.$$

Between the forms A, B, and C simple explicit mappings B $\to$ A and A $\to$ C (thus also B $\to$ C) exist, and their spectra are equivalent. Therefore, with appropriate transformations, the boundary-value problem can be brought into the form C, which can also be solved by non-shooting methods. Physically relevant are also problems with the boundary conditions in the form

$$\frac{y'}{y}(0) = \cot\phi\,, \qquad \frac{y'}{y}(1) = \alpha\lambda + \beta - \sum_n \frac{\gamma_n}{\lambda - \delta_n} = f(\lambda)\,, \qquad (9.104)$$

where $\phi \in [0, \pi)$. A transformation exists for such problems that generates a boundary-value problem with different functions $q(x) \to \hat{q}(x)$ and $f(\lambda) \to \hat{f}(\lambda)$, such that $\hat{f}(\lambda)$ either contains less terms than $f(\lambda)$, or these terms contain fewer singularities [26]. By repeated use of this transformation, the problem can be converted to the standard Sturm–Liouville problem whose eigenvalues are almost identical to those of the problem with the condition (9.104). Similar transformations exist for boundary conditions in the form

$$\frac{y'}{y}(1) = \alpha\lambda^2 + \beta\lambda + \gamma$$

see [27]. See also Chap. 7 of [17].

9.8 Singular Sturm–Liouville Problems ★

So far we have only discussed regular Sturm–Liouville problems of the form (9.84), which we rewrite in terms of the linear differential operator L:

$$Ly(x) = \lambda y(x)\,, \quad L = \frac{1}{w(x)}\left[-\frac{\mathrm{d}}{\mathrm{d}x}\left(p(x)\frac{\mathrm{d}}{\mathrm{d}x}\right) + q(x)\right]\,, \quad a < x < b\,.$$
$$(9.105)$$

This problem becomes singular if one or both endpoints become singular. This means that either (a, b) is an infinite interval—say, $[0, \infty]$ or $[-\infty, \infty]$—or any of the functions $1/p, q$, or w is non-integrable in the neighborhood of an endpoint, that is

$$\int_a \left[\frac{1}{p(x)} + |q(x)| + w(x)\right] \mathrm{d}x = \infty\,.$$

(Here p and w are assumed to be positive, with possible exception at a finite number of points.) We are interested in solutions of singular variants of (9.105) that are square-integrable with respect to the weight function w, thus $y \in L^2_w(a, b)$ or

$$\int\limits_a^b |y(x)|^2 w(x)\, \mathrm{d}x < \infty .$$

(For continuous functions this is equivalent to being square-integrable at both end-points.) Note that singularity is a *local property*: it depends only on how the coefficient functions behave in the vicinity of the point in question.

The standard approach to a singular problem is to *regularize* it. This can be accomplished either by truncating the nominal domain to a finite interval (that is, by replacing the problematic endpoint with a nearby point) or by altering the coefficient functions slightly. The resulting regular problem is then solved by using one of the methods described in Sect. 9.7.

A physicist's prototype example of a singular Sturm–Liouville problem is the radial part of the Schrödinger equation for the hydrogen atom,

$$- y''(x) + \left[-\frac{1}{x} + \frac{l(l+1)}{x^2} \right] y(x) = \lambda y(x) , \qquad x \in [0, \infty) , \qquad (9.106)$$

whose exact eigenvalues for $l = 1$ are $\lambda_k = -1/(2k+4)^2$, $k = 0, 1, 2, \ldots$ (or principal quantum numbers $n = 2, 3, 4, \ldots$). We truncate $[a, b) = [0, \infty)$ to, say, $[a_*, b_*] = [0.0001, 100]$, enforce the Dirichlet boundary conditions $y(a_*) = y(b_*) = 0$, and solve the corresponding regular boundary-value problem to desired precision. A typical result for the $n = 4$ eigenfunction should look like the black curve in Fig. 9.15 (right) and the first four eigenvalues should match the exact values very well, as shown in Fig. 9.15 (left). For higher k, however, the eigenfunctions protrude farther and farther away from the origin and we will be forced to manually increase b_* in order to properly capture their asymptotic fall-off: see the $k = 12$ ($n = 14$) case in Fig. 9.15 (left and right, blue symbols and curve) where at least $b_* \approx 1000$ is needed.

One would imagine that by trial and error, with physical intuition and perhaps some idea on the asymptotic behavior of the eigenvalues or the eigenfunctions, a suitable regularization can be established empirically such that the determination of the eigenvalues and eigenfunctions may be deemed accurate enough and reliable. This is true; the best solvers, however, do this automatically, and the purpose of this Section is to provide an overview of the considerations that appear in the solution process. We follow the exposition in Chaps. 7 and 8 of [20], omitting the strictly mathematical details.

9.8.1   The "Shapes" of Spectra

For a regular Sturm–Liouville problem and for any eigenvalue $\lambda \in \mathbb{C}$ (or $\mathbb{R}$) there are two possibilities: either λ is in the resolvent set, meaning that

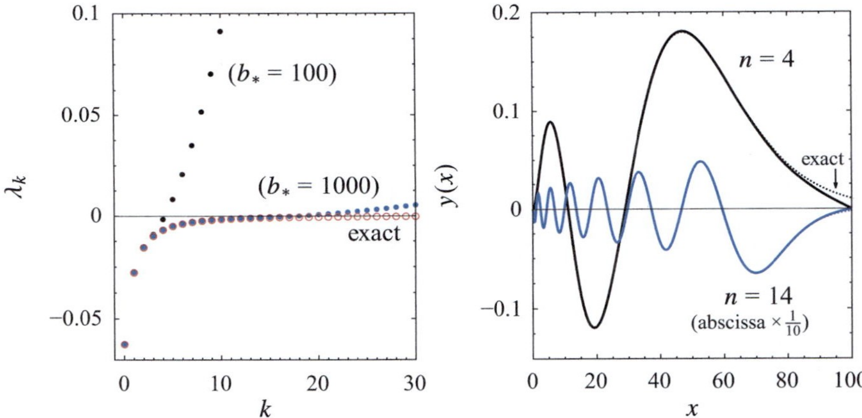

Fig. 9.15 Solving Eq. (9.106), the Schrödinger equation for the hydrogen atom with the electron in the p-wave, as a regular Sturm–Liouville problem on a truncated domain $[0.0001, b_*]$. [LEFT] The approximate eigenvalues obtained with $b_* = 100$ and $b_* = 1000$, respectively. Higher eigenvalues can only be extracted if b_* is large enough. [RIGHT] The approximate eigenfunctions for $n = 4$ and $n = 14$. (Note change of scale.)

$$(L - \lambda)v = f$$

has a unique solution satisfying the boundary conditions for any square-integrable f, or λ is in the *point spectrum* σ_{p} (comprising the eigenvalues), meaning that there exist nontrivial solutions of

$$(L - \lambda)u = 0 . \tag{9.107}$$

Whereas the spectrum of a regular problem always consists of a sequence of distinct eigenvalues tending monotonously to $+\infty$ (see Sect. 9.7.2), the spectrum of a singular problem is a closed infinite subset of $\mathbb{R}$ which can assume several characteristic "shapes". In particular, a singular problem may possess a continuous (part of the) spectrum, σ_{c}, implying that there is only the trivial solution of (9.107), so that $(L - \lambda)$ is a bijective map, yet $(L - \lambda)^{-1}$ is not a bounded operator on $L^2_w(a, b)$. The forms of commonly encountered spectra are illustrated in Fig. 9.16.

Substantial information on the spectrum and the corresponding eigenfunctions can be obtained by examining the properties of the singular endpoints, based on two criteria discussed below. An endpoint is either of the *limit-point* (LP) type or the *limit-circle* (LC) type, and this distinction is independent of λ. Additionally, an endpoint may be *oscillatory* (O) or *non-oscillatory* (N), which may depend on λ. Provided that the coefficient functions themselves do not oscillate infinitely near the endpoints, singular endpoints have been shown to fall into five basic categories [29]:

LCN: LC and non-oscillatory for all real λ;
LCO: LC and oscillatory for all real λ;
LPN: LP and non-oscillatory for all real λ;

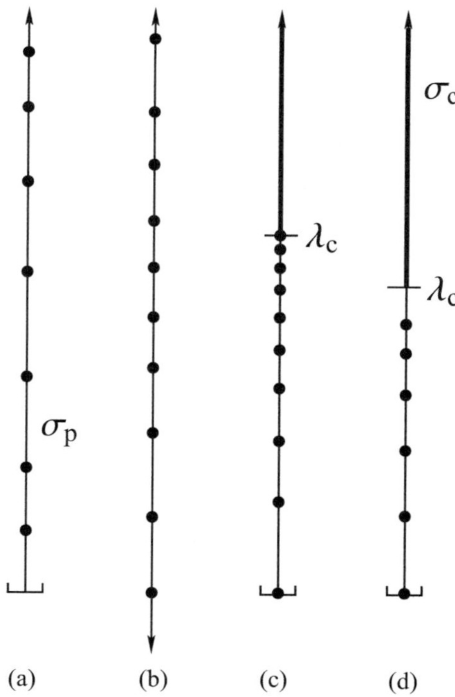

Fig. 9.16 Typical "shapes" of spectra encountered in singular Sturm–Liouville problems. (a) Pure point spectrum, with eigenvalues λ_k forming an infinite sequence bounded from below, with $+\infty$ the only accumulation point. (b) Sequence of λ_k unbounded in both directions, characteristic for LCO problems. (c) Infinite sequence of eigenvalues, bounded from below, with a single finite accumulation point λ_c such that all $\lambda \geq \lambda_\mathrm{c}$ are in the continuous part of the spectrum, σ_c. (d) A finite (possibly empty) sequence of eigenvalues strictly smaller than λ_c such that all $\lambda \geq \lambda_\mathrm{c}$ are in σ_c

LPN/O: LP and non-oscillatory for $\lambda < \lambda_\mathrm{c}$, oscillatory for $\lambda > \lambda_\mathrm{c}$;

LPO: LP and oscillatory for all λ.

If both endpoints of a singular Sturm–Liouville problem are of the LC type (or regular), the spectrum is purely discrete and unbounded from above. It is also unbounded from below if and only if either endpoint is oscillatory. On the other hand, if at least one point is of the LP type, then any of the cases listed above can occur. The spectrum is then purely discrete (bounded from below and unbounded from above) if and only if each LP endpoint is non-oscillatory for all λ [30].

9.8.2 Classification of Singular Endpoints

9.8.2.1 Limit-Point or Limit-Circle

The limit-point/limit-circle (LP/LC) classification allows us to determine the number of square-integrable solutions of (9.105) at a singular endpoint. For a given $\lambda \in \mathbb{C}$ the set of solutions is a two-dimensional subspace of functions. It is not clear that this subspace contains *any* non-trivial function that is square-integrable at an endpoint: consider, for instance, the problem $-y''(x) = \lambda y(x)$ for $x \in [0, \infty)$ with a real $\lambda > 0$, whose solution, $y(x) \propto \sin(\sqrt{\lambda}x)$, is not square-integrable. However, it has been proven that for either endpoint and for any non-real λ (that is, a λ with a non-zero imaginary part) there is at least one non-zero solution of (9.105) that is square-integrable at that point [31]. The theorem identifies two possible cases, both of which are independent of the particular value of λ chosen:

LC: all solutions of (9.105) are square-integrable at the endpoint; or
LP: only one non-zero solution (up to a scalar factor) is square-integrable
    at the endpoint.

In the LC case the same also holds true for all real λ. In the LP case it holds for all real λ outside σ_c. (Also note that a regular endpoint is automatically LC.)

The character of an endpoint (call it e) can be inferred from the behavior of the ratios $m(x) = -y(x)/(py')(x)$ as $y(x)$ ranges over all non-trivial solutions of (9.105) at a fixed λ. We fix c and d in (a, b) and look at the values of $m(c)$ as $m(d)$ varies over $\mathbb{R}$ and as d approaches either a or b. The solution $y(x)$ has the form $y(x) = \theta(x) + m(c)\phi(x)$, where θ and ϕ are the two independent solutions satisfying the conditions

$$\theta(c) = 0, \quad (p\theta')(c) = -1, \quad \phi(c) = 1, \quad (p\phi')(c) = 0. \tag{9.108}$$

Denoting $A = \theta(d)$, $B = (p\theta')(d)$, $C = \phi(d)$, $D = (p\phi')(d)$, one can demonstrate that $m(c)$ traces a circle (call it C_d) as $m(d)$ varies over $\mathbb{R}$. The center of this circle is given by

$$\mu_d = \frac{AD^* - BC^*}{C^*D - CD^*}, \tag{9.109}$$

while its radius is

$$r_d = \frac{|AD - BC|}{|C^*D - CD^*|}, \tag{9.110}$$

which, by resorting to Green's identity (see Sect. 7 of [20]), can be expressed by

$$r_d = \left[2 \operatorname{Im} \lambda \int_c^d |\phi(x)|^2 w(x)\, \mathrm{d}x \right]^{-1}. \tag{9.111}$$

In the limit $d \to e$ the consecutive circles C_d shrink either to a limiting circle C_e (the "Weyl circle") or a limiting point m_e, each of these features occupying the opposite halves of the complex plane, as illustrated by the Example below. These two cases classify an endpoint $e = a$ or b as either LC or LP. In terms of the circle radius, the LC case occurs if $\int_c^d |\phi(x)|^2 w(x)\, dx$ converges, while the LP case occurs if this integral diverges as $d \to e$.

Example As an illustration we construct the Weyl circles for the hydrogen equation (9.106) in the case of zero orbital angular momentum,

$$- y'' - (1/x)y = \lambda y \quad \text{on } [a, b] = [0, \infty].$$

We choose $\lambda = i$ and take $c = 1$ as the departure point, then solve the differential equation for θ and ϕ as an initial-value problem on $[c, d]$ with initial conditions given by (9.108) and d approaching the endpoints. When $d \to a$, the consecutive circles specified by Eqs. (9.109) and (9.110) converge to a limiting circle in the lower half-plane; in the limit $d \to b$, the circles shrink to a point in the upper half-plane, as shown in Fig. 9.17. Thus the problem is LC at $x = 0$ and LP at $x = \infty$.   ◁

This is a purely geometric consideration. Formally, the limit-point vs. limit-circle assignment is covered by the **LP/LC classification theorem** [28]: Let the Sturm–Liouville differential operator L be defined on $(0, b)$ with $p(x) = w(x) = 1$ (Liouville normal form, Eq. (9.86)). If there is a $C \in \mathbb{R}$ such that

$$q(x) \geq C + \frac{3}{4}\frac{1}{x^2} \quad \text{for } x \text{ close to } 0 , \tag{9.112}$$

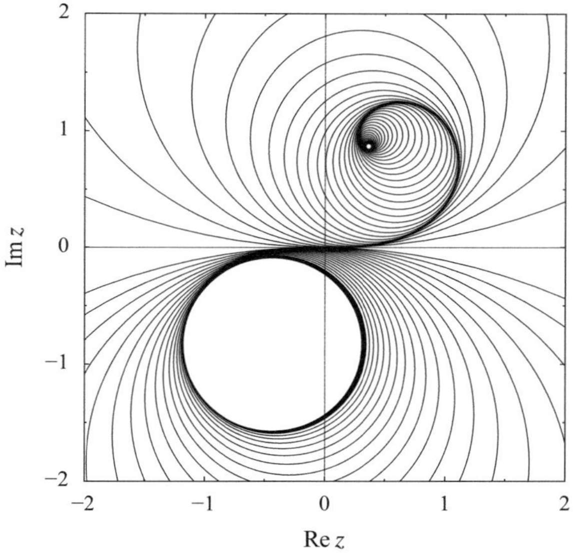

Fig. 9.17 Weyl circles for the hydrogen equation with a purely Coulomb potential. A limiting circle forming when $x \to 0$ and a circle shrinking to a point when $x \to \infty$ show that the problem is of the LC type at $x = 0$ and of the LP type at $x = \infty$

then L is in the LP case at 0. On the other hand, if there is an $\varepsilon > 0$ such that

$$|q(x)| \le \left(\frac{3}{4} - \varepsilon\right) \frac{1}{x^2} \quad \text{for } x \text{ close to } 0 , \tag{9.113}$$

then L is in the LC case at 0.

Example Consider the Bessel eigenvalue equation

$$-\left(x y'(x)\right)' + \frac{v^2}{x} y(x) = \lambda x y(x) , \quad x \in (0, b) ,$$

with b finite and $v \ge 0$. Thus $p(x) = x$, $q(x) = v^2/x$ and $w(x) = x$, and by substitution $y(x) = u(x)/\sqrt{x}$ this can be recast in Liouville normal form

$$-u''(x) + q(x)u(x) = \lambda u(x) , \quad q(x) = \frac{v^2 - 1/4}{x^2} .$$

Equation (9.112) asks for C such that $(v^2 - 1/4)/x^2 \ge C + 3/(4x^2)$ or $v^2 \ge Cx^2 + 1$ for x close to 0. For $v \ge 1$, any $C \ge 0$ will do, thus we are in the LP case for $v \ge 1$. Similarly, Eq. (9.113) calls for ε such that $|(v^2 - 1/4)/x^2| \le (3/4 - \varepsilon)/x^2$ when x approaches 0. For $0 \le v < 1$ this inequality is satisfied by $0^+ \le \varepsilon \le 1/2$, thus the problem is in the LC case for $0 \le v < 1$. The endpoint $x = b$ is regular.   ◁

Numerical LP/LC endpoint classification In state-of-the-art packages for singular Sturm–Liouville problems the endpoint analysis is performed automatically based on user-specified coefficient functions. Below we list the algorithm implemented in the SLEDGE package [32, 33] that classifies an endpoint as limit-point (LP) or limit-circle (LC) type.

Let us assume without loss of generality that the singular endpoint is $x = 0$. Other cases can be reduced to this one by a substitution $x' = x - x_0$ for finite x_0 or $x' = 1/x$ for endpoints at infinity, with appropriate modification of the equation itself. One then approximates the functions $f = p$, q and w by power functions, such that

$$f(x) = c_f |x|^{E_f} \left(1 + \mathcal{O}\left(|x|^{E'_f}\right)\right) \quad \text{as } x \to 0 ,$$

where c_f, E_f and E'_f are constants, with $E'_f > 0$. A sequence of points $\{x_n\}$ is devised that approaches zero (for instance, $x_0 = 1$, $x_{n+1} = x_n/1.1$ for $n = 0, 1, 2, \ldots, 100$), and E_f and c_f are read off from the asymptotics of the expressions

$$E_f = \frac{\log[f(x_{n+1})/f(x_n)]}{\log[x_{n+1}/x_n]} ,$$

$$c_f = f(x_n)/|x_n|^{E_f} .$$

In this manner the parameters E_p, c_p, E_q, c_q, E_w and c_w are estimated for $p(x), q(x)$ and $w(x)$, respectively. From these we calculate

$$\eta = (E_w - E_p + 2)/2 \,, \quad \tau = (E_p + E_w)/(4\eta) \,, \quad E_{q\mathrm{N}} = (E_q - E_w)/\eta$$

and

$$C_1 = \frac{c_q}{c_w} \left(|\eta| \sqrt{c_p/c_w} \right)^{E_{q\mathrm{N}}} \,, \quad C_2 = \tau(\tau - 1) \,, \quad C_3 = \frac{c_p}{c_w} \left(\frac{E_p + E_w}{4} \right)^2 \,.$$

Here $E_{q\mathrm{N}}$ is the power E_q of the coefficient function $q(x)$ of the Liouville normal form of the problem. We then have the following theorem [30]:

- If $\eta > 0$ the problem is of the LP type at $x = 0$ if any of the following hold:

$$\begin{aligned}
c_q &> 0 & \text{and} \quad E_{q\mathrm{N}} &< -2 \,; \\
C_1 + C_2 &> 3/4 & \text{and} \quad E_{q\mathrm{N}} &= -2 \,; \\
C_2 &> 3/4 & \text{and} \quad E_{q\mathrm{N}} &> -2 \,.
\end{aligned}$$

The cases ($C_1 + C_2 = 3/4$ and $E_{q\mathrm{N}} = -2$) and ($C_2 = 3/4$ and $E_{q\mathrm{N}} > -2$) are indeterminate, meaning that there is insufficient information to make the classification.

It is of the LC type at $x = 0$ if any of the following hold:

$$\begin{aligned}
c_q &< 0 & \text{and} \quad E_{q\mathrm{N}} &< -2 \,; \\
C_1 + C_2 &< 3/4 & \text{and} \quad E_{q\mathrm{N}} &= -2 \,; \\
C_2 &< 3/4 & \text{and} \quad E_{q\mathrm{N}} &> -2 \,.
\end{aligned}$$

- If $\eta < 0$ the problem is of the LP type at $x = 0$ if either of the following hold:

$$\begin{aligned}
c_q &> 0 \text{ and } E_{q\mathrm{N}} > 2 \,; \\
E_{q\mathrm{N}} &\leq 2 \,.
\end{aligned}$$

It is of the LC type at $x = 0$ if $c_q < 0$ and $E_{q\mathrm{N}} > 2$.
- If $\eta = 0$ the problem is of the LP type at $x = 0$ if either of the following hold:

$$\begin{aligned}
c_q &> 0 \text{ and } E_q - E_w < 0 \,; \\
E_q - E_w &\geq 0 \,.
\end{aligned}$$

It is of the LC type at $x = 0$ if $c_q < 0$ and $E_q - E_w < 0$.

Note on the form of boundary conditions In problems of the LC type some care is needed when imposing the boundary conditions. Let us define the modified Wronski's determinant as $[f, g](x) = p(x)\left[f(x)g'^*(x) - f'(x)g^*(x) \right]$. The appropriate form of the boundary condition at an LC-type endpoint e is

$$[y, \cos(\gamma)f - \sin(\gamma)g](e) = 0$$

for some γ. One way to choose f and g is to take two linearly independent solutions of the differential equation for some arbitrary λ. If only one endpoint is LC, one such boundary conditions is needed; if both endpoints are LC, two such conditions must be imposed. See [20] for details.

Example (Example 7.12 from [20].) The equation

$$- \left(x^{1-\alpha-\beta}y'(x)\right)' - \alpha\beta x^{-1-\alpha-\beta}y(x) = \lambda y(x), \quad x \in (0, 1],$$

has solutions $y(x) = x^\alpha$ and $y(x) = x^\beta$ when $\lambda = 0$. If $\alpha, \beta > -1/2$ the endpoint $x = 0$ is of the LC type. Assuming that α and β are distinct, two independent boundary conditions at $x = 0$ are

$$[y, x^\alpha](0) = \lim_{x \to 0}\left(-\alpha x^{-\beta}y(x) + x^{1-\beta}y'(x)\right) = 0,$$

$$[y, x^\beta](0) = \lim_{x \to 0}\left(-\beta x^{-\alpha}y(x) + x^{1-\alpha}y'(x)\right) = 0,$$

and any boundary condition at $x = 0$ is a combination of these two. ◁

Example (Problem 7.19 from [20].) The equation

$$- y''(x) + q(x)y(x) = \lambda y(x), \quad q(x) = -\mathrm{e}^x,$$

has an LC-type oscillatory endpoint at $+\infty$. In order to establish the appropriate boundary condition, we choose $\lambda = 0$ and use the lowest-order solution in the WKB approximation described in Sect. 1.3.6, which has the form

$$y(x) = [\omega(x)]^{-1/2}\mathrm{e}^{\pm i\,\Omega(x)}.$$

Here $\omega(x) = \sqrt{-q(x)}$ and $\Omega(x) = \int \omega(t)\,\mathrm{d}t$. This procedure yields damped oscillatory approximate solutions of the form

$$y(x) \approx \mathrm{e}^{-x/4}\left[\cos\left(2\mathrm{e}^{x/2}\right) \pm i\sin\left(2\mathrm{e}^{x/2}\right)\right],$$

The functions above, as well as $f(x) = \mathrm{e}^{-x/4}\cos(2\mathrm{e}^{x/2})$, $g(x) = \mathrm{e}^{-x/4}\sin(2\mathrm{e}^{x/2})$ or simply $\mathrm{e}^{-x/4}\sin(2\mathrm{e}^{x/2} + \delta)$, where δ is a phase shift, are exact solutions of the differential equation for $\lambda = -1/16$, and any linear combination of them generates all valid boundary conditions $[y, f](\infty) = 0$ and $[y, g](\infty) = 0$ as δ varies. ◁

9.8.2.2 Oscillatory or Non-oscillatory

An endpoint e is oscillatory (for a given real value of λ) if all solutions of (9.105) have infinitely many zeros in any neighborhood of e; otherwise it is non-oscillatory. One can show that [28]:

- At an LC endpoint, the solutions are either oscillatory for all (real) λ or oscillatory for no λ.
- At an LP endpoint, suppose that for each λ, $\lambda w(x) - q(x)$ does not change sign in some neighborhood of the endpoint. Then the solutions are either oscillatory for all (real) λ or oscillatory for no λ, or there is a λ_c such that the solutions are non-oscillatory for $\lambda < \lambda_c$ and oscillatory for $\lambda > \lambda_c$, in which case all eigenvalues are $\leq \lambda_c$.

The problems that are oscillatory for all λ are those whose spectra are unbounded from below. If at least one endpoint is of the LCO type and neither is of the LPO or LPN/O type the spectrum consists of a sequence $\{\lambda_k\}_{k\in\mathbb{Z}}$ tending to $\pm\infty$ as $k \to \pm\infty$: see Fig. 9.16b.

Numerical O/N endpoint classification With the same definitions as in the LP/LC classification criteria listed above, we have the following theorem [30]:

- If $\eta > 0$ the problem is non-oscillatory for all λ at $x = 0$ if any of the following hold:

$$
\begin{aligned}
c_q &> 0 \quad &\text{and} \quad E_{q\mathrm{N}} &< -2 \,; \\
C_1 + C_2 &> -1/4 \quad &\text{and} \quad E_{q\mathrm{N}} &= -2 \,; \\
C_2 &> -1/4 \quad &\text{and} \quad E_{q\mathrm{N}} &> -2 \,.
\end{aligned}
$$

The cases ($C_1 + C_2 = -1/4$ and $E_{q\mathrm{N}} = -2$) and ($C_2 = -1/4$ and $E_{q\mathrm{N}} > -2$) are indeterminate.

It is oscillatory for all λ at $x = 0$ if either of the following hold:

$$
\begin{aligned}
c_q &< 0 \quad &\text{and} \quad E_{q\mathrm{N}} &< -2 \,; \\
C_1 + C_2 &< -1/4 \quad &\text{and} \quad E_{q\mathrm{N}} &= -2 \,.
\end{aligned}
$$

- If $\eta < 0$ the problem is non-oscillatory for all λ at $x = 0$ if $c_q > 0$ and $E_{q\mathrm{N}} > 0$. It is oscillatory for all λ if $c_q < 0$ and $E_{q\mathrm{N}} > 0$. It is non-oscillatory in $(-\infty, \lambda_c)$ and oscillatory in (λ_c, ∞) if either of the following hold:

$$
\begin{aligned}
E_{q\mathrm{N}} &= 0 \quad (\text{then } \lambda_c = c_q/c_w) \,; \\
E_{q\mathrm{N}} &< 0 \quad (\text{then } \lambda_c = 0) \,.
\end{aligned}
$$

- If $\eta = 0$ the problem is non-oscillatory for all λ at $x = 0$ if $c_q > 0$ and $E_q - E_w < 0$. It is oscillatory for all λ if $c_q < 0$ and $E_q - E_w < 0$. It is non-oscillatory in $(-\infty, \lambda_c)$ and oscillatory in (λ_c, ∞) if any of the following hold:

$$
\begin{aligned}
E_q - E_w &= 0 \quad (\text{then } \lambda_c = c_q/c_w + C_3) \,; \\
E_q - E_w &< 0 \quad (\text{then } \lambda_c = C_3) \,; \\
c_q &= 0 \quad (\text{then } \lambda_c = C_3) \,.
\end{aligned}
$$

Example Consider the endpoint $b = \infty$ of the hydrogen equation (9.106) with $l = 1$. By substitution $x \mapsto 1/x$ the coefficient functions become $p(x) = x^2$, $q(x) = 2 - 1/x$ and $w(x) = 1/x^2$. Applying the above procedure with x_n approaching zero we get $\eta = -1$ and $E_{q\mathrm{N}} = -2$, resulting in the non-oscillatory assignment in $(-\infty, \lambda_\mathrm{c})$ and oscillatory assignment in $(\lambda_\mathrm{c}, \infty)$ with $\lambda_c = 0$, as expected for the energies of the hydrogen atom: we have negative energies $\lambda_k = -1/(2k + 4)^2$ of the bound states bounded from below and tending to zero for $k \to \infty$ plus the positive energies of the continuum states, as illustrated by Fig. 9.16c.   ◁

9.8.3 *Titchmarsh–Weyl m-Function and Spectral Density*

Two powerful tools exist that help us to characterize the nature of the spectrum of a given singular eigenvalue problem. One is the Titchmarsh–Weyl m-function [34]; its main benefit is that the spectral properties of a given problem can be inferred from its limiting behavior for λ near the real axis. The other is the spectral density function [35]; it allows us to understand how the approximating regularized problem (as solved numerically) converges to the exact singular problem that may have some continuous spectrum. Both of these tools provide us with a deep insight into the features of the spectrum and the asymptotics of the eigenfunction expansions.

Firstly let us consider the problem

$$- y''(x) + q(x)y(x) = \lambda y(x) \,, \quad x \in [0, \infty) \,, \tag{9.114}$$

where the $x = 0$ endpoint is *regular*, q is real-valued and continuous on $[0, \infty)$. Let the solutions $\theta, \phi : [0, \infty) \times \mathbb{C} \to \mathbb{C}$ of (9.114) be defined by the initial conditions

$$\theta(0, \lambda) = \sin \alpha \,, \quad \theta'(0, \lambda) = -\cos \alpha \,, \quad \phi(0, \lambda) = \cos \alpha \,, \quad \phi'(0, \lambda) = \sin \alpha \,.$$

for some $\alpha \in [0, \pi)$ and for all $\lambda \in \mathbb{C}$. (These conditions are generalizations of those given in Eq. (9.108).) It has been shown [36] that there exists an analytic function (the Titchmarsh–Weyl m-function, also called m-coefficient) with the properties: m is regular on $\mathbb{C} \setminus \mathbb{R}$; $[m(\lambda)]^* = m(\lambda^*)$ for all $\lambda \in \mathbb{C} \setminus \mathbb{R}$; $\operatorname{Im} m(\lambda) > 0$ for all λ with $\operatorname{Im} \lambda > 0$, and vice-versa; the solutions $y(x, \lambda)$ of (9.114) defined by

$$y(x, \lambda) = \theta(x, \lambda) + m_\alpha(\lambda)\phi(x, \lambda) \,, \quad x \in [0, \infty) \,, \tag{9.115}$$

satisfy

$$\int_0^\infty |y(x, \lambda)|^2 \mathrm{d}x = \frac{\operatorname{Im} m_\alpha(\lambda)}{\operatorname{Im} \lambda} < \infty \quad \forall \lambda \in \mathbb{C} \,,$$

that is, they are square-integrable. By writing the functions (9.115) at the origin,

$$y(0, \lambda) = \quad \sin \alpha + m_\alpha(\lambda) \cos \alpha \,,$$
$$y'(0, \lambda) = -\cos \alpha + m_\alpha(\lambda) \sin \alpha \,,$$

we see that $m_\alpha(\lambda)$ is easy to extract. Typically one considers two cases, with $\alpha = 0$ and $\alpha = \pi/2$. For $\alpha = 0$ we have $\theta(0, \lambda) = \phi'(0, \lambda) = 0$, $\theta'(0, \lambda) = -1$ and $\phi(0, \lambda) = 1$, while for $\alpha = \pi/2$ we have $\theta(\pi/2, \lambda) = \phi'(\pi/2, \lambda) = 1$ and $\phi(\pi/2, \lambda) = \theta'(\pi/2, \lambda) = 0$, resulting in

$$m_0(\lambda) = -\frac{y(0, \lambda)}{y'(0, \lambda)} = -\frac{1}{m_{\pi/2}(\lambda)} \,.$$

The positions of the poles of $m_\alpha(\lambda)$ are the eigenvalues. Analytic calculations of $m_\alpha(\lambda)$ exist only for very specific (and simple) functional forms of $q(x)$, but the nice thing is that $m_\alpha(\lambda)$ can be computed numerically [37–39]. There are two options:

1. For a chosen λ, start at a large x-value, $x = X$, set the initial conditions $y(X, \lambda) = 0$ and $y'(X, \lambda) = 1$ and (numerically) integrate (9.114) *backwards* to $x = 0$, calculating the ratio $-y(0, \lambda)/y'(0, \lambda)$. As X is increased this ratio tends towards its asymptotic value $m_0(\lambda)$.
2. Start at $x = 0$, using two functions u and v, and set the initial conditions $u(0, \lambda) = v'(0, \lambda) = 0$ and $v(0, \lambda) = u'(0, \lambda) = 1$. (Numerically) integrate (9.114) *forwards* to $x = X$ and form the ratio $-v(X, \lambda)/u(X, \lambda)$. As X is increased this ratio tends towards $m_0(\lambda)$.

In principle these options are equivalent, but in practice one method may be better than the other for a particular value of λ. In deciding whether a large enough X has been chosen for the backward integration (or set for the forward integration), one can check the convergence of the Weyl circle radius given by Eq. (9.111).

Example As an illustration, we show the calculation of the m-function in the case of $q(x) = x^2$, corresponding to the linear harmonic oscillator problem on $[0, \infty)$ in quantum mechanics. In this particular case both m_0 and $m_{\pi/2}$ can be computed analytically [39]. We have

$$m_0(\lambda) = \frac{\Gamma\left(\frac{1}{4} - \frac{1}{4}\lambda\right)}{2\Gamma\left(\frac{3}{4} - \frac{1}{4}\lambda\right)} \,, \tag{9.116}$$

which is a meromorphic function with simple poles at $\lambda_k = 4k + 1$ ($k = 0, 1, 2, \ldots$). The positions of these poles on the real axis are the eigenvalues of the oscillator problem with Neumann boundary conditions at the regular endpoint ($x = 0$), corresponding to the even solutions (Hermite polynomials of degree $2k$ times Gaussians). In turn, with $\alpha = \pi/2$ we have

$$m_{\pi/2}(\lambda) = -\frac{1}{m_0(\lambda)} = -\frac{2\Gamma\left(\frac{3}{4} - \frac{1}{4}\lambda\right)}{\Gamma\left(\frac{1}{4} - \frac{1}{4}\lambda\right)} \,. \tag{9.117}$$

Now the poles are at $\lambda_k = 4k + 3$ $(k = 0, 1, 2, \ldots)$: these are the eigenvalues of the oscillator problem with Dirichlet boundary conditions at $x = 0$, corresponding to the odd solutions (Hermite polynomials of degree $2k + 1$ times Gaussians). The results of the numerical calculation of $m_0(\lambda)$ and $m_{\pi/2}(\lambda)$ according to the procedure outlined above are shown in Fig. 9.18.   ◁

In order to introduce the concept of spectral density, we closely follow the standard presentation in [35] and start with an eigenvalue problem of the form

$$- \big(p(x)y'(x)\big)' + q(x)y(x) = \lambda w(x)y(x) \,, \quad a < x < \infty \,, \tag{9.118}$$

with the boundary condition at $x = a$ that depends on the character of that endpoint. If $x = a$ is a regular endpoint, the condition is

$$a_1 y(a) - a_2 (py')(a) = \lambda\big(a_3 y(a) - a_4 (py')(a)\big) \tag{9.119}$$

with either $a_3 = a_4 = 0$ and a_1, a_2 not both zero, or $a_3 a_2 - a_4 a_1 > 0$. If, on the other hand, $x = a$ is singular and is non-oscillatory for all λ, we set

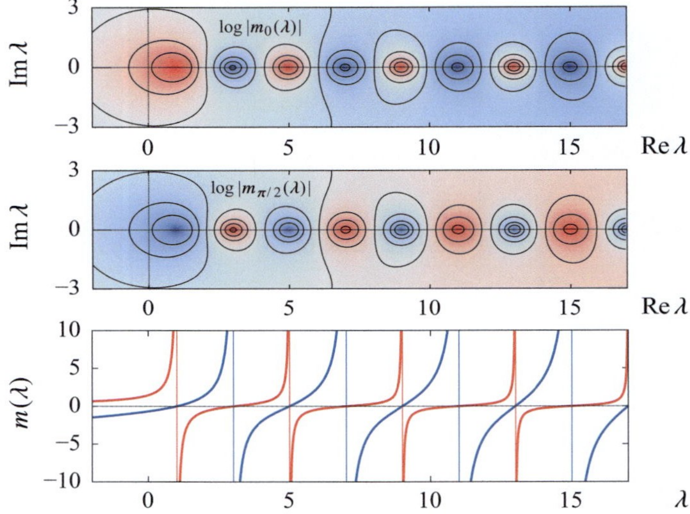

Fig. 9.18 The Titchmarsh–Weyl m-function for the Schrödinger equation with $q(x) = x^2$ (harmonic oscillator) for $x \in [0, \infty)$. [TOP] Contour plot of $m_0(\lambda)$ in the vicinity of the real axis, with poles at $\lambda = 1, 5, 9, \ldots$ [CENTER] Plot of $m_{\pi/2}(\lambda)$, with poles at $\lambda = 3, 7, 11, \ldots$ [BOTTOM] Plot of $m_0(\lambda)$ (red curves) and $m_{\pi/2}(\lambda)$ (blue curves) along the real axis

$$\lim_{x \to a} p(x)\big(y(x, \lambda)v'(x) - y'(x, \lambda)v(x)\big) = 0 \,, \tag{9.120}$$

where v is the principal solution of (9.118) for any fixed real value of λ. To define the singular spectral function associated with the problem (9.118)–(9.119) we assume the point $x = a$ to be regular and choose $a_3 = a_4 = 0$. We study the problem on the finite interval $[a, b]$, where $b \in (a, \infty)$, and ultimately let b tend to infinity. At the right endpoint, we set the Dirichlet boundary condition

$$u(b) = 0 \,. \tag{9.121}$$

Let us denote the eigenvalues and eigenfunctions of the regular eigenvalue problem (9.118)–(9.121) by $\lambda_{k,b}$ and $\phi(x, \lambda_{k,b})$, respectively, where $k = 0, 1, 2, \ldots$ and the eigenvalues are ordered as $\lambda_{0,b} < \lambda_{1,b} < \lambda_{2,b} < \cdots$. Let $H = L_w^2(a, b)$ denote the Hilbert space with the norm (A.2) on $[a, b]$. Each $f \in H$ has an expansion in eigenfunctions of the regular problem specified by Eqs. (9.118)–(9.121) of the form

$$f(x) = \sum_{k=0}^{\infty} c_k \phi(x, \lambda_{k,b}) \,,$$

where

$$c_k = \frac{1}{\|\phi(\cdot, \lambda_{k,b})\|^2} \int_a^b f(x)\phi(x, \lambda_{k,b})w(x)\, \mathrm{d}x \,.$$

Alternatively, this expansion can be written as a Riemann–Stieltjes integral

$$f(x) = \int_{-\infty}^{\infty} T_b[f](\lambda)\phi(x, \lambda)\, \mathrm{d}\rho^{(b)}(\lambda) \,,$$

in which

$$T_b[f](\lambda) = \int_a^b f(x)\phi(x, \lambda)w(x)\, \mathrm{d}x$$

and $\rho^{(b)}(\lambda)$ is the step function integrator defined for all real λ by

$$\rho^{(b)}(\lambda) = \sum_{\lambda_{k,b} \leq \lambda} \|\phi(\cdot, \lambda_{k,b})\|^{-2} \,.$$

The step spectral function $\rho^{(b)}$ for the problem on the finite interval $[a, b]$ is monotone non-decreasing and right-continuous at the eigenvalues.

To obtain the eigenfunction expansion for the *singular* problem (9.118)–(9.119) on $[0, \infty)$, assuming that the endpoint at infinity is of the LP type, we define the spectral density function as the limit

$$\rho(\lambda) = \lim_{b \to \infty} \rho^{(b)}(\lambda)$$

for all λ. The function ρ is continuous (or right-continuous at points of discontinuity) and satisfies

$$\int_{-\infty}^{\infty} \frac{\mathrm{d}\rho(\lambda)}{1 + \lambda^2} < \infty .$$

Then we have the expansion

$$f(x) = \int_{-\infty}^{\infty} T[f](\lambda)\phi(x, \lambda)\,\mathrm{d}\rho(\lambda) , \tag{9.122}$$

where

$$T[f](\lambda) = \int_{a}^{\infty} f(x)\phi(x, \lambda)w(x)\,\mathrm{d}x .$$

The m-function and the spectral density function (both taken at the same α) are closely related. If $m(\lambda)$ is analytic in the upper complex half-plane, maps the upper half-plane onto itself and satisfies the condition (Im $\lambda > 0 \Rightarrow \mathrm{Im}\, m(\lambda) > 0$), then $m(\lambda)$ can be represented in the form

$$m(\lambda) = C_1 + C_2\lambda + \int_{-\infty}^{\infty} \left[\frac{1}{t - \lambda} - \frac{t}{1 + t^2} \right] \mathrm{d}\rho(t) ,$$

where C_1 and $C_2 \geq 0$ are real constants. Given $\rho(\lambda)$ we define the *spectral density* as $\mathrm{d}\rho(\lambda)/\mathrm{d}\lambda$, which is given almost everywhere by

$$\frac{\mathrm{d}\rho(\lambda)}{\mathrm{d}\lambda} = \lim_{\varepsilon \to 0^+} \frac{1}{\pi} \mathrm{Im}\big[m(\lambda + \mathrm{i}\,\varepsilon) \big] , \tag{9.123}$$

hence

$$\rho(\lambda) = \lim_{\varepsilon \to 0^+} \int_{-\infty}^{\lambda} \frac{1}{\pi} \mathrm{Im}\big[m(\mu + \mathrm{i}\,\varepsilon) \big]\,\mathrm{d}\mu .$$

Spectral density defined in this manner provides a complete description of both the discrete and the continuous part of the spectrum; in quantum mechanics, it can be understood as the local probability density for the energy of the system. In a handful of cases, ρ can be computed analytically, but in general it has to be estimated numerically. At present the only Sturm–Liouville code with the capability to calculate it is SLEDGE [40]; also refer to Refs. [41–43] for state-of-the-art updates.

Example As an illustration we present the spectral density function corresponding to the Schrödinger equation for the hydrogen atom with zero orbital angular momentum (pure Coulomb potential):

$$-y''(x) + q(x)y(x) = \lambda y(x), \quad q(x) = -\frac{1}{x}, \quad x \in [0, \infty).$$

In addition to $b = \infty$ this also requires us handle the singular $x = 0$ endpoint, which is of the LC type and non-oscillatory; we use the boundary condition (9.120) dictating $y(0) = 0$. The spectral density function is given by [35]

$$\rho(\lambda) = \begin{cases} \displaystyle\sum_{\lambda_k < \lambda} \frac{1}{2(k+1)^3} & ; \ \lambda \leq 0, \\[2ex] \frac{1}{2}\zeta(3) \approx 0.6010284516 & ; \ \lambda = 0, \\[2ex] \rho(0) + \displaystyle\int_0^\lambda \left[1 - e^{-\pi/\sqrt{\mu}}\right]^{-1} d\mu & ; \ \lambda > 0, \end{cases}$$

where ζ is the Riemann zeta function, and the expansion of any f in terms of eigenfunctions then has the form (9.122) with

$$\phi(x, \lambda) = x + \mathcal{O}(x^2) \quad \forall \lambda \in \mathbb{C}.$$

The contour plot of the corresponding m-function [44] is shown in Fig. 9.19 (top). Note the poles in the vicinity of the negative real axis which represent the eigenvalues $\{\lambda_k\}$ and comprise the discrete (point) spectrum. The center panel shows an estimate of the spectral density obtained by choosing $\varepsilon = 0.0001$ to evaluate the limit in (9.123). Note the spikes corresponding to the eigenvalues and the smooth contribution at $\lambda \geq 0$ corresponding to the continuous part of the spectrum. The bottom panel shows the spectral density function. ◁

Spectral characterization of Sturm–Liouville problems with both endpoints singular has also been explored for more general potentials of the form

$$q(x) = \frac{q_0}{x^2} + \frac{q_1}{x} + \sum_{n=0}^{\infty} q_{n+2}x^n, \quad x \in (0, \infty),$$

assuming $\lim_{x \to \infty} q(x) = 0$, wih q_n real, $-1/4 \leq q_0 \leq \infty$, and q_0, q_1 not both zero; see [43]. In addition, spectral formalism has been developed for singular Sturm–Liouville problems representing linear Hamiltonian systems of the form $J\,y'(x) = [\lambda A(x) + B(x)]y(x) + A(x)f(x)$, where A and B are $2n \times 2n$ matrices, y and f

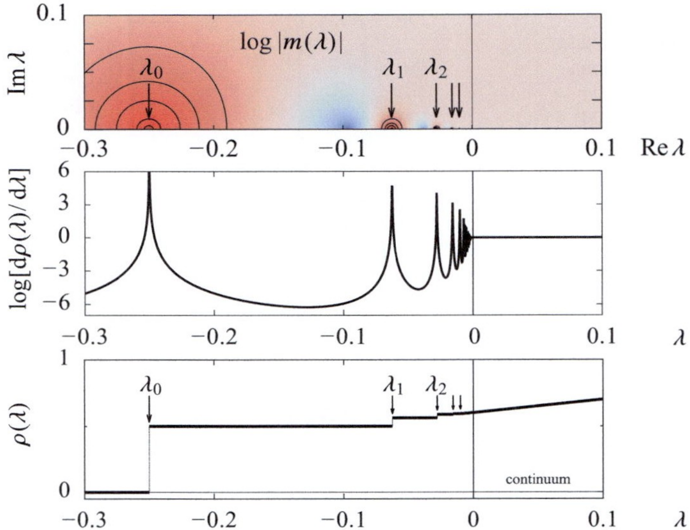

Fig. 9.19 Spectral properties of the Schrödinger equation for the hydrogen atom in the ground state. [TOP] The m-function in the complex plane with the poles near the negative real axis corresponding to the eigenvalues. [CENTER] An approximation of the spectral density obtained by evaluating $\mathrm{Im}[m(\lambda + \mathrm{i}\,0.0001)]/\pi$. [BOTTOM] The spectral density function with jumps at the eigenvalues and a smooth rise in the continuum

are $2n \times 1$ vector functions and $J = ((0, -I_n), (I_n, 0))$. In this case the $m(\lambda)$ is replaced by its matrix generalization, $M(\lambda)$. For details see [45–49].

9.8.4 Resonances

In particular settings the Schrödinger equation with potentials that involve barriers allow for quasi-bound states (resonances), which may be sharply localized and therefore loosely identified as "eigenvalues". The way to locate them is to vary λ and observe the phase shift $\delta(\lambda)$ of the asymptotic wave-function

$$y(x) \sim y_0 \sin\!\big(\sqrt{\lambda}\,x + \delta(\lambda)\big)$$

as $x \to \infty$, then calculate the time delay by taking the derivative $\tau(\lambda) = \mathrm{d}\delta/\mathrm{d}\lambda$— think of the point of inflection of the phase shift in a forced damped pendulum. The location of the maximum of the time delay is the center of the resonance, and this is associated with an "eigenvalue". Figure 9.20 shows an example on $[0, \infty)$ for a rectangular barrier with

$$q(x) = \begin{cases} 1 \; ; \; \alpha < x < \beta \, , \\ 0 \; ; \; \text{otherwise} \, . \end{cases} \tag{9.124}$$

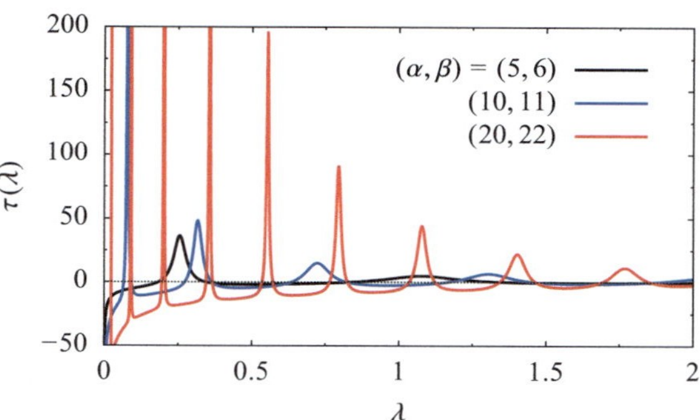

Fig. 9.20 Time delays calculated from the phase shifts of wave-functions passing the rectangular potential barrier (9.124) with $(\alpha, \beta) = (5, 6)$, $(10, 11)$ and $(20, 22)$

and initial conditions $y(0) = 0$, $y'(0) = 1$. With appropriate bracketing of the peaks, Sturm–Liouville solvers discussed previously can be used to pinpoint the precise values. For an introduction see Chap. 10 in [20].

9.8.5 Available Software

Several popular and robust software packages of comparable performance are available for solving singular (as well as regular) Sturm–Liouville problems in an automatic fashion: the stand-alone package SLEIGN2 [50] based on [51–53]; the SL02F program based on [54–56]; the D02KAF, D02KDF, and D02KEF routines incorporated in the NAG library [57]; as well as the SLEDGE package [32, 33, 58, 59]. All these solvers are included in SLTSTPAK [60, 61], a comprehensive test suite that also includes 60 characteristic Sturm–Liouville problems most frequently encountered in physics. A more recent development is MATSLISE [62], a MATLAB package customized for an efficient solution the Schrödinger equation.

9.9 Problems

9.9.1 Gelfand–Bratu Equation

One of the problems encountered in chemical kinetics involving diffusion with exothermic reactions is the Gelfand–Bratu boundary-value problem

$$y'' = -\delta\, e^y , \qquad 0 < x < 1 ,$$

with boundary conditions $y(0) = y(1) = 0$. For $0 < \delta < \delta_c \approx 3.51$ this problem has two solutions, while for $\delta > \delta_c$ there are no real solutions. In the physically relevant domain the two exact solutions are given by

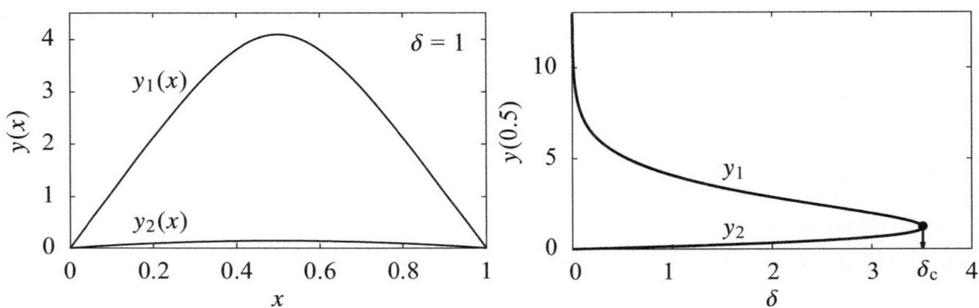

Fig. 9.21 Gelfand–Bratu equation. [LEFT] Two possible exact solutions at $\delta = 1$. [RIGHT] The dependence of the value $y(0.5)$ on the parameter δ (bifurcation diagram)

$$y(x) = -2\log\left[\frac{\cosh\xi(1-2x)}{\cosh\xi}\right],$$

where ξ is one of the two solutions of $\cosh\xi = \sqrt{8/\delta}\,\xi$.

$\odot$ Solve the problem by using difference methods described in Sect. 9.1. Start with $\delta = 1$: in this case the two exact solutions are defined by the parameters $\xi_1 \approx 0.379291149762689$ and $\xi_2 \approx 2.734675693030527$ (Fig. 9.21 (left)). Use the explicit iterative method with an appropriate convergence parameter ω. The equation tells us that the solution should be concave on $[0, 1]$, so choose the initial approximation accordingly, e.g. $y(x) = \mu x(1-x)$, where $\mu = 1$ or $\mu = 16$. Change the parameter δ and note the value of the smaller and larger of the two solutions at $x = 0.5$. By increasing δ you should observe two curves that ultimately meet at $\delta = \delta_{\rm c}$ (bifurcation diagram, Fig. 9.21 (right)).

$\oplus$ Solve the problem by using the Newton's method (9.16) and (9.17).

9.9.2 *Measles Epidemic*

In a simple epidemiological model [2] of a measles epidemic spreading in a population of size P we assume that any person in the population belongs to one of four groups: at time t there are $S(t)$ susceptible persons; $F(t)$ infectives; $L(t)$ latent persons (infected but with clinical signs as yet unpronounced); and $I(t)$ immunes. So at any time $t \in [0, 1]$ we have

$$S(t) + F(t) + L(t) + I(t) = P\,.$$

We use dimensionless quantities $y_1 = S/P$, $y_2 = L/P$, $y_3 = F/P$. The dynamics of the infection is described by a system of non-linear differential equations,

$$\dot{y}_1 = \mu - \beta(t)y_1 y_3 = f_1(t, \boldsymbol{y}),$$
$$\dot{y}_2 = \beta(t)y_1 y_3 - y_2/\lambda = f_2(t, \boldsymbol{y}),$$
$$\dot{y}_3 = y_2/\lambda - y_3/\eta = f_3(t, \boldsymbol{y}),$$

where $\boldsymbol{y} = (y_1, y_2, y_3)^{\mathrm{T}}$ and $\beta(t) = \beta_0(1 + \cos 2\pi t)$. The spreading infection is periodic: the boundary condition is $\boldsymbol{y}(1) = \boldsymbol{y}(0)$.

⊙ This non-linear problem has the form (9.19) and you can solve it on $t \in [0, 1]$ by using Newton's method described in Sect. 9.2. Use $y_1(t) = 0.1$, $y_2(t) = y_3(t) = 0.005(1 - \cos 2\pi t)$ as the initial approximation for the iteration. Choose a reasonably large mesh size N, as the division to N intervals implies a $3(N + 1) \times 3(N + 1)$ Jacobi matrix. Use the parameters $\mu = 0.02$, $\lambda = 0.0279$, $\eta = 0.01$, and $\beta_0 = 1575$.

9.9.3 *Diffusion-Reaction Kinetics in a Catalytic Pellet*

An important problem in chemical engineering is the computation of diffusion and reaction kinetics in a small porous sphere (a pellet). Assume that within the pellet with radius R a substance (platinum) is distributed in order to catalyze the dehydrogenation of cyclohexane [63]. The kinetics in spherical geometry in the region $0 \leq r \leq R$ is given by the boundary-value problem

$$D\left[\frac{1}{r^2}\frac{\mathrm{d}}{\mathrm{d}r}\left(r^2\frac{\mathrm{d}c}{\mathrm{d}r}\right)\right] = k\mathcal{R}(c), \qquad \left.\frac{\mathrm{d}c}{\mathrm{d}r}\right|_{r=0} = 0, \qquad c\big|_{r=R} = c_0,$$

where c is the concentration of cyclohexane, D is the diffusion constant, k is the reaction rate (all at given temperature), while the function $\mathcal{R}(c)$ defines the dependence on concentration. By substituting $r/R \to r$ and $c(r)/c(R) \to c(r)$ we write the problem in dimensionless form

$$\frac{\mathrm{d}^2 c}{\mathrm{d}r^2} = \Phi^2\frac{\mathcal{R}(c)}{c_0} - \frac{2}{r}\frac{\mathrm{d}c}{\mathrm{d}r} = f(r, c, c'), \qquad \left.\frac{\mathrm{d}c}{\mathrm{d}r}\right|_{r=0} = 0, \qquad c\big|_{r=1} = 1,$$

where $\Phi^2 = kR^2/D$.

⊙ If $\mathcal{R}(c) = c$ the equation is linear. At temperature 700 K the realistic parameter values are $k = 4/\mathrm{s}^{-1}$ and $D = 0.05\,\mathrm{cm}^2/\mathrm{s}$, so that for $R = 2.5\,\mathrm{mm}$ we obtain $\Phi \approx 2.236$. The exact solution is

$$c(r) = \frac{\sinh \Phi r}{r \sinh \Phi}$$

(see Fig. 9.22 (left)). Solve the problem by using Newton's method from Sect. 9.3.3 (even though the problem is linear). You are solving the initial-value problems

$$\begin{pmatrix} u \\ v \end{pmatrix}' = \begin{pmatrix} v \\ \Phi^2 u - \frac{2}{r}v \end{pmatrix}, \qquad \begin{pmatrix} \xi \\ \eta \end{pmatrix}' = \begin{pmatrix} \eta \\ \Phi^2\xi - \frac{2}{r}\eta \end{pmatrix},$$

with boundary conditions $u(0) = s$, $v(0) = 0$, $\xi(0) = 1$, and $\eta(0) = 0$. Newton's iteration (9.44) involves functions $\phi(s) = u(1; s) - 1$ and $\phi'(s) = \xi(1; s)$.

We are also interested in the average reaction rate within the pellet and the average rate computed at the surface:

$$
E = \left[\int_0^R \mathcal{R}\left(c(r)\right) r^2 \, dr \right] \left[\int_0^R \mathcal{R}(c_0) \, r^2 \, dr \right]^{-1} .
$$

The quantity E is a measure for the efficiency of the catalytic process: if mass transfer does not restrict the reaction rate, the average rate is the same in the interior of the pellet and at the surface, and then $E = 1$; if diffusion is significant, the mass transfer reduces the reaction rate within the pellet, indicated by $E < 1$. We integrate the differential equation by parts and obtain

$$
D \int_0^R \left[\frac{1}{r^2} \frac{d}{dr} \left(r^2 \frac{dc}{dr} \right) \right] r^2 \, dr = k \int_0^R \mathcal{R}(c) \, r^2 \, dr = D R^2 \frac{dc}{dr} \bigg|_{r=R} ,
$$

so that

$$
E = \left[k \mathcal{R}(c_0) R^3 \right]^{-1} 3 R^2 D \frac{dc}{dr} \bigg|_{r=R} .
$$

In dimensionless quantities you are therefore seeking $E = (3/\Phi^2)(dc/dr)|_{r=1}$, while the value from the exact solution is

$$
E = \frac{3}{\Phi} \left[\frac{1}{\tanh \Phi} - \frac{1}{\Phi} \right] \approx 0.7727 .
$$

$\oplus$ Solve the pellet problem with $\mathcal{R}(c) = c^2$ by using Newton's method for non-linear problems described in Sect. 9.3.3 With the same boundary conditions as before, you are now solving the coupled initial-value problems

$$
\begin{pmatrix} u \\ v \end{pmatrix}' = \begin{pmatrix} v \\ \Phi^2 u^2 - \frac{2}{r} v \end{pmatrix} , \qquad \begin{pmatrix} \xi \\ \eta \end{pmatrix}' = \begin{pmatrix} \eta \\ 2\Phi^2 u \xi - \frac{2}{r} \eta \end{pmatrix} .
$$

9.9.4 Deflection of a Beam with Inhomogeneous Elastic Modulus

Under certain assumptions, small vertical deflections of a freely supported beam can be described by the fourth-order boundary-value problem [2]

$$\left(x^3 u''\right)'' = 1 \,,$$
$$u(1) = u''(1) = u(2) = u''(2) = 0 \,,$$

on the interval $x \in [1, 2]$. The exact solution is

$$u(x) = \frac{10 \log 2 - 3}{4}(1 - x) + \frac{1}{2}\left[\frac{1}{x} + (3 + x)\log x - x\right] \,.$$

⊙ Solve this problem by using the collocation method for scalar boundary-value problems of higher orders, as described in Sect. 9.5.2. Use the uniform mesh (9.2) and Gauss collocation of orders $K = 4$ (see Table 9.2) and $K = 6$ (compute the corresponding canonical parameters ρ_k on your own). How crude a mesh (what N) can you afford that the error of the numerical solution is still acceptable?

⊕ Transform this problem to a set of first-order equations and solve it by shooting or by using the difference method.

9.9.5 A Boundary-Layer Problem

Here we discuss the boundary-value problem (adapted from [64]):

$$\varepsilon y'' - x^2 y' - y = 0 \,, \qquad 0 < x < 1 \,, \tag{9.125}$$

with boundary conditions $y(0) = y(1) = 1$ and parameter $\varepsilon = 0.01$. An exact solution exists that involves hypergeometric functions, but it can not be written in a simple closed form. The main feature of this problem is the rapid change of the solution in the vicinity of the right endpoint $x = 1$ (see Fig. 9.22 (right)). Such a region is known as the *boundary layer* and it requires particular care in order to avoid the potential instabilities in the numerical solution.

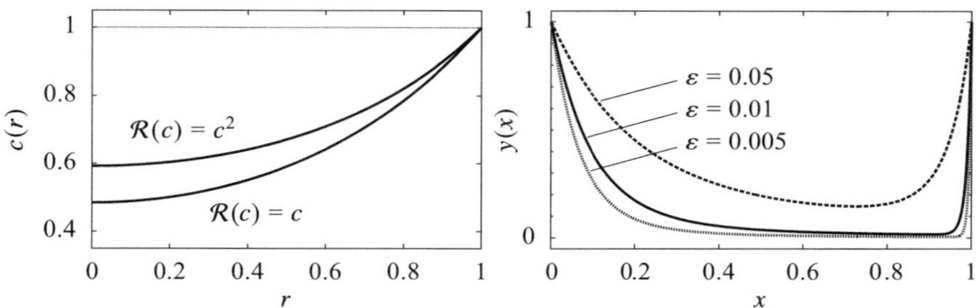

Fig. 9.22 Solutions of Problems 9.9.3 and 9.9.5. [LEFT] Concentration of cyclohexane in the catalytic pellet as a function of radius, for two different functions $\mathcal{R}(c)$ determining the kinetics rate. [RIGHT] The boundary-layer problem (Eq. (9.125)) for three values of the parameter ε

$\odot$ Solve the problem by using the usual difference method described in Sect. 9.1 (Eq. (9.9)), where you can also find the stability criterion for the size of the mesh spacing h. You may also blindly divide the interval $[0, 1]$ to $N = 10, 20, 50$, or 100 subintervals, and observe the corresponding solutions.

Use the same method to study the boundary-layer behavior in the problem

$$\varepsilon y'' - y' + 1 = 0 , \qquad 0 < x < 1 , \tag{9.126}$$

with boundary conditions $y(0) = 1$ and $y(1) = 3$. The exact solution is $y(x) = 1 + x + (e^{x/\varepsilon} - 1)/(e^{1/\varepsilon} - 1)$. Solve the same problem by keeping the central difference for the second derivative, but using the one-sided formula (9.4) for the first derivative. Write the difference equations in matrix form and solve the resulting equation for $N = 10, 20$, and 40. Discuss the solutions with $\varepsilon = 0.1$ and 0.01. Compare all solutions. How does the error change when N increases? Which method is preferable if ε is reduced still further?

$\oplus$ Solve the problems (9.125) and (9.126) by using the collocation method with cubic B-splines, ending in Eq. (9.65). How do the solutions depend on the number of points (or the number of basis functions) on the interval?

Recall that the key step of the collocation method is the requirement that the remainder $\mathcal{R}(x) = y'' + p(x)y' + q(x)y - r(x)$ is zero at the collocation points. There is another way of finding the zero of the remainder: we construct the *least-square error function*

$$E = \int_a^b [\mathcal{R}(x)]^2 \, \mathrm{d}x ,$$

which, of course, depends on the coefficients a_j that enter the expression through the expansion (9.64). We find the zero of $\mathcal{R}$ when we minimize E with respect to a_j, i.e. by requiring $\partial E / \partial a_j = 0$. We obtain the matrix system

$$\sum_{j=0}^{N} \alpha_{ij} a_j = \beta_i ,$$

where

$$\alpha_{ij} = \int_a^b \chi_i(x)\chi_j(x) \, \mathrm{d}x , \qquad \beta_i = \int_a^b \chi_i(x)r(x) \, \mathrm{d}x ,$$

and $\chi_i(x) = \phi_i''(x) + p(x)\phi_i(x)'' + q(x)\phi_i(x)$. This method is more stable since the error function E in some sense averages the integrand over the interval, while standard collocation "checks" only the agreement at the collocation points. The price we pay for this is an increase in the matrix bandwidth.

9.9.6 *Small Oscillations of an Inhomogeneous String*

Small transverse deflections $U(x, t)$ of a string with linear mass density ρ are described by the equation

$$\rho(x)\frac{\partial^2 U}{\partial t^2}(x, t) = \frac{\partial}{\partial x}\left(T(x)\frac{\partial U}{\partial x}(x, t)\right) + F(x, t),$$

where T is the longitudinal internal tension and F is the external force per unit length. At the endpoints of $[0, L]$ the string is fixed, so $U(0, t) = U(L, t) = 0$. We are interested in the oscillations of an inhomogeneous string, which we seek by the ansatz $U(x, t) = u(x)\exp(i\omega t)$. The eigenmodes are given by the equation

$$-\omega^2\rho(x)u(x) = \frac{\partial}{\partial x}\left(T(x)\frac{\partial u}{\partial x}(x)\right). \tag{9.127}$$

⊙ Calculate the eigenfunctions and eigenvalues of the string with the density

$$\rho(x) = \begin{cases} \rho_0 \; ; \; x \in [0, L/2], \\ \rho_1 \; ; \; x \in (L/2, L]. \end{cases}$$

Solve the problem by using the difference method: divide the interval $[0, L]$ to N subintervals with endpoints $x_j = jh$, $h = L/N$, and use this mesh to discretize the string deflections $\boldsymbol{u} = (u_1, u_2, \ldots, u_{N-1})^{\mathrm{T}}$, the first and second derivative in Eq. (9.127), as well as ρ and T. You obtain a system of difference equations

$$-(h\omega)^2\rho_j u_j = T_j(u_{j+1} - 2u_j + u_{j-1}) + \frac{1}{4}(T_{j+1} - T_{j-1})(u_{j+1} - u_{j-1}),$$

valid for $1 \leq j \leq N - 1$. Initially assume constant string tension, $T = $ const. By including the boundary conditions $u_0 = u_N = 0$, the system can be written in matrix form $A\boldsymbol{u} = (h\omega)^2\boldsymbol{\rho}\,\boldsymbol{u}$, where A is symmetric, or

$$\rho^{-1/2}A\rho^{-1/2}\boldsymbol{v} = (h\omega)^2\boldsymbol{v},$$

where $\boldsymbol{v} = \rho^{-1/2}\boldsymbol{u}$ and ρ is the diagonal density matrix $\rho = \mathrm{diag}\,(\rho_j)_{j=1}^{N-1}$. Use methods from Sect. 4.6 to solve the resulting matrix eigenvalue problem.

Solve this problem by shooting: choose an energy $E = \omega^2$ and solve Eq. (9.127) as an initial-value problem by using an integrator of your choice (for example, RK4), starting at $x = 0$ with initial conditions $u(0) = 0$ and $u'(0) = 1$ towards $x = L$. Change E until the boundary condition $u(L) = 0$ is satisfied to desired precision. This gives you just one solution, but further eigenmodes can be found in the same manner by realizing that the sequence of eigenvalues follows the number of nodes of u on the interval.

Discuss the solutions obtained by the difference and shooting methods, and compare both to the exact solution (two modes corresponding to the segments with ρ_0 and ρ_1 joined smoothly at the junction). Plot the computed eigenvalues as a function of the ratio ρ_1/ρ_0.

⊕ Use the difference method to study the oscillation of the string under constant tension, $T = \mathrm{const}$, but with a random density distribution,

$$\rho_j = a w_j \left[\frac{1}{N} \sum_{n=1}^{N-1} w_n \right]^{-1} , \qquad w_n = \chi_n + 1 ,$$

for $a = 0, 1, 10,$ and 100. Here χ_n is a random variable uniformly distributed on $[0, 1]$. Analyze the form of the eigenmodes for different N and plot the most appealing ones. At chosen a and N, compute the center-of-mass of the eigenmode, $M(\boldsymbol{u})$, and its dispersion, $S(\boldsymbol{u})$,

$$M(\boldsymbol{u}) = \sum_{j=1}^{N-1} j\, u_j , \qquad S(\boldsymbol{u}) = \sqrt{ \sum_{j=1}^{N-1} j^2\, u_j^2 - M(\mathbf{u})^2 } .$$

9.9.7 *One-Dimensional Schrödinger Equation*

In this Problem we are solving the stationary Schrödinger equation

$$-\frac{\hbar^2}{2m} \frac{\mathrm{d}^2 \psi}{\mathrm{d}x^2} + V(x)\psi = E\psi$$

in an infinite potential well of width a, with the potential

$$V = \begin{cases} 0 & ; \ |x| < a/2 , \\ \infty & ; \ |x| \ge a/2 , \end{cases}$$

and in a finite well $(V(|x| \ge a/2) = V_0)$. The exact solutions for both cases can be found in any textbook on quantum mechanics. Here we wish to compute the eigenfunctions ψ and eigenenergies E numerically.

⊙ First use the difference method to solve the infinite well problem. Divide the interval $[-a/2, a/2]$ with N points $(x_j = -a/2 + j(a/N))$ and discretize the differential equation accordingly. In dimensionless form you get

$$\frac{\psi_{i-1} - 2\psi_i + \psi_{i+1}}{h^2} + E\psi_i = 0$$

or $\psi_{i-1} - (2 - \lambda)\psi_i + \psi_{i+1} = 0$, where $\lambda = Eh^2 = k^2 h^2$. You should also discretize the boundary conditions at $x = -a/2$ and $x = a/2$ which, in general (and for the finite well) are of the mixed type,

$$c_1 \psi_0 + c_2 \frac{\psi_1 - \psi_{-1}}{2h} = 0 \,,$$

$$d_1 \psi_N + d_2 \frac{\psi_{N+1} - \psi_{N-1}}{2h} = 0 \,,$$

while for the infinite well we clearly have $\psi_0 = \psi_N = 0$. Both cases lead to tridiagonal systems of equations

$$A\psi = \lambda\psi$$

for the eigenvectors $\psi = (\psi_0, \psi_1, \ldots, \psi_N)^{\mathrm{T}}$ and eigenvalues λ, which can be solved by using methods from Sect. 4.6. Determine a few lowest eigenfunctions and eigenvalues! What plays a larger role in the error: the approximation of the second derivative by the central difference or the graininess of the interval (the finite dimension of the matrix being diagonalized)?

Solve the problem by shooting: start with a "cosine" ($\psi(0) = 1$, $\psi'(0) = 0$) or "sine" ($\psi(0) = 0$, $\psi'(0) = 1$) initial condition at the origin, select a value of E, and solve the initial-value problem by integrating the differential equation to $x = a/2$. Check the validity of the boundary condition $\psi(a/2) = 0$. Change E (and/or ψ') until the boundary condition is fulfilled to required precision. This procedure gives you both the odd and even eigenfunctions with the corresponding eigenvalues.

$\oplus$ Use shooting to solve the finite well problem. Note: the spectrum is finite! Only the boundary conditions change; the condition at $x = a/2$ is given by the requirement that the wave-function is smooth at the boundary, and has the form

$$c_1 \psi(a/2) + c_2 \psi'(a/2) = 0 \,.$$

9.9.8   A Fourth-Order Eigenvalue Problem

Difference and shooting methods are also suitable for the solution of higher-order eigenvalue problems: think of them as Sturm–Liouville problems but involving higher derivatives. An example of such a problem is

$$\frac{\mathrm{d}^4 y}{\mathrm{d}x^4} - [\lambda q(x) - r(x)]\, y(x) = 0 \,, \qquad x \in [a, b] \,, \tag{9.128}$$

with boundary conditions

$$y(a) = y(b) = y'(a) = y'(b) = 0 \,.$$

We approximate the differential equation by the difference scheme on a discrete mesh (9.2). As usual we use u_j to denote the approximation of $y(x_j)$, as well as $q_j =$

$q(x_j)$ and $r_j = r(x_j)$. We approximate the fourth derivative at x_j, $j = 2, 3, \ldots, N - 2$, by using Eq. 25.3.28 from [8],

$$\frac{d^4 y}{dx^4}\bigg|_{x=x_j} \approx \frac{u_{j-2} - 4u_{j-1} + 6u_j - 4u_{j+1} + u_{j+2}}{h^4} .$$

We write the boundary conditions as $u(a) = u_0 = 0$, $u(b) = u_N = 0$, and

$$u'(a) \approx \frac{u_1 - u_{-1}}{2h} = 0 , \qquad u'(b) \approx \frac{u_{N+1} - u_{N-1}}{2h} = 0 .$$

At the left endpoint ($x = x_0 = a$) we use

$$\frac{d^4 y}{dx^4}\bigg|_{x=x_1} \approx \frac{-4u_0 + 7u_1 - 4u_2 + u_3 - 2hu'(a)}{h^4} ,$$

and analogously at $x = x_N = b$. By using the difference equations at $j = 1$ and $j = N - 1$ and the boundary conditions for the first derivative, we eliminate u_{-1} and u_{N+1}, ending up with a system of equations for the solution vector $\boldsymbol{u} = (u_1, u_2, \ldots, u_{N-1})^{\mathrm{T}}$ which has the form of a generalized matrix eigenvalue problem,

$$\left(A + h^4 R\right) \boldsymbol{u} = \Lambda h^4 Q \boldsymbol{u} . \tag{9.129}$$

Here $R = \mathrm{diag}\,(r_1, r_2, \ldots r_{N-1})$ and $Q = \mathrm{diag}\,(q_1, q_2, \ldots q_{N-1})$, while A is penta-diagonal, with the structure

$$A_{1,1} = A_{N-1,N-1} = 7 ,$$
$$A_{i,i} = 6 , \; i = 2, 3, \ldots, N - 2 ,$$
$$A_{i,i\pm1} = -4 , \; i = 1, 2, \ldots, N - 2 \;\; \text{or} \;\; i = 2, 3, \ldots, N - 1 ,$$
$$A_{i,i\pm2} = 1 , \; i = 1, 2, \ldots, N - 3 \;\; \text{or} \;\; i = 3, 4, \ldots, N - 1 .$$

⊙ Solve the problem (9.128) by using the difference method described above: solve (9.129) on $[a, b] = [1, e]$, with the coefficient functions $r(x) = 0$ and $q(x) = 1/x^4$. Find some of the lowest eigenvalues. The exact values of the first five are $\lambda_1 = 531.836459064$, $\lambda_2 = 3919.16273370$, $\lambda_3 = 14865.4293298$, $\lambda_4 = 40373.3097672$, $\lambda_5 = 89795.9545147$ (precise to 12 decimal places). The method is second order, but it can be turned into a fourth-order method if a seven-point difference is used for the fourth derivative. When the boundary conditions are accounted for, we obtain a heptadiagonal matrix given in [65].

⊕ Solve this problem by shooting, and compare the results to those obtained in the first part. Use the paper [66] for further reference.

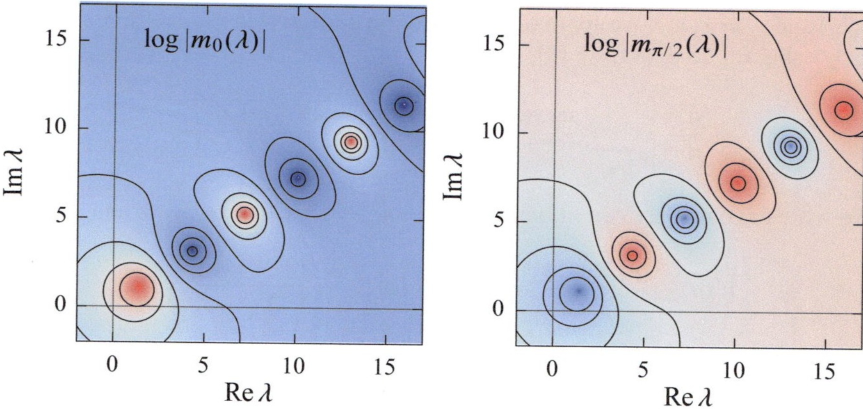

Fig. 9.23 Contour plots of the functions $m_0(\lambda)$ and $m_{\pi/2}(\lambda)$ for the harmonic oscillator with a complex potential, as specified by Eq. (9.130)

9.9.9 Harmonic Oscillator with a Complex Potential

Consider the Schrödinger equation for the linear harmonic oscillator with a complex constant in the potential:

$$- y''(x) + cx^2 y(x) = \lambda y(x) , \quad x \in [0, \infty) , \quad c = 1 + \mathrm{i}4 . \tag{9.130}$$

Because c is complex, the eigenvalues move into the complex plane as well [67, 68]. (This is a generalization of the Example on page 618.)

⊙ Visualize the spectrum of this problem by calculating the Titchmarsh–Weyl m-function with the procedure described on page 618. Apply Neumann and Dirichlet boundary conditions at the origin, i.e. $y'(0) = 0$ and $y(0) = 0$. The exact values are given by Eqs. (9.116) and (9.117) in which the constant 2 in the numerator and denominator, respectively, is replaced by $2c^{1/4}$. Plot the m-functions in the complex plane (see Fig. 9.23) and extract the corresponding eigenvalues from their poles: confirm that these are at $\lambda_k = \sqrt{c}(4k + 1)$, $k = 0, 1, 2, \ldots$ in the Neumann case, and at $\lambda_k = \sqrt{c}(4k + 3)$, $k = 0, 1, 2, \ldots$ in the Dirichlet case [69, 70]. ◁

References

1. H.B. Keller, *Numerical Methods for Two-Point Boundary-Value Problems* (Blaisdell Publishing Company, Waltham, 1968)
2. U.M. Ascher, R.M.M. Mattheij, R.D. Russell, *Numerical Solution of Boundary Value Problems for Ordinary Differential Equations* (SIAM, Philadelphia, 1995). See also U. Ascher, R.D. Russell, SIAM Review **23**, 238 (1981)

3. E. Isaacson, H.B. Keller, *Analysis of Numerical Methods* (John Wiley & Sons, New York, 1966)
4. P. Henrici, *Discrete Variable Methods in Ordinary Differential Equations* (John Wiley & Sons, New York, 1962)
5. P. Henrici, *Error Propagation for Difference Methods* (John Wiley & Sons, New York, 1963)
6. E.W. Larsen, J.E. Morel, W.F. Miller, Asymptotic solutions of numerical transport problems in optically thick, diffusive regimes. J. Comput. Phys. **69**, 283 (1987)
7. E.W. Larsen, J.E. Morel, Asymptotic solutions of numerical transport problems in optically thick, diffusive regimes II. J. Comput. Phys. **83**, 212 (1989)
8. M. Abramowitz, I.A. Stegun, *Handbook of mathematical functions*, 10th edn. (Dover Publications, Mineola, 1972)
9. D. Knoll, J. Morel, L. Margolin, M. Shashkov, Physically motivated discretization methods. Los Alamos Sci. **29**, 188 (2005)
10. U. Ascher, J. Christiansen, R.D. Russell, Collocation software for boundary-value ODEs. ACM Trans. Math. Softw. **7**, 209 (1981); G. Bader, U. Ascher, A new basis implementation for a mixed order boundary value ODE solver. SIAM J. Sci. Stat. Comput. **8**, 483 (1987)
11. J.R. Cash, M.H. Wright, A deferred correction method for nonlinear two-point boundary value problems: implementation and numerical evaluation. SIAM J. Sci. Stat. Comput. **12**, 971 (1991)
12. W.H. Enright, P.H. Muir, Runge-Kutta software with defect control for boundary value ODEs. SIAM J. Sci. Comput. **17**, 479 (1996)
13. L.F. Shampine, P.H. Muir, H. Xu, A user-friendly fortran BVP solver. J. Numer. Anal. Indust. Appl. Math. **1**, 201 (2006)
14. J.E. Flaherty, *Finite Element Analysis, CSCI, MATH 6860 Lecture Notes* (Rensselaer Polytechnic Institute, Troy, 2000)
15. W.O. Amrein, A.M. Hinz, D.B. Pearson (eds.), *Sturm-Liouville Theory (Past and present* (Birkhäuser Verlag, Basel, 2005)
16. A. Zettl, *Sturm-Liouville Theory* (AMS, Providence, 2005)
17. A. Zettl, *Recent Developments in Sturm-Liouville Theory* (De Gruyter, Berlin, 2021)
18. C.T. Fulton, S.A. Pruess, Eigenvalue and eigenfunction asymptotics for regular Sturm-Liouville problems. J. Math. Anal. Appl. **188**, 297 (1994)
19. Q. Kong, A. Zettl, Eigenvalues of regular Sturm-Liouville problems. J. Diff. Equ. **131**, 1 (1996)
20. J.D. Pryce, *Numerical Solution of Sturm-Liouville Problems* (Clarendon Press, Oxford, 1993)
21. S. Pruess, Estimating the eigenvalues of Sturm-Liouville problems by approximating the differential equation. SIAM J. Numer. Anal. **10**, 55 (1973)
22. J. Walter, Regular eigenvalue problems with eigenvalue parameter in the boundary condition. Math. Z. **133**, 301 (1973)
23. C.T. Fulton, Two-point boundary value problems with eigenparameter contained in the boundary conditions. Proc. R. Soc. Edinb. **77A**, 293 (1977); C.T. Fulton, Singular eigenvalue problems with eigenvalue parameter contained in the boundary conditions. Proc. R. Soc. Edinb. **87A**, 1 (1980)
24. D. Hinton, An expansion theorem for an eigenvalue problem with eigenvalue parameter in the boundary condition. Quart. J. Math. Oxford **2**, 33 (1979); D. Hinton, Eigenfunction expansions for a singular eigenvalue problem with eigenparameter in the boundary condition. SIAM J. Math. Anal. **12**, 572 (1981)
25. P.A. Binding, P.J. Browne, B.A. Watson, Transformations between Sturm-Liouville problems with eigenvalue dependent and independent boundary conditions. Bull. London Math. Soc. **33**, 749 (2001)
26. P.A. Binding, P.J. Browne, B.A. Watson, Sturm–Liouville problems with boundary conditions rationally dependent on the eigenparameter, I. Proc. Edinb. Math. Soc. **45**, 631 (2002); the inverse problem is discussed in P.A. Binding, P.J. Browne, B.A. Watson, Sturm–Liouville problems with boundary conditions rationally dependent on the eigenparameter, II. J. Comput. Appl. Math. **148**, 147 (2002)
27. W.J. Code, P.J. Browne, Sturm-Liouville problems with boundary conditions depending quadratically on the eigenparameter. J. Math. Anal. Appl. **309**, 729 (2005)

28. J. Weidmann, *Spectral Theory of Ordinary Differential Operators Lecture Notes in Mathematics*, vol. 1258 (Springer-Verlag, Berlin, 1987)
29. J.D. Pryce, Classical and vector Sturm-Liouville problems: recent advances in singular-point analysis and shooting-type algorithms. J. Comput. Appl. Math. **50**, 455 (1994)
30. N. Dunford, J.T. Schwarz, *Linear Operators (Spectral theory)* (John Wiley Interscience, New York, Part II, 1988)
31. H. Weyl, Über gewöhnliche Differentialgleichungen mit Singularitäten und die zugehörigen Entwicklungen willkürlicher Funktionen. Math. Ann. **68**, 220 (1910)
32. S. Pruess, C.T. Fulton, Y. Xie, An asymptotic numerical method for a class of singular Sturm-Liouville problems. SIAM J. Numer. Anal. **32**, 1658 (1995)
33. C.T. Fulton, S. Pruess, Y. Xie, The automatic classification of Sturm-Liouville problems. J. Appl. Math. Comput. **124**, 149 (2005)
34. W.N. Everitt, A personal history of the m-coefficient. J. Comput. Appl. Math. **171**, 185 (2004)
35. C.T. Fulton, S. Pruess, The computation of spectral density functions for Sturm-Liouville problems involving simple continuous spectra. ACM Trans. Math. Softw. **24**, 107 (1998)
36. E.C. Titchmarsh, On expansions in eigenfunctions (IV). Quart. J. Math. Oxford **12**, 33 (1941)
37. B.M. Brown, V.G. Kirby, J.D. Pryce, Numerical determination of the Titchmarsh-Weyl m-coefficient and its applications to HELP inequalities. Proc. R. Soc. Lond. **426**, 167 (1989)
38. B.M. Brown, V.G. Kirby, J.D. Pryce, A numerical method for the determination of the Titchmarsh-Weyl m-coefficient. Proc. R. Soc. Lond. **435**, 535 (1991)
39. M.R.M. Witwit, N.A. Gordon, J.P. Killingbeck, Numerical computation and analysis of the Titchmars-Weyl $m_\alpha(\lambda)$ function for some simple potentials. J. Comput. Appl. Math. **106**, 131 (1999)
40. C. Fulton, D. Pearson, S. Pruess, Computing the spectral functions for singular Sturm-Liouville problems. J. Comput. Appl. Math. **176**, 131 (2005)
41. C. Fulton, D. Pearson, S. Pruess, Efficient calculation of spectral density functions for specific classes of singular Sturm-Liouville problems. J. Comput. Appl. Math. **212**, 150 (2008)
42. C. Fulton, D. Pearson, S. Pruess, New characterization of spectral density functions for singular Sturm-Liouville problems. J. Comput. Appl. Math. **212**, 194 (2008)
43. C. Fulton, D. Pearson, S. Pruess, Estimating spectral density functions for Sturm-Liouville problems with two singular endpoints. IMA J. Numer. Anal. **34**, 609 (2014)
44. C.T. Fulton, Titchmarsh-Weyl m-functions for second-order Sturm-Liouville problems with two singular endpoints. Math. Nachr. **281**, 1418 (2008)
45. D.B. Hinton, J.K. Shaw, On Titchmarsh-Weyl $M(\lambda)$-functions for linear Hamiltonian systems. J. Differ. Equ. **40**, 316 (1981)
46. D.B. Hinton, J.K. Shaw, Hamiltonian systems of limit point or limit circle type with both endpoints singular. J. Differ. Equ. **50**, 444 (1983)
47. D.B. Hinton, J.K. Shaw, On boundary value problems for Hamiltonian systems with two singular points. SIAM J. Math. Anal. **15**, 272 (1984)
48. D.B. Hinton, J.K. Shaw, On the spectrum of a singular Hamiltonian system. Quaest. Math. **5**, 29 (1982)
49. D.B. Hinton, J.K. Shaw, On the spectrum of a singular Hamiltonian system. II. Quaest. Math. **10**, 1 (1986)
50. P.B. Bailey, W.N. Everitt, A. Zettl, Algorithm 810: the SLEIGN2 Sturm–Liouville code. ACM Trans. Math. Softw. **27**, 143 (2001); the code can be downloaded from: https://www.niu.edu/math/sl2/index.shtml
51. P.B. Bailey, B.S. Garbow, H.G. Kaper, A. Zettl, Eigenvalue and eigenfunction computations for Sturm-Liouville problems. ACM Trans. Math. Softw. **17**, 491 (1991)
52. P.B. Bailey, B.S. Garbow, H.G. Kaper, A. Zettl, Algorithm 700: A FORTRAN software package for Sturm-Liouville problems. ACM Trans. Math. Softw. **17**, 500 (1991)
53. P.B. Bailey, M.K. Gordon, L.F. Shampine, Automatic solution of the Sturm-Liouville problem. ACM Trans. Math. Softw. **4**, 193 (1978)
54. M. Marletta, J.D. Pryce, LCNO Sturm-Liouville problems: computational difficulties and examples. Numer. Math. **69**, 303 (1995)

55. J.D. Pryce, M. Marletta, A new multi-purpose software package for Schrödinger equation and Sturm-Liouville computations. Comput. Phys. Commun. **62**, 42 (1991)
56. M. Marletta, J.D. Pryce, Automatic solution of Sturm-Liouville problems using the Pruess method. J. Comput. Appl. Math **39**, 57 (1992)
57. Numerical Algorithms Group. http://www.nag.com
58. S. Pruess, C.T. Fulton, Mathematical software for Sturm-Liouville problems. ACM Trans. Math. Softw. **19**, 360 (1993)
59. M.S.P. Eastham, C.T. Fulton, S. Pruess, Using the SLEDGE package on Sturm-Liouville problems having nonempty essential spectra. ACM Trans. Math. Softw. **22**, 423 (1996)
60. J.D. Pryce, A test package for Sturm-Liouville solvers. ACM Trans. Math. Softw. **25**, 21 (1999)
61. J.D. Pryce, Algorithm 789: SLTSTPAK: a test package for Sturm-Liouville solvers. ACM Trans. Math. Softw. **25**, 58 (1999)
62. V. Ledoux, M. Van Daele, G. Vanden Berghe, MATSLISE: a MATLAB package for the numerical solution of Sturm–Liouville and Schrödinger equations. ACM Trans. Math. Softw. **31**, 532 (2005)
63. M.E. Davis, *Numerical Methods and Modeling for Chemical Engineers* (John Wiley & Sons, New York, 1984)
64. M.H. Holmes, *Introduction to Numerical Methods in Differential Equations* (Springer-Verlag, New York, 2007)
65. G. Vanden Berghe, M. Van Daele, H. De Meyer, A five-diagonal finite difference method based on mixed-type interpolation for computing eigenvalues of fourth-order two-point boundary-value problems. J. Comput. Appl. Math. **41**, 359 (1992)
66. D.J. Jones, Use of a shooting method to compute eigenvalues of fourth-order two-point boundary value problems. J. Comput. Appl. Math. **47**, 395 (1993)
67. B.M. Brown, D.K.R. McCormack, W.D. Evans, M. Plum, On the spectrum of second-order differential operators with complex coefficients. Proc. R. Soc. Lond. A **455**, 1235 (1999)
68. A.A. Abramov, A. Aslanyan, E.B. Davies, Bounds on complex eigenvalues and resonances. J. Phys. A: Math. Gen. **34**, 57 (2001)
69. E.B. Davies, Pseudo-spectra, the harmonic oscillator and complex resonances. Proc. R. Soc. Lond. A **455**, 585 (1999)
70. L. Greenberg, M. Marletta, Numerical solution of non-self-adjoint Sturm-Liouville problems and related systems. SIAM J. Numer. Anal. **38**, 1800 (2001)

Chapter 10
Difference Methods for One-Dimensional PDE

Abstract Finite-difference methods for one-dimensional partial differential equations are introduced by first identifying classes of equations upon which suitable discretizations are constructed. It is shown how parabolic equations and their boundary conditions are discretized such that a desired local order of error is achieved and that the discretization is consistent and yields a stable and convergent solution scheme. Convergence criteria are established for a variety of explicit and implicit difference schemes. Difference schemes for hyperbolic equations are introduced from the standpoint of the Courant–Friedrichs–Lewy criterion, dispersion and dissipation. Various techniques for non-linear equations and equations of mixed type are given, including high-resolution schemes for equations that can be expressed in terms of conservation laws. The Problems include the (parabolic) diffusion and (hyperbolic) advection equation, Burgers equation, the shock-tube problem, Korteweg–de Vries equation, and the non-stationary linear and cubic Schrödinger equations.

Solving partial differential equations (PDE) is so crucial to mathematical physics that we devote three chapters to it. The solution methods largely depend on the type of the PDE. The type of a general system of linear first-order equations

$$A\boldsymbol{v}_x + B\boldsymbol{v}_y = \boldsymbol{c}\,,$$

where $\boldsymbol{v} = (v_1, v_2, \ldots, v_M)^{\mathrm{T}}$ is the solution vector, $\boldsymbol{c} = (c_1, c_2, \ldots, c_M)^{\mathrm{T}}$ is the vector of inhomogeneous terms, and A and B are $M \times M$ matrices, is determined by the zeros of the characteristic polynomial $\rho(\lambda) = \det(A - \lambda B)$. The system is *hyperbolic* if $\rho(\lambda)$ has precisely M distinct zeros, or if it has M real zeros and the system $(A - \lambda B)\boldsymbol{u} = \boldsymbol{0}$ possesses precisely M linearly independent solutions. The system is *parabolic* if $\rho(\lambda)$ has M real zeros but $(A - \lambda B)\boldsymbol{u} = \boldsymbol{0}$ does not have M linearly independent solutions. The system without real zeros of $\rho(\lambda)$ is *elliptic*. (The classification of systems with mixed real and complex zeros of $\rho(\lambda)$ is more involved.) For second-order linear PDE of the general form

$$Lv = av_{xx} + bv_{xy} + cv_{yy} + dv_x + ev_y + fv = Q\,, \tag{10.1}$$

where we seek the scalar solution $v(x, y)$ on some domain on the (x, y)-plane at given initial and boundary conditions, the classification is simpler: it is determined already by the discriminant $\mathcal{D} = b^2 - 4ac$.

Hyperbolic equations PDE of the hyperbolic type ($\mathcal{D} > 0$) describe time-dependent conservative processes that do not evolve towards a stationary state, for example, the propagation of waves. We will learn the basic numerical approaches on the case of the first-order advection equation

$$v_t \pm c v_x = 0$$

and the second-order wave equation

$$v_{tt} = c^2 v_{xx} \ .$$

We will be interested in the dissipation of the numerical solution (diminishing magnitude of the solution or its Fourier components) and its dispersion (different propagation velocity of Fourier components with different wave-vectors). Among other, we use high-order non-linear hyperbolic equations to model atmospheric phenomena and solve complex aerodynamic problems.

Parabolic equations The equations of the parabolic type ($\mathcal{D} = 0$) describe time-dependent dissipative processes that evolve to the stationary final state. We will study the fundamentals on the case of the diffusion equation

$$v_t = D \nabla^2 v$$

in one, two, and three spatial dimensions, which we use to model the diffusion of substances or heat in matter. We shall pay some attention to peculiarities that occur in the presence of non-linearities.

Elliptic equations The equations of the elliptic type ($\mathcal{D} < 0$) describe systems in equilibrium or stationary states, and their solutions (in the context of variational calculus) usually maximize or minimize the integrals representing the energy of the system. Among the best known is the Poisson equation

$$\nabla^2 v = \rho \ ,$$

which lends itself for the study of stationary heat currents or slow curl-free currents of inviscid and incompressible fluids. Of course, we encounter it time and again in the studies of electrostatics and gravitation.

The classification of PDE is not merely academic. Frequently we meet equations that have different types in different parts of the definition domain. An example is the Euler-Tricomi equation

$$v_{xx} = x v_{yy} \ ,$$

which is relevant for the study of motion of bodies in matter with velocities that approach the speed of sound in that matter. This equation is hyperbolic at $x > 0$, parabolic at $x = 0$, and elliptic at $x < 0$. For mixed-type equations, the numerical methods often need to be tailored uniquely. Further problems appear in non-linear PDE, as in the Burgers equation used to describe the propagation of shock waves,

$$v_t + v v_x = D v_{xx} ,$$

where the dominant regime (hyperbolic or parabolic) is determined by the parameter D. On the other hand, the numerical method itself may leave its footprint in the character of the solution: this danger lurks in difference schemes for hyperbolic equations, where parabolic components often steal their way into the solution (see Sect. 10.7).

In this Chapter we discuss *difference methods* for one-dimensional PDE (one space coordinate). Retracing the footsteps of Ref. [1], we learn the basics by studying the parabolic problem $v_t - D v_{xx} = Q$ in Sects. 10.1–10.5. As for ordinary differential equations, solving PDE by difference schemes requires us to grasp three basic concepts: the *consistency* of the difference scheme with the differential equation, the *stability* of the scheme, and the *convergence* of the numerical solution to the true solution. Later we discuss hyperbolic and elliptic PDE.

10.1 Discretization of the Differential Equation

Here we set up difference methods for parabolic PDE involving two independent variables t and x that most often represent the time and space coordinates. Our prototype problem is the linear diffusion equation

$$v_t - D v_{xx} = Q . \tag{10.2}$$

We shall distinguish pure initial-value problems, e.g. on $x \in \mathbb{R}$ with the initial condition $v(x, 0) = f(x)$, and initial-boundary-value problems, e.g. on $x \in [0, 1]$ with the same initial condition and additional boundary conditions. An example of the former is the time dependence of the temperature of a thin, infinitely long wire; an example of the latter is a thin rod of finite length with both ends held at constant temperature.

Initial-boundary-value problems are more relevant. We discretize each of the continuous variables separately (see Fig. 10.1) by forming a two-dimensional mesh along which we trace the time evolution of the initial condition. We also have to discretize the initial and boundary conditions (see Sect. 10.2). A unique solution of Eq. (10.2) exists for pure Dirichlet or mixed Dirichlet–Neumann boundary conditions. For pure Neumann conditions the solution is determined only up to an additive constant: if $v(x, y)$ solves the equation, so does $v(x, y) + C$.

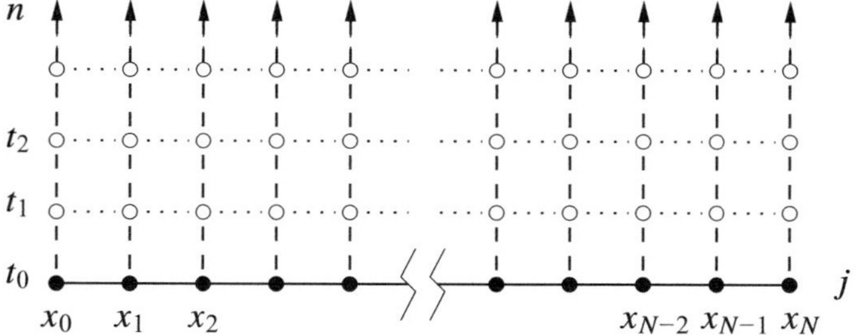

Fig. 10.1 The uniform mesh for the solution of initial-boundary-value problems for evolution (parabolic and hyperbolic) PDE in one temporal and one spatial dimension on the interval $[x_0, x_N]$. The discretization Δt and $\Delta x = (x_N - x_0)/N$ determines the mesh points $(x_j, t_n) = (j\Delta x, n\Delta t)$, and at each of them the true solution v_j^n and its approximation u_j^n. We arrange the solutions in the vector $\boldsymbol{u}^n = (u_0^n, u_1^n, \ldots, u_N^n)^{\mathrm{T}}$

On this mesh we construct various finite differences, for example

$$\left.\frac{\partial v}{\partial x}\right|_{(j\Delta x, n\Delta t)} = \frac{v_{j+1}^n - v_j^n}{\Delta x} + O(\Delta x)\,,$$

$$\left.\frac{\partial v}{\partial x}\right|_{(j\Delta x, n\Delta t)} = \frac{v_j^n - v_{j-1}^n}{\Delta x} + O(\Delta x)\,,$$

$$\left.\frac{\partial v}{\partial x}\right|_{(j\Delta x, n\Delta t)} = \frac{v_{j+1}^n - v_{j-1}^n}{2\Delta x} + O(\Delta x^2)\,, \tag{10.3}$$

$$\left.\frac{\partial v}{\partial t}\right|_{(j\Delta x, n\Delta t)} = \frac{v_j^{n+1} - v_j^n}{\Delta t} + O(\Delta t)\,, \tag{10.4}$$

$$\left.\frac{\partial v}{\partial t}\right|_{(j\Delta x, n\Delta t)} = \frac{v_j^{n+1} - v_j^{n-1}}{2\Delta t} + O(\Delta t^2)\,, \tag{10.5}$$

$$\left.\frac{\partial^2 v}{\partial x^2}\right|_{(j\Delta x, n\Delta t)} = \frac{v_{j+1}^n - 2v_j^n + v_{j-1}^n}{\Delta x^2} + O(\Delta x^2)\,, \tag{10.6}$$

$$\left.\frac{\partial^2 v}{\partial t^2}\right|_{(j\Delta x, n\Delta t)} = \frac{v_j^{n+1} - 2v_j^n + v_j^{n-1}}{\Delta t^2} + O(\Delta t^2)\,. \tag{10.7}$$

(Note that expressions (10.4)–(10.7) tell us how well the individual parts of the difference schemes approximate the exact partial derivatives in the differential equation, but this has only an indirect relation to how well the solution u of the difference scheme approximates the true solution v of the differential equation.) In writing down the differences we use the shorthand notation

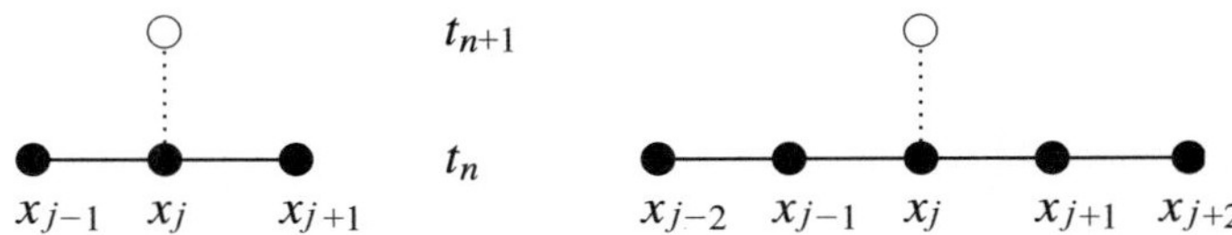

Fig. 10.2   [LEFT] The FTCS scheme with the spatial difference $\Delta_2^{(x)}$ defined by Eq. (10.11). [RIGHT] The FTCS scheme with the difference $\Delta_4^{(x)}$ (Eq. (10.12)) for which two additional (numerical) boundary conditions at the points x_{-1} and x_{N+1} are needed

$$\Delta_+^{(x)} u_j = u_{j+1}^n - u_j^n \, , \tag{10.8}$$

$$\Delta_-^{(x)} u_j = u_j^n - u_{j-1}^n \, , \tag{10.9}$$

$$\Delta_0^{(x)} u_j = u_{j+1}^n - u_{j-1}^n \, , \tag{10.10}$$

$$\Delta_2^{(x)} u_j = u_{j+1}^n - 2u_j^n + u_{j-1}^n \, , \tag{10.11}$$

occasionally also

$$\Delta_4^{(x)} u_j = -\frac{1}{12} u_{j+2}^n + \frac{4}{3} u_{j+1}^n - \frac{5}{2} u_j^n + \frac{4}{3} u_{j-1}^n - \frac{1}{12} u_{j-2}^n \, . \tag{10.12}$$

We can therefore discretize the one-dimensional linear diffusion Eq. (10.2) as

$$u_j^{n+1} = u_j^n + r \left(u_{j+1}^n - 2u_j^n + u_{j-1}^n \right) + \Delta t \, q_j^n \, , \tag{10.13}$$

where we have denoted $r = D\Delta t/\Delta x^2$ and $q_j^n = Q(j\Delta x, n\Delta t)$. This gives us an explicit method of order $O(\Delta t) + O(\Delta x^2)$ which advances one step in time by using the symmetric spatial difference $\Delta_2^{(x)}$, thus we name it *Forward-Time, Centered-Space* (FTCS, Fig. 10.2 (left)). The scheme is stable for $0 < r \leq 1/2$ (we will say more on stability in Sect. 10.5). This method is just one of the possible explicit *two-level* difference methods of the form $\boldsymbol{u}^{n+1} = F\boldsymbol{u}^n + \Delta t \, \boldsymbol{q}^n$.

10.2   Discretization of Initial and Boundary Conditions

A consistent discretization of the initial and boundary conditions is essential for a correct numerical solution of PDE. If either of these conditions are discretized to an order lower than the order of the difference scheme, the solution also has a lower order. (In general, the order of the numerical solution does not exceed the lowest order of any discrete approximation in the formulation of the problem.) The importance of boundary conditions for the solution in the interior of the definition domain is illustrated in the following and in Fig. 10.3.

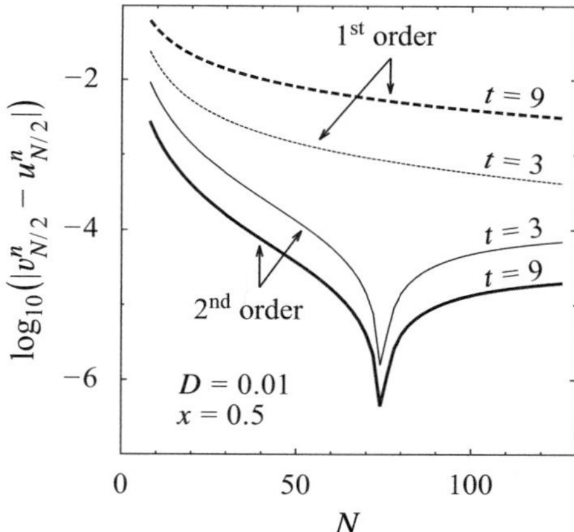

Fig. 10.3 The error of the numerical solution of $v_t = Dv_{xx}$ by the FTCS scheme on $x \in [0, 1]$ for $t > 0$ with the initial condition $v(x, 0) = \sin^2 \pi x$ and boundary conditions $v_x(0, t) = v_x(1, t) = 0$, and $D = 0.01$. Shown is the error at $x = 0.5$ in dependence of the discretization ($\Delta x = 1/N$) for two different treatments of the boundary conditions: to first order (10.17) or to second order (10.18), at two different times $t = n\Delta t$

In the scheme for the one-dimensional diffusion equation, the easiest boundary conditions to implement are Dirichlet: imagine a rod with a prescribed temperature dependence at its endpoints ($x_0 = 0$ and $x_N = 1$), for example, due to heating and cooling, independent of the inhomogeneous term Q. We discretize the initial condition $v(x, 0) = f(x)$ as

$$u_j^0 = f(j\Delta x), \qquad j = 0, 1, \ldots, N, \tag{10.14}$$

and the boundary conditions $v(0, t) = a(t)$ and $v(1, t) = b(t)$ as

$$u_0^{n+1} = a((n + 1)\Delta t) = a^{n+1}, \qquad n = 0, 1, \ldots, N, \tag{10.15}$$
$$u_N^{n+1} = b((n + 1)\Delta t) = b^{n+1}, \qquad n = 0, 1, \ldots, N. \tag{10.16}$$

Homogeneous Dirichlet conditions ($u_0^n = u_N^n = 0$) are a special case: they describe a rod whose ends are in contact with a heat reservoir at zero temperature.

For one complete time step in the FTCS scheme both (10.15) and (10.16) are essential. Namely, we use Eq. (10.13) at $n = 0$ to compute the values u_j^1 for $j = 1, 2, \ldots, N - 1$ from the initial condition (10.14), but not for $j = 0$ and $j = N$, since in this case the second difference at the right of (10.13) would reach for indices outside of the definition domain ($j = -1$ and $j = N + 1$, Fig. 10.2 (left)). In one time step we therefore map $N + 1$ values in the space coordinate to the next $N + 1$ values: the scheme (10.13) contains $N - 1$ equations, and the missing equations are (10.15) and (10.16).

We face a different situation with Neumann boundary conditions which fix the derivative at the endpoints, like $v_x(0, t) = c(t)$, $v_x(1, t) = d(t)$. Such conditions prescribe the heat current density from the rod or into it. (The homogeneous condition $c(t) = 0$ means that the rod is thermally isolated at the left end.) At $x = x_0 = 0$ we get, to first order,

$$\frac{u_1^{n+1} - u_0^{n+1}}{\Delta x} = c^{n+1} , \tag{10.17}$$

thus $u_0^{n+1} = u_1^{n+1} - \Delta x\, c^{n+1}$ (and analogously at $x = x_N = 1$). Since the FTCS scheme is second-order in the space coordinate, such a crude approximation may spoil the complete numerical solution (see Fig. 10.3). It is preferable to use the symmetric difference

$$\frac{u_1^{n+1} - u_{-1}^{n+1}}{2\Delta x} = c^{n+1} ,$$

but this formula reaches to a *ghost point* $x_{-1} = x_0 - \Delta x$ outside of the mesh. This additional point implies that we are dealing with $N + 2$ unknowns, while having only $N + 1$ equations at our disposal. But we can write the FTCS scheme at $(0, n + 1)$ by including the symmetric difference in the form $u_{-1}^n = u_1^n - 2\Delta x\, c^n$, and thereby eliminate the point x_{-1}. At $x = 0$ we get

$$\begin{aligned}
u_0^{n+1} &= u_0^n + r\left(u_1^n - 2u_0^n + u_{-1}^n\right) \\
&= u_0^n + 2r\left(u_1^n - u_0^n\right) - 2r\Delta x\, c^n
\end{aligned} \tag{10.18}$$

(and similarly at $x = 1$). Equation (10.13) and two boundary conditions therefore again constitute $N + 1$ equations for $N + 1$ variables, and all discrete approximations remain consistently of second order in the space coordinate. The advantage of the symmetric difference becomes clear from the example in Fig. 10.3.

10.3 Consistency $\star$

We define the consistency of the differential equation and the corresponding difference scheme in analogy to the ordinary differential equations (Chap. 8). The equation $Lv = Q$ and the scheme $L_j^n u_j^n = q_j^n$ are *point-wise consistent* if for each function $\phi(x, t)$ (even for the true solution v) at the point (x, t) we have

$$(L\phi - Q)_j^n - \left[L_j^n \phi(j\Delta x, n\Delta t) - q_j^n\right] \to 0 ,$$

when $\Delta x \to 0$, $\Delta t \to 0$, and $(j\Delta x, (n + 1)\Delta t) \to (x, t)$. Instead of the point-wise consistency, we sometimes define a more strict *consistency in norm*: the equation that is first-order in time and the explicit two-level scheme

$$\boldsymbol{u}^{n+1} = F\boldsymbol{u}^n + \Delta t\boldsymbol{q}^n \tag{10.19}$$

are consistent when the true solution of the equation v satisfies

$$v^{n+1} = Fv^n + \Delta t q^n + \Delta t \tau^n ,$$

where the discretization error $\|\tau^n\| \to 0$ when $\Delta x \to 0$, $\Delta t \to 0$ (the vector u^n is defined in the caption to Fig. 10.1, and v^n has the components $v(j\Delta x, n\Delta t)$). We say that the difference scheme is of order (p, q) when $\|\tau^n\| = O(\Delta x^p) + O(\Delta t^q)$.

Up to which order in time and space coordinate is the FTCS scheme consistent with the diffusion equation? We Taylor-expand the expressions

$$v_j^{n+1} = v_j^n + \Delta t (v_t)_j^n + \frac{\Delta t^2}{2}(v_{tt})_j^n + O(\Delta t^3) ,$$

$$v_{j\pm 1}^n = v_j^n \pm \Delta x (v_x)_j^n + \frac{\Delta x^2}{2}(v_{xx})_j^n \pm \frac{\Delta x^3}{6}(v_{xxx})_j^n$$
$$+ \frac{\Delta x^4}{24}(v_{xxxx})_j^n \pm \frac{\Delta x^5}{120}(v_{xxxxx})_j^n + O(\Delta x^6) ,$$

and compute the difference between the differential equation and its difference approximation (10.13). We have $v_t = D v_{xx}$ and $v_{tt} = D(v_{xx})_t = D^2 v_{xxxx}$, whence

$$[v_t - D v_{xx}]_j^n - \left[\frac{\Delta_+^{(t)}}{\Delta t} - D\frac{\Delta_2^{(x)}}{\Delta x^2}\right] v_j^n = \frac{D}{2}\left[\frac{\Delta x^2}{6} - D\Delta t\right](v_{xxxx})_j^n + O(\Delta t^2) + O(\Delta x^4) .$$

We see that with a general $r = D\Delta t / \Delta x^2$, the equation and its difference approximation are point-wise consistent to order $O(\Delta t) + O(\Delta x^2)$, while in the special case $r = 1/6$ they are consistent even to $O(\Delta t^2) + O(\Delta x^4)$ (see Fig. 10.4). The local discretization error in the FTCS scheme can therefore be strongly reduced if we fix $\Delta t = \Delta x^2/(6D)$ once the spatial discretization Δx has been decided for. Alas, with such short steps Δt, the solution evolves very slowly: it takes $6N^2$ steps for it to diffuse from one end of the interval to another.

Consistency in norm requires us to distinguish between initial-value and initial-boundary-value problems. In the "sup"-norm (Appendix A) the initial-value problem for the diffusion equation on $x \in \mathbb{R}$ for $t > 0$ and the FTCS scheme are consistent to order $O(\Delta t) + O(\Delta x^2)$, if v_{tt} and v_{xxxx} are uniformly bounded on $\mathbb{R} \times [0, t_0]$ for some $t_0 > t$. If we can restrict $(\|v_{tt}^n\|_{2,\Delta x})^2 < A < \infty$ and $(\|v_{xxxx}^n\|_{2,\Delta x})^2 < B < \infty$, they are also consistent in the $l_{2,\Delta x}$ norm [1].

In initial-boundary-value problems, the order of consistency depends on the discretization of the boundary conditions. For homogeneous Dirichlet conditions, the FTCS scheme on $[0, 1]$ is point-wise consistent to order $O(\Delta t) + O(\Delta x^2)$ as well as in the "sup" and $l_{2,\Delta x}$ norms (Fig. 10.4). Things are different if we have, say, a homogeneous Dirichlet condition $u_N^{n+1} = 0$ at $x = 1$, and a homogeneous Neumann condition at $x = 1$. If the latter is discretized to second order (as in (10.18) with $c^n = 0$), the scheme and the equation are consistent only to order $O(\Delta t) + O(\Delta x)$ in "sup" and $l_{2,\Delta x}$ norms: one order less in spite of the effort! Worse still, discretiz-

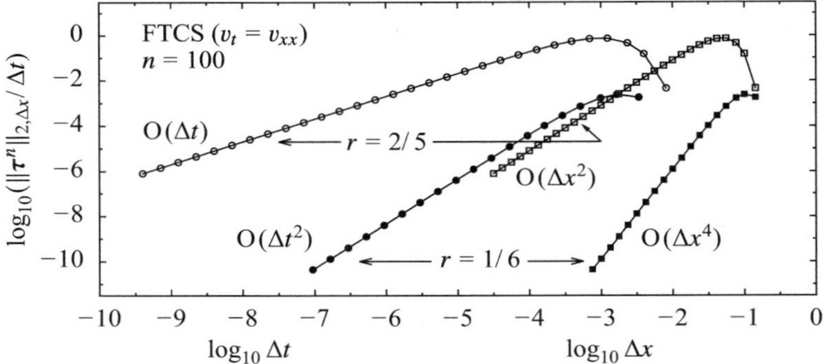

Fig. 10.4 The order of consistency (and convergence) in the $\|\cdot\|_{2,\Delta x}$ norm, of the FTCS scheme for the diffusion equation $v_t = Dv_{xx}$ on $[0, 1]$ with the initial condition $v(x, 0) = \sin \pi x$ and Dirichlet boundary conditions $v(0, t) = v(1, t) = 0$. Shown are two pairs of curves for $r = 2/5$ (asymptotic order $O(\Delta t) + O(\Delta x^2)$) and $r = 1/6$ (asymptotic order $O(\Delta t^2) + O(\Delta x^4)$). The figure for $\|\cdot\|_\infty$ closely resembles this one

ing it to first order (as in (10.17) with $c^n = 0$), we fail to attain even first-order consistency, since the local discretization error becomes [1]

$$\tau_1^n = -\frac{D}{2}\,(v_{xx})_1^n + O(\Delta t) + O(\Delta x)\,.$$

We can avoid this problem by shifting the mesh for $\Delta x/2$ (*offset grid*): we discretize the spatial coordinate as

$$x_j = (j - 1)\Delta x + \Delta x/2\,, \qquad j = 0, 1, \ldots, N\,,$$

hence $x_0 = -\Delta x/2$ and $x_1 = \Delta x/2$. This gives us a more natural mesh for the homogeneous Neumann boundary condition $u_0^{n+1} = u_1^{n+1}$ at $x = 0$, and makes the scheme consistent with the equation to order $O(\Delta t) + O(\Delta x)$.

10.4 Implicit Schemes

Similarly as in ordinary differential equations (Chap. 8), the stability properties of difference schemes for PDE improve when we resort to implicit methods. Instead of computing the second difference in the space coordinate and the source term at time $n\Delta t$ (as in FTCS), we compute it at time $(n + 1)\Delta t$, yielding the implicit two-level *Backward-Time, Centered-Space* (BTCS) scheme of order $O(\Delta t) + O(\Delta x^2)$,

$$u_j^{n+1} = u_j^n + r\left(u_{j+1}^{n+1} - 2u_j^{n+1} + u_{j-1}^{n+1}\right) + \Delta t q_j^{n+1}\,, \tag{10.20}$$

which is absolutely stable for any r. Similarly, by using the averages of the BTCS and FTCS expressions at the right-hand side of the equation,

$$u_j^{n+1} = u_j^n + \frac{r}{2}\,\Delta_2^{(x)}\left(u_j^{n+1} + u_j^n\right) + \frac{\Delta t}{2}\left(q_j^{n+1} + q_j^n\right)\,, \tag{10.21}$$

we have formed the Crank–Nicolson method of order $O(\Delta t^2) + O(\Delta x^2)$, which is absolutely stable. The notion of "absolute stability" should be taken with a grain of salt: see Sect. 10.5, then the discussion in Appendix I, Eq. (I.4).

What is the most sensible way to implement the *temporal* discretization of the differential equation? We should be particularly cautious in the study of reaction-diffusion processes described by the equations of the form

$$v_t = Dv_{xx} + bv\,,$$

especially in cases where the characteristic time scale of the diffusion part of the process, $\tau_d = L^2/D$, and the scale of the reaction part, $\tau_r = |1/b|$, are very different. (Here L is the distance on which the solution gradient changes substantially.) Three obvious ways to perform an implicit discretization of this equation are:

$$\frac{u_j^\star - u_j^n}{\Delta t} = bu_j^\star\,, \qquad \frac{u_j^{n+1} - u_j^\star}{\Delta t} - D\Delta_2^{(x)}u_j^{n+1} = 0\,,$$

$$\frac{u_j^{n+1} - u_j^n}{\Delta t} - D\Delta_2^{(x)}u_j^{n+1} = bu_j^{n+1}\,,$$

$$\frac{u_j^{n+1} - u_j^n}{\Delta t} - D\Delta_2^{(x)}\tfrac{1}{2}\left[u_j^{n+1} + u_j^n\right] = b\,\tfrac{1}{2}\left[u_j^{n+1} + u_j^n\right]\,.$$

Here we do not discuss the nuances of these schemes, but the reader should be aware that there are important distinctions between their solutions [2].

10.5   Stability and Convergence ★

The consistency of the difference scheme $L_j^n u_j^n = q_j^n$ with the differential equation $Lv = Q$ by itself does not guarantee that the solution of the difference equation converges to the solution of the differential equation in the sense

$$\|u^{n+1} - v^{n+1}\| = O(\Delta x^p) + O(\Delta t^q)\,.$$

For convergence, the scheme also needs to be *stable*. As in the initial-value problems for ordinary differential equations, this means that small errors in the initial condition map to small errors in the solution.

The concepts of consistency, stability, and convergence are linked by the Lax theorem [1]. It tells us that the two-level scheme (10.19) which is consistent with the differential equation to order $O(\Delta t^q) + O(\Delta x^p)$ (in some norm) and is stable in this norm, is also convergent in the same order. We should understand this as a warning. We can envision many discretizations and apparently promising difference schemes for any given PDE, but only some of them are stable.

The difference scheme (10.19) is stable in the norm $\| \cdot \|$ when such constants $\Delta x_0 > 0$, $\Delta t_0 > 0$, $\Lambda \geq 0$, and $\beta \geq 0$ exist that

$$\|u^{n+1}\| \leq \Lambda e^{\beta t}\|u^0\| \tag{10.22}$$

for $0 < \Delta x \leq \Delta x_0$, $0 < \Delta t \leq \Delta t_0$, and $0 \leq (n+1)\Delta t = t$. (The inhomogeneous term $\Delta t q^n$ contributes only to the discretization error $\Delta t \tau^n$, so the definition used here applies both to homogeneous and inhomogeneous schemes. For the remainder of this Section we set $q = 0$.) Therefore, the solutions are allowed to grow, but not faster than exponentially. Let us stress the message of Eq. (10.22): a stable solution may grow with time, but not with the number of time steps.

10.5.1 Initial-Value Problems

For the stability analysis of difference schemes for initial-value problems with PDE we rely on the discrete Fourier transformation. (By using this tool we already wade into spectral methods for PDE discussed in Chap. 12.) Following [1] we merge the explicit scheme (FTCS), the implicit scheme (BTCS), and the Crank–Nicolson (CN) implicit scheme for the diffusion equation to a single expression

$$
\begin{aligned}
- ar u_{j-1}^{n+1} &+ (1 + 2ar)u_j^{n+1} - ar u_{j+1}^{n+1} \\
&= (1-a)r u_{j-1}^n + [1 - 2(1-a)r]\, u_j^n + (1-a)r u_{j+1}^n .
\end{aligned} \tag{10.23}
$$

The value $a = 0$ corresponds to FTCS, the value $a = 1$ to BTCS, while $a = 1/2$ gives the Crank–Nicolson scheme. By using the discrete Fourier transformation $\widetilde{u}^n(\xi) = \sum_j e^{-ij\xi} u_j^n$ we get

$$
\begin{aligned}
- ar e^{-i\xi} \widetilde{u}^{n+1}(\xi) &+ (1 + 2ar)\widetilde{u}^{n+1}(\xi) - ar e^{i\xi}\widetilde{u}^{n+1}(\xi) \\
&= (1-a)r e^{-i\xi}\widetilde{u}^n(\xi)\, [1 - 2(1-a)r]\,\widetilde{u}^n(\xi)(1-a)r e^{i\xi}\widetilde{u}^n(\xi)
\end{aligned}
$$

or

$$
\left(1 + 4ar\sin^2\frac{\xi}{2}\right)\widetilde{u}^{n+1}(\xi) = \left(1 - 4(1-a)r\sin^2\frac{\xi}{2}\right)\widetilde{u}^n(\xi) .
$$

We write this as $\widetilde{u}^{n+1}(\xi) = \rho(\xi)\widetilde{u}^n(\xi)$, whence we read off the ratio between the transform at time $(n+1)\Delta t$ and the transform at $n\Delta t$. The ratio

$$\rho(\xi) = \frac{1 - 4(1-a)r\sin^2\frac{\xi}{2}}{1 + 4ar\sin^2\frac{\xi}{2}} \qquad (10.24)$$

is called the *symbol* of the difference scheme. We can repeat this step for all subsequent times, hence $\widetilde{u}^{n+1}(\xi) = [\rho(\xi)]^{n+1}\widetilde{u}^0(\xi)$. Obviously the scheme is stable by definition (10.22) if its symbol is bounded at least as

$$|\rho(\xi)| \le 1 + C\Delta t \le e^{C\Delta t} \qquad (10.25)$$

(Neumann criterion). The one single numerical value of the symbol elegantly "warns" us that during the time evolution one of the Fourier components is spinning out of control. For the FTCS, BTCS, and CN schemes we establish the stability criterion by seeking the maxima of $\partial\rho(\xi)/\partial\xi = 0$. For $a \ge 1/2$ the scheme (10.23) is unconditionally stable, while for $a < 1/2$ it is conditionally stable, with the condition

$$r = D\frac{\Delta t}{\Delta x^2} \le \frac{1}{2(1-2a)} \, . \qquad (10.26)$$

The stability regions of the FTCS and Crank–Nicolson schemes in dependence of the Fourier parameter ξ for different values of r are shown in Fig. 10.5. We will shed a different light on the limiting case of $r = 1/2$ for the FTCS scheme in the discussion of dispersion and dissipation (Sect. 10.9).

Does the FTCS method remain stable if a convection term is added to the diffusion equation, $v_t + cv_x = Dv_{xx}$ with the parameter $c < 0$? The corresponding scheme is

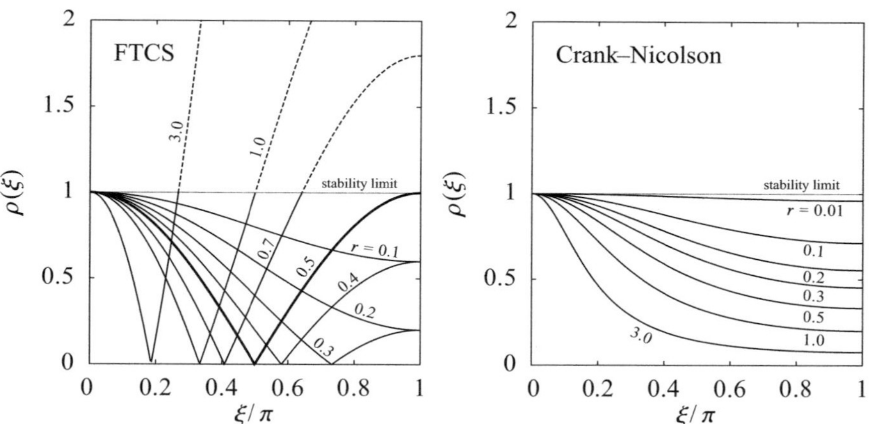

Fig. 10.5 [LEFT] The absolute value of the symbol $\rho(\xi)$ of the conditionally stable FTCS scheme and [RIGHT] of the absolutely stable Crank–Nicolson scheme for the diffusion equation, as a function of the Fourier parameter ξ at different $r = D\Delta t/\Delta x^2$. The stability limit and the symbols in the unstable region are denoted by dotted lines

$$u_j^{n+1} = u_j^n + r\Delta_2^{(x)}u_j^n - \frac{R}{2}\Delta_0^{(x)}u_j^n ,$$

where $r = D\Delta t/\Delta x^2$ and $R = c\Delta t/\Delta x$. If $r^2 \geq R^2/4$, the parabolic character of the equation is more pronounced; in the opposite case, its hyperbolic nature prevails. The symbol of the scheme $\rho(\xi) = (1 - 2r) + 2r\cos\xi - iR\sin\xi$ is complex, with the absolute value

$$|\rho(\xi)|^2 = (1 - 2r)^2 + R^2 + 4r(1 - 2r)\cos\xi + (4r^2 - R^2)\cos^2\xi .$$

A short calculation [1] shows that we may again choose $\Lambda = 1$, $\beta = C = 0$, and thus bound $|\rho(\xi)| \leq 1$ and achieve conditional stability, with the condition

$$R^2/2 \leq r \leq 1/2 \tag{10.27}$$

(check it). What about the diffusion equation in the form $v_t = Dv_{xx} + bv$ with the parameter $b > 0$? The FTCS scheme with all terms included is

$$u_j^{n+1} = u_j^n + r\Delta_2^{(x)}u_j^n + b\Delta t u_j^n ,$$

and has the symbol

$$\rho(\xi) = \left(1 - 4r\sin^2\frac{\xi}{2}\right) + b\Delta t .$$

Now we must choose $\beta = C \neq 0$ and allow the solution to grow with time, although slower than exponentially. Whenever terms with zeroth derivatives are added, one may expect similar behavior. The stability criterion with $\beta \neq 0$ only ensures that the numerical solution converges to the exact solution when $\Delta t \to 0$ and $\Delta x \to 0$. It does *not* guarantee that this occurs for *any* Δt and Δx.

10.5.2 Initial-Boundary-Value Problems

The stability of schemes for initial-boundary-value problems can be approached by using two tools: operator norms and Fourier transformation. Any two-level scheme for the homogeneous equation ($q = 0$) can be written in the form (10.19). This also implies to implicit schemes $F_1 u^{n+1} = F_2 u^n$ with invertible matrices F_1, since then $u^{n+1} = (F_1^{-1}F_2)u^n = Fu^n$. The criterion (10.22) implies that we must have $\|F^{n+1}\| \leq \Lambda e^{\beta t}$ for stability. The spectral radius of an arbitrary matrix A is bounded as $\rho(A) \leq \|A\|$ (Appendix A.4), so we must ensure that $\rho(F) \leq 1 + C\Delta t$, in analogy to the initial-value problem (10.25).

We insert an illustration that serves as an example for the Problems in Sect. 10.12. Let us write the FTCS scheme by using a spatial discretization at $N + 1$ points for the homogeneous equation $v_t = Dv_{xx}$ with the initial condition $v(x, 0) = f(x)$ (10.14) and Dirichlet boundary conditions $v(0, t) = 0$ (10.15) and $v(1, t) = 0$ (10.16). In matrix form the scheme becomes $\boldsymbol{u}^{n+1} = F\boldsymbol{u}^n$ where $\boldsymbol{u} = (u_1, u_2, \ldots, u_{N-1})^\mathrm{T}$ is the solution vector and F is a symmetric tridiagonal matrix of the form $T(a, b, c)$ (A.8) with $a = r, b = 1 - 2r, c = r$. (Check it! Why this equation does not include values at $j = 0$ and $j = N$?) The eigenvalues of F are $\lambda_j = 1 - 2r + 2r \cos(j\pi/N) = 1 - 4r \sin^2(j\pi/2N)$. We will attain stability if the spectral radius (the maximum eigenvalue) can be bounded as

$$\max \left| 1 - 4r \sin^2 \frac{j\pi}{2N} \right| \le 1 \,,$$

which translates to

$$0 \le r \le \left[2 \sin^2 \frac{(N - 1)\pi}{2N} \right]^{-1} . \tag{10.28}$$

(Convince yourself that to compute the spectral radius you need the eigenvalue at $j = N - 1$, not $j = 1$!) What happens if we replace the Dirichlet boundary condition at $x = 0$ by a Neumann condition $v_x(0, t) = 0$ in the form (10.17)? Only the extreme upper left matrix element changes from $1 - 2r$ to $1 - r$. The matrix of the scheme then becomes $I - rT_{N_1 D}$, where $T_{N_1 D}$ has the form (A.11). The eigenvalues of the matrix $I - rT_{N_1 D}$ are $\lambda_j = 1 - r[2 - 2\cos((2j - 1)\pi/(2N - 1))] = 1 - 4r \sin^2((2j - 1)\pi/2(2N - 1))$, and we have stability under the condition

$$0 \le r \le \left[2 \sin^2 \frac{(2N - 3)\pi}{2(2N - 1)} \right]^{-1} .$$

With an ever finer spatial discretization (increasing N) the upper limit of r for stability in both cases approaches the value of $1/2$ *from above*. In both cases we may therefore simply require $0 \le r \le 1/2$. Nevertheless, these bounds should not lull the reader into the mistaken belief that this is the only stability criterion that ever appears in relation to the diffusion equation.

In the second approach we apply the Fourier transformation and assign to the approximate solution at each mesh point a *discrete Fourier mode*

$$u_j^n = \xi^n \mathrm{e}^{\mathrm{i}jp\pi \Delta x} \,, \tag{10.29}$$

where the superscript n in ξ^n indeed means the nth power. We seek the solution of the difference equations in the form

$$u^n(x) = \sum_p c_p^n \mathrm{e}^{\mathrm{i}p\pi x} \,.$$

Let us derive the criterion for the stability of the FTCS scheme for the equation $v_t = Dv_{xx}$. We insert the terms (10.29) in (10.23) with $a = 0$, yielding

$$\xi^{n+1}e^{ipj\pi\Delta x} = \xi^n e^{ipj\pi\Delta x}(re^{-ip\pi\Delta x} + (1 - 2r) + re^{ip\pi\Delta x})$$

or $\xi = 1 - 4r\sin^2(p\pi\Delta x/2)$. We have stability if no Fourier mode (in magnitude) exceeds unity, i.e.

$$|\xi| = \left| 1 - 4r\sin^2\frac{p\pi\Delta x}{2} \right| \leq 1 .$$

The ξ is the *amplification factor* and the corresponding inequality is the *discrete Neumann criterion*. (As an exercise, compute ξ for the general scheme (10.23).) We have derived a requirement that is the same as (10.28) if we set $\Delta x = 1/N$, even though we have not mentioned the boundary conditions. Namely, the stability of the scheme for an initial-value problem is just a necessary condition for the stability of the scheme for an initial-boundary-value problem.

There is yet a third way to approach stability that treats the boundary conditions properly and relies on the Gershgorin's theorem (see e.g. [1]). This path can be chosen in the cases when the Fourier approach is difficult, for example, when dealing with mixed boundary conditions

$$v(0, t) - g_0 v_x(0, t) = f_0 ,$$
$$v(1, t) + g_1 v_x(1, t) = f_1 ,$$

with constant f_j and $g_j \geq 0$. We just give the final result: the *necessary* condition for stability is

$$r \leq \min\left\{ \frac{1}{2 + \Delta x/g_0}, \frac{1}{2 + \Delta x/g_1} \right\} < \frac{1}{2} .$$

A consistent inclusion of boundary conditions therefore narrows the range of r for which the scheme for an initial-boundary-value problem is stable.

10.6 Higher Order Schemes

One might imagine that the order of the FTCS scheme for (10.2) in the time variable could be improved by using the symmetric difference (10.5) instead of (10.4). Indeed, this gives us the explicit *leapfrog* method of order $O(\Delta t^2) + O(\Delta x^2)$,

$$u_j^{n+1} = u_j^{n-1} + 2r\left(u_{j+1}^n - 2u_j^n + u_{j-1}^n \right) + \Delta t q_j^n . \tag{10.30}$$

This is a three-level scheme (Fig. 10.6): we compute the values at time $(n + 1)\Delta t$ from those at times $n\Delta t$ and $(n - 1)\Delta t$. But unfortunately this scheme is unstable regardless of r (Sect. 10.5).

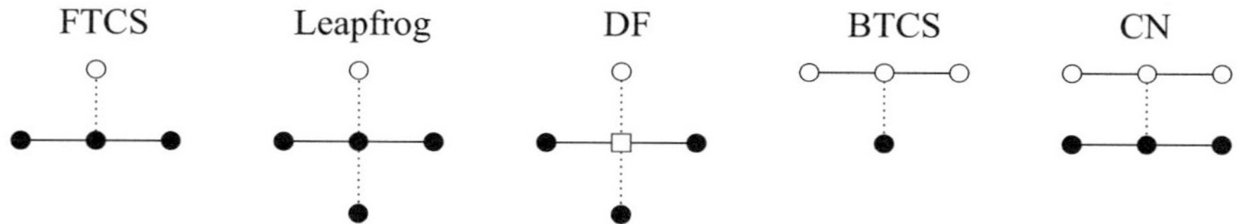

Fig. 10.6 Two-level (FTCS, BTCS, CN) and three-level (Leapfrog, DF) difference schemes for the diffusion equation $v_t - Dv_{xx} = Q$. (Compare to Fig. 10.2.) The properties of the depicted schemes are listed in Table 10.1. In the DF scheme the value of u at $(j\Delta x, n\Delta t)$ does not appear (empty square)

The leapfrog method can be stabilized by replacing u_j^n by its time average $\frac{1}{2}(u_j^{n+1} + u_j^{n-1})$. This gives us the explicit three-level Dufort–Frankel (DF) scheme

$$u_j^{n+1} = \frac{2r}{1+2r}\left(u_{j+1}^n + u_{j-1}^n\right) + \frac{1-2r}{1+2r}\,u_j^{n-1} + \frac{1}{1+2r}\,\Delta t q_j^n\,, \tag{10.31}$$

which is unconditionally stable, but only conditionally consistent (and thus only conditionally convergent). Convergence is ensured if r is constant [1].

On the other hand, we can increase the order of the FTCS scheme in the space variable by replacing the three-point difference by a five-point formula,

$$u_j^{n+1} = u_j^n + r\left(-\frac{1}{12}u_{j+2}^n + \frac{4}{3}u_{j+1}^n - \frac{5}{2}u_j^n + \frac{4}{3}u_{j-1}^n - \frac{1}{12}u_{j-2}^n\right) + \Delta t q_j^n\,. \tag{10.32}$$

The resulting scheme is of order $O(\Delta t) + O(\Delta x^4)$ and remains stable, but additional boundary conditions must be specified at the points x_{-1} and x_{N+1} that lie outside of the definition domain (Fig. 10.2 (right)). In this case we prescribe *numerical boundary conditions*. We solve the diffusion equation with homogeneous Dirichlet condition to this order if we use $u_{-1} = u_{N+1} = 0$ (solution vanishes at the ghost points) or $u_{-1}^n - 2u_0^n + u_1^n = u_{N+1}^n - 2u_N^n + u_{N-1}^n = 0$ (the second derivatives at the endpoints of the rod are zero).

We have described just a few representative explicit schemes for the solution of the one-dimensional diffusion equation. By Taylor expansions and the method of undetermined coefficients it is possible to construct many other schemes [1]. As in ordinary differential equations (Chap. 8), an alternative is offered by the implicit schemes (Table 10.1).

It is difficult to formulate a general advice when to use a high-order scheme. Now and then high-order schemes are a good choice, but if the numerical cost allows it, a finer discretization might be preferable to increasing the order. Table 10.1 summarizes the basic properties of the methods for the solution of (10.2).

Table 10.1 Select explicit (E) and implicit (I) difference schemes for the solution of the one-dimensional diffusion equation $v_t - Dv_{xx} = Q$, where $r = D\Delta t/\Delta x^2$

Scheme	Type	Order of error	Stability
FTCS (10.13)	E	$O(\Delta t) + O(\Delta x^2)$	$0 < r \le 1/2$
Leapfrog (10.30)	E	$O(\Delta t^2) + O(\Delta x^2)$	Unstable
Dufort–Frankel (10.31)	E	$O(\Delta t^2) + O(\Delta x^2)$	Stable
FTCS5 (10.32)	E	$O(\Delta t) + O(\Delta x^4)$	$0 < r \le 3/8$
BTCS (10.20)	I	$O(\Delta t) + O(\Delta x^2)$	$\forall r > 0$
Crank–Nicolson (10.21)	I	$O(\Delta t^2) + O(\Delta x^2)$	$\forall r > 0$

10.7 Hyperbolic Equations

In this Section we discuss hyperbolic equations of first order,

$$av_x + bv_y = c ,$$

where a, b, and c may in general be functions of x, y, and v (but not v_x or v_y), and of second order,

$$av_{xx} + bv_{xy} + cv_{yy} + d = 0 ,$$

where a, b, c, and d may be functions of x, y, v, v_x, and v_y (but not v_{xx}, v_{xy}, or v_{yy}). (The reader will reassign $x \to t$ and $y \to x$ if needed.) Note that the character of the equation may change on the domain: for example, the equation

$$yv_{xx} + xv_{xy} + yv_{yy} = F(x, y, v, v_x, v_y) ,$$

is hyperbolic for $|x| > 2|y|$, parabolic for $|x| = 2|y|$, and elliptic for $|x| < 2|y|$.

Characteristics A familiar property of the mentioned quasi-linear hyperbolic equations are the *characteristic curves* or *characteristics*: they represent directions defined at each point of the domain, along which the solution is independent of the partial derivatives in other directions. The method of solving hyperbolic equations by characteristics excels in the propagation of discontinuities: the sequence of characteristics is a natural mesh for them. But this method becomes cumbersome when the equations become more complex [3]. In the following we therefore seek the solutions by difference methods. We shall return to the problem of discontinuities in Sect. 10.11.

Properties of solutions In this Section and later in Sect. 10.10 we show that the scalar hyperbolic equation

$$v_t + cv_x = 0 \tag{10.33}$$

with a constant value of c is an appropriate model equation for the study of one-dimensional hyperbolic problems. Namely, the related matrix problem

$$\boldsymbol{v}_t = A\boldsymbol{v}_x$$

can be decoupled to a set of equations of the form (10.33) by diagonalizing A. Its exact solution $v(x, t) = f(x - ct)$ is known: for an arbitrary function f the equation describes the quantity v (for example, a compression in the wave) "carried" by the flow *without change in shape* (no dissipation) along the x-axis with velocity c. In parabolic PDE (diffusion equation) dissipation is contained already in the equation, and that beneficially influences the stability of the corresponding difference schemes. For hyperbolic equations we also desire numerical solutions that maintain an undamped propagation of waves and remain stable. In addition, *dispersion* can be induced by a carelessly designed difference scheme: this occurs when different Fourier components of the solution propagate with different velocities, causing the solution to decay or become distorted.

10.7.1   Explicit Schemes

We start with a naive approximation of the initial-value problem for Eq. (10.33). At lowest order $O(\Delta t) + O(\Delta x)$ we discretize the equation by approximating the spatial derivative by $v_x \approx (\Delta_+^{(x)} u_j)/\Delta x$ (*Forward-Time, Forward-Space, FTFS*) or by $v_x \approx (\Delta_-^{(x)} u_j)/\Delta x$ (*Forward-Time, Backward-Space, FTBS*), resulting in

$$u_j^{n+1} = u_j^n - R(u_{j+1}^n - u_j^n) , \qquad (10.34)$$

$$\text{or} \quad u_j^{n+1} = u_j^n - R(u_j^n - u_{j-1}^n) , \qquad (10.35)$$

where $R = c\Delta t/\Delta x$. The FTFS scheme for $c < 0$ is conditionally stable, with the condition $|R| \leq 1$ (and unstable for $c > 0$), while the FTBS scheme is conditionally stable for $c > 0$, with the condition $0 \leq R \leq 1$ (and unstable for $c < 0$, see also Sect. 10.5). This symmetry between the direction of spatial differencing and stability is related to the behavior of the solution near the characteristics: in the FTFS scheme with $c < 0$ (or FTBS with $c > 0$) the solution for $|R| \leq 1$ (or $0 \leq R \leq 1$) moves along the characteristics towards the stable solution; in the opposite cases it moves away from them.

Courant–Friedrichs–Lewy criterion We have discussed the stability of difference schemes for parabolic PDE in Sect. 10.5. The stability conditions pertaining to schemes for hyperbolic PDE are not as easy to formulate. When parabolic schemes, either explicit or implicit, are written in matrix form, we obtain symmetric matrices that allow us to derive necessary *and* sufficient conditions for stability in terms of spectral radii. The corresponding matrices for hyperbolic PDE are often non-symmetric, and we can obtain only necessary conditions. In the cases where we

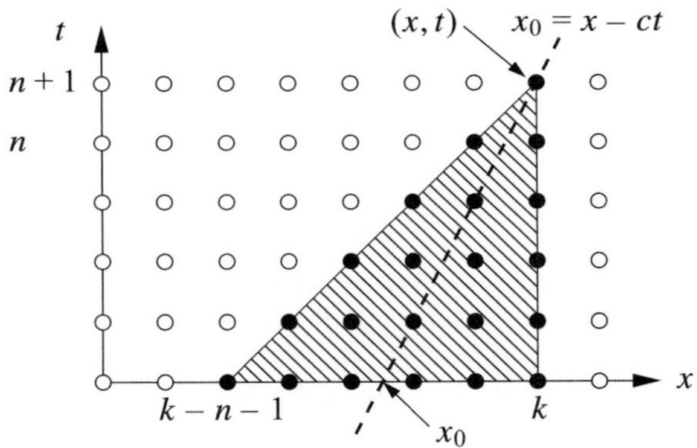

Fig. 10.7 Analytic and numerical domain of dependence of the point (x, t) for the hyperbolic equation $v_t + cv_x = 0$ with the initial condition $v(x, 0) = f(x)$. The analytic domain is the point x_0 (where the characteristic intersects the $t = 0$ axis) and is contained in the numerical domain of the FTBS scheme (10.35), which is $[(k - n - 1)\Delta x, k\Delta x]$

are unable to find necessary and sufficient conditions by means of Neumann (discrete Fourier) analysis, we can rely on the Courant–Friedrichs–Lewy (CFL) criterion which serves as a good orientation [1].

The CFL criterion relates the *analytic and numerical domains of dependence*. The analytic domain of dependence of the point (x, t) for the equation $v_t + cv_x = 0$ with the initial condition $v(x, 0) = f(x)$ is the set of points on which the solution of the equation at (x, t) depends. This set includes a single point $x_0 = x - ct$: the solution at (x, t) depends exclusively on the value of the function f at x_0. The numerical domain of dependence is the set of points on the x-axis from which the difference scheme transports the initial condition to the final value at $(x, t) = (k\Delta x, (n + 1)\Delta t)$. Figure 10.7 shows the numerical domain of dependence for the FTBS scheme (10.35): it is the interval $[(k - n - 1)\Delta x, k\Delta x]$ at $t = 0$. All past time steps are contained in the triangle defined by this interval and the apex at (x, t).

The partial differential equation and the *explicit* difference scheme for it satisfy the CFL criterion when the analytic domain for this equation is contained in the numerical domain of the corresponding scheme. But fulfilling the CFL requirement is just a *necessary* condition for stability. If we use the FTBS scheme for $v_t + cv_x = 0$, the analytic domain is $x_0 = x - ct = k\Delta x - c(n + 1)\Delta t = [k - R(n + 1)]\Delta x$, and it is contained in the numerical domain precisely when

$$(k - n - 1)\Delta x \le [k - R(n + 1)]\Delta x \le k\Delta x \,.$$

The inequalities hold true if $0 \le R \le 1$. In a similar way the necessary conditions for stability of other schemes for hyperbolic PDE can be derived. The results are summarized in Table 10.2. Implicit schemes are absolutely stable and the CFL criterion is not useful for them.

Increasing the order As in the leapfrog method for the diffusion equation, we can use the centered difference $\Delta_0^{(x)}$ (FTCS) instead of the one-side differences $\Delta_+^{(x)}$ or $\Delta_-^{(x)}$ to gain one order in the space coordinate,

$$u_j^{n+1} = u_j^n - \frac{R}{2}\left(u_{j+1}^n - u_{j-1}^n\right) . \tag{10.36}$$

But recall that the stability condition of the scheme for $v_t + cv_x = Dv_{xx}$ was (10.27), while here we have $r = 0$: the scheme (10.36) is therefore unstable.

How do we attain a higher order and yet maintain stability? Since $v_t = -cv_x$ and $v_{tt} = (-cv_x)_t = -(cv_t)_x = c^2 v_{xx}$, we resort to the Taylor expansion

$$v_j^{n+1} = v_j^n + (v_t)_j^n \Delta t + (v_{tt})_j^n \frac{\Delta t^2}{2} + O(\Delta t^3)$$

$$= v_j^n + (-cv_x)_j^n \Delta t + (c^2 v_{xx})_j^n \frac{\Delta t^2}{2} + O(\Delta t^3) .$$

The approximations $v_x \approx (\Delta_0^{(x)} u_j^n)/\Delta x$ and $v_{xx} \approx (\Delta_2^{(x)} u_j^n)/\Delta x^2$ give us the *linear Lax–Wendroff scheme*

$$u_j^{n+1} = u_j^n - \frac{R}{2}\Delta_0^{(x)} u_j^n + \frac{R^2}{2}\Delta_2^{(x)} u_j^n , \tag{10.37}$$

with the error of order $O(\Delta t^2) + O(\Delta x^2)$. (Beware that in this book and in literature there are other schemes that carry the same name.) From the symbol $|\rho(\xi)|^2 = 1 - 4R^2 \sin^2(\xi/2) + 4R^4 \sin^4(\xi/2)$ we infer the stability criterion $|R| \leq 1$. We have sinned a bit. If we read the scheme backwards, the equation

$$v_t + cv_x = \frac{c^2 \Delta t}{2} v_{xx} .$$

emerges: a diffusive (dissipative) term has appeared at the right-hand side of the equation that damps the solution for all $t > 0$. The consequences will be discussed in Sect. 10.9. Let us also mention the Lax–Friedrichs scheme

$$u_j^{n+1} = \frac{1}{2}\left(u_{j+1}^n + u_{j-1}^n\right) - \frac{R}{2}\Delta_0^{(x)} u_j^n , \tag{10.38}$$

which is conditionally stable ($|R| \leq 1$) and consistent to $O(\Delta t) + O(\Delta x^2/\Delta t)$. Both schemes will become more familiar in Problem (10.12.2) and will be useful for the solution of PDE that can be expressed in conservative form (Sect. 10.11).

10.7.2 Implicit Schemes

If hyperbolic PDE in some specific regimes (e.g. aerodynamic computations involving high Reynolds numbers) were solved by explicit schemes, prohibitively short time steps would be required to ensure stability. Implicit schemes have much better stability properties and become a viable option in such cases. Here we only mention the simplest ones [1]. It is easy to derive the implicit versions of the FTFS (10.34), FTBS (10.35), FTCS (10.36), and Lax–Wendroff schemes (10.37): all one has to do is to evaluate the spatial derivative at time $(n + 1)$ instead of n. The corresponding quartet of implicit schemes is

$$(1 - R)u_j^{n+1} + Ru_{j+1}^{n+1} = u_j^n , \qquad (10.39)$$

$$- Ru_{j-1}^{n+1} + (1 + R)u_j^{n+1} = u_j^n , \qquad (10.40)$$

$$- \frac{R}{2}u_{j-1}^{n+1} + u_j^{n+1} + \frac{R}{2}u_{j+1}^{n+1} = u_j^n , \qquad (10.41)$$

$$\left(-\frac{R^2}{2} - \frac{R}{2}\right) u_{j-1}^{n+1} + \left(1 + R^2\right) u_j^{n+1} + \left(-\frac{R^2}{2} + \frac{R}{2}\right) u_{j+1}^{n+1} = u_j^n . \qquad (10.42)$$

If we approximate the term cv_x by the central difference $\Delta_0^{(x)}$ averaged between $n\,\Delta t$ and $(n + 1)\Delta t$, we obtain the Crank–Nicolson scheme

$$- \frac{R}{4}u_{j-1}^{n+1} + u_j^{n+1} + \frac{R}{4}u_{j+1}^{n+1} = \frac{R}{4}u_{j-1}^n + u_j^n - \frac{R}{4}u_{j+1}^n , \qquad (10.43)$$

with the symbol $\rho(\xi) = (1 - (\mathrm{i}R/2)\sin\xi)/(1 + (\mathrm{i}R/2)\sin\xi)$, for which obviously $|\rho(\xi)| \equiv 1$. The BTCS, implicit Lax–Wendroff, and Crank–Nicolson schemes are stable regardless of the sign of c, which makes us comfortable in the cases where c is a function of independent variables (when c changes throughout the definition domain). The main properties of the explicit and implicit schemes for the one-dimensional hyperbolic equation $v_t + cv_x = 0$ are summarized in Table 10.2.

10.7.3 Wave Equation

The one-dimensional second-order wave equation

$$v_{tt} = c^2 v_{xx} , \qquad x \in [0, 1] , \qquad c > 0 , \qquad (10.44)$$

with the initial conditions $v(x, 0) = f(x)$ and $v_t(x, 0) = g(x)$ is important enough to merit its own Subsection. It is easy to derive a simple explicit difference scheme for it: we approximate the time derivative at the left as (10.7) and the spatial derivative at the right as (10.6), and obtain

Table 10.2 Consistency and stability properties of the explicit (E) and implicit (I) difference schemes for solving the one-dimensional hyperbolic problem $v_t + cv_x = 0$ with the parameter $R = c\Delta t / \Delta x$

Scheme	Type	Order of error	Stability
FTFS (10.34)	E	$O(\Delta t) + O(\Delta x)$	$-1 \leq R \leq 0$
FTBS (10.35)	E	$O(\Delta t) + O(\Delta x)$	$0 \leq R \leq 1$
FTCS (10.36)	E	$O(\Delta t) + O(\Delta x^2)$	Unstable
Lax–Wendroff (10.37)	E	$O(\Delta t^2) + O(\Delta x^2)$	$|R| \leq 1$
Lax–Friedrichs (10.38)	E	$O(\Delta t) + O(\Delta x^2/\Delta t)$	$|R| \leq 1$
BTFS (10.39)	I	$O(\Delta t) + O(\Delta x)$	$R \leq 0$
BTBS (10.40)	I	$O(\Delta t) + O(\Delta x)$	$R \geq 0$
BTCS (10.41)	I	$O(\Delta t) + O(\Delta x^2)$	$\forall R$
Lax–Wendroff (10.42)	I	$O(\Delta t^2) + O(\Delta x^2)$	$\forall R$
Crank–Nicolson (10.43)	I	$O(\Delta t^2) + O(\Delta x^2)$	$\forall R$

$$u_j^{n+1} = R^2 \left(u_{j+1}^n + u_{j-1}^n \right) + 2 \left(1 - R^2 \right) u_j^n - u_j^{n-1} , \qquad (10.45)$$

where $R = c\Delta t / \Delta x$. The scheme (10.45) is stable for $R \leq 1$ and has three levels: we need the solutions at $(n - 1)\Delta t$ and $n\Delta t$ to compute the solution at $(n + 1)\Delta t$. At $t = 0$ the solution is given by the initial condition, $u_j^0 = f(j\Delta x) = f_j$. But the solution in the first time step $t = \Delta t$—the so-called *initialization scheme*—depends on the discretization of the remaining initial and boundary conditions. For Dirichlet boundary conditions $v(0, t) = a(t)$ and $v(1, t) = b(t)$ we use the approximation $v_t(x, 0) = g(j\Delta x) = g_j \approx (u_j^1 - u_j^0)/\Delta t$ at $j = 0, 1, \ldots, N$ to express

$$u_j^1 = u_j^0 + \Delta t g(j\Delta x) = f_j + \Delta t g_j .$$

One can show that with such initialization the scheme (10.45) is convergent only to order $O(\Delta t) + O(\Delta x^2)$ [1]. The symmetric approximation for the time derivative $g_j \approx (u_j^1 - u_j^{-1})/(2\Delta t)$ offers a better initialization

$$u_j^1 = \frac{R^2}{2} \left(f_{j+1} + f_{j-1} \right) + \left(1 - R^2 \right) f_j + \Delta t g_j$$

(derive this as an exercise). This improvement in the initialization makes the scheme (10.45) convergent at order $O(\Delta t^2) + O(\Delta x^2)$, as expected from it.

For the wave equation, we could also use the three-level implicit scheme

$$\frac{1}{\Delta t^2} \Delta_2^{(t)} u_j^n = \frac{1}{\Delta x^2} \left[\frac{1}{4} \Delta_2^{(x)} u_j^{n+1} + \frac{1}{2} \Delta_2^{(x)} u_j^n + \frac{1}{4} \Delta_2^{(x)} u_j^{n-1} \right] , \qquad (10.46)$$

which is of order $O(\Delta t^2) + O(\Delta x^2)$ and is absolutely stable for all $\Delta t/\Delta x > 0$. It requires us to solve a tridiagonal system of equations at each time step.

In Sect. 10.10 and Problem 10.12.4 we show yet another attractive option, namely how we can use the substitutions $v_1 = cv_x$ and $v_2 = v_t$ to translate the solution of (10.44) to the solution of the system of two first-order hyperbolic equations

$$\begin{pmatrix} v_1 \\ v_2 \end{pmatrix}_t = \begin{pmatrix} 0 & c \\ c & 0 \end{pmatrix} \begin{pmatrix} v_1 \\ v_2 \end{pmatrix}_x .$$

Diagonalizing this system reveals the eigenvalues $\pm c$ corresponding to the decoupled scalar equations $(V_1)_t + c(V_1)_x = 0$ and $(V_2)_t - c(V_2)_x = 0$.

10.8 Non-linear Equations and Equations of Mixed Type ⋆

So far we have studied the solution methods for PDE and the criteria for consistency, stability, and convergence of difference schemes only in the case of linear equations. But real life abounds in non-linear equations involving time and space derivatives of higher orders. Here we illustrate some typical approaches to such problems.

Non-linear diffusion equation Let us discuss the non-linear equation

$$v_t = (D(v)v_x)_x \ , \qquad D(v) = \frac{1 + av^2}{1 + bv^2} \ , \qquad a > b > 0 \ , \tag{10.47}$$

which is used to describe diffusion in porous substances or heat transfer in matter with a temperature-dependent coefficient of thermal conductivity. (This equation is the topic of Problem 10.12.1. An even more interesting case, the two-dimensional growth of biofilms on substrates, is treated in Problem 11.9.2.) We denote $\mathcal{V} = D(v)v_x$ and approximate the time derivative to first order, $v_t(x, t) \approx (u_j^{n+1} - u_j^n)/\Delta t$, and the spatial derivative $\mathcal{V}_x$ by the symmetric difference

$$\mathcal{V}_x(x, t) \approx (\mathcal{V}_{j+1/2}^n - \mathcal{V}_{j-1/2}^n)/\Delta x \ ,$$

where we use the average of the function $D(v)$ in both terms of the numerator and the one-sided difference for the time derivative. We obtain

$$\mathcal{V}_{j+1/2}^n \approx \frac{1}{2} \left[D\left(u_{j+1}^n\right) + D\left(u_j^n\right) \right] \frac{u_{j+1}^n - u_j^n}{\Delta x} \ , \tag{10.48}$$

$$\mathcal{V}_{j-1/2}^n \approx \frac{1}{2} \left[D\left(u_j^n\right) + D\left(u_{j-1}^n\right) \right] \frac{u_j^n - u_{j-1}^n}{\Delta x} \ , \tag{10.49}$$

which ultimately gives us the scheme

$$\frac{u_j^{n+1} - u_j^n}{\Delta t} = \frac{\mathcal{V}_{j+1/2}^n - \mathcal{V}_{j-1/2}^n}{\Delta x} \, .$$

The discretization itself was easy, but when we wish to establish whether this scheme is stable, we must make the key—and risky—step. If we assume that the solution v is smooth, $D(v)$ near some point $(x_\star, t_\star)$ may be replaced by the constant value $D_\star = D(u(x_\star, t_\star))$. The solution near that point is then determined by the linear diffusion equation $v_t = D_\star v_{xx}$ or the FTCS scheme (10.13), which is stable if $D_\star \Delta t / \Delta x^2 \le 1/2$ (see Eq. (10.26)). The function $D(v)$ is bounded, $D(v) \le a/b$, so the linearized scheme is conditionally stable, with the condition $\Delta t \le (b/a)(\Delta x^2/2)$.

Burgers equation Linearization by local "freezing" of the numerical solution works surprisingly well even in problems with stronger non-linearities. But the choice of the appropriate difference scheme is always followed by the difficult question of its stability. Further problems may appear in equations of mixed types, where several properties of the solution interlace. A typical example is the Burgers equation

$$v_t + v v_x = D v_{xx} \, , \qquad D = \text{const} \, ,$$

that is used in models of gas dynamics, acoustic phenomena like shock-waves, and turbulence. The equation occupies and important place in the hierarchy of the approximations of the Navier–Stokes equation, in particular because it is one of the few non-linear PDE with known analytic solutions (at various boundary conditions). It is therefore a benchmark problem for numerical methods for non-linear PDE. The Burgers equation is of mixed parabolic-hyperbolic type. For small values of D (negligible diffusive term $D v_{xx}$) the equation is strongly hyperbolic and the solutions tend to evolve to propagating discontinuities that are hard to resolve numerically. For large D (relatively small non-linear advection term $v v_x$) its parabolic character is more pronounced.

The non-linearities in the Burgers equation are seemingly easy to handle by an explicit scheme: we attempt

$$u_j^{n+1} = u_j^n - \frac{R}{2} u_j^n \, \Delta_0^{(x)} u_j^n + r \Delta_2^{(x)} u_j^n \, , \tag{10.50}$$

where $R = \Delta t / \Delta x$ and $r = D \Delta t / \Delta x^2$. But for small D the scheme is unstable even if we choose $r \le 1/2$ as instructed by (10.26). Stability is restored if we resort to the implicit scheme (BTCS), which we obtain by evaluating the last two terms on the right-hand side of (10.50) at time $(n + 1)\Delta t$ instead of at $n\Delta t$,

$$u_j^{n+1} = u_j^n - \frac{R}{2} u_j^{n+1} \, \Delta_0^{(x)} u_j^{n+1} + r \Delta_2^{(x)} u_j^{n+1} \, , \tag{10.51}$$

but this gives birth to a new problem: at each time step we need to solve a system of non-linear equations! The most direct approach, also frequently used in practice, is solving (10.51) by Newton's method. Let us assume Dirichlet boundary conditions $u_0 = u_N = 0$ so that the complete solution fits into the vector $\boldsymbol{u} = (u_1, u_2, \ldots, u_{N-1})^{\mathrm{T}} = (u_1^{n+1}, u_2^{n+1}, \ldots, u_{N-1}^{n+1})^{\mathrm{T}}$. We write the system (10.51) in the form $\boldsymbol{f}(\boldsymbol{u})$, where $\boldsymbol{f} : \mathbb{R}^{N-1} \to \mathbb{R}^{N-1}$, with the components

$$f_j(\boldsymbol{u}) = u_j + \frac{R}{2} u_j \left(u_{j+1} - u_{j-1} \right) - r \left(u_{j+1} - 2u_j + u_{j-1} \right) - u_j^n \, ,$$

where $j = 1, 2, \ldots, N - 1$. The Jacobi matrix $[J(\boldsymbol{u})]_{jk} = \partial f_j / \partial u_k$ is tridiagonal,

$$J(\boldsymbol{u}) = \begin{pmatrix} b_1 & c_1 & & & \\ a_2 & b_2 & c_2 & & \\ & a_3 & b_3 & c_3 & \\ & & \ddots & \ddots & \ddots \\ & & & a_{N-1} & b_{N-1} \end{pmatrix} ,$$

with the matrix elements

$$a_j = -\frac{R}{2} u_j - r \, , \qquad b_j = 1 + \frac{R}{2} \left(u_{j+1} - u_{j-1} \right) + 2r \, , \qquad c_j = \frac{R}{2} u_j - r \, .$$

Starting with an appropriate initial approximation (e.g. $\boldsymbol{u}^{n+1} = \boldsymbol{u}^n$) we iterate

$$J(\boldsymbol{u}) \Delta \boldsymbol{u} = -\boldsymbol{f}(\boldsymbol{u}) \, ,$$
$$\boldsymbol{u} = \boldsymbol{u} + \Delta \boldsymbol{u} \, ,$$

until $\Delta \boldsymbol{u}$ in some norm, e.g. (A.5), drops below the specified tolerance. When the iteration has converged, we have only accomplished a single time step and mapped the solution $\boldsymbol{u}^n$ to $\boldsymbol{u}^{n+1}$.

If solving the non-linear system appears to be too big of a nuisance, we may again try some linearization. We can do this by *lagging* one part of the non-linear term, for example, the zeroth derivative (function value),

$$u_j^{n+1} = u_j^n - \frac{R}{2} u_j^n \, \Delta_0^{(x)} u_j^{n+1} + r \Delta_2^{(x)} u_j^{n+1} \, . \tag{10.52}$$

This leads to a system of *linear* equations of the form $F \boldsymbol{u}^{n+1} = \boldsymbol{u}^n$ for the solution $\boldsymbol{u}^{n+1}$, with a tridiagonal coefficient matrix F.

The non-linearity in the scheme can also be circumvented by expanding the factors in the non-linear term to first order,

$$u_j^{n+1} = u_j^n + \delta_j \, . \tag{10.53}$$

When we insert this in (10.51) and neglect all terms quadratic in δ, we again obtain a system of *linear* equations for the correction $\boldsymbol{\delta} = (\delta_1, \delta_2, \ldots, \delta_{N-1})^{\mathrm{T}}$,

$$
\left(\frac{R}{2} u_j^n - r\right) \delta_{j+1} + \left(1 + 2r + \frac{R}{2} \Delta_0^{(x)} u_j^n\right) \delta_j + \left(-\frac{R}{2} u_j^n - r\right) \delta_{j-1}
$$
$$
= -\frac{R}{2} u_j^n \, \Delta_0^{(x)} u_j^n + r \, \Delta_2^{(x)} u_j^n \,. \tag{10.54}
$$

We solve this tridiagonal system for $\boldsymbol{\delta}$ by using the current solution $\boldsymbol{u}^n$, which appears at both sides of the equation, and compute $\boldsymbol{u}^{n+1}$ by Eq. (10.53).

The numerical solution of non-linear PDE is full of pitfalls and it is nearly impossible to formulate a set of common instructions. In most cases, the stability criteria developed for linear problems are not applicable; we have only limited control over dispersion and dissipation (Sect. 10.9); without approximations, we are forced to solve systems of non-linear equations. With almost every new equation we learn from the start. Problem 10.12.5 helps us to realize how the simple methods described above work (or fail) in practice in the case of the Burgers equation. Problem 10.12.8 is devoted to the related Korteweg–de Vries equation, and Problem 10.12.10 to the cubic Schrödinger equation.

10.9   Dispersion and Dissipation ⋆

Exact solutions of hyperbolic and parabolic one-dimensional PDE can be written as the sum of the terms $v(x, t) = \widehat{v} \exp[\mathrm{i}(\omega t + kx)]$. Such functions satisfy the chosen PDE only if a particular analytic connection—a dispersion relation—exists between the frequency of the wave ω and the wave vector $k = 2\pi/\lambda$.

Examples (Rephrased from Ref. [1] whose definitions and notation we adopt in this Section.) In the advection equation, $v_t \pm cv_x = 0$, the dispersion relation holds only if $\omega = \mp kc$, with the solution components $v(x, t) = \widehat{v} \exp[\mathrm{i}(kx - \omega t)]$: such a wave propagates with the velocity $\mp c = \omega/k$ that does not depend on the frequency, and with a constant amplitude.

The equation $v_t \pm cv_{xxx} = 0$ has the dispersion relation $\omega = k^3 c$, and the solution is $v(x, t) = \widehat{v} \exp[\mathrm{i}(kx + k^3 ct)]$: waves travel without dissipation with velocities $\pm k^2 c$ in directions opposite to those of the first-order equation, but we see dispersion: components with different wave vectors propagate with different velocities.

The dispersion relation for the diffusion equation $v_t = Dv_{xx}$ is $\omega = \mathrm{i}Dk^2$, and the solution components are $v(x, t) = \widehat{v} \exp(-Dk^2 t) \exp(\mathrm{i}kx)$: such "waves" do not propagate and their amplitude diminishes with time.   ◁

What about dispersion and dissipation in the *difference schemes* for PDE? In the usual discretization $x = j\Delta x$ and $t = n\Delta t$, the solution components are

$$
u_j^n = \widehat{u} \, \mathrm{e}^{\mathrm{i}(kj\Delta x + \omega n\Delta t)} \,. \tag{10.55}
$$

The component with the highest frequency in the solution corresponds to the term $\exp[2\pi i(N/2)j\,\Delta x]$, thus the interval $0 \le k\Delta x \le \pi$ contains the information on all components in the Fourier representation of the solution. We separate the real and imaginary parts of the dispersion relation $\omega = \omega(k)$ such that $\omega = \alpha(k) + i\beta(k)$, thus $e^{i\omega t} = e^{i\alpha t}e^{-\beta t}$, and

$$u_j^n = \widehat{u}\,\left(e^{-\beta\Delta t}\right)^n\, e^{ik[j\Delta x - (-\alpha/k)n\Delta t]}\,.$$

The discrete dispersion relation is intimately related to the symbol of the difference scheme, $e^{i\omega\Delta t} = \rho(k\Delta x)$ (Sect. 10.5). The real part,

$$\alpha\Delta t = \arctan\left[\frac{\operatorname{Im}\rho(k\Delta x)}{\operatorname{Re}\rho(k\Delta x)}\right]\,,$$

determines the propagation and dispersion, while the imaginary part,

$$\beta\Delta t = -\log|\rho(k\Delta x)|\,,$$

determines the dissipation of the solution. This immediately allows us to classify the solutions of a given difference scheme:

$$
\begin{aligned}
\alpha = 0 \text{ for all } k \quad &\Longrightarrow \text{ solution does not propagate,}\\
\alpha \ne 0 \text{ for some } k \quad &\Longrightarrow \text{ solution propagates with velocity } -\alpha/k,\\
d^2\alpha/dk^2 \ne 0 \quad &\Longrightarrow \text{ scheme is dispersive,}\\
\beta > 0 \text{ for some } k \quad &\Longrightarrow \text{ scheme is dissipative,}\\
\beta < 0 \text{ for some } k \quad &\Longrightarrow \text{ solution diverges (scheme unstable),}\\
\beta = 0 \text{ for all } k \quad &\Longrightarrow \text{ scheme is non-dissipative.}
\end{aligned}
$$

Example Let us inspect the dispersion and dissipation properties of the FTCS scheme (10.13) for the diffusion equation. The symbol of the scheme is given by (10.24) where $a = 0$ and $\xi = k\Delta x$, thus $\exp(i\omega\Delta t) = 1 - 4r\sin^2(k\Delta x/2)$, and hence

$$\alpha = 0\,,\qquad \omega = i\beta = -\frac{i}{\Delta t}\log\left|1 - 4r\sin^2\frac{k\Delta x}{2}\right|\,.$$

The solution u_j^n therefore does not propagate ($\alpha = 0$). We now see Eq. (10.26) in a different light from the viewpoint of the parameter β: if $r > 1/2$, then $\beta < 0$, so the factor $(\exp(-\beta\Delta t))^n$ diverges and the scheme is unstable. But if $r \le 1/2$, we have $\beta \ge 0$, and $(\exp(-\beta\Delta t))^n$ reduces the amplitude of the solution. All Fourier components fade out, only the one with $\beta = 0$ survives, and this one determines the asymptotics. The dissipation of the solution of the difference scheme follows the physical dissipation dictated by the diffusion equation.   ◁

Example We repeat the exercise for the FTFS scheme (10.34) for the hyperbolic advection equation $v_t + cv_x = 0$ with $c < 0$ and $R = c\Delta t/\Delta x$. When we insert the Fourier expansion (10.55) in the scheme, we obtain the dispersion relation $\exp(i\omega\Delta t) = 1 + R - R\cos k\Delta x - iR\sin k\Delta x$, whence we read off

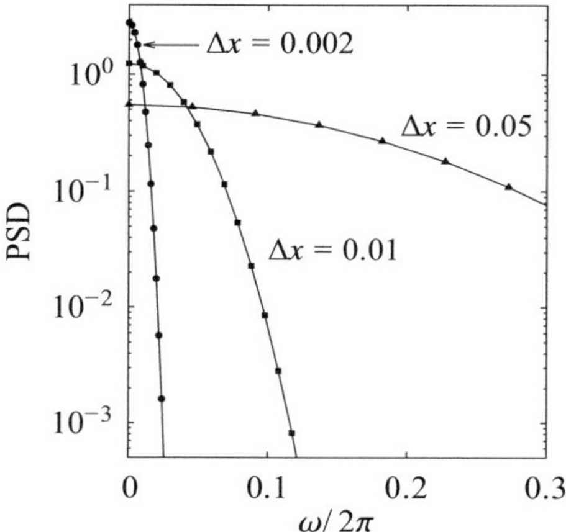

Fig. 10.8 Power spectral density of the function $v_j^0 = \sin^{40} \pi j \Delta x$ (as an initial condition for the equation $v_t + c v_x = 0$) with spatial discretizations $\Delta x = 0.05$, $\Delta x = 0.01$, and $\Delta x = 0.002$. Only the lowest Fourier components are relevant at small Δx

$$\alpha = -\frac{1}{\Delta t} \arctan\left[\frac{R \sin k \Delta x}{1 + 2R \sin^2 \frac{k \Delta x}{2}} \right], \tag{10.56}$$

$$\beta = -\frac{1}{2\Delta t} \log\left[(1 + R)^2 - 2R(1 + R) \cos k \Delta x + R^2 \right]. \tag{10.57}$$

Since α is non-linear in k, we have $d^2\alpha/dk^2 \neq 0$ and the scheme (10.34) is therefore always dispersive. If $\beta < 0$, the scheme is unstable. If $\beta > 0$, the scheme is stable for $|R| \leq 1$ and then all solution components with $k \neq 0$ decay, while the $k = 0$ component neither grows nor decays. The scheme is dissipative, except for $R = -1$, when $\beta = 0$ for all k. It is impossible to completely remove dissipation from (10.34), but it can be limited by choosing a sufficiently small Δx. Namely, by reducing Δx we shift the solution components to lower frequencies, where dissipation is less pronounced (see Fig. 10.8 and [1]).

From the dispersion relations (10.56) and (10.57) we see that high-frequency components of the numerical solution (small λ or $k\Delta x$ near π) propagate with velocity $-\alpha/k = 0$ if $R > -1/2$, or with velocity $-\alpha/k = c/R$ if $R < -1/2$. Low-frequency components (large λ or $k\Delta x \ll 1$) are more interesting: if $|R| < 1/2$, they propagate slower than the components of the exact solution, and faster if $1/2 < |R| < 1$. But sometimes there is a relief [1]: the components with the most "wrong" velocities are also the most strongly damped. If the underlying problem (differential equation) or the corresponding difference scheme contain dissipation, it often hides the effects of dispersion. Figure 10.9 (left) shows the error of the propagation velocity $c - (-\alpha/k)$ in dependence of R for the FTFS scheme. The error is smallest near the stability limit ($R = -1$).   ◁

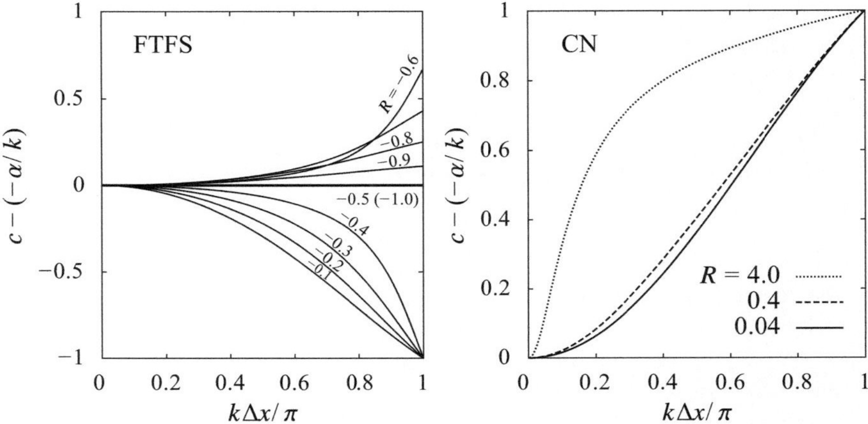

Fig. 10.9 [LEFT] Error of propagation velocity $c - (-\alpha/k)$ as a function of R in the FTFS scheme. The error is smallest near the stability limit ($R = -1$). [RIGHT] Error of propagation velocity in the Crank–Nicolson scheme. By further decrease of R the curves become barely distinguishable from the curve corresponding to $R = 0.04$

A similar analysis can be performed for the Crank–Nicolson scheme. Its symbol is $|\rho(k\Delta x| = 1$, so it is non-dissipative, but is has strong dispersion: as shown in Fig. 10.9 (right), the phase error can be reduced somewhat by decreasing Δx, but even this works only for low-frequency components. This means that solutions with broad Fourier spectra are hard to represent on the mesh. In general, Δx must be sufficiently small for high-frequency components to become irrelevant for the solution. See also Problem 10.12.2.

10.10 Systems of Linear Hyperbolic and Parabolic PDE ⋆

In this Section we discuss difference schemes for some classes of systems of one-dimensional hyperbolic and parabolic PDE. We adopt the definitions of consistency, stability, and convergence with obvious generalizations from our treatment of the scalar cases (Sects. 10.3 and 10.5). By the Lax theorem, consistent stable schemes are also convergent. Stability is again probed by the discrete Fourier transformation (page 649): the role of the symbol of the scheme is now taken by the *amplification matrix* $G(\xi)$ measuring the growth of the solution's Fourier components in a given time step, $\widetilde{\boldsymbol{u}}^{n+1}(\xi) = G(\xi)\widetilde{\boldsymbol{u}}^n = G^{n+1}(\xi)\widetilde{\boldsymbol{u}}^0$. In the following we give the stability criteria of the basic schemes without proof.

A general system of one-dimensional linear hyperbolic PDE with constant coefficients can be written as

$$\boldsymbol{v}_t = A\boldsymbol{v}_x , \qquad \boldsymbol{v}(x,0) = \boldsymbol{v}_0(x) , \qquad x \in \mathbb{R}, \quad t > 0 , \tag{10.58}$$

where $v = (v_1, v_2, \ldots, v_M)^{\mathrm{T}}$ is the vector of unknowns and A is a $M \times M$ matrix. Let us assume that A is diagonalizable, so that a matrix S exists such that the similarity transformation $D = SAS^{-1}$ diagonalizes A, hence $Sv_t = SAv_x = SAS^{-1}Sv_x = DSv_x$, and S^{-1} is the matrix containing the eigenvectors of A. The new vector $V = Sv$ can be used to translate the original system of coupled equations for the *primitive variables* v to the diagonal system $V_t = DV_x$, i.e.

$$(V_m)_t - \lambda_m (V_m)_x = 0 , \qquad m = 1, 2, \ldots, M , \tag{10.59}$$

for the *characteristic variables* V, where λ_m are the eigenvalues of A. The individual solutions of the decoupled system obviously solve the scalar problem (10.33), $V_m(x, t) = V_{0m}(x + \lambda_m t)$, where $V_0 = (V_{01}, V_{02}, \ldots, V_{0M})^{\mathrm{T}} = Sv_0$ is the initial condition for (10.59). This looks like a set of M waves propagating with phase velocities λ_m along the positive or negative x-axis, depending on the sign of the eigenvalues λ_m. The solution of the original problem (10.58) is

$$v(x, t) = S^{-1}V = S^{-1} \begin{pmatrix} V_{01}(x + \lambda_1 t) \\ V_{02}(x + \lambda_2 t) \\ \vdots \\ V_{0M}(x + \lambda_M t) \end{pmatrix} .$$

Hyperbolic systems For hyperbolic systems of the form

$$v_t = Av_x + B_0 v , \qquad x \in \mathbb{R} , \quad t > 0 , \tag{10.60}$$

where A and B_0 are real constant $M \times M$ matrices, and A is diagonalizable (with eigenvalues λ_m), two explicit schemes are immediately at hand:

$$u_j^{n+1} = RAu_{j+1}^n + (I - RA)u_j^n + \Delta t B_0 u_j^n , \tag{10.61}$$

$$u_j^{n+1} = (I + RA)u_j^n - RAu_{j-1}^n + \Delta t B_0 u_j^n , \tag{10.62}$$

where $R = \Delta t / \Delta x$. The scheme (10.61) is stable in the $l_{2,\Delta x}$-norm if $\lambda_m \geq 0$ and $\lambda_m R \leq 1$ for each $m \in \{1, 2, \ldots, M\}$. In the same sense the scheme (10.62) is stable exactly when all $\lambda_m \leq 0$ and $|\lambda_m| R \leq 1$. For the system (10.60) in which the eigenvalues of A have different signs, we would like to use some "mixture" of the methods (10.61) and (10.62). We achieve this by *flux splitting* which is a popular trick in systems of non-linear PDE. We arrange the eigenvalues along the diagonal of the diagonal matrix $D = SAS^{-1}$ in decreasing order $\lambda_1 \geq \lambda_2 \geq \ldots \geq \lambda_M$ and construct matrices D_+ and D_- containing only the part of D with positive or negative eigenvalues, respectively. Then we use the matrices $A_+ = S^{-1}D_+S$ and $A_- = S^{-1}D_-S$ to form the split scheme

$$u_j^{n+1} = u_j^n + RA_+\Delta_+^{(x)}u_j^n + RA_-\Delta_-^{(x)}u_j^n + \Delta t B_0 u_j^n , \tag{10.63}$$

where $\Delta_+^{(x)}$ and $\Delta_-^{(x)}$ are defined by Eqs. (10.8) and (10.9). There are also schemes in which we do not need to pay attention to the sign of the eigenvalues of A, for example, the Lax–Wendroff scheme

$$\boldsymbol{u}_j^{n+1} = \boldsymbol{u}_j^n + \frac{R}{2}\, A\, \Delta_0^{(x)} \boldsymbol{u}_j^n + \frac{R^2}{2}\, A^2\, \Delta_2^{(x)} \boldsymbol{u}_j^n + \Delta t\, B_0 \boldsymbol{u}_j^n \,, \tag{10.64}$$

which is of order $O(\Delta t^2) + O(\Delta x^2)$. It is stable if $|\lambda_m| R \leq 1$. The matrix A should be evaluated at time $n\Delta t$, so $A = A(\boldsymbol{u}^n)$, $A^2 = [A(\boldsymbol{u}^n)]^2$.

Parabolic systems For parabolic systems of the form

$$\boldsymbol{v}_t = B_2 \boldsymbol{v}_{xx} + B_0 \boldsymbol{v} \,, \qquad x \in \mathbb{R}\,, \quad t > 0\,,$$

we could suggest the explicit scheme

$$\boldsymbol{u}_j^{n+1} = \boldsymbol{u}_j^n + r\, B_2\, \Delta_2^{(x)} \boldsymbol{u}_j^n + \Delta t\, B_0 \boldsymbol{u}_j^n \,,$$

where $r = \Delta t / \Delta x^2$ (the diffusion constant D, which we kept throughout the scalar case, can be absorbed in the matrix B_2). If B_2 is a real, symmetric, and positive definite $M \times M$ matrix (with eigenvalues λ_m), and B_0 is bounded, the above scheme is stable precisely when $r\lambda_m \leq 1/2$ for each m.

Mixed-type systems We obtain a much more general class of systems of PDE if the diffusion (parabolic) and advection (hyperbolic) terms are included simultaneously,

$$\boldsymbol{v}_t = B_2 \boldsymbol{v}_{xx} + B_1 \boldsymbol{v}_x + B_0 \boldsymbol{v} \,, \qquad x \in \mathbb{R}\,, \quad t > 0\,, \tag{10.65}$$

where B_0, B_1, and B_2 are real constant matrices and B_2 is positive definite. In such cases we should devise a sensible difference scheme on the basis of the relative magnitudes of the eigenvalues of B_1 and B_2. If the eigenvalues of B_2 are substantially larger than the eigenvalues of B_1, the parabolic nature of the equations will prevail over the hyperbolic, and the scheme of the form

$$\boldsymbol{u}_j^{n+1} = \boldsymbol{u}_j^n + r\, B_2\, \Delta_2^{(x)} \boldsymbol{u}_j^n + \frac{R}{2}\, B_1\, \Delta_0^{(x)} \boldsymbol{u}_j^n + \Delta t\, B_0\, \boldsymbol{u}_j^n \tag{10.66}$$

will behave decently (will be stable). In the opposite case, the hyperbolic character dominates and the scheme will become unstable, just as (10.36) was unstable for scalar equations. We face an even greater confusion if some eigenvalues of B_2 are smaller than those of B_1, while some are larger: in such instances the whole system must be carefully decoupled and components with the same regime rejoined. Systems of this type are too complicated to be discussed in this book. Yet a simplified variant of the parabolic-hyperbolic system,

$$\boldsymbol{v}_t = \boldsymbol{v}_{xx} + B_1 \boldsymbol{v}_x + B_0 \boldsymbol{v} \,, \qquad x \in \mathbb{R}\,, \quad t > 0\,,$$

where B_1 and B_0 are real constant matrices, and where B_1 has M non-degenerate eigenvalues, is manageable and still interesting enough: a good scheme for it is (10.66) where one sets $B_2 = I$. It is stable if for all eigenvalues λ_m of B_1 it holds that $\lambda_m^2 R^2/2 \le r \le 1/2$. If for some m the opposite inequality $r < \lambda_m^2 R^2/2$ is fulfilled (dominating hyperbolic term), it is better to differentiate the term $B_1 v_x$ "hyperbolically" (Table 10.2).

We do not discuss implicit schemes for systems of PDE. (Their main advantage are the improved stability properties: many are absolutely stable.) We also disregard systems of PDE with more than one spatial dimension. However, the problems in solving such systems are not immense; further reading can be found in Ref. [1].

10.11   Conservation Laws and High-Resolution Schemes ⋆

Section 10.9 taught us that seemingly well-designed difference schemes that are both consistent and stable do not perform satisfactorily for certain classes of PDEs: the deficiencies can usually be spotted in the suspicious behavior of the solution near discontinuities. Figure 10.10 shows the solution of $v_t - v_x = 0$ at $t = 1.0$ by using three difference schemes: regardless of the method, dispersion and dissipation distort the solutions already within a single period.

Nevertheless, by using a special approach to solve the PDEs expressed in conservative forms, it is possible to resolve and monitor the discontinuities evolving from the initial conditions, as well as their propagation velocities. In this Section we discuss the solution of initial-boundary-value problems for PDEs that can be written in

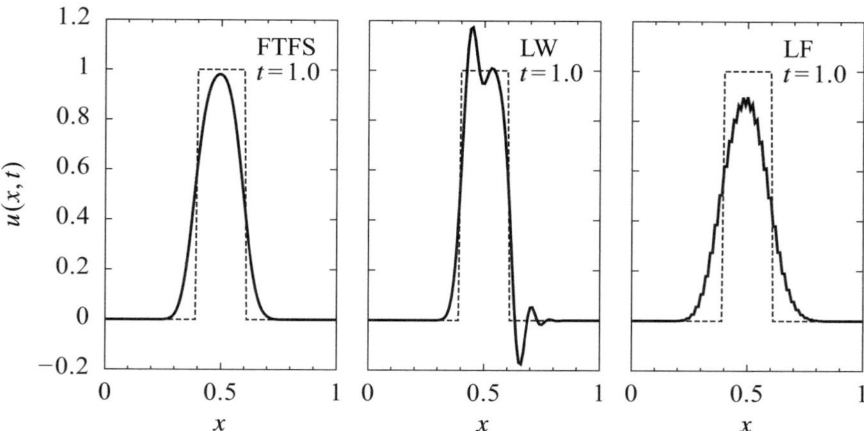

Fig. 10.10   The numerical solution of the hyperbolic problem $v_t - v_x = 0$ with periodic boundary conditions $v(0, t) = v(1, t)$ and the "step" initial condition (dashed line). Shown are the solutions by [LEFT] the FTFS scheme, [CENTER] the Lax–Wendroff scheme and [RIGHT] the Lax–Friedrichs scheme at time $t = 1.0$

the form of a conservation law for some quantity (e.g. mass, momentum, or energy). For such equations one can devise consistent, stable schemes with low dissipation and dispersion [4].

Initially we discuss conservation laws in the form

$$v_t + [F(v)]_x = 0 , \tag{10.67}$$

where we assume that the function $F : \mathbb{R}^N \to \mathbb{R}^N$ is convex ($\partial^2 F_j / \partial v_k^2 \geq 0$). A scalar example of such a law with $F(v) = v^2/2$ is the Burgers equation without diffusion,

$$v_t + v v_x = v_t + \left(\tfrac{1}{2} v^2\right)_x = 0 .$$

The corresponding scheme is at hand: we integrate (10.67) from $x_{j-1/2}$ to $x_{j+1/2}$ over x and from $t_n = n\Delta t$ to $t_{n+1} = (n+1)\Delta t$ over t, resulting in

$$\Delta x \left(v_j^{n+1} - v_j^n\right) + \left[\int_{t_n}^{t_{n+1}} F\left(v(x_{j+1/2}, t)\right) \, dt - \int_{t_n}^{t_{n+1}} F\left(v(x_{j-1/2}, t)\right) \, dt \right] = 0 ,$$

which has an obvious physical interpretation: the difference in the amount of "matter" entering the volume $[t_n, t_{n+1}] \times [x_{j-1/2}, x_{j+1/2}]$ or exiting it (through the boundary $t = t_n$ or $t = t_{n+1}$) is balanced by the amount of "matter" entering or exiting through $x = x_{j-1/2}$ or $x = x_{j+1/2}$. Such an equation can be approximated by the difference scheme

$$u_j^{n+1} = u_j^n - R \left[h_{j+1/2}^n - h_{j-1/2}^n\right] , \tag{10.68}$$

where $R = \Delta t / \Delta x$. In this approach, the discrete approximations for the fluxes are given in terms of the *numerical flux functions*

$$h_{j+1/2}^n = h(u_{k-p}^n, \dots, u_{k+q}^n) , \tag{10.69}$$

$$h_{j-1/2}^n = h(u_{k-p-1}^n, \dots, u_{k+q-1}^n) , \tag{10.70}$$

which in general depend on values at $p + q + 1$ points. The scheme (10.68) is consistent with the conservation law (10.67) when $h(v, \dots, v) = F(v)$.

Difference schemes that can be written as (10.68) with the flux functions (10.69) and (10.70) are called *conservative*. But this transcription by itself does not guarantee any of the appealing properties advertised in the introduction! A non-linear generalization of the linear Lax–Wendroff scheme (10.37),

$$u_j^{n+1} = u_j^n - \frac{R}{2} \Delta_0^{(x)} F_j^n + \frac{R^2}{2} \left[J\left(u_{j+1/2}^n\right) \Delta_+^{(x)} F_j^n - J\left(u_{j-1/2}^n\right) \Delta_-^{(x)} F_j^n \right] , \tag{10.71}$$

where $[J(u)]_{ab} = \partial F_a / \partial u_b$, can be cast in conservative form if we write

$$h_{j+1/2}^n = \frac{1}{2} \left[F_{j+1}^n + F_j^n\right] - \frac{R}{2} J\left(u_{j+1/2}^n\right) \Delta_+^{(x)} F_j^n .$$

Similarly we write the expression for $h^n_{j-1/2}$, we just replace j by $j-1$ everywhere. Since we do not know the solution at $x_{j+1/2}$ and $x_{j-1/2}$ we approximate

$$J\left(u^n_{j+1/2}\right) \approx J\left(\tfrac{1}{2}\left(u^n_{j+1} + u^n_j\right)\right), \qquad J\left(u^n_{j-1/2}\right) \approx J\left(\tfrac{1}{2}\left(u^n_j + u^n_{j-1}\right)\right).$$

Thus, the scheme is conservative and consistent with the conservation law (10.67), since $h(u, u) = F(u)$, but its solution are still influenced by dispersion. The Lax–Friedrichs scheme

$$u^{n+1}_j = \frac{1}{2}\left(u^n_{j+1} + u^n_{j-1}\right) - \frac{R}{2}\Delta^{(x)}_0 F^n_j,$$

which corresponds to the flux functions

$$h^n_{j+1/2} = \frac{1}{2}\left(F^n_{j+1} + F^n_j\right) - \frac{1}{2R}\left(u^n_{j+1} - u^n_j\right), \qquad h^n_{j-1/2} = h^n_{j+1/2}[j \leftrightarrow j-1],$$

is also conservative and consistent, but strong dissipation is present in its solutions.

10.11.1   High-Resolution Schemes: TVD Versus ENO

Dispersion and dissipation of the solutions of difference methods used to approximate PDEs in the form of conservation laws can be harnessed by using *high-resolution schemes*. Henceforth we discuss only scalar problems

$$v_t + [F(v)]_x = 0. \tag{10.72}$$

High-resolution schemes can also be formulated for *systems* of linear or non-linear equations in the form (10.67), but we restrict the discussion to the scalar case which allows us to develop all key concepts. Further reading can be found in [5–7].

The first requirement that we impose on the scheme is that the solution in subsequent time steps changes less and less, in the sense

$$\sum_{j=-\infty}^{\infty}\left|\Delta^{(x)}_+ u^{n+1}_j\right| \le \sum_{j=-\infty}^{\infty}\left|\Delta^{(x)}_+ u^n_j\right| \qquad \forall n \ge 0. \tag{10.73}$$

Schemes that fulfill (10.73) are known as *total variation diminishing* (TVD), and their typical representatives are the *flux-limiter methods*. In these methods we include a special "smoothing" function in the scheme in order to introduce artificial dissipation that prevents non-physical oscillations of the solutions near discontinuities. Their upgrade are the *slope-limiter methods* [8–13] in which one attempts to prevent too large solution slopes. Another option are the *modified flux methods* where one does not modify the numerical flux function h but rather the function F appearing in the conservation law. Here we get to know the first two types of schemes. (Note, however, that the listed approaches are not mutually exclusive and that hybrid schemes also exist.)

Sometimes the TVD requirement (10.73) is relaxed such that

$$\sum_{j=-\infty}^{\infty} \left| \Delta_+^{(x)} u_j^{n+1} \right| \le \sum_{j=-\infty}^{\infty} \left| \Delta_+^{(x)} u_j^n \right| + O(\Delta x^p) \qquad \forall n \ge 0 \tag{10.74}$$

and for some p. If the solution produced by the scheme satisfies (10.74), the scheme is said to be *essentially non-oscillatory* (ENO). One of the reasons that the TVD requirement is relaxed to ENO (or something less restrictive than TVD) is to improve accuracy: without such a relaxation, the most one can expect from a TVD scheme is second order accuracy away from the extreme points of the solution (and merely first order accuracy in their vicinity): see [14] and Sect. 9.7.2 of Ref. [4].

10.11.2   High-Resolution Flux-Limiter Schemes

The numerical flux functions of high-resolution flux-limiter schemes are constructed [4] as a combination of the flux function ^{L}h belonging to a low-order scheme, and the function ^{H}h belonging to a high-order scheme:

$$h_{j+1/2}^n = {}^{L}h_{j+1/2}^n + \phi_j^n \left[{}^{H}h_{j+1/2}^n - {}^{L}h_{j+1/2}^n \right] .$$

This allows us to enjoy the best of the two worlds: the admixture of the low-order scheme introduces dissipation into the composite scheme, helping us to reduce unwanted oscillations, while the admixture of the high-order scheme improves the accuracy. The ultimate success depends on the choice of the *flux limiter function* ϕ_j^n, which is the weight of the anti-diffusive part of the flux,

$$^{H}h_{j+1/2}^n - {}^{L}h_{j+1/2}^n .$$

The limiter function to which we entrust the smoothing of the solution is written as

$$\phi_j^n = \phi(\theta_j^n) ,$$

where θ_j^n is the *smoothing parameter* that can be computed from various components of the current solution.

In order for a flux-limiter scheme to be consistent with the conservation law to second order and possibly to be TVD as well, the function $\phi(\theta)$ for positive θ must satisfy the conditions

$$0 \le \frac{\phi(\theta)}{\theta} \le 2 , \qquad 0 \le \phi(\theta) \le 2 .$$

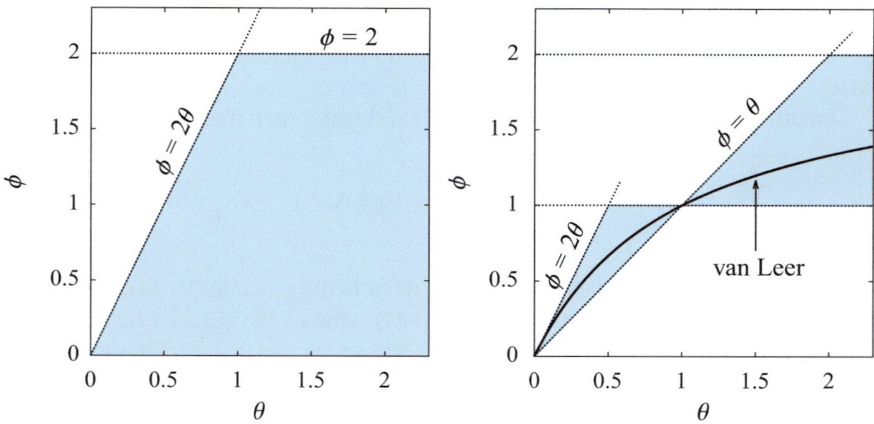

Fig. 10.11 [LEFT] The TVD region of the flux-limiter $\phi(\theta)$ in a high-resolution scheme for the scalar conservation law $v_t + [F(v)]_x = 0$. [RIGHT] Optimal region with the van Leer limiter (Eq. (10.82)) drawn in. The schemes that utilize flux limiters that pass the point $(1, 1)$ are consistent with the conservation law to second order

Fig. 10.11 (left) shows the allowed region for the function $\phi(\theta)$. This basic requirement is met by many functions (see [15] for a comprehensive review and comparison). The most frequently used are

$$\text{minmod :} \quad \phi(\theta) = \max\{0, \min\{\theta, 1\}\}\,, \tag{10.75}$$

$$\text{MC :} \quad \phi(\theta) = \max\{0, \min\{2\theta, (1+\theta)/2, 2\}\}\,, \tag{10.76}$$

$$\text{ospre :} \quad \phi(\theta) = \tfrac{3}{2}(\theta^2 + \theta)/(\theta^2 + \theta + 1)\,, \tag{10.77}$$

$$\text{superbee :} \quad \phi(\theta) = \max\{0, \min\{2\theta, 1\}, \min\{\theta, 2\}\}\,, \tag{10.78}$$

$$\text{Sweby :} \quad \phi(\theta) = \max\{0, \min\{b\theta, 1\}, \min\{\theta, b\}\}\,, \tag{10.79}$$

$$\text{UMIST :} \quad \phi(\theta) = \max\{0, \min\{2\theta, (1+3\theta)/4, (3+\theta)/4, 2\}\}\,, \tag{10.80}$$

$$\text{van Albada :} \quad \phi(\theta) = (\theta^2 + \theta)/(\theta^2 + 1)\,, \tag{10.81}$$

$$\text{van Leer :} \quad \phi(\theta) = (|\theta| + \theta)/(|\theta| + 1)\,, \tag{10.82}$$

(The Sweby limiter is just a generalization of the superbee limiter to $1 \leq b \leq 2$.) The functions whose graphs remain in the shaded area shown in Fig. 10.11 (right) are preferable [16]. All forms enumerated above go through this region and also have the property $\phi(1) = 1$ which is crucial for second-order consistency [4]. Moreover, all of these limiters possess the symmetry

$$\frac{\phi(\theta)}{\theta} = \phi\left(\frac{1}{\theta}\right),$$

which is also physically relevant: a high-resolution scheme with a symmetric flux-limiter function handles increasing and decreasing spatial gradients equivalently.

10.11.2.1 Linear Problem $v_t + cv_x = 0$

To form $^{\mathrm{L}}h$ for the scalar problem $v_t + cv_x = 0$, one uses a scheme that is insensitive to the sign of c, while to form $^{\mathrm{H}}h$, one takes the Lax–Wendroff scheme:

$$^{\mathrm{L}}h^n_{j+1/2} = \tfrac{1}{2}c\left(u^n_j + u^n_{j+1}\right) - \tfrac{1}{2}|c|\left(u^n_{j+1} - u^n_j\right) , \qquad (10.83)$$

$$^{\mathrm{H}}h^n_{j+1/2} = cu^n_j + \tfrac{1}{2}c(1 - cR)\Delta^{(x)}_+ u^n_j .$$

The method with the flux function (10.83) is known as the *upwind* scheme since it always differentiates in the direction opposite to the sense of advection: for $c < 0$ it turns into the FTFS scheme, and for $c > 0$ into the FTBS scheme. The complete flux function is then

$$h^n_{j+1/2} = {}^{\mathrm{L}}h^n_{j+1/2} + \tfrac{1}{2}\,\phi^n_j c\left(\mathrm{sign}(c) - cR\right)\Delta^{(x)}_+ u^n_j , \qquad (10.84)$$

and the corresponding high-resolution scheme becomes

$$u^{n+1}_j = u^n_j - \tfrac{1}{2}cR\Delta^{(x)}_0 u^n_j + \tfrac{1}{2}\,|c|R\Delta^{(x)}_2 u^n_j$$
$$- \tfrac{1}{2}\,cR\left(\mathrm{sign}(c) - cR\right)\left(\phi^n_j \Delta^{(x)}_+ u^n_j - \phi^n_{j-1}\Delta^{(x)}_- u^n_j\right) , \qquad (10.85)$$

where $R = \Delta t/\Delta x$. (Note that in this Section the velocity c has not been absorbed into R for practical reasons.) This scheme is consistent, second-order accurate away from zeros of v_x, and TVD. The smoothing parameters are computed as

$$\theta^n_j = \begin{cases} \Delta^{(x)}_- u^n_j \big/ \Delta^{(x)}_+ u^n_j \; ; & c > 0 , \\[2mm] \Delta^{(x)}_+ u^n_{j+1} \big/ \Delta^{(x)}_+ u^n_j \; ; & c < 0 . \end{cases}$$

Note that the differences in these formulas access points outside of the definition domain, and need to be handled separately. Trivial flux-limiter function ϕ do not bring anything new: $\phi = 0$ means just the FTFS or FTBS schemes, while $\phi = 1$ reproduces the Lax-Wendroff scheme. We therefore use a function from the set (10.75)–(10.82). How this works in practice is illustrated by Problem 10.12.2.

10.11.2.2 Non-linear Conservation Laws of the Form $v_t + [F(v)]_x = 0$

A non-linear conservation law obviously requires a non-linear high-resolution scheme. By analogy with the linear problem we use the upwind scheme for $^{\mathrm{L}}h$ and the non-linear Lax-Wendroff scheme to construct $^{\mathrm{H}}h$:

$$^{\mathrm{L}}h^n_{j+1/2} = \frac{1}{2}\left(F^n_j + F^n_{j+1}\right) - \frac{1}{2}\left|a^n_{j+1/2}\right| \Delta^{(x)}_+ u^n_j \,, \tag{10.86}$$

$$^{\mathrm{H}}h^n_{j+1/2} = \frac{1}{2}\left(F^n_j + F^n_{j+1}\right) - \frac{R}{2}\left(a^n_{j+1/2}\right)^2 \Delta^{(x)}_+ u^n_j \,.$$

The justification for this particular choice and many additional details on their properties can be found in Ref. [4]. We end up with the numerical flux function

$$h^n_{j+1/2} = \,^{\mathrm{L}}h^n_{j+1/2} + \frac{1}{2}\phi^n_j \left|a^n_{j+1/2}\right| \left(1 - R\left|a^n_{j+1/2}\right|\right)\Delta^{(x)}_+ u^n_j \,, \tag{10.87}$$

where $\phi^n_j = \phi(\theta^n_j)$,

$$a^n_{j+1/2} = \begin{cases} \Delta^{(x)}_+ F^n_j \big/ \Delta^{(x)}_+ u^n_j \;;\; \Delta^{(x)}_+ u^n_j \neq 0 \,, \\[6pt] F'(u^n_j) \;;\; \Delta^{(x)}_+ u^n_j = 0 \,, \end{cases} \tag{10.88}$$

and

$$\theta^n_j = \begin{cases} \Delta^{(x)}_- u^n_j \big/ \Delta^{(x)}_+ u^n_j \;;\; a^n_{j+1/2} > 0 \,, \\[6pt] \Delta^{(x)}_+ u^n_{j+1} \big/ \Delta^{(x)}_+ u^n_j \;;\; a^n_{j+1/2} < 0 \,. \end{cases}$$

10.11.3   High-Resolution Slope-Limiter Schemes

In slope-limiter schemes the solutions are sought in the form of piecewise linear functions as illustrated by Fig. 10.12: on the interval $[x_{j-1/2}, x_{j+1/2}]$ ("the cell") the approximate solution at time $t_n = n\Delta t$ has the form $\overline{u}^n_j = u^n_j + \sigma^n_j(x - x_j)$, where σ^n_j is an appropriate slope by which the velocity changes over the interval. The way this local slope is determined differs depending on whether a linear or a non-linear problem is being considered.

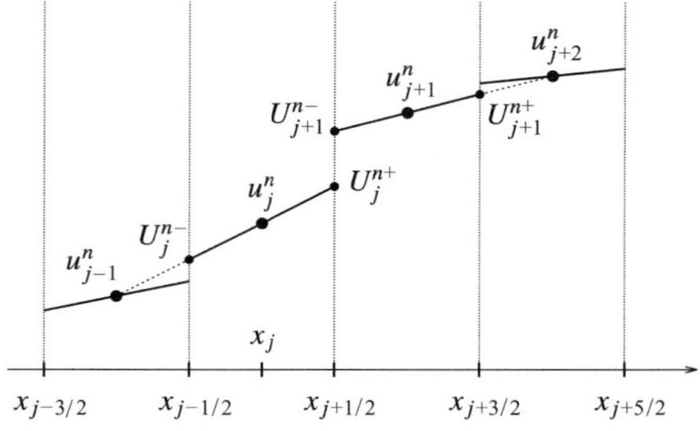

Fig. 10.12 A part of the piecewise linear approximate solution of $v_t + [F(v)]_x = 0$ with the mid-cell velocities u^n_j and edge-of-cell (sometimes denoted as "left" and "right") velocities $U^\pm_j$, with the dashed lines indicating the slopes

10.11.3.1   Linear Problem $v_t + cv_x = 0$

In the case of the linear advection equation $v_t + cv_x = 0$ which can be treated exactly we obtain the conservative difference scheme

$$
u^n_{j+1} = \begin{cases} u^n_j - cR\Delta^{(x)}_- u^n_j - \frac{1}{2}cR(1-cR)\big(\Delta^{(x)}_- \sigma^n_j\big)\Delta x \; ; \; c > 0 \,, \\[2mm] u^n_j - cR\Delta^{(x)}_+ u^n_j + \frac{1}{2}cR(1+cR)\big(\Delta^{(x)}_+ \sigma^n_j\big)\Delta x \; ; \; c < 0 \,. \end{cases}
$$

The CFL stability condition requires $|c|R \leq \frac{1}{2}$. As before the scheme can be expressed by its most important component, the numerical flux function

$$
h^n_{j+1/2} = \begin{cases} cu^n_j + \frac{1}{2}c(1-cR)\sigma^n_j \Delta x \quad\;\; ; \; c > 0 \,, \\[2mm] cu^n_{j+1} - \frac{1}{2}c(1+cR)\sigma^n_{j+1}\Delta x \; ; \; c > 0 \,. \end{cases}
$$

The cases $c > 0$ and $c < 0$ can be merged into a single expression

$$
h^n_{j+1/2} = \frac{1}{2}c\big(u^n_j + u^n_{j+1}\big) - \frac{1}{2}|c|\Delta^{(x)}_+ u^n_j + \frac{c}{2}\big(\text{sign}(c) - cR\big)\sigma^n_{j+\ell}\Delta x \,, \qquad (10.89)
$$

where $\ell = 0$ if $c > 0$ and $\ell = 1$ if $c < 0$. It is a straightforward exercise to show that if we define the slope as

$$
\sigma^n_j = \frac{\Delta^{(x)}_+ u^n_j}{\Delta x} = \frac{u^n_{j+1} - u^n_j}{\Delta x} \,,
$$

plugging (10.89) into the scalar version of (10.68) results in the linear Lax-Wendroff scheme (10.37) if c is reabsorbed into R—and the lesson is that one may easily have bad luck and end up with a second-order scheme which fails to be TVD. In order to attain the desired TVD property, we must put some sort of constraint on the slopes, in contrast to the flux-limiter schemes where we were obliged to restrict the fluxes. A popular slope limiter is

$$
\sigma^n_j = \frac{1}{\Delta x}\,\text{minmod}\big\{\Delta^{(x)}_+ u^n_j, \Delta^{(x)}_- u^n_j\big\} \,,
$$

where $\text{minmod}\{a, b\} = \frac{1}{2}\big(\text{sign}(a) + \text{sign}(b)\big)\min\{|a|, |b|\}$. Other popular dedicated functional forms exist, yet it is instructive to note that the flux-limiter schemes can be considered as special cases of the slope-limiter schemes if the slopes are defined as

$$
\sigma^n_j = \phi^n_j \frac{u^n_{j+1} - u^n_j}{\Delta x} \,.
$$

With this choice, for instance, the scheme (10.89) reduces to (10.84), and any flux-limiter formula from the set (10.75)–(10.82) is suitable for smoothing purposes.

10.11.3.2   Non-linear Conservation Laws of the Form $v_t + [F(v)]_x = 0$

The derivation of the slope-limiter schemes for non-linear equations of the form $v_t + [F(v)]_x = 0$ is more involving as the integration of the conservation law over a given cell that results in the numerical flux function can only be performed in an approximative manner. Just as in the linear case one resorts to a piecewise linear approximate solution such that the velocities at the endpoints of a given cell are related to the mid-cell velocities by

$$U_j^{n\pm} = u_j^n \pm \frac{1}{2}\sigma_j^n \Delta x \,,$$

as shown in Fig. 10.12. One replaces the exact flux function F by the approximate flux function G, and the numerical flux function $h_{j+1/2}^n$ by its approximation, $\tilde{h}_{j+1/2}^n$. In addition, since a TVD scheme is desired, one must ensure that the total variation of the piecewise linear approximation is less than or equal to the total variation of the point-wise approximation (see Sect. 9.7 of Ref. [4] as well as [17] for details).

We define G as the piecewise linear approximation that interpolates F at the four points $U_j^{n\pm}$ and $U_{j+1}^{n\pm}$, and calculate its slope across the cell j:

$$G_j'^n = \begin{cases} [F(U_j^{n+}) - F(U_j^{n-})]/[U_j^{n+} - U_j^{n-}] \; ; \; \sigma_j^n \neq 0 \,, \\ F'(u_j^n) \qquad\qquad\qquad\qquad\qquad\quad\; ; \; \sigma_j^n = 0 \,. \end{cases}$$

The function G, then, has the form

$$G(u)^n = \begin{cases} F(U_j^{n+}) + (u - U_j^{n+})G_j'^n \quad\;\; ; \; u \in [U_j^{n-}, U_j^{n+}] \,, \\ F(U_{j+1}^{n-}) + (u - U_{j+1}^{n-})G_{j+1}'^n \; ; \; u \in [U_{j+1}^{n-}, U_{j+1}^{n+}] \,. \end{cases}$$

This allows us to devise a scheme that depends on whether the considered spatial region contains a sonic point or not. A *sonic point* corresponds to the value of u at which the characteristic speed is zero, such that $F'(u_{j-1}^n) < 0 < F'(u_j^n)$, that is, u_s defines the point where F' changes sign, $F'(u_s) = 0$. This definition translates to the approximate derivative G' as well, and we end up with the following recipe.

The case $G_j'^n G_{j+1}'^n > 0$ (no sonic point in the region) The approximate numerical flux function is given by

$$\tilde{h}_{j+1/2}^n = \begin{cases} F(U_j^{n+}) - \frac{1}{2}\sigma_j^n \Delta t (G_j'^n)^2 \qquad ; \; F' > 0 \,, \\ F(U_{j+1}^{n-}) - \frac{1}{2}\sigma_{j+1}^n \Delta t (G_{j+1}'^n)^2 \; ; \; F' < 0 \,. \end{cases}$$

The case $G''_j G''_{j+1} \leq 0$ (there is a sonic point in the region) We define

$$
G''^n_{j+1/2} = \begin{cases} \left[F(U^{n-}_{j+1}) - F(U^{n+}_j) \right] / \left[U^{n-}_{j+1} - U^{n+}_j \right] \; ; \; U^{n-}_{j+1} \neq U^{n+}_j \; , \\ F'(U^{n+}_j) \qquad\qquad\qquad\qquad\qquad\qquad ; \; U^{n-}_{j+1} = U^{n+}_j \; , \end{cases}
$$

and consider two options. (1) If $G''_j > 0$, $G''_{j+1/2}(U^{n-}_{j+1} - U^{n+}_j) = 0$ and $G''_{j+1} < 0$, the approximate numerical flux function is

$$
\tilde{h}^n_{j+1/2} = \begin{cases} F(U^{n+}_j) - \frac{1}{2}\sigma^n_j \Delta t \left(G''^n_j \right)^2 \qquad ; \; \sigma^n_j \left(G''^n_j \right)^2 \geq \sigma^n_{j+1} \left(G''^n_{j+1} \right)^2 \; , \\ F(U^{n-}_{j+1}) - \frac{1}{2}\sigma^n_{j+1} \Delta t \left(G''^n_{j+1} \right)^2 \; ; \; \text{otherwise} \; , \end{cases}
$$

otherwise, it is

$$
\tilde{h}^n_{j+1/2} = \begin{cases} F(U^{n+}_j) - \frac{1}{2}\sigma^n_j \Delta t \left(G''^n_j \right)^2 \qquad ; \; G''^n_j \geq 0 \text{ and } G''^n_{j+1/2} \geq 0 \; , \\ F(U^{n-}_{j+1}) - \frac{1}{2}\sigma^n_{j+1} \Delta t \left(G''^n_{j+1} \right)^2 \; ; \; G''^n_{j+1/2} \leq 0 \text{ and } G''^n_{j+1} \leq 0 \; , \\ F(v_0) \qquad\qquad\qquad\qquad\qquad\;\; ; \; G''^n_j < 0 \text{ and } G''^n_{j+1} > 0 \; , \end{cases}
$$

where $v_0 = \min\left\{ \max\left\{ U^{n+}_j, u_s \right\}, U^{n-}_{j+1} \right\}$.

10.11.3.3   Kurganov–Tadmor Second Order Scheme

A very popular and robust scheme for the solution of non-linear conservation laws,

$$
v_t + [F(v)]_x = 0 \; , \tag{10.90}
$$

as well as convection-diffusion equations of the form

$$
v_t + [F(v)]_x = [Q(v, v_x)]_x \; , \tag{10.91}
$$

is the Kurganov–Tadmor scheme [18]. As before, v is a vector of conserved quantities, $F(v)$ is the non-linear convection flux, and $Q(v, v_x)$ is the dissipation (diffusive) flux. In the following, we write down the scheme for the scalar case in one space dimension; the formulas for two space dimensions are given in Sect. 11.3.3. The extension to higher dimensionalities of v (systems of scalar conservation laws), is straightforward as it proceeds component-wise (just replace $u \rightarrow u$).

We begin by describing the *fully discrete* (in both space and time) version of the scheme for (10.90) which is second-order accurate in the spatial domain. Firstly, the mid-cell velocities are calculated as

$$
u^{(n+1/2)-}_{j+1/2} = u^{n-}_{j+1/2} - \frac{\Delta t}{2} F\left(u^{n-}_{j+1/2} \right)_x \; , \quad u^{n-}_{j+1/2} = u^n_j + \Delta x (u_x)^n_j \left(\frac{1}{2} - R a^n_{j+1/2} \right) \; ,
$$

$$
u^{(n+1/2)+}_{j+1/2} = u^{n+}_{j+1/2} - \frac{\Delta t}{2} F\left(u^{n+}_{j+1/2} \right)_x \; , \quad u^{n+}_{j+1/2} = u^n_{j+1} - \Delta x (u_x)^n_{j+1} \left(\frac{1}{2} - R a^n_{j+1/2} \right) \; ,
$$

where $R = \Delta t / \Delta x$. The $+/-$ signs next to the time superscript denote the velocities at the left and right cell edge, in analogy to Fig. 10.12—although spatial differencing in the Kurganov–Tadmor scheme does not proceed in the same manner. The exact spatial derivatives, $v_x(x_{j+1/2}, t_{n+1})$, are approximated by

$$(u_x)_{j+1/2}^{n+1} = \frac{2}{\Delta x}\,\text{minmod}\left\{ \frac{w_{j+1}^{n+1} - w_{j+1/2}^{n+1}}{1 + R\big(a_{j+1/2}^n - a_{j+3/2}^n\big)}, \frac{w_{j+1/2}^{n+1} - w_j^{n+1}}{1 + R\big(a_{j+1/2}^n - a_{j-1/2}^n\big)} \right\},$$

where $\text{minmod}\{\alpha, \beta\} = \frac{1}{2}\big(\text{sign}(\alpha) + \text{sign}(\beta)\big)\min\{|\alpha|, |\beta|\}$. The quantities $a_{j\ldots}^n$ are the local maximal speeds which in the general (vector) case may be determined from the spectral radius of the Jacobian $\partial F / \partial u$:

$$a_{j+1/2}^n = \max\left\{ \rho\left(\frac{\partial F}{\partial u}\big(u_{j+1/2}^{n-}\big)\right),\ \rho\left(\frac{\partial F}{\partial u}\big(u_{j+1/2}^{n+}\big)\right)\right\}. \tag{10.92}$$

(In the scalar case this is no obstacle: in the Burgers equation, for instance, we have $F(u) = u^2/2$, and the maximum local speed is simply $a = \max_u F'(u) = \max_u u$.) All this information is then used to calculate the cell averages at time t_{n+1}:

$$\begin{aligned}
w_{j+1/2}^{n+1} &= \frac{u_j^n + u_{j+1}^n}{2} + \frac{\Delta x - a_{j+1/2}^n \Delta t}{4}\left((u_x)_j^n - (u_x)_{j+1}^n\right) \\
&\quad - \frac{1}{2a_{j+1/2}^n}\left[F\left(u_{j+1/2}^{(n+1/2)+}\right) - F\left(u_{j+1/2}^{(n+1/2)-}\right)\right], \\
w_j^{n+1} &= u_j^n + \frac{\Delta t}{2}\left(a_{j-1/2}^n - a_{j+1/2}^n\right)(u_x)_j^n \\
&\quad - \frac{R}{1 - R\big(a_{j-1/2}^n + a_{j+1/2}^n\big)}\left[F\left(u_{j+1/2}^{(n+1/2)-}\right) - F\left(u_{j-1/2}^{(n+1/2)+}\right)\right].
\end{aligned}$$

Finally, everything is put in place by using

$$\begin{aligned}
u_j^{n+1} &= Ra_{j-1/2}^n w_{j-1/2}^{n+1} + \left[1 - R\left(a_{j-1/2}^n + a_{j+1/2}^n\right)\right]w_j^{n+1} + Ra_{j+1/2}^n w_{j+1/2}^{n+1} \\
&\quad + \frac{\Delta x}{2}\left[\left(Ra_{j-1/2}^n\right)^2 (u_x)_{j-1/2}^{n+1} - \left(Ra_{j+1/2}^n\right)^2 (u_x)_{j+1/2}^{n+1}\right]. \tag{10.93}
\end{aligned}$$

The fully discrete scheme (10.93) can be turned into its semi-discrete version by taking the $\Delta t \to 0$ limit. We obtain

$$\begin{aligned}
\frac{du_j}{dt} &= -\frac{1}{2\Delta x}\left[\left(F\big(u_{j+1/2}^+(t)\big) + F\big(u_{j+1/2}^-(t)\big)\right) - \left(F\big(u_{j-1/2}^+(t)\big) + F\big(u_{j-1/2}^-(t)\big)\right)\right] \\
&\quad + \frac{1}{2\Delta x}\left[a_{j+1/2}(t)\big(u_{j+1/2}^+(t) - u_{j+1/2}^-(t)\big) - a_{j-1/2}(t)\big(u_{j-1/2}^+(t) - u_{j-1/2}^-(t)\big)\right].
\end{aligned}$$

This can be recast in a compact conservative form

$$\frac{\mathrm{d}u_j}{\mathrm{d}t} = -\frac{h_{j+1/2}(t) - h_{j-1/2}(t)}{\Delta x} \,, \tag{10.94}$$

where

$$h_{j+1/2}(t) = \frac{F\big(u^+_{j+1/2}(t)\big) + F\big(u^-_{j+1/2}(t)\big)}{2} - \frac{a_{j+1/2}(t)}{2}\left(u^+_{j+1/2}(t) - u^-_{j+1/2}(t)\right)$$

is the numerical flux function. The intermediate velocities $u^{\pm}_{j+1/2}$ are given by

$$\begin{aligned} u^+_{j+1/2}(t) &= u_{j+1}(t) - \frac{\Delta x}{2}(u_x)_{j+1}(t) \,, \\ u^-_{j+1/2}(t) &= u_j(t) + \frac{\Delta x}{2}(u_x)_j(t) \,, \end{aligned} \tag{10.95}$$

while the local speeds are still determined by (10.92), with the obvious change of dropping the superscript n and furnishing the velocities by the (t) argument.

Such semi-discrete schemes are useful, for instance, in strong stability preserving (SSP) methods for time integration presented in Sect. 12.6.1. In SSP methods the integration from $t_n \to t_{n+1}$ proceeds in a convex combination of explicit-Euler steps. Denoting the right-hand side of (10.94) for a general u by

$$L(u) = -\frac{h_{j+1/2}(u) - h_{j-1/2}(u)}{\Delta x} \,,$$

the complete time step is typically executed by an s-stage Runge–Kutta method:

$$\begin{aligned} u^{(1)} &= u^n + \Delta t\, L\big(u^n\big) \,, \\ u^{(l+1)} &= \eta_l u^n + (1 - \eta_l)\big(u^{(l)} + \Delta t\, L\big(u^{(l)}\big)\big) \,, \quad l = 1, 2, \ldots, s-1 \,, \\ u^{n+1} &= u^{(s)} \,. \end{aligned}$$

In the two-stage (second order in Δt) modified Euler method, for instance, we have $\eta_1 = \frac{1}{2}$, while in the three-stage (third order in Δt) which goes under the brand-name SSPRK(3, 3) we take $\eta_1 = \frac{3}{4}$ and $\eta_2 = \frac{1}{3}$. Other SSP variants are presented in Sect. 12.6.1.

Example As an illustration we solve the one-dimensional inviscid Burgers equation $v_t + (v^2/2)_x = 0$ with the initial condition $v(x, 0) = \sin x$ and periodic boundary condition. We use the semi-discrete version (10.94) of the Kurganov–Tadmor scheme, which allows us to use any ODE solver; we pick the SSPRK(3,3) mentioned above. The solution develops a shock discontinuity at critical time $t_c = 1$, which the difference scheme effortlessly resolves: see Fig. 10.13. Compare this solution to the solutions obtained by other techniques in Problem 10.12.5! ◁

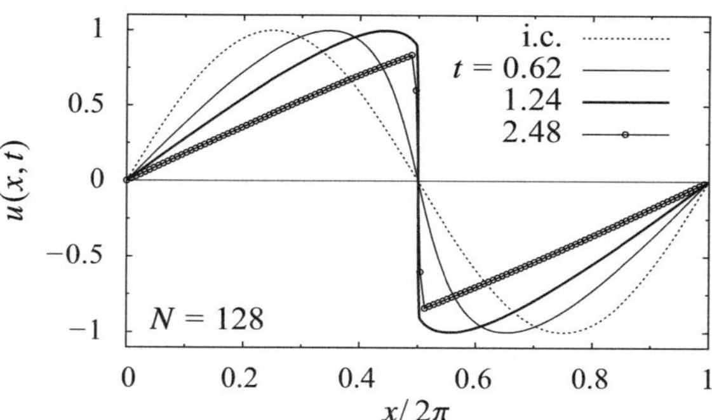

Fig. 10.13 Solution of the inviscid Burgers equation with periodic boundary conditions and smooth initial data by using the Kurganov–Tadmor scheme with $\theta = 1$ in the slope limiter for the spatial part and the SSPRK(3,3) method for time integration. The solution has been computed with $\Delta x = \Delta t = 0.0495$

Due to spatial differencing performed in a symmetric ("central") fashion (see Sect. 2 of Ref. [18]) the Kurganov–Tadmor scheme has several advantages with respect to high-resolution schemes based on upwind discretizations discussed previously. Its numerical viscosity—loosely speaking, the property of the scheme causing sharp solution features to flatten out and lose shape — is of the order $O\big(\Delta x^3 (a(u)u_{xxx})_x\big)$ as opposed to, for instance, the Lax–Friedrichs scheme where it is $O\big(\Delta x^2/\Delta t\big)$. It is also easy to implement, in particular because the computation of the numerical derivatives, $(u_x)_j$, may be carried out component-wise. A desirable way to perform this important step is to use some sort of a generalized minmod limiter, say,

$$(u_x)_j^n = \text{minmod} \left\{ \theta \frac{u_j^n - u_{j-1}^n}{\Delta x}, \; \frac{u_{j+1}^n - u_{j-1}^n}{2\Delta x}, \; \theta \frac{u_{j+1}^n - u_j^n}{\Delta x} \right\}, \quad 1 \le \theta \le 2,$$

where

$$\text{minmod}\{z_1, z_2, z_3, \ldots\} = \begin{cases} \min_j\{z_j\} \; ; \; z_j > 0 \;\; \forall j, \\ \max_j\{z_j\} \; ; \; z_j < 0 \;\; \forall j, \\ 0 \qquad\qquad ; \; \text{otherwise}. \end{cases}$$

In this case the scheme (10.94) is TVD if

$$\left| \frac{(u_x)_{j\pm1/2}}{\Delta u_{j\pm1/2}} \le 2 \right.$$

The conservative form (10.94) is also ideally suited to introduce the additional parabolic diffusion term as in (10.91). The semi-discrete scheme then assumes the form

$$\frac{\mathrm{d}u_j}{\mathrm{d}t} = -\frac{h_{j+1/2}(t) - h_{j-1/2}(t)}{\Delta x} + \frac{p_{j+1/2}(t) - p_{j-1/2}(t)}{\Delta x}, \tag{10.96}$$

where the diffusion flux may be discretized as

$$
p_{j+1/2}(t) = \frac{1}{2}\left[Q\left(u_j(t), \frac{u_{j+1}(t) - u_j(t)}{\Delta x}\right) + Q\left(u_{j+1}(t), \frac{u_{j+1}(t) - u_j(t)}{\Delta x}\right)\right].
$$

An example of treating the combined advection-diffusion problem by the Kurganov–Tadmor scheme is provided by Problem 10.12.6.

10.11.3.4 Kurganov–Levy Third Order Scheme

In central-differencing approaches the order of the scheme depends on the degree of the polynomial that is used to (re)construct the velocity profile across the neighboring cells, computed from the cell averages $\bar{u}_j^n$ of the approximate solution. With a piecewise linear reconstruction and a second-order quadrature in time, we obtain the second-order scheme just described. To obtain a third-order central scheme, a piecewise parabolic reconstruction must be used, in conjunction with a more accurate time quadrature (for instance, Simpson). The net result is that the third-order scheme can be written exactly as in Eq. (10.94), the only difference being the calculation of intermediate velocities: instead of (10.95) one has

$$
\begin{aligned}
u_{j+1/2}^{+}(t) &= A_{j+1}(t) - \frac{\Delta x}{2} B_{j+1} + \frac{(\Delta x)^2}{8} C_{j+1}, \\
u_{j+1/2}^{-}(t) &= A_{j}(t) + \frac{\Delta x}{2} B_{j} + \frac{(\Delta x)^2}{8} C_{j},
\end{aligned}
\tag{10.97}
$$

where A_j, B_j and C_j depend on the chosen reconstruction method. Any piecewise quadratic reconstruction can be exploited to produce the $O(1)$–$O(\Delta x)$–$O(\Delta x^2)$ hierarchy of (10.97). The diffusive term may be added precisely as in (10.96), with an appropriate (fourth-order) discretization of the diffusion flux: see [19] for details.

10.11.4 High-Order Essentially Non-oscillatory (ENO) Schemes

As indicated in Sect. 10.11.1 it is difficult to construct TVD schemes with more than first-order accuracy near the solution extrema, and one way to attain higher spatial accuracy is to relax the TVD requirement (10.73) to (10.74); see also [20, 21]. In the ENO approach [22], the difference scheme is permitted to produce a certain amount of spurious oscillations, yet the $O(1)$ Gibbs-like phenomena near the discontinuities are "essentially" prevented—hence the name of the technique. The main idea behind the ENO approximations is to adaptively avoid the cell containing the discontinuity (or a sharp gradient) within the whole stencil, the size of which depends on the spatial order one aims to achieve. The weighted ENO (WENO) schemes attain even better smoothness by a weighted combination of the information from several candidate stencils. The WENO research field is lively; in the following we present only a handful of the most popular, well established and reliable fifth-order semi-discrete schemes for conservation laws of the form (10.91).

Weighted Essentially Non-oscillatory (WENO) Schemes

The schemes for (10.91) are still written in the usual conservative form (10.94), only the approximations of the fluxes are modified. In fifth-order methods, which apparently enjoy the highest popularity, the approximate numerical flux functions assume the form

$$h_{j+1/2}(t) = \sum_{m=0}^{2} \omega_m h_{j+1/2}^{(m)}(t) \,. \tag{10.98}$$

To ensure numerical stability all fluxes F, be they exact or approximate, are split into the negative and positive parts, such that

$$F(u) = F^+(u) + F^-(u) \,,$$

where $F^+(u)$ and $F^-(u)$ satisfy $\mathrm{d}F^+(u)/\mathrm{d}u \geq 0$ and $\mathrm{d}F^-(u)/\mathrm{d}u \leq 0$. A standard way to enforce this is the Lax–Friedrichs splitting

$$F^{\pm}(u) = \frac{1}{2}\bigl(F(u) \pm \mu u\bigr) \,,$$

where $\mu = \max_u |F'(u)|$, and the maximum is taken over the whole relevant range of u. (In the trivial case of the linear advection equation $u_t + cu_x = 0$, for instance, we have $F(u) = cu$, hence $F'(u) = c$, $\mu = c$ and $F^+(u) = cu$, $F^-(u) = 0$.) For other types of flux splittings refer to Ref. [23]. In the formulas below we only show how $h_{j+1/2}^+$ is constructed and drop the '+' sign for clarity; the negative part $h_{j+1/2}^-$ is symmetric to the positive part with respect to $x_{j+1/2}$. We also suppress the explicit dependence on t. In the original scheme by Jiang and Shu [23], labeled by WENO–JS5, the flux in (10.98) is composed by the weighted sum of

$$h_{j+1/2}^{(0)} = \frac{1}{6}\bigl(2F_{j-2} - 7F_{j-1} + 11F_j\bigr) \,,$$

$$h_{j+1/2}^{(1)} = \frac{1}{6}\bigl(-F_{j-1} + 5F_j + 2F_{j+1}\bigr) \,,$$

$$h_{j+1/2}^{(2)} = \frac{1}{6}\bigl(2F_j + 5F_{j+1} - F_{j+2}\bigr) \,,$$

where $F_j = F(x_j, t)$ and ω_p are the weights given by

$$\omega_m = \frac{\alpha_m}{\sum_{n=0}^{2} \alpha_n}, \quad m = 0, 1, 2, \tag{10.99}$$

where

$$\alpha_m = \frac{d_m}{(\varepsilon + \beta_m)^2}$$

and

$$d_0 = 1/10, \ d_1 = 6/10, \ d_2 = 3/10.$$

The parameter ε ($0 < \varepsilon \ll 1$, typically 10^{-6} or less) is introduced to prevent the denominator of α_n from becoming zero. The quantities

$$\begin{aligned}
\beta_0 &= \frac{13}{12}\left(F_{j-2} - 2F_{j-1} + F_j\right)^2 + \frac{1}{4}\left(F_{j-2} - 4F_{j-1} + 3F_j\right)^2, \\
\beta_1 &= \frac{13}{12}\left(F_{j-1} - 2F_j + F_{j+1}\right)^2 + \frac{1}{4}\left(F_{j+1} - F_{j-1}\right)^2, \\
\beta_2 &= \frac{13}{12}\left(F_j - 2F_{j+1} + F_{j+2}\right)^2 + \frac{1}{4}\left(3F_j - 4F_{j+1} + F_{j+2}\right)^2,
\end{aligned} \tag{10.100}$$

are the *smoothness indicators*, which measure the relative smoothness of the function $u(x)$ within the stencil; a larger β_n indicates less smoothness.

In general the classic WENO–JS5 scheme attains only third-order accuracy near smooth extrema or at critical points. This deficiency has been cured by using a mapping function for ω_m such that the original Jiang–Shu weights are used as a first approximation and mapped to more accurate values, resulting in a fifth-order scheme known as WENO–M5 [24]. The mapping function is

$$g_m(\omega) = \frac{\omega\left(d_m + d_m^2 - 3d_m\omega + \omega^2\right)}{d_m^2 + \omega(1 - 2d_m)},$$

and the weights are defined as

$$\omega_m^{\mathrm{M}} = \frac{\alpha_m^{\mathrm{M}}}{\sum_{n=0}^{2} \alpha_n^{\mathrm{M}}}, \quad \alpha_m^{\mathrm{M}} = g_m\left(\omega_m^{\mathrm{JS}}\right), \quad m = 0, 1, 2,$$

where ω_m^{JS} are calculated by using Eq. (10.99).

Another scheme which ensures less dissipation and higher resolution than the classical WENO–JS5 scheme has been developed, and as far as can be inferred from the literature it has become somewhat of a standard: it is known as WENO–Z5 [25, 26]. Since the weight mapping of the M–scheme is time-consuming, the idea of the Z–scheme is to increase the order of the smoothness indicators instead. These are defined as

$$\beta_m^Z = \frac{\beta_m + \varepsilon}{\beta_m + \tau_5 + \varepsilon} \, , \quad m = 0, 1, 2 \, ,$$

where β_m are calculated from (10.100) and

$$\tau_5 = |\beta_0 - \beta_2|$$

is an additional global smoothness indicator (evaluated across the complete five-point stencil). The corresponding WENO–Z5 weights are then given by

$$\omega_m^Z = \frac{\alpha_m^Z}{\sum_{n=0}^{2} \alpha_n^Z} \, , \quad \alpha_m^Z = \frac{d_m}{\beta_m^Z} = d_m \left[1 + \left(\frac{\tau_5}{\beta_m + \varepsilon} \right)^p \right] \, , \quad m = 0, 1, 2 \, .$$

Here $p \geq 1$ is a parameter that can be used to keep the order of the scheme at critical points as high as possible: see [26] for details. In order to better capture parts of the solution with curvature features, the WENO–Z5+ scheme [27] may be applied in which the weights are slightly modified by using

$$\alpha_m^{Z+} = d_m \left[1 + \left(\frac{\tau_5}{\beta_m + \varepsilon} \right)^p + \eta \frac{\beta_m}{\tau_5 + \varepsilon} \right] \, , \quad \eta \approx (\Delta x)^{2/3} \, .$$

See [28, 29] for further information on how schemes of higher orders (7, 9, and 11) are constructed. Many other variants of WENO schemes exist: see e.g. [30, 31].

Example Consider the linear advection equation

$$v_t + v_x = 0 \, , \quad x \in [-1, 1] \, , \tag{10.101}$$

with periodic boundary condition and the initial condition given by

$$v(x, t) = \begin{cases} \frac{1}{6}\big(G(x; \beta, z - \delta) + 4G(x; \beta, z) + G(x; \beta, z + \delta)\big) & ; \ x \in [-0.8, -0.6] \, , \\ 1 & ; \ x \in [-0.4, -0.2] \, , \\ 1 - 10|x - 0.1| & ; \ x \in [0, 0.2] \, , \\ \frac{1}{6}\big(E(x; \alpha, a - \delta) + 4E(x; \alpha, a) + E(x; \alpha, a + \delta)\big) & ; \ x \in [0.4, 0.6] \, , \\ 0 & ; \ \text{otherwise} \, , \end{cases}$$

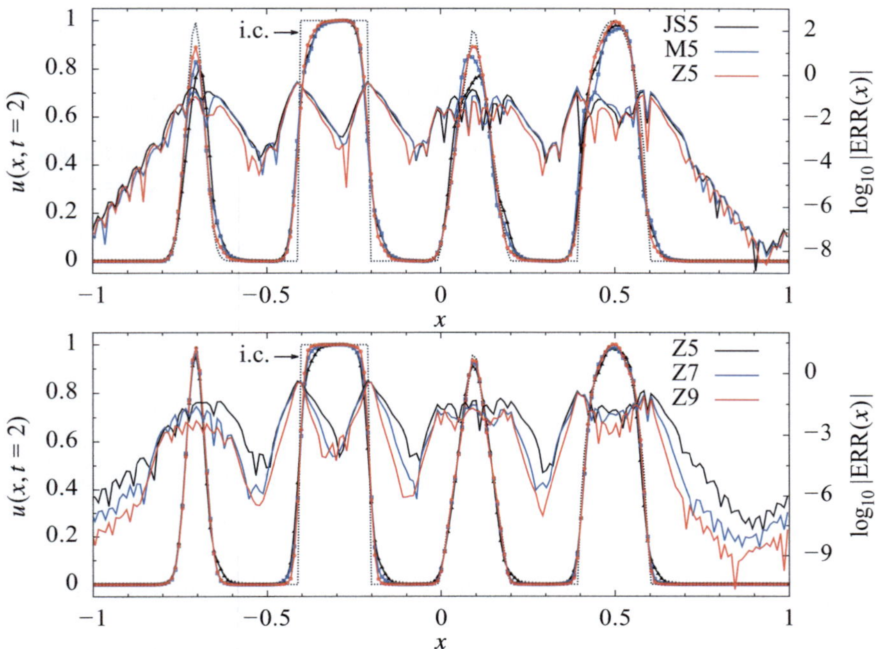

Fig. 10.14 [Top] Solution of the linear advection equation (10.101) by using fifth-order WENO–JS, –M and –Z spatial schemes on a 200-point grid, coupled to third-order time integration. Shown are the solution values (left ordinate) and the point-wise errors (right ordinate) after a full period (at $t = 2$). The parameter $p = 2$ was used in the WENO–Z5 scheme. [Bottom] Solutions by using the WENO–Z scheme of order 5, 7 and 9, with the step size adjusted as $\Delta t = (\Delta x)^{q/3}$

where

$$G(x; \beta, z) = e^{-\beta(x-z)^2} ,$$
$$E(x; \alpha, a) = \left[\max\{1 - \alpha^2(x - a)^2, 0\}\right]^{1/2} ,$$

$z = -0.7$, $\delta = 0.005$, $\beta = \log 2/(36\delta^2)$, $\alpha = 10$ and $a = 0.5$. This popular benchmark problem discussed in detail in Ref. [28] involves several shapes—a Gaussian, a square wave, a sharp triangle and a half-ellipse. Although the transport equation is simple and linear, this velocity profile represents a very stringent test because it contains a combination of functions that are not smooth and functions that are smooth but sharply peaked, and the scheme therefore may have difficulties to advect all of them with sufficient fidelity.

Figure 10.14 (top) shows the solutions by using the WENO–JS5, WENO–M5 and WENO–Z5 schemes on a mesh with $N = 200$ points and $\varepsilon = 10^{-6}$, integrated up to $t = 2$ (a full period) by a third-order Runge–Kutta scheme with $\Delta t = \Delta x$. Coupling high-spatial-order schemes to a low-order time integrator is not doing them a favor;

to roughly compensate for this deficiency one can adjust the temporal step size by rescaling it as

$$\Delta t = \left(\Delta x\right)^{q/3} ,$$

where q is the spatial order. Accordingly, Fig. 10.14 (bottom) shows the solutions after a full period by using the WENO–Z schemes of order 5, 7 and 9 with $\Delta t = (\Delta x)^{5/3}$, $(\Delta x)^{7/3}$ and $(\Delta x)^{9/3}$, respectively.   ◁

10.12   Problems

10.12.1   Diffusion Equation

Use several difference methods to solve the linear diffusion equation

$$v_t = D v_{xx} , \qquad x \in [0, a] , \qquad t > 0 , \qquad D > 0 ,$$

with specified initial and boundary conditions. The equation describes the heat transfer in an infinite slab of thickness a (or along a thin insulated wire).

⊙   Discretize the diffusion equation on the time-space mesh as described in Sect. 10.1. Solve the equation by using the explicit scheme (10.13) and pay attention to stability: keep $0 < r = D\Delta t/\Delta x^2 < 1/2$. Compare the solution at $r = 1/6$ to solutions at different r. The initial condition is: temperature zero everywhere except between $x = 0.2a$ and $x = 0.4a$, where the slab is at constant non-zero temperature T_0. At time $t = 0$ we switch on additional heating between $x = 0.5a$ and $x = 0.75a$, with a heat release rate of $5T_0\lambda/a^2$. Plot the temperature profile $T(x, t) = v(x, t)$ at dimensionless times $Dt/a^2 = 0(0.3)3$!

Solve the diffusion equation by the implicit scheme (10.20) or the Crank–Nicolson scheme (10.21). Do you observe more freedom in choosing r? How does the choice of r influence the precision? What happens to the discrete "energy"

$$E^n = \Delta x \sum_{j=0}^{N} \left(u_j^n\right)^2 ,$$

when you use the difference schemes (in stable and unstable regimes)?

⊕   Use the explicit scheme with local "freezing" of the function $D(v)$ described in Sect. 10.8 to solve the non-linear diffusion equation

$$v_t = (D(v) v_x)_x , \qquad D(v) = \frac{1 + 3v^2}{1 + v^2} , \qquad x \in [0, 1] ,$$

with the initial condition $v(x, 0) = \sin 3\pi x$ and boundary conditions $v(0, t) = v(1, t) = 0$. Use the discretization $N = 50$ ($\Delta x = 0.02$) and $\Delta t = 0.00005$, and monitor the solution until $T = 0.1$. By changing Δx and Δt confirm the stability criterion. Does the stability criterion remain valid for a different initial condition, e.g. $v(x, 0) = 100\, x(1 - x)|x - 0.5|$? Write down and solve the corresponding implicit scheme.

10.12.2 Initial-Boundary Value Problem for $v_t + cv_x = 0$

We would like to solve numerically the first-order (hyperbolic) advection equation

$$v_t + cv_x = 0\,, \qquad c = \pm 1\,,$$

on $x \in [0, 1]$ with the initial condition $v(x, 0) = \sin^{80} \pi x$ and periodic boundary condition $v(0, t) = v(1, t)$.

$\odot$ Solve the equation $v_t + v_x = 0$ by using the FTBS scheme (10.35) and the Lax–Wendroff scheme (10.37) with the discretization $N = 20$ and time step $\Delta t = 0.01$. Plot the solutions at $t = 0.0, 0.1, 0.2, 0.4$, and 1.0. Repeat the exercise with $N = 100$ and time step $\Delta t = 0.002$. Solve the equation $v_t - v_x = 0$ with the same discretization, but use the FTFS scheme (10.34) instead of FTBS. What happens to the amplitudes and phases of the solutions? (Compare them to exact solutions at chosen times.) Discuss the solutions in the cases when the parameter R is near the stability limit (Table 10.2).

Repeat the exercise by using the Crank–Nicolson scheme (10.43). Pay attention to the correct treatment of the boundary conditions (periodicity) when you rewrite the system of difference equations in matrix form $F_1 u^{n+1} = r^n$. (The matrix is cyclic tridiagonal.) Do you observe any peculiarities or limitations regarding the parameter R? (See Fig. 10.9.) Compare the amplitudes and phases of the solutions to those from the first part of the Problem. What happens when discretization is refined to, say, $\Delta x = 0.001$, $\Delta t = 0.0002$?

$\oplus$ We put the difference schemes for hyperbolic PDE to a harsher test by including discontinuous initial conditions. We discuss the case $v_t - v_x = 0$ with a periodic boundary condition $v(0, t) = v(1, t)$ and a "step" initial condition

$$v(x, 0) = \begin{cases} 1\;; & 0.4 \le x \le 0.6\,, \\ 0\;; & \text{otherwise}\,. \end{cases}$$

Use the same discretization as in the first part of the Problem and compute the solution by using the schemes: FTFS (10.34), Lax–Wendroff (10.37), Lax–Friedrichs (10.38), and Crank–Nicolson (10.43): the results should look like Fig. 10.10.

Finally, use the high-resolution scheme (10.85) with the flux limiter function (10.82) and (10.78). If the formulas access the points x_{-1} or x_{N+1}, apply correctly the periodic boundary condition. If the expression for θ has a zero in the denominator, set $\theta = 0$. Use $(\Delta x, \Delta t) = (100, 0.002)$ and $(1000, 0.0002)$.

10.12.3   Dirichlet Problem for a System of Non-linear Hyperbolic PDE

We are interested in the solutions of a hyperbolic system (example from Ref. [1])

$$\boldsymbol{v}_t = A\boldsymbol{v}_x \,, \qquad x \in [0, 1] \,, \quad t > 0 \,,$$

with the matrix

$$A = \frac{1}{3}\begin{pmatrix} 5 & 3 & 5 \\ -1 & -3 & -7 \\ 1 & -3 & 1 \end{pmatrix} \,.$$

$\odot$ Compute the matrices S and S^{-1} occurring in the similarity transformation $D = SAS^{-1}$ that diagonalizes A. Then solve the system by using the Lax–Wendroff scheme (10.64), with the initial condition

$$\boldsymbol{v}(x, 0) = \left(\cos 4\pi x, \;\; (1 - x)\cos \pi x, \;\; e^{-x}\right)^{\mathrm{T}}$$

and discretization $N = 100$ ($\Delta x = 0.01$). From the eigenvalues of A it is clear that three boundary conditions may be assigned to this system, two at $x = 1$ and one at $x = 0$. We choose $v_1(1, t) = 1$, $v_2(1, t) = \sin 2t$, and $v_3(0, t) = \cos 2t$. Elsewhere, we prescribe numerical boundary conditions $u_{10}^n = u_{11}^n$, $u_{20}^n = u_{21}^n$, and $u_{3N}^n = u_{3N-1}^n$. Compute the solutions up to $t = 1.0$. Repeat the computation with the schemes (10.61) and (10.62) in the regime where they are expected to be stable.

Redo the calculation by using the initial condition

$$\boldsymbol{v}(x, 0) = \left(\sin \frac{\pi}{2}x, \;\; \cos \pi x, \;\; xe^{1-x}\right)^{\mathrm{T}}$$

and boundary conditions $v_1(0, t) = \sin 2t$, $v_1(1, t) = 1$, and $v_3(1, t) = \cos 2t$. Take $u_{20}^n = u_{21}^n$, $u_{2N}^n = u_{2N-1}^n$, and $u_{30}^n = u_{31}^n$ as numerical boundary conditions.

$\oplus$ Repeat the exercise by using the flux-splitting scheme (10.63) without changing the spatial discretization or the length of the time step. Compare the results to those obtained in the first part of the Problem.

10.12.4   Second-Order and Fourth-Order Wave Equations

This Problem [3] deals with the one-dimensional second-order wave equation

$$v_{tt} = c^2 v_{xx} \,, \qquad x \in [0, L] \,, \qquad t > 0 \,, \qquad c > 0 \,,$$

describing the oscillations of a thin light string of length L with specified initial and boundary conditions. By substitution $x/L \to x$, $v/L \to v$, and $cT/L \to t$ we bring the equation to dimensionless form. The initial conditions are

$$v(x, 0) = \tfrac{1}{2} x(1 - x) , \qquad v_t(x, 0) = 0 ,$$

and the boundary solutions are $v(0, t) = v(1, t) = 0$. The exact solution is

$$v(x, t) = \frac{2}{\pi^3} \sum_{k=1}^{\infty} \frac{1}{k^3} (1 - \cos k\pi) \cos k\pi t \, \sin k\pi x .$$

$\odot$ Solve the wave equation by using the explicit difference scheme (10.45). Compute the deflections u_j^n after five oscillation periods. Pay attention to stability, which is ensured if $R = c\Delta t / \Delta x \le 1$. How do the results depend on the initialization scheme (Sect. 10.7.3)? Compare the numerical and analytic solutions, and find a good criterion for stopping the summation in the exact solution. Test the implicit method (10.46) too.

Solve the Problem by treating the equation $v_{tt} = v_{xx}$ as a system of first-order hyperbolic equations. By denoting $p = u_x$ and $q = u_t$, it assumes a simple form $p_x - q_t = 0, q_x - p_t = 0$. To solve it, first use the difference scheme

$$\frac{p_{j+1}^n - p_{j-1}^n}{2\Delta x} = \frac{q_j^{n+1} - \tfrac{1}{2}\left(q_{j+1}^n + q_{j-1}^n\right)}{\Delta t} ,$$

$$\frac{q_{j+1}^n - q_{j-1}^n}{2\Delta x} = \frac{p_j^{n+1} - \tfrac{1}{2}\left(p_{j+1}^n + p_{j-1}^n\right)}{\Delta t} .$$

Along $t = 0$ we have $p = u_x = \tfrac{1}{2}(1 - 2x)$ and $q = u_t = 0$. Along $x = 0$ we have $u = 0$, thus $q = 0$ and $q_t = p_x = 0$. This provides the constraints for points outside of the mesh: $p_{-1}^n = p_1^n$, $p_{N+1}^n = p_{N-1}^n$, $q_{-1}^n = -q_1^n$, and $q_{N+1}^n = -q_{N-1}^n$. The solution at arbitrary time are the vectors p and q. The deflections can then be computed by simple quadrature, e.g. $\Delta u \approx \tfrac{1}{2}(p_j + p_{j+1})\Delta x$. Also solve the problem by the Lax–Wendroff method (10.37) in the form

$$p_j^{n+1} = p_j^n + \frac{R}{2}\left[q_{j+1}^n - q_{j-1}^n\right] + \frac{R^2}{2}\left[p_{j+1}^n - 2p_j^n + p_{j-1}^n\right] ,$$

$$q_j^{n+1} = q_j^n + \frac{R}{2}\left[p_{j+1}^n - p_{j-1}^n\right] + \frac{R^2}{2}\left[q_{j+1}^n - 2q_j^n + q_{j-1}^n\right] .$$

Carefully determine the requirements at the boundary points x_0 and x_N as dictated by the scheme and the boundary conditions! What differences between the methods do you observe?

$\oplus$ How would you approach the solution of the fourth-order wave equation

$$\left(c_1^2 \frac{\partial^2}{\partial x^2} - \frac{\partial^2}{\partial t^2}\right) \left(c_2^2 \frac{\partial^2}{\partial x^2} - \frac{\partial^2}{\partial t^2}\right) v = 0$$

with specified initial conditions for $v(x, 0)$, $v_t(x, 0)$, $v_{tt}(x, 0)$, and $v_{ttt}(x, 0)$ on $-\infty < x < \infty$? Such an equation describes the propagation of long shallow waves in an infinite channel with two fluids that do not mix [32]. In equilibrium (and at large distances from the waves) the fluid with density ρ_2 and height h_2 is at the bottom, and the fluid with density $\rho_1 < \rho_2$ and height h_1 on the top. The parameters are related by

$$c_1^2 + c_2^2 = g(h_1 + h_2)$$

and

$$c_1^2 c_2^2 = g^2 h_1 h_2 (1 - \rho_1/\rho_2) .$$

The solution $v(x, t)$ measures the deviation from the equilibrium height $h_1 + h_2$.

10.12.5   Burgers Equation

The one-dimensional Burgers equation

$$v_t + v v_x = D v_{xx}$$

is widely used in modeling gas dynamics, acoustic phenomena, and turbulence. This is a mixed-type equation: for small values of D it is hyperbolic, while for large D its parabolic nature dominates. In this Problem we get to know the basic approaches to solving non-linear partial differential equations.

⊙   Solve the Burgers equation on $x \in [0, 1]$ with the initial condition $v(x, 0) = \sin 2\pi x$ and boundary conditions $v(0, t) = v(1, t) = 0$. Use several methods, e.g. the explicit one (Eq. (10.50)), lagging the non-linear term (Eq. (10.52)), or by expanding the current solution around the solution at the previous time step (Eq. (10.54)). Use the discretization $N = 100$, $\Delta t = 0.007$, and the parameter $D = 0.1$. Compute the solution up to $T \approx 0.2$. Gradually decrease D. What do you observe? Try the discretization $N = 1000$, $\Delta t = 0.0007$ and reduce D down to $D \approx 0.0001$. Compute the solution at each time step of the implicit scheme (10.51) directly (without linearization) by using the multi-dimensional Newton's method described in Sect. 10.8.

Use the generalization of the scheme for the linear problem $v_t + c v_x = D v_{xx}$, with the usual notations $R = \Delta t / \Delta x$ and $r = D \Delta t / \Delta x^2$,

$$u_j^{n+1} = u_j^n - \frac{R}{2} u_j^n \Delta_0^{(x)} u_j^n + \left(r + \frac{R^2}{2} \left(u_j^n \right)^2 \right) \Delta_2^{(x)} u_j^n .$$

In addition, try out the "mixed" scheme

$$u_j^{n+1} = u_j^n - R u_j^n \left(\mathcal{D}^n u_j^n \right) + r \Delta_2^{(x)} u_j^n , \tag{10.102}$$

in which the diffusion term Dv_{xx} is approximated by the central difference, while the first difference in the non-linear term depends on the sign of the solution,

$$\left(\mathcal{D}^n u_j^n\right) = \begin{cases} u_{j+1}^n - u_j^n \; ; \; u_j^n < 0 \, , \\ u_j^n - u_{j-1}^n \; ; \; u_j^n \geq 0 \, . \end{cases}$$

$\oplus$ Solve the simplified equation $v_t + vv_x = 0$ (i.e. $D = 0$) for the same initial and boundary conditions by using the high-resolution scheme (10.87) with the flux-limiters (10.82) and (10.78). Compare the solutions. The exact solution is

$$v(x, t) = -2 \sum_{k=1}^{\infty} \frac{J_k(-2\pi kt)}{2\pi kt} \, \sin(2\pi kx) \, ,$$

but it is only valid until $1/(2\pi)$ when the shock-wave "breaks" (Fig. 10.15). Repeat the exercise on $x \in [-1, 1]$ with a discontinuous initial condition

$$v(x, 0) = \begin{cases} 1.0 \; ; \; x \leq 0 \, , \\ 0.5 \; ; \; x > 0 \, . \end{cases}$$

See also Example on page 681.

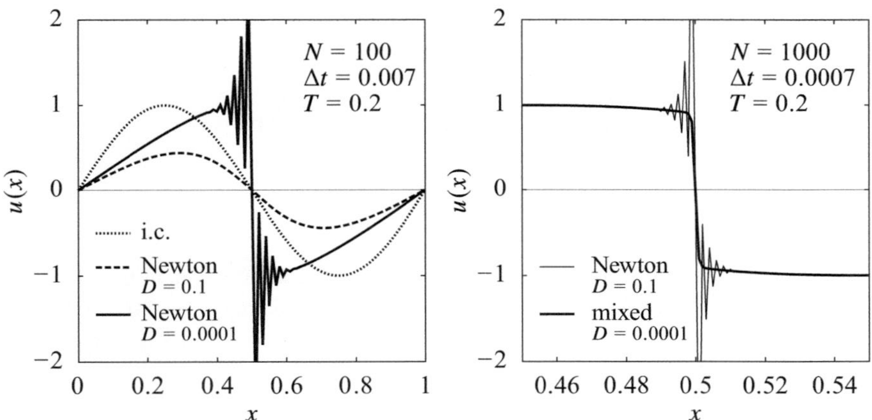

Fig. 10.15 Solutions of the Burgers equation. [LEFT] Initial condition (i.c.) and the solution by Newton's method with $D = 0.1$ and $D = 0.0001$ on a mesh with 100 subintervals. At small D a discontinuity forms that the scheme is unable to resolve. [RIGHT] Solution by Newton's method and by the mixed scheme (10.102) with $D = 0.0001$ by using a 10-times finer mesh and 10-times shorter time step (zoom-in on the discontinuity)

10.12.6 Buckley–Leverett Equation

Consider the scalar Buckley–Leverett equation [33]

$$v_t + [F(v)]_x = \varepsilon\big(D(v)v_x\big)_x \,, \quad \varepsilon D(v) \geq 0 \,, \tag{10.103}$$

which is a prototype model for oil-reservoir simulations involving two-phase flow: it describes the displacement of oil from the reservoir by either water or gas. In general the equation is of mixed convection-diffusion type, but typically $D(v)$ is chosen to vanish at specific values of v, rendering the equation degenerate parabolic.

$\odot$ Solve Eq. (10.103) for $x \in [0, 1]$ by using the Kurganov–Tadmor scheme described in Sect. 10.11.3.3 with

$$F(v) = \frac{v^2}{v^2 + (1 - v)^2}$$

and

$$D(v) = 4v(1 - v) \,. \tag{10.104}$$

Set $\varepsilon = 0.01$ and use the initial condition

$$v(x, 0) = \begin{cases} 1 - 3x \; ; \; 0 \leq x \leq \frac{1}{3} \,, \\ 0 \qquad ; \; \frac{1}{3} < x \leq 1 \,, \end{cases}$$

while the boundary value $v(0, t) = 1$ should remain fixed. The solution should look like the one shown in Fig. 10.16 (left).

$\oplus$ Gravitational effects may be added into the model by using

$$F(v) = \frac{v^2}{v^2 + (1 - v)^2}\big(1 - 5(1 - v)^2\big) \,,$$

but keeping the same functional form for the diffusion coefficient as in Eq. (10.104). (The equation is slightly more intricate in this case as one needs to handle the flux where $F'(v)$ changes sign.) Solve the Buckley–Leverett problem including gravitational effects with the initial condition

$$v(x, 0) = \begin{cases} 0 \; ; \; 0 \leq x < 1 - \frac{1}{\sqrt{2}} \,, \\ 0 \; ; \; 1 - \frac{1}{\sqrt{2}} \leq x \leq 1 \,, \end{cases}$$

and the boundary value $v(1, t) = 1$ fixed: compare your result to the one presented in Fig. 10.16 (right).

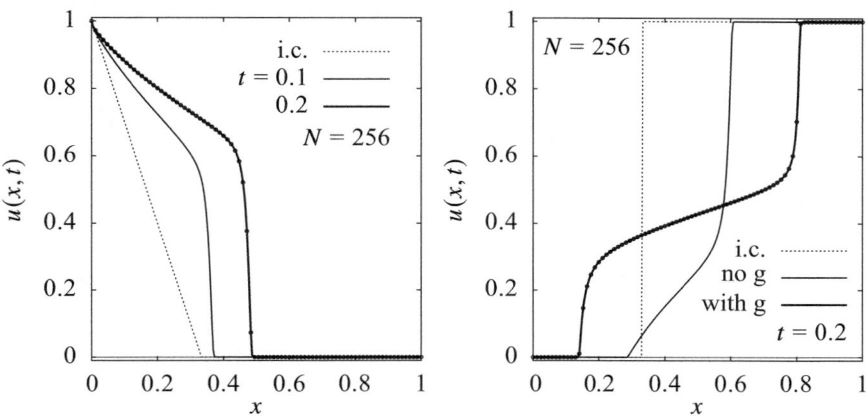

Fig. 10.16 Solutions of the Buckley–Leverett equation on a 256-point grid on [0, 1]. [LEFT] The first part of the Problem (no gravitational effects). The initial condition (i.c.) and the solutions at $t = 0.1$ and $t = 0.2$ are shown. (In the latter case only every third point is drawn for reasons of clarity.) [RIGHT] The second part of the Problem with the modified flux function modeling the gravitational effects ("with g") and with gravitational effects excluded ("no g"), drawn at $t = 0.2$

10.12.7 The Shock-Tube Problem

Here we study the dynamics of a gas in an infinite tube, where a barrier at $x = 0$ initially separates two regions with different pressures and densities. We remove the barrier and study the time dependence of the density, pressure, and velocity of the gas. The system is described by three coupled non-linear PDE [1]:

$$\rho_t + (\rho v)_x = 0\,,$$
$$(\rho v)_t + (\rho v^2 + p)_x = 0\,,$$
$$E_t + [v(E + p)]_x = 0\,,$$

where ρ is the density, p is the pressure, v is the velocity of the gas, and E its total energy. The pressure and energy are related by the equation of state $p = (\kappa - 1)(E - \rho v^2/2)$, where $\kappa = c_{\mathrm{p}}/c_{\mathrm{v}} = 1.4$ is the specific heat ratio. For the solution we use the variables ρ, the momentum density $m = \rho v$, and E. The system can then be written in matrix form $\boldsymbol{v}_t = A\boldsymbol{v}_x$, where $\boldsymbol{v} = (\rho, m, E)^{\mathrm{T}}$ and

$$A = \begin{pmatrix} 0 & -1 & 0 \\[2ex] -\dfrac{\kappa - 3}{2}\dfrac{m^2}{\rho^2} & (\kappa - 3)\dfrac{m}{\rho} & -(\kappa - 1) \\[2ex] \dfrac{m}{\rho^2}\left(\kappa E - (\kappa - 1)\dfrac{m^2}{\rho}\right) & -\dfrac{1}{\rho}\left(\kappa E - \dfrac{3(\kappa - 1)}{2}\dfrac{m^2}{\rho}\right) & -\kappa\dfrac{m}{\rho} \end{pmatrix}\,.$$

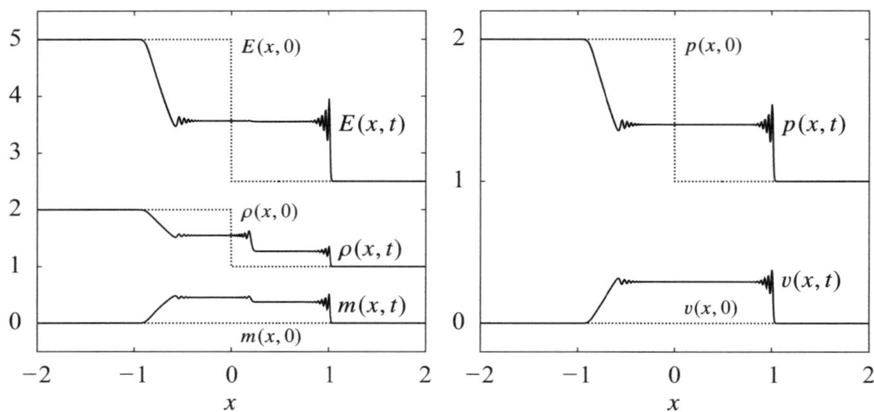

Fig. 10.17 Solution of the shock-tube problem by the non-linear Lax-Wendroff scheme with $\Delta t = 0.00125$ on a mesh with $N = 800$ subintervals. [LEFT] Spatial dependence of E, ρ, and m at $t = 0.75$. [RIGHT] Spatial dependence of p and v at $t = 0.75$

$\odot$ Solve the shock-tube problem on $x \in [-2, 2]$ with the boundary conditions $\rho_0 = \rho_1$, $m_0 = m_1$ and $E_0 = E_1$ at $x = -2$, $\rho_N = \rho_{N-1}$, $m_N = m_{N-1}$ and $E_N = E_{N-1}$ at $x = 2$, and the initial conditions

$$v(x, 0) = 0, \qquad \rho(x, 0) = p(x, 0) = \begin{cases} 2 \,; & -2 \leq x < 0 \,, \\ 1 \,; & 0 \leq x \leq 2 \,. \end{cases}$$

Compute the initial condition for E from those for p, ρ, and v, while the initial condition for $m = \rho v$ is $m(x, 0) = 0$ (since $v(x, 0) = 0$). Use the linearized Lax-Wendroff method (10.64) with the discretization $N = 400$ ($\Delta x = 0.01$) and compute the solution in steps of $\Delta t = 0.0025$ until $t = 1.0$.

$\oplus$ Use the non-linear variant of the Lax–Wendroff scheme (10.71) with the same discretization and compute the solution to $t = 1.0$. Pay attention to the signs in the term proportional to R: while (10.64) is written for systems of the form $\boldsymbol{v}_t = A \boldsymbol{v}_x$, Eq. (10.71) applies to $\boldsymbol{v}_t + [\boldsymbol{F}(\boldsymbol{v})]_x = F'(\boldsymbol{v})\boldsymbol{v}_x = 0$. Here you should therefore use $A = F'(\boldsymbol{v})$ and $\boldsymbol{F} = (\rho v, \rho v^2 + p, v(E + p))^{\mathrm{T}}$. Compare the propagation velocities of the shock-waves. Do the results change if you extend the domain to $x \in [-8, 8]$ and trace the solution to $t = 4.0$? What about if you refine the discretization to $\Delta x = 0.001$, $\Delta t = 0.00025$? See also Fig. 10.17.

10.12.8   Korteweg–de Vries Equation

Korteweg–de Vries equation [34] is a non-linear evolution equation useful in studying solitary waves (solitons). Solitons are waves that propagate with constant velocity and without change in form. In its original version, $v_t + cvv_x + \mu v_{xxx} = 0$, the equation is appropriate for the description of one-dimensional waves with small amplitudes in dispersive systems, like waves in shallow water or spilling over a submerged obstacle. The generalized Korteweg–de Vries equation

$$v_t + \frac{\beta}{\alpha + 1}\left(v^{\alpha+1}\right)_x + \mu v_{xxx} = 0$$

with constant α, β, and μ has a much broader applicability [35]. Physically most relevant are the cases $\alpha = 1$ and $\alpha = 2$. In this Problem we analyze the simplest possibilities for the numerical solution of the generalized equation [36, 37].

⊙ We discretize the non-linear term $(v^{\alpha+1})_x$ and the dispersive term v_{xxx} at time $(n+1)\Delta t$ as

$$\left. \left(v^{\alpha+1}\right)_x \right|_{x=x_j} \approx \frac{1}{2\Delta x}\left[\left(u_{j+1}^{n+1}\right)^\alpha u_{j+1}^{n+1} - \left(u_{j-1}^{n+1}\right)^\alpha u_{j-1}^{n+1}\right],$$

$$\left. v_{xxx} \right|_{x=x_j} \approx \frac{1}{2\Delta x^3}\left[-u_{j-2}^{n+1} + 2u_{j-1}^{n+1} - 2u_{j+1}^{n+1} + u_{j+2}^{n+1}\right].$$

The equation is then linearized by using the approximation $\left(u_{j\pm1}^{n+1}\right)^\alpha \approx \left(u_{j\pm1}^{n}\right)^\alpha$. This discretization yields a (diagonally dominant) system of linear equations

$$-\frac{\mu\Delta t}{2\Delta x^3}u_{j-2}^{n+1} + \left(\frac{\mu\Delta t}{\Delta x^3} - \frac{\beta\Delta t\left(u_{j-1}^n\right)^\alpha}{2\Delta x(\alpha+1)}\right)u_{j-1}^{n+1} + u_j^{n+1}$$

$$+ \left(-\frac{\mu\Delta t}{\Delta x^3} + \frac{\beta\Delta t\left(u_{j+1}^n\right)^\alpha}{2\Delta x(\alpha+1)}\right)u_{j+1}^{n+1} + \frac{\mu\Delta t}{2\Delta x^3}u_{j+2}^{n+1} = u_j^n,$$

which has to be solved at each time step. (LAPACK/BLAS libraries contain optimized routines SGBSV (single) and DGBSV (double precision) for the solution of linear systems with banded matrices.) Use $\alpha = 1$, $\beta = -6$, $\mu = 1$, $\Delta x = 0.1$, and $\Delta t = 0.001$ with the initial condition $v(x,0) = -2k_x^2\,\mathrm{sech}^2 k_x x$, where $k_x = 0.7$. The corresponding exact solution is

$$v(x,t) = \left(\frac{2k_x^2(\alpha+1)(\alpha+2)}{\beta\alpha^2}\right)^{1/\alpha}\mathrm{sech}^{2/\alpha}k_x\left(x - \frac{4k_x^2}{\alpha}t\right).$$

Solve the equation numerically on the interval $x \in [-8, 10]$ with the boundary conditions $u(-8,t) = u(10,t) = 0$. Monitor the time evolution of the solution from the initial condition up to $t = 1.0$. Also discuss the cases $\alpha = 2$, $\alpha = 3$, and $\alpha = 4$. What is the influence of the (approximate) boundary conditions on the numerical error?

$\oplus$ A slightly different discretization can be obtained by averaging the non-linear and dispersion terms at times $n\Delta t$ and $(n+1)\Delta t$,

$$u_t + \frac{\beta}{\alpha+1}\frac{1}{2}\left[\left(u_j^{n+1}\right)^{\alpha+1} + \left(u_j^n\right)^{\alpha+1}\right]_x + \mu\frac{1}{2}\left[u_j^{n+1} + u_j^n\right]_{xxx} = 0 .$$

The equation can now be linearized by Taylor-expanding the expressions $\left(u_{j\pm1}^{n+1}\right)^{\alpha+1}$ up to first order in time. This results in an implicit scheme which requires us to solve a pentadiagonal matrix system at each time step. Compare its solutions to those from the first part of the Problem.

10.12.9   *Non-stationary Schrödinger Equation*

The one-dimensional non-stationary Schrödinger equation

$$\left(i\hbar\frac{\partial}{\partial t} - H\right)\psi(x,t) = 0 , \qquad \psi(x,t_0) = \phi(x) ,$$

is the basic tool for a non-relativistic description of the time evolution of quantum states or wave packets in different potentials. Here we only discuss time-independent Hamiltonian operators

$$H = -\frac{\hbar^2}{2m}\frac{\partial^2}{\partial x^2} + V(x) .$$

(Time evolution involving time-dependent Hamiltonians is discussed in [38]). The evolution of the state $\psi(x,t)$ to $\psi(x,t+\Delta t)$ is described by the approximation

$$\psi(x,t+\Delta t) = e^{-iH\Delta t/\hbar}\psi(x,t) \approx \frac{1-\frac{1}{2}iH\Delta t/\hbar}{1+\frac{1}{2}iH\Delta t/\hbar}\psi(x,t) + O(\Delta t^3) ,$$

which is unitary. We compute the time evolution on the mesh $x_j = x_0 + j\Delta x$ ($j = 0, 1, \ldots, N_x$) at times $t_n = n\Delta t$ ($n = 0, 1, \ldots, N_t$), approximating the spatial derivative by $\psi''(x) \approx (\psi(x+\Delta x) - 2\psi(x) + \psi(x-\Delta x))/\Delta x^2$. The values of the wave-function and the potential at the mesh points at time t_n are denoted by $\psi(x_j, t_n) = \psi_j^n$ and $V(x_j) = V_j$, respectively, resulting in the equations

$$\psi_j^{n+1} - \frac{i\hbar\Delta t}{4m\Delta x^2}\left[\psi_{j+1}^{n+1} - 2\psi_j^{n+1} + \psi_{j-1}^{n+1}\right] + \frac{i\Delta t}{2\hbar}V_j\psi_j^{n+1}$$
$$= \psi_j^n + \frac{i\hbar\Delta t}{4m\Delta x^2}\left[\psi_{j+1}^n - 2\psi_j^n + \psi_{j-1}^n\right] - \frac{i\Delta t}{2\hbar}V_j\psi_j^n .$$

We arrange the values of the wave-function at x_j in the vector $\boldsymbol{\Psi}^n = \left(\psi_0^n, \ldots, \psi_{N_x}^n\right)^{\mathrm{T}}$ and rewrite the system in matrix form

$$
A\boldsymbol{\Psi}^{n+1} = A^*\boldsymbol{\Psi}^n, \qquad A =
\begin{pmatrix}
d_0 & a & & & & \\
a & d_1 & a & & & \\
& a & d_2 & a & & \\
& & \ddots & \ddots & \ddots & \\
& & & a & d_{N_x-1} & a \\
& & & & a & d_{N_x}
\end{pmatrix},
$$

where

$$
b = \frac{i\hbar\Delta t}{2m\Delta x^2}, \qquad a = -\frac{b}{2}, \qquad d_j = 1 + b + \frac{i\Delta t}{2\hbar}V_j.
$$

The matrix A and vector $\boldsymbol{\Psi}$ are complex, so the computation is best done in complex floating-point arithmetic. With the above approximation of the second derivative (which is of order $O(\Delta x^2)$) we obtain a tridiagonal matrix. Had we used a difference of higher order, we would have obtained a banded matrix, but this would also improve the spatial precision: see [39, 40].

$\odot$ Compute the time evolution of the initial state

$$
\psi(x, 0) = \sqrt{\frac{\alpha}{\sqrt{\pi}}}\, e^{-\alpha^2(x-\lambda)^2/2}
$$

in the harmonic potential $V(x) = \frac{1}{2}kx^2$, where $\alpha = (mk/\hbar^2)^{1/4}$, $\omega = \sqrt{k/m}$. Set $\hbar = m = 1$ and use $\omega = 0.2$, $\lambda = 10$. Place the spatial mesh on the interval $[x_0, x_{N_x}] = [-40, 40]$ with $N_x = 300$. The exact solution is the coherent state

$$
\psi(x, t) = \sqrt{\frac{\alpha}{\sqrt{\pi}}}\exp\left[-\frac{1}{2}\left(\xi - \xi_\lambda\cos\omega t\right)^2 - i\left(\frac{\omega t}{2} + \xi\xi_\lambda\sin\omega t - \frac{1}{4}\xi_\lambda^2\sin 2\omega t\right)\right],
$$

where $\xi = \alpha x$, $\xi_\lambda = \alpha\lambda$. In the chosen units the period is $T = 10\pi$.

$\oplus$ Compute the time evolution of the Gaussian wave packet

$$
\psi(x, 0) = (2\pi\sigma_0^2)^{-1/4}e^{ik_0(x-\lambda)}e^{-(x-\lambda)^2/(2\sigma_0)^2}
$$

in potential-free space ($V(x) = 0$). Set $\hbar = 1$, $m = 1/2$, $\sigma_0 = 1/20$, $k_0 = 50\pi$, $\lambda = 0.25$. The spatial interval is $[x_0, x_{N_x}] = [-0.5, 1.5]$ with $N_x = 2000$, and we choose $\Delta t = 2\Delta x^2$. Compute the solution until the center of the packet reaches $x \approx 0.75$. The exact solution is

$$
\psi(x, t) = \frac{(2\pi\sigma_0^2)^{-1/4}}{\sqrt{1 + i\hbar t/(2m\sigma_0^2)}}\exp\left[\frac{-(x-\lambda)^2/(2\sigma_0)^2 + ik_0(x-\lambda) - i\hbar k_0^2 t/(2m)}{1 + i\hbar t/(2m\sigma_0^2)}\right].
$$

Section 1.2.2 tells you how the precision in the time variable can be improved.

10.12.10 Non-stationary Cubic Schrödinger Equation

The non-linear Schrödinger equation with a cubic non-linearity,

$$
\mathrm{i}\frac{\partial \psi}{\partial t} + \frac{\partial^2 \psi}{\partial x^2} + q|\psi|^2\psi = 0 \,,
$$

where q is a parameter [41], is used to describe numerous physical processes, like propagation of solitons in optical fibers, propagation of plasma waves, and even Bose–Einstein condensates. Splitting $\psi(x, t) = u(x, t) + \mathrm{i}v(x, t)$ turns the equation in a system of coupled non-linear PDE:

$$
\frac{\partial u}{\partial t} + \frac{\partial^2 v}{\partial x^2} + q\left(u^2 + v^2\right)v = 0 \,, \tag{10.105}
$$

$$
\frac{\partial v}{\partial t} - \frac{\partial^2 u}{\partial x^2} - q\left(u^2 + v^2\right)u = 0 \,. \tag{10.106}
$$

We are interested in the solutions of this system on $x \in [0, 2\pi]$ with the periodic boundary condition $\psi(x) = \psi(x + 2\pi)$. The spatial derivatives are approximated by differences of the form (10.6) with the discretization $x_j = j\Delta x$, $j = 0, 1, \ldots, N$, where $\Delta x = 2\pi/N$. The solution vectors are $u = (u_1, u_2, \ldots, u_N)^{\mathrm{T}}$, $v = (v_1, v_2, \ldots, v_N)^{\mathrm{T}}$. When the boundary condition is accounted for, Eqs. (10.105) and (10.106) can be written in canonical Hamiltonian form

$$
\frac{\mathrm{d}}{\mathrm{d}t}\begin{pmatrix} u \\ v \end{pmatrix} = \begin{pmatrix} 0 & -1 \\ 1 & 0 \end{pmatrix}\begin{pmatrix} \partial H/\partial u \\ \partial H/\partial v \end{pmatrix} = J^{-1}\begin{pmatrix} \partial H/\partial u \\ \partial H/\partial v \end{pmatrix} \,,
$$

where $\partial H/\partial u = (\partial H/\partial u_1, \partial H/\partial u_2, \ldots \partial H/\partial u_N)^{\mathrm{T}}$ (analogously for $\partial H/\partial v$), and

$$
H(u, v) = \frac{1}{2}\left(u^{\mathrm{T}}, v^{\mathrm{T}}\right)\begin{pmatrix} F & 0 \\ 0 & F \end{pmatrix}\begin{pmatrix} u \\ v \end{pmatrix} + \frac{q}{4}\sum_{j=1}^{N}\left(u_j^2 + v_j^2\right)^2 \,,
$$

where

$$
F = \frac{1}{\Delta x^2}\begin{pmatrix} -2 & 1 & & & 1 \\ 1 & -2 & 1 & & \\ & \ddots & \ddots & \ddots & \\ & & 1 & -2 & 1 \\ 1 & & & 1 & -2 \end{pmatrix} \,.
$$

One therefore needs to solve the system of equations $\dot{z} = J^{-1}\nabla H = J^{-1}(\partial H/\partial z)$, where $z = (u^{\mathrm{T}}, v^{\mathrm{T}})^{\mathrm{T}}$. This is best accomplished by the implicit symplectic Euler integrator

$$
z^{n+1} = z^n + \Delta t\, J^{-1}(\nabla H)_{z=\frac{1}{2}(z^{n+1}+z^n)} \,, \tag{10.107}
$$

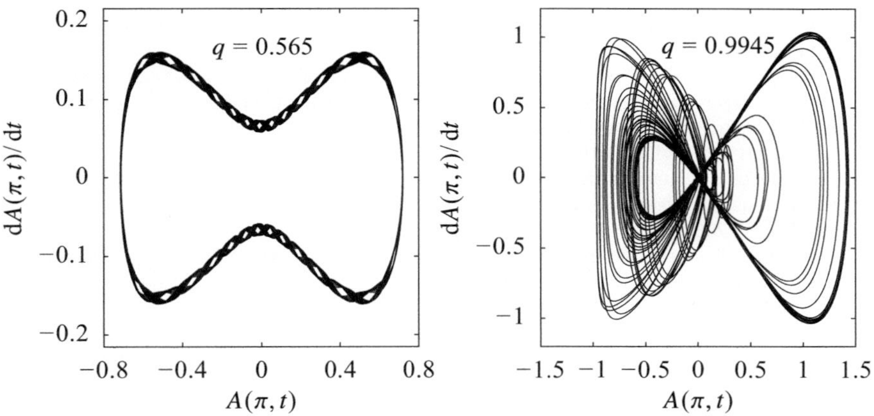

Fig. 10.18 Phase diagrams in solving the cubic Schrödinger equation where the solution at two values of the non-linearity parameter q is computed until time T. Shown is the amplitude $A(\pi, t) = |\psi(\pi, t)| - 1$ versus its time derivative $dA(\pi, t)/dt$. [LEFT] $q = 0.565$, $T = 400$. [RIGHT] $q = 0.9945$, $T = 1000$

where the solution at each time step is computed iteratively.

⊙ By using (10.107) solve the non-linear Schrödinger equation with the initial condition $\psi(x, 0) = 1 + \varepsilon e^{i\theta} \cos x$, where $\varepsilon = 0.1$ and $\theta = 45.18°$. Use the time step $\Delta t = 0.001$ and compute the solutions at a given parameter q until time T. Use the pairs of parameters $(q, T) = (0.01, 400)$, $(0.325, 400)$, $(0.565, 400)$, $(0.9825, 400)$, $(0.9945, 1000)$, $(1.0115, 1000)$, $(1.315, 1000)$, and $(1.630, 1000)$. Plot $A(\pi, t) = |\psi(\pi, t)| - 1$ and $A_t(\pi, t) = dA(\pi, t)/dt$ as functions of t, and the phase diagram $A_t(\pi, t)$ versus $A(\pi, t)$ (see examples in Fig. 10.18). You can use a simple approximation to compute the derivative A_t, e.g. $A_t \approx (A^{n+1} - A^n)/\Delta t$. Plot the spectrum of the signals A and A_t (since you will be dealing with samples of several hundred thousand points, use FFT). For the integration (10.107) set the tolerance for the difference of the current and previous iteration to less than $\approx 10^{-13}$. Otherwise the numerical errors in A and A_t will be too large.

⊕ Plot the regions $|\psi(x, t)|^2 = \text{const}$. When computing with parameters q at which you have observed irregular behavior, you may restrict the plots to the part of the interval $[0, T]$ with the most interesting structure.

References

1. J.W. Thomas, *Numerical Partial Differential Equations: Finite Difference Methods, Springer Texts in Applied Mathematics*, vol. 22 (Springer-Verlag, Berlin, 1998)
2. D. Knoll, J. Morel, L. Margolin, M. Shashkov, Physically motivated discretization methods. Los Alamos Sci. **29**, 188 (2005)

3. G.D. Smith, *Numerical Solution of Partial Differential Equations* (Oxford University Press, Oxford, 2003)
4. J.W. Thomas, *Numerical Partial Differential Equations: Conservation Laws and Elliptic Equations, Springer Texts in Applied Mathematics*, vol. 33 (Springer-Verlag, Berlin, 1999)
5. E. Godlewski, P.-A. Raviart, *Numerical Approximations of Hyperbolic Systems of Conservation Laws* (Springer-Verlag, Berlin, 1996)
6. R.J. LeVeque, *Numerical Methods for Conservation Laws* (Birkhäuser Verlag, Basel, 1990)
7. R.J. Leveque, *Finite-Volume Methods for Hyperbolic Problems* (Cambridge University Press, Cambridge, 2004)
8. B. Van Leer, Towards the ultimate conservative difference scheme I. The quest of monotonicity. Springer Lecture Notes Phys. **18**, 163 (1973)
9. B. Van Leer, Towards the ultimate conservative difference scheme II. Monotonicity and conservation combined in a second order scheme. J. Comput. Phys. **14**, 361 (1974)
10. B. Van Leer, Towards the ultimate conservative difference scheme III. Upstream-centered? Nite-difference schemes for ideal compressible? Ow. J. Comput. Phys. **23**, 263 (1977)
11. B. Van Leer, *Towards the ultimate conservative difference scheme IV. A new approach to numerical convection.* J. Comput. Phys. **23**, 276 (1977)
12. B. Van Leer, Towards the ultimate conservative difference scheme V. A second order sequel to Godunov's method. J. Comput. Phys. **32**, 101 (1979)
13. B. Van Leer, Towards the ultimate conservative difference scheme. J. Comput. Phys. **135**, 229 (1997)
14. S. Osher, Convergence of generalized MUSCL schemes. SIAM J. Numer. Anal. **22**, 947 (1985)
15. F. Kemm, A comparative study of TVD limiters—well-known limiters and an introduction of new ones. Int. J. Numer. Meth. Fluids **67**, 404 (2011)
16. P.K. Sweby, High resolution schemes using flux limiters for hyperbolic conservation laws. SIAM J. Numer. Anal. **21**, 995 (1984)
17. J.B. Goodman, R.J. LeVeque, A geometric approach to high resolution TVD schemes. SIAM J. Numer. Anal. **25**, 268 (1988)
18. A. Kurganov, E. Tadmor, New high-resolution schemes for nonlinear conservation laws and convection-diffusion equations. J. Comput. Phys. **160**, 241 (2000)
19. A. Kurganov, D. Levy, A third-order semidiscrete central scheme for conservation laws and convection-diffusion equations. SIAM J. Sci. Comput. **22**, 1461 (2000)
20. C. Brehm, M.F. Barad, J.A. Housman, C.C. Kiris, A comparison of higher-order finite-difference shock capturing schemes. Comput. Fluids **122**, 184 (2015)
21. H.Q. Yang, A.J. Przekwas, A comparative study of advanced shock-capturing schemes applied to Burgers' equation. J. Comput. Phys. **102**, 139 (1992)
22. C.-W. Shu, S. Osher, Efficient implementation of essentially non-oscillatory shock-capturing schemes. J. Comput. Phys. **77**, 439 (1988); C.-W. Shu, S. Osher, Efficient implementation of essentially non-oscillatory shock-capturing schemes, II. J. Comput. Phys. **83**, 32 (1989)
23. G.-S. Jiang, C.-W. Shu, Efficient implementation of weighted ENO schemes. J. Comput. Phys. **126**, 202 (1996)
24. A.K. Henrick, T.D. Aslam, J.M. Powers, Mapped weighted essentially non-oscillatory schemes: achieving optimal order near critical points. J. Comput. Phys. **207**, 542 (2005)
25. R. Borges, M. Carmona, B. Costa, W.S. Don, An improved weighted essentially non-oscillatory scheme for hyperbolic conservation laws. J. Comput. Phys. **227**, 3191 (2008)
26. M. Castro, B. Costa, W.S. Don, High order weighted essentially non-oscillatory WENO-Z schemes for hyperbolic conservation laws. J. Comput. Phys. **230**, 1766 (2011)
27. F. Acker, R. B. de R. Borges, B. Costa, An improved WENO–Z scheme. J. Comput. Phys. **313**, 726 (2016)
28. D.S. Balsara, C.-W. Shu, Monotonicity preserving weighted essentially non-oscillatory schemes with increasingly high order of accuracy. J. Comput. Phys. **160**, 405 (2000)
29. G.A. Gerolymos, D. Sénéchal, I. Vallet, Very-high-order WENO schemes. J. Comput. Phys. **228**, 8481 (2009)

30. S. Rathan, G. Naga Raju, A modified fifth-order WENO scheme for hyperbolic conservation laws. Comput. Math. Appl. **75**, 1531 (2018)
31. D. Barreto, R. B. de R. Borges, B. Costa, S. Santos, *New weighting strategies for WENO schemes*, arXiv:2311.09332 [math.NA]
32. D. Eisen, On the numerical analysis of a fourth order wave equation. SIAM J. Numer. Anal. **4**, 457 (1967)
33. S.E. Buckley, M.C. Leverett, Mechanism of fluid displacement in sands. Trans. AIME **146**, 107 (1942)
34. D.J. Korteweg, G. de Vries, On the change of form of long waves advancing in a rectangular channel and on a new type of long stationary wave. Phil. Mag. Series 5 **39**, 422 (1895)
35. B. Fornberg, G.B. Whitham, *A numerical and theoretical study of certain nonlinear wave phenomena*. Phil. Trans. R. Soc. London, Series A **289**, 373 (1978)
36. X. Lai, Q. Cao, Some finite difference methods for a kind of GKdV equations. Commun. Numer. Meth. Eng. **23**, 179 (2007)
37. Q. Cao, K. Djidjeli, W.G. Price, E.H. Twizell, Computational methods for some non-linear wave equations. J. Eng. Math. **35**, 323 (1999)
38. K. Kormann, S. Holmgren, O. Karlsson, Accurate time propagation for the Schrödinger equation with an explicitly time-dependent Hamiltonian. J. Chem. Phys. **128**, 184101 (2008); see also C. Lubich, Integrators for quantum dynamics, in *Quantum simulations of complex many-body systems: from theory to algorithm*, vol. 10, ed. J. Grotendorst, D. Marx, A. Muramatsu, NIC Series (John von Neumann Institute for Computing, Jülich, 2002), p. 459
39. W. van Dijk, F.M. Toyama, Accurate numerical solutions of the time-dependent Schrödinger equation. Phys. Rev. E **75**, 036707 (2007)
40. T.N. Truong et al., A comparative study of time dependent quantum mechanical wave packet evolution methods. J. Chem. Phys. **96**, 2077 (1992)
41. X. Liu, P. Ding, Dynamic properties of cubic nonlinear Schrödinger equation with varying nonlinear parameter. J. Phys. A: Math. Gen. **37**, 1589 (2004)

Chapter 11
Difference Methods for PDE in Two or More Dimensions

Abstract Finite-difference methods for partial differential equations in two dimensions are presented by first handling the basic (parabolic) diffusion equation in two dimensions by explicit and implicit difference schemes, allowing us to introduce the corresponding stability criteria. The treatment of alternating direction implicit (ADI) schemes is supplemented by the extension to three space dimensions. Several schemes for solutions of hyperbolic equations are listed. Solving elliptic equations (Poisson and Laplace) by classical relaxation methods (Jacobi, Gauss-Seidel, SOR, SSOR) is discussed, as well as high-resolution schemes for hyperbolic equations. The need to respect the underlying nature of the physical problems is emphasized by solving the two-dimensional diffusion and Poisson equations in polar coordinates with various boundary conditions. The boundary element method, the finite element method, and a mesh-free method based on radial basis functions are presented. In addition to many variants of Laplace and Poisson equations, the Problems include several instances of the diffusion equation, the boundary-element treatment of potential flow, as well as non-linear cases like the Buckley–Leverett equation and the equation describing the growth of bacterial biofilms.

11.1 Parabolic and Hyperbolic PDE

The basic concepts of difference methods for PDE in several dimensions are readily adopted from the discussion of one-dimensional initial-boundary-value problems (Chap. 10). We are seeking consistent, stable difference schemes, and corresponding discretizations of the initial and boundary conditions by which we obtain convergence of the numerical solution to the exact solution of the differential equation. Except in Eqs. (11.21) and (11.22) we restrict the discussion to PDE with time dependence in two spatial coordinates.

11.1.1   Parabolic Equations

The basic model problem for two-dimensional parabolic PDE is the initial-boundary-value problem for the diffusion equation with an inhomogeneous term,

$$v_t = D\nabla^2 v + Q = D\left(v_{xx} + v_{yy}\right) + Q(x, y, t)\,,$$
$$v(x, y, 0) = f(x, y)\,,$$

which we solve for $t > 0$ in a region with a simple geometry, e.g. on a square $(x, y) \in [0, 1] \times [0, 1]$. More complicated geometries are discussed in Sects. 11.4, 11.5 and 11.6. Let us ignore the issue of boundary conditions for the moment. By analogy with the one-dimensional case (Fig. 10.1) we discretize the time axis and both spatial axes.

The exact solution $v(x_j, y_k, t_n) = v_{jk}^n$ at $(j\Delta x, k\Delta y)$ and time $n\Delta t$ corresponds to the numerical solution u_{jk}^n, where $j = 0, 1, \ldots, N_x$ and $k = 0, 1, \ldots, N_y$ (Fig. 11.1 (left), the time axis is perpendicular to the page). The most obvious book-keeping choice for the vector of approximate solutions $\boldsymbol{u}$ is

$$\boldsymbol{u}^n = \left(u_{00}^n, u_{10}^n, \ldots, u_{N_x0}^n, u_{01}^n, u_{11}^n, \ldots, u_{N_x1}^n, \ldots, u_{0N_y}^n, u_{1N_y}^n, \ldots, u_{N_xN_y}^n\right)^{\mathrm{T}}.$$

$$(11.1)$$

Next, we adapt the formulas for the differences to two dimensions. The obvious generalizations of the approximation for the second derivative (10.6) are

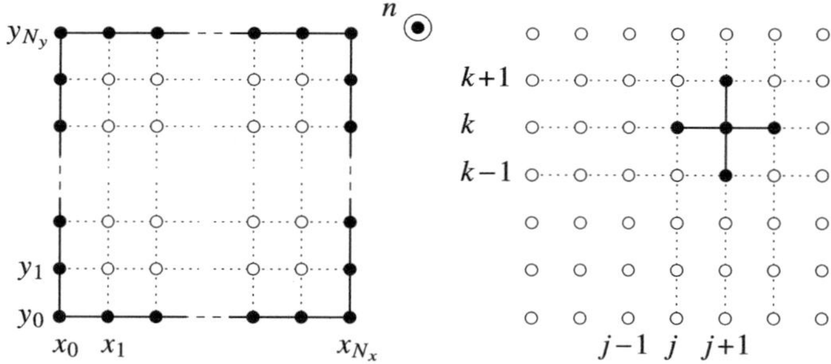

Fig. 11.1 [LEFT] The uniform mesh for the solution of the initial-boundary-value problem for a PDE with one time and two space dimensions on $[x_0, x_{N_x}] \times [y_0, y_{N_y}] = [0, 1] \times [0, 1]$. The discretization Δt, $\Delta x = 1/N_x$, $\Delta y = 1/N_y$ defines the mesh points $(x_j, y_k, t_n) = (j\Delta x, k\Delta y, n\Delta t)$ and in each of them the exact solution v_{jk}^n and its approximate value u_{jk}^n. [RIGHT] The stencil of mesh points for the evaluation of $(v_{xx} + v_{yy})_{jk}$

$$\frac{\partial^2 v}{\partial x^2}\bigg|_{(j\Delta x, k\Delta y, n\Delta t)} = \frac{v_{j+1k}^n - 2v_{jk}^n + v_{j-1k}^n}{\Delta x^2} + O(\Delta x^2)\,, \tag{11.2}$$

$$\frac{\partial^2 v}{\partial y^2}\bigg|_{(j\Delta x, k\Delta y, n\Delta t)} = \frac{v_{jk+1}^n - 2v_{jk}^n + v_{jk-1}^n}{\Delta y^2} + O(\Delta y^2)\,, \tag{11.3}$$

which need to be evaluated in the characteristic *stencil* of mesh points (Fig. 11.1 (right)). Instead of one auxiliary notation (10.11) we now have two,

$$\Delta_2^{(x)} = u_{j+1k}^n - 2u_{jk} + u_{j-1k}^n\,, \qquad \Delta_2^{(y)} = u_{jk+1}^n - 2u_{jk} + u_{jk-1}^n\,. \tag{11.4}$$

11.1.2 Explicit Scheme

The two-dimensional generalization of the explicit FTCS scheme Eq. (10.13) is

$$u_{jk}^{n+1} = u_{jk}^n + \left(r_x\Delta_2^{(x)} + r_y\Delta_2^{(y)}\right)u_{jk}^n + \Delta t\, q_{jk}^n\,, \tag{11.5}$$

where $r_x = D\Delta t/\Delta x^2$, $r_y = D\Delta t/\Delta y^2$, and $Q(x_j, y_k, t_n) = q_{jk}^n$. By analogy to the one-dimensional variant we conclude that the discretization error of the scheme is $O(\Delta t) + O(\Delta x^2) + O(\Delta y^2)$, and that the scheme can be written in matrix form

$$\boldsymbol{u}^{n+1} = F\boldsymbol{u}^n + \boldsymbol{r}^n + \Delta t\, \boldsymbol{q}^n\,.$$

The inhomogeneous term (the last two terms at the right) has two contributions. The vector $\boldsymbol{r}^n$ contains the requirements at the boundary: for example, for homogeneous Dirichlet boundary conditions, the value u_{jk} on all boundaries of the region $[x_0, x_{N_x}] \times [y_0, y_{N_y}]$ is zero, and then $\boldsymbol{r}^n = 0$. The second part, the vector $\Delta t\, \boldsymbol{q}^n$, describes the sources. In practical problems the reader will have to figure out the structure of these terms for the chosen scheme and boundary conditions.

Let us inspect the typical cases. Dirichlet boundary conditions (the values of the function v specified on all four sides of the square) are expressed as

$$u_{0k}^n = f(0, k\Delta y, n\Delta t)\,, \tag{11.6}$$

$$u_{N_x k}^n = f(1, k\Delta y, n\Delta t)\,, \tag{11.7}$$

$$u_{j0}^n = f(j\Delta x, 0, n\Delta t)\,, \tag{11.8}$$

$$u_{jN_y}^n = f(j\Delta x, 1, n\Delta t)\,. \tag{11.9}$$

In this case the solution can be represented by a shorter vector $\boldsymbol{u}$, since all values at $j = 0$, $j = N_x$, $k = 0$, and $k = N_y$ are fixed by the boundary conditions. Instead of with (11.1) we work with a $(N_x - 1)(N_y - 1)$-dimensional vector

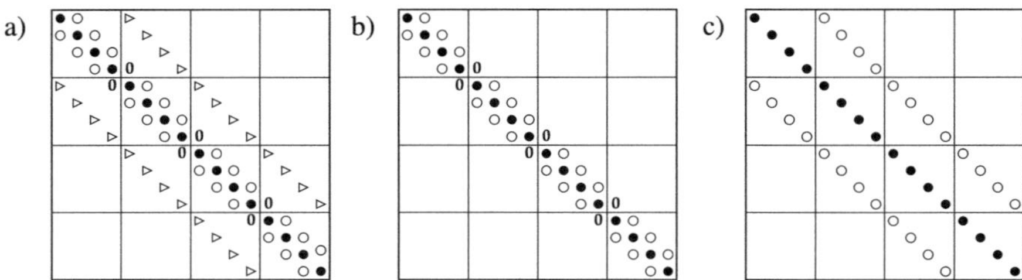

Fig. 11.2 Common structures of matrices appearing in explicit ($u^{n+1} = Fu^n$) and implicit schemes ($F_1 u^{n+1} = Fu^n$) for the solution of two-dimensional diffusion equation. The values of the elements denoted by ● / ○ / ▷ depend on the difference method (see text). Shown are matrices for the case $N_x = N_y = 5$

$$u^n = \left(u^n_{11}, u^n_{21}, \ldots, u^n_{N_x-11}, \ldots, u^n_{1N_y-1}, u^n_{2N_y-1}, \ldots, u^n_{N_x-1N_y-1} \right)^{\mathrm{T}} . \qquad (11.10)$$

The matrix F therefore has the size $(N_x - 1)(N_y - 1) \times (N_x - 1)(N_y - 1)$. If $N_x = N_y$ (which is not obligatory) it has the form shown in Fig. 11.2 (case a)), with the elements ● $= 1 - 2r_x - 2r_y$, ○ $= r_x$, and ▷ $= r_y$. Here we realize the practical constraints brought about by the additional space dimension: even a modest spatial discretization in $N_x = N_y = 100$ points in each coordinate implies manipulations of $(N_x - 1)(N_y - 1) \times (N_x - 1)(N_y - 1) \approx 10000 \times 10000$ matrices without simple structure (e.g. not tridiagonal). This becomes relevant in implicit schemes where we need to solve matrix systems of such sizes.

For Dirichlet boundary conditions we use the discrete Fourier transformation (by analogy to the one-dimensional case) to derive the sufficient condition for stability of (11.5). We extend the discrete Fourier mode (10.29) to two dimensions,

$$u^n_{jk} = \xi^n e^{ijp\pi \Delta x} e^{ikq\pi \Delta y} ,$$

insert it into the homogeneous part of the difference equation (11.5), and divide the resulting equation by ξ^{n+1}, whence

$$\xi = 1 - 4r_x \sin^2 \frac{p\pi \Delta x}{2} - 4r_y \sin^2 \frac{q\pi \Delta y}{2} .$$

The discrete Neumann criterion $|\xi| \le 1$ means

$$r_x + r_y \le 1/2 .$$

The stability condition for the two-dimensional scheme (11.5) is therefore more restrictive than in the one-dimensional case (first row of Table 10.1). If we use the uniform mesh, $\Delta x = \Delta y$, such a condition requires $r_x = r_y \le 1/4$ and forces us to take a twice smaller Δt (or $\sqrt{2}$-times larger Δx) to ensure stability.

Neumann conditions, as usual, require special care. If we have three Dirichlet conditions, e.g. $v(1, y, t) = v(x, 0, t) = v(x, 1, t) = 0$, and one Neumann, $v_x(0, y, t) = g(y, t)$, the joint order of the difference scheme depends on the order of the approximation for the Neumann condition. If the latter is discretized to first order, as in (10.17),

$$u_{1k}^n = u_{0k}^n + \Delta x g(k \Delta y, n \Delta t) , \tag{11.11}$$

the FTCS scheme is only of order $O(\Delta x)$. We retrieve the order $O(\Delta x^2)$ if the Neumann condition is discretized to second order. Confirm that in this case

$$u_{0k}^{n+1} = u_{0k}^n + 2r_x(u_{1k}^n - u_{0k}^n) - 2r_x \Delta x g(k \Delta y, n \Delta t) + r_y \Delta_2^{(y)} u_{0k}^n , \tag{11.12}$$

by analogy to Eq. (10.18)! The stability condition for the scheme with Neumann conditions is just a necessary one, since the mapping matrix is not symmetric and we need to evaluate its spectral radius by Gershgorin's theorem [1].

11.1.3 Crank–Nicolson Scheme

The two-dimensional generalization of the Crank–Nicolson scheme Eq. (10.21) is

$$\left(1 - \frac{r_x}{2} \Delta_2^{(x)} - \frac{r_y}{2} \Delta_2^{(y)}\right) u_{jk}^{n+1} = \left(1 + \frac{r_x}{2} \Delta_2^{(x)} + \frac{r_y}{2} \Delta_2^{(y)}\right) u_{jk}^n$$
$$+ \frac{\Delta t}{2} \left(q_{jk}^n + q_{jk}^{n+1}\right) . \tag{11.13}$$

The scheme has order $O(\Delta t^2) + O(\Delta x^2) + O(\Delta y^2)$, and can be written as

$$F_1 u^{n+1} = F u^n + r^n + \frac{\Delta t}{2} \left(q^n + q^{n+1}\right) ,$$

where (for Dirichlet conditions (11.6)–(11.9)) the matrix F_1 has the form as in Fig. 11.2 (case a)), with the elements $\bullet = 1 + r_x + r_y$, $\circ = -r_x/2$, and $\triangleright = -r_y/2$, while F has the same form except that all signs of r_x and r_y are flipped. The term r^n includes the boundary conditions at $j = 0$, $j = N_x$, $k = 0$, and $k = N_y$ that are not included in F_1 or F. The discrete Neumann criterion for stability is again derived by Fourier analysis as for the explicit scheme. For the symbol

$$\xi = \frac{1 - 2r_x \sin^2 p\pi \Delta x/2 - 2r_y \sin^2 q\pi \Delta y/2}{1 + 2r_x \sin^2 p\pi \Delta x/2 + 2r_y \sin^2 q\pi \Delta y/2}$$

we obviously get $|\xi| \leq 1$ for all non-negative r_x and r_y, so the two-dimensional Crank-Nicolson scheme is absolutely stable (the conclusion applies to Dirichlet and Neumann boundary conditions). As an exercise, generalize the implicit scheme (10.20) to two dimensions and check that its amplification factor is

$$\xi = \frac{1}{1 + 4r_x \sin^2 p\pi\,\Delta x/2 + 4r_y \sin^2 q\pi\,\Delta y/2}\,.$$

The two-dimensional scheme is absolutely stable since $|\xi| \le 1$.

11.1.4   Alternating Direction Implicit Schemes

When implicit schemes discussed above (either in two or three space dimensions) are written in a matrix representation, the matrices become sparse. If the matrices are very large, the numerical solution of the corresponding matrix system may become too costly, even if one resorts to iterative methods (e.g. Gauss–Seidel, Jacobi, SOR, or multi-grid; see Sects. 11.2 and 11.4). The troublesome solution of such systems can be circumvented by *Alternating Direction Implicit* (ADI) schemes. In a two-dimensional ADI scheme we compute the spatial derivatives in one dimension explicitly, and implicitly in the other (or vice-versa). This enables us to benefit from the good stability properties of implicit schemes, while keeping the tridiagonal structure of the matrix system.

Peaceman–Rachford scheme A typical ADI method is the Peaceman–Rachford scheme: in the first half of the time step Δt we compute the spatial derivative implicitly in coordinate x and explicitly in coordinate y, and vice-versa in the second half of Δt. This can be expressed as a two-level scheme

$$\left(1 - \frac{r_x}{2}\,\Delta_2^{(x)}\right) u_{jk}^{n+\frac{1}{2}} = \left(1 + \frac{r_y}{2}\,\Delta_2^{(y)}\right) u_{jk}^n + \frac{\Delta t}{2}\,q_{jk}^n\,, \tag{11.14}$$

$$\left(1 - \frac{r_y}{2}\,\Delta_2^{(y)}\right) u_{jk}^{n+1} = \left(1 + \frac{r_x}{2}\,\Delta_2^{(x)}\right) u_{jk}^{n+\frac{1}{2}} + \frac{\Delta t}{2}\,q_{jk}^{n+1} \tag{11.15}$$

which is absolutely stable. (Optionally, the q_{jk} terms in the inhomogeneous part may be evaluated at the same time, $(n + 1/2)\Delta t$.) Moreover, the order of spatial derivatives in Eqs. (11.14) and (11.15) can also be reversed: in this case we simply rewrite $x \longleftrightarrow y$. The scheme is of order $O(\Delta t^2) + O(\Delta x^2) + O(\Delta y^2)$, but only if boundary conditions are treated consistently. For example, if we have a Dirichlet condition $v(0, y, t) = f(y, t)$ at $x = 0$ $(j = 0)$, we do not spoil the nominal order of the scheme if the boundary condition at the intermediate time $(n + \frac{1}{2})\Delta t$ is computed as

$$u_{0k}^{n+\frac{1}{2}} = \frac{1}{2}\left(1 - \frac{r_y}{2}\,\Delta_2^{(y)}\right) f(0, k\Delta y, (n+1)\Delta t)$$

$$+ \frac{1}{2}\left(1 + \frac{r_y}{2}\,\Delta_2^{(y)}\right) f(0, k\Delta y, n\Delta t)\,. \tag{11.16}$$

If the condition is time-independent, it clearly simplifies to $u_{0k}^{n+\frac{1}{2}} = f(0, k\Delta y)$. For the corresponding condition at $x = 1$ ($j = N_x$) we compute $f(1, k\Delta y, \ldots)$ instead of $f(0, k\Delta y, \ldots)$; the discretization of the analogous conditions at $y = 0$ ($k = 0$) and $y = 1$ ($k = N_y$) should be done by the reader as an exercise. (See Problem 11.9.1 which also involves Neumann boundary conditions.)

The steps (11.14) and (11.15) can be written in matrix form

$$F_1 u^{n+\frac{1}{2}} = F u^n + r_y^n + r_x^{n+\frac{1}{2}} + \frac{\Delta t}{2} q^n , \tag{11.17}$$

$$F_1' u^{n+1} = F' u^{n+\frac{1}{2}} + r_x^{n+\frac{1}{2}} + r_y^{n+1} + \frac{\Delta t}{2} q^{n+1} . \tag{11.18}$$

The terms r_x and r_y contain the boundary conditions along x and y at times $n\Delta t$ and $(n + 1/2)\Delta t$. The matrix F_1 is tridiagonal, as in Fig. 11.2 (case b)), with the elements $\bullet = 1 + r_x, \circ = -r_x/2$, while the matrix F has the form shown in Fig. 11.2 (case c)), with the elements $\bullet = 1 - r_y, \circ = r_y/2$. We get F_1' from F by replacing $r_y \to -r_y$, and F' from F_1 by replacing $r_x \to -r_x$.

In each time step (11.17)–(11.18) we need to solve an $(N_x - 1)(N_y - 1) \times (N_x - 1)(N_y - 1)$ matrix system in which F_1' is not tridiagonal. But by reordering the solution components as $u_{jk} \to u_{kj}$, the system (11.18) becomes tridiagonal. The whole procedure then decouples to the solution of $N_y - 1$ systems of size $(N_x - 1) \times (N_x - 1)$ in the first time step, and $N_x - 1$ systems of size $(N_y - 1) \times (N_y - 1)$ in the second step:

$$F_x u_k^{n+\frac{1}{2}} = r_k + \frac{\Delta t}{2} q_k^n , \qquad k = 1, 2, \ldots, N_y - 1 , \tag{11.19}$$

$$F_y u_j^{n+1} = r_j + \frac{\Delta t}{2} q_j^{n+1} , \qquad j = 1, 2, \ldots, N_x - 1 , \tag{11.20}$$

where the matrix F_x (or F_y) is tridiagonal, with the values $1 + r_x$ (or $1 + r_y$) on the main diagonal and $-r_x/2$ (or $-r_y/2$) on the subdiagonals. In the solution vectors u of dimension $(N_x - 1)$ (or $(N_y - 1)$) we should pay attention to the ordering of components,

$$u_k = (u_{1k}, u_{2k}, \ldots, u_{N_x-1\,k})^{\mathrm{T}} ,$$
$$u_j = (u_{j1}, u_{j2}, \ldots, u_{jN_y-1})^{\mathrm{T}} ,$$

and analogously for q_k and q_j. The vectors containing the boundary conditions are

$$
\boldsymbol{r}_k = \begin{pmatrix} \left(1+\frac{r_y}{2}\Delta_2^{(y)}\right)u_{1k}^n + \frac{r_x}{2}u_{0k}^{n+\frac{1}{2}} \\ \left(1+\frac{r_y}{2}\Delta_2^{(y)}\right)u_{2k}^n \\ \vdots \\ \left(1+\frac{r_y}{2}\Delta_2^{(y)}\right)u_{N_x-2k}^n \\ \left(1+\frac{r_y}{2}\Delta_2^{(y)}\right)u_{N_x-1k}^n + \frac{r_x}{2}u_{N_xk}^{n+\frac{1}{2}} \end{pmatrix}, \quad
\boldsymbol{r}_j = \begin{pmatrix} \left(1+\frac{r_x}{2}\Delta_2^{(x)}\right)u_{j1}^{n+\frac{1}{2}} + \frac{r_y}{2}u_{j0}^{n+1} \\ \left(1+\frac{r_x}{2}\Delta_2^{(x)}\right)u_{j2}^{n+\frac{1}{2}} \\ \vdots \\ \left(1+\frac{r_x}{2}\Delta_2^{(x)}\right)u_{jN_y-2}^{n+\frac{1}{2}} \\ \left(1+\frac{r_x}{2}\Delta_2^{(x)}\right)u_{jN_y-1}^{n+\frac{1}{2}} + \frac{r_y}{2}u_{jN_y}^{n+1} \end{pmatrix}.
$$

The columns $\boldsymbol{r}_k$ (dimension $N_x - 1$) and $\boldsymbol{r}_j$ (dimension $N_y - 1$) contain expressions evaluated at times $n\Delta t$, $(n + 1/2)\Delta t$, and $(n + 1)\Delta t$: they are either boundary conditions u_{0k} and u_{N_xk}, computed at intermediate time $(n + 1/2)\Delta t$ by using Eq. (11.16), boundary conditions u_{j0} and u_{jN_y} at time $(n + 1)\Delta t$, or the solution values u_{jk} that we know from the previous step.

The key advantage of the decomposition of large matrix systems (11.17) and (11.18) to a multitude of smaller ones becomes clear at large N_x and N_y (solving systems with large matrices F_1 and F_1'). The subsystems of equations in (11.19) and (11.20) are independent: the solution of Eq. (11.19) is valid at $y_k = k\Delta y$ and it does not depend on values with other indices k, except those on the right hand-side of the equation, which is known from the previous time step. (Analogously for Eq. (11.20) at $x_j = j\Delta x$.) This decomposition can be exploited in vector and parallel-processing implementations.

D'Yakonov form The Peaceman–Rachford scheme can be expressed in a form that leads to a more efficient numerical implementation [1],

$$
\left[1 - \frac{r_x}{2}\Delta_2^{(x)}\right]u_{jk}^\star = \left[1 + \frac{r_x}{2}\Delta_2^{(x)}\right]\left[1 + \frac{r_y}{2}\Delta_2^{(y)}\right]u_{jk}^n + \frac{\Delta t}{2}\left(q_{jk}^n + q_{jk}^{n+1}\right),
$$
$$
\left[1 - \frac{r_y}{2}\Delta_2^{(y)}\right]u_{jk}^{n+1} = u_{jk}^\star .
$$

Its main feature is that the second step does not involve the inhomogeneous term q. Of course we still have to compute the boundary conditions corresponding to the equation for the intermediate value $u_{jk}^\star$, but we must be careful not to spoil the order of the methods by discretizing these conditions inappropriately. For example, if we have a Dirichlet boundary condition at $x = 0$, the overall order of the scheme will remain $O(\Delta t^2) + O(\Delta x^2) + O(\Delta y^2)$ if we evaluate the boundary condition as

$$
u_{0k}^\star = -\frac{r_y}{2}f(0, (k + 1)\Delta y, (n + 1)\Delta t) + (1 + r_y)f(0, k\Delta y, (n + 1)\Delta t)
$$
$$
- \frac{r_y}{2}f(0, (k - 1)\Delta y, (n + 1)\Delta t)
$$

(and similarly for other domain boundaries). In Problem 11.9.1 you can find the expressions for Neumann boundary conditions to   first and second order.

11.1.5 Three Space Dimensions

Dealing with PDE in one time and three space dimensions is beyond our scope, but by using our knowledge accumulated so far it is easy to write at least the simplest schemes. An explicit scheme for the three-dimensional diffusion equation in Cartesian coordinates, $v_t = D\nabla^2 v + Q(x, y, z, t)$, is at hand: when we organize the numerical approximations u^n_{jkl} of the exact solution $v(x_j, y_k, z_l, t_n) = v(j\Delta x, k\Delta y, l\Delta z, n\Delta t)$ in a vector $\boldsymbol{u}^n$ of dimension $(N_x + 1) \times (N_y + 1) \times (N_z + 1)$, a simple analogy to one dimension (10.13) and two dimensions (11.5) gives

$$u^{n+1}_{jkl} = u^n_{jkl} + \left(r_x \Delta_2^{(x)} + r_y \Delta_2^{(y)} + r_z \Delta_2^{(z)}\right) u^n_{jkl} + \Delta t\, Q^n_{jkl}\,, \tag{11.21}$$

with the error of order $O(\Delta t) + O(\Delta x^2) + O(\Delta y^2) + O(\Delta z^2)$. This scheme is conditionally stable, with the condition $r_x + r_y + r_z \le 1/2$. Similarly, we generalize (11.13) to the absolutely stable three-dimensional Crank–Nicolson scheme

$$\left(1 - \frac{r_x}{2} \Delta_2^{(x)} - \frac{r_y}{2} \Delta_2^{(y)} - \frac{r_z}{2} \Delta_2^{(z)}\right) u^{n+1}_{jkl}$$
$$= \left(1 + \frac{r_x}{2} \Delta_2^{(x)} + \frac{r_y}{2} \Delta_2^{(y)} + \frac{r_z}{2} \Delta_2^{(z)}\right) u^n_{jkl} + \frac{\Delta t}{2} \left(q^n_{jkl} + q^{n+1}_{jkl}\right)\,, \tag{11.22}$$

which is of order $O(\Delta t^2) + O(\Delta x^2) + O(\Delta y^2) + O(\Delta z^2)$. At each time step this scheme requires us to solve a matrix equation of the form $F_1 \boldsymbol{u}^{n+1} = F\boldsymbol{u}^n + \cdots$ with $(N_x - 1)(N_y - 1)(N_z - 1) \times (N_x - 1)(N_y - 1)(N_z - 1)$ matrices F_1 and F. Obviously, even a coarse discretization may lead to practical (memory) obstacles. If we insist on difference methods for parabolic PDE, we also resort to ADI schemes in three dimensions, as the corresponding matrices have a simpler structure.

11.1.6 Hyperbolic Equations

We start our study of two-dimensional hyperbolic PDE by the initial-value problem

$$\begin{aligned} v_t + cv_x + dv_y &= 0\,, \\ v(x, y, 0) &= f(x, y)\,. \end{aligned} \tag{11.23}$$

This problem has the exact solution $v(x, y, t) = f(x - ct, y - dt)$ which tells us that at $t > 0$ the initial condition propagates without change of amplitude or phase with velocity c along the x-axis and velocity d along the y-axis. In the following we describe three classes of corresponding difference schemes.

11.1.7   Explicit Schemes

Again, the analogy to one-dimensional discussion allows us to augment the schemes (10.34), (10.35), and (10.36) by the additional dimension:

$$u_{jk}^{n+1} = \left(1 - R_x \Delta_+^{(x)} - R_y \Delta_+^{(y)}\right) u_{jk}^n \,, \tag{11.24}$$

$$u_{jk}^{n+1} = \left(1 - R_x \Delta_-^{(x)} - R_y \Delta_-^{(y)}\right) u_{jk}^n \,, \tag{11.25}$$

$$u_{jk}^{n+1} = \left(1 - \frac{R_x}{2} \Delta_0^{(x)} - \frac{R_y}{2} \Delta_0^{(y)}\right) u_{jk}^n \,, \tag{11.26}$$

where $R_x = c\Delta t/\Delta x$ and $R_y = d\Delta t/\Delta y$. The FTFS scheme (11.24) is consistent to order $O(\Delta t) + O(\Delta x) + O(\Delta y)$, and its symbol is $\rho(\xi, \eta) = 1 - R_x[\exp(i\xi) - 1] - R_y[\exp(i\eta) - 1]$. From the conditions $|\rho(\pm\pi, 0)| \le 1$ and $|\rho(0, \pm\pi)| \le 1$ (at the points where $|\rho|^2$ has the extrema) we obtain the conditions $-1 \le R_x \le 0$ and $-1 \le R_y \le 0$, while the condition $|\rho(\pm\pi, \pm\pi)| \le 1$ yields the final stability criterion $-1 \le R_x + R_y \le 0$. The FTBS scheme (11.25) has the same order as FTFS, and its symbol $\rho(\xi, \eta) = 1 - R_x[1 - \exp(-i\xi)] - R_y[1 - \exp(-i\eta)]$ implies the stability criterion $0 \le R_x + R_y \le 1$ with $0 \le R_x \le 1$ and $0 \le R_y \le 1$. (Check this as an exercise and use the discussion accompanying (10.23)!) The FTCS scheme (11.26) is absolutely unstable, just as (10.36) has been.

11.1.8   Schemes for Equations in the Form of Conservation Laws

The analogy to the one-dimensional case (10.72) also leads directly to two-dimensional difference schemes for hyperbolic differential equations that are written in the form of a conservation law

$$v_t + [F(v)]_x + [G(v)]_y = 0 \,. \tag{11.27}$$

But recall the discussion in Sect. 10.11! The conservation law formulation by itself does not protect us from the unwanted effects of the difference scheme like dissipation and dispersion (Fig. 10.10), but it is a natural step on the path towards high-resolution schemes. In spite of this warning, we use the template (10.68) to construct the scheme

$$u_{jk}^{n+1} = u_{jk}^n - R_x \left[h_{j+1/2\,k}^{(x)n} - h_{j-1/2\,k}^{(x)n}\right] - R_y \left[h_{j\,k+1/2}^{(y)n} - h_{j\,k-1/2}^{(y)n}\right] \,, \tag{11.28}$$

where $R_x = \Delta t/\Delta x$ and $R_y = \Delta t/\Delta y$, and where $h^{(x)}$ and $h^{(y)}$ are numerical flux functions in the x and y directions. For the linear conservative problem of the form $v_t + cv_x + dv_y = 0$ the flux functions may be simply

$$h^{(x)n}_{j+1/2\,k} = c u^n_{jk}\,, \qquad h^{(y)n}_{j\,k+1/2} = d u^n_{jk}\,. \tag{11.29}$$

For a general non-linear conservative problem (11.27) the upwind scheme (10.86) is generalized by introducing the flux functions

$$h^{(x)n}_{j+1/2\,k} = \frac{1}{2}\left[F^n_{jk} + F^n_{j+1\,k}\right] - \frac{1}{2}\left|a^n_{j+1/2\,k}\right|\Delta^{(x)}_+ u^n_{jk}\,,$$

$$h^{(y)n}_{j\,k+1/2} = \frac{1}{2}\left[G^n_{jk} + G^n_{jk+1}\right] - \frac{1}{2}\left|a^n_{jk+1/2}\right|\Delta^{(y)}_+ u^n_{jk}\,,$$

except that the expressions $a^n_{j+1/2\,k}$ and $a^n_{jk+1/2}$ have to be computed such that in the former case we use the x coordinate (index j) as in Eq. (10.88), while in the latter we use the y coordinate (index k) analogously. The split (two-step) Lax–Wendroff scheme

$$u^{n+1/2}_{jk} = u^n_{jk} - R_x \Delta^{(x)}_-\left[h^{(x)n}_{j+1/2\,k}\right]\,, \tag{11.30}$$

$$u^{n+1}_{jk} = u^{n+1/2}_{jk} - R_y \Delta^{(y)}_-\left[h^{(y)n+1/2}_{jk+1/2}\right]\,, \tag{11.31}$$

is also very useful. It exploits the flux functions

$$h^{(x)n}_{j+1/2\,k} = c u^n_{jk} + \frac{1}{2}c(1 - cR_x)\Delta^{(x)}_+ u^n_{jk}\,, \tag{11.32}$$

$$h^{(y)n+1/2}_{j\,k+1/2} = d u^{n+1/2}_{jk} + \frac{1}{2}d(1 - dR_y)\Delta^{(y)}_+ u^{n+1/2}_{jk}\,. \tag{11.33}$$

Dissipation and dispersion effects in difference schemes become apparent already in linear problems with discontinuous initial conditions: see Fig. 11.3 and compare it to Fig. 10.10! This comparison clearly demonstrates the need to build methods capable of resolving discontinuities. You will encounter the two-dimensional upwind scheme and the split Lax–Wendroff scheme in Problem 11.9.4, accompanied by high-resolution schemes from Sect. 11.3.

11.1.9   Implicit and ADI Schemes

Good stability properties of implicit methods represent a significant advantage that should certainly be exploited in schemes for two-dimensional hyperbolic PDE. For example, the two most obvious absolutely stable schemes for problems of the form (11.23), where we define $R_x = c\Delta t/\Delta x$ and $R_y = d\Delta t/\Delta y$, are the BTCS scheme

$$\left(1 + \frac{R_x}{2}\Delta^{(x)}_0 + \frac{R_y}{2}\Delta^{(y)}_0\right)u^{n+1}_{jk} = u^n_{jk}\,,$$

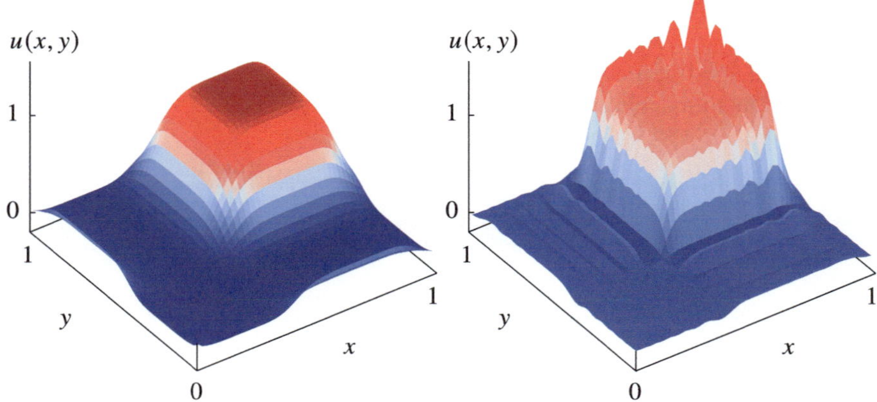

Fig. 11.3 The numerical solution of the hyperbolic problem $v_t + v_x + v_y = 0$ on $(x, y) \in [0, 1] \times [0, 1]$ with the initial condition $v(x, y, 0) = 1$ for $(x, y) \in [\frac{1}{4}, \frac{3}{4}] \times [\frac{1}{4}, \frac{3}{4}]$ at time $t = 0.2$. [LEFT] Dissipation of the solution in the upwind scheme. [RIGHT] Strong dispersion of the solution in the split Lax–Wendroff scheme. The discretization in both cases is $N_x = N_y = 48$ and $\Delta t = 0.002$. Compare to Fig. 11.5

which is convergent to order $O(\Delta t) + O(\Delta x^2) + O(\Delta y^2)$, and the Crank–Nicolson scheme

$$\left(1 + \frac{R_x}{4}\Delta_0^{(x)} + \frac{R_y}{4}\Delta_0^{(y)}\right) u_{jk}^{n+1} = \left(1 - \frac{R_x}{4}\Delta_0^{(x)} - \frac{R_y}{4}\Delta_0^{(y)}\right) u_{jk}^{n} \,,$$

which is of order $O(\Delta t^2) + O(\Delta x^2) + O(\Delta y^2)$.

Similarly to their application to parabolic equations (page 710), alternating direction implicit (ADI) methods allow us to avoid sparse matrices on the left-hand sides of the matrix equations that express the scheme in matrix notation. A typical representative is the absolutely stable Beam–Warming scheme of order $O(\Delta t^2) + O(\Delta x^2) + O(\Delta y^2)$,

$$\left(1 + \frac{R_x}{4}\Delta_0^{(x)}\right) u_{jk}^{\star} = \left(1 - \frac{R_x}{4}\Delta_0^{(x)}\right)\left(1 - \frac{R_y}{4}\Delta_0^{(y)}\right) u_{jk}^{n} \,,$$

$$\left(1 + \frac{R_y}{4}\Delta_0^{(y)}\right) u_{jk}^{n+1} = u_{jk}^{\star} \,.$$

11.2 Elliptic PDE

Numerical solution of elliptic PDE is fundamentally different from the solution of parabolic and hyperbolic equations. While in the latter two cases we seek the time dependence of the solution on some spatial definition domain, the solution of elliptic problems is a time-independent function that solves the given PDE on its domain with specified boundary conditions. By far the most well-known elliptic PDE is the Poisson equation $-\nabla^2 v = Q$.

11.2.1 Dirichlet Boundary Conditions

The basic model problem with Dirichlet boundary conditions is the two-dimensional Poisson equation "on the square":

$$
\begin{aligned}
-\nabla^2 v &= Q(x, y), \quad (x, y) \in R = [0, 1] \times [0, 1], \\
v(x, y) &= f(x, y), \quad (x, y) \in \partial R.
\end{aligned}
\tag{11.34}
$$

We discretize the equation along both space axes as shown in Fig. 11.1—we just ignore the time axis (index n). When we approximate the Laplace operator by the finite differences (11.2) and (11.3), the problem (11.34) becomes

$$
-\frac{1}{\Delta x^2} \Delta_2^{(x)} u_{jk} - \frac{1}{\Delta y^2} \Delta_2^{(y)} u_{jk} = q_{jk},
\tag{11.35}
$$

with the boundary conditions on four sides of the square

$$
\begin{aligned}
u_{0k} &= f_{0k}, \quad u_{N_x k} = f_{N_x k}, \quad k = 0, 1, \ldots, N_y, \\
u_{j0} &= f_{j0}, \quad u_{jN_y} = f_{jN_y}, \quad j = 0, 1, \ldots, N_x.
\end{aligned}
\tag{11.36}
$$

In all Problems we will maintain $N_x = N_y = N$ ($\Delta x = \Delta y$) for simplicity; this will not deprive us of any essential ingredient.

We shall also encounter a much more general elliptic PDE,

$$
a v_{xx} + c v_{yy} + d v_x + e v_y + f v = Q,
\tag{11.37}
$$

where a, c, d, e, f, and Q may be functions of x and y (we require only $a < 0$, $c < 0$, and $f > 0$). We discretize it as

$$
\frac{a}{\Delta x^2} \Delta_2^{(x)} u_{jk} + \frac{c}{\Delta y^2} \Delta_2^{(y)} u_{jk} + \frac{d}{2\Delta x} \Delta_0^{(x)} u_{jk} + \frac{e}{2\Delta y} \Delta_0^{(y)} u_{jk} + f u_{jk} = q_{jk}.
$$

From the programming viewpoint—following the advice of Ref. [2]—it is worthwhile to prepare the ground for such a general scheme, and write it in the form

$$
\alpha_{jk}^1 u_{j+1k} + \alpha_{jk}^2 u_{j-1k} + \alpha_{jk}^3 u_{jk+1} + \alpha_{jk}^4 u_{jk-1} - \alpha_{jk}^0 u_{jk} = q_{jk},
\tag{11.38}
$$

where the coefficients depend on the location (indices j and k).

We arrange the solution in a $(N_x + 1)(N_y + 1)$-dimensional vector of the form (11.1), we just drop the time index n. But due to the Dirichlet boundary conditions a shorter, $(N_x - 1)(N_y - 1)$-dimensional, vector (11.10) that does not include the components (11.36), suffices. The problems (11.34) and (11.37) can then be written as

$$A\boldsymbol{u} = \boldsymbol{q} \, ,$$

where A is a $(N_x - 1)(N_y - 1) \times (N_x - 1)(N_y - 1)$ matrix. By comparing the coefficients in Eqs. (11.35) and (11.38) we find

$$\alpha_{jk}^0 = -2\left(\frac{1}{\Delta x^2} + \frac{1}{\Delta y^2}\right) \, , \quad \alpha_{jk}^1 = \alpha_{jk}^2 = -\frac{1}{\Delta x^2} \, , \quad \alpha_{jk}^3 = \alpha_{jk}^4 = -\frac{1}{\Delta y^2} \, .$$

$$\tag{11.39}$$

The matrix A corresponding to problem (11.34) therefore has the form as in Fig. 11.2 (case a)) with the elements $\cdot = 2(1/\Delta x^2 + 1/\Delta y^2)$, $\circ = -1/\Delta x^2$, and $\triangleright = -1/\Delta y^2$. The symmetric matrix A is strictly positive definite, thus invertible, and the problem (11.35) has a unique solution. On the other hand, the matrix A corresponding to problem (11.37) is not symmetric in general. It is invertible only if it is strictly diagonally dominant, i.e. when $|a_{jj}| > \sum_{k=1,k\neq j}^{N-1} |a_{jk}|$ for each $j = 1, 2, \ldots, N - 1$. This requirement is met by fulfilling the conditions

$$0 < \Delta x < -\frac{2a_{jk}}{|d_{jk}|} \, , \qquad 0 < \Delta y < -\frac{2c_{jk}}{|e_{jk}|} \, .$$

The discrete scheme (11.35) is consistent with the differential equation (11.34) to order $O(\Delta x^2) + O(\Delta y^2)$. If we assume that the solution of the equation, v, is four times continuously differentiable with respect to x and y on $R^0 = R \backslash \partial R$ (the definition domain minus the boundary), it can be shown that the scheme is convergent to the same order,

$$\|\boldsymbol{v} - \boldsymbol{u}\|_\infty \leq C \left(\Delta x^2 + \Delta y^2\right) \|\partial^4 v\|_{\infty 0} \, . \tag{11.40}$$

Here $\|\partial^4 v\|_{\infty 0} = \sup\{|\partial^4 v(x, y)/\partial x^p \partial y^q| : p + q = 4 ; 0 \leq p, q \leq 4 ; (x, y) \in R^0\}$. The scheme for problem (11.37) described above has the same order of convergence.

11.2.2 Neumann Boundary Conditions

The basic elliptic model problem involving Neumann boundary conditions is the two-dimensional Poisson equation "on the square":

$$-\nabla^2 v = Q(x, y) \, , \quad (x, y) \in R = [0, 1] \times [0, 1] \, ,$$
$$\frac{\partial v}{\partial n}(x, y) = g(x, y) \, , \quad (x, y) \in \partial R \, .$$

We discretize the operator side of the equation as in the case of Dirichlet boundary conditions, and the boundary conditions to first or second order (compare to Eqs. (11.11) or (11.12) for the two-dimensional diffusion equation). Most of the numerical methods described below for Dirichlet problems can be applied to Neumann problems equally well. A solution example is given in Problem 11.9.3.

11.2.3 Mixed Boundary Conditions

Difference schemes for two-dimensional elliptic PDE with mixed boundary conditions $\alpha v + \beta(\partial v/\partial n) = h(x, y)$ for $(x, y) \in \partial R$ become somewhat cumbersome even in simple geometries. Such problems do not offer major new insights and we do not discuss them here. Details can be found in Ref. [2], Sect. 10.8.

11.2.4 Iterative Solution Methods: Relaxation

Several classes of methods are available to solve the system $A\boldsymbol{u} = \boldsymbol{q}$. In *direct methods* we solve the system directly, by "inverting" A. These classical techniques have been presented in Sects. 4.2 and 4.3. In *iterative methods* we obtain the answer after a sequence of steps in which the current approximate solution $\boldsymbol{w}$ (by some convergence criterion) becomes increasingly similar to the final numerical solution $\boldsymbol{u}$. The main representatives of this approach are the *relaxation methods*—also going under the name of *residual correction methods*—in which we strive, in each step, to reduce the *algebraic error* $\boldsymbol{e} = \boldsymbol{u} - \boldsymbol{w}$ and the *residual error* $\boldsymbol{r} = \boldsymbol{q} - A\boldsymbol{w}$.

In relaxation methods we approximate the matrix A^{-1} and define the iteration

$$\boldsymbol{w}^{n+1} = \boldsymbol{w}^n + B\boldsymbol{r}^n , \qquad n = 0, 1, \dots ,$$

where $\boldsymbol{r}^n = \boldsymbol{q} - A\boldsymbol{w}^n$ and B is some approximation of A^{-1}. On purpose, we have chosen n to denote the iteration index, just as in methods for parabolic and hyperbolic PDE that indeed contain the dependence on time. The errors $\boldsymbol{e}^n$ and $\boldsymbol{r}^n$ are related by the equation $A\boldsymbol{e}^n = A(\boldsymbol{u} - \boldsymbol{w}^n) = \boldsymbol{q} - A\boldsymbol{w}^n = \boldsymbol{r}^n$, so in "time" we iterate the solution as $\boldsymbol{w}^{n+1} = \boldsymbol{w}^n + BA\boldsymbol{e}^n$. The changing of the algebraic error $\boldsymbol{e}^{n+1} = (I - BA)\boldsymbol{e}^n$ is determined by the *iteration matrix* (also called the *error propagation matrix*) $R = I - BA$. We split the matrix A to the lower-triangular, diagonal, and upper-triangular part, $A = L + D + U$. Various relaxation schemes differ by the choice of the matrix B (Table 11.1).

Jacobi method In the Jacobi method we iterate $\boldsymbol{w}^{n+1} = D^{-1}(\boldsymbol{q} - (L + U)\boldsymbol{w}^n)$. For the Dirichlet problem (11.35) this means

Table 11.1 Forms of the matrix B in residual correction methods for iterative solution of the system $Au = q$. The iteration matrix is $I - BA$, where $A = L + D + U$, and ω is a free parameter

Method	Matrix B
Jacobi	D^{-1}
Gauss–Seidel	$(L + D)^{-1}$
SOR	$\omega(D + \omega L)^{-1}$
SSOR	$\omega(2 - \omega)(D + \omega U)^{-1}D(D + \omega L)^{-1}$

$$u_{jk}^{n+1} = -\frac{1}{\alpha_{jk}^0} \left[q_{jk} - \alpha_{jk}^1 u_{j+1k}^n - \alpha_{jk}^2 u_{j-1k}^n - \alpha_{jk}^3 u_{jk+1}^n - \alpha_{jk}^4 u_{jk-1}^n \right]$$

$$= \frac{1}{d} \left[q_{jk} + \frac{1}{\Delta x^2} \left(u_{j+1k}^n + u_{j-1k}^n \right) + \frac{1}{\Delta y^2} \left(u_{jk+1}^n + u_{jk-1}^n \right) \right] , \quad (11.41)$$

where α_{jk} are given by Eq. (11.39) and $d = -\alpha_{jk}^0 = 2(1/\Delta x^2 + 1/\Delta y^2)$. Equation (11.41) applies at $j = 1, 2, \ldots, N_x - 1$ and $k = 1, 2, \ldots, N_y - 1$, while the remaining values are specified by the boundary conditions. At each iteration step, we therefore replace the solution at (j, k) by the average of the solutions at four neighboring points (Fig. 11.1 (right)), and admix the source term q.

A common recommendation for implementing the method [2] is to code it along the first, not the second line of Eq. (11.41), even though such luxury appears to be redundant in such a simple method. This arms us for the attack on more general equations with space-dependent coefficients, as in (11.37), and on other types of boundary conditions. The coefficients $\alpha^{0,1,2,3,4}$ may be matrices (in this case relocate α_{jk}^0 to the left of the equation, in front of u_{jk}^{n+1}). The same advice applies to other relaxation methods.

Gauss–Seidel method In the Gauss–Seidel method we acquire consecutive approximations of the solution by the iteration $w^{n+1} = (L + D)^{-1}(q - Uw^n)$. For the problem (11.35) this can be written as

$$u_{jk}^{n+1} = -\frac{1}{\alpha_{jk}^0} \left[q_{jk} - \alpha_{jk}^1 u_{j+1k}^n - \alpha_{jk}^2 \underline{u_{j-1k}^{n+1}} - \alpha_{jk}^3 u_{jk+1}^n - \alpha_{jk}^4 \underline{u_{jk-1}^{n+1}} \right]$$

$$= \frac{1}{d} \left[q_{jk} + \frac{1}{\Delta x^2} \left(u_{j+1k}^n + \underline{u_{j-1k}^{n+1}} \right) + \frac{1}{\Delta y^2} \left(u_{jk+1}^n + \underline{u_{jk-1}^{n+1}} \right) \right] . \quad (11.42)$$

This iteration differs from the Jacobi only in the underlined terms. Even though the iteration index $n + 1$ appears on both sides of the equation, the scheme is explicit: the values u_{j-1k}^{n+1} and u_{jk-1}^{n+1} in the loops with increasing j and k are always computed just before they are actually needed. The averaging over the neighbors of the point (j, k) takes place in the current iteration step.

SOR and SSOR methods A tremendous improvement of the convergence speed can be obtained by the *successive over-relaxation* (SOR) scheme. In the first part of the iteration we perform one Gauss–Seidel step, and then form the average of the values from this step and the previously computed values:

$$u^{\star}_{jk} = -\frac{1}{\alpha^0_{jk}} \left[q_{jk} - \alpha^1_{jk}u^n_{j+1k} - \alpha^2_{jk}u^{n+1}_{j-1k} - \alpha^3_{jk}u^n_{jk+1} - \alpha^4_{jk}u^{n+1}_{jk-1} \right] ,$$

$$u^{n+1}_{jk} = u^n_{jk} + \omega \left(u^{\star}_{jk} - u^n_{jk} \right) ,$$

where

$$0 < \omega < 2$$

is a free parameter; outside of this range the method does not converge. If ω is chosen carefully, the number of necessary iteration steps drops dramatically (Fig. 11.4). At $\omega = 1$ the SOR scheme is equivalent to the Gauss–Seidel scheme. The choice of the optimal ω is described in the following. The SOR scheme can also be written in symmetrized form (SSOR), which combs the discrete mesh in two directions, and is less sensitive to the choice of optimal ω. It can be further accelerated by a trick due to Chebyshev; we eschew these finesses [2, 3].

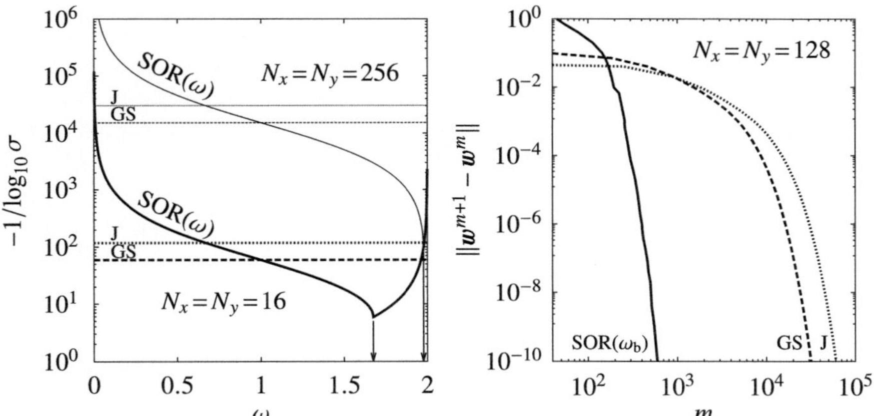

Fig. 11.4 [LEFT] The spectral radius σ of Jacobi, Gauss–Seidel, and SOR iteration matrices for the solution of the problem (11.35) on $N_x = N_y = 16$ and $N_x = N_y = 256$ meshes. Shown is the number of iteration steps needed to improve the precision of the result by one digit (the ordinate corresponds to Eq. (11.43) with $\zeta = 0.1$). The vertical arrows denote the optimal values of the parameter $\omega = \omega_{\rm b}$ for the SOR method (1.67351 and 1.97575), where the number of steps drops dramatically. [RIGHT] The number of steps m needed to achieve the chosen precision, measured by (11.46). In both figures we see the typical double convergence speed of the Gauss–Seidel method compared to the Jacobi method, and the much faster convergence of the SOR method when $\omega_{\rm b}$ is optimal

Example A residual correction method converges for any initial approximation $\boldsymbol{w}^0$ precisely when the spectral radius of its iteration matrix R is strictly less than one, $\sigma(R) < 1$ [4]. Yet we wish the method not only to converge, but to converge quickly! The speed of convergence is roughly determined by the largest eigenvalue λ_1 of the iteration matrix R: to reduce the algebraic error $\boldsymbol{e}$ by a factor $\zeta = \|\boldsymbol{e}^{n+m}\|/\|\boldsymbol{e}^n\|$, we must iterate

$$m \approx \frac{\log \zeta}{\log \sigma(R)} = \frac{\log \zeta}{\log |\lambda_1|} \tag{11.43}$$

times. The number of iterations m may be very large if $|\lambda_1| \approx 1$, or if there are several eigenvalues for which $|\lambda_j/\lambda_1| < 1$ while $|\lambda_j/\lambda_1| \approx 1$. For example, the spectral radius $\sigma(R_{\mathrm{J}})$ of the Jacobi iteration matrix $R_{\mathrm{J}} = I - D^{-1}A$ for the scheme (11.35) can be computed exactly:

$$\sigma(R_{\mathrm{J}}) = \frac{2}{d}\left(\frac{1}{\Delta x^2}\cos\frac{\pi}{N_x} + \frac{1}{\Delta y^2}\cos\frac{\pi}{N_y}\right) < 1 . \tag{11.44}$$

For $N_x = N_y = 1000$ ($\Delta x = \Delta y = 1/1000$) we get $\sigma(R_{\mathrm{J}}) = 0.999995$, which means that in order to improve the precision of the result by one digit we should make $-1/\log\sigma(R_{\mathrm{J}}) \approx 466000$ iterations! The Gauss–Seidel iteration does not fare much better: its speed of convergence is namely just double that of Jacobi since $\sigma(R_{\mathrm{GS}}) = \sigma(R_{\mathrm{J}})^2$.   ◁

Choosing the parameter ω in the SOR method By appropriately choosing the parameter ω in the SOR method we try to minimize the spectral radius of the corresponding iteration matrix (Table 11.1). Determining the optimal value ω_{b} is crucial for improving the speed of convergence (Fig. 11.4). In rare cases the spectral radius of the Jacobi iteration matrix can be computed analytically, e.g. as in Eq. (11.44). Then the optimal ω can be computed as

$$\omega_{\mathrm{b}} = \frac{2}{1 + \sqrt{1 - \overline{\lambda}_{\mathrm{J}}^2}} , \tag{11.45}$$

where $\overline{\lambda}_{\mathrm{J}} = \sigma(R_{\mathrm{J}})$. If $\sigma(R_{\mathrm{J}})$ can not be computed, one may resort to the procedure given below (and justified in detail in Sect. 10.5.12 of Ref. [2]).

We start by choosing a value of ω to initialize the SOR scheme. After a large number of iteration steps the ratio of differences of consecutive approximations becomes almost equal to the largest eigenvalue of the SOR iteration matrix:

$$n \gg 1 \implies \frac{w_j^{n+1} - w_j^n}{w_j^n - w_j^{n-1}} \approx \lambda_{1,\omega} .$$

(The expression applies to the jth component of the approximate solution vector $\boldsymbol{w}$ at three consecutive iteration steps $n - 1$, n, and $n + 1$; in practice we may indeed monitor just one component j or a set of different components.) We use this value to compute the approximation of the largest eigenvalue of the Jacobi iteration matrix,

$$\bar{\lambda}_J \approx \frac{1 - \omega - \lambda_{1,\omega}}{\omega\sqrt{\lambda_{1,\omega}}} \, ,$$

which can finally be used in Eq. (11.45). If we happen to choose $\omega > \omega_b$ (which, unfortunately, we only realize once we try), the procedure does not work. One must keep on trying until $\omega \leq \omega_b$.

The convergence of iterative methods should be monitored in order to achieve the desired precision. The iteration may be terminated when the norm of the difference between the subsequent approximations,

$$\| \boldsymbol{w}^{n+1} - \boldsymbol{w}^n \| \, , \tag{11.46}$$

or the norm of the current residual error,

$$\| \boldsymbol{q} - A\boldsymbol{w}^n \| \, , \tag{11.47}$$

drop below the specified tolerance. For safety, we often check both convergences in various norms, e.g. l_2, $l_{2,\Delta x}$, or sup-norm. It makes no sense to choose a tolerance smaller than the discretization error (11.40) since this is just wasting CPU time. Because the error constants C of $\|\partial^4 v\|_{\infty 0}$ in (11.40) are not known, we may estimate the appropriate tolerance by trial-and-error, e.g. with analytically solvable cases. Further advice is offered by Ref. [2] in Sect. 10.5.4.

Finally, what is it that actually "relaxes" in relaxation methods? We may see the solution of $-\nabla^2 v = Q$ as the solution of $v_t = \nabla^2 v + Q$ at very long times: the initial temperature distribution $v(x, y, 0)$ relaxes to the equilibrium solution at $t \to \infty$ where $v_t = 0$. In other words, relaxation methods are, at least in principle, also applicable to non-stationary parabolic problems, since two-step difference schemes can be expressed in the form $F_1 \boldsymbol{u}^{n+1} = F\boldsymbol{u}^n + \Delta t \boldsymbol{q}$. At each time step we are therefore again solving the matrix equation $A\boldsymbol{u} = \boldsymbol{q}$. Yet simple iterative relaxation methods are much slower than alternating gradient implicit methods. Moreover, iterative methods are not suitable for the solution of hyperbolic equations $v_t + av_x + bv_x = v_{xx} + v_{yy}$, as   the structure of the corresponding matrices becomes too complicated.

11.2.5 *Iterative Solution Methods: Conjugate Gradients*

Relaxation methods like the SOR and its improved versions (symmetrized SOR with Chebyshev acceleration) are useful only if near-optimal convergence parameters can be determined accurately and effectively. This problem can be avoided by using *conjugate gradient (CG) methods* that belong to a broader class of Krylov subspace

Table 11.2 Direct (D) and iterative (I) methods for solving the matrix problem $Au = q$ following from the discretization of the Poisson equation on a $N \times N$ mesh. The numerical cost (number of arithmetic operations and amount of memory) is given in terms of the *order of magnitude* in the power $n = N^2$: the actual cost in iterative methods depends on the criterion used to terminate the iteration

Method	Type	Number of operations	Memory
Jacobi	I	n^2	n
Gauss–Seidel	I	n^2	n
Conjugate gradients	I	$n^{3/2}$	n
SOR (SSOR+Chebyshev)	I	$n^{3/2}$ $(n^{5/4})$	n
FFT	D	$n \log n$	n
Block cyclic reduction	D	$n \log n$	n
Multi-grid	I	n	n

methods [5]. In these methods we do not solve the equation $Au = q$ but rather seek the minimum of the scalar function $F(u) = \frac{1}{2}u^{\mathrm{T}}Au - q^{\mathrm{T}}u$ or the root of $F'(u) = Au - q$ by using algorithms of steepest descent (to the minimum). Naive formulations of CG methods are not significantly better than classical relaxation methods (see Table 11.2), whereas the *preconditioned CG methods* truly out-perform them. Krylov subspace methods are discussed separately in Sect. 4.3.2.

11.3 High-Resolution Schemes $\star$

At the end of Sect. 11.1 we stopped mid-way in difference schemes for two-dimensional hyperbolic problems that can be written in the form of a conservation law $v_t + [F(v)]_x + [G(v)]_y = 0$ or their linear advective simplifications, $v_t + cv_x + dv_y = 0$. In two dimensions we encounter difficulties closely resembling those from one space dimension (Fig. 10.10): especially for discontinuous initial conditions we observe dissipation and strong dispersion in the time evolution of the solutions. We tame them (for example) by using flux limiter methods from Sect. 10.11.1. In the following we present the baseline schemes for the linear advection problem as well as the robust and popular Kurganov–Tadmor scheme for the non-linear problem.

11.3.1 Two Basic Schemes

A simple high-resolution scheme for $v_t + cv_x + dv_y = 0$ can be constructed if Eq. (11.28) is equipped by the numerical flux functions

$$
h_{j+1/2\,k}^{(x)n} = cu_{jk}^n + \phi_{jk}^{(x)n}\frac{1}{2}\,c(1 - cR_x)\Delta_+^{(x)}u_{jk}^n \,,
\tag{11.48}
$$

$$
h_{j\,k+1/2}^{(y)n} = du_{jk}^n + \phi_{jk}^{(y)n}\frac{1}{2}\,d(1 - dR_y)\Delta_+^{(y)}u_{jk}^n \,,
\tag{11.49}
$$

where

$$
\phi_{jk}^{(x)n} = \phi^{(x)}\left(\theta_{jk}^{(x)n}\right) \,, \qquad \theta_{jk}^{(x)n} = \left(\Delta_-^{(x)}u_{jk}^n\right)\Big/\left(\Delta_+^{(x)}u_{jk}^n\right) \,,
$$

$$
\phi_{jk}^{(y)n} = \phi^{(y)}\left(\theta_{jk}^{(y)n}\right) \,, \qquad \theta_{jk}^{(y)n} = \left(\Delta_-^{(y)}u_{jk}^n\right)\Big/\left(\Delta_+^{(y)}u_{jk}^n\right) \,.
$$

These are obvious generalizations of Eqs. (11.32) and (11.33) with the limiter functions $\phi^{(x)}$ and $\phi^{(y)}$, which we choose for each direction separately (x or y), and according to the physical character of the problem, from the set (10.75)–(10.82). We should pay attention to zeros of the denominators in θ_{jk} and to cases when j or k reach outside of the definition domain (boundary condition). An example of the solution of a hyperbolic problem $v_t + v_x + v_y = 0$ on $(x, y) \in [0, 1] \times [0, 1]$ with this scheme is shown in Fig. 11.5.

We obtain a high-resolution split scheme of the Lax–Wendroff type by including the flux functions (11.48) and (11.49) with the chosen flux limiter functions into the expressions (11.30) and (11.31).

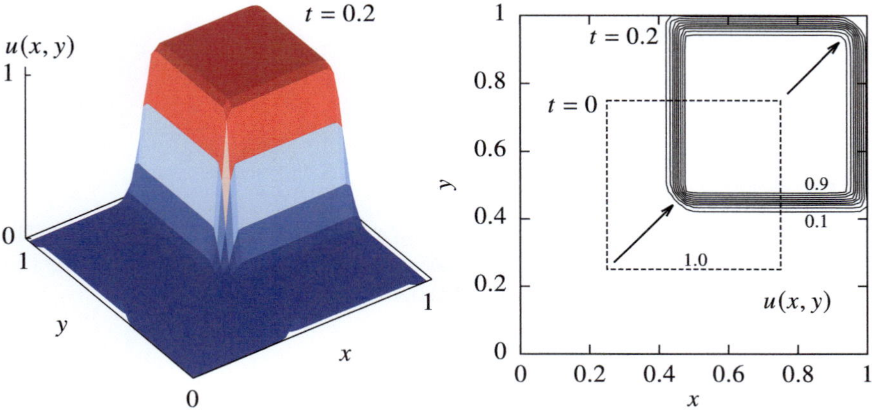

Fig. 11.5 Numerical solution of the linear hyperbolic problem $v_t + v_x + v_y = 0$ on$(x, y) \in$ $[0, 1] \times [0, 1]$ with same conditions as in Fig. 11.3. [LEFT] Solution by (11.48) and (11.49). [RIGHT] Solution isolines. The initial condition (dashed square) moves to the top right corner after $t = 0.2$ (a fifth of the period). In comparison to Fig. 11.3(left and right) we observe significantly less dissipation and dispersion

11.3.2   Zalesak–Smolarkiewicz Scheme

In Problem 11.9.4 we test several difference methods to solve $v_t + cv_x + dv_y = 0$ with constant c and d, and discontinuous initial conditions. In this test we notice that both high-resolution schemes for linear problems discussed above fail with only a modest change in the coefficients c and d, for example, by allowing them to depend on x and y,

$$v_t + c(x, y)v_x + d(x, y)v_y = 0 . \tag{11.50}$$

One of the reasons for this failure is the one-dimensional character of the limiter functions built into the two-dimensional scheme: the limiting of flux in one direction is decoupled from the limiting in the other. For transport problems of the form (11.50) one should therefore seek high-resolution schemes that limit the flux two-dimensionally. Here we describe the Zalesak–Smolarkiewicz method [6] in the notation of Ref. [2]. The landmark feature of this method are its two components: a low-order scheme (L) and a high-order (H) scheme. Its second feature is that at each point (j, k) we try to introduce enough anti-diffusive flux into the solution u_{jk} such that u_{jk}^{n+1} is still bounded by some values $u_{jk}^{\min}$ and $u_{jk}^{\max}$ to be determined on the fly. The low-order component,

$$^{\mathrm{L}}u_{jk}^n = u_{jk}^n - R_x \Delta_-^{(x)} \left[{}^{\mathrm{L}}h_{j+1/2\,k}^{(x)n} \right] - R_y \Delta_-^{(y)} \left[{}^{\mathrm{L}}h_{j\,k+1/2}^{(y)n} \right] ,$$

contains the upwind scheme

$$^{\mathrm{L}}h_{j+1/2\,k}^{(x)n} = \frac{1}{2}c \left(u_{jk}^n + u_{j+1\,k}^n \right) - \frac{1}{2}|c|\Delta_+^{(x)}u_{jk}^n ,$$

$$^{\mathrm{L}}h_{j\,k+1/2}^{(y)n} = \frac{1}{2}d \left(u_{jk}^n + u_{j\,k+1}^n \right) - \frac{1}{2}|d|\Delta_+^{(y)}u_{jk}^n .$$

At all j and k, only the values u_{jk} at time $n\Delta t$ are needed for the computation. The high-order component is constructed in two steps. In the first step, we adopt the flux functions of the non-split Lax–Wendroff scheme,

$$^{\mathrm{H}}h_{j+1/2\,k}^{(x)n} = cu_{jk}^n + \frac{1}{2}c(1 - cR_x)\Delta_+^{(x)}u_{jk}^n ,$$

$$^{\mathrm{H}}h_{j\,k+1/2}^{(y)n} = du_{jk}^n + \frac{1}{2}d(1 - dR_y)\Delta_+^{(y)}u_{jk}^n ,$$

while in the second step, we use the differences

$$^{\mathrm{D}}h_{j+1/2\,k}^{(x)n} = {}^{\mathrm{H}}h_{j+1/2\,k}^{(x)n} - {}^{\mathrm{L}}h_{j+1/2\,k}^{(x)n} , \qquad {}^{\mathrm{D}}h_{j\,k+1/2}^{(y)n} = {}^{\mathrm{H}}h_{j\,k+1/2}^{(y)n} - {}^{\mathrm{L}}h_{j\,k+1/2}^{(y)n} ,$$

to form the solution at time $(n + 1)\Delta t$:

$$u_{jk}^{n+1} = {}^{\mathrm{L}}u_{jk}^{n} - R_x \Delta_{-}^{(x)} \left[\phi_{jk}^{(x)n}\, {}^{\mathrm{D}}h_{j+1/2\,k}^{(x)n} \right] - R_y \Delta_{-}^{(y)} \left[\phi_{jk}^{(y)n}\, {}^{\mathrm{D}}h_{j\,k+1/2}^{(y)n} \right] .$$

We define the limiter functions ϕ in the form

$$\phi_{jk}^{(x)n} = \begin{cases} \min\left\{ b_{j+1\,k}^{+}, b_{jk}^{-} \right\} & ;\ {}^{\mathrm{D}}h_{j+1/2\,k}^{(x)n} \geq 0 \,, \\[2mm] \min\left\{ b_{jk}^{+}, b_{j+1\,k}^{-} \right\} & ;\ {}^{\mathrm{D}}h_{j+1/2\,k}^{(x)n} < 0 \,, \end{cases}$$

$$\phi_{jk}^{(y)n} = \begin{cases} \min\left\{ b_{j\,k+1}^{+}, b_{jk}^{-} \right\} & ;\ {}^{\mathrm{D}}h_{j\,k+1/2}^{(y)n} \geq 0 \,, \\[2mm] \min\left\{ b_{jk}^{+}, b_{j\,k+1}^{-} \right\} & ;\ {}^{\mathrm{D}}h_{j\,k+1/2}^{(y)n} < 0 \,, \end{cases}$$

where b_{jk}^{+} is the smallest upper limit for the factor multiplying the anti-diffusive flux a_{jk}^{+} entering the mesh point (j, k), while b_{jk}^{-} is the largest upper limit for the factor multiplying the corresponding exiting flux a_{jk}^{-}:

$$b_{jk}^{+} = \begin{cases} \min\left\{ 1, \left(u_{jk}^{\max} - u_{jk}^{\mathrm{L}} \right) \Big/ a_{jk}^{+} \right\} & ;\ a_{jk}^{+} > 0 \,, \\[2mm] 0 & ;\ a_{jk}^{+} = 0 \,, \end{cases}$$

$$b_{jk}^{-} = \begin{cases} \min\left\{ 1, \left(u_{jk}^{\mathrm{L}} - u_{jk}^{\min} \right) \Big/ a_{jk}^{-} \right\} & ;\ a_{jk}^{-} > 0 \,, \\[2mm] 0 & ;\ a_{jk}^{-} = 0 \,. \end{cases}$$

The anti-diffusive fluxes are defined by

$$\begin{aligned} a_{jk}^{+} &= \max\left\{ 0, {}^{\mathrm{D}}h_{j-1/2\,k}^{(x)n} \right\} - \min\left\{ 0, {}^{\mathrm{D}}h_{j+1/2\,k}^{(x)n} \right\} \\[2mm] &\quad + \max\left\{ 0, {}^{\mathrm{D}}h_{j\,k-1/2}^{(y)n} \right\} - \min\left\{ 0, {}^{\mathrm{D}}h_{j\,k+1/2}^{(y)n} \right\} , \\[3mm] a_{jk}^{-} &= \max\left\{ 0, {}^{\mathrm{D}}h_{j+1/2\,k}^{(x)n} \right\} - \min\left\{ 0, {}^{\mathrm{D}}h_{j-1/2\,k}^{(x)n} \right\} \\[2mm] &\quad + \max\left\{ 0, {}^{\mathrm{D}}h_{j\,k+1/2}^{(y)n} \right\} - \min\left\{ 0, {}^{\mathrm{D}}h_{j\,k-1/2}^{(y)n} \right\} . \end{aligned}$$

Only $u_{jk}^{\min}$ and $u_{jk}^{\max}$ were left to be determined. We choose

$$u_{jk}^{\min} = \min\left\{ u_{j+1\,k}, u_{j\,k+1}, u_{jk}, u_{j-1\,k}, u_{j\,k-1}, u_{j+1\,k}^{\mathrm{L}}, u_{j\,k+1}^{\mathrm{L}}, u_{jk}^{\mathrm{L}}, u_{j-1\,k}^{\mathrm{L}}, u_{j\,k-1}^{\mathrm{L}} \right\} ,$$

$$u_{jk}^{\max} = \max\left\{ u_{j+1\,k}, u_{j\,k+1}, u_{jk}, u_{j-1\,k}, u_{j\,k-1}, u_{j+1\,k}^{\mathrm{L}}, u_{j\,k+1}^{\mathrm{L}}, u_{jk}^{\mathrm{L}}, u_{j-1\,k}^{\mathrm{L}}, u_{j\,k-1}^{\mathrm{L}} \right\} .$$

This choice embodies the two-dimensional nature of the method, as $u_{jk}^{\min}$ and $u_{jk}^{\max}$ at each point (j, k) couple the values from the four neighboring points.

11.3.3 Kurganov–Tadmor Scheme

The Kurganov–Tadmor scheme [7] introduced in the one-dimensional context in Sect. 10.11.3.3 is easily generalized to two or three spatial dimensions. Below we give the formulas for the two-dimensional version of the problem (10.91),

$$v_t + [F(v)]_x + [G(v)]_y = [Q^x(v, v_x, v_y)]_x + [Q^y(v, v_x, v_y)]_y .$$

For reasons of clarity, we write down the formulas for the scalar case ($v \rightarrow v$, $F \rightarrow F$, and so on). The extension to higher dimensionalities of v, F, ... is straightforward as it is done component-wise. The two-dimensional version of the scalar second-order semi-discrete scheme (10.94) becomes

$$\frac{\mathrm{d}u_{jk}}{\mathrm{d}t} = -\frac{h^x_{j+1/2k}(t) - h^x_{j-1/2k}(t)}{\Delta x} - \frac{h^y_{jk+1/2}(t) - h^y_{jk-1/2}(t)}{\Delta y}$$
$$+\frac{p^x_{j+1/2k}(t) - p^x_{j-1/2k}(t)}{\Delta x} + \frac{p^y_{jk+1/2}(t) - p^y_{jk-1/2}(t)}{\Delta y},$$

where

$$h^x_{j+1/2k}(t) = \frac{F\left(u^+_{j+1/2k}(t)\right) + F\left(u^-_{j+1/2k}(t)\right)}{2} - \frac{a^x_{j+1/2k}(t)}{2}\left(u^+_{j+1/2k}(t) - u^-_{j+1/2k}(t)\right),$$
$$h^y_{jk+1/2}(t) = \frac{G\left(u^+_{jk+1/2}(t)\right) + G\left(u^-_{jk+1/2}(t)\right)}{2} - \frac{a^y_{jk+1/2}(t)}{2}\left(u^+_{jk+1/2}(t) - u^-_{jk+1/2}(t)\right),$$

are the numerical convection fluxes along x and y, while

$$p^x_{j+1/2k}(t) = \frac{1}{2}\left[Q^x\left(u_{jk}(t), \frac{u_{j+1k}(t) - u_{jk}(t)}{\Delta x}, (u_y)_{jk}(t)\right)\right.$$
$$\left. + Q^x\left(u_{j+1k}(t), \frac{u_{j+1k}(t) - u_{jk}(t)}{\Delta x}, (u_y)_{j+1k}(t)\right)\right],$$
$$p^y_{jk+1/2}(t) = \frac{1}{2}\left[Q^y\left(u_{jk}(t), (u_x)_{jk}(t), \frac{u_{jk+1}(t) - u_{jk}(t)}{\Delta y}\right)\right.$$
$$\left. + Q^y\left(u_{jk+1}(t), (u_x)_{jk+1}(t), \frac{u_{jk+1}(t) - u_{jk}(t)}{\Delta y}\right)\right],$$

are the corresponding numerical diffusion fluxes. The local maximum speeds may be evaluated as in Eq. (10.92), hence

$$a^x_{j+1/2k}(t) = \max_{\pm} \rho\left(\frac{\partial F}{\partial u}(u^\pm_{j+1/2k}(t))\right), \quad a^y_{jk+1/2}(t) = \max_{\pm} \rho\left(\frac{\partial G}{\partial u}(u^\pm_{j+1/2k}(t))\right),$$

for the general (non-scalar) case. By analogy to Eq. (10.95) the intermediate velocities are given by

$$u^{\pm}_{j+1/2k}(t) = u_{j+1/2\pm1/2k}(t) \mp \frac{\Delta x}{2}\left(u_x\right)_{j+1/2\pm1/2k}(t) \,,$$

$$u^{\pm}_{jk+1/2}(t) = u_{jk+1/2\pm1/2}(t) \mp \frac{\Delta y}{2}\left(u_y\right)_{jk+1/2\pm1/2}(t) \,.$$

The third-order version of this scheme is described in Ref. [8].

Example (Adapted Example 11 from Ref. [7].) We apply the Kurganov–Tadmor scheme to the two-dimensional Burgers equation involving diffusive terms,

$$v_t + \left(v^2\right)_x + \left(v^2\right)_y = \varepsilon\big(D(v)v_x\big)_x + \varepsilon\big(D(v)v_y\big)_y \,, \quad (x, y) \in [-1.5, 1.5]^2 \,,$$

where $\varepsilon = 0.1$. We consider two cases, $D(v) = 0$ (a purely hyperbolic problem), and

$$D(v) = \begin{cases} 0 \,; \; |v| \le 0.25 \,, \\ 1 \,; \; |v| > 0.25 \,. \end{cases}$$

This is a discontinuous function, rendering the problem hyperbolic when $|v| \le 0.25$ and parabolic otherwise. Let the boundary conditions be periodic, while the initial condition is

$$v(x, 0) = \begin{cases} -1 \,; \; (x - 0.5)^2 + (y - 0.5)^2 \le 0.4^2 \,, \\ 1 \,; \; (x + 0.5)^2 + (y + 0.5)^2 \le 0.4^2 \,, \\ 0 \,; \; \text{otherwise} \,. \end{cases}$$

The numerical solutions calculated on a 60×60 spatial grid up to $t = 0.5$ with $\Delta t = 0.1\Delta x$ are shown in Fig. 11.6 for the purely hyperbolic case (left panel) and the diffusive case (right panel). Note how well the scheme behaves even with such a modest number of grid points. See also Problem 11.9.5. ◁

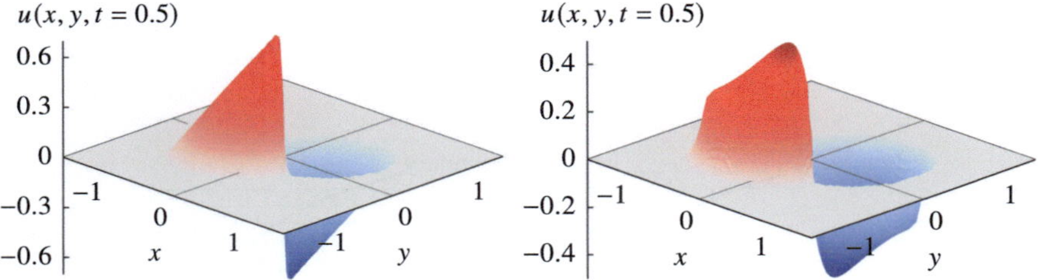

Fig. 11.6 The solution of the two-dimensional Burgers equation on a 60×60 grid at $t = 0.5$ by using the Kurganov–Tadmor scheme with $\theta = 1$ in the slope limiter. [LEFT] The purely hyperbolic case (no diffusion). [RIGHT] The case with diffusion

11.4   Physically Motivated Discretizations

In each difference scheme for PDE or systems of PDE, regardless of the dimension-ality, we use some procedure to discretize the space coordinates (and the time axis in evolution problems). Of course, for numerous realistic (two- and three-dimensional) physics problems, the basic Cartesian layout with square or rectangular domains is not the natural environment. In exceptional cases we may apply coordinate trans-formations to convert the problems with PDE in non-Cartesian frames to Cartesian ones, but most often the suitable transformation is very difficult to find. Computa-tions in planar polar coordinates, for example, also introduce peculiarities not seen in Cartesian coordinates.

Another option is to solve the equation on the existing (geometrically irregular) domain: in this case the main obstacle is the proper implementation of the boundary conditions in parts of the domain boundary bypassing the mesh points, as shown in Fig. 11.7 (left). For example, if we wish to compute the derivatives u_x and u_{xx} at C, we may use a non-uniform mesh with $\Delta x_j = x_j - x_{j-1}$ and $\Delta x_{j+1} = x_{j+1} - x_j$ near the boundary, and use the differences evaluated at the points L, C, and R' (instead of at L, C, and R):

$$(u_x)_j = \frac{u_{j+1} - u_{j-1}}{\Delta x_{j+1} + \Delta x_j} + O\left(\frac{\Delta x_{j+1}^2}{\Delta x_{j+1} + \Delta x_j}\right) + O\left(\frac{\Delta x_j^2}{\Delta x_{j+1} + \Delta x_j}\right) , \quad (11.51)$$

$$(u_{xx})_j = 2\,\frac{\Delta x_j u_{j+1} - (\Delta x_{j+1} + \Delta x_j)u_j + \Delta x_{j+1} u_{j-1}}{\Delta x_{j+1}\Delta x_j(\Delta x_{j+1} + \Delta x_j)}$$
$$+ O\left(\frac{\Delta x_{j+1}^2}{\Delta x_{j+1} + \Delta x_j}\right) + O\left(\frac{\Delta x_j^2}{\Delta x_{j+1} + \Delta x_j}\right) . \quad (11.52)$$

We may also make the mesh denser locally, or superpose a finer mesh on a coarser mesh, as in Fig. 11.7 (right), but such a procedure is hard to automate; dedicated programs exist for this task. Details on constructing overlapping discrete meshes can be found in Ref. [9, 10] in the context of multi-grid methods (see also Sect. 11.8).

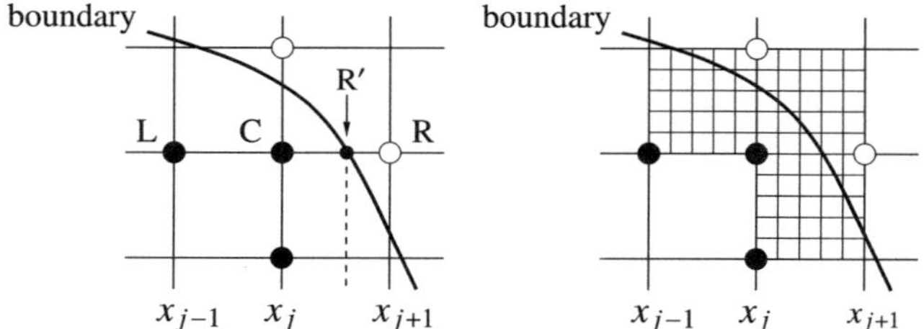

Fig. 11.7 [LEFT] Computing the approximations of u_x and u_{xx} in the vicinity of the domain boundary bypassing the mesh points. In the first order of the mesh spacings we use Eqs. (11.51) and (11.52). [RIGHT] Making the discrete mesh denser

11.4.1 *Two-Dimensional Diffusion Equation in Polar Coordinates*

Let us translate the difference scheme for the two-dimensional diffusion equation in Cartesian coordinates to polar coordinates. We are solving the equation

$$v_t = D \left(\frac{1}{r}(r v_r)_r + \frac{1}{r^2} v_{\theta\theta} \right) + Q(r, \theta, t) \tag{11.53}$$

on $(r, \theta) \in [0, 1] \times [0, 2\pi]$ for $t > 0$, with the initial and boundary conditions

$$v(r, \theta, 0) = f(r, \theta), \quad 0 \le \theta < 2\pi, \quad 0 \le r \le 1,$$
$$v(1, \theta, t) = g(\theta, t), \quad 0 \le \theta < 2\pi, \quad t > 0.$$

These conditions are supplemented by the periodicity requirement

$$v(r, 0, t) = v(r, 2\pi, t).$$

As in Sect. 11.1 we count the time steps as $t = n\Delta t$, and segment the radial and angular coordinates uniformly:

$$r_j = j\Delta r, \quad \Delta r = 1/N_r, \quad j = 0, 1, \ldots, N_r,$$
$$\theta_k = k\Delta\theta, \quad \Delta\theta = 2\pi/N_\theta, \quad k = 0, 1, \ldots, N_\theta,$$

thus $r_0 = 0, r_{N_r} = 1, \theta_0 = 0, \theta_{N_\theta} = 2\pi$ (Fig. 11.8 (left)). Since the values at $j = 0$ do not depend on k, we abbreviate $u_{0k}^n = u_0^n$, and the continuity in θ implies $u_{jN_\theta}^n = u_{j0}^n$.

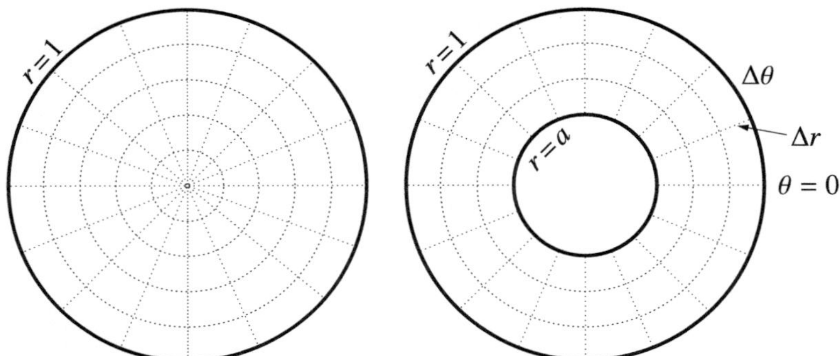

Fig. 11.8 [LEFT] The mesh for the solution of diffusion and Poisson equations in polar coordinates on $[r_0, r_{N_r}] \times [\theta_0, \theta_{N_\theta}] = [0, 1] \times [0, 2\pi]$. The discretization $\Delta\theta = 2\pi/N_\theta$, $\Delta r = (r_{N_r} - r_0)/N_r$, and Δt defines the mesh points $(r_j, \theta_k, t_n) = (r_0 + j\Delta r, k\Delta\theta, n\Delta t)$, and at each of them the true solution $v_{jk}^n = v(r_j, \theta_k, t_n)$ and its approximation u_{jk}^n. [RIGHT] The mesh for the Poisson problem on $[a, 1] \times [0, 2\pi]$

We replace the partial derivatives by the corresponding differences:

$$(v_t)_{jk}^n \approx \frac{u_{jk}^{n+1} - u_{jk}^n}{\Delta t} \, ,$$

$$\left(\frac{1}{r^2} \, v_{\theta\theta} \right)_{jk}^n \approx \frac{1}{r_j^2} \frac{1}{\Delta\theta^2} \Delta_2^{(\theta)} u_{jk}^n \, ,$$

$$\left(\frac{1}{r} \, (r v_r)_r \right)_{jk}^n \approx \frac{1}{r_j} \frac{1}{\Delta r^2} \left[r_{j+\frac{1}{2}} \left(u_{j+1k}^n - u_{jk}^n \right) - r_{j-\frac{1}{2}} \left(u_{jk}^n - u_{j-1k}^n \right) \right] \, ,$$

where $\Delta_2^{(\theta)}$ is defined by (11.4), where the role of y is now played by θ. We should be aware of the approximation of $(r v_r)_r / r$, in which the radial coordinate appears outside of the mesh points (where it is known): it turns out that such a difference is more precise than the discretization of the equivalent expression $v_r / r + v_{rr}$. When all terms are joined, an explicit difference scheme emerges,

$$u_{jk}^{n+1} = u_{jk}^n + \frac{1}{r_j} \frac{D\Delta t}{\Delta r^2} \left[r_{j+\frac{1}{2}} \left(u_{j+1k}^n - u_{jk}^n \right) - r_{j-\frac{1}{2}} \left(u_{jk}^n - u_{j-1k}^n \right) \right]$$
$$+ \frac{1}{r_j^2} \frac{D\Delta t}{\Delta\theta^2} \left[u_{jk+1}^n - 2u_{jk}^n - \underline{u_{jk-1}^n} \right] + \Delta t q_{jk}^n \, , \tag{11.54}$$

which applies at $j = 1, 2, \ldots, N_r - 1$ and $k = 1, 2, \ldots, N_\theta - 1$. For $k = 0$ the underlined term should be understood as $u_{jN_\theta-1}^n$ (periodicity). We obtain the missing difference equation at $j = 0$ by integrating the differential equation (11.53),

$$\iiint\limits_{r \, \theta \, t} v_t \, d\Omega = D \iiint \left(\frac{1}{r} \, (r v_r)_r + \frac{1}{r^2} \, v_{\theta\theta} \right) d\Omega + \iiint Q(r, \theta, t) \, d\Omega \, ,$$

on the time interval $t \in [t_n, t_{n+1}]$ and space intervals ("control volume") $r \in [0, \Delta r/2]$ and $\theta \in [0, 2\pi]$. (We have denoted $d\Omega = r \, dr \, d\theta \, dt$.) In the term containing the spatial derivatives we use the divergence theorem

$$\int\limits_0^{\Delta r/2} \left(\frac{1}{r} \, (r v_r)_r + \frac{1}{r^2} \, v_{\theta\theta} \right) r \, dr = \frac{\Delta r}{2} \, v_r \left(\frac{\Delta r}{2}, \theta, t \right) \, ,$$

and use the trapezoidal formula for the remaining integrals, whence

$$\pi \frac{\Delta r^2}{4} \left(u_0^{n+1} - u_0^n \right) \approx D \frac{\Delta r}{2} \Delta t \sum_{k=0}^{N_\theta-1} \frac{u_{1k}^n - u_0^n}{\Delta r} \Delta\theta + \pi \frac{\Delta r^2 \Delta t}{4} \, q_0^n \, .$$

We then use $\sum_k u_0^n = N_\theta u_0^n = (2\pi/\Delta\theta) u_0^n$, which ultimately results in

$$u_0^{n+1} = \left(1 - \frac{4D\Delta t}{\Delta r^2}\right) u_0^n + \frac{2D\Delta\theta\,\Delta t}{\pi\,\Delta r^2} \sum_{k=0}^{N_\theta - 1} u_{1k}^n + \Delta t q_0^n \,. \tag{11.55}$$

The complete scheme is given by Eqs. (11.54) and (11.55). Polar coordinates are no hindrance for implicit difference schemes, like the pure implicit or Crank–Nicolson by analogy to Eq. (11.13). The resulting matrix systems for solution vectors feature matrices with relatively ugly structures. However, these matrices can be transformed to triangular forms; details are given in Ref. [1].

11.4.2 Two-Dimensional Poisson Equation in Polar Coordinates

When two-dimensional elliptic PDE are solved in polar coordinates, further peculiarities appear. We discuss two model problems [2]: in the first, we solve the Poisson equation on the annulus,

$$-\nabla^2 v = Q(r, \theta) \,,$$

where $a < r < 1$ and $0 \le \theta \le 2\pi$ (Fig. 11.8 (right)), with Dirichlet boundary conditions

$$v(a, \theta) = f_1(\theta)\,, \quad 0 \le \theta < 2\pi\,, \quad a < r < 1\,,$$
$$v(1, \theta) = f_2(\theta)\,, \quad 0 \le \theta < 2\pi\,;$$

in the second, we include the origin (thus $a = 0$), so that the first initial condition (11.56) does not apply. In both problems, the spatial part of the difference operator is the same as for the diffusion equation (11.53), so precisely that discretization can be adopted. We span the mesh (r_j, θ_k) on the annulus such that $r_0 = a, r_{N_r} = 1$, $\theta_0 = 0$, $\theta_{N_\theta} = 2\pi$, $\Delta r = (1 - a)/N_r$, and $\Delta\theta = 2\pi/N_\theta$; if the origin is included, we have $r_0 = 0$ and $\Delta r = 1/N_r$. For either of the problems, this boils down to

$$-\frac{1}{\Delta r^2}\frac{1}{r_j}\left[r_{j+1/2}\left(u_{j+1k} - u_{jk}\right) - r_{j-1/2}\left(u_{jk} - u_{j-1k}\right)\right] - \frac{1}{\Delta\theta^2}\frac{1}{r_j^2}\,\underline{\Delta_2^{(\theta)} u_{jk}} = q_{jk}\,,$$

where $q_{jk} = Q(r_j, \theta_k)$ for $j = 1, 2, \ldots, N_r - 1$ and $k = 1, 2, \ldots, N_\theta - 1$. When $k = 0$, the scheme accesses the point u_{j-1} in the underlined term; periodicity in θ comes to rescue by setting $u_{j-1} = u_{jN_\theta - 1}$. The remaining components of the solution are given by the boundary conditions. In the annulus problem we have

$$u_{jN_\theta} = u_{j0}\,, \qquad j = 0, 1, \ldots, N_r\,,$$
$$u_{0k} = f_1(\theta_k)\,, \quad k = 0, 1, \ldots, N_\theta\,,$$
$$u_{N_r k} = f_2(\theta_k)\,, \quad k = 0, 1, \ldots, N_\theta\,.$$

In the problem that includes the origin we must be careful about—the origin: the values u_{0k} and q_{0k} actually can not depend on k, thus we abbreviate $u_{0k} = u_0$ and $q_{0k} = q_0$. The boundary conditions then become

$$
\begin{aligned}
u_{jN_\theta} &= u_{j0} , & j &= 0, 1, \ldots, N_r , \\
u_{N_r k} &= f_2(\theta_k) , & k &= 0, 1, \ldots, N_\theta ,
\end{aligned}
$$

and

$$
\frac{4}{\Delta r^2} u_0 - \frac{2\Delta\theta}{\pi \Delta r^2} \sum_{k=0}^{N_\theta-1} u_{1k} = q_0 .
$$

11.5 Boundary Element Method ⋆

The *boundary element method* (BEM) allows us to solve PDE in non-trivial geometries in which the discretization of the interior of the definition domain is difficult, while it is relatively easy to express (at least in some approximation) the boundary conditions on the boundaries of this domain. This is what makes the BEM method so appealing: just by using the information from the boundaries we solve the problem on the whole domain.

Here the basic outline of BEM is presented, following closely Ref. [11]. As an example we discuss the two-dimensional Laplace equation

$$
\frac{\partial^2 \phi}{\partial x^2} + \frac{\partial^2 \phi}{\partial y^2} = 0 , \tag{11.56}
$$

which we try to solve in the xy-plane in the domain R bounded by the piecewise smooth closed curve C. Along the individual segments C_i of C, either Dirichlet or Neumann boundary conditions are specified:

$$
\begin{aligned}
\phi(x, y) &= f_i(x, y) , & (x, y) &\in C_i , \\
\frac{\partial\phi(x, y)}{\partial n} &= f_j(x, y) , & (x, y) &\in C_j ,
\end{aligned}
$$

as shown in Fig. 11.9 (left). The normal derivative $\partial\phi/\partial n = n_x(\partial\phi/\partial x) + n_y(\partial\phi/\partial y)$ is defined by the components of the unit normal vector $\boldsymbol{n} = (n_x, n_y)^{\mathrm{T}}$, which points away from the domain R. A physical example of such a problem is the stationary state of heat conduction in isotropic matter. The solution $\phi(x, y)$ describes the distribution of temperature in the domain R with the boundary C, some pieces of which are held at constant temperature, while the others experience a constant heat flux $-\lambda(\partial\phi/\partial n)$.

The particular solution of (11.56) is $\phi(x, y) = A \log \sqrt{x^2 + y^2} + B$ for $(x, y) \neq (0, 0)$. We choose $A = 1/(2\pi)$, $B = 0$, and move the origin from $(0, 0)$ to (ξ, η).

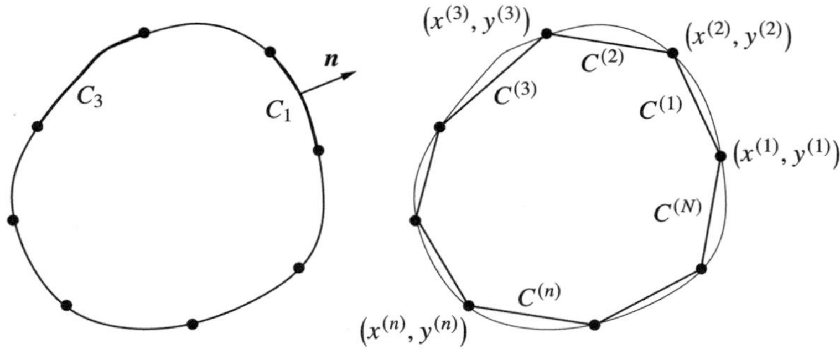

Fig. 11.9 [LEFT] The domain R with the smooth boundary C on which we solve the two-dimensional Poisson equation. As an example, we have Dirichlet boundary conditions on segment C_3 and Neumann boundary conditions on C_1. [RIGHT] The approximation of the boundary C by the inscribed polygon with sides $C^{(n)}$

This results in the *fundamental solution* of the Laplace equation

$$\Phi(x, y; \xi, \eta) = \frac{1}{4\pi} \log\left[(x - \xi)^2 + (y - \eta)^2\right] ,$$

which is defined everywhere except at (ξ, η). Gauss theorem for vector functions, $\int_C \mathbf{V} \cdot \mathbf{n}\, ds(x, y) = \int_R \nabla \cdot \mathbf{V}\, dx\, dy$, can be applied to show that for any two solutions ϕ_1 and ϕ_2 of Eq. (11.56), we have

$$\int_C [\phi_2(\partial\phi_1/\partial n) - \phi_1(\partial\phi_2/\partial n)]\, ds(x, y) = 0 .$$

We choose $\phi_1 = \Phi(x, y; \xi, \eta)$ and $\phi_2 = \phi(x, y)$, where $\phi(x, y)$ is the desired solution of Eq. (11.56). A few basic tricks of complex analysis are then needed to connect the desired solution and the fundamental solution by the *boundary* integral equation

$$\lambda(\xi, \eta)\phi(\xi, \eta) = \int_C \left[\phi(x, y)\frac{\partial}{\partial n}\Phi(x, y; \xi, \eta) - \Phi(x, y; \xi, \eta)\frac{\partial}{\partial n}\phi(x, y)\right] ds ,$$

$$(11.57)$$

where $\lambda(\xi, \eta) = 1/2$. In the boundary element method, the solution of the basic problem in the interior of the domain R is obtained by solving this integral equation *on the boundary C*. The form of the integral equation remains the same in all parts of the domain of (11.56), only the parameter $\lambda(\xi, \eta)$ changes:

$$\lambda(\xi, \eta) = \begin{cases} 0 \; ; \; (\xi, \eta) \notin R \cup C , \\ \frac{1}{2} \; ; \; (\xi, \eta) \text{ on smooth part of } C , \\ 1 \; ; \; (\xi, \eta) \in R . \end{cases}$$

We approximate the boundary C by a polygon with N sides, as shown in Fig. 11.9 (right), such that

$$C \approx C^{(1)} \cup C^{(2)} \cup \cdots \cup C^{(N)} \,.$$

Each side $C^{(n)}$ is a segment between the points $(x^{(n)}, y^{(n)})$ and $(x^{(n+1)}, y^{(n+1)})$. We assume that the values of the functions and their derivatives are constant along individual sides $C^{(n)}$, i.e.

$$\phi(x, y) \approx v^{(n)} \,, \qquad \frac{\partial \phi(x, y)}{\partial n} \approx d^{(n)} \,, \qquad (x, y) \in C^{(n)} \,, \qquad n = 1, 2, \ldots, N \,,$$

where $v^{(n)}$ is the value of ϕ and $d^{(n)}$ is the value of $\partial \phi / \partial n$ in the middle of $C^{(n)}$. Now the integral equation (11.57) can be approximately written as

$$\lambda(\xi, \eta) \phi(\xi, \eta) \approx \sum_{n=1}^{N} \left[v^{(n)} \mathcal{D}^{(n)}(\xi, \eta) - d^{(n)} \mathcal{V}^{(n)}(\xi, \eta) \right] \,, \tag{11.58}$$

where

$$\mathcal{V}^{(n)}(\xi, \eta) = \int_{C^{(n)}} \Phi(x, y; \xi, \eta) \, \mathrm{d}s(x, y) \,, \tag{11.59}$$

$$\mathcal{D}^{(n)}(\xi, \eta) = \int_{C^{(n)}} \frac{\partial \Phi}{\partial n}(x, y; \xi, \eta) \, \mathrm{d}s(x, y) \,. \tag{11.60}$$

For some index n, the boundary condition defines either the value $v^{(n)}$ or the derivative $d^{(n)}$ (but not both), so there are N unknowns at the right-hand side of Eq. (11.58). We choose (ξ, η) to lie in the middle of the sides $C^{(n)}$ and obtain

$$\frac{1}{2} v^{(m)} = \sum_{n=1}^{N} \left[v^{(n)} \mathcal{D}^{(n)}\left(\overline{x}^{(m)}, \overline{y}^{(m)}\right) - d^{(n)} \mathcal{V}^{(n)}\left(\overline{x}^{(m)}, \overline{y}^{(m)}\right) \right] \,, \qquad m = 1, 2, \ldots, N \,,$$

where $\left(\overline{x}^{(m)}, \overline{y}^{(m)}\right)$ is the midpoint of $C^{(m)}$. In Eq. (11.58) we have used $\lambda = 1/2$, since all midpoints lie on the smooth parts of the approximate boundary C. This is a system of N linear equation, which can be written as

$$\sum_{n=1}^{N} a_{mn} z_n = \sum_{n=1}^{N} b_{mn} \,, \qquad m = 1, 2, \ldots, N \,, \tag{11.61}$$

where

$$a_{mn} = -\mathcal{V}^{(n)}\left(\overline{x}^{(m)}, \overline{y}^{(m)}\right),$$
$$b_{mn} = v^{(n)}\left[-\mathcal{D}^{(n)}\left(\overline{x}^{(m)}, \overline{y}^{(m)}\right) + \tfrac{1}{2}\delta_{m,n}\right],$$
$$z_n = d^{(n)},$$

if the boundary condition on $C^{(n)}$ prescribes the value ϕ, or

$$a_{mn} = \mathcal{D}^{(n)}\left(\overline{x}^{(m)}, \overline{y}^{(m)}\right) - \tfrac{1}{2}\delta_{m,n},$$
$$b_{mn} = d^{(n)}\mathcal{V}^{(n)}\left(\overline{x}^{(m)}, \overline{y}^{(m)}\right),$$
$$z_n = v^{(n)},$$

if the boundary condition on $C^{(n)}$ prescribes the derivative $\partial\phi/\partial n$. When this system is solved, each component z_n contains precisely the missing information from $C^{(n)}$ that was not expressed by the boundary condition: if the derivative was specified, Eq. (11.61) provides the value of the function, and vice-versa. We end up with N values of ϕ and N values of $\partial\phi/\partial n$ on N segments of C. Finally, the solution in the interior of R is obtained by Eq. (11.58), in which we set $\lambda = 1$,

$$\phi(\xi, \eta) \approx \sum_{n=1}^{N}\left[v^{(n)}\mathcal{D}^{(n)}(\xi, \eta) - d^{(n)}\mathcal{V}^{(n)}(\xi, \eta)\right], \qquad (\xi, \eta) \in R.$$

The elegance of the method has thus been clearly revealed: we obtain the solution on the whole domain R by manipulating information from its boundary C. In constructing $\mathcal{V}(\xi, \eta)$ and $\mathcal{D}(\xi, \eta)$, the line integrals (11.59) and (11.60) need to be computed along individual segments $C^{(n)}$. The segments are parameterized as

$$x(t) = x^{(n)} - tl^{(n)}n_y^{(n)}, \qquad y(t) = y^{(n)} + tl^{(n)}n_x^{(n)}, \qquad 0 \le t \le 1. \qquad (11.62)$$

Here $l^{(n)}$ is the length of $C^{(n)}$, while the unit vector $\boldsymbol{n}^{(n)} = (n_x^{(n)}, n_y^{(n)})^{\mathrm{T}}$ perpendicular to this segment and pointing away from R has the components

$$n_x^{(n)} = \frac{1}{l^{(n)}}\left(y^{(n+1)} - y^{(n)}\right), \qquad n_y^{(n)} = \frac{1}{l^{(n)}}\left(x^{(n)} - x^{(n+1)}\right).$$

We define

$$A^{(n)} = \left(l^{(n)}\right)^2,$$
$$B^{(n)}(\xi, \eta) = 2l^{(n)}\left(-n_y^{(n)}\left(x^{(n)} - \xi\right) + n_x^{(n)}\left(y^{(n)} - \eta\right)\right),$$
$$C^{(n)}(\xi, \eta) = \left(x^{(n)} - \xi\right)^2 + \left(y^{(n)} - \eta\right)^2.$$

In the parameterization (11.62), the quantity $4A^{(n)}C^{(n)} - (B^{(n)})^2$ is non-negative,

$$F^{(n)} \equiv 4A^{(n)}C^{(n)}(\xi, \eta) - \left[B^{(n)}(\xi, \eta)\right]^2 \ge 0, \qquad \forall(\xi, \eta).$$

Elementary integration [11] brings us to the final expressions for the line integrals $\mathcal{V}^{(n)}$ and $\mathcal{D}^{(n)}$, which are distinguished according to the value of $F^{(n)}$. If $F^{(n)} > 0$, we evaluate

$$
\mathcal{V}^{(n)} = \frac{l^{(n)}}{4\pi} \left\{ 2\left(\log l^{(n)} - 1\right) - \frac{B^{(n)}}{2A^{(n)}} \log\left|\frac{C^{(n)}}{A^{(n)}}\right| + \left[1 + \frac{B^{(n)}}{2A^{(n)}}\right] \log\left|1 + \frac{B^{(n)}}{A^{(n)}} + \frac{C^{(n)}}{A^{(n)}}\right| \right.
$$
$$
\left. + \frac{\sqrt{F^{(n)}}}{A^{(n)}} \left[\arctan \frac{2A^{(n)} + B^{(n)}}{\sqrt{F^{(n)}}} - \arctan \frac{B^{(n)}}{\sqrt{F^{(n)}}}\right] \right\} ,
$$
$$
\mathcal{D}^{(n)} = \frac{l^{(n)}\left[n_x^{(n)}\left(x^{(n)} - \xi\right) + n_y^{(n)}\left(y^{(n)} - \eta\right)\right]}{\pi \sqrt{F^{(n)}}} \left[\arctan \frac{2A^{(n)} + B^{(n)}}{\sqrt{F^{(n)}}} - \arctan \frac{B^{(n)}}{\sqrt{F^{(n)}}}\right] ,
$$

On the other hand, if $F^{(n)} = 0$, we need to compute

$$
\mathcal{V}^{(n)} = \frac{l^{(n)}}{2\pi} \left\{ \log l^{(n)} + \left[1 + \frac{B^{(n)}}{2A^{(n)}}\right] \log\left|1 + \frac{B^{(n)}}{2A^{(n)}}\right| - \frac{B^{(n)}}{2A^{(n)}} \log\left|\frac{B^{(n)}}{2A^{(n)}}\right| - 1 \right\} ,
$$
$$
\mathcal{D}^{(n)} = 0 .
$$

The boundary element method is also applicable to non-stationary problems: time integration is performed separately. A superb introduction is given in Ref. [11]. See also Problems 11.9.9 and 11.9.10.

11.6   Finite Element Method ★

Finite element methods (FEM) were conceived already in the late 1940s [12], yet they remain the basic tool of engineers and natural scientists, especially for problems in elastomechanics, hydro- and aero-dynamics in complex geometries. In difference methods we use finite differences to approximate the differential equation, i.e. its space and time derivatives, and impose boundary conditions appropriately. But in complex geometries this is very hard to accomplish.

In the finite element approach, we find the integral of the differential equation on its definition domain, and thereby express the equation in its variational form. The domain is then divided into smaller units (finite elements) on which the solution of the equation is approximated by a linear combination of some basis functions. The individual elements may be positioned over the domain quite freely (Fig. 11.10), and this liberty represents the main charm and strength of the method. Finally, the variational integral is computed by summing the contributions from all elements; we end up with a system of algebraic equations for the coefficients multiplying the basis functions in the solution expansion.

Here we present the essence of FEM. For greater clarity, the basic concepts are introduced in one space dimension, while in applications the method is overwhelm-

Fig. 11.10 Positioning the finite elements for the problem of propagating acoustic waves in the geometry of a "snail". Such complex geometries are virtually unmanageable by classical difference methods. Figure courtesy of M. Melenk and S. Langdon

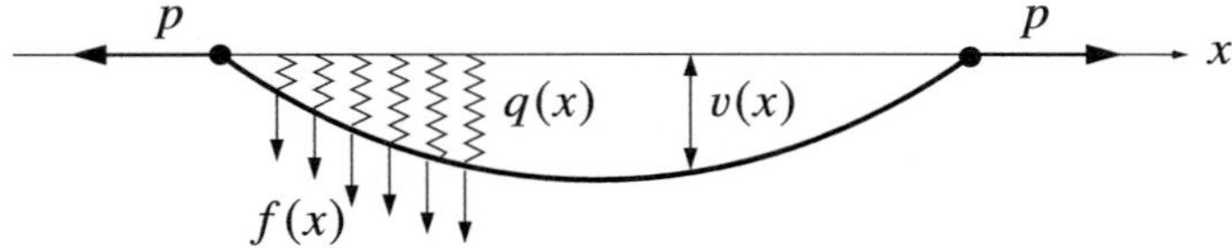

Fig. 11.11 The classical boundary-value problem illustrating the one-dimensional finite-element method: transverse deflections of a cable under non-uniform load

ingly used in two and three dimensions. Further reading can be found in the textbooks [13, 14] which we follow closely. See also [15].

11.6.1 *One Space Dimension*

The finite element method started to bloom in construction problems, so it is fair to introduce it by a civil-engineering example: we are interested in the transverse deflections $v(x)$ of a support cable pulled at its ends by the force p. There is another force (per unit length) acting in the transverse direction, $f(x)$, and the cable is held in equilibrium by springs with elastic modulus $q(x)$ (Fig. 11.11).

The deflection $v(x)$ from the horizontal is the solution of the boundary-value problem

$$ - \big(p(x)v'(x)\big)' + q(x)v(x) = f(x) , \qquad x \in [0, 1] , \tag{11.63} $$

with boundary conditions $v(0) = v(1) = 0$. For simplicity we assume constant $p > 0$ and $q \geq 0$. We have shown in Sect. 9.6 (see Eq. 9.82) that seeking the solution of this problem is equivalent to determining the function v satisfying the equation

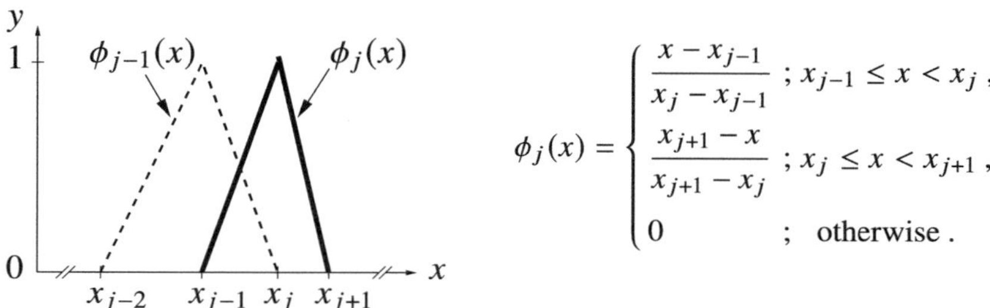

$$\phi_j(x) = \begin{cases} \dfrac{x - x_{j-1}}{x_j - x_{j-1}} & ; x_{j-1} \leq x < x_j , \\[2ex] \dfrac{x_{j+1} - x}{x_{j+1} - x_j} & ; x_j \leq x < x_{j+1} , \\[2ex] 0 & ; \text{ otherwise} . \end{cases}$$

Fig. 11.12 Basis functions for the one-dimensional finite element method. The function ϕ_j is non-zero only on the interval $(x_{j-1}, x_j] \cup [x_j, x_{j+1})$, and $\phi_j(x_k) = \delta_{j,k}$

$$A(w, v) = \int_0^1 \left[w' p v' + w q v \right] \mathrm{d}x = \int_0^1 w f \, \mathrm{d}x = \langle w, f \rangle$$

for any weight function w.

We divide the interval $[0, 1]$ to N (not necessarily uniform) subintervals such that $0 = x_0 < x_1 < x_2 < \ldots < x_{N-1} < x_N = 1$. Each subinterval $[x_{j-1}, x_j]$ with length $h_j = x_j - x_{j-1}$ is called the *finite element*. Over the adjacent elements $[x_{j-1}, x_j]$ and $[x_j, x_{j+1}]$ we span the basis function ϕ_j with a peak at x_j and the *nodes* (zeros) at x_{j-1} and x_{j+1} (Fig. 11.12). We use piecewise linear basis functions here; quadratic and cubic functions are also widely used.

The approximate solution is expanded in terms of the basis functions ϕ_j:

$$u(x) = \sum_{j=1}^{N-1} c_j \phi_j(x) . \tag{11.64}$$

Since $\phi_j(x_k) = \delta_{j,k}$, we get $u(x_k) = \sum_j c_j \delta_{j,k} = c_k$. The coefficients c_k are therefore equal to the values of u at the interior nodes. For Dirichlet boundary conditions, $u(x_0) = c_0$ and $u(x_N) = c_N$, so we may include the contributions at the boundary nodes in Eq. (11.64) if we set $c_0 = c_N = 0$. The weight function is expanded similarly,

$$\widetilde{w}(x) = \sum_{j=1}^{N-1} d_j \psi_j(x) .$$

We stay in the Galerkin framework, where the basis functions ϕ_j for the solution u are the same as the basis functions ψ_j for the weight function $\widetilde{w}$.

The coefficients c_k in the expansion (11.64) are computed by solving the variational equation (9.83), where the sum runs over all finite elements. The equation to be solved is

$$\sum_{j=1}^{N} \left[A_j(\widetilde{w}, u) - \langle \widetilde{w}, f \rangle_j \right] = 0 \quad \forall \widetilde{w} , \tag{11.65}$$

where u is the approximate solution (11.64), $\widetilde{w}$ is the weight function, and f is the right-hand side of Eq. (11.63). The subscript j denotes the element $[x_{j-1}, x_j]$ and runs from 1 to N, since all finite elements from $[x_0, x_1]$ through $[x_{N-1}, x_N]$ have to be considered. We split the expression for A_j in two parts,

$$A_j(\widetilde{w}, u) = A_j^{\mathrm{S}}(\widetilde{w}, u) + A_j^{\mathrm{M}}(\widetilde{w}, u) = \int_{x_{j-1}}^{x_j} p\widetilde{w}'u'\,\mathrm{d}x + \int_{x_{j-1}}^{x_j} q\widetilde{w}u\,\mathrm{d}x \ . \tag{11.66}$$

Stiffness matrix, mass matrix, and load vector In engineering problems the first term of (11.66) corresponds to the internal energy of the body (due to compression or expansion), and the second term to external influences (e.g. due to the change of potential energy under load). Over the element $[x_{j-1}, x_j]$, the approximate solution and the weight function have the forms

$$u(x) = c_{j-1}\phi_{j-1}(x) + c_j\phi_j(x) = (c_{j-1}, c_j)\begin{pmatrix}\phi_{j-1}(x) \\ \phi_j(x)\end{pmatrix} ,$$

$$\widetilde{w}(x) = d_{j-1}\phi_{j-1}(x) + d_j\phi_j(x) = (d_{j-1}, d_j)\begin{pmatrix}\phi_{j-1}(x) \\ \phi_j(x)\end{pmatrix} ,$$

with the derivatives

$$u'(x) = (c_{j-1}, c_j)\begin{pmatrix}-1/h_j \\ 1/h_j\end{pmatrix} , \qquad \widetilde{w}'(x) = (d_{j-1}, d_j)\begin{pmatrix}-1/h_j \\ 1/h_j\end{pmatrix} ,$$

where $h_j = x_j - x_{j-1}$. From these expressions one finds

$$A_j^{\mathrm{S}}(\widetilde{w}, u) = (d_{j-1}, d_j)\left[\int_{x_{j-1}}^{x_j}\begin{pmatrix}1/h_j^2 & -1/h_j^2 \\ -1/h_j^2 & 1/h_j^2\end{pmatrix} p(x)\,\mathrm{d}x\right]\begin{pmatrix}c_{j-1} \\ c_j\end{pmatrix} ,$$

$$A_j^{\mathrm{M}}(\widetilde{w}, u) = (d_{j-1}, d_j)\left[\int_{x_{j-1}}^{x_j}\begin{pmatrix}\phi_{j-1}^2 & \phi_{j-1}\phi_j \\ \phi_{j-1}\phi_j & \phi_j^2\end{pmatrix} q(x)\,\mathrm{d}x\right]\begin{pmatrix}c_{j-1} \\ c_j\end{pmatrix} .$$

For constant p and q, as assumed for (11.63), only elementary integrals are involved (trivial integration in A_j^{S} and integration of quadratic functions in A_j^{M}). In this case we get

$$A_j^{\mathrm{S}}(\widetilde{w}, u) = (d_{j-1}, d_j)\,S_j\begin{pmatrix}c_{j-1} \\ c_j\end{pmatrix} , \qquad S_j = \frac{p}{h_j}\begin{pmatrix}1 & -1 \\ -1 & 1\end{pmatrix} , \tag{11.67}$$

$$A_j^{\mathrm{M}}(\widetilde{w}, u) = (d_{j-1}, d_j)\,M_j\begin{pmatrix}c_{j-1} \\ c_j\end{pmatrix} , \qquad M_j = \frac{qh_j}{6}\begin{pmatrix}2 & 1 \\ 1 & 2\end{pmatrix} . \tag{11.68}$$

The matrix S_j is known as the *element (local) stiffness matrix*, and the matrix M_j is the *element (local) mass matrix*.

In general, the integral $\langle \widetilde{w}, f \rangle_j$ in Eq. (11.65) (the scalar product of functions $\widetilde{w}$ and f on the jth finite element) is computed numerically. But a good approximation can be obtained if f on $[x_{j-1}, x_j]$ is replaced by a piecewise linear function, $f(x) \approx f_{j-1}\phi_{j-1}(x) + f_j\phi_j(x)$. Then

$$\langle \widetilde{w}, f \rangle_j \approx \left(d_{j-1}, d_j \right) \boldsymbol{g}_j, \qquad \boldsymbol{g}_j = \frac{h_j}{6} \begin{pmatrix} 2f_{j-1} + f_j \\ f_{j-1} + 2f_j \end{pmatrix}. \tag{11.69}$$

The vector $\boldsymbol{g}_j$ is the *element (local) load vector*, as it describes the external force or loading upon the system.

Assembly The next step is the *assembly* of the element matrices into the global stiffness matrix by summing the contributions of all elements, and adding all element load vectors into the global load vector. To simplify, we consider a uniform mesh, $h_j = h = 1/N$. We sum the contributions (11.67), (11.68), and (11.69) over all elements $1, 2, \ldots, N$, where we consider the boundary condition $c_0 = d_0 = 0$ in the terms involving the element $j = 0$, and $c_N = d_N = 0$ in those involving $j = N$. We collect the coefficients c_j and d_j for the interior points of $[x_0, x_N]$ in the vectors $\boldsymbol{c} = (c_1, c_2, \ldots, c_{N-1})^{\mathrm{T}}$ and $\boldsymbol{d} = (d_1, d_2, \ldots, d_{N-1})^{\mathrm{T}}$, and put the components $\langle \widetilde{w}, f \rangle_j$ in the global load vector $\boldsymbol{g} = (g_1, g_2, \ldots, g_{N-1})^{\mathrm{T}}$. The equation with the sum over the elements,

$$\sum_{j=1}^{N} \left[A_j^{\mathrm{S}}(\widetilde{w}, u) + A_j^{\mathrm{M}}(\widetilde{w}, u) \right] = \sum_{j=1}^{N} \langle \widetilde{w}, f \rangle_j,$$

can then be written in matrix form $\boldsymbol{d}^{\mathrm{T}} \left[(S + M)\, \boldsymbol{c} - \boldsymbol{g} \right] = 0$, where

$$S = \frac{p}{h} \begin{pmatrix} 2 & -1 & & & \\ -1 & 2 & -1 & & \\ & \ddots & \ddots & \ddots & \\ & & -1 & 2 & -1 \\ & & & -1 & 2 \end{pmatrix}, \qquad M = \frac{qh}{6} \begin{pmatrix} 4 & 1 & & & \\ 1 & 4 & 1 & & \\ & \ddots & \ddots & \ddots & \\ & & 1 & 4 & 1 \\ & & & 1 & 4 \end{pmatrix},$$

are the $(N-1) \times (N-1)$ global stiffness and mass matrices, and

$$\boldsymbol{g} = \frac{h}{6} \begin{pmatrix} f_0 + 4f_1 + f_2 \\ f_1 + 4f_2 + f_3 \\ \vdots \\ f_{N-2} + 4f_{N-1} + f_N \end{pmatrix}$$

is the global load vector of dimension $N - 1$ (in the linear approximation on each element). This equation must hold true for any d, so the coefficients c_j of the expansion (11.64) for $j = 1, 2, \ldots, N - 1$ are obtained by solving the system

$$(S + M)c = g \,.$$

11.6.2   Two Space Dimensions

In two dimensions the problem (11.63) is generalized to a planar region R,

$$-\frac{\partial}{\partial x}\left(p(x, y)\frac{\partial v}{\partial x}\right) - \frac{\partial}{\partial y}\left(p(x, y)\frac{\partial v}{\partial y}\right) + q(x, y)v = f(x, y)\,, \qquad (x, y) \in R\,,$$

with the boundary condition $v(x, y) = \alpha(x, y)$ for $(x, y) \in \partial R$. (A problem with a non-homogeneous condition $\alpha \neq 0$ can be turned into the problem with $\alpha = 0$ by shifting $\hat{v} = v - \alpha$.) The variational formulation of the problem is analogous to the one-dimensional case; things start to become complicated when we attempt to divide (partition) the region R to finite elements.

In one dimension the finite elements are intervals; in a planar region the role of the elements is taken by various geometric shapes. Most frequently one uses (not necessarily equilateral or isosceles) triangles, rarely quadrilaterals (Fig. 11.13). The partitioning of the region is known as triangulation (or quadrangulation). For complex geometries (see Fig. 11.10) it almost amounts to art. Effective triangulation is accomplished by dedicated commercial programs (see e.g. Ref. [16]). Among the most well known is the Delaunay triangulation [17]. A good triangulation guarantees that none of the interior angles in the triangles is too small (it maximizes the smallest used angles), thereby ensuring numerical stability.

On the chosen triangulation the corresponding basis functions are defined. Here we restrict the discussion to piecewise linear functions of x and y. The basis function on the nodes of the triangle $x_j = (x_j, y_j)^{\mathrm{T}}$, $x_{j+1} = (x_{j+1}, y_{j+1})^{\mathrm{T}}$, and $x_{j+2} = (x_{j+2}, y_{j+2})^{\mathrm{T}}$, with an apex at x_j (Fig. 11.14 a)) has the form

$$\phi_j(x, y) = \left[\det\begin{pmatrix} 1 & x_j & y_j \\ 1 & x_{j+1} & y_{j+1} \\ 1 & x_{j+2} & y_{j+2} \end{pmatrix}\right]^{-1} \cdot \det\begin{pmatrix} 1 & x & y \\ 1 & x_{j+1} & y_{j+1} \\ 1 & x_{j+2} & y_{j+2} \end{pmatrix}\,. \qquad (11.70)$$

The function ϕ_j is non-zero only over the triangle defined by the nodes x_j, x_{j+1}, and x_{j+2}. It also holds that

$$\phi_j(x_k, y_k) = \delta_{j,k}\,.$$

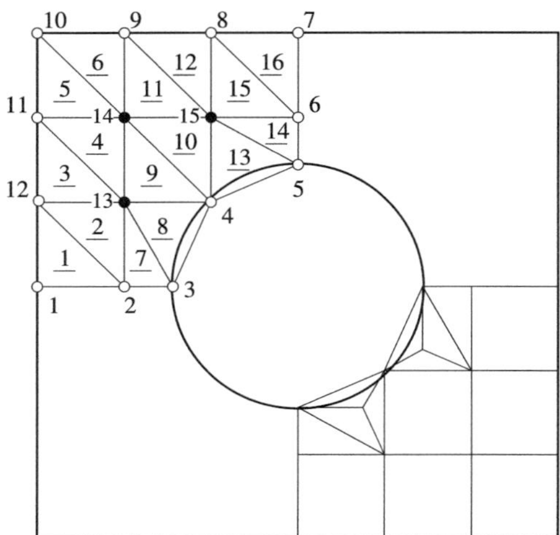

Fig. 11.13 Triangulation for the finite element method in the geometry of a square with a cut-out circle. Top left: a coarse mesh of triangular elements with the enumeration of nodes and elements. Bottom right: in a part of the region we are allowed to use other shapes (e.g. squares) or refine the mesh in physically more interesting regions

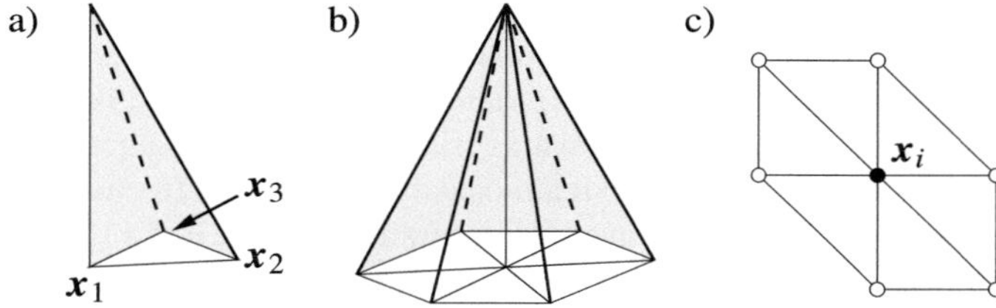

Fig. 11.14 Three-node element functions for the finite element method on the plane: **a** The basic function above the triangular element with the value 1 at the node x_1 and values 0 at the side nodes x_2 and x_3; **b** The basis functions above six elements; **c** The neighborhood of the node x_i as used in Problem 11.9.8

Note that Eq. (11.70) expresses only the intersection of the planes defining the three side surfaces of the pyramid. For each pair of x and y one needs to check whether they lie inside the triangle defined by the three nodes.

Assembly When assembling the elements of local stiffness and mass matrices in the corresponding global matrices, one must distinguish between the local indices of nodes $m, n \in \{1, 2, 3\}$, and the global indices of nodes $j, k \in \{1, 2, \ldots, N_{\mathcal{N}}\}$ and elements $t \in \{1, 2, \ldots, N_{\mathcal{T}}\}$. Global indices label the whole triangulation with $N_{\mathcal{N}}$ nodes and $N_{\mathcal{T}}$ elements (triangles). In the variational requirement we sum over all triangles,

$$\sum_{t=1}^{N_{\mathcal{T}}} A^{(t)}(\widetilde{w}, u) = \sum_{t=1}^{N_{\mathcal{T}}} \langle \widetilde{w}, f \rangle^{(t)} \,,$$

where t is the label of the triangle T_t. The surface integrals

$$A^{(t)}(\widetilde{w}, u) = \int_{T_t} \left[p \, (\nabla \widehat{w})^{\mathrm{T}} \cdot \nabla u + q \, \widetilde{w} \, u \right] \mathrm{d}x \, \mathrm{d}y \,, \qquad \langle \widetilde{w}, f \rangle^{(t)} = \int_{T_t} \widetilde{w} f \, \mathrm{d}x \, \mathrm{d}y \,,$$

are computed on each triangle T_t separately. This is relatively easily done in the Galerkin form of the method with piecewise linear basis functions, where $\phi_j = \psi_j$, as we need only the expressions for the functions ϕ_j (11.70) and their derivatives

$$\nabla \phi_j(x, y) = \frac{1}{2\,|T|} \begin{pmatrix} y_{j+1} - y_{j+2} \\ x_{j+2} - x_{j+1} \end{pmatrix} \,,$$

where $|T| = \frac{1}{2}|(x_2 - x_1)(y_3 - y_1) - (y_2 - y_1)(x_3 - x_1)|$ is the surface area of the triangle. The local contributions of triplets of nodes to the stiffness matrix and the corresponding components of the local load vector are then

$$A_{mn}^{(t)}(\widetilde{w}, u) = \int_{T_t} \left[p \, (\nabla \phi_m)^{\mathrm{T}} \nabla \phi_n + q \, \phi_m \phi_n \right] \mathrm{d}x \, \mathrm{d}y \,,$$

$$\langle \widetilde{w}, f \rangle_m^{(t)} = \int_{T_t} \phi_m f \, \mathrm{d}x \, \mathrm{d}y \,.$$

Example How local matrices and vectors are assembled in the global matrix and global vector is best explained by an example. To make the discussion easier, we set $p = 1$ and $q = 0$ (we are solving the Poisson equation $-\nabla^2 v = f$). In this case the surface integrals over the triangles T_t are simple,

$$A_{mn}^{(t)} = \frac{1}{4\,|T|} \, (y_{m+1} - y_{m+2} \,, \; x_{m+2} - x_{m+1}) \begin{pmatrix} y_{n+1} - y_{n+2} \\ x_{n+2} - x_{n+1} \end{pmatrix} \,,$$

$$\langle \widetilde{w}, f \rangle_m^{(t)} \approx \frac{1}{6} \det \begin{pmatrix} x_{m+1} - x_m & x_{m+2} - x_m \\ y_{m+1} - y_m & y_{m+2} - y_m \end{pmatrix} f(x_{\mathrm{T}}, y_{\mathrm{T}}) \,,$$

where $(x_{\mathrm{T}}, y_{\mathrm{T}})$ are the coordinates of the center of mass of the triangle T_t.

As an illustration of the method, the top left portion of Fig. 11.13 shows a coarse triangulation of a square region with a cut-out circle. This problem (adopted from Ref. [14]) is still manageable by classical difference methods, but we use it here to serve as a typical example. We arrange the nodes in the matrix

$$
\mathcal{N} = \begin{pmatrix} 1\,2 & 3\,4 & 5 \cdots 11 & 12 & 13 & 14 & 15 \\ 0\,1 & 1.59\,2 & 3 \cdots 0 & 0 & 1 & 1 & 2 \\ 0\,0 & 0\,1 & 1.41 \cdots 2 & 1 & 1 & 2 & 2 \end{pmatrix}^{\mathsf{T}}
$$

that includes the triplets {index of node, x coordinate, y coordinate}: here we specify the actual locations of the nodes in the planar region. The matrix

$$
\mathcal{T} = \begin{pmatrix} \underline{1} & \underline{2} & \underline{3} & \underline{4} & \underline{5} \cdots & \underline{12} & \underline{13} & \underline{14} & \underline{15} & \underline{16} \\ 1 & 2 & 11 & 11 & 10 \cdots & 8 & 4 & 5 & 6 & 6 \\ 2 & 13 & 12 & 13 & 11 \cdots & 9 & 5 & 6 & 8 & 7 \\ 12 & 12 & 13 & 14 & 14 \cdots & 15 & 15 & 15 & 15 & 8 \end{pmatrix}^{\mathsf{T}}
$$

has the entries {index of element (underlined), index of node 1, index 2, index 3} relating the elements to the global indices of their nodes. The connection between the space of physical coordinates and the space of elements must be provided in our own code by some mapping between the sets $\mathcal{N}$ and $\mathcal{T}$. It is recommendable that the main loop in the code runs over the triangles $t = 1, 2, \ldots, N_{\mathcal{T}}$, not over the nodes, which would also be possible. Namely, it turns out that fewer surface integrals (evaluated numerically in general) need to be computed by using the element loop than by looping over the nodes [14]. The components of the global stiffness matrix S and the global load vector $\boldsymbol{g}$ are obtained by summing

```
loop over                    t = 1, 2, ..., N_T
```
$$
S_{\mathcal{T}(t,m),\mathcal{T}(t,n)} \mathrel{+}= A^{(t)}_{mn} \qquad (m, n = 1, 2, 3)
$$
$$
g_{\mathcal{T}(t,m)} \mathrel{+}= \langle \widetilde{w}, f \rangle^{(t)}_m \qquad (m = 1, 2, 3)
$$
```
end loop
```

By solving $S\boldsymbol{c} = \boldsymbol{g}$ (as in the one-dimensional case) we finally obtain the vector $\boldsymbol{c}$ containing the expansion coefficients of the solution, $u(x, y) = \sum_j c_j \phi_j(x, y)$.

Try not to confuse the labelings of the nodes and the elements, since in general the number of nodes $N_{\mathcal{N}}$ is different from the number of elements $N_{\mathcal{T}}$. While the program loops over the elements t, the values are inserted in the global matrix, and the corresponding load vector according to the labeling of nodes. Compare Fig. 11.14 c) to Fig. 11.19: around the node $\boldsymbol{x}_i$ (in global enumeration) six triangles are arranged, and they form the support for the basis function shown in Fig. 11.14 b). Check your understanding in Problem 11.9.8! ◁

We have merely traced the very first steps of FEM. A genuine expert use of the method only just starts here: non-uniform planar or spatial meshes are devised; basis functions are realized as splines of different degrees so that specific requirements at the boundaries between the elements are met; the mesh can be made denser during the computation if an increase in local precision is called for; discontinuities may be incorporated; non-stationary problems can be implemented; a road opens towards non-linear problems with a multitude of initial and boundary conditions. The finite element method is already a part of modern numerical packages like Matlab: nice pedagogical examples that can serve as a good vantage point for further study are given in Ref. [18]. A myriad of commercial program packages is available for a more demanding use of the finite element method [19]. For an excitingly propulsive free version, see [20, 21].

11.7 Mimetic Discretizations ⋆

A powerful tool for solving PDE in complex geometries are the *mimetic (or compatible) discretizations* [22] that attempt to mimic the properties of the physical problem as closely as possible. A nice example is the diffusion in strongly heterogeneous and non-isotropic media described by the equation $v_t = \nabla \cdot D(v)\nabla v + Q$. The space is divided in "logically rectangular" convex cells onto which scalar and vector fields are attached (Fig. 11.15 (left)).

But the key step is the discretization of differential operators, like the diffusion operator $\nabla \cdot D(v)\nabla$ for the problem mentioned above [24, 25]. The discretization should, at least to some order, respect the symmetry and conservative properties of the underlying equation, for example, the conservation of mass, momentum, or energy (in studies of fluid flows) or Maxwell's equations (in electro-magnetic problems). Ultimately, the problems are translated to systems of algebraic equations and solved by preconditioned matrix relaxation methods.

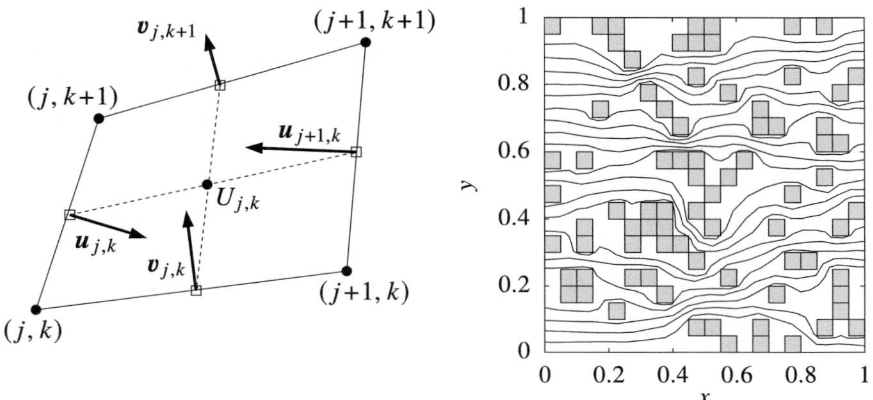

Fig. 11.15 [LEFT] A "logically rectangular" cell in a planar mimetic discretization. The scalar fields $U_{j,k}$ are defined at the cell centers, and the vector fields u and v are given by the components of vectors perpendicular to the cell sides. [RIGHT] An example of solving the problem of a fluid flowing through a random arrangement of shale blocks in sand. The shale occupies 20 % of the total surface area (figure adapted from [23].)

A detailed presentation is beyond the scope of this book, but we let them lurk at the horizon due to their flexibility and development in the recent years. An introduction is given in Ref. [22] and the mathematical background in Refs. [26, 27].

11.8   Multi-grid and Mesh-Free Methods ★

Two further unique approaches to solving PDE in complex geometries should be mentioned. In the *multi-grid* approach the differential problem is discretized on grids (meshes) of varying coarseness. The method is applied on ever finer meshes, and the resulting sequence of solutions converges very rapidly: even in complex geometries and with non-trivial boundary conditions, the multi-grid methods are among the fastest on the market, but each application of the method needs to be tailored carefully to the demands of the individual problem. Multi-grid methods were originally developed for boundary problems involving elliptic PDE, but they can be used to solve other types for PDE as well. An excellent introduction is given by Ref. [28]; further reading is offered by Refs. [29, 30].

The *mesh-free methods* liberate us from the shackles of discretization: we express the approximate solution as a function spanned over a set of nodes in the definition domain of the differential problem; there is no need for these nodes to be systematically related in any manner (as e.g. in the finite element method). A certain algorithm leads to a system of equations relating the solution to the information at the nodes and on the domain boundaries. Mesh-free methods excel in problems calling for a large degree of flexibility, for example, in situations with non-fixed boundaries between regions with different physical properties. Examples include phase changes (boundaries between fluid and solid phases of alloys) or mechanical defects (cracks in materials). Especially in three space dimensions, mesh-free methods compete successfully with finite element methods where adaptive triangulation may become too costly. We recommend [31, 32] for further reading. In the following we describe the popular mesh-free method based on *radial basis functions*.

11.8.1   A Mesh-Free Method Based on Radial Basis Functions

Here we explain the basic idea of the radial basis functions (RBF) method by solving the Poisson equation on a square with Dirichlet boundary conditions,

$$\nabla^2 v = Q(x, y), \quad (x, y) \in R = [0, 1] \times [0, 1],$$
$$v(x, y) = f(x, y), \quad (x, y) \in \partial R,$$

On the definition domain we choose N points $\{(x_i, y_i)\}_{i=1}^{N}$, of which N_1 are in the interior of the domain, while the remaining $N_2 = N - N_1$ are on its boundary. We seek the approximate solution u in the form

$$u(x, y) = \sum_{j=1}^{N} a_j \phi_j(x, y) \,, \tag{11.71}$$

where ϕ_j are the basis functions. A possible choice for the form of the basis functions is

$$\phi_j(x, y) = \sqrt{(x - x_j)^2 + (y - y_j)^2 + c^2} = \sqrt{r_j^2 + c^2} \,, \tag{11.72}$$

where c is an adjustable parameter [33]. For such functions we have

$$\frac{\partial^2 \phi_j}{\partial x^2} = \frac{(y - y_j)^2 + c^2}{\left(r_j^2 + c^2\right)^{3/2}} \,, \qquad \frac{\partial^2 \phi_j}{\partial y^2} = \frac{(x - x_j)^2 + c^2}{\left(r_j^2 + c^2\right)^{3/2}} \,.$$

At the interior points we insert the ansatz for the solution u in the differential equation, while at the boundary points we insert it into the boundary conditions. This results in a system of linear equations for the coefficients a_j, represented by a $N \times N$ matrix:

$$\sum_{j=1}^{N} \left(\frac{\partial^2 \phi_j}{\partial x^2} + \frac{\partial^2 \phi_j}{\partial y^2} \right) (x_i, y_i)\, a_j = Q(x_i, y_i) \,, \quad i = 1, 2, \ldots, N_1 \,,$$

$$\sum_{j=1}^{N} \phi_j(x_i, y_i)\, a_j = f(x_i, y_i) \,, \quad i = N_1 + 1, N_1 + 2, \ldots, N \,.$$

In the variables x and y, the basis functions (11.72) have a radial symmetry around the collocation points (x_j, y_j) which gave the method its name; but a large set of functions possessing this property is in wide use, for example,

$$\sqrt{r_j^2 + c^2} \,, \qquad \frac{1}{\sqrt{r_j^2 + c^2}} \,, \qquad \frac{1}{r_j^2 + c^2} \,, \qquad \mathrm{e}^{-(cr_j)^2} \,.$$

Radial basis function methods [34] are suitable for both time-dependent and time-independent problems. Instructive case studies of parabolic problems (diffusion equation) can be found in Ref. [35] and for hyperbolic problems (wave equation, Burgers equation) in Ref. [36]. Examples of solving elliptic problems (Poisson equation) are discussed in Ref. [37].

Example By using the mesh-free method utilizing radial basis functions [33] we solve the Poisson equation on the unit square,

$$- \nabla^2 v = Q , \qquad (x, y) \in [0, 1] \times [0, 1] ,$$

where $Q(x, y) = -13 \, e^{-2x+3y}$, with Dirichlet boundary conditions

$$v(0, y) = e^{3y} , \qquad v(1, y) = e^{-2+3y} , \qquad v(x, 0) = e^{-2x} , \qquad v(x, 1) = e^{-2x+3} .$$

The exact solution is

$$v(x, y) = e^{-2x+3y} .$$

On the definition domain we randomly choose $N_1 = 16$ collocation points on the boundary of the square (four on each side), and $N_2 = 24$ points in its interior (Fig. 11.16 (left)). In total, we have $N_1 + N_2 = 40$ collocation points, each corresponding to a basis function of the form (11.72), where we choose $c = 0.5$. When the ansatz (11.71) at the collocation points is inserted in the differential equation and in the equations specifying the boundary conditions, we obtain a system of equations for the expansion coefficients a_j, as described on page 749. We define the absolute error of the numerical solution as

$$\mathrm{ERR} = \left[\sum_{i=1}^{N} (u(x_i, y_i) - v(x_i, y_i))^2 \right]^{1/2} .$$

For $c = 0.5$ this error amounts to $\mathrm{ERR} \approx 0.91$ (Fig. 11.16 (right)).

The parameter c defines the scale of the radial basis function, determining whether the function is more spike-like or rather smooth. Obviously, the precision of the

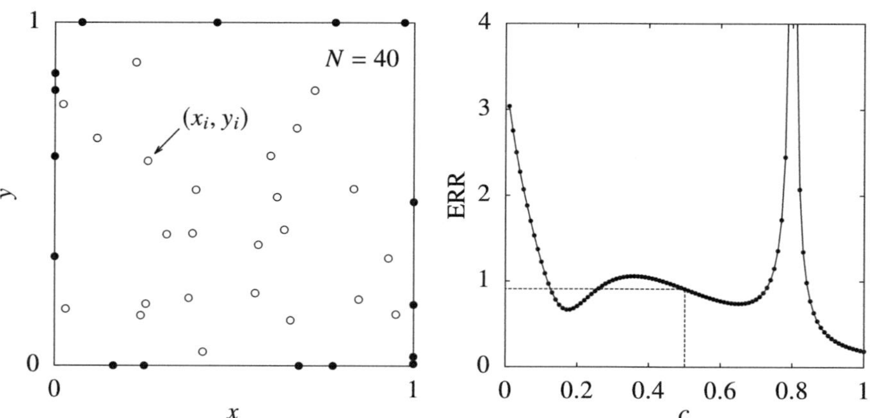

Fig. 11.16 Solving the Poisson equation on the square by using a mesh-free method based on radial basis functions. [LEFT] A random distribution of 16 points on the boundaries of the square (symbols ·) and 24 points in its interior (symbols o). [RIGHT] The error of the solution as a function of the parameter c

solution can be strongly influenced by changing c. Figure 11.16 (right) shows the error ERR in dependence of c: at certain values of c, the error may even explode! Some general instructions about the choice of the optimal c are given in Ref. [38]. ◁

11.9 Problems

11.9.1 Two-Dimensional Diffusion Equation

The initial-boundary-value problem for the inhomogeneous diffusion equation

$$v_t = D\left(v_{xx} + v_{yy}\right) + Q$$

on $(x, y) \in [0, 1] \times [0, 1]$ is the prototype parabolic problem in two dimensions that can be most easily solved by difference methods (Sect. 11.1). Particular attention should be paid to the consistent treatment of the boundary conditions, so that the order of the difference scheme is not spoiled.

⊙ By using the explicit (FTCS) scheme in two dimensions (11.5), solve the following problems with Dirichlet boundary conditions (examples from Ref. [1]):

(1)

$$v(x, y, 0) = \sin(\pi x)\,\sin(2\pi y)\,,$$
$$v(0, y, t) = v(1, y, t) = v(x, 0, t) = v(x, 1, t) = Q(x, y, t) = 0\,;$$

compute the solutions at times $0 \le t \le 1$ (choose a few representative instants) with step $\Delta t = 0.0005$ (0.001), spatial discretization $N_x = N_y = 20$ and $D = 1.0$. What happens if the discretization in the x coordinate is twice as coarse?

(2)

$$v(x, y, 0) = 0\,,$$
$$v(0, y, t) = v(1, y, t) = v(x, 0, t) = v(x, 1, t) = 0\,,$$
$$Q(x, y, t) = \sin(2\pi x)\,\sin(4\pi y)\,\sin t\,;$$

compute the solution at times $0 \le t \le 2$ with the discretization $N_x = N_y = 20$, time step $\Delta t = 0.0005$ (or $N_x = N_y = 100$ and $\Delta t = 0.00002$), and parameter $D = 0.5$.

(3)

$$v(x, y, 0) = 0\,,$$
$$v(0, y, t) = \sin(\pi y)\,\sin t\,, \quad v(x, 0, t) = \sin(\pi x)\,\sin t\,,$$
$$v(1, y, t) = v(x, 1, t) = Q(x, y, t) = 0\,;$$

use $N_x = N_y = 20$, $\Delta t = 0.001$, $D = 1.0$, and compute the solution at $0 \leq t \leq 10$. Compare the numerical result to the analytic one. What can you say about the stability of the schemes? Solve the three problems listed above by using the Crank–Nicolson scheme (11.13).

$\bigoplus$ We are also interested in the solutions of the homogeneous equation $(Q(x, y, t) = 0)$, with a Neumann boundary condition on one side of the square and Dirichlet conditions on the remaining three sides. A consistent inclusion of the boundary conditions is of key importance for the convergence of the chosen difference scheme. Discuss the problem

$$v(x, y, 0) = \sin(\pi x)\, \sin(2\pi y)\,,$$
$$v(0, y, t) = v(x, 0, t) = v(x, 1, t) = 0\,,$$
$$v_x(1, y, t) = g^{\mathrm{N}}(y, t) = -\pi\, \sin(2\pi y)\, \mathrm{e}^{-5D\pi^2 t}\,.$$

We implement the Neumann conditions at $x = 1$ for the explicit scheme by analogy to Eqs. (11.11) or (11.12) that apply at $x = 0$. At time $(n + 1)\Delta t$ we get, to first order in the space variable, $u_{N_x k}^{n+1} = u_{N_x-1 k}^{n+1} + \Delta x\, g^{\mathrm{N}}(k\Delta y, (n + 1)\Delta t)$, which is already the missing equation for the value at $j = N_x$. To second order, we write the Neumann condition at time $n\Delta t$ as $u_{N_x+1 k}^{n} = u_{N_x-1 k}^{n} + 2\Delta x\, g^{\mathrm{N}}(k\Delta y, n\Delta t)$. We insert the value $u_{N_x+1 k}^{n}$ in the difference scheme at $j = N_x$, whence we again obtain the missing equation at $j = N_x$,

$$u_{N_x k}^{n+1} = (1 - 2r_x)u_{N_x k}^{n} + 2r_x u_{N_x-1 k}^{n} + r_y \Delta_2^{(y)} u_{N_x k}^{n} + 2r_x \Delta x\, g^{\mathrm{N}}(k\Delta y, n\Delta t)\,.$$

For the Peaceman–Rachford scheme, the first-order Neumann condition is imposed in the same manner. The second-order condition is easier to compute if the order of spatial derivatives in the scheme is interchanged, i.e. by just substituting $x \longleftrightarrow y$ in Eqs. (11.14) and (11.15). As in the implicit case, the condition is inserted in the scheme at $j = N_x$, but at time $(n + 1)\Delta t$. We obtain

$$u_{N_x k}^{n+1} = \frac{r_x}{1+r_x} u_{N_x-1 k}^{n+1} + \frac{1}{1+r_x}\left(1 + \frac{r_y}{2}\Delta_2^{(y)}\right) u_{N_x k}^{n+\frac{1}{2}} + \frac{r_x}{1+r_x}\Delta x\, g^{\mathrm{N}}(k\Delta y, (n + 1)\Delta t)\,.$$

In the D'Yakonov variant of the scheme, the Neumann condition needs to be translated into a condition for $u^{\star}$. Check that for the first-order Neumann condition this means

$$u_{N_x k}^{\star} = u_{N_x-1 k}^{\star} + \Delta x\left(1 - \frac{r_x}{2}\Delta_2^{(x)}\right) g^{\mathrm{N}}(k\Delta y, (n + 1)\Delta t)\,,$$

while for the second-order Neumann condition one should use

$$u^{\star}_{N_x k} = \frac{r_y}{1+r_y} u^{\star}_{N_x-1k} + \frac{1}{1+r_y}\left(1 + \frac{r_y}{2}\Delta_2^{(y)}\right)\left(1 + \frac{r_x}{2}\Delta_2^{(x)}\right)u^n_{N_x k}$$
$$+ \frac{r_y}{1+r_y}\,\Delta x\, g^{\mathrm{N}}(k\Delta y, (n+1)\Delta t)\,.$$

Compute the solution by using the explicit scheme (11.5) at times $0 \le t \le 1$ with the step size $\Delta t = 0.001$ and discretization $N_x = N_y = 10$, and then with $\Delta t = 0.0005$, $N_x = N_y = 20$ and $\Delta t = 0.0001$, $N_x = N_y = 40$. Use $D = 1.0$. Repeat the exercise with the Peaceman-Rachford and D'Yakonov scheme with $\Delta t = 0.01$. In all cases use first- and second-order approximations for the Neumann boundary condition. Compare the analytic and numerical solutions.

11.9.2 Non-linear Diffusion Equation

This problem acquaints us with a method of solving a diffusion-reaction problem from the field of biotechnology: we are interested in the formation and growth of bacterial biofilms on a nutritious substrate [39, 40]. The density of the created biomass $v(x, y, t)$ is determined by the partial differential equation

$$v_t = \nabla\,[D(v)\nabla v] + \kappa v\,, \qquad (x, y) \in [0, 1] \times [0, 0.3]\,,$$

where

$$D(v) = \delta\frac{v^b}{(1-v)^a}\,, \qquad 0 < \delta \ll 1 \le a, b\,.$$

The quantity κ drives the rate of biofilm formation: a constant $\kappa > 0$ implies unlimited nutrients. In this case the bacterial culture spreads until it attains a homogeneous distribution (Fig. 11.17).

We monitor the approximate density of the biofilm $u(j\Delta x, k\Delta y, n\Delta t)$ on the uniform mesh with $\Delta x = 1/N_x$ and $\Delta y = 0.3/N_y$. We discretize the time derivative to second order, $v_t \approx (u^{n+1}_{jk} - u^n_{jk})/\Delta t$, while the reaction term κv is evaluated at time $(n+1)\Delta t$. For the diffusion term we apply the difference formula

$$\nabla\,[D(v)\nabla v] \approx \frac{1}{2\Delta x^2}\left\{\left[D\left(u^n_{jk}\right) + D\left(u^n_{j+1k}\right)\right]\left[u^{n+1}_{j+1k} - u^{n+1}_{jk}\right]\right.$$
$$\left[D\left(u^n_{jk}\right) + D\left(u^n_{j-1k}\right)\right]\left[u^{n+1}_{j-1k} - u^{n+1}_{jk}\right]\Big\}$$
$$+ \frac{1}{2\Delta y^2}\left\{\left[D\left(u^n_{jk}\right) + D\left(u^n_{j\,k+1}\right)\right]\left[u^{n+1}_{jk+1} - u^{n+1}_{jk}\right]\right.$$
$$\left.\left[D\left(u^n_{jk}\right) + D\left(u^n_{jk-1}\right)\right]\left[u^{n+1}_{jk-1} - u^{n+1}_{jk}\right]\right\}\,.$$

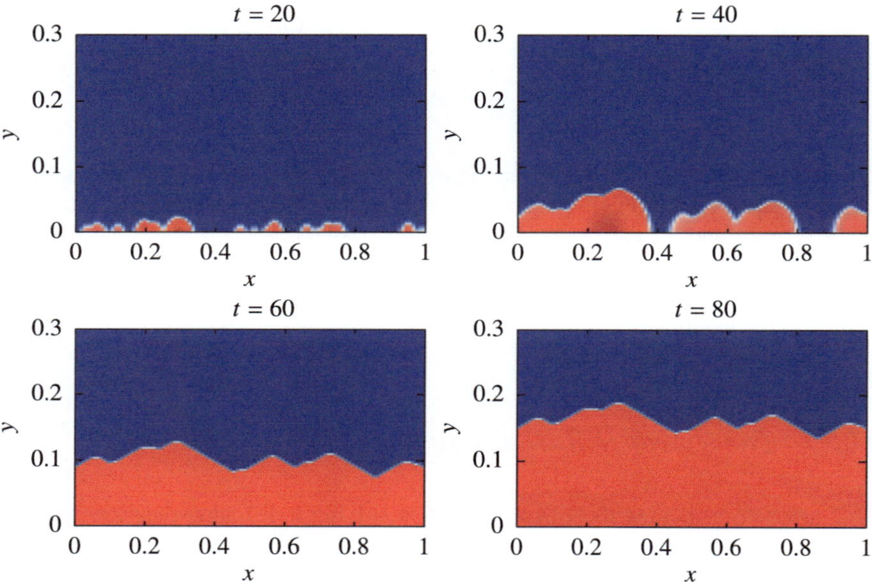

Fig. 11.17 Formation of the bacterial layer from an initial random distribution at $y = 0$. An unlimited supply of nutrients is available in the model, so the blotches grow, merge, and gradually expand to a homogeneous distribution of maximum density

If D is constant, this formula turns into the usual discretization of $\nabla^2 v$ in Cartesian coordinates Eqs. (11.2) and (11.3). Beware that the diffusion coefficients depend on the current values of u at times $n\,\Delta t$, while the values of the functions at the right-hand side are given at $(n + 1)\Delta t$. The difference scheme is therefore implicit and does not suffer from stability problems.

⊙ Carefully join the terms v_t, $\nabla\,[D(v)\nabla v]$, and κv to form an implicit difference scheme that can be written in matrix form as

$$A\boldsymbol{u}^{n+1} = \boldsymbol{u}^n \; .$$

Here $\boldsymbol{u}^n$ is the solution vector at time $n\,\Delta t$ with the coordinates indexed as in (11.1). Solve the problem by using the parameters $a = b = 4$, $\kappa = 0.1$, and $\delta = 0.1$. Start with the spatial discretization $N_x = N_y = 60$ and the time step $\Delta t = 1$. Attention: even with such a coarse discretization, the matrix A has the size $(N_x + 1)(N_y + 1) \times (N_x + 1)(N_y + 1) = 3721 \times 3721$. Fortunately it is banded: only the diagonal, the first sub- and super-diagonal, and the $(N_x + 1)$th sub- and super-diagonals have non-zero entries, so the computation is much faster if one resorts to methods for banded matrices, e.g. DGBSV from the LAPACK library (see Sect. 4.2.3).

The initial condition is N_i randomly distributed values of the density between 0 and 1 that appear at time $t = 0$ along $y = 0$ in $N_i \approx N_x/2$ intervals of the mesh; everywhere else the density should be zero. Impose Neumann boundary conditions

$v_x(x=0) = v_x(x=1) = v_y(y=0) = 0$ on three boundaries of the domain, and a Dirichlet condition $v(y=0.3) = 0$ on the fourth boundary. Compute the solution until $t \approx 50$. If the speed of your computer (or the algorithm to solve the banded system) allows it, gradually increase N_x and N_y.

11.9.3 Two-Dimensional Poisson Equation

We discuss the elliptic boundary problem (adapted from Ref. [2])

$$\nabla^2 v = 2\pi^2 (\sin(\pi x)\cos(\pi y) + \cos(\pi x)\sin(\pi y))\, e^{\pi(x+y)}$$

on $(x, y) \in R = [0, 1] \times [0, 1]$ with Dirichlet boundary condition

$$v = 0, \quad (x, y) \in \partial R,$$

that has the exact solution $v(x, y) = \sin(\pi x)\sin(\pi y)\, e^{\pi(x+y)}$.

   ⊙   Discretize the problem on the unit square $(x, y) \in [0, 1] \times [0, 1]$ with a uniform mesh $N_x = N_y = 128$ and solve it by using the Jacobi Eq. (11.41) and Gauss–Seidel Eq. (11.42) iteration. Terminate the iteration when the value of the sup-norm (A.4) or the energy norm (A.6) of the difference between subsequent solution vectors (11.46) or the residual error (11.47) drops below a certain value, say, 10^{-6}. How does the number of iteration steps in both methods change when you refine the discretization (increase $N_x = N_y = N$)?

   ⊕   On the domain $(x, y) \in R = [0, 1] \times [0, 1]$, solve the following boundary problem

$$\nabla^2 v = e^{x+y} = Q(x, y)$$

with Neumann boundary conditions $(\partial v / \partial n)(x, y) = g(x, y)$, in detail:

$$\frac{\partial v}{\partial x}(0, y) = \tfrac{1}{2}\, e^y, \qquad \frac{\partial v}{\partial y}(x, 0) = \tfrac{1}{2}\, e^x,$$

$$\frac{\partial v}{\partial x}(1, y) = \tfrac{1}{2}\, e^{1+y}, \qquad \frac{\partial v}{\partial y}(x, 1) = \tfrac{1}{2}\, e^{x+1}.$$

The operator side of the equation is discretized as in Eq. (11.35), while the boundary conditions are approximated at first order,

$$u_{1k} = u_{0k} + \Delta x\, g_{0k}, \qquad u_{j1} = u_{j0} + \Delta y\, g_{j0},$$

$$u_{N_x k} = u_{N_x-1 k} + \Delta x\, g_{N_x k}, \quad u_{jN_y} = u_{jN_y-1} + \Delta y\, g_{jN_y},$$

where $j = 1, 2, \ldots, N_x - 1$ and $k = 1, 2, \ldots, N_y - 1$. We rewrite the complete system of equations (including the boundary conditions) as $A\boldsymbol{u} = \boldsymbol{q}$, where $\boldsymbol{u}$ is the solution vector of the form (11.1) that does not contain components $j = 0$, $j = N_x$, $k = 0$, and $k = N_y$ (its dimension is $(N_x - 1)(N_y - 1)$). The matrix of the system,

$$
A = \begin{pmatrix}
T_1 & -Y & 0 & & \\
-Y & T & -Y & 0 & \\
& \ddots & \ddots & \ddots & \\
& 0 & -Y & T & -Y \\
& & 0 & -Y & T_1
\end{pmatrix},
$$

is symmetric block-tridiagonal with $(N_y - 1) \times (N_y - 1)$ blocks, where the non-zero blocks lie only along the diagonal and the first sub- and super-diagonal. The matrices T_1, T, and Y have size $(N_x - 1) \times (N_x - 1)$. By abbreviating $a = 1/\Delta x^2$ and $b = 1/\Delta y^2$, the matrices T_1 and T are

$$
T_1 = \begin{pmatrix}
a + b & -a & 0 & & \\
-a & 2a + b & -a & 0 & \\
& \ddots & \ddots & \ddots & \\
& 0 & -a & 2a + b & -a \\
& & 0 & -a & a + b
\end{pmatrix}
$$

and

$$
T = \begin{pmatrix}
a + 2b & -a & 0 & & \\
-a & 2a + 2b & -a & 0 & \\
& \ddots & \ddots & \ddots & \\
& 0 & -a & 2a + 2b & -a \\
& & 0 & -a & a + 2b
\end{pmatrix},
$$

while $Y = (1/\Delta y^2)I$. The right-hand side of the equation is a vector of dimension $(N_x - 1) \times (N_y - 1)$,

$$
q = Q + b_x + b_y ,
$$

with three contributions: the source term and two terms picking up those parts of the expressions for boundary conditions that have not been absorbed in the matrix A. The components are arranged in the usual "$j - k$" ordering:

$$
Q = \left(Q_{11}, Q_{21}, \ldots, Q_{N_x-1\,1},\; Q_{12}, Q_{22}, \ldots,\; Q_{1N_y-1}, Q_{2N_y-1}, \ldots, Q_{N_x-1\,N_y-1} \right)^{\mathrm{T}} ,
$$

$$
b_x = \frac{1}{\Delta x} \left(g_{01}, 0, \ldots, 0, g_{N_x 1},\; g_{02}, 0, \ldots,\; g_{0N_y-1}, 0, \ldots, 0, g_{N_x N_y-1} \right)^{\mathrm{T}} ,
$$

$$
b_y = \frac{1}{\Delta y} \left(g_{10}, g_{20}, \ldots, 0, g_{N_x-1\,0},\; 0, 0, \ldots,\; g_{1N_y}, g_{2N_y}, \ldots, 0, g_{N_x-1\,N_y} \right)^{\mathrm{T}} .
$$

Discretize the problem on a reasonably fine uniform mesh with N_x, $N_y \approx 100$. Solve the matrix equation by the Gauss–Seidel and SOR method. Because the boundary conditions have been discretized only to first order, the convergence of the solution will also be just of order $O(\Delta x) + O(\Delta y)$. Use the convergence tolerance of $\approx 10^{-6}$ in both schemes. The problem can be simplified by setting $N_x = N_y = N$, since in this case the value of the optimal relaxation parameter ω for the SOR scheme can be determined by

$$\omega_b = \frac{2}{1 + \sqrt{1 - \frac{1}{4}\left(1 + \cos\frac{\pi}{N}\right)^2}} \, .$$

11.9.4 High-Resolution Schemes for the Advection Equation

This Problem (taken from Ref. [2]) involves the two-dimensional advection equation

$$v_t + c(x, y)v_x + d(x, y)v_y = 0 \, , \qquad (x, y) \in [0, 1] \times [0, 1] \, ,$$

with the initial condition

$$v(x, y, 0) = f(x, y) = \begin{cases} 1 \; ; \; (x, y) \in \left[\frac{1}{4}, \frac{3}{4}\right] \times \left[\frac{1}{4}, \frac{3}{4}\right] \, , \\ 0 \; ; \; \text{otherwise} \, . \end{cases}$$

The exact solution $v(x, y, t) = f(x - ct, y - dt)$ at constant c and d does not bother about the discontinuity in the initial condition: it just advances the initial "jump" in time. In contrast, propagation of discontinuities in difference schemes is plagued by typical annoyances discussed in this Problem.

⊙ Set $c = d = 1$ and impose periodic boundary conditions

$$v(0, y, t) = v(1, y, t) \, , \quad y \in [0, 1] \, ,$$
$$v(x, 0, t) = v(x, 1, t) \, , \quad x \in [0, 1] \, .$$

Use the scheme (11.28) with the flux functions (11.29) and discretization $N_x = N_y = 100$ ($\Delta x = 0.01$). Compute the solution until $t = 1.0$ in time steps of $\Delta t = 0.002$. In same conditions, apply the split Lax–Wendroff method Eqs. (11.30) and (11.31) with the flux functions (11.32) and (11.33). Compute the solution until $t = 0.2$. Finally, solve the problem by using the flux functions (11.48) and (11.49); once in the scheme (11.28), and once in the split Lax–Wendroff scheme. Compare the solutions and describe their behavior.

⊕ For all schemes mentioned above, non-constant coefficients c and d are a much harder nut to crack. You can see that by running the programs from the first part of this Problem with a pair of seemingly harmless coefficient functions

$$c(x, y) = \sqrt{2}\left(y - \tfrac{1}{2}\right) \, , \qquad d(x, y) = \sqrt{2}\left(x - \tfrac{1}{2}\right) \, .$$

This part of the Problem can be solved well by the high-resolution Zalesak–Smolarkiewicz method described in Sect. 11.3.

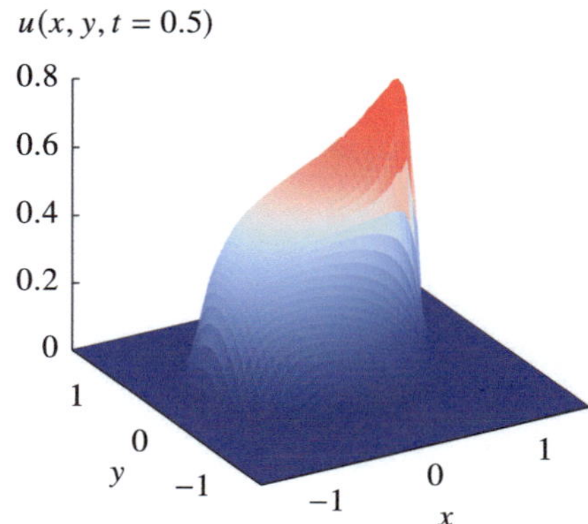

Fig. 11.18 Solution of the two-dimensional Buckley–Leverett equation (11.73) at $t = 0.5$ with the initial condition (11.74) by using the Kurganov–Tadmor scheme with $\theta = 1$ in the slope limiter

11.9.5   Two-Dimensional Buckley–Leverett Equation

We consider the two-dimensional version of the Buckley–Leverett equation (10.103),

$$v_t + [F(v)]_x + [G(v)]_y = \varepsilon\left(v_{xx} + v_{yy}\right), \quad (x, y) \in [-1.5, 1.5]^2, \qquad (11.73)$$

where $\varepsilon = 0.01$. The flux functions are

$$F(v) = \frac{v^2}{v^2 + (1 - v)^2},$$
$$G(v) = F(v)\left(1 - 5(1 - v)^2\right).$$

(Gravitational effects are included in the y-direction.)

  $\odot$ Solve (11.73) by using the two-dimensional Kurganov–Tadmor scheme discussed in Sect. 11.3.3. Use a 100×100 spatial grid to calculate the solution up to $t = 0.5$ with periodic boundary conditions and the initial condition

$$v(x, y, 0) = \begin{cases} 1 \; ; \; x^2 + y^2 < 0.5, \\ 0 \; ; \; \text{otherwise}. \end{cases} \qquad (11.74)$$

Compare your results to Fig. 11.18.

11.9.6   Two-Dimensional Diffusion Equation in Polar Coordinates

Here we try to solve the diffusion equation [2]

$$v_t = \frac{1}{r}\,(r v_r)_r + \frac{1}{r^2}\,v_{\theta\theta}\,, \qquad v = v(r, \theta, t)\,,$$

on $(r, \theta) \in [0, 1] \times [0, 2\pi]$ with pairs of initial and boundary conditions

$$v(r, \theta, 0) = 0 , \qquad v(1, \theta, t) = \sin(4\theta) \sin t ,$$

or

$$v(r, \theta, 0) = (1 - r^2) \sin(2\theta) , \qquad v(1, \theta, t) = 0 .$$

$\odot$ Use the explicit scheme (11.54) and (11.55) with $N_r = 20$, $N_\theta = 32$, $\Delta t = 0.001$. For the first condition pair, compute the solution at $t = 0.1, 0.5, 1.5, 3.0$, and 6.0; for the other pair, compute it at $t = 0.1, 0.5$, and 1.0.

11.9.7 Two-Dimensional Poisson Equation in Polar Coordinates

We would like to solve the Poisson equation in planar polar coordinates

$$- \nabla^2 v = Q ,$$

by considering in two geometries (an annulus and a disk including the origin) and different boundary conditions. (This problem is adapted from Ref. [2].)

$\odot$ In the annular geometry (Fig. 11.8 (right)), let $a = 0.1$ and $Q(r, \theta) = 0$, and impose boundary conditions $f_1(\theta) = \sin 2\theta$ and $f_2(\theta) = \sin 3\theta$ for $\theta \in [0, 2\pi]$. Solve the problem by using the difference scheme described in Sect. 11.4.2. Use the discretization $N_r = N_\theta = 20$ and solve the resulting matrix system by using the Gauss–Seidel method. Repeat the exercise by using a finer discretization $N_r = N_\theta = 100$ with the Gauss–Seidel and the optimal SOR method. In addition, use SOR with $N_r = N_\theta = 100$, $a = 0.5$, $Q(r, \theta) = \exp(r) \sin 2\pi\theta$, and boundary conditions $f_1(\theta) = \sin 4\theta$, $f_2(\theta) = \sin 3\theta$ for $\theta \in [0, 2\pi]$. Examine the difference between the subsequent solutions (11.46) or the residual (11.47) to ascertain convergence.

$\oplus$ Solve the Poisson equation in the disk geometry (Fig. 11.8 (left)). Discuss the case $Q(r, \theta) = 0$ with the boundary condition $f_2(\theta) = \sin 2\theta$ for $\theta \in [0, 2\pi]$. Use the Jacobi and SOR schemes with $N_r = N_\theta = 20$. Additionally, use the SOR method to solve the problem with $Q(r, \theta) = \cos(\pi r) \cos(2\pi\theta)$ and the boundary condition $f_2(\theta) = \sin 4\theta$ on the mesh $N_r = N_\theta = 100$. Use the quantities (11.46) and (11.47) when determining when to stop the iteration.

11.9.8 Finite Element Method

We would like to use the finite element method to solve the Poisson equation

$$-\nabla^2 v = \frac{\partial^2 v}{\partial x^2} + \frac{\partial^2 v}{\partial y^2} = f(x, y), \qquad (x, y) \in R = [0, 1] \times [0, 1],$$

with $f(x, y) = 8\pi^2 \sin(2\pi x) \sin(2\pi y)$ and homogeneous Dirichlet boundary conditions on all sides of R. The analytic solution is $v(x, y) = \sin(2\pi x) \sin(2\pi y)$. (Example adapted from Ref. [14].)

⊙　First, use uniform triangulation with the lengths of intervals h on each axis (Fig. 11.19), in which you span piecewise continuous basis functions on the individual elements, as described in Sect. 11.6. Set $h = 0.1, 0.01$, and 0.001. In all cases, compute the error

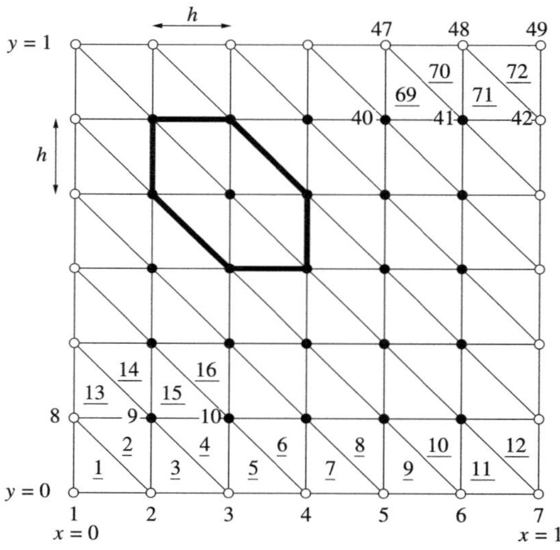

Fig. 11.19 Regular triangulation for solving the Poisson equation on the square by the finite element method. The indices of the elements are underlined; the indices of the nodes are not. (Only one of the many options is drawn.)

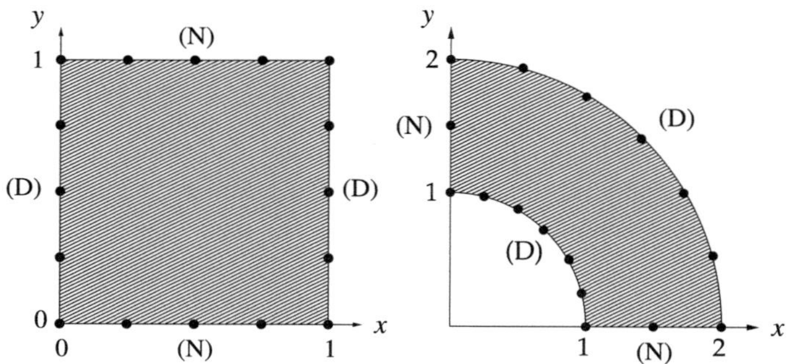

Fig. 11.20 [LEFT] Square geometry for the solution of the planar Laplace equation. The letters at the boundaries denote the type of the boundary condition: (D) Dirichlet, (N) Neumann. [RIGHT] The geometry of a section of an annulus

$$\left[\int_R |\nabla v - \nabla u|^2 \, \mathrm{d}x \right]^{1/2} .$$

⊕ Solve the problem by using a different enumeration of nodes and/or elements. By doing this, you obtain a different stiffness matrix and a different load vector. Is the matrix sparse? How does this influence the speed at which the matrix system can be solved?

11.9.9 Boundary Element Method for the Two-Dimensional Laplace Equation

In this example (adapted from Ref. [11]) we are solving the two-dimensional Laplace equation

$$\frac{\partial^2 \phi}{\partial x^2} + \frac{\partial^2 \phi}{\partial y^2} = 0$$

in the geometry of a square (see Fig. 11.20 (left)).

Along $x = 0$ and $x = 1$, impose Dirichlet boundary conditions, while along $y = 0$ and $y = 1$, impose Neumann conditions:

$$\phi = 0 \text{ on boundary } x = 0 \text{ for } 0 < y < 1 ,$$
$$\phi = \cos(\pi y) \text{ on boundary } x = 1 \text{ for } 0 < y < 1 ,$$
$$\frac{\partial \phi}{\partial n} = 0 \text{ on boundaries } y = 0 \text{ and } y = 1 \text{ for } 0 < x < 1 .$$

The exact solution is

$$\phi(x, y) = \frac{\sinh(\pi x) \, \cos(\pi y)}{\sinh \pi} .$$

⊙ Use the boundary element method to solve the Laplace equation on the square with the conditions specified above. Divide each side of the square into N_1 boundary elements, so that the boundary points are at

$$
\begin{aligned}
\left(x^{(n)}, y^{(n)} \right) &= \left((n-1)l, 0 \right) && \text{(bottom boundary)} , \\
\left(x^{(N_1+n)}, y^{(N_1+n)} \right) &= \left(1, (n-1)l \right) && \text{(right boundary)} , \\
\left(x^{(2N_1+n)}, y^{(2N_1+n)} \right) &= \left(1 - (n-1)l, 1 \right) && \text{(top boundary)} , \\
\left(x^{(3N_1+n)}, y^{(3N_1+n)} \right) &= \left(0, 1 - (n-1)l \right) && \text{(left boundary)} ,
\end{aligned}
$$

for $n = 1, 2, \ldots, N_1$, where $l = 1/N_1$ is the length of each element. We have a total of $N = 4N_1$ boundary elements, and clearly

$$\left(x^{(N+1)}, y^{(N+1)}\right) = \left(x^{(1)}, y^{(1)}\right).$$

⊕　Use the boundary element method to solve the Laplace equation in the geometry shown in Fig. 11.20 (right). Impose Neumann conditions on the straight boundaries of the domain, and Dirichlet conditions on the circular arcs:

$$\frac{\partial \phi}{\partial n} = 0 \quad \text{on boundary } x = 0 \text{ for } 1 < y < 2\,,$$

$$\frac{\partial \phi}{\partial n} = 0 \quad \text{on boundary } y = 0 \text{ for } 1 < x < 2\,,$$

$$\phi = \cos\left(4 \arctan \frac{y}{x}\right) \quad \text{on arc } x^2 + y^2 = 1 \text{ for } x, y > 0\,,$$

$$\phi = 3 \cos\left(4 \arctan \frac{y}{x}\right) \quad \text{on arc } x^2 + y^2 = 4 \text{ for } x, y > 0\,.$$

The exact solution is

$$\phi(x, y) = \left\{\frac{16}{85}\left[r^4 - \frac{1}{r^4}\right] - \frac{16}{255}\left[\frac{r^4}{16} - \frac{16}{r^4}\right]\right\} \cos\left(4 \arctan \frac{y}{x}\right)\,,$$

where $r^2 = x^2 + y^2$. Divide the straight sections of the boundary into N_1, the exterior arc into $8N_1$, and the interior arc into $2N_1$ elements, so that the total number of the boundary elements is $N = 12N_1$:

$$\left(x^{(n)}, y^{(n)}\right) = \left(1 + \frac{n-1}{N_1}, 0\right), \qquad\qquad n = 1, 2, \ldots, N_1\,,$$

$$\left(x^{(N_1+n)}, y^{(N_1+n)}\right) = \left(2 \cos \frac{(n-1)\pi}{16N_1}, 2 \sin \frac{(n-1)\pi}{16N_1}\right), \quad n = 1, 2, \ldots, 8N_1\,,$$

$$\left(x^{(9N_1+n)}, y^{(9N_1+n)}\right) = \left(0, 2 - \frac{n-1}{N_1}\right), \qquad n = 1, 2, \ldots, N_1\,,$$

$$\left(x^{(10N_1+n)}, y^{(10N_1+n)}\right) = \left(\sin \frac{(n-1)\pi}{4N_1}, \cos \frac{(n-1)\pi}{4N_1}\right), \quad n = 1, 2, \ldots, 2N_1\,.$$

11.9.10　Boundary Element Method for Potential Flow

The boundary element method is also highly suitable for the calculation of flow of ideal fluids around solid bodies. In the case of potential (irrotational) flow, i.e. when the velocity can be expressed as a gradient of a scalar function, $\boldsymbol{u} = \nabla\phi$, the very same ϕ that represents the electrostatic potential in the Laplace equation can be used to generate the velocities. In two dimensions, the integral over a line segment (a "panel") of length l aligned with the x-axis is

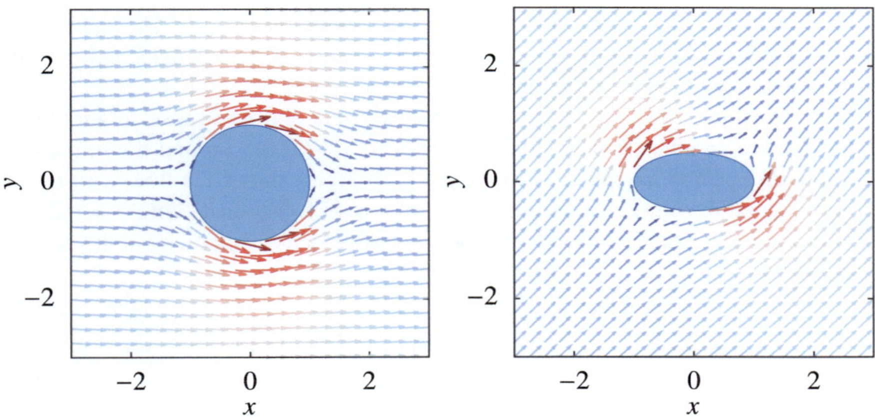

Fig. 11.21 Velocity vector field corresponding to potential flow around [LEFT] a unit circle and around [RIGHT] an ellipse, calculated by the BEM method employing 32 panels. The direction of the incoming flow is $\boldsymbol{u}_\infty = (1, 0)^{\mathrm{T}}$ and $\boldsymbol{u}_\infty = (1, 1)^{\mathrm{T}}$, respectively

$$\phi(x, y) = \frac{1}{4\pi} \int\limits_{-l/2}^{+l/2} \log\big((x - \xi)^2 + y^2\big)\, \mathrm{d}\xi$$

$$= \frac{1}{2\pi}\left[-l + y\left(\arctan\frac{x_-}{y} - \arctan\frac{x_+}{y} \right) + \frac{x_-}{2}\log(x_-^2 + y^2) - \frac{x_+}{2}\log(x_+^2 + y^2) \right],$$

where $x_\pm = x \mp l/2$. The parallel (tangential) and perpendicular (normal) velocity components in the frame of the individual panel can be calculated as

$$u_\parallel = \frac{\partial\phi}{\partial x} = \log\frac{x_-^2 + y^2}{x_+^2 + y^2}, \quad u_\perp = \frac{\partial\phi}{\partial y} = \frac{1}{2\pi}\left(\arctan\frac{x_-}{y} - \arctan\frac{x_+}{y} \right).$$

$$(11.75)$$

Given the velocity of the flow far away from the body, $\boldsymbol{u}_\infty$, we wish to calculate the "velocity sources" σ_j—analogous to charge densities in the electrostatic case—on each panel such that they cancel the corresponding component of $\boldsymbol{u}_\infty$ and the total normal velocity $u_{i\perp}$ on that panel vanishes: no fluid is allowed to enter or exit the body. This consideration results in the system of linear equations

$$u_{\perp i} = \sum_{j=1}^{N} U_{ij}^\perp \sigma_j + u_{\infty\perp i} \equiv 0 \quad \forall i\,,$$

in short,

$$U^\perp \boldsymbol{\sigma} = -\boldsymbol{u}_{\infty\perp}\,.$$

Here $U^\perp$ is the matrix whose elements are the normal velocity components produced by the ith panel on the location of the jth panel ($j \neq i$), obtained by rotation, while the diagonal elements are given by the $x, y \to 0$ limits of (11.75). The right hand-side $\boldsymbol{u}_{\infty\perp}$ contains the normal projections of $\boldsymbol{u}_\infty$ on individual panels, and $\boldsymbol{\sigma}$ is the vector containing the desired intensities of "velocity sources".

$\odot$ Calculate the potential flow around a circle and an ellipse. Figure 11.21 (left) shows a sample result for the flow with $\boldsymbol{u}_\infty = (1, 0)^\mathrm{T}$ around the unit circle represented by $N = 32$ panels, while Fig. 11.21 (right) shows the flow with $\boldsymbol{u}_\infty = (1, 1)^\mathrm{T}$ around the ellipse represented by $x = \cos\phi$, $y = b\sin\phi$, with $b = 0.5$ and $\phi \in [0, 2\pi]$. The exact solution for the tangential velocity component at the surface of the ellipse is

$$u_\parallel(x, y) = u_\infty \frac{(1 + b)y}{\sqrt{y^2 + b^4 x^2}} \, .$$

References

1. J.W. Thomas, *Numerical Partial Differential Equations: Finite Difference Methods, Springer Texts in Applied Mathematics*, vol. 22 (Springer-Verlag, Berlin, 1998)
2. J.W. Thomas, *Numerical Partial Differential Equations: Conservation Laws and Elliptic Equations, Springer Texts in Applied Mathematics*, vol. 33 (Springer-Verlag, Berlin, 1999)
3. W.H. Press, B.P. Flannery, S.A. Teukolsky, W.T. Vetterling, *Numerical Recipes: The Art of Scientific Computing*, 3 rd edn. (Cambridge University Press, Cambridge, 2007). See also The Equivalent Handbooks in Fortran, Pascal and C, as Well as http://numerical.recipes
4. R.A. Horn, C.R. Johnson, *Matrix Analysis* (Cambridge University Press, Cambridge, 1985)
5. J.W. Demmel, *Applied Numerical Linear Algebra* (SIAM, Philadelphia, 1997)
6. S.T. Zalesak, Fully multidimensional flux-corrected transport algorithms for fluids. J. Comp. Phys. **31**, 335 (1979); P.K. Smolarkiewicz, A fully multidimensional positive definite advection transport algorithm with small implicit diffusion, J. Comp. Phys. **54**, 325 (1984)
7. A. Kurganov, E. Tadmor, New high-resolution schemes for nonlinear conservation laws and convection-diffusion equations. J. Comput. Phys. **160**, 241 (2000)
8. A. Kurganov, D. Levy, A third-order semidiscrete central scheme for conservation laws and convection-diffusion equations. SIAM J. Sci. Comput. **22**, 1461 (2000)
9. N.A. Peterson, An algorithm for assembling overlapping grid systems. SIAM J. Sci. Comput. **20**, 1995 (1999)
10. W.D. Henshaw, On multigrid for overlapping grids. SIAM J. Sci. Comput. **26**, 1547 (2005)
11. W.T. Ang, *A Beginner's Course in Boundary Element Methods* (Universal Publishers, Boca Raton, 2007)
12. W.K. Liu, S. Li, H.S. Park, Eighty years of the finite element method: birth, evolution, and future. Arch. Comput. Meth. Eng. **29**, 4431 (2022)
13. J.E. Flaherty, *Finite Element Analysis, CSCI, MATH 6860 Lecture Notes* (Rensselaer Polytechnic Institute, Troy, 2000)
14. Z. Chen, *Finite Element Methods and their Applications* (Springer-Verlag, Berlin, 2005)
15. M.S. Gockenbach, *Understanding and Implementing the Finite Element Method* (SIAM, Philadelphia, 2006)
16. Computational Geometry Algorithms Library, https://www.cgal.org/
17. M. de Berg, O. Cheong, M. van Kreveld, M. Overmars, *Computational Geometry: Algorithms and Applications*, 3rd edn (Springer-Verlag, Berlin, 2008). See also J.R. Shewchuk, Delaunay refinement algorithms for triangular mesh generation. Comp. Geom. **22**, 21 (2002)
18. J. Alberty, C. Carstensen, S.A. Funken, Remarks around 50 lines of MATLAB: short finite element implementation. Num. Algor. **20**, 117 (1999); J. Alberty, C. Carstensen, S.A. Funken, R. Klose, MATLAB implementation of the finite element method in elasticity. Computing **69**, 239 (2002)

19. https://en.wikipedia.org/wiki/List_of_finite_element_software_packages
20. F. Hecht, New development in freefem++. J. Numer. Math. **20**, 251 (2012)
21. http://www.freefem.org
22. D. Knoll, J. Morel, L. Margolin, M. Shashkov, Physically motivated discretization methods. Los Alamos Sci. **29**, 188 (2005)
23. M. Shashkov, S. Steinberg, Solving diffusion equations with rough coefficients on rough grids. J. Comp. Phys. **129**, 383 (1996); J. Hyman, M. Shashkov, S. Steinberg, The numerical solution of diffusion problems in strongly heterogeneous non-isotropic materials. J. Comp. Phys. **132**, 130 (1997); J. Hyman, M. Shashkov, Mimetic discretizations for Maxwell's equations. J. Comp. Phys. **151**, 881 (1999)
24. J. Hyman, J. Morel, M. Shashkov, S. Steinberg, Mimetic finite difference methods for diffusion equations. Comput. Geosci. **6**, 333 (2002)
25. Y. Kuznetsov, K. Lipnikov, M. Shashkov, The mimetic finite difference method on polygonal meshes for diffusion-type problems. Comput. Geosci. **8**, 301 (2004)
26. P. Bochev, J. Hyman, Principles of mimetic discretizations of differential operators, in *Compatible Spatial Discretizations, The IMA Volumes in Mathematics and its Applications*, ed. by D.N. Arnold, P.B. Bochev, R.B. Lehoucq, R.A. Nicolaides, M. Shashkov, vol. 142 (Springer-Verlag, Berlin, 2006), p. 89
27. K. Lipnikov, G. Manzini, M. Shashkov, Mimetic finite difference method. J. Comp. Phys. **257**, 1163 (2014)
28. Multiple Authors: Special Issue of Computing in Science and Engineering (2006)
29. P. Wesseling, *An Introduction to Multigrid Methods* (R. T. Edwards, Philadelphia, 2004)
30. W.L. Briggs, H. van Emden, S.F. McCormick, *A Multigrid Tutorial*, 2nd edn (SIAM, Philadelphia, 2000)
31. S. Li, W.K. Liu, *Mesh-Free Particle Methods* (Springer-Verlag, Berlin, 2004)
32. G.R. Liu, *Mesh-Free Methods: Moving Beyond the Finite-Element Method* (CRC Press, Boca Raton, 2003)
33. E.J. Kansa, Multiquadrics—a scattered data approximation scheme with applications to computational fluid dynamics, I: surface approximations and partial derivative estimates. Comp. Math. Appl. **19**, 127 (1990); II: Solutions of parabolic, hyperbolic and elliptic partial differential equations. Comp. Math. Appl. **19**, 147 (1990)
34. N. Flyer, B. Fornberg, Radial basis functions: developments and applications to planetary scale flows. Comput. Fluids **46**, 23 (2011)
35. M. Tatari, M. Dehghan, A method for solving partial differential equations via radial basis functions: application to the heat equation. Eng. Anal. Boundary Elem. **34**, 206 (2010)
36. E.J. Kansa, Exact explicit time integration of hyperbolic partial differential equations with mesh free radial basis functions. Eng. Anal. Boundary Elem. **31**, 577 (2007)
37. E. Larsson, B. Fornberg, A numerical study of some radial basis function based solution methods for elliptic PDEs. Comp. Math. Appl. **46**, 891 (2003)
38. C.-S. Huang, H.-D. Yen, A.H.-D. Cheng, On the increasingly flat radial basis function and optimal shape parameter for the solution of elliptic PDEs. Eng. Anal. Boundary Elem. **34**, 802 (2010)
39. H.J. Eberl, L. Demaret, A finite difference scheme for a degenerate diffusion equation arising in microbial biology. Electron. J. Diff. Equations **2007**, 77 (2007)
40. I. Klapper, J. Dockery, Mathematical description of microbial biofilms. SIAM Rev. **52**, 221 (2010)

Chapter 12
Spectral Methods for ODE and PDE

Abstract The representation of spatial derivatives is at the heart of spectral methods for partial differential equations, so the main kinds (Fourier, Chebyshev, Legendre, Laguerre, Hermite) are analyzed at the outset, together with efficient means to compute them. Galerkin methods involving all three classes of basis functions are discussed for both stationary (Helmholtz equation) and non-stationary (advection equation) problems. Tau methods are also applicable to both types of problems and are presented next. Due to their straightforward implementation of boundary conditions, they offer an exciting alternative to Galerkin approaches. Separate Sections are devoted to collocation methods and efficient means to solve non-linear equations in the spectral framework. Strong stability preserving time integration methods are discussed, and spectral methods for ODE and PDE posed on semi-infinite and infinite definition domains are introduced. Examples and Problems include the Galerkin method for the advection, diffusion and Poisson equations (Poiseuille flow), the tau method for the Poisson equation, the diffusion equation in collocation approaches, and the Burgers equation.

In finite difference methods the exact solution v of the differential equation is approximated by low-order polynomials interpolating v at several nearby mesh points. For example, Eq. (10.3) is an approximation of the derivative at x_j obtained by parabolic interpolation between x_{j-1}, x_j, and x_{j+1}. The solution on the whole interval is constructed by superposing many such overlapping polynomials as the weighted sum of the function values at the interpolation points.

In spectral methods [1] the approach is from the global standpoint: we approximate the solution *on the whole interval* by a *single* high-degree polynomial (or a linear combination of basis functions) and establish conditions at which this polynomial approximates the exact solution as closely as possible. Different spectral methods exploit different classes of polynomials or basis functions and different ways of realizing these conditions.

Linear Stationary Problems

Spectral methods represent a subclass of the *methods of weighted residuals*. Initially we discuss linear PDE (e.g. of the form (10.1)) on the domain Ω, with the boundary condition specified at the domain boundary $\partial\Omega$:

© The Author(s), under exclusive license to Springer Nature Switzerland AG 2025 767
S. Širca and M. Horvat, *Computational Methods in Physics*, Graduate Texts in Physics,
https://doi.org/10.1007/978-3-031-68566-8_12

$$Lv = Q \quad \text{in } \Omega,$$
$$Bv = 0 \quad \text{on } \partial\Omega, \tag{12.1}$$

where L and B are linear operators. It is easier to grasp the basic ideas of spectral methods in the context of mappings between vector spaces. The operator L in (12.1) acts in Hilbert space X, the space of real or complex functions defined on Ω and square-integrable with respect to a continuous positive weight function. We present the operator L in its discrete form L_N, which is defined on $X_N \subset X$ and maps to X. At the heart of all spectral methods is a condition for the *spectral approximation* $u^N \in X_N$ or for the *residual* $R = L_N u^N - Q$. We require that the projection of the residual from a space $Z \subseteq X$ to a subspace $Y_N \subset Z$ is zero,

$$P_N\left(L_N u^N - Q\right) = 0. \tag{12.2}$$

The operator P_N is an orthogonal projector. With the scalar product $\langle u, v \rangle_N$ in the space Y_N this means $\langle z - P_N z, v \rangle_N = 0$ for $\forall v \in Y_N$. Under this assumption the basic spectral requirement can be written in its *variational form*

$$\left\langle L_N u^N - Q,\, v \right\rangle_N = 0 \quad \forall v \in Y_N. \tag{12.3}$$

The way in which the residual R is minimized depends on the choice of spaces X_N, Y_N, and the form of the projector P_N. Clearly X_N and Y_N should have equal dimensions for the solution of (12.2) to be unique, but in general these spaces need not be identical. In Galerkin and collocation methods for problems involving Dirichlet boundary conditions, usually $X_N = Y_N$, while in tau methods $X_N \neq Y_N$. The choice of the vector spaces and the definition of the scalar product (i.e. the projection of the differential equation to the corresponding subspace) depends on the nature of the problem and on the individual method.

In one dimension the spectral solution is sought in the form of a finite sum

$$u^N(x) = \sum_k u_k \phi_k(x), \tag{12.4}$$

where ϕ_k are the *trial* (or *expansion*, or *approximating*) *functions* that form the basis of X_N. The range of the index k depends on N and on the type of the method. In addition to the trial functions, we also utilize *test* or *weight functions* ψ_n, which are used to minimize the residual $R = L_N u^N - Q$ in the sense of the scalar product (12.3), thus

$$\langle \psi_n, R \rangle = 0 \quad \forall n. \tag{12.5}$$

In Galerkin and tau methods, the trial and test functions are equal ($\psi_n = \phi_n$), but in the Galerkin approach we choose them such that they fulfill boundary conditions by themselves; in tau methods, boundary conditions are imposed by supplemental equations. In collocation methods, the test functions are the delta-"functions" $\psi_n = \delta(x - x_n)$ applied at the collocation points x_n.

Linear Evolution Problems

In the analysis of linear evolution problems

$$
\begin{aligned}
v_t + Lv &= Q \quad \text{in } \Omega \times (0, \infty) , \\
Bv &= 0 \quad \text{on } \partial\Omega \times (0, \infty) , \\
v &= v_0 \quad \text{in } \Omega \text{ at } t = 0 ,
\end{aligned}
\tag{12.6}
$$

the spectral approximation is usually realized in the spatial part only. We treat time as a separate variable and compute the time evolution by using the methods for initial-value problems: see Chap. 8 as well as Sect. 12.6.

The spectral approximation u^N for $t \geq 0$ is a continuously differentiable function with values in X_N. At $t = 0$ it satisfies the initial condition, while at $t > 0$ it satisfies

$$
P_N \left(\frac{\mathrm{d}u^N}{\mathrm{d}t} + L_N u^N - Q \right) = 0 ,
$$

or, in variational form,

$$
\left\langle \frac{\mathrm{d}u^N}{\mathrm{d}t} + L_N u^N - Q, \, v \right\rangle_N = 0 \quad \forall v \in Y_N ,
\tag{12.7}
$$

which is analogous to Eq. (12.3) for stationary problems. We shall see that the conditions (12.5) translate to systems of algebraic equations, and conditions (12.7) to systems of differential equations.

Comparison of Difference and Spectral Methods

Trial functions distinguish difference and finite element methods (Chaps. 10 and 11) from spectral methods. In the former two classes, they are overlapping low-degree local polynomials, thus more suitable for irregular geometries. In spectral methods, they are infinitely differentiable global functions; in this book we exploit trigonometric functions, Chebyshev and Legendre polynomials, Laguerre and Hermite functions, as well as coordinate-mapped Chebyshev polynomials. We apply these methods to problems in regular geometries, although they have also invaded the field of complex geometries [2].

12.1 Spectral Representation of Spatial Derivatives

In spectral methods the spatial derivatives can be computed either in configuration (physical) or in transform space, where the computation becomes very simple.

12.1.1 Fourier Spectral Derivatives

The Fourier series (5.8) corresponds to the Fourier series for the derivative of u,

$$Su' = \sum_{k=-\infty}^{\infty} \mathrm{i}k\,\widehat{u}_k\phi_k\,,$$

which is known as the spectral (Fourier–Galerkin) derivative. Deriving with respect to x in configuration space is equivalent to multiplying each Fourier coefficient by $\mathrm{i}k$ in transform space. (Recall the momentum operator $\boldsymbol{p} = \hbar\boldsymbol{k} = -\mathrm{i}\hbar\nabla$ in non-relativistic quantum mechanics.) The derivation and the truncation of the series ($Su \to S_N u$) commute, thus $(S_N u)' = S_N u'$. We obtain ever higher derivatives by multiplying the coefficients repeatedly by $\mathrm{i}k$,

$$\frac{\mathrm{d}^m u(x)}{\mathrm{d}x^m} \quad\longleftrightarrow\quad (\mathrm{i}k)^m\widehat{u}_k\,. \tag{12.8}$$

In the discrete case, the derivative is computed by using the function values $u_j = u(x_j)$ given at the Fourier collocation points

$$x_j = 2\pi j/N\,, \qquad j = 0, 1, \ldots, N-1\,, \qquad N \text{ even}\,.$$

First, we compute the discrete coefficients $\widetilde{u}_k$, multiply them by $\mathrm{i}k$, and transform back to configuration space by using (5.12). The approximation for the derivative at x_j is then

$$(\mathcal{D}_N u)_j = \sum_{k=-N/2}^{N/2-1} \widetilde{u}_k^{(1)}\, \mathrm{e}^{2\mathrm{i}jk\pi/N}\,, \qquad j = 0, 1, \ldots, N-1\,,$$

where

$$\widetilde{u}_k^{(1)} = \mathrm{i}k\widetilde{u}_k = \frac{\mathrm{i}k}{N}\sum_{j=0}^{N-1} u(x_j)\,\mathrm{e}^{-2\mathrm{i}jk\pi/N}\,, \qquad k = -N/2, -N/2+1, \ldots, N/2-1\,.$$

The $(\mathcal{D}_N u)$ is called the *Fourier collocation (or interpolation) derivative*, as the values $(\mathcal{D}_N u)_j$ are equal to the derivatives of the discrete Fourier transform at the mesh points. By changing the order of summation over k and j, the Fourier derivative can be expressed as a matrix multiplication

$$(\mathcal{D}_N u)_l = \sum_{k=-N/2}^{N/2-1}\left(\frac{\mathrm{i}k}{N}\sum_{j=0}^{N-1} u(x_j)\,\mathrm{e}^{-2\pi\mathrm{i}jk/N}\right)\mathrm{e}^{2\pi\mathrm{i}kl/N} = \sum_{j=0}^{N-1}\left(D_N^{(1)}\right)_{lj} u_j\,, \tag{12.9}$$

where

$$\left(D_N^{(1)}\right)_{lj} = \frac{1}{N} \sum_{k=-N/2}^{N/2-1} ik\,e^{2ik(l-j)\pi/N} \,. \tag{12.10}$$

If u is a real function, the term with $k = -N/2$ makes a purely imaginary contribution to the sum. In other words, if the Fourier coefficient $\widehat{u}_{-N/2}$ has a non-zero imaginary component, u^N is not a real function. As a rule, we therefore drop the term with $k = -N/2$, i.e. we set $\widehat{u}_{-N/2} \equiv 0$. This asymmetry between the lower and upper limit for k and the mentioned artifact originate in numerous implementations of the FFT calling for meshes with even numbers of points. The sum in (12.10) can then be computed analytically, yielding [3]

$$\left(D_N^{(1)}\right)_{lj} = \begin{cases} 0 & ; \, l = j \,, \\ \dfrac{1}{2}\,(-1)^{l+j}\cot\left(\dfrac{(l-j)\pi}{N}\right) & ; \, l \neq j \,. \end{cases} \tag{12.11}$$

An analogous procedure gives the matrix corresponding to the second derivative,

$$\left(D_N^{(2)}\right)_{lj} = \begin{cases} -\dfrac{1}{12}\,(N-1)(N-2) & ; \, l = j \,, \\ \dfrac{1}{4}\,(-1)^{l+j}N + \dfrac{1}{2}\,(-1)^{l+j+1}\sin^{-2}\left(\dfrac{(l-j)\pi}{N}\right) & ; \, l \neq j \,. \end{cases} \tag{12.12}$$

This matrix can be used to compute the values of the second derivative at the collocation points by simple multiplication,

$$(\mathcal{D}_N^2 u)_l = \sum_{j=0}^{N-1} \left(D_N^{(2)}\right)_{lj} u_j \,. \tag{12.13}$$

Example One can get a feel for the beauty of spectral derivation in the case of a simple function $u(x) = \exp(\sin x)$ which is periodic on $[0, 2\pi]$. We compute its derivative first by using finite differences, then by using (12.9). The fourth-order difference $u'(x_j) \approx [u(x_{j-2}) - 8u(x_{j-1}) + 8u(x_{j+1}) - u(x_{j+2})]/(12h)$ corresponds to a pentadiagonal circulant matrix. The error with respect to the exact solution $\|(u^N)' - u'\|$ decreases as $\mathcal{O}(N^{-4})$ (Fig. 12.1 (left)).

All non-diagonal elements of the matrix $D_N^{(1)}$ in Eq. (12.11) are non-zero, but the error drops to round-off level already at very small N. For sufficiently smooth functions the error decreases as $\mathcal{O}(N^{-m})$ for *any* m. This dramatic fall-off that gave the spectral methods their name is known as *spectral convergence*. Do not expect it for functions with discontinuities! For such functions—regardless of N—typical oscillations occur throughout the domain, in particular at its edges (Gibbs phenomenon). An example for the function $u(x) = \sin(x/2)$ with a discontinuity in the derivative $u'(x) = \frac{1}{2}\cos(x/2)$ is shown in Fig. 12.1 (right). ◁

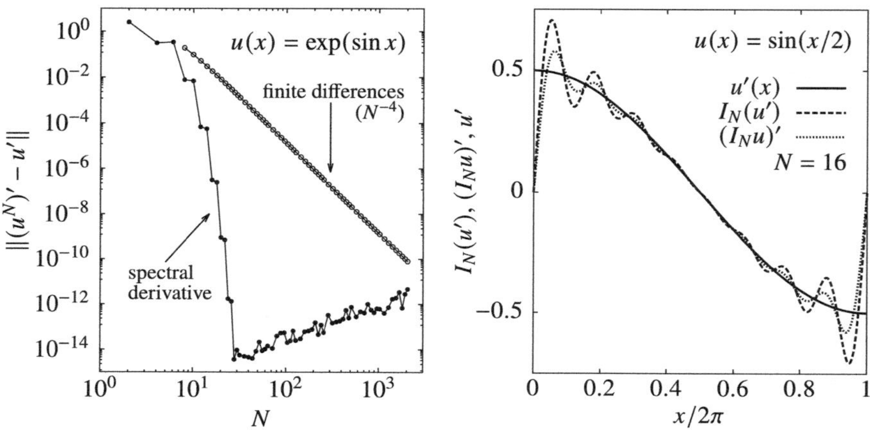

Fig. 12.1 [LEFT] The precision of the derivative of the function $u(x) = \exp(\sin x)$ on $[0, 2\pi]$. Finite difference methods converge as $\mathcal{O}(N^{-m})$, where m is the order of the method (here $m = 4$). The spectral derivative converges faster than any power N^{-m}. [RIGHT] The spectral derivatives of $\sin(x/2)$ compared to the exact derivative. Due to the discontinuities at the boundaries Gibbs oscillations appear. Here I_N denotes the interpolants at N points (see Eq. (5.13))

The collocation derivative (or second derivative) can be computed naively, by multiplying the $N \times N$ matrix and the N-dimensional vector u_j according to (12.9) and (12.13), which takes $2N^2$ operations. There is a much faster way. First we use the inverse FFT to compute the interior sums, multiply the computed Fourier coefficients by ik (or $(ik)^2$ for the second derivative), and finally use FFT to compute the exterior sum. For a real function u the total cost is $N(5 \log_2 N - 5)$ operations if the classic FFT is used (see pp. 253–254 and Fig. 5.7 (right)).

We have written the matrix representations of Fourier derivatives for an even number of mesh points N. The formulas for odd N are given in [4].

12.1.2 Legendre Spectral Derivatives

Spectral derivatives of functions expanded in terms of orthogonal polynomials can also be computed in either physical space or transform space. In transform space, the first and second derivatives of a function with the Legendre expansion (5.33) are

$$u'(x) = \sum_{k=0}^{\infty} \widehat{u}_k^{(1)} P_k(x) , \quad \widehat{u}_k^{(1)} = (2k + 1) \sum_{\substack{p=k+1 \\ p+k \text{ odd}}}^{\infty} \widehat{u}_p ,$$

$$[-4pt]u''(x) = \sum_{k=0}^{\infty} \widehat{u}_k^{(2)} P_k(x) , \quad \widehat{u}_k^{(2)} = \left(k + \frac{1}{2}\right) \sum_{\substack{p=k+2 \\ p+k \text{ even}}}^{\infty} \left[p(p+1) - k(k+1)\right]\widehat{u}_p . \tag{12.14}$$

These formulas follow from the recurrence relations for Legendre polynomials [5]. The finite sum $(P_N u)'$ corresponding to the first derivative is known as the *Legendre–Galerkin derivative*. In contrast to the Fourier expansion, this derivative and the truncation of the sum do not commute, thus $(P_N u)' \neq P_{N-1} u'$. Asymptotically $P_{N-1} u'$ is a better approximation of u' than $(P_N u)'$ [3].

If the values of a function are given at the Gauss, Gauss–Radau, or Gauss–Lobatto quadrature nodes (see Eqs. (5.34), (5.35), and (5.37)), it can be differentiated in physical space by differentiating the interpolation polynomial $I_N u$ of degree N and evaluating it at these points. The *Legendre collocation derivative* is defined as the derivative of the discrete finite series for the function u,

$$\mathcal{D}_N u = (I_N u)' , \tag{12.15}$$

and is a polynomial of degree $N - 1$. We compute the collocation derivative $(\mathcal{D}_N u)_l = (\mathcal{D}_N u)(x_l)$ at $x_0, x_1, \ldots, x_N$ from the values $u(x_j)$ as

$$(\mathcal{D}_N u)_l = \sum_{j=0}^{N} \left(D_N^{(1)} \right)_{lj} u(x_j) , \qquad l = 0, 1, \ldots, N . \tag{12.16}$$

The matrix elements $(D_N^{(1)})_{lj}$ can be computed explicitly for all three types of quadrature [3]. Most frequently, the Gauss–Lobatto variant (5.36) is used, for which

$$\left(D_N^{(1)} \right)_{lj} = \begin{cases} -\dfrac{N(N+1)}{4} & ; \; j = l = 0 , \\[2mm] 0 & ; \; 1 \leq j = l \leq N - 1 , \\[2mm] \dfrac{N(N+1)}{4} & ; \; j = l = N , \\[2mm] \dfrac{P_N(x_l)}{P_N(x_j)} \dfrac{1}{x_l - x_j} & ; \; j \neq l . \end{cases}$$

The matrix corresponding to the second derivative is

$$\left(D_N^{(2)} \right)_{lj} = \begin{cases} \dfrac{N(N+1)(N^2 + N - 2)}{24} & ; \; j = l = 0 , \; j = l = N , \\[2mm] \dfrac{(-1)^N}{P_N(x_j)} \dfrac{N(N+1)(1+x_j) - 4}{2(1+x_j)^2} & ; \; l = 0 , \; 1 \leq j \leq N , \\[2mm] \dfrac{1}{P_N(x_j)} \dfrac{N(N+1)(1-x_j) - 4}{2(1-x_j)^2} & ; \; l = N , \; 0 \leq j \leq N - 1 , \\[2mm] \dfrac{1}{3} \dfrac{P_N''(x_j)}{P_N(x_j)} & ; \; 1 \leq j = l \leq N - 1 , \\[2mm] -\dfrac{P_N(x_l)}{P_N(x_j)} \dfrac{2}{(x_l - x_j)^2} & ; \; \begin{matrix} 1 \leq l \leq N - 1 , \\ 0 \leq j \leq N , \end{matrix} \; j \neq l . \end{cases}$$

A fast Legendre transform is not known, so in order to compute the Legendre spectral derivative, one should use the matrix multiplication (12.16).

12.1.3 Chebyshev Spectral Derivatives

In transform space, the first and second Chebyshev–Galerkin derivative of the function with the expansion (5.43) can be computed by using the formulas

$$u'(x) = \sum_{k=0}^{\infty} \widehat{u}_k^{(1)} T_k(x) \,, \qquad \widehat{u}_k^{(1)} = \frac{2}{c_k} \sum_{\substack{p=k+1 \\ p+k \text{ odd}}}^{\infty} p\,\widehat{u}_p \,, \tag{12.17}$$

$$u''(x) = \sum_{k=0}^{\infty} \widehat{u}_k^{(2)} T_k(x) \,, \qquad \widehat{u}_k^{(2)} = \frac{1}{c_k} \sum_{\substack{p=k+2 \\ p+k \text{ even}}}^{\infty} p(p^2 - k^2)\,\widehat{u}_p \,. \tag{12.18}$$

(The formulas for the third and fourth derivative are listed in [4].) The expansion coefficients of the derivative u' and of the function u are related by

$$c_k \widehat{u}_k^{(1)} = \widehat{u}_{k+2}^{(1)} + 2(k + 1)\widehat{u}_{k+1} \,, \qquad k = 0, 1, \ldots, N - 1 \,. \tag{12.19}$$

This allows us to efficiently differentiate a degree-N polynomial in transform space: since $\widehat{u}_k^{(1)} = 0$ for all $k \geq N$, the expansion coefficients of u' can be computed from the expansion of u by Eq. (12.19). The generalization of this relation is

$$c_k \widehat{u}_k^{(q)} = \widehat{u}_{k+2}^{(q)} + 2(k + 1)\widehat{u}_{k+1}^{(q-1)} \,, \qquad k \geq 0 \,.$$

The Chebyshev interpolation derivative in the sense of (12.15) can be represented in matrix form appearing in (12.16) just as in the case of Legendre polynomials. Here we list only the matrices corresponding to the first and second derivative for the Gauss–Lobatto nodes [3]. The matrix for the first derivative is

$$\left(D_N^{(1)}\right)_{lj} = \begin{cases} \dfrac{1}{6}\left(2N^2 + 1\right) \,; & j = l = 0 \,, \\[2ex] -\dfrac{1}{2}\dfrac{x_j}{1 - x_j^2} \,; & 1 \leq j = l \leq N - 1 \,, \\[2ex] -\dfrac{1}{6}\left(2N^2 + 1\right) \,; & j = l = N \,, \\[2ex] \dfrac{\overline{c}_l}{\overline{c}_j}\dfrac{(-1)^{l+j}}{x_l - x_j} \,; & j \neq l \,. \end{cases} \tag{12.20}$$

The matrix corresponding to the second derivative is

$$
\left(D_N^{(2)}\right)_{lj} =
\begin{cases}
\dfrac{1}{15}\left(N^4 - 1\right) & ;\ j = l = 0,\ j = l = N,\\[2ex]
\dfrac{2}{3}\dfrac{(-1)^j}{\bar{c}_j}\dfrac{(2N^2 + 1)(1 - x_j) - 6}{(1 - x_j)^2} & ;\ l = 0,\ 1 \le j \le N,\\[2ex]
\dfrac{2}{3}\dfrac{(-1)^{j+N}}{\bar{c}_j}\dfrac{(2N^2 + 1)(1 + x_j) - 6}{(1 + x_j)^2} & ;\ l = N,\ 0 \le j \le N - 1,\\[2ex]
-\dfrac{(N^2 - 1)(1 - x_l^2) + 3}{3(1 - x_l^2)^2} & ;\ 1 \le j = l \le N - 1,\\[2ex]
\dfrac{(-1)^{j+l}}{\bar{c}_j}\dfrac{x_l^2 + x_j x_l - 2}{(1 - x_l^2)(x_j - x_l)^2} & ;\ \begin{array}{l} 1 \le l \le N - 1,\\ 0 \le j \le N, \end{array}\ j \ne l.
\end{cases}
\tag{12.21}
$$

(The coefficients $\bar{c}_i$ are defined below Eq. (5.46).) In computing the denominators $x_l - x_j$ and $1 - x_j^2$ for large N round-off errors may occur due to subtraction of almost equal values. The following expressions that can be derived from the definition of Chebyshev nodes, $x_j = \cos(\pi j / N)$, are more stable:

$$
\frac{1}{1 - x_j^2} \longrightarrow 1/\sin^2(j\pi/N),
$$

$$
\frac{1}{x_l - x_j} \longrightarrow -\frac{1}{2}\frac{1}{\sin\big[(l + j)\pi/(2N)\big]\sin\big[(l - j)\pi/(2N)\big]},
$$

For further hints on improving the precision of matrix differentiation see [6].

As with Legendre polynomials, the Chebyshev series truncation and interpolation do not commute with differentiation. This implies that asymptotically $P_{N-1}u'$ and $I_{N-1}u'$ are better approximations of u' than $(P_N u)'$ or $(I_N u)'$.

12.1.4 Computing the Chebyshev Spectral Derivative by Fourier Transformation

The Chebyshev collocation spectral derivative can be computed efficiently by using the fast Fourier transformation. We see this from the structure of the interpolation polynomial in dependence on x or θ, where $x = \cos\theta$. In the following we restrict the discussion to the Chebyshev–Gauss–Lobatto collocation (5.45). The polynomial

$$
p(x) = \sum_{n=0}^{N} a_n T_n(x), \qquad x \in [-1, 1],
\tag{12.22}
$$

interpolates an arbitrary function f on $[-1, 1]$ at Chebyshev collocation points $x_j = \cos(\pi j / N)$, where $j = 0, 1, \ldots, N$. On the other hand, the polynomial

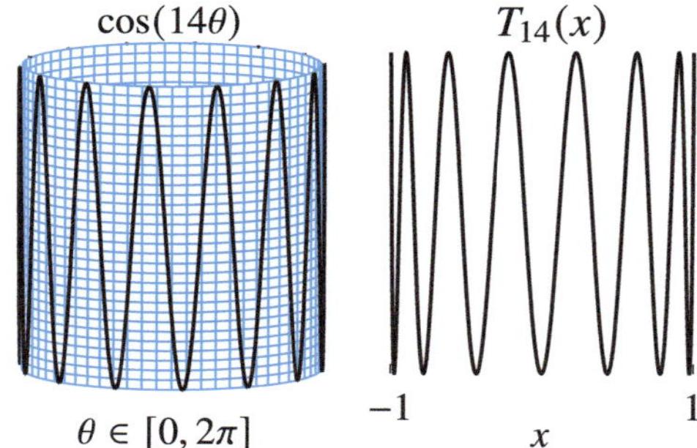

Fig. 12.2 The cosine function $\cos(14\theta)$ wrapped around a cylinder (left) and its planar projection, the Chebyshev polynomial $T_{14}(x)$ (right)

$$P(\theta) = \sum_{n=0}^{N} a_n \cos(n\theta) , \qquad \theta \in \mathbb{R} , \tag{12.23}$$

interpolates an arbitrary even and 2π-periodic function F at equidistant points $\theta_j = \pi j/N$. With the transformation $x = \cos\theta$ we have $F(\theta) = f(x) = f(\cos\theta)$, which implies $P(\theta) = p(x) = p(\cos\theta)$. The Chebyshev transform (12.22) in x is therefore nothing but the Fourier (cosine) transform (12.23) in θ. In other words, if one "wraps" the cosine function $\cos(n\theta)$ around a half-cylinder with unit radius and looks at it from the side, the Chebyshev polynomial T_n emerges (Fig. 12.2).

The procedure is [7]: compute the derivative of the Chebyshev interpolation polynomial p of the function f by first finding the trigonometric interpolation polynomial P of the corresponding function F; compute the derivative in Fourier space; compute the inverse Fourier transform on the uniform mesh; finally, display the obtained values on the mesh of Chebyshev collocation points.

We write the values u_j at the Chebyshev collocation points $x_j = \cos(\pi j/N)$, $j = 0, 1, \ldots, N$, as an array $U = \{U_1, U_2, \ldots, U_{2N}\}$ of dimension $2N$,

$$U_{j+1} = u_j , \quad j = 0, 1, \ldots, N ,$$

$$U_{2N-j+1} = u_j , \quad j = 1, 2, \ldots, N - 1 .$$

By fast Fourier transformation we calculate

$$\widehat{U}_k = \frac{\pi}{N} \sum_{j=1}^{2N} \mathrm{e}^{-\mathrm{i}k\theta_j} U_j , \qquad k = -N + 1, -N + 2, \ldots, N .$$

The polynomial P depends on the transformed values $\widehat{U}_k$ and has the form

$$P(\theta) = \frac{1}{2\pi} \sum_{k=-N+1}^{N} \mathrm{e}^{\mathrm{i}k\theta} \widehat{U}_k = \sum_{n=0}^{N} a_n \cos(n\theta) .$$

At this point we actually execute the differentiation by using (12.8) at $m = 1$, thus

$$\widehat{V}_k = \begin{cases} ik\,\widehat{U}_k \; ; \; k = -N+1, -N+2, \ldots, N-1 \, , \\ \quad 0 \; ; \; k = N \, , \end{cases} \tag{12.24}$$

and compute the inverse Fourier transform

$$W_j = \frac{1}{2\pi} \sum_{k=-N+1}^{N} e^{ik\theta_j}\,\widehat{V}_k \, , \qquad j = 1, 2, \ldots, 2N \, .$$

This yields the derivative of the trigonometric interpolation polynomial P on the equidistant mesh θ_j. In the last step the derivative of the interpolation polynomial p is computed on the mesh x_j. This is accomplished by using the chain rule

$$p'(x) = \frac{dp}{dx} = \frac{dP}{d\theta}\frac{d\theta}{dx} = -\frac{P'(\theta)}{\sin\theta} = \frac{\sum_{n=0}^{N} a_n n \sin(n\theta)}{\sqrt{1-x^2}} \, .$$

Denote $p'(x_j) = w_j$ and $P'(\theta_j) = W_j$. At the interior collocation points we get

$$w_j = -\frac{W_j}{\sqrt{1-x_j^2}} \, , \qquad j = 1, 2, \ldots, N-1 \, . \tag{12.25}$$

Note that at $x_0 = 1$ ($\theta = 0$) and $x_N = -1$ ($\theta = \pi$), the expressions for w_0 and w_N are of the form $0/0$ and must be harnessed by the l'Hôpital rule, resulting in

$$w_0 = \frac{1}{\pi}\left[\sum_{n=1}^{N-1} n^2\widehat{U}_n + \frac{N^2}{2}\widehat{U}_N\right] , \qquad w_N = \frac{1}{\pi}\left[\sum_{n=1}^{N-1}(-1)^{n+1}n^2\widehat{U}_n - \frac{N^2}{2}\widehat{U}_N\right] .$$

Example Compute the Chebyshev spectral derivative of the function

$$u(x) = (x^2 - 1)\exp(x^3) \, ,$$

given at the Chebyshev nodes $x_j = \cos(\pi j/N)$, $j = 0, 1, \ldots, N$ where its values are determined by the interpolation polynomial (12.22); see Fig. 12.3 (left). The derivatives at x_j can be computed simply by multiplying u_j by the matrix $D_N^{(1)}$ of Eq. (12.20), but a more elegant way to the derivatives $p'(x_j) = w_j$ leads through Eq. (12.25) and two additional equations for w_0 and w_N by using FFT. Figure 12.3 (right) shows the absolute error of both methods.   ◁

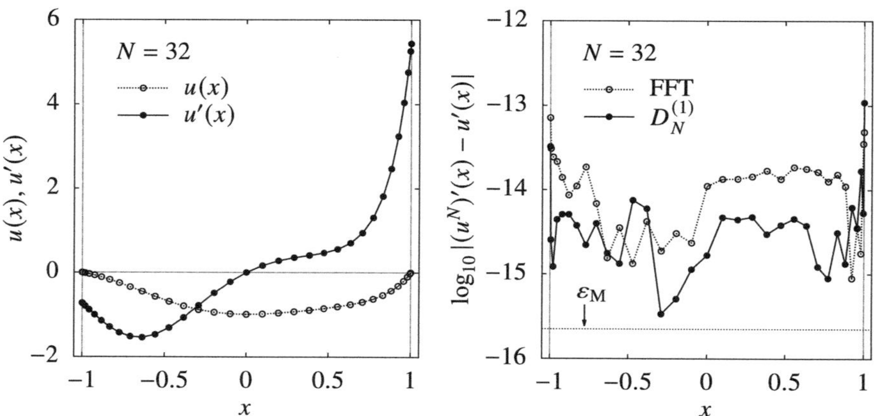

Fig. 12.3 [LEFT] The graphs of $u(x) = (x^2 - 1)\exp(x^3)$ and its derivative $u'(x)$. The symbols o and
• denote their values at the Chebyshev points $x_j = \cos(\pi j/N)$, $j = 0, 1, \ldots, N$, where $N = 32$.
[RIGHT] The absolute error of the derivative, computed via multiplication by the matrix (12.20) and
according to (12.25) via FFT

Higher derivatives can be computed similarly. To compute the mth derivative in
Fourier space, step (12.24) should be repeated m-times, as shown in (12.8). We define

$$\widehat{V}_k^{(m)} = (\mathrm{i}k)^m\,\widehat{U}_k$$

and set $\widehat{V}_{-N} = 0$, if m is odd. Let us compute the second derivative:

$$p''(x) = \frac{\mathrm{d}P}{\mathrm{d}\theta}\frac{\mathrm{d}^2\theta}{\mathrm{d}x^2} + \frac{\mathrm{d}^2 P}{\mathrm{d}\theta^2}\left(\frac{\mathrm{d}\theta}{\mathrm{d}x}\right)^2 = -\frac{x}{(1-x^2)^{3/2}}\,P'(\theta) + \frac{1}{1-x^2}\,P''(\theta)\,.$$

By denoting $p''(x_j) = w_j^{(2)}$ and $P''(\theta_j) = W_j^{(2)}$ we obtain at the interior points

$$w_j^{(2)} = -\frac{x_j}{(1-x_j^2)^{3/2}}\,W_j + \frac{1}{1-x_j^2}\,W_j^{(2)}\,, \qquad j = 1, 2, \ldots, N-1\,,$$

while at the boundary points (check this as an exercise) we get

$$w_0^{(2)} = \frac{1}{3\pi}\left[\sum_{n=1}^{N-1}(n^4 - n^2)\widehat{U}_n + \frac{N^4 - N^2}{2}\,\widehat{U}_N\right],$$

$$w_N^{(2)} = \frac{1}{3\pi}\left[\sum_{n=1}^{N-1}(-1)^n(n^4 - n^2)\widehat{U}_n + \frac{N^4 - N^2}{2}\,\widehat{U}_N\right].$$

12.1.5   *Laguerre Spectral Derivatives*

Spectral methods for problems on the semi-infinite domain $[0, \infty]$ call for Laguerre polynomials and Laguerre functions as the basis sets. The matrix for the first spectral derivative corresponding to Laguerre–Gauss–Radau nodes is [8, 9]

$$
\left(D_N^{(1)}\right)_{lj} = \begin{cases}
-(N+1)/2 & ; \, j = l = 0 \,, \\
0 & ; \, j = l \neq 0 \,, \\
\dfrac{\widehat{L}_{N+1}(x_l)}{\widehat{L}_{N+1}(x_j)} \dfrac{1}{x_l - x_j} & ; \, l \neq j \,,
\end{cases}
\tag{12.26}
$$

where $\widehat{L}_k(x)$ are the Laguerre functions defined in (5.50) and $\{x_j\}_{j=0}^N$ are the nodes: $x_0 = 0$, while $x_1, x_2, \ldots, x_N$ are the roots of the polynomial L'_{N+1}. (For alternatives and Laguerre–Gauss nodes see Table 3.1.) Note that $x_N = \mathcal{O}(N)$ as $N \to \infty$. Higher derivatives are obtained by raising (12.26) to the appropriate power,

$$
D_N^{(m)} = \left(D_N^{(1)}\right)^m \,, \quad m \geq 2 \,.
\tag{12.27}
$$

Fig. 12.4 illustrates first-order and second-order Laguerre spectral differentiation of the function $u(x) = xe^{-x}$ for $x \in [0, \infty]$ with $N = 5$, 10 and 15 by showing the pointwise errors at the nodes with respect to the exact derivatives.

Because the interval $[0, \infty]$ can be mapped onto itself by the change of variable $x = a\xi$, where a is any positive real number, the chain rule

$$
\frac{\mathrm{d}f}{\mathrm{d}\xi} = a\frac{\mathrm{d}f}{\mathrm{d}x} \,, \quad \frac{\mathrm{d}^2 f}{\mathrm{d}\xi^2} = a^2\frac{\mathrm{d}^2 f}{\mathrm{d}x^2} \,, \ldots
$$

Fig. 12.4 Pointwise error of the first (black points and curves) and second (blue points and curves) Laguerre spectral derivative of the function $u(x) = xe^{-x}$, for $N = 5$, 10 and 15. (The values at the nodes have been connected by straight lines, not by the interpolant

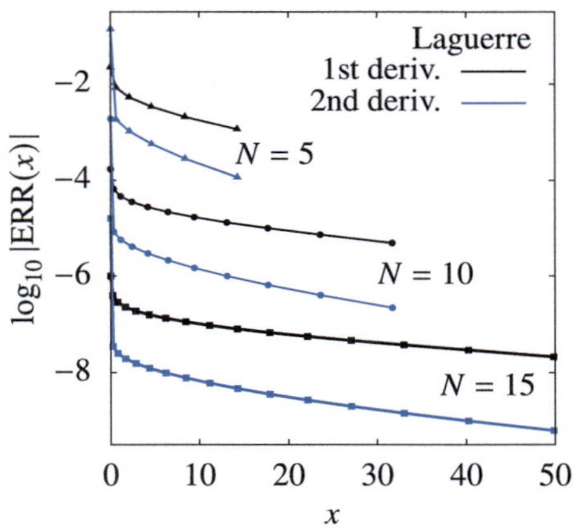

tells us that upon rescaling $D_N^{(1)}$ (for $a = 1$) should be multiplied by a, $D_N^{(2)}$ by a^2, and so on, while the nodes must be rescaled as $x_j \to \xi_j = x_j/a$. Thus Laguerre differentiation is exact for functions of the form

$$e^{-ax/2} p(x)\,, \tag{12.28}$$

where p is any polynomial of degree at most $N - 1$. The freedom offered by the scaling parameter a may be put to good use in the optimization of differencing: this is exploited in the standard MATLAB Differentiation Matrix Suite [10].

12.1.6 Hermite Spectral Derivatives

Spectral methods for problems on the whole real axis $[-\infty, \infty]$ are based on Hermite polynomials and Hermite functions as the basis sets. The matrix corresponding to the first spectral derivative with Gauss quadrature nodes is [11]

$$\left(D_N^{(1)}\right)_{lj} = \begin{cases} 0 & ; \ j = l\,, \\[2ex] \dfrac{\widehat{H}_N(x_l)}{\widehat{H}_N(x_j)} \dfrac{1}{x_l - x_j} & ; \ l \neq j\,, \end{cases} \tag{12.29}$$

where $\widehat{H}_k(x)$ are the Hermite functions defined by (5.54) and $\{x_j\}_{j=0}^N$ are the roots of the polynomial H_{N+1}. Note that $-x_0 = x_N = \mathcal{O}(\sqrt{N})$ as $N \to \infty$. In contrast to Laguerre spectral differentiation, matrices for Hermite derivatives of higher orders ($m \geq 2$) are *not* obtained by simple successive application of (12.29):

$$D_N^{(m)} \neq \left(D_N^{(1)}\right)^m ! \tag{12.30}$$

Instead, they may be constructed in the following manner [11]. We first devise a spectral derivative matrix not for the function f that we wish to differentiate but rather for the function

$$g(x) = e^{x^2/2} f(x)\,.$$

This is accomplished by the matrix

$$D = HQH^{-1}\,,$$

where

$$H_{ij} = \begin{cases} H_N(x_j) & ; \ i = j\,, \\ 0 & ; \ i \neq j\,, \end{cases} \qquad Q_{ij} = \begin{cases} x_i & ; \ i = j\,, \\ 1/(x_i - x_j) & ; \ i \neq j\,. \end{cases}$$

To obtain the differentiation matrices for the original function f we differentiate

$$f'(x) = e^{-x^2/2}\big(g'(x) - xg(x)\big)\,,$$
$$f''(x) = e^{-x^2/2}\big(g''(x) - 2xg'(x) + (x^2 - 1)g(x)\big)\,,$$

and so on, whence we infer that the subsequent derivatives of f sampled at the nodes $\{x_j\}$ and organized into $f = (f(x_0), f(x_1), \ldots, f(x_N))^{\mathrm{T}}$ may be computed by

$$f' = E[D - X]E^{-1}f\,, \tag{12.31}$$
$$f'' = E[D^2 - 2XD + (X^2 - I)]E^{-1}f\,, \tag{12.32}$$

and in an analogous fashion for higher orders. Here $X = \mathrm{diag}(x_j)$, $E = \mathrm{diag}(e^{-x_j^2/2})$, and I is the identity matrix. (Note that the computation of all inverses in the formulas above is trivial.) From (12.31) we read off

$$D_N^{(1)} = E[D - X]E^{-1}\,,$$

yielding precisely (12.29), while (12.32) gives us

$$D_N^{(2)} = E[D^2 - 2XD + (X^2 - I)]E^{-1}\,. \tag{12.33}$$

Since the expression $D^2 - 2XD + X^2 - I = (D - X)^2 - I$ is just one $-I$ shy of a complete square, we clearly have (12.30). Figure 12.5 illustrates the first-order and second-order Hermite spectral differentiation of the function

$$u(x) = \frac{x - 1}{x^2 + 1}\, e^{-x^2/3 + x - 1}\,, \quad x \in [\infty, \infty]\,, \tag{12.34}$$

with $N = 10, 20$ and 30, and shows the pointwise errors at the nodes with respect to the exact values of the derivatives.

Because the interval $[-\infty, \infty]$ can be mapped onto itself by the change of variable $x = a\xi$, where a is any positive real, we can rescale the nodes and the derivative matrices in the same manner as in Laguerre spectral differentiation described in Sect. 12.1.5. In contrast to (12.28) Hermite spectral differentiation is exact for functions of the form

$$e^{-(ax)^2/2}p(x)\,, \tag{12.35}$$

where p is any polynomial of degree at most $N - 1$.

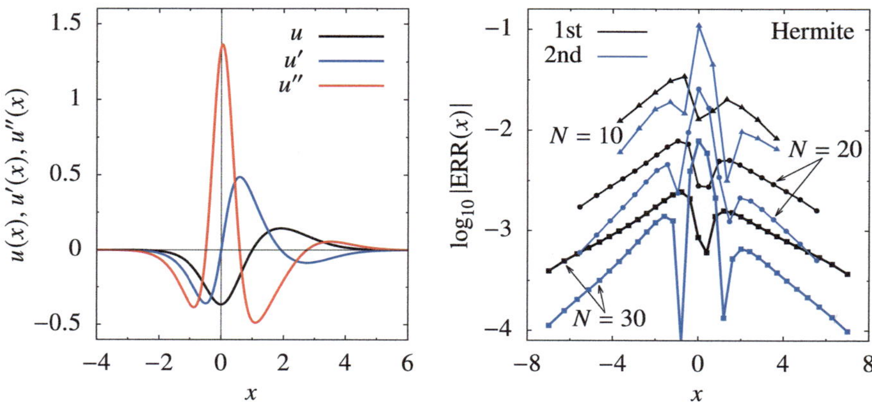

Fig. 12.5 [LEFT] The function (12.34) and its first and second derivative. Note the asymmetric nature of the chosen function and the increasingly steep slopes of its derivatives. [RIGHT] Error of the first (black points and curves) and second (blue points and curves) Hermite spectral derivatives of (12.34), for $N = 10, 20$ and 30

12.2   Galerkin Methods

In Galerkin methods for stationary problems of the form (12.1) the residual R is minimized by requiring (12.5). This means

$$\langle \psi_n, L_N u^N - Q \rangle = \Big\langle \psi_n, L_N \sum_k u_k \phi_k - Q \Big\rangle = \sum_k u_k \langle \psi_n, L_N \phi_k \rangle - \langle \psi_n, Q \rangle = 0 \,.$$

In the classical Galerkin methods, the trial and test functions are the same, so $\psi_n = \phi_n$, and the equations above become

$$\sum_k L_{nk} u_k = \langle \phi_n, Q \rangle \,, \qquad n = 0, 1, \ldots, N,$$

where $L_{nk} = \langle \phi_n, L_N \phi_k \rangle$. By solving this linear system we get $N + 1$ coefficients u_k of the spectral expansion (12.4) of the function u^N. We have not specified the range of the index k. For example, depending on the test function, it may run from $-N/2$ to $N/2 - 1$ (trigonometric polynomials) or from 0 to N (orthogonal polynomials).

12.2.1   Fourier–Galerkin

The classic model stationary problem is the one-dimensional Helmholtz equation

$$-\frac{\mathrm{d}^2 v}{\mathrm{d}x^2} + \lambda v = Q \,, \qquad x \in (0, 2\pi) \,, \tag{12.36}$$

where $\lambda \geq 0$. We seek solutions that are periodic on the interval $[0, 2\pi]$. The equation is of the form $Lv = Q$, where $L = -\mathrm{d}^2/\mathrm{d}x^2 + \lambda$ is a linear differential operator. This is not a partial differential equation as it only contains the spatial derivative, but we use it to learn the basics of Galerkin methods. Later on we add another spatial dimension and time dependence.

First let us construct the spectral solution in the space of trigonometric polynomials. The trial space X_N is spanned on a set of such polynomials (basis functions) ϕ_k of degree $\leq N/2$,

$$\left\{ \phi_k(x) = \mathrm{e}^{\mathrm{i}kx} \; ; \; -N/2 \leq k \leq N/2 - 1 \right\}.$$

The spectral solution $u = u^N$ will then have the form of a finite Fourier series,

$$u(x) = u^N(x) = \sum_{k=-N/2}^{N/2-1} \widehat{u}_k \phi_k(x) = \sum_{k=-N/2}^{N/2-1} \widehat{u}_k \mathrm{e}^{\mathrm{i}kx} . \tag{12.37}$$

We interpret L as L_N, and insert the solution ansatz in the Galerkin condition

$$\left\langle \psi_n, L_N u^N \right\rangle = \langle \psi_n, Q \rangle , \tag{12.38}$$

where

$$\psi_n(x) = \frac{1}{2\pi} \, \mathrm{e}^{-\mathrm{i}nx} \tag{12.39}$$

are the test functions. The trial and the test functions are orthogonal,

$$\langle \psi_n, \phi_k \rangle = \int_0^{2\pi} \psi_n(x) \phi_k^*(x) \, \mathrm{d}x = \delta_{k,n} . \tag{12.40}$$

From the system (12.38) one therefore reaps only the simple relation

$$\widehat{u}_k = \frac{\widehat{Q}_k}{k^2 + \lambda} , \qquad -\frac{N}{2} \leq k \leq \frac{N}{2} - 1 , \tag{12.41}$$

where

$$\widehat{Q}_k = \frac{1}{2\pi} \int_0^{2\pi} Q(x) \, \mathrm{e}^{-\mathrm{i}kx} . \tag{12.42}$$

The value of $\widehat{u}_0$ is arbitrary when $\lambda = 0$ (a non-zero value means just a translation of the solution along the dependent axis). In order to exploit the standard Fourier transformation with even N, we set $\widehat{Q}_{-N/2} = 0$ and thus damp only the highest Fourier component that has no match in $\widehat{Q}_{N/2}$ (as argued on page 770).

We have described the *Fourier–Galerkin method*. It is a "Fourier" method because the trial and test functions are trigonometric polynomials; it is also a "Galerkin" method because the minimization of the residual occurs in the characteristic manner (12.38). For certain functions Q the Fourier coefficients (12.42) can be computed analytically; otherwise numerical integration must be used. An example is discussed in Problem 12.9.1.

12.2.2   Legendre–Galerkin

The Fourier–Galerkin method is applicable to the problem (12.36) with periodic solutions. For the same problem with homogeneous Dirichlet conditions,

$$-\frac{\mathrm{d}^2 v}{\mathrm{d}x^2} + \lambda v = Q\,, \qquad x \in (-1, 1)\,, \qquad v(-1) = v(1) = 0\,, \tag{12.43}$$

the Legendre polynomials also offer a very natural basis. (More general guidelines on choosing the basis functions can be found in Sect. 12.7.) Following [3] one can seek the spectral solution on the interval $[-1, 1]$ in the form

$$u(x) = \sum_{k=2}^{N} \breve{u}_k \phi_k(x)\,, \tag{12.44}$$

where the basis functions are

$$\phi_k(x) = \begin{cases} P_0(x) - P_k(x)\,; & k \text{ even and } \geq 2\,, \\ P_1(x) - P_k(x)\,; & k \text{ odd and } \geq 3 \end{cases} \tag{12.45}$$

(see Fig. 12.6 (left and center)).

Legendre polynomials have the property $P_k(\pm 1) = (\pm 1)^k$, so the particular choice of the basis functions (12.45) satisfies the boundary conditions $u(-1) = u(1) = 0$, even though ϕ_k are *not* orthogonal. Because the spectral solution is expanded in terms of the functions ϕ_k instead of the polynomials P_k, we use the notation $\breve{u}_k$ for the expansion coefficients instead of the usual $\widehat{u}_k$ as in (5.33). We take the test and trial functions to be identical, $\psi_k = \phi_k$. The Galerkin condition then becomes

$$-\left\langle \frac{\mathrm{d}^2 u}{\mathrm{d}x^2}, \phi_l \right\rangle + \lambda \langle u, \phi_l \rangle = \langle Q, \phi_l \rangle = b_l\,, \qquad l = 2, 3, \ldots, N\,, \tag{12.46}$$

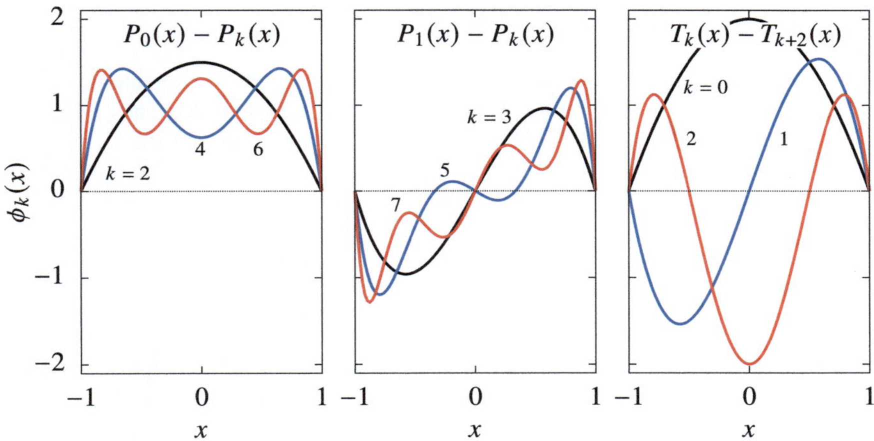

Fig. 12.6 [LEFT, CENTER] The basis functions (12.45) for the Legendre–Galerkin method with $k \in \{2, 4, 6\}$ and $k \in \{3, 5, 7\}$. [RIGHT] The basis functions (12.50) for the Chebyshev–Galerkin method with $k \in \{0, 1, 2\}$

where the scalar products should be understood as

$$\langle u, v \rangle = \int\limits_{-1}^{1} u(x) v(x) \, \mathrm{d}x$$

(Eq. (A.1) with the Legendre weight $w(x) = 1$). After integrating by parts and taking into account the boundary conditions, one obtains

$$-\left\langle \frac{\mathrm{d}^2 u}{\mathrm{d}x^2}, \phi_l \right\rangle = \underbrace{-\left[\phi_l(x) \frac{\mathrm{d}u}{\mathrm{d}x} \right]\Big|_{-1}^{1}}_{0} + \left\langle \frac{\mathrm{d}u}{\mathrm{d}x}, \frac{\mathrm{d}\phi_l}{\mathrm{d}x} \right\rangle .$$

We write the expansion coefficients $\breve{u}_k$ and the expressions for b_l as vectors $\breve{\boldsymbol{u}} = (\breve{u}_2, \breve{u}_3, \ldots, \breve{u}_N)^{\mathrm{T}}$ and $\boldsymbol{b} = (b_2, b_3, \ldots, b_N)^{\mathrm{T}}$. When the solution ansatz is inserted in (12.46), the system can be written as a matrix equation for $\breve{\boldsymbol{u}}$:

$$(K + \lambda M)\breve{\boldsymbol{u}} = \boldsymbol{b} , \tag{12.47}$$

where the matrix elements of K and M are

$$K_{lk} = \left\langle \frac{\mathrm{d}\phi_k}{\mathrm{d}x}, \frac{\mathrm{d}\phi_l}{\mathrm{d}x} \right\rangle = \begin{cases} \min(k,l)\,[\min(k,l)+1] & ; \ k,l \ \text{even and} \geq 2, \\ \min(k,l)\,[\min(k,l)+1] - 2 & ; \ k,l \ \text{odd and} \geq 3, \\ 0 & ; \ \text{otherwise}, \end{cases}$$

$$M_{lk} = \langle \phi_k, \phi_l \rangle = \begin{cases} 2 + \dfrac{2}{2k+1}\,\delta_{k,l} & ; \ k,l \ \text{even and} \geq 2, \\[2mm] \dfrac{2}{3} + \dfrac{2}{2k+1}\,\delta_{k,l} & ; \ k,l \ \text{odd and} \geq 3, \\[2mm] 0 & ; \ \text{otherwise}. \end{cases}$$

(Compute the elements K_{kl} and M_{kl} as an exercise. You only need the basic properties of Legendre polynomials.) Through solving (12.47) we obtain the Legendre–Galerkin spectral approximation of u by Eq. (12.44).

The matrix $K + \lambda M$ is sparse (empty subdiagonals interchange with the full), so the solution of the system (12.47) may be time-consuming and prone to round-off errors. With a different choice of the basis functions,

$$\phi_k(x) = \frac{1}{\sqrt{4k+6}}\left[P_k(x) - P_{k+2}(x)\right], \qquad k \geq 0, \tag{12.48}$$

that fulfill the boundary conditions, and the solution ansatz

$$u(x) = \sum_{k=0}^{N-2} \breve{u}_k \phi_k(x) \tag{12.49}$$

we obtain a tridiagonal system. For details see [2] and Problem 12.9.1.

12.2.3 Chebyshev–Galerkin

Legendre polynomials are not the only option to solve problems of the form (12.43). They can be also treated by Chebyshev polynomials. The spectral approximation is written in the form (12.49), with the basis functions

$$\phi_k(x) = T_k(x) - T_{k+2}(x), \qquad k = 0, 1, \ldots, N-2, \tag{12.50}$$

which again are not orthogonal, but they do fulfill the boundary conditions $u(-1) = u(1) = 0$ due to the property $T_k(\pm 1) = (\pm 1)^k$ (Fig. 12.6 (right)). The Galerkin condition now reads

$$-\left\langle \frac{\mathrm{d}^2 u}{\mathrm{d}x^2}, \phi_l \right\rangle_w + \lambda\,\langle u, \phi_l \rangle_w = \langle Q, \phi_l \rangle_w = b_l, \qquad l = 0, 1, \ldots, N-2,$$

where the scalar products involve the Chebyshev weight:

$$\langle u, v \rangle_w = \int_{-1}^{1} u(x)v(x)w(x)\,dx\,, \qquad w(x) = \frac{1}{\sqrt{1-x^2}}\,.$$

When the expansion coefficients $\breve{u}_k$ and the expressions for b_l at the right-hand side of the Galerkin equation are arranged as vectors $\breve{u} = (\breve{u}_0, \breve{u}_1, \ldots, \breve{u}_{N-2})^{\mathrm{T}}$ and $b = (b_0, b_1, \ldots, b_{N-2})^{\mathrm{T}}$, we again obtain a matrix system of the form (12.47), where

$$K_{lk} = \begin{cases} 2\pi(l+1)(l+2) & ;\ k = l\,, \\ 4\pi(l+1) & ;\ k = l+2,\ l+4,\ l+6,\ \ldots\,, \\ 0 & ;\ k < l \ \text{or}\ k+l = \text{odd}\,, \end{cases}$$

$$M_{lk} = \begin{cases} \frac{\pi}{2}(c_l+1) & ;\ k = l\,, \\ -\frac{\pi}{2} & ;\ k = l \pm 2\,, \\ 0 & ;\ \text{otherwise}\,, \end{cases}$$

where $c_0 = 2$ and $c_k = 1$ for $k \geq 1$. In this case the matrix system is non-symmetric (K is upper triangular and M is tridiagonal). The b_l can be computed numerically, and they can be expressed in terms of the usual Chebyshev coefficients,

$$b_l = \frac{\pi}{2}\left(c_l\widehat{Q}_l - c_{l+2}\widehat{Q}_{l+2}\right)\,, \qquad l = 0, 1, \ldots, N-2\,.$$

Once the coefficients $\breve{u}_k$ are known, the solution is given by the sum (12.49), or else we transform the coefficients,

$$\widehat{u}_k = \begin{cases} \breve{u}_k & ;\ k = 0, 1\,, \\ \breve{u}_k - \breve{u}_{k-2} & ;\ k = 2, 3, \ldots, N-2\,, \end{cases}$$

and apply the expansion in terms of the Chebyshev polynomials,

$$u(x) = \sum_{k=0}^{N-2} \widehat{u}_k T_k(x)\,.$$

Fig. 12.8 shows the error of the Chebyshev–Galerkin method for the problem with the function $Q(x) = (16\pi^2 + \lambda)\sin(4\pi x)$ in comparison to other methods.

12.2.4 Two Space Dimensions

Extending the Galerkin method for our chosen model problem to two space dimensions is straightforward. The Helmholtz equation (12.36) becomes

$$-\nabla^2 v + \lambda v = Q\,, \qquad x \in \Omega = (0, 2\pi) \times (0, 2\pi)\,,$$

and instead of (12.37) the spectral solution is sought in the form

$$u(x, y) = \sum_{kl} \widehat{u}_{kl}\, e^{i(kx+ly)}\,.$$

The test functions, the scalar product (12.40), and the Galerkin condition (12.38) should be generalized to two dimensions as well. Instead of (12.41) we get

$$\widehat{u}_{kl} = \frac{\widehat{Q}_{kl}}{k^2 + l^2 + \lambda}\,, \qquad k, l = -N/2, -N/2 + 1, \ldots, N/2 - 1\,,$$

with the coefficients

$$\widehat{Q}_{kl} = \int_{\Omega} Q(x, y)\, e^{-i(kx+ly)}\, dx\, dy\,.$$

To solve the equation $-\nabla^2 v + \lambda v = Q$ with Dirichlet boundary conditions (two-dimensional generalization of Eq. (12.43)) by Legendre-Galerkin or Chebyshev-Galerkin method, the basis functions (12.45) and (12.50) are used in product,

$$\phi_{kl}(x, y) = \phi_k(x)\phi_l(y)\,.$$

Only the book-keeping becomes slightly more complicated: the expansion coefficients are organized as $\approx N \times N$-dimensional vectors, and the matrices K and M of the corresponding matrix systems blow up to sizes $\approx N^2 \times N^2$. (Recall the vector (11.1) that holds the two-dimensional finite-difference solution.)

12.2.5 Non-stationary Problems

The spectral approach changes substantially when time dependence is included. Following [3], the basic model problem is the linear hyperbolic equation

$$\frac{\partial v}{\partial t} - \frac{\partial v}{\partial x} = 0\,, \qquad x \in (0, 2\pi)\,, \qquad t > 0\,,$$

with a periodic boundary condition $v(0, t) = v(2\pi, t)$ for each t, and the initial condition $v(x, 0) = f(x)$. The ansatz for the solution is

$$u(x, t) = \sum_{k=-N/2}^{N/2-1} \widehat{u}_k(t)\, \phi_k(x)\,, \tag{12.51}$$

where the basis functions $\phi_k(x) = e^{ikx}$ depend on the space coordinate while the expansion coefficients depend on time. We keep (12.39) as test functions.

In evolution problems of the form (12.6) the residual R is minimized by the variational condition (12.7). The Galerkin condition gives

$$\frac{1}{2\pi} \int_0^{2\pi} \left[\left(\frac{\partial}{\partial t} - \frac{\partial}{\partial x} \right) \sum_{k=-N/2}^{N/2-1} \widehat{u}_k(t)\, e^{ikx} \right] e^{-inx}\, dx = 0\,.$$

When the terms $\widehat{u}_k(t)\, e^{ikx}$ are differentiated with respect to t and x, this becomes

$$\frac{1}{2\pi} \int_0^{2\pi} \left[\sum_{k=-N/2}^{N/2-1} \left(\frac{d\widehat{u}_k}{dt} - ik\widehat{u}_k \right) e^{ikx} \right] e^{-inx}\, dx = 0\,.$$

The left-hand side of the equation can be integrated analytically. In fact, due to the orthogonality (12.40), the integration is trivial, yielding a system of differential equations for the coefficients $\widehat{u}_k(t)$,

$$\frac{d\widehat{u}_k}{dt} = ik\widehat{u}_k\,, \qquad -N/2 \le k \le N/2 - 1\,, \tag{12.52}$$

with the initial conditions

$$\widehat{u}_k(0) = \frac{1}{2\pi} \int_0^{2\pi} f(x)\, e^{-ikx}\, dx\,. \tag{12.53}$$

The system (12.52) can be solved by some method for initial-value problems described in Chap. 8. The complete solution at time t is given by the sum (12.51).

How the Fourier–Galerkin method works in practice can be experienced in Problem 12.9.2. In the case of a Dirichlet boundary condition (the equation is only allowed one boundary condition) the expansion in terms of Chebyshev polynomials can be used (the Chebyshev–Galerkin method).

Example In this example of solving time-dependent problems by the Fourier method, an important advantage of spectral approaches over the finite difference methods emerges. We seek the solution of the equation

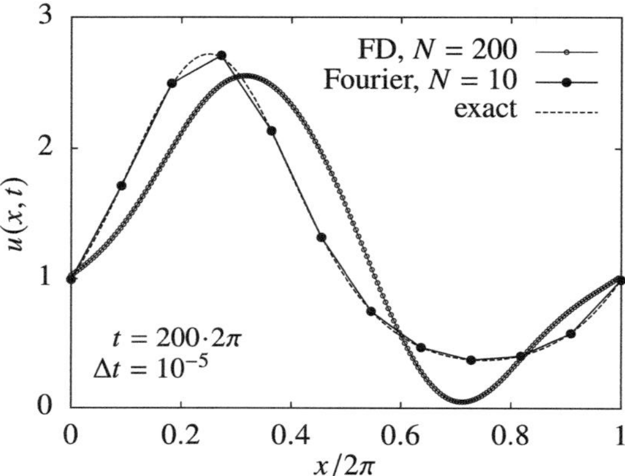

Fig. 12.7 The numerical solution of the equation $v_t = v_x$ with the initial condition $v(x, 0) = \exp(\sin x)$ and periodic boundary conditions. Shown is the solution at $t = 400\pi$ obtained by the explicit Euler method. The finite difference solution (FD) develops typical phase errors in spite of the relatively fine discretization. The solution by the Fourier spectral method exhibits no such error even at very modest N

$$v_t - v_x = 0 \,, \qquad x \in (0, 2\pi) \,, \qquad t > 0 \,,$$

with the boundary condition $v(0, t) = v(2\pi, t)$ and the initial condition $v(x, 0) = \exp(\sin x)$, after long times. The exact solution is $v(x, t) = \exp(\sin(x + t))$. In a finite difference approach, one writes the equation in terms of *local* differences,

$$\frac{u_j^{n+1} - u_j^n}{\Delta t} = \frac{u_{j+1}^n - u_{j-1}^n}{2\Delta x} \,,$$

as in Eqs. (10.7) and (10.3), and then finds the solution at consecutive times $n\,\Delta t$ by, say, the explicit Euler method (8.5). The solution at time 400π on a mesh with $N = 200$ points with time step $\Delta t = 10^{-5}$ is shown in Fig. 12.7 (left). The solution develops an error in both amplitude and phase.

A computation on a very coarse mesh with $N = 10$ using the Fourier collocation derivative (12.9), which approximates the derivative by a *global* sum of basis functions, gives the result shown in Fig. 12.7 (right), all other parameters unchanged. (With $N = 200$ the differences between the numerical and exact solution would no longer be noticeable on this scale.) ◁

If the differential equations are linear, adding terms or increasing the order of derivatives causes only minor complications. For example, the solution of the convection-diffusion problem

$$\frac{\partial v}{\partial t} = c\frac{\partial v}{\partial x} + D\frac{\partial^2 v}{\partial x^2} \,, \qquad x \in (0, 2\pi) \,, \qquad t > 0 \,,$$

with a smooth initial condition $v(x, 0) = f(x)$ and periodic boundary condition, can again be sought in the form (12.51). Instead of the system (12.52) we get

$$\frac{\mathrm{d}\widehat{u}_k}{\mathrm{d}t} = \left(\mathrm{i}kc - k^2 D\right)\widehat{u}_k, \qquad -N/2 \le k \le N/2 - 1, \tag{12.54}$$

with the initial condition (12.53). The $c = 0$ case is discussed in Problem 12.9.3.

Fourier–Galerkin methods work best for linear evolution problems or problems with constant coefficients. They are less suitable for non-linear or variable-coefficient problems, as the resulting systems of differential equations for the evolution coefficients become too complicated, if they can be written down compactly at all. (If you doubt that, write the corresponding systems for the equations $v_t = v v_x$ and $v_t = \mathrm{e}^v v_x$, see also Sect. 12.5 and [4]).

Problem 12.9.3 also teaches us about the usefulness of Galerkin method using orthogonal polynomials. Solving evolution problems by the Legendre–Galerkin or Chebyshev–Galerkin method leads to a matrix system of ordinary differential equations for the evolution coefficients a of the form

$$M\frac{\mathrm{d}a(t)}{\mathrm{d}t} = Sa(t),$$

where the matrix M depends on the chosen basis (and is easy to compute), while S depends on the nature of the differential problem. All basis functions must satisfy the boundary conditions, time-dependent or not. A more flexible treatment of the boundary conditions is offered by tau methods (Sect. 12.3).

12.3 Tau Methods

Tau methods are representatives of weighted residual methods, in which solving the differential equation leads to the minimization (12.5) as in Galerkin methods. But in contrast to Galerkin methods, the test functions in tau methods do not satisfy the boundary conditions of the differential problem; additional equations are needed to impose boundary conditions. (Slightly different interpretations of tau methods can be found in literature; see e.g. Chap. 21 in [12].)

The relatively simple implementation of the boundary conditions—even time-dependent ones—is a major advantage of tau methods. Still, each problem requires us to derive a distinct system of equations for the expansion coefficients of the spectral solution. Tau methods are optimally suited for linear elliptic problems with Chebyshev or Legendre polynomials as test functions.

12.3.1　Stationary Problems

To explain the method we stick to the model problem (12.43) and use Chebyshev polynomials. The expansion of an arbitrary function u on $x \in [-1, 1]$ has the form (5.43). When we insert this expansion at finite N in Eq. (12.5), consider the expression for u'' (Eq. (12.18)) and orthogonality (5.25), we get

$$-\widehat{u}_k^{(2)} + \lambda\widehat{u}_k = \widehat{Q}_k\,, \qquad k = 0, 1, \ldots, N - 2\,. \tag{12.55}$$

The boundary conditions are expressed by the additional equations

$$u(-1) = 0 \Longrightarrow \sum_{k=0}^{N} \widehat{u}_k T_k(-1) = \sum_{k=0}^{N} \widehat{u}_k(-1)^k = 0\,,$$

$$u(1) = 0 \Longrightarrow \sum_{k=0}^{N} \widehat{u}_k T_k(1) = \sum_{k=0}^{N} \widehat{u}_k(1)^k = 0\,,$$

where we have used $T_k(\pm 1) = (\pm 1)^k$. These equations can also be written as

$$\sum_{\substack{k=0 \\ k \text{ even}}}^{N} \widehat{u}_k = 0\,, \qquad \sum_{\substack{k=1 \\ k \text{ odd}}}^{N} \widehat{u}_k = 0\,. \tag{12.56}$$

This is typical of tau methods: when the minimization condition is imposed, the residual $Lu - Q$ is projected on a $(N - N_b)$-dimensional space, where N_b is the number of boundary conditions (here $N_b = 2$). The polynomial approximation u has degree N, but N_b degrees of freedom are reserved for the boundary conditions. The coefficients $\widehat{u}_k^{(2)}$ in (12.55) can be expressed by (12.18), yielding

$$-\frac{1}{c_k} \sum_{\substack{p=k+2 \\ p+k \text{ even}}}^{N} p(p^2 - k^2)\widehat{u}_p + \lambda\widehat{u}_k = \widehat{Q}_k\,, \qquad k = 0, 1, \ldots, N - 2\,, \tag{12.57}$$

where $c_0 = 2$ and $c_k = 1$ for $k \geq 1$, and

$$\widehat{Q}_k = \frac{2}{\pi c_k} \int_{-1}^{1} Q(x) T_k(x) \frac{1}{\sqrt{1 - x^2}}\, \mathrm{d}x\,.$$

Equations (12.57) and (12.56) constitute a matrix system for the coefficients $\widehat{u}_k$. The matrix is upper triangular (with additional rows for the boundary condition) and can be ill-conditioned if λ is small. The method to convert the system to a more robust quasi-tridiagonal form can be found in [3]. The error of the Chebyshev tau method for the Helmholtz problem with $Q(x) = (16\pi^2 + \lambda) \sin(4\pi x)$ is shown in Fig. 12.8.

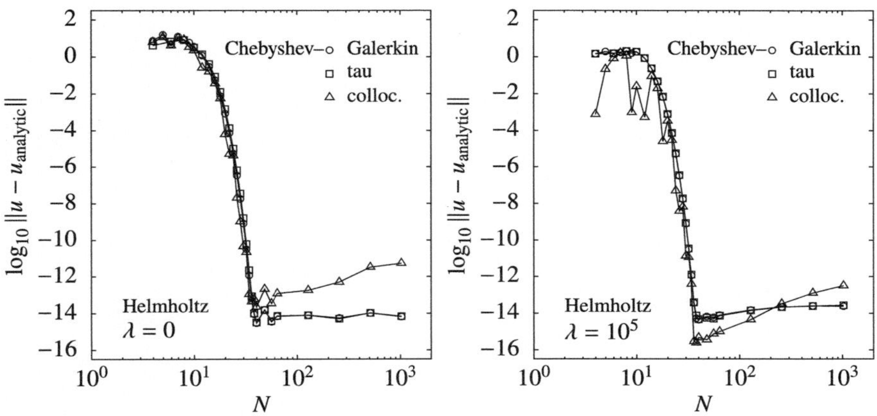

Fig. 12.8 The error of the numerical solution of the Helmholtz problem (12.43) in dependence of the number of basis functions in Chebyshev–Galerkin and Chebyshev tau methods, and in dependence of the number of collocation points in Chebyshev collocation. The function Q corresponds to the exact solution $v(x) = \sin(4\pi x)$. [LEFT] $\lambda = 0$ (Poisson equation). [RIGHT] $\lambda = 10^5$ (a typical value encountered in computations of incompressible flows)

More general (mixed) boundary conditions

$$\alpha_1 v(-1) + \beta_1 v_x(-1) = c_- \,,$$
$$\alpha_2 v(1) + \beta_2 v_x(1) = c_+$$

can be incorporated in the tau method quite unobtrusively: the system of equations for the expansion coefficients $\widehat{u}_k$ for $k = 0, 1, \ldots, N - 2$ is still given by Eq. (12.57), only the last two rows of the matrix change: since

$$T_k(\pm 1) = (\pm 1)^k \,, \qquad T_k'(\pm 1) = (\pm 1)^{k+1} k^2 \,,$$

Equations (12.56) are replaced by

$$\sum_{k=0}^{N} \widehat{u}_k \left(\alpha_1 (-1)^k + \beta_1 (-1)^{k+1} k^2 \right) = c_- \,,$$
$$\sum_{k=0}^{N} \widehat{u}_k \left(\alpha_2 + \beta_2 k^2 \right) = c_+ \,.$$

12.3.2　Non-stationary Problems

Tau methods easily accommodate non-stationary problem as well: time dependence is assigned to the expansion coefficients. As an example we discuss the Chebyshev tau method for the diffusion equation

$$v_t = v_{xx}\,, \qquad x \in (-1, 1)\,, \qquad t > 0\,,$$

with mixed boundary conditions

$$\alpha_1 v(-1, t) + \beta_1 v_x(-1, t) = c_-(t)\,, \tag{12.58}$$

$$\alpha_2 v(1, t) + \beta_2 v_x(1, t) = c_+(t)\,, \tag{12.59}$$

and the initial condition $v(x, 0) = f(x)$. We seek the solution in the form

$$u(x, t) = \sum_{k=0}^{N} \widehat{u}_k(t) T_k(x)\,. \tag{12.60}$$

By Eq. (12.7) we require that the residual $R = u_t - u_{xx}$ is orthogonal to all functions from the space of Chebyshev polynomials of degree $N - 2$, so

$$\frac{2}{\pi c_k} \int_{-1}^{1} \Big(u_t(x, t) - u_{xx}(x, t)\Big) T_k(x) \frac{1}{\sqrt{1 - x^2}}\, \mathrm{d}x = 0\,, \qquad k = 0, 1, \ldots, N - 2\,.$$

The coefficients of the second space derivative are again given by Eq. (12.18), whence a system of linear differential equations for the expansion coefficients emerges,

$$\frac{\mathrm{d}\widehat{u}_k}{\mathrm{d}t} = \frac{1}{c_k} \sum_{\substack{p=k+2 \\ p+k \text{ even}}}^{N} p(p^2 - k^2)\widehat{u}_p(t)\,, \qquad k = 0, 1, \ldots, N - 2\,. \tag{12.61}$$

The initial conditions are

$$\widehat{u}_k(0) = \frac{2}{\pi c_k} \int_{-1}^{1} f(x) T_k(x) \frac{1}{\sqrt{1 - x^2}}\, \mathrm{d}x\,.$$

The system (12.61) yields the first $N - 1$ solution coefficients at time t, while the remaining coefficients $\widehat{u}_{N-1}(t)$ and $\widehat{u}_N(t)$ are given by two supplementary equations for the boundary conditions (12.58) and (12.59), in which again the properties $T_k(\pm 1) = (\pm 1)^k$ and $T_k'(\pm 1) = k^2(\pm 1)^{k+1}$ are exploited:

$$\sum_{k=0}^{N} \widehat{u}_k(t) \left(\alpha_1(-1)^k + \beta_1(-1)^{k+1}k^2 \right) = c_-(t) \,,$$

$$\sum_{k=0}^{N} \widehat{u}_k(t) \left(\alpha_2 + \beta_2 k^2 \right) = c_+(t) \,.$$

The boundary conditions therefore remain detached from the representation of the differential problem. Apart from the time evolution of the coefficients, the method is no harder than in the stationary case.

12.4 Collocation Methods

In collocation (also called pseudospectral) methods—see comment to [13]—the solution u is represented by the values $u(x_j) = u_j$ at specific *collocation points* or *nodes*. In Fourier collocation methods for periodic problems on the interval $[0, 2\pi]$, the nodes are the Fourier points (5.10). In methods with Legendre or Chebyshev polynomials, the location of nodes on the interval $[-1, 1]$ depends on the type of collocation. For example, the points (5.36) are the nodes of Legendre-Gauss-Lobatto collocation, while the points (5.45) are the nodes of Chebyshev-Gauss-Lobatto collocation (see Fig. 12.9). Between the nodes we span the interpolation polynomial $I_N u$, and represent the derivatives of u by exact derivatives of this polynomial.

12.4.1 Stationary Problems

We first consider one-dimensional stationary problems, which can be formulated as differential equations of the form (12.1), where Ω is an interval $[a, b]$ an L is a linear differential operator, while v and Q are functions. The collocation method amounts to expanding the approximate solution u as in (12.4) and applying delta-"functions" in the test condition (12.5) evaluated by the scalar product with the appropriate weight function.

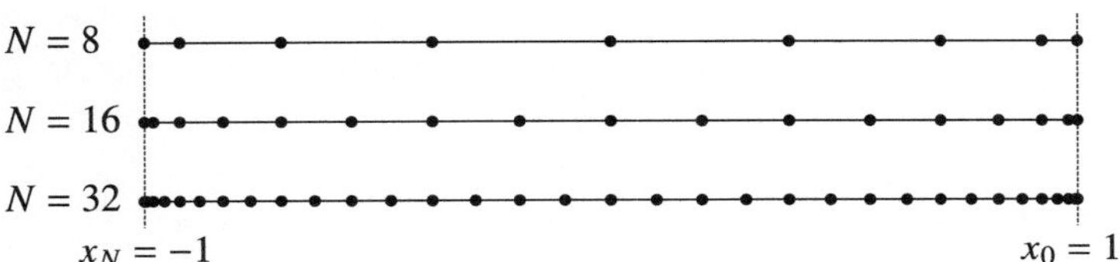

Fig. 12.9 Chebyshev–Gauss–Lobatto nodes (5.45) on the interval $[x_N, x_0] = [-1, 1]$ for $N = 8$, 16, and 32. (See also Fig. 5.12.)

Let us keep the Helmholtz periodic problem (12.36) as our baseline example, closely following [13]. The Fourier collocation solution u is given by the equations

$$\left[-\frac{\mathrm{d}^2 u}{\mathrm{d}x^2} + \lambda u - Q\right]\bigg|_{x=x_j} = 0, \qquad j = 0, 1, \ldots, N - 1, \tag{12.62}$$

where N is even and x_j are the Fourier collocation points (5.10) with the indices running from 0 to $N - 1$. This is equivalent to the requirement for the minimization of the residual (12.5) with $L_N = L = -\mathrm{d}^2/\mathrm{d}x^2 + \lambda$, test functions $\psi_n(x) = \delta(x - x_n)$, and the usual scalar product on $[0, 2\pi]$. When the discrete Fourier transform (5.12) is inserted in the equations above, we get

$$k^2 \widetilde{u}_k + \lambda \widetilde{u}_k = \widetilde{Q}_k, \qquad -\frac{N}{2} \leq k \leq \frac{N}{2} - 1,$$

and

$$\widetilde{Q}_k = \frac{1}{N} \sum_{j=0}^{N-1} Q(x_j) \, \mathrm{e}^{-ikx_j},$$

which closely resembles (but is not identical) to Eqs. (12.41) and (12.42). In practice, we evaluate Q at the Fourier collocation points x_j and use the transformation (5.11), in FFT form, to compute the coefficients $\widetilde{Q}_k$, whence we obtain $\widetilde{u}_k = \widetilde{Q}_k/(k^2 + \lambda)$. Finally, the transformation (5.12) is applied, again as FFT, to turn $\widetilde{u}_k$ into the function values $u(x_j)$ which represent our final solution. If we wish to know the values $u(x)$ on the subintervals *between* the collocation points, the complete Lagrange interpolation polynomial can be plotted.

The problem with Dirichlet boundary conditions can be handled just as efficiently. Consider the case of Chebyshev collocation with Gauss–Lobatto nodes. The form of the collocation condition (12.62) remains unchanged, but x_j are now given by Eq. (5.45), resulting in the system of equations

$$\sum_{k=0}^{N} \left[-\left(D_N^{(2)}\right)_{jk} + \lambda \delta_{j,k}\right] u(x_j) = Q(x_j),$$

where the matrix $D_N^{(2)}$ represents the Chebyshev collocation second derivative (see Eq. (12.21)). If the solution u and the right-hand sides of the equation Q at the collocation points are arranged in the vectors

$$\boldsymbol{u} = \left(u_0, u_1, \ldots, u_N\right)^{\mathrm{T}}, \qquad \boldsymbol{Q} = \left(Q_0, Q_1, \ldots, Q_N\right)^{\mathrm{T}},$$

with components $u_j = u(x_j)$ and $Q_j = Q(x_j)$, respectively, we obtain, in short,

$$Z_N\left[-D_N^{(2)} + \lambda I\right]\boldsymbol{u} = \boldsymbol{Q}.$$

Dirichlet boundary conditions are imposed by setting the first and last column of the system matrix to zero: this is accomplished symbolically by the matrix Z_N in the above expression. This is a typical feature of collocation methods: one sacrifices as many collocation nodes as needed to accommodate the boundary conditions. We solve the system by classical methods (the system matrix is sparse). An alternative to matrix multiplication leads through Chebyshev transformation and FFT (Sect. 12.1.3). The error of the Chebyshev collocation method for the Helmholtz problem with $Q(x) = (16\pi^2 + \lambda)\sin(4\pi x)$ is shown in Fig. 12.8.

In analogous fashion the collocation method with differentiation matrices can be applied to any linear differential equation $Lv = Q$, where the differential operator has the structure

$$L = \sum_m L_m(x)\frac{\mathrm{d}^m}{\mathrm{d}x^m} \ .$$

(For non-linear equations see Sect. 12.5.) Given the collocation points (nodes) x_j corresponding to the chosen orthogonal polynomials (or other orthogonal basis functions) and type of quadrature, the differential equation is translated to a system of linear equations

$$A\boldsymbol{u} = \boldsymbol{Q} \ , \tag{12.63}$$

which, by components, is given by

$$L_0(x_j)u_j + L_1(x_j)\sum_{k=0}^{N}\left(D_N^{(1)}\right)_{jk}u_k + L_2(x_j)\sum_{k=0}^{N}\left(D_N^{(2)}\right)_{jk}u_k + \cdots = Q_j \ ,$$

where $j = 0, 1, \ldots, N$. The system can be solved by standard methods. Depending on the basis functions used, however, the system matrix A may or may not possess any favorable properties: it may be full, non-symmetric, and so on. A general approach to the construction of such collocation methods (sometimes also called "discrete ordinate methods") is described in Ref. [14].

Alternatively, a pseudospectral discretization of the linear differential equation based on orthogonal polynomials or other basis functions ϕ_k can also be devised by starting from the discrete expansion of the solution u in the manner of (5.27), so that

$$u(x_j) = \sum_{k=0}^{N}\tilde{u}_k\phi_k(x_j) \ , \quad j = 0, 1, \ldots, N \ .$$

(It is easy to check that in the Fourier case, collocation and pseudospectral methods are algebraically identical.) This results in the system of linear equations for the expansion coefficients that can be written as:

$$A\tilde{\boldsymbol{u}} = \boldsymbol{Q} \ , \tag{12.64}$$

where

$$A_{jk} = \left(L\phi_k(x)\right)\big|_{x=x_j} \; ; \quad j, k = 0, 1, \ldots, N \; ,$$
$$Q_j = Q(x_j) \, , \qquad\quad ; \quad j = 0, 1, \ldots, N \; .$$

(Recall that in the Galerkin approach discussed in Sect. 12.2 the structure of the system (12.64) remains the same, while A and $\boldsymbol{Q}$ are given by

$$A_{jk} = \langle\phi_j, L\phi_k\rangle_w \; ; \quad j, k = 0, 1, \ldots, N \; ,$$
$$Q_j = \langle\phi_j, Q\rangle_w \, , \; ; \quad j = 0, 1, \ldots, N \; ,$$

and the inner products must be taken with the suitable weight w.) Eigenvalue problems may be solved in a similar manner; illustrations of the collocation method on $[0, \infty]$ and $[-\infty, \infty]$ are given in Sects. 12.7.4 and 12.7.7, respectively.

12.4.2   Non-stationary Problems

A typical non-stationary periodic problem is the advection-diffusion equation

$$v_t = c(x)v_x + D(x)v_{xx} \, , \qquad x \in (0, 2\pi) \, , \qquad t > 0 \, ,$$

with the initial condition $v(x, 0) = f(x)$ and the boundary condition $v(0, t) = v(2\pi, t)$. The Fourier collocation solution can be sought in the form [4]

$$u(x, t) = \sum_{j=0}^{N-1} u(x_j, t)g_j(x) \, , \tag{12.65}$$

where g_j is the Lagrange interpolation polynomial with the property $g_j(x_i) = \delta_{i,j}$ (see Eq. (5.15) and Fig. 5.2). As in the stationary case we require that the residual $u_t - cu_x - Du_{xx}$ is zero at the Fourier collocation points (5.10). This requirement leads to the system of differential equations

$$\frac{\mathrm{d}u(x_j, t)}{\mathrm{d}t} = c(x_j)\, I_N \frac{\partial}{\partial x} \left[I_N u(x_j, t)\right] + D(x_j)\, I_N \frac{\partial^2}{\partial x^2} \left[I_N u(x_j, t)\right]$$
$$= \sum_{k=0}^{N-1} \left[c(x_j) \left(D_N^{(1)}\right)_{jk} + D(x_j) \left(D_N^{(2)}\right)_{jk}\right] u(x_k, t) \, ,$$

describing the time evolution of $u(x_j, t)$. The first row of the equation fulfills the promises of the introduction to this Section: the derivatives of a function are represented by exact derivatives of its interpolation polynomial. The second row contains the instructions on how to solve the system. The derivatives at the right can be evaluated by direct multiplication of matrices (12.11) and (12.12) by the vector $\boldsymbol{u}(t)$, but one is better off by using the discrete Fourier transformation: we first transform the values $u(x_j, t)$ to Fourier space, multiply the obtained coefficients $\widetilde{u}_k(t)$ by ik and $(ik)^2$ (for first and second derivative, respectively), and map the computed products back to configuration space, where we ultimately multiply them by $c(x_j)$ or $D(x_j)$, and the right-hand side is ready to go. Finally, we follow the time evolution of the values $u(x_j, t)$ with the initial condition

$$u(x_j, 0) = f(x_j)$$

by using methods for initial-value problems for systems of ordinary differential equations. The examples of solving the diffusion equation (the above problem with $c = 0$) can be found in Problems 12.9.6 and 12.9.7. Collocation methods are ideally suited for the solution of non-linear problems: this is discussed in Sect. 12.5.

12.4.3 Spectral Elements: Collocation with B-Splines

In the ansatz for the collocation solution one is not limited to trigonometric functions or orthogonal polynomials. A special class of methods is based on approximating the solution by a sum of localized low-degree polynomials. Instead of global, local functions can be used that are defined only on specific parts of the domain, and zero elsewhere. We are referring to *spectral element methods* [15]. For the lowest polynomial degrees (linear functions) they are identical to finite element methods described in Sect. 11.6.

In the following we present a method in which the spectral elements are cubic *B-splines*. Because the solution approximation is expressed by values at characteristic points (maxima) of these polynomial splines, we in fact still witness "collocation". As an example, again take the diffusion equation $v_t = Dv_{xx}$ on the interval $x \in [0, L]$ with Dirichlet boundary conditions $v(0, t) = v(L, t) = 0$ (see also Problem 12.9.6). The solution is sought in the form [16]

$$u(x, t) = \sum_{k=-1}^{N+1} a_k(t) B_k(x) \,, \tag{12.66}$$

where B_j is the cubic spline centered at $x = x_j$ on the mesh $x_j = j\Delta x$ ($j = 0, 1, \ldots, N$ and $\Delta x = L/N$). This special class of basis functions already appeared in Chap. 9 on scalar boundary-value problems with ordinary differential equations (see Eq. (9.64), Fig. 9.9, and Sect. 9.5 in general). We require that the expansion (12.66) satisfies the differential equation and the boundary conditions. Inserting the expansion (12.66) in the differential equation and evaluating the result at $x = x_j$, we get

$$\sum_{k=-1}^{N+1} \frac{\mathrm{d}a_k(t)}{\mathrm{d}t} B_k(x_j) = D \sum_{k=-1}^{N+1} a_k(t) B_k''(x_j) , \qquad j = 0, 1, \dots, N ,$$

where $'$ denotes the derivative with respect to x. When properties of B-splines are accounted for, one ends up with a system of differential equations for the coefficients $a_j(t)$:

$$\frac{\mathrm{d}}{\mathrm{d}t} \left[a_{j-1}(t) + 4a_j(t) + a_{j+1}(t) \right] = \frac{6D}{\Delta x^2} \left(a_{j-1}(t) - 2a_j(t) + a_{j+1}(t) \right) ,$$

where $j = 0, 1, \dots, N$. From the boundary condition at $x = 0$ we determine $a_{-1} = -4a_0 - a_1$. When the above equation is used at $j = 0$, it also follows that $a_0 = 0$ and $a_{-1} = -a_1$. Similarly, at $x = L$ we see that $a_N = 0$ and $a_{N+1} = -a_{N-1}$, and end up with a system of linear equations

$$A \frac{\mathrm{d}\boldsymbol{a}}{\mathrm{d}t} = B\boldsymbol{a} ,$$

where $\boldsymbol{a}(t) = (a_1(t), a_2(t), \dots a_{N-1}(t))^{\mathrm{T}}$ and

$$A = \begin{pmatrix} 4 & 1 & & & \\ 1 & 4 & 1 & & \\ & \ddots & \ddots & \ddots & \\ & & 1 & 4 & 1 \\ & & & 1 & 4 \end{pmatrix} , \qquad B = \frac{6D}{\Delta x^2} \begin{pmatrix} -2 & 1 & & & \\ 1 & -2 & 1 & & \\ & \ddots & \ddots & \ddots & \\ & & 1 & -2 & 1 \\ & & & 1 & -2 \end{pmatrix} .$$

The initial condition for the differential equation is $u(x_j, 0) = g(x_j)$, so the initial approximation for the collocation approximation is

$$A\boldsymbol{a}(0) = 6\boldsymbol{g} ,$$

where $\boldsymbol{g} = (g(x_1), g(x_2), \dots, g(x_{N-1}))^{\mathrm{T}}$. The coefficients $a_k(t)$ can then be evolved in time by using some method discussed in Chap. 8. The solution at arbitrary time and location is given by the sum (12.66).

12.5 Non-linear Equations

The spectral expansion coefficients in Galerkin and tau methods result from solving systems of algebraic equations and systems of differential equations, respectively. For differential equations with general boundary conditions, non-constant coefficients, and in particular for non-linear equations, these systems become cumbersome or even impossible to write down (e.g. Fourier–Galerkin treatment of $v_t = \mathrm{e}^v v_x$).

Non-linear problems are therefore most frequently solved by collocation (pseudospectral) methods. Galerkin and tau methods do turn out to be effective for certain non-linear problems, although complications appear that are not present when solving linear problems.

Burgers equation In the following we describe the typical approaches to non-linear problems in the case of the Burgers equation

$$Lv = \frac{\partial v}{\partial t} + v\,\frac{\partial v}{\partial x} - D\,\frac{\partial^2 v}{\partial x^2} = 0\,, \qquad x \in \Omega\,, \qquad t > 0\,,$$

where D is a positive constant and $v(x,0) = f(x)$ is the initial condition. We are interested in periodic solutions on $\Omega = [0, 2\pi]$ and in non-periodic solutions with Dirichlet boundary conditions on $\Omega = [-1, 1]$. For the periodic problem both Fourier–Galerkin and Fourier collocation methods are suitable; for the Dirichlet problem we will use the Chebyshev tau and collocation methods.

Fourier–Galerkin method In this method the approximation u of the exact solution v has the form (12.51). When the Galerkin condition $\langle Lu, \mathrm{e}^{-ikx}\rangle = 0$ is enforced and the equation for the Fourier coefficients (5.7) is used, we obtain a system of differential equations for the coefficients $\widehat{u}_k(t)$:

$$\frac{\mathrm{d}\widehat{u}_k}{\mathrm{d}t} + \left(\widehat{u\,\frac{\partial u}{\partial x}}\right)_k + Dk^2\widehat{u}_k = 0\,, \qquad -N/2 \le k \le N/2 - 1\,. \tag{12.67}$$

The corresponding initial conditions are

$$\widehat{u}_k(0) = \frac{1}{2\pi} \int\limits_0^{2\pi} f(x)\,\mathrm{e}^{-ikx}\,\mathrm{d}x\,. \tag{12.68}$$

The diffusion term $-Dv_{xx}$ is linear and turns to $Dk^2\widehat{u}_k$ in transform space, as expected from Eq. (12.8), while the non-linear advection term vv_x generates the characteristic "compound coefficient"

$$\left(\widehat{u\,\frac{\partial u}{\partial x}}\right)_k = \frac{1}{2\pi} \int\limits_0^{2\pi} u\,\frac{\partial u}{\partial x}\,\mathrm{e}^{-ikx}\,\mathrm{d}x\,. \tag{12.69}$$

This is a special case of the more general coefficient

$$(\widehat{uv})_k = \frac{1}{2\pi} \int\limits_0^{2\pi} u(x)\,v(x)\,\mathrm{e}^{-ikx}\,\mathrm{d}x\,,$$

where u and v are trigonometric polynomials. By using discrete Fourier expansion and the orthogonality of basis functions, one immediately realizes

$$(\widehat{uv})_k = \frac{1}{2\pi} \int_0^{2\pi} \sum_l \widehat{u}_l\, \mathrm{e}^{ilx} \sum_m \widehat{u}_m\, \mathrm{e}^{imx} \mathrm{e}^{-ikx} = \sum_{l,m} \widehat{u}_l\, \widehat{u}_m\, \delta_{l+m,k} = \sum_{l+m=k} \widehat{u}_l\, \widehat{v}_m \ ,$$

which is nothing but a convolution of discrete functions. A naive evaluation of this formula requires $\mathcal{O}(N^2)$ operations, while a computation exploiting FFT requires only $\mathcal{O}(N \log_2 N)$ (see Sects. 7.6 and 3.4 in [3]). This logarithmic discount is particularly relevant in two- and three-dimensional problems.

Another trap is hidden in the terms with non-constant coefficients in the differential equation (e.g. $a(x)v_x$) or in the non-linear terms (e.g. vv_x). If high Fourier modes are present in the expansions of the functions a and v, the terms $a(x)v_x$ and vv_x correspond to even higher modes, which may result in aliasing effects (Sect. 5.2.3). For example, if $v(x) = \sin kx$, the non-linear term becomes $vv_x = (k/2)\sin 2kx$: we witness frequency doubling. Aliasing can be removed by two relatively simple tricks described in Sect. 3.4 of [3].

Chebyshev tau method The tau method involving Chebyshev polynomials can be applied to the Burgers equation with Dirichlet boundary conditions $v(-1, t) = v_-(t)$ and $v(1, t) = v_+(t)$, and the initial condition $v(x, 0) = f(x)$. The solution is sought in the form (12.60). We require that the expression $u_t + uu_x - Du_{xx}$ is orthogonal to all test functions T_k, thus

$$\int_{-1}^{1} \left(\frac{\partial u}{\partial t} + u\,\frac{\partial u}{\partial x} - D\,\frac{\partial^2 u}{\partial x^2} \right) T_k(x)\, \frac{1}{\sqrt{1-x^2}}\, \mathrm{d}x = 0\ , \qquad k = 0, 1, \ldots, N-2\ .$$

Think: why does the index k run only up to $N - 2$? We obtain a system of linear differential equations for the expansion coefficients,

$$\frac{\mathrm{d}\widehat{u}_k}{\mathrm{d}t} + \left(\widehat{u\,\frac{\partial u}{\partial x}} \right)_k - D\widehat{u}_k^{(2)} = 0\ , \qquad k = 0, 1, \ldots, N-2\ ,$$

where $\widehat{u}_k^{(2)}$ is given by (12.18). This system for $N - 1$ coefficients is supplemented by the relations expressing the boundary conditions

$$\sum_{k=0}^{N} \widehat{u}_k(t) = v_-(t)\ , \qquad \sum_{k=0}^{N} (-1)^k \widehat{u}_k(t) = v_+(t)\ ,$$

as well as by the initial condition for all $N + 1$ coefficients,

$$\widehat{u}_k(0) = \frac{2}{\pi c_k} \int\limits_{-1}^{1} f(x)\, T_k(x)\, \frac{1}{\sqrt{1-x^2}}\, dx\,, \qquad k = 0, 1, \ldots, N\,.$$

The contribution of the non-linear term,

$$\left(\widehat{u\, \frac{\partial u}{\partial x}}\right)_k = \frac{2}{\pi c_k} \int\limits_{-1}^{1} \left(u\, \frac{\partial u}{\partial x}\right) T_k(x)\, \frac{1}{\sqrt{1-x^2}}\, dx\,,$$

is again a special case of an expression that is evaluated as the convolution sum,

$$(\widehat{uv})_k = \frac{2}{\pi c_k} \int\limits_{-1}^{1} u(x)\, v(x)\, T_k(x)\, \frac{1}{\sqrt{1-x^2}}\, dx = \frac{1}{2} \sum_{l+m=k} \widehat{u}_l\, \widehat{v}_m + \sum_{|l-m|=k} \widehat{u}_l\, \widehat{v}_m\,.$$

Effective means to compute this sum and protect against aliasing are discussed by [3] in Sect. 3.4. With some effort (and by resorting to recurrence relations for Chebyshev polynomials) even the non-linear term can be expressed with the coefficients $\widehat{u}_k^{(1)}$ from Eq. (12.17), and thus the evaluation of convolutions can be avoided: see Sect. 7.2 in [4].

Fourier collocation The collocation solution of the Burgers equation with the initial condition $v(x, 0) = f(x)$ and periodic boundary conditions is represented by the values at the collocation points $x_j = 2\pi j/N$, where $j = 0, 1, \ldots, N-1$, and by its interpolation (12.65) if needed. The collocation requirement is

$$\left[\frac{\partial u}{\partial t} + u\, \frac{\partial u}{\partial x} - D\, \frac{\partial^2 u}{\partial x^2}\right]\bigg|_{x=x_j} = 0\,, \qquad j = 0, 1, \ldots, N-1\,,$$

while the initial conditions are $u(x_j, 0) = f(x)$. If the solution components are arranged in the vector $\boldsymbol{u} = (u_0(t), u_1(t), \ldots, u_{N-1}(t))^{\mathrm{T}}$, the collocation condition translates to a system of equations

$$\frac{d\boldsymbol{u}}{dt} + \boldsymbol{u} \otimes D_N^{(1)} \boldsymbol{u} - D D_N^{(2)} \boldsymbol{u} = \boldsymbol{0}\,,$$

where $D_N^{(1)}$ and $D_N^{(2)}$ are the Fourier collocation derivative matrices (Eqs. (12.11) and (12.12)). The corresponding contributions can be computed directly by matrix multiplication, or by transformation to Fourier space, multiplication by (ik) or $(-k^2)$, followed by the inverse transformation (see Sect. 12.1.1).

Chebyshev collocation In a similar manner one can tackle the Burgers equation with Dirichlet boundary conditions by Chebyshev collocation. We represent the solution by the values at the Gauss–Lobatto collocation points (5.45) and require that the differential equation is fulfilled at those points,

$$\left[\frac{\partial u}{\partial t} + u\,\frac{\partial u}{\partial x} - D\,\frac{\partial^2 u}{\partial x^2} \right]\Bigg|_{x=x_j} = 0\,, \qquad j = 1, 2, \ldots, N-1\,.$$

This requirement is supplemented by the boundary conditions $u(x_0, t) = u_+(t)$ and $u(x_N, t) = u_-(t)$, and the initial condition

$$u(x_j, 0) = f(x_j)\,, \qquad j = 0, 1, \ldots, N\,.$$

When the solution is arranged in the vector $\boldsymbol{u} = (u_0(t), u_1(t), \ldots, u_N(t))^{\mathrm{T}}$, the collocation condition gives the system of equations

$$Z_N \left(\frac{\mathrm{d}\boldsymbol{u}}{\mathrm{d}t} + \boldsymbol{u} \otimes D_N^{(1)} \boldsymbol{u} - D D_N^{(2)} \boldsymbol{u} \right) = \boldsymbol{0}\,,$$

where $D_N^{(1)}$ and $D_N^{(2)}$ are the first and second Chebyshev collocation derivative matrices (see Eqs. (12.20) and (12.21)). Here Z_N is the matrix that, upon multiplication with a vector, sets its first and last component to zero.

12.6   Time Integration $\star$

Regardless of the chosen approach all spectral representations of the spatial operator in a partial differential equation result in a system of ordinary differential equations of the form

$$\frac{\mathrm{d}\boldsymbol{u}}{\mathrm{d}t} = \boldsymbol{L}\big(\boldsymbol{u}(x, t), x, t\big)\,, \tag{12.70}$$

where in general $\boldsymbol{L}$ is a non-linear function of $\boldsymbol{u}$. We know from Chap. 8 (see in particular Sect. 8.5) that the integration of such systems calls for a stability analysis that is usually performed on the linearized (or locally linear) version of the problem,

$$\frac{\mathrm{d}\boldsymbol{u}}{\mathrm{d}t} = L\boldsymbol{u}(x, t)\,, \tag{12.71}$$

where $\boldsymbol{u}$ is an $M(N-1)$-dimensional vector containing the $N+1$ expansion coefficients (in the case of a Galerkin or tau method) or the values of $\boldsymbol{u}$ at the grid points (in the case of a collocation method), and L is an $M(N-1) \times M(N-1)$ matrix. The exact solution of (12.71) is given by $\boldsymbol{u}(x, t) = \mathrm{e}^{Lt}\boldsymbol{u}(x, 0)$, yet even though the matrix exponential can be calculated—say, by the method described in Appendix A.9—the exact time propagation is usually replaced by a single-step finite-difference approximation

$$\boldsymbol{u}^{n+1} = S(L, \Delta t)\boldsymbol{u}^n\,,$$

where $S(L, \Delta t)$ is some approximation of e^{Lt}, typically Taylor or Padé of chosen order. The choice of the time step size Δt used in the difference schemes for such systems is closely related to the spectrum (the set of eigenvalues) of matrices representing the spectral derivatives contained in L. A discrete temporal scheme is said to be *strongly stable* when $\|[S(L, \Delta t)]^n\|_{L^2_w} \le B(\Delta t)$, where $\| \cdot \|_{L^2_w}$ is the induced matrix norm of the kind (A.7) with the appropriate weight function w in the corresponding vector norm. A sufficient condition for strong stability is [4]

$$\|S(L, \Delta t)\|_{L^2_w} \le 1 + b\Delta t$$

for some bounded b and sufficiently small Δt, although in practice a more restrictive condition with $b = 0$ may be preferred. The eigenvalues of $S(L, \Delta t)$ provide conditions for stability which, however, are only necessary in general. In the special case when S is a normal matrix ($S^T S = S S^T$), strong stability is ensured if

$$\rho(S) \le 1 + b\Delta t \, . \tag{12.72}$$

The explicit form of L and its spectrum depend on the type of the spectral method. The spectrum of the matrix (12.11) for the Fourier collocation derivative is

$$\{i(-N/2 + 1), \ldots, -i, 0, i, \ldots, i(N/2 - 1)\} \, ,$$

while the spectrum of the matrix (12.12) for the second derivative is

$$\left\{ (-N/2 + 1)^2, \ldots, -1, 0, 1, \ldots, (N/2 - 1)^2 \right\} \, .$$

The spectral radius for the mth derivative is

$$\rho\left(D_N^{(m)} \right) = \max \left| \lambda^{(m)} \right| = \left(\frac{N}{2} - 1 \right)^m = \mathcal{O}(N^m) \, . \tag{12.73}$$

Since the differentiation matrices in the Fourier collocation method are all normal, discrete stability for a linear problem is guaranteed if $\rho(S)$ can be bounded as in (12.72). Moreover, from (12.73) we read off that the step size Δt needed for a stable temporal integration of Fourier approximations of advection problems (terms with v_x) scales as $\Delta t \sim 1/N$, and for diffusion problems (terms with v_{xx}) as $\Delta t \sim 1/N^2$: an increase of N, the number of Fourier basis functions (or collocation points), requires a corresponding decrease of Δt. The restriction in the Fourier–Galerkin method is essentially the same.

The limitations in spectral methods based on orthogonal polynomials are more severe: for explicit integration of Chebyshev or Legendre approximation of advection problems the step size must scale as $\Delta t \sim 1/N^2$, while for diffusion problems $\Delta t \sim 1/N^4$. This behavior can be seen in Fig. 12.10 (left). It shows the spectrum of the differential operator d/dx, i.e. the set of eigenvalues of the matrix $D_N^{(1)}$ for the Chebyshev collocation derivative with Gauss–Lobatto nodes (Eq. (12.20)) and

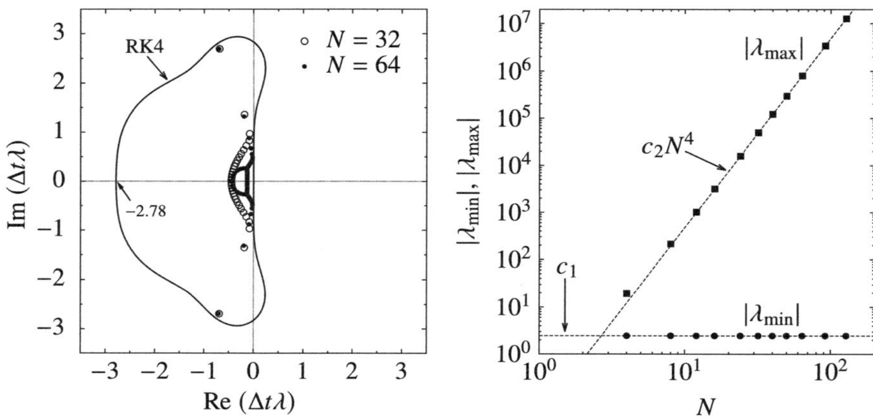

Fig. 12.10 [LEFT] The rescaled spectrum of the differential operator d/dx in Chebyshev–Gauss–Lobatto collocation, embedded in the absolute stability region of the RK4 method. [RIGHT] Minimal and maximal eigenvalues of the operator $-\mathrm{d}^2/\mathrm{d}x^2$

boundary condition $u(x_0) = 0$ (the first row and first column in the matrix are zero). The eigenvalues are rescaled such that the maximum eigenvalue of $D_N^{(1)}$ remains in the absolute stability region of the RK4 method (see also Fig. 8.3). These run-away eigenvalues that asymptotically behave as $\lambda \sim \mathcal{O}(N^2)$ force us to advance in very short time steps Δt. Similarly, Fig. 12.10 (right) shows the eigenvalues of the differential operator $-\mathrm{d}^2/\mathrm{d}x^2$ represented by the Chebyshev collocation second derivative matrix $-D_N^{(2)}$. Its eigenvalues are real and positive. They are bounded by $0 < c_1 \leq \lambda \leq c_2 N^4$, so they asymptotically increase as $\lambda \sim \mathcal{O}(N^4)$.

We know from Sect. 10.7.1 that explicit time integration schemes for problems of the form (12.70) are inherently limited by an unphysical, purely computational phenomenon known as the Courant–Friedrichs–Lewy (CFL) instability: roughly speaking, the maximum time step $(\Delta t)_{\mathrm{max}}$ for stable integration is going to be of the same order of magnitude as the time scale for advection (or wave propagation, or diffusion, or any other mechanism) across the smallest distance $(\Delta x)_{\mathrm{min}}$ between two grid points, so that, for instance, $c_{\mathrm{max}}(\Delta t)_{\mathrm{max}}/(\Delta x)_{\mathrm{min}} = \mathcal{O}(1)$ for advection and $(\Delta t)_{\mathrm{max}}/\nu(\Delta x)_{\mathrm{min}}^2 = \mathcal{O}(1)$ for diffusion.

Such severe restrictions on the time step in spectral methods therefore call for either explicit or implicit integrators of orders high enough to exhibit good stability properties. These tools are known as *strong stability preserving methods*.

12.6.1 *Strong Stability Preserving Methods*

Strong stability preserving (SSP) methods are a tool to solve certain classes of large systems of ordinary differential equations of the form (12.70), most commonly those arising from the spatial discretization of hyperbolic PDEs. Many of the ODE integrators discussed in Chap. 8 can be used to solve such systems of equations, but if the solutions are known or presumed to exhibit discontinuities—see, for instance, Figs. 10.15 and 10.17—one is obliged to think of a reliable way of propagating them in time with step sizes as large as possible and without losing stability.

Suppose, then, that a hyperbolic conservation law of the form $\boldsymbol{u}_t + [\boldsymbol{F}(\boldsymbol{u})]_x = \boldsymbol{0}$ has been rewritten as (12.70), where $\boldsymbol{L}$ represents the spatial discretization such that $\boldsymbol{L}(\boldsymbol{u}) = -[\boldsymbol{F}(\boldsymbol{u})]_x$. We are interested in strongly stable solutions, meaning that in some norm $\|\cdot\|$ the sequence $\{\boldsymbol{u}^n\}$ satisfies $\|\boldsymbol{u}^{n+1}\| \le \|\boldsymbol{u}^n\|$ for all $n \ge 0$. Let us assume that the discretization $\boldsymbol{L}$ has the property that when it is combined with the first-order explicit Euler (EE) method,

$$\boldsymbol{u}^{n+1} = \boldsymbol{u}^n + \Delta t\, \boldsymbol{L}(\boldsymbol{u}^n)\,,$$

then, for a sufficiently small time step dictated by the CFL condition, $\Delta t \le (\Delta t)_{\mathrm{EE}}$, the solution is total variation diminishing (TVD), behaving in the manner defined by Eq. (10.73). One way to preserve this feature is to refine the spatial discretization and resort to high-resolution schemes of Sect. 10.11: they retain the simplicity of EE time-stepping and still preserve the nonlinear stability properties for approximating discontinuous solutions. Often, however, high-order temporal propagation is required, and in this case there is no guarantee that a spatial discretization which is strongly stable for a nonlinear problem under EE integration will retain this property when coupled with a linearly stable higher order temporal discretization. The objective of SSP discretizations is precisely to maintain nonlinear stability (the TVD feature) while attaining higher order accuracy in time, possibly with a modified CFL condition

$$\Delta t \le C(\Delta t)_{\mathrm{EE}}\,, \tag{12.74}$$

where C is known as the CFL coefficient.

Stark reminder Let us illustrate the issues at hand by an example where we use a sophisticated spatial discretization and a high-order time-propagation scheme, and still end up with instabilities. We consider the simplest case of a *linear* one-dimensional advection equation $v_t + v_x = 0$ for $x \in [0, 2]$ with the initial condition

$$v(x, 0) = \begin{cases} 1\,; & 0 \le x < 0.5 \text{ or } 1.5 < x \le 2\,, \\ 0\,; & 0.5 \le x \le 1.5\,, \end{cases} \tag{12.75}$$

as shown in Fig. 12.11, and periodic boundary condition. Thus $[F(v)]_x = v_x$, and the spatial discretization is furnished by the fifth-order WENO–JS5 scheme of Eq. (10.99), while the time propagation is accomplished by a 5-stage, third-order Runge–Kutta method RK(5,3) given by

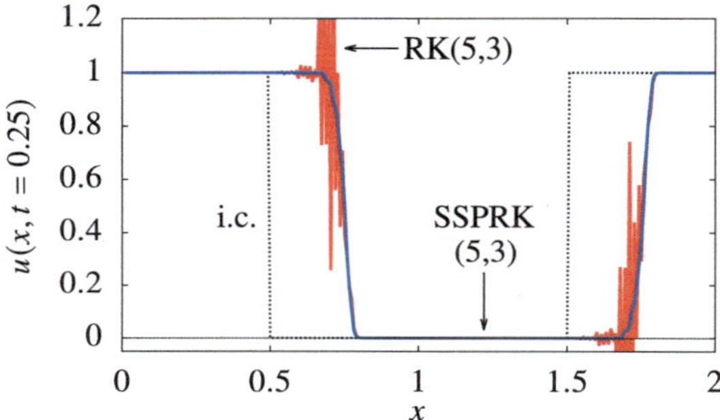

Fig. 12.11 The solution of $v_t + v_x = 0$ at $t = 0.25$ obtained by using the WENO–JS5 spatial discretization ($\Delta x = 0.0078$) and two time evolutions (step size $\Delta t = 2\Delta x$): a "standard" five-stage, third order Runge–Kutta (red), and a corresponding strong stability preserving scheme (blue)

$$u^{(1)} = u^n + \frac{1}{7}\Delta t L(u^n),$$

$$u^{(2)} = u^n + \frac{3}{16}\Delta t L(u^{(1)}),$$

$$u^{(3)} = u^n + \frac{1}{3}\Delta t L(u^{(2)}),$$

$$u^{(4)} = u^n + \frac{2}{3}\Delta t L(u^{(3)}),$$

$$u^{n+1} = u^{(5)} = u^n + \frac{1}{4}\Delta t L(u^n) + \frac{3}{4}\Delta t L(u^{(4)}).$$

Figure 12.11 (red curve) shows the solution at $t = 0.25$ obtained by taking a very large step size, $\Delta t = 2\Delta x$, exhibiting strong oscillations. If, on the other hand, an SSP version of the integrator of the same order and with the same number of stages is used—the SSPRK(5,3) described below—the time evolution remains stable (blue curve). Note that the potential increase in the complexity of a SSP scheme with respect to a non-SSP scheme is in fact more than compensated by being able to use larger step sizes; in other words, SSP methods may reduce the overall computational cost when the step size is limited by stability considerations.   ◁

SSP methods [18–21] are devised along the same lines as the ODE solvers of Chap. 8: they can be explicit or implicit, single-step or multi-step [22, 23], with fixed or adaptive step sizes [24]. In the following we discuss explicit SSP methods of the Runge–Kutta type that can be written in the form [25]

$$\boldsymbol{u}^{(0)} = \boldsymbol{u}^n,$$

$$\boldsymbol{u}^{(i)} = \sum_{k=0}^{i-1}\left(\alpha_{ik}\boldsymbol{u}^{(k)} + \beta_{ik}\Delta t \boldsymbol{L}(\boldsymbol{u}^{(k)})\right), \quad i = 1, 2, \ldots, s, \qquad (12.76)$$

$$\boldsymbol{u}^{n+1} = \boldsymbol{u}^{(s)},$$

where s is the number of stages and all $\alpha_{ik} \geq 0$, $\beta_{ik} \geq 0$ (compare this to the construction in Eq. (8.9)). If the solutions obtained by the EE method are strongly stable under the CFL restriction $\Delta t \leq (\Delta t)_{EE}$, then the Runge–Kutta method (12.76) is SSP under the restriction (12.74) with

$$C = \min_{i,k} \alpha_{ik}/\beta_{ik} \,. \tag{12.77}$$

The ultimate goal of an SSP method is to make C as high as possible, and the scheme that attains the highest C for a given order in Δt and a given number of stages is called *optimal*.

SSPRK(2,2) An optimal second-order, two-stage Runge–Kutta method is

$$u^{(1)} = u^n + \Delta t L(u^n) \,,$$
$$u^{n+1} = \frac{1}{2}u^n + \frac{1}{2}u^{(1)} + \frac{1}{2}\Delta t L(u^{(1)}) \,.$$

SSPRK(3,3) An optimal third-order, three-stage Runge–Kutta formula, also known as the Shu–Osher scheme, is

$$u^{(1)} = u^n + \Delta t L(u^n) \,,$$
$$u^{(2)} = \frac{3}{4}u^n + \frac{1}{4}u^{(1)} + \frac{1}{4}\Delta t L(u^{(1)})$$
$$u^{n+1} = \frac{1}{3}u^n + \frac{2}{3}u^{(2)} + \frac{2}{3}\Delta t L(u^{(2)}) \,,$$

and is probably the most commonly used SSP Runge–Kutta method. Both of these methods have $C = 1$, which permits the same step size as the EE scheme would permit, but because two or three stages (evaluations of L) are required, the computational cost roughly doubles or triples with respect to EE. It is therefore better to define the *effective SSP coefficient* as $C_{\text{eff}} = C/s$, where s is the number of stages. The methods SSPRK(2,2) and SSPRK(3,3) have $C_{\text{eff}} = 1/2$ and $1/3$, respectively.

For a given order of accuracy, optimal SSPRK methods with more stages typically have larger C_{eff}, and this is what we may keep in mind when matching a spatial discretization of a chosen order to a time integration scheme of its own order, pondering between the additional cost of increasing the order and the gain of a larger permissible step size.

SSPRK(5,3) The following five-stage, third-order scheme:

$$u^{(1)} = u^n + 0.37726891511710\,\Delta t L(u^n)\,,$$

$$u^{(2)} = u^{(1)} + 0.37726891511710\,\Delta t L(u^{(1)})\,,$$

$$u^{(3)} = 0.56656131914033\,u^n$$
$$+0.43343868085967\,u^{(2)} + 0.16352294089771\,\Delta t L(u^{(2)})\,,$$

$$u^{(4)} = 0.09299483444413\,u^n$$
$$+0.00002090369620\,u^{(1)} + 0.90698426185967\,u^{(3)}$$
$$+0.00071997378654\,\Delta t L(u^n) + 0.34217696850008\,\Delta t L(u^{(3)})\,,$$

$$u^{(4)} = 0.00736132260920\,u^n + 0.20127980325145\,u^{(1)} + 0.00182955389682\,u^{(2)}$$
$$+0.78952932024253\,u^{(4)} + 0.00277719819460\,\Delta t L(u^n)$$
$$+0.00001567934613\,\Delta t L(u^{(1)}) + 0.29786487010104\,\Delta t L(u^{(4)})\,,$$

for instance, has $C_{\text{eff}} \approx 2.65/5 \approx 0.53$ [26]. (For an alternative set of optimal coefficients with the same C_{eff} see Table 1 of [27].) This scheme was used in the Example in Fig. 12.11.

In large-scale computations—in particular in three dimensions—the problem of storage is of utmost importance. Williamson schemes [28] in their simplest form require only two units of storage: the schemes take the form

$$v^{(i)} = A_i v^{(i-1)} + \Delta t L(u^{(i-1)})\,,$$
$$u^{(i)} = u^{(i-1)} + B_i v^{(i)}\,,$$

with i indexing the stages ($i = 1, 2, \ldots, s$) and where $u^{(0)} = u^n$, $u^{n+1} = u^{(s)}$ and $A_1 = 0$. A popular representative of such schemes is given below.

Williamson(5,3) A numerically optimal five-stage, third-order low-storage SSPRK scheme (see Table 6 of [27]) is

$$u^{(0)} = u^n\,,$$
$$v^{(1)} = \Delta t L(u^{(0)})\,,$$
$$u^{(1)} = u^{(0)} + 0.713497331193829\,v^{(1)}\,,$$
$$v^{(2)} = -4.344339134485095\,v^{(1)} + \Delta t L(u^{(1)})\,,$$
$$u^{(2)} = u^{(1)} + 0.133505249805329\,v^{(2)}\,,$$
$$v^{(3)} = \Delta t L(u^{(2)})\,,$$
$$u^{(3)} = u^{(2)} + 0.713497331193929\,v^{(3)}\,,$$
$$v^{(4)} = -3.770024161386381\,v^{(3)} + \Delta t L(u^{(3)})\,,$$
$$u^{(4)} = u^{(3)} + 0.149579395628565\,v^{(4)}\,,$$
$$v^{(5)} = -0.046347284573284\,v^{(4)} + \Delta t L(u^{(4)})\,,$$
$$u^{n+1} = u^{(5)} = u^{(4)} + 0.384471116121269\,v^{(5)}\,.$$

This scheme has $C_{\text{eff}} \approx 1.402/5 \approx 0.280$ and is only slightly less efficient (more expensive) than the SSPRK(3,3) method which has $C_{\text{eff}} = 1/3$. This added computational cost is acceptable when storage represents a critical requirement.

SSPRK(5,4) No four-stage, fourth-order SSPRK methods with positive C exist, but fourth-order scheme with five stages do exist, and a popular representative [27] is

$$
\begin{aligned}
u^{(1)} &= u^n + 0.391752226571890\,\Delta t L\!\left(u^n\right), \\
u^{(2)} &= 0.444370493651235\,u^n \\
&\quad +0.555629506348765\,u^{(1)} + 0.368410593050371\,\Delta t L\!\left(u^{(1)}\right), \\
u^{(3)} &= 0.620101851488403\,u^n \\
&\quad +0.379898148511597\,u^{(2)} + 0.251891774271694\,\Delta t L\!\left(u^{(2)}\right), \\
u^{(4)} &= 0.178079954393132\,u^n \\
&\quad +0.821920045606868\,u^{(3)} + 0.544974750228521\,\Delta t L\!\left(u^{(3)}\right), \\
u^{n+1} &= 0.517231671970585\,u^{(2)} \\
&\quad +0.096059710526147\,u^{(3)} + 0.063692468666290\,\Delta t L\!\left(u^{(3)}\right) \\
&\quad +0.386708617503269\,u^{(4)} + 0.226007483236906\,\Delta t L\!\left(u^{(4)}\right).
\end{aligned}
$$

SSPRK methods of higher orders exists, but are much less popular. The reason is that any SSPRK method of order $p > 4$ with non-zero C must have at least one negative β_{ik}, although this complication can be overcome by rewriting (12.77) as

$$
C = \min_{i,k} \alpha_{ik}/|\beta_{ik}|
$$

and replacing $\beta_{ik} L(\cdot)$ by $\beta_{ik}\widetilde{L}(\cdot)$ in the scheme whenever β_{ik} is negative. Here $\widetilde{L}(\cdot)$ approximates the same spatial derivatives as $L(\cdot)$ but is assumed to be stable for Euler's method applied *backwards* in time, i.e. by downwinding instead of upwinding: see Eq. (10.83) and [29] for details.

Remark Instructive graphical representations of the spectral properties of differential operators can be found in Sect. 4.3 of [3], while further insights into time propagation can be gained from Chaps. 9, 13 and 14 of [12]. An important point of contact between matrices, discrete and continuous differential operators is established through the concept of *pseudospectra*: see Sect. 4.8 and [17].

12.7   Semi-infinite and Infinite Domains ⋆

If spectral methods are used to solve problems with periodic solutions on $[0, 2\pi]$ or non-periodic solutions on $[-1, 1]$, the choice of basis functions is simple. Boyd [12] (see Chap. 1 in that book) advises:

1. When in doubt, use Chebyshev polynomials unless the solution is spatially periodic, in which case an ordinary Fourier series is better.
2. Unless you're sure another set of basis functions is better, use Chebyshev polynomials.
3. Unless you're really, really sure that another set of basis functions is better, use Chebyshev polynomials.

This is an unforgettable memorandum, but how do we deal with problems posed on semi-infinite and infinite domains, with $y \in [0, \infty]$ and $y \in [-\infty, \infty]$?[1] We have three options. The first one—obvious and crude, and applicable only to rapidly decaying functions—is domain truncation: replace $\pm\infty$ by $\pm L$, where L is "large", and solve the problem with the basis functions for the finite domain; we have chosen this approach e.g. in the treatment of singular Sturm–Liouville problems in Sect. 9.8. The second possibility is to use basis functions inherent to the semi-infinite and infinite domain, namely Laguerre and Hermite polynomials, respectively, as well as Laguerre and Hermite functions based on them—see Sects. 5.3.3 and 5.3.5. One may exploit these basis functions in the same group of methods that we have presented with respect to the trigonometric functions, Legendre and Chebyshev polynomials: Galerkin, tau and collocation. The third option is to resort to coordinate transformations.

12.7.1 Domain Truncation

The domain truncation error may be defined as $E_{\mathrm{dt}}(L) = \max_{|y| \geq L}\{v(y)\}$, where v is the true solution. For most real-world problems, the error of the numerical solution u_N expanded in terms of N basis functions will behave as $\max_{|y|=L}\{v(y) - u_N(y)\} \sim \mathcal{O}\big(E_{\mathrm{dt}}(L)\big)$. Once L is fixed while N is sent to infinity, however, the error does not converge to zero but rather to $E_{\mathrm{dt}}(L)$—thus in order to improve accuracy, both the L and N must be increased simultaneously [12].

12.7.2 Expansions in Terms of Laguerre Functions on $[0, \infty]$

The basis functions intrinsically suitable for solving ODE and PDE on the semi-infinite domain are the Laguerre functions, i.e. Laguerre polynomials multiplied by an exponential: see Eq. (5.50). These functions solve the radial part of the Schrödinger equation with the Coulomb potential, and they are the staple diet in quantum physics and chemistry. Let us show how they can be used in a pseudospectral method to solve the boundary-value problem

---

[1] In this Section y is the variable spanning the (semi)-infinite domain, while x and t will be used for finite domains $[-1, 1]$ and $[0, \pi]$, respectively.

$$y^2 v_{yy} + y v_y - \left(1 + y^2\right) v = 0 \,, \quad y \in [1, \infty] \,,$$
$$v(1) = K_1(1) \approx 0.601907 \,, \tag{12.78}$$
$$v(\infty) = 0 \,,$$

where K_1 is the modified Bessel function of the second kind. The exact solution of (12.78) is $v(y) = K_1(y)$, which decays exponentially,

$$K_1(y) \sim \sqrt{\frac{\pi}{2y}} e^{-y} \left(1 + \mathcal{O}\left(\frac{1}{y}\right)\right) \quad \text{as } y \to \infty \,,$$

so that the boundary condition at infinity is natural.

Upon pseudospectral discretization the differential equation posed on $y \in [0, \infty]$ is translated into a matrix problem of the form (12.64), where

$$A_{jk} = y_j^2 L_k''(y_j) + y_j L_k'(y_j) - \left(1 + y_j^2\right) L_k(y_j) \,, \tag{12.79}$$
$$Q_j = 0 \,.$$

Here $j = 0, 1, \ldots, N - 1$ and $k = 0, 1, \ldots, N$, while $'$ denotes the derivative with respect to y. The last row of the matrix ($j = N$) is reserved to enforce the boundary condition at $y = 1$, thus the remaining entries are $A_{Nk} = L_k(1)$ and $Q_N = K_1(1)$. Because the boundary condition is posed at $y = 1$ and requires one collocation node for itself, the set of $N + 1$ collocation nodes $\{y_j\}_{j=0}^N$ consists of the origin, $y_0 = 1$, plus the N zeros of L'_{N+1} offset by y_0 (Laguerre–Gauss–Lobatto). Once the expansion coefficients $\widetilde{u}_k$ are obtained by solving (12.64), the solution is given by

$$u(y) = \sum_{k=0}^{N} \widetilde{u}_k L_k(y) \,.$$

We may also solve (12.78) by using the collocation method employing differentiation matrices acting directly on the components of the solution $\boldsymbol{u}$. The collocation solution with the components u_j can then be obtained by solving the system of linear equations of the form (12.63):

$$y_j^2 \sum_{k=0}^{N} D_{jk}^{(2)} u_k + y_j \sum_{k=0}^{N} D_{jk}^{(1)} u_k - \left(1 + y_j^2\right) u_j = 0 \,, \quad j = 1, 2, \ldots, N \,,$$

where $D_{jk}^{(1)}$ and $D_{jk}^{(2)}$ are the Laguerre spectral derivative matrices: see Eqs. (12.26) and (12.27). We keep the first row ($j = 0$) of the system matrix A for the boundary condition, such that $A_{00} u_0 = 1\, u_0 = K_1(1)$. The errors of the collocation solution for $N = 10, 20$ and 30 with respect to the exact solution are shown in Fig. 12.12.

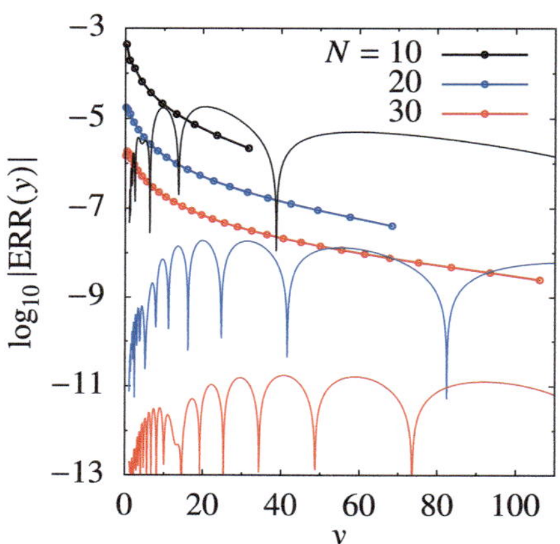

Fig. 12.12 Errors of the solutions of (12.78) by the Laguerre collocation method, for $N = 10, 20$ and 30. For clarity the values at the nodes have been connected by lines (not by the interpolation polynomial). Also shown are the errors obtained by using the Chebyshev *TL* functions discussed below. Note that in the latter case *optimal values of the scaling parameter L* have been used

12.7.3 Coordinate Mappings for $[a, b] \leftrightarrow [0, \infty]$

For functions that decay algebraically rather than exponentially at infinity (or asymptotically decay to a constant) direct Laguerre expansions typically converge too slowly, indicating a need for different basis functions. By coordinate transformations applied to the arguments of functions from a given basis set, new basis sets can be obtained which possess better asymptotic properties and for which the derivatives may be computed much easier and faster. The most effective approach on the $[0, \infty]$ domain is to exploit the mappings (5.51) yielding the *rational Chebyshev functions TL* of Eq. (5.52). Upon the mapping $y = L \cot^2(t/2)$ the rational-Chebyshev series in $TL_k(y)$ are converted to a Fourier cosine series in $\cos(kt)$, and accordingly the derivatives in y are translated to derivatives in t:

$$\frac{d}{dy} = -\frac{1}{L}\Big[\sin^2(t/2)\tan(t/2)\Big]\frac{d}{dt}\,.$$

Higher derivatives $(d^2/dy^2, d^3/dy^3, \ldots)$ needed to express the $u_y, u_{yy}, \ldots$ of the original problem posed on $[0, \infty]$ by the $u_t, u_{tt}, \ldots$ of the problem posed on $[0, \pi]$, are obtained by iterating this basic rule. The first four relations are

$$u_y = -\big(S^3/(LC)\big)u_t\,, \tag{12.80}$$

$$u_{yy} = \big(S^5/(2L^2C^3)\big)\big[2CS\,u_{tt} + (3 - 2S^2)u_t\big]\,, \tag{12.81}$$

$$u_{yyy} = \big(S^7/(4L^3C^5)\big)\big[(4S^4 - 4S^2)u_{ttt}$$
$$+ (12CS^3 - 18CS)u_{tt} - (8S^4 - 20S^2 + 15)u_t\big]\,,$$

$$u_{yyyy} = \big(S^9/(8L^4C^7)\big)\big[(8CS^3 - 8CS^5)u_{tttt} + (48S^6 - 120S^4 + 72S^2)u_{ttt}$$
$$+ (88CS^5 - 232CS^3 + 174CS)u_{tt} - (48S^6 - 168S^4 + 210S^2 - 105)u_t\big]\,,$$

where $S \equiv \sin(t/2)$ and $C \equiv \cos(t/2)$. The expressions for the fifth and sixth derivative are given in [30].

Example Let us illustrate the use of rational Chebyshev functions by again solving the problem (12.78) in the pseudospectral manner. By complete analogy to the discretization (12.79) now the differential equation in terms of $y \in [0, \infty]$ morphs into the matrix problem (12.64) constructed around the TL basis set,

$$A_{jk} = y_j^2\, TL_k''(y_j) + y_j TL_k'(y_j) - \left(1 + y_j^2\right) TL_k(y_j)\,, \quad Q_j = 0\,.$$

where $j = 0, 1, \ldots, N - 1$ and $k = 0, 1, \ldots, N$, and $'$ denotes the derivative with respect to y. Note that in the actual code the derivatives in y are most efficiently computed in terms of the t variable with equidistant values, by using the transcriptions (12.80) and (12.81)! We may again allocate the last row of the matrix ($j = N$) to enforce the boundary condition at $y = 1$, thus $A_{Nk} = TL_k(1)$ and $Q_N = K_1(1)$. The $y_0 = 1$ node is used to pin down the value at the origin, while the others are

$$y_j = y_0 + L \cot^2(t_j/2)\,, \quad t_j = (2j + 1)\pi/(2N)\,, \quad j = 0, 1, \ldots, N - 1\,.$$

(Otherwise, the t_j's would be specified by $t_j = (2j + 1)\pi/(2N + 2))$, $j = 0, 1, \ldots,$ N—see Table 3.1.) Once the coefficients $\widetilde{u}_k$ are obtained, the solution is given by

$$u(y) = \sum_{k=0}^{N} \widetilde{u}_k TL_k(y)\,.$$

Fig. 12.13 (left) shows the errors of the solution for $N = 10, 20$ and 30, calculated with the scaling parameter set to 1 or to the optimal L in each case. Figure 12.13 (right) shows the error as a function of L, whence the optimal values of L can be established. The corresponding minimized errors are shown in the left panel. See also Fig. 12.12 where they are compared to the Laguerre collocation method.   ◁

12.7.4 Eigenvalue Problems on $[0, \infty]$

Eigenvalue problems on the semi-infinite domain can be solved by spectral techniques in a completely analogous fashion as those not involving eigenvalues. (For a treatment of singular Sturm–Liouville eigenvalue problems by means of non-spectral methods on truncated domains see Sect. 9.8.) For example, a collocation method based on the TL basis functions for an eigenvalue problem of the form

$$L_2(y)v_{yy} + L_1(y)v_y + L_0(y)v = \lambda\big[M_2(y)v_{yy} + M_1(y)v_y + M_0(y)v\big]\,, \quad y \in [0, \infty]\,,$$

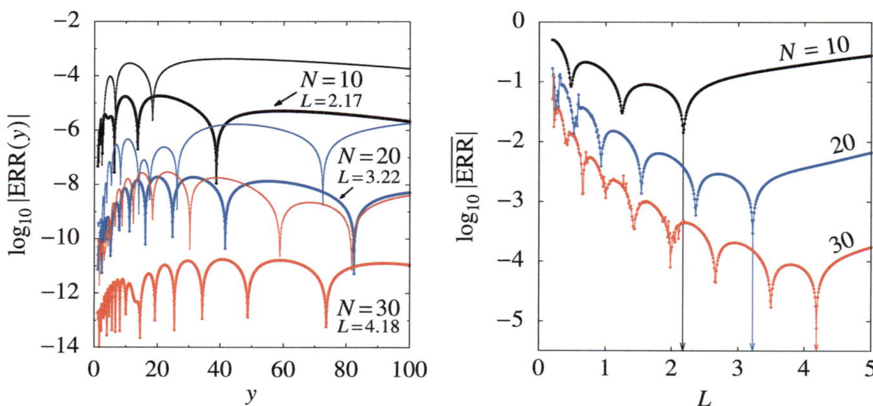

Fig. 12.13 [LEFT] Pointwise errors of the numerical solution of (12.78) by the pseudospectral method involving Chebyshev *TL* functions, for $N = 10$, 20 and 30 (thick curves), with optimal values of the scaling parameter L inferred from the right panel. The corresponding errors for $L = 1$ are drawn with thin curves. [RIGHT] The average error over the interval spanned by the collocation points, evaluated by a simple trapezoidal formula. The arrows point to the optimal values of L

with the boundary condition

$$a_1 v_y + a_0 v + \lambda(b_1 v_y + b_0 v) = 0, \quad y = 0,$$

where λ is the eigenvalue, can be pseudospectrally discretized into a system of linear equations representable as a generalized eigenvalue problem

$$A\widetilde{u} = \lambda B\widetilde{u}, \tag{12.82}$$

where $\widetilde{u}$ are the expansion coefficients of the (eigen)solution; compare this to (12.64). The elements of the matrices A and B are given by

$$A_{jk} = L_2(y_j)TL_k''(y_j) + L_1(y_j)TL_k'(y_j) + L_0(y_j)TL_k(y_j),$$
$$B_{jk} = M_2(y_j)TL_k''(y_j) + M_1(y_j)TL_k'(y_j) + M_0(y_j)TL_k(y_j),$$

where $j = 0, 1, \ldots, N - 1$ and $k = 0, 1, \ldots, N$. As in the preceding Example, the last rows of A and B should be allocated for the boundary condition, thus

$$A_{Nk} = a_1 TL_k'(0) + a_0 TL_k(0),$$
$$B_{Nk} = b_1 TL_k'(0) + b_0 TL_k(0).$$

Equivalent formulas apply to any other basis function we may choose, e.g. Laguerre: simply replace TL_k by L_k. Further reading can be found in [31–36].

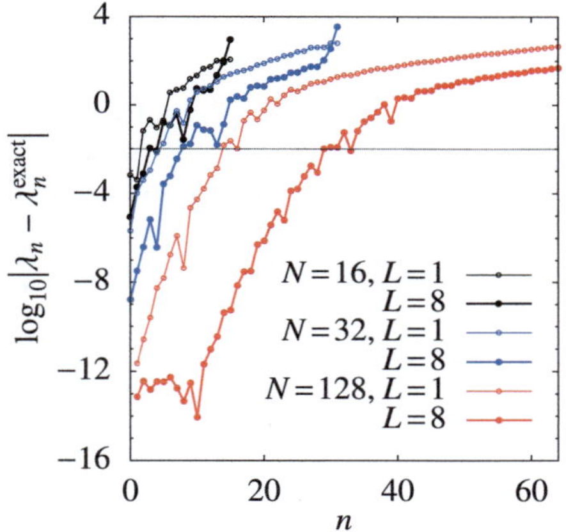

Fig. 12.14 Errors of the eigenvalues of the Whittaker's problem (12.83) for three different N and two different scaling parameters L. The horizontal line represents approximately two digits of precision, corresponding to about three quarters of eigenvalues obtained with $L = 8$ at any given N being "not good"

Example To illustrate the method we solve the Whittaker's eigenvalue problem

$$v_{yy} + \left[-1/4 + (\lambda + 1)/y\right]v = 0 , \quad y \in [0, \infty] , \tag{12.83}$$

where λ is the eigenvalue. For $\lambda = -1$ the exact solution is $v(y) = \exp(-y/2)$, while for non-negative λ the exact solutions are

$$v(y) = y e^{-y/2} L_n^1(y) , \quad \lambda_n = n , \quad n = 0, 1, 2, \dots ,$$

where L_n^1 is the generalized Laguerre polynomial. Figure 12.14 shows the errors of the eigenvalues for the pseudospectral solution with $N = 16$, 32 and 128, with the scaling parameters $L = 1$ and 8 in each case. The errors rapidly increase with n; arbitrarily proclaiming a given eigenvalue as "good" if its error is less than about two decimal places (horizontal line in the Figure) about a quarter of the eigenvalues are "good" when $L = 8$. This is a feature of spectral methods, regardless of whether finite or non-finite domains are involved; see [12].   ◁

12.7.5 *Expansions in Terms of Hermite Functions on* $[-\infty, \infty]$

To solve ODE and PDE on the infinite domain the intrinsically suitable basis functions are either the $\sin(ky)/y$ functions [37, 38], which we shall not discuss in this book, or the Hermite functions, i.e. Hermite polynomials multiplied by an exponential: see Eq. (5.54). These functions solve the linear harmonic oscillator eigenvalue problem in quantum mechanics, but to illustrate their functionality in spectral methods, let us solve the one-dimensional diffusion equation by Hermite collocation. We shall apply the collocation method to the boundary-value problem

$$v_t = v_{yy} , \quad y \in [-\infty, \infty] ,$$
$$v(\pm\infty, t) = 0 ,$$
$$v(y, 0) = (2\pi)^{-1/2} \exp(-y^2/2) . \tag{12.84}$$

The second Hermite collocation derivative matrix acting on the components $\boldsymbol{u}_j$ of the numerical solution $\boldsymbol{u}$ is given by (12.33), and all one needs to do is integrate the system of ordinary differential equations

$$\frac{\mathrm{d}u_j}{\mathrm{d}t} = \sum_{k=0}^{N} \left(D_N^{(2)}\right)_{jk} u_k , \quad j = 0, 1, \ldots, N ,$$

where $u_j = u(y_j)$ and $\{y_j\}_{j=0}^{N}$ are the Hermite collocation nodes (the zeros of H_{N+1}). The exact solution is

$$v(y, t) = \frac{1}{\sqrt{2\pi(2t + 1)}} \exp\left(-\frac{y^2}{2(2t + 1)}\right) .$$

Fig. 12.15 (left) shows the results for $N = 10$ and 20 at times $t = 1$ and 4 by using the explicit Euler scheme with step size $\Delta t = 0.01$. Figure 12.15 (right) shows the pointwise errors at the collocation nodes as a function of N at $t = 4$. See also [39].

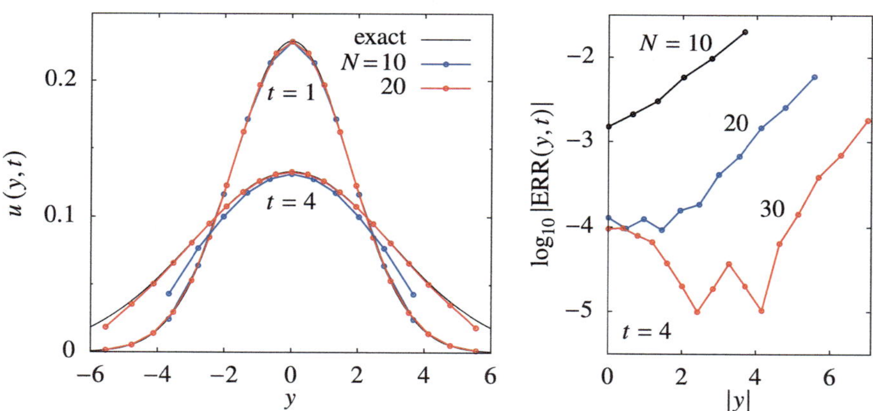

Fig. 12.15 The solution of the diffusion equation by the Hermite collocation method. [LEFT] The solutions at $t = 1$ and 4 for $N = 10$ and 20, compared to the exact solution. [RIGHT] Pointwise errors at the collocation nodes for $N = 10, 20$ and 30 at $t = 4$

12.7.6 Coordinate Mappings for $[a, b] \leftrightarrow [-\infty, \infty]$

By the same token as discussed in Sect. 12.7.3, solutions that decay algebraically rather than exponentially at $\pm\infty$ (or asymptotically decay to a constant) the convergence of Hermite expansions may be too slow, indicating a need for different basis functions. On the infinite domain $[-\infty, \infty]$ the best approach is to resort to the mapping (5.55) yielding the *rational Chebyshev functions TB* of Eq. (5.56). Upon the mapping $y = L \cot(t)$ the rational-Chebyshev series in $TB_k(y)$ are converted to a Fourier cosine series in $\cos(kt)$, and accordingly the derivatives in y are translated to derivatives in t:

$$\frac{\mathrm{d}}{\mathrm{d}y} = -\frac{\sin^2 t}{L}\frac{\mathrm{d}}{\mathrm{d}t}\,.$$

The higher derivatives $(\mathrm{d}^2/\mathrm{d}y^2, \mathrm{d}^3/\mathrm{d}y^3, \ldots)$ needed to express the $u_y, u_{yy}, \ldots$ of the original problem posed on $[-\infty, \infty]$ by the $u_t, u_{tt}, \ldots$ of the problem posed on $[0, \pi]$, are obtained by iterating this basic rule. The first four relations are

$$\begin{aligned}
u_y &= -\left(S^2/L\right)u_t\,,\\
u_{yy} &= \left(S^3/L^2\right)\left[S\,u_{tt} + 2\,C\,u_t\right],\\
u_{yyy} &= \left(S^4/L^3\right)\left[-S^2\,u_{ttt} - 6CS\,u_{tt} + \left(8S^2 - 6\right)u_t\right],\\
u_{yyyy} &= \left(S^5/L^4\right)\left[S^3 u_{tttt} + 12CS^2 u_{ttt} + \left(36S - 44S^2\right)u_{tt} + \left(24C - 48CS^2\right)u_t\right],
\end{aligned}$$

(12.85)

where $S \equiv \sin(t)$ and $C \equiv \cos(t)$. The expressions for the fifth and sixth derivative are given in [40].

Example Let us consider the boundary-value problem

$$v_{yy} - y^2 v = y\,,\quad y \in [-\infty, \infty]\,,$$

(12.86)

with the boundary conditions $v(\pm\infty) = 0$—and since the solution is anti-symmetric about the origin, $v(0) = 0$. This equation describes the north-south current of the steady part of the "Yoshida jet", a wind-driven strong equatorial jet in the ocean: see Chap. 9 of [41]. The exact solution is known [42] but it either requires the calculation of Bessel and Struve functions with fractional indices and brings about cancellations of large numbers, or it consists of an infinite Hermite series with a very slow convergence because the decay of the solution at $|y| \to \infty$ is only algebraic, $v(y) \sim -(1/y)\left(1 + \mathcal{O}(1/y^4)\right)$ as $|y| \to \infty$, causing the coefficients of the Hermite expansion to decrease only as $\mathcal{O}\left(k^{-3/4}\right)$! In contrast, the pseudospectral method based on the TB functions performs well. To obtain the expansion coefficients of the series

$$u(y) = \sum_{k=0}^{N} \widetilde{u}_{2k+1} TB_{2k+1}(y)\,,$$

in which we should keep kept *only the odd-order basis functions* due to the anti-symmetric nature of the solution, we need to solve the matrix problem (12.64) constructed around the *TB* basis set. The corresponding matrix elements and the elements of the right hand-side vector are given by

$$A_{jk} = TB_k''(y_j) - y_j^2\, TB_k(y_j)\,, \quad Q_j = y_j\,,$$

where $j = 0, 1, \ldots, N$ and $k = 1, 3, \ldots, 2N + 1$, while $'$ denotes the derivative with respect to y. No modifications of the matrix are needed due to the boundary conditions, since they are enforced naturally. As we have seen in the case of *TL* expansions, in the actual code the second derivative in y is best evaluated in the t variable by using (12.85). The optimal collocation nodes are

$$y_j = L \cot(t)\,, \quad t_j = (j + 1)\pi/(2(N + 2))\,, \quad j = 0, 1, \ldots, N\,.$$

This choice is important: because of the anti-symmetry of the solution the basis is "halved" to include only the odd-order TB_k's, hence the collocation points should be evenly spaced on $[\pi/2, \pi]$ rather than $[0, \pi]$.

   Figure 12.16 (left) shows the solution for $N = 2$ (three basis functions TB_1, TB_3 and TB_5) and $N = 3$ (with the additional TB_7), computed with $L = 3$. Figure 12.16 (right) shows the error as a function of y for $N = 4, 7$ and 10. Note that even better results can be obtained by using SB_k basis functions, which are images of $\sin(kt)$ instead of $\cos(kt)$ under the map $y = L \cot(t)$: see [40] for details.   ◁

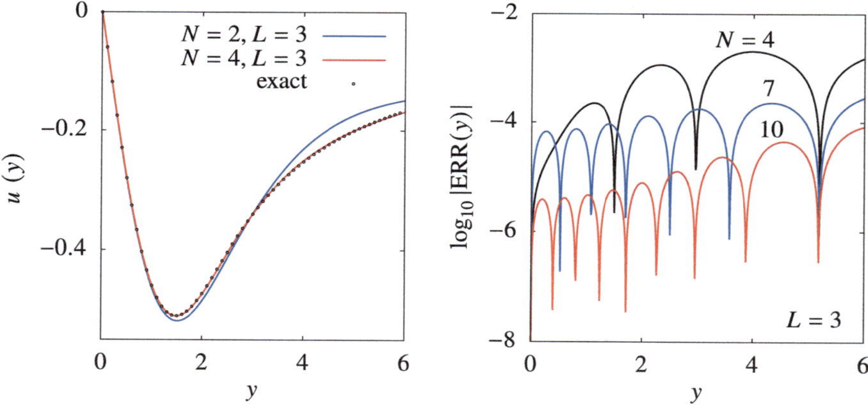

Fig. 12.16 The solution of (12.86) by the pseudospectral method based on rational Chebyshev functions TB_k. [LEFT] The solution for $N = 2$ and 4 with the scaling parameter $L = 3$. [RIGHT] The errors of the solution for $N = 4, 7$ and 10

12.7.7 Eigenvalue Problems on $[-\infty, \infty]$

We could discuss the eigenvalue problems on the infinite domain by repeating almost verbatim the approach of Sect. 12.7.4. To give the exposition a slightly different slant, however, we pick a specific quantum-mechanical problem and describe the Hermite collocation approach which yields the eigenvalue analogue of (12.63) and manipulates directly the solution components u_j at the collocation nodes y_j; note that this approach differs from the collocation based on expansion coefficients $\widetilde{u}_k$, which resulted in (12.82). To illustrate the method, we choose the Schrödinger equation for the anharmonic oscillator with the usual quadratic plus a sextic term in the potential:

$$- v''(y) + \left(y^2 + \beta y^6\right)v(y) = Ev , \quad y \in [-\infty, \infty] , \tag{12.87}$$

where $\beta = 0.2$.

The collocation method amounts to the requirement that the polynomial interpolant matches the solution of the differential equation at the collocation nodes. Denoting $p(y) = \left(y^2 + \beta y^6\right)$ the continuous problem (12.87) translates to the discretized problem

$$- \left(D_N^{(2)}\right)u + \operatorname{diag}\left(p(y_j)\right)u = Eu , \tag{12.88}$$

where the collocation nodes $\{y_j\}_{j=0}^{N}$ are the zeros of H_{N+1}, and $D_N^{(2)}$ is the second-order Hermite spectral differentiation matrix given by (12.33). To compute the approximate eigenvalues and eigenfunctions, we must diagonalize the coefficient matrix in (12.88), i.e. solve the eigenvalue problem $Au = Eu$ with

$$A = -\left(D_N^{(2)}\right) + \operatorname{diag}\left(p(y_j)\right) . \tag{12.89}$$

Suppose that we are interested in the lowest eigenvalue. The exact value is [44]

$$E_0(\text{exact}) = 1.173889345125433315298177395 .$$

Taking $N = 16$ or $N = 32$ and diagonalizing A with a direct application of $D_N^{(2)}$ gives $E_0 \approx 1.17394$ or $E_0 = 1.1738892$, correct to about 4 and 7 digits, respectively. But we can do much better than that! Recall from (12.35) that we have the liberty simultaneously to rescale the coordinates (hence the nodes) and differentiation matrices by the scaling parameter a, and indeed by optimizing a we can dramatically improve the accuracy, as shown in Fig. 12.17: with $N = 32$ and $a = 2$, for instance, we obtain $E_0 \approx 1.17388934512546$, correct to about 14 digits—assuming that the diagonalization code was at least as accurate. For details see [43].

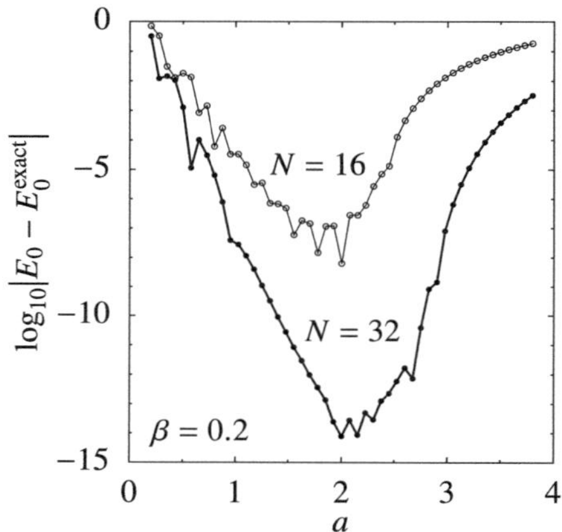

Fig. 12.17 The error of the ground-state energy of the sextic anharmonic oscillator described by the Schrödinger equation (12.87), obtained by the spectral collocation method with $N = 16$ and 32, as a function of the scaling parameter a involved in the evaluation of the second-derivative. With optimal a nearly machine precision can be achieved

12.8   Complex Geometries ⋆

In this Chapter we discussed almost exclusively one-dimensional problems. We encountered two space dimensions only with the Fourier–Galerkin method in Sect. 12.2.4, where the extension to multiple dimensions (from the interval to the square or cube) was conceptually simple. In Problem 12.9.5 one can obtain a good two-dimensional impression while solving the Poisson equation by the Legendre tau method. Cylindrical and spherical geometries, as well as all non-trivial geometries appearing in advanced physics problems require greater involvement and are beyond the scope of this book. Spectral methods for solving ODEs and PDEs in cylindrical and spherical geometries are introduced in Chap. 18 of [12] and Chap. 6 of [13], but our Problem 12.9.4 may also serve as an appetizer. The book [7] gives appealing two-dimensional examples solved by MATLAB. As a very complete textbook on spectral methods in complex geometries we recommend [2].

12.9   Problems

12.9.1   Galerkin Methods for the Helmholtz Equation

This Problem is devoted to solving the one-dimensional Helmholtz equation

$$-\frac{\mathrm{d}^2 v}{\mathrm{d}x^2} + \lambda v = Q, \qquad \lambda \geq 0,$$

by Galerkin spectral methods. The solution of the problem on $x \in [0, 2\pi]$ with the periodic boundary condition $v(0, t) = v(2\pi, t)$ can be expanded in terms of trigonometric polynomials, while the solution of the problem on $x \in [-1, 1]$ with homogeneous Dirichlet conditions $v(-1, t) = v(1, t) = 0$ can be constructed as a linear combination of Legendre or Chebyshev polynomials.

⊙ Solve the periodic problem by using the Fourier–Galerkin method described in Sect. 12.2.1, with the function

$$Q(x) = 4(\pi - x)\cos x + [(1 + \lambda)(\pi - x)^2 - 2]\sin x\ .$$

The exact solution is $v(x) = (\pi - x)^2 \sin x$ and does not depend on λ. Determine the Fourier coefficients (12.42) by analytic or numerical integration. Is the method of integration relevant for the precision of the final result if N is relatively small, for example, $N = 8$ or $N = 16$?

⊙ Solve the Dirichlet version of the problem with $Q(x) = x\sin(\pi x)$ by using the Legendre–Galerkin method from Sect. 12.2.2 and the Chebyshev–Galerkin method from Sect. 12.2.3. Try to solve the Legendre-Galerkin part by using the optimized basis functions (12.48).

12.9.2 Galerkin Methods for the Advection Equation

The one-dimensional linear hyperbolic equation

$$\frac{\partial v}{\partial t} - \frac{\partial v}{\partial x} = 0\ , \qquad x \in (0, 2\pi)\ ,$$

with the periodic boundary condition $v(0, t) = v(2\pi, t)$ and specified initial condition $v(x, 0)$ is a benchmark problem for spectral methods (e.g. Fourier–Galerkin) for non-stationary problems. The Galerkin condition with the approximation (12.51) leads to an uncoupled system of ordinary differential equations (12.52) that can be solved by methods for initial-value problems.

⊙ Apply the Fourier–Galerkin method to solve the problem outlined above, with the initial condition $v(x, 0) = \sin(\pi \cos x)$. The exact solution [3] is

$$v(x, t) = \sin[\pi \cos(x + t)]\ ,$$

and it corresponds to the Fourier expansion

$$v(x, t) = \sum_{k=-\infty}^{\infty} \widehat{v}_k(t)\, e^{ikx}\ , \qquad \widehat{v}_k(t) = \sin\left(\frac{k\pi}{2}\right) J_k(\pi)\, e^{ikt}\ .$$

Of course, the spectral approximation will have a finite N, e.g. $N = 16$. Compare the numerical and exact solutions at long times (large multiple of 2π). Determine the speed of convergence of the numerical solution to the exact one when N increases. How important is the choice of the integration method? What role does the time step size play? Solve the problem by using some difference method and establish a comparison similar to that in Fig. 12.7!

⊙ Use the Chebyshev–Galerkin method to solve the problem

$$\frac{\partial v}{\partial t} - \frac{\partial v}{\partial x} = 0 \,, \qquad x \in (-1, 1) \,,$$

with the boundary condition $v(1, t) = 0$ and initial condition $v(x, 0) = \sin \pi x$. This time the spectral solution can be cast in the form [4]

$$u(x, t) = \sum_{k=1}^{N} a_k(t)\phi_k(x) \,,$$

where the basis functions are $\phi_k(x) = T_k(x) - 1$ (since $\phi_0(x) = 0$, the summation starts at $k = 1$). According to Galerkin we require the residual $R = u_t - u_x$ to be orthogonal to ϕ_n in the sense of the scalar product

$$\langle R, \phi_n \rangle_w = \frac{2}{\pi} \int_{-1}^{1} (u_t - u_x)\phi_n(x)\frac{1}{\sqrt{1 - x^2}}\, dx = 0 \,, \qquad n = 1, 2, \ldots, N \,.$$

When orthogonality relations are applied, the following system of differential equations emerges:

$$\sum_{k=1}^{N} M_{nk}\frac{da_k(t)}{dt} = \sum_{k=1}^{N} S_{nk}a_k(t) \,, \qquad n = 1, 2, \ldots, N \,,$$

where the coefficient matrices are

$$M_{nk} = \frac{2}{\pi} \int_{-1}^{1} \phi_n(x)\phi_k(x)\frac{1}{\sqrt{1 - x^2}}\, dx = 2 + \delta_{n,k} \,,$$

$$S_{nk} = \frac{2}{\pi} \int_{-1}^{1} \phi_n(x)\phi_k'(x)\frac{1}{\sqrt{1 - x^2}}\, dx = 2k \sum_{\substack{p=0 \\ p+k \ \text{odd}}}^{k-1} (\delta_{n,p} - \delta_{0,p}) \,.$$

When the expansion coefficients are arranged in the vector $\boldsymbol{a} = (a_1, a_2, \ldots, a_N)^{\mathsf{T}}$ the system of differential equations can be written in matrix form

$$\frac{d\boldsymbol{a}(t)}{dt} = M^{-1}S\boldsymbol{a}(t) \,, \qquad a_k(0) = \frac{2}{\pi} \int_{-1}^{1} f(x)\phi_k(x)\frac{1}{\sqrt{1 - x^2}}\, dx \,.$$

Solve it by using standard methods for initial-value problems.

12.9.3 *Galerkin Method for the Diffusion Equation*

Check your understanding of Galerkin methods by solving the diffusion equation

$$v_t = D v_{xx} , \qquad x \in (0, 2\pi) ,$$

with boundary condition $v(0, t) = v(2\pi, t)$ and initial condition $v(x, 0) = f(x)$.

⊙ Using the ansatz (12.51) in the Galerkin condition leads to the system (12.54) (with the simplification $c = 0$) that should be solved with the initial condition (12.53). Solve it by using the initial condition $f(x) = x^2(2\pi - x)^2$ and $D = 1.0$. Plot the solution at selected times $t \in [0, 10]$.

⊙ To solve the diffusion equation on $x \in (-1, 1)$ with Dirichlet boundary conditions $v(-1, t) = v(1, t) = 0$, we choose the Legendre–Galerkin approach. The solution ansatz is

$$u(x, t) = \sum_{n=1}^{N-1} a_n(t)\phi_n(x) , \tag{12.90}$$

with the basis functions $\phi_n(x) = P_{n+1}(x) - P_{n-1}(x)$, where $n \geq 1$. We require the residual $u_t - u_{xx}$ to be orthogonal to all functions ϕ_k, hence

$$\frac{2k+1}{2} \int_{-1}^{1} \left(\frac{da_n}{dt} \phi_n(x) - a_n \frac{d^2\phi_n(x)}{dx^2} \right) \phi_k(x)\, dx = 0 , \qquad k = 1, 2, \ldots, N-1 .$$

By considering the orthogonality of Legendre polynomials, we get the system

$$\sum_{n=1}^{N-1} M_{kn} \frac{da_n(t)}{dt} = \sum_{n=1}^{N-1} S_{kn} a_n(t) , \qquad k = 1, 2, \ldots, N-1 ,$$

with the matrices

$$M_{kn} = \frac{2k+1}{2} \int_{-1}^{1} \phi_k(x)\phi_n(x)\, dx ,$$

$$S_{kn} = \frac{2k+1}{2} \int_{-1}^{1} \phi_k(x) \frac{d^2\phi_n(x)}{dx^2}\, dx .$$

When the expansion coefficients are arranged in the vector $\boldsymbol{a} = (a_1, a_2, \ldots, a_{N-1})^{\mathrm{T}}$, this system of differential equations can be written in matrix form

$$\frac{d\boldsymbol{a}(t)}{dt} = M^{-1} S \boldsymbol{a}(t) ,$$

and solved by standard methods for initial-value problems. The initial condition vector can be computed from the expansion of $u(x, 0) = f(x)$ in terms of the functions ϕ_n as in (12.90), which leads to the recurrence

$$a_{n+1}(0) = a_{n-1}(0) - \frac{2n+1}{2} \int_{-1}^{1} f(x) P_n(x) \, dx \,, \qquad a_{-1}(0) = a_0(0) = 0 \,.$$

Solve this part of the Problem by using $f(x) = (1 - x)^2 \cos^4(\pi x/2)$.

12.9.4   Galerkin Method for the Poisson Equation: Poiseuille Law

The flow of a viscous fluid in a long straight pipe driven by a pressure gradient p' is described by the Poisson equation $\nabla^2 v = -p'/\eta$, where v is the longitudinal velocity component that depends only on the coordinates in the plane of the pipe's cross-section, and η is the viscosity. The fluid velocity at the pipe's walls is zero. The flux is given by the Poiseuille law

$$\Phi = \int v \, dS = \frac{C}{8\pi} \frac{S^2}{\eta} p' \,,$$

where the coefficient C depends on the shape of the cross-section. For a circular pipe, $C = 1$; but what is the coefficient for a *semi-circular* pipe with the radius R? Changing to variables $x = r/R$ and $u = v\eta/(p'R^2)$, the problem becomes

$$\nabla^2 u(x, \varphi) = -1, \qquad u(x = 1, \varphi) = u(x, 0) = u(x, \varphi = \pi) = 0 \,,$$

and the flux coefficient is given by

$$C = 8\pi \int_{0}^{1} \int_{0}^{\pi} \frac{u(x, \varphi) \, x \, dx \, d\varphi}{(\pi/2)^2} \,.$$

The eigenfunctions of the Poisson equation for the semi-circular geometry are

$$g_{ms}(x, \varphi) = J_{2m+1}(y_{ms}x) \, \sin(2m + 1)\varphi \,, \tag{12.91}$$

where y_{ms} is the sth zero of the $(2m + 1)$th Bessel function. The functions g_{01} and g_{12} are shown as examples in Fig. 12.18.

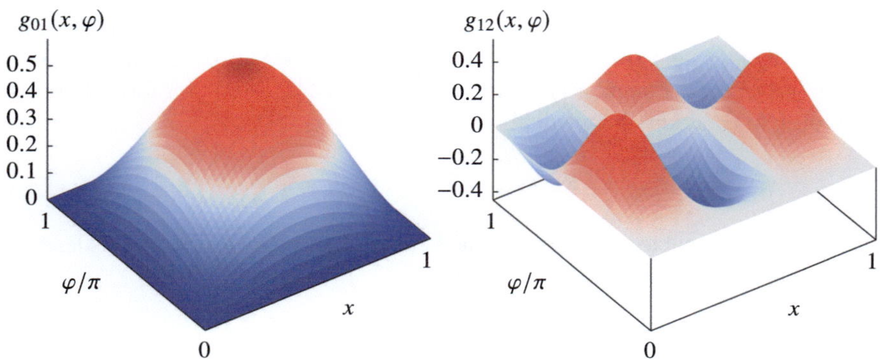

Fig. 12.18 Examples of eigenfunctions (12.91) for the Poisson equation in semi-circular geometry. [LEFT] The function $g_{01}(x, \varphi)$. [RIGHT] The function $g_{12}(x, \varphi)$

By using these functions the solution can be written as [45]

$$u(x, \varphi) = \sum_{ms} \frac{A_{ms}\, g_{ms}(x, \varphi)}{y_{ms}^2}, \qquad m = 0, 1, 2, \ldots, \qquad s = 1, 2, \ldots,$$

where

$$A_{ms} = \frac{\langle 1, g_{ms} \rangle}{\langle g_{ms}, g_{ms} \rangle},$$

and the scalar product is $\langle u, v \rangle = \iint u(x, \varphi) v(x, \varphi) x \, \mathrm{d}x \, \mathrm{d}\varphi$. The flux coefficient is then

$$C = 8 \sum_{ms} \left[\frac{8}{\pi} \frac{I_{ms}}{(2m+1)\, y_{ms}\, J_{2m+2}(y_{ms})} \right]^2.$$

To compute the integral $I_{ms} = \frac{1}{2}(2m+1)\langle 1, g_{ms} \rangle$ use the relation

$$\int_0^1 x\, J_{2m+1}(y_{ms} x)\, \mathrm{d}x = \frac{4(2m+1)}{y_{ms}} \sum_{k=m+1}^{\infty} \frac{k\, J_{2k}(y_{ms})}{(4k^2 - 1)}.$$

$\odot$ Had we not known Bessel functions (and had we been ignorant of their eigenfunction property), we could appeal to Galerkin to find an approximate solution. We write the solution as a linear combination of trial functions

$$u(x, \varphi) = \sum_{k} a_k \phi_k. \tag{12.92}$$

The functions ϕ_k need not be orthogonal, but they should satisfy the boundary conditions so that they are also satisfied by the sum (12.92). We keep the exact functions $\sin(2m+1)\varphi$ for the angular part, while we take $x^{2m+1}(1-x)^n$ for the radial part. According to Galerkin, the residual

$$R(x, \varphi) = \nabla^2 u(x, \varphi) + 1$$

should be orthogonal to all test functions $\psi_j = \phi_j$, hence $\langle R, \psi_j \rangle = 0$. This leads to the system of equations for the coefficients a_i,

$$\sum_j A_{ij} a_j = b_i \, , \qquad A_{ij} = \langle \psi_i, \nabla^2 \psi_j \rangle \, , \qquad b_i = \langle -1, \psi_i \rangle \, ,$$

and the flux coefficient is

$$C = -\frac{32}{\pi} \sum_i \sum_j b_i A_{ij}^{-1} b_j \, .$$

The indices i and j are "double": both run through the complete set of m and n. Due to orthogonality in m the matrix A has a block structure. How does the precision of C depend on the number of included angular and radial terms?

$\odot$ Try to compute the coefficient C for a rectangular pipe.

12.9.5   *Legendre Tau Method for the Poisson Equation*

Tau methods are well suited for solving the two-dimensional Poisson equation

$$- \nabla^2 v = Q$$

on the square $(x, y) \in [-1, 1] \times [-1, 1]$ with Dirichlet boundary conditions

$$v(x, -1) = B_1(x) \, , \quad v(x, 1) = B_2(x) \, ,$$
$$v(-1, y) = B_3(y) \, , \quad v(1, y) = B_4(y) \, .$$

Both Chebyshev and Legendre polynomials can be used as trial functions, even if they do not satisfy the boundary conditions: in tau methods they are established by separate equations. In this Problem the two-dimensional trial functions are products of Legendre polynomials

$$\phi_{jk}(x, y) = P_j(x) P_k(y) \, , \qquad j, k = 0, 1, \ldots, N \, ,$$

and we seek the solution in the form

$$u(x, y) = \sum_{j=0}^{N} \sum_{k=0}^{N} a_{jk} \phi_{jk}(x, y) \,. \tag{12.93}$$

In this case we need one set of test functions for the representation of the differential equation, and another set to impose the boundary conditions. The test functions for the equation are

$$\psi_{jk}(x, y) = \tilde{P}_j(x)\tilde{P}_k(y) \,, \qquad j, k = 0, 1, \ldots N - 2 \,,$$

where

$$\tilde{P}_k(x) = \left(k + \frac{1}{2}\right) P_k(x) \,,$$

(the factor $k + 1/2$ originates in the relation $(k + \frac{1}{2}) \int_{-1}^{1} P_j(x) P_k(x) \, dx = \delta_{j,k}$). The test functions for the boundary conditions depend only on one variable,

$$\psi_j^{(1)}(x) = \psi_j^{(2)}(x) = \tilde{P}_j(x) \,, \qquad j = 0, 1, \ldots N \,,$$
$$\psi_k^{(3)}(y) = \psi_k^{(4)}(y) = \tilde{P}_k(y) \,, \qquad k = 0, 1, \ldots N \,.$$

From the minimization condition for the residual we obtain

$$\int_{-1}^{-1} \int_{-1}^{-1} \left(\frac{\partial^2 u}{\partial x^2} + \frac{\partial^2 u}{\partial y^2} + Q(x, y)\right) \psi_{jk}(x, y) \, dx \, dy = 0 \,, \quad j, k = 0, 1, \ldots N - 2 \,,$$

$$\int_{-1}^{-1} B_1(x)\psi_j^{(1)}(x) \, dx = \int_{-1}^{-1} B_2(x)\psi_j^{(2)}(x) \, dx = 0 \,, \quad j = 0, 1, \ldots, N \,,$$

$$\int_{-1}^{-1} B_3(y)\psi_k^{(3)}(x) \, dy = \int_{-1}^{-1} B_4(y)\psi_k^{(4)}(x) \, dy = 0 \,, \quad k = 0, 1, \ldots, N \,.$$

With the chosen expansion of u the integrals can be computed analytically [2]. For the part pertaining to the differential equation one gets

$$-\left[a_{jk}^{[2,0]} + a_{jk}^{[0,2]}\right] = Q_{jk} \,, \qquad j, k = 0, 1, \ldots N - 2 \,, \tag{12.94}$$

where

$$a_{jk}^{[2,0]} = \left(j + \frac{1}{2}\right) \sum_{\substack{p=j+2 \\ p+j \text{ even}}}^{N} [p(p+1) - j(j+1)] \, a_{pk} \,, \qquad (12.95)$$

$$a_{jk}^{[0,2]} = \left(k + \frac{1}{2}\right) \sum_{\substack{q=k+2 \\ q+k \text{ even}}}^{N} [q(q+1) - k(k+1)] \, a_{jq} \,,$$

$$Q_{jk} = \int_{-1}^{-1} \int_{-1}^{-1} Q(x, y) \psi_{jk}(x, y) \, dx \, dy \,. \qquad (12.96)$$

This is a generalization of the coefficients $\widehat{u}_k^{(2)}$ from Eq. (12.14) to two dimensions: the second derivative in x (and zeroth in y) corresponds to the coefficients (12.95); the second derivative in y (and zeroth in x) corresponds to (12.96). Both include linear combinations of coefficients in (12.93), and Eq. (12.94) is the system for these coefficients. Boundary conditions generate constraints supplementing the system of equations, rendering the final solution unique:

$$\sum_{k=0}^{N} a_{jk} = \sum_{k=0}^{N} (-1)^k a_{jk} = 0 \,, \qquad j = 0, 1, \ldots, N \,,$$

$$\sum_{j=0}^{N} a_{jk} = \sum_{j=0}^{N} (-1)^j a_{jk} = 0 \,, \qquad k = 0, 1, \ldots, N \,.$$

⊙   Solve the Dirichlet problem for the Poisson equation with the Legendre tau method described above, by using $Q(x, y) = 2\pi^2 \sin(\pi x) \sin(\pi y)$. The exact solution is $u(x, y) = \sin(\pi x) \sin(\pi y)$.

12.9.6   *Collocation Methods for the Diffusion Equation I*

In this Problem we are solving the one-dimensional diffusion equation that describes the temperature profile $T(x, t)$ in a homogeneous infinite slab of thickness L:

$$\frac{\partial T}{\partial t} = D \frac{\partial^2 T}{\partial x^2} + \frac{Q}{\rho c} \,, \qquad 0 < x < L \,, \qquad D = \frac{\lambda}{\rho c} \,.$$

For the time being we do not consider additional heat sources, $Q(x, t) = 0$. The temperature at an arbitrary point x at time t is given by the Fourier series

$$T(x, t) \approx \sum_{k=0}^{N-1} \widehat{T}_k(t) e^{2\pi i f_k x} ,$$

where $f_k = k/L$. When this ansatz is inserted in the Galerkin requirement, we get the evolution equation for the Fourier coefficients:

$$\frac{d\widehat{T}_k(t)}{dt} = D\left(-4\pi^2 f_k^2\right) \widehat{T}_k(t) .$$

The explicit Euler method is used to advance in time,

$$\widehat{T}_k(t + \Delta t) = \widehat{T}_k(t) + \Delta t\, D(-4\pi^2 f_k^2)\widehat{T}_k(t) .$$

The temperature profile $T(x, t)$ at an arbitrary time can be computed by using the inverse Fourier transformation.

$\odot$ Use the Fourier method to compute the time evolution of the temperature profile with the initial condition $T(x, 0) = \sin^2(\pi x/L)$. By using the Fourier transformation you get the initial condition for the evolution equation. Pay attention to the stability of the Euler difference scheme: at any time step maintain

$$\left| \frac{\widehat{T}_k(t + \Delta t)}{\widehat{T}_k(t)} \right| = \left| 1 + \Delta t D(-4\pi^2 f_k^2) \right| < 1 .$$

The discretization also requires some care: in FFT one should keep $f_k < f_{\text{Nyquist}}$ for each k (see Eq. (5.6)).

In addition, solve the Dirichlet problem: the boundary conditions are $T(0, t) = T(L, t) = 0$, while at time zero the slab should have the ambient temperature T_0 everywhere, except between $0.2L$ and $0.4L$ where it has been heated up to temperature $T_1 > T_0$. Moreover, at time zero we switch on a heater between $0.5L$ and $0.75L$ with the power density of $5T_0\lambda/L^2$. The suitable eigenfunctions are the sine functions with multiples of half-waves on the interval. FFT implies an expansion in terms of sines *and* cosines, but it can still be used with Dirichlet boundary conditions if the function to be transformed is extended to an odd function on the interval $[-L, L]$.

$\odot$ Solve the problem by using the collocation method with B-splines discussed in Sect. 12.4.3. Use the initial condition $T(x, 0) = g(x) = \sin(\pi x/L)$ and homogeneous Dirichlet boundary conditions $T(0, t) = T(L, t) = 0$. Evolve the expansion coefficients $a_j(t)$ in time by using the explicit Euler method: the initial condition for the vector of coefficients $\boldsymbol{a}$ is $A\boldsymbol{a}^0 = 6\boldsymbol{g}$. At subsequent times $n\Delta t$ we get

$$\boldsymbol{a}^{n+1} = \boldsymbol{a}^n + \Delta t A^{-1} B \boldsymbol{a}^n = (I + \Delta t A^{-1} B)\boldsymbol{a}^n .$$

The time profile at any later time can be computed by evaluating the sum (12.66). The implicit method is more stable, but at every step it requires you to solve

$$\left(A - \frac{\Delta t}{2} B \right) a^{n+1} = \left(A + \frac{\Delta t}{2} B \right) a^n .$$

12.9.7 Collocation Methods for the Diffusion Equation II

The homogeneous Dirichlet problem for the diffusion equation on $x \in [-1, 1]$,

$$\frac{\partial v}{\partial t} = \frac{\partial^2 v}{\partial x^2} , \qquad v(-1, t) = v(1, t) = 0 ,$$

and initial condition $v(x, 0) = f(x)$ can be efficiently solved by collocation methods with orthogonal polynomials. The first choice for the trial functions are the Chebyshev polynomials, $\phi_k = T_k$, and the collocation solution acquires the form

$$u(x, t) = \sum_{k=0}^{N} a_k(t) T_k(x) = \sum_{j=0}^{N} u_j(t) l_j(x) , \tag{12.97}$$

where $l_j(x)$ are the polynomials (5.47) interpolating at the Chebyshev–Gauss–Lobatto collocation points $x_j = \cos(j\pi/N)$ (see Eq. (5.45)). Their property is $l_j(x_k) = \delta_{j,k}$. Since the complete solution is represented as the weighted sum of the values at the nodes, we have denoted $u_j(t) = u(x_j, t)$.

One requires that the differential equation $u_t - u_{xx} = 0$ is fulfilled exactly at the collocation points [4],

$$\left[\frac{\partial u}{\partial t} - \frac{\partial^2 u}{\partial x^2} \right]\Bigg|_{x=x_j} = 0 , \qquad j = 1, 2, \ldots, N - 1 ,$$

while the boundary conditions are $u(x_0, t) = u(x_N, t) = 0$. By using the expressions for the (second) Chebyshev collocation derivative (see Eq. (12.21)) one obtains a system of differential equations for the values $u_j(t)$ at the nodes,

$$\frac{d u_j(t)}{dt} = \sum_{l=0}^{N} \left(D_N^{(2)} \right)_{jl} u_l(t) , \qquad j = 1, 2, \ldots, N - 1 , \tag{12.98}$$

that should be solved with the initial conditions

$$u_k(0) = u(x_k, 0) = f(x_k) , \qquad k = 0, 1, \ldots, N .$$

$\odot$ Solve the Dirichlet problem for the diffusion equation by using the Chebyshev collocation method with Gauss–Lobatto nodes. The initial condition is $v(x, 0) = f(x) = \sin \pi x$. The exact solution is $v(x, t) = e^{-\pi^2 t} \sin \pi x$ and it corresponds to the expansion

$$v(x, t) = \sum_{k=0}^{\infty} b_k(t) T_k(x) , \qquad b_k(t) = \frac{2}{c_k} \sin \left(\frac{k\pi}{2} \right) J_k(\pi) e^{-\pi^2 t} ,$$

where $c_0 = 2$ and $c_k = 1$ for $k \geq 1$.

$\odot$ Collocation is not restricted to Dirichlet boundary conditions. This part of the Problem deals with the solution of the diffusion equation with more general boundary conditions

$$\alpha_1 v(-1, t) - \beta_1 v_x(-1, t) = g(t) ,$$
$$\alpha_2 v(1, t) + \beta_2 v_x(1, t) = h(t) .$$

The solution again has the form (12.97), but Chebyshev polynomials should be substituted by Legendre polynomials, so for the Gauss–Lobatto nodes the corresponding interpolation polynomials are (5.39). By requiring that the differential equation is fulfilled at the collocation points, we obtain an equation like (12.98), except that the collocation derivative matrix is given by the expression on page 773. The boundary conditions are embodied by two additional equations:

$$\alpha_1 u_0(t) - \beta_1 \sum_{j=0}^{N} \left(D_N^{(1)} \right)_{0j} u_j(t) = g(t) ,$$

$$\alpha_2 u_N(t) + \beta_2 \sum_{j=0}^{N} \left(D_N^{(1)} \right)_{Nj} u_j(t) = h(t) .$$

A system of two equations for the remaining unknowns $u_0(t)$ and $u_N(t)$ follows:

$$\left[\alpha_1 - \beta_1 \left(D_N^{(1)} \right)_{00} \right] u_0(t) - \beta_1 \left(D_N^{(1)} \right)_{0N} u_N(t) = g(t) + \beta_1 \sum_{j=1}^{N-1} \left(D_N^{(1)} \right)_{0j} u_j(t) ,$$

$$\beta_2 \left(D_N^{(1)} \right)_{N0} u_0(t) + \left[\alpha_2 + \beta_2 \left(D_N^{(1)} \right)_{NN} \right] u_N(t) = h(t) - \beta_2 \sum_{j=1}^{N-1} \left(D_N^{(1)} \right)_{Nj} u_j(t) .$$

12.9.8 Burgers Equation

An nice example of a non-linear advection-diffusion problem that can be solved by almost all spectral methods in this Chapter is the Burgers equation

$$v_t + vv_x - Dv_{xx} = 0$$

with the parameter $D > 0$ and initial condition $v(x, 0) = f(x)$. We are interested in the periodic solutions on the interval $[0, 2\pi]$ and non-periodic solutions on $[-1, 1]$ computed by using the methods discussed in Sect. 12.5.

$\odot$　Use the Fourier–Galerkin method described on page 801 to solve the Burgers equation with periodic solutions on $[0, 2\pi]$. In computing the "compound coefficient" (12.69), recall the simple relation (12.8): the system of differential equations for the expansion coefficients $\widehat{u}_k(t)$ (12.67) then becomes

$$\frac{d\widehat{u}_k}{dt} + i \sum_{m+l=k} m\widehat{u}_m\widehat{u}_l + Dk^2\widehat{u}_k = 0, \qquad -N/2 \le k \le N/2 - 1,$$

with the initial conditions (12.68). The exact solution is given by

$$v(x, t) = c + v_b(x - ct, t + t_0),$$

where

$$v_b(x, t) = -2D\frac{1}{\phi_b}\frac{\partial\phi_b}{\partial x}, \qquad \phi_b(x, t) = \frac{1}{\sqrt{4\pi Dt}}\sum_{n=-\infty}^{\infty} e^{-(x-2\pi n)^2/(4Dt)}.$$

Use $N = 16, 32$, and 64 with $D = 0.2, c = 4$, and $t_0 = 1$. The initial condition is the exact solution at $t = 0$. Compute the solution until $t = \pi/8$ (Fig. 12.19 might serve as an example). Solve the problem by Fourier collocation as well.

$\odot$　Use the Chebyshev tau and Chebyshev collocation methods to solve the Burgers equation on $[-1, 1]$ with homogeneous Dirichlet boundary conditions $v(-1, t) = v(1, t) = 0$. The exact solution in this case is

$$v(x, t) = 4\pi D\left[\sum_{n=1}^{\infty} na_n e^{-n^2\pi^2 Dt}\sin n\pi x\right]\left[a_0 + 2\sum_{n=1}^{\infty} a_n e^{-n^2\pi^2 Dt}\cos n\pi x\right]^{-1},$$

where the coefficients a_n are given by the modified Bessel functions of the first kind as $a_n = (-1)^n I_n(1/(2\pi D))$. The initial condition is the exact solution at time zero. Use $D = 0.1$ and compute with $N = 16, 32$, and 64 in both Chebyshev tau and collocation methods. Compute the solutions until $t = 1$.

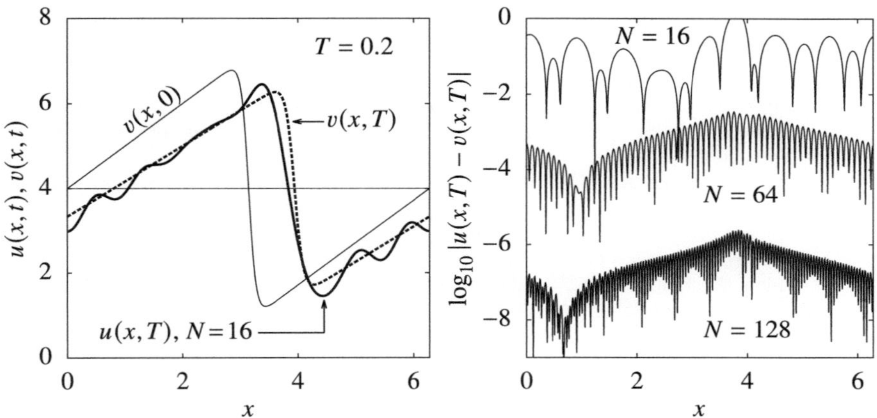

Fig. 12.19 Solution of the Burgers equation with periodic boundary conditions by the Fourier–Galerkin method. [LEFT] The initial condition and the comparison of numerical ($N = 16$) and exact solution at $T = 0.2$. [RIGHT] The error of the numerical solution along $[0, 2\pi]$ for different approximation levels (different number of basis functions N)

References

1. D. Gottlieb, S.A. Orszag, *Numerical Analysis of Spectral Methods: Theory and Applications* (SIAM, Philadelphia, 1987)
2. C. Canuto, M.Y. Hussaini, A. Quarteroni, T.A. Zang, *Spectral Methods. Evolution to Complex Geometries and Applications to Fluid Mechanics* (Springer-Verlag, Berlin, 2007)
3. C. Canuto, M.Y. Hussaini, A. Quarteroni, T.A. Zang, *Spectral Methods. Fundamentals in Single Domains* (Springer-Verlag, Berlin, 2006)
4. J. Hesthaven, S. Gottlieb, D. Gottlieb, *Spectral Methods for Time-Dependent Problems* (Cambridge University Press, Cambridge, 2007)
5. M. Abramowitz, I.A. Stegun, in *Handbook of Mathematical Functions*, 10th edn. (Dover Publications, Mineola, 1972)
6. R. Baltensperger, M.R. Trummer, Spectral differencing with a twist. SIAM J. Sci. Comput. **24**, 1465 (2002)
7. L.N. Trefethen, *Spectral Methods in Matlab* (SIAM, Philadelphia, 2000)
8. J. Shen, T. Tang, *Spectral and High-Order Methods with Applications* (Science Press, Beijing, 2006)
9. D. Funaro, *Polynomial Approximation of Differential Equations* (Springer-Verlag, Berlin, 1992)
10. J.A.C. Weideman, S.C. Reddy, A MATLAB differentiation matrix suite. ACM Trans. Math. Softw. **26**, 465 (2000); the corresponding MATLAB software package is available at http://www.mathworks.com/matlabcentral/fileexchange/29-dmsuite
11. J.A.C. Weideman, The eigenvalues of Hermite and rational spectral differentiation matrices. Numer. Math. **61**, 409 (1992)
12. J.P. Boyd, in *Chebyshev and Fourier Spectral Methods*, 2nd edn (Dover Publications, Mineola, 2000); available at http://www-personal.engin.umich.edu/~jpboyd
13. B. Fornberg, in *A Practical Guide to Pseudospectral Methods* (Cambridge University Press, Cambridge, 1998). The concepts of "collocation" and "pseudospectral" methods are synonymous in our context, while a part of the expert community distinguishes between them and states that the only true pseudospectral methods are those involving global basis functions; see Section 3.5 in [3] and Section 3.1 in [12]

14. B. Shizgal, R. Blackmore, A discrete ordinate method of solution of linear boundary value and eigenvalue problems. J. Comput. Phys. **55**, 313 (1984)
15. D. Funaro, Lecture notes in computational science and engineering, in*Spectral Elements for Transport-Dominated Equations*, vol 1 (Springer-Verlag, Heidelberg, 1997)
16. M.H. Holmes, *Introduction to Numerical Methods in Differential Equations* (Springer-Verlag, New York, 2007)
17. L.N. Trefethen, M. Embree, *Spectra and Pseudospectra* (Princeton University Press, Princeton, 2005)
18. S. Gottlieb, D. Ketcheson, C.-W. Shu, *Strong Stability Preserving Runge-Kutta and Multistep Time Discretizations* (World Scientific, Singapore, 2011)
19. S. Gottlieb, D.I. Ketcheson, C.-W. Shu, High order strong stability preserving time discretizations. J. Sci. Comput. **38**, 251 (2009)
20. S. Gottlieb, On high order strong stability preserving Runge-Kutta and multi step time discretizations. J. Sci. Comput. **25**, 105 (2005)
21. A.J. Christlieb, S. Gottlieb, Z. Grant, D.C. Seal, Explicit strong stability preserving multistage two-derivative time-stepping schemes. J. Sci. Comput. **68**, 914 (2016); see also *Erratum*, J. Sci. Comput. **68**, 943 (2016)
22. C. Bresten, S. Gottlieb, Z. Grant, D. Higgs, D.I. Ketcheson, A. Németh, Explicit strong stability preserving multistep Runge-Kutta methods. Math. Comput. **86**, 747 (2017)
23. G. Izzo, Z. Jackiewicz, Strong stability preserving Runge-Kutta and linear multistep methods. Bull. Iran. Math. Soc. **48**, 4029 (2022)
24. Y. Hadjimichael, D.I. Ketcheson, L. Lóczi, A. Németh, Strong stability preserving explicit linear multistep methods with variable step size. SIAM J. Numer. Anal. **54**, 2799 (2016)
25. S. Gottlieb, C.-W. Shu, E. Tadmor, Strong stability-preserving high-order time discretizations. SIAM Rev. **43**, 89 (2001)
26. R.J. Spiteri, S.J. Ruuth, A new class of optimal high-order strong-stability-preserving time discretization methods. SIAM J. Numer. Anal. **40**, 469 (2002)
27. S.J. Ruuth, Global optimization of explicit strong-stability-preserving Runge-Kutta methods. Math. Comput. **75**, 183 (2006)
28. J.H. Williamson, Low-storage Runge-Kutta schemes. J. Comput. Phys. **35**, 48 (1980)
29. S.J. Ruuth, R.J. Spiteri, High-order strong-stability-preserving Runge-Kutta methods with downwind-biased spatial discretizations. SIAM J. Numer. Anal. **42**, 974 (2004)
30. J.P. Boyd, Orthogonal rational functions on a semi-infinite interval. J. Comput. Phys. **70**, 63 (1987)
31. V. Iranzo, A. Falqués, Some spectral approximations for differential equations in unbounded domains. Comp. Methods Appl. Mech. Eng. **98**, 105 (1992)
32. J. Shen, Stable and efficient spectral methods in unbounded domains using Laguerre functions. SIAM J. Numer. Anal. **38**, 1113 (2000)
33. B. Guo, J. Shen, Z. Wang, Chebyshev rational spectral and pseudospectral methods on a semi-infinite interval. Int. J. Numer. Meth. Eng. **53**, 65 (2002)
34. J.P. Boyd, C. Rangan, P.H. Bucksbaum, Pseudospectral methods on a semi-infinite interval with application to the hydrogen atom: a comparison of the mapped Fourier-sine method with Laguerre series and rational Chebyshev expansions. J. Comput. Phys. **188**, 56 (2003)
35. J.P. Boyd, Rational Chebyshev series for the Thomas-Fermi function: endpoint signularities and spectral methods. J. Comput. Appl. Math. **244**, 90 (2013)
36. J.P. Boyd, Five themes in Chebyshev spectral methods applied to the regularized Charney eigenproblem. Comput. Math. Appl. **71**, 1227 (2016)
37. F. Stenger, Numerical methods based on Whittaker cardinal, or sinc functions. SIAM Rev. **23**, 165 (1981)
38. K.M. McArthur, K.L. Bowers, J. Lund, The sinc method in multiple space dimensions: model problems. Numer. Math. **56**, 789 (1990)
39. D. Funaro, O. Kavian, Approximation of some diffusion evolution equations in unbounded domains by Hermite functions. Math. Comp. **57**, 597 (1991)

40. J.P. Boyd, Spectral methods using rational basis functions on an infinite interval. J. Comput. Phys. **69**, 112 (1987)
41. J.P. Boyd, *Dynamics of the Equatorial Ocean* (Springer-Verlag, Berlin, 2018)
42. J.P. Boyd, Five ways to solve the Yoshida jet problem of wind-driven equatorial flow. Dyn. Atmos. Oceans **83**, 1 (2018)
43. J.A.C. Weideman, Spectral differentiation matrices for the numerical solution of Schrödinger's equation. J. Phys. A: Math. Gen. **39**, 10229 (2006)
44. H. Meißner, E.O. Steinborn, Quartic, sextic, and octic anharmonic oscillators: precise energies of ground state and excited states by an iterative method based on the generalized Bloch equation. Phys. Rev. A **56**, 1189 (1997)
45. I. Kuščer, A. Kodre, H. Neunzert, *Mathematik in Physik und Technik* (Springer-Verlag, Berlin, 1993)

Chapter 13
Inverse and Ill-Posed Problems ★

Abstract Solving inverse and ill-posed problems is a common yet one of the most challenging tasks faced by a physicist. In this Chapter we discuss the basic properties of linear and non-linear operators involved in solving direct ("forward") problems and their inverse ("backward") counterparts, the reasons for their ill-posedness, and measures that can be taken to control and stabilize the inversion procedure by regularization. In the three core Sections we discuss topics that are presumably of greatest interest to physicists: the solution of Fredholm and Volterra integral equations of the first and second kind, the solution of inverse Sturm-Liouville problems, i. e. the reconstruction of the operators from their spectra and known symmetry properties, and the solution of retrospective and source/coefficient-recovery problems for partial differential equations. We present methods of phase retrieval based on the data providing only the intensities. Examples and Problems include stable deconvolution of noisy signals, image deblurring techniques, the inverse Radon transformation, and identifying the shape and refraction index of inhomogeneous domains by inverting the far-field pattern of scattered plane waves.

When we evaluate the expression $f = Au$, where u and f are vectors and A is a matrix, we solve a *direct* or *forward* problem. Given A we can precisely calculate f for any u. The only danger we may anticipate in a computer is the one of over- or underflow, or perhaps loss of precision. Solving a *backward* or *inverse* problem is a step in the opposite direction: we either wish to reconstruct u from f, knowing A—the so-called *inverse reconstruction problem*—or learn something about A from given u and f, which is known as the *inverse identification problem* because we are identifying the parameters of the underlying "model"—the presumed linear mapping embodied by the matrix A. In either case trouble is brewing: A may not be invertible, and even if it is, it may be ill-conditioned. The difficulties could be compounded by the data f being noisy, or A depending on u, introducing nonlinearity.

One could say that everything we ever do in physics research—identifying inputs to the system based on its outputs or assessing models—amounts to solving inverse problems in the above sense [1–4]. Most often such problems are *ill-posed*. A problem is considered to be *well-posed* if its solution exists, if this solution is unique, and if the solution depends smoothly on the data. If any of these three qualifiers is absent, the problem is said to be ill-posed.

S. Širca and M. Horvat, *Computational Methods in Physics*, Graduate Texts in Physics,
https://doi.org/10.1007/978-3-031-68566-8_13

Example Consider the linear equation $A\boldsymbol{u} = \boldsymbol{f}$ with

$$
A = \begin{pmatrix} 1 & 0 \\ 0 & 10^{-3} \end{pmatrix}, \qquad \boldsymbol{f} = \begin{pmatrix} 1 \\ 10^{-3} \end{pmatrix},
$$

whose unique solution is $\boldsymbol{u} = A^{-1}\boldsymbol{f} = (1, 1)^{\mathrm{T}}$. Let us modify the second element of $\boldsymbol{f}$ by $\delta = 10^{-3}$, such that $\boldsymbol{f}^{\delta} = \boldsymbol{f} + (0, \delta)^{\mathrm{T}} = (1, 2\cdot 10^{-3})^{\mathrm{T}}$. The solution $\boldsymbol{u}^{\delta}$ of the perturbed problem, $A\boldsymbol{u}^{\delta} = \boldsymbol{f}^{\delta}$, is $\boldsymbol{u}^{\delta} = (1, 2)^{\mathrm{T}}$, a large change in the result! Indeed, the ratio of the norm of the solution difference, $\boldsymbol{u} - \boldsymbol{u}^{\delta}$, to the norm of the perturbation, $\boldsymbol{f} - \boldsymbol{f}^{\delta}$, is

$$
r = \frac{\|\boldsymbol{u} - \boldsymbol{u}^{\delta}\|_2}{\|\boldsymbol{f} - \boldsymbol{f}^{\delta}\|_2} = 1000 \, .
$$

Now suppose we change the matrix A by introducing a parameter α as

$$
A_{\alpha} = \begin{pmatrix} 1 & 0 \\ 0 & (1 - \alpha)^{-3} 10^{-3} \end{pmatrix},
$$

and choose $\alpha = 0.2$. By keeping the same perturbation as before, the solution of $A_{\alpha}\boldsymbol{u}^{\delta} = \boldsymbol{f}^{\delta}$ becomes $\boldsymbol{u}^{\delta} = (1, 1.024)^{\mathrm{T}}$, with the error ratio of just $r = 24$. By guesswork we quickly find that taking $\alpha = 0.2063$ yields $u^{\delta} \approx (1, 0.999998)^{\mathrm{T}}$ and $r \approx 0.002$! In this obviously ill-posed problem, a particular choice of the functional form of the correction and a magical value of its parameter α dramatically reduce the reconstruction error. ◁

Example Figure 13.1 (left) shows a discrete square-wave signal $\boldsymbol{u}$ sampled at $N = 100$ points spaced $h = 1/N$ apart, its smoothly smeared version $A\boldsymbol{u}$ by using the matrix A with the elements

$$
A_{ij} = \frac{h}{\sqrt{2\pi}\,\gamma} \exp\left[-\frac{\big((i - j)h\big)^2}{2\gamma^2} \right], \qquad i, j = 1, 2, \ldots, 100 \, , \tag{13.1}
$$

where $\gamma = 0.03$, as well as the signal $\boldsymbol{f}^{\delta} = A\boldsymbol{u} + \boldsymbol{\delta}$ contaminated by seemingly harmless normally distributed noise, $\delta_i \sim 0.2 N(0, \sigma^2)$ with $\sigma = 0.2$.

The right panel shows a naive reconstruction of $\boldsymbol{u}$ by calculating $A^{-1}\boldsymbol{f}^{\delta}$. Since A is ill-conditioned, the attempt fails miserably. The problem obviously needs to be modified in some way in order to deliver a stable solution. ◁

Finding a suitable way to modify the matrix A such that its inversion is as insensitive to data perturbation as possible, rendering the inverse problem well-posed, is called *regularization* and is at the heart of successful solutions of inverse problems. In the present Chapter this strategy will be applied to a much broader class of linear and non-linear operators.

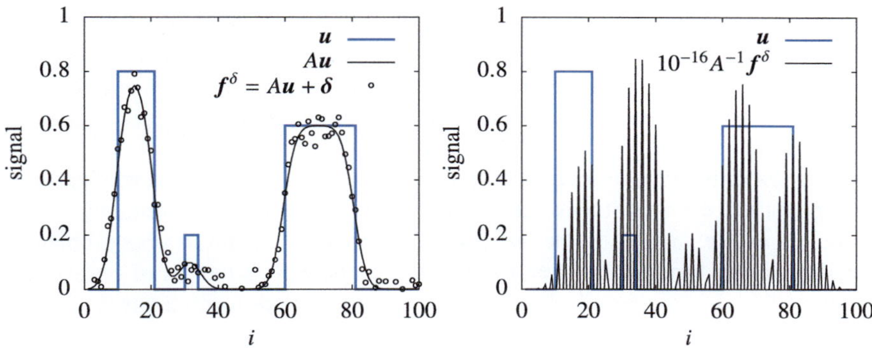

Fig. 13.1 [LEFT] The signal u, its smeared version Au and the final signal with added Gaussian noise. [RIGHT] The naive reconstruction of u by solving the equation $Au = f^\delta$ for u. The recovered solution is utterly useless: in the figure it has been multiplied by 10^{-16} (!) to fit into the scale of the plot

13.1   Classification of Inverse Problems

Each inverse problem is peculiar in its own way, and there is no universally accepted classification. One possible categorization is based on which segment of the quantities characterizing the direct problem is retrieved in solving its inverse. A direct problem is formulated by specifying the equation of the physical process, the domain Ω in which the process is studied, the initial conditions (for non-stationary processes) and the boundary conditions on the domain boundary $\partial\Omega$. As an illustration [5], consider the wave equation

$$u_{tt} - c^2(x)\left[\nabla^2 u - \nabla \log \rho(x) \cdot \nabla u\right] = h(x, t)\,, \quad x \in \Omega \subset \mathbb{R}^n\,, \quad t > 0\,,$$

where $u(x, t)$ is the pressure, $c(x)$ is the velocity of sound in the inhomogeneous medium with density $\rho(x)$, and $h(x, t)$ is the source term. Let the solution satisfy the initial conditions $u(x, 0) = \phi(x)$ and $u_t(x, 0) = \psi(x)$ as well as the boundary condition $(\partial u/\partial n)|_{\partial\Omega} = g(x, t)$.

Suppose now that we have some knowledge about the values of u at the boundary, $u|_{\partial\Omega} = f(x, t)$, the so-called *data of the inverse problem*. Various inverse problems can be classified based on which piece of the model is unknown. If we are required to determine the initial conditions, i.e. the functions ϕ and ψ, we are referring to a *retrospective* problem. If we need to establish the function g, we are about to solve an *inverse boundary-value problem*. Sometimes the initial conditions ϕ and ψ are unknown, while the boundary conditions and the supplemental information (f and g, respectively) are specified only on a part of the boundary and we need to find the solution $u(x, t)$. Such inverse problems are called *extension problems* as we are trying to extend the solution to the interior of the domain. If we are trying to determine the source function $h(x, t)$, we are solving a so-called *inverse source problem.* If we wish to determine the function $c(x)$, we are facing an *inverse coefficient problem.* The latter, for instance, is a paradigmatic problem in seismology [6, 7]: the determination of the properties of the Earth's crust and mantle from the measured pressure waves at the surface and their known source, the epicenter of the earthquake.

Often, as in the study of wave-propagation in the medium or scattering of particles on a potential, the amplitudes can be written in terms of their harmonic content as $u(x, t) = \widetilde{u}(x; \omega)\, e^{i\omega t}$. If the additional information for the inverse problem is specified in the form $\widetilde{u}(x; \omega_i) = f_i(x)$, $x \in X$, where X is the set of observation points and $\{\omega_i\}$ is the set of observed frequencies, we are formulating an *inverse scattering problem.* Just as important are the *inverse spectral* (or *isospectral*) *problems* in which we are seeking the coefficient functions based on some knowledge on the eigenvalues λ_i and/or eigenvectors u_i of the corresponding differential operator, for instance,

$$\nabla^2 u - \nabla \log \rho \cdot \nabla u = \lambda u \,.$$

Inverse problems can also be classified quite naturally according to the dimensionality of the process (or its model) itself and of the quantities being estimated [8]. The model may be of finite dimension (for instance, a system of n equations and m unknowns) or of infinite dimension (like an initial-value problem for a PDE). The quantities one wishes to reconstruct may also be of finite dimension (a limited set of scalar parameters of the model) or infinite dimension (for example, the coefficient functions in a PDE).

In this Section we explain the basic concepts of inverse problems involving bounded linear operators A acting between Hilbert spaces $\mathcal{P}$ and $\mathcal{Q}$, i.e. problems of the form

$$Au = f \,, \quad u \in \mathcal{P}\,, \quad f \in \mathcal{Q}\,, \quad A \in \mathcal{L}(\mathcal{P}, \mathcal{Q})\,, \tag{13.2}$$

where we are trying to determine u based on f. Matrices A establishing the mappings of the form $f = Au$, where u and f are vectors, for instance, will represent just a subfamily of such operators; the general exposition will also apply to other types of A, like integral or differential operators. The domain, the null-space (kernel) and the range of A will be denoted by $\mathcal{D}(A)$, $\mathcal{N}(A)$ and $\mathcal{R}(A)$, respectively. The adjoint operator of A is denoted by A^* and defined by

$$\langle Au, v \rangle_\mathcal{Q} = \langle u, A^*v \rangle_\mathcal{P}\,, \tag{13.3}$$

relating the inner products in the spaces $\mathcal{P}$ and $\mathcal{Q}$.

Example If A is a real matrix, we have simply $\langle Au, v \rangle = (Au)^\mathrm{T}v = v^\mathrm{T}Au = (A^\mathrm{T}v)^\mathrm{T}u = u^\mathrm{T}(A^\mathrm{T}v) = \langle u, A^\mathrm{T}v \rangle$, thus $A^* = A^\mathrm{T}$. For complex A, $A^* = A^\dagger$, the inner products being evaluated in an analogous fashion.   ◁

Example The integral operator A defined by

$$(Au)(y) = \int_a^b k(y, x)u(x)\,\mathrm{d}x = f(y)\,, \quad y \in (c, d)\,, \quad u \in L^2(a, b)\,,$$

where $k \in L^2\big((c, d) \times (a, b)\big)$, is bounded and linear in u. Its adjoint is given by

$$(A^*v)(y) = \int_c^d k(x, y)\,v(x)\,\mathrm{d}x\,, \quad y \in (a, b)\,, \quad v \in L^2(c, d)\,.$$

Indeed $\int_c^d v(y)\,\mathrm{d}y \int_a^b k(y, x)u(x)\,\mathrm{d}x = \int_a^b u(y)\,\mathrm{d}y \int_c^d k(x, y)v(x)\,\mathrm{d}x$, confirming the relation (13.3). $\lhd$

Example Consider the acceleration of a unit mass moving under the influence of the external force $f(t)$ acting during the interval $t \in [0, 1]$, with the initial conditions $u(0) = \dot{u}(0) = 0$. The equation of motion, $\ddot{u} = f(t)$, can be written as $Au(t) = f(t)$, where $A = \mathrm{d}^2/\mathrm{d}t^2$. What is the adjoint of A?

The relation (13.3) can be rewritten as $\langle Au, v \rangle - \langle u, A^*v \rangle = 0$ or

$$\int_0^1 \big[\ddot{u}(t)v(t) - u(t)A^*v(t)\big]\,\mathrm{d}t = 0\,.$$

This will obviously only work if $A^* = A$, since then

$$\int_0^1 \big[\ddot{u}(t)v(t) - u(t)\ddot{v}(t)\big]\,\mathrm{d}t = \int_0^1 \frac{\mathrm{d}}{\mathrm{d}t}\big[\dot{u}v - u\dot{v}\big]\,\mathrm{d}t = \big[\dot{u}v - u\dot{v}\big]\Big|_0^1 = 0\,.$$

Because $u(0) = \dot{u}(0) = 0$, the last expression gives $\dot{u}(1)v(1) - u(1)\dot{v}(1) = 0$. Since $u(1)$ and $\dot{u}(1)$ were left unspecified, this must apply to *any* $u(1)$ and $\dot{u}(1)$, which can only be fulfilled by $v(1) = \dot{v}(1) = 0$. The fully specified adjoint of A, operating on v, is therefore $A^*v = \mathrm{d}^2v/\mathrm{d}t^2$, $v(1) = \dot{v}(1) = 0$. $\lhd$

13.2 Generalized Solutions of $Au = f$

The key issue with any inverse problem of the form (13.2) is that its solution does not exist if $f \notin \mathcal{R}(A)$. This can easily happen as in general there is no guarantee that in the presence of noise f will remain in $\mathcal{R}(A)$. In the framework of linear algebra (see Sect. 4.5 and Appendix A.7) we circumvented this obstacle by trying to find u that minimizes the residual $Au - f$ in some norm: in Euclidean norm this amounted to finding the least-squares solution of $Au = f$, and this particular solution could be obtained by solving the normal equation (4.15). However, for $\mathcal{N}(A) \neq \{0\}$ there are infinitely many solutions that minimize $\|Au - f\|_Q$. The usual strategy is to pick the one with the smallest norm. For matrix problems, this was accomplished in Sect. 4.5.3 by means of the generalized (Moore–Penrose) inverse of A.

For general bounded linear operators, we call $u \in \mathcal{P}$ a least-squares solution of the inverse problem (13.2) if

$$\|Au - f\|_Q \le \|A\tilde{u} - f\|_Q \quad \forall \tilde{u} \in \mathcal{P}\,.$$

It can be shown [9] that the set of least-squares solutions is non-empty if and only if $f \in \mathcal{R}(A) \oplus \mathcal{R}(A)^{\perp}$ (see Fig. A.1), and that for arbitrary f the solution need not exist unless $\mathcal{R}(A)$ is closed. If the solution does exist, however, it can be obtained by solving the normal equation

$$A^*Au = A^*f\,,$$

where A^* is the adjoint of A. The solution $u^+ \in \mathcal{P}$ that satisfies

$$\|u^+\|_{\mathcal{P}} \le \|u\|_{\mathcal{P}} \quad \text{for all least} - \text{squares solutions } u$$

is called the *minimum-norm* solution of the inverse problem (13.2). If a least-squares solution exists, the minimum-norm solution is unique and can be computed by the generalized (Moore–Penrose) inverse of A,

$$u^+ = A^+f\,,$$

introduced in exact analogy to the finite-dimensional case (see A.12 and the subsequent definition). We will construct A^+ in Sect. 13.2.2.

13.2.1   Compact Operators and Their Singular Value Decomposition

Let $A \in \mathcal{L}(\mathcal{P}, Q)$. Then A is said to be *compact* if the image of a bounded sequence $\{u_j\}_{j\in\mathbb{N}} \subset \mathcal{P}$ contains a convergent subsequence $\{Au_{j_k}\}_{k\in\mathbb{N}} \subset Q$. Nearly all linear inverse problems involve compact operators, which can be seen as the infinite-dimensional analogues to ill-conditioned matrices. Namely, if the range of a compact operator A is infinite-dimensional, its Moore–Penrose inverse A^+ is discontinuous [10, 11]. Compactness is the main source of numerical instabilities (ill-posedness) in infinite-dimensional inverse problems.

The best way to understand the misbehavior of compact linear operators and manipulate it in order to stabilize their inverses leads through their spectral properties. If A is a self-adjoint operator, i.e. $A = A^*$ and $\langle Au, v \rangle_Q = \langle u, Av \rangle_{\mathcal{P}}$ for all $u \in \mathcal{P}$ and $v \in Q$, one may use its spectral decomposition

$$Au = \sum_{i=1}^{\infty} \lambda_i \langle u, u_i \rangle_{\mathcal{P}} u_i \, , \quad \forall u \in \mathcal{P} \, , \tag{13.4}$$

where λ_i are the non-zero real eigenvalues and $\{u_i\}$ is the set of corresponding orthogonal eigenvectors. If A is not self-adjoint, the spectral analysis in terms of eigenvalues and eigenvectors must be replaced by singular values and singular vectors. The singular system can be established by making a detour through operators $A^*A : \mathcal{P} \to \mathcal{P}$ and $AA^* : Q \to Q$, both of which are compact, non-negative and self-adjoint—hence by (13.4) their spectral representations are

$$A^*Au = \sum_{i=1}^{\infty} \mu_i \langle u, u_i \rangle_{\mathcal{P}} u_i \, , \quad \forall u \in \mathcal{P} \, ,$$

and

$$AA^*v = \sum_{i=1}^{\infty} \mu_i \langle v, v_i \rangle_{Q} v_i \, , \quad \forall v \in Q \, .$$

Due to orthogonality of u_i and v_i we have $A^*Au_i = \mu_i u_i$ and $AA^*v_i = \mu_i v_i$, and the square roots of the eigenvalues of A^*A and AA^*, $\sigma_i = \sqrt{\mu_i}$, are the singular values of A and A^*. The singular value decompositions of A and A^* are given by

$$Au = \sum_{i=1}^{\infty} \sigma_i \langle u, u_i \rangle_{\mathcal{P}} v_i \, , \quad u \in \mathcal{P} \, , \tag{13.5}$$

and

$$A^*v = \sum_{i=1}^{\infty} \sigma_i \langle v, v_i \rangle_{Q} u_i \, , \quad v \in Q \, ,$$

where σ_i are ordered as $\sigma_1 \geq \sigma_2 \geq \cdots > 0$, the *right singular vectors* $\{u_i\}_{i \in \mathbb{N}} \subset \mathcal{P}$ are an orthonormal basis of $\mathcal{N}(A)^{\perp}$, and the *left singular vectors* $\{v_i\}_{i \in \mathbb{N}} \subset Q$ are an orthonormal basis of $\mathcal{R}(A)$, such that

$$Au_i = \sigma_i v_i \, , \quad A^*v_i = \sigma_i u_i \, . \tag{13.6}$$

The construction of a self-adjoint operator A^*A or AA^* from A and A^* that individually do not possess this property sometimes allows for a relatively easy determination of the singular system of A and A^*, as illustrated by the following Example.

Example (Adapted from [12], Ex. 2.4.1.) Let $\mathcal{P} = Q = L^2(0, \pi)$ and let the linear integral operator A be defined by

$$(Au)(x) = \int_{x}^{\pi} u(t) \, dt \, , \quad x \in (0, \pi) \, , \quad u \in \mathcal{P} \, .$$

It is easy to see that the relation $\langle Au, v \rangle_Q = \langle u, A^*v \rangle_{\mathcal{P}}$ is satisfied when the adjoint of A operates as

$$(A^*v)(x) = \int_0^x v(t)\, dt \,, \quad x \in (0, \pi) \,, \quad v \in \mathcal{P} \,.$$

Neither A nor A^* are self-adjoint, but A^*A is: it is given by

$$(A^*Au)(x) = \int_0^x dt \int_t^\pi u(s)\, ds \tag{13.7}$$

and has positive eigenvalues $\mu_i = \sigma_i^2$ which can be determined from the equation

$$(A^*Au)(x) = \sigma^2 u(x) \,. \tag{13.8}$$

Differentiating both sides twice with respect to x leads to the second-order differential equation $-u''(x) = \lambda u(x)$, $\lambda = 1/\sigma^2$, calling for two boundary conditions. The first one is obtained by using $x = 0$ in (13.8), yielding $u(0) = 0$ by (13.7). Differentiating both sides of (13.8) with respect to x and evaluating the result at $x = \pi$ we get $u'(\pi) = 0$. We have thus transcribed (13.8) to the classical Sturm–Liouville problem

$$- u''(x) = \lambda u(x) \,, \quad x \in (0, \pi) \,, \quad u(0) = u'(\pi) = 0 \,,$$

whose eigenvalues are $\lambda_i = (i - 1/2)^2$, $i \in \mathbb{N}$, while its normalized eigenvectors are $u_i(x) = \sqrt{2/\pi}\,\sin(\sqrt{\lambda_i}\,x)$. Therefore the eigenvalues $\sigma_i^2 = 1/\lambda_i$ and the corresponding eigenvectors of A^*A are

$$\sigma_i^2 = \frac{1}{(i - 1/2)^2} \,, \quad u_i(x) = \sqrt{\frac{2}{\pi}}\,\sin\left[\left(i - \tfrac{1}{2}\right)x\right] \,, \quad i \in \mathbb{N} \,.$$

In turn, the singular values and the right and left singular vectors of A are

$$\sigma_i = \frac{1}{i - 1/2} \,, \quad u_i(x) = \sqrt{\frac{2}{\pi}}\,\sin\left[\left(i - \tfrac{1}{2}\right)x\right] \,, \quad v_i(x) = \sqrt{\frac{2}{\pi}}\,\cos\left[\left(i - \tfrac{1}{2}\right)x\right] \,,$$

all of which can be checked explicitly by evaluating (13.8) and (13.6).   ◁

13.2.2 The Pseudoinverse of Compact Operators

The singular value decomposition (13.5) of a compact linear operator A can now be used to construct its Moore–Penrose inverse, which is given by

$$A^+ f = \sum_{i=1}^{\infty} \frac{1}{\sigma_i} \langle f, v_i \rangle_Q u_i \, , \quad f \in \mathcal{D}(A^+) \, . \tag{13.9}$$

If this expansion converges (see condition (13.11) below), $u^+ = A^+ f$ is the minimum-norm solution of $Au = f$, while the set of all least-squares solutions is $u^+ + \mathcal{N}(A)$ [11]. Furthermore, u^+ satisfies the normal equation

$$A^* A u = \sum_{i=1}^{\infty} \sigma_i \langle f, v_i \rangle_Q u_i = A^* f \, , \tag{13.10}$$

which is an alternative way of computing the inverse. It is useful to know that the minimum-norm solution exists only if

$$\sum_{i=1}^{\infty} \frac{1}{\sigma_i^2} \left| \langle f, v_i \rangle_Q \right|^2 < \infty \, . \tag{13.11}$$

This is known as the Picard criterion and imposes a condition on the decay of the expansion coefficients of the pseudoinverse (13.9). In particular, if the singular vectors are harmonic basis functions, the inner products $\langle f, v_i \rangle$ are simply the Fourier coefficients of f, and the Picard criterion is a statement about the relative importance of low-frequency and high-frequency components of f, i.e. its smoothness. Note that $\sigma_i \to 0$ as $i \to \infty$, hence the expansion coefficients must decay to zero faster than σ_i in order to ensure convergence.

The $1/\sigma_i$ dependence of the individual terms in (13.9) is the key source of ill-posedness of inverse problems. Assume that f is contaminated by noise with the functional form v_j and amplitude $\delta > 0$, resulting in the right-hand side of the form $f^\delta = f + \delta v_j$, and let u_δ^+ denote the minimal-norm solution of $Au = f^\delta$. Then

$$\left\| u^+ - u_\delta^+ \right\|_{\mathcal{P}} = \left\| A^+ f - A^+ f^\delta \right\|_{\mathcal{P}} = \delta \left\| A^+ v_j \right\|_{\mathcal{P}} = \frac{\delta}{\sigma_j} \to \infty \quad \text{as} \quad j \to \infty \, ,$$

where we have exploited the orthonormality of u_i and v_i. At fixed j, therefore, the amplification of noise during the inversion depends on the magnitude of σ_j. A faster decay of the singular values implies a stronger amplification of noise.

13.3 Regularization of Linear Inverse Problems

The fast decay of the singular values of A which primarily causes the inverse problem
to be ill-posed can be counteracted by *regularization*. The idea is to introduce a
positive real parameter α and replace the $1/\sigma$ factor in (13.9) by a bounded function
g_α which attenuates the relative importance of small σ and at the same time satisfies
$\lim_{\alpha \to 0} g_\alpha(\sigma) = 1/\sigma$. The operator Γ_α defined by

$$\Gamma_\alpha f = \sum_{i=1}^{\infty} g_\alpha(\sigma_i)\langle f, v_i\rangle_Q u_i \tag{13.12}$$

is then called a *regularization* of A^+—and thus of the inverse problem. Ideally,
eliminating the contribution of small singular values makes the inversion insensitive
to the presence of noise in f. The desired effect is illustrated in Fig. 13.2.

Regardless of the form of $g_\alpha(\sigma)$ regularization implies an artificial modification
of the operator's spectral properties, introducing an *approximation error* in addition
to the *data error* caused by the noisiness of f. This can be seen by evaluating the
error of the regularized solution of the noisy problem, $u_\alpha^\delta = \Gamma_\alpha f^\delta$, with respect to
the minimal-norm solution of the noise-free problem, $u^+ = A^+ f$, and splitting it as

$$\begin{aligned}
\|u_\alpha^\delta - u^+\|_{\mathcal{P}} &\leq \|\Gamma_\alpha f^\delta - \Gamma_\alpha f\|_{\mathcal{P}} + \|\Gamma_\alpha f - u^+\|_{\mathcal{P}} \\
&\leq \underbrace{\delta\|\Gamma_\alpha\|_{\mathcal{L}(Q,\mathcal{P})}}_{E_{\text{data}}(\alpha)} + \underbrace{\|\Gamma_\alpha f - A^+ f\|_{\mathcal{P}}}_{E_{\text{approx}}(\alpha)} = E_{\text{total}}(\alpha)\,.
\end{aligned} \tag{13.13}$$

The first term depends on the level of noise δ and does not stay bounded as $\alpha \to 0$.
The second term does vanish in the limit $\alpha \to 0$ but grows rapidly when one tries to
"over-regularize" the problem, thus removing it far from its original setting. In plain

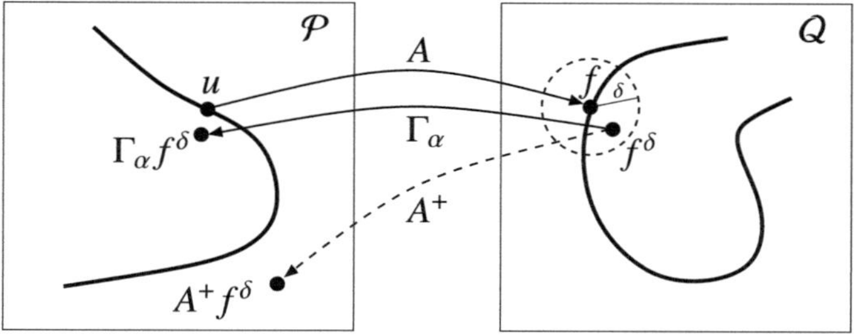

Fig. 13.2 Regularization of the inverse problem. When A^+ is applied to a noisy right-hand side of
$Au = f$ (indicated by f^δ residing within a δ-neighborhood of f), the result will typically be far
away from the desired minimum-norm solution u^+ (dashed arrow). The regularized operator Γ_α
maps f^δ much closer to the true solution and remains in its vicinity for a wide range of α and even
in the presence of noise

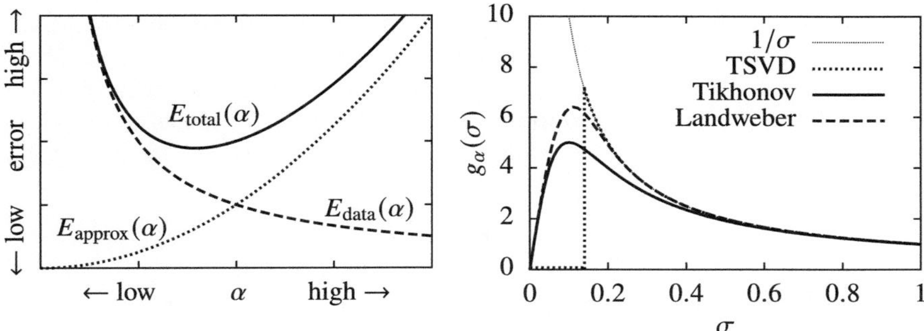

Fig. 13.3 [LEFT] The error of the regularized inverse (Eq. (13.13)) with contributions from the noisy data and the error of the approximation. [RIGHT] Some standard choices for the regularization function $g_\alpha(\sigma)$: TSVD (here with $\alpha = 0.14$), Tikhonov (Eq. (13.15), here with $\alpha = 0.01$), and Landweber (Eq. (13.22), here with $\alpha = 0.01$, $\tau = 1$)

words, the solutions with $\alpha \to 0$ are useless because regularization is insufficient and the noise wins, while the solutions at large α are useless because they come from solving a problem that bears little resemblance to the original one. A sound approach is to try to find an optimal α where the total error is minimal: see Figs. 13.3 (left) and 13.4 (top left). In the following, we present a few of the simplest regularization techniques.

13.3.1 Truncated Singular Value Decomposition

The simplest way to suppress small singular values is to truncate the expansion (13.12), i.e. discard all $1/\sigma$ that fall below the threshold specified by α,

$$g_\alpha(\sigma) = \begin{cases} 1/\sigma & ;\ \sigma \geq \alpha\,, \\ 0 & ;\ \sigma < \alpha\,. \end{cases}$$

Thus the truncated SVD (TSVD) solution is given by

$$u_\alpha = \Gamma_\alpha f = \sum_{\sigma_i \geq \alpha} \frac{1}{\sigma_i} \langle f, v_i \rangle_Q\, u_i\,. \tag{13.14}$$

If the data error can be bounded as $\| f - f^\delta \|_Q \leq \delta$, the TSVD results in the error estimate $\| Au_\alpha - Au_\alpha^\delta \|_Q \leq \delta$ and, more importantly,

$$\left\| u_\alpha - u_\alpha^\delta \right\|_{\mathcal{P}} \leq \delta/\alpha\,.$$

This inequality tells us that higher noise implies that more and more singular values need to be suppressed if the solution error is to be kept below a desired value. Moreover, the choice of α is not arbitrary as it may not lead to a convergent regularization: in particular, α may be assumed to depend on δ alone, on f^δ alone, or on both δ and f^δ. The methods by which the optimal α can be determined will be discussed in Sect. 13.3.7.

13.3.2 Tikhonov Regularization

Classic Tikhonov regularization uses the function

$$g_\alpha(\sigma) = \frac{\sigma}{\sigma^2 + \alpha} , \tag{13.15}$$

which generates the regularized inverse

$$u_\alpha = \Gamma_\alpha f = \sum_{i=1}^{\infty} \frac{\sigma_i}{\sigma_i^2 + \alpha} \langle f, v_i \rangle_Q \, u_i . \tag{13.16}$$

Assuming that α depends only on δ and that $\lim_{\delta \to 0} \delta/(2\sqrt{\alpha(\delta)}) = 0$, the Tikhonov regularization brings about the error estimate

$$\left\| u_\alpha - u_\alpha^\delta \right\|_{\mathcal{P}} \le \delta \Big/ \left(2\sqrt{\alpha(\delta)} \right) . \tag{13.17}$$

Example Let us revisit the Example of Fig. 13.1 and try to regularize the ill-conditioned inverse by using the Tikhonov technique. We are dealing with an algebraic problem $Au = f^\delta$ with the matrix $A \in \mathbb{R}^{n \times n} = \mathcal{P} = Q$ given by Eq. (13.1), $u \in \mathbb{R}^n$ and $f^\delta \in \mathbb{R}^n$, $n = 100$. The singular value decomposition of A,

$$A = Q\Sigma P^{\mathrm{T}} ,$$

yields the singular values σ_i (diagonal of Σ) and the singular vectors $u_i \in \mathbb{R}^n$ (columns of P) and $v_i \in \mathbb{R}^n$ (columns of Q), so the regularized inverse (13.16) applied to the noisy data f^δ becomes

$$u_\alpha^\delta = \Gamma_\alpha f^\delta = \sum_{i=1}^{n} \frac{\sigma_i}{\sigma_i^2 + \alpha} \left(f^\delta \cdot v_i \right) u_i . \tag{13.18}$$

Figure 13.4 shows the regularized solutions u_α^δ (top right and bottom panels) for three choices of α. For illustration we can assume that we know the exact solution u and take the norm of the error $\|u_\alpha^\delta - u\|_2$ as a measure of the quality of the regularized

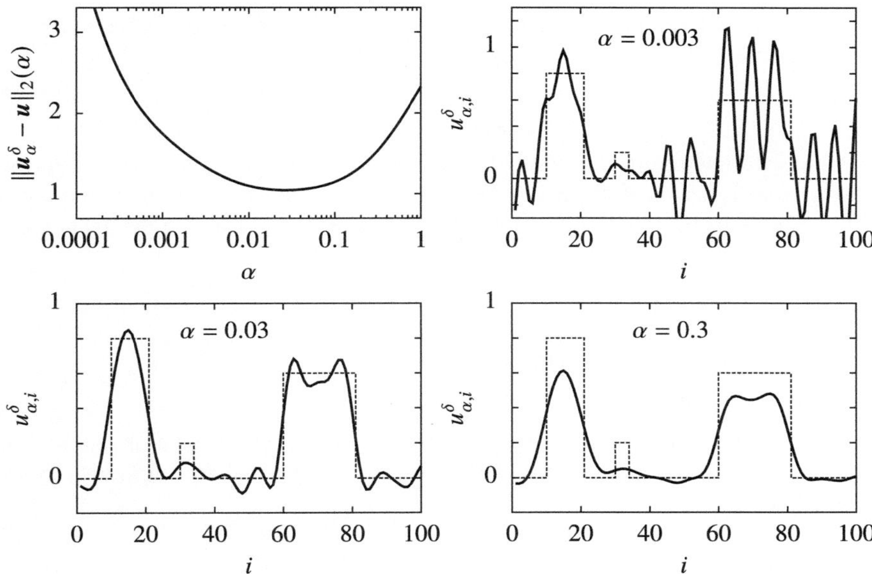

Fig. 13.4 [TOP LEFT] The norm of the reconstructed solution error and [TOP RIGHT, BOTTOM] the regularized solutions (13.18) for three values of α

inverse. It is shown in Fig. 13.4 (top left). The optimal $\alpha \approx 0.03$ corresponds to the minimum of the error. We shall have another look at how to stabilize the deconvolution in Sect. 13.3.6.   ◁

Tikhonov regularization has an extremely important advantage: it can be computed without explicit knowledge of the singular values and singular vectors. This can be seen by evaluating the decomposition (13.16) of $(A^*A + \alpha I)u_\alpha$ instead of u_α:

$$(A^*A + \alpha I)u_\alpha = \sum_{i=1}^{\infty} \frac{\sigma_i}{\sigma_i^2 + \alpha} \langle f, v_i \rangle_Q A^*Au_i + \sum_{i=1}^{\infty} \frac{\alpha\sigma_i}{\sigma_i^2 + \alpha} \langle f, v_i \rangle_Q u_i$$

$$= \sum_{i=1}^{\infty} \frac{\sigma_i(\sigma_i^2 + \alpha)}{\sigma_i^2 + \alpha} \langle f, v_i \rangle_Q u_i = \sum_{i=1}^{\infty} \sigma_i \langle f, v_i \rangle_Q u_i = A^*f,$$

where we have used (13.6) to evaluate $A^*Au_i = \sigma_i A^*v_i = \sigma_i^2 u_i$. In short, the Tikhonov-regularized solution can be obtained by solving the equation

$$(A^*A + \alpha I)u_\alpha = A^*f \tag{13.19}$$

for u_α, which is a well-posed linear problem involving a self-adjoint, positive definite operator. In other words, an ill-posed equation involving A has been replaced by a "nearby" well-posed equation involving A^*A.

Note that Tikhonov regularization may also be understood as the minimization of the functional

$$J_\alpha(u) = \frac{1}{2}\|Au - f\|_Q^2 + \frac{\alpha}{2}\|u\|_{\mathcal{P}}^2 \tag{13.20}$$

by which we attempt to minimize the residual $Au - f$ (i.e. solve $Au = f$ in the least-squares sense) with the added "penalty term" intended to minimize the norm of u (i.e. find the minimum-norm solution) [13]. For $f \in Q$ the u_α given by (13.16) is the unique global minimizer of $J_\alpha(u)$.

13.3.3   Landweber Iteration

Another regularization method can be devised by rewriting the equation $Au = f$ in the form $u = (I - \tau A^*A)u + \tau A^* f$ for some $\tau > 0$ and interpreting it as a fixed-point mapping

$$u^{(k+1)} = \left(I - \tau A^*A\right)u^{(k)} + \tau A^* f , \qquad k = 0, 1, 2, \ldots \tag{13.21}$$

with $u^{(0)} = 0$, known as Landweber iteration [13]. When it converges to the final u, this very u solves (13.10). Note that τ is not yet the true regularization parameter but an auxiliary scalar which must fulfill certain requirements for the iteration to converge:

$$0 < \tau < \frac{2}{\|A\|_{\mathcal{L}(\mathcal{P},Q)}^2} \qquad \text{or} \qquad 0 < \tau\sigma_i < 2 \;\; \forall i ,$$

where σ_i are the singular values of A. If, however, the iteration index k is interpreted as the inverse of the regularization parameter, $k = 1/\alpha$, Landweber iteration is equivalent to a regularization method with the weight function

$$g_\alpha(\sigma) = \frac{1}{\sigma}\left[1 - \left(1 - \tau\sigma^2\right)^{1/\alpha}\right] , \tag{13.22}$$

that is,

$$u_\alpha = \Gamma_\alpha f = \sum_{i=1}^{\infty} \frac{1}{\sigma_i}\left[1 - \left(1 - \tau\sigma_i^2\right)^{1/\alpha}\right]\langle f, v_i\rangle_Q\, u_i .$$

For $f = Au^+$ (noise-free data) the Landweber iterates satisfy the error estimate

$$\left\|u^{(k)} - u^+\right\|_{\mathcal{U}} = O\left(1/\sqrt{k}\right) = O\left(\sqrt{\alpha}\right) .$$

Assuming that a $w \in Q$ exists such that $u^+ = A^*w$, one can also derive the error estimate for the solution u_δ corresponding to noisy data with $\|f^\delta - f\|_Q \leq \delta$:

$$\left\| u_\delta^{(k)} - u^+ \right\|_{\mathcal{P}} \le \tau k \delta \| A \|_{\mathcal{L}(\mathcal{P},\mathcal{Q})} + \frac{\| w \|_{\mathcal{Q}}}{\sqrt{\tau(2k-1)}} \, ,$$

where $f = Au^+$. Hence the "step size" τ should not be too large and, obviously, k may not be pushed all the way to infinity.

If $A \in \mathbb{R}^{m \times n}$ with $m \ge n$ and $\mathrm{rank}(A) = n$, an alternative stopping criterion for the iteration can be established if one assumes that for some $\delta > 0$ we have

$$\left\| f - f^\delta \right\|_2 + \left\| Au - f \right\|_2 \le \delta \, , \qquad \left\| f^\delta \right\|_2 > \tau \delta \, ,$$

where $\tau > 1$ and $u^+ = A^+ f$ is the pseudoinverse computed with exact data, i.e. the solution of $A^* A u = A^* f$. Then there exists a smallest iteration index k such that

$$\left\| Au^{(k)} - f^\delta \right\|_2 \le \tau \delta < \left\| Au^{(k-1)} - f^\delta \right\|_2 \, ,$$

which can be used to terminate the iteration. At this k, the estimate

$$\left\| u^{(k)} - u^+ \right\|_2 \le \frac{\tau + 1}{\sigma_n} \delta$$

applies, where σ_n is the smallest singular value of A [6].

Example (Adapted from [13].) Consider the integral equation

$$\int_0^1 (1 + xt)\mathrm{e}^{xt} u(t) \, \mathrm{d}t = \mathrm{e}^x = f(x) \, , \qquad 0 \le x \le 1 \, , \tag{13.23}$$

with the unique exact solution $u_{\mathrm{e}}(x) = 1$. This is a representative of Fredholm integral equations of the first kind which will be treated in more detail in Sect. 13.4. The operator $A : L^2(0, 1) \to L^2(0, 1)$ corresponding to this problem,

$$(Au)(x) = \int_0^1 (1 + xt)\mathrm{e}^{xt} u(t) \, \mathrm{d}t \, ,$$

is self-adjoint, $A^* = A$. We discretize the problem by setting up a uniform mesh with $x_i = i/N, i = 0, 1, \ldots, N$, with N even, and evaluate the integral by Simpson's formula, yielding the discrete version of the operator:

$$(Au)(x_i) = \sum_{j=0}^{N} w_j (1 + x_i x_j)\, \mathrm{e}^{x_i x_j} u(x_j) \, , \qquad w_j = \begin{cases} 1/(3N) \; ; \; j = 0 \text{ or } 1 \, , \\ 4/(3N) \; ; \; j = 1, 3, \ldots, N-1 \, , \\ 2/(3N) \; ; \; j = 2, 4, \ldots, N-2 \, , \end{cases}$$

where w_j are the Simpson quadrature weights. The operator A is now represented by a $(N+1) \times (N+1)$ matrix, while the solution and the right-hand side are $(N+1)$-dimensional vectors, $\boldsymbol{u} = (u_0, u_1, \ldots, u_N)^{\mathrm{T}}$ and $\boldsymbol{f} = (f_0, f_1, \ldots, f_N)^{\mathrm{T}}$ with $u_i = u(x_i)$ and $f_i = f(x_i)$. We perturb $\boldsymbol{f} \to \boldsymbol{f}^\delta$ by uniformly distributed random numbers $r_i \sim U(-1/2, 1/2)$, forming the components $f_i^\delta = f_i + r_i$ and normalizing the perturbation such that $\left\| \boldsymbol{f} - \boldsymbol{f}^\delta \right\|_{2,\Delta x} = \delta$. (Note that we are obliged to use the "energy" norm (A.6) with $\Delta x = 1/N$ in order to eliminate the effect of discretization on the magnitude of the error.)

Let us first solve the problem by using Tikhonov regularization. Since $A^* = A$, Tikhonov's equation (13.19) becomes

$$\alpha\, \boldsymbol{u}_\alpha^\delta + A^2 \boldsymbol{u}_\alpha^\delta = A \boldsymbol{f}^\delta$$

and we solve it many times at fixed δ and chosen α in order to reduce the effect of statistical fluctuations in $\boldsymbol{f}^\delta$ on $\boldsymbol{u}_\alpha^\delta$. The average errors of the results with $N = 16$ for a variety of δ and a broad span of α are shown in Fig. 13.5 (left).

At fixed δ the error shows the characteristic initial drop-off when α is being reduced, down to the optimum beyond which the error starts to increase again. Of course, a part of this rise stems from the discretization error of the Simpson formula itself. But the effects of different N become visible only at very low α.

Next we solve the discretized problem by the Landweber method (13.21) with $\tau = 0.5$, again using $N = 16$. The error of the solution changes during the iteration but ultimately stabilizes, as shown in Fig. 13.5 (right) for several perturbations δ. Note that the errors are comparable in size to the ones obtained by Tikhonov's method, and remain stable even at relatively large δ.   ◁

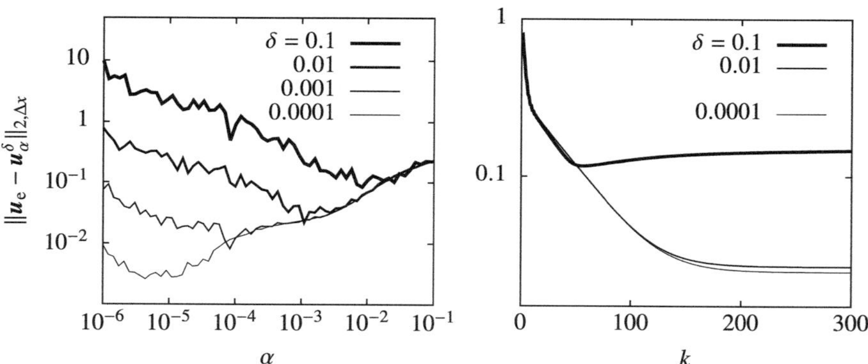

Fig. 13.5 [LEFT] The solution error of the discretized and Tikhonov-regularized problem (13.23) as a function of α for several noise levels δ. [RIGHT] The evolution of the solution error during the Landweber iteration with $\tau = 0.5$ for different δ

13.3.4   Conjugate Gradients Method Applied to the Normal Equation

Linear inverse problems involving matrices can also be solved by resorting to the conjugate gradient (CG) method (see page 185). The idea is to minimize the norm $\|Au - f^\delta\|_2$ by applying the conjugate gradient algorithm to the normal equation with perturbed data, $A^*Au = A^*f^\delta$. We assume that $A \in \mathbb{R}^{m \times n}$, $m \geq n$, with $\mathrm{rank}(A) = n$, while $u \in \mathbb{R}^n$ and $b \in \mathbb{R}^m$. The iterative CG algorithm consists of picking an initial solution guess (which can be zero) and an initial descent direction,

$$u^{(0)} = 0, \qquad r^{(0)} = f^\delta, \qquad p^{(0)} = A^*r^{(0)},$$

then repeating

$$
\begin{aligned}
\alpha_k &= \left\|A^*r^{(k-1)}\right\|_2^2 \Big/ \left\|Ap^{(k-1)}\right\|_2^2, \\
u^{(k)} &= u^{(k-1)} + \alpha_k p^{(k-1)}, \\
r^{(k)} &= r^{(k-1)} - \alpha_k A p^{(k-1)}, \\
\beta_k &= \left\|A^*r^{(k)}\right\|_2^2 \Big/ \left\|A^*r^{(k-1)}\right\|_2^2, \\
p^{(k)} &= A^*r^{(k)} + \beta_k p^{(k-1)},
\end{aligned}
\tag{13.24}
$$

for $k = 1, 2, 3, \ldots$ until convergence is achieved. Convergence can be controlled by the following estimates. Assume that for some known $\delta > 0$ one can establish the bound

$$\left\|f - f^\delta\right\|_2 + \left\|Au^+ - f\right\|_2 \leq \delta < \left\|f^\delta\right\|_2,$$

where $u^+ = A^+f$ is the pseudoinverse computed with exact data. Then a smallest integer k exists at which the subsequent iterates satisfy

$$\left\|Au^{(k)} - f^\delta\right\|_2 \leq \delta < \left\|Au^{(k-1)} - f^\delta\right\|_2, \tag{13.25}$$

which can be used as a stopping criterion for the iteration. At this very k, we have

$$\left\|u^{(k)} - u^+\right\|_2 \leq \frac{2\delta}{\sigma_n},$$

where σ_n is the smallest eigenvalue of A. Most importantly, the CG algorithm has a peculiar property of being self-regularizing, at least to some extent: see Appendix D of [6]. No regularization parameter needs to be introduced! While the algorithm implicitly minimizes the functional $J(u) = \|Au - f^\delta\|$, it can also be set up to minimize (13.20), which corresponds to Tikhonov regularization—but it may happen that such explicit regularization is redundant.

As stressed in Sect. 4.3.3, the convergence of the CG algorithm can be improved by preconditioning, in which the normal system $A^*Au = A^*f^\delta$ is replaced by an equivalent system

$$M^{-1}A^*A(M^*)^{-1}M^*u = M^{-1}A^*f^\delta$$

with a suitably chosen preconditioning matrix M.

The above procedure has been spelled out with the $\|\cdot\|_2$ norm and the usual inner (scalar) product, but in fact it can be applied to any other norm and extended to quite general Banach spaces: see, for instance, [14–17].

Example (Adapted from [6].) Seismic tomography is a method of exploration of Earth's interior, measuring the sound waves produced at the surface and reflected from the depths with unknown properties characterized by *acoustic impedance*. In a simple picture we may imagine plane waves propagating through horizontally stratified medium. The process can be described by an integral equation

$$f(t) = -\int_0^{X_0} g(t - 2s)u(s)\,\mathrm{d}s\,, \qquad X_0 = \tfrac{1}{2}\,,$$

where $g(t)$ is the pressure applied at time t on the surface, $f(t)$ is the vertical velocity at the surface recorded by geophones (seismogram), and $u(t)$ is a measure of deviation of acoustic impedance from a constant value. We take

$$g(t) = G(5t - 1)\,, \qquad G(z) = (1 - 2\pi^2 z^2)\,\mathrm{e}^{-\pi^2 z^2}\,,$$

and pretend for the moment that u (indexed by 'e' for 'exact') is known,

$$u_\mathrm{e}(t) = 2t(1 - 2t)\,, \quad t \in \left[0, \tfrac{1}{2}\right].$$

With given $g(t)$ and $u_\mathrm{e}(t)$ the integral can be calculated analytically:

$$f(t) = \frac{1}{500\,\pi^{5/2}}\left[5\sqrt{\pi}\left(\mathrm{e}^{-A^2(t)} + \mathrm{e}^{-B^2(t)}\right) + \mathrm{erf}\big(A(t)\big) - \mathrm{erf}\big(B(t)\big)\right]\,,$$

where $A(t) = \pi(1 - 5t)$ and $B(t) = \pi(6 - 5t)$. The functions $g(t)$ and $f(t)$ are shown in Fig. 13.6.

The inverse problem is to find u from a realistic, noisy seismogram f^δ. In order to use the conjugate gradients method operating with matrices and vectors, we discretize the problem on two different uniform grids, one on which to capture the measured signal, $t_i = iT_0/m$, $T_0 = 1$, $i = 1, 2, \ldots, m$, and one on which to represent the approximate solution, $\tau_j = jX_0/n$, $X_0 = 1/2$, $j = 1, 2, \ldots, n$. The latter grid is used to expand the solution in terms of piecewise-linear "hat" functions defined in Fig. 11.12:

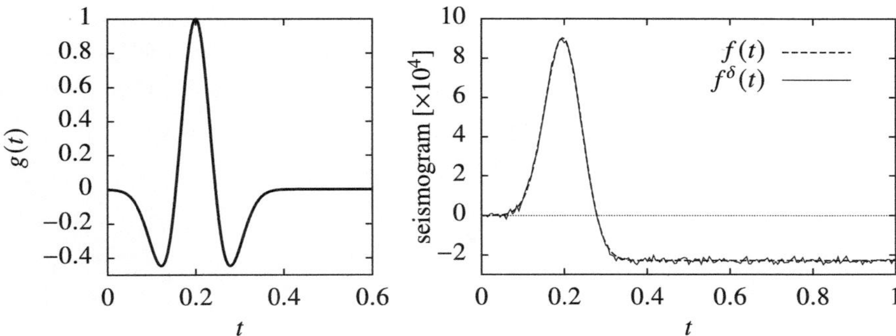

Fig. 13.6 Typical signals involved in linearized acoustic tomography. [LEFT] Source signal. [RIGHT] Ideal and realistic (noisy) seismograms $f(t)$ and $f^\delta(t)$

$$\widetilde{u}(t) = \sum_{j=1}^{n} u_j \phi_j(t) \,.$$

Organizing the components of f and the expansion coefficients u_j as vectors

$$f = \left(f_1, f_2, \ldots, f_m\right)^{\mathrm{T}} \,, \qquad u = \left(u_1, u_2, \ldots, u_n\right)^{\mathrm{T}} \,,$$

leads to a system of linear equations $Au = f$ for the expansion coefficients with a $m \times n$ matrix A whose elements are given by

$$A_{ij} = -\int_{0}^{X_0} g(t_i - 2\,s)\phi_j(s)\,\mathrm{d}s \,,$$

where $i = 1, 2, \ldots, m$ and $j = 1, 2, \ldots, n$. We simulate the measured seismogram (see Fig. 13.6 (right)) by adding normally distributed noise with zero mean and $\sigma = 10^{-5}$ to its individual components,

$$f_i^\delta = f_i + R_i \,, \qquad R_i \sim N(0, \sigma^2) \,, \tag{13.26}$$

such that $\delta = \|f - f^\delta\|_2 \approx \sqrt{m}\sigma$, and solve the equation $Au^\delta = f^\delta$ by iteration (13.24), using (13.25) as the stopping criterion. The resulting approximate solution

$$\widetilde{u}^\delta(t) = \sum_{j=1}^{n} u_j^\delta \phi_j(t)$$

for $m = 200$ and $n = 100$ is shown in Fig. 13.7 (left) together with the presumed exact solution $u_{\mathrm{e}}(t)$, while Fig. 13.7 (right) shows the evolution of two error estimates during the iteration. The reconstructed u exhibits wiggles that persist even if the

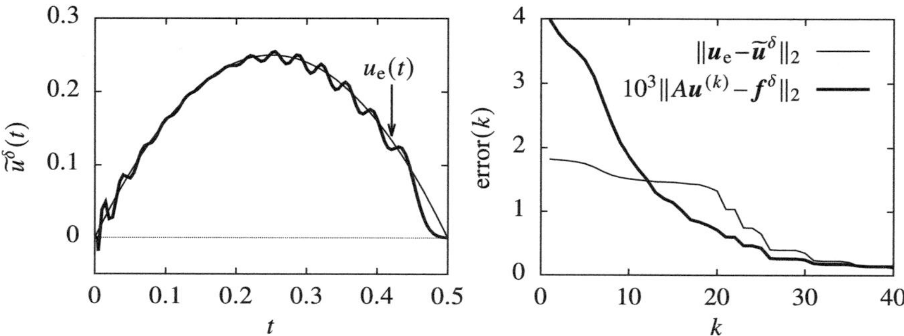

Fig. 13.7 [LEFT] The acoustic impedance $\widetilde{u}^{\delta}$ reconstructed from the noisy seismogram, compared to the presumed ideal profile u_{e}. [RIGHT] The evolution of two error norms during the conjugate gradients iteration

discretization is further refined. A specific coordinate transformation, corresponding to a particular factorization of A, as well as explicit regularization, help to smoothen them out: see [6] for details.   ◁

13.3.5 Variational Regularization

We have seen that Tikhonov regularization is equivalent to the minimization of (13.20). But obviously one could think of many other functionals of the form

$$J_{\alpha}(u) = \frac{1}{2}\|Au - f\|^2 + \alpha R(u) \,, \tag{13.27}$$

where $\alpha R(u)$ is the "penalty" part of the functional intended to impose some additional condition on the solution u and thereby enforce regularization. The idea is to punish not the excessive norm of u but another property of u, like its smoothness or entropy content. The methods by which we are seeking the regularized solution u_{α} as the global minimum of an optimization problem—that is, a global minimizer of (13.27)—are known as *variational regularization techniques*, and are suitable in both linear and non-linear inverse problems. Among the most popular choices of the regularization functional are

$$R(u) = \frac{1}{2}\|u\|^2 + \frac{1}{2}\|\nabla u\|^2 \,, \tag{13.28}$$

often used in image processing to control rapid intensity variations at object edges, the *total variational regularization* using

$$R(u) = \int \|\nabla u(x)\| \, dx \, ,$$

as well as the *maximum entropy regularization*

$$R(u) = \int u(x) \log u(x) \, dx$$

commonly used in the solution of inverse problems where u is assumed to be a properly normalized probability density function. For details see Chap. 8 of [18] as well as [19, 20].

13.3.6 Regularization of Fourier-Transform-Based Deconvolution

Convolution equations of the form

$$f(y) = \int_{-a}^{a} c(y - x)u(x) \, dx = (c * u)(y) \, , \qquad y \in [-a, a] \, , \tag{13.29}$$

in which the kernel function c is known, can be solved for u by using the Fourier transformation. With c and f discretized on a suitable grid, we calculate the discrete N-point Fourier transforms $C = \mathcal{F}_N[c]$ and $F = \mathcal{F}_N[f]$, form their ratio $U = F/C$, and take its inverse Fourier transformation:

$$u = \mathcal{F}_N^{-1}\left[\frac{\mathcal{F}_N[f]}{\mathcal{F}_N[c]}\right] = \mathcal{F}_N^{-1}\left[\frac{F_k}{C_k}\right] \, , \tag{13.30}$$

where the division in the square brackets is to be performed component-wise. This is the inverse procedure of the "forward" convolution problem discussed in Sect. 5.2.6. However, if f is noisy—even to the slightest extent—its high-frequency components will become greatly amplified during the inversion process, usually rendering the resulting u useless. The remedy is to modify the denominator by preventing its components from reaching small values or becoming zero.

Following the approach of [6] we suppose that a discrete version of u is to be recovered at equidistant points $x_j = jh$, $j \in \Omega = \{-N/2, \ldots, N/2 - 1\}$ with N even, and $h = 2a/N$. We expand the approximations of both u and f in terms of the "hat" functions,

$$\widetilde{u}(x) = \sum_{j\in\Omega} u_j \phi(x/h - j) \, , \qquad \widetilde{f}(x) = \sum_{j\in\Omega} f_j \phi(x/h - j) \, , \tag{13.31}$$

where $u_j = u(x_j)$, $f_j = f(x_j)$ and

$$
\phi(z) = \begin{cases} 1 + z \, ; & -1 \le z \le 0 \, , \\ 1 - z \, ; & 0 \le z \le 1 \, , \\ 0 & ; \quad \text{otherwise} \, . \end{cases}
$$

The coefficients f_j are known since the convoluted signal has been sampled at x_j's; the u_j are to be determined. Since one usually performs the deconvolution with a certain idea about the functional form of the kernel c, we further assume that its Fourier transform at points $y = k/(2a)$,

$$
\widehat{c}(y) = \int_{-\infty}^{\infty} c(x) \, e^{-2\pi i x y} dx \, , \qquad C_k = \widehat{c}\left(\frac{k}{2a}\right) \, , \qquad k \in \mathbb{Z} \, ,
$$

can be computed accurately enough. The advantage of the expansions (13.31) is that their Fourier transforms U and F can be calculated exactly:

$$
U(y) = h \left(\frac{\sin(\pi h y)}{\pi h y} \right)^2 \sum_{j \in \Omega} u_j \, e^{-2\pi i j h y} \, ,
$$

and analogously for $F(y)$. (Of course the sums can still be evaluated by an appropriate fast Fourier algorithm.) For $y = k/(2a)$, $k \in \mathbb{Z}$, we obtain

$$
U\left(\frac{k}{2a}\right) = \sigma_k U_k \, , \qquad F\left(\frac{k}{2a}\right) = \sigma_k F_k \, , \qquad \sigma_k = 2a \left(\frac{\sin(\pi k / N)}{\pi k / N} \right)^2 \, ,
$$

where

$$
U_k = \frac{1}{N} \sum_{j \in \Omega} u_j \, e^{-2\pi i j k / N} \, , \qquad F_k = \frac{1}{N} \sum_{j \in \Omega} f_j \, e^{-2\pi i j k / N} \, .
$$

A realistic f_j will be perturbed to f_j^δ, causing noisy transforms F_k^δ. To avoid the instability in the deconvolution, the denominator of the F_k/C_k ratio is modified before transforming it back to the array of the expansion coefficients of $\widetilde{u}^\delta$,

$$
u^\delta = \mathcal{F}_N^{-1}\left[U^\delta\right] \, , \qquad U_k^\delta = \frac{F_k^\delta C_k^*}{|C_k|^2 + \alpha} \, , \qquad k \in \Omega \, ,
$$

where $*$ denotes complex conjugation and $\alpha > 0$ is the regularization parameter; $\alpha = 0$ corresponds to the unregularized case (13.30). The final solution for the deconvoluted function is then

$$
\widetilde{u}^\delta(x) = \sum_{j \in \Omega} u_j^\delta \phi(x/h - j) \, .
$$

The remaining task is to determine the optimal α for the given noise level δ assumed to be present in f. One approach is to insert $\widetilde{u}^\delta(x)$ back into the original convolution (13.29) and sample the output, resulting in samples of the "simulated" effect f'^δ, then finding the maximum α at which

$$\sum_{j \in \Omega} \left| f_j'^\delta - f_j^\delta \right|^2 \leq \delta^2 \tag{13.32}$$

is satisfied. Such α can be found by solving the equation

$$N \sum_{k \in \Omega} \left(\frac{\sigma_k}{2a} \right)^2 \left| C_k^\delta \right|^2 \left(\frac{1}{1 + \beta \left| C_k^\delta \right|^2} \right)^2 = \delta^2$$

for β and taking $\alpha = 1/\beta$ [6].

13.3.7 Determination of the Optimal Regularization Parameter

In the few examples above we have seen that a sensible regularized solution could be obtained with a regularization parameter α that was "optimal" in the sense that it minimized the reconstruction error measured with respect to the exact solution— recall, for instance, Figs. 13.4 (top left) and 13.5 (left). But in reality we do not know the exact solution—otherwise we would not be solving the inverse problem in the first place! A more general strategy is called for.

There are three distinct approaches to choosing the appropriate regularization parameter, depending on what we know about the noise on the right-hand side of the equation $Au = f$. *Prior* parameter choices depend only on the amount δ of noise that contaminates f. *Posterior* choices depend on both δ and f^δ. *Heuristic* determinations of α depend only on f^δ—which indeed may often be the only available quantity.

Prior Parameter Choice Rules

It can be shown that for any regularization Γ_α there exists a prior parameter choice rule with $\alpha = \alpha(\delta)$ which makes the regularization convergent, i.e. the norms of regularized inverses stay bounded. This happens if and only if

$$\lim_{\delta \to 0} \alpha(\delta) = 0 \quad \text{and} \quad \lim_{\delta \to 0} \delta \| \Gamma_{\alpha(\delta)} \|_{\mathcal{L}(Q,\mathcal{P})} = 0 \,.$$

A rule conforming to these requirements implies that $\alpha(\delta)$ should not converge to zero too rapidly with respect to δ—see, for instance, the condition (13.17) derived for the Tikhonov case—often motivating choices of the form $\alpha(\delta) = \delta^p$. Finding an optimal power p, however, may require additional information on u^+ in terms of the so-called source conditions [9]. Posterior choice rules described next have the advantage that this additional knowledge is not needed. Instead, they rely on the Morozov discrepancy principle.

Posterior Rules and Morozov's Discrepancy Principle

Assume that $f = Au + \varepsilon$ where ε is noise with magnitude $\|\varepsilon\|_Q \leq \delta > 0$. We may claim that the regularized inverse $\Gamma_\alpha f$ introduced in (13.12) represents an acceptable reconstruction if $\|A\Gamma_\alpha f - f\|_Q \leq \delta$. For example, if the components of the vector $\varepsilon \in \mathbb{R}^n$ are normally distributed random numbers $\varepsilon_i \sim N(0, \sigma^2)$, the expected value of the error level is $\overline{\|\varepsilon\|_2} = \sqrt{n}\sigma$, therefore we can take $\delta = \sqrt{n}\sigma$—just as it has been done in (13.26). Morozov's discrepancy principle then suggests to determine $\alpha(\delta, f^\delta) > 0$ such that the requirement

$$\left\| A\Gamma_{\alpha(\delta, f^\delta)} f^\delta - f^\delta \right\|_Q \leq \eta\delta \tag{13.33}$$

is fulfilled for some fixed constant $\eta > 1$. So the recipe is: take the noisy right-hand side f^δ from Q, find its regularized inverse in $\mathcal{P}$ by means of $\Gamma_{\alpha(\delta, f^\delta)}$, map it back to Q by A, and evaluate the norm of the resulting difference to see if it remains below a desired value. An estimate of this kind has been enforced in Eq. (13.32). In general, a posterior regularization strategy can be devised by picking a null sequence $\{\alpha_i\}_{i \in \mathbb{N}}$ and iteratively computing $u_{\alpha_i} = \Gamma_{\alpha_i} f^\delta$ for $i \in \{1, 2, \ldots, i^*\}$ until $u_{\alpha_{i*}}$ satisfies (13.33).

Heuristic Approaches

Heuristic determinations of the regularization parameter do not depend on the knowledge of the noise level δ. They require only the noisy f^δ as input [9]. One option is to take the first n elements of a null sequence $\{\alpha_i\}_{i \in \mathbb{N}}$ and choose the optimal $\alpha(f^\delta) = \alpha_{i*}$ such that

$$\alpha_{i*} = \underset{1 \leq i < n}{\arg\min} \ \|u_{\alpha_{i+1}} - u_{\alpha_i}\|_{\mathcal{P}} \,,$$

which is known as the *quasi-optimality principle*. We may also follow the Morozov approach—start with a noisy f^δ, perform the inversion to yield u_α, and map it back—but since δ is presumed to be unknown, there is nothing against which we could calibrate the resulting discrepancy $Au_\alpha - f^\delta$, hence we choose $\alpha(f^\delta)$ such that

$$\alpha(f^\delta) = \underset{\alpha > 0}{\arg\min} \ \frac{1}{\sqrt{\alpha}} \left\| Au_\alpha - f^\delta \right\|_Q \,,$$

which is known as the Hanke–Raus rule. Alternatively, we choose $\alpha(f^\delta)$ such that

$$\alpha(f^\delta) = \underset{\alpha > 0}{\arg\min} \ \|u_\alpha\|_{\mathcal{P}} \left\| Au_\alpha - f^\delta \right\|_Q \,.$$

This is the formal expression of the so-called "L-curve method" which can be applied in a very practical manner. We plot the points

$$\left(\log \|u_\alpha\|_{\mathcal{P}}, \ \log \|Au_\alpha - f^\delta\|_Q \right)$$

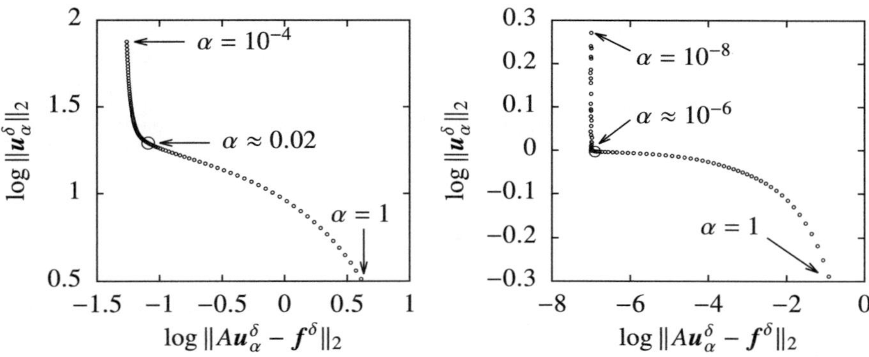

Fig. 13.8 Determination of the optimal regularization parameter α (denoted by a larger circle) by the L-curve method [LEFT] for the signal reconstruction problem posed in Fig. 13.1, yielding $\alpha \approx 0.02$ (compare to Fig. 13.4 (top left)) and [RIGHT] for the Fredholm problem (13.23) with $\delta = 0.001$ (compare to Fig. 13.5 (left))

as a function of α for a wide range of α. These points usually lie on a curve resembling the letter L, giving the method its name. The optimal parameter α is chosen to match a place near the knee of the curve—the approximate balance point of the two concurrent errors. The precise placement of the optimal point is a matter of debate; a popular choice is to take the point of maximum curvature. Figure 13.8 illustrates the L-curve method for two previously encountered examples of regularized inversion.

Caveat Heuristic choices of the regularization parameter do have an important theoretical drawback: it can be proven [9] that in the infinite-dimensional case they can not lead to convergent regularizations. However, when applied to discretized inverse problems, heuristic parameter choices do work in practice, in particular if additional assumptions on the perturbed data are made—for instance, $f^\delta \notin \mathcal{D}(A^+)$.

13.3.8 *Regularization of Non-linear Inverse Problems*

Regularization of inverse problems involving non-linear operators in the corresponding direct problems,

$$A(u) = f \ ,$$

is more difficult than in the linear case [21]. By analogy to Eq. (13.27) a regularized inverse of a non-linear problem is typically still attempted in the variational (Tikhonov) sense by trying to minimize a functional like

$$J_\alpha(u) = \frac{1}{2}\|A(u) - f\|_Q^2 + \alpha R(u) \tag{13.34}$$

with the second (penalty) term either in the form (13.28) or $R(u) = \|u - u^*\|$, where u^* is an approximation of the unknown solution u. Due to non-linearity such functionals may not have unique minimizers. A detailed description is beyond the scope of this book. The interested reader is referred to [22–25] as well as Chap. 4 of [6].

13.4 Solution of Integral Equations

Integral equations abound in physics and other natural sciences [26, 27]. They appear in four basic varieties, depending on the placement of the unknown function with respect to the integral sign and on the integration limits. Fredholm integral equations of the first kind have the form

$$\int_a^b K\big(x, t, q(t)\big)\, \mathrm{d}t = f(x)\,, \qquad x \in [a, b]\,, \tag{13.35}$$

where $K\big(x, t, q(t)\big)$ is the kernel and $q(x)$ is the function we wish to determine. Volterra integral equations of the first kind are a special case of Fredholm equations with a floating upper integration limit,

$$\int_a^x K\big(x, t, q(t)\big)\, \mathrm{d}t = f(x)\,, \qquad x \in [a, T]\,. \tag{13.36}$$

If the unknown function appears also outside of the integral, we are referring to Fredholm equations of the second kind,

$$q(x) = f(x) + \int_a^b K\big(x, t, q(t)\big)\, \mathrm{d}t\,, \qquad x \in [a, b]\,, \tag{13.37}$$

and Volterra equations of the second kind,

$$q(x) = f(x) + \int_a^x K\big(x, t, q(t)\big)\, \mathrm{d}t\,, \qquad x \in [a, T]\,. \tag{13.38}$$

If the integrand can be written in the form $K\big(x, t, q(t)\big) = k(x, t)q(t)$, all the above equations become linear in q, and we stick to this special case in the following unless noted otherwise. Disentangling q from the kernel does simplify some solution techniques, but the main computational problems appear when the isolated q is absent: while inverse problems with equations of the second kind are well-posed, those involving equations of the first kind are ill-posed.

13.4.1 Fredholm Equations of the Second Kind

If f is continuous on $[a, b]$ and K is continuous on $[a, b] \times [a, b]$, a linear Fredholm equation of the second kind (13.37) has a unique solution q which is also continuous on $[a, b]$.

There are three straightforward approaches to solving such equations numerically. The obvious one is to replace the integral by a quadrature formula over a grid $\{t_i\}$, $i = 0, 1, \ldots, n$, and an analogous discretization in the x variable; the grids need not be uniform. We obtain

$$q(x_i) = f(x_i) + \sum_{j=0}^{n} w_j k(x_i, t_j) q(t_j) , \qquad i = 0, 1, \ldots, n , \tag{13.39}$$

where w_j are the quadrature weights. With $f(x_i) = f_i$, $k(x_i, t_j) = k_{ij}$ and $q(x_j) = q_j$ this leads to a system of linear equations for q_j,

$$\sum_{j=0}^{n} \left(\delta_{i,j} - w_j k_{ij} \right) q_j = f_i , \qquad i = 0, 1, \ldots, n . \tag{13.40}$$

The corresponding continuous approximant is given by

$$\widetilde{q}_n(x) = f(x) + \sum_{j=0}^{n} w_j k(x, t_j) .$$

The second option is to replace the kernel $k(x, t)$ with its separable approximation of the form $k(x, t) \approx \sum_{i=0}^{n} \phi_i(x) \psi_i(t)$, where the functions ϕ_i are linearly independent. The solution of the resulting equation,

$$q(x) = f(x) + \int_{a}^{b} \sum_{i=0}^{n} \phi_i(x) \psi_i(t) q(t) \, dt ,$$

is given by the expansion

$$\widetilde{q}_n(x) = f(x) + \sum_{i=0}^{n} a_i \phi_i(x) ,$$

with coefficients $a_i = \int_{a}^{b} \widetilde{q}_n(t) \psi_i(t) \, dt$ to be determined. Inserting $\widetilde{q}_n$ in (13.37) and comparing the coefficients of individual ϕ_i we obtain a system of linear equations for the a_i's:

$$\sum_{j=0}^{n} \left(\delta_{i,j} - A_{ij} \right) a_j = F_i , \qquad i = 0, 2, \ldots, n ,$$

where

$$A_{ij} = \int_a^b \psi_i(t)\phi_j(t)\,dt \,, \qquad F_i = \int_a^b \psi_i(t)f(t)\,dt \,.$$

Finally, Fredholm equations of the second kind can also be solved efficiently by using the Picard method, which is nothing but the fixed-point iteration

$$q^{(n+1)}(x) = f(x) + \int_a^b k(x,t)q^{(n)}(t)\,dt \,, \qquad n = 0, 1, 2, \ldots \,, \tag{13.41}$$

where n is the iteration index and $q^{(0)} = f(x)$ is the starting point. The process will converge as long as $|\int_a^b k(\cdot\,,t)\,dt| < 1$. An iteration of this type is also the preferred method for solving non-linear Fredholm integral equations.

Example The equation

$$q(x) = 1 + x\int_0^1 tq(t)\,dt \,, \qquad 0 \le x \le 1 \,, \tag{13.42}$$

is a linear Fredholm integral equation of the second kind for $q(x)$ with $f(x) = 1$ and a separable kernel $k(x,t) = xt$. Its exact solution is $q_e(x) = 1 + 3x/4$. We solve the problem by using the iteration (13.41) and the trapezoidal formula to evaluate the integrals. Figure 13.9 shows the absolute error of the solution over the interval $[0, 1]$ for three lengths of the iteration sequence.

When the iteration converges, the error stops diminishing. This is a consequence of the particular choice of the trapezoidal formula: more sophisticated quadratures would lead to smaller errors. ◁

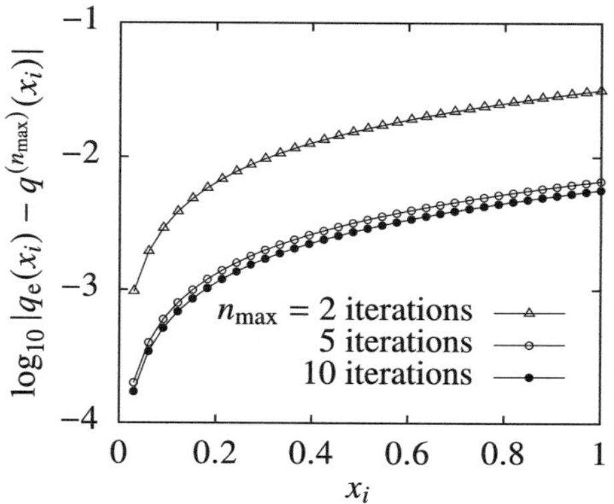

Fig. 13.9 The absolute error of the numerical solution of (13.42) obtained by iteration (13.41) and 21-point trapezoidal rule for the integrals

13.4.2 *Volterra Equations of the Second Kind*

If $k(x, t)$ is continuous for $a \leq t \leq x \leq T$ and f is continuous on $a \leq x \leq T$, a linear Volterra equation of the second kind (13.38) possesses a unique continuous solution $q(x)$ on $a \leq x \leq T$. Beware that although there is only *one* continuous solution, there may be discontinuous solutions as well—see p. 31 of [28] for an example.

Volterra equations of the second kind can be solved by any method of the previous Subsection; the only difference is the floating upper limit of the integral over t. One could in principle bypass this restriction by extending the kernel function such that $k(x, t) = 0$ for $t > x$, thereby converting a Volterra problem into a Fredholm problem, but introducing such asymmetric kernels is not recommended as certain segments of theoretical support are lost.

Following the quadrature approach, we now have

$$q(x_i) = f(x_i) + \sum_{j=0}^{i} w_j k(x_i, t_j) q(t_j) \,, \qquad i = 0, 1, \ldots, n \,,$$

instead of (13.39), while (13.40) is replaced by the triangular system

$$q_0 = f_0 \,, \qquad q_i - \sum_{j=0}^{i} w_j k_{ij} q_j = f_i \,, \qquad i = 1, 2, \ldots, n \,,$$

where we have used the fact that $q(a) = f(a)$.

The Picard method now reads

$$q^{(n+1)}(x) = f(x) + \int_{a}^{x} k(x, t) q^{(n)}(t) \, \mathrm{d}t \,, \qquad n = 0, 1, 2, \ldots \,,$$

and may again be initialized with $q^{(0)} = f(x)$. An iteration of this type can also be used to solve non-linear Volterra problems of the second kind,

$$q^{(n+1)}(x) = f(x) + \int_{a}^{x} K\left(x, t, q^{(n)}(t)\right) \, \mathrm{d}t \,, \qquad n = 0, 1, 2, \ldots \,,$$

but with certain restrictions on the kernel. If K satisfies a Lipschitz condition of the form

$$|K(x, t, u) - K(x, t, v)| \leq \Lambda |u - v| \,,$$

then (13.38) still has a unique continuous solution for all finite T. Even if the Lipschitz inequality is not fulfilled or if it is satisfied in a relaxed form

$$|K(x, t, u) - K(x, t, v)| \leq \lambda(x, t)|u - v| \,,$$

where $\lambda(x, t)$ is square-integrable, the existence and uniqueness of the solution are guaranteed at least locally, i.e. in the neighborhood of the exact solution. For highly stable Runge–Kutta methods for Volterra equations of the second kind see [29]. Further theoretical details are given in Chaps. 3, 4 and 7 of [28].

13.4.3 Fredholm Equations of the First Kind

Inverse problems involving Fredholm integral equations of the first kind

$$(Aq)(x) = \int_a^b k(x, t)q(t)\,\mathrm{d}t = f(x)\,, \qquad a \leq x \leq b\,, \tag{13.43}$$

(as well as their Volterra counterparts) are ill-posed [30]. To appreciate this, consider the trivial example when both k and f are constant functions. Solving (13.43) for q means finding the curve from the area under the curve, which obviously translates to non-uniqueness, a landmark of ill-posedness.

If the kernel $k \in L^2\big((a, b) \times (a, b)\big)$ is real and symmetric, $k(x, t) = k(t, x)$, and if $f, q \in L^2(a, b)$, there exists a complete set of orthonormal eigenfunctions ϕ_n and the corresponding eigenvalues λ_n of the integral operator A ($|\lambda_{n+1}| \leq |\lambda_n|$, $\lambda_n \in \mathbb{R}$) such that

$$(A\phi_n)(x) = \int_a^b k(x, t)\phi_n(t)\,\mathrm{d}t = \lambda_n\phi_n(x)\,, \qquad n \in \mathbb{N}\,.$$

The kernel of A may be degenerate: this means that a n_0 exists beyond which all eigenvalues are zero, i.e. $\lambda_n \neq 0$ for $1 \leq n \leq n_0$ and $\lambda_n = 0$ for all $n > n_0$. If the kernel is degenerate, (13.43) has either no solution or infinitely many solutions, depending on f. One can show [31] that it has a solution only if f is a linear combination of the form $f(x) = \sum_{n=1}^{n_0} f_n\phi_n(x)$.

If the data of the inverse problem is noisy—denoted by f^δ as usual—(13.43) must be regularized. A possible method in the case of symmetric kernels is

$$\alpha q(x) + \int_a^b k(x, t)q(t)\,\mathrm{d}t = f^\delta(x)\,, \tag{13.44}$$

where $\alpha > 0$ is the regularization parameter. If we assume that all eigenvalues of A are positive and that an exact solution q_e of (13.43) exists for some f, then there exist $\alpha_*, \delta_* > 0$ such that for all $\alpha \in (0, \alpha_*)$, $\delta \in (0, \delta_*)$ and for any f^δ satisfying the bound $\|f - f^\delta\| \leq \delta$, Eq. (13.44) has a solution q_α^δ satisfying

$$\left\| q_\alpha^\delta - q_e \right\| \le \frac{\delta}{\alpha} + \omega(\alpha) ,$$

where $\lim_{\alpha\to 0} \omega(\alpha) = 0$. Such regularization works for both non-degenerate and degenerate symmetric kernels. If the kernel is asymmetric, regularization requires an intermediate step via the adjoint operator A^* defined as

$$(A^* p)(x) = \int_a^b k(t, x) p(t) \, dt .$$

Applying A^* to $Aq = f^\delta$ yields the integral operator equation with a symmetric kernel, $A^*Aq = A^* f^\delta$, which can be regularized in the usual manner as

$$\left(A^*A + \alpha I\right)q = A^* f^\delta . \tag{13.45}$$

The solution of this equation converges to a quasi-solution of (13.43) as α and δ tend to zero at matched rates [32].

For another type of regularization method based on the truncation of the eigen-function expansions instead of the Tikhonov-like approach outlined above see [33]. For maximum-entropy regularization see [34, 35].

Example Let us demonstrate the regularization approach (13.44) by solving

$$2x = \int_0^1 xtq(t) \, dt , \qquad 0 \le x \le 1 ,$$

which is a Fredholm equation of the first kind with $k(x, t) = xt$ and $f(x) = 2x$ (assumed to be exact for clarity). Regularization results in an equation of the second kind

$$q_\alpha(x) = \frac{2}{\alpha} x - \frac{1}{\alpha} \int_0^1 xtq_\alpha(t) \, dt ,$$

which we solve by iteration initialized with $q_\alpha^{(0)} = 0$. The consecutive iterates are

$$q_\alpha^{(0)} = 0 , \qquad q_\alpha^{(1)}(x) = \frac{2}{\alpha} x ,$$

$$q_\alpha^{(2)}(x) = \frac{2}{\alpha} x - \frac{2}{3\alpha^2} x ,$$

$$q_\alpha^{(3)}(x) = \frac{2}{\alpha} x - \frac{2}{3\alpha^2} x + \frac{2}{9\alpha^3} x ,$$

$$q_\alpha^{(4)}(x) = \frac{2}{\alpha} x - \frac{2}{3\alpha^2} x + \frac{2}{9\alpha^3} x - \frac{2}{27\alpha^4} x ,$$

and so on. The regularized solution is therefore

$$q_\alpha(x) = \frac{2}{\alpha}\, x \left(1 - \frac{1}{3\alpha} + \frac{1}{9\alpha^2} - \frac{1}{27\alpha^3} + \cdots \right) = \frac{2}{\alpha}\left(\frac{3\alpha}{1 + 3\alpha}\right) x\,.$$

The $\alpha \to 0$ limit gives the final answer $q(x) = \lim_{\alpha\to 0} q_\alpha(x) = 6x$. Note that this solution is not unique: another one is given by $q(x) = 4x^2 + 5x^3$. $\quad\lhd$

If the kernel is simple and the integral is easy to evaluate, such analytic procedure is quite useful [36]. Otherwise, the most straightforward way to calculate the regularized solution of (13.43) is by iteration of the form

$$q^{(n+1)}(x) = q^{(n)}(x) + \alpha \left[f^\delta(x) - \int_a^b k(x, t) q^{(n)}(t)\, \mathrm{d}t \right]\,, \qquad n = 0, 1, \ldots,$$

with $\alpha > 0$ and $q^{(0)} = 0$ [37]. It is applicable also for non-linear Fredholm equations, but it may be quite slow to converge due to the usual trade-off between α and δ involved in the desired precision of the solution.

Example (Adapted from [38].) Let us solve the Fredholm equation of the first kind

$$(Aq)(x) = \int_a^b k(x, t) q(t)\, \mathrm{d}t = f^\delta(x)\,, \qquad c \le x \le d\,,$$

with

$$k(x, t) = \frac{1}{x + t}\,, \qquad f^\delta(x) = \frac{1}{x} \log\left(\frac{1 + x/a}{1 + x/b}\right) + \mathrm{noise}(\delta)\,,$$

on the intervals $[a, b] = [c, d] = [1, 5]$. The exact solution for noiseless f is $q_\mathrm{e}(x) = 1/x$.

We will use the regularization (13.45) and discretize the problem in a form similar to the one used in Example (13.23), but we will construct a more general penalty part of the functional (13.27):

$$J_\alpha(u) = \frac{1}{2}\|Aq - f\|^2 + \alpha\left(a_0\|q - \widehat{q}\|^2 + a_1\|q'\|^2 + a_2\|q''\|^2\right)\,.$$

The idea is to impose regularization by punishing not just the fluctuations of q (or the differences with respect to some guess of the solution, $\widehat{q}$) but also the excessive first or second derivatives. These ingredients of the regularization can be controlled by the parameters a_0, a_1 and a_2.

Let $f^\delta(x)$ be given at N points (not necessarily equidistant), such that $c \le x_1 < x_2 < \ldots < x_N \le d$, with $f_i^\delta = f^\delta(x_i)$. The integration interval $[a, b]$ is split into N

subintervals $[t_{j-1}, t_j]$, $j = 1, 2, \ldots, N$, such that $a = t_0 < t_1 < \ldots < t_N = b$. For simplicity, we choose uniform grids $x_i = c + (i - 1)\Delta x$ with $\Delta x = (d - c)/(N - 1)$, $i = 1, 2, \ldots, N$ and $t_j = a + j\Delta t$ with $\Delta t = (b - a)/N$, $j = 0, 1, \ldots, N$, and arrange the discretized functions in vectors

$$\boldsymbol{f}^{\delta} = \left(f_1^{\delta}, f_2^{\delta}, \ldots, f_N^{\delta} \right)^{\mathrm{T}}, \qquad \boldsymbol{q} = \left(q_1, q_2, \ldots, q_N \right)^{\mathrm{T}},$$

and similarly for $\widehat{\boldsymbol{q}}$, if we happen to know it. We approximate the integrals at given x_i by using the repeated mid-point rule:

$$(Aq)(x_i) = \int_a^b k(x_i, t) q(t)\, \mathrm{d}t = \sum_{j=1}^{N} \int_{t_{j-1}}^{t_j} k(x_i, t) q(t)\, \mathrm{d}t \approx \sum_{j=1}^{N} K_{ij} q_j ,$$

where $K_{ij} = (t_j - t_{j-1}) k(x_i, \bar{t}_j)$ and $\bar{t}_j = (t_{j-1} + t_j)/2$. The task, then, is to minimize the functional

$$J_{\alpha}(\boldsymbol{q}) = \sum_{i=1}^{N} \left(\sum_{j=1}^{N} K_{ij} q_j - f_j^{\delta} \right)^2 + \alpha \left\{ a_0 \sum_{j=1}^{N} \left(q_j - \widehat{q}_j \right)^2 \right. $$
$$\left. + a_1 \sum_{j=1}^{N-1} \left(q_{j+1} - q_j \right)^2 + a_2 \sum_{j=2}^{N-1} \left(q_{j+1} - 2q_j + q_{j-1} \right)^2 \right\}$$

for all vectors $\boldsymbol{q}$. The requirements $\partial J_{\alpha}/\partial q_j = 0$ for $j = 1, 2, \ldots, N$ lead to a symmetric system of linear equations

$$\left[K^{\mathrm{T}} K + \alpha \left(a_0 H_0 + a_1 H_1 + a_2 H_2 \right) \right] \boldsymbol{q} = K^{\mathrm{T}} \boldsymbol{f}^{\delta} + \alpha a_0 \widehat{\boldsymbol{q}} , \tag{13.46}$$

where $K = [K_{ij}]$, $H_0 = I_N$ is the $N \times N$ identity matrix, and H_1 and H_2 are given by

$$H_1 = \begin{pmatrix} 1 & -1 & & & & 0 \\ -1 & 2 & -1 & & & \\ & -1 & 2 & -1 & & \\ & & \ddots & \ddots & \ddots & \\ & & & -1 & 2 & -1 \\ 0 & & & & -1 & 1 \end{pmatrix}_{N \times N}$$

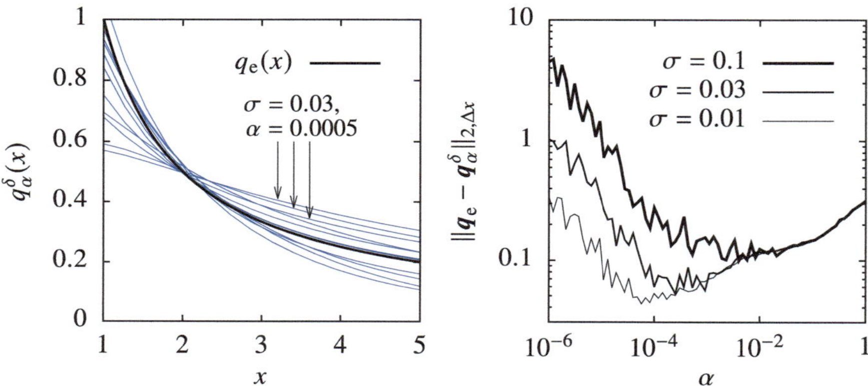

Fig. 13.10 [LEFT] Solutions of (13.46) for $\alpha = 0.0005$ and ten different realizations of noise in f^δ. [RIGHT] The absolute errors for various noise levels σ as a function of α, each averaged over 100 solutions. Compare to Fig. 13.5 (left)

and

$$
H_2 = \begin{pmatrix}
1 & -2 & 1 & & & & & & 0 \\
-2 & 5 & -4 & 1 & & & & & \\
1 & -4 & 6 & -4 & 1 & & & & \\
& 1 & -4 & 6 & -4 & 1 & & & \\
& & \ddots & \ddots & \ddots & \ddots & \ddots & & \\
& & & & 1 & -4 & 6 & -4 & 1 \\
& & & & & 1 & -4 & 5 & -2 \\
0 & & & & & & 1 & -2 & 1
\end{pmatrix}_{N \times N} .
$$

The noise on f is modeled as $f_i^\delta = f_i(1 + \delta)$, where $\delta \sim N(0, \sigma^2)$, $\sigma = 0.03$. By solving (13.46) with the chosen regularization parameter α we obtain the discretized solution $q = q_\alpha^\delta$. The results for ten different realizations of f^δ are shown in Fig. 13.10 (left) for $N = 16$, $\widehat{q} = 0$, $a_0 = 1$, $a_1 = a_2 = 0$ and $\alpha = 0.0005$.

Figure 13.10 (right) shows the absolute error of the solution as a function of α. The value of $\alpha = 0.0005$ in the left panel roughly corresponds to the minimum of the $\sigma = 0.03$ curve.   ◁

13.4.4   *Volterra Equations of the First Kind, Smooth Kernels*

Linear Volterra integral equations of the first kind,

$$
\int_a^x k(x, t) q(t)\, \mathrm{d}t = f(x)\,, \qquad a \le x \le T\,, \tag{13.47}
$$

can be solved by the standard methods described in Sect. 13.4.1, e.g. quadrature, expansion in terms of basis functions or iteration. For instance, any quadrature approximation will have the form

$$\sum_{j=0}^{i} w_j k(x_i, t_j) q(t_j) = f(x_i) , \qquad i = 0, 1, \ldots, n ,$$

and will yield a triangular system of equations for $q_j = q(t_j)$; see Chap. 9 of [28] for a variety of further options. However, just as their Fredholm counterparts, Volterra equations of the first kind are ill-posed—so in principle any of these methods will work, but will be highly sensitive to the level of noise in f.

As we have suggested before, (13.47) can formally be rewritten as a Fredholm equation of the first kind by pushing the upper integration limit to T and extending the domain of the kernel function such that $k(x, t) = 0$ for $x < t \leq T$. But the variable upper limit allows us to differentiate the equation with respect to x, giving

$$k(x, x)q(x) + \int_a^x \frac{\partial k}{\partial x}(x, t)q(t) \, dt = f'(x) .$$

If $k(x, x) \neq 0$ for all $x \in [a, T]$—this is essential!—we can divide the above equation by $k(x, x)$, resulting in a Volterra equation of the second kind. Also note that if k and q are to be bounded, one *must* have $f(a) = 0$, otherwise no solution of (13.47) exists. Although we have thus apparently converted an ill-posed problem into a well-posed one, the advantage is difficult to exploit in practice. The main reason is again the noise on the right-hand side: in reality we have f^δ instead of f, often in tabular form, and evaluating numerical derivatives of f^δ will likely lead to large errors in the solution.

Some sort of regularization is needed, and the classical variant, by analogy to (13.44), is

$$\alpha q(x) + \int_a^x k(x, t)q(t) \, dt = f^\delta(x) . \tag{13.48}$$

Assume that f and f^δ are continuous and satisfy the estimate $\| f - f^\delta \| \leq \delta$, that an exact solution q_e of (13.47) exists such that $q_e(a+) = 0$, and that k, $\partial k/\partial x$ and $\partial^2 k/\partial x^2$ are continuous. Then there exist positive real numbers α_*, δ_* and C such that, for $\alpha \in (0, \alpha_*)$ and $\delta \in (0, \delta_*)$, Eq. (13.48) has a unique continuous solution q_α^δ that satisfies [31]

$$\left\| q_\alpha^\delta - q_e \right\| \leq C \left(\alpha + \frac{\delta}{\alpha} \right) .$$

This regularization is also applicable to non-linear Volterra equations of the first kind. A broader survey of appropriate regularization methods is offered by [39].

13.4.5   *Volterra Equations of the First Kind, Unbounded Kernels*

In various areas of physics we encounter Volterra equations of the first kind with unbounded kernels, many of which are covered by the basic form

$$\int_0^x \frac{q(t)}{(x-t)^p}\, dt = f(x)\,, \qquad 0 \le x \le T\,, \qquad 0 < p < 1\,, \tag{13.49}$$

known as Abel's equation. If f is continuous for $0 < x \le T$ and $x^n f(x) \to C$ as $x \to 0$, where $C \ne 0$ and $n < p$, then (13.49) has an explicit solution:

$$q(x) = \frac{\sin p\pi}{\pi} \frac{d}{dx} \int_0^x \frac{f(t)}{(x-t)^{1-p}}\, dt\,, \qquad 0 < x \le T\,.$$

(If $f(0) = 0$, remove d/dx and replace $f(t)$ by $f'(t)$ in the above formula.) This solution is continuous and satisfies

$$q(x) = \big[C + o(1)\big]\frac{\Gamma(1-n)}{\Gamma(1-p)\Gamma(p-n)}\, x^{p-n-1}$$

as $x \to 0$. Generalized Abel's equation

$$\int_0^x \frac{k(x,t)}{(x-t)^p}\, q(t)\, dt = f(x)\,, \tag{13.50}$$

where $k(x,t)$ is a smooth function satisfying $k(x,x) \ne 0$ for all x, also has a unique continuous solution. A transformation exists that turns (13.50) into an equation of the first kind with a smooth kernel (see p. 75 of [28]), but that does not circumvent its ill-posedness.

13.4.6   *The Radon Transformation and Its Inverse*

The Radon problem is the problem of reconstructing a function from the values of its line integrals along different directions [40–42]. This task is at the heart of modern diagnostic tools like X-ray computerized tomography (CT) where two-dimensional planar sections of a three-dimensional body are scanned by X-rays emitted from various directions in the plane of the section, as shown in Fig. 13.11 (left). We wish to reconstruct the absorption coefficient $q(x, y)$ from the measured attenuated intensity of the rays transmitted through the body along a specific line L,

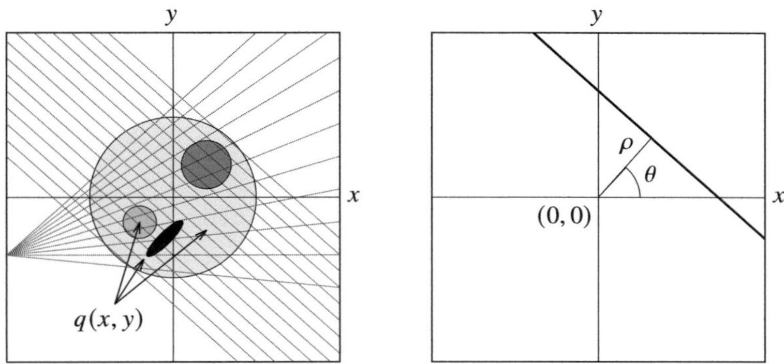

Fig. 13.11 [LEFT] X-ray scanning of a (x, y) section of a three-dimensional inhomogeneous body with absorption coefficient $q(x, y)$. [RIGHT] The definition of parameters ρ and θ corresponding to the line $x \cos \theta + y \sin \theta = \rho = \text{const}$.

$$\left.\frac{j_{\text{out}}}{j_{\text{in}}}\right|_L = \exp\left(-\int_L q(x, y)\, dl\right).$$

The family of lines $L(\rho, \theta)$ in the (x, y) plane is defined by

$$x \cos \theta + y \sin \theta = \rho\,, \qquad 0 \le \theta < \pi\,, \qquad -\rho_{\max} \le \rho \le \rho_{\max}\,,$$

where ρ is the shortest distance from the origin to the line and θ is the angle between the abscissa and the perpendicular segment connecting the origin and the line (Fig. 13.11 (right)). Any line from this family can be represented in parametric form $x = \rho \cos \theta - s \sin \theta$ and $y = \rho \sin \theta + s \cos \theta$ with $s \in \mathbb{R}$. The Radon transform of $q(x, y)$ is the integral of $q(x, y)$ along a given line,

$$f(\rho, \theta) = \int_{L(\rho,\theta)} q(x, y)\, dl = \int_{-\infty}^{\infty} q\big(\rho \cos \theta - s \sin \theta, \rho \sin \theta + s \cos \theta\big)\, ds\,.$$

Note that for certain functions q the transform does not exist, for instance, $q(x, y) = \text{const} \ne 0$—although in practice, of course, we are investigating neither infinite homogeneous bodies nor bodies with singular densities.

When examining a body in X-ray tomography it is the physical process of scanning and measuring the transmitted intensities itself that executes the mathematics of "forward" Radon transformation, which is a straightforward, well-posed procedure. The result $f(\rho, \theta)$, either in continuous or discrete form, is called the *sinogram*. Except for very simple geometries and functional forms of q (see the following Example), the true challenge is the inversion of the Radon transformation, i.e. the unfolding or "back-projection" of the sinogram, which is an ill-posed and numerically intensive problem.

Example Consider the Radon transformation of a continuous radially symmetric function vanishing outside of the disk of radius R centered at the origin, denoted by the largest circle in Fig. 13.11 (left) without the insets. Let $q(x, y) = q_0\big((x^2 + y^2)^{1/2}\big) = q_0(r)$ for $r \leq R$, and $q_0(r) = 0$ for $r > R$. Then

$$f(\rho, \theta) = \int_{-\infty}^{\infty} q\big(\rho \cos\theta - s \sin\theta, \; \rho \sin\theta + s \cos\theta\big) \, \mathrm{d}s \tag{13.51}$$

$$= \int_{-\sqrt{R^2-\rho^2}}^{\sqrt{R^2-\rho^2}} q_0\left(\sqrt{\rho^2 + s^2}\right) \mathrm{d}s = 2 \int_0^{\sqrt{R^2-\rho^2}} q_0\left(\sqrt{\rho^2 + s^2}\right) \mathrm{d}s \; ,$$

where $|\rho| \leq R$. Note that the dependence on θ has disappeared, so the Radon transform in this particular geometry becomes a function of ρ only. The change of variables $z = (\rho^2 + s^2)^{1/2}$ gives

$$f(\rho) = \int_{\rho}^{R} \frac{2 q_0(z) z}{\sqrt{z^2 - \rho^2}} \, \mathrm{d}z \; , \qquad 0 \leq \rho \leq R \; ,$$

which is a generalized Abel equation with an integrable singularity. Its solution (see p. 132 of [31] for the derivation) is

$$q_0(z) = -\frac{1}{\pi z} \frac{\mathrm{d}}{\mathrm{d}z} \int_z^R \rho \sqrt{\rho^2 - z^2} \, f(\rho) \, \mathrm{d}\rho \; .$$

The radially symmetric function $q_0(r)$ is therefore uniquely determined by its Radon transform.   ◁

Fourier-Based Inversion

One of the standard methods of inverting the continuous Radon transformation is based on the Fourier slice theorem [43] and involves one one-dimensional and one two-dimensional Fourier transformation. Given the forward Radon transform $f(\rho, \theta)$ we first calculate its Fourier transform

$$F(k_x, k_y) = F(\nu \cos\theta, \nu \sin\theta) = \int_{-\infty}^{\infty} f(\rho, \theta) \, \mathrm{e}^{-2\pi \mathrm{i} \rho \nu} \, \mathrm{d}\nu \; ,$$

where $k_x = \nu \cos\theta$ and $k_y = \nu \sin\theta$ are the so-called polar frequency parameters. The function q is then obtained by Fourier-transforming F,

$$q(x, y) = \int\limits_{-\infty}^{\infty} \int\limits_{-\infty}^{\infty} F(k_x, k_y)\, e^{2\pi i(k_x x + k_y y)}\, dk_x\, dk_y .$$

Filtered Back-Projection

The second standard approach is *filtered back-projection* [42]:

$$q(x, y) = \int\limits_{0}^{\pi} \left[\int\limits_{-\infty}^{\infty} \Phi(\nu) \left(\int\limits_{-\infty}^{\infty} f(\sigma, \theta)\, e^{-2\pi i\sigma\nu}\, d\sigma \right) e^{2\pi i\nu\rho(x,y,\theta)} \right] d\theta , \qquad (13.52)$$

where $\rho(x, y, \theta) = x\cos\theta + y\sin\theta$. Here the Fourier transform of f (the innermost integral in the above expression) is filtered by the filter $\Phi(\nu)$ before being transformed back to configuration space and integrated over θ to yield the desired q. The basic form of the filter is the low-pass $\Phi(\nu) = |\nu|$, known as the ramp filter, but other forms are in use that also suppress the high-frequency part in order to improve the signal-to-noise ratio, for instance

$$\Phi(\nu) = |\nu|\frac{1}{2}\sin\left(\frac{\pi\nu}{2\nu_{\min}}\right) \Big/ \left(\frac{\pi\nu}{2\nu_{\min}}\right) \quad \text{or} \quad \Phi(\nu) = |\nu|\left(1 + \cos\frac{\pi\nu}{\nu_{\min}}\right) ,$$

where for $\nu_{\min} < |\nu| < \nu_{\max}$ the filter is set to zero. Filtering can also be enforced after the back-projection—see Sect. 7.3 of [44]. In practice these formulas are evaluated in discrete form; as three-fold integrals or sums are involved, the numerical cost of a naive implementation is $O(N^3)$, where N is the typical size of the discretization in any variable (usually they are all comparable). See Problem 13.8.4. For information on fast versions of the Radon transformation and its inverse (typically $O(N^2 \log N)$) consult [45] and references therein.

13.5 Inverse Sturm–Liouville Problems

Let us motivate this Section by the paper [46] in which Mark Kac asked whether one can "*hear the shape of a drum*". In other words, can we uniquely determine the *shape* of the drum membrane if we measure its eigenfrequencies? To play a drum means to solve the wave equation $u_{tt} = \nabla^2 u$ on the domain (drum) Ω by the ansatz $u(x, y, t) = \phi(x, y)T(t)$, whence

$$T_{tt}/T = \nabla^2\phi/\phi = \text{const} = \lambda .$$

The displacement of the membrane from equilibrium is $u(x, t) = \phi(x, y)\sin\sqrt{\lambda}t$, where $\phi(x, y)$ is the solution of the two-dimensional eigenvalue problem

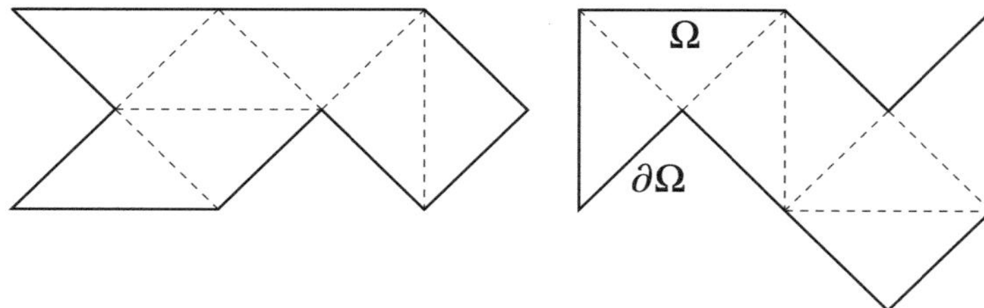

Fig. 13.12 Two geometrically different drum membranes (isospectral domains) with identical discrete sets of eigenfrequencies

$$-\nabla^2\phi = \lambda\phi \quad \text{on membrane } \Omega \,,$$
$$\phi = 0 \quad \text{on membrane boundary } \partial\Omega \,,$$

and where the eigenvalues λ_i are the squares of the drum's eigenfrequencies. The inverse problem consists of asking whether the geometry of the domain Ω is uniquely determined by the discrete set of eigenvalues (the spectrum). The non-trivial answer is: it is not. One can show that the observed eigenvalues do determine some properties of Ω, like its surface area, circumference, and connectivity [46], but geometrically different domains can be found with precisely the same spectrum, including possible degeneracies [47, 48]. We call such domains (and inverse problems) *isospectral*. Figure 13.12 shows the example of two membranes with identical spectra.

Similarly, one could introduce a coefficient function $c(x, y)$ in the wave equation to represent the position-dependent speed of sound, and try to infer the functional form of c from the measured spectrum. Problems of this kind are the topic of this Section. To keep the presentation within the scope of this book, the discussion will be restricted to one-dimensional isospectral problems involving regular Sturm–Liouville operators on a finite interval.

13.5.1 Information Needed to Recover $q(x)$

The classical inverse Sturm–Liouville problem consists of reconstructing the potential function $q(x)$ from the solution of the direct problem

$$\left.\begin{aligned}
-y''(x) + q(x)y(x) &= \lambda y(x) \,, \quad 0 < x < 1 \,, \\
y'(0) - hy(0) &= 0 \,, \\
y'(1) + Hy(1) &= 0 \,,
\end{aligned}\right\} \tag{13.53}$$

by knowing something about its eigenvalues λ_n, the properties of its eigenfunctions $y_n = y(x, \lambda_n)$, or the symmetries of q. Note that the above boundary conditions translate to $\alpha_1 = h$, $\alpha_2 = 1$ (or $h = \alpha_1/\alpha_2$) and $\beta_1 = H$, $\beta_2 = 1$ (or $H = \beta_1/\beta_2$) with $B = 1$ in the notation of (9.86) and (9.87).

Even if one knows the complete spectrum $\{\lambda_n\}_{n=0}^{\infty}$ of (13.53), this is insufficient to recover q; some additional information must be provided. For instance, if one knows in advance that q is symmetric about the midpoint of the interval, $q(x) = q(1 - x)$, and that $h = H$, q can be recovered uniquely from the spectrum [49]. Similarly, if the values of q are given over at least a half of the interval, knowing the spectrum allows for a unique determination of the remaining portion of q [50]. The third option for the recovery of q is that we know the spectrum as well as the eigenfunction norms, usually defined by

$$\rho_n = \|y_n\|^2 / y_n^2(0) \quad \text{(finite } h) \qquad \text{or} \qquad \rho_n = \|y_n\|^2 / y_n'^2(0) \quad (h = \infty) \,,$$

where $\| \cdot \| = \| \cdot \|_{L^2(0,1)}$. The sets $\{\lambda_n\}_{n=0}^{\infty}$ and $\{\rho_n\}_{n=0}^{\infty}$ constitute the so-called *spectral data* of the Sturm–Liouville problem. Yet another possibility to uniquely reconstruct q is to specify *two* spectra of related Sturm–Liouville problems differing only in their boundary conditions, for instance, the spectrum $\{\lambda_n\}_{n=0}^{\infty}$ from (13.53) and the spectrum $\{\mu_n\}_{n=0}^{\infty}$ from the same problem but with the constant H replaced by $H' \neq H$ [51].

13.5.2 The Gelfand–Levitan Method

Gelfand and Levitan [52] have shown that the Sturm–Liouville problem for which the spectral data $\{\lambda_n\}_{n=0}^{\infty}$ and $\{\rho_n\}_{n=0}^{\infty}$ are known can be inverted analytically, recovering q as well as the boundary-value parameters h and H. Here we sketch the main steps of the method [53] for the direct problem in the form (13.53). Other types of boundary conditions are treated similarly; see, for instance, [54] or Chap. 4 of [13]. One starts with a transformation property for the eigenfunctions $y(x, \lambda)$ of the operator $-\mathrm{d}^2/\mathrm{d}x^2 + q(x)$, namely

$$y(x, \lambda) = \cos(\sqrt{\lambda}\, x) + \int_0^x G(x, t) \cos(\sqrt{\lambda}\, t)\, \mathrm{d}t \,, \tag{13.54}$$

where $G(x, t)$ is a real-valued function satisfying

$$G(x, x) = h + \frac{1}{2} \int_0^x q(t)\, \mathrm{d}t \,. \tag{13.55}$$

The function $G(x, t)$ must be found by solving the Gelfand–Levitan equation

$$G(x, t) + F(x, t) + \int_0^x F(u, t) G(x, u)\, \mathrm{d}u = 0 , \qquad 0 < t < x < 1 , \qquad (13.56)$$

where

$$F(x, t) = \sum_{n=0}^{\infty} \left(\frac{\cos\left(\sqrt{\lambda_n}\, x\right) \cos\left(\sqrt{\lambda_n}\, t\right)}{\rho_n} - \frac{\cos n\pi x \cos n\pi t}{\rho_n^0} \right) \qquad (13.57)$$

and $\rho_n^0 = (1 + \delta_{n,0})/2$. Once $G(x, t)$ is found, the potential function can be recovered from $G(x, x)$ by calculating

$$q(x) = 2 \frac{\mathrm{d}}{\mathrm{d}x} G(x, x) .$$

From (13.55) we also get one unknown parameter of the boundary condition,

$$h = G(0, 0) ,$$

while the second one is retrieved from the asymptotic expansion of the eigenvalues. According to (9.88) we have

$$\sqrt{\lambda_n} = n\pi + \frac{C_1}{n\pi} + O\left(\frac{1}{n^3}\right) , \qquad C_1 = h + H + \frac{1}{2} \int_0^1 q(t)\, \mathrm{d}t .$$

Hence

$$H = C_1 - h - \int_0^1 q(t)\, \mathrm{d}t = C_1 - G(1, 1) .$$

Example (Based on Example 1.5.1 of [53].) Take $\lambda_n = (n\pi)^2$ ($n \geq 0$), the normalizations $\rho_n = 1/2$ ($n \geq 1$) and a positive $\rho_0 \neq 1/2$. Set $r = 1/\rho_0 - 1$. By (13.57) we obtain $F(x, t) = r$. Solving (13.56) we find

$$G(x, t) = -\frac{r}{1 + rx} .$$

The potential is then given by

$$q(x) = 2 \frac{\mathrm{d}}{\mathrm{d}x} G(x, x) = \frac{2r^2}{(1 + rx)^2} ,$$

while the boundary condition constants are

$$h = G(0,0) = -r \,, \qquad H = -G(1,1) = \frac{r}{1+r} = r\rho_0 \,,$$

since we infer from the asymptotics of λ_n that $C_1 = 0$. Finally, formula (13.54) yields the eigenfunctions

$$y(x,\lambda) = \cos(\sqrt{\lambda}\,x) - \frac{r}{1+rx}\,\frac{\sin(\sqrt{\lambda}\,x)}{\sqrt{\lambda}}\,.$$

It is easy to check that they are indeed normalized to

$$\rho_n = \frac{1}{y^2(0,\lambda_n)} \int_0^1 y^2(x,\lambda_n)\,\mathrm{d}x = \frac{1}{2}\,, \qquad n \geq 1\,,$$

while in the $n = 0$ case we retrieve the chosen ρ_0. $\triangleleft$

While the Gelfand–Levitan method is suitable for specific analytic studies, its real-life usefulness is limited as usually neither the full spectrum nor the normalization constants are known. The method also has a very high operation count: assuming a discretization $(x_i, t_j) = (i/N, j/N)$ with $0 \leq j \leq i \leq N$ the computational cost of solving (13.56) is $O(N^4)$. In addition, the data function $F(x,t)$ itself is often very difficult to evaluate: the weights ρ_n and ρ_n^0 may fall off rapidly when $n \to \infty$ so that large cancellation errors may occur in the evaluation of (13.57). In the following, we list a few alternative approaches.

13.5.3   Successive Approximations

In this Subsection we describe the procedure of recovering q in the particular case when two finite spectra of two direct Sturm–Liouville problems (13.53) with different boundary conditions are available [55]: one with $h = \infty$ and $H = \infty$ (pure Dirichlet case) with the spectrum $\{\lambda_n\}$, and one with $H' = 0$ (homogeneous Neumann condition at $x = 1$), giving the spectrum $\{\mu_n\}$. For other types of boundary conditions, only minor changes are required; see Sect. 3.9 of [56]. The vantage point is that for a given q there exists a function $K(x,t) = K(x,t;q)$ that solves the characteristic boundary-value problem

$$K_{tt} - K_{xx} + q(x)K = 0\,, \quad 0 \leq |t| \leq x \leq 1\,,$$

$$K(x,\pm x) = \pm\frac{1}{2}\int_0^x q(s)\,\mathrm{d}s\,, \quad 0 \leq x \leq 1\,.$$

First we must determine two functions of a single argument, $K(1, t)$ and $K_x(1, t)$. This is done by solving the equation

$$\int_0^1 K(1, t) \sin\left(\sqrt{\lambda_n}\, t\right) \mathrm{d}t = -\sin\sqrt{\lambda_n} \tag{13.58}$$

for $K(1, t)$ and the equation

$$\int_0^1 K_x(1, t) \sin\left(\sqrt{\mu_n}\, t\right) \mathrm{d}t = -\sqrt{\mu_n} \cos\sqrt{\mu_n} - \frac{1}{2} \sin\sqrt{\mu_n} \int_0^1 q(s)\, \mathrm{d}s \tag{13.59}$$

for $K_x(1, t)$. These tasks are readily accomplished as λ_n, μ_n and $\int_0^1 q(s)\, \mathrm{d}s$ are known from the asymptotic forms (see (9.92) and the square of (9.90))

$$\lambda_n = (n + 1)^2 \pi^2 + \int_0^1 q(s)\, \mathrm{d}s + O\left(\frac{1}{n^2}\right) ,$$

$$\mu_n = (n + 1/2)^2 \pi^2 + \int_0^1 q(s)\, \mathrm{d}s + O\left(\frac{1}{n^2}\right) .$$

For instance, if $K(1, t) \equiv f(t)$ is represented in terms of some basis functions $\{\phi_k\}_{k=1}^N$ suitable for the boundary conditions of the problem,

$$f(t) = \sum_{k=1}^N f_k \phi_k(t) ,$$

the vector of its expansion coefficients $\boldsymbol{f} = (f_1, f_2, \ldots, f_N)^\mathrm{T}$ is obtained by solving the equation

$$A\boldsymbol{f} = \boldsymbol{b} ,$$

where A is a $N \times N$ matrix with the elements

$$A_{nk} = \int_0^1 \phi_k(t) \sin\left(\sqrt{\lambda_n}\, t\right) \mathrm{d}t ,$$

while the components of the right-hand side $\boldsymbol{b} = (b_1, b_2, \ldots, b_N)^\mathrm{T}$ are given by

$$b_n = -\sin\sqrt{\lambda_n} .$$

One can then formulate an iterative algorithm as follows [55]. For a given $q(x)$, let $u = u(x, t; q)$ solve the boundary-value problem

$$\begin{aligned}
u_{tt} - u_{xx} + q(x)u = 0, \quad 0 \le |t| \le x \le 1, \\
\left.\begin{array}{l} u(1, t) = K(1, t) \\ u_x(1, t) = K_x(1, t) \end{array}\right\} \quad -1 \le t \le 1.
\end{aligned} \tag{13.60}$$

As demonstrated in [57] this problem is most easily solved by rewriting it in the form of a system of two PDE of first order and integrating it along the characteristics $x + t = \text{const}$ and $x - t = \text{const}$. Pick an initial guess $q^{(0)}(x)$ and iterate

$$q^{(i+1)}(x) = 2\frac{\mathrm{d}}{\mathrm{d}x} u\big(x, x; q^{(i)}\big), \qquad i = 0, 1, \dots,$$

until convergence. Note that the problem (13.60) needs to be solved at each step of the iteration. An equivalent form of the iteration is

$$q^{(i+1)}(x) = G(x) - 2 \int_x^1 q^{(i)}(z) u\big(z, 2x - z; q^{(i)}\big) \, \mathrm{d}z,$$

where

$$\begin{aligned}
G(x) &= 2\big[G_1(2x - 1) + G_2(2x - 1)\big], \\
G_1(t) &= K_t(1, t; q), \\
G_2(t) &= K_x(1, t; q),
\end{aligned}$$

are also evaluated at $q = q^{(i)}$.

13.5.4 Quasi-Newton Method

For a given $q(x)$, let $v = v(x, t; q)$ be the solution of the characteristic initial-value problem

$$v_{tt} - v_{xx} + q(x)v = 0, \quad 0 \le |t| \le x \le 1,$$

$$v(x, \pm x) = \pm\frac{1}{2} \int_0^x q(s) \, \mathrm{d}s, \quad 0 \le x \le 1.$$

An alternative iterative method to retrieve q is then [58, 59]

$$q^{(i+1)}(x) = q^{(i)}(x) + G(x) - v_t\big(1, 2x - 1; q^{(i)}\big) - v_x\big(1, 2x - 1; q^{(i)}\big).$$

See also [60, 61].

13.5.5 Shooting Method

An appealing and easily programmable iterative method to solve the inverse Sturm–Liouville problem with a variety of boundary conditions, properties of the available spectrum, and assumptions about the potential function q has been proposed in [62]. The idea is to use the shooting method to solve the initial-value problem yielding $y_n(x)$ for each eigenvalue λ_n,

$$-y_n''(x) + q(x)y_n(x) = \lambda_n y_n(x) \, ,$$
$$y_n(0) = 1 \, , \quad y_n'(0) = h \, , \tag{13.61}$$

complemented by a solution of an auxiliary initial-value problem

$$-\widehat{y}_{nk}''(x) + q(x)\widehat{y}_{nk}(x) = \lambda_n \widehat{y}_{nk}(x) - \phi_k(x)y_n(x) \, ,$$
$$\widehat{y}_{nk}(0) = 0 \, , \quad \widehat{y}_{nk}'(0) = 0 \, , \tag{13.62}$$

where $q(x)$ is updated during each iteration of the above pair of equations and $\{\phi_k(x)\}_{k=1}^{B}$ are the basis functions used in its expansion. The problem (13.62) controls the Fréchet derivatives $\widehat{y}_{jk}$ of $y_n(x)$ with respect to q in the direction $\phi_k(x)$. In the following we shall consider the recovery of the potential function q which is assumed to be symmetric on [0, 1], thus $q(x) = q(1 - x)$, from a subset $\{\lambda_n\}_{n=1}^{N}$ of a single spectrum. An example of retrieving a non-symmetric q from two spectra corresponding to two Sturm–Liouville problems with different boundary conditions is given in Problem 13.8.5. Several other cases which ensure unique reconstruction of q are discussed in [62]. For an alternative approach to the regularized recovery of symmetric potentials by resorting to pseudospectral methods see [63].

In (13.61) the initial conditions at the left point ($x = 0$) correspond to the general boundary condition $y'(0) - hy(0) = 0$ of the underlying direct Sturm–Liouville problem. If the boundary condition at $x = 0$ is homogeneous Dirichlet, we simply set $y_n(0) = 0$ and normalize by $y_n'(0) = 1$. Let the general boundary condition at the right point ($x = 1$) be denoted by

$$\mathcal{B}[y] \equiv y'(1) + Hy(1) = 0 \, .$$

The idea is to solve (13.61) by shooting and trying to capture this final condition $\mathcal{B}[y(\cdot\,; q, \lambda_n)] = 0$ for each $n = 1, 2, \ldots, N$ with an increasingly better knowledge on $q(x)$. Given its ith approximation $q^{(i)}$ we update

$$\mathcal{B}\left[y_n(\cdot\,; q^{(i)}) + \left.\frac{\partial y_n}{\partial q_k}\right|_{q=q^{(i)}} \Delta q^{(i)} \right] = 0 \, .$$

Since $\mathcal{B}$ is linear, this amounts to solving

$$\mathcal{B}\left[\widehat{y}_{nk}(\cdot\,; q^{(i)}) \cdot \Delta q^{(i)} \right] = -\mathcal{B}\left[y_n(\cdot\,; q^{(i)}) \right]$$

at each step of the iteration. Recall that we are trying to recover a symmetric q, hence we must set $H = h$ also in the target boundary condition at $x = 1$. The iterative algorithm, then, is

> **Input**: subset of spectrum $\{\lambda_n\}_{n=1}^N$
> set $\mathcal{B}[y] = y'(1) + Hy(1)$ // $H = h$ for symmetric q
> set $q^{(0)}(x)$ // can be zero
> **for** $i = 1$ **step** 1 **to** $i_{\max}$ **do**
>> **for** $n = 1$ **step** 1 **to** N **do**
>>> solve initial-value problem (13.61) for y_n
>>> $b_n = \mathcal{B}[y_n]$
>>> **for** $k = 1$ **step** 1 **to** B **do**
>>>> solve initial-value problem (13.62) for $\widehat{y}_{nk}$
>>>> $A_{nk} = \mathcal{B}[\widehat{y}_{nk}]$
>>> **end**
>> **end**
>> solve $A_{nk}c_k = -b_n$
>> $\Delta q(x) = \sum_{k=1}^B c_k \phi_k(x)$
>> $q^{(i)}(x) = q^{(i-1)}(x) + \Delta q(x)$
> **end**
> **Output**: reconstructed approximate $q(x)$

Note that no additional information on q is needed except the fact that it is symmetric about $x = 1/2$. In particular, we do not have to provide the average potential $\int_0^1 q(t)\, dt$ that some methods require on input.

Example (Adapted from [62].) Consider a Sturm–Liouville problem with homogeneous Dirichlet boundary conditions at both $x = 0$ ($h = \infty$, hence $y_n(0) = 0$ and $y'(0) = 1$) and $x = 1$ ($H = \infty$, thus $\mathcal{B}[y] = y[1]$) with the symmetric potential

$$q(x) = 1 - \exp\left(-20(x - 0.5)^2\right)$$

shown by the thin full curve in Fig. 13.13 (left). The first three ($N = 3$) eigenvalues, computed by the SLEIGN2 code [64], are

$$\{\lambda_n\}_{n=1}^3 = \left\{10.2302265,\ 40.1367683,\ 89.4264374\right\}.$$

The accuracy of this calculation is crucial! Let us take three ($B = 3$) basis functions for the expansion of the updates $\Delta q^{(i)}$, namely

$$\{\phi_k(x)\}_{k=1}^B = \left\{1,\ \cos(2k\pi x),\ \cos(4k\pi x)\right\}.$$

One is free to use other types of basis functions, for instance, linear or cubic splines, but in general they should be matched to the anticipated symmetry properties of q.

The reconstructed potential after a single iteration ($i_{\max} = 1$) and after two iterations ($i_{\max} = 2$) is shown by the dashed and thick full curve in Fig. 13.13 (left),

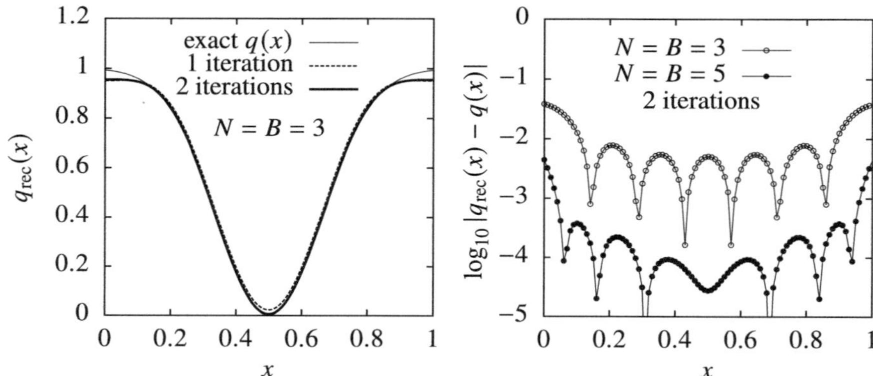

Fig. 13.13 Recovering a symmetric potential $q(x)$ of a Sturm–Liouville problem on $[0, 1]$ with homogeneous Dirichlet boundary conditions from its partial spectrum. [LEFT] Reconstruction from the lowest three eigenvalues by using an expansion in terms of three basis functions. [RIGHT] Error of the reconstruction after two iterations with $N = B = 3$ and $N = B = 5$

respectively. Figure 13.13 (right) shows the corresponding absolute errors. The algorithm converges very quickly, and increasing $i_{\max}$ does not eliminate the residual disagreement observed near the edges of the domain.

Choosing $N = B = 3$ and $N = B = 5$ in the above example makes the matrix A square, letting us exploit as much information as possible and keep A well conditioned. In practice, the size of the spectrum is usually beyond our control: we are forced to live with whatever low N we have available. If, however, we happen to have a larger set of N presumably noisy eigenvalues and opt for a relatively short expansion basis, $B < N$, the problem can still be solved in the least-squares sense: just prior to solving the system $A_{nk}c_k = -b_n$, perform the singular value decomposition $A = U\Sigma V^{\mathrm{T}}$, filter or zero out small singular values σ_i contained in Σ, and recombine to obtain the whittled-down A. This is instrumental not only in solving a non-square system, but primarily in weeding out the otherwise irrelevant components that cause the inversion instabilities. ◁

Other types of inverse Sturm–Liouville problems For the solution of inverse Sturm–Liouville problems in the so-called impedance form

$$- \big(a(x)y'(x)\big)' = \lambda a(x)y(x) \, ,$$

where the task is to recover $a(x)$ from spectral data, see [65, 66]. Inverse Sturm–Liouville problems in which the eigenvalues are contained in the boundary conditions are discussed in [67].

13.6   Inverse Problems for Partial Differential Equations

In this Section we discuss inverse problems for a very modest sample of hyperbolic, parabolic and elliptic partial differential equations. For further details consult [12, 13, 31].

13.6.1   Retrospective Inverse Problem for the Heat Equation

Often we wish to reconstruct the temperature distribution in a body at a certain time $t \in (0, T)$ from the temperature profile measured at $t = T$. As we are asking for a solution at earlier times, we are referring to a *retrospective inverse problem*. For instance, in studying heat diffusion in a bar of length L with a coordinate-dependent thermal diffusivity $D(x)$ and homogeneous Dirichlet boundary conditions we must find $f(x)$ appearing in

$$\left.\begin{aligned}
u_t &= \big(D(x)u_x\big)_x \,, \quad x \in (0, L)\,, \ t \in (0, T]\,, \\
u(0, t) &= u(L, t) = 0\,, \quad t \in (0, T)\,, \\
u(x, 0) &= f(x)\,, \quad x \in (0, L)\,,
\end{aligned}\right\} \tag{13.63}$$

from the measured temperature distribution $u_T(x)$ at the final time $t = T$,

$$u_T(x) = u(x, T)\,, \qquad x \in (0, L)\,. \tag{13.64}$$

The operator Φ corresponding to the problem $\Phi f = u_T$ posed by (13.63) and (13.64) is self-adjoint and has the property

$$(\Phi \phi_n)(x) = \kappa_n \phi_n(x)\,, \qquad \kappa_n = e^{-\lambda_n T}\,, \qquad n = 1, 2, 3, \dots\,,$$

where $\{\lambda_n\}_{n=1}^{\infty}$ and $\{\phi_n\}_{n=1}^{\infty}$ are the eigenvalues and orthonormal eigenvectors of the eigenproblem

$$-\big(D(x)\phi'(x)\big)' = \lambda \phi(x)\,, \qquad \phi(0) = \phi(L) = 0\,.$$

Because Φ is self-adjoint, its left and right singular vectors are the same, ϕ_n, and its singular values are $\sigma_n = \kappa_n$. In principle f can then be obtained from the series

$$f(x) = \sum_{n=1}^{\infty} e^{\lambda_n T} u_{T,n} \phi_n(x)\,,$$

where $u_{T,n} = \langle u_T, \phi_n \rangle$ are the Fourier coefficients of $u_T(x)$. The exponential growth of the terms in T, however, renders the inverse problem severely ill-posed. Indeed, the above solution exists if and only if $u_{T,n}$ decrease rapidly enough, such that

$$\sum_{n=1}^{\infty} e^{\lambda_n T} u_{T,n}^2 < \infty \,,$$

therefore some sort of regularization is imperative.

Example For simplicity, let us set $D(x) = 1$ and $L = 1$ so that the problem (13.63) and (13.64) consists of finding $f(x) = u(x, 0)$ from $u_T(x) = u(x, T)$ for $x \in (0, 1)$. The eigenvalues and normalized eigenfunctions of the homogeneous Dirichlet eigenvalue problem $-\phi''(x) = \lambda\phi(x)$, $\phi(0) = \phi(1) = 0$, are

$$\lambda_n = (n\pi)^2 \,, \qquad \phi_n(x) = \sqrt{2}\sin(n\pi x) \,, \qquad n = 1, 2, 3, \ldots$$

Now assume that the data u_T is noisy, such that $\|u_T^\delta - u_T\| \leq \delta$, and see whether a good way to find a regularized inverse would be to truncate the SVD expansion of Φ by keeping just its singular values $\sigma_n = \kappa_n = \exp(-n^2\pi^2 T)$ up to some fixed $n = N$. As in (13.14)—be alert to the unfortunate change of lettering between u and f—we write

$$f(x) \approx \left(\Gamma^N u_T^\delta\right)(x) = \sum_{n=1}^{N} \frac{1}{\kappa_n} \langle u_T^\delta, \phi_n\rangle \phi_n = \sum_{n=1}^{N} e^{n^2\pi^2 T} u_{T,n}^\delta \phi_n(x) \,,$$

where N depends on the noise level δ. The error of this approximation is [12]

$$\left\|\Gamma^N u_T^\delta - \Phi^+ u_T\right\|_{L^2(0,1)}^2 \leq 2\,e^{2N^2\pi^2 T}\delta^2 + 2\sum_{n=N+1}^{\infty} e^{2n^2\pi^2 T} u_{T,n}^2 \,,$$

where the two terms correspond to the data error and the approximation error, respectively, as in (13.13). Just by looking at the first term we see that we are fighting an uphill battle: at any realistic noise level δ the solution error will be unacceptably large already for N on the order of one, except for very small T. From a reverse perspective: if the solution error is to remain below a practicable value and we wish to keep at least a few Fourier coefficients in order to capture the correct functional form of f, the error on u_T must be minute.

One does not fare much better by resorting to Tikhonov regularization as in (13.16). The Nth partial sum of the approximate inverse becomes

$$\left(\Gamma_\alpha^N u_T^\delta\right)(x) = \sum_{n=1}^{N} \frac{\kappa_n^2}{\kappa_n^2 + \alpha} \frac{1}{\kappa_n} \langle u_T^\delta, \phi_n\rangle \phi_n = \sum_{n=1}^{N} \frac{e^{n^2\pi^2 T}}{1 + \alpha\,e^{2n^2\pi^2 T}} u_{T,n}^\delta \phi_n(x) \,,$$

where formally both N and α depend on δ, and the solution error is

$$\left\|\Gamma_\alpha^N u_T^\delta - f\right\|_{L^2(0,1)}^2 \leq \frac{\delta^2}{\alpha} + \frac{4\alpha}{(1+\sqrt{\alpha})^2} \sum_{n=1}^{N} e^{2n^2\pi^2 T} u_{T,n}^2 + 2\sum_{n=N+1}^{\infty} e^{2n^2\pi^2 T} u_{T,n}^2 \,.$$

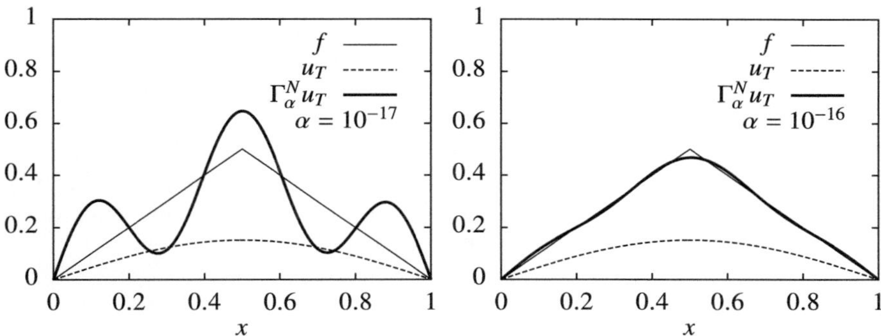

Fig. 13.14 Backtracing the solution $u_T(x) = u(x, T)$ at $T = 0.1$ of the heat equation to its triangular initial condition $u(x, 0) = f(x)$ by using Tikhonov regularization and the first three nonzero Fourier coefficients ($n = 1, 3$ and 5). [LEFT] Under-regularized solution by using $\alpha = 10^{-17}$. [RIGHT] The solution by using $\alpha = 10^{-16}$

Obviously this is not helpful either: even with moderate α one can again afford only the lowest Fourier coefficients because of the exponential growth of the sums. This severely limits our capability to recover the initial temperature profile except over very small time spans T. See Fig. 13.14.   ◁

Lattès–Lions regularization A different approach to the retrospective problem for the heat equation has been suggested by Lattès and Lions [68]. Let us reverse the direction of time in (13.63) and (13.64) by substituting $t \to T - t$, and augment the differential operator by a fourth-order derivative term,

$$\left.\begin{array}{l} u_t = -u_{xx} - \alpha u_{xxxx} \,, \\ u(0, t) = u(1, t) = u_{xx}(0, t) = u_{xx}(1, t) = 0 \,, \\ u(x, 0) = f(x) \,, \end{array}\right\} \qquad (13.65)$$

where $0 \le t \le T$ and we set $L = 1$ henceforth. Thus $f(x)$ has become the present (time-zero) temperature profile which we wish to trace to positive times, now lying in the past. The u_{xxxx} term, together with the additional boundary conditions on u_{xx} in (13.65), provides the regularization which is controlled by the parameter $\alpha > 0$. The regularized solution is given by

$$u_\alpha(x, t) = \sum_{n=1}^{\infty} f_n \exp\left\{ (n\pi)^2 \left[1 - \alpha (n\pi)^2 \right] t \right\} \sin(n\pi x) \,,$$

where f_n are the Fourier coefficients of f. Assuming that $\|u(\cdot, T)\| \le C$ the error of the approximation can be estimated as [31]

$$
\|u_\alpha(\cdot, t) - u(\cdot, t)\| = \left(\int_0^1 \big(u_\alpha(x, t) - u(x, t)\big)\, \mathrm{d}x \right)^{1/2} \le \frac{4Ct}{(T - t)^2}\, \alpha\, \mathrm{e}^{-2} \, .
$$

Example Let the initial condition be $f(x) = u(x, 0) = \sin(\pi x)$, shown by the thin curves labeled $T = 0$ in Fig. 13.15. We wish to propagate this initial profile to $T = 0.2$, with the temperatures going up. As f is equal to the first eigenfunction of the Dirichlet problem $-\phi'' = \lambda\phi$, the exact solution is simply

$$
u(x, T) = \mathrm{e}^{\pi^2 T} u(x, 0) \, .
$$

We use ten Fourier components ($n = 1, 2, \ldots, 10$) to expand f and u_α. Since only f_1 is non-zero, we randomly perturb the remaining nine as $f_n = 0 + R_n$, where $R_n \sim N(0, \varepsilon)$ and $\varepsilon = 10^{-7}$. A hundred solutions backtraced by using $\alpha = 0.003$ as the regularization parameter are shown in Fig. 13.15 (left).

Another hundred solutions backtraced by using a slightly larger value of $\alpha = 0.0035$ are shown in Fig. 13.15 (right). Note the extreme sensitivity of the problem and the explosion of the error estimate when $t \to T$!   ◁

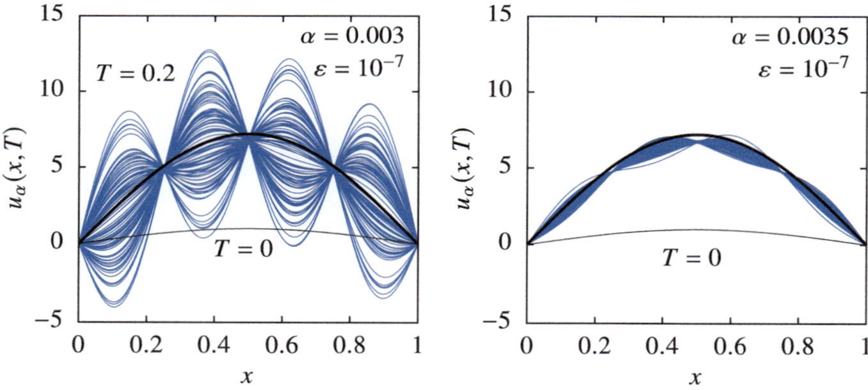

Fig. 13.15 Lattès–Lions regularization of the retrospective problem for the heat equation. The sinusoidal distribution $u(x, 0) = f(x)$ at $T = 0$ is traced *back* to $T = 0.2$. [LEFT] Hundred solutions whose initial conditions were randomly perturbed as described in the text. Regularization parameter $\alpha = 0.003$. [RIGHT] Same but with $\alpha = 0.0035$

Finding the initial condition from an additional measurement at a point Consider the well-posed initial-boundary-value problem for the one-dimensional heat equation [31]

$$
\left.
\begin{aligned}
u_t &= u_{xx}\,, \quad x \in (0, 1)\,,\ t \in (0, T)\,, \\
u(x, 0) &= q(x)\,, \quad x \in (0, 1)\,, \\
u_x(0, t) &= u_x(1, t) = 0\,, \quad t \in (0, T)\,,
\end{aligned}
\right\}
\tag{13.66}
$$

where $q(x)$ is given. Now assume that $q(x)$ is unknown, but we possess an additional piece of information, namely the time dependence of the temperature measured at a specific point $x_0 \in [0, 1]$:

$$u(x_0, t) = f(t), \qquad t \in (t_0, t_1), \qquad 0 < t_0 < t_1 < T. \tag{13.67}$$

The recovery of q with the help of auxiliary knowledge on f is the inverse problem to (13.66). The solution to the direct problem can be expressed as a Fourier series

$$u(x, t) = \frac{1}{\sqrt{2}} q_0 + \sqrt{2} \sum_{n=1}^{\infty} q_n \, e^{-n^2\pi^2 t} \cos(n\pi x),$$

where

$$q_n = \sqrt{2} \int_0^1 q(z) \cos(n\pi z)\, dz, \qquad n = 0, 1, 2, \ldots$$

Substituting the expansion of $u(x, t)$ into (13.67) and evaluating it at $x = x_0$, we obtain a Fredholm integral equation of the first kind for q,

$$\int_0^1 K(t, x_0, z) q(z)\, dz = f(t), \qquad t \in (t_0, t_1),$$

with the kernel

$$K(t, x_0, z) = 1 + 2 \sum_{n=1}^{\infty} e^{-n^2\pi^2 t} \cos(n\pi x_0) \cos(n\pi z).$$

13.6.2 Reconstructing the Source Term in the Heat Equation

When the heat equation involves source terms, the most common inverse problem is to recover the position dependence of the source function. To be specific, consider a heat-conduction problem in which the spatial and temporal dependence of the source are separated,

$$\left.\begin{aligned}
u_t &= u_{xx} + q_1(x)q_2(t), \quad x \in (0, 1),\ t \in (0, T), \\
u(x, 0) &= 0, \quad x \in (0, 1), \\
u(0, t) &= u_x(1, t) = 0, \quad t \in (0, T], \\
u_T(x) &= u(x, T) \quad x \in (0, 1).
\end{aligned}\right\} \tag{13.68}$$

We wish to recover $q_1(x)$ from the measured temperature distribution $u_T(x)$ at $t = T$ under the assumption that $q_2(t)$ is known. If, as in Sect. 13.6.1, the problem (13.68) is framed as the action of a differential operator $\Phi q_1 = u_T$, the eigensystem of Φ is again defined by $(\Phi \phi_n)(x) = \kappa_n \phi_n(x)$, where $\{\phi_n\}_{n=1}^{\infty}$ are the orthonormal eigenvectors of the Dirichlet–Neumann eigenvalue problem

$$-\phi''(x) = \lambda \phi(x) , \qquad \phi(0) = \phi'(1) = 0 .$$

The eigenvalues $\{\lambda_n\}_{n=1}^{\infty}$ corresponding to individual ϕ_n then specify the eigenvalues of Φ itself:

$$\kappa_n = \int_0^T e^{-\lambda_n(T-t)} q_2(t) \, dt , \qquad n = 1, 2, 3, \ldots \tag{13.69}$$

Since Φ is self-adjoint (see Sect. 3.1 of [12]), the left and right singular vectors of Φ are again the same, ϕ_n, and its singular values are $\sigma_n = \kappa_n$. One can therefore recover q_1 from u_T by the Tikhonov-regularized inverse of the form

$$q_1(x) \approx \left(\Gamma_\alpha^N u_T^\delta \right)(x) = \sum_{n=1}^{N} \frac{\kappa_n^2}{\kappa_n^2 + \alpha \, \kappa_n} \frac{1}{\kappa_n} \left\langle u_T^\delta, \phi_n \right\rangle \phi_n(x) . \tag{13.70}$$

The absolute error of this Nth partial sum is given by

$$\left\| \Gamma_\alpha^N u_T^\delta - q_1 \right\|_{L^2(0,1)}^2 \leq \frac{\delta^2}{\alpha} + \frac{4\alpha}{(1 + \sqrt{\alpha})^2} \sum_{n=1}^{N} \frac{u_{T,n}^2}{\kappa_n^2} + 2 \sum_{n=N+1}^{\infty} \frac{u_{T,n}^2}{\kappa_n^2} , \tag{13.71}$$

where $u_{T,n} = \langle u_T, \phi_n \rangle$. See Problem 13.8.6.

Recovering the source from an additional measurement at a point Consider a heat-transfer problem of the form (setting $D(x) = 1$ for simplicity)

$$\left. \begin{aligned} u_t &= u_{xx} + q_1(x) q_2(t) , & x &\in (0, 1) , \ t \in (0, T] , \\ u(x, 0) &= 0 , & x &\in [0, 1] , \\ u_x(0, t) &= u_x(1, t) = 0 , & t &\in (0, T] , \end{aligned} \right\} \tag{13.72}$$

in which instead of the final temperature distribution $u(x, T)$ an additional measurement is available at a fixed point $x_0 \in [0, 1]$ at all times:

$$u(x_0, t) = f(t) , \qquad t \in [0, T] . \tag{13.73}$$

Now $q_1(x)$ is assumed to be known and we wish to recover the time-dependent part of the source, $q_2(t)$. The solution of the direct problem (13.72) is

$$u(x, t) = q_{1,0} \int_0^t q_2(\tau)\, d\tau + \sqrt{2} \sum_{n=1}^{\infty} q_{1,n} \cos(n\pi x) \int_0^t q_2(\tau)\, e^{-n^2\pi^2(t-\tau)}\, d\tau \, ,$$

where

$$q_{1,0} = \int_0^1 q_1(z)\, dz \, , \qquad q_{1,n} = \sqrt{2} \int_0^1 q_1(z) \cos(n\pi z)\, dz \, .$$

Inserting this expansion into (13.73) we obtain

$$q_{1,0} \int_0^t q_2(\tau)\, d\tau + \sqrt{2} \sum_{n=1}^{\infty} q_{1,n} \cos(n\pi x_0) \int_0^t q_2(\tau)\, e^{-n^2\pi^2(t-\tau)}\, d\tau = f(t) \, .$$

Assuming continuity of q_1 and q_2 we may change the order of summation and integration, which leads to a Volterra equation of the first kind for q_2,

$$\int_0^t K(t, \tau) q_2(\tau)\, d\tau = f(t) \, , \qquad t \in [0, T] \, ,$$

with the kernel

$$K(t, \tau) = q_{1,0} + \sqrt{2} \sum_{n=1}^{\infty} q_{1,n}\, e^{-n^2\pi^2(t-\tau)} \cos(n\pi x_0) \, .$$

If $q_1(x_0) = 0$ the solution of the inverse problem (13.72) is not necessarily unique. This can be seen by choosing $x_0 = 1/2$ and letting q_1 be odd with respect to x_0, that is, $q_1(x) = -q_1(x_0 - x)$. The solution $q_2(t)$ *is* unique if q_1 satisfies $q_1 \in C^4[0, 1]$, $q_1(x_0) \neq 0$ and $q_1'(+0) = q_1'(1 - 0) = 0$, and if f satisfies $f \in C^4[0, T]$ and $f(+0) = 0$ [31].

13.6.3 *Inverse Source Problems for the Wave Equation*

Consider the wave equation describing the transverse displacements along a unit-length string attached at both ends, with given initial conditions for the displacement and velocity,

$$u_{tt} = c^2(x) u_{xx} + q_1(x) q_2(t) \, , \quad x \in (0, 1) \, , \ t \in (0, T) \, ,$$
$$u(x, 0) = f(x) \, , \quad u_t(x, 0) = g(x) \, , \quad x \in (0, 1) \, ,$$
$$u(0, t) = u(1, t) = 0 \, , \quad t \in (0, T) \, ,$$

Assuming that $c(x)$ and the temporal dependence of the source term, $q_2(t)$, are known, one might think that providing additional information on the displacement at some later time $t = T$, e.g.

$$u_T(x) = u(x, T), \qquad x \in (0, 1), \tag{13.74}$$

would allow us to recover the spatial load $q_1(x)$, just as we have done in Sect. 13.6.2 for the heat equation. However, it can be shown that the final-state over-determination in the form (13.74) allows for a unique reconstruction of $q_1(x)$ only if T is not a rational number, which is impossible to ensure in practice. The analogous kind of over-determination in which the velocity distribution $v_T(x) = u_t(x, T)$ is known, has the same insurmountable condition: see p. 95 of [12] for details. In the following, we shall restrict the discussion to the cases where final over-determination is not needed.

Determining the purely temporal part of the source Consider the initial-value problem

$$\left.\begin{aligned}
u_{tt} &= u_{xx} + \rho(x, t)q(t), \quad x, t > 0, \\
u(x, 0) &= u_t(x, 0) = 0, \quad x > 0, \\
u_x(0, t) &= 0, \quad t > 0,
\end{aligned}\right\} \tag{13.75}$$

with the missing boundary condition provided by the trace of the function $u(x, t)$ measured at $x = 0$,

$$u(0, t) = f(t), \qquad t \geq 0. \tag{13.76}$$

This is the data of the inverse problem that must fulfill the consistency condition $f(0) = f'(0) = 0$. Assuming that $\rho(x, t)$ is known and that $\rho(0, t) \neq 0$ for all $t \in [0, T]$, it is then easy to show by extending the problem to negative x that $q(t)$ satisfies a Volterra equation of the second kind,

$$q(t) + \int_0^t K(t - \tau, \tau)q(\tau)\, d\tau = F(t), \qquad t \in [0, T], \tag{13.77}$$

where

$$K(x, \cdot) = \frac{\rho_x(x, \cdot)}{\rho(0, t)}, \qquad F(t) = \frac{f''(t)}{\rho(0, t)}.$$

This integral equation is best solved by iteration

$$q^{(n)}(t) = F(t) - \int_0^t K(t - \tau, \tau)\, q^{(n-1)}(\tau)\, d\tau, \qquad n = 1, 2, \ldots,$$

with the initial guess $q^{(0)}(t) = F(t)$, or by any other method discussed in Sect. 13.4. Note that even though this integral equation has a unique solution and converges rapidly, there is no way to overcome the burden of (numerically) evaluating the second derivative of the data function f. Do not despair: refer to Sect. 3.6.

Example Let us demonstrate the reconstruction of the purely temporal part of the source in the wave equation (13.75) by choosing

$$\rho(x, t) = 1 + x^2 t, \qquad q(t) = \sin t .$$

Our task is first to solve the direct (initial-value) problem with the known $q(t)$, sample the resulting $u(x, t)$ at $x = 0$ to obtain the function $u(0, t) = f(t)$, construct $F(t)$ by taking the numerical second derivative of $f(t)$ and, finally, handle the inverse problem by solving (13.77) for q, ending up with the reconstructed $q_{\text{rec}}(t)$.

The solution $u(x, t)$ of the direct problem calculated on the domain $(x, t) \in [0, T] \times [0, T]$ with $T = 20$ is shown in Fig. 13.16 (top left). This calculation must be as precise as possible in order to achieve sufficient smoothness of $F(t)$. The solution in the Figure has been calculated by the NDSolve routine of MATHEMATICA to a precision of six digits and interpolated, so that the second derivative of f was performed on the interpolant. The resulting $F(t)$ is shown by the smooth curve in Fig. 13.16 (top right).

The integral equation (13.77) for q is then solved by iteration with the initial guess $q^{(0)} = F(t)$. The reconstructed potential after 10, 20, 50 and 100 iterations is shown in Fig. 13.16 (bottom left). The corresponding errors with respect to the exact source are shown in the bottom right panel.

Reader's homework: perturb $F(t)$ in some manner—obtaining something like the thin noisy trace in Fig. 13.16 (top right)—and repeat the exercise. Study the sensitivity of the reconstruction to the level of noise introduced at other stages of the procedure. ◁

Determining the purely spatial part of the source The treatment of the inverse problem in which the purely *spatial* component of the source needs to be recovered, i.e. the inverse of the initial-value problem

$$\begin{aligned}
u_{tt} &= u_{xx} + \rho(x, t)q(x) , \quad x, t > 0 , \\
u(x, 0) &= u_t(x, 0) = 0 , \quad x > 0 , \\
u_x(0, t) &= 0 , \quad t > 0 ,
\end{aligned}$$

for which the additional data is provided as in (13.76), proceeds in an analogous manner. If $\rho(x, 0) \neq 0$ for all $x \in [0, T]$ and $f(0) = f'(0) = 0$, the unique $q(x)$ can again be obtained by solving a Volterra equation of the second kind

$$q(x) + \int_0^x K(\xi, x - \xi)q(\xi)\, d\xi = F(x) , \qquad x \in [0, T] ,$$

where

$$K(\cdot, t) = \frac{\rho_t(\cdot, t)}{\rho(x, 0)} , \qquad F(x) = \frac{f''(x)}{\rho(x, 0)} .$$

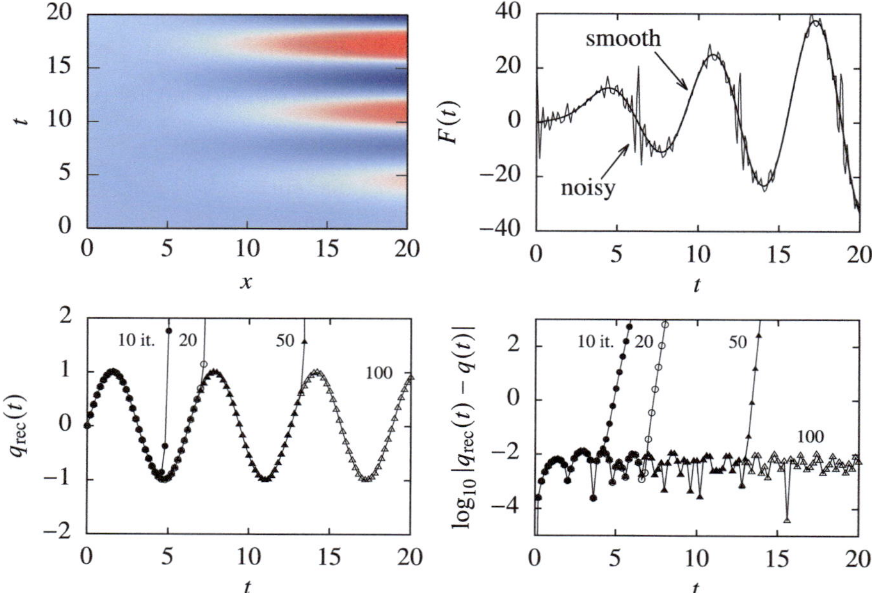

Fig. 13.16 Recovering the purely temporal part of the source term in the wave equation. [TOP LEFT] The solution $u(x, t)$ on the domain $(x, t) \in [0, 20] \times [0, 20]$. [TOP RIGHT] The normalized second derivative of $f(t)$. [BOTTOM LEFT] The reconstructed temporal part of the source. [BOTTOM RIGHT] The error of the reconstruction

13.6.4   Recovering the Potential in the String Equation

Inverse problems requiring the determination of the coefficient function $q(x)$ in hyperbolic equations of the type

$$u_{tt} = u_{xx} - q(x)u + F(x, t), \qquad x \in \mathbb{R},\ t > 0,$$

with initial conditions

$$u(x, 0) = \phi(x), \qquad u_t(x, 0) = \psi(x), \qquad x \in \mathbb{R},$$

are discussed in Sect. 10.1.1 of [31]. If $F(x, t)$ is known and additional data is available, namely

$$u(x_0, t) = f(t), \qquad u_x(x_0, t) = g(t), \qquad t \geq 0,$$

where $x_0 \in \mathbb{R}$ is a fixed point, the inverse problem of recovering q (the potential of the "string equation") has a unique solution. The recovery involves the solution of a system of three coupled non-linear integral equations of the second kind on the triangular domain bounded by the x-axis and the characteristics $x - t = x_0 - t_0$

and $x + t = x_0 + t_0$. The recovery of q in the particular case that $F(x, t) = 0$ also results in a system of integral equations: such inverse problems with various initial and boundary conditions are discussed in Sect. 4.2 of [12] and Sect. 10.1.2 of [31].

13.6.5 Inverse Scattering Problem

Reconstructing the properties of an object from waves scattered off its surface or transmitted through its interior is a canonical problem in quantum mechanics, particle and nuclear physics, electromagnetic theory, acoustics and many other fields of science and engineering. The principles are the same regardless of the character and frequencies of the waves.

The Direct Problem

To remain focused let us consider planar acoustic waves of constant angular frequency ω arriving with velocity c_0 (wave number $k = \omega/c_0$) from a distant source and impinging on an inhomogeneous medium (scatterer) Ω as shown in Fig. 13.17 (left). If the deviation p of the pressure from its static value is assumed to be periodic in time and can be factorized as $p(x, t) = \mathrm{Re}[u(x)\exp(-i\,\omega t)]$, its spatial part satisfies the Helmholtz equation

$$\nabla^2 u(x) + k^2 n(x)u(x) = 0 \,, \tag{13.78}$$

where

$$n(x) = \frac{c_0^2}{c^2(x)}\left(1 + i\,\frac{\beta}{\omega}\right)\,, \qquad \mathrm{Re}\,n(x) \ge 0\,, \quad \mathrm{Im}\,n(x) \ge 0\,,$$

is the complex index of refraction. (Note the unconventional definition through the *squares* of velocities.) Here $c(x)$ is the propagation velocity in the inhomogeneous part of the medium and β the absorption parameter. (In scattering of particle projectiles on target nuclei, the real and imaginary part of n represent the interaction and absorption part of the scattering potential.) It is assumed that beyond a sphere $K[0, R]$ of sufficiently large radius R centered on the scatterer one has $c(x) = c_0$ and $\beta = 0$.

The incident field has the form

$$u^{\mathrm{i}}(x) = \mathrm{e}^{ik\widehat{\theta}\cdot x} \,,$$

where $\widehat{\theta}$ is the unit vector specifying the direction of its propagation, and satisfies the unperturbed Helmholtz equation $\nabla^2 u^{\mathrm{i}} + k^2 u^{\mathrm{i}} = 0$. The inhomogeneous medium modifies the incident field and produces the scattered field u^{s}. The total field $u = u^{\mathrm{i}} + u^{\mathrm{s}}$ satisfies (13.78), which implies

$$\nabla^2 u^{\mathrm{s}} + k^2 n u^{\mathrm{s}} = k^2 (1 - n)u^{\mathrm{i}} \,.$$

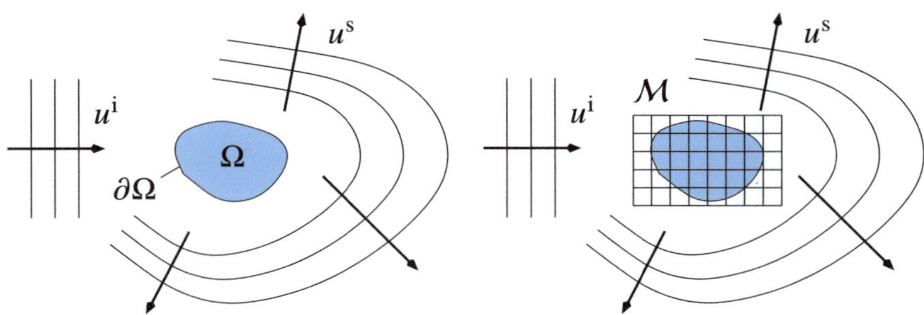

Fig. 13.17 Scattering of acoustic waves in an inhomogeneous medium. [LEFT] Depiction of incident plane waves and scattered waves. [RIGHT] The mesh $\mathcal{M}$ enclosing the scatterer used in the sampling method

Regardless of the details of the process, in three dimensions the scattered wave becomes a spherical wave far away from the inhomogeneity, and satisfies the Sommerfeld radiation condition

$$\lim_{r \to \infty} r \left(\frac{\partial u^{\mathrm{s}}}{\partial r} - \mathrm{i} k u^{\mathrm{s}} \right) = 0 \,,$$

where $r = |\boldsymbol{x}|$. The total field has the asymptotic behavior [69]

$$u(\boldsymbol{x}) = u^{\mathrm{i}}(\boldsymbol{x}) + \frac{\mathrm{e}^{\mathrm{i} k r}}{r} u_\infty(\widehat{\boldsymbol{x}}; \widehat{\boldsymbol{\theta}}) + O \left(\frac{1}{r^2} \right) \,, \qquad r \to \infty \,,$$

where u_∞ is the *scattering amplitude* or the *far-field pattern* of the scattered field u^{s}, observed in the direction $\widehat{\boldsymbol{x}}$. In two-dimensional scattering (polar coordinates in $\mathbb{R}^2$) the radiation condition reads

$$\lim_{r \to \infty} \sqrt{r} \left(\frac{\partial u^{\mathrm{s}}}{\partial r} - \mathrm{i} k u^{\mathrm{s}} \right) = 0 \,,$$

while the total asymptotic field has the form

$$u(r, \phi) = u^{\mathrm{i}}(\boldsymbol{x}) + \frac{\mathrm{e}^{\mathrm{i} k r}}{\sqrt{r}} u_\infty(\phi; \theta) + O \left(\frac{1}{r^{3/2}} \right) \,, \qquad r \to \infty \,.$$

To solve the direct problem in either case we need to determine the total field u satisfying the Helmholtz equation such that $u^{\mathrm{s}} = u - u^{\mathrm{i}}$ fulfills the radiation condition. The basic inverse problem consists of reconstructing the shape of the scatterer, i.e. its approximate boundary $\partial \Omega$ from the (noisy) far-field pattern measured at fixed k for a variety of $\widehat{\boldsymbol{x}}$ and $\widehat{\boldsymbol{\theta}}$ on the unit sphere (in three dimensions) or for multiple ϕ and θ

(in two dimensions). A much more ambitious goal is to determine also the index of refraction $n(x)$ in Ω [70].

Determining the Shape of the Scatterer

There are many ways to determine the shape of the scatterer from the far-field pattern; a state-of-the-art review is given in Refs. [71, 72]. A simple two-dimensional sampling method, which we offer as illustration, has been proposed in Ref. [73]. No assumption on the precise nature of the scatterer is needed, i.e. whether or not the waves penetrate into Ω and, if they do not, what type of boundary conditions are met by the total field at $\partial\Omega$. The procedure is as follows.

Construct a rectangular mesh $\mathcal{M}$ of points (x_i, y_j) which is known to contain the scatterer, as illustrated in Fig. 13.17 (right). If you have no idea about its size, just make the mesh large enough to be beyond reasonable doubt. The nodes of this mesh are specified by the vector $\rho = (\rho \cos \beta, \rho \sin \beta)^{\mathrm{T}}$, where $\rho^2 = x^2 + y^2$ and $\beta = \arctan(x, y)$. For each ρ solve the *far-field equation*

$$\int_{-\pi}^{\pi} u_\infty(\phi; \theta) g(\theta; \rho)\, \mathrm{d}\theta = \mathrm{e}^{-\mathrm{i} k\rho(\phi-\beta)}, \qquad \phi \in [-\pi, \pi], \tag{13.79}$$

for $g = g(\theta; \rho)$. This is a Fredholm integral equation of the first kind, best solved in discretized form, as any realistic far-field pattern $u_\infty(\phi; \theta)$ will also be supplied only at discrete incident angles θ and observation angles ϕ. As u_∞ will most certainly be noisy, some sort of regularization is needed, for instance, constraining $\|g'\|$. Note, however, that the inverse problem at hand is peculiar in the sense that the forward operator itself, i.e. the kernel u_∞, is noisy in practice, while the right-hand side that usually caused the loss of stability is presumed to be exactly known (the incoming wave).

Having found $g = g(\cdot\,; \rho)$ in this manner, the outline of $\partial\Omega$ is roughly given by those points $\rho \in \mathcal{M}$ where

$$\|g(\cdot\,; \rho)\|^2_{L^2(-\pi,\pi)} = \int_{-\pi}^{\pi} \left|g(\theta; \rho)\right|^2 \mathrm{d}\theta \tag{13.80}$$

achieves its maximum. In fact, as ρ approaches the boundary of the scatterer from within, i.e. $\rho \to \partial\Omega$, $\rho \in \Omega$, the norm $\|g(\cdot\,; \rho)\|$ becomes unbounded, but typically a node of $\mathcal{M}$ will not lie precisely on $\partial\Omega$.

Example Consider the scattering of electromagnetic plane waves with $k = 1$ in the (x, y) plane by an infinite cylinder of radius $R = 1$ oriented along the z-axis. For a Dirichlet boundary condition $u = 0$ on the unit circle $\partial\Omega$, the far-field pattern can be calculated analytically:

$$u_\infty(\phi; \theta) = -\mathrm{e}^{-\mathrm{i}\pi/4}\sqrt{\frac{2}{\pi k}} \sum_{n=-\infty}^{\infty} \frac{J_n(kR)}{H_n^{(1)}(kR)} \, \mathrm{e}^{\mathrm{i} n(\phi-\theta)}, \tag{13.81}$$

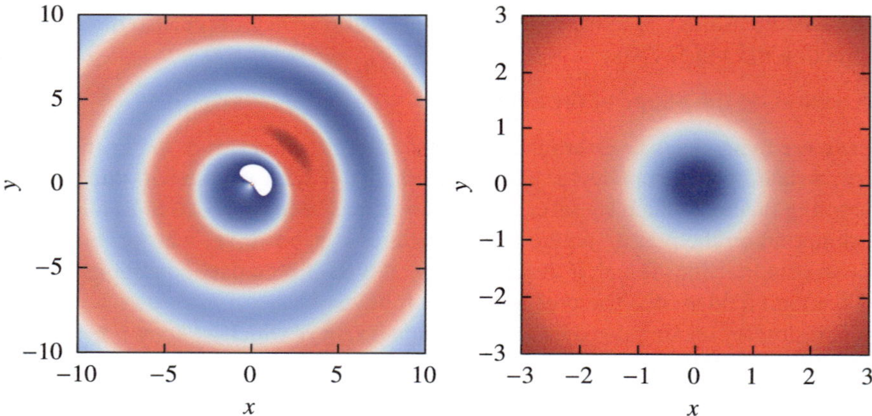

Fig. 13.18 Reconstructing the border of the scattering region from the far-field pattern. [LEFT] Contour plot of the scattered field u^s (not valid near the origin) for $\theta = \pi/4$. [RIGHT] Contour plot of the norm $\log \| g(\cdot\,; \boldsymbol{\rho}) \|$

where J_n and $H_n^{(1)}$ are the Bessel and Hankel functions of the first kind, respectively. The scattered field $u^s(r, \phi; \theta) = (\mathrm{e}^{\mathrm{i}kr}/\sqrt{r})\, u_\infty(\phi; \theta)$ corresponding to this pattern with the incident wave impinging under $\theta = \pi/4$ is shown in Fig. 13.18 (left).

Figure 13.18 (right) shows the norms (13.80) obtained by solving Eq. (13.79) with the noise-free pattern (13.81) uniformly sampled over $[-\pi, \pi]$ in 32 points in both ϕ and θ, and with the uniform mesh $\mathcal{M} = [-3, 3] \times [-3, 3]$ consisting of 128×128 sampling points for the vector $\boldsymbol{\rho}$. The norms change rapidly when $\boldsymbol{\rho}$ approaches the boundary of Ω.

Additional exercise: add noise to the far-field pattern and repeat the above procedure. For an alternative approach see Problem 13.8.7 and [74]. Redo the calculation for acoustic waves scattering off a solid cylinder, in which case the far-field pattern is also known: see [75]. ◁

An improved version of the method by using Morozov's discrepancy principle is described in [76]. The three-dimensional upgrades involving scattering of acoustic and electromagnetic waves are given in [77, 78], respectively. For objections voiced against the sampling approach in general see Sect. 12 of [79]. See also Sects. 3, 4 and 5 of [3] as well as [80].

Determining the Index of Refraction

Determining the nature of the inhomogeneity in the scattering region, specifically its index of refraction $n(\boldsymbol{x})$, is a harder task than finding just its support. In the following we quote the recipe for a regularized, generally applicable three-dimensional method based on Newton's iteration, developed in Sect. 6.6 of [13] under the assumption that the incident waves are *plane waves*. Instead of n the method yields the *contrast* defined as $m(\boldsymbol{x}) = n(\boldsymbol{x}) - 1$.

Assuming that $n(x) = 1$ outside of the sphere $K = K[0, R]$ with radius R, define the *volume potential* with density ϕ as

$$(V\phi)(x) = \int_K \frac{e^{ik|x-y|}}{4\pi|x-y|}\,\phi(y)\,\mathrm{d}^3 y\,, \qquad x \in K\,,$$

and the integral operator W by

$$(W\psi)(\widehat{x}) = \frac{k^2}{4\pi}\int_K \psi(y)\,e^{-ik\widehat{x}\cdot y}\,\mathrm{d}^3 y\,, \qquad \widehat{x} \in S^2\,,$$

where S^2 is the unit sphere. Define $N \in \mathbb{N}$ to be the largest integer not exceeding $2k/\sqrt{3}$, and the set of grid points

$$\mathcal{Z}_N = \left\{ j \in \mathbb{Z}^3 : |j_n| \le N,\ n = 1, 2, 3 \right\}.$$

For every $j \in \mathcal{Z}_N$, fix the unit vectors $\widehat{x}_j$ and $\widehat{\theta}_j$ such that $j = k(\widehat{x}_j - \widehat{\theta}_j)$. Define the operator

$$(L\rho)(x) = -\frac{1}{2\pi^2 k^2} \sum_{j \in \mathcal{Z}_N} \rho(\widehat{x}_j, \widehat{\theta}_j)\,e^{ij\cdot x}$$

and follow the iterative procedure

1. Set $m^{(0)} = 0$, $u^{(0)} = u^{\mathrm{i}}$ and $\ell = 0$.
2. Calculate

$$\mu = L\left[u_\infty - W\left(m^{(\ell)} u^{(\ell)}\right)\right]\,,$$
$$v = u^{\mathrm{i}} - u^{(\ell)} - k^2\,V\left(m^{(\ell)} u^{(\ell)}\right) - k^2\,V\left(\mu u^{\mathrm{i}}\right)\,.$$

3. Set

$$m^{(\ell+1)} = m^{(\ell)} + \mu\,, \qquad u^{(\ell+1)} = u^{(\ell)} + v\,,$$

replace ℓ by $\ell + 1$, and continue with step 2 until convergence at final ℓ_*. The resulting index of refraction is then given by $n = m^{(\ell_*)} + 1$.

Note that u^{i}, $u^{(\ell)}$ and v are functions of x and $\widehat{\theta}$, u_∞ depends on $\widehat{x}$ and $\widehat{\theta}$, while m and μ are functions of r only.

13.7 Phase Retrieval

Phase retrieval (also known as object reconstruction or signal recovery) is the process of recovering both the modulus and the phase of a complex-valued input signal from measurements that provide only the intensities [81, 82]. In experiments with coherent diffractive imaging, for instance, one attempts to determine the electric field E_{obj} at an object with typical transverse dimension a from the intensities I_{diff} provided by the observed diffraction pattern. In the far-field approximation, that is, at distances L such that $\lambda L/a^2 \gg 1$, these entities are related by Fourier transformation:

$$I_{\text{diff}}(x, y) \propto \left| \mathcal{F}[E_{\text{obj}}] \left(\frac{x}{\lambda L}, \frac{y}{\lambda L} \right) \right|^2 .$$

More generally, we are dealing with a quantity $f(x) = |f(x)|\, e^{i\eta(x)}$, where x is a d-dimensional spatial coordinate (most often $d = 1$ or $d = 2$), and its Fourier transform $F = \mathcal{F}[f]$ with $F(k) = |F(k)|\, e^{i\psi(k)}$, where k is a d-dimensional wave-number coordinate. Phase retrieval consists of establishing the phase $\eta(x)$ from only the magnitude $|F|$. Virtually all applications of phase retrieval in physics, however, deal with discrete signals. In the one-dimensional case, we have

$$F = \mathcal{F}_N[f] , \quad f_j = |f_j|\, e^{i\eta_j} , \quad F_k = |F_k|\, e^{i\psi_k} , \tag{13.82}$$

where $j = 0, 1, \ldots, N - 1$ and $k = 0, 1, \ldots, N - 1$, and the task at hand is: find the f_j's (modulus *and* phase), knowing only the magnitudes of the corresponding Fourier transforms, the $|F_k|$'s (or their squares). Note that Fourier phase recovery is a special case of the more general problem in which we are seeking $x \in \mathbb{C}^N$, given the measurements

$$b = |Ax|^2 \in \mathbb{R}^M \tag{13.83}$$

and assuming that the matrix $A \in \mathbb{C}^{M \times N}$ is known. (Here $|\cdot|$ denotes the element-wise modulus operator.) In discrete one-dimensional phase retrieval, for example, A is the Fourier matrix with the elements $A_{jk} = e^{-i2\pi jk/N}$. Alternatively, the problem can be formulated as

$$\begin{aligned} &\text{minimize } \|Ax - y\|_2^2 \\ &\text{subject to } |y| = b , \end{aligned} \tag{13.84}$$

where we optimize with respect to both x and y. This is the at the heart of iterative algorithms discussed below: one step of the iteration minimizes the quadratic error, the subsequent one renormalizes the modulus in order to satisfy the constraint.

In general the recovery of a signal from the modulus of its Fourier transform alone is not unique. There are trivial ambiguities which one should always be aware of and may appear as bogus features in a reconstruction procedure if used indiscriminately: the Fourier magnitude $|F|$ is invariant with respect to global phase shifts, $f_j \to f_j e^{i\phi}$, conjugate inversions, $f_j \to f_{-j}^*$, as well as spatial offsets, $f_j \to f_{j+j_0}$.

There are also nontrivial ambiguities which depend on the dimensionality d of the problem. In one dimension, for instance, there is no uniqueness: one can devise many different signals with the same Fourier magnitude. This is clear from Eq. (13.83) if, say, $M = N$: solving it implies finding $2N$ (real and imaginary) components of x by exploiting just N equations, which is clearly ill-posed. There is no uniqueness even if the support of the signal to be reconstructed—an additional piece of information that effectively increases the number of equations—is known.

For $d \geq 2$, however, real d-dimensional signal with support $N = (N_1, N_2, \ldots, N_d)$, meaning that $f_{n_1, n_2, \ldots, n_d} = 0$ whenever $n_k < 0$ or $n_k \geq N_k$ for $k = 1, 2, \ldots, d$, is uniquely specified by the magnitude of its continuous Fourier transform, up to the trivial ambiguities listed above. An equally important finding pertains to purely discrete f and F: in contrast to the one-dimensional case, for $d \geq 2$ unique retrieval of f from $|F|$ *can* be attained, and the key trick is to extend the domain of f in each of the d dimensions, so-called *oversampling*. If the f to be reconstructed, say, the $N \times N$ image of a teddy-bear in the upper left panel of Fig. 13.19, is (assumed to be) embedded in a wider space, $M \times M$, where $M \geq 2N - 1$, knowing the corresponding $M \times M$ points of $|F|$ is sufficient to guarantee uniqueness. Practice shows that for typical images, the $M \geq 2N - 1$ condition (which needs to be fulfilled in all dimensions) is too severe: even $M \geq 2^{1/d} N$ seems to be sufficient in most cases [83, 84]. In general, any additional information either in the real-space or in the Fourier-space domain will augment the system of equations (13.83) with supplementary conditions, thus improve the conditioning of the problem and perhaps render the inversion unique. Some such constraints will be mentioned during the description of the algorithms below; for details, see Table 1 of [81] and references therein, as well as [85–87].

13.7.1 Alternating-Projection Algorithms

The most popular methods of phase retrieval are based on iterative algorithms performing successive projections onto object and Fourier domains. The simplest of them is the Gerchberg–Saxton algorithm [88] which can be used in its original form if *both* intensities, $|f_j|$ and $|F_k|$, are known, as, for instance, in electron microscopy where the magnitude of the complex signal corresponding to the object can be measured along with its diffraction pattern. (In the following, single indices j and k are used: generalizations to ≥ 2 dimensions should be obvious.) One proceeds as follows: 1) Fourier-transform an estimate of the object, g_j; 2) replace the modulus of the resulting transform, $|G_k|$, by the measured modulus $|F_k|$, but keep the phase; this results in an approximation of the Fourier transform, G'_k; 3) calculate the inverse Fourier transform of G'_k to obtain g'_j, an intermediate representation of the object; 4) replace the modulus of g'_j with the measured object modulus $|f_j|$ to obtain a new estimate of the object. So one pursues the following iterative procedure:

$$
\begin{aligned}
G_k^{(\nu)} &= \mathcal{F}_N\!\left[g_j^{(\nu)}\right] = \left|G_k^{(\nu)}\right| e^{i\phi_j^{(\nu)}}, \\
G_k'^{(\nu)} &= |F_k|\, e^{i\phi_j^{(\nu)}}, \\
g_j'^{(\nu)} &= \mathcal{F}_N^{-1}\!\left[G_k'^{(\nu)}\right] = \left|g_j'^{(\nu)}\right| e^{i\theta_k'^{(\nu)}}, \\
g_j^{(\nu+1)} &= |f_j|\, e^{i\theta_k'^{(\nu)}} = |f_j|\, e^{i\theta_k^{(\nu+1)}},
\end{aligned}
\tag{13.85}
$$

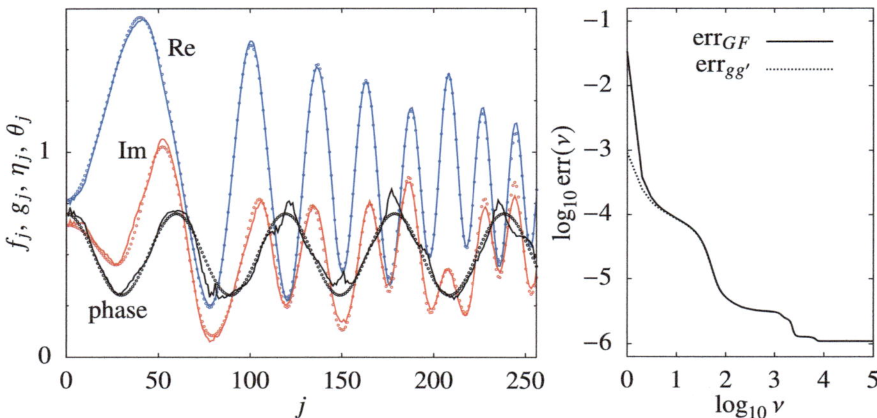

Fig. 13.19 Phase retrieval of a one-dimensional signal with $N = 256$ samples by the Gerchberg–Saxton algorithm. [LEFT] Real part, imaginary part and the phase of the true signal f (symbols) and the corresponding quantities of its approximation g (full lines). [RIGHT] Typical behavior of the errors (13.87) and (13.88) as a function of the iteration index

for $\nu = 0, 1, 2, \ldots$. Here g, θ, G' and ϕ are estimates of f, η, F and ψ, respectively: see Eq. (13.82). The loop may be started either by an initial guess of g_j, which typically boils down to a rough outline of the assumed $|f_j|$ and randomly distributed phases, $\phi_j \sim U(-\pi, \pi)$, or by the corresponding guesses in the Fourier domain. A one-dimensional example is illustrated in Fig. 13.19. Note that the algorithm tends to stagnate in local minima (extended plateaux of the error seen in the right panel) from which it may or may not recover.

Most often the magnitudes $|f_j|$ are unknown, as, for example, in astronomy where the observed objects are optically incoherent sources, meaning that the phases are random and the only measurable quantity is the intensity of light received from them. There is no '$|f|$' to calibrate against, just '$|F|$'. However, the fact that the signal is non-negative (the so-called "non-negativity constraint") together with at least a partial knowledge of its size (the "support constraint") can be and should be exploited in the algorithm in order to achieve a satisfactory reconstruction. To make it work, we replace its fourth step by

$$g_j^{(\nu+1)} = \begin{cases} g_j'^{(\nu)} & ; \ j \in \mathcal{S}, \\ 0 & ; \ j \notin \mathcal{S}, \end{cases} \tag{13.86}$$

where $\mathcal{S}$ is the set of points at which $g_j'^{(\nu)}$ satisfies the object-domain constraints. (In the teddy-bear reconstruction shown in Fig. 13.21, $\mathcal{S}$ is the complement of the blacked-out region of the top left panel.)

In general, alternating-projection algorithms operate in the manner schematically shown in Fig. 13.20, and the more prior information one has on f or F, the better chance one has in reconstructing the desired signal uniquely and efficiently.

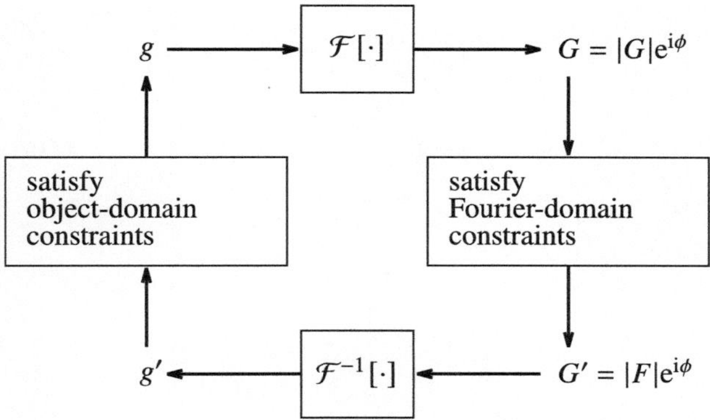

Fig. 13.20 Mode of operation of alternating-projection algorithms for phase retrieval. Successive projections onto object and Fourier domains are performed, while imposing various constraints

The error of the reconstruction can be defined in many ways. In the object domain, for instance, we may compare the successive estimates of the object,

$$\mathrm{err}_{gg'}(\nu) = \sum_j \left| g_j^{(\nu+1)} - g_j'^{(\nu)} \right|^2 . \tag{13.87}$$

For the problem of two intensity measurements, the right-hand side can be expressed as

$$\sum_j \left[|f_j| - |g_j'^{(\nu)}| \right]^2 ,$$

while for the problem with a single intensity measurement (where $|f|$ is unknown) it amounts to

$$\sum_{j \notin S} |g_j'^{(\nu)}|^2 .$$

A good measure of the error in the Fourier domain, which can be used in both the single-intensity and two-intensity measurement case, is

$$\mathrm{err}_{GF}(\nu) = \sum_k \left| G_k^{(\nu)} - G_k'^{(\nu)} \right|^2 = \sum_k \left[|G_k^{(\nu)}| - |F_k| \right]^2 . \tag{13.88}$$

A typical behavior of these two error estimates in the Gerchberg–Saxton algorithm is shown in the left panel of Fig. 13.19. The application of this method in (two-dimensional) digital holography is discussed in [89–91].

The Gerchberg–Saxton algorithm is extremely simple to code and use, but its stagnation problems and slow, jumpy convergence make it difficult to apply in multi-dimensional settings. It has been shown [92, 93] that another modification of the

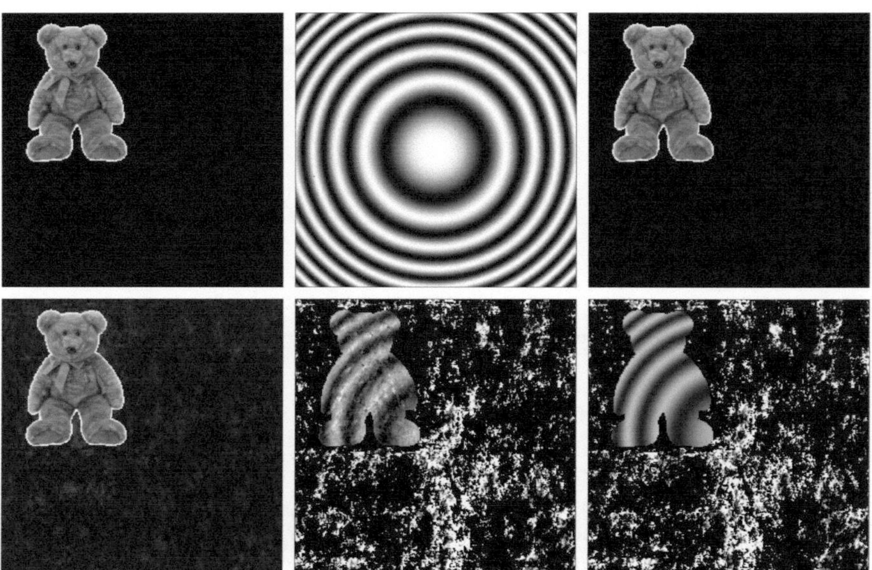

Fig. 13.21 Phase retrieval (object reconstruction) by the hybrid input-output algorithm initialized by a random distribution of phases. [TOP LEFT] The original (oversampled) image of an object (signal) with known support. [TOP CENTER] The true phase. [TOP RIGHT] The Fourier transform of the object, the departure point for the reconstruction. [BOTTOM LEFT] Reconstructed image after 200 iterations. [BOTTOM CENTER AND RIGHT] Reconstructed phase after 20 and 200 iterations, respectively

fourth step of the algorithm in Eq. (13.85), which supersedes the one proposed in Eq. (13.86), makes the reconstruction much more robust and quickly converging. By writing

$$
g_j^{(v+1)} = \begin{cases} g_j^{\prime(v)} & ; \ j \in \mathcal{S}, \\ g_j^{(v)} - \beta g_j^{\prime(v)} & ; \ j \notin \mathcal{S}, \end{cases} \tag{13.89}
$$

where β is a feedback parameter (usually $0.5 < \beta < 1$ works well), one obtains the so-called *[Fienup's] hybrid input-output* (HIO) algorithm, which currently appears to be the most popular simple method of phase retrieval in problems with a single intensity measurement. An example of its use in a two-dimensional setting is shown in Fig. 13.21. Note that in this example the magnitudes $|F|$ of the Fourier transform are the only quantities needed for the full recovery of f.

The HIO method has its own deficiencies: the support must be predefined, and the algorithm may perform poorly in the presence of high noise in the diffraction intensity, $|F|$, and produce oscillating images. An obvious solution to the problem of noisiness is offered by a noise-robust version of the HIO algorithm [94] in which the step (13.89) is replaced by

$$g_j^{(\nu+1)} = \begin{cases} g_j^{\prime(\nu)} & ; \ j \in S \,, \\ g_j^{(\nu)} - \beta g_j^{\prime(\nu)} & ; \ j \notin S \ \& \ \left| g_j^{\prime(\nu)} \right| > 3\sigma_{\text{noise}} \,, \\ 0 & ; \ j \notin S \ \& \ \left| g_j^{\prime(\nu)} \right| \le 3\sigma_{\text{noise}} \,. \end{cases}$$

The two constraints imposed at points outside the support reduce the sensitivity to noise along the "3σ-rule" of robust statistics discussed on page 318.

A similar approach is taken by the *oversampling smoothness* algorithm [95] in which a Gaussian filter is applied to the intermediate image in the complement of the support. In this method, the step (13.89) is replaced by

$$g_j^{\prime\prime(\nu)} = \begin{cases} g_j^{\prime(\nu)} & ; \ j \in S \,, \\ g_j^{(\nu)} - \beta g_j^{\prime(\nu)} & ; \ j \notin S \,, \end{cases}$$

$$g_j^{(\nu+1)} = \begin{cases} g_j^{\prime\prime(\nu)} & ; \ j \in S \,, \\ \mathcal{F}_N[G_k^{\prime\prime(\nu)} W_k] & ; \ j \notin S \,, \end{cases}$$

where $G'' = \mathcal{F}_N[g'']$ and W_k is a (discretized) Gaussian function in k-space whose variance is made to decrease during the iteration.

Most recently, it has been observed that a small modification of the standard HIO algorithm which makes $g^{(\nu)}$ a continuous function both within and outside the support, results in excellent object recovery by using the non-negativity constraint *alone*, even in the presence of high noise in the Fourier magnitudes [96]. This approach works particularly well for objects with non-convex and non-centrosymmetric support (as in Fig. 13.21)—although the main benefit of the method is the true support need not be specified in advance! In the *continuous* hybrid input-output (CHIO) method the fourth step of the standard HIO algorithm is replaced by

$$g_j^{(\nu+1)} = \begin{cases} g_j^{\prime(\nu)} & ; \ \alpha g_j^{(\nu)} \le g_j^{\prime(\nu)} \ \& \ j \in S \,, \\ g_j^{(\nu)} - \left(\frac{1-\alpha}{\alpha}\right) g_j^{\prime(\nu)} & ; \ 0 \le g_j^{\prime(\nu)} \le \alpha g_j^{(\nu)} \ \& \ j \in S \,, \\ g_j^{(\nu)} - \beta g_j^{\prime(\nu)} & ; \ \text{otherwise} \,, \end{cases} \tag{13.90}$$

where α is an additional parameter. Stable convergence is usually attained when $\alpha < 1/(1 + \beta)$. Note that the non-negativity and support constraints are explicitly separated in the second line of (13.90), and the surprising finding of [96] is that often the latter constraint, which may be difficult to specify in practice, is redundant.

13.7.2 Semidefinite Programming Algorithms

An alternative approach to phase retrieval leads through the realization that the set of quadratic equations (13.83) can be rewritten as

$$|a_k^\dagger \cdot x|^2 = b_k \,, \quad k = 1, 2, \ldots, M \,, \tag{13.91}$$

and that the unknown vector $x \in \mathbb{C}^N$ to be determined from it can be embedded in higher-dimensional space by transforming $x \mapsto X = xx^\dagger \in \mathbb{C}^{N \times N}$, resulting in a system *linear* in X. This procedure is called *lifting*: quadratic measurements of x are "lifted" and interpreted as linear measurements involving a rank-1 positive-semidefinite matrix X. So instead of (13.91) one has

$$|a_k^\dagger \cdot x|^2 = \mathrm{tr}(x^\dagger a_k a_k^\dagger x) = \mathrm{tr}(a_k a_k^\dagger x x^\dagger) = \mathrm{tr}(A_k X) = b_k \,,$$

where $A_k = a_k a_k^\dagger$. The phase retrieval problem, then, is rephrased as

$$
\begin{array}{ll}
\text{find} \quad X & \\
\text{subject to} \quad \mathrm{tr}(A_k X) = b_k \,, & \qquad
\begin{array}{l}
\text{minimize} \quad \mathrm{rank}(X) \\
\text{subject to} \quad \mathrm{tr}(A_k X) = b_k \,, \\
\qquad\qquad X \geq 0 \,.
\end{array}
\end{array}
\tag{13.92}
$$

(with the left problem also requiring $\mathrm{rank}(X) = 1$, $X \geq 0$, and $\Leftrightarrow$ between the two)

This rank-minimization problem is NP-hard, but the constraints in (13.92) are *linear* in X, hence a simplification has been suggested in which the minimum rank objective is replaced by the minimization of the trace: one then needs to solve

$$
\begin{array}{l}
\text{minimize} \quad \mathrm{tr}(X) \\
\text{subject to} \quad \mathrm{tr}(A_k X) = b_k \,, \\
\qquad\qquad X \geq 0 \,.
\end{array}
\tag{13.93}
$$

This represents a semidefinite programming problem [97] with convex constraints that can be handled by well established solvers: see, for example, Refs. [99–101]. Solving (13.93) is at the heart of the PhaseLift method [102]. Since the estimate $\widehat{X}$ of X found in this manner does not necessarily have low rank, one usually resorts to keeping only the largest rank-1 component of its spectral decomposition: expand X as

$$\widehat{X} = \sum_{i=1}^{N} \lambda_i u_i u_i^\dagger \,, \quad \lambda_1 \geq \lambda_2 \geq \cdots \geq \lambda_N \geq 0 \,,$$

where u_i are mutually orthogonal, and set

$$\widehat{x} = \sqrt{\lambda_1} u_1 \,.$$

It has been shown [103] that this method allows for an exact recovery of any complex vector $x \in \mathbb{C}^N$ from $O(N)$ quadratic equations of the form (13.91). The downside is that the problem becomes highly overdetermined ($N \times N$ Hermitian matrices compared to $O(N)$ real data points), and the computational complexity increases rapidly with N. Consider, for instance, an $N = 128 \times 128 = 16384$ pixel image that becomes embedded in an $N \times N$ matrix X; standard convex relaxation techniques then call for an optimization on $N \times N \approx 2.7 \cdot 10^8$-dimensional space.

What factor in front of N is implied by $O(N)$ such that the reconstructed x is unique depends on how the diffraction data is obtained. PhaseLift (as well as other similar techniques) benefit substantially from so-called structured illumination: multiple diffraction patterns are acquired, either in an actual experiment or by simulation, each of them providing a different "snapshot" of the object. This can be achieved, for example, by modulating the incoming beam probing the object, say, by a diffraction grating; by oblique illuminations of the object at different angles; by illuminating overlapping portions of the object (ptychography); or by placing a mask or a phase plate just after the object. Variability in the illumination is encoded by a specific weight, so that instead of the original signal components x_j one has the weighted components $w_j x_j$, and the algorithm is applied to the union of such data sets.

In the PhaseLift approach a low-rank solution for X can be obtained even more accurately and reliably if a weight matrix $W^{(\nu)}$ is introduced in (13.93) and the algorithm is made iterative [102]: choose $\varepsilon > 0$, start with $W^{(0)} = I$ as in the unweighted problem, and iterate

$$
\begin{aligned}
\text{for } \nu = 0, 1, 2, \ldots \quad \text{minimize} \quad & \mathrm{tr}(W^{(\nu)} X) \\
\text{subject to} \quad & \mathrm{tr}(A_k X) = b_k \,, \\
& X \geq 0 \,,
\end{aligned}
$$

where the weight matrix is updated as

$$
W^{(\nu+1)} = \left(X^{(\nu)} - \varepsilon I \right)^{-1} .
$$

(The weight matrix for the next step of the loop is approximately the inverse of the current solution estimate.)

An alternative strategy involving semidefinite solution techniques is to solve the phase retrieval problem by explicitly separating the amplitude and phase variables, and only optimizing in terms of the latter. This is the essence of the PhaseCut method [104]. The original retrieval problem

$$
\begin{aligned}
&\text{find } X \\
&\text{subject to } |Ax| = b \,,
\end{aligned}
\tag{13.94}
$$

where $A \in \mathbb{C}^{M \times N}$, $x \in \mathbb{C}^N$ and $b \in \mathbb{R}^M$, can be rewritten as a minimization

$$
\min_{\substack{u, |u_k|=1 \\ x}} \| Ax - \mathrm{diag}(b)u \|_2^2 \,,
$$

where one optimizes over both x and phase vectors $u \in \mathbb{C}^M$, the components of which satisfy $|u_k| = 1$. The x-portion of this minimization is a standard least-squares problem and can be solved explicitly by setting $x = A^\dagger \mathrm{diag}(b)u$, resulting in the reduced problem $\min_{u, |u_k|=1} \| AA^\dagger \mathrm{diag}(b)u - \mathrm{diag}(b)u \|_2^2$ which only involves u. Then the phase retrieval problem (13.94) can be restated as

$$\text{minimize } \boldsymbol{u}^{\dagger} H \boldsymbol{u}$$
$$\text{with } H = \mathrm{diag}(\boldsymbol{b})\big(I - AA^{\dagger}\big)\mathrm{diag}(\boldsymbol{b}) \qquad (13.95)$$
$$\text{subject to } |u_k| = 1 , \quad k = 1, 2, \ldots, M ,$$

where the Hermitian matrix H is positive-semidefinite. Solving the problem (13.95), which is non-convex in $\boldsymbol{u}$, implies the same forbidding computational demands as the ancestor to PhaseLift, Eq. (13.92), and one again resorts to "lifting" and replacing rank-minimization by trace-minimization. Setting $U = \boldsymbol{u}\boldsymbol{u}^{\dagger} \in \mathbb{C}^{M \times M}$, the PhaseCut method can be stated as

$$\text{minimize } \mathrm{tr}(UH)$$
$$\text{subject to } \boldsymbol{d}(U) = \mathbf{1} ,$$
$$U \geq 0 ,$$

where $\boldsymbol{d}(U)$ denotes the vector containing the diagonal elements of U. For details see [105].

Semidefinite programming approaches to phase retrieval are known to be robust and are broadly used. Note, however, that even the venerable Gerchberg–Saxton's and Fienup's algorithms can be made quite competitive: a carefully devised *resampling* variant of these two alternating-projection techniques has been shown to converge geometrically and to perform just as well as the "lifting" methods outlined above: see, for example, the AltMinPhase method described in [106].

Regardless of the approach, phase retrieval calls for either additional prior information or additional magnitude-only measurements to attain uniqueness and stability. Recently, great progress has been made by taking into account that the signals representing the objects may be sparse ("the sparsity prior") and that additional magnitude information can be obtained from overlapping short-time Fourier transforms reminiscent of the structured illumination idea: see [107] and references therein for an overview of such approaches. For a comprehensive MATLAB library of phase recovery routines see [108].

13.8   Problems

13.8.1   Image Deblurring

Reconstruction of blurred images is the example of Eq. (13.1) extended to two dimensions. Let a square $n \times n$ grayscale image be represented by a real matrix $U \in \mathbb{R}^{n \times n}$. If A is the blurring operator, $F = AU$ is the blurred image. There are many methods to deblur an image [109]. Blurring occurs when the picture is recorded—due to poor optics, atmospheric disturbances, electromagnetic or thermal noise, and so on—and implies a convolution of U (the undistorted signal) with the blurring kernel which we typically do not know. In deblurring, by which we wish to reconstruct the original image, a specific form of the kernel must therefore be assumed from the beginning.

A common guess is the Gaussian kernel, a matrix representation of a bivariate normal *point spread function* (PSF) which smears each pixel into a weighted average of itself and its neighbors within a radius r, for instance

$$A_2 = \begin{pmatrix} 0.00297 & 0.01331 & 0.02194 & 0.01331 & 0.00297 \\ 0.01331 & 0.05963 & 0.09832 & 0.05963 & 0.01331 \\ 0.02194 & 0.09832 & 0.16210 & 0.09832 & 0.02194 \\ 0.01331 & 0.05963 & 0.09832 & 0.05963 & 0.01331 \\ 0.00297 & 0.01331 & 0.02194 & 0.01331 & 0.00297 \end{pmatrix} \tag{13.96}$$

for $r = 2$. (Typically $r \ll n$.) Blurring means sliding this $(2r + 1) \times (2r + 1)$ matrix across all pixels of the image and keeping track of the cumulative intensity for each pixel. This amounts to two-dimensional convolution, while deblurring, the inverse problem, implies deconvolution, i.e. "taking out" the kernel from the distorted image. Both are most easily accomplished by using the FFT.

$\odot$ Download the image `teddy200.pgm` ($n \times n = 200 \times 200$ matrix U) from the book's web page (Fig. 13.22 (left)). Take the elementary blur matrix as above (or construct it with a different blur radius r) and pad it with zeros to match the size of U. Cyclically translate the entries of the extended A along its diagonal such that A_{11} contains the central weight of the PSF (16.21 % in the above example) while the values overflowing the top left corner reappear from the bottom right corner inwards. Calculate the two-dimensional Fourier transforms $\mathcal{F}[U]$ and $\mathcal{F}[A]$, multiply them component-wise—not as a matrix product!—then calculate the inverse Fourier transformation of the product matrix:

$$F = \mathcal{F}^{-1}\big[\mathcal{F}[A]\mathcal{F}[U]\big] \,.$$

The resulting F should be a blurred image similar to the one shown in the second panel of Fig. 13.22. Deconvolution is achieved by the inverse procedure: calculate the two-dimensional Fourier transform $\mathcal{F}[F]$ of F, divide it component-wise by $\mathcal{F}[A]$, and take the inverse transformation of the ratio. This yields the reconstructed image

$$U_{\text{rec}} = \mathcal{F}^{-1}\big[\mathcal{F}[F]/\mathcal{F}[A]\big] \,.$$

The Fourier transformation and its inverse are bijective, hence the reconstructed image, shown in the third panel of Fig. 13.22 is equal to the original up to round-off errors. Now add noise to the $\mathcal{F}[F]/\mathcal{F}[A]$ ratio before taking the Fourier inverse. How does that influence the deblurring? Study the level of noise you can add before the image becomes unrecognizable. Regularize the inverse by analogy to one-dimensional deconvolution discussed in Sect. 13.3.6.

$\oplus$ Images can also be blurred and deblurred by using linear algebra without resorting to FFT. If the blurs in horizontal and vertical directions are independent, as is the case with Gaussian kernels, two-dimensional convolution reduces to two one-dimensional convolutions. For example, if b is the original and c is the distorted vector, Gaussian blurring with $r = 2$ is accomplished by

Fig. 13.22 Image blurring and deblurring by Fourier transformation. [LEFT] Original image. [LEFT CENTER] Blurred image. [RIGHT CENTER] Noise-free reconstruction. [RIGHT] Reconstruction from a noisy blurred image

$$c = Ab\,, \qquad A = \begin{pmatrix} \ddots & \ddots & \ddots & \ddots & \ddots & \ddots & & \\ 0 & k_1 & k_2 & k_3 & k_4 & k_5 & 0 & \cdots \\ 0 & 0 & k_1 & k_2 & k_3 & k_4 & k_5 & 0 & \cdots \\ & \ddots & \ddots & \ddots & \ddots & \ddots & \ddots \end{pmatrix},$$

where k_i are the elements of $\mathbf{k} = (0.05449,\ 0.24420,\ 0.40262,\ 0.24420,\ 0.05449)^{\mathrm{T}}$ obtained by summing (13.96) over all rows. The corners of the blurring matrices depend on our decision on how to handle the elements of $\mathbf{b}$ that do not exist and $\mathbf{k}$ wants to pick up, e.g. b_0 and b_{-1} as well as b_{n+1} and b_{n+2} in our example. The usual assumption is that $b_i = b_1$ for $i < 1$ and $b_i = b_n$ for $i > n$ (the first and last element are repeated for the whole extension of the kernel). In this case, the upper left corner of the blurring matrix has the form

$$A = \begin{pmatrix} k_1 + k_2 + k_3 & k_4 & k_5 & & & \\ & k_1 + k_2 & k_3 & k_4 & k_5 & \\ & & k_1 & k_2 & k_3 & k_4 & k_5 & 0 & \cdots \\ & & 0 & k_1 & k_2 & k_3 & k_4 & k_5 & 0 & \cdots \\ & & & & \ddots & \ddots & \ddots & \ddots & \ddots & \ddots \end{pmatrix}.$$

The other frequently used options are setting all auxiliary elements of $\mathbf{b}$ to zero, i.e. $b_i = 0$ for all $i < 1$ and $i > n$, or to enforce periodic boundary conditions $b_0 = b_n$, $b_{-1} = b_{n-1}$, and so on. The choice of boundary conditions leads to different structures of A (Toeplitz, circulant, ...) [110, 111] and therefore to different fast algorithms for the inversion. The SVD approach described next works with any type of boundary conditions.

Turning from vectors to matrices (two-dimensional grayscale images), if $U \in \mathbb{R}^{m \times n}$ is the original image and $F \in \mathbb{R}^{m \times n}$ is the blurred image, and blurring the columns is independent of the blurring of the rows, U and F are related by

$$F = A_{\mathrm{c}} U A_{\mathrm{r}}^{\mathrm{T}}\,,$$

where $A_c \in \mathbb{R}^{m \times m}$ blurs the columns of U and $A_r \in \mathbb{R}^{n \times n}$ blurs its rows. A naive inversion, $U = A_c^{-1} F A_r^{-T}$, usually does not work due to matrices being ill-conditioned—try it! The idea, then, is to use singular value decomposition (SVD) to regularize the matrix inversion by controlling the behavior of the singular values. For example, if we decompose A_c and A_r as

$$A_c = U_c \Sigma_c V_c^T, \qquad A_r = U_r \Sigma_r V_r^T \tag{13.97}$$

(see Sect. 4.5.2), the SVD expansions of their inverses are given by

$$(A_c)_k^{-1} = \sum_{i=1}^{k} \frac{v_{ci} u_{ci}^T}{\sigma_{ci}} , \qquad (A_r)_k^{-1} = \sum_{i=1}^{k} \frac{v_{ri} u_{ri}^T}{\sigma_{ri}} , \qquad k \leq \min(m, n) ,$$

where u_{ci}, v_{ci}, u_{ri} and v_{ri} are the ith columns of U_c, V_c, U_r and V_r, respectively, and σ_{ci} and σ_{ri} are the singular values contained in Σ_c and Σ_r. The order-k reconstructed image can then be obtained by calculating

$$U_k = (A_c)_k^{-1} F (A_r)_k^{-T} . \tag{13.98}$$

Download the image of a blurred quote (Fig. 13.23) from the book's web page. The text has been blurred by using a Gaussian kernel with $r = 20$. Find the minimum k in the truncated SVD expansion (13.98) to make the text readable. See also Fig. 4.7.

A better way to proceed is to keep all singular values but control the reconstruction by Tikhonov regularization. We exploit (13.97) but use the weights (13.15) for each singular value contained in Σ_c and Σ_r. This can be done by forming diagonal matrices Φ_c and Φ_r with the elements

$$(\Phi_c)_{ii} = \frac{\sigma_{ci}^2}{\sigma_{ci}^2 + \alpha} , \qquad (\Phi_r)_{ii} = \frac{\sigma_{ri}^2}{\sigma_{ri}^2 + \alpha} ,$$

where $\alpha > 0$ is the regularization parameter. The reconstructed image is given by

$$U = V_c \Phi_c \Sigma_c^{-1} U_c^T F U_r \Phi_r \Sigma_r^{-1} V_r^T .$$

Fig. 13.23 A grayscale image with 1979×313 pixels blurred by a Gaussian kernel with $r = 20$

13.8.2   *Gravitational Pull of Circularly Distributed Mass*

(Adapted from [27].) Suppose that mass is distributed on a ring of radius $1/2$ centered at the origin, with density $q(\theta)$, where θ is the polar angle. At points on the coplanar, concentric circle of radius 1 the radial component of the gravitational force $f(\phi)$ at polar angle ϕ is measured—see Fig. 13.24 (left) and imagine a satellite orbiting an inhomogeneous planetary ring.

By the cosine law the square of the distance between an infinitesimal mass element at angle θ on the inner ring and a point on the outer ring at angle ϕ is given by $r^2 = 5/4 - \cos(\phi - \theta)$. By the same law, the angle ψ between the radial vector originating from the point experiencing the pull and the vector from the attracted mass to the mass element $q(\theta)\mathrm{d}\theta$ on the inner ring satisfies $\cos\psi = (2 - \cos(\phi - \theta))/(2r)$. The total radial force on a point at polar angle ϕ on the outer circle is then

$$f(\phi) = G \int_0^{2\pi} \frac{2 - \cos(\phi - \theta)}{(5 - 4\cos(\phi - \theta))^{3/2}} \, q(\theta) \, \mathrm{d}\theta \, , \tag{13.99}$$

where G is the gravitational constant. The inverse problem of determining the mass distribution q from the measured forces f amounts to solving a Fredholm integral equation of the first kind which can be approached by methods of Sect. 13.4.3.

⊙   Choose an exact density of the inner-ring mass distribution, for instance

$$q_{\mathrm{e}}(\theta) = 0.02 + 0.15\big[0.4\, g(\theta; \mu_1, \sigma) + 0.9\, g(\theta; \mu_2, \sigma) + 0.6\, g(\theta; \mu_3, \sigma)\big] \, ,$$

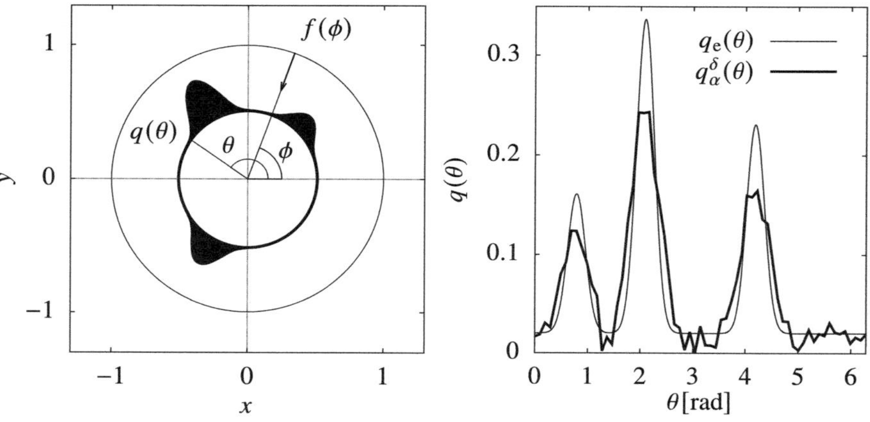

Fig. 13.24 [LEFT] An inhomogeneous mass distribution with density $q(\theta)$ on the inner ring exerts a net gravitational pull $f(\phi)$ on a satellite orbiting on a concentric ring twice as large. [RIGHT] The exact input mass density q_{e} and a typical reconstructed density $q_\alpha^\delta(\theta)$

where $g(\theta; \mu, \sigma) = (2\pi\sigma^2)^{-1/2} \exp(-(x - \mu)^2/(2\sigma^2))$ is the Gaussian function. Take $\mu_1 = \pi/4$, $\mu_2 = 2\pi/3$, $\mu_3 = 4\pi/3$ and $\sigma = 0.17$. The $q_e(\theta)$ corresponding to this parameter set is shown in Fig. 13.24 (left). Numerically calculate the integral (13.99) as precisely as possible to obtain the almost exact $f(\phi)$, then solve the inverse problem of recovering $q(\theta)$.

Add noise to f before reconstructing q, for instance, $f_i \to f_i + R_i$ for each component i of the discretized function, where R_i is a normally distributed random number with zero mean and $\sigma = 0.005$. Regularize the problem. How much noise on f can you afford to introduce?

13.8.3 Polymer Sedimentation in a Centrifuge

One method of determining the distribution of molecular weights in the chemical composition of polymers is to use high-speed centrifuges. A polymer solute is placed in a cell and centrifuged until an equilibrium between sedimentation and diffusion is reached. The distribution of the molecular weights $q(w)$ then corresponds to a certain concentration profile $f(r)$ in the radial direction, which can be measured by optical means. These two distributions are related by Fujita's equation [112] which is a Fredholm equation of the first kind

$$f(r) = f_0 \int_0^\infty \frac{\lambda w \exp(-\lambda wr)}{1 - \exp(-\lambda w)} \, q(w) \, \mathrm{d}w \, ,$$

where r is a radial measure related to the distance from the rotation center ($r = 1$ at the meniscus and $r = 0$ at the sample bottom), λ is a parameter proportional to the square of the rotation frequency, and f_0 is the initial concentration of the polymer in the solute.

⊙ First choose an exact function $q_e(w)$, then numerically integrate Fujita's equation to obtain as precise $f(r)$ as possible, and finally solve the inverse problem of reconstructing $q(w)$ by using the method summarized in Eq. (13.46), with or without adding noise to f.

The practical range of w does not extend to infinity, thus $f(w) = 0$ beyond some $w = w_{\max}$. The interval for w can therefore be restricted to $[0, w_{\max}]$ and further rescaled to $[0, 1]$ by using a dimensionless variable $x = w/w_{\max}$. Try the following normalized unimodal and bimodal test cases suggested by [113]:

$$q(x) = 30\, x^2(1-x)^2 , \qquad 0 \le x \le 1 ,$$

$$\text{and} \quad q(x) = q_1(x) + q_2(x) ,$$

$$q_1(x) = \begin{cases} 210.625\, x^2(x-0.6)^2 , & 0 \le x \le 0.6 , \\ 0 , & 0.6 < x \le 1 , \end{cases}$$

$$q_2(x) = \begin{cases} 0 , & 0 \le x < 0.5 , \\ 435.9(x-0.5)^2(1-x)^2 & 0.5 \le x \le 1 , \end{cases}$$

or devise different, less smooth initial distributions. See also [114].

13.8.4 Discrete Radon Transformation and Its Inverse

The purpose of this Problem is to get acquainted with the basics of the discrete Radon transformation and its inverse. The planar coordinates x and y and the transform-space coordinates ρ and θ can be discretized as

$$\begin{aligned} x_m &= x_{\min} + m\Delta x , & m &= 0, 1, \ldots, M-1 , \\ y_n &= y_{\min} + n\Delta y , & n &= 0, 1, \ldots, N-1 , \\ \rho_r &= \rho_{\min} + r\Delta\rho , & r &= 0, 1, \ldots, R-1 , \\ \theta_t &= \theta_{\min} + t\Delta\theta , & t &= 0, 1, \ldots, T-1 , \end{aligned}$$

where

$$x_{\min} = -\tfrac{1}{2}(M-1)\,\Delta x , \qquad y_{\min} = -\tfrac{1}{2}(N-1)\,\Delta y , \qquad \rho_{\min} = -\tfrac{1}{2}(R-1)\,\Delta\rho ,$$

with $x_{\max} = -x_{\min}$, $y_{\max} = -y_{\min}$, $\rho_{\max} = -\rho_{\min}$, $\theta_{\min} = 0$ and $\Delta\theta = \frac{\pi}{T}$. A straightforward discrete approximation of (13.51) is

$$f(\rho_r, \theta_t) \approx \Delta s \sum_{j=0}^{J-1} q\big(\rho_r \cos\theta_t - s_j \sin\theta_t,\ \rho_r \sin\theta_t + s_j \cos\theta t\big) , \qquad (13.100)$$

with an appropriate discretization of s, for instance $s_j = s_{\min} + j\Delta s$ with $s_{\min} = -s_{\max} = -((J-1)/2)\,\Delta s$. The correspondingly simple version of the discrete inverse Radon transformation—in the so-called *nearest neighbor approximation* [44]—is

$$q(x_m, y_n) \approx \Delta\theta \sum_{t=0}^{T-1} f\big(\rho_{[r_*]}, \theta_t\big) , \qquad (13.101)$$

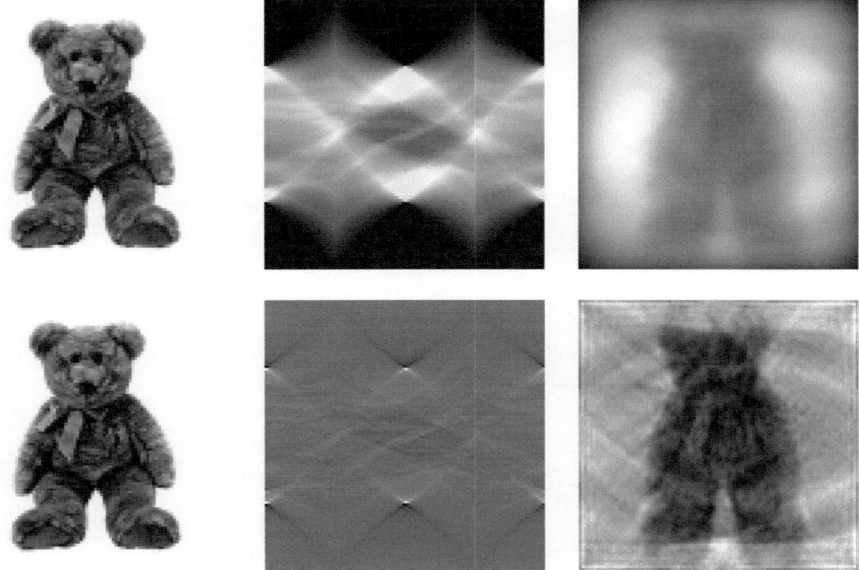

Fig. 13.25 The discrete Radon transform of a 100×100 grayscale image and its inverse [TOP ROW] without filtering (original, transform and reconstructed image) and [BOTTOM ROW] with filtering in the frequency domain by using (13.52) with $\Phi(\nu) = |\nu|$

where

$$r_*(m, n; t) = \frac{x_m \cos \theta_t + y_n \sin \theta_t - \rho_{\min}}{\Delta \rho} \tag{13.102}$$

and $[\cdot]$ denotes rounding to the nearest integer.

⊙ Download the image `teddy100.pgm` (a 100×100 matrix representing $q(x_m, y_n)$) from the book's web page, see Fig. 13.25 (top left). Calculate the discrete Radon transformation of the image by using (13.100) and its inverse by (13.101). Use $M = N = R = T = 100$, $\Delta x = \Delta y = 1$ and $\Delta \rho = 2$. Choose a discretization for the variable s. The obtained sinogram should be similar to the one shown in Fig. 13.25 (top center) and the image reconstructed by (13.101) to the rather fuzzy top right panel.

⊕ The drawback of the naive form (13.100) of the forward transformation is that for a given value of j, the points $(x, y) = (\rho_r \cos \theta_t - s_j \sin \theta_t, \rho_r \sin \theta_t + s_j \cos \theta_t)$ almost never coincide with the samples (pixel values) in the assumed image $q(x_m, y_n)$. In principle, one could alleviate this problem by interpolation to the desired position, but two-dimensional interpolation tends to increase noise. A better forward transform, incorporating only one-dimensional interpolation, reads [44]

$$\sin\theta_t \le \frac{1}{\sqrt{2}} \;:\; f(\rho_r,\theta_t) \approx \frac{\Delta x}{|\cos\theta_t|} \sum_{n=0}^{M-1} q\big(x_{[\alpha n+\beta]},\, y_n\big)\,, \tag{13.103}$$

$$\alpha = -\tan\theta_t\,, \qquad \beta = \frac{\rho_r - x_{\min}(\cos\theta_t + \sin\theta_t)}{\Delta x \cos\theta_t}\,,$$

$$\sin\theta_t > \frac{1}{\sqrt{2}} \;:\; f(\rho_r,\theta_t) \approx \frac{\Delta x}{|\sin\theta_t|} \sum_{m=0}^{M-1} q\big(x_m,\, y_{[\alpha m+\beta]}\big)\,, \tag{13.104}$$

$$\alpha = -\cot\theta_t\,, \qquad \beta = \frac{\rho_r - x_{\min}(\cos\theta_t + \sin\theta_t)}{\Delta x \sin\theta_t}\,,$$

where we have again assumed $M = N$ (square image) and $\Delta x = \Delta y$. The corresponding linearly interpolated discrete inverse transform is

$$q(x_m, y_n) \approx \Delta\theta \sum_{t=0}^{T-1} (1-w) f(\rho_{r_l},\theta_t) + w f(\rho_{r_l+1},\theta_t)\,, \tag{13.105}$$

where $r_l = \lfloor r^*(m,n;t)\rfloor$, $w = r^* - r_l$, and r^* is again given by (13.102).

⊕ Repeat the exercise by calculating the forward transform through (13.103) and (13.104), but apply a filter $\Phi(\nu)$ on f prior to calculating its inverse transform by (13.105), i.e. construct a discrete version of filtered back-projection (13.52). The pseudo-code for the filtering part is listed in Algorithm 8.1 of [44]. You should obtain much better signal-to-noise ratios even with relatively coarse grids—see Fig. 13.25 (bottom). Use different filters, study the dependence on $\Delta\rho$ and the refinement of the discretization in ρ and θ.

13.8.5 Reconstruction of the Potential from Two Sturm–Liouville Spectra

The iterative shooting method for the solution of inverse Sturm–Liouville problems described in Sect. 13.5.5 allows for other types of boundary conditions and arbitrary forms of the potential function q. For instance, let us explore the possibility of reconstructing

$$q(x) = (x - 0.3)^2 + \frac{1}{1 + \mathrm{e}^{100(x-0.3)}} + \frac{1}{1 + \mathrm{e}^{100(x-0.7)}}\,,$$

based on the knowledge of spectra from two different Sturm–Liouville problems on $[0, 1]$ with this potential, one with homogeneous Dirichlet boundary conditions at both ends, and another one with a homogeneous Dirichlet condition at $x = 0$ and a homogeneous Neumann condition at $x = 1$.

⊙ To solve this problem split the inner loops of the algorithm on page 885 in two and let the number of known eigenvalues equal the number of the basis functions used for the expansion of q. For instance, the first three eigenvalues of the Dirichlet–Dirichlet (DD) and Dirichlet–Neumann (DN) problems are

$$\{\lambda_n^{DD}\}_{n=1}^3 = \{10.2386770,\ 40.2787933,\ 89.6048813\},$$
$$\{\lambda_n^{DN}\}_{n=1}^3 = \{3.25992084,\ 22.9471359,\ 62.4128532\},$$

which can be arranged in a set of $N = 3 + 3 = 6$ values used in the n-loop of the algorithm:

$$\lambda_n \in \left\{\lambda_1^{DD},\ \lambda_2^{DD},\ \lambda_3^{DD},\ \lambda_1^{DN},\ \lambda_2^{DN},\ \lambda_3^{DN}\right\}.$$

For the DD case the boundary conditions at $x = 1$ are evaluated as $\mathcal{B}[y_n] = y_n(1)$ in determining b_n and $\mathcal{B}[\widehat{y}_{nk}] = \widehat{y}_{nk}(1)$ in finding A_{nk}, while in the DN case they are evaluated with the corresponding derivatives $y_n'(1)$ and $\widehat{y}_{nk}'(1)$. Choose $B = B_c + B_s = 3 + 3 = 6$ trigonometric basis functions, B_c of which are cosines and B_s are sines, for instance,

$$\phi_k(x) \in \left\{\underbrace{1,\ \cos(2\pi x),\ \cos(4\pi x)}_{B_c},\ \underbrace{\sin(2\pi x),\ \sin(4\pi x),\ \sin(6\pi x)}_{B_s}\right\},$$

or

$$\phi_k(x) \in \left\{\underbrace{1,\ \cos(\pi x),\ \cos(2\pi x)}_{B_c},\ \underbrace{\sin(\pi x),\ \sin(2\pi x),\ \sin(3\pi x)}_{B_s}\right\}. \tag{13.106}$$

You should observe something similar to Fig. 13.26 (left). Increase the size of the spectrum subset to, say, $N = 20$, and the basis to $B_c + B_s = 10 + 10 = 20$, and try to reproduce Fig. 13.26 (right).

⊕ Which of the two basis sets suggested above is more suitable? Does the ordering of the basis functions matter? Check this by reshuffling the cosines and sines. Would a different type of basis functions like linear "hat" functions or cubic splines perform better? Study the sensitivity of the reconstructed q to perturbations in the two spectra. (For example, use two methods or two tools to compute λ_n^{DD} and λ_n^{ND} and compare the reconstructed q in each case.) Choose a different potential and repeat the exercise. What happens if the potential is discontinuous?

13.8.6 Identifying the Source Term in the Heat Equation

Here we learn how to reconstruct the spatial part of the source, $q_1(x)$, in the initial-boundary-value problem for the heat equation posed in the form (13.68). Let

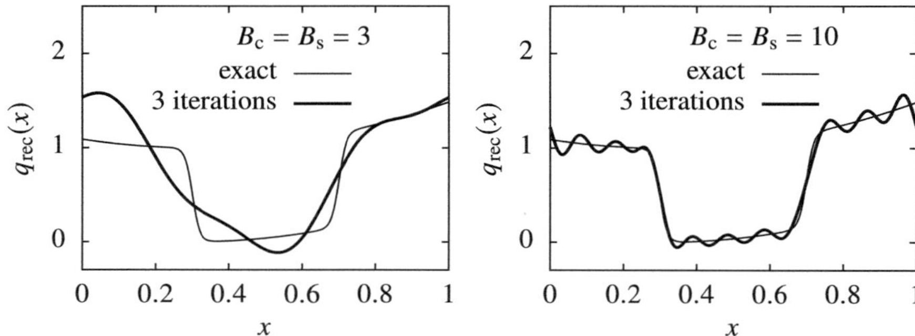

Fig. 13.26 Recovering the potential from the partial spectra of two Sturm–Liouville problems on [0, 1] with different boundary conditions. [LEFT] Reconstruction from the lowest three eigenvalues of the DD problem and the lowest three eigenvalues of the DN problem, by using the expansion in terms of 6 basis functions from the set (13.106). [RIGHT] Same but with a set of 20 eigenvalues and 20 basis functions

$$q_2(t) = \frac{\pi^4}{4}\left[1 + e^{-\pi^2 t/2}\right].$$

If we take $q_1(x) = \sin(\pi x/2)$, the exact solution of the direct problem is

$$u(x, t) = \sin\left(\frac{\pi x}{2}\right)\left[1 - e^{-\pi^2 t/2}\right],$$

with the temperature profile at $t = T$ given by $u_T = u(x, T)$.

⊙   Use the truncated SVD expansion with Tikhonov regularization as given in Eq. (13.70) to recover q_1 from u_T at $T = 1$. Study the n-dependence of the singular values κ_n given by formula (13.69) to get a feeling for the maximum feasible N. Compute the unregularized ($\alpha = 0$) solution at different N and the regularized solutions at fixed N and different values of the regularization parameter, for instance $\alpha = 10^{-5}$, 10^{-3} and 10^{-1}. Perturb the noise-free temperature distribution such that

$$\left\| u_T - u_T^\delta \right\|_{L^2(0,1)} \leq \delta \,.$$

You can do this either by modifying the values of the function given on a grid, or by smearing its Fourier coefficients. See Fig. 13.27.

⊕   Analyze the structure of the error estimate (13.71). Change the formalism to accommodate thermal diffusivities $D(x) \neq 1$ and in particular $D(x) \neq$ const. Moreover, generalize the problem such that the spatial and temporal parts of the source are not decoupled, i.e. solve the reconstruction problem in the general case

$$u_t = \big(D(x)u_x\big)_x + q_1(x)q_2(x, t)\,.$$

See Sect. 3.1 of [12] for details.

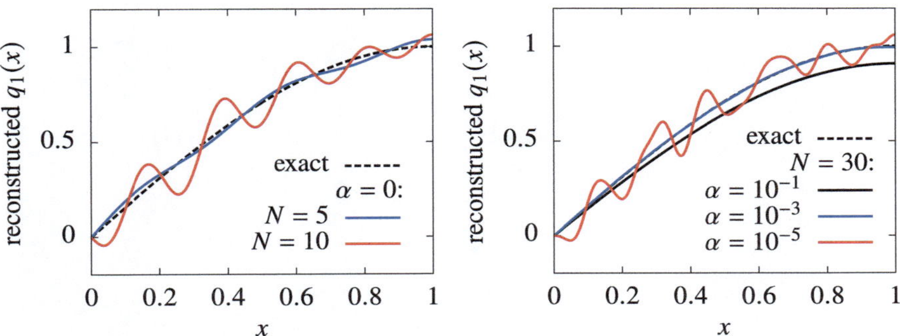

Fig. 13.27 Recovering the spatial part of the source in the heat equation. [LEFT] Reconstructed $q_1(x)$ without regularization, for different N and with Gaussian noise at the level $\delta \approx 10^{-4}$ added to u_T. [RIGHT] Reconstruction of $q_1(x)$ at fixed N but different values of the regularization parameter: $\alpha = 10^{-5}$ (insufficient regularization), $\alpha = 10^{-3}$ (indistinguishable from the exact dependence) and $\alpha = 10^{-1}$ (over-regularized solution with good qualitative behavior but wrong magnitude)

$\bigoplus$ Repeat the exercise with chosen $D(x)$, $q_1(x)$ and $q_2(t)$ for which you do not know the exact solution. Use the methods of Chaps. 10 or 12 to solve the direct problem, yielding $u(x, t)$. Perturb this solution at $t = T$ and reconstruct $q_1(x)$ by using the previous procedure.

13.8.7 *Reconstructing the Scatterer from the Far-Field Pattern*

Let us reconsider the Example on page 899 illustrating the recovery of the scatterer's shape from the far-field pattern u_∞ of the outgoing wave. The approach described below (see p. 248 of [3] for details) is based on the singular-value decomposition of the matrix encoding the scattering amplitude. This allows us to control the instabilities due to noise in u_∞ by eliminating the redundant singular values.

$\odot$ Suppose that we know the approximate values of the far-field pattern $u_\infty(\phi_i, \theta_j)$ measured along uniformly distributed directions specified by the observation and incident angles: $\phi_i = 2\pi i/N, \theta_j = 2\pi j/N, i, j = 1, 2, \ldots, N$. If the index of refraction n of the scatterer is assumed to be real, arrange the discrete values of u_∞ into a $N \times N$ matrix with the elements $A_{ij} = u_\infty(\phi_i, \theta_j)$. If, however, n is assumed to be complex, i.e. it is expected to contain the absorption term, set up a matrix with the elements

$$A_{ij} = \frac{1}{2}\left(u_\infty(\phi_i, \theta_j) + \overline{u}_\infty(\phi_i, \theta_j)\right),$$

where $^-$ denotes complex conjugation. (Our specific example deals with an impenetrable cylinder with a real n.) Perform the singular value decomposition of A,

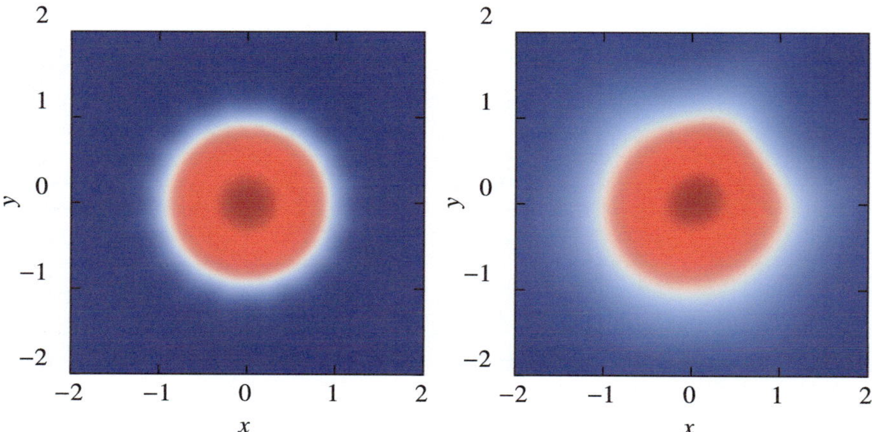

Fig. 13.28 Reconstructing the border of the scatterer (assumed to be contained in the domain $[-2, 2] \times [-2, 2]$ and embedded in an $N \times N = 128 \times 128$ mesh of sampling points) from the far-field pattern by using its singular value decomposition truncated to 16 terms. [LEFT] Contour plot of the function $\Phi(\boldsymbol{\rho})$ calculated from the exact u_∞. [RIGHT] The same, but with the coefficients of u_∞ perturbed by $1/10$ of standardized normal random deviates

$$A = U \Sigma V^\dagger , \qquad \Sigma = \mathrm{diag}(\sigma_1, \sigma_2, \ldots, \sigma_N) .$$

For each vector $\boldsymbol{\rho} = (\rho_x, \rho_y)$ on the mesh $\mathcal{M}$ enclosing the scatterer calculate the expansion coefficients with respect to the columns of V by

$$a_m^{(\rho)} = \sum_{j=1}^{N} V_{jm} \exp\left[-\mathrm{i}\, m\left(\rho_x \cos \theta_j + \rho_y \sin \theta_j\right)\right] , \qquad m = 1, 2, \ldots, N .$$

For each $\boldsymbol{\rho}$ compute

$$\Phi(\boldsymbol{\rho}) = \left[\sum_{m=1}^{N} \left| a_m^{(\rho)} \right|^2 \Big/ \sigma_m \right]^{-1} .$$

The values of Φ should be much smaller for $\boldsymbol{\rho} \notin \Omega$ than for those lying within Ω. If need be, regularization can be introduced at this stage by removing or filtering the unwanted σ_m from the denominator.

Figure 13.28 (left) shows the reconstruction of the scatterer based on the analytic u_∞ given in Eq. (13.81) sampled at $N = 16$ points in both ϕ and θ, for $k = R = 1$. The contour plot of the function Φ reveals the circular outline of the scatterer. The right panel shows the reconstructed shape with the nth coefficient of u_∞ perturbed by $0.1\, R_n$, where R_n is a standardized normal random deviate.

References

1. J.B. Keller, Inverse problems. Am. Math. Monthly **83**, 107 (1976)
2. A.N. Tikhonov, V. Arsenin, *Solutions of ill-posed problems* (John Wiley & Sons, New York, 1977)
3. A. G. Ramm, *Inverse problems. Mathematical and analytical techniques with applications to engineering* (Springer Science+Business Media, New York, 2005)
4. J.L. Mueller, S. Siltanen, *Linear and nonlinear inverse problems with practical applications* (SIAM, Philadelphia, 2012)
5. S.I. Kabanikhin, Definitions and examples of inverse and ill-posed problems. J. Inv. Ill-Posed Problems **16**, 317 (2008)
6. M. Richter, *Inverse problems. Basics, theory and applications in geophysics* (Springer International Publishing AG, Cham, 2016)
7. A.B. Weglein et al., Inverse scattering series and inverse seismic exploration. Inverse Problems **19**, R27 (2003)
8. F. D. M. Neto, A. J. da S. Neto, *An introduction to inverse problems with applications* (Springer-Verlag, Berlin, 2013)
9. M. Benning, M. J. Ehrhardt, *Inverse problems in imaging*, Lecture Notes (Cambridge University, 2016)
10. A. Ben-Israel, T. N. E. Greville, *Generalized inverses. Theory and applications* 2nd edition (Springer-Verlag, New York, 2003)
11. H.W. Engl, M. Hanke, A. Neubauer, *Regularization of inverse problems* (Kluwer Academic Publishers, Dordrecht, 1996)
12. A.H. Hasanoğlu, V.G. Romanov, *Introduction to inverse problem for differential equations* (Springer International Publishing AG, Cham, 2017)
13. A. Kirsch, *An introduction to the mathematical theory of inverse problems,* 2nd edition (Springer Science+Business Media, New York, 2011)
14. I. Stein, Conjugate gradient methods in Banach spaces. Nonlin. Anal. **63**, e2621 (2005)
15. F. Heber, F. Schöpfer, T. Schuster, Acceleration of sequential subspace optimization in Banach spaces by orthogonal search directions. J. Comput. Appl. Math. **345**, 1 (2019)
16. F. Lenti, F. Nunziata, C. Estatico, M. Migliaccio, Conjugate gradient method in Hilbert and Banach spaces to enhance the spatial resolution of radiometer data. IEEE Trans. Geosci. Rem. Sens. **54**, 397 (2016)
17. C. Estatico, S. Gratton, F. Lenti, D. Titley-Peloquin, A conjugate gradient like method for p-norm minimization in functional spaces. Numer. Math. **137**, 895 (2017)
18. C.R. Vogel, *Computational methods for inverse problems* (SIAM, Philadelphia, 2002)
19. A. Tarantola, *Inverse problem theory and methods for model parameter estimation* (SIAM, Philadelphia, 2005)
20. L. Tenorio, Statistical regularization of inverse problems. SIAM Review **43**, 347 (2001)
21. A. N. Tikhonov, A. S. Leonov, A. G. Yagola, *Nonlinear ill-posed problems*, Vols. I and II (Chapman and Hall, London, 1998)
22. H.W. Engl, K. Kunisch, A. Neubauer, Convergence rates for Tikhonov regularization of nonlinear ill-posed problems. Inverse Problems **5**, 523 (1989)
23. A. Neubauer, Tikhonov regularization for non-linear ill-posed problems: optimal convergence rates and finite-dimensional approximation. Inverse Problems **5**, 541 (1989)
24. S.S. Pereverzyev, R. Pinnau, N. Siedow, Regularized fixed-point iterations for nonlinear inverse problems. Inverse Problems **22**, 1 (2006)
25. M. Bachmayr, M. Burger, Iterative total variation schemes for nonlinear inverse problems. Inverse Problems **25**, 105004 (2009)
26. A.-M. Wazwaz, *A first course in integral equations,* 2nd edition (World Scientific, Singapore, 2015)
27. C. W. Groetsch, *Integral equations of the first kind, inverse problems and regularization: a crash course*, J. Phys.: Conf. Ser. **73**, 012001 (2007)

28. P. Linz, *Analytical and numerical methods for Volterra equations* (SIAM, Philadelphia, 1985)
29. G. Izzo, E. Russo, C. Chiapparelli, Highly stable Runga-Kutta methods for Volterra integral equations. Appl. Num. Math. **62**, 1002 (2012)
30. P.C. Hansen, Numerical tools for analysis and solution of Fredholm equations of the first kind. Inverse Problems **8**, 849 (1992)
31. S.I. Kabanikhin, *Inverse and ill-posed problems* (Theory and applications (De Gruyter, Berlin, 2012)
32. C.W. Groetsch, *The theory of Tikhonov regularization for Fredholm equations of the first kind* (Pitman Publishing, London, 1984)
33. E. de Micheli, N. Magnoli, G.A. Viano, On the regularization of Fredholm integral equations of the first kind. SIAM J. Math. Anal. **29**, 855 (1998)
34. P.P.B. Eggermont, Maximum entropy regularization for Fredholm integral equations of the first kind. SIAM J. Math. Anal. **24**, 1557 (1993)
35. U. Amato, W. Hughes, Maximum entropy regularization of Fredholm integral equations of the first kind. Inverse Problems **7**, 793 (1991)
36. A.-M. Wazwaz, The regularization method for Fredholm integral equations of the first kind. Comp. Math. Appl. **61**, 2981 (2011)
37. V. M. Fridman, *Method of successive approximations for Fredholm integral equations of the first kind*, Russian version in Usp. Mat. Nauk, Tom XI, 1(67), 233 (1956)
38. H.J.J. Riele, A program for solving first kind Fredholm equations by means of regularization. Comp. Phys. Comm. **36**, 423 (1985)
39. P. K. Lamm, *A survey of regularization methods for first kind Volterra equations*, in: D. Colton, H. W. Engl, A. K. Louis, J. R. McLaughlin, W. Rundell (eds.), *Surveys on solution methods for inverse problems*, p. 53 (Springer-Verlag, Wien, 2000)
40. J. Radon, *Über die Bestimmung von Funktionen durch ihre Integralwerte längs gewisser Mannigfaltigkeiten*, Berichte über die Verhandlungen der Königlich-Sächsischen Gesellschaft der Wissenschaften zu Leipzig, Math.-Phys. Klasse, **69**, 262 (1917)
41. F. Natterer, *The mathematics of computerized tomography* (SIAM, Philadelphia, 2001)
42. S. Helgason, *Integral geometry and Radon transforms* (Springer Science+Business Media, New York, 2010)
43. R.M. Mersereau, Recovering multidimensional signals from their projections. Computer Graphics and Image Processing **1**, 179 (1973)
44. P. A. Toft, J. A. Sørensen, *The Radon transform — theory and implementation*, PhD Thesis (Technical University of Denmark, Lyngby, 1996)
45. W.H. Press, Discrete Radon transform has an exact, fast inverse and generalizes to operations other than sums along lines. Proc. Nat. Acad. Sci. **103**, 19249 (2006)
46. M. Kac, *Can one hear the shape of a drum?*, Am. Math. Monthly **73**, No. 4, p. 1 (1966)
47. C. Gordon, D. Webb, S. Wolpert, *One cannot hear the shape of the drum*, Bulletin Am. Math. Soc. **27**, No. 1, p. 134 (1992)
48. S. J. Chapman, *Drums that sound the same*, Am. Math. Monthly **102**, No. 2, p. 124 (1995)
49. G. Borg, Eine Umkehrung der Sturm-Liouville Eigenwertaufgabe. Acta Math. **76**, 1 (1946)
50. H. Hochstadt, B. Lieberman, An inverse Sturm-Liouville problem with mixed given data. SIAM J. Appl. Math. **34**, 676 (1976)
51. N. Levinson, The inverse Sturm-Liouville Problem. Math. Tidsskr. B **25**, 25 (1949)
52. I. M. Gelfand, B. M. Levitan, *On the determination of a differential equation from its spectral function*, Amer. Math. Soc. Transl. (2) **1**, 253 (1955)
53. G. Freiling, V. Yurko, *Inverse Sturm-Liouville problems and their applications* (Nova Science Publishers, New York, 2001)
54. J.R. McLaughlin, Analytical methods for recovering coefficients in differential equations from spectral data. SIAM Review **28**, 53 (1986)
55. W. Rundell, P.E. Sacks, Reconstruction techniques for classical inverse Sturm-Liouville problems. Math. Comp. **58**, 161 (1992)
56. K. Chadan, D. Colton, L. Päivärinta, W. Rundell, *An introduction to inverse scattering and inverse spectral problems* (SIAM, Philadelphia, 1997)

57. M. C. Drignei, *Inverse Sturm–Liouville problems using multiple spectra* (Iowa State University, 2008). Accessible at https://doi.org/10.31274/rtd-180813-16905
58. M.C. Drignei, A numerical method for solving a Goursat-Cauchy boundary value problem. Appl. Math. Comp. **220**, 123 (2013)
59. M.C. Drignei, A Newton-type method for solving an inverse Sturm-Liouville problem. Inv. Prob. Sci. Eng. **23**, 851 (2015)
60. P.E. Sacks, An iterative method for the inverse Dirichlet problem. Inverse Problems **4**, 1055 (1988)
61. V. Barcilon, Iterative solution of the inverse Sturm-Liouville problem. J. Math. Phys. **15**, 429 (1974)
62. B.D. Lowe, M. Pilant, W. Rundell, The recovery of potentials from finite spectral data. SIAM J. Math. Anal. **23**, 482 (1992)
63. H. Altundağ, C. Böckmann, H. Taşeli, Inverse Sturm-Liouville problems with pseudospectral methods. Int. J. Comp. Math. **92**, 1373 (2015)
64. P. B. Bailey, W. N. Everitt, A. Zettl, *Algorithm 810: The SLEIGN2 Sturm–Liouville code*, ACM Trans. Math. Softw. **27**, 143 (2001); the code can be downloaded from https://www.niu.edu/math/sl2/index.shtml
65. W. Rundell, P.E. Sacks, The reconstruction of Sturm-Liouville operators. Inverse Problems **8**, 457 (1992)
66. L.-E. Andersson, *Algorithms for solving inverse eigenvalue problems for Sturm–Liouville equations*, in: P. C. Sabatier (ed.), *Inverse problems in action*, p. 138 (Springer-Verlag, Berlin, 1990)
67. C.M. McCarthy, W. Rundell, Eigenparameter dependent inverse Sturm-Liouville problems. Num. Func. Anal. Optim. **24**, 85 (2003)
68. R. Lattès, J.-L. Lions, *Méthode de quasi-réversibilité et applications* (Dunod, Paris, 1967)
69. D. Colton, R. Kress, *Inverse acoustic and electromagnetic scattering theory*, 3rd edition (Springer-Verlag, New York, 2013)
70. D. Colton, J. Coyle, P. Monk, Recent developments in inverse acoustic scattering theory. SIAM Review **42**, 369 (2000)
71. R. Potthast, A survey on sampling and probe methods for inverse problems. Inverse Problems **22**, R1 (2006)
72. G. Nakamura, R. Potthast, *Inverse modeling* (IOP Publishing, Bristol, 2015)
73. D. Colton, A. Kirsch, A simple method for solving inverse scattering problems in the resonance region. Inverse Problems **12**, 383 (1996)
74. A. Kirsch, Factorization of the far-field operator for the inhomogeneous medium case and an application in inverse scattering theory. Inverse Problems **15**, 413 (1999)
75. J.J. Faran, Sound scattering by solid cylinders and spheres. J. Acoust. Soc. Am. **23**, 405 (1951)
76. D. Colton, M. Piana, R. Potthast, A simple method using Morozov's discrepancy principle for solving inverse scattering problems. Inverse Problems **13**, 1477 (1997)
77. D. Colton, K. Giebermann, P. Monk, A regularized sampling method for solving three-dimensional inverse scattering problems. SIAM J. Sci. Comput. **21**, 2316 (2000)
78. D. Colton, H. Haddar, P. Monk, The linear sampling method for solving the electromagnetic inverse scattering problem. SIAM J. Sci. Comput. **24**, 719 (2002)
79. A.G. Ramm, S. Gutman, Numerical solution of obstacle scattering problems. Int. J. Appl. Math. Mech. **1**, 1 (2005)
80. D. Colton, R. Kress, *Integral equation methods in scattering theory* (SIAM, Philadelphia, 2013)
81. Y. Shechtman, Y.C. Eldar, O. Cohen, H.N. Chapman, J. Miao, M. Segev, Phase retrieval with applications to optical imaging: a contemporary overview. IEEE Signal Process. Mag. **32**, 87 (2015)
82. J. Dong, L. Valzania, A. Maillard, T.-A. Pham, S. Gigan, M. Unser, Phase retrieval: from computational imaging to machine learning: a tutorial. IEEE Signal Process. Mag. **40**, 45 (2023)

83. T. Latychevskaia, Iterative phase retrieval in coherent diffractive imaging: practical issues. Appl. Opt. **57**, 7187 (2018)

84. T. Latychevskaia, Phase retrieval methods applied to coherent imaging. Adv. Imaging Electron. Phys. **218**, 1 (2021)

85. T. Bendory, R. Beinert, Y. C. Eldar, *Fourier phase retrieval: uniqueness and algorithms*, in: H. Boche et al. (eds.), *Compressed sensing and its applications*, p. 55 (Birkhäuser, Cham, 2017)

86. P. Grohs, S. Koppensteiner, M. Rathmair, Phase retrieval: uniqueness and stability. SIAM Review **62**, 301 (2020)

87. A. Fannjiang, T. Strohmer, The numerics of phase retrieval. Acta Numer. **29**, 125 (2020)

88. R.W. Gerchberg, W.O. Saxton, A practical algorithm for the determination of phase from image and diffraction plane pictures. Optik **35**, 237 (1972)

89. T. Latychevskaia, Reconstruction of missing information in diffractive patterns and holograms by iterative phase retrieval. Opt. Commun. **452**, 56 (2019)

90. T. Latychevskaia, Iterative phase retrieval for digital holography: tutorial. J. Opt. Soc. Am. A **36**, D31 (2019)

91. F. Momey, L. Denis, T. Olivier, C. Fournier, From Fienup's phase retrieval techniques to regularized inversion for in-line holography: tutorial. J. Opt. Soc. Am. A **36**, D62 (2019)

92. J.R. Fienup, Reconstruction of an object from the modulus of its Fourier transform. Opt. Lett. **3**, 27 (1978)

93. J.R. Fienup, Phase retrieval algorithms: a comparison. Appl. Opt. **21**, 2758 (1982)

94. A.V. Martin et al., Noise-robust coherent diffractive imaging with a single diffraction pattern. Optics Express **20**, 16650 (2012)

95. J.A. Rodriguez, R. Xu, C.-C. Chen, Y. Zhou, J. Miao, Oversampling smoothness: an effective algorithm for phase retrieval of noisy diffraction intensities. J. Appl. Cryst. **46**, 312 (2013)

96. Y. Tian, J.R. Fienup, Phase retrieval with only a nonnegativity constraint. Opt. Lett. **48**, 135 (2023)

97. L. Vandenberghe, S. Boyd, Semidefinite programming. SIAM Review **38**, 49 (1996)

98. N.J. Walkington, Nesterov's method for convex optimization. SIAM Review **65**, 539 (2023)

99. Y. Nesterov, *Lectures on convex optimization* (Springer-Verlag, Cham, 2018)

100. S.R. Becker, E.J. Candès, M.C. Grant, Templates for convex cone problems with applications to sparse signal recovery. Math. Prog. Comp. **3**, 165 (2011)

101. S. Boyd, L. Vandenberghe, *Convex optimization* (Cambridge University Press, Cambridge, 2009)

102. E.J. Candès, Y.C. Eldar, T. Strohmer, V. Voroninski, Phase retrieval via matrix completion. SIAM Review **57**, 225 (2015)

103. E.J. Candès, X. Li, Solving quadratic equations via PhaseLift when there are about as many equations and unknowns. Found. Comput. Math. **14**, 1017 (2014)

104. F. Fogel, I. Waldspurger, A. d'Aspremont, Phase retrieval for imaging problems. Math. Prog. Comp. **8**, 311 (2016)

105. I. Waldspurger, A. d'Aspremont, S. Mallat, Phase recovery, MaxCut and complex semidefinite programming. Math. Program. A **149**, 47 (2015)

106. P. Netrapalli, P. Jain, S. Sanghavi, Phase retrieval using alternating minimization. IEEE Trans. Signal Process. **63**, 4814 (2015)

107. K. Jaganathan, Y. C. Eldar, B. Hassibi, *Phase retrieval: an overview of recent developments*, in: A. Stern (ed.), *Optical compressive imaging*, p. 264 (CRC Press, Boca-Raton, 2016)

108. R. Chandra, Z. Zhong, J. Hontz, V. McCulloch, C. Studer, T. Goldstein, *PhasePack: a phase retrieval library*, 51st Asilomar Conference on Signals, Systems and Computers, Pacific Grove, CA, USA, 2017, pp. 1617. See also https://doi.org/10.48550/arXiv.1711.10175

109. A. Buades, B. Coll, J.M. Morel, Image denoising. A new nonlocal principle, SIAM Review **52**, 113 (2010)

110. P.C. Hansen, J.G. Nagy, D.P. O'Leary, *Deblurring images: matrices, spectra, and filtering* (SIAM, Philadelphia, 2006)

111. J.G. Nagy, K. Palmer, L. Perrone, Iterative methods for image deblurring: a Matlab object-oriented approach. Num. Alg. **36**, 73 (2004)
112. H. Fujita, *Mathematical theory of sedimentation analysis* (Academic Press, New York, 1962)
113. J.M. Jeffcoate, A. Wragg, The numerical solution of Fujita's integral equation to provide estimates of molecular weight distributions. J. Polymer Science **21**, 123 (1983)
114. J. T. Marti, *Numerical solution of Fujita's equation*, in: G. Hämmerlin et al. (eds.), *Improperly posed problems and their numerical treatment*, p. 179 (Springer-Verlag, Basel, 1983)

Appendix A
Mathematical Tools

A.1 Asymptotic Notation

In asymptotic analysis we use several sets of functions that acquire their special meaning in the limit of a real parameter $x \to a$, where a is finite or infinite.

Symbol $O(\cdot)$ The $O(f)$ symbol denotes the set of functions $g \in O(f)$ which, in the limit $x \to a$, are bounded from above in magnitude as $|g(x)| \le C|f(x)|$, with a finite positive constant C. For a function g we thus have

$$\lim_{x \to a} \frac{|g(x)|}{|f(x)|} \in (0, C] \, .$$

We say that the function g is *asymptotically bounded from above* by the function f up to a constant factor.

Symbol $o(\cdot)$ The $o(f)$ symbol denotes the set of functions $g \in o(f)$, where f dominates over g such that $|g(x)| < C|f(x)|$ for each $C > 0$, when $x \to a$. So for a function $g \in o(f)$ we have

$$\lim_{x \to a} \frac{|g(x)|}{|f(x)|} = 0 \, .$$

We say that the function f is *asymptotically dominant* with respect to g. One has:

$$o(f) + o(f) \subseteq o(f) \, , \quad o(f)o(g) \subseteq o(fg) \, , \quad o(o(f)) \subseteq o(f) \, , \quad o(f) \subset O(f) \, .$$

Symbol $\Theta(\cdot)$ The $\Theta(f)$ symbol denotes the set of functions $g \in \Theta(f)$ which, in the limit $x \to a$, are bounded as $C_1|f(x)| \le |g(x)| \le C_2|f(x)|$ for two positive constants C_1 and C_2. Then for $g \in \Theta(f)$ we have

$$\lim_{x \to a} \frac{|g(x)|}{|f(x)|} \in [C_1, C_2] \, ,$$

© The Editor(s) (if applicable) and The Author(s), under exclusive license to Springer Nature Switzerland AG 2025

S. Širca and M. Horvat, *Computational Methods in Physics*, Graduate Texts in Physics, https://doi.org/10.1007/978-3-031-68566-8

i.e. f *asymptotically tightly bounds* g, which we also denote by $g \asymp f$.

Symbol $(\cdot \sim \cdot)$ The $\sim f(x)$ symbol denotes the set of functions $g \sim f$ that are asymptotically equivalent to the function f, so

$$\lim_{x \to a} \frac{f(x)}{g(x)} = 1 \,.$$

These symbols represent sets of functions, but they are often used to express the relations among their elements. For example, $g \in O(f)$ is interpreted as $g = O(f)$.

A.2   The Norms in Spaces $L^p(\Omega)$ and $L_w^p(\Omega)$, $1 \le p \le \infty$

Over a bounded open domain $\Omega \in \mathbb{R}^d$ ($d = 1, 2, 3$) we define the space $L_w^p(\Omega)$ of measurable functions $u : \Omega \to \mathbb{R}$ which are Lebesgue-integrable on Ω, i.e. for which $\int_\Omega |u(x)|^p \, w(x) \, dx < \infty$, where w is a continuous, strictly positive and integrable weight function on Ω. The space $L_w^p(\Omega)$ for $1 \le p < \infty$, furnished with the norm

$$\|u\|_{L_w^p(\Omega)} = \left(\int_\Omega |u(x)|^p \, w(x) \, dx \right)^{1/p} ,$$

is a Banach space. (Most often we encounter one-dimensional cases $\Omega = [a, b]$, e.g. $\Omega = [0, 2\pi]$ in Fourier analysis, or $\Omega = [-1, 1]$ with orthogonal polynomials. The simplifications $\Omega \to [a, b]$ and $x \to x$ are trivial.) Among these, the space $L_w^2(\Omega)$ is virtually ubiquitous: it is also a Hilbert space with the scalar product

$$\langle u, v \rangle_w = \int_\Omega u(x) v(x) w(x) \, dx \tag{A.1}$$

and the induced weighted norm

$$\|u\|_{L_w^2(\Omega)} = \left(\int_\Omega |u(x)|^2 \, w(x) \, dx \right)^{1/2} . \tag{A.2}$$

The corresponding expressions for the spaces $L^p(\Omega)$ are obtained simply by dropping all occurrences of the weight function w. We also encounter spaces of complex measurable functions $u : \Omega \to \mathbb{C}$. In this case, the expressions remain as they are, except that $|\cdot|$ stands for the modulus of the complex number instead of the usual absolute value; for example, in $L^2(a, b)$, the scalar product and the norm are

$$\langle u, v \rangle_2 = \int_a^b u(x) v^*(x) \, dx \, , \qquad \|u\|_2 = \int_a^b |u(x)|^2 \, dx \, . \tag{A.3}$$

(We drop the index 2 when there is no danger of confusion with another norm.) In the family of spaces $L^p(\Omega)$, the extreme case $p = \infty$ represents a Banach space of measurable functions $u : \Omega \to \mathbb{R}$ such that $|u|$ is bounded outside of a set with measure zero, and is furnished with the norm

$$\|u\|_{L^\infty(\Omega)} = \sup_{x \in \Omega} |u(x)| = M \, .$$

In other words, M is the smallest real number for which $|u(x)| \leq M$ holds true outside of a set with measure zero.

A.3 Discrete Vector Norms

We are mostly working with n-dimensional vectors $u \in \mathbb{R}^n$ and square matrices $A \in \mathbb{R}^{n \times n}$. Several vector, matrix, and operator norms can be defined (see also Sect. A.4). The most important class of vector norms are the p-norms

$$\|u\|_p = \left(\sum_{j=1}^n |u_j|^p \right)^{1/p} , \qquad 1 \leq p < \infty \, .$$

The unit balls ("circles") with $n = 2$ for the norms $\| \cdot \|_1$, $\| \cdot \|_2$, and $\| \cdot \|_p$ for some large p are the first three shapes in this Figure:

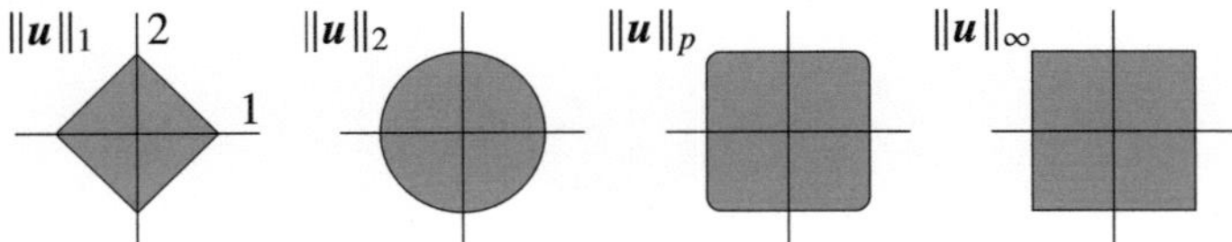

A commonly used norm is the "max"-norm $\|u\|_\infty = \max_{1 \leq j \leq n} |u_j|$, for which the unit "circle" is shown by the fourth shape. In the numerical methods for partial differential equations, where the space axes are discretized in $N + 1$ points (for example, x_j with indices from $j = 0$ to $j = N$), the "max"-norm appears in two disguises. One of them includes all points, while the other excludes the boundary points:

$$\|u\|_\infty = \max_{0 \leq j \leq N} |u_j| \, , \qquad \|u\|_{\infty 0} = \max_{1 \leq j \leq N-1} |u_j| \, . \tag{A.4}$$

Various inequalities exist among the vector norms. They can be used in establishing the hierarchy of the estimates of differences, errors, and remainders in virtually all numerical problems. Among others, the following relations hold:

$$\|u\|_2 \le \|u\|_1 \le \sqrt{n}\|u\|_2 \,, \quad \|u\|_\infty \le \|u\|_2 \le \sqrt{n}\|u\|_\infty \,, \quad \|u\|_\infty \le \|u\|_1 \le n\|u\|_\infty \,.$$

Of all p-norms, by far the most widely used is the Euclidean (l_2):

$$\|u\|_2 = \sqrt{\sum_{j=1}^{n} |u_j|^2} \,. \tag{A.5}$$

(If the subscript 2 is missing, it is usually precisely 2 that is meant.) If we compute the norm by naively coding the expression (A.5) in a computer program, about a half of all representable numbers in floating-point arithmetic will cause an overflow or an underflow. This is how LAPACK's experts circumvent this problem (the DNRM2 routine from the level-1 BLAS library) [1]:

> **Input**: Values $\{u_j\}_{j=1}^n$
> $t = 0; s = 1;$
> **for** $i = 1$ **step** 1 **to** n **do**
> > **if** ($u_j \ne 0$) **then**
> > > **if** ($|u_j| > t$) **then**
> > > > $s = 1 + s(t/u_j)^2;$
> > > > $t = |u_j|;$
> > >
> > > **else**
> > > > $s = s + (u_j/t)^2;$
> > >
> > > **end**
> >
> > **end**
>
> **end**
> **Output**: $\|u\|_2 = t\sqrt{s}$

In the Chapters on differential equations and inverse methods we use the so-called "energy" ($l_{2,\Delta x}$) norm, which takes into account the discretization Δx in one of the coordinates,

$$\|u\|_{2,\Delta x} = \sqrt{\sum_{j=1}^{n} |u_j|^2 \Delta x} \,. \tag{A.6}$$

Namely, the usual l_2-norm is poorly suited to distinguish among different discretizations of the function values. Why? A vector describing the discrete approximation of a function defined on the whole real axis, $u_{\Delta x} = (\dots, u(-\Delta x), u(0), u(\Delta x), \dots)^{\mathrm{T}}$, has the l_2-norm of $\|u_{\Delta x}\|_2$. The approximation $u_{\Delta x/2}$ on a mesh twice as dense (with a spacing of $\Delta x/2$) has twice as many components as $u_{\Delta x}$, and its l_2-norm is $\|u_{\Delta x/2}\|_2 \approx \sqrt{2}\|u_{\Delta x}\|_2$, thus for any smooth function $\|u_{\Delta x}\|_2 \to \infty$ if $\Delta x \to 0$. In such cases it is better to use the $l_{2,\Delta x}$-norm, which possesses all formal properties of the l_2-norm, and which approximates (A.3) for square-integrable functions.

The generalization to vectors from spaces $\mathbb{R}^n \oplus \mathbb{R}^n \oplus \cdots$ is straightforward. For example, for $u \in \mathbb{R}^n \oplus \mathbb{R}^n$ we have

$$\|u\|_{2,\Delta x} = \sqrt{\sum_{j=1}^{n}\sum_{k=1}^{n} |u_{jk}|^2 \Delta x \, \Delta y} \ .$$

In infinitely-dimensional spaces the sums run from $-\infty$ to $+\infty$. Such examples can be found in discussing initial problems for ordinary and partial differential equations, where $u = (\ldots, u_{-1}, u_0, u_1, \ldots)^{\mathrm{T}}$. In such cases the norm (A.6) is computed as

$$\|u\|_{2,\Delta x} = \sqrt{\sum_{j=-\infty}^{\infty} \|u_j\|_2 \, \Delta x} \ .$$

A.4 Matrix and Operator Norms

Matrix norms can be introduced by analogy with the vector norms. We are primarily dealing with real square $n \times n$ matrices. A commonly used norm is the "max"-norm

$$\|A\|_{\max} = \max_{ij} |A_{ij}| \ ,$$

but this "norm" does not satisfy one of the mathematical requirements for the norm: namely, the inequality $\|AB\|_{\max} \leq \|A\|_{\max} \|B\|_{\max}$ is not always fulfilled. The Euclidean norm

$$\|A\|_{\mathrm{F}} = \sqrt{\sum_{ij} |A_{ij}|^2} \ ,$$

also known as the Frobenius or Hilbert–Schmidt norm, is a "genuine" matrix norm. We also use the norm $\|A\|_1$, which measures the largest column sum of the absolute values of the matrix elements, and the norm $\|A\|_\infty$, which measures the largest row sum of the absolute values:

$$\|A\|_1 = \max_{j} \sum_{i} |A_{ij}| \ , \qquad \|A\|_\infty = \max_{i} \sum_{j} |A_{ij}| \ .$$

Note that $\|A\|_1 = \|A^{\mathrm{T}}\|_\infty$. The operator norms

$$\|A\| = \max_{u \neq 0} \frac{\|Au\|}{\|u\|} \tag{A.7}$$

(which are also matrix norms) are defined according to the underlying vector norm, and are thus also known as the subordinate or induced norms. In general, operator norms are difficult to compute. The operator norm corresponding to the Euclidean vector norm is

$$\|A\|_2 = \max_{u \neq 0} \frac{\|Au\|_2}{\|u\|_2} = \sqrt{\lambda_{\max}} \, ,$$

where $\lambda_{\max}$ is the largest eigenvalue of the matrix $A^{\mathrm{T}} A$. Various relations exist among different matrix norms, for example

$$\|A\|_2 \leq \|A\|_{\mathrm{F}} \leq \sqrt{n}\,\|A\|_2 \leq \|A\|_\infty \, , \qquad \frac{1}{n}\|A\|_\infty \leq \|A\|_1 \leq n\|A\|_\infty \leq n^{3/2}\|A\|_2 \, .$$

The spectral radius of the matrix A with the eigenvalues $\lambda_1, \ldots, \lambda_s$ is defined as

$$\rho(A) = \max_{1 \leq k \leq s} \{\, |\lambda_k(A)| \,\} \, .$$

For Hermitian matrices the Euclidean matrix norm is equal to the spectral radius of the matrix, $\rho(A) = \|A\|_2$, while in general $\rho(A) \leq \|A\|_2$. These considerations apply equally well to vectors in $\mathbb{C}^n$ and matrices in $\mathbb{C}^{n \times n}$: the absolute value signs denote the magnitudes of the complex numbers, while for matrices A^{T} (transposition) should be substituted by $A^\dagger$ (transposition and complex conjugation).

A.5   Eigenvalues of Tridiagonal Matrices

Here we list the eigenvalues λ_s and eigenvectors $\boldsymbol{v}_s = (v_1, v_2, \ldots, v_k, \ldots, v_N)_s^{\mathrm{T}}$ of some special $N \times N$ matrices appearing in problems with partial differential equations. The eigenvalues of a tridiagonal matrix

$$T(a, b, c) = \begin{pmatrix} b & c & 0 & & \\ a & b & c & 0 & \\ & \ddots & \ddots & \ddots & \\ & & 0 & a & b & c \\ & & & 0 & a & b \end{pmatrix} \tag{A.8}$$

are

$$\lambda_s = b + 2c\sqrt{\frac{a}{c}}\, \cos\frac{s\pi}{N+1} \, , \qquad s = 1, 2, \ldots, N \, , \tag{A.9}$$

while the corresponding components of the eigenvectors are

$$(\boldsymbol{v}_s)_k = 2\left(\sqrt{\frac{a}{c}}\right)^k \sin\frac{ks\pi}{N+1} \, , \qquad k = 1, 2, \ldots, N \, . \tag{A.10}$$

On page 652 we also refer to the tridiagonal symmetric matrix

$$
T_{N_1 D} = \begin{pmatrix} 1 & -1 & 0 & & & \\ -1 & 2 & -1 & 0 & & \\ & & \ddots & \ddots & \ddots & \\ & & 0 & -1 & 2 & -1 \\ & & & 0 & -1 & 2 \end{pmatrix} , \tag{A.11}
$$

with eigenvalues

$$
\lambda_s = 2 - 2 \cos \frac{(2s - 1)\pi}{2N + 1} , \qquad s = 1, 2, \ldots, N ,
$$

and eigenvector components

$$
(\boldsymbol{v}_s)_k = \cos \frac{(2k - 1)(2s - 1)\pi}{2(2N + 1)} , \qquad k = 1, 2, \ldots, N .
$$

A.6 Singular Values of X and Eigenvalues of $X^{\mathrm{T}} X$ and $X X^{\mathrm{T}}$

The singular values and the corresponding singular vectors of the $n \times m$ matrix X of rank r are closely related to the eigenvalues of the products $X^{\mathrm{T}} X$ and $X X^{\mathrm{T}}$. Within this textbook, this connection is exploited in multivariate statistical methods, where it is sometimes preferable to use the covariance matrix $X^{\mathrm{T}} X$ rather than the data matrix X (Sects. 6.7, 6.8, 6.11, and 6.12). Let

$$
X = U \Sigma V^{\mathrm{T}} , \qquad U \in \mathbb{R}^{n \times r} , \qquad \Sigma = \mathrm{diag}(\sigma_1, \sigma_2, \ldots, \sigma_r) \in \mathbb{R}^{r \times r} , \qquad V \in \mathbb{R}^{m \times r} ,
$$

be the singular value decomposition of the matrix X. Then the following holds:

1. The product matrix $X^{\mathrm{T}} X \in \mathbb{R}^{m \times m}$ is symmetric and has r positive eigenvalues $\{\sigma_1^2, \sigma_2^2, \ldots, \sigma_r^2\}$, with the corresponding eigenvectors $\{\boldsymbol{v}_1, \boldsymbol{v}_2, \ldots, \boldsymbol{v}_r\}$, and $m - r$ zero eigenvalues. The singular values $\{\sigma_1, \sigma_2, \ldots, \sigma_r\}$ of X are therefore just the positive square roots of the positive eigenvalues of $X^{\mathrm{T}} X$, and the columns of V are the corresponding eigenvectors.
2. The product matrix $X X^{\mathrm{T}} \in \mathbb{R}^{n \times n}$ is symmetric and has r positive eigenvalues $\{\sigma_1^2, \sigma_2^2, \ldots, \sigma_r^2\}$, with the corresponding eigenvectors $\{\boldsymbol{u}_1, \boldsymbol{u}_2, \ldots, \boldsymbol{u}_r\}$, and $n - r$ zero eigenvalues. The singular values $\{\sigma_1, \sigma_2, \ldots, \sigma_r\}$ of X are therefore just the positive square roots of the positive eigenvalues of $X X^{\mathrm{T}}$, and the columns of U are the corresponding eigenvectors.

This implies that the positive eigenvalues of the matrices $X^{\mathrm{T}} X$ or $X X^{\mathrm{T}}$ are equal, and the eigenvectors of these matrix products are related by

$$u_i = \frac{1}{\sigma_i} X v_i , \qquad v_i = \frac{1}{\sigma_i} X^{\mathrm{T}} u_i , \qquad i = 1, 2, \ldots, r ,$$

or, in matrix form,

$$U = X V \Sigma^{-1} , \qquad V = X^{\mathrm{T}} U \Sigma^{-1} .$$

A.7 The Generalized (Moore–Penrose) Inverse of a Matrix

Let $A \in \mathbb{C}^{m \times n}$ be a general complex matrix representing the linear transformation $A x = b$ with $x \in \mathbb{C}^n = \mathcal{P}$ and $b \in \mathbb{C}^m = \mathcal{Q}$, see Fig. A.1. If $m = n$ and A is non-singular, the solution $x = A^{-1} b$ exists and is unique. If $m \neq n$, there may be no solution at all, there can be a single solution, but also multiple or even infinitely many solutions, depending on b and on the rank of A.

To devise the generalized inverse of A, we first define a restriction of A to $\mathcal{N}(A)^\perp$, which we denote by $\widetilde{A}$:

$$\widetilde{A} = A|_{\mathcal{N}(A)^\perp} : \mathcal{N}(A)^\perp \to \mathcal{R}(A) ,$$

so that $\widetilde{A}$ represents a bijective (one-to-one) mapping between $\mathcal{N}(A)^\perp$ and $\mathcal{R}(A)$, the bottom two boxes in the Figure. Now we need an inverse mapping between the *whole* $\mathcal{Q}$ and the *whole* $\mathcal{P}$. This is accomplished by the Moore–Penrose inverse

$$A^+ : \mathcal{D}(A^+) \to \mathcal{P} \tag{A.12}$$

defined as the unique linear extension of $\widetilde{A}^{-1}$ to $\mathcal{R}(A) \oplus \mathcal{R}(A)^\perp = \mathcal{D}(A^+)$ such that $A^+ b = 0$ for $b \in \mathcal{R}(A)^\perp$, hence $\mathcal{N}(A^+) = \mathcal{R}(A)^\perp$.

In the following Examples we describe two methods to compute A^+, namely by explicit construction of the appropriate bases and by exploiting the singular value decomposition. For alternative representations, consult [2].

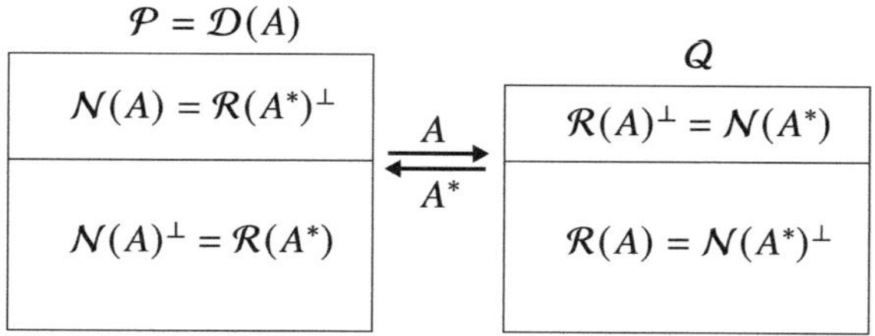

Fig. A.1 The domain $\mathcal{P} = \mathcal{D}(A)$, the null-space (kernel) $\mathcal{N}(A)$ and the range $\mathcal{R}(A)$ of a linear operator A acting between spaces $\mathcal{P}$ and $\mathcal{Q}$, together with the subspaces involving its adjoint A^* (transpose in $\mathbb{R}$, Hermitian conjugate in $\mathbb{C}$), such that $\mathcal{P} = \mathcal{N}(A) \oplus \mathcal{R}(A^*)$ and $\mathcal{Q} = \mathcal{N}(A^*) \oplus \mathcal{R}(A)$. Bijective mapping exists between $\mathcal{N}(A)^\perp$ and $\mathcal{R}(A)$

Example Let

$$A = \begin{pmatrix} 1 & 1 & 2 \\ 0 & 2 & 2 \\ 1 & 0 & 1 \\ 1 & 0 & 1 \end{pmatrix}. \tag{A.13}$$

Then $\mathcal{R}(A^*)$ is spanned by the columns of $A^* = A^T$ or the rows of A, i.e. by the vectors $(1, 1, 2)^T$, $(0, 2, 2)^T$ and $(1, 0, 1)^T$, all of which can be expressed as linear combinations of $e_1 = (1, 0, 1)^T$ and $e_2 = (0, 1, 1)^T$ that form the basis of $\mathcal{R}(A^*)$. Since $Ae_1 = (3, 2, 2, 2)^T \equiv f_1$ and $Ae_2 = (3, 4, 1, 1)^T \equiv f_2$, we have

$$A^+ f_1 = e_1, \quad A^+ f_2 = e_2. \tag{A.14}$$

Now we must compute the basis for $\mathcal{R}(A)^\perp = \mathcal{N}(A^*)$, i.e. find the vectors x that A^* maps to zero. Solving the equation $A^* x = 0$ yields $x = x_3 f_3 + x_4 f_4$, where $f_3 = (-1, 1/2, 1, 0)^T$, $f_4 = (-1, 1/2, 0, 1)^T$ and x_3 and x_4 are arbitrary real constants. This implies

$$A^+ f_3 = A^+ f_4 = (0, 0, 0)^T. \tag{A.15}$$

The intermediate results (A.14) and (A.15) taken together give

$$A^+ \begin{pmatrix} \overset{f_1\downarrow}{3} & \overset{f_2\downarrow}{3} & \overset{f_3\downarrow}{-1} & \overset{f_4\downarrow}{-1} \\ 2 & 4 & 1/2 & 1/2 \\ 2 & 1 & 1 & 0 \\ 2 & 1 & 0 & 1 \end{pmatrix} = \begin{pmatrix} \overset{e_1\downarrow}{1} & \overset{e_2\downarrow}{0} & \overset{0\downarrow}{0} & \overset{0\downarrow}{0} \\ 0 & 1 & 0 & 0 \\ 1 & 1 & 0 & 0 \end{pmatrix}$$

and therefore

$$A^+ = \begin{pmatrix} 1/7 & -5/21 & 11/42 & 11/42 \\ 0 & 1/3 & -1/6 & -1/6 \\ 1/7 & 2/21 & 2/21 & 2/21 \end{pmatrix}. \tag{A.16}$$

One can further check that the pseudoinverse A^+ satisfies the relations $AA^+A = A$, $A^+AA^+ = A^+$, $(AA^+)^* = AA^+$ and $(A^+A)^* = A^+A$.   ◁

Example Alternatively, the pseudoinverse of (A.13) can be found by using its singular value decomposition $A = U\Sigma V^T$ as in (4.22), yielding

$$U \approx \begin{pmatrix} -0.6212 & 0.2067 & 0.7559 & 0 \\ -0.6719 & -0.6369 & -0.3780 & 0 \\ -0.2852 & 0.5252 & -0.3780 & -0.7071 \\ -0.2852 & 0.5252 & -0.3780 & 0.7071 \end{pmatrix},$$

$$\Sigma \approx \begin{pmatrix} 3.9045 & & 0 & 0 \\ & 0 & 1.6598 & 0 \\ & 0 & & 0 & 0 \\ & 0 & & 0 & 0 \end{pmatrix} \implies \Sigma_1 \approx \begin{pmatrix} 3.9045 & 0 \\ & 0 & 1.6598 \end{pmatrix},$$

$$V \approx \begin{pmatrix} -0.3052 & 0.7573 & -0.5774 \\ -0.5033 & -0.6430 & -0.5774 \\ -0.8084 & 0.1144 & 0.5774 \end{pmatrix},$$

hence, by (4.24),

$$A^+ = V \begin{pmatrix} \Sigma_1^{-1} & 0_{2 \times 2} \\ 0_{1 \times 2} & 0_{1 \times 2} \end{pmatrix} U^{\mathrm{T}} = \begin{pmatrix} 0.1429 & -0.2381 & 0.2619 & 0.2619 \\ 0 & 0.3333 & -0.1667 & -0.1667 \\ 0.1429 & 0.0952 & 0.0952 & 0.0952 \end{pmatrix}.$$

Within rounding error this is equal to (A.16). $\triangleleft$

A.8   The "Square Root" of a Matrix

The "square root" of a symmetric positive semi-definite matrix $A \in \mathbb{R}^{n \times n}$ is computed by first finding its Cholesky decomposition $A = GG^{\mathrm{T}}$, then computing the singular value decomposition $G = U\Sigma V^{\mathrm{T}}$ of the matrix G, and finally form $X = U\Sigma U^{\mathrm{T}}$. For the matrix X obtained in this way we have $X^2 = A$, and thus $A^{1/2} = X$ or $A^{-1/2} = X^{-1}$ [3].

A.9   The Matrix Exponential

The calculation of the exponential function of a real or complex $n \times n$ matrix A,

$$e^A = I + A + \frac{A^2}{2!} + \frac{A^3}{3!} + \cdots, \tag{A.17}$$

is at the heart of many scientific problems represented by systems of linear ordinary differential equations of the form

$$\dot{x}(t) = Ax(t) + Bu(t), \quad t \geq 0,$$

with the initial condition $x(0) = x_0$. (See, for example, Eqs. (1.17) and (8.17) and Appendix I.) The exact solution of such problems is

$$x(t) = e^{tA} x_0 + \int_0^t e^{(t-s)A} B u(s)\, ds \ . \tag{A.18}$$

Even though the expansion (A.17) formally converges for any square matrix A, the straightforward summation of the Taylor series is obviously not the proper strategy to pursue. Rather, the standard approach to computing $\exp(A)$ exploits the relation

$$e^A = \left(e^{A/\sigma}\right)^\sigma$$

with some scaling factor $\sigma = 2^s$ $(s = 1, 2, \ldots)$ which reduces $\|A/\sigma\|_1$ to a value small enough that the scaled matrix can be accurately approximated by its Padé approximant. The latter is usually taken in its diagonal form, $[m/m](z) = P_m(z)/Q_m(z) = R_m(z)$, where P and Q are polynomials of degree m (see Eq. (1.11)), hence

$$e^{A/2^s} \approx R_m\left(A/2^s\right) , \quad e^A \approx \left[R_m\left(A/2^s\right)\right]^{2^s} \ .$$

The challenge is to find the right balance between the depth of "downscaling" (the power s) and the order m of the Padé approximant without wasting computational time and losing precision [4, 5]. It can be shown that if $\|A\| \leq 2^{s-1}$ then the Padé approximation implies

$$\left[R_m\left(A/2^s\right)\right]^{2^s} = e^{A+E}$$

with the error bound

$$\frac{\|E\|}{\|A\|} \leq 8 \left(\frac{\|A\|}{2^s}\right)^{2m} \left(\frac{(m!)^2}{(2m)!(2m+1)!}\right) \ .$$

Here we give an all-purpose implementation devised with $m = 13$ [6]. Due to the peculiar property of the exponential, $P_m(A) = Q_m(-A)$, calculating $R_m(A)$ from the equation $Q_m(A) R_m(A) = P_m(A)$ amounts to solving the matrix equation

$$(-U + V)R = U + V \tag{A.19}$$

for R, with

$$U = A\left[A_6(b_{13}A_6 + b_{11}A_4 + b_9 A_2) + b_7 A_6 + b_5 A_4 + b_3 A_2 + b_1 I\right] ,$$
$$V = A_6(b_{12}A_6 + b_{10}A_4 + b_8 A_2) + b_6 A_6 + b_4 A_4 + b_2 A_2 + b_0 I ,$$
$$A_2 = A^2 , \quad A_4 = A_2^2 , \quad A_6 = A_2 A_4 ,$$

and the coefficients

$$b_0 = 64764752532480000 \,,\ b_1 = 32382376266240000 \,,$$
$$b_2 = 7771770303897600 \,,\ b_3 = 1187353796428800 \,,$$
$$b_4 = 129060195264000 \,,\ b_5 = 10559470521600 \,,$$
$$b_6 = 670442572800 \,,\ b_7 = 33522128640 \,,\ b_8 = 1323241920 \,,$$
$$b_9 = 40840800 \,,\ b_{10} = 960960 \,,\ b_{11} = 16380 \,,\ b_{12} = 182 \,,\ b_{13} = 1 \,.$$

The algorithm is:

Input: matrix $A \in \mathbb{C}^{n \times n}$

$\mu = \mathrm{tr}(A)/n$;

$A = A - \mu I$;

$A = D^{-1} A D$; // optional balancing: if it fails to reduce $\|A\|_1$, use $D = I$

$\theta_{13} = 5.371920351148152$;

$s = \lceil \log_2(\|A\|_1 / \theta_{13}) \rceil$;

$A = A/2^s$;

solve $(-U + V)R = U + V$ for R by using Eq. (A.19);

$X = R^{2^s}$;

$X = e^\mu D X D^{-1}$;

Output: $X \approx \exp(A)$

This method is at the bottom of the `expm` routine in MATLAB and `MatrixExp` of MATHEMATICA. To avoid unnecessary downscaling and squaring, Padé approximations of degrees lower than $m = 13$ (e. g. $m = 9, 7, 5$ or 3) may be used if $\|A\|_1$ after the $A = A - \mu I$ step and balancing turns out to be small enough. Typically, $m = 9$ is sufficient when $\|A\|_1 \lesssim \theta_9 \approx 2.098$, $m = 7$ when $\|A\|_1 \lesssim \theta_7 \approx 0.9504$, $m = 5$ when $\|A\|_1 \lesssim \theta_5 \approx 0.2539$, and $m = 3$ when $\|A\|_1 \lesssim \theta_3 \approx 0.1496$ [7]. See also [8].

There are at least nineteen other ways to compute $\exp(A)$, several of which make use of specific structures of A, all with their own peculiar advantages and disadvantages [4], yet the scaling-and-squaring method described above appears to possess the most benefits.

In applications involving Markov chains, the solution $x(t)$ of (A.18) with $B = 0$ embodies the probability distribution describing the states of the chain and must therefore satisfy the usual probability requirements, that is, $x_i \in [0, 1]$ for all i and $\sum_i x_i = 1$. For calculations of $\exp(tA)$ with these supplementary restrictions—in particular in the context of sparse matrices A frequently occurring in Markov processes—see [9, 10].

A.10 Computation of Arbitrary Matrix Functions

To compute a general function $f(A)$ of an $n \times n$ matrix A we exploit the property $f(X^{-1}AX) = X^{-1}f(A)X$, which is valid for any nonsingular matrix X. If A is diagonalizable, such that $A = XDX^{-1}$ with $D = \mathrm{diag}(\lambda_1, \lambda_2, \ldots, \lambda_n)$, then

$$f(A) = Xf(D)X^{-1} = X \, \mathrm{diag}\big(f(\lambda_1), f(\lambda_2), \ldots, f(\lambda_n)\big) X^{-1} \, .$$

This works best for Hermitian, symmetric or, more generally, normal matrices (for which $AA^{\mathrm{T}} = A^{\mathrm{T}}A$ or $AA^{\dagger} = A^{\dagger}A$) where X can be chosen to be unitary. For general forms of A, however, the diagonalizing matrix X can be ill-conditioned, resulting in inaccurate evaluation in floating-point arithmetic. In such cases one resorts to similarity transformations $A = XUX^{\dagger}$ where X is unitary and U is upper triangular. Then

$$f(A) = Xf(U)X^{\dagger} \, ,$$

and the problem is reduced to computing a function of a triangular matrix. A stable and best general-purpose method to compute $f(U)$ is provided by the Schur–Parlett algorithm [11], which is implemented in the `funm` routine in MATLAB and `MatrixFunc` of MATHEMATICA.

References

1. *LAPACK, The Linear Algebra PACKage*, http://netlib.org/lapack
2. J.C.A. Barata, M.S. Hussein, *The Moore–Penrose pseudoinverse: a tutorial review of the theory*, Braz. J. Phys. **42**, 146 (2012)
3. G.H. Golub, C.F. Van Loan, *Matrix Computations*, 3rd edn. (Johns Hopkins University Press, Baltimore, 1996)
4. C. Moler, C. van Loan, *Nineteen dubious ways to compute the exponential of a matrix, twenty-five years later*, SIAM Rev. **45**, 3 (2003)
5. N.J. Higham, *Functions of matrices: theory and computation*, (SIAM, Philadelphia, 2005)
6. N.J. Higham, *The scaling and squaring method for the matrix exponential revisited*, SIAM Rev. **51**, 747 (2009)
7. A.H. Al-Mohy, N.J. Higham, *A new scaling and squaring method for the matrix exponential*, SIAM J. Matrix Anal. Appl. **31**, 970 (2009)
8. S. Güttel, Y. Nakatsukasa, *Scaled and squared subdiagonal Padé approximation for the matrix exponential*, SIAM J. Matrix Anal. Appl. **37**, 145 (2016)
9. R.B. Sidje, EXPOKIT: *a software package for computing matrix exponentials*, ACM Trans. Math. Softw. **24**, 130 (1998)
10. R.B. Sidje, W.J. Stewart, *A numerical study of large sparse matrix exponentials arising in Markov chains*, Comput. Stat. Data Anal. **29**, 345 (1999)
11. P.I. Davies, N.J. Higham, *A Schur–Parlett algorithm for computing matrix functions*, SIAM J. Matrix Anal. Appl. **25**, 464 (2003)

Appendix B
Standard Numerical Data Types

In serious computer programming, memory manipulations are needed (e.g. in dynamic allocation). Moreover, the characteristics of data structures should be well matched to the compiler in order to optimize execution speed (e.g. by aligning the structures to the boundaries of 32–bit regions on 32–bit architectures). To be able to do this, we should know how real and integer numbers are stored.

B.1 Real Numbers in Floating-Point Arithmetic

The IEEE 754–2008 standard for floating-point arithmetic was devised by the IEEE [1]. This set of rules prescribes the computer representation of numbers in single (32 bits) and double precision (64 bits) as well as the rules for arithmetic operations upon these types of numbers (algorithmic prescriptions for addition, subtraction, multiplication, division, and taking the square root). In the following we discuss the numbers in the *little-endian* notation (see Sect. B.2).

Single precision (C/C++ `float`) The single-precision floating-point representation allows us to represent normal numbers in the range of $\approx 10^{\pm 38}$. According to the IEEE standard, a 32–bit word (4 bytes) is sufficient for the complete representation: it can be written as a sequence of 32 bits from 31 to 0 from left to right. The bit S at the extreme left determines the sign of the whole number (the *sign bit*). It is followed by 8 *exponent bits* E specifying the exponent of the number, and finally by 23 bits representing the *mantissa* or the *fraction* F:

1	8	23

S	E E E E E E E E	F F F F F F F F F F F F F F F F F F F F F F F

31 30   23 22   0

| 0 | 0 1 1 1 1 1 0 0 | 0 1 0 0 0 0 0 0 0 0 0 0 0 0 0 0 0 0 0 0 0 0 0 | = 0.15625 |
| 1 | 1 0 0 0 0 1 0 1 | 1 1 0 1 1 0 1 0 1 0 0 0 0 0 0 0 0 0 0 0 0 0 0 | = − 118.625 |

© The Editor(s) (if applicable) and The Author(s), under exclusive license to Springer Nature Switzerland AG 2025

S. Širca and M. Horvat, *Computational Methods in Physics*, Graduate Texts in Physics, https://doi.org/10.1007/978-3-031-68566-8

The numerical value of V can be determined by using the scheme

$$
\begin{aligned}
E = 255 \text{ and } F \neq 0 &\implies V = \text{NaN } (\textit{Not a Number})\,, \\
E = 255 \text{ and } F = 0 \text{ and } S = 1 & \quad V = -\infty\,, \\
E = 255 \text{ and } F = 0 \text{ and } S = 0 & \quad V = +\infty\,, \\
0 < E < 255 & \quad V = (-1)^S\, 2^{E-127}\, 1.F_{10}\,, \\
E = 0 \text{ and } F \neq 0 & \quad V = (-1)^S\, 2^{-126}\, 0.F_{10}\,, \\
E = 0 \text{ and } F = 0 \text{ and } S = 1 & \quad V = -0\,, \\
E = 0 \text{ and } F = 0 \text{ and } S = 0 & \quad V = +0\,.
\end{aligned}
$$

The values $0.F_{10}$ and $1.F_{10}$ are obtained by taking the negative powers of 2 corresponding to each bit of F to the right of the decimal point, and summing them to form the decimal value, e.g. $1.011_{10} = 1 + 0 \cdot 2^{-1} + 1 \cdot 2^{-2} + 1 \cdot 2^{-3} = 1.875$ (obviously always $0 \leq 0.F_{10} < 1$ and $1 \leq 1.F_{10} < 2$). Two examples are shown in the Figure: in the first case $S = 0$, $E = 124$, and $F = 01$ (thus $1.F_{10} = 1.25$) and we obtain $V = (-1)^0 2^{124-127} 1.25 = 0.15625$; in the second case $S = 1$, $E = 133$, and $F = 110110101$ (thus $1.F_{10} = 1.85351562$) and therefore $V = (-1)^1 2^{133-127} 1.85351562 = -118.625$. The zeroth (rightmost) bit is known as the *least significant bit* (LSB), in contrast to the *most significant bit* (MSB) at the extreme left: if the LSB changes, the whole value V changes least. The notation $S = 0$, $E = 0$, and just LSB $= 1$ therefore corresponds to the smallest representable positive number in single-precision, $V = 2^{-126} 2^{-23} = 2^{-149} \approx 1.4 \cdot 10^{-45}$.

Double precision (C/C++ `double`) The representation of a double-precision floating-point number by the IEEE standard requires a 64–bit word (8 bytes) which can be written as a sequence of 64 bits from 63 to 0 from left to right. This packaging allows us to represent normal numbers in the range of $\approx 10^{\pm308}$. By analogy to the single-precision representation, the leftmost bit S determines the sign of the whole number, followed by 11 exponent bits and 52 bits for the mantissa:

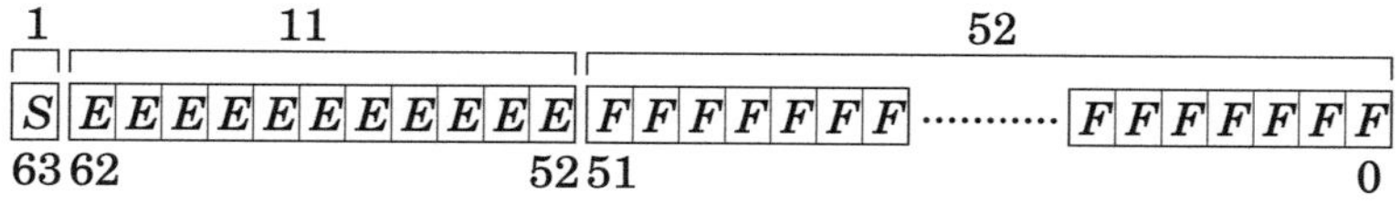

$$
\begin{array}{c|ccccccccccc|ccccccc}
1 & & & & & 11 & & & & & & & & & & 52 & & & \\
\hline
S & E & E & E & E & E & E & E & E & E & E & E & F & F & F & F & F & F & \cdots & F & F & F & F & F & F \\
\hline
63\,62 & & & & & & & & & & 52\,51 & & & & & & & & & & & & & 0
\end{array}
$$

The numerical value of V is determined as before,

$$
\begin{aligned}
E = 2047 \text{ and } F \neq 0 &\implies V = \text{NaN } (\textit{Not a Number})\,, \\
E = 2047 \text{ and } F = 0 \text{ and } S = 1 & \quad V = -\infty\,, \\
E = 2047 \text{ and } F = 0 \text{ and } S = 0 & \quad V = +\infty\,, \\
0 < E < 2047 & \quad V = (-1)^S\, 2^{E-1023}\, 1.F_{10}\,, \\
E = 0 \text{ and } F \neq 0 & \quad V = (-1)^S\, 2^{-1022}\, 0.F_{10}\,, \\
E = 0 \text{ and } F = 0 \text{ and } S = 1 & \quad V = -0\,, \\
E = 0 \text{ and } F = 0 \text{ and } S = 0 & \quad V = +0\,.
\end{aligned}
$$

Quadruple precision (C/C++ `long double`) In exceptional cases we need quadruple-precision numbers requiring 96 bits of memory (1 sign bit, 31 exponent bits, and 64 bits for the mantissa). This data type enables us to represent normal numbers in the range of $\approx 10^{\pm 4932}$. If our algorithm does not work in single or double precision, we should not resort blindly to quadruple precision: rather, discover the deficiency of the algorithm or find the error in the code.

Exceptions in floating-point arithmetic A computer code performing operations in floating-point arithmetic may generate certain types of critical errors which should be made known to the user. Locating such errors is made easier by tracing *exceptions*. The IEEE 745 standard defines several critical errors and the corresponding exceptions that these errors should trigger. The most commonly encountered exceptions occurring in computations by a real number R, are

- `FE_DIVBYZERO`: division by zero ($R/0$),
- `FE_INVALID`: invalid conversion of data to R (its floating-point form),
- `FE_OVERFLOW`: $|R|$ exceeds the maximum value (range overflow),
- `FE_UNDERFLOW`: $|R|$ is smaller than the smallest non-zero value (range underflow),
- `FE_INEXACT`: inexact conversion of data to R.

Triggering a floating-point exception is not automatic. In a program written in C/C++, exceptions must be enabled explicitly: for example, the exceptions `FE_DIVBYZERO`, `FE_INVALID`, or `FE_OVERFLOW`) are activated by the command

```
#include <fenv.h>
feenableexcept(FE_DIVBYZERO | FE_INVALID | FE_OVERFLOW);
```

and are deactivated by the command `fedisableexcept()`. If an exception is triggered, the operating system terminates the program with the SIGFPE signal (*floating-point exception*). See `exceptions.cc` on the book's web page.

Normal and subnormal numbers and their limits The numbers V corresponding to "non-extreme" E (in the ranges $0 < E < 255$ or $0 < E < 2047$) are known as *normal* or *normalized*, since their mantissa has the form $1.F_{10}$. In the case $E = 0$ and $F \neq 0$ (mantissa in the form $0.F_{10}$) we are dealing with *subnormal* or *denormalized* numbers. Both types of numbers appear in a variety of applications. A simple C++ code may be used to establish their upper and lower limits: see program `limits.cc` on the book's web page.

Combining Types With Different Precisions

On all computer architectures, computation with more precise data types is more time- and memory-consuming than computing with less precise ones. By knowing their inherent limitations, however, various data types can be sometimes combined such that the precision of the final result is dictated by the more precise type, yet the computation proceeds relatively faster.

Let us discuss such a case in which `float` and `double` data types are combined. Suppose we have a real number x of type `double` and we wish to compute the value of a function f at x. We represent x as a sum of the nearest number s of type `float` and the remainder ε of type `double`:

$$x(\text{double}) = s(\text{float}) + \varepsilon(\text{double}) .$$

In double-precision floating-point arithmetic we have $1 \oplus \varepsilon^3 = 1$, so that certain functions can be evaluated faster and sometimes even more precisely. In the case of the exponential function, $f(x) = \exp(x) = \exp(s)\exp(\varepsilon)$, we can use the power expansion $\exp(\varepsilon) \approx 1 + \varepsilon + \varepsilon^2/2$, since ε is small and the third order of the expansion does not contribute anything at the specified precision. In the case of $f(x) = \exp(x^2)$, we use the expansion $x^2 = s^2 + \varepsilon(x + s)$. Similar conclusions follow in the evaluation of polynomials of low degrees or smooth functions f that can be approximated by $f(x) \approx f(s) + f'(s)\varepsilon + f''(s)\varepsilon^2/2$, if one can ascertain that $f(s)$ is computed to sufficient precision.

The speed of the operations with different numerical types strongly depends on the computer architecture and the programming language. In the order of increasing numerical cost, addition is followed by subtraction (about the same cost as addition), multiplication, and division (about the same cost as multiplication). The speed of the algorithm is determined mostly by the number of multiplications and divisions. Multiplication and division of `double` types are about twice as time-consuming as multiplication and division of `float` types.

B.2 Integer Numbers

The representation of integer numbers is more transparent than the representation of floating-point numbers. The most frequently used types in programming languages related to C/C++ are `char` (8 bits in memory), `short int` (at least 16 bits), and `int`, where "int" is an abbreviation for "integer". According to the C language standard, the type `int` has the same size as a typical processor register and hence occupies 32 (64, 128, ...) bits on 32 (64, 128, ...) bit architectures. The binary representation of an integer with p bits $b_i \in \{0, 1\}$ further depends on an additional declaration qualifier. If the variable is `unsigned`, it describes a non-negative number of the form

$$b_p 2^p + b_{p-1} 2^{p-1} + \ldots + b_1 2^1 + b_0 2^0 = (b_p, b_{p-1}, \ldots, b_0)_2 ,$$

which is represented by the vector of bits $(b_p, b_{p-1}, \ldots, b_0)$. If it is declared as `signed`, it can stand for either positive or negative numbers of the form

$$\pm (b_{p-1}, b_{p-2}, \ldots, b_0)_2 .$$

A positive integer can be represented by the vector of bits $(0, b_{p-1}, \ldots, b_0)$, i.e. by a non-negative number

$$x = (0, b_{p-1}, \ldots, b_0)_2 ,$$

while a negative integer can be represented by

$$(1, \neg b_{p-1}, \ldots, \neg b_0)_2 + 1 \,,$$

known as the *two's binary complement* of x. The leftmost bit (MSB) is the *sign bit*. This rather unusual notation for negative integers allows for a faster summation of positive and negative numbers. The 32–bit computer architecture permits the following integer ranges:

```
                 unsigned           signed
    char         0...255            −128...127
    short int    0...65535          −32768...32767
    int          0...4294967295     −2147483648...2147483647
```

Two further types, `long int` and `long long int`, are also in use. Their implementation depends on the compiler, computer architecture, and even the operating system: on UNIX and Linux systems, see system variable `__WORDSIZE`) and other declarations in the file `/usr/include/limits.h`.

An even more important characteristic of the representation of integers is the ordering of bits within a byte or word. In *big-endian* ordering the most significant bit (MSB) is stored at the lowest memory address. In *little-endian* ordering the least significant bit (LSB) is stored first. The distinction between *big-endian* and *little-endian* may apply even at the level of words, not just at the level of bits: the number 1025 in integer type of length 4 bytes can be represented as 00000000|00000000|00000100|00000001, but this sequence of bits will be stored differently in the two orderings:

```
    address     big-endian      little-endian
    00          00000000        00000001
    01          00000000        00000100
    10          00000100        00000000
    11          00000001        00000000
```

Most modern computers use *little-endian* ordering. Problems may appear when one attempts to port the programs between architectures with different orderings. This is known as the "NUXI problem": the word "UNIX", written in two two-byte words, is stored in memory as "UNIX" on *big-endian* systems (in the sense of ordering bytes within a word), or as "NUXI" on *little-endian* systems. Note that within an individual byte, even the bits can be stored in one ordering or another. Moreover, on some systems the order of significance of bits in a byte can be opposite to the order of significance of bytes in a word! The code `biglittle.cc` on the book's web page can be used to check the endianness of your machine.

B.3   (Almost) Arbitrary Precision

Computations in number theory, experimental mathematics, cryptography and numerous physical applications [2] sometimes require arbitrary precision in integer or floating-point arithmetic. Standard libraries exist for this purpose. An important library for integer arithmetic is BigDigits [3], where the interval $[L_{min}, L_{max}]$ of integers is practically unbounded. Of course, with increasing interval length $L = L_{max} - L_{min} + 1$, memory consumption also increases, and amounts to $\lceil \log_2 L \rceil$ bits in the optimal case. In order to ensure good portability, well-defined use of memory, and transparency of computations, we sometimes also use the character representation of numbers. In this representation, numbers in some base B (e.g. decimal, $B = 10$) are stored as arrays of characters, e.g. the number 123 is stored as the array "123". Such storage requires $8\lceil \log_B L \rceil$ bits (which is a lot). Another important library for integer arithmetic in arbitrary precision is NTL [4].

Arbitrary integer and floating-point arithmetic are implemented in the GNU Multi-Precision (GMP) library, *"the fastest bignum library on the planet"*) [5]. Implementations in various precisions (`double-double precision`, `quad-double precision`, and arbitrary precision) for Fortran77, Fortran90, and C++ are listed in [6]. For multiple-precision implementations of BLAS and LAPACK see [7].

References

1. IEEE Standard 754–2008 for binary floating-point arithmetic, IEEE Standards Association, 2008; available at http://standards.ieee.org/ieee/754/4211
2. D.H. Bailey, J.M. Borwein, High-precision arithmetic in mathematical physics. Math. **3**, 337 (2015)
3. BigDigits multiple-precision arithmetic, http://www.di-mgt.com.au/bigdigits.html
4. NTL: A Library for doing Number Theory, http://shoup.net/ntl
5. GNU Multi Precision (GMP), free library for arbitrary precision arithmetic, available at http://gmplib.org. See also L. Fosse et al., *MPFR: a multiple-precision binary floating-point library with correct rounding*, ACM Trans. Math. Softw. **33**, 13 (2007)
6. http://www.davidhbailey.com/dhbsoftware
7. http://mplapack.sourceforge.net, http://github.com/nakatamaho/mplapack

Appendix C
Generation of Pseudorandom Numbers

Generation of *pseudorandom* numbers is a deterministic process of computing sequences appearing to be random. The degree of their randomness is established by special statistical tests. *Random number generators* are used in a variety of computations, and individual applications may require these generators to possess specific statistical properties. In the following, generation of pseudorandom numbers is called simply *drawing*, and the generated numbers are called *random numbers*. The mathematical basis of random numbers is presented in [1], the analysis of generators from the mathematical perspective is given by [2], and from the viewpoint of theoretical computing by [3]. Details on the use of generators can be found in [4–6].

C.1 Uniform Generators: From Integers to Reals

To generate random numbers with a uniform probability distribution, we use *uniform generators*. A discrete integer random variable $X \in \mathbb{Z}_m = [0, m - 1] \subset \mathbb{N}_0$ is distributed uniformly if it assumes any value on the interval with equal probability, thus

$$\text{Prob}(X = i) = \frac{1}{m}, \qquad i = 0, 1, \ldots, m - 1 .$$

A continuous real random variable $X \in [a, b] \subset \mathbb{R}$ is distributed uniformly if its probability density is

$$p(x) = \begin{cases} (b - a)^{-1} & ; \ x \in [a, b] , \\ 0 & ; \ \text{otherwise} . \end{cases}$$

We denote the uniform distribution on the real or integer axis by $U(a, b)$.

A good uniform generator that yields the sequence of numbers $\{x_i\}$ is expected to generate *uncorrelated sequences*: this means that the vectors of subsequences $(x_i, x_{i+1}, \ldots, x_{k+i})$ are as weakly correlated as possible, for any k. The sequence

S. Širca and M. Horvat, *Computational Methods in Physics*, Graduate Texts in Physics, https://doi.org/10.1007/978-3-031-68566-8

$\{x_i\}$ should also have a *long period* (it should not repeat itself for as long as possible) and should be *uniform and unbiased* (equal number of points should fall into spaces of equal sizes). In other words, we require a uniform distribution of points $(x_i, x_{i+1}, \ldots, x_{k+i-1})$ in a hypercube with dimension k as large as possible; this is known as *serial uniformity of the sequence*. Moreover, the generator should be numerically efficient [7].

Most modern uniform random generators are based on integer arithmetic. Such generators typically return numbers with equal probabilities on the interval $[0, m - 1]$, where $m = 2^{32}$ or 2^{64}. Uniform generators are standard components of general libraries and tools, e.g. the command `rand()` in MATLAB and C/C++, `gsl_rng_rand` in GSL, or `Random[]` in MATHEMATICA. Random integers $x_k \in \mathbb{Z}_m$ returned by the generator can be converted to uniformly distributed random real numbers y_k from $U(0, 1)$ by using the transformations [2]:

$$
\begin{aligned}
y_k &= x_k/m &&\text{approximately uniform in } [0, 1)\,, \\
y_k &= x_k/(m - 1) &&[0, 1]\,, \\
y_k &= (x_k + 1)/m &&(0, 1]\,, \\
y_k &= (x_k + 1/2)/m &&(0, 1)\,.
\end{aligned}
$$

The real numbers obtained in this manner have $b = \log_2 m$ random most significant bits. Frequently this is not enough (and, at any rate, less than what is supported by the mantissa of the real data type). But there is another way of converting x_k to y_k. In real arithmetic with a n-bit mantissa, the precision of a number is 2^{-n} with $n > b$. An approximation of a random real number y on the interval $[0, 1)$ with all bits random can be obtained by independently drawing the integers $\{x_k \in \mathbb{Z}_m\}_{k=1}^h$ and using the formula

$$
y = x_1 m^{-1} + x_2 m^{-2} + \cdots + x_h m^{-h}\,,
$$

where h is chosen such that

$$
(h - 1)b < n < (h + 1)b\,.
$$

In the case of the 32-bit integer random generator `int32()` and the real data type `double`, the right-hand side of this formula is (see Chap. 7 in [4]):

```
|2.32830643653869629E-10 * (int32() + 2.32830643653869629E-10*int32())|
```

C.2   Transformations Between Distributions

A generator of random numbers with an arbitrary discrete or continuous distribution can be obtained by transforming the numbers returned by a uniform generator. Here we present the most frequently encountered transformations.

C.2.1   Discrete Distribution

Let X be a random variable that can assume n distinct values. An example is the interval of integers $I = [1, n]$, where X assumes each value with a probability $p_i = \text{Prob}(X = i)$ for $i \in I$ and $\sum_{i \in I} p_i = 1$ holds true. The probabilities p_i represent a discrete probability distribution with the cumulative distribution

$$P_i = \text{Prob}(X \le i) = \sum_{j \le i} p_j \quad \text{for} \quad i \in I \cup \{0\} \,.$$

A discrete random variable X can be expressed by a continuous random variable U from $U(0, 1)$ by inverting the cumulative distribution, by using the formula $X = P^{-1}(U)$: this means that $X = i$ precisely when $P_{i-1} \le U < P_i$.

This statement can be turned into a method of generating numbers, where for each drawn number u from $U(0, 1)$ we find an interval $R_i = [P_i, P_{i+1})$ such that $u \in R_i$, and then i is the sought value of the random variable X. Finding the corresponding interval can be accomplished linearly in $O(n)$ operations at worst, or in $O(\log n)$ by using search trees [8].

In [6] and [9] other methods of generation of random numbers according to a discrete distribution are presented. Here we mention the Walker's alias method [10] with the computational cost of $O(1)$ (independent of n) and memory requirement of just $O(n)$ [11]. In this method, any discrete random variable $X \in [1, n]$ with the probability distribution $\{p_i\}_{i=1}^n$ can be expressed as a random variable $Y \in [1, n]$ with n two-point probability distributions

$$\text{Prob}(Y = y_{ij}) = q_{ij} \,, \qquad i = 1, 2, \ldots, n \,, \qquad j = 1, 2 \,,$$

where $q_{i1} + q_{i2} = 1$, and we denote $y_{i1} = i$, $y_{i2} = L_i$, $q_{i1} = F_i$ and $q_{i2} = 1 - F_i$. This transformation of random variables is not unique. We draw the value of the random variable X by using the algorithm

Input: Constants F_i and L_i for $i = 1, 2, \ldots, n$.
Draw u from $U(0, n)$;
$i = \lceil u \rceil$;
$v = i - u$;
if ($v > F_i$) **then**
$\quad | \quad i = L_i$;
end
Output: i is one realization of the discrete random variable X with the
$\qquad\qquad$ distribution p_i.

The constants F_i and L_i depend on the distribution p_i and are determined below. The constant F_i represents the limit value to which we compare the randomly chosen value u from $U(0, 1)$, while L_i is the alias into which i is transformed if the comparison fails. The F_i and L_i can be determined by using the procedure

Input: Discrete distribution given by vector $p = (p_i)_{i=1}^n$.
Define vectors $F = (F_i)_{i=1}^n$ and $L = (L_i)_{i=1}^n$.
for $i = 1, 2, \ldots, n$ **do**
    | $F_i = np_i$;
    | $L_i = 0$;
end
Define sets $G = \{i : F_i > 1\}$ and $S = \{i : F_i < 1\}$.
if ($|S| = 0$) **then**
    | All p_i are equal to $1/n$, thus terminate the routine, since the values of the
    | random variable X can be drawn from the uniform distribution.
end
repeat
    | Choose an element from G (denoted by k) and an element from S (denoted
    | by j).
    | $L_j = k$;   // alias assigned to element j
    | $S = S \backslash \{j\}$;
    | $F_k = F_k - (1 - F_j)$;   // change kth limit value
    | **if** ($F_k < 1$) **then**
    | | $G = G \backslash \{k\}$;
    | | $S = S \cup \{k\}$;
    | **end**
until ($|S| > 0$)
Output: Vectors of limit values F and aliases L for distribution p.

Here $|\cdot|$ denotes cardinality (the number of elements in the set). An illustration of the Walker algorithm in action is given in Fig. C.1. Walker's method is implemented in the GSL library (`gsl_ran_discrete*`) and the R project (`sample[·]`).

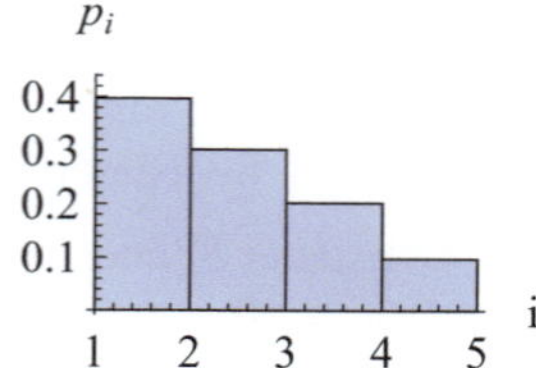

step	L	F	G	S
1	$(0, 0, 0, 0)$	$(1.6, 1.2, 0.8, 0.4)$	$\{1, 2\}$	$\{3, 4\}$
2	$(0, 0, 1, 0)$	$(1.4, 1.2, 0.8, 0.4)$	$\{1, 2\}$	$\{4\}$
3	$(0, 0, 1, 1)$	$(0.8, 1.2, 0.8, 0.4)$	$\{2\}$	$\{1\}$
4	$(2, 0, 1, 1)$	$(0.8, 1.0, 0.8, 0.4)$	$\{2\}$	$\{\}$

Fig. C.1 The sequence of vectors of constants L and F, and the sets G and S during the Walker's algorithm for the discrete distribution $p = (p_i)_{i=1}^4 = (0.4, 0.3, 0.2, 0.1)$ of numbers $\{1, 2, 3, 4\}$. Step 1 represents the status of (L, F, G, S) prior to entering the loop

C.2.2   *Continuous Distribution*

If a real random variable X is distributed according to the continuous probability density p, the probability of an event $X \in A = [a, b]$ is given by the integral

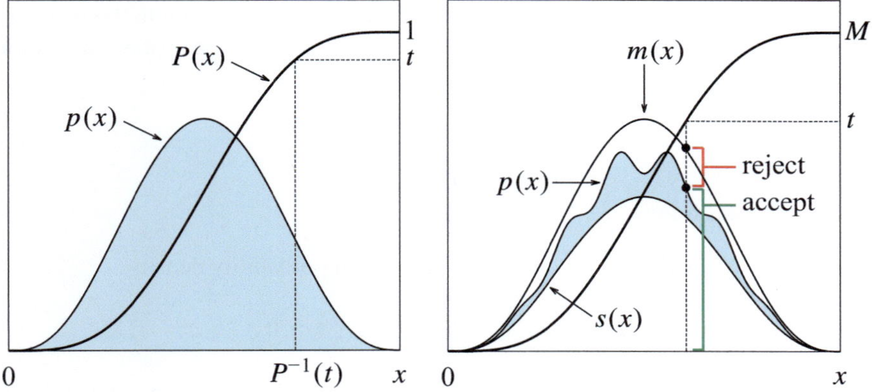

Fig. C.2 Generation of random numbers [LEFT] by the inverse method; [RIGHT] by the rejection method

$$\mathrm{Prob}(X \in A) = \int_A p(t)\,\mathrm{d}t\,,$$

The function p is integrable and usually at least piecewise continuous. The cumulative distribution function is

$$P(x) = \int_{-\infty}^{x} p(t)\,\mathrm{d}t\,, \qquad 0 \le P(x) \le 1\,.$$

Random values of the variable X can be obtained by using uniform generators and transformations between distributions of various variables. In the following we present the most commonly used methods.

The inverse method Assume that X is a random variable with the probability density $p(x)$ and U is a variable distributed uniformly according to $U(0, 1)$. Then X can be expressed as

$$X = P^{-1}(U)$$

(Fig. C.2 (left)). This method of generating the values of X is useful if the inverse of the cumulative distribution function $P^{-1}(y)$ is easy to compute.

If a random variable $X \in \Omega$ has the probability density p_X and h is a differentiable function with the inverse $g = h^{-1}$, the random variable $Y = h(X)$ has the probability density

$$p_Y(y) = \int_{\Omega} \delta(h(t) - y)p_X(t)\,\mathrm{d}t = p_X(g(y))\left|g'(y)\right|\,.$$

If independent random variables X and Y are distributed according to the densities p_X and p_Y, the probability density of the sum $Z = X + Y$ is given by the convolution of p_X and p_Y,

$$p_Z(z) = \int_{\mathbb{R}} p_X(t - z) p_Y(t)\, dt \ .$$

A few examples follow.

- A variable X distributed according to the Weibull probability density

$$p(t) = k t^{k-1} \exp(-t^k)\,, \qquad t > 0\,,$$

can be obtained by the transformation $X = (-\log U)^{1/k}$. We exclude the value 0 from the domain of U, so that X is always bounded. A special case of the Weibull density is the exponential density $p(t) = \exp(-t)$.

- If U_1 and U_2 are independent random variables distributed according to $U(0, 1)$, the Box–Muller transformation [12]

$$X_1 = \sqrt{-2 \log U_1}\ \cos(2\pi U_2)\,, \qquad X_2 = \sqrt{-2 \log U_1}\ \sin(2\pi U_2)\,,$$

yields independent random variables X_1 and X_2 distributed according to the normal distribution $N(0, 1)$. The variables U_1 and U_2 define the length $R = \sqrt{-2 \log U_1}$ and angle $\theta = 2\pi U_2$ of the two-dimensional vector $(X_1, X_2)^{\mathrm{T}}$. The costly computation of trigonometric functions can be avoided in the Marsaglia implementation of the Box–Muller method (see [3], Chap. 7, algorithm P):

repeat
> Independently draw u_1 and u_2 from $U(0, 1)$.
> $v = 2(u_1, u_2)^{\mathrm{T}} - (1, 1)^{\mathrm{T}}$;
> $s = |v|^2$;

until $(s \geq 1 \vee s \neq 0)$
$(x_1, x_2)^{\mathrm{T}} = \sqrt{-2 \log(s)/s}\ v$;
Output: Realization of two independent random variables (x_1, x_2) distributed
according to $N(0, 1)$.

On the average, the drawn vector v covers the unit circle, while approximately $1 - \pi/4 \approx 21.5\,\%$ of generated points are discarded: for one pair (x_1, x_2) we draw $2/(\pi/4) \approx 2.54$ uniformly distributed numbers. This algorithm is suitable for the generation of complex normally distributed numbers $x_1 + \mathrm{i}\,x_2$ for the construction of random matrices from the GUE (Sect. 4.9.2).

- The values x of a random vector $X \in \mathbb{R}^d$ distributed according to the multivariate normal probability density

$$p(x) = \frac{1}{(2\pi)^{d/2}(\det \Sigma)^{1/2}} \exp\left(-\frac{1}{2}(x - \mu)^{\mathrm{T}} \Sigma^{-1}(x - \mu)\right)$$

with the mean $\boldsymbol{\mu}$ and correlation matrix Σ can be generated by independently drawing d components of the vector $\boldsymbol{y} = (y_1, y_2, \ldots, y_d)^{\mathrm{T}}$ according to the standard normal distribution $N(0, 1)$, and computing

$$x = Ly + \boldsymbol{\mu} \,,$$

where L is the lower-triangular $d \times d$ matrix from the Cholesky decomposition of the correlation matrix, $\Sigma = LL^{\mathrm{T}}$.

• The values of a random variable X with the probability density in the form of a symmetric trapezoid

$$p(x) = \frac{1}{ab} \begin{cases} (a+b)/2 - |t| & ; \ 2|t| \in [|a-b|, (a+b)] \,, \\ \min(a, b) & ; \ 2|t| \in [0, |a-b|] \,, \\ 0 & ; \ \text{otherwise} \,, \end{cases}$$

can be obtained by combining the values of two random variables U_1 and U_2 uniformly distributed on the unit interval $[0, 1]$, by using the formula

$$X = a \left(U_1 - \frac{1}{2} \right) + b \left(U_2 - \frac{1}{2} \right) \,.$$

• The points $x = (x_1, x_2, \ldots, x_d)^{\mathrm{T}} \in \mathbb{R}^d$ that are uniformly distributed over the $(d-1)$-dimensional real sphere $S_{d-1} \in \mathbb{R}^d$ can be generated [3] by independently drawing the components of the vector $\boldsymbol{y} = (y_1, y_2, \ldots, y_d)^{\mathrm{T}}$ with probability density $N(0, 1)$, and normalizing it: $x_i = y_i / \|\boldsymbol{y}\|_2$, where $\|\boldsymbol{y}\|_2^2 = \sum_{i=1}^{d} y_i^2$.

• The points $x = (x_1, x_2, \ldots, x_d)^{\mathrm{T}}, x_i > 0$, which are uniformly distributed over the plane defined by $\sum_{i=1}^{d} a_i x_i = b$ with positive real constants a_i and b, are generated by independently drawing d components of the vector $\boldsymbol{y} = (y_1, y_2, \ldots, y_d)^{\mathrm{T}}$ with exponential probability density $p(y) = \exp(-y)$, and computing [13]

$$S = \sum_{i=1}^{d} a_i y_i \,, \qquad x_i = \frac{b}{S} a_i y_i \,.$$

Rejection method Suppose a function m exists that bounds the probability density function p from above as tightly as possible (Fig. C.2 (right)), and that the integral of m is $M = \int_{\mathbb{R}} m(x)\mathrm{d}x$. If we possess an efficient method to generate random numbers distributed according to $\frac{1}{M} m$ and the computation of p is too costly, we introduce a function $s \geq 0$ that bounds p as tightly as possible from below, and follow the algorithm

Input: Probability density p, functions m and s.
repeat
> Draw x with probability density $\frac{1}{M} m(x)$ and u with density $U(0, 1)$.
> **if** ($s(x) > u\, m(x)$) **then**
> | Terminate the loop.
> **end**

until ($p(x) < u\, m(x)$)
Output: x is the value of one realization of the random variable X.

The function s increases the rate at which the values are accepted, which is known as *squeezing*. If we do not know the function s, we simply remove the **if** conditional statement from the loop. We discard $\approx 1 - \frac{1}{M}$ drawn pairs (x, v), so the algorithm becomes the most efficient when M is close to 1. If m is a piecewise constant function, the Walker's method described previously can be used to draw the values according to $\frac{1}{M} m$. Some examples follow.

- The Cauchy–Lorentz (Breit–Wigner) distribution

$$p(x) = \frac{1}{\pi(1 + x^2)} \tag{C.1}$$

is a frequent occurrence: it describes shapes of spectral lines and nuclear resonances in quantum physics, or resonance curves in classical forced oscillations. The corresponding cumulative distribution function P and its inverse P^{-1} are

$$P(x) = \frac{1}{\pi} \arctan x + \frac{1}{2}, \qquad P^{-1}(t) = \tan\left[\pi\left(t - \frac{1}{2}\right)\right]. \tag{C.2}$$

The values of the variable X distributed according to (C.1) could be generated by the inverse method where in Eq. (C.2) one would use a random variable U distributed uniformly on $[-\pi/2, \pi/2]$, and compute $X = \tan U$. Instead of the costly evaluation of $\tan U = \sin U / \cos U$, we prefer to see the value of X as the ratio of the projections of a point within a circle onto the x- and y-axis. These points are uniformly distributed over the angles. We use the algorithm

repeat
> | Draw u_1 from $U(-1, 1)$ and u_2 from $U(0, 1)$.

until ($u_1^2 + u_2^2 > 1 \vee u_2 = 0$)
$x = u_1/u_2$;
Output: x is a realization of a Cauchy-distributed random variable.

- Assume that a random variable X has the probability density p defined on the interval $[s, s + h]$ such that

$$\int_s^{s+h} p(x)\, dx = \int_0^h p(x + s)\, dx = 1 ,$$

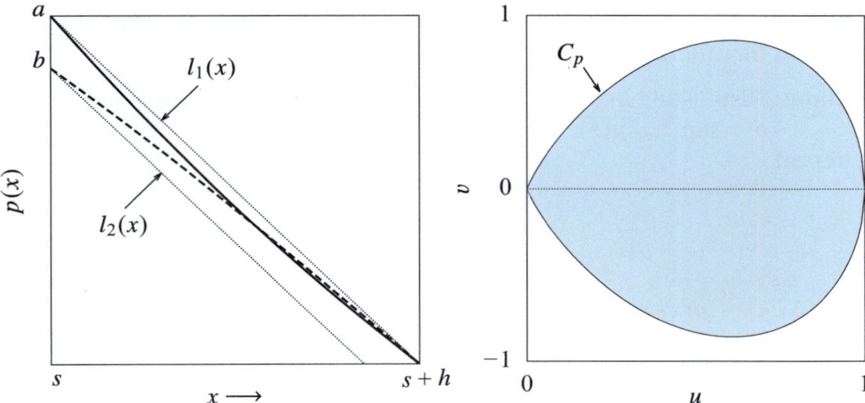

Fig. C.3 [LEFT] An almost linear portion of the probability density $p(x)$ on the interval $[s, s + h]$, which is bounded from above and below by straight lines (C.3). [RIGHT] The region C_p where points are accepted in the method of uniform deviates, for the standard normal distribution with the probability density $p(x) = (2\pi)^{-1/2} \exp(-x^2/2)$. The boundary of the region is defined by $v = \pm 2u\sqrt{-\log u}$

and that it is bounded from above and below by the straight lines

$$l_1(x) = a - \frac{b}{h}(x - s) , \qquad l_2(x) = b - \frac{b}{h}(x - s) , \qquad (C.3)$$

as shown in Fig. C.3 (left). The values of the variable X can be drawn by using the following algorithm [3] based on the rejection method:

Input: Distribution p, interval $[s, s + h]$, constants a and b.
repeat
    Independently draw u and v from $U(0, 1)$.
    if ($u > v$) **then**
      | Swap u and v.   // implies $u \leq v$
    end
    $x = s + hu$;
    if ($v \leq a/b$) **then**
      | Terminate the loop.
    end
until ($v > u + p(x)/b$)
Output: $x \in [s, s + h]$ is a realization of a random variable distributed
        according to p.

Method of the ratio of uniform deviates Assume that the random variable X has the probability density p, to which we assign the region

$$C_p = \left\{ (u, v) : 0 \leq u \leq \sqrt{p(v/u)} \right\} .$$

If (U, V) is a random variable uniformly distributed over C_p, we have $X = V/U$ [14]. This transformation is embodied in the algorithm

> **Input**: Distribution p, constants $a = \sup_x \sqrt{p(x)}$, $b = \inf_x x\sqrt{p(x)}$, and
> $\qquad c = \sup_x x\sqrt{p(x)}$.
> **repeat**
> $\quad\mid$ Independently draw u and v from $U(0, 1)$.
> $\quad\mid$ $u_1 = au$;
> $\quad\mid$ $u_2 = b + (c - b)v$;
> $\quad\mid$ $x = v/u$;
> **until** ($u^2 \le p(x)$)
> **Output**: x is the value of one realization of the random variable distributed
> $\qquad$ according to p.

This algorithm is a variation of the rejection method, and can be optimized similarly. An example of optimized generation of random numbers from $N(0, 1)$ is described in [15]: the region C_p where the points are accepted is shown in Fig. C.3 (right). This method requires ≈ 2.74 draws from $U(0, 1)$ to generate one number from $N(0, 1)$ [4], which is worse than in the Box–Muller method.

C.3 Using and Testing Random Number Generators

There are many random number generators branded as "good" or "best" by specialists. But every generator has its own deficiencies, some of them very specific. If we (as non-specialists) must invoke a generator very frequently in our code, we may be guided by the following advice [3, 5, 16, 17].

Choose only generators that were created and tested by the experts in the field. The code should be terse and preferably based on integer arithmetic in favor of greater speed. Choose a generator with the largest period and best serial uniformity for as many dimensions as possible. Preferably use generators that are accessible in source code, since modern compilers allow us to link short code fragments in the main program by using inline functions. Prior to its use, study the statistical property of a generator and determine whether any of its known deficiencies may jeopardize the validity of your results. Individual rounds of computation should be performed by using different generators and different initialization seeds.

The quality of random number generators can be checked by statistically founded *batteries of empirical tests*. The best known batteries can be found in the classical collection due to Knuth [3], in the slightly more restrictive set DIEHARD by Marsaglia [18], in the NIST Statistical Test Suite [19, 20], and in the large set TESTU01 established by L'Ecuyer [21]. See also Appendices A–C in [22].

C.4 Random Permutations

Problems with discrete and finite sets of solutions often require us to handle *permutations*. Permutations are familiar from combinatorics [23] and group theory [24], and also play a central role in quantum mechanics of many-body systems. A permutation π of n elements on n enumerated places is written as

$$\pi \begin{pmatrix} 1 & 2 & \dots & n \\ k_1 & k_2 & \dots & k_n \end{pmatrix} ,$$

indicating a reordering of the elements: the ith element goes in the k_ith place. Since the elements are merely reshuffled,

$$\{k_1, k_2, \dots, k_n\} = \{1, 2, \dots, n\} .$$

Numerical analysis of combinatorial problems calls for a generator of permutations, which can be deterministic or stochastic (random). If the deterministic procedure is not tailored to the problem such that the desired set of permutations is obtained after a predefined number of generations, all permutations must be calculated in order to attain reliable results. Here we present two simple and popular varieties; a review is given in [25].

C.4.1 Recursive Generation of Permutations

In the Heap's method [26] all permutations of length n are found recursively. The final result consists of $n!$ different arrangements of numbers from the set $\{1, 2, \dots, n\}$ into the vector $v \in \{1, 2, \dots, n\}^n$, for instance

$$\{1, 2, 3\} \rightarrow v \in \big\{\{1, 2, 3\}, \ \{1, 3, 2\}, \ \{2, 1, 3\}, \ \{2, 3, 1\}, \ \{3, 1, 2\}, \ \{3, 2, 1\}\big\} .$$

Initially we set $v_i = 0$ for $i = 1, 2, \dots, n$. The initial zeros indicate that no place (component) of v has been visited yet. In each step of the recurrence we seek the kth unvisited place and assign it the value l that no other place had been assigned previously, and increment l by one. The procedure is repeated recursively until no more unvisited places are available. In the process, l sequentially exhausts all values from 1 to n. The algorithm is:

Procedure Heap(v, k, L)
Input: component k in vector v, assigned the value $l + 1$
$l = L + 1$;
$v_k = l$;
if ($l = n$) **then**
  | A unique permutation of v has been found, store it or print it
else
  | Initiate the visit of other, unvisited components ($v_i = 0$)
  | **forall the** i for which $v_i = 0$ **do**
  | | Heap(v, i, l);
  | **end**
end
$v_k = 0$;

In the main program the procedure is invoked by Heap($\{0, 0, \ldots, 0\}, 1, -1$). The computational cost of retrieving all permutations is $O(n \cdot n!)$.

C.4.2 Stochastic Generation of Permutations

If the length of the permutations, n, is too large, the deterministic procedure does not permit the calculation of all $n!$ permutations. In such cases one resorts to stochastic methods. Typically we desire that each new permutation has the same probability as all the others, $1/n!$, and that it is uncorrelated to all previously generated permutations. This can be achieved by an algorithm based on Fisher–Yates mixing [3]. In spite of its old age (1938) it remains the most reliable and most general procedure to generate random permutations. Durstenfeld's implementation [27], in which an existing input permutation v is cleverly mixed in n steps, is:

Input: permutation v with length n
for $i = 1$ **to** $n - 1$ **do**
  | $j = \text{rand}(i, n)$;
  | Swap v_i and v_j.
end
Output: random permutation of v

The algorithm utilizes the function $\text{rand}(a, b)$ providing the value of a pseudorandom variable uniformly distributed on $[a, b]$. The computational cost of an individual random permutation is $O(n)$.

References

1. L. Devroye, *Non-Uniform Random Variate Generation* (Springer-Verlag, Berlin, 1986)
2. P. L'Ecuyer, Random number generation, in *The Handbook of Computational Statistics*, ed. by J.E. Gentle, W. Haerdle, Y. Mori (eds.), 2nd edn (Springer-Verlag, Berlin, 2004), p. 35. See also P. L'Ecuyer, in Uniform random number

generators and Non-uniform random variate generation, in *International Encyclopedia of Statistical Science*, ed. by M. Lovric (Springer-Verlag, Berlin, 2011)

3. D. Knuth, *The Art of Computer Programming, Volume 2: Seminumerical Algorithms*, 3rd edn (Addison-Wesley Professional, Reading, 1998)
4. W.H. Press, B.P. Flannery, S.A. Teukolsky, W.T. Vetterling, *Numerical Recipes: The Art of Scientific Computing*, 3rd edn (Cambridge University Press, Cambridge, 2007). See also the equivalent handbooks in Fortran, Pascal and C, as well as http://numerical.recipes
5. H. Niederreiter, *Random Number Generation and Quasi-Monte Carlo Methods* (SIAM, Philadelphia, 1992)
6. J. E. Gentle, *Random Number Generation and Monte Carlo Methods* (Springer-Verlag, Berlin, 2003)
7. P. L'Ecuyer, Uniform random number generation. Ann. Operations Res. **53**, 77 (1994)
8. T.H. Cormen, C.E. Leiserson, R.L. Rivest, C. Stein, *Introduction to Algorithms*, 3rd edn (MIT Press, Cambridge, 2009)
9. A.V. Peterson Jr., R.A. Kronmal, On mixture methods for the computer generation of random variables, Amer. Statistician **36**, 184 (1982)
10. A.J. Walker, An efficient method for generating discrete random variables with general distributions. ACM Trans. Math. Softw. **3**, 253 (1977)
11. R.A. Kronmal, A.V. Peterson, On the alias method for generating random variables from a discrete distribution. Amer. Statis. **33**, 214 (1979)
12. G.E.P. Box, M.E. Muller, A note on the generation of random normal deviates. Ann. Math. Statist. **29**, 610 (1958)
13. M. Horvat, The ensemble of random Markov matrices. J. Stat. Mech. **2009**, P07005 (2009)
14. A.J. Kinderman, J.F. Monahan, Computer generation of random variables using the ratio of uniform deviates. ACM Trans. Math. Softw. **3**, 257 (1977)
15. J. L. Leva, A fast normal random number generator. ACM Trans. Math. Softw. **18**, 449 (1992)
16. S. Tezuka, *Uniform Random Numbers: Theory and Practice* (Kluwer Academic Publishers, Norwell, 1995)
17. O. Goldreich, *Modern Cryptography, Probabilistic Proofs and Pseudo-Randomness* (Springer-Verlag, Berlin, 1999)
18. G. Marsaglia, *DIEHARD Battery of Tests of Randomness*, further developed by R. G. Brown: see https://webhome.phy.duke.edu/~rgb/General/dieharder.php
19. A.L. Rukhin, Testing randomness: a suite of statistical procedures. Theor. Probab. Applic. **45**, 111 (2001)
20. https://csrc.nist.gov/projects/random-bit-generation
21. P.L'Ecuyer, R. Simard, TestU01: a C library for empirical testing of random number generators. ACM Trans. Math. Softw. **33**, art. 22 (2007). The C source codes and binaries are available at http://simul.iro.umontreal.ca/testu01/tu01.html
22. J. C. Collins, *Testing, selection, and implementation of random number generators*, Army Research Laboratory, Report ARL-TR-4498 (2008)

23. M. Bóna, *Combinatorics of Permutations* (Chapman & Hall/CRC, Boca Raton, 2004)
24. M. Hamermesh, *Group Theory and Its Application to Physical Problems* (Dover Publications, New York, 1989)
25. D. Knuth, *The Art of Computer Programming, Volume 4A: Combinatorial algorithms, Part 1* (Addison-Wesley, Upper Saddle River, 2011)
26. A. Levitin, *Introduction to the Design and Analysis of Algorithms* (Addison-Wesley, Reading, 2003)
27. R. Durstenfeld, Algorithm 235: random permutation. Comm. ACM **7**, 420 (1964)

Appendix D
Dual Unscented Kalman Filter

Simultaneous estimation of both the states x_n and parameters θ_n appearing in nonlinear models—the so-called *adaptive nonlinear filtering* introduced in Sect. 7.8.6—can be accomplished by means of a dual unscented Kalman filter [1, 2] based on the following dynamic equations:

$$x_{n+1} = f_n(x_n, u_n, q_{x,n}; \theta_n) \,, \tag{D.1}$$

$$z_n = h_n(x_n, r_{x,n}; \theta_n) \,, \tag{D.2}$$

$$\theta_{n+1} = \theta_n + q_{\theta,n} \,,$$

$$d_{n+1} = h_n \left[f_n(x_n, u_n, q_{x,n}; \theta_n), r_{x,n}; \theta_n \right] + r_{\theta,n} \,,$$

where

$$q_x \sim N(\overline{q}_x, Q_x) \,, \quad r_x \sim N(\overline{r}_x, R_x) \,, \quad q_\theta \sim N(\overline{q}_\theta, Q_\theta) \,, \quad r_\theta \sim N(\overline{r}_\theta, R_\theta)$$

are normally distributed noise terms with known or estimated averages and covariance matrices, while P_x and P_θ are the corresponding state and parameter covariances. The state part of the filter operates with the augmented M_x-dimensional state vector x^{a} and the augmented block-diagonal covariance matrix P^{a}:

$$x^{\mathrm{a}} = \left(x^{\mathrm{T}}, q_x^{\mathrm{T}}, r_x^{\mathrm{T}}\right)^{\mathrm{T}} \,, \quad P^{\mathrm{a}} = \mathrm{diag}(P_x, Q_x, R_x) \,.$$

Initializing the filter with

$$\hat{x}_0^{\mathrm{a}+} = \left(\hat{x}_0^{+\mathrm{T}}, \mathbf{0}^{\mathrm{T}}, \mathbf{0}^{\mathrm{T}}\right)^{\mathrm{T}} , \quad P_0^{\mathrm{a}+} = \mathrm{diag}(P_{x,0}^+, Q_{x,0}, R_{x,0}) \,,$$

$$\hat{\theta}_0^+ = \langle \theta_0 \rangle \,, \qquad\qquad P_{\theta,0}^+ = \langle (\theta_0 - \hat{\theta}_0^+)(\theta_0 - \hat{\theta}_0^+)^{\mathrm{T}} \rangle \,,$$

where $\hat{x}_0^{+\mathrm{T}} = \langle x_0 \rangle$, one proceeds (for $n = 1, 2, \ldots$) as follows.

First, the parameter estimate and covariances are updated:

© The Editor(s) (if applicable) and The Author(s), under exclusive license to Springer Nature Switzerland AG 2025

S. Širca and M. Horvat, *Computational Methods in Physics*, Graduate Texts in Physics, https://doi.org/10.1007/978-3-031-68566-8

$$\hat{\boldsymbol{\theta}}_n^- = \hat{\boldsymbol{\theta}}_{n-1}^+ \,,$$
$$P_{\theta,n}^- = P_{\theta,n-1}^+ + Q_\theta \,.$$

We then form an $M_x \times (2M_x + 1)$ matrix $\mathcal{X}^{\text{a}}$ whose $2M_x + 1$ columns are the M_x-dimensional sigma vectors $(\mathcal{X}^{\text{a}})_i$:

$$(\mathcal{X}_{n-1}^{\text{a}})_i = \begin{cases} \hat{\boldsymbol{x}}_{n-1}^{\text{a}+} & ; \quad i = 0 \,, \\ \hat{\boldsymbol{x}}_{n-1}^{\text{a}+} + \sqrt{M_x + \mu_x} \left(\sqrt{P_{n-1}^{\text{a}+}}\right)_i & ; \quad i = 1, 2, \ldots, M_x \,, \\ \hat{\boldsymbol{x}}_{n-1}^{\text{a}+} - \sqrt{M_x + \mu_x} \left(\sqrt{P_{n-1}^{\text{a}+}}\right)_{i-M_x} & ; \quad i = M_x + 1, M_x + 2, \ldots, 2M_x \,, \end{cases}$$

where $M_x = \dim(\boldsymbol{x}^{\text{a}})$. Note that the $\mathcal{X}^{\text{a}}$ matrix itself also consists of its state and noise parts, possessing the structure $\mathcal{X}^{\text{a}} = \left((\mathcal{X}^x)^{\text{T}}, (\mathcal{X}^q)^{\text{T}}, (\mathcal{X}^r)^{\text{T}}\right)^{\text{T}}$. Then the $2M_x + 1$ sigma points constructed around the augmented state are transported through the nonlinearity, adopting the present estimate of the parameters:

$$(\mathcal{X}_n^{x-})_i = \boldsymbol{f}_{n-1} \left((\mathcal{X}_{n-1}^{x+})_i, \boldsymbol{u}_{n-1}, (\mathcal{X}_{n-1}^q)_i; \hat{\boldsymbol{\theta}}_n^-\right) \,, \qquad i = 0, 1, \ldots, 2M_x \,.$$

The prior state and covariance estimates are then calculated by a weighted average of the x-components of the resulting vectors:

$$\hat{\boldsymbol{x}}_n^- = \sum_{i=0}^{2M_x} w_{x,i}^{(\text{m})} (\mathcal{X}_n^{x-})_i \,,$$

$$P_n^- = \sum_{i=0}^{2M_x} w_{x,i}^{(\text{c})} \left[(\mathcal{X}_n^{x-})_i - \hat{\boldsymbol{x}}_n^-\right] \left[(\mathcal{X}_n^{x-})_i - \hat{\boldsymbol{x}}_n^-\right]^{\text{T}} \,.$$

The scaling parameter μ_x and the weights for the means and the covariances are given by the formulas (7.44), (7.46) and (7.47), respectively, with $M = M_x, \mu = \mu_x$, $\alpha = \alpha_x$ and $\beta = \beta_x$. For the parameter part of the filter we need $M = M_\theta$ ($\neq M_x$ in general), and also the constants μ_θ, α_θ and β_θ may be chosen differently. Next, we construct the sigma points for the model parameters:

$$(\mathcal{X}_n^\theta)_i = \begin{cases} \hat{\boldsymbol{\theta}}_n^- & ; \quad i = 0 \,, \\ \hat{\boldsymbol{\theta}}_n^- + \sqrt{M_\theta + \mu_\theta} \left(\sqrt{P_{\theta,n}^-}\right)_i & ; \quad i = 1, 2, \ldots, M_\theta \,, \\ \hat{\boldsymbol{\theta}}_n^- - \sqrt{M_\theta + \mu_\theta} \left(\sqrt{P_{\theta,n}^-}\right)_{i-M_\theta} & ; \quad i = M_\theta + 1, M_\theta + 2, \ldots, 2M_\theta \,, \end{cases}$$

where $M_\theta = \dim(\boldsymbol{\theta})$. Now these sigma points are carried through the measurement model as if this were an observation based on $\hat{\boldsymbol{\theta}}_n^-$ and its sigma-point neighbors:

$$(\mathcal{D}_n)_i = \boldsymbol{h}_n \left[\boldsymbol{f}_{n-1} \left(\hat{\boldsymbol{x}}_{n-1}^+, \boldsymbol{u}_{n-1}, \overline{\boldsymbol{q}}_x; (\mathcal{X}_n^\theta)_i\right), \overline{\boldsymbol{r}}_x; (\mathcal{X}_n^\theta)_i\right] \,,$$

allowing us to form the output estimate of the parameter filter,

$$\boldsymbol{d}_n = \sum_{i=0}^{2M_\theta} w_{\theta,i}^{(\mathrm{m})} (\mathcal{D}_n)_i \, ,$$

followed by the output estimate of the state filter,

$$(\mathcal{Z}_n)_i = \boldsymbol{h}_n \left((\mathcal{X}_{n-1}^{x-})_i, (\mathcal{X}_{n-1}^{r})_i ; \hat{\boldsymbol{\theta}}_n^- \right) \, , \qquad i = 0, 1, \ldots, 2M_x \, ,$$

$$\hat{\boldsymbol{z}}_n = \sum_{i=0}^{2M_x} w_{x,i}^{(\mathrm{m})} (\mathcal{Z}_n)_i \, .$$

We compute

$$P_{xz,n} = \sum_{i=0}^{2M_x} w_{x,i}^{(\mathrm{c})} \left[(\mathcal{X}_n^{x-})_i - \hat{\boldsymbol{x}}_n^- \right] \left[(\mathcal{Z}_n)_i - \hat{\boldsymbol{z}}_n \right]^{\mathrm{T}} \, ,$$

$$P_{zz,n} = \sum_{i=0}^{2M_x} w_{x,i}^{(\mathrm{c})} \left[(\mathcal{Z}_n)_i - \hat{\boldsymbol{z}}_n \right] \left[(\mathcal{Z}_n)_i - \hat{\boldsymbol{z}}_n \right]^{\mathrm{T}} \, ,$$

which allows us to evaluate the Kalman gain matrix for the states,

$$K_{x,n} = P_{xz,n} P_{zz,n}^{-1} \, ,$$

and, by analogy,

$$P_{\theta d,n} = \sum_{i=0}^{2M_\theta} w_{\theta,i}^{(\mathrm{c})} \left[(\mathcal{X}_n^{\theta})_i - \hat{\boldsymbol{\theta}}_n^- \right] \left[(\mathcal{D}_n)_i - \hat{\boldsymbol{d}}_n \right]^{\mathrm{T}} \, ,$$

$$P_{dd,n} = \sum_{i=0}^{2M_\theta} w_{\theta,i}^{(\mathrm{c})} \left[(\mathcal{D}_n)_i - \hat{\boldsymbol{d}}_n \right] \left[(\mathcal{D}_n)_i - \hat{\boldsymbol{d}}_n \right]^{\mathrm{T}} \, ,$$

leading to the Kalman gain matrix for the parameters,

$$K_{\theta,n} = P_{\theta d,n} P_{dd,n}^{-1} \, .$$

Finally, we calculate the posterior state and covariance estimates,

$$\hat{\boldsymbol{x}}_n^+ = \hat{\boldsymbol{x}}_n^- + K_{x,n} \left(\boldsymbol{z}_n - \hat{\boldsymbol{z}}_n \right) \, ,$$

$$P_{x,n}^+ = P_{x,n}^- - K_{x,n} P_{zz,n} K_{x,n}^{\mathrm{T}} \, ,$$

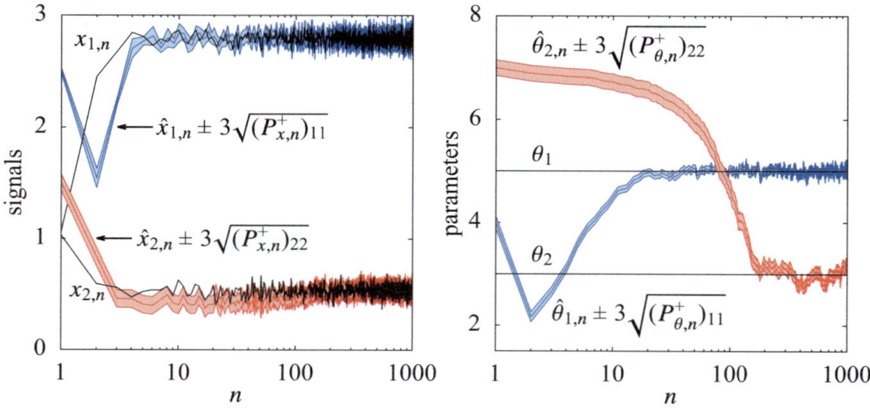

Fig. D.1 Dual nonlinear estimation with the unscented Kalman filter. [LEFT] State estimates and their uncertainties. [RIGHT] Parameter estimates and their uncertainties. Fitting the last 80 % of the points gives $\theta_1 = 5.000 \pm 0.004$ and $\theta_2 = 2.988 \pm 0.015$

followed by the posterior estimates of the parameters and their corresponding covariances,

$$\hat{\boldsymbol{\theta}}^+_n = \hat{\boldsymbol{\theta}}^-_n + K_{\theta,n}\left(z_n - \hat{\boldsymbol{d}}_n\right) ,$$

$$P^+_{\theta,n} = P^-_{\theta,n} - K_{\theta,n} P_{dd,n} K^{\mathrm{T}}_{\theta,n} .$$

This closes the $n = 1, 2, \ldots$ loop.

Example Consider a two-component system represented by the model equation of the form (D.1) with the function

$$\boldsymbol{f}(\boldsymbol{x}, \boldsymbol{q}_x; \boldsymbol{\theta}) = \begin{pmatrix} \sqrt{\theta_1 + x_1} \\ 1/\sqrt{\theta_2 + x_2} \end{pmatrix} + \begin{pmatrix} q_{x,1} \\ q_{x,2} \end{pmatrix} ,$$

supplemented by the measurement equation of the form (D.2), where

$$\boldsymbol{h}(\boldsymbol{x}, \boldsymbol{r}_x) = \begin{pmatrix} x_1^2 \\ x_2^2 \end{pmatrix} + \begin{pmatrix} r_{x,1} \\ r_{x,2} \end{pmatrix} .$$

Let the true value of the parameters be $\boldsymbol{\theta} = (\theta_1, \theta_2)^{\mathrm{T}} = (5, 3)^{\mathrm{T}}$ and the initial value of the true signal be $\boldsymbol{x}_0 = (1, 1)^{\mathrm{T}}$. The true states $\boldsymbol{x}_n$ and the measurements z_n were distorted by normal noise with zero mean and covariance matrices $Q_x = R_x = \mathrm{diag}(0.05^2, 0.05^2)$. Running the dual filter described above with the initial conditions $\hat{\boldsymbol{x}}^+_0 = (2.5, 1.5)^{\mathrm{T}}, \hat{\boldsymbol{\theta}}^+_0 = (4, 7)^{\mathrm{T}}, P^+_{x,0} = \mathrm{diag}(0.03^2, 0.03^2)$, and $P^+_{\theta,0} = \mathrm{diag}(0.05^2, 0.05^2)$ yields the simultaneous estimation of the states and parameters as shown in Fig. D.1. Reader, try to reduce the wiggles in $\hat{\boldsymbol{x}}_n$ and $\hat{\boldsymbol{\theta}}_n$!   ◁

References

1. E. A. Wan, A. T. Nelson, Dual extended Kalman filter methods, in *Kalman Filtering and Neural Networks*, ed. by S. Haykin (John Wiley & Sons, New York, 2001), pp. 123–173
2. E.A. Wan, R. van der Merwe, A. T. Nelson, Dual estimation and the unscented transformation. Adv. Neural Inform. Processing Syst. **12**, 666 (1999)

Appendix E
Computation of Poincaré Maps

We describe a simple method devised by Hénon [1] allowing us to compute Poincaré maps in dynamical systems. The trick at the heart of the method is very different from the interpolation approach based on the dense-output capability of ODE solvers (see pages 477 and 482), although both achieve the same goal.

Consider an autonomous dynamical system $\boldsymbol{y}' = \boldsymbol{f}(\boldsymbol{y})$ defined by a system of N (coupled) differential equations:

$$\frac{\mathrm{d}y_1}{\mathrm{d}x} = f_1(y_1, y_2, \ldots, y_N) ,$$
$$\vdots$$
$$\frac{\mathrm{d}y_N}{\mathrm{d}x} = f_N(y_1, y_2, \ldots, y_N) . \tag{E.1}$$

When the system evolves, the solution traces a trajectory in an N-dimensional phase space $(y_1, y_2, \ldots, y_N)$. We are interested in the successive intersections of the trajectory with a *surface of section* Σ, which in general is an $(N-1)$-dimensional subset of the phase space, defined by an equation of the form

$$S(y_1, y_2, \ldots, y_N) = 0 . \tag{E.2}$$

The dynamical system (E.1) defines a mapping of Σ on itself, known as the Poincaré map. The problem at hand, then, amounts to following the points on the trajectory until a change of sign of S is detected. Interpolation between the two relevant sets of point is one way to find the approximate surface of section; Hénon's method is an excellent alternative which guarantees that one trajectory point lies *exactly* on Σ.

We discuss first the case where (E.2) has the simple form $y_i - c = 0$ which, by a permutation of coordinates, can be brought to

$$y_N - c = 0 . \tag{E.3}$$

© The Editor(s) (if applicable) and The Author(s), under exclusive license to Springer Nature Switzerland AG 2025
S. Širca and M. Horvat, *Computational Methods in Physics*, Graduate Texts in Physics,
https://doi.org/10.1007/978-3-031-68566-8

The key realization is that y_N in this equation is a dependent variable. By rearranging the system (E.1) such that y_N acts as the *independent* variable, the problem of finding the surface of section at the point of passage translates to making a step not in x, but rather in y_N such that (E.3) is exactly fulfilled. The rearranged system has the form

$$
\begin{aligned}
\frac{\mathrm{d}y_1}{\mathrm{d}y_N} &= \frac{f_1}{f_N}\,, \\
&\ \ \vdots \\
\frac{\mathrm{d}y_{N-1}}{\mathrm{d}y_N} &= \frac{f_1}{f_N}\,, \\
\frac{\mathrm{d}x}{\mathrm{d}y_N} &= \frac{1}{f_N}\,.
\end{aligned}
\tag{E.4}
$$

The procedure is the following. We trace the solution by integrating (E.1) until we detect a change of sign of $S = x_N - c$. We then use the last calculated trajectory point as the initial condition for (E.4) and integrate *for a single step*, taking

$$
\Delta y_N = -S
\tag{E.5}
$$

as the step size. This brings us *exactly* on the surface of section. After having noted the coordinates of the point, we revert to the system (E.1), and continue.

Both modalities can be brought under a single roof by defining ξ as the current independent variable, writing $R = \mathrm{d}x/\mathrm{d}\xi$, and recasting the system as

$$
\begin{aligned}
\frac{\mathrm{d}y_1}{\mathrm{d}\xi} &= Rf_1\,, \\
&\ \ \vdots \\
\frac{\mathrm{d}y_N}{\mathrm{d}\xi} &= Rf_N\,, \\
\frac{\mathrm{d}x}{\mathrm{d}y_N} &= R\,,
\end{aligned}
$$

where $R = 1$ for "forward" integration and $R = 1/f_N$ for "backward" integration. Some care is needed in multi-stage methods, of course; applying the $R = 1/f_N$ recipe in the four stages of the RK4 method, for instance, means

$$
\begin{aligned}
k_1 &= R_1 f(y)\,, \\
k_2 &= R_2 f\left(y + \tfrac{1}{2}\Delta\xi\, k_1\right), \\
k_3 &= R_3 f\left(y + \tfrac{1}{2}\Delta\xi\, k_2\right), \\
k_4 &= R_4 f\left(y + \Delta\xi\, k_3\right),
\end{aligned}
$$

where $R_1 = 1/f_N(y)$, $R_2 = 1/f_N\left(y + \tfrac{1}{2}\Delta\xi\, k_1\right)$, $R_3 = 1/f_N\left(y + \tfrac{1}{2}\Delta\xi\, k_2\right)$ and $R_4 = 1/f_N\left(y + \Delta\xi\, k_3\right)$.

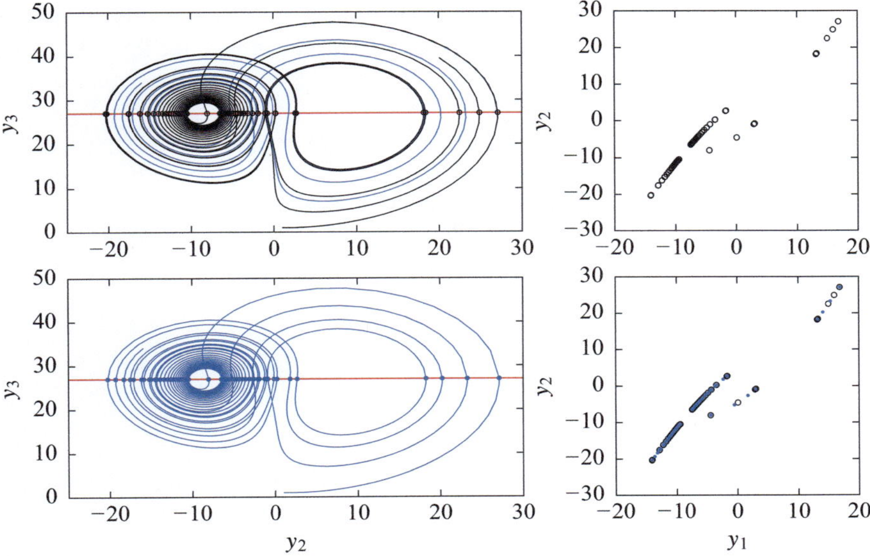

Fig. E.1 Poincaré maps for the Lorenz system. [TOP LEFT] The (y_2, y_3) plane of the phase space showing the intersections with the line $y_3 = 27$. Black curve: trajectory with Hénon's method applied; blue curve: direct integration of the system (E.1). [TOP RIGHT] Points on the surface of section (y_1, y_2) defined by the condition $y_3 = 27$, found by using Hénon's method. [BOTTOM LEFT] Same as above, but showing the trajectory when the improved version of the algorithm is used (see text). Now the intersections with the line $y_3 = 27$ are located on the correct trajectory. [BOTTOM RIGHT] Same as above, but showing how the correct intersections (blue filled circles) have shifted with respect to the less precise ones (black empty circles)

To illustrate the method we integrate the Lorenz system (8.71) with parameters $\sigma = 10$, $b = 8/3$, $r = 28$ and initial conditions $(2, 1, 1)$, and seek for points on the surface of section defined by $y_3 - 27 = 0$ ($y_3 = Z$ in the notation of (8.71)). Figure E.1 (top left) shows the (y_2, y_3) plane of the phase space with the trajectory in black and the locations of the intersections with the line $y_3 = 27$. The right panel shows the corresponding points in the (y_1, y_2) plane.

Note that if we had continuously integrated the system without trying to locate the intersections, the solution would follow the trajectory drawn in blue: it differs from the black trajectory because at each crossover of the surface of section some of the points of the solution, not only the relevant decision coordinate, are obtained by backward integration, resulting in a different trajectory when forward integration is resumed. This deficiency can be cured by a simple augmentation of the Hénon's classic procedure: immediately after crossing the surface of section, store the values of all variables; perform the backward step (E.5); reset the values of all stored variables, and resume the integration [2]. The trajectory and the Poincaré map for the Lorenz system with the same parameter set as before are shown in Fig. E.1 (bottom left and right).

Both the standard and the improved Hénon's technique can be easily adapted to seeking surfaces of section defined by a more general equation of the form (E.2). We introduce an additional variable

$$y_{N+1} = S(y_1, y_2, \ldots, y_N),$$

and add to the system the corresponding differential equation

$$\frac{dy_{N+1}}{dx} = f_{N+1}(y_1, y_2, \ldots, y_N),$$

where

$$f_{N+1}(\boldsymbol{y}) = \sum_{i=1}^{N} f_i(\boldsymbol{y}) \frac{\partial S(\boldsymbol{y})}{\partial y_i}.$$

We obtain a new system of $N + 1$ first-order equations, with the surface of section defined by

$$y_{N+1} = 0.$$

This is of the form (E.3), and the above procedure can be applied.

Finally, to directly quote Hénon's original paper, "the technique described here is also of interest when some of the functions in (E.1), or some of their low-order derivatives, are discontinuous across an $(N - 1)$-dimensional subset ob phase space defined by an equation

$$R(y_1, y_2, \ldots, y_N) = 0. \tag{E.6}$$

The use of a high-order integration algorithm can then produce large errors whenever the »surface of discontinuity« (E.6) is crossed, because the algorithm implicitly assumes that all derivatives exist up to its order. The only remedy is to use an algorithm with independent time steps and to ensure that all points of intersection of the trajectory with the surface of discontinuity are integration points. Within each integration step, there is then no discontinuity and the algorithm gives accurate results" [1].

References

1. M. Hénon, On the numerical computation of Poincaré maps. Phys. D **5**, 412 (1982)
2. P. Palaniyandi, On computing Poincaré map by Hénon method. Chaos Solit. Fractals **39**, 1877 (2009)

Appendix F
Fixed Points and Stability ★

F.1 Linear Stability

In Chapter 8 we discuss autonomous systems of ordinary differential equations with specified initial conditions:

$$\boldsymbol{y}' = \boldsymbol{f}(\boldsymbol{y}), \qquad \boldsymbol{y}(0) = \boldsymbol{y}_0 \in \mathbb{R}^M, \qquad x \geq x_0. \tag{F.1}$$

In explicit one-step difference methods we approximate the system (F.1) by the discrete mapping

$$\boldsymbol{y}_{n+1} = \boldsymbol{F}(\boldsymbol{y}_n), \tag{F.2}$$

where $\boldsymbol{F}$ is a known function (e.g. $\boldsymbol{F}(\boldsymbol{y}) = \boldsymbol{y} + h\boldsymbol{f}(\boldsymbol{y})$ in the explicit Euler scheme). When analyzing the stability of the solutions of (F.1) one would like to know whether two solutions of the equation that are nearby at some time, remain nearby at later times, or perhaps even asymptotically approach each other. We define two types of stability visualized in Fig. F.1. (The reader can easily generalize these definitions to the mapping (F.2).)

DEFINITION. The solution $\boldsymbol{y}^*$ corresponding to (F.1) is locally attractive (*linearly stable*) *in Lyapunov sense* if for some $\varepsilon > 0$ such $\delta > 0$ exists that for any other solution z of (F.1) that satisfies $\|\boldsymbol{y}^*(x_0) - z(x_0)\| < \delta$, also $\|\boldsymbol{y}^*(x) - z(x)\| < \varepsilon$ holds true for $x > x_0$.

DEFINITION. The solution $\boldsymbol{y}^*$ corresponding to (F.1) is *asymptotically stable* if it is Lyapunov stable and for any other solution z of (F.1) such $\beta > 0$ exists that $\lim_{x \to \infty} \|\boldsymbol{y}^*(x) - z(x)\| = 0$ when $\|\boldsymbol{y}^*(x_0) - z(x_0)\| < \beta$.

Suppose that the true solution (flow) $\boldsymbol{y}$ approaches a constant value $\boldsymbol{y}^*$ when $x \to \infty$. Then clearly $\boldsymbol{f}(\boldsymbol{y}^*) = 0$, and $\boldsymbol{y}^*$ is called the *fixed point* of the flow $\boldsymbol{y}$ corresponding to (F.1). In the case of a mapping the fixed point is the point $\boldsymbol{y}^*$ for

© The Editor(s) (if applicable) and The Author(s), under exclusive license to Springer Nature Switzerland AG 2025

S. Širca and M. Horvat, *Computational Methods in Physics*, Graduate Texts in Physics, https://doi.org/10.1007/978-3-031-68566-8

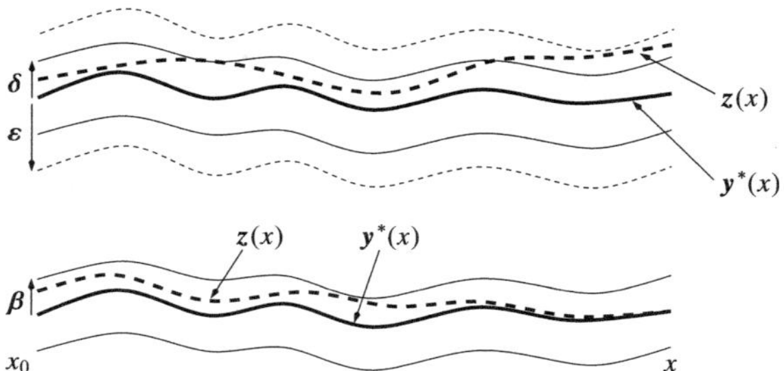

Fig. F.1 Geometric illustration of stability in the [TOP] Lyapunov sense and [BOTTOM] asymptotic stability

which $F(y^*) = y^*$. The study of stability of differential equations and difference schemes focuses primarily on the stability near the fixed points.

In the vicinity of the fixed point y^* we denote $y = y^* + \delta$ when dealing with a differential equation, or $y_n = y^* + \delta_n$ when we think in terms of a discrete mapping. Then Eq. (F.1) can be linearized,

$$\delta' = y' = f(y) \approx \underbrace{f(y^*)}_{0} + \left.\frac{\partial f}{\partial y}\right|_{y=y^*} \delta \,,$$

and similarly for the mapping (F.2),

$$y^* + \delta_{n+1} = F(y_n) \approx \underbrace{F(y^*)}_{y^*} + \left.\frac{\partial F}{\partial y}\right|_{y=y^*} \delta_n \,.$$

In the following we shift the fixed point to the origin, so $\delta \to y$ for equations and $\delta_n \to y_n$ for mappings. Then we have, to first order,

$$y' = Ay \,, \qquad A_{ij} = \left[\left.\frac{\partial f}{\partial y}\right|_{y=0}\right]_{ij} \,, \tag{F.3}$$

$$y_{n+1} = Ay_n \,, \qquad A_{ij} = \left[\left.\frac{\partial F}{\partial y}\right|_{y=0}\right]_{ij} \,. \tag{F.4}$$

The behavior at large x for equations (or large n for mappings) is determined by the eigenvalues of the Jacobi matrix $\partial f / \partial y$ (or $\partial F / \partial y$). The system (F.3) is stable in the Lyapunov sense precisely when all roots of the equation

$$\det(\lambda I - A) = a_0 \lambda^n + a_1 \lambda^{n-1} + \cdots + a_{n-1}\lambda + a_n = 0$$

satisfy $\mathrm{Re}\,\lambda_i \leq 0$, and all multiple roots satisfy $\mathrm{Re}\,\lambda_i < 0$. A similar argument applies to the asymptotic stability of the linearized system (F.4): the sufficient condition for it is that the spectral radius of the Jacobi matrix for the mapping (F.2) is bounded,

$$\rho\left(\left.\frac{\partial \mathbf{F}}{\partial \mathbf{y}}\right|_{\mathbf{y}=\mathbf{y}^*}\right) < 1 \,.$$

The Routh–Hurwitz criterion [1] allows us to determine whether the eigenvalues satisfy the stability condition from the coefficients a_i alone, i.e. without actually solving the characteristic equation. Since the computation of the characteristic polynomial for large matrices is difficult and the search for zeros is plagued by round-off errors, it is recommendable to use dedicated, numerically stable algorithms to find the eigenvalues.

If the matrix elements of A are not constant (non-autonomous systems), as in

$$\mathbf{y}' = A(x)\mathbf{y} \,,$$

and we wish to examine their asymptotic stability, it *does not suffice* to check the criteria $\mathrm{Re}\,\lambda_i \leq 0$ for simple roots or $\mathrm{Re}\,\lambda_i < 0$ for multiple roots of the characteristic equation. But specific cases exist [1] for which these checks are sufficient. Asymptotic stability is ensured if $A_{ii}(x) < 0$ and $A(x)$ is diagonally dominant; or if a decomposition $A(x) = B + C(x)$ exists, where B is a constant matrix and all its eigenvalues satisfy $\mathrm{Re}\,\lambda_i < 0$ and $\|C(x)\| < \varepsilon$.

Example Examine the stability of the system of differential equations

$$y_1' = \tfrac{1}{3}(y_1 - y_2)(1 - y_1 - y_2) \,, \tag{F.5}$$

$$y_2' = y_1(2 - y_2) \,. \tag{F.6}$$

Linearize the system, compute the eigenvalues and eigenvectors for the problem $\mathbf{y}' = A\mathbf{y}$, and confirm that the system has four fixed points in the (y_1, y_2) plane: an unstable focus $(0, 0)$, a saddle $(0, 1)$, an unstable knot $(2, 2)$, and an unstable point $(-1, 2)$ with the eigenspace spanned by the vectors whose y_2-components are zero. Based on (F.3), classify the linear stability (or instability) in the vicinity of fixed points. Figure F.2 shows the numerical solution of the system. Reproduce it by randomly selecting initial conditions $(y_1(0), y_2(0))$ near the fixed points $(\pm 1, \pm 1)$, and run the integration to $x \approx 0.5$. Use the same case to check the stability of the explicit Euler difference scheme. Such studies aid us, on a case-by-case basis, in deciding to what extent linear measures of stability can be trusted in non-linear systems. ◁

F.2 Spurious Fixed Points

Are the fixed points of the chosen difference scheme the same as the fixed points of the differential equation it approximates? We examine the stability of the (non-linear) logistic equation

$$y' = \lambda y(1 - y) \,, \qquad y(0) = y_0 \,,$$

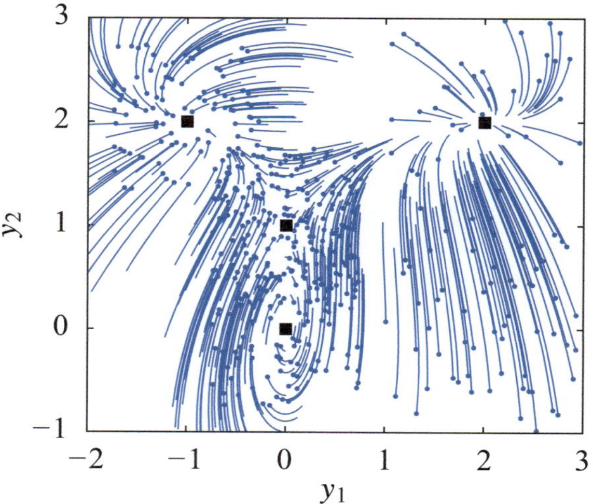

Fig. F.2 The numerical solution (flow) of the non-linear system (F.5)-(F.6), with 500 randomly chosen initial conditions (small filled circles), integrated to $x = 0.7$ by the adaptive RK4 method. The fixed points are denoted by filled squares

with $\lambda \in \mathbb{C}$ and of the corresponding difference approximations for it by using the criteria of *linear* stability [2]. The exact solution of the equation is

$$y(x) = \frac{y_0}{y_0 + (1 - y_0)\mathrm{e}^{-\lambda x}} \, ,$$

and the stable fixed points are obviously $y^* = 0$ for $\mathrm{Re}\,\lambda < 0$ and $y^* = 1$ for $\mathrm{Re}\,\lambda > 0$. The Jacobi matrix in one dimension is just the derivative $\partial f/\partial y = \lambda - 2\lambda y$, so at the fixed points $[\partial f/\partial y](0) = \lambda$ or $[\partial f/\partial y](1) = -\lambda$. The fixed point $y^* = 0$ is linearly stable when $\mathrm{Re}\,\lambda < 0$, and the fixed point $y^* = 1$ is linearly stable if $\mathrm{Re}\,\lambda > 0$.

The corresponding explicit Euler scheme is

$$y_{n+1} = y_n + h\lambda y_n(1 - y_n) = F(y_n) \, ,$$

with the same fixed points as before, $y^* = 0$ and $y^* = 1$. But now the Jacobi derivative is $\partial F/\partial y = 1 + h\lambda - 2h\lambda y$, thus $[\partial F/\partial y](0) = 1 + h\lambda$ and $[\partial F/\partial y](1) = 1 - h\lambda$. The fixed point $y^* = 0$ corresponds to the absolute stability region which is exactly the same as in the linear case (the interior of the circle denoted by $p = 1$ in Fig. 8.3). The point $y^* = 1$ corresponds to this circle mirrored across the imaginary axis. Let us look one order higher, to the improved Euler method (8.6). When we write the difference scheme

$$y_{n+1} = y_n + hf(y_n + (h/2)f(y_n)) \, ,$$

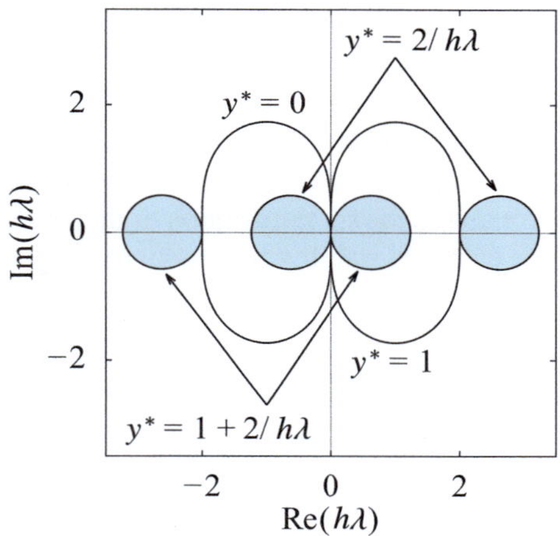

Fig. F.3 Stability regions with respect to genuine and spurious fixed points for the improved Euler method (second-order explicit RK method). See also Fig. 8.3

for the logistic equation, we get

$$
y_{n+1} = y_n + h\lambda \left[y_n + \frac{h\lambda}{2} y_n (1 - y_n) \right] \left[1 - y_n - \frac{h\lambda}{2} y_n (1 - y_n) \right]
$$

$$
= y_n + h\lambda y_n (1 - y_n) \left[1 + \frac{h\lambda}{2} - \frac{h\lambda}{2} y_n \right] \left[1 - \frac{h\lambda}{2} y_n \right] .
$$

This expression reveals *four* fixed points: the familiar $y^* = 0$ and $y^* = 1$, as well as two new ones, $2/h\lambda$ and $1 + 2/h\lambda$! The fixed points that are not seen in the differential equation, but appear in the corresponding difference scheme, are called *spurious* or *ghost points*. Characteristically, spurious fixed points depend on the step size h. When one computes $\partial F / \partial y$, finding the stability region of different fixed points translates to a search of the set of points $h\lambda$ in the complex plane satisfying

$$
\left| 1 \pm h\lambda \pm \frac{(h\lambda)^2}{2} \right| < 1 ,
$$

similar to what we have done in Sect. 8.5. The fixed point $y^* = 0$ corresponds to the interior of the shape denoted by $p = 2$ in Fig. 8.3 (and its mirror image corresponds to $y^* = 1$). To each of the spurious fixed point corresponds a pair of smaller stability regions: the complete pattern of stability regions for all four fixed points is shown in Fig. F.3.

F.3   Non-linear Stability

When we studied a non-linear system by using methods of linear stability we learned that one should not apply linear measures to non-linear systems. The stability region of a linear discrete mapping used to approximate a linear problem is not necessarily the same as the stability region of the same scheme applied to a non-linear problem. The theory of stability for non-linear systems requires us to understand the concepts of *contractivity* and *convergence*.

DEFINITION. If $\widetilde{y}$ and y are any solutions of the system $y' = f(x, y)$ corresponding to different initial conditions, the system is *contractive* if

$$\|\widetilde{y}(x_2) - y(x_2)\| \le \|\widetilde{y}(x_1) - y(x_1)\|$$

for any pair (x_1, x_2) from the interval $a \le x_1 \le x_2 \le b$. The corresponding difference scheme is contractive if

$$\|\widetilde{y}_{n+1} - y_{n+1}\| \le \|\widetilde{y}_n - y_n\| \ .$$

Linear problems Contractivity is easy to understand for the linear problem

$$y' = Ay \ .$$

What is the norm of the numerical solution $y_{n+1} = S(hA)y_n$ if S is the stability function of the applied difference method? Since $\|y_{n+1}\| \le \|S(hA)\| \, \|y_n\|$, we have contractivity if $\|S(hA)\| \le 1$. For functions $S(z)$ that ensure stability throughout the complex half-plane (A-stability, Sect. 8.8), we have contractivity when, in the Euclidean norm, Re $\langle y, Ay \rangle \le 0$ for each y [3].

Semi-linear problems For semi-linear systems of the form

$$y' = Ay + g(x, y) \, ,$$

where $g(x, y)$ is a small non-linear perturbation, we can resort to similar measures as in the linear case. If all eigenvalues of A satisfy Re $\lambda_i < 0$, and if for any $\varepsilon > 0$ a $\delta > 0$ exists such that $\|g(x, y)\| \le \varepsilon \|y\|$ for $\|y\| < \delta$ and $x \ge x_0$, then the origin is asymptotically stable in the Lyapunov sense. But the parameter ε must be sufficiently small. This means that in non-linear systems, one may expect absolute stability to occur only in the vicinity of stable fixed points. Contractivity is achieved when

$$\langle y, Ay \rangle \le \mu \|y\|^2 \, , \qquad \|g(x, y) - g(x, z)\| \le \Lambda \|y - z\| \, ,$$

where μ and Λ are constants that must satisfy $\mu + \Lambda \le 0$ (with Λ small). Similar conditions apply to numerical solutions of difference schemes, e.g. the implicit Runge–Kutta methods [3].

Non-linear problems The study of stability of "genuine" non-linear systems

$$y' = f(x, y)$$

rests upon one-sided Lipschitz conditions of the form

$$\mathrm{Re}\,\langle f(x, y) - f(x, z),\ y - z \rangle \leq \nu \|y - z\|^2, \tag{F.7}$$

where ν is the one-sided Lipschitz constant of f (compare Eq. (F.7) to Eq. (8.4)). When $\nu \leq 0$, the distance between any two solutions of the equation $y' = f(x, y)$ is a non-increasing function of x. We wish the numerical solution to possess the same property, e.g. when using implicit Runge–Kutta methods. For this purpose we define B-stability and B-convergence, which we refer to in Sect. 8.8.

DEFINITION. A method of the Runge–Kutta type is B-stable if from the contractivity condition $\mathrm{Re}\,\langle f(x, y) - f(x, z),\ y - z \rangle \leq 0$ for all $h \geq 0$, we can conclude

$$\|\widetilde{y}_{n+1} - y_{n+1}\| \leq \|\widetilde{y}_n - y_n\|,$$

where y_{n+1} (or $\widetilde{y}_{n+1}$) are the numerical solutions obtained from y_n (or $\widetilde{y}_n$) in one step. (B-stability implies A-stability, see [3]).

DEFINITION. A method of the Runge–Kutta type is B-convergent of order r for the non-linear problem $y' = f(x, y)$ satisfying the one-sided Lipschitz condition (F.7), if the global error can be estimated as

$$\|y_n - y(x_n)\| \leq h^r \gamma(x_n - x_0, \nu) \max_j \max_x \|y^{(j)}(x)\|$$

for $h\nu \leq \alpha$, where $h = \max h_i$, $1 \leq j \leq l$, and $a \leq x \leq b$. The function γ and the parameter α depend only on the method.

For stiff problems, the measures of convergence have to be adapted. Namely, if we use A-stable methods to solve large systems of stiff non-linear differential equations, some may yield unstable results. Besides, it may happen that the precision of the solution has no connection to the order of the method used. A decrease of the order of convergence occurs [4], i.e. the order of the method for a stiff problem with large λ becomes much smaller than the order for a non-stiff problem. Through B-convergence one can also introduce global error estimates that do not depend on the degree of stiffness [5]. The implicit method Radau 5 (page 492) is B-stable and B-convergent of order 3 [3].

References

1. E. Hairer, S.P. Nørsett, G. Wanner, *Solving Ordinary Differential Equations I; Nonstiff Problems*, Springer Series in Computational Mathematics, Vol. 8 (Springer-Verlag, Berlin, 2000)
2. J. H. E. Cartwright, O. Piro, The dynamics of Runge–Kutta methods. Int. J. Bifurcation and Chaos **2**, 427 (1992)
3. E. Hairer, G. Wanner, *Solving Ordinary Differential Equations II; Stiff and Differential-Algebraic Problems*, Springer Series in Computational Mathematics, Vol. 14 (Springer-Verlag, Berlin, 2004)
4. A. Prothero, A. Robinson, On the stability and accuracy of one-step methods for solving stiff systems of ordinary differential equations. Math. Comput. **28**, 145 (1974)
5. R. Frank, J. Schneid, C.W. Ueberhuber, The concept of B-Convergence. SIAM J. Numer. Anal. **18**, 753 (1981)

Appendix G
Regularization in Orbital Mechanics

In calculations of trajectories of astrophysical objects—e.g. highly eccentric planetary orbits or motion of stars in large stellar clusters—at some instant two bodies may face an encounter with a negligible impact parameter (that is, a very small orbital angular momentum), resulting in a large acceleration due to the singularity of the equation of motion $\ddot{r} = -GM/r^2$ for $r \to 0$. The most extreme case is the head-on collision in which the impact parameter is zero: in this case one can show that $r(t) \propto |t - t_{\mathrm{c}}|^{2/3}$ in the vicinity of $r = 0$, where t_{c} is the collision time, hence any adaptive step size integrator will be forced to take ever shorter time steps as $r \to 0$, thus dramatically increasing the number of force evaluations, slowing down the whole N-body integration and potentially spoiling its accuracy: see Fig. G.1. This problem can be avoided either by a transformation to a coordinate system in which the equation of motion becomes non-singular and can be handled by standard integrators, or by keeping the singularity but constructing an integration scheme that is insensitive to it. The procedure is called *regularization* [1].

G.1 Burdet–Heggie Regularization

The simplest approach consists in transforming the time variable [2, 3]. We write the equation of motion for the perturbed two-body problem as

$$\ddot{r} = -\frac{GM}{r^3} r + g \,, \tag{G.1}$$

where $g = -\nabla \Phi$ is the acceleration due to the gravitational field Φ of the remaining $N - 2$ bodies in the calculation (or other sources). We rescale the time variable as

$$\mathrm{d}t = r \, \mathrm{d}\tau \,.$$

S. Širca and M. Horvat, *Computational Methods in Physics*, Graduate Texts in Physics, https://doi.org/10.1007/978-3-031-68566-8

Time-stepping in fictitious time τ allows for a finer representation of the dynamics where r is small (e.g. near the periapsis), leading to larger accuracy.

Denoting the derivative with respect to τ by a prime we find

$$\dot{\boldsymbol{r}} = \frac{\mathrm{d}\boldsymbol{r}}{\mathrm{d}\tau}\frac{\mathrm{d}\tau}{\mathrm{d}t} = \frac{\boldsymbol{r}'}{r}\,, \qquad \ddot{\boldsymbol{r}} = \frac{\mathrm{d}}{\mathrm{d}\tau}\left(\frac{\boldsymbol{r}'}{r}\right)\frac{\mathrm{d}\tau}{\mathrm{d}t} = \frac{\boldsymbol{r}''}{r^2} - \frac{\boldsymbol{r}\cdot\boldsymbol{r}'}{r^4}\boldsymbol{r}'\,.$$

Inserting the expression for $\ddot{\boldsymbol{r}}$ in (G.1) yields $\boldsymbol{r}'' = (\boldsymbol{r}\cdot\boldsymbol{r}')\boldsymbol{r}'/r^2 - GM\boldsymbol{r}/r + r^2\boldsymbol{g}$. Using the Laplace–Runge–Lenz vector (i.e. the eccentricity vector) given by

$$\boldsymbol{e} = \dot{\boldsymbol{r}}\times(\boldsymbol{r}\times\dot{\boldsymbol{r}}) - GM\frac{\boldsymbol{r}}{r} = \frac{|\boldsymbol{r}'|^2}{r^2}\boldsymbol{r} - \frac{\boldsymbol{r}\cdot\boldsymbol{r}'}{r^2}\boldsymbol{r}' - GM\frac{\boldsymbol{r}}{r}$$

we can eliminate the $(\boldsymbol{r}\cdot\boldsymbol{r}')\boldsymbol{r}'$ term, hence $\boldsymbol{r}'' = |\boldsymbol{r}'|^2\boldsymbol{r}/r^2 - 2GM\boldsymbol{r}/r - \boldsymbol{e} + r^2\boldsymbol{g}$. The energy corresponding to the orbit is

$$E = \frac{v^2}{2} - \frac{GM}{r} + \Phi = \frac{|\boldsymbol{r}'|^2}{2r^2} - \frac{GM}{r} + \Phi\,,$$

so the $|\boldsymbol{r}'|^2$ term can also be eliminated, and we obtain

$$\boldsymbol{r}'' = 2E\boldsymbol{r} - \boldsymbol{e} - \nabla\left(r^2\Phi\right)\,. \tag{G.2}$$

This is the regularized equation of motion from which the singularity has disappeared: $\boldsymbol{r}(\tau)$ is a smooth function, even in the case of a central collision. The price one has to pay is an additional equation that relates the fictitious time τ to the physical time t, plus evolution equations for the energy and the eccentricity vector:

$$t' = r\,,$$
$$E' = r\frac{\partial\Phi}{\partial t}\,,$$
$$\boldsymbol{e}' = (\boldsymbol{r}\cdot\nabla\Phi)\,\boldsymbol{r}' + \left(\boldsymbol{r}\cdot\boldsymbol{r}'\right)\nabla\Phi - 2\left(\boldsymbol{r}'\cdot\nabla\Phi\right)\boldsymbol{r}\,.$$

When the external potential vanishes, E is just the Keplerian energy $v^2/2 - GM/r$ which is constant, and so is the eccentricity vector $\boldsymbol{e}$, thus $E' = 0$, $\boldsymbol{e}' = \boldsymbol{0}$. In this case Eq. (G.2) describes the motion of a linear harmonic oscillator with frequency $\sqrt{-2E}$ experiencing a constant force $-\boldsymbol{e}$.

The "RK, BH" curve in Fig. G.1 shows the relative energy error from integrating a single pericenter passage of a Keplerian orbit with eccentricity $\varepsilon = 0.99$ by using an embedded Runge–Kutta method of order 4(5), with Burdet–Heggie regularization applied: the number of force evaluations drops by an order of magnitude for a given energy error (and the error drops by six orders of magnitude for a given number of force evaluations)—compare to the "RK, U" result obtained by the unregularized calculation. The "BH" curve in Fig. G.2 shows the energy error for an orbit with

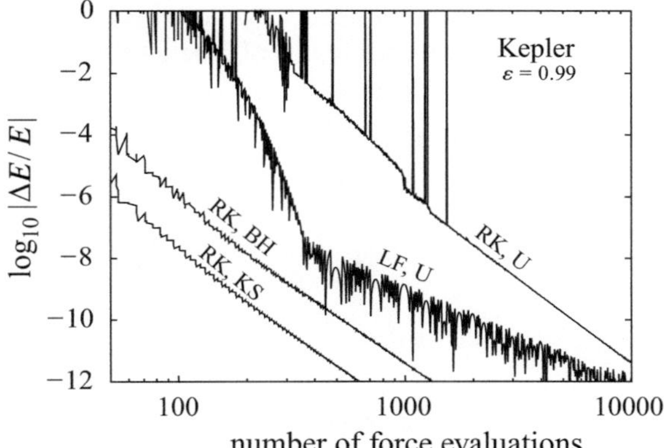

Fig. G.1 Relative energy error from integrating one pericenter passage of a Keplerian orbit with eccentricity $\varepsilon = 0.99$, as a function of the number of force evaluations. RK: using the adaptive step size Runge–Kutta method; LF: using a leapfrog scheme with $h = \Delta t \propto r$; U: unregularized; BH: with Burdet–Heggie regularization (Sect. G.1); KS: with Kustaanheimo–Stiefel regularization (Sect. G.2)

an even higher eccentricity ($\varepsilon = 0.999$) as a function of time: with the symplectic Euler scheme the error oscillates but does not increase with time—compare to the "Hermite" curve obtained by the standard fourth-order Hermite integrator.

G.2 Kustaanheimo–Stiefel Regularization

In contrast to the Burdet–Heggie approach, Kustaanheimo–Stiefel regularization [4] involves the transformation of both the time and space coordinates. Defining the vector $\boldsymbol{u} = (u_1, u_2, u_3, u_4)^{\mathrm{T}}$ and trivially extending the physical position vector to four dimensions, $\boldsymbol{r} = (x, y, z, 0)^{\mathrm{T}}$, we consider the transformation

$$\boldsymbol{r} = L(\boldsymbol{u})\boldsymbol{u} \,, \quad L(\boldsymbol{u}) = \begin{pmatrix} u_1 & -u_2 & -u_3 & u_4 \\ u_2 & u_1 & -u_4 & -u_3 \\ u_3 & u_4 & u_1 & u_2 \\ u_4 & -u_3 & u_2 & -u_1 \end{pmatrix} \,,$$

or, by components,

$$x = u_1^2 - u_2^2 - u_3^2 + u_4^2 \,, \quad y = 2(u_1 u_2 - u_3 u_4) \,, \quad z = 2(u_1 u_3 + u_2 u_4) \,, \quad \text{(G.3)}$$

with $r = |\boldsymbol{u}|^2 = u_1^2 + u_2^2 + u_3^2 + u_4^2$. Given that the transformation is not bijective, we have some liberty in defining its inverse. This is needed, among other things, to translate the initial conditions for $\boldsymbol{r}$ to the initial conditions for $\boldsymbol{u}$, so some care is called for regarding the sign of the coordinates. For instance, if $x \geq 0$, one can show that $u_1^2 + u_4^2 = \frac{1}{2}(r + x)$, and if we choose $u_4 = 0$, the inverse relations are

$$u_1 = \sqrt{(r + x)/2}\,, \quad u_2 = \frac{y}{2u_1}\,, \quad u_3 = \frac{z}{2u_1} \quad (x \geq 0)\,.$$

Alternatively, if $x < 0$, it holds that $u_2^2 + u_3^2 = \frac{1}{2}(r - x)$, hence by setting $u_3 = 0$ the inverses become

$$u_2 = \sqrt{(r - x)/2}\,, \quad u_1 = \frac{y}{2u_2}\,, \quad u_4 = \frac{z}{2u_2} \quad (x < 0)\,.$$

One may also use, for example, the mapping

$$\begin{aligned}
u_1 &= \sqrt{(r + x)/2}\cos\alpha\,, \quad u_4 = \sqrt{(r + x)/2}\sin\alpha\,, \\
u_2 &= \frac{yu_1 + zu_4}{r + x}\,, \qquad\qquad u_3 = \frac{zu_1 - yu_4}{r + x}\,,
\end{aligned} \tag{G.4}$$

where α is an arbitrary parameter and $x \geq 0$.

Rescaling the time as $\mathrm{d}t = r\,\mathrm{d}\tau$ and denoting the derivatives with respect to τ by primes, thus $t' = r = |\boldsymbol{u}|^2$, the transformation from the velocity $\boldsymbol{u}' = (u_1', u_2', u_3', u_4')^{\mathrm{T}}$ in transform space to the corresponding physical velocity $\dot{\boldsymbol{r}} = (\dot{x}, \dot{y}, \dot{z}, 0)^{\mathrm{T}}$ is

$$\dot{\boldsymbol{r}} = 2\,L(\boldsymbol{u})\dot{\boldsymbol{u}} = 2\,L(\boldsymbol{u})\boldsymbol{u}'/|\boldsymbol{u}|^2\,,$$

or, by components:

$$\begin{aligned}
\dot{x} &= 2(u_1 u_1' - u_2 u_2' - u_3 u_3' + u_4 u_4')/|\boldsymbol{u}|^2\,, \\
\dot{y} &= 2(u_1' u_2 + u_1 u_2' - u_3' u_4 - u_3 u_4')/|\boldsymbol{u}|^2\,, \\
\dot{z} &= 2(u_1' u_3 + u_1 u_3' + u_2' u_4 + u_2 u_4')/|\boldsymbol{u}|^2\,.
\end{aligned} \tag{G.5}$$

If the constraint $u_1' u_4 - u_1 u_4' - u_2' u_3 + u_2 u_3' = 0$ is imposed, it is easy to check that $v^2 = |\dot{\boldsymbol{r}}|^2 = \dot{x}^2 + \dot{y}^2 + \dot{z}^2 = |2L(\boldsymbol{u})\dot{\boldsymbol{u}}|^2 = 4|\boldsymbol{u}|^2|\dot{\boldsymbol{u}}|^2 = 4(u_1'^2 + u_2'^2 + u_3'^2 + u_4'^2)/|\boldsymbol{u}|^2$. In devising the inverse mapping we exploit the r-orthogonality of the transformation matrix, $L^{-1}(\boldsymbol{u}) = L^{\mathrm{T}}(\boldsymbol{u})/r$ or $L^{\mathrm{T}}(\boldsymbol{u})L(\boldsymbol{u}) = rI$. It follows that

$$\boldsymbol{u}' = \tfrac{1}{2}L^{\mathrm{T}}(\boldsymbol{u})\dot{\boldsymbol{r}}\,,$$

that is,

$$\begin{aligned}
u_1' &= (u_1\dot{x} + u_2\dot{y} + u_3\dot{z})/2\,, \\
u_2' &= (-u_2\dot{x} + u_1\dot{y} + u_4\dot{z})/2\,, \\
u_3' &= (-u_3\dot{x} - u_4\dot{y} + u_1\dot{z})/2\,, \\
u_4' &= (u_4\dot{x} - u_3\dot{y} + u_2\dot{z})/2\,.
\end{aligned} \tag{G.6}$$

The energy corresponding to the orbit is

$$E = \frac{v^2}{2} - \frac{GM}{r} + \Phi = 2\frac{|\boldsymbol{u}'|^2}{|\boldsymbol{u}|^2} - \frac{GM}{|\boldsymbol{u}|^2} + \Phi \,.$$

Taking into account that $L'(\boldsymbol{u}) = L(\boldsymbol{u}')$, the regularized equation of motion and the auxiliary evolution equations for t and E' are readily derived [5]:

$$\boldsymbol{u}'' = \frac{1}{2}E\boldsymbol{u} - \frac{1}{4}\frac{\partial}{\partial\boldsymbol{u}}\left(|\boldsymbol{u}|^2\Phi\right) , \tag{G.7}$$

$$t' = |\boldsymbol{u}|^2 ,$$

$$E' = |\boldsymbol{u}|^2\frac{\partial\Phi}{\partial t} \,.$$

The initial conditions for $\boldsymbol{r}$ and $\dot{\boldsymbol{r}}$ must first be translated into the initial conditions for $\boldsymbol{u}$ and $\boldsymbol{u}'$ by using Eqs. (G.4) and (G.6), respectively, while after the evolution, the transformed quantities $\boldsymbol{u}(\tau)$ and $\boldsymbol{u}'(\tau)$ must be converted back to the physical $\boldsymbol{r}(t)$ and $\dot{\boldsymbol{r}}(t)$ by using Eqs. (G.3) and (G.5), respectively.

In the absence of the external field, the regularized equation of motion (G.7) again describes a linear harmonic oscillator with frequency $\sqrt{-E/2}$, which is a factor of two smaller than the frequency $\sqrt{-2E}$ pertaining to Eq. (G.2), and because the frequency is reduced, the integrator is able to follow the orbit more accurately. As an illustration, the "KS" curve in Fig. G.2 shows the relative energy error of a Keplerian orbit with eccentricity $\varepsilon = 0.999$ traced by (G.7) over many periods. (An embedded Runge–Kutta 4(5) method with a tolerance of 10^{-11} has been used.) See also the "RK, KS" curve in Fig. G.1 and compare it to the unregularized case ("RK, U") and to the integration involving the Burdet–Heggie regularization ("RK, BH").

G.3   Algorithmic Regularization

In the special case of second-order differential equations of the form

$$\ddot{\boldsymbol{r}} = \boldsymbol{F}(\boldsymbol{r}) \tag{G.8}$$

a powerful regularization procedure has been devised which involves only the time transformation, yet produces regular results even though the singularity in the force— as, for instance, in (G.1)—remains in place, and even for collision orbits [6, 7]. The transformation of time is taken to be more general,

$$d\tau = \Omega(\boldsymbol{r})\,dt \,,$$

where Ω is an arbitrary positive function. Introducing the auxiliary quantity $W = \Omega$ and denoting the derivatives with respect to τ by primes the equation of motion (G.8),

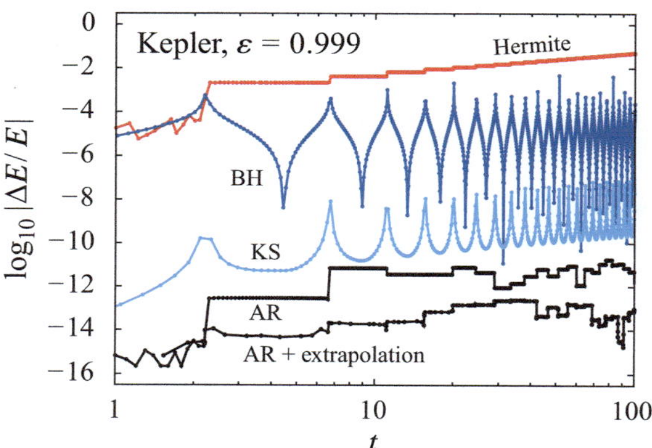

Fig. G.2 Relative energy error from integrating many repetitions of a Keplerian orbit with eccentricity $\varepsilon = 0.999$. Hermite: unregularized calculation with the Hermite integrator of Sect. 8.9.5; BH (with symplectic Euler), KS: as in Fig. G.1; AR: algorithmic regularization using the time-transformed leapfrog integrator of Sect. G.3, without and with Bulirsch–Stoer extrapolation

equivalent to the set $\dot{\boldsymbol{r}} = \boldsymbol{v}$, $\dot{\boldsymbol{v}} = \boldsymbol{F}(\boldsymbol{r})$, can be recast as $\boldsymbol{r}' = \boldsymbol{v}/W$, $t' = 1/W$, $\boldsymbol{v}' = \boldsymbol{F}/\Omega$. If the value of W is obtained from the differential equation $\dot{W} = \boldsymbol{v} \cdot (\partial \Omega / \partial \boldsymbol{r})$ instead of using $W = \Omega$ directly, the equations of motion become $\boldsymbol{r}' = \boldsymbol{v}/W$, $t' = 1/W$, $\boldsymbol{v}' = \boldsymbol{F}(\boldsymbol{r})/\Omega(\boldsymbol{r})$ and $W' = (\boldsymbol{v}/\Omega(\boldsymbol{r})) \cdot \partial \Omega / \partial \boldsymbol{r}$. This set is then transcribed into a time-transformed leapfrog (TTL) scheme of the form [8, 9]

$$\boldsymbol{r}_{1/2} = \boldsymbol{r}_0 + \frac{h}{2} \frac{\boldsymbol{v}_0}{W_0},$$

$$t_{1/2} = t_0 + \frac{h}{2} \frac{1}{W_0},$$

$$\boldsymbol{v}_1 = \boldsymbol{v}_0 + h \frac{\boldsymbol{F}(\boldsymbol{r}_{1/2})}{\Omega(\boldsymbol{r}_{1/2})},$$

$$W_1 = W_0 + h \frac{\boldsymbol{v}_0 + \boldsymbol{v}_1}{2\Omega(\boldsymbol{r}_{1/2})} \cdot \frac{\partial \Omega(\boldsymbol{r}_{1/2})}{\partial \boldsymbol{r}_{1/2}},$$

$$\boldsymbol{r}_1 = \boldsymbol{r}_{1/2} + \frac{h}{2} \frac{\boldsymbol{v}_1}{W_1},$$

$$t_1 = t_{1/2} + \frac{h}{2} \frac{1}{W_1},$$

where the subscripts 0 and 1 denote the values at the beginning and end of a single time step $h = \Delta \tau$.

As an example, the "AR" curve in Fig. G.2 shows the energy error from integrating a high-eccentricity Kepler orbit by using the TTL scheme with step size $h = 0.01$ and $\Omega(\boldsymbol{r}) = 1/r$. If an extrapolation method is used, the accuracy can be improved even further: this is demonstrated by the "AR + extrapolation" curve in Fig. G.2 obtained by plugging the TTL stepper into the extrapolation scheme of the Bulirsch–Stoer type (see Sect. 8.6).

References

1. J. Roa, *Regularization in Orbital Mechanics* (De Gruyter, Berlin, 2017)
2. C. A. Burdet, Theory of Kepler motion: the general perturbed two-body problem. Z. Angew. Math. Phys. **19**, 345 (1968)
3. D. C. Heggie, Regularization using a time-transformation only, in *Recent Advances in Dynamical Astronomy*, ed. B.D. Tapley, V. Szebehely (D. Reidel Publishing Company, Dordrecht, 1973), p. 34
4. P. Kustaanheimo, E. Stiefel, Perturbation theory of Kepler motion based on spinor regularization. J. Reine Angew. Math. **218**, 204 (1965)
5. S. Tremaine, *Dynamics of Planetary Systems* (Princeton University Press, Princeton, 2023)
6. S. Mikkola, K. Tanikawa, *Algorithmic regularization of the few-body problem*, Mon. Not. R. Astron. Soc. **310**, 745 (1999)
7. S. Mikkola, K. Tanikawa, Explicit Symplectic Algorithms for Time-Transformed Hamiltonians. Celest. Mech. Dyn. Astr. **74**, 287 (1999)
8. S. Mikkola, S. Aarseth, A time-transformed leapfrog scheme. Celest. Mech. Dyn. Astr. **84**, 343 (2002)
9. S. Mikkola, Regular algorithms for the few-body problem. Lect. Notes Phys. **760**, 31 (2008)

Appendix H
Construction of Symplectic Integrators ⋆

To study time evolution of dynamical systems, an operator formalism can be developed [1] that leads to a special family of integrators. We have seen in Chap. 8 that the evolution amounts to solving the system of differential equations

$$\dot{\boldsymbol{y}} = \boldsymbol{f}(\boldsymbol{y})$$

over some phase space χ, where $\boldsymbol{y}$ is a vector in this space, and $\boldsymbol{f}$ is a vector of scalar functions[1]. The solutions of this equations are the trajectories traced by $\boldsymbol{y}$ in time t, and described by the mapping $M^t(\boldsymbol{y})$. In general, we would like to learn about the dynamics of some variable (observable) of the system $\boldsymbol{\phi}(\boldsymbol{y})$, which can be very simple (for example, the "position" $\boldsymbol{\phi}(\boldsymbol{y}) = \boldsymbol{y}$). The time dynamics is defined by the map M, $\boldsymbol{\phi}^t(\boldsymbol{y}) = \boldsymbol{\phi}(M^t(\boldsymbol{y}))$, thus

$$\dot{\boldsymbol{\phi}} = (\boldsymbol{f} \cdot \nabla)\boldsymbol{\phi} \, .$$

The linear differential operator $\boldsymbol{f} \cdot \nabla = D$ is called the *dynamics generator* and represents the derivative in the direction tangential to the trajectory at the point $\boldsymbol{y}$ (Fig. H.1). The equation above can be formally integrated, yielding

$$\boldsymbol{\phi}^t(\boldsymbol{y}) = \exp(t D)\boldsymbol{\phi}(\boldsymbol{y}) = \boldsymbol{\phi}(M^t(\boldsymbol{y})) \, , \tag{H.1}$$

where the exponential acts in the operator sense. This equation embodies the *evolution operator* or *integrator* describing the time evolution of $\boldsymbol{\phi}$.

Hamiltonian systems Let us apply the procedure outlined above to Hamiltonian systems with the Hamiltonian function $H(\boldsymbol{p}, \boldsymbol{q})$, where $(\boldsymbol{p}, \boldsymbol{q}) \in \chi = \mathbb{R}^{2n}$. The dynamics (and the trajectory) is governed by the Hamiltonian equations

---

[1] In this Appendix and in Sect. 8.11 we use the pair of variables $(t, \boldsymbol{y})$ instead of $(x, \boldsymbol{y})$, while we express $\boldsymbol{y}$ by the Hamiltonian canonical variables $\boldsymbol{p}$ and $\boldsymbol{q}$.

989

S. Širca and M. Horvat, *Computational Methods in Physics*, Graduate Texts in Physics, https://doi.org/10.1007/978-3-031-68566-8

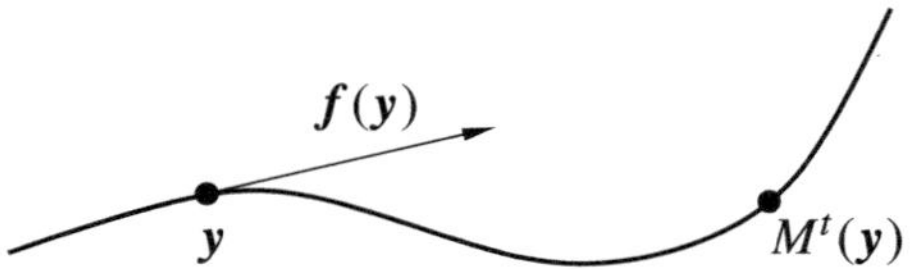

Fig. H.1 The trajectory of a dynamical system is described by the mapping $M^t(y)$ which corresponds to the solution of a system of differential equations $\dot{y} = f(y)$

$$\dot{y} = J^{-1}\nabla H(y)\,, \qquad J = \begin{pmatrix} 0 & 1 \\ -1 & 0 \end{pmatrix}\,, \qquad \nabla = \begin{pmatrix} \nabla_p \\ \nabla_q \end{pmatrix}\,,$$

while the time evolution of an arbitrary observable ϕ is given by the Liouville equation

$$\dot{\phi} = \{\phi, H\}\,,$$

where $\{A, B\} = (\partial_q A)(\partial_p B) - (\partial_q B)(\partial_p A)$ is the Poisson bracket. We use the notation $D_H = \{\cdot, H\}$ to write the equation in the form $\dot{\phi} = D_H \phi$ and formally integrate it over an infinitesimally short time $t = \tau$:

$$\phi^\tau(y) = \exp(\tau D_H)\phi(y) = \exp\!\Big(-\tau(\nabla H\, J)\nabla\Big)\phi(y) = \phi(M^\tau(y))\,.$$

The exponential operator in this equation is unitary in the function space $\phi : \chi \to \mathbb{C}^n$, thus the norm $\int_\chi |\phi|^2\, dy$ along the solution of the system remains constant. In the case $\phi(y) = y = (p, q)$ this means that the mapping from the point (p, q) in phase space at $t = 0$ to the point $(\widetilde{p}, \widetilde{q})$ at a later time along the solution preserves the volume in phase space, i.e. the form $\sum dp \wedge dq = \sum d\widetilde{p} \wedge d\widetilde{q}$: the mapping is *symplectic* [2]. Moreover, it preserves the energy, $H(p, q) = H(\widetilde{p}, \widetilde{q})$.

In practice we rarely sum more than just a few terms in the exponential series

$$\phi^\tau(y) = \sum_{n=0}^N \frac{\tau^n}{n!}\, D_H^n\, \phi(y) + O(\tau^{N+1})\,,$$

where $D_H^n \phi = \{\{\cdots\{\phi, H\}, \cdots, H\}, H\}$. The truncated series still preserves the energy and possible other constants (integrals) of motion $I_k(p, q)$, for which $\{H, I_k\} = 0$ applies, but the truncated operator is no longer unitary and violates the symplectic nature of the dynamics. Nevertheless, for short times τ even the finite series provides a good description.

Symplectic integrators As we have shown in Sect. 8.11, the traditional integrators (for example, Euler or Runge–Kutta methods) in general do not preserve the symplectic structure, nor the energy. In the following we chart a path to a special class of explicit symplectic integrators with which we attempt to fulfill at least one of these requirements. We focus on solving the equations of motion for Hamiltonian systems,

$$\dot{q} = \frac{\partial H}{\partial p} \,, \qquad \dot{p} = -\frac{\partial H}{\partial q} \,,$$

where the Hamiltonian is separable (the kinetic energy depending only on p and the potential only on q), thus

$$H(p, q) = T(p) + V(q) \,.$$

(For greater clarity, we discuss only the scalar case and write ∂_p, ∂_q instead of ∇_p, ∇_q.) When the equations of motion are rewritten in the form $\dot{y} = \{y, H\}$, the dynamics of the system is described by

$$y(\tau) = \exp(\tau D_H)y(0) = \exp(\tau(D_T + D_V))y(0) \,. \tag{H.2}$$

The exponential operators composed of partial generators $D_T = (\partial_p T)\partial_q$ and $D_V = -(\partial_q V)\partial_p$, which in general do not commute, have a particularly appealing property: they generate shifts in the canonical variables,

$$\exp(\tau D_T)\phi(q, p) = \phi(q + \tau(\partial_p T), p) = \phi(S_\tau(y)) \,,$$
$$\exp(\tau D_V)\phi(q, p) = \phi(q, p - \tau(\partial_q V)) = \phi(C_\tau(y)) \,,$$

which we denote as new mappings $S_\tau, C_\tau : \chi \to \chi$. (This should be familiar from quantum mechanics: the linear momentum operator $\boldsymbol{p} = -i\hbar\nabla_x$ is the generator of infinitesimal translations, while the orbital angular momentum operator $\boldsymbol{L} = \boldsymbol{r} \times \boldsymbol{p}$ is the generator of infinitesimal rotations.) We approximate the exact time evolution operator by the product of the exponential operators

$$\exp(\tau(D_T + D_V)) = \prod_{i=1}^{k} \exp(c_i \tau D_T) \exp(d_i \tau D_V) + O(\tau^{r+1}) \,, \tag{H.3}$$

where c_i and d_i are real numbers, and r is an integer determining the quality of the approximation and the order of the integrator. We have thus replaced the full evolution operator by a compositum of shifts in the phase space:

$$e^{\tau D_H}\phi(y) = e^{c_1 \tau D_T} e^{d_1 \tau D_V} \cdots e^{c_k \tau D_T} e^{d_k \tau D_V}\phi(y) + O(\tau^{r+1})$$
$$= \phi\Big(C_{d_k \tau} \circ S_{c_k \tau} \circ \cdots \circ C_{d_1 \tau} \circ S_{c_1 \tau}(y)\Big) + O(\tau^{r+1}) \,.$$

By comparison to Eq. (H.1), we infer that we have thus also obtained an approximation for the true mapping $M^t(y)$ of the dynamical system,

$$M^\tau = C_{d_k \tau} \circ S_{c_k \tau} \circ \cdots \circ C_{d_1 \tau} \circ S_{c_1 \tau} + O(\tau^{r+1}) \,.$$

The scheme for the integrator is now at hand [3]. When expression (H.3) is inserted in (H.2), we obtain a sequence of simpler mappings

$$q_i = q_{i-1} + \tau c_i \left(\frac{\partial T}{\partial p}\right)_{p=p_{i-1}} \,, \qquad p_i = p_{i-1} - \tau d_i \left(\frac{\partial V}{\partial q}\right)_{q=q_i} \,, \tag{H.4}$$

which map the initial configuration $y(0) = (p_0, q_0)$ in k steps into the final configuration $y(\tau) = (p_k, q_k)$. This sequence of mappings is symplectic, since it is composed of smaller steps which are all symplectic by themselves. We now know how to integrate the equation of motion; the main challenge is how to find k as small as possible (as few operations as possible) with the order of the integrator r as large as possible.

In principle, the procedure is straightforward: we expand the left-hand side of (H.3) in powers of τ and compare the coefficients of the corresponding powers of τ at the right (up to including order k). We end up with a system of non-linear algebraic equations for the unknowns c_i and d_i, which is easily solvable only for small k. In the simplest non-trivial case we obtain $k = 1$, $c_1 = d_1 = 1$, therefore

$$\exp(\tau D_T) \exp(\tau D_V) = \exp(\tau D_{\widetilde{H}_1}) \,. \tag{H.5}$$

For small τ, the left-hand side of the equation represents a simple iterative scheme which requires a single reference to q and p for one step in time τ according to (H.4). The right-hand side can be computed by using the Baker–Campbell–Hausdorff (BCH) formula describing the decomposition of a product of exponential functions of two *non-commuting* operators A and B,

$$\exp(A) \exp(B) = \exp(C) \,,$$

where

$$C = A + B + \frac{1}{2}[A, B] + \frac{1}{12}\Big([A, [A, B]] + [B, [B, A]]\Big) + \frac{1}{24}[A, [B, [B, A]]] + \cdots$$

and $[A, B] = AB - BA$. At the right of (H.5) we then obtain

$$\begin{aligned}
D_{\widetilde{H}_1} = D_T + D_V + \frac{\tau}{2}[D_T, D_V] + \frac{\tau^2}{12}\Big([D_T, [D_T, D_V]] + [D_V, [D_V, D_T]]\Big) \\
+ \frac{\tau^3}{24}[D_T, [D_V, [D_V, D_T]]] + \cdots \,,
\end{aligned}$$

and it follows that

$$\begin{aligned}
\widetilde{H}_1 = T + V + \frac{\tau}{2}\{V, T\} + \frac{\tau^2}{12}\Big(\{\{T, V\}, V\} + \{\{V, T\}, T\}\Big) \\
+ \frac{\tau^3}{12}\Big(\{\{\{T, V\}, V\}, T\}\Big) + \cdots \,. \tag{H.6}
\end{aligned}$$

(Note that a change in the signs occurs when passing from the usual commutators to Poisson brackets of odd orders, $[D_T, D_V] \rightarrow -\{T, V\}$, $[D_T, [D_V, [D_T, D_V]]] \rightarrow -\{\{\{T, V\}, T\}, V\}$, and so on. For details see [4].) We see that the expansion (H.6) has given us a first-order symplectic integrator preserving the Hamiltonian structure of the system, but which evolves according to a slightly different Hamiltonian operator $\widetilde{H}_1$. Only when $\tau \rightarrow 0$ do we recover the original Hamiltonian dynamics governed by $H = T + V$.

A second-order integrator We obtain an explicit symplectic integrator of the second order involving two stages ($k = 2$) by using the constants

$$c_1 = c_2 = \frac{1}{2}, \qquad d_1 = 1, \qquad d_2 = 0.$$

Let us write down the formulas for this integrator explicitly and use the notation of Chapter 8: instead of τ we write h, and replace the index i by n. The solutions at time t_n are p_n and q_n. The solution at the next time $t_{n+1} = t_n + h$ is obtained by computing

$$q^* = q_n + hc_1 \left(\frac{\partial T}{\partial p} \right)_{p=p_n},$$

$$p^* = p_n - hd_1 \left(\frac{\partial V}{\partial q} \right)_{q=q^*},$$

$$q_{n+1} = q^* + hc_2 \left(\frac{\partial T}{\partial p} \right)_{p=p^*},$$

$$p_{n+1} = p^* - hd_2 \left(\frac{\partial V}{\partial q} \right)_{q=q_{n+1}},$$

where q^* and p^* are auxiliary variables. In integrators of higher orders, there are even more such variables which form a chain-like progression to consecutive parts of an individual time step. Because $d_2 = 0$, only a single evaluation of the force $F = -\partial_q V$ is needed. This integrator corresponds to the decomposition

$$\exp\left(\tfrac{1}{2}h D_T\right) \exp(h D_V) \exp\left(\tfrac{1}{2}h D_T\right) = \exp(h D_{\widetilde{H}_2}),$$

whence it follows that

$$\widetilde{H}_2 = T + V + \frac{h^2}{12} \left(\{\{T, V\}, V\} - \frac{1}{2}\{\{V, T\}, T\} \right) + O(h^4).$$

A fourth-order integrator A very popular fourth-order symplectic integrator [5] requiring four stages ($k = 4$) has the parameters

$$c_1 = c_4 = \frac{1}{2(2 - 2^{1/3})}\,, \qquad c_2 = c_3 = \frac{1 - 2^{1/3}}{2(2 - 2^{1/3})}\,, \tag{H.7}$$

$$d_1 = d_3 = \frac{1}{2 - 2^{1/3}}\,, \qquad d_2 = -\frac{2^{1/3}}{2 - 2^{1/3}}\,, \qquad d_4 = 0\,, \tag{H.8}$$

and has the highest order for which the coefficients could still be determined analytically. Again, the last coefficient is $d_4 = 0$: at each time step h only three evaluations of the force $F = -\partial_q V$ are needed, one less than, for example, in the standard RK4 method calling for four evaluations. The coefficients (H.7) and (H.8) do not solve the system of equations for c_i and d_i at fourth order uniquely: they can be further optimized with respect to a certain aspect of the numerical problem, for example, for optimal conservation of energy [6].

A sixth-order integrator The path to practically useful symplectic integrators also leads through the BCH formula, but the procedure of determining the coefficients c_i and d_i becomes more complicated and they can no longer be given analytically. The coefficients (in double precision) for the sixth-order integrator with eight stages ($k = 8$) are

$$
\begin{aligned}
c_1 = c_8 &= 0.392256805238780\,, \\
c_2 = c_7 &= 0.510043411918458\,, \\
c_3 = c_6 &= -0.471053385409758\,, \\
c_4 = c_5 &= 0.06875316825251980\,, \\
d_1 = d_7 &= 0.784513610477560\,, \\
d_2 = d_6 &= 0.235573213359357\,, \\
d_3 = d_5 &= -1.17767998417887\,, \\
d_4 &= 1.31518632068391\,, \\
d_8 &= 0\,.
\end{aligned}
$$

An eighth-order integrator requiring sixteen stages is described in [7].

References

1. A.J. Dragt, J.M. Finn, Lie series and invariant functions for analytic symplectic maps. J. Math. Phys. **17**, 2215 (1976)
2. V.I. Arnol'd, *Mathematische Methoden der klassischen Mechanik* (VEB Deutscher Verlag der Wissenschaften, Berlin, 1988)
3. H. Yoshida, Recent progress in the theory and application of symplectic integrators. Celest. Mech. Dyn. Astr. **56**, 27 (1993); H. Kinoshita, H. Yoshida, H. Nakai, Symplectic integrators and their application to dynamical astronomy. Celest. Mech. Dyn. Astr. **50**, 59 (1991)
4. S.R. Scuro, S.A. Chin, Forward symplectic integrators and the long-time phase error in periodic motions. Phys. Rev. E **71**, 056703 (2005)
5. E. Forest, R. D. Ruth, Fourth-order symplectic integration. Phys. D **43**, 105 (1990)

6. D. Donnelly, E. Rogers, Symplectic integrators: an introduction. Am. J. Phys. **73**, 938 (2005)
7. H. Yoshida, Construction of higher order symplectic integrators. Phys. Lett. A **150**, 262 (1990)

Appendix I
Transforming PDEs to Systems of ODEs

Partial differential equations can be transformed to systems of ordinary differential equations. This conversion reveals some new possibilities to construct specific difference schemes, yet this process is not without obstacles [1].

I.1 Diffusion Equation

The first example is the one-dimensional linear diffusion equation $v_t = Dv_{xx}$ with the initial condition $v(x, 0) = f(x)$ and homogeneous Dirichlet boundary conditions at $x = 0$ and $x = 1$. We approximate the time derivative at time t by the central difference,

$$\frac{\mathrm{d}v(x, t)}{\mathrm{d}t} = \frac{D}{\Delta x^2}\Big(v(x - \Delta x, t) - 2v(x, t) + v(x + \Delta x, t)\Big) + O(\Delta x^2) \,,$$

and do this at each $x_j = j\Delta x$. We obtain a system of ordinary differential equations, which can be written in matrix form

$$\frac{\mathrm{d}\boldsymbol{u}(t)}{\mathrm{d}t} = A\,\boldsymbol{u}(t) + \boldsymbol{b}(t) \,. \tag{I.1}$$

Here $\boldsymbol{u}(t) = (u_1, u_2, \ldots, u_{N-1})^\mathrm{T}$ is the solution vector, while the boundary conditions are contained in the vector $\boldsymbol{b}(t) = (D/\Delta x^2)(u_0(t), 0, \ldots, 0, u_N(t))^\mathrm{T}$. The matrix of the system is

$$A = \frac{D}{\Delta x^2}\begin{pmatrix} -2 & 1 & & & \\ 1 & -2 & 1 & & \\ & \ddots & \ddots & \ddots & \\ & & 1 & -2 & 1 \\ & & & 1 & -2 \end{pmatrix} \,.$$

© The Editor(s) (if applicable) and The Author(s), under exclusive license to Springer Nature Switzerland AG 2025
S. Širca and M. Horvat, *Computational Methods in Physics*, Graduate Texts in Physics,
https://doi.org/10.1007/978-3-031-68566-8

If the boundary conditions $\boldsymbol{b}$ do not depend on time, the solution of the system (I.1) with the initial condition $\boldsymbol{u}(0) = (f_1, f_2, \ldots, f_{N-1})^{\mathrm{T}} = \boldsymbol{f}$ has the form $\boldsymbol{u}(t) = -A^{-1}\boldsymbol{b} + \exp(tA)\,(\boldsymbol{f} + A^{-1}\boldsymbol{b})$. The solutions at times t and $t + \Delta t$ are then related by the evolution equation

$$\boldsymbol{u}(t + \Delta t) = -A^{-1}\boldsymbol{b} + \exp(\Delta t A)\left(\boldsymbol{u}(t) + A^{-1}\boldsymbol{b}\right) ,$$

which becomes even simpler in the case of homogeneous Dirichlet boundary conditions ($\boldsymbol{b} = \boldsymbol{0}$).

We can expand the exponential function with a matrix argument in the Taylor series $\exp(\Delta t A) = I + \Delta t A + \frac{1}{2}\Delta t^2 A^2 + \cdots$ or use a Padé approximation. The latter can be used to approximate the exponential as a ratio of two polynomials (rational approximation)

$$\exp(X) = \underbrace{\frac{1 + p_1 X + p_2 X^2 + \cdots + p_L X^L}{1 + q_1 X + q_2 X^2 + \cdots + q_M X^M}}_{[L/M](X)} + c_{L+M+1} X^{L+M+1} + O(X^{L+M+2}) ,$$

where p_l, q_m, and c are real coefficients that can be determined uniquely. The function $[L/M](X)$ is the Padé approximation of order (L, M) of the function $\exp(X)$ with the leading error term $c_{L+M+1} X^{L+M+1}$ (see also Sect. 1.2.2). Table I.1 shows a few lowest orders of the Padé approximations.

Table I.1 Padé approximations $[L/M]$ of order (L, M) of the function $\exp(X)$, and their leading error terms. If X is a matrix, the denominators of expressions $[L/M]$ should be understood in the sense of matrix inverses: we write $\boldsymbol{u}(t + \Delta t) = (A/B)\boldsymbol{u}(t)$, while we actually solve the system $B\boldsymbol{u}(t + \Delta t) = A\boldsymbol{u}(t)$

(L, M)	$[L/M](X)$	Error
(0,1)	$1/(1 - X)$	$-\frac{1}{2}X^2$
(1,0)	$1 + X$	$\frac{1}{2}X^2$
(0,2)	$1\big/\left(1 - X + \frac{1}{2}X^2\right)$	$\frac{1}{6}X^3$
(1,1)	$\left(1 + \frac{1}{2}X\right)\big/\left(1 - \frac{1}{2}X\right)$	$-\frac{1}{12}X^3$
(2,0)	$1 + X + \frac{1}{2}X^2$	$\frac{1}{6}X^3$
(1,2)	$\left(1 + \frac{1}{3}X\right)\big/\left(1 - \frac{2}{3}X + \frac{1}{6}X^2\right)$	$\frac{1}{72}X^4$
(2,1)	$\left(1 + \frac{2}{3}X + \frac{1}{6}X^2\right)\big/\left(1 - \frac{1}{3}X\right)$	$-\frac{1}{72}X^4$
(2,2)	$\left(1 + \frac{1}{2}X + \frac{1}{12}X^2\right)\big/\left(1 - \frac{1}{2}X + \frac{1}{12}X^2\right)$	$\frac{1}{720}X^5$

In difference schemes for partial differential equations, X are matrices: how one should interpret "matrix division" in the expression for $\exp(X)$, will become clear instantly. The Padé approximation $[1/0]$ means

$$u(t + \Delta t) = (I + \Delta t A)\, u(t)\,, \tag{I.2}$$

which is nothing but the well-known explicit FTCS scheme. In the case of implicit schemes where $M \geq 1$, the denominators in Table I.1 should be understood in the sense of matrix inverses. The $[0/1]$ approximation is the implicit BTCS scheme,

$$(I - \Delta t A)u(t + \Delta t) = u(t)\,,$$

while the $[1/1]$ approximation is the Crank–Nicolson scheme,

$$\left(I - \tfrac{1}{2}\Delta t A\right) u(t + \Delta t) = \left(I + \tfrac{1}{2}\Delta t A\right) u(t)\,. \tag{I.3}$$

The compact notation hides a trap shaking our conviction that from the viewpoint of stability (the value of the parameter $r = D\Delta t/\Delta x^2$), we may safely rely on the Crank–Nicolson scheme. The difference scheme with the approximation of order (L, M) maps the initial condition $u(0)$ to $u(t_n) = ([L/M](\Delta t A))^n u(0)$ at time t_n. The non-degenerate eigenvalues λ_s of the matrix A (A.9) correspond to the independent eigenvectors v_s (A.10), which we can use to expand the initial approximation as $u(0) = \sum_s c_s v_s$. Since $A v_s = \lambda_s v_s$, we also have $F(A)v_s = F(\lambda_s)v_s$, thus

$$u(t_n) = \sum_{s=1}^{N-1} c_s ([L/M](\Delta t \lambda_s))^n\, v_s\,.$$

The solution remains stable, $\lim_{n\to\infty} u(t_n) = \mathbf{0}$, if $|[L/M](\Delta t \lambda_s)| < 1$ for each s. This is clearly true for the Crank–Nicolson scheme, since $\lambda_s < 0$ for all s and the growth factor is $\mu_s = [1/1](\Delta t \lambda_s) = (1 + \Delta t \lambda_s)/(1 - \Delta t \lambda_s) < 1$. Yet $[1/1](\Delta t \lambda_s)$ may come close to the value -1 for large $\Delta t \lambda_s = -4r \sin^2(s\pi/2N)$, i.e. when r is large and $s\pi/2N \approx \pi/2$ (both N and s large). This can occur when there are discontinuities in the initial condition or between the initial and the boundary condition, and may cause oscillations of the solution in the vicinity of such discontinuities. Figure I.1 shows an example.

Oscillations originate in the "high-frequency" terms $c_{N-1}v_{N-1}, c_{N-2}v_{N-2}, \ldots,$ which become $c_{N-1}\mu_{N-1}^n v_{N-1}, c_{N-2}\mu_{N-2}^n v_{N-2}, \ldots$ in the nth time step. These terms alternate in sign when n increases, and are only slowly damped. Still, it can be shown [1] that such instabilities can be harnessed with a sufficiently small time step

$$\Delta t \lesssim \Delta x/\pi\,. \tag{I.4}$$

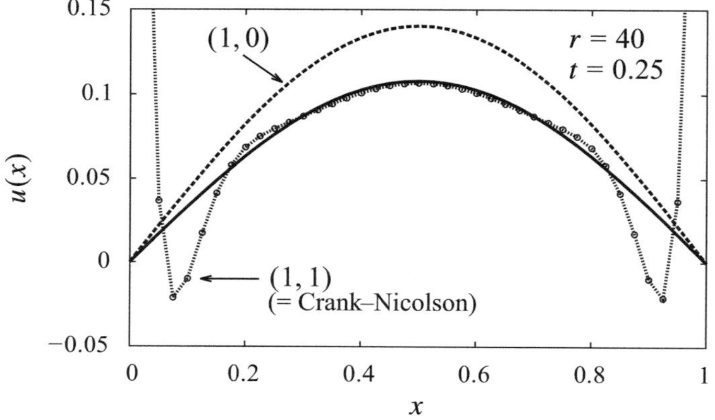

Fig. I.1 Numerical solution of the diffusion equation $v_t = D v_{xx}$ on $0 \leq x \leq 1$ with the initial condition $v(x, 0) = 1$ and boundary conditions $v(0, t) = v(1, t) = 0$. (Example adapted from [1].) Shown are the exact solution (full curve) at $t = 0.25$, the corresponding numerical solution ($D = 1$, $\Delta x = \Delta t = 0.025$, $r = 40$) by the Padé approximation of order $(1, 0)$ (Eq. (I.2)), and the solution of order $(1, 1)$ (Crank–Nicolson, Eq. (I.3)), in which typical oscillations appear in the vicinity of the discontinuity between the initial and boundary condition. For the example shown in this figure we have set $\Delta t = 0.025 > \Delta x / \pi = 0.008$, in disagreement with the requirement (I.4)

I.2 Advection Equation

As an example of hyperbolic equations, we discuss $v_t + c v_x = 0$ with the parameter $c > 0$ at $x \geq 0$, the initial condition $v(x, 0) = f(x)$, and the boundary condition $v(0, t) = b(t)$. According to the sign of c it is sensible to discretize the spatial derivative as $v_x(x, t) \approx (v(x, t) - v(x - \Delta x, t))/\Delta x$. As in the preceding Section, we establish an ordinary differential equation at each point $x_j = j \Delta x$,

$$\frac{\mathrm{d}v(x, t)}{\mathrm{d}t} = -\frac{c}{\Delta x}\Big(v(x, t) - v(x - \Delta x, t)\Big) + O(\Delta x) \,,$$

so that the whole domain can be covered by the matrix system of the form

$$\frac{\mathrm{d}\boldsymbol{u}(t)}{\mathrm{d}t} = -c B\, \boldsymbol{u}(t) + c \boldsymbol{b}(t) \,,$$

where $\boldsymbol{u}(t) = (u_1, u_2, \ldots, u_N)^{\mathrm{T}}$ is the solution vector, the boundary condition is contained in the vector $\boldsymbol{b}(t) = (1/\Delta x)(b(t), 0, \ldots, 0)^{\mathrm{T}}$, and B is bidiagonal,

$$
B = \frac{1}{\Delta x}
\begin{pmatrix}
1 & 0 & & & \\
-1 & 1 & 0 & & \\
 & \ddots & \ddots & \ddots & \\
 & & -1 & 1 & 0 \\
 & & & -1 & 1
\end{pmatrix} .
$$

The solution of this system is

$$
\boldsymbol{u}(t) = B^{-1}\boldsymbol{b}(t) + \exp(-c\Delta t\, B)(\boldsymbol{f} - B^{-1}\boldsymbol{b}(t)) ,
$$

In subsequent time steps we therefore obtain

$$
\boldsymbol{u}(t + \Delta t) = B^{-1}\boldsymbol{b}(t) + \exp(-c\Delta t\, B)(\boldsymbol{u}(t) - B^{-1}\boldsymbol{b}(t)) .
$$

The exponential function $\exp(-c\Delta t\, B)$ can now be replaced by a Padé approximation. For example, the order $(0, 1)$ approximation represents the implicit scheme $(I + c\Delta t\, B)\boldsymbol{u}(t + \Delta t) = \boldsymbol{u}(t) + c\Delta t\, \boldsymbol{b}(t)$, which can be transformed into explicit form. At the points x_j, $j = 1, 2, \ldots, N$, the solution is

$$
u_j^{n+1} = \frac{1}{1 + cR}\left(cR u_{j-1}^{n+1} + u_j^n\right) ,
$$

where $R = \Delta t/\Delta x$ (check this as an exercise). At order $(1, 1)$, one again retrieves the Crank–Nicolson scheme. We may also increase the precision of the schemes by using a three-point difference approximation for the spatial derivative,

$$
\frac{dv(x, t)}{dt} = -\frac{c}{2\Delta x}\left(3v(x, t) - 4v(x - \Delta x, t) + v(x - 2\Delta x, t)\right) + O(\Delta x^2) .
$$

In this case, the matrix system becomes

$$
\frac{d\boldsymbol{u}(t)}{dt} = -\tfrac{1}{2}cB\,\boldsymbol{u}(t) + \tfrac{1}{2}c\boldsymbol{b}(t) , \tag{I.5}
$$

the boundary condition is contained in $\boldsymbol{b}(t) = (1/\Delta x)(2b(t), -b(t), 0, \ldots, 0)^{\mathrm{T}}$, and the matrix B is

$$
B = \frac{1}{\Delta x}
\begin{pmatrix}
2 & & & & \\
-4 & 3 & & & \\
1 & -4 & 3 & & \\
 & \ddots & \ddots & \ddots & \\
 & & 1 & -4 & 3 \\
 & & & 1 & -4 & 3
\end{pmatrix} .
$$

The solution of this system is $\boldsymbol{u}(t) = B^{-1}\boldsymbol{b}(t) + \exp(-\frac{1}{2}c\Delta t B)(\boldsymbol{f} - B^{-1}\boldsymbol{b}(t))$. In subsequent time steps we obtain

$$\boldsymbol{u}(t + \Delta t) = B^{-1}\boldsymbol{b}(t) + \exp(-\tfrac{1}{2}c\Delta t B)(\boldsymbol{u}(t) - B^{-1}\boldsymbol{b}(t)) \,.$$

The $(0, 1)$ Padé approximation results in the stable implicit scheme

$$(I + \tfrac{1}{2}c\Delta t B)\boldsymbol{u}(t + \Delta t) - \tfrac{1}{2}c\Delta t \boldsymbol{b}(t + \Delta t) = \boldsymbol{u}(t) \,,$$

which can also be rewritten in explicit form.

An unpleasant surprise is in store at the very end [1]. If we use the symmetric difference for the spatial derivative,

$$\frac{\partial v(x, t)}{\partial x} \approx \frac{v(x + \Delta x, t) - v(x - \Delta x, t)}{2\Delta x} \,, \tag{I.6}$$

we obtain the matrix system of the same form as before (Eq. (I.5)), with the matrix

$$B = \frac{1}{\Delta x}\begin{pmatrix} 0 & 1 & & & \\ -1 & 0 & 1 & & \\ & \ddots & \ddots & \ddots & \\ & & -1 & 0 & 1 \\ & & & -1 & 0 \end{pmatrix}$$

and vector $\boldsymbol{b}(t) = (1/\Delta x)(b(t), 0, 0, \ldots, -u_{N+1}(t))^{\mathrm{T}}$ containing the boundary condition. Due to the symmetric difference the eigenvalues of B are complex. The solution $\boldsymbol{u}(t)$ therefore includes oscillatory terms that ruin the solution (Fig. I.2).

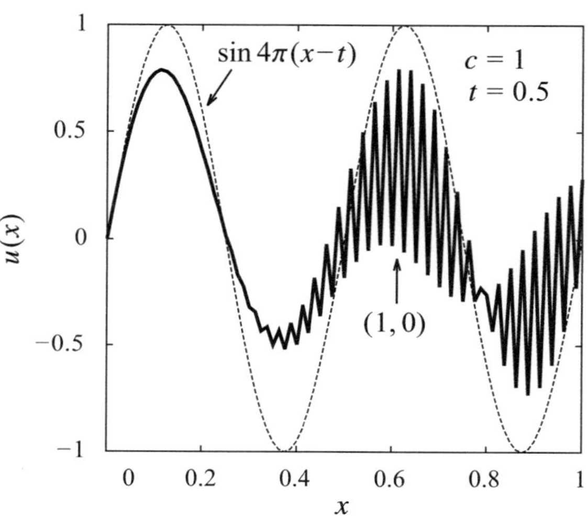

Fig. I.2 Solution of $v_t + cv_x = 0$ $(c = 1)$ on $0 \le x \le 1$ with the initial condition $v(x, 0) = \sin 4\pi x$, at $t = 0.5$. Shown are the exact solution $v(x, t) = \sin 4\pi(x - t)$ and the numerical solution $(\Delta x = 0.0125, \Delta t = 0.025)$ in the $(1, 0)$ Padé approximation. The symmetric approximation for v_x (Eq. (I.6)) causes strong oscillations

Reference

1. G.D. Smith, *Numerical Solution of Partial Differential Equations* (Oxford University Press, Oxford, 2003)

Appendix J
Numerical Libraries, Auxiliary Tools, and Languages

J.1 Important Numerical Libraries

A detailed explanation of individual methods and algorithms is neither the only nor the main purpose of this textbook: we are primarily interested in the mathematical and physical background of a problem rather than the details of the numerical method used to solve it. In such cases we build the skeleton of the program code, but to efficiently and reliably manage the core of the computations, we resort to specialized collection of numerical routines called *libraries*.

Linear algebra packages The routines for the manipulation of vectors and matrices (for example, to compute scalar product of vectors, multiply matrices, solve systems of equations, $Ax = b$, or eigenvalue problems, $Ax = \lambda x$) should never be coded by ourselves; this would be like reinventing the wheel. Almost invariably our program, even if literally adopted from a textbook on numerical methods, will be inferior to a carefully optimized algorithm from a dedicated library.

The BLAS (Basic Linear Algebra Subprograms) library [1] contains basic routines for vector and matrix operations used by more general libraries. At level 1 of BLAS we find routines for scalar, vector, and vector-vector operations; level 2 contains routines for matrix-vector operations; level 3 allows for matrix-matrix manipulations.

The basic building blocks of BLAS are used by numerous commercial software packages as well as by higher-level libraries, the most famous of which is LAPACK (Linear Algebra PACKage) [2]. It contains a collection of subprograms to solve systems of linear equations, eigenvalue problems, and problems involving singular value decomposition. Most of the routines have been prepared in real and complex versions (vectors and matrices), in single- and double-precision floating-point arithmetic. The library is well tuned to dense and banded matrices, while it is less appropriate for use with large general sparse matrices. To solve matrix problems with dense and banded matrices on multi-processor systems (clusters) with distributed memory, a scalable version SCALAPACK is available [3]. The same family of libraries also encompasses packages for the solution of eigenvalue problems with very large sparse

S. Širca and M. Horvat, *Computational Methods in Physics*, Graduate Texts in Physics, https://doi.org/10.1007/978-3-031-68566-8

matrices ARPACK [4], for direct solution of sparse systems (CAPSS, MFACT), as well as preconditioning routines used in iterative methods for large sparse systems (PARPRE). See also [5].

Libraries for specific processor architectures The original versions of LAPACK and BLAS libraries have been developed in Fortran77, but both have been ported for use with Fortran95, C/C++, and Java. They are incorporated in numerous distributions of Linux/UNIX systems. If we wish faster libraries, they need to be tuned to the individual processor architecture. For the BLAS library and a subset of routines from the LAPACK library, this can be done by the ATLAS (Automatically Tuned Linear Algebra Software) package [6]. This tool compiles the libraries and adapts them to the chosen computer platform. Precompiled optimized codes are readily available for most processors.

Perhaps the most widely spread and powerful among them is the Intel oneAPI Math Kernel Library (INTEL ONEMKL; formerly Intel Math Kernel Library or INTEL MKL) [7] offering highly optimized mathematical routines for a broad spectrum of scientific, engineering, and financial applications with interfaces to Fortran and C. It contains BLAS/LAPACK, SCALAPACK, and LINPACK, as well as the routines for direct and iterative solution of systems with sparse matrices. It performs particularly well in algorithms for direct solution of systems with large symmetric sparse matrices (of sizes 10000×10000 or more) [8]. INTEL ONEMKL also includes a set of (up to seven-dimensional) Fourier transformations for digital signal analysis, image processing, and solving partial differential equations. Moreover, the INTEL ONEMKL package contains well tuned vector implementations of computationally intensive elementary mathematical functions (above all, trigonometric and hyperbolic, where INTEL ONEMKL boasts several times larger speeds than those attained in standard versions). The INTEL ONEMKL package is available for Windows, Linux, and Mac OS operating systems.

For AMD processors we may use the AMD Optimizing CPU Libraries (AOCL) [9], the successors of the AMD Core Math Library, ACML. In addition to the optimized BLAS, LAPACK and SCALAPACK routines, this package offers a set of routines for fast Fourier transformations, random number generation, compression and cryptography.

Dedicated Fourier transformation packages The "working horse" for the discrete Fourier transformation (especially for very long data arrays) is the fast Fourier transform FFTW3 [10] (Fastest Fourier Transform in the West). It is written in C and is well portable between different architectures and operating systems. (In one form or another, FFTW3 is implemented in most libraries for specific architectures mentioned above.) One of the advantages of FFTW3 is the capability to adapt to the hardware used, allowing for improvement in speed (see Fig. 5.7). The method is of order $O(N \log N)$ for any N, in contrast to numerous other implementations that are restricted to a subset of sizes N (e.g. to powers of two, $N = 2^n$) or that become of order $O(N^2)$ for specific values of N (e.g. when N is prime). While the majority of competitive FFT libraries is limited to one-, two-, and three-dimensional transformations, FFTW3 can be applied in arbitrary dimensions.

Packages for Sturm–Liouville eigenvalue problems Several sophisticated packages exist for solving Sturm–Liouville boundary-value problems for ODEs involving eigenvalues and eigenfunctions, both in the regular (Sect. 9.7) and singular case (Sect. 9.8). These are SLEIGN2 [11] based on [12, 13, 14]; the SL02F program based on [15, 16, 17]; and the D02KAF, D02KDF, and D02KEF routines incorporated in the NAG library [18] (see description below). These packages use the shooting method based on the Prüfer transformation (Sect. 9.7.5) and can be applied to regular and singular Sturm–Liouville problems with separated or unseparated boundary conditions, as well as for the approximation of the continuous spectrum in the case of singular problems. Another state-of-the-art package, which is the only one that provides completely automatic endpoint classification in singular problems, is SLEDGE [19, 20, 21, 22]. All solvers listed above are included in SLTSTPAK [23, 24], a comprehensive test suite that also includes 60 characteristic Sturm–Liouville problems most frequently encountered in physics. For regular problems the code SLCPM12 [25] has been developed that is less general than the above all-purpose codes, yet precise and fast. A recent development is MATSLISE [26], a MATLAB package customized for an efficient solution of the Schrödinger equation.

Large general libraries The collections of routines mentioned above are specialized and highly optimized for a particular kind of numerical problems or algorithms. We therefore use them with a very specific purpose or goal. For everyday use, much more general and extensive libraries are available.

The longest tradition among them is claimed by the NAG (Numerical Analysis Group) corpus [18]. The NAG library is probably the largest commercial collection of mathematical and statistical algorithms and corresponding routines (more than 1000) for programming in C/C++, Fortran, Java, Python and MATLAB. It includes subprograms for optimization and minimization, the solution of ordinary and partial differential equations, finding zeros of non-linear equations, solving dense, banded, and sparse matrix systems and corresponding eigenvalue problems, for special functions, for approximation, fitting and interpolation, as well as for numerical integration. It also offers routines for random number generation, correlation and regression analysis, for multivariate methods, variance analysis and time series analysis.

A similar collection, NUMERICAL RECIPES (NR) [27], is a practical tool for everyday numerical computations that is not extremely intensive. For decades, physicists considered the NR library (in versions for Pascal, Fortran77, Fortran90, and C) to be the first choice for numerical work, while the latest (third) edition for C++ (NR3E) became commercial, with relatively severe copyright restrictions in academia. This may be one of the reasons that many users of Linux/UNIX systems prefer the open-code GSL (GNU Scientific Library) [28]. The NR3E collection also offers an interface to MATLAB.

The BOOST project [29] is a kind of a library of libraries. It contains a large collection of algorithms and tools for general structured and object-oriented programming, numerous data types, structures, classes, mathematical routines, and input-output tools. Components of Boost are gradually being integrated into standard C++ libraries.

J.2 Choosing the Programming Language

Fortran More than six decades have passed since its first steps, but the FORTRAN (The IBM FORmula TRANslating System) language in its newer versions like Fortran77 and Fortran95 remains the most widely spread and used language in high-performance computing, such as the study of fluid dynamics, weather predictions, finite element method analyses, theoretical physical chemistry, simulations of nuclear explosions, and similar numerically intensive applications. The latest revision of the language is Fortran 2023 [30].

In Fortran basic scientific and engineering ideas can be expressed very naturally and directly. Excellent compilers exist, coupled to an exceptionally rich assortment of libraries and packages. In modern versions of Fortran most of the constructs offered by C++ and Java can be realized, including pointers, dynamical memory allocation, and classes, with the exception of inheritance (relating general and particular properties of classes) and dynamical polymorphism (using equal names for abstract referencing to families of data types or routines) [31]. The standard Linux/UNIX compiler is `gfortran` [32], which is a component of the GNU Compiler Collection `gcc`. Commercial versions of Fortran compilers for high-performance computing (also for Windows and Mac OS) are offered by Intel and NAG.

C/C++ The C and C++ languages are closely related. The syntactically and semantically relatively simple language C [33] was developed in 1972 to cater to the needs of low-level programming, allowing for direct access to memory and other hardware components. A code written in C is easily portable among different architectures and operating systems, so C can be found everywhere, from supercomputers to drivers steering the lowest levels of electronic circuits.

The C++ language [34] was designed as a broad, object-oriented extension of the C language whose nature is markedly procedural, with the added option of *exception handling*. Another important enrichment is the possibility of dynamical resource control (memory allocation and de-allocation, opening and closing files, connecting to databases and libraries depending on the context (*scoped resource management*). The C++ language derives part of its strength from generic programming: algorithms can be written in a form that does not depend on the type of objects occurring in the algorithm. It also benefits from a rich standard library facilitating the work with arrays, iterators, input-output operations, algorithms of general and numerical types, and diagnostic tools. With few exceptions, the C language is a subset of C++, so a valid code written in C is valid in C++ as well (but not the other way around). The standard open-code compiler for C/C++ on Linux/UNIX systems is `gcc` [35], while an excellent commercial compiler is offered by Intel.

Java Similar to C++, Java [36] was developed for object-oriented programming, originally in the context of network computing and the desire for easy portability. Yet syntactically it is compatible with neither C++ or C (although it resembles them in many ways). Perhaps the most essential benefit of using Java is the Java Runtime Environment (JRE), an environment in which user application or libraries actually run. An application in Java "communicates" with the physical computer or the actual operating system by means of the *Java Virtual Machine* (JVM). The virtual machine represents a certain abstraction of the actual system. The program in Java therefore need not be written for each architecture or operating system separately, since to do this, one would have to know their numerous details and peculiarities: we write the program only once, and then it can be embedded in any platform on which a JRE is running.

Python As opposed to the three compiled languages outlined above Python [37] is an interpreted language: there are no explicit compilation and linking stages. The interpreter can be used interactively, which makes it easy to write test code in the process of bottom-up program development. Python is a high-level language, and there is broad consensus that it is simple to use and learn, and at the same time very efficient: the high-level data types allow one to express complex operations in a very compact manner and, as opposed to Fortran, C++/C and Java, no variable or argument declarations are necessary. Algorithms for optimization, integration, interpolation, eigenvalue problems, algebraic equations, differential equations, statistics and many other classes of problems are provided by SciPy [38], a wrapper around highly-optimized implementations of algorithms written in Fortran and C++/C.

The TIOBE index At the time of finishing this edition (June 2024), the "popularity share" of the above programming languages, as measured by the TIOBE Index [39], has been: Python = 15 % (top place), C++ = 10 % (second place), C = 9 % (third place), Java = 8 % (fourth place) and Fortran = 1.5 % (place 10).

References

1. L. S. Blackford et al., BLAS, Basic linear algebra subprograms. ACM Trans. Math. Softw. **28**, 135 (2002); J. Dongarra, Basic linear algebra subprograms technical forum standard. Int. J. High Perf. Comp. Appl. **16**, 1 (2002) and **16**, 115 (2002). The reference (unoptimized) library is available at http://www.netlib. org/blas

2. *LAPACK, The Linear Algebra PACKage*, http://www.netlib.org/lapack (version for Fortran77), `../lapack95` (version for Fortran95), `../clapack` (version for C), `../lapack++` (version for C++), `../java/f2j` (version for Java)

3. SCALAPACK, http://www.netlib.org/scalapack

4. https://github.com/opencollab/arpack-ng

5. J.J. Dongarra, V. Eijkhout, Numerical linear algebra algorithms and software. J. Comput. Appl. Math. **123**, 489 (2000)

6. ATLAS, R.C. Whaley, A. Petitet, J.J. Dongarra, Automated empirical optimization of software and the ATLAS project. Parallel Comp. **27**, 3 (2001); the latest repository is https://github.com/math-atlas/math-atlas

7. Intel Math Kernel Library, https://software.intel.com/mkl

8. N.I.M. Gould, J.A. Scott, Y. Hu, A numerical evaluation of sparse direct solvers for the solution of large sparse symmetric linear systems of equations. ACM Trans. Math. Softw. **33**, Art. 10 (2007)

9. https://www.amd.com/en/developer/aocl.html

10. M. Frigo, S.G. Johnson, The design and implementation of FFTW3. Proc. IEEE **93**, 216 (2005); see also http://www.fftw.org

11. P. B. Bailey, W. N. Everitt, A. Zettl, Algorithm 810: The SLEIGN2 Sturm–Liouville code. ACM Trans. Math. Softw. **27**, 143 (2001); the code can be downloaded from https://www.niu.edu/math/sl2/index.shtml

12. P.B. Bailey, B.S. Garbow, H.G. Kaper, A. Zettl, Eigenvalue and eigenfunction computations for Sturm–Liouville problems. ACM Trans. Math. Softw. **17**, 491 (1991)

13. P.B. Bailey, B. S. Garbow, H.G. Kaper, A. Zettl, Algorithm 700: a FORTRAN software package for Sturm–Liouville problems. ACM Trans. Math. Softw. **17**, 500 (1991)

14. P.B. Bailey, M.K. Gordon, L.F. Shampine, Automatic solution of the Sturm–Liouville problem. ACM Trans. Math. Softw. **4**, 193 (1978)

15. M. Marletta, J. D. Pryce, LCNO Sturm–Liouville problems: computational difficulties and examples. Numer. Math. **69**, 303 (1995)

16. J. D. Pryce, M. Marletta, A new multi-purpose software package for Schrödinger equation and Sturm–Liouville computations. Comput. Phys. Commun. **62**, 42 (1991)

17. M. Marletta, J.D. Pryce, Automatic solution of Sturm–Liouville problems using the Pruess method. J. Comput. Appl. Math **39**, 57 (1992)

18. Numerical Algorithms Group, http://www.nag.com

19. C.T. Fulton, S. Pruess, Y. Xie, The automatic classification of Sturm–Liouville problems. J. Appl. Math. Comput. **124**, 149 (2005)

20. S. Pruess, C. T. Fulton, Y. Xie, sl An asymptotic numerical method for a class of singular Sturm–Liouville problems. SIAM J. Numer. Anal. **32**, 1658 (1995)

21. S. Pruess, C. T. Fulton, Mathematical software for Sturm–Liouville problems. ACM Trans. Math. Softw. **19**, 360 (1993)

22. M.S.P. Eastham, C.T. Fulton, S. Pruess, Using the SLEDGE package on Sturm–Liouville problems having nonempty essential spectra. ACM Trans. Math. Softw. **22**, 423 (1996)

23. J.D. Pryce, A test package for Sturm–Liouville solvers. ACM Trans. Math. Softw. **25**, 21 (1999)

24. J.D. Pryce, Algorithm 789: SLTSTPAK: A test package for Sturm–Liouville solvers. ACM Trans. Math. Softw. **25**, 58 (1999)

25. L.G. Ixaru, H. De Meyer, G. Vanden Berghe, SLCPM12—a program for solving regular Sturm–Liouville problems. Comput. Phys. Commun. **118**, 259 (1999)

26. V. Ledoux, M. Van Daele, G. Vanden Berghe, MATSLISE: a MATLAB package for the numerical solution of Sturm–Liouville and Schrödinger equations. ACM Trans. Math. Softw. **31**, 532 (2005)

27. W.H. Press, B.P. Flannery, S.A. Teukolsky, W.T. Vetterling, *Numerical Recipes: The Art of Scientific Computing*, 3rd edn (Cambridge University Press, Cambridge, 2007). See also the equivalent handbooks in Fortran, Pascal and C, as well as http://numerical.recipes

28. GSL (GNU Scientific Library), http://www.gnu.org/software/gsl

29. BOOST C++ Libraries, http://www.boost.org

30. https://fortran-lang.org/
31. J. Reid, *The future of Fortran*, Computing in Science and Engineering, Jul/Aug 2003, p. 59; V.K. Decyk, C.D. Norton, H.J. Gardner, *Why Fortran?*, ibid., Jul/Aug 2007, p. 68
32. GNU Fortran Compiler, http://gcc.gnu.org/wiki/GFortran
33. B.W. Kernighan, DM. Ritchie, *C Programming Language*, 2nd edn (ANSI C) (Prentice Hall, Englewood Cliffs 1988); the current official language standard is ISO/IEC 9899:1999, also adopted by ANSI; see also http://www.open-std.org/jtc1/sc22/wg14
34. B. Stroustrup, *The C++ Programming Language*, 3rd edn (Addison-Wesley, Reading, 2000); N.M. Josuttis, *The C++ Standard Library: A Tutorial and Reference* (Addison-Wesley, Reading 1999); S. Prata, *C++ Primer Plus*, 5th edn (Sams Publishing (Pearson Education), Indianapolis 2004); the current official language standard is ISO/IEC 14882:2003, see also http://www.open-std.org/jtc1/sc22/wg21
35. GNU Project C/C++ Compiler, http://gcc.gnu.org
36. Java Platform (Standard edition), http://java.sun.com/javase
37. https://www.python.org/
38. https://scipy.org/
39. https://www.tiobe.com/tiobe-index/

Index

A

$A(\alpha)$-stability, 511
Abel's equation, 874
Abel summation, 49
Absolute convergence, 34
Absolute stability, 484–485
Acceleration of convergence, 38–41, 81
Acoustic impedance, 856
Action function, 29
Adams–Bashforth–Moulton method, 499
Adams method, 497
Adaptive estimation, 445–446, 963–966
Adaptive step size, 480, 483
Admissibility condition, 292
Advection equation, 757, 1000
Affine invariance, 77
Affinity matrix, 370
Airy functions, 55
Aitken's Δ^2 method, 40, 81
Algebraic multiplicity, 198
Algorithm
  alternating-projection, 903
  Arnoldi's, 209–210
  Bartels–Stewart, 188
  Buchberger's, 95
  Burg's, 427
  Cooley–Tukey, 254
  Coppersmith–Winograd, 174
  distillation, 38
  divide-and-conquer, 254
  Durstenfeld's, 960
  EM, 368
  Euclid's, 88
  Faugère's (F_5), 95
  FFT, 254

  for M-estimate of location, 322
  for M-estimate of scale, 323
  Gerchberg–Saxton, 903–905
  hierarchical clustering, 363
  Horner's, 83
  hybrid input-output, 905–907
  Kahan's, 37
  Lanczos', 210–212
  Linz's, 37
  PhaseCut, 909–910
  PhaseLift, 907–909
  QR, 199
  QZ, 206
  Remez', 8
  Ruffini's, 83
  Strassen's, 174
  Wynn's, 13
Algorithmic regularization, 985–987
Aliasing, 251
Almost optimal approximation, 9
Alternating series, 41, 59
AMD Optimizing CPU Libraries (AOCL),
  1006
Amplification factor, 653
Amplification matrix, 667
Amplitude function, 29
Analysis
  asymptotic, 17–34
  auto-regressive, 424–432
  canonical correlation, 375–377
  cluster, 361–371
  factor, 377–382
  linear discriminant, 371–373
  logistic discriminant, 374
  of errors in solving $Ax = b$, 177